Immunobiology of HLA

Bo Dupont
Editor

Immunobiology of HLA

Volume II

Immunogenetics and Histocompatibility

With 188 Illustrations, 8 in Full Color

Springer-Verlag
New York Berlin Heidelberg
London Paris Tokyo Hong Kong

Bo Dupont
Member, Sloan-Kettering Institute for Cancer Research;
Professor of Immunology
Cornell University Graduate School of Medical Science
New York, New York 10021, USA

Library of Congress Cataloging-in-Publication Data
Immunobiology of HLA / Bo Dupont, editor.
p. cm.
Proceedings of the Tenth International Histocompatibility Workshop and Conference, held in Princeton, N.J. and New York City, Nov. 12–23, 1987.
Dedicated to the memory of Ruggero Ceppellini.
Contents: v. 1. Histocompatibility testing 1987 — v. 2. Immunogenetics and histocompatibility.
Includes bibliographies and indexes.
1. Histocompatibility testing—Congresses. 2. HLA histocompatibility antigens—Congresses. I. Dupont, Bo. II. Ceppellini, Ruggero. III. International Histocompatibility Workshop and Conference (10th : 1987 : Princeton, N.J. and New York, N.Y.)
[DNLM: 1. Histocompatibility Testing—congresses. 2. HLA Antigens—immunology—congresses. WO 680 I3237 1987]
QR187.H57I45 1989
616.07′9—dc19
DNLM/DLC

88-39761

Typeset by Publishers Service, Bozeman, Montana.
Printed and bound by Arcata Graphics/Halliday, West Hanover, Massachusetts.
Printed in the United States of America.

9 8 7 6 5 4 3 2 1

ISBN 0-387-96826-1 Springer-Verlag New York Berlin Heidelberg
ISBN 3-540-96826-1 Springer-Verlag Berlin Heidelberg New York

Preface to Volume II

This publication contains six invited review papers covering major topics on the molecular biology of HLA (Part 1) and 253 original contributions (Part 2) selected from the 540 abstracts presented at the Histocompatibility Conferences in New York City, November 18–23, 1987.

The format of organizing an international conference on immunogenetics and histocompatibility in conjunction with the international histocompatibility workshop occurred for the first time in 1980 at the Eighth International Histocompatibility Workshop in Los Angeles. This arrangement was repeated in 1984 when the Ninth Workshop meeting took place in Munich and the conference in Vienna. The back-to-back organization of histocompatibility workshop and histocompatibility conference takes advantage of the fact that investigators with primary interest and active participation in the histocompatibility workshops constitute only a subset of the scientific community involved in research on the major histocompatibility complex in man. Conversely, many immunogeneticists, biochemists, and molecular biologists who are not engaged in histocompatibility testing are keenly interested in and wish to communicate and interact with their colleagues engaged in the application of immunogenetics.

The scope of the histocompatibility and immunogenetic conference was further expanded on this occasion because it was possible to arrange the 13th annual meeting of the American Society for Histocompatibility and Immunogenetics (ASHI) immediately following the international conference. The scientific programs for these two meetings were coordinated and planned jointly, allowing a comprehensive coverage of many timely issues.

The organization and planning of these two conferences required close collaboration between the Council and Program Committee of ASHI and the Organizing Committee of the International Histocompatibility Workshop. I wish to express my thanks for the productive collaboration with Herbert A. Perkins, Chairman, 13th Annual Meeting; Rene Duquesnoy, President, ASHI; Peter Parham, Scientific Program, ASHI; Adriana Zeevi, Abstract Committee, ASHI; and David Eckels, Educational Program, ASHI.

The attendance at the two conferences in New York City far exceeded our expectations. More than 1,400 active participants attended the meetings, and we reached the limit of what the hotel conference center could accommodate. It was indeed a very exciting experience to see the vibrant and engaged participation of colleagues who have been active in this field for more than 25 years as well as the many students, fellows, and other younger investigators.

This publication is designed with the intent of providing a handbook on the "state of the art" primarily concerning molecular biology and immunogenetics of the HLA system. The review papers included in Part 1 were prepared during the spring of 1988 and the authors of the papers in Part 2 have been requested to update their manuscripts to reflect the changes made by the WHO Nomenclature Committee. It is my hope that this publication will be useful as an easily accessible reference source for several years to come. The references in each paper are provided with full titles for this purpose.

New York, New York, USA

Bo Dupont
Chairman
Tenth International Histocompatibility
Workshop and Conference

Contents of Volume II

Section III Class I Histocompatibility Antigens: Transfection and Gene Expression

Section IV Class I Histocompatibility Antigens: Regulation

Section V Class I Histocompatibility Antigens: New Class I, TL/Qa-Like, and Others

Section VI Class I Histocompatibility Antigens: HLA-DRB1

Section VIII Class II Histocompatibility Antigens: HLA-DQ

Section XIII Antigen Presentation and HLA Restriction

Section XIV HLA and Disease Associations

Section XVIII Clinical Transplantation

Section XXI Cell Surface Antigens Other than HLA

Contents of Volume I

David Rush, Pascale Perrier, John Marwick, Gen-yao Ye, Derek Middleton, Robert A. Good, Jill K. Benson, C. Bernard Carpenter, Rick Belcher, John Lin, and Michael Riley

Antigen Society #2 Report (A25, A26, AW34, AW43, and AW66)
V. Joysey, E.D. du Toit, R.W. Martell, D.G. Taljaard, B. Crichton, T. Doran, Y. Morishima, P. Wheeler, V. Lazda, S.M. Tait, and A. Cambon-Thomsen

Antigen Society #3 Report (AW19 and A32/Aw74)
C.B. Carpenter, P. Dyer, H. Festenstein, J. Moraes, J. Gagne, D.J. McCloskey, J. Marwick, D. Tate, T. Barbalho, M. Beach, M. Lima, P. Mackintosh, M. Moraes, and L. Ratner

Antigen Society #4 Report (A2, AW68, AW69 and A28)
J.M. Kreisler, M. Morellini, M.E. Moreno, W.F. Wakeling, S. Marsh, L.J. Kennedy, P. Hall, S.C. Knight, B. Tait, C. Vermylen, Y.S. Kao, P. Saunders, K.W.K. Cheung, C. Giraudo, N. Prigoshin, S. Trabace, A. Salerno, L. Salvaneschi, A. Amoroso, and H. Mervart

Antigen Society #5 Report (B5, B51, Bw52, B18, B35, Bw53)
Clara Gorodezky, Pablo Rubinstein, Bruce Barger, Carol Carpenter, James Wolf, Rhonda Andersen, Josefina Flores, Frances E. Ward, Bento Arce-Gomez, Marc de Bruyere, Benjamin Bradley, Sanking Cheng, Donald E. Cross, Maria DiLonardo, Philippe Dyer, Leonardo Fainboim, Luis F. Garcia, Maria Gerbase de Lima, Emilio Haas, Eugene Heise, Ram Kakaiya, Andrija Kastelan, Pauline Mackintosh, Helena Mervart, Norma Prigoshin, Ma Carmen Samaniego, Argimiro Suarez, Marlis Schroeder, Paul Terasaki, Luis Teran, and Kimiyoshi Tsuji

Antigen Society #6 Report (B21, B49, Bw50, BN21, B12, B44, B45)
R.C. Williams, S.N. Chen, D.K. Gill, J.T. Lane, J.E. McAuley, R. Strothman, and K.K. Mittal

Antigen Society #7 Report (B8, w59, B14, w64, w65)
G. Mueller-Eckhardt, M.L. Schroeder, C. Darke, R. Khan, R. Wank, G. Dejour, R. Fauchet, F. Christiansen, E. Du Toit, H. Mervart, F. Schunter, A. Stratton, and G. Sudmeyer

Antigen Society #8 Report (Bw70, Bw71, Bw72)
Godfrey J. Laundy, Marilyn S. Pollack, Martin G. Guttridge, and Peter T. Klouda

Antigen Society #9 Report (Bw46 and the Subgroups of B15)
E.D. Albert, D. Chandanayingyong, J.S. Thompson, T. Zhao, M.G. Hammond, S. Naito, and G.M. Sierp

Antigen Society #10 Report (Bw57 and Bw58)
S.F. Goldmann, E. Ivaskova, H.J. von Keyserlingk, H. Mervart, E. Yang, V. Lenhard, S.H. Chan, M. Minev, E. Gyodi, G. Petranyi, S. Nyulassy, C. Brautbar, C. Jersild, S. Chen, and G. Dunston

Antigen Society #11 Report (B37, B38, B39, and Bw67)
G.M. Troup, L. Fainboim, S. Naito, A.H. Mota, and E.A. Palavecino

Antigen Society #12 Report (Bw54, Bw55, Bw56 and Bw42)
S. Sekiguchi, H. Neumeyer, N. Kashiwagi, K. Tsuji, K. Kobayashi, Y. Konoeda, M. Ohkubo, M. Atoh, K. Tokunaga, A. Yagita, H. Inoko, R. Fong, H. Mervart, Y. Paik, P. Reekers, M. Hammond, E. Du Toit, and E. Call

Joint Reports: T-Cell Recognition of HLA Class I Molecules

Joint Reports: Structural Polymorphism of the HLA Class II Alpha and Beta Chains

DR5 (DRw11, DRw12): 2-D Gel Patterns
Alejandro Madrigal, Ronald Bontrop, Maria Cristina Mazzilli, Cecile Kaplan, Kim Wong, Nan Knitter-Jack, Gerald Nepom, Marie-Christine Morel, Joan Lindsay, Dominique Charron, and Manfred Ballas

DRw6 (DRw13, DRw14): 2-D Gel Patterns
Dominique Charron, Pierre Hermans, Manfred Ballas, Zacharie Brahmi, Ernette du Toit, Robert Martell, Laurine Bow, and Ronald Bontrop

DR7: 2-D Gel Patterns
Robert Winchester, Erin Matthews, Francois Amesland, Maria Christina Mazzilli, Nelson Fernandez, Bernadette Kienzle, and Robert W. Knowles

DRw8: 2-D Gel Patterns
Lydia Kulova, Ronald Bontrop, Soo Young Yang, and Hiroo Maeda

DR9: 2-D Gel Patterns
Renée Fauchet, Odile Bouhallier, Dominique J. Charron, Robert Winchester, Erin Matthews, Bernadette Kienzle, and Robert W. Knowles

DRw52: 2-D Gel Patterns
Ronald Bontrop, Manfred Ballas, Alejandro Madrigal, Pierre Hermans, Susan Fuggle, Nel Otting, Christine Carter, Steven Sacks, Cecile Kaplan, Marie-Christine Morel, Susan F. Radka, Barbara Nepom, Laurine Bow, Ernette Du Toit, Stephanie Crafford, Nan Knitter-Jack, Gerald Nepom, Maria Christina Mazzilli, Zacharie Brahmi, and Dominique Charron

DRw53: 2-D Gel Patterns
Robert W. Knowles, Bernadette Kienzle, Robert Winchester, Erin Matthews, Odile Bouhallier, Renee Fauchet, William H. Marshall, Dimitri Monos, Francoise Amesland, Maria Christina Mazzilli, Marie-Anne Henry, Pascale Perrier, Cora Dejelo, Nelson Fernandez, Barbara Nepom, Nan Knitter-Jack, Gerald Nepom, Malek Kamoun, Chester Zmijewski, and Jerry Lanchbury

DQw1 (DQw5, DQw6): 2-D Gel Patterns
Cecile Kaplan, Marie-Christine Morel, Pierre Hermans, Jean Tkaczuk, Takako Naskatsuji, Nelson Fernandez, Nigel Rust, Christine Carter, Susan Fuggle, Corazon Dejelo, William H. Marshall, Kim Wong, Manfred Ballas, Zacharie Brahmi, Ernette Du Toit, Robert Martell, Laurine Bow, Lydia Kulova, Robert Winchester, Ronald Bontrop, and Dominique Charron

DQw2: 2-D Gel Patterns
Susan Radka, Susan Fuggle, Maria Christina Mazzilli, Christine Carter, Nigel Rust, Robert Winchester, Erin Matthews, Cecile Kaplan, Barbara Nepom, Francoise Amesland, Nelson Fernandez, Ernette Du Toit, Stephanie Crafford, Marie-Christine Morel, and Robert W. Knowles

DQw3 (DQw7, DQw8, DQw9): 2-D Gel Patterns
Christina Mazzilli, Enrico Gandini, Jerry Lanchbury, Alejandro Madrigal, Manfred Ballas, Susan Fuggle, Christine Carter, Nigel Rust, Marie-Anne Henry, Pascale Perrier, William H. Marshall, Dimitri Monos, Cora Dejelo, Cecile Kaplan, Marie-Christine Morel, Kim Wong, Pierre Hermans, Robert Winchester, Erin Matthews, Francoise Amesland, Odile Bouhallier, Renee Fauchet, Barbara Nepom,

Joint Reports: T-Cell Recognition of HLA Class II Molecules

RFLP Standardization Reports

RFLP Summary Reports

Color Plates

Color Plate I

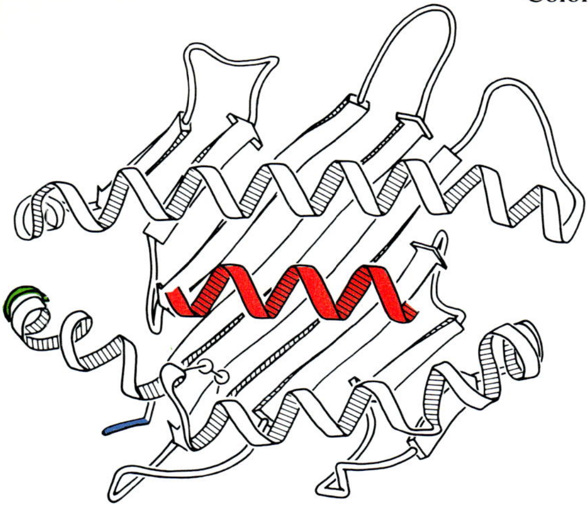

Figure 1. A schematic view of the α_1 and α_2 domains of HLA-A2 as seen from the viewpoint of a T-cell receptor. The β strands are shown as broad arrows pointing from amino-terminus to carboxy-terminus. The α helices are shown as helical ribbons. The amino terminus of the α_1 domain is the blue region at the bottom of the figure, and the carboxy-terminus of the α_2 domain is the dark green region at the left of the figure. The four β strands to the left and the top α helix are from α_1; the four β strands to the right and the bottom α helix are from α_2. The red helix at the center indicates how an α-helical peptide of 15 amino acids might sit in the peptide binding site. Adapted from Figure 2B of Bjorkman et al. (4). For further discussion, see text page 11.

```
α1 domain          ▼              ▼                    ▼                ▼  ▼▼▼ ▼▼▼  ▼ ▼▼  ▼
           1        10        20        30        40        50        60        70        80        90
consensus  GSHSMRYFYTSVSRPGRGEPRFIAVGYVDDTQFVRFDSDAASPRMEPRAPWIEQEGPEYWDRETQIVKAQSQTDRESLRTLRGYYNQSEA
A2.1       --------F---------------------------------Q------------------G--RK---H---H-VD-G-----------
A2.2Y      ------------------------------------------R------------------G--RK---H---H-VD-G-----------
A2.2F      --------F---------------------------------R------------------G--RK---H---H-VD-G-----------
A2.3       --------F---------------------------------Q------------------G--RK---H---H-VD-G-----------
A2.4a      ------------------------------------------Q------------------G--RK---H---H-VD-G-----------
A2.4b      --------F---------------------------------Q------------------G--RK---H---H-VD-G-----------
Aw69       ------------------------------------------Q-------------------N-RN---------VD-G-----------
Aw68.2     -----------M------------------------------Q-------------------N-RN---------VD-G-----------
Aw68.1     ------------------------------------------Q-------------------N-RN---------VD-G-----------
A3.1       --------F---------------------------------Q------------------Q--RN---------VD-G-----------
A3.2       --------F---------------------------------Q------------------Q--RN---------VD-G-----------
A11        ------------------------------------------Q------------------Q--RN---------VD-G----------D
A1         --------F---------------------------------QK-----------------Q--RNM--H-----AN-G----------D
A32        --------F---------------------------------Q------------------Q--RN---H---------IALR-------
Aw24       --------S---------------------------------Q------------------E--GK---H------N--IALR-------
Bw58       ----------AM--------------------------------T----------------G--RNM--SA--Y--N--IALR-------
B27.1      --------H--------------T-------L------------E---------------------C--KA-----D----LR-------
B27.2      --------H--------------T-------L------------E---------------------C--KA-----N--IALR-------
B27.3      --------H--------------T-------L------------E---------------------C--KA----------LR-------
B27.4      --------H--------------T-------L------------E---------------------C--KA----------LR-------
B27f       --------H--------------T-------L------------E---------------------C--KA--Y--N---ALR-------
B44.1      ----------AM-----------T-------L--------T---K---------------------S-TNT--Y--N---AAR-------
B44.2      ----------AM-----------T-------L--------T---K---------------------S-TNT--Y--N---ALR-------
B13        ----------AM-----------T----------------T---A---------------------S-TNT--Y--N---ALR-------
Bw47       ----------AM-----------T-------L--------T---K---------------------S-TNT--Y--D----LR-------
Bw65       ----------A------------S--------------------E----------------N---C-TNT--------N----------
B14        -----------------------S--------------------E----------------N---C-TNT--------N----------
B18        --------H--------------S-----G--------------T----------------N---S-TNT--Y-----N----------
B40*       --------H--------------T-------L--------T---K---------------------S-TNT--Y-----N----------
Bw41       --------H-AM-----------T-------L--------T---K---------------------S-TNT--Y-----N----------
Bw60       --------H-AM-----------T-------L--------T---K---------------------S-TNT--Y-----N----------
B8         --------D-AM-----------S--------------------E----------------N---F-TNT--------N----------
Bw42       -----------------------S--------------------E----------------N---Y---A--------N----------
B7.1       -----------------------S--------------------E----------------N---Y---A--------N----------
B7.2       -----------------------S--------------------E----------------N---Y---A--------N----------
Cw1        C----K--F--------------S--------------------G------V------------KY-R-A----V---N----------
Cw2.1      C---------A----S----H-----------------------G---GR-V------------KY-R-A----VN--K----------
Cw2.2      C---------A----S----H-----------------------G------V------------KY-R-A----VN--K----------
Cw3        --------C-A---------H------------------DE---G------V-RK---------KY-P-A----V---N----------
consensus  GSHSMRYFYTSVSRPGRGEPRFIAVGYVDDTQFVRFDSDAASPRMEPRAPWIEQEGPEYWDRETQIVKAQSQTDRESLRTLRGYYNQSEA
           1        10        20        30        40        50        60        70        80        90
```

Figure 3. The protein sequence of 15 HLA-A, 20 HLA-B, and 4 HLA-C molecules in the α_1 domain (above), the α_2 domain (color plate II), and the α_3 domain (color plate III). Each sequence is compared with a consensus constructed from the most frequent amino acid at each position. Dashes indicate identity with the consensus. The residue at position 116 in HLA-B7.2 is unknown, but is different from Y found in HLA-B7.1(5). It is therefore indicated with an X. The figure is derived from Figure 1 of Parham et al. (81).

Residues with high variability (see Fig. 4) are given by ▼. Locus-specific residues are colored yellow. Residues found in only one molecule or subtypes of a family are colored pink, those shared by small groups of distinct molecules are colored blue. These pink and blue residues with restricted distribution are good candidates for involvement in the highly specific epitopes seen by the "best" typing reagents. They are colored to aid in further prediction of alloantigenic epitopes. Residues critical for the Bw4 and Bw6 epitopes are colored green. For further discussion, see pages 12 to 14.

Color Plate II

α2 domain

```
                ▼ ▼                ▼ ▼                                     ▼      ▼
                   100       110       120       130       140       150       160       170       180
consensus  GSHTLQRMYGCDVGPDGRLLRGYHQYAYDGKDYIALNEDLRSWTAADTAAQITQRKWEAARVAEQLRAYLEGTCVEWLRRYLENGKETLQRA
A2.1       ----V---------S-W-F-----------------K----------M---T-KH-----H------------------------------T
A2.2Y      --------------S-W-F-----------------K----------M---T-KH-----H----W-------------------------T
A2.2F      --------------S-W-F-----------------K----------M---T-KH-----H----W-------------------------T
A2.3       ----V---------S-W-F-----------------K----------M---T-KH---T-HE---W-------------------------T
A2.4a      ----V---------S-W-F-----------------K----------M---T-KH-----H------------------------------T
A2.4b      ----V---C-----S-W-F-----------------K----------M---T-KH-----H------------------------------T
Aw69       ----V---------S-W-F-----------------K----------M---T-KH-----H------------------------------T
Aw68.2     ----I-------------F-----------------K----------M---T-KH-----H----W-------------------------T
Aw68.1     ----I-M-------S---F----R-D----------K----------M---T-KH-----H----W-------------------------T
A3.1       ----I-I-------S---F----R-D---------------------M-----K------HE--------D--------------------T
A3.2       ----I-I-------S---F----R-D---------------------M-----K------H----Q----D--------------------T
A11        ----I-I-----------F----R-D---------------------M-----K------HA---Q------R------------------T
A1         ----I-I-----------F----R-D---------------------M-----K-----VHA---R-V----R--DG--------------T
A32        ----I-M----------------Q-D---------------------M-------------------------------------------T
Aw24       ------M-F-----S---F-----------------K----------M-----K------H----Q---------DG---------------
Bw58       ---II-------L---------HD-S--------------S-------------------------------L-------------------
B27.1      ------N------------------D--------------S-------------------------------E-------------------
B27.2      ------N------------------D--------------S-------------------------------E-------------------
B27.3      ------N------------------D--------------S--------------------E----------E-------------------
B27.4      ------N----------------D----------------S--------------------E----------E-------------------
B27f       ------N------------------D--------------S-------------------------------E-------------------
B44.1      ---II------------------D-D--------------S------------------------D------L---S---------------
B44.2      ---II------------------D-D--------------S------------------------D------L---S---------------
B13        ----W-T-----L---------HN-L--------------S-------------L-----------------E-------------------
Bw47       --------F----------------D--------------S-------------------------------E-------------------
Bw65       ------W----------------N-F--------------S--------------------E------------------H-----------
B14        ------W----------------N-F--------------S--------------------E------------------H-----------
B18        ----------------------HD-S--------------S---------------------------------------H-----------
B40*       ------S---------------HN------------------------------------------------E-------------------
Bw41       ----W-----------------HN-----------------------------------------D--------------------D--E--
Bw60       ----------------------HN----------------------------S---L---------------E-------------DK-E--
B8         ------S---------------HN-----------------------------------------D--------------------D--E--
Bw42       ------S---------------HN-----------------------------------------D--------------------D--E--
B7.1       ------S---------------HD-------------------------------------E---R------E-------------DK-E--
B7.2       ------S---------------HD-X-----------------------------------E---R------E-------------DK-E--
Cw1        ------W-C---L----------D-------------------------------------E---R---------------------S----
Cw2.1      ------------L----------D-S-----------------------------------E---W------E--------------K----
Cw2.2      ------------L----------D-S-----------------------------------E---W------E-------------------
Cw3        ---II------------------D-H--------------------N--------------E----------L---------K-------G-
consensus  GSHTLQRMYGCDVGPDGRLLRGYHQYAYDGKDYIALNEDLRSWTAADTAAQITQRKWEAARVAEQLRAYLEGTCVEWLRRYLENGKETLQRA
                   100       110       120       130       140       150       160       170       180
```

Figure 3. *(Continued).*

Color Plate III

α3 domain

```
                 190       200       210       220       230       240       250       260       270
consensus  DPPKTHVTHHPISDHEATLRCWALGFYPAEITLTWQRDGEDQTQDTELVETRPAGDRTFQKWAAVVVPSGEEQRYTCHVQHEGLPKPLTLRW
A2.1       -A----M---AV------------S-------------------------------G-------------Q---------------------
A2.2Y      -A----M---AV------------S-------------------------------G-------------Q---------------------
A2.2F      -A----M---AV------------S-------------------------------G-------------Q---------------------
A2.3       -A----M---AV------------S-------------------------------G-------------Q---------------------
A2.4a      -A----M---AV------------S-------------------------------G-------------Q---------------------
A2.4b      -A----M---AV------------S-------------------------------G-------------Q---------------------
Aw69       -A----M---AV------------S-------------------------------G-------------Q---------------------
Aw68.2     -A----M---AV------------S-------------------------------G-----V-------Q---------------------
Aw68.1     -A----M---AV------------S-------------------------------G-----V-------Q---------------------
A3.1       ------M-------------------------------------------------G-----------------------------------
A3.2       ------M-------------------------------------------------G-----------------------------------
A11        ------M-------------------------------------------------G-----------------------------------
A1         ------M-------------------------------------------------G-----------------------------------
A32        -A----M---AV------------S-------------------------------G------S------Q---------------------
Aw24       ------M-------------------------------------------------G-----------------------------------
Bw58       -----------V--------------------------------------------------------------------------------
B27.1      --------------------------------------------------------------------------------------------
B27.2      --------------------------------------------------------------------------------------------
B27.3      --------------------------------------------------------------------------------------------
B27.4      --------------------------------------------------------------------------------------------
B27f       --------------------------------------------------------------------------------------------
B44.1      ----------------------S---------------------------------------------------------------------
B44.2      ----------------V---------------------------------------------------------------------------
B13        --------------------------------------------------------------------------------------------
Bw47       --------------------------------------------------------------------------------------------
Bw65       --------------------------------------------------------------------------------------------
B14        --------------------------------------------------------------------------------------------
B18        --------------------------------------------------------------------------------------------
B40*       --------------------------------------------------------------------------------------------
Bw41       --------------------------------------------------------------------------------------------
Bw60       --------------------------------------------------------------------------------------------
B8         --------------------------------------------------------------------------------------------
Bw42       --------------------------------------------------------------------------------------------
B7.1       --------------------------------------------------------------------------------------------
B7.2       --------------------------------------------------------------------------------------------
Cw1        EH---------V------------------------W-------------------G--------M-------------------E------
Cw2.1      EH---------V----------------T---------------------------G----------------------------E------
Cw2.2      EH---------V----------------T---------------------------G----------------------------E------
Cw3        EH---------V------------------------W-------------------G----------------------------E------
consensus  DPPKTHVTHHPISDHEATLRCWALGFYPAEITLTWQRDGEDQTQDTELVETRPAGDRTFQKWAAVVVPSGEEQRYTCHVQHEGLPKPLTLRW
                 190       200       210       220       230       240       250       260       270
```

Figure 3. *(Continued).*

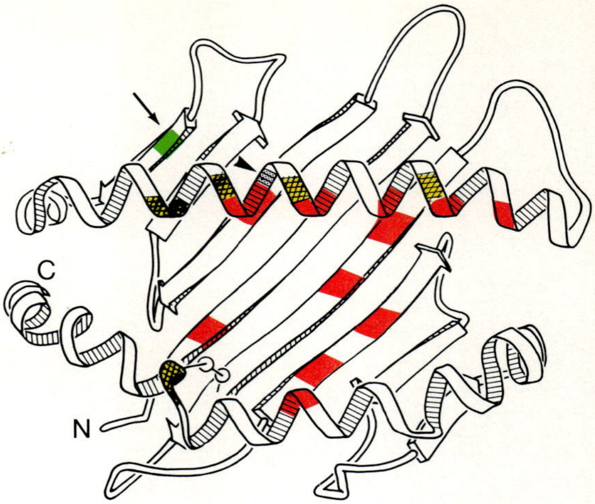

Figure 5. Position of residues having high variability in the structure of HLA-A2. The highlighted residues are those of high variability, as designated by ▼ in Figures 3 and 4. The schematic structure of the α_1 and α_2 domains is as in Figure 1. Solid areas are those residues with side chains pointing into the peptide binding site and potentially involved in contacting bound peptide. Yellow areas are those residues with side chains pointing up from the helices and potentially involved in contacting the T-cell receptor. The side chain of residue 71 (indicated by the arrowhead) in the α_1 helix points away from the peptide site, and residue 45 (green area) is not directly involved in the binding site. For further discussion, see text page 16.

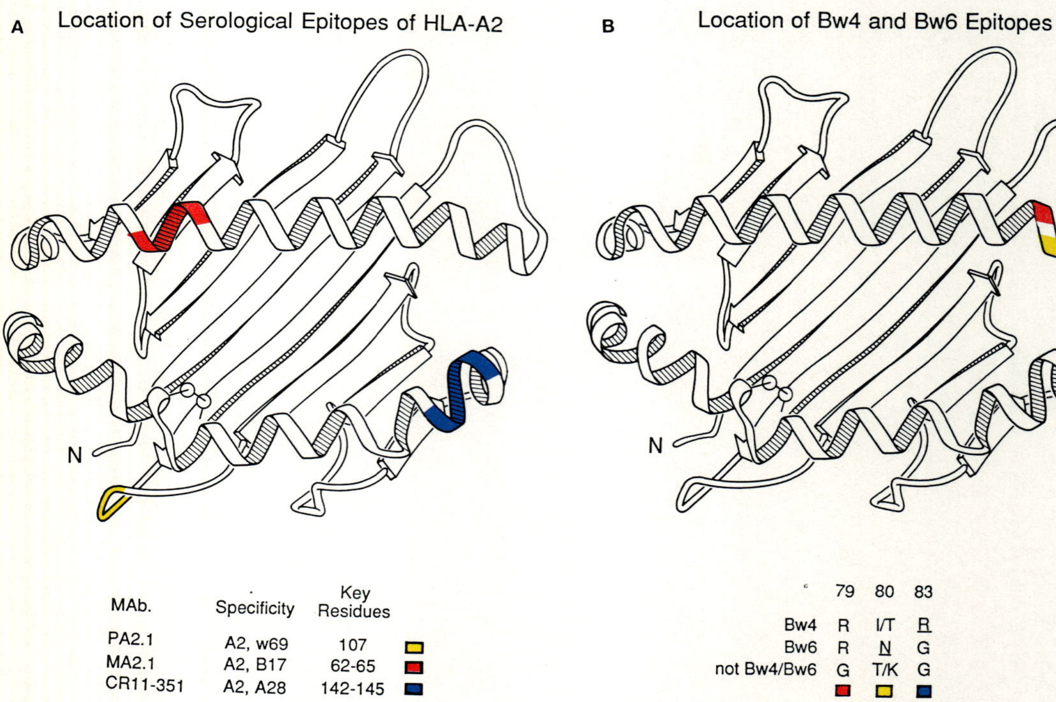

Figure 6. Mapping serologic epitopes. The format and labels in this schematic diagram of the α_1 and α_2 domains of the class I molecule are as in Figure 1. Residues involved in the epitopes are highlighted, and their respective epitopes are given in the keys to each panel. Panel A: The monoclonal antibody defined epitopes of HLA-A2. Panel B: Localization of the Bw4 and Bw6 epitopes. For further discussion, see text page 17.

Color Plate V

```
SIGNAL PEPTIDE
               -24                      -1
  A2.1          -a-------v----------q---
  A2.4a         -a-------v----------q---
  B27.1         ---t-----l---w--v-------
  Bw58          ---t----vl---w--v-------
  Bw46          ---t----vl--------------
  Cw1/3       mctklsggnlsrkhfpkea-------
  Cw1           ---------i--------------
  Cw3           ---------i--------------
  Cw2.1         ---------i--------------
  Consensus     MRVMAPRTL-LLLSGALALTETWA

ALPHA 1 DOMAIN
                1                            30                             60                           90
  A2.1          --------f-s-------------------------------q-m----------------g--r-v-ahs--h--d-gt----------
  A2.4a         ----------s-------------------------------q-m----------------g--r-v-ahs--h--d-gt----------
  B27.1         --------h-s------------t-------l------------e--------------------ic-ak-----ed--t-lr-------
  Bw58          -----------m--------------------------------t----------------g--rnm-as---y-en--ialr-------
  Bw46          -----------m--------------------------------ma----------------------r---------------------
  Cw1/3         c---------------------------------q----------------v----------------r---------------------
  Cw1           c----k--f-s------------s---------------------------v----------------r---------------------
  Cw3           --------c-----------h------------------de----------v-rk-------------p---------------------
  Cw2.1         c--------------s----h---------------------------gr-v----------------r-------n--k----------
  Consensus     GSHSMRYFYTAVSRPGRGEPRFIAVGYVDDTQFVRFDSDAASPRGEPRAPWIEQEGPEYWDRETQKYK-QAQTDRVSLRNLRGYYNQSEA

ALPHA 2 DOMAIN
                91                           120                            150                           180
  A2.1          ----v---------s-w-f----h-y----------k----------m---t-kh-----hv-----------------------------t
  A2.4a         ----v---------s-w-f----h-y----------k----------m---t-kh-----hv-----------------------------t
  B27.1         ------n----------------h-d--------------s--------------------v----------e-------------------
  Bw58          ---ii-------l---------h--s--------------s--------------------v----------l-------------------
  Bw46          ----------------------h--s--------------s--------------------e---w------l-------------------
  Cw1/3         ------------l----------n-f-----------------------------------t------------------------k-----
  Cw1           ------w-c---l------------y-----------------------------------e---r---------------------s----
  Cw3           ---ii--------------------h--------------------n--------------e----------l---------k-------g-
  Cw2.1         ------------l------------s-----------------------------------e---w------e--------------k----
  Consensus     GSHTLQRMYGCDVGPDGRLLRGYDQ-AYDGKDYIALNEDLRSWTAADTAAQITQRKWEAAR-AEQLRAYLEGTCVEWLRRYLENGKETLQRA

ALPHA 3 DOMAIN
                183                          212                            242                           272
  A2.1          -a----m---a-------------s---------------------------------------------q---------------------
  A2.4a         -a----m---a-------------s---------------------------------------------q---------------------
  B27.1         -p---------i--------------------------------------------r-----------------------------------
  Bw58          -p------------------------------------------------------r-----------------------------------
  Bw46          -p---------i--------------------------------------------r-----------------------------------
  Cw1/3         e------------------------------------------------------------------------------------e------
  Cw1           e-----------------------------------w----------------------------m-------------------e------
  Cw3           e-----------------------------------w------------------------------------------------e------
  Cw2.1         e---------------------------t--------------------------------------------------------e------
  Consensus     DHPKTHVTHHPVSDHEATLRCWALGFYPAEITLTWQRDGEDQTQDTELVETRPAGDGTFQKWAAVVVPSGEEQRYTCHVQHEGLPKPLTLRW

TRANSMEMBRANE DOMAIN
                275                           305      314
  A2.1          -------------i---vlfga-it -----a--w-----
  A2.4a         -------------i---vlfga-it -----a--w-----
  B27.1         -----s-v----------------- -----a--------
  Bw58          -----s------------------- -----t--------
  Bw46          -----s------------------- -----t--------
  Cw1/3         g---------------------lav-----m---------
  Cw1           ----------------------lav---------------
  Cw3           ----------------------lav---------------
  Cw2.1         ----------------------lav---------------
  Consensus     EPSSQPTIPIVGIVAGLAVLAVVVILGAVVAVVMCRRKSS

CYTOPLASMIC DOMAIN
                315                        342
  A2.1          dr-------------------------v*
  A2.4a         dr-------------------------v*
  B27.1         -----------c--------------*...
  Bw58          --------------------------*...
  Bw46          --------------------------*...
  Cw1/3         ------c------n-------e--i---*
  Cw1           ------c------n-------e--i---*
  Cw3           ------c------n-------e--i---*
  Cw2.1         ------c------n-------e--i---*
  Consensus     GGKGGSYSQAASSDSAQGSDVSLTACKA-
```

Figure 11. Gene conversion between HLA-B and HLA-C in an oriental haplotype. Class I HLA-encoding cDNA clones were isolated from a cell line typed homozygous HLA-A2, HLA-Bw46, and HLA-Cw1/3, and were kindly provided by Dr. Malcolm Simons. The specificity designated Cw1/3 is identical to Cw3.3 and CSH1 and is now called Cw11 (114). Three distinct cDNA were obtained and sequenced. The molecule typed as HLA-A2 has the sequence corresponding to the HLA-A2.4α subtype. The sequences are compared with those of HLA-A2.1, HLA-B27.1, HLA-Bw58, HLA-Cw1, Cw3, and Cw2.1. The region 53–96 from which an exchange from HLA-Cw1/3 into an acceptor B locus allele to give HLA-Bw46 is indicated in red. The consensus is derived from the nine molecules in the figure. The HLA-C locus molecules have an extra amino acid in the transmembrane domain (59,74). Termination codons are indicated by an asterisk. For further discussion, see text page 29.

Part 1

Summary Reports

HLA 1987*

Walter F. Bodmer

The year of the Workshop, 1987, was the 50th anniversary of the discovery of H2 by Peter Gorer. His work in the mouse was stimulated by the great population geneticist J.B.S. Haldane, leading to an approach more like that taken up later in the human field. What would he think now of the HLA system—the human analogue of his mouse H2 system?

This Workshop Summary is my fourth out of the last five Workshops since 1972. The nine previous Workshops, starting from the first organized by Bernard Amos in 1964, encapsulate the history of the development of the HLA field from a small collaboration, with workers gathered together in a single laboratory, to the mammoth collaboration of the present Workshop. Serology remains the starting point of the system, leading to the definition of its specificities. The products we now study in such detail at the molecular level came from simple serology and statistics, leading to data clusters and the establishment of antigen types.

The major principle of the Workshops is that we do together what no single laboratory could do on its own. Another important principle of such collaboration is that we must get as much out of it, each in our separate laboratories, as we put into it. Thus, we exchange cells and sera that no single laboratory can collect, and acquire an international typed cell panel, defined reagents, and an enormous and valuable body of data.

This Workshop effectively started at the end of the last one in 1984. Organizing a Workshop has become an increasing problem and an enormous effort was put into the organization of this Workshop by Bo Dupont and his colleagues. Perhaps only those who have themselves organized a Workshop can fully appreciate the effort that is involved. Now that the DNA genie has come out of its box, the high technology of molecular genetics makes the organization more complex than ever before.

The HLA Map

The genetic map, as defined by both conventional and molecular genetics, is the best starting point for an overview of the HLA region. (The version shown in Fig. 1 incorporates the decisions made by the Nomenclature Committee following the Workshop.) There are a number of features worth emphasizing. DOβ and DNα (formerly DZα) do not yet have any partners. The map based on pulse-field gel electrophoresis is a map of actual physical distances and shows the region to be about 3.5 $\times 10^6$ bases long just like *E. coli*. This map was first predicted some time ago based on crude genetic recombination data at a time when there were many fewer genes. There are some quite large gaps in the map. Are there extra genes to be found in these gaps, or have we nearly saturated the region? Even, say, 75% coverage would still leave a number of new genes to be discovered. What, for example, about the very large gap that is now clear between A and C? The ideas of recombination hot spots and corresponding patterns of linkage disequilibrium that were beginning to be discerned at the last Workshop are on the whole confirmed. However, the distance between A and C of nearly 1% recombination fraction is now consistent with the physical distance, without the need to postulate a hot spot there.

The interesting cluster of tumor necrosis factor (TNF) genes, which are involved in many functions, is newly placed in the region. Their function as a form of lymphokine can, as in the case of the complement genes, explain in evolutionary terms why they are there. Thus, an interaction between their function and the immune response control functions of the other products may have led to an advantage for inserting them in the region through, for example, some form of coordinated control of activity. However, the case for any such interaction accounting for the presence of the 21 hydroxylase (21OH) genes is still hard to understand. They are not uninteresting, but do not fit into a discussion of the functional significance of the HLA class I and II products. Perhaps, as suggested before, the 21OH genes just hitchhiked into the region attached to the complement genes, so that one should look near related complement genes in other genetic regions for possible common ancestors of the 21OH genes.

The search is still on for the human equivalents of TL and QA. There are some hints of their presence, both from the Workshop serology and from the results described by Harry Orr and others, indicating the presence of three potentially functional genes—6, 6.2 (now HLA-E), and 5.4 lying between A and C. Only one of these genes is apparently expressed on the cell surface, and none have the classic tissue distribution expected from the distribution of the mouse TL determinants. The mouse apparently has many more functional genes in the corresponding part of the H2 region. However, the mouse equivalents of the human CD1 genes cloned by Milstein and his colleagues and assigned to chromosome 1, which were initially thought to be the human equivalents of TL because of their similar tissue distribution, have not yet been identified. Why is there this difference between man and mouse? One of the most obvious differences between these species is that mice have tails, but it seems unlikely that the function of the TL and QA genes in the mouse H2 region has anything

*This text is an edited version of the Final Summary talk as given at the Conference in New York, the only change being the use of the Nomenclature established by the WHO Committee, which met and reported after the Workshop and Conference.

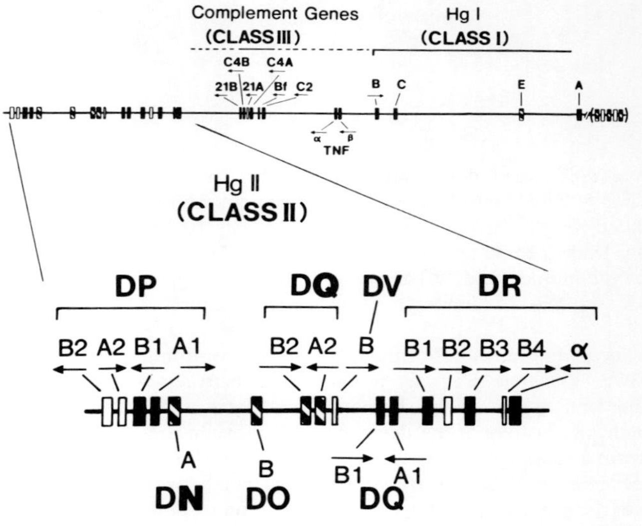

Figure 1. Map of the HLA region based on a general conseneus from many different laboratories, together with pulse-field gel electrophoretic data. It is drawn to scale, with the total length being approximately 3.5×10^6 base pairs. In the HLA D region expanded map, full shading signifies the gene is expressed, no shading that it is a pseudogene, and cross-hatching that it is potentially expressable but so far no expression has been detected. The arrows indicate the known orientations of the gene with respect to expression. I am grateful to John Trowsdale for providing this version of the map.

to do with tails. Might these genes be additional restriction elements analogous to the other class I products? This may even be the case for the human CD1 products, in spite of their substantial difference in sequence from the HLA class I products. Perhaps at times a positive immune response is disadvantageous and so leads to selection against the expression of a gene in the major histocompatibility region. Thus, from time to time, in addition to the positive advantages of an immune response leading to the selection for new variants and sometimes new genes, there would be a counterbalancing negative selection leading to the functional extinction of genes in the region, thus contributing to the difference between man and mouse with respect to the human equivalents of TL and perhaps QA. The CD1 determinants are specifically expressed on Langerhans and other similar histiocytic cells, perhaps suggesting that if they have a function analogous to the class I determinants, this may be specifically related to the functional requirements of particular cell types.

Some argue that patterns of linkage disequilibrium suggest the existence of "extended" haplotypes, in which alleles from several loci spread throughout the HLA region appear to be maintained together on a given chromosome. Daniel Cohen, for example, pointed out a constellation of differences that seemed particularly to be associated with DR1-containing haplotypes. One explanation for such a clustering of genes on an extended haplotype is that selection has pushed the haplotype into the population comparatively quickly, so that there has not been enough time to disrupt the association between its constituent alleles. The second, and contrasting, explanation is that such haplotypes contain sequences that tend to suppress genetic recombination. However, there is no evidence, for example, that the common A1- and B8-containing haplotypes give rise to recombinants with a frequency that is different from the average. So far, in this Workshop, comparatively few recombinant families have been analyzed. For the future, it might be worth doing a form of case control study of recombinants to ask the question: Are there particular haplotypes that are under- or overrepresented among observed recombinants relative to their frequency in the population, and does that therefore provide part of the explanation for observed patterns of linkage disequilibrium? That, perhaps, is a collaborative project that could be taken up in the next Workshop.

Nomenclature

There is no convenient name for the products of the HLA region analogous, for example, to immunoglobulins or hemaglobins. This is because the products were defined by the genes as, for example, was also the case for the A B O blood groups. As a convenient parallel to the immunoglobulins, I have suggested the name *histoglobulins* for the products. Then, for example, one can use symbols HgI and HgII for the class I and class II products to distinguish them from the mixed bag of other products coded for by genes in the middle of the HLA region including 210H, complement components, and TNF, for which there is no rationale for attaching the name class III. In the Workshops, it is particularly the genetics, variability, and function of the histoglobulins that are being studed.

The Nomenclature Committee started its meeting immediately after the Workshop, and had to come to grips with the problem of naming the genes in the HgII region. There are some simple principles for nomenclature that have been widely accepted by the human gene mappers, and which have been applied to the nomenclature of the HLA region for many years. First among these is that Roman, not Greek, letters are used for gene names. Thus, as shown in Fig. 1, the alpha-chain genes in the D region are associated with the letter A and the beta-chain genes with B. Following the Nomenclature Committee, it was decided that the numbers distinguishing different A or B genes within a subregion should be placed after the letter to provide a consistent hierarchical nomenclature. There is still a problem that one cannot always be sure that the collection of genes in different haplotypes is precisely homologous. Thus, what may be called B1 in one haplotype may not be the same gene in sequence in another haplotype. This appears to be the case for DR2, as compared with other haplotypes containing, for example, DR3, 5, or 6.

Restriction Fragment Length Polymorphism

For the first time, this Workshop included a systematic study of DNA-based restriction fragment length polymorphisms (RFLP). A key feature of the success of this study was the effectiveness of the approach to standardization, using a least-squares fit of a formula proposed by Southern to describe the relationship between migration distance in a gel and molecular weight. By using a good range of molecular weight controls, and through repetition of different probe-restriction enzyme combinations, it was possible to obtain remarkably precise estimates of the molecular weights of particular fragments, enabling differences as small as 50 base pairs to be clearly identified. It is interesting to see the parallel between the early days of clustering of serum reactions and the current interest in the clustering of RFLP patterns. The presence and absence of a fragment are just like the presence and absence of a serum reaction, and thus the same 2×2 analyses and clustering approaches can be applied as first done by Jon van Rood more than 25 years ago. Some of the pairwise correlation coefficients within clusters are very high (more than 0.8 or 0.9), and such clusters can easily be used for the definition of apparent types, just as in the case of serum reactions. The equivalents of DQW1, DRW52, and DRW53 can easily be seen, as we first pointed out some years ago.

Certain patterns stand out clearly and distinctively. Thus, the allelic sets corresponding to the DQA2 and B2 (formerly DXα and DXβ) genes are clearly identified, as is their relative lack of linkage disequilibrium with the DR and DQ serological types. There is, as expected, strong linkage disequilibrium between the patterns for DQA2 and DQB2 and an occasional surprising association with DR, such as that with a subset of DR3 that distinguishes the B18 from the B8-associated DR3 haplotypes. Since to date there has been no evidence for expression of the DQA2 and B2 products, these are results that could not have been obtained by serology. Nevertheless, it is important to be aware that in spite of the apparent similarity to serological typing, the underlying basis for the RFLP clusters is very different. They are the products of a given haplotype cut by different restriction enzymes, and so must be correlated with the restriction maps of different haplotypes. Such maps must be obtained from a genomic analysis and the use of genomic clones. Most of the restriction fragment differences lie outside the coding regions and probably also the control regions, and so few, if any, of the restriction fragment differences are themselves likely to be of any functional relevance. It is through their association by linkage disequilibrium with functionally relevant variation which may not yet have been identified by other approaches, that the RFLP clusters will be of functional interest. Functionally relevant variation is eventually more likely to be identified by serology, by T cell clones or by the level of antigen expression and, in due course, by its relevance to immune response differences, disease associations, and transplantation.

The strong association, due to linkage disequilibrium, between certain RFLP clusters and serological types means that it may be possible to do a form of typing using RFLP analysis. However, it is important not to be circular in this analysis and not to extrapolate from the Caucasoids, who are so much studied, to other populations where the patterns of association between RFLP clusters and serological types may be quite different. Nevertheless, there are clearly situations where an RFLP typing can be very useful. This applies, for example, where the source of tissue, as may often happen, is not suitable for a serological analysis or when, as in the case of the DP typing, the cellular approach is technically demanding and limiting and the range of monoclonal antibodies available for serological analysis is limited. In that case, as Julia Bodmer, her colleagues, and others have shown, a combination of RFLP and monoclonal antibody typing can be very effective. Undoubtedly, in such a situation, the use of oligonucleotides specific to the sequence differences between expressed products will soon replace the more conventional RFLP analysis.

Having defined in great detail the patterns of RFLP variation in a central panel of cells in this Workshop, perhaps the next Workshop should be devoted to extending the range of variation studied to a wider collection of cells from different ethnic, disease, and family groups, and working with a smaller number of highly selected enzyme-probe combinations.

HgI (HLA-ABC) Serology

Serology is still, in many ways, the mainstay of the definition of new specificies because it can be so easily applied to the study of very large numbers of individuals. Thus, in this Workshop, between 10,000 and 15,000 cells were typed, and a tremendous job was done by

Edgar Milford and his colleagues in assembling this huge body of data. There are relatively few new splits in the A, B, and C serology, suggesting that we are beginning to reach a saturation point with respect to the definition of new HgI determinants. I suspect that a plot of the accumulation of new specificities against time would indicate that we are beginning to approach an asymptote, especially in terms of the amount of variability that is explained rather than the number of specific variants.

One-dimensional isoelectric focusing (IEF) gel electrophoresis, introduced into this Workshop for the first time, made a major contribution to the distinction between HgI variants at the biochemical level. The patterns of charge differences were clear and reproducible. The precise pI values for the variants can be calculated using a similar least-square approach to that used for calculating the molecular weights of RFLP fragments, with the simplification that in this case, the pI appears to be linearly related to migration distance over a wide range of values.

In spite of the clarity with which HgI variants are defined using the IEF technique, there is a need to be cautious in using the technique as a basis for assigning new names to specificities. The gene products are ultimately defined by the amino acid sequence, and this will only give rise to IEF variants if there are detectable charge differences. Two different products could easily have the same charge and not be distinguished by the IEF technique. Since sequencing at the DNA level is now such a straightforward procedure, the Nomenclature Committee decided that the naming of alleles should be based on known and confirmed sequence data, while retaining, for obvious reasons of convenience and continuity, the definition of specificities identified by serological and cellular techniques. Clearly, the biochemical approaches may identify new variations not yet seen using the serological or cellular techniques and these, when confirmed by sequencing, will take their place in the new nomenclature.

HgII (HLA-D) Serology

The HLA D region, HgII serology, has matured considerably since the last Workshop. Several new specificities have been defined, helped considerably by the availability of monoclonal antibodies that can be used reproducibly in conventional cytotoxicity assays. Most of the HgII serological determinants can now be defined with monoclonal antibodies, especially if one uses a combination of cytotoxicity and enzyme-linked immunosorbent assays (ELISA). This clearly points to the serology of the future.

Biochemical analysis of the HgII products using two-dimensional gel immunoelectrophoresis made a further significant contribution to the definition of these products and their variants. In the last Workshop, the main and important contribution of two-dimensional gel analysis was to the definition of the set of products. In this Workshop, following the dedication to detail and standardization of the techniques by Bob Knowles and his colleagues, the results were sufficiently reproducible to help in the definition of new variants that could be correlated with serology, T cell determinants, and amino acid sequence. Thus, the lack of variability of the DRB4 product, DRW53, was clearly seen as established by the constancy of the DNA sequence in different DR4-, 7-, and 9-bearing haplotypes. Intriguing variation in the products from DR5-bearing haplotypes was demonstrated, suggesting unexpected gene duplication or perhaps even formation of different products by differential mRNA splicing.

A particular achievement of this Workshop has been the combined analysis of the core cell lines by serological, cellular, biochemical, and molecular techniques. Thus, the target of the T-cell clones could be clearly identified by blocking with monoclonal antibodies, the specificities of which had been clearly established by serological and biochemical analysis.

Monoclonal Antibodies. Monoclonal antibodies were first mentioned 10 years ago in the Oxford Workshop with reference to the use of the now well known HgI monoclonal antibody W6/32 for biochemical studies, giving just a hint of the future of monoclonal antibody-based serology. In the next Workshop in 1980 in Los Angeles, there were two monoclonal antibodies contributed by Ruggero Ceppellini (who we were sorry to miss for the first time at this Workshop) and Massimo Trucco. By the time of the 9th Workshop in 1984, more than 50 monoclonal antibodies were studied using the cytotoxicity assay, but with the technical problems of carryover and other related issues not yet solved. In this Workshop, not only were there many more monoclonal antibodies studied using the cytotoxicity assay, but additionally a micro-ELISA technique developed by Genetic Systems was introduced. As already emphasized, many of the specificities, especially of the HLA D region, can now be well defined using monoclonal antibodies. An aim for the future must be to obtain a set of monoclonal antibodies that can cover the full range of specificities using either a cytotoxicity or a micro-ELISA technique. Current micro-ELISA techniques work well on cell lines and peripheral blood T cells, but still need a boost in sensitivity before they can be used reliably for typing peripheral blood B cells. Improving the sensitivity and reliability of micro-ELISA techniques for monoclonal antibodies should perhaps be another goal for the next Workshop.

Those who have been involved for some time in making monoclonal antibodies know the frustrations of getting good and specific reagents. Now, however, new and powerful techniques are becoming available which can focus the immune system of the mouse onto sequences of special interest. These depend on the use of transfectants, mainly of L cells or other mouse cell lines, namely cells into which have been inserted one or more defined human genes in such a way that they are expressed normally on the cell surface and thus can be used for serological, biochemical, and functional analyses. These transfected cell lines are extremely valuable reagents, not only for the characterization of monoclonal antibodies, but more especially for their production and for the analysis of function. This is because individual human products can be expressed on a mouse background in the absence, not only of other human products, but also of mouse products. Thus, in our laboratory, Austin et al. (1) showed that L-cell transfectants carrying the DP4 gene reacted as expected with monomorphic monoclonal antibodies and were able to present antigen in a functionally effective way. Subsequently, Heyes et al. (2) used these transfectants to make a polymorphic anti-DP monoclonal antibody, DP11, with specificity DPw2+DPw4, and which was used in this Workshop. By immunizing C3H mice, from which the L cells are derived, the only product seen by an immunized mouse should be that introduced by transfection, thus focusing the mouse's immune system on the transfected human product. More recently, we have shown that this antibody, though it reacts with DP α-chains, is not polymorphic on these chains when tested by Western blotting on the separated α-chain. This is consistent with the fact that DPw2 and DPw4 α-chains do not differ in sequence, whereas the corresponding β chains do (Young and Trowsdale, personal communication). The polymorphism detected by the DP11 antibody must therefore be with respect to a site on the α-chain that is only revealed or made accessible when this is complexed with the DPw2 or DPw4 β-chains (3). This is another example of the type of interaction between α- and β-chains discussed extensively by Ron Germain. Another polymorphic antibody, ILR1, made by Lee Nadler but which was not used in the Workshop, detects DPw2,3 and probably the CP63 determinant as well as DR5. With these two antibodies, four of the DP types can readily be identified serologically in most cases. Because of the limited differences in amino acid sequence between the DP β-chains, the specific amino acid most probably recognized by ILR1 can be identified as 58, which, in terms of the beautiful structure for A2 produced by Bjorkman, Wiley, Strominger, and others, is at the beginning of the α-helices. Similarly, more recent studies with DR7 expressing L-cell transfectants have clarified the difference between monoclonal antibodies that react with DR7 and with DRW53, and have identified monoclonal antibodies previously thought to be monomorphic as having the reactivity all but DR7 with respect to DR, while being monomorphic with respect to DP (4). These results confirm the common serological assumption that the same specificity may occur on quite different gene products. If there is linkage disequilibrium between the corresponding alleles, then there may be a considerable confusion between cross-reaction and association due to linkage disequilibrium. This is, for example, the situation with respect to the association between the reactivity of monoclonal antibody 17-3-3 and DRW53. This association must be due to linkage disequilibrium, since the 17-3-3 antibody reacts with the DR7 product in transfectants, but not with DRW53. Further interesting examples of the sharing of determinants between different gene products involving DRW52 and DR3 have been clearly described by Bernard Mach and his colleagues.

More specific focusing of the mouse immune response on to the polymorphic human determinant can be achieved by transfecting L cells not with the whole human gene product, but with human-mouse chimeric products. For example, if just a small part of a polymorphic human sequence is inserted into the appropriate position in the mouse molecule, then all the mouse immune system should detect is this short sequence against a mouse background. Structural homology between the products of the two species should ensure that the inserted human sequence takes up its correct configuration when replacing the corresponding sequence of the mouse product. This approach is now being used by Jill Maddox in Julia Bodmer's laboratory as the basis of a more directed approach to the production of monoclonal antibodies against particular polymorphic determinants.

A further interesting example of the sorting out of specificities using transfectants comes from the analysis by John Trowsdale and his colleagues (personal communications) of the two β-chains expressed from DR2-bearing haplotypes. These have been cloned separately from a homozygous cell line and individually transfected into mouse L cells. Though there are amino acid sequence differences between the two products, they are not distinguishable using any of the available monomorphic or polymorphic DR monoclonal antibodies. Nevertheless, they can be distinguished in functional studies, where it has been shown that L-cell transfectants expressing one of the chains can present a particular antigen to a T-cell clone, while the other cannot. The fact that L cells expressing only the DR product can function in this way, without a requirement for accessory molecules such as LFA1, suggests that the switching on of D region products by gamma-interferon in, for example, epithelial cells may be the first stage in antigen stimulation and immune response. Only at the second stage, when the response is amplified, do the highly specialized antigen-presenting cells, such as macrophages and dendritic and other histiocytic cells that have the full range of related products (such as LFA1) and so are able to function more efficiently, take over. The first stage of the immune response may be very important and account for the ability of gamma-interferon and other similar products to switch on HgII expression in virtually any cell type.

Definition and Origin of Serological Specificities

The original basis for the definition of a serological specificity was the coordinate reactivity of a set of sera with very closely related patterns of activity. Such serological specificities corresponded effectively to alleles at the various loci of the HLA region and defined the genetics of the HLA system. Every now and then, a new "split" was discovered, showing that what was previously thought to be a single specificity was in fact a mixture of two related specificities, sometimes perhaps one broader than the other. Such was the history, for example, of A2 and A28 and of the numerous splits of B15. Recall also the cross-reacting pairs, for example, the splits of B12, namely B44 which is BW4-associated, while B45 is BW6-associated. More than 20 years ago, we suggested that new alleles could be derived from old ones by recombination and that this could explain patterns of cross-reaction and associations with BW4 and BW6. These ideas came directly from R.A. Fisher's celebrated interpretation of the Rhesus blood groups. Now, the molecular genetics show the correctness of these ideas based on classical approaches.

The marvelous collections of sequence data can now provide the ultimate explanation of the serological patterns. This is most clearly shown by Peter Parham's analysis of the relationships between A2, A28, AW68, and AW69. What now really is A2, the good old Mach? In conventional polyclonal serology, every serum is a mixture of antibodies. If you mix a suitable set of monoclonal antibodies, you reconstruct a polyclonal antiserum. The reactivity of the antiserum corresponds to the lowest common denominator of the reactivity of its constituent antibodies. It is this common element, for example, the A2 in the mixture of monoclonal antibodies reacting to A2 and AW69, and A2 and B17, which defines the specificity of a serum. The averaging process is taken still further when clustering the reactions of several polyclonal antisera. One classic serological specificity can therefore be divided into many different combinations by the patterns of reactions of individual monoclonal antibodies.

Should we give the different epitopes corresponding to single monoclonal antibodies a separate name? An allele is best defined, as agreed by the Nomenclature Committee, by the amino acid sequence of its expressed product. The extent to which small regional differences in sequence are given a separate identity is almost a matter of individual choice and convenience. But the Nomenclature Committee has chosen not to name these regional epitopes and this is perhaps appropriate, as the *reductio ad absurdum* of this approach would be to name separately each amino acid in the sequence.

Defining an allele by the amino acid sequence of the expressed product may not be enough. There can be nucleotide base pair changes that do not give rise to amino acid substitutions. There may also be changes outside the coding region that influence levels and tissue patterns of expression. The Nomenclature Committee leaves open the possibility that such allele "splits" can be identified by adding further digits to the name, thus providing a hierarchical numbering that clearly identifies in common those alleles with the same amino acid sequence.

The secret of the difference between BW4 and BW6, the first alleles of the HLA system, is now revealed by their sequence. It is just a small difference in a part of the alpha-helix. The BW4, BW6 difference appears to be very old since, as shown many years ago, it can be picked up by polymorphic antibodies in other species. Why has this difference been maintained for such a long time? As already mentioned, there are many pairs of B locus alleles which cross-react and associate, respectively, with BW4 and BW6. One of these can be produced from the other by recombination, as already emphasized. Perhaps that small change at the end of a segment of the alpha-helix interacts with other changes to give functionally significant differences in immune response. The distinctive contribution of the BW4 and BW6 sequences in this respect may thus explain why they have been maintained together in populations for such a long time.

The amino acid sequence data of the HgI products assembled by Peter Parham give rise to a histogram of the frequency with which pairs of alleles differ with respect to a given number of amino acid substitutions. This distribution is trimodal. Cross-reacting combinations (such as A2 and A28) have the least difference, differences between alleles of the same locus come next, and finally come differences between alleles at different loci. Sequence data thus confirm the notion of cross-reacting groups, and of the characteristic "A-ness" or "B-ness" attached to the products of the A and B loci. This is in contrast to the situation in the mouse, where the distinction between K, D, and L is not so clear. It has always seemed to me that it is this difference, which leads to more cross-reactivity between the products of the loci in the mouse than in man, which made it more difficult to see the separation of the H2-K, D and L loci in the mouse than of the HLA A, B, and later C loci in man. the problem was accentuated by the use of inbred lines of mice, which give extreme patterns of linkage disequilibrium, in contrast to the outbred human population, in which linkage disequilibrium may be comparatively rapidly dissipated. If the HLA system had been worked out just in the North of Scotland, then it might have been very difficult to distinguish A1 from B8, as they are so often associated in that population. Indeed, that was a problem for workers in the United Kingdom in the initial distinction between A1 and B8, since their sera so often contained antibodies to both specificities. Polyclonal antisera are often likely to contain antibodies to products of different loci associated in the population by linkage disequilibrium.

There are many ways in which parts of the sequence from one gene can be transferred to another, in addition to recombination between homologous alleles. These mechanisms are often loosely called "gene conversion." This is a term that was, however, originally confined to transfer of sequences from one allele to its homologue in a heterozygote. The mechanisms that transfer sequences from an allele at one locus to one at another are non-homologous events, which could be analogous to a classic gene conversion or which could also, for example, be intrachromosomal unequal double cross-overs. Such transfers are, of course, the basis for the existence of serological determinants that are shared between different gene products. If the transfer is produced by an intrachromosomal exchange, then initially there is complete linkage disequilibrium between the presence of the specificities in the products of loci that may be adjacent or very close together on the same chromosome. The closer together the genes are, the more persistent will be the linkage disequilibrium generated by the initial event. As pointed out before, this is the probable origin of linkage disequilibrium between cross-reacting determinants on different locus products, giving rise to a serological confusion that may not be resolved until the products are studied separately in transfectants against a neutral background.

There is no reason to suppose that linkage disequilibrium is constrained by the boundaries between genes. It is largely determined by the position of recombinational hot spots, and thus the association of serological determinants within a gene is as much linkage disequilibrium as that between genes, though the former is naturally in general expected to be much stronger.

Allelic Combinations and Conformational Effects

It has been clearly shown by Ron Germain that different allelic combinations in the mouse may vary in their efficiency of association and that this could be due to conformational effects connected with different specific interactions between particular α- and β-chains. This possibility was predicted at the time of the last Workshop, following the discovery of considerable polymorphism for both the DQ α- and β-chains. This led to the suggestion that the association of insulin-dependent diabetes with DR3/DR4 heterozygotes may actually be due to associations between DQ α- and β-chains determined by DQ alleles associated by linkage disequilibrium with DR3 and DR4. Indeed, that was the basis for the initial suggestion that the IDDM association may in fact be with DQ rather than DR, as is now widely believed to be the case. A double heterozygote for DQ α and β variation can form two parental combinations of products and two cross-combinations, namely involving DQ α, for example, determined by one chromosome and DQ β determined by the homologous chromosome. The latter may give rise to novel determinants, specific to the heterozygotes, as has often been pointed out. What has been shown by Ron Germain is that in some cases in the mouse at least, these cross-combinations do not function very well. The complement to that is that the determinants that do function well together may be held together on the same chromosome by linkage disequilibrium and may have been maintained in that way by natural selection. That may, of course, account for strong linkage disequilibrium between particular DQ α- and β-chain determinants. For example, some people have argued that the DQW1 specificity is in the α-chain, while others suggest it is in the β-chain. The answer could well be that it is in both. Polyclonal antisera may contain mixtures of antibodies, some of which react against determinants coded for by the DQA1 gene on a DQW1 haplotype, while others react with the product of the DQB1 gene. Thus, one of the best monoclonal antibodies produced in Julia Bodmer's laboratory to DQW1 clearly reacts in Western blots with the β-chain, while others have suggested that their monoclonal antibodies may react with the DQ α-chain.

Structure and Function of the HLA Products and Peptide Binding

The most frequently shown picture at the Symposium following the Workshop was the beautiful three-dimensional structure of the A2 product established by Bjorkman, Wiley, Strominger, and their colleagues. Now, this provides the basis not only for predicting the structure of HgI, but also HgII products, and relating these structures to their function in interacting with the T-cell receptor and antigen. My interpretation of this interaction is simply that the HLA products cannot interact with the T receptor unless antigen is in its place attached to the T-receptor idiotype where it belongs. Then, with this triple interaction between T receptor, antigen, and HLA product established, effecting of function, such as killing or stimulation to divide, can take place. The logic of this implies that association between the T receptor and its antigen must implicitly lead to a conformation change in the T receptor that enables the binding between HLA product and T receptor to be effective. This is, of course, entirely analogous to the situation for antigen-antibody-complement complexes. Complement only associates effectively with antibody when antibody and antigen are associated.

How does this triple interaction relate to the interesting and exciting ideas, now widely supported, on the role of specific binding of peptides, the products of antigen processing, to HLA products? The dominant theme is that peptide binding to the HLA product is a *necessary* part of the immune response, and that is a dogma which I believe should be seriously questioned. The issue is not whether peptide interactions with HLA products happen, nor whether they may be important for particular immune responses, especially those with a genetic basis. The question is: Are they a normal and necessary part of the immune response? I would like to raise three major problems for this dogma.

The first is that the ultimate specificity for antigen recognition must lie with the T receptor. The T receptor has the somatic variability, with all the complex and elegant mechanisms that have evolved for its generation. Specificity is equivalent to affinity. Thus, ultimately the higher affinity has to be with the T receptor rather than with HLA product, otherwise there would be no need for the T receptor. Of course, the required affinity may be achieved by the triple combination of T-receptor antigen and HLA product. If that is the case, however, it does seem odd that the primary association of all antigens is with the molecule that must ultimately have the much lower affinity with antigens.

The second point is that it seems to be commonly assumed that T cells can only see peptides, namely only recognize processed proteins. But there are clear indications in the literature that T cells can recognize conformational determinants and nonprotein haptens. Surely, they must be able to recognize sugar differences, and I am sure this will be found if enough care and effort are put into the search.

The third point, which may be the most critical from a practical point of view, is that I believe that the in vitro cellular assays are incomplete. The majority of antigen presentation most probably, especially if the antigen is processed, takes place in the lymph nodes. The late Bill Ford, who was both an anatomist and an immunologist, wrote very effectively on the role of the lymphoid anatomy. He said "the lymph node is the control post for the immune system. . . . , it's a site of confluence of lymph-borne antigen and blood-borne lymphocytes." The anatomy of the lymphoid system must be taken into account when analyzing the immune response. It cannot be forgotten that antigen presentation in lymph nodes involves other cells, such as dendritic and endothelial cells, and interactions with the extracellular matrix and other products of other cells that contribute to the effectiveness of cellular interactions. If these other interactions are left out, as they are in all current in vitro cellular immunoassays, these assays may be intrinsically insensitive. They may identify only high avidity cellular interactions, which, in the absence of the essential third component, may be particularly those interactions in which peptide does bind to the HLA product. These may be the situations in which HLA variation often does lead to immune response differences. This gives a very different view of the nature of the immune response in which peptide interaction with the HLA product is not necessary, but may certainly sometimes be important. The challenge is to develop assays that pick up a wider range of responses than what I believe at the moment may be being identified, namely just the tip of the avidity iceberg in T-cell interactions. Perhaps cellular immunoassays should be done in lymph node skeletons that parallel the thymus skeletons, which have been pioneered by John Owen for the study of thymus ontogeny in relation to lymphocyte function. Such an approach may give the increased sensitivity, which could show up a wide range of unexpected T-cell responses. It is interesting that when Peter Parham changed an extremely conservative amino acid in the HgI "RFDS" sequence, it had no effect on the in vitro T cell response. Perhaps that sequence is somehow involved in binding to the missing component from current in vitro cellular assays.

Such interactions must, of course, also play a role in the function of the thymus and thymic education. Is the interaction between T cells in the thymus and epithelial or other cellular components mainly to throw out T cells carrying receptor sequences that do not function properly with self Hg products? It seems unlikely that interactions in the thymus are mainly for the development of tolerance. That may happen anywhere in the body where immune responses can take place. Are there, for example, particular LFA-ICAM like interactions specific to the thymus that mediate the sort of triggering that determines that only functionally effective T receptor-carrying T lymphocytes are passed into the general circulation?

Are we now finally beginning to understand the significance of antigen processing? A cell seems to leave its decomposed garbage on its face, or its behind, depending on your viewpoint. Some of the breakdown products of disposed proteins simply stick to the outside surface of the cell. This gives the outside surface a signature of its internal components, displayed for the immune system to see. That is why, as shown by Alain Townsend and others, T cells can respond to internal cellular proteins such as the proteins of the flu core. This very important result means that it is very hard for a virus to hide from the immune system inside a cell by not giving rise to products that are expressed on the surface, because its internal products will be recognized from their decomposed remnants on the outside of the cell. This view of T-cell recognition enormously widens the range of potential T-cell responses to cellular products. That may be particularly important in T-cell responses to tumors, where internal changes such as in the ras and myc oncogenes may be recognized by T cells and contribute, admittedly perhaps at a late stage in tumor progression, to T-cell responses to tumors.

Disease Associations

Narcolepsy is now the most striking association between HLA and disease. I have been told that a way to distinguish ordinary sleepiness from narcolepsy is to point out that it is the difference between the audience being asleep having been bored by the lecturer, as compared with the lecturer going to sleep. Disease associations were not a major focus in this Workshop. Outside it, and as reported in the Symposium, there has been some excitement in the amino acid 57 position change identified by Hugh McDevitt, his colleagues, and others. The picture is, however, not entirely clear since, for example, when DR1 is involved in IDDM it appears to be only in DR4/DR1 heterozygotes (6). Perhaps that amino acid position change is necessary, but not sufficient for the disease association. To determine the complete functional basis for a disease association, one needs appropriate animal models. These may now be available, since it has been shown that one can create models for human autoimmune disease in transgenic mice by humanizing the mouse's immune system. Eventually, we should be able to satisfy the equivalent of Koch's postulates for the role of a particular sequence by showing that when the sequence is there, it gives rise to the disease in the model, and not otherwise.

Vaccination programs should be monitored with respect to immune response differences. This applies particularly to immunization programs for the hepatitis virus and HIV, once these are started. Immune response differences may, after all, answer the question as to the basis for differences in the expression of the disease in infected individuals. Perhaps for a future Workshop, there are areas where collaboration is appropriate. No doubt treatments will be developed using monoclonal antibodies to the HLA products. There are already good mouse models and these must be transferred to primate systems where the anti-human reagents that will be used in people can be tested. Some disease variation, as in forms of the "bare lymphocyte syndrome," may be connected with defects in the regulation of expression. At the cellular level, such defects may be relevant to stages of cancer progression when cells are selected, because of immune responses, to lose their expression of HgI or HgII determinants.

The Future

The fundamental challenge for the future is clearly to understand function in relation to structure, but such analysis does not fit easily into the framework of collaboration in an international workshop. There remain, however, many very practical aims that do provide a good rationale for continued collaboration in the definition of variability, specificity, and function in the HLA region in different populations. The importance of HLA matching for transplantation was again emphasized by Jean Dausset and Paul Terasaki. Notwithstanding the successful use of cyclosporin A, matching does matter, not only for kidneys but also for hearts, livers, eyes, and bone marrow. We need to know whether DP matching matters, or whether A, B, C, DR- and DQ-matching, which is easier because of linkage disequilibrium, is enough. How important are minor differences between cross-reacting determinants? If unrelated bone marrow grafts are going to work, much collaboration is needed to find suitable matched types. The search for the best way to match, therefore, remains an important component of future Workshop collaboration.

We should be very grateful that our Japanese colleagues have accepted the challenge of the next Workshop. We must thank them very much and welcome their decision. During the coming months and years, they will be seeking advice from all of us on the direction for the future, and this must be suggested by the analysis of this Workshop and Conference. The definition of specificity is still paramount. Monoclonal antibodies, especially using ELISA assays, are likely to remain the most convenient technical option, though perhaps the technology of DNA typing using oligonucleotides will improve to be competitive. The practical applications of transplantation, as already emphasized, also need collaboration. Perhaps, in addition, we should return to population studies and cross the population bridge from Europe via the Soviet Union and China, which makes such studies possible. Then we can look at the apparent intriguing gradient of European type specificities from North to South in China. Is that telling us that the Mongols are after all not Mongoloids, but Caucasoids? The functional studies, as already pointed out, will proceed apace, but cannot really be an effective part of the Workshop. What is a relevant part of the Workshop for function is to accumulate the data base of sequences and correlate that information with the serology.

As we approach the end of this Workshop, we have clearly passed another milestone in the development of the HLA system. This Workshop has had major new components which have provided an enormous challenge that has been met. It has produced those most valuable reagents, the core cell lines, which must now be the best typed cells in the world. It has produced standards for electrophoresis and restriction fragment length polymorphism analysis and has defined new antigens. All this has been a mammoth task for which we must thank Bo Dupont and his colleagues, Edgar, Edmond, Neal, Maryann, Sally, Soo Young, Daniel, Bob, John, Janet and many others for making it a success.

In the tradition of previous Workshops that I have summarized, I finish with a poem by Julia Bodmer, which is a summary of my summary (it can be sung to the Eton Boat Song).

> Where are we now, Mac,
> Thirty years on we forgetfully wonder
> What it was like to define HLA
> Three drops of white cells and three drops of serum
> Gave Lukes and Gillespie 4B and 4A.
> Ten Workshops later, we really are moving
> We've sequenced and crystallized much HLA
> We can open the molecule and view many epitopes
> And guess at their function, by position and lay
> So armed with this knowledge we go to the battle
> With AIDS and diseases which plague us today
> Match grafts where they matter and alter receptors
> And change the expression of our old HLA.

References*

1. Austin P, Trowsdale J, Rudd C, Bodmer W, Feldmann M, Lamb J. Functional expression of HLA-DP genes transfected into mouse fibroblasts. Nature 1985;313(5997):61–64.
2. Heyes J, Austin P, Bodmer J, et al. Monoclonal antibodies to HLA-DP transfected mouse L cells. Proc Natl Acad Sci USA 1986;83(10):3417–3412.
3. Bodmer J, Bodmer W, Heyes J, et al. Identification of HLA-DP polymorphism with DPα and β probes and monoclonal antibodies—correlation with PLT defined types. Proc Natl Acad Sci USA 1987;84:4596–4600.
4. Young JAT, Wilkinson D, Bodmer WF, Trowsdale J. Sequence and evolution of HLA-DR7 and DRw53 associated β chain genes. Proc Natl Acad Sci USA 1987;84:4929–4933.
5. Wilkinson D, De Vries RRP, Madrigal A, et al. Analysis of HLA-DR glycoproteins by DNA-mediated gene transfer. J Exp Med 1988;167:1442–1458.
6. Winearls B, Bodmer J, Bodmer WF, et al. A family study of the association between insulin dependent diabetes mellitus, autoantibodies and the HLA system. Tissue Antigens 1984; 24:234–246.

Author Affiliation

Imperial Cancer Research Fund, Lincoln's Inn Fields, London, England

---

*The major references for this summary consist of all the other papers in this Immunobiology of HLA, Volume II: Immunogenetics and Histocompatibility (this volume). Only those specific to work from our laboratory are listed here.

HLA-A,B,C: Patterns of Polymorphism in Peptide-Binding Proteins

P. Parham, D.A. Lawlor, R.D. Salter, C.E. Lomen, P.J. Bjorkman, and P.D. Ennis

Specific Immune Recognition: The Prevailing Paradigms

It is now over 50 years since Gorer's discovery of H-2 (1,2) and 30 years since Dausset's description of the first HLA specificity – HLA-A2 (3). Just prior to this conference, that same molecule provided the first three-dimensional structure of a transplantation antigen (4,5), enabling the synthesis and rationalization of a large body of information. The same period also witnessed the passing of Peter Medawar (6) and John Humphrey (7), pioneers in transplantation immunology and the genetics of immune responsiveness.

Developments in this field have commonly been made through the juxtaposition of contrasting results or hypotheses derived from pairs of apparently similar systems. One pair is simply that of species, man or mouse, HLA or H-2, and their similarities and differences. Then there are class I and class II antigens, helper T cells versus cytotoxic T cells, CD4 and CD8. The debates these engender and the differences they create may not be fortuitous since there is a need to simplify complex questions into simple either/or hypotheses. Well designed experiments, we are told, are those that test a hypothesis, give a simple yes no answer, and distinguish between two alternatives.

A classic confrontation has been between humoral and cellular immunology. Silverstein (8) has argued that much of the debate arose from the political animosities of the late 19th century. The immunologic product of this competition is still with us. Long after it has become clear that cells are the basis for all immune responses, tired texts continue to embrace their readers with humoral and cellular arms of the immune response, thus paying homage to the Franco-Prussian War.

Despite such forces of reaction, a consensus concerning the molecular nature of specific immune recognition has been formed (9–12). The basis for immune specificity lies with the two classes of lymphocyte – B cells and T cells – each expressing a variable clonal receptor. Triggering of these receptors by foreign antigen initiates the immune response. In the case of the B cell, this receptor is a membrane-bound form of the antibody the cell is destined to secrete (13).

For T cells, there is no antigen-specific secreted product and the receptor is the once elusive but now familiar T-cell receptor (a heterodimeric protein formally resembling a membrane-associated Fab fragment) (11,14–16). Although the structure and genetic basis for diversity in antibodies and T-cell receptors are similar, there is a fundamental difference in their interaction with antigens. Antibodies can interact with conformational or sequential determinants of peptides or proteins and are not dependent on other components for stabilization of these interactions. In contrast, T-cell receptors only interact with sequential determinants. These have been generally found on short peptides of 10 to 20 amino acids in length, resulting from degradation or processing of proteins (9,10,12,17–19). The interaction of antigen with the T-cell receptor takes place between two membranes, one containing the receptor and the other containing the antigenic peptide bound to a glycoprotein encoded by the major histocompatibility complex (MHC). The MHC molecule is said to present antigen to the T cell. The requirement of T cells to see antigen bound to, "presented," or "in the context of" MHC molecules is known as MHC restriction, and the MHC molecule involved in response to a particular peptide or epitope is called the restriction element (20).

The MHC glycoproteins come in two varieties – class I and class II – and much of this two-volume publication is dedicated to describing the details of their differences and polymorphism. Despite many distinguishing features, these two classes of molecule have much in common including a structure with two immunoglobulin-like and two polymorphic domains (21,22). Modeling of class II primary structures into the three-dimensional structure of HLA-A2 suggests that the basic folding pattern for class I and II will be the same (4). The force of this argument based on structural similarity is strengthened by the recent appreciation that class I and class II restriction both involve the binding and presentation of processed proteins to T lymphocytes (23).

These cells also come in two varieties. Class I-restricted T cells express the CD8 glycoprotein and are mostly of the cytotoxic phenotype. Class II present to T cells expressing the CD4 glycoprotein and are generally of the helper variety (24). The correlation of class I with CD8 and class II with CD4 led to the hypothesis that these pairs of molecules interact with each other (25). Binding studies between transfected cells that over-express CD4 and target cells expressing class II molecules demonstrate this to be the case for the class II/CD4 pair (26).

Although we have stated that MHC glycoproteins are peptide-binding proteins, this has only been directly demonstrated for class II molecules (18). However, two lines of evidence argue the case for class I. The first comes from experiments in which cells incubated with peptides can then be lysed by cytotoxic T cells specific for an epitope contained in the peptide (28). The interpretation of these experiments being that peptide binds to the appropriate class I molecule, thus forming the complex recognized by the T cell. A second set of arguments derives from the crystal structure of HLA-A2 (4,5). The α_1 and α_2 domains combine to form a structure that is consistent with a peptide-binding function (Fig. 1). Two α helices form the sides of a long deep

Immunobiology of HLA
Volume II

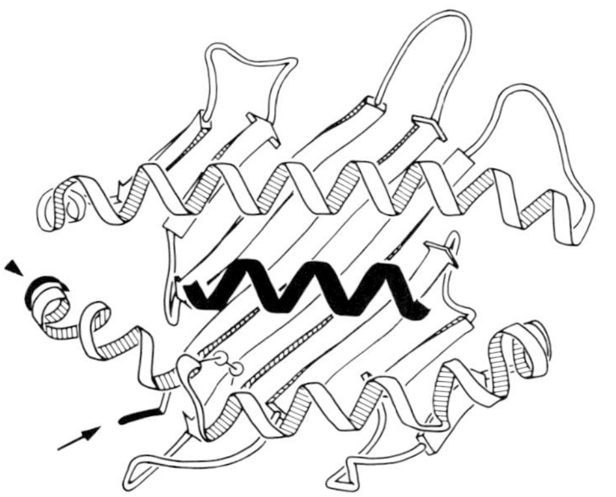

Figure 1. A schematic view of the α_1 and α_2 domains of HLA-A2 as seen from the viewpoint of a T-cell receptor. The β strands are shown as broad arrows pointing from amino-terminus to carboxy-terminus. The α helices are shown as helical ribbons. The amino terminus of the α_1 domain is the darkened region at the bottom of the figure (indicated by an arrow) and the carboxy-terminus of the α_2 domain is the darkened region at the left of the figure (indicated by an arrowhead). The four β strands to the left and the top α helix are from α_1; the four β strands to the right and the bottom α helix are from α_2. The darkened helix at the center indicates how an α-helical peptide of 15 amino acids might sit in the peptide binding site. Adapted from Figure 2B of Bjorkman et al. (4). The full color illustration of this figure appears in Color Plate I at the beginning of this volume.

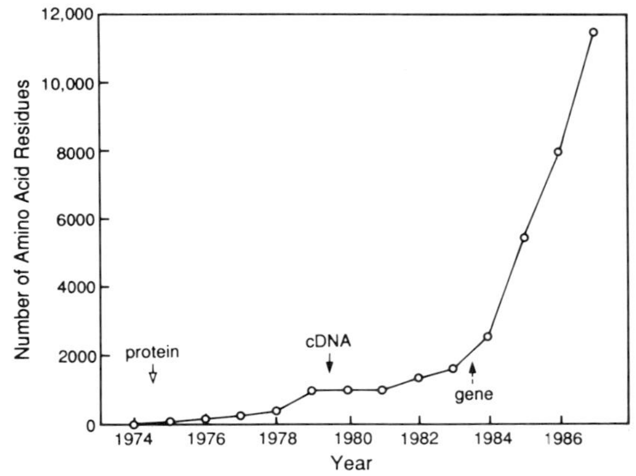

Figure 2. The accumulation of protein sequence information for HLA-A,B,C molecules. Results from direct determination and the predicted protein sequences from nucleotide sequences have been combined. Only nucleotide sequences for which the gene product is known have been included. The approximate times at which protein, cDNA, and genomic sequencing were first reported are indicated by arrows. Multiple independent sequences of the same allele are included only once.

groove, the floor of which is formed by a plane of antiparallel β pleated sheet. The dimensions of this groove are 25×10 Å, sufficient to accommodate peptides in α helical or more extended structures. Of critical importance was the discovery of material within this site that could not be accounted for by the known amino acid sequences of the HLA-A2 heavy chain (29) and β_2-microglobulin (30). The structure of this material was ill-defined, suggesting it represents a heterogeneous mixture of bound peptides.

Why do these peptide-binding proteins cause rejection of transplanted tissues? Two properties of class I molecules contribute to their role as transplantation antigens. First is their central involvement in forming the ligands that stimulate and act as targets for cytotoxic T lymphocytes, cells that are involved in graft rejection. Second is their high polymorphism, which ensures that unrelated individuals will invariably differ in their class I molecules.

The differences between the class I molecules are powerful stimulators of cytotoxic T cells and this can be explained in terms of the structure. The geometry and topology of the peptide-binding groove of HLA-A2 indicate that when the functional complex between peptide, MHC, and T-cell receptor is formed, the combining site of the T-cell receptor will interact extensively with the helices of the α_1 and α_2 domains of the MHC molecule as well as with the bound peptide (5). Many of the differences between class I molecules arise from amino acid substitutions at residues within the helices that may affect interactions with T-cell receptors. The T cells in an individual have been selected to both interact with, and yet be tolerant to, the class I antigens of self. To those T cells, the class I antigens of another individual may resemble the combination of self class I binding a foreign antigenic peptide. The allogeneic class I molecule is thus sufficiently similar to self to act as a restriction element, yet has enough differences to also be seen as foreign. This combination of factors may explain the power of the allogeneic response (31).

Determination of Primary Structures for HLA-A,B,C

Class I MHC molecules were first described in mice as a series of polymorphic cell surface antigens encoded by a single genetic locus (H-2) and providing the major immunologic barrier to transplantation (1,32). The principal method used to describe HLA, the human counterpart, has been that of serology, commencing with the studies of Dausset, Payne, van Rood, and their colleagues (3,33,34). Over the years and through the agency of the International Workshops, the polymorphism of class I HLA molecules has been defined. It is built up at two levels. First there are three different loci; HLA-A,B,C; second there are multiple alleles at each locus. After the Ninth Workshop, 19 distinct antigenic specificities were defined at the HLA-A locus, 37 at the HLA-B locus, and 8 at the HLA-C locus (35). These numbers underestimate the total number of alleles because many molecules that type identically with alloantisera are distinguished on the basis of other assays (36–41) .

Once the unusual character of MHC polymorphism was appreciated, an understanding of its chemical and

```
α1 domain          ▼              ▼                    ▼                ▼  ▼▼▼ ▼▼▼  ▼ ▼▼  ▼
           1        10        20        30        40        50        60        70        80        90
consensus  GSHSMRYFYTSVSRPGRGEPRFIAVGYVDDTQFVRFDSDAASPRMEPRAPWIEQEGPEYWDRETQIVKAQSQTDRESLRTLRGYYNQSEA
A2.1       --------F---------------------------------Q------------------G--RK---H---H-VD-G-----------
A2.2Y      ------------------------------------------R------------------G--RK---H---H-VD-G-----------
A2.2F      --------F---------------------------------R------------------G--RK---H---H-VD-G-----------
A2.3       --------F---------------------------------Q------------------G--RK---H---H-VD-G-----------
A2.4a      ------------------------------------------Q------------------G--RK---H---H-VD-G-----------
A2.4b      --------F---------------------------------Q------------------G--RK---H---H-VD-G-----------
Aw69       ------------------------------------------Q-------------------N-RN---------VD-G-----------
Aw68.2     -----------M------------------------------Q-------------------N-RN---------VD-G-----------
Aw68.1     ------------------------------------------Q-------------------N-RN---------VD-G-----------
A3.1       --------F---------------------------------Q------------------Q--RN---------VD-G-----------
A3.2       --------F---------------------------------Q------------------Q--RN---------VD-G-----------
A11        ------------------------------------------Q------------------Q--RN---------VD-G----------D
A1         --------F---------------------------------QK-----------------Q--RNM--H-----AN-G----------D
A32        --------F---------------------------------Q------------------Q--RN---H---------IALR-------
Aw24       --------S---------------------------------Q------------------E--GK---H------N--IALR-------
Bw58       ----------AM--------------------------------T----------------G--RNM--SA--Y--N--IALR-------
B27.1      --------H--------------T-------L------------E---------------------C--KA-----D----LR-------
B27.2      --------H--------------T-------L------------E---------------------C--KA-----N--IALR-------
B27.3      --------H--------------T-------L------------E---------------------C--KA----------LR-------
B27.4      --------H--------------T-------L------------E---------------------C--KA----------LR-------
B27f       --------H--------------T-------L------------E---------------------C--KA--Y--N---ALR-------
B44.1      ----------AM-----------T-------L--------T---K---------------------S-TNT--Y--N---AAR-------
B44.2      ----------AM-----------T-------L--------T---K---------------------S-TNT--Y--N---ALR-------
B13        ----------AM-----------T-----------------T----A--------------------S-TNT--Y--N---ALR-------
Bw47       ----------AM-----------T-------L--------T---K---------------------S-TNT--Y--D----LR-------
Bw65       ----------A------------S---------------------E-----------------N---C-TNT--------N----------
B14        -----------------------S---------------------E-----------------N---C-TNT--------N----------
B18        --------H--------------S-----G--------------T-----------------N---S-TNT--Y-----N----------
B40*       --------H--------------T-------L--------T---K---------------------S-TNT--Y-----N----------
Bw41       --------H-AM-----------T-------L--------T---K---------------------S-TNT--Y-----N----------
Bw60       --------H-AM-----------T-------L--------T---K---------------------S-TNT--Y-----N----------
B8         --------D-AM-----------S---------------------E-----------------N---F-TNT--------N----------
Bw42       -----------------------S---------------------E-----------------N---Y---A--------N----------
B7.1       -----------------------S---------------------E-----------------N---Y---A--------N----------
B7.2       -----------------------S---------------------E-----------------N---Y---A--------N----------
Cw1        C----K--F--------------S---------------------G------V-------------KY-R-A----V---N----------
Cw2.1      C---------A----S----H-----------------------G---GR-V-------------KY-R-A----VN--K----------
Cw2.2      C---------A----S----H-----------------------G------V-------------KY-R-A----VN--K----------
Cw3        --------C-A---------H-------------------DE---G------V-RK----------KY-P-A----V---N----------
consensus  GSHSMRYFYTSVSRPGRGEPRFIAVGYVDDTQFVRFDSDAASPRMEPRAPWIEQEGPEYWDRETQIVKAQSQTDRESLRTLRGYYNQSEA
           1        10        20        30        40        50        60        70        80        90
```

Figure 3. The protein sequence of 15 HLA-A, 20 HLA-B, and 4 HLA-C molecules in the α_1 domain (this page), the α_2 domain (page 13), and the α_3 domain (page 14). Each sequence is compared with a consensus constructed from the most frequent amino acid at each position. Dashes indicate identity with the consensus. The residue at position 116 in HLA-B7.2 is unknown, but is different from Y found in HLA-B7.1 (5). It is therefore indicated with an X. The figure is derived from Figure 1 of Parham et al. (81).

Residues with high variability (see Fig. 4) are given by ▼ Locus-specific residues are colored yellow. Residues found in only one molecule or subtypes of a family are colored pink, those shared by small groups of distinct molecules are colored blue. These pink and blue residues with restricted distribution are good candidates for involvement in the highly specific epitopes seen by the "best" typing reagents. They are colored to aid in further prediction of alloantigenic epitopes. Residues critical for the Bw4 and Bw6 epitopes are colored green. The actual color figures appear in Color Plates I to III at the beginning of this volume.

genetic basis was sought. This necessitated determination of the primary structures of the HLA-A,B,C heavy chain glycoproteins, and preferably of the genes that encode them.

In 1975, the amino terminal residues of three preparations of papain-solubilized HLA antigens were determined and shown to be glycine (42). This initiated the determination of the primary structure of HLA-A,B,C molecules, the progress of which is shown in Fig. 2. Amino-terminal sequences were determined in a number of laboratories using a variety of innovative methodologies (43–46). However, it was conventional protein chemistry of mass-isolated, papain-solubilized material that pushed the analysis beyond this initial stage. By 1979, sequences for the extracellular domains (α_1, α_2, and α_3) of two HLA preparations were complete (47,48). One was of HLA-B7 from the JY lymphoid cell line (47) and the other was of a mixture, "comprising a multitude of allelic forms," isolated from pooled spleens and peripheral blood lymphocytes (48). Subsequently, this approach was also used to determine a complete sequence of HLA-B27 (49) and most of the sequence for HLA-A2 (50), -A28 (51), and -Bw60 (52).

In choosing pairs of serologically cross-reactive molecules (A2/A28 and B7/Bw60) for study, it was hoped that the number of substitutions would be small and

```
α2 domain

                  ▼ ▼                ▼ ▼                                      ▼      ▼
                   100       110       120       130       140       150       160       170       180
consensus  GSHTLQRMYGCDVGPDGRLLRGYHQYAYDGKDYIALNEDLRSWTAADTAAQITQRKWEAARVAEQLRAYLEGTCVEWLRRYLENGKETLQRA
A2.1       ----V---------S-W-F-----------------K----------M---T-KH-----H------------------------------T
A2.2Y      --------------S-W-F-----------------K----------M---T-KH-----H----W-------------------------T
A2.2F      --------------S-W-F-----------------K----------M---T-KH-----H----W-------------------------T
A2.3       ----V---------S-W-F-----------------K----------M---T-KH---T-HE---W-------------------------T
A2.4a      ----V---------S-W-F-----------------K----------M---T-KH-----H------------------------------T
A2.4b      ----V---C-----S-W-F-----------------K----------M---T-KH-----H------------------------------T
Aw69       ----V---------S-W-F-----------------K----------M---T-KH-----H------------------------------T
Aw68.2     ----I-------------F-----------------K----------M---T-KH-----H----W-------------------------T
Aw68.1     ----I-M-------S---F----R-D----------K----------M---T-KH-----H----W-------------------------T
A3.1       ----I-I-------S---F----R-D---------------------M-----K------HE--------D--------------------T
A3.2       ----I-I-------S---F----R-D---------------------M-----K------H----Q----D--------------------T
A11        ----I-I-----------F----R-D---------------------M-----K------HA---Q------R------------------T
A1         ----I-I-----------F----R-D---------------------M-----K-----VHA---R-V----R--DG--------------T
A32        ----I-M----------------Q-D---------------------M-------------------------------------------T
Aw24       ------M-F-----S---F-----------------K----------M-----K------H----Q---------DG---------------
Bw58       ---II-------L---------HD-S--------------S-------------------------------L-------------------
B27.1      ------N------------------D--------------S-------------------------------E-------------------
B27.2      ------N------------------D--------------S-------------------------------E-------------------
B27.3      ------N------------------D--------------S--------------------E----------E-------------------
B27.4      ------N----------------D----------------S--------------------E----------E-------------------
B27f       ------N------------------D--------------S-------------------------------E-------------------
B44.1      ---II------------------D-D--------------S------------------------D------L---S---------------
B44.2      ---II------------------D-D--------------S------------------------D------L---S---------------
B13        ----W-T-----L---------HN-L--------------S-------------L-----------------E-------------------
Bw47       --------F----------------D--------------S-------------------------------E-------------------
Bw65       ------W----------------N-F--------------S--------------------E------------------H-----------
B14        ------W----------------N-F--------------S--------------------E------------------H-----------
B18        ----------------------HD-S--------------S---------------------------------------H-----------
B40*       ------S---------------HN------------------------------------------------E-------------------
Bw41       ----W-----------------HN-----------------------------------------D--------------------D--E--
Bw60       ----------------------HN----------------------------S---L---------------E-------------DK-E--
B8         ------S---------------HN-----------------------------------------D--------------------D--E--
Bw42       ------S---------------HN-----------------------------------------D--------------------D--E--
B7.1       ------S---------------HD-------------------------------------E---R------E-------------DK-E--
B7.2       ------S---------------HD-X-----------------------------------E---R------E-------------DK-E--
Cw1        ------W-C---L----------D-------------------------------------E---R---------------------S----
Cw2.1      ------------L----------D-S-----------------------------------E---W------E-------------K----
Cw2.2      ------------L----------D-S-----------------------------------E---W------E-------------------
Cw3        ---II------------------D-H--------------------N--------------E----------L---------K-------G-
consensus  GSHTLQRMYGCDVGPDGRLLRGYHQYAYDGKDYIALNEDLRSWTAADTAAQITQRKWEAARVAEQLRAYLEGTCVEWLRRYLENGKETLQRA
                   100       110       120       130       140       150       160       170       180
```

Figure 3. *(Continued).*

would allow precise identification of alloantigenic epitopes. This reasoning proved false, with 13 substitutions between A2 and A28 and 19 between B7 and Bw60. However, more closely related groups of molecules became available with the discovery that serologically identical HLA-A,B,C molecules could be subdivided by cytotoxic T cells (36–39). The existence of subtypes of HLA-A,B,C specificities has proved quite general, although most extensively documented for HLA-A2 and -B27. Once the amino acid sequence for one subtype was complete, it became possible to determine the structures of other subtypes by comparative peptide mapping and sequencing of the difference peptides (53–56).

For examination of the genetics and evolution of HLA-A,B,C polymorphism, the protein approach is limited in that it is relatively slow and does not give direct information on the genetic material. With the development of recombinant DNA methods, it became possible to take a direct approach. The first cDNA clones were isolated in 1980 (57,58). However, the first sequence of a functionally identified molecule was not published until 4 years later (59,60). The reasons for this hiatus, which can be seen in Fig. 2, are as follows. First, the human MHC was found to contain many homologous genes that cross-hybridized with HLA-A,B,C (61,62). Thus, it became imperative to express isolated DNA clones and determine by serology the identity of the protein products. This proved difficult with cDNAs clones, many of which were incomplete, and the study of Arnot et al. (63) revealed the potential for confusion. Attention subsequently turned to the isolation of complete genomic clones. Jordan and colleagues first analyzed HLA-Cw3 (59) and HLA-A3 (60) genes, which were identified on the basis of transfection and absorption analysis with alloantisera (64). Koller and Orr used deletion mutants of the LCL.721 cell line to identify the HLA-A2.1 gene, a result confirmed by transfection and analysis with monoclonal antibodies (29). As the number of gene sequences increased, it became possible to refine procedures. Locus-specific probes were used to map genes

```
α3 domain

                  190       200       210       220       230       240       250       260       270
 consensus DPPKTHVTHHPISDHEATLRCWALGFYPAEITLTWQRDGEDQTQDTELVETRPAGDRTFQKWAAVVVPSGEEQRYTCHVQHEGLPKPLTLRW
      A2.1 -A----M---AV------------S------------------------------G---------------Q--------------------
     A2.2Y -A----M---AV------------S------------------------------G---------------Q--------------------
     A2.2F -A----M---AV------------S------------------------------G---------------Q--------------------
      A2.3 -A----M---AV------------S------------------------------G---------------Q--------------------
     A2.4a -A----M---AV------------S------------------------------G---------------Q--------------------
     A2.4b -A----M---AV------------S------------------------------G---------------Q--------------------
      Aw69 -A----M---AV------------S------------------------------G---------------Q--------------------
    Aw68.2 -A----M---AV------------S------------------------------G------V--------Q--------------------
    Aw68.1 -A----M---AV------------S------------------------------G------V--------Q--------------------
      A3.1 ------M------------------------------------------------G------------------------------------
      A3.2 ------M------------------------------------------------G------------------------------------
       A11 ------M------------------------------------------------G------------------------------------
        A1 ------M------------------------------------------------G------------------------------------
       A32 -A----M---AV------------S------------------------------G-------S-------Q--------------------
      Aw24 ------M------------------------------------------------G------------------------------------
      Bw58 -----------V--------------------------------------------------------------------------------
     B27.1 --------------------------------------------------------------------------------------------
     B27.2 --------------------------------------------------------------------------------------------
     B27.3 --------------------------------------------------------------------------------------------
     B27.4 --------------------------------------------------------------------------------------------
      B27f --------------------------------------------------------------------------------------------
     B44.1 ----------------------S---------------------------------------------------------------------
     B44.2 ----------------V---------------------------------------------------------------------------
       B13 --------------------------------------------------------------------------------------------
      Bw47 --------------------------------------------------------------------------------------------
      Bw65 --------------------------------------------------------------------------------------------
       B14 --------------------------------------------------------------------------------------------
       B18 --------------------------------------------------------------------------------------------
      B40* --------------------------------------------------------------------------------------------
      Bw41 --------------------------------------------------------------------------------------------
      Bw60 --------------------------------------------------------------------------------------------
        B8 --------------------------------------------------------------------------------------------
      Bw42 --------------------------------------------------------------------------------------------
      B7.1 --------------------------------------------------------------------------------------------
      B7.2 --------------------------------------------------------------------------------------------
       Cw1 EH---------V-----------------------W-------------------G--------M-------------------E-------
     Cw2.1 EH---------V---------------T---------------------------G----------------------------E-------
     Cw2.2 EH---------V---------------T---------------------------G----------------------------E-------
       Cw3 EH---------V-----------------------W-------------------G----------------------------E-------
 consensus DPPKTHVTHHPISDHEATLRCWALGFYPAEITLTWQRDGEDQTQDTELVETRPAGDRTFQKWAAVVVPSGEEQRYTCHVQHEGLPKPLTLRW
                  190       200       210       220       230       240       250       260       270
```

Figure 3. *(Continued).*

into particular restriction fragments (65–67), permitting the construction of partial libraries from complete genomic digests (68,69). This decreased the complexity of libraries, which could then be selectively screened with the locus-specific probes. For example, genes encoding HLA-A11 and subtypes of HLA-A3 are uniquely found on a 15·5-kb BamHI fragment, and this property was used in their isolation (69,70). Although complete sequences were determined for the first genes isolated (29,59,60,71–74), subsequent interest has focused on the exons (68–70,75–81). For this reason, oligonucleotides have been developed, which specifically prime sequencing of the exons (68,69). Application of these more streamlined procedures has resulted in a significant increase in the rate of sequence accumulation (Fig. 2).

The strategy we have used over the last 3 years is as follows. Peripheral blood lymphocytes of individuals expressing HLA types of interest are transformed with Epstein-Barr virus to give B-cell lines. DNA is prepared and digested with the restriction enzyme EcoRI, which produces individual fragments of 6 to 8 kb containing individual intact HLA-A,B,C genes. These fragments are size-selected on sucrose gradients and used to construct partial libraries in λ phage vectors (68). The libraries are screened with a complete HLA-B7 cDNA probe (61), which hybridizes to all class I genes and with two locus-specific probes—one that hybridizes to HLA-A locus genes and the other that hybridizes to HLA-B and HLA-C locus genes (65,66). Clones that hybridize to the B7 cDNA probe and to one of the locus-specific probes are selected for further study. Restriction mapping allows determination of the number of unique clones and representatives of each type are analyzed by transfection. In most studies, we have used mouse L cells as the recipients for HLA genes that are cotransfected with thymidine kinase DNA. Lines of stable transfectants are obtained and tested for expression of HLA-A,B,C products by cell-binding assay using a panel of monoclonal antibodies (71). In this system, the HLA heavy chain is required to associate with murine β_2-microglobulin, which is an unfavorable reaction for certain allelic products (82). A second disadvantage is that the conventional HLA-typing assay cannot be used to type these cells.

To eliminate these problems, we have begun to use a mutant human B-cell line as the recipient in transfection analysis (83,84). This cell (called Hmy2 CIR) has lost expression of HLA-A,B and possibly HLA-C. A low level of binding with monomorphic anti-HLA-A,B,C reagents such as W6/32 is obtained and may be due to HLA-C or to a nonclassic class I gene such as those described by Orr and colleagues (85). Before transfection into Hmy2 CIR, the HLA-A,B,C genes are subcloned into the EBV-based vector pHEBO, which has the EBV origin of replication and the Hygromycin resistance gene (86). Transfectants are selected for resistance to Hygromycin, and stable lines are obtained in about 3 weeks. These cells contain functional human β_2-microglobulin genes, thus avoiding the potential complications of forming hybrids between human heavy chains and mouse β_2-microglobulin (82). Transfectants in Hmy2 CIR can be reliably typed for HLA using modifications that are commonly used for B-lymphoblastoid cell lines. In a recent experiment, we made transfectants expressing genes, which from the HLA type of the donor, were thought to be HLA-B13, B38, B49, and B51. These were coded and sent to Dr. Frances E. Ward (Duke University), who established their identity by HLA typing. This approach will be particularly useful for identifying genes coding for molecules for which there are no diagnostic monoclonal antibodies.

With improved methods for making cDNA libraries, it has been easier to obtain full length cDNA clones that can be transfected and their products analyzed by serology (77). It is also true that as the data base increases, the ability to assign identity on the basis of sequence alone improves. Kottmann et al. (87) isolated a cDNA derived from a cell homozygous for HLA-A2, -B44, -Cw5 and assigned it as B44 on the basis of hybridization with locus-specific probes and sequence comparisons. Szöts et al. (88) identified a cDNA encoding HLA-B27 by comparison of the predicted protein sequence with that of the B27 protein (49).

We have obtained cDNA clones containing full coding regions for HLA-A2.2Y (77), HLA-A2.4a, HLA-Bw46, and HLA-Cw11 a molecule that reacts with HLA-Cw1- and HLA-Cw3-specific sera. This approach has also been successful in obtaining full-length cDNA clones for class I MHC molecules of cattle (89) and the chimpanzee. Sequence analysis, whether of cDNA or genomic clones, involves the use of specific oligonucleotide primers. This obviates the need to fragment and subclone and reduces the chance for error or confusion. For genomic clones, a pair of oligonucleotides provides sequencing of each exon, whereas for cDNA, conserved sequences at intervals of $\sim$200 bp throughout the sequence were selected. It need hardly be emphasized that in these studies, it is the small number of nucleotide substitutions between highly homologous sequences that is important. Thus, there is no acceptable error rate in sequencing, and it is critical that complete sequences on both strands be performed.

Subtypes Defined by Sequence

The application of various analytical methods has subdivided the types defined by serologic HLA typing (36–41). Subtypes defined by cytotoxic T cells have been sequenced, and they have proven to be the products of novel alleles (53–56,69,77,79). It becomes relevant to ask the question whether additional heterogeneity would be found by randomly sequencing genes that were typed by other methods as being identical. This would be the only way of identifying alleles defined by silent substitutions. We have studied three molecules typed as HLA-Aw69, two molecules typed as HLA-Aw68.2, and two molecules typed as HLA-B13, and in each case, identical sequences were obtained (68,77,80,81). In addition, four separate laboratories have sequenced HLA-B27.1 and obtained identical results (49,73,76,88–90). These results suggest there is not a massive amount of hidden heterogeneity.

On the other hand, there are examples where heterogeneity has been found (Fig. 3), although in each case, the gene products were not well characterized by isoelectric focusing (91) or with cytotoxic T cells (36). Two molecules typed as B14 and one subtyped for Bw65 showed one nucleotide substitution giving a coding change at position 11. In this case, the serologic subtype of the second molecule was not known and it may represent Bw64. The HLA-B44 gene sequenced in this laboratory (B44.2) differed by five nucleotide substitutions and three amino acids from that (B44.1) reported by Kottmann et al. (87). Two molecules typed as Cw2, one sequenced in this laboratory (Cw2.2) and one (Cw2.1) by Ploegh and colleagues (74), showed seven nucleotide substitutions and three amino acid changes, one in the leader peptide, the others at positions 49 and 50 in α_1. Differences were also observed in two clones encoding Cw1 (74,88). This result may reflect the relatively crude state of typing for HLA-C, and one expects that sequencing may be the best method for improving this state of affairs.

The Pattern of Polymorphic Variation in HLA-A,B,C

Comparison of the sequences of HLA-A,B,C molecules shows a complex pattern of amino acid substitutions that is fundamentally different from that encountered with other large families of polymorphic proteins such as the hemoglobin variants (Fig. 3) (81). In those families, low frequency variants differ from a predominant wild type sequence by only one or two amino acid substitutions, and the wild type sequence is clearly the common ancestor of the variants. In contrast, there are no predominant alleles at the HLA-A,B, and C loci, and all sequences differ from the consensus by multiple differences.

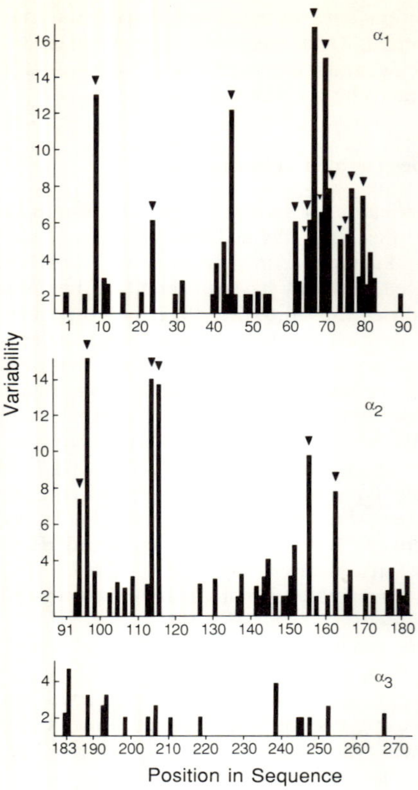

Figure 4. Variability plot of 39 HLA-A,B,C sequences. Separate plots of the α_1, α_2, and α_3 domains are shown. Variability is defined as the number of different amino acids found at a particular position divided by the frequency of the most common amino acids (92). Positions with variability greater than or equal to 5.0 (high variability residues) are indicated by (▼). The sequences from Figure 3 were used in this analysis (81).

In the three extracellular domains (α_1, α_2, and α_3) comprising that part of the molecule which interacts with T cells, we find positions of polymorphism spread throughout the sequence (Fig. 3). A method to analyze complex patterns of sequence variation is the variability plot of Wu and Kabat (92). This plot illustrates that the number of positions showing variation and the magnitude of that variation are significantly lower in the α_3 domain compared with that in the α_1 and α_2 domains. On this basis, we have divided the positions of polymorphism into two groups. Positions of high variability have a variability greater than that found in α_3 (≥ 5.0), and positions of low variability have a variability comparable with that found in α_3 (<5.0).

A total of 20 residues with high variability are found. Although the α_1 and α_2 domains have homologous structures (4), their complement of residues with high variability is different. For each domain, the amino-terminal half comprises the region of β strands, and the carboxy-terminal half the α-helical region. The helical region of the α_1 domain has 11 residues of high variability, compared with 2 in the α_2 domain. In contrast, there is somewhat greater variability in the β strand region of α_2 compared with α_1 (Figs. 3, 4, and 5).

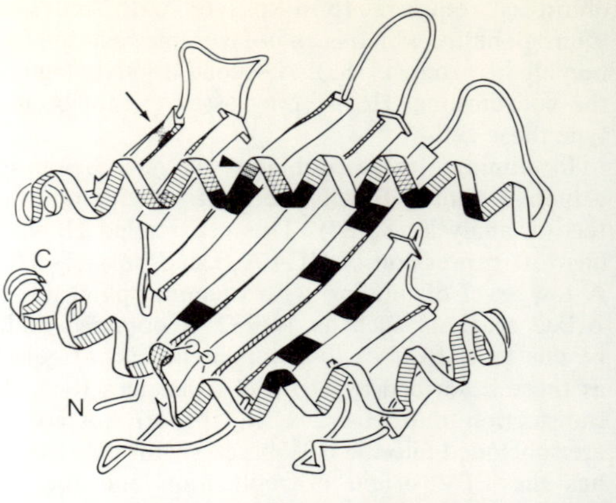

Figure 5. Position of residues having high variability in the structure of HLA-A2. The highlighted residues are those of high variability, as designated by ▼ in Figures 3 and 4. The schematic structure of the α_1 and α_2 domains is as in Figure 1. Solid areas are those residues with side chains pointing into the peptide binding site and potentially involved in contacting bound peptide. Cross-hatched areas are those residues with side chains pointing up from the helices and potentially involved in contacting the T-cell receptor. The side chain of residue 71 (arrowhead) in the α_1 helix points away from the peptide site, and residue 45 (indicated by the arrow), is not directly involved in the binding site. The full color illustration of this figure appears in Color Plate IV at the beginning of this volume.

In their analysis of the crystal structure of HLA-A2, Bjorkman et al. (5) identified residues that could potentially interact with bound peptide (solid area in Fig. 5) or the T-cell receptor (cross-hatched area in Fig. 5). Eighteen of the 20 residues with high variability fall into this category, the exceptions being residue 45 (area indicated by an arrow in Fig. 5), which is on a strand external to the binding site, and residue 71 (dotted area in Fig. 5), which is on an α helix with the side chain pointing away from the binding site. It is worth noting that neither of these two positions shows variability amongst HLA-A locus molecules, and that the HLA-A2 crystal structure may not be giving an accurate picture of the role of these side chains in HLA-B and C locus molecules.

Thus, we see that residues with high variability are likely to be directly involved in interactions with antigenic peptides and T-cell receptors, and that their variation may serve to modify these interactions. In essence, this explains the function and selective advantage of HLA-A,B,C polymorphism.

In addition to the 20 positions with high variability, there are 71 positions with low variability. These are found throughout the structure, often at positions quite distant from the peptide binding site. These substitutions may influence the interactions with peptides and T-cell receptors by indirect conformational perturbations. Alternatively, they may represent neutral muta-

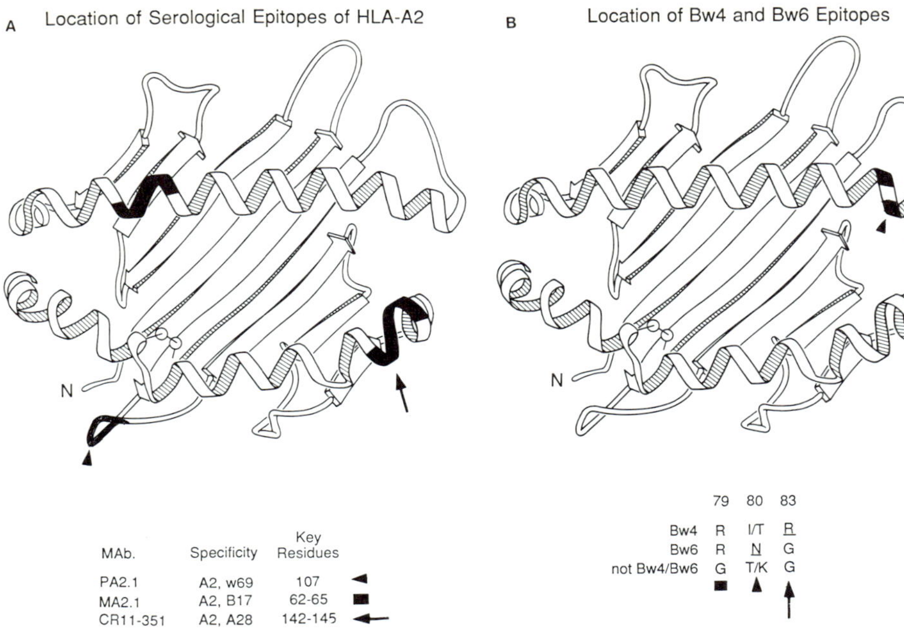

Figure 6. Mapping serologic epitopes. The format and labels in this schematic diagram of the α_1 and α_2 domains of the class I molecule are as in Figure 1. Residues involved in the epitopes are highlighted, and their respective epitopes are given in the keys to each panel. Panel A: The monoclonal antibody defined epitopes of HLA-A2. Panel B: Localization of the Bw4 and Bw6 epitopes. The full color illustration of this figure can be found in Color Plate IV at the beginning of this volume.

tions of no functional import. Evidence supporting the feasibility of the first hypothesis is the observation that a single mutation in the α_3 domain of H-2D^d obliterates its reactivity with cytotoxic T cells (93).

Mapping Serologic Epitopes

Almost all of the epitopes recognized by alloantisera and monoclonal antibodies are dependent on the three-dimensional conformation of HLA-A,B,C molecules. Few antibodies made against the native molecule react with the separated heavy chains or peptides derived from HLA-A,B,C sequences. Only in the case of β_2-microglobulin have cross-reactivities with peptides been successfully used to map two epitopes to residues 45 (94) and 84 (95). It is therefore surprising that simple comparison of linear sequences has been remarkably accurate in identifying residues critical for alloantigenic epitopes (71,77,78). The strategy is to compare sequences of molecules and search for patterns of polymorphic substitution that correlate with the distribution of particular epitopes. We shall briefly describe some of the epitopes mapped and some predictions for where further epitopes may be found.

HLA-A2. In this manner, three epitopes of HLA-A2 have been mapped. The epitope shared by A2 and B17 to residues 62 to 65 of the α_1 helix, the epitope shared by A2 and A28 to residues 142 to 145 of the α_2 helix, and the epitope shared by A2 and Aw69 to residue 107 on a tight turn between two strands of β structure (Fig. 6A). In each case, these assignments have been substantiated by site-directed mutagenesis (84,96–98). None of the three epitopes is specific to HLA-A2, and no monoclonal antibody with such a specificity has so far been obtained. However, the combination of all three epitopes does uniquely define A2. The same may also be true for alloantisera in which a typing reaction depends on the activity of a combination of antibodies.

HLA-Bw4/Bw6. Of particular interest for many years have been the Bw4 and Bw6 epitopes originally described as 4a and 4b by Jon Van Rood and the first antigens of the HLA-B series (34). Every HLA-B locus molecule has either the Bw4 or Bw6 epitope (35). In addition, the HLA-A locus molecules HLA-Aw24 and HLA-A32 have the Bw4 epitope (99), and HLA-Cw3 is cross-reactive with Bw6 (100). As part of an analysis of subtypes of HLA-A2, Nick Holmes planned to isolate the gene for HLA-A2.2Y from Andrew McMichael's lymphocytes. On probing with an HLA-A locus-specific probe, he also obtained the other HLA-A locus allele, which was HLA-A32. In due course, this was sequenced and provided us with the second HLA-A locus molecule that was Bw4-positive (75) as Bertrand Jordan's group had already isolated HLA-Aw24 (72). Comparison of these two sequences with other HLA-A locus molecules that are negative for Bw4 identified a sequence (RIALR) between residues 79 and 83 that was specifically shared by these two molecules. Examination of the sequences from the B locus showed that variations on this sequence were only found with Bw4-associated molecules. A different sequence (RNLRG) was found in Bw6-associated molecules. The assignment of Bw4 and Bw6 to the region encompassing residues 79 to 83 was originally made on the basis of 16 sequences (75). All subsequent sequences have confirmed the correlation and, with greater information, we have identified asparagine at position 80 as critical for Bw6 and arginine at position 83 for Bw4 (Fig. 6B). These epitopes are formed by the carboxy-terminal region of the long α helix of the α_1 domain. Their apparent diallelism at the B locus clearly stems from their being derived from alternate sequences in the same part of the molecule.

The identity of the Bw4 sequences (residues 79 to 83) of HLA-Aw24, A32, and Bw58 compared with some variability of Bw4 sequences in other HLA-B molecules suggests that a recent mutational event moved the Bw4 sequence from the B locus into the A locus.

Certain pairs of serologically similar HLA-B locus molecules (such as B44 and B45, Bw62 and Bw63, Bw49 and Bw50) are generally distinguished on the basis of their segregation of the Bw4 and Bw6 epitopes (35). These pairs of alleles may have arisen from genetic exchange of the Bw4/Bw6 encoding region between alleles at the B locus, predicting that the major differences between such pairs of molecules will include residues 79 to 83. Clear evidence for this type of exchange can be seen in the known sequences (Fig. 3). For example, residues 62 to 76 of the variable α_1 helix are identical in B44.2 and B13 molecules that are Bw4-positive and in B40*, B41, and Bw60 molecules that are Bw6-positive. From residues 77 to 83, B44.2 and B13 have an identical Bw4 sequence, which differs from the Bw6 sequence shared by B40*, B41 and Bw60.

Since the Bw4 epitope is found on HLA-Aw24 and HLA-A32, why has Bw6 not been found on HLA-A locus molecules? One possibility was that a nonfunctional molecule would be made if such a genetic exchange took place. This now seems unlikely because in vitro mutagenesis by Jesus Santos-Aguado and coworkers has shown that Bw6 reactivity can be simply introduced into HLA-A2 (98).

HLA-B27, Ankylosing Spondylitis, and a Disease-Specific Epitope? Susceptibility to many human diseases has been associated with HLA type (101). In cases such as congenital adrenal hyperplasia (102,103), the HLA allele is merely a marker for a closely linked gene. However, the direct involvement of MHC molecules is implicated in other diseases, many of which are autoimmune. How this occurs is readily rationalized in terms of the structure and function of MHC molecules. A predisposition for generating a T-cell response against a particular self-peptide could arise by having an MHC molecule that either selects for such T cells in thymic ontogeny, has a particularly high affinity for the self-peptide, binds an exogenous peptide that mimics the self-peptide, or some combination of these three.

Although the majority of disease associations are with class II alleles, the most dramatic is that of HLA-B27 with ankylosing spondylitis, Reiters syndrome, and other rheumatoid disorders (104). Ninety percent of patients with ankylosing spondylitis type for B27, whereas the overall frequency is 10% in Caucasians and considerably less in other populations. For this reason, many investigations of class I structure and function have focused on the HLA-B27 specificity and its many subtypes (39,49,56,73,76,88,90). A complete sequence for 271 residues of the B27.1 heavy chain was first obtained from papain-solubilized protein by Lopez de Castro and colleagues (49). With the sole difference of glutamic acid instead of glutamine at position 242, probably due to misassignment in the protein analysis, this sequence proved identical to that coded by the six B27.1 nucleotide sequences that were subsequently obtained (73,76, 88,90). In one study, the sequence of the gene from a patient with ankylosing spondylitis was shown to be the same as that from a healthy individual (76). The hypothesis that a subtype of HLA-B27 would yield an even higher association with disease was widely embraced and stimulated the subdivision of B27 into at least seven subtypes (105). Nucleotide sequences for two subtypes (B27.1, B27.2) (73) and protein sequences for three additional subtypes (B27.3, B27.4, B27.f) (56) have been determined. However, the disease appears equally correlated with all subtypes, leading to the conclusion that it is a shared feature of this family of molecules that is responsible for the disease association.

Substitutions that are exclusive to the B27 family are at two positions of high variability—lysine at position 70 in α_1, and asparagine at position 97 in α_2. The side chains of both residues point into the peptide-binding site, are close together, and should influence the nature of bound peptide.

Previous comparisons of B27 focused attention on the serine at position 131 and cysteine at position 67 (49,88). Position 131 is found on a turn between two of the β strands and is well away from the putative sites of interaction with peptides and T cells. The serine, which was initially found in B27 and is preserved in all subtypes, has now appeared in B44.1, B44.2, B13, B47, B14, Bw65, and B18 and is thus an unlikely candidate for the disease-specific structure of B27. In contrast, cysteine 67 is at a position of high variability, points into the peptide binding site, and is of more restricted distribution, being common to subtypes of B27 and B14. Ezquerra et al. (49) suggested that the reactive sulfhydryl of cysteine 67 "may endow the molecule with a potential for establishing intermolecular disulfide bonds with other surface components." With a knowledge of the class I structure, one can now suggest that this residue may establish covalent bonds and, thus, extremely stable complexes with peptides or other molecules that are found in the binding site. Cysteine 67 is close to lysine 70 and asparagine 97 (which are not found in B14) and perhaps a combination of these three residues, which so far are found in all B27 subtypes and in no other HLA-A,B,C molecules, is responsible for the high correlation with susceptibility to ankylosing spondylitis. These residues also provide the best candidates for forming the serologic determinants that define the B27 family (Fig. 3).

HLA-B7/Bw42. HLA-Bw42, a molecule found in African blacks, is cross-reactive with B7 (106). There is a clear basis for this cross-reactivity in the primary structure—Bw42 has an α_1 domain that is identical to B7, an α_2 domain that is identical to B8, and an α_3 domain that is shared by all three (81). This suggests that the α_1 domain is responsible for the epitope defined by the monoclonal antibody BB7.1, which exclusively reacts with B7 and Bw42 (106). There are no residues in α_1 that are uniquely found in the sequence shared by B7 and Bw42. However, a unique combination of residues is found in the helical region from residues 63 to 83. The Bw6 determinant of these molecules is known to involve residues 79 to 83, and this epitope is topographically separate from that defined by the BB7.1 antibody (107). Thus, the BB7.1 epitope is likely to involve residues 63 to 78. Given that Bw42 is identical to B8 in α_2, it is surprising that cross-reactions resulting from this relationship have not been found.

The structure suggests that Bw42 was derived from a homologous recombination between allelic B7 and B8 genes. Because Bw42 is only found in Africans, this recombination is likely to have occurred subsequent to the separation of blacks from Caucasians and orientals. However, it is also of interest that Bw42 is closest to the consensus sequence for HLA-A,B,C molecules, having only 12 differences (Fig. 3). This might indicate that Bw42 is older than B7 and B8, in which case those two genes would have been derived by separate recombinations between Bw42 and other B locus alleles.

HLA-B7/B27/Bw42. No single substitution is exclusive to this group of molecules. However, the sequence AQTDRE at residues 71 to 76 is exclusive to Bw42 and subtypes of B7 and B27 (Fig. 3). Thus, this central part of the α_1 helix provides the best candidate for this epitope, which would be flanked on the amino-terminal side by B7, Bw42-specific epitopes, and on the carboxy-terminal side by Bw6. This alignment is consistent with topographical mapping and the blocking of the BB7.1 antibody (B7, Bw42-specific) by the ME1 antibody (108), which recognizes an epitope shared by B7, B27, and Bw42 (106).

HLA-B7/B40. Cross-reactivity between B7 and B40 with alloantisera has been known for some time, and this determined the choice of B7 and Bw60 as the first B locus molecules to be sequenced (47,52). The monoclonal antibody MB40.2 defines an epitope shared by B7, Bw60, Bw61, and B41 and also identified a novel subtype of B40 (B40*) that lacked the MB40.2 epitope (106,109). Sequencing of B40* showed it differed from Bw60 at eight scattered residues throughout the α_1 and α_2 domains (78). Of interest is the identity between Bw60 and B40* in the α helical region of the α_1 domain. It is presumably this region that is responsible for the alloantigenic determinants that are shared by these molecules.

Comparison of sequences for B7, Bw60, and B40* showed that the pattern of substitutions at positions 177 to 180 uniquely correlated with the presence of the MB40.2 epitope (Fig. 3). MB40.2-positive molecules had DKLE, and MB40.2-negative molecules had ETLQ (78). Subsequent to that analysis, 14 additional HLA-A,B,C sequences have been determined (81). Eleven MB40.2-negative molecules have either ETLQ, ESLQ, or EKLQ at positions 177 to 180. Bw41 and Bw42 are both MB40.2-positive and have the sequence DTLE, suggesting that lysine 178 is not critical for the epitope. This has been independently established by site-directed mutagenesis (97), and is also suggested by the sequence of EKLQ in the MB40.2-negative molecule Cw2.1. The one exception that does not fit the simple correlation is HLA-B8, which is negative for the epitope but has the sequence DTLE. One explanation is that substitutions in B8 at another part of the sequence are responsible for the loss of the epitope. As B8 is identical to Bw42 in α_2, an effect such as this would have to derive from the α_1 domain.

The only residues that distinguish B8 from B7, Bw41, Bw42, and Bw60 in the α_1 domain are the unique aspartic acid at position 9 and phenylalanine at position 67, which is shared by B8 and Bw51 (sequence not shown). Neither of these residues is particularly close to position 177 to 180 in the three-dimensional structure, and both point into the peptide-binding site. Thus, it is not clear how they would alter the epitope formed by residues 177 to 180, unless perhaps by altering the nature of bound peptides.

HLA-Bw60/Bw41. The basis for epitopes shared by these molecules is clearly the α_1 domain, which is identical in Bw60 and Bw41. The α helical region is also identical to B40*. In α_2, Bw41 differs from Bw60 by six dispersed substitutions, but is very similar to B8, from which it differs by two substitutions at positions 95 and 97 (Fig. 3). This indicates that a homologous recombination between Bw60 and B8 or a closely related gene may have been involved in the generation of Bw41. Again as seen for B7, Bw42, and B8, differences in the α_2 domain are serologically dominated by similarities in the α_1 domain.

HLA-B8/B14. Although B8 is identical to Bw42 and similar to Bw41 in the α_2 domain, it is apparent that differences in α_1 play a serologically more dominant role and this group of molecules has not been grouped on the basis of serologic cross-reactivities. In contrast, cross-reactions between B8 and B14 are documented (110), and these molecules are predictably more related in the

α_1 domain than in the α_2 domain. Two B14 molecules have been sequenced – one typed as Bw65 and the other as B14 without further splitting. They represent distinct subtypes, differing by an amino acid substitution at position 11. B8 differs from Bw65 by three substitutions (positions 9, 12, and 67) and from B14 by four substitutions (positions 9, 11, 12, and 67) in the α_1 domain. The sequence of TNTQTDRES between residues 69 and 77 of the helical region is uniquely shared by these three molecules. By comparison, there is no comparable sequence in the α_2 domain, where B8 differs by eight dispersed substitutions from B14 and Bw65.

HLA-A1/A3/A11. A3 and A11 provide one of the oldest and best documented examples of cross-reactivity between HLA molecules. This has been analyzed at the molecular level by Cowan et al. (70). In the α_1 domain, A11 differs from the two subtypes of A3 by two substitutions of Y for F at position 9 and D for A at position 90. This substitution is in the β strand and involves a residue of high variability at the bottom of the peptide-binding site. The α helical region, which is implicated in many of the serologic determinants, is identical in all three molecules. In α_2, A11 differs from A3.1 by five residues and A3.2 by four residues (Fig. 3). In comparison with 21 HLA-A,B,C sequences, Cowan et al. (70) identified glutamine 62 as the only residue uniquely shared by A3 and A11. They hypothesized that "it is likely that glutamine 62 is involved in the formation of an epitope recognized by antibodies cross-reactive on HLA-A11 and HLA-A3." Since then, A32, a molecule that is not cross-reactive with A3 or A11, has been shown to have glutamine 62 (75), indicating that this residue may be necessary but not sufficient to give the cross-reactive epitope. The sequence of A32 in the carboxy-terminal part of the α_1 helix is quite different from A3 and A11, indicating the importance of the combination of residues throughout this region, which so far is unique to A11 and A3.

A1 cross-reacts with A3 and A11 and shares glutamine 62. Three residues – aspartic acid 90, alanine 152, and arginine 163 – are only found in A1 and A11, and these residues are likely to contribute to A1/A11 cross-reactive epitopes.

Residue 62 has a side chain that points up from the peptide-binding site and is accessible for interaction with antibodies (5). The region from residues 62 to 65 forms an epitope shared by A2 and B17 (Fig. 6A), as well as being implicated in the cross-reactions between A3, A11, and A1. HLA-Aw24 has unique residues at position 62 (E) and 65 (G), which provide good candidates for involvement in the alloantigenic determinants that define this molecule. Position 66 in Aw24 is lysine, which is shared with subtypes of A2 and no other HLA-A,B molecules. This residue may therefore be critical for the cross-reactivity between A2 and A9 (Aw23 and Aw24). Lysine 66 is also present in HLA-C locus molecules, but the surrounding sequences are quite different from those found in A2 and Aw24.

Prediction of Immunologic Epitopes: Alloantigenicity Plots

Certain HLA-A,B,C molecules are more immunogenic and have more alloantigenic determinants than others. Such differences are important in assessments of histocompatibility, and it would be useful to have ways of predicting the epitopes that would stimulate immune responses in transplant situations. To do this, one must ask what factors might contribute to the formation of an alloantigenic epitope?

1. Having an amino acid that is not found in many other allelic products. In the ideal case, the residue will be unique and could therefore give rise to an allelic-specific epitope.
2. Having a low variability so that differences are more readily distinguished.
3. Having the "unique" amino acid be on the surface of the protein with its side chain accessible to the solvent.
4. Having the amino acid be in the α_1 or α_2 domains, as we have evidence (95) that the proximity of the membrane to the α_3 domain renders it relatively inaccessible to antibodies and presumably to alloreactive T cells.

The Wu and Kabat (92) plot has facilitated analysis of sequence variability in families of immunologic molecules. Perhaps plots measuring alloantigenicity could contribute to understanding the antigenic properties of HLA-A,B,C molecules and their capacity to stimulate allogeneic responses. As an initial step towards this goal, we have defined a parameter of alloantigenicity (or A) that is computed for a particular residue i at a given position χ in the sequence:

$$A_{\chi}^{i} = \frac{(N\text{-}n)}{(n)} \times \frac{(m)}{(N)} \times \frac{(a)}{(10)} ,$$

where N is the total number of sequences analyzed, n is the number of sequences that have the residue i at position χ, m is the number of sequences having the most frequent residue apart from i, and a is the percentage accessibility of the side chain of the residue at position χ in the HLA-A2 structure.

In defining this parameter, we have taken into account the following premises:

1. We assume an individual is tolerant to the sequences of his or her own class I molecules, and it is the sequence differences between allogeneic and self class I antigens that cause stimulation. Rare sequences are therefore more likely to be stimulatory than are common sequences, and the rarer the sequence, the more specific will be the resulting immune response. The term N-n/n thus weights for residues that are found with lower frequencies.
2. We assume that stronger and more specific immune responses will be made in situations where all class I antigens of the responder have an identical residue which is different to that of the stimulating molecule

to weight in favor of positions where there is a predominant amino acid that is different from the molecule under consideration. This is represented by term m/N.

3. We assume that in general, substitutions at residues on the surface of the class I molecule will have a greater effect on the immunogenicity and antigenicity than residues that are buried in the structure. This is accounted for by the a/10 term. Note this does not eliminate the involvement of conformational changes caused by internal substitution, but weights in favor of surface changes.
4. The analysis takes no account of whether a peptide is bound to the class I molecule. Likewise, weighting based on the type of the side chain has not been made because the known alloantigenic epitopes show that residues with small side chains, for example glycine 62 in HLA-A2 and -B17 (Fig. 6A, 71), can be important. The phenotypic frequencies of HLA-A,B,C molecules are also not involved.

Table 1 lists the values of alloantigenicity for the polymorphic positions found in the 39 sequences in Fig. 3. The maximum possible value is 370. From these values, plots of alloantigenicity can be made for any 1 of the 39 molecules. One should not always expect the regions of highest alloantigenicity to correspond to epitopes defined by typing alloantisera because the choice of alloantisera is extremely selective. In particular, there has been a selection against sera that identify public or broadly cross-reactive epitopes.

How well does the alloantigenicity plot reflect known serologic properties? HLA-A1 is serologically well defined and has a number of unique features that result in 8 residues having an alloantigenicity in excess of 40 (positions 44, 76, 90, 150, 158, 163, 166, and 167) (Fig. 7A). Examination of the sequence suggests that positions 44, 76, 150, and 157 could contribute to A1-specific determinants, positions 90, 152, and 163 to epitopes shared by A1 and A11, and positions 166 and 167 to cross-reactive epitopes of A1 and Aw24. In contrast to A1, the molecule HLA-B40* has been elusive to serologic definition, being identified by its lack of reaction with certain monoclonal antibodies (109). No positions in the B40* sequence have alloantigenicity greater than 40, the highest values being 36.3 at position 41 and 13.8 at residue 113 (Fig. 7A). At both these positions, the residues are shared with other members of the B40 cross-reacting group (Fig. 3).

The major peak of alloantigenicity for HLA-B7 is at residues 177 to 180, where an epitope shared by B7, Bw60, Bw41, and Bw42 has been localized (Fig. 7B). Values of alloantigenicity will generally be lower for molecules such as HLA-A2 and HLA-B27, where a number of closely related sequences are included in the analysis. Biases introduced in this manner will decrease as a more comprehensive description of the polymorphism is achieved. For HLA-A2, the maximum alloantigenicity is 20.6 (Fig. 7C). However, the three regions of the molecule that we know contribute to serologic epitopes (positions 62 to 65), (positions 105 to 109), (positions 141 to 146) do feature as peaks of alloantigenicity (Figs. 6A and 7C). In addition, residues at positions 43 and 127 show high values and may contribute to additional epitopes. There are also residues in the α_3 domain that appear as peaks of alloantigenicity, and their possible involvement should be born in mind. However, such effects may be small or indirect because of the potential inaccessibility of this region of the molecule at the cell surface. Plots of the sequences for subtypes HLA-A2.2 and HLA-A2.3 defined by cytotoxic T cells show significant peaks of alloantigenicity and suggest that in appropriate responder stimulator combinations, alloantisera or monoclonal antibodies defining these molecules could be obtained.

This analysis is based on a position-by-position approach and will not give weight to rare combinations of more frequent residues. Such patterns of sequence are not uncommon as is clearly seen in the helical region of the α_1 domain. For example, the epitope shared by B7 and Bw42 is likely to result from substitutions at residues 62 to 79, although alloantigenicity values are low (Fig. 7B). These molecules are identical throughout this region, and only differ from the consensus at positions 63, 67, and 71. However, the combination of substitutions at these three positions is unique to B7 and Bw42, although asparagine 63 is shared with 8 other molecules, tyrosine 67 with 5 other molecules, and alanine 71 with 11 other molecules (Fig. 3).

The role of bound peptides in forming alloantigenic epitopes is an important question that is addressed by little data. The side chains that point into the site at many positions of high variability are relatively inaccessible and are negatively weighted in the calculation of alloantigenicity. However, if the normal state of the cell surface HLA-A,B,C molecules is to have peptide bound, then it is likely, given the architecture of the molecule, that the peptide will form part of or affect the conformation of alloantigenic determinants. As the nature of the bound peptide will itself be determined by the side chain forming the binding site, it is possible that substitutions at these positions may have detectable effects on alloantigenic determinants.

Locus-Specific Characteristics of HLA-A,B,C

Comparison of the nucleotide sequences for HLA-A,B,C genes has revealed locus-specific differences, especially at the 3′ end (29,72,74,88). This is more apparent at the nucleotide level than from the protein sequences (72). In the extracellular domains (α_1 α_2 α_3), there are 183 positions that show conservation between all HLA-A,B,C molecules, compared with only 6 that show locus-specific segregation of different amino acids (Fig. 3, ref 81). In fact, there is no position, as yet, at which all molecules of a particular locus have an identical amino acid with all loci being distinct. As more sequences have been obtained, the number of locus-specific residues has

Table 1. Alloantigenicity Values

Position	Alloantigenicity of Residues at the Position													
	α_1 Domain													
1+	G	0	C	11										
6	R	0	K	196										
Δ 9	Y	0	F	0	H	0	S	2	D	2	C	2		
11	S	0	A	0										
12	V	0	M	4										
16+	G	0	S	18										
21	R	0	H	38										
Δ 24	A	0	T	0	S	1								
30	D	0	G	100										
32	Q	0	L	2										
40	A	0	D	44										
41	A	0	T	36	E	302								
43	P	1	Q	8	R	75								
44	R	0	K	344										
Δ 45	M	0	E	0	K	1	G	1	T	2				
46	E	0	A	204										
49	A	0	G	22										
50	P	0	R	333										
52	I	0	V	1										
54	Q	0	R	222										
55	E	0	K	41										
Δ 62+	R	0	G	3	Q	5	E	25						
63	E	0	N	2										
Δ 65	Q	2	R	8	G	182								
Δ 66	I	2	N	6	K	8								
Δ 67	V	0	S	1	C	1	Y	1	M	4	F	8		
Δ 69	A	1	T	11	R	50	P	159						
Δ 70	Q	1	N	2	H	2	K	4	S	24				
Δ 71	S	1	A	2	T	2								
Δ 74	D	0	Y	2	H	3								
Δ 76	E	2	V	5	A	133								
Δ 77	S	2	D	3	N	4								
79+	R	0	G	1										
Δ 80	T	1	N	6	I	22	K	46						
81	L	0	A	2										
82	R	1	L	9	A	137								
83+	G	0	R	2										
90	A	0	D	176										
	α_2 Domain													
94	T	0	I	0										
Δ 95	L	0	I	1	V	1	W	4						
Δ 97	R	0	N	7	S	7	I	8	W	12	M	12	T	37
99	Y	0	C	50	F	50								
103	V	0	L	3										
105	P	1	S	16										
107	G	0	W	19										
109	L	1	F	3										
113	Y	1	H	14										
Δ 114	H	0	D	0	N	1	R	1	Q	7				
Δ 116	Y	1	D	1	S	5	F	11	L	23	H	23		
127	N	1	K	14										
131	R	0	S	2										
137	D	0	N	156										
138	T	2	M	9										
142	I	0	T	14										
143	T	0	S	67										
144	Q	1	K	4										
145	R	1	H	21	L	235								
147	W	0	L	122										
149	A	0	T	281										

Table 1. *(Continued)*

Position	Alloantigenicity of Residues at the Position							
	α_2 Domain							
150	A	0	V	322				
151	R	1	H	6				
152	V	1	E	6	A	50		
Δ 156	L	0	W	6	D	8	R	11	Q	15
158	A	0	V	326				
161	E	0	D	137				
Δ 163	T	2	E	5	L	23	R	49
166	E	0	D	125				
167	W	0	G	106	S	106		
171	Y	0	H	4				
173	E	0	K	233				
177	E	0	D	29				
178	T	0	K	92	S	290		
180	Q	0	E	12				
181	R	0	G	270				
182	A	1	T	7				
	α_3 Domain							
183	D	0	E	49				
184	P	0	A	4	H	13		
189	V	0	M	1				
193	P	1	A	19				
194	I	1	V	5				
199	A	0	V	30				
205	A	0	S	4				
207	G	1	S	9				
211	A	0	T	61				
219	R	0	W	93				
239	R	0	G	1				
245	A	0	V	4				
246	A	0	S	89				
248	V	0	M	170				
253	E	0	Q	6				
268	K	0	E	53				

Values of alloantigenicity, to the nearest integer, are given for all positions of polymorphism in the 39 sequences shown in Figure 3. The column at the left gives the position in the sequence, and positions with high variability are indicated by Δ. Subsequent columns to the right give each amino acid found at that position and the corresponding values of alloantigenicity. The alloantigenicity of a residue at position χ is A^i_χ and is given by the equation:

$$A^i_\chi = \left(\frac{N-n}{n}\right)\left(\frac{m}{N}\right)\left(\frac{a}{10}\right)$$

where N is the total number of sequences analyzed, n is the number of sequences having residue i at position χ, m is the number of sequences having the most frequent residue excluding i, and a is the percent accessibility of the side chain at position χ as determined from unrefined coordinates of the HLA-A2 structure. For positions in HLA-A2 where there is a glycine residue (indicated by +), the percent accessibility has been arbitrarily assigned as 10.

decreased to the current number of six (positions 52, 138, 183, 239, and 268). At each of these positions, the products of two of the loci are identical, and those of the third are distinct. Methionines 138 and 189 are A locus-specific, arginine 239 is B locus-specific, and valine 52 and glutamic acids 183 and 268 are C locus-specific (Fig. 3).

Segregation of polymorphic differences between HLA-A,B,C loci is found. There are positions where one locus is conserved and another is highly variable. For example, at position 45, all A locus molecules have methionine, all C locus molecules have glycine, and four different residues (TEMK) are found in B locus molecules. There are sequences that are common at one locus and absent in another. For example, the sequence TNT at positions 69 to 71 is found in 11 of 20 B locus sequences, but is absent in HLA-A and HLA-C locus molecules (Fig. 3). Similarly, the combination of serine 105 with phenylalanine at position 109 is commonly found in HLA-A locus molecules, yet these residues are always proline and leucine in HLA-B and HLA-C locus molecules.

This segregation of polymorphism effectively adds to the degree of locus identity, as can be seen by comparing

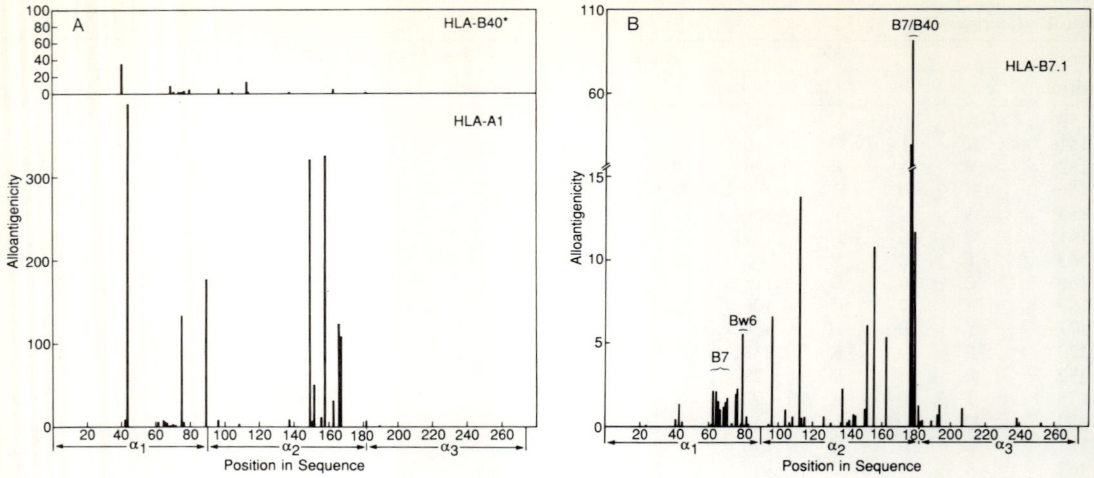

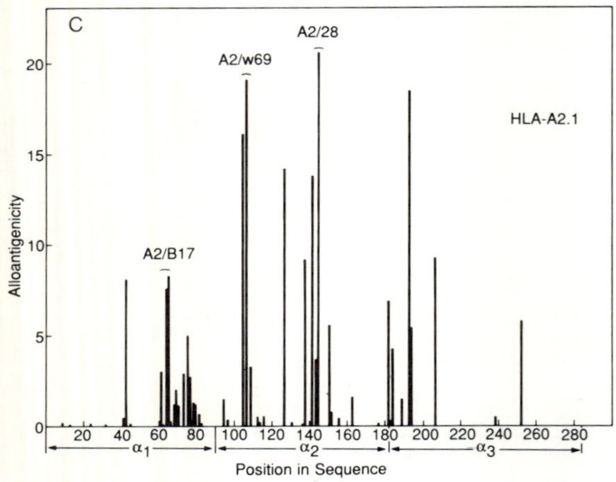

Figure 7. Alloantigenicity plots of HLA-A,B,C molecules. The plots were calculated according to the formula given in the text and Table 1. Panel A: Comparison of an antigenically well defined molecule, HLA-A1, and a poorly defined molecule, HLA-B40*. Panel B: Alloantigenicity plot for HLA-B7.1. Residues involved in B7-specific epitopes, the Bw6 epitope, and the epitope shared by B7 and Bw60 are indicated by the brackets. Panel C: Alloantigenicity plot for HLA-A2.1. Residues involved in the epitopes shared with B17, Aw69, and A28 are indicated by the brackets. See also Figure 6A.

the sequences in a pairwise fashion and calculating the number of amino acid differences. A frequency distribution of the values obtained from this calculation shows a trimodal distribution (Fig. 8A) in which comparisons between products of the same locus are readily distinguished from comparisons between products of different loci. This demonstrates that products of one locus have more in common with each other than with products of another locus. The distribution given by intralocus comparisons gives two peaks – one at lower values, which represents the comparison of subtypes and is dominated by the A2/A28 and B27 families, and the other at higher values representing the comparison between members of different cross-reacting groups. When the interlocus comparison is broken down into its three components (Fig. 8B), one sees that B and C locus products are mosts closely related, and that A and C locus products are most disparate. This is consistent with their close chromosomal linkage and the hypothesis that HLA-B and HLA-C represent the products of a more recent duplication.

Similar results are obtained from a comparison of nucleotide sequences encoding the α_1, α_2, and α_3 domains (Fig. 9). This shows that alleles of one locus are more closely related to alleles of that same locus than to alleles of a different locus.

Of practical consequence is that it should now be possible to assign unidentified sequences to particular loci. To test this idea, we shall consider the sequence of HLA-B49 that was completed subsequent to the analysis shown in Figs. 3 to 9. Pairwise comparison of B49 with 11 HLA-A alleles gives a range of 73 to 93 nucleotide differences (Table 2), clearly in the distribution of interlocus differences (Fig. 9). Comparison with four HLA-C alleles gives a range of 63 to 69 nucleotide differences (Table 2), again in the range for interlocus differences (Fig. 9). In contrast, comparison with 15 HLA-B alleles gives differences in the range 18 to 48 (Table 2), which is consistent with an intralocus distribution (Fig. 9). These results unequivocally demonstrate that the novel sequence is from an allele of the HLA-B locus, and confirmation is

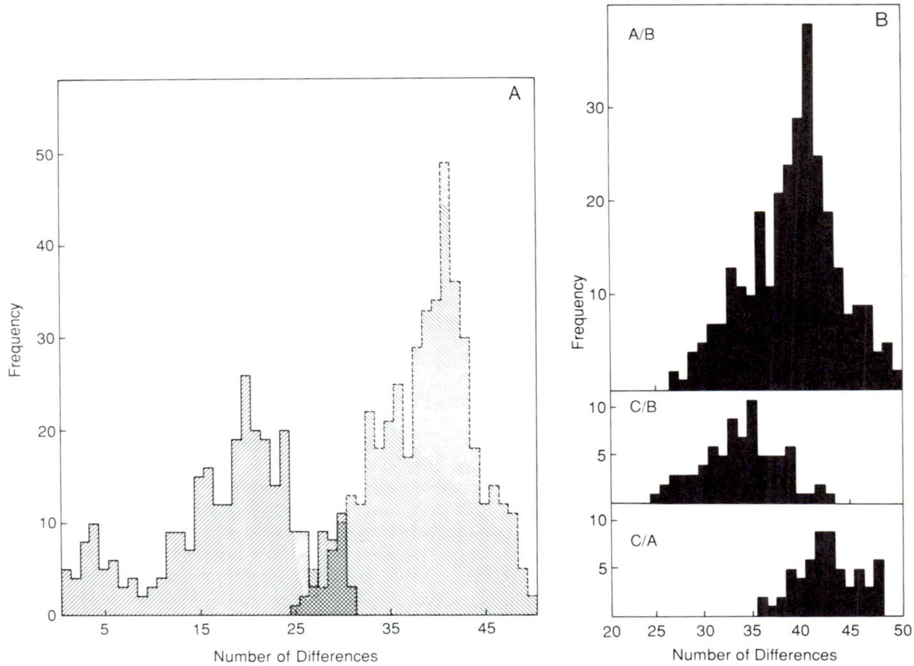

Figure 8. Pairwise companions of the protein sequences shown in Figure 3 were made to determine the number of amino acid differences. These values were then plotted as a frequency distribution (panel A). The distribution obtained by comparing proteins of the same locus is given by ▨; that obtained by comparison of proteins from different loci is given by ▧. In panel B, the comparison of proteins from different loci is separated into the three components. A/B compares A locus molecules to B locus molecules, C/B compares C locus molecules to B locus molecules, and C/A compares C locus molecules to A locus molecules.

provided by a characteristic pattern of residues at the six locus-specific positions.

This form of analysis can also be extended to class I molecules of different species (Fig. 10). Comparison of the available sequences from human, mouse, rabbit, miniature swine, cow, and syrian hamster show that interspecies comparisons are readily distinguished from intraspecies comparisons. When class I genes from two different species are compared, the number of substitutions varies between 100 and 200, whereas for comparisons within a species, the number is less than 100 (Fig. 11). Similar calculations taking two species at a time show that of these six species, man and mouse are the most disparate. The six species so far analyzed are phylogenetically quite distant. Clear-cut distinctions between xenogenic and allogeneic comparisons of class I molecules may not be the case for more closely related species such as mouse and rat or human and chimpanzee.

Evolution of HLA-A,B,C Genes

For most proteins exhibiting genetic polymorphism, there is a predominant "wild type" allele and a number of variants that are present in lower frequencies. This is not the case for class I and II MHC genes; many alleles are present at significant frequency and they are related by complex substitutions. Tracing the evolution of class I MHC molecules is thus a challenging problem.

Once the complexity of substitutions between class I molecules was appreciated, it became logical to first focus on alleles that by immunologic criteria were closely related. In the mouse, the H-2K^b allele and a series of spontaneous mutants provided a system in which limited changes correlated with functional differences (112). In most K^b mutants, the changes involve a localized segment having multiple nucleotide substitutions. An example is the bml mutant, which has seven nucleotide substitutions, leading to changes in amino acids 152, 155, and 156. Such changes are unlikely to occur through multiple point substitutions. That alleles of other class I genes on the K^b haplotype have the new sequences found in the mutants supports a model in which an exchange of a short segment of DNA occurred between the two cis-oriented genes. In the case of the bml mutant, the Q10 gene would have donated a sequence including codons 149 to 156 to the K^b gene. This process has been called gene conversion, genetic exchange, or nonreciprocal genetic exchange (112).

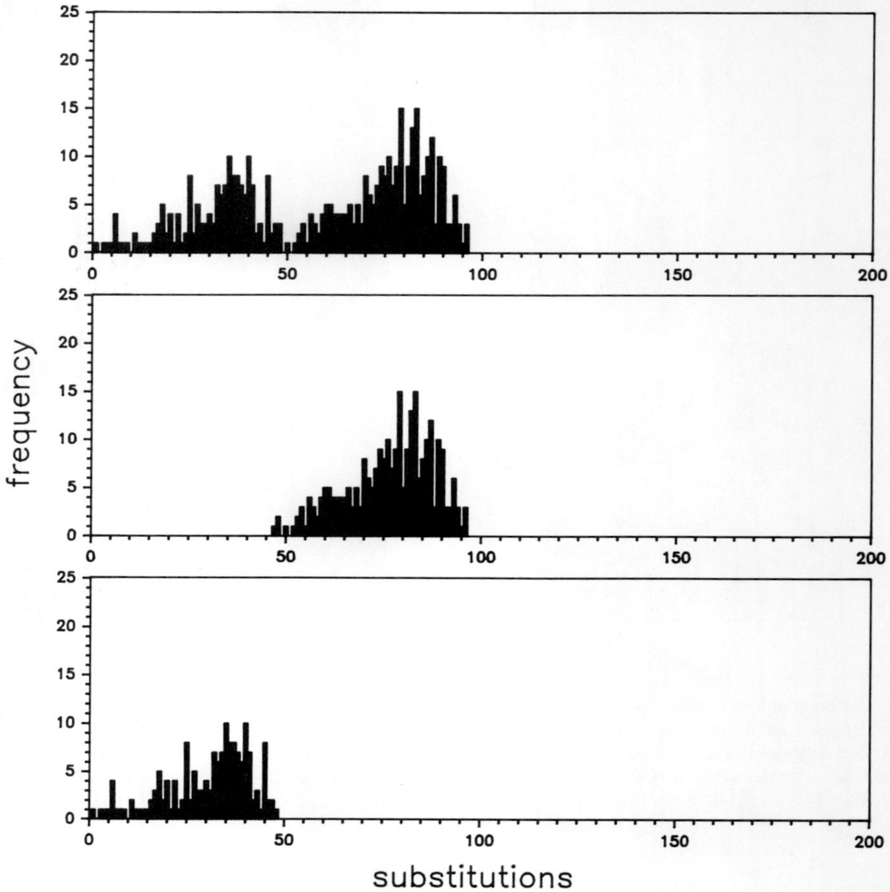

Figure 9. Histograms of nucleotide substitutions between HLA-A, HLA-B, and HLA-C. Pairwise comparisons of the nucleotide sequences encoding the α_1, α_2, and α_3 domains of the 31 HLA-A,B,C molecules were made to determine the number of nucleotide substitutions. The top panel shows the frequency distribution of these differences, the middle panel shows only the comparison between alleles of different loci, and the bottom panel shows only the comparison between alleles of the same loci.

Table 2. Pairwise Comparison of HLA-B49 with HLA-A,B,C Alleles

Allele	Number of Nucleotide Substitutions from HLA-B49	Range	Allele	Number of Nucleotide Substitutions from HLA-B49	Range
HLA-A2.1	93		B44.2	23	
A2.2Y	93		B13	18	
A2.3	94		Bw47	25	
Aw69	90	73-95	Bw65	33	18-48
Aw68.2	89		B14	34	
Aw68.1	95		B18	36	
A3	80		B40*	27	
A11	79		Bw41	23	
A1	85		Bw60	32	
A32	76		B8	41	
Aw24	73		Bw42	48	
HLA-Bw58	33		HLA-Cw1	66	
B27.1	39		Cw2.1	66	
B27.2	35		Cw2.2	63	63-69
B44.1	26		Cw3	69	

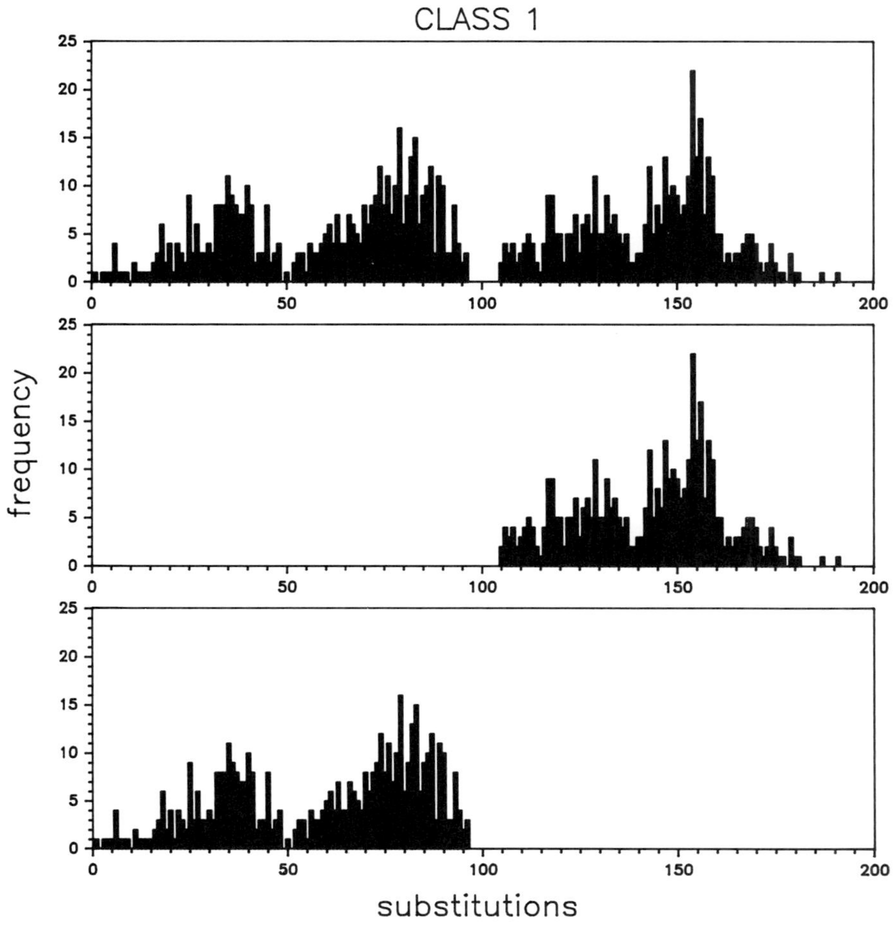

Figure 10. Comparison of nucleotide sequences encoding the α_1, α_2, and α_3 domains of class I molecules from different species. The number of nucleotide differences in all pairwise comparisons of 31 human, 6 mouse, 2 cow, 1 rabbit, 2 miniature swine, and 1 syrian hamster class I molecules is plotted as a frequency distribution. The top panel shows all comparisons, the middle panel shows only comparisons between pairs of molecules from different species, and the bottom panel shows only comparisons between pairs of molecules from the same species. The sequences and primary citations are listed in references 89 and 111.

In examination of HLA-A,B,C sequences, a search has been made for patterns similar to those found in the bm mutants. For example, Seemann et al. (73) found that subtypes HLA-B27.1 and -B27.2 differed by four substitutions in a stretch of 14 nucleotides that led to substitutions at amino acids 77, 80, and 81. They noticed that HLA-Aw24 had a sequence similar to that of B27.2 in this region, and speculated that a gene conversion between B27.1 and Aw24 or a related gene could have given rise to B27.2. However, HLA-Bw58 has the identical sequence to Aw24 (71), and an alternative hypothesis is that the exchange occurred between HLA-B27.1 and -Bw58 alleles.

The analysis shown in Figs. 8 and 9 has implications for the evolution of HLA-A,B,C genes. The result that alleles of one locus are more closely related to each other than to alleles of another locus shows that the exchange of genetic material between loci has been less important to the diversification of HLA-A,B,C than has exchange between the alleles of a locus. The situation is clearly different from that found for the H-$2K^b$ mutants, where most exchange is between loci (112).

What mechanisms have contributed to the diversification of HLA-A,B,C genes? Point mutations must serve to introduce new polymorphisms, examples being the single substitution in codon 11 between B14 and Bw65 or the substitution in codon 43 that produces the unique arginine residue in HLA-A2.2 (Fig. 3). Inspection of the sequences suggests that single and double recombination between alleles have been common events that assort elements of sequence into new combinations. A number of simple cases in which various molecules are related by

```
SIGNAL PEPTIDE
              -24                   -1
A2.1          -a-------v----------q---
A2.4a         -a-------v----------q---
B27.1         ---t-----l---w--v-------
Bw58          ---t----vl---w--v-------
Bw46          ---t----vl--------------
Cw1/3       mctklsggnlsrkhfpkea-------
Cw1           ---------i--------------
Cw3           ---------i--------------
Cw2.1         ---------i--------------
Consensus     MRVMAPRTL-LLLSGALALTETWA

ALPHA 1 DOMAIN
              1                           30                            60                            90
A2.1          --------f-s-------------------------------q-m----------------g--r-v-ahs--h--d-gt----------
A2.4a         ----------s-------------------------------q-m----------------g--r-v-ahs--h--d-gt----------
B27.1         --------h-s------------t-------l------------e--------------------ic-ak-----ed--t-lr-------
Bw58          -----------m--------------------------------t----------------g--rnm-as---y-en--ialr-------
Bw46          -----------m--------------------------------ma----------------------r---------------------
Cw1/3         c---------------------------------q----------------v----------------r---------------------
Cw1           c----k--f-s------------s---------------------------v----------------r---------------------
Cw3           --------c-----------h------------------de----------v-rk-------------p---------------------
Cw2.1         c--------------s----h---------------------------gr-v----------------r-------n--k----------
Consensus     GSHSMRYFYTAVSRPGRGEPRFIAVGYVDDTQFVRFDSDAASPRGEPRAPWIEQEGPEYWDRETQKYK-QAQTDRVSLRNLRGYYNQSEA

ALPHA 2 DOMAIN
              91                           120                            150                          180
A2.1          ----v---------s-w-f----h-y----------k----------m---t-kh-----hv-----------------------------t
A2.4a         ----v---------s-w-f----h-y----------k----------m---t-kh-----hv-----------------------------t
B27.1         ------n----------------h-d--------------s--------------------v----------e-------------------
Bw58          ---ii-------l---------h--s--------------s--------------------v----------l-------------------
Bw46          ----------------------h--s--------------s--------------------e---w------l-------------------
Cw1/3         ------------l----------n-f-----------------------------------t------------------------k-----
Cw1           ------w-c---l------------y-----------------------------------e---r---------------------s----
Cw3           ---ii--------------------h--------------------n--------------e----------l---------k-------g-
Cw2.1         ------------l------------s-----------------------------------e---w------e--------------k----
Consensus     GSHTLQRMYGCDVGPDGRLLRGYDQ-AYDGKDYIALNEDLRSWTAADTAAQITQRKWEAAR-AEQLRAYLEGTCVEWLRRYLENGKETLQRA

ALPHA 3 DOMAIN
              183                          212                            242                          272
A2.1          -a----m---a-------------s---------------------------------------------q---------------------
A2.4a         -a----m---a-------------s---------------------------------------------q---------------------
B27.1         -p---------i--------------------------------------------r-----------------------------------
Bw58          -p------------------------------------------------------r-----------------------------------
Bw46          -p---------i--------------------------------------------r-----------------------------------
Cw1/3         e------------------------------------------------------------------------------------e------
Cw1           e-----------------------------------w----------------------------m-------------------e------
Cw3           e-----------------------------------w------------------------------------------------e------
Cw2.1         e---------------------------t--------------------------------------------------------e------
Consensus     DHPKTHVTHHPVSDHEATLRCWALGFYPAEITLTWQRDGEDQTQDTELVETRPAGDGTFQKWAAVVVPSGEEQRYTCHVQHEGLPKPLTLRW

TRANSMEMBRANE DOMAIN
              275                           305     314
A2.1          -------------i---vlfga-it -----a--w-----
A2.4a         -------------i---vlfga-it -----a--w-----
B27.1         -----s-v----------------- -----a--------
Bw58          -----s------------------- -----t--------
Bw46          -----s------------------- -----t--------
Cw1/3         g---------------------lav----m----------
Cw1           ----------------------lav---------------
Cw3           ----------------------lav---------------
Cw2.1         ----------------------lav---------------
Consensus     EPSSQPTIPIVGIVAGLAVLAVVVILGAVVAVVMCRRKSS

CYTOPLASMIC DOMAIN
              315                        342
A2.1          dr-------------------------v*
A2.4a         dr-------------------------v*
B27.1         -----------c-------------*...
Bw58          -------------------------*...
Bw46          -------------------------*...
Cw1/3         ------c------n------e--i----*
Cw1           ------c------n------e--i----*
Cw3           ------c------n------e--i----*
Cw2.1         ------c------n------e--i----*
Consensus     GGKGGSYSQAASSDSAQGSDVSLTACKA-
```

Figure 11. Gene conversion between HLA-B and HLA-C in an oriental haplotype. Class I HLA-encoding cDNA clones were isolated from a cell line typed homozygous HLA-A2, HLA-Bw46, and HLA-Cw1/3, and were kindly provided by Dr. Malcolm Simons. The specificity designated Cw1/3 is identical to Cw3.3 and CSH1 and is now called Cw11 (114). Three distinct cDNA were obtained and sequenced. The molecule typed as HLA-A2 has the sequence corresponding to the HLA-A2.4α subtype. The sequences are compared with those of HLA-A2.1, HLA-B27.1, HLA-Bw58, HLA-Cw1, Cw3, and Cw2.1. The region 53–96 from which an exchange from HLA-Cw1/3 into an acceptor B locus allele to give HLA-Bw46 is shaded. The consensus is derived from the nine molecules in the figure. The HLA-C locus molecules have an extra amino acid in the transmembrane domain (59,74). Termination codons are indicated by an asterisk. The full color illustration of this figure appears in Color Plate V at the beginning of this volume.

single events can be found. HLA-Aw69 was probably formed by a recombination between HLA-A2.1 and HLA-Aw68.1 genes (68). Likewise HLA-Bw42 has the α_1 domain of HLA-B7 and the α_2 domain of HLA-B8, and probably results from a recombination between these two genes (81). A single recombination between HLA-A2.1 and HLA-A2.2Y would have been sufficient to generate the HLA-A2.4 and HLA-A2.2F alleles (77). Other putative pathways between current alleles involve multiple events. HLA-Bw41 is identical to HLA-Bw60 in the α_1 domain, and in α_2 it differs from HLA-B8 by two juxtaposed substitutions at positions 95 and 97 (81). Thus, HLA-Bw41 could have derived from HLA-Bw60 and HLA-B8 by a single recombination to generate the exon shuffle. This intermediate could then have undergone either a double recombination, in which a short sequence in the region of residues 95 and 97 was exchanged, or a double point mutation. More complex patterns of interallelic exchange can be discerned, especially in the α_1 domain of HLA-B locus molecules where a striking pathwork quilt motif is seen.

A critical problem is analyzing the events that give rise to new alleles in humans is that no example of the formation of a new allele has been found either in family studies or in analysis of cell lines in tissue culture. Identification of the actual donor and acceptor genes contributing to exchange events has therefore not been possible. We are thus limited to examination of alleles that are fixed in the contemporary population, and can say nothing about the relative frequency of exchange events between alleles of the same locus compared with alleles of different loci. Our analysis does not rule out a role for exchange events between loci in the diversification of HLA-A,B,C. In fact, this seems a likely mechanism by which the sequence encoding the Bw4 epitope (residues 79 to 83) was introduced into the HLA-A locus, and a similar argument can be made for sharing of the sequence (residues 62 to 65) responsible for the MA2.1 epitope by HLA-A2 and HLA-B17 (71).

Preliminary analysis of an oriental haplotype provides good evidence for a nonreciprocal genetic exchange between the HLA-B and HLA-C loci. This haplotype has been of interest because serologic analysis suggested it had a duplicated HLA-C locus with both Cw1 and Cw3 alleles (113). Subsequent analysis has shown that this serologic pattern is due to Cw11 a novel HLA-C molecule previously called Cw3.3, CSH1, or Cw1/3, which is invariably linked with HLA-Bw46 (114). Sequence analysis of cDNA from a cell line homozygous for this haplotype shows a unique C locus molecule (Fig. 11). This putative Cw1/3 molecule has a sequence with all the characteristics of a C locus molecule, although the 5′ end of the clone including part of the signal peptide has an unusual sequence, which may represent a cloning artifact. A second clone has many of the features expected of a B locus molecule, and is putatively assigned as coding for HLA-Bw46. However, in the region from residues 53 to 96, it has a characteristic C locus sequence that is identical to that found in the linked Cw1/3 molecule and in Cw1.

Thus, we hypothesize that HLA-Bw46 was formed by donation of all or part of the region encoding residues 53 to 96 from HLA-Cw1/3 to a linked HLA-B gene. A potential acceptor has not yet been found. If this was a relatively recent event, it could explain why HLA-Bw46 is always found in association with HLA-Cw1/3, insufficient time having elapsed to recombine HLA-Bw46 away from HLA-Cw1/3. However, it would not explain why the older gene HLA-Cw1/3 is always found with HLA-Bw46. This apparent total linkage disequilibrium could result from a serologic problem derived from the existence of a common helix in the α_1 domains of both the HLA-B and HLA-C locus molecules. This helix is often found to be serologically dominant, and the characteristics of HLA-Cw1/3 may be dependent on reactivity with both molecules, in which case the phenotype of the HLA-Cw1/3 molecule in the absence of HLA-Bw46 would not be picked up.

Conclusions

The sequences of a significant number of HLA-A,B,C molecules have been obtained. In combination with the crystallographic structure of HLA-A2, they provide a unique base of structural information for understanding the function and polymorphism of MHC molecules. Patterns of polymorphic substitution are becoming clear, and methods for locating epitopes recognized by alloantibodies are established. In certain instances, the evolutionary relationships between alleles can be inferred, and the divergence of alleles at the HLA-A,B, and C loci revealed (81). The evolutionary heritage of class I genes in man and mouse appears different, and one would like to know why. Lifespan, lifestyle, and chance, in addition to differences in potential pathogens, may all play a role.

Indirect evidence strongly argues that the function of HLA-A,B,C molecules is to bind antigenic peptides and present them to cytotoxic T lymphocytes. The peptide-combining site must be capable of binding peptides with many different primary sequences derived from a multitude of proteins. Polymorphism in MHC permits alterations in the specificity of the combining site and the repertoire of peptides bound. Multiple codominant loci and their extraordinary polymorphism both serve to increase the number of distinct class I molecules expressed by an individual, thereby expanding the number of foreign antigens that can be the target of a T-cell response. Although persuasive, these proposals are speculative, as is much of the interpretation of the crystallographic structure of HLA-A2 (5), and experimental tests of these hypotheses are critical.

To understand the specificity of the combining site and the influence of polymorphic differences, it is imperative that assays for measuring the binding of peptide to HLA-A,B,C be developed. Attempts so far, and in contrast to class II MHC molecules (18), have been unsuccessful. These efforts are hindered by the paucity of information on T-cell epitopes and their corresponding restriction elements. To date, the epitopes for only three HLA-A,B,C-restricted responses have been defined (19). A much larger data base is essential, and its production would be a worthwhile goal for subsequent workshops.

Despite 30 years of investigation, our understanding of the numbers of alleles at the HLA-A,B, and C loci, and their frequency, and geographic distribution is still very incomplete. With refinement of technique, many of the defined antigens are being split into a multitude of subtypes representing the products of different alleles. For a number of reasons, research efforts have been heavily focused on HLA-A2 and HLA-B27; in both cases, between 5 and 10 subtypes are at varying stages of characterization and the supply is not exhausted. If only partly generalized, one must begin to think in terms of hundreds of alleles.

Certain subtypes of HLA-A2 and HLA-B27 are specific to oriental populations, and there are likely to be many more population-specific alleles. The patterns of substitution in HLA-A,B,C sequences suggests that the polymorphism has been accumulating over a long period of time, starting long before the origin of modern Homo sapiens and the separation of races (77). Thus, one would expect that more recently evolved polymorphisms with population specificity would also involve the most subtle distinctions and be most difficult to detect. Extensive studies are still restricted to a select few of the world's populations, increasing the chance that many regional subtypes remain undetected. It is 16 years since an international workshop focused on the diversity of human populations and that involved alloantisera detecting a small number of antigens. Further population studies are important and are ideally suited as the focus of future workshops.

During the past 10 years, there has been relatively little emphasis on the development of serologic methods that will refine the definition of HLA-A,B,C to the point where there is a congruency between antigens, molecules, and alleles. In part, this can be attributed to the discovery of class II molecules, their complexity, and more prominent associations with disease. There also exists uncertainty as to the value of more splits of HLA-A,B,C for transplantation matching, especially when it is unclear which differences are important and perfectly clear that the ability to match decreases with the number of types. However, greater understanding of the genetics, evolution, and functional role of class I MHC polymorphism is dependent on further definition of alleles, their structures, frequencies, and distribution. This in turn should lead to an understanding of the allogeneic response, the hierarchy of alloantigenic epitopes, and quantitative assessments of match based on better HLA typing.

References

1. Gorer PA. The detection of antigenic differences in mouse erythrocytes by the employment of immune sera. Br J Exp Pathol 1936;17:42–50.
2. Carlson GA. Fifty years of H-2 research. Immunology Today 1987;8:317–319.
3. Dausset J. Iso-leuco-anticorps. Acta Haematol 1958;20: 156–166.
4. Bjorkman PJ, Saper MA, Samraoui B, Bennett WS, Strominger JL, Wiley DC. Structure of the human class I histocompatibility antigen, HLA-A2. Nature 1987;329:506–512.
5. Bjorkman PJ, Saper MA, Samraoui B, Bennett WS, Strominger JL, Wiley DC. The foreign antigen binding site and T cell recognition regions of class I histocompatibility antigens. Nature 1987;329:512–518.
6. Simpson E. Sir Peter Medawar. Immunology Today 1988; 9:4–6.
7. Askonas BA, Klaus GGB. John H. Humphrey (1915–1987). Immunology Today 1988;9:100–101.
8. Silverstein AM. Cellular versus humoral immunity: determinants and consequences of an epic 19th century battle. Cell Immunol 1979;48:208–221.
9. Livingstone AM, Fathman CG. The structure of T-cell epitopes. Ann Rev Immunol 1987;5:477–501.
10. Braciale TJ, Morrison LA, Sweetser MT, Sambrook J, Gething M-J, Braciale VL. Antigen presentation pathways to class I and class II MHC-restricted T lymphocytes. Immunol Rev 1987;98:95–114.
11. Marrack P, Kappler J. The T cell receptor. Science 1987; 238:1073–1079.
12. Allen PM. Antigen processing at the molecular level. Immunology Today 1987;8:270–273.
13. Rogers J, Early P, Carter C, et al. Two mRNAs with different 3′ ends encode membrane-bound and secreted forms of immunoglobulin mu chain. Cell 1980;20:303–312.
14. Marrack P, Kappler J. The antigen-specific major histocompatibility complex-restricted receptor on T cells. Advances Immunol 1986;38:1–30.
15. Allison JP, Lanier LL. Structure, function and serology of the T-cell antigen receptor complex. Ann Rev Immunol 1987;5:503–540.
16. Toyonaga B, Mak TW. Genes of the T-cell antigen receptor in normal and malignant T cells Ann Rev Immunol 1987;5:585–620.
17. Berzofsky JA, Cease KB, Cornette JL, et al. Protein antigenic structure recognized by T cells: potential applications to vaccine design. Immunol Rev 1987;98:9–52.
18. Allen PM, Babbit BP, Unanue ER. T-cell recognition of lysozyme: the biochemical basis of presentation. Immunol Rev 1987;98:171–187.
19. Rothbard JB, Taylor WR. A sequence common to T cell epitopes. EMBO J 1988;7:93–100.
20. Matzinger P, Zamoyska R. A beginner's guide to major histocompatibility complex function. Nature 1982;297: 628.
21. Ploegh HL, Orr HT, Strominger JL. Major histocompatility antigens: the human (HLA-A,B,C) and murine (H-2K, H-2D) class I molecules. Cell 1981;24:287–299.
22. Kaufman JF, Auffray C, Korman AJ, Shackelford DA, Strominger JL. The class II molecules of the human and murine major histocompatibility complex. Cell 1984; 36:1–13.
23. Germain RN. The ins and outs of antigen processing and presentation. Nature 1986;322:687–689.
24. Littman DR. The structure of the CD4 and CD8 genes. Ann Rev Immunol 1987;5:561–584.
25. Swain SL. Significance of Lyt phenotypes: Lyt2 antibodies block activities of T cells that recognize class I major histocompatibility complex antigens regardless of their function. Proc Natl Acad Sci USA 1981;78:7101–7105.
26. Doyle C, Strominger JL. Interaction between CD4 and

class II MHC molecules mediates cell adhesion. Nature 1987;330:256–259.
27. Babbit BP, Allen PM, Matsueda G, Haber E, Unanue ER. Binding of immunogenic peptides to Ia histocompatibility molecules. Nature 1985;317:359–361.
28. Townsend ARM, Rothbard J, Gotch FM, Bahadur G, Wraith D, McMichael AJ. The epitopes of influenza nucleoprotein recognized by cytotoxic T lymphocytes can be defined with short synthetic peptides. Cell 1986;44: 959–968.
29. Koller BH, Orr HT. Cloning and complete sequence of an HLA-A2 gene: analysis of two HLA-A alleles at the nucleotide level. J Immunol 1985;13:2727–2733.
30. Suggs SV, Wallace RB, Hirose T, Kawashima EH, Itakura K. Use of synthetic oligonucleotides as hybridization probes: isolation of clones cDNA sequences for human beta$_2$-microglobulin. Proc Natl Acad Sci USA 1981;78:6613–6617.
31. Matzinger P, Bevan MJ. Why do so many lymphocytes respond to major histocompatibility antigens? Cell Immunol 1977;29:1–5.
32. Gorer PA, Lyman S, Snell GD. Studies on the genetic and antigenic basis of tumour transplantation. Linkage between a histocompatibility gene and "fused" in mice. Proc Roy Soc B 1948;135:499–505.
33. Payne R, Hackel E. Inheritance of human leukocyte antigens. Am J Human Genet 1961;13:306–319.
34. van Rood JJ, van Leeuwen A. Leukocyte grouping. A method and its application. J Clin Invest 1963;42:1382–1390.
35. Bodmer WF, Albert E, Bodmer JG, et al. Nomenclature for factors of the HLA system 1984. In: Albert ED, Baur MP, Mayr WR (eds.). Histocompatibility Testing 1984. Berlin, Heidelberg: Springer-Verlag, 1984.
36. Biddison WE, Ward FE, Shearer GM, Shaw S. The self determinants recognized by human virus immune T cells can be distinguished from the serologically defined HLA antigens. J Immunol 1980;124:548–552.
37. Spits H, Breuning M, Ivanyi P, Russo C, de Vries JE. In vitro-isolated human cytotoxic T-lymphocyte clones detect variations in serologically defined HLA antigens. Immunogenetics 1982;16:503–512.
38. van der Poel JJ, Molders H, Thompson A, Ploegh HL. Definition of four HLA-A2 subtypes by CML typing and biochemical analysis. Immunogenetics 1983;17:609–621.
39. Toubert A, Gomard G, Grumet FC, Amor B, Muller JY, Levy JP. Identification of several functional subgroups of HLA-B27 by restriction of the activity of antiviral T killer lymphocytes. Immunogenetics 1984; 20:513–525.
40. Neefjes JJ, Breur-Vriesendorp BS, van Seventer GA, Ivanyi P, Ploegh HL. An improved biochemical method for the analysis of HLA-class I antigens. Definition of new HLA-class I subtypes. Human Immunol 1986;16: 169–181.
41. Kennedy LJ, Wallace LE, Madrigal JA, Rickinson AB, Bodmer JG. New HLA-A2 variations defined by monoclonal antibodies and cytotoxic T lymphocytes. Immunogenetics 1987;26:155–160.
42. Parham P, Terhorst C, Herrmann H, Humphreys RE, Waterfield MD, Strominger JL. Immunological and chemical purity of papain-solubilized HLA antigens. Proc Natl Acad Sci USA 1975;72:1594–1598.
43. Apella E, Tanigaki N, Fairwell T, Pressman D. Partial amino acid sequences of the heavy chains of human HLA histocompatibility antigens. Biochem Biophys Res Commun 1976;71:286–292.
44. Ballou B, McKean DJ, Freedlender EF, Smithies O. HLA membrane antigens: sequencing by intrinsic radioactivity. Proc Natl Acad Sci USA 1976;73:4487–4491.
45. Bridgen J, Snary D, Crumpton MJ, Barnstable C, Goodfellow P, Bodmer WF. Isolation and N-terminal amino acid sequence of membrane bound human HLA-A and HLA-B antigens. Nature 1976;261:200–205.
46. Terhorst C, Parham P, Mann DL, Strominger JL. Structure of HLA antigens: Amino-acid and carbohydrate compositions and NH_2-terminal sequences of four antigen preparations. Proc Natl Acad Sci USA 1976;73:910–914.
47. Orr HT, Lopez de Castro JA, Lancet D, Strominger JL. Complete amino acid sequence of a papain-solubilized human histocompatibility antigen, HLA-B7. 2. Sequence determination and search for homologies. Biochemistry 1979;18:5711–5720.
48. Tragardh L, Rask L, Wiman K, Peterson PA. Primary structure of pooled, papain-solubilized HLA-A,B and C antigens. Scand J Immunol 1979;10:597–600.
49. Ezquerra A, Bragado R, Vega MA, Strominger JL, Woody J, Lopez de Castro JA. Primary structure of papain-solubilized human histocompatibility antigen HLA-B27. Biochemistry 1985;24:1733–1741.
50. Orr HT, Lopez de Castro JA, Parham P, Ploegh HL, Strominger JL. Comparison of amino acid sequences of two human histocompatibility antigens, HLA-A2 and HLA-B7: location of putative alloantigenic sites. Proc Natl Acad Sci USA 1979;76:4395–4399.
51. Lopez de Castro JA, Strominger JL, Strong DM, Orr HT. Structure of crossreactive human histocompatibility antigens HLA-A28 and HLA-A2: possible implications for the generation of HLA polymorphism. Proc Natl Acad Sci USA 1982;79:3813–3817.
52. Lopez de Castro JA, Bragado R, Strong DM, Strominger JL. Primary structure of papain-solubilized human histocompatibility antigen HLA-B40 (-Bw60). An outline of alloantigenic determinants. Biochemistry 1983;22:3961–3969.
53. Krangel MS, Taketani S, Biddison WE, Strong DM, Strominger JL. Comparative structural analysis of HLA-A2 antigens distinguishable by cytotoxic T lymphocytes: variants M7 and DR1. Biochemistry 1982;21:6313–6321.
54. Taketani S, Krangel MS, Spits H, de Vries J, Strominger JL. Structural analysis of an HLA-B7 antigen variant detected by cytotoxic T lymphocytes. J Immunol 1984; 133:816–821.
55. van Schravendijk MR, Biddison WE, Berger AE, Coligan JE. Comparative structural analysis of HLA-A3 antigens distinguishable by cytotoxic T lymphocytes: variant E1. J Immunol 1985;134:410–416.
56. Rojos S, Aparicio P, Choo SY, Hansen JA, Lopez de Castro JA. Structural analysis of an HLA-B27 population variant, B27f. Multiple patterns of amino acid changes within a single polypeptide segment generates polymorphism in HLA-B27. J Immunol 1987;139:831–836.
57. Ploegh HL, Orr HT, Strominger JL. Molecular cloning of a human histocompatibility antigen cDNA fragment. Proc Natl Acad Sci USA 1980;77:6081–6085.
58. Sood AK, Pereira D, Weissman SM. Isolation and partial nucleotide sequence of a cDNA clone for human histocompatibility antigen HLA-B by use of a oligodeoxynucleotide primer. Proc Natl Acad Sci USA 1981;78: 616–620.

59. Sodoyer R, Damotte M, Delovitch TL, Trucy J, Jordan BR, Strachan T. Complete nucleotide sequence of a gene encoding a functional human class I histocompatibility antigen (HLA-CW3). EMBO J 1984;3:879–885.
60. Strachan T, Sodoyer R, Damotte M, Jordan BR. Complete nucleotide sequence of a functional class I HLA gene, HLA-A3: implications for the evolution of HLA genes. EMBO J 1984;3:887–894.
61. Biro PA, Pan J, Sood AK, Kole R, Weissman SM. Sequences of human repetitive DNA, non-alpha-globin genes, and major histocompatibility locus genes (part III), the major histocompatibility complex. Cold Spring Harbor Symp Quant Biol 1983;47:1079–1086.
62. Koller BH, Geraghty D, Orr HT, Shimuzu Y, De Mars R. Organization of the human class I major histocompatibility complex genes. Immunol Res 1987;6:1–10.
63. Arnot D, Lillie JW, Auffray C, Kappes D, Strominger JL. Inter-locus and intra-allelic polymorphisms of class I antigen gene mRNA. Immunogenetics 1984;20:237–252.
64. Lemonnier FA, Malissen M, Golstein P, et al. Expression of human class I histocompatibility antigens at the surface of DNA transformed mouse L cells. Immunogenetics 1982;16:355–361.
65. Koller BH, Sidwell B, De Mars R, Orr HT. Isolation of HLA locus-specific DNA probes from the 3′-untranslated region. Proc Natl Acad Sci USA 1984;81:5175–5178.
66. Coppin HL, Denny DW Jr, Weissman SM, McDevitt HO. HLA-B locus polymorphism: Studies with a specific hybridization probe. Proc Natl Acad Sci USA 1985;82:8614–8618.
67. Strachan T, Dodge AB, Smillie D, et al. An HLA-C specific DNA probe. Immunogenetics 1986;23:115–120.
68. Holmes N, Parham P. Exon shuffling in vivo can generate novel class I MHC antigens. EMBO J 1985;4:2849–2854.
69. Cowan EP, Jordan BR, Coligan JE. Molecular cloning and DNA sequence analysis of genes encoding cytotoxic T lymphocyte-defined HLA-A3 subtypes: the E1 subtype. J Immunol 1985;135:2835–2841.
70. Cowan EP, Jelachich ML, Biddison WE, Coligan JE. DNA sequence of HLA-A11: remarkable homology with HLA-A3 allows identification of residues involved in epitopes recognized by antibodies and T cells. Immunogenetics 1987;25:241–250.
71. Ways JP, Coppin H, Parham P. The complete primary structure of HLA-Bw58. J Biol Chem 1985;260:11924–11933.
72. N'Guyen C, Sodoyer R, Trucy J, Strachan T, Jordan BR. The HLA-AW24 gene: sequence, surroundings and comparison with the HLA-A2 and HLA-A3 genes. Immunogenetics 1985;21:479–489.
73. Seemann GHA, Rein RS, Brown CS, Ploegh HL. Gene conversion-like mechanisms may generate polymorphism in human class I gene. EMBO J 1986;5:547–552.
74. Gussow D, Rein RS, Meyer I, et al. Isolation, expression and the primary structure of HLA-Cw1 and HLA-Cw2 genes: evolutionary aspects. Immunogenetics 1987;25: 313–322.
75. Wan AM, Ennis P, Parham P, Holmes N. The primary structure of HLA-A32 suggests a region involved in formation of Bw4/Bw6 epitopes. J Immunol 1986;137: 3671–3674.
76. Coppin HL, McDevitt HO. Absence of polymorphism between HLA-B27 genomic exon sequences isolated from normal donors and ankylosing spondylitis patients. J Immunol 1986;137:2169–2172.
77. Holmes N, Ennis P, Wan AM, Denney DW, Parham P. Multiple genetic mechanisms have contributed to the generation of the HLA-A2/A28 family of class I MHC molecules. J Immunol 1987;139:936–941.
78. Ways JP, Lawlor DA, Wan AM, Parham P. A transposable epitope of HLA-B7,B40 molecules. Immunogenetics 1987;25:323–328.
79. Mattson DH, Handy DE, Bradley DA, Coligan JE, Cowan EP, Biddison WE. DNA sequences of the genes that encode CTL-defined HLA-A2 variations M7 and DK1. Immunogenetics 1987;26:190–192.
80. Zemmour J, Ennis PD, Parham P, Dupont B. Comparison of the structure of HLA-Bw47 to HLA-B13 and its relationship to 21-hydroxylase deficiency. Immunogenetics 1988;27:281–287.
81. Parham P, Lomen CE, Lawlor DA, et al. The nature of polymorphism in HLA-A,B,C molecules. Proc Natl Acad Sci USA 1988;85:4005–4009.
82. Rein RS, Seemann GHA, Neefjes JJ, Hochstenbach FMH, Stam NJ, Ploegh HL. Association with beta$_2$-microglobulin controls the expression of transfected human class I genes. J Immunol 1987;138:1178–1183.
83. Storkus W, Howell DN, Salter RD, Dawson JR, Cresswell P. NK susceptibility varies inversely with target cell class I HLA antigen expression. J Immunol 1987;138:1657–1659.
84. Salter RD, Clayberger C, Lomen CE, Krensky AM, Parham P. In vitro mutagenesis at a single residue introduces B and T cell epitopes into a class I HLA molecule. J Exp Med 1987;166:283–288.
85. Shimizu Y, Geraghty DE, Koller BH, Orr HT, De Mars R. Transfer and expression of three cloned human non-HLA-A,B,C class I major histocompatibility complex genes in mutant lymphoblastoid cells. Proc Natl Acad Sci USA 1988;85:227–231.
86. Sugden B, Marsh K, Yates J. A vector that replicates on a plasmid and can be efficiently selected in B lymphoblasts transformed by Epstein-Barr virus. Mol Cell Biol 1985;5:410–413.
87. Kottmann AH, Seeman GHA, Guessow HD, Roos MH. DNA sequence of the coding region of the HLA-B44 gene. Immunogenetics 1986;23:396–400.
88. Szöts H, Riethmuller G, Weiss E, Meo T. Complete sequence of HLA-B27 cDNA identified through the characterization of structural markers unique to the HLA-A,-B and -C allelic series. Proc Natl Acad Sci USA 1986;83:1428–1432.
89. Ennis PD, Jackson AP, Parham P. Molecular cloning of bovine class I MHC cDNA. J Immunol 1988;141:642–651.
90. Weiss EH, Kron W, Dorner C, Lang M, Riethmuller G. Organization, sequence and expression of the HLA-B27 gene: a molecular approach to analyze HLA and disease associations. Immunobiology 1985;170:367–380.
91. Yang SY, Morishima Y, Collins NC, et al. Comparison of one-dimensional IEF patterns for serologically detectable HLA-A and -B allotypes. Immunogenetics 1984;19: 217–231.
92. Wu TT, Kabat E. An analysis of the sequences of the variable regions of Bence Jones proteins and myeloma light chains and their implications for antibody complementarity. J Exp Med 1970;132:211–250.
93. Potter TA, Bluestone JA, Rajan TV. A single amino acid substitution in the alpha$_3$ domain of an H-2 class I molecule abrogates reactivity with CTL. J Exp Med 1987;166:556–566.
94. Parham P, Androlewicz MJ, Holmes NJ, Rothenberg BE. Arginine 45 is a major part of the antigenic determinant of

humans $beta_2$-microglobulin recognized by mouse monoclonal antibody BBM.1. J Biol Chem 1983;258:6179–6186..
95. Parham P, Lutz PM, Cresswell P. Localization of an epitope of beta-2 microglobulin that is shared with other members of the immunoglobulin superfamily. Immunogenetics 1987;26:323–326.
96. Santos-Aguado J, Biro PA, Fuhrmann U, Strominger JL, Barbosa JA. Amino acid sequences in the $alpha_1$ domain and not glycosylation are important in HLA-A2/$Beta_2$-microglobulin association and cell surface expression. Mol Cell Biol 1987;7:982–990.
97. Barbosa JA, Santos-Aguado J, Mentzer SJ, Strominger JL, Burakoff SJ, Biro PA. Site-directed mutagenesis of class I HLA genes. J Exp Med 1987;166:1329–1350.
98. Santos-Aguado J, Yunis J, Milford JE, Yunis JE, Strominger JL. Molecular characterization of the serologic recognition sites in the HLA-A2 molecule. Abstract 295. Tenth International Histocompatibility Conference, 1987.
99. Kostyu DD, Cresswell P, Amos DB. A public HLA antigen associated with HLA-A9, Aw32 and Bw4. Immunogenetics 1980;10:433–442.
100. Layet C, Delovitch T, Ferrier P, Caillol DH, Jordan BR, Lemonnier FA. Expression of an HLA-Bw6-related specificity by the HLA-Cw3 molecule. Immunogenetics 1985;21:469–478.
101. Tiwari JL, Terasaki PI. HLA and Disease Associations. New York: Springer-Verlag, 1985.
102. Dupont B, Oberfield SE, Smithwick EM, Lee TD, Levine LS. Close genetic linkage between HLA and congenital adrenal hyperplasia (21-hydroxylase deficiency). Lancet 1977;2:1309–1311.
103. White PC, New MI, Dupont B. HLA-linked congenital adrenal hyperplasia results from a defective gene encoding a cytochrome P-450 specific for steroid 21 hydroxylation. Proc Natl Acad Sci USA 1984;81:7505–7509.
104. Brewerton DA, Caffrey M, Hart FD, James DCO, Nicholls A, Sturrock RD. Ankylosing spondylitis and HL-B27. Lancet 1973;1:904–907.
105. Choo SY, Seyfried C, Hansen JA, Nepom GT. Tryptic peptide mapping identifies structural heterogeneity among six variants of HLA-B27. Immunogenetics 1986; 23:409–412.
106. Parham P, Antonelli P, Herzenberg LA, Kipps T, Fuller A, Ward FE. Further studies on the epitopes of HLA-B7 defined by murine monoclonal antibodies. Hum Immunol 1985;15:44–67.
107. Parham P. Antigenic determinants of the B7 molecule; Bw6 and B7 specific determinants are spatially separate. Immunogenetics 1983;18:1–16.
108. Ellis SA, Taylor C, McMichael A. Recognition of HLA-B27 and related antigens by monoclonal antibody. Hum Immunol 1982;5:49–59.
109. Parham P. Monoclonal antibodies against two separate alloantigenic sites of HLA-B40. Immunogenetics 1981; 13:509–527.
110. Tongio MM, Mitsuisjhi Y, Hauptmann G, Mayer S. In Histocompatibility Testing 1984. Albert ED, Baur MP, Mayr MW (eds.). Berlin, Heidelberg: Springer-Verlag, 1984:138–139.
111. Maloy WL. Comparison of the primary structure of class I molecules. Immunol Res 1987;6:11–29.
112. Nathenson SG, Geliebter J, Pfaffenbach GM, Zeff RA. Murine major histocompatibility complex class-I mutants: molecular analysis and structure-function implications. Ann Rev Immunol 1986;4:471–502.
113. Goodfellow PN, Payne RO. Serological studies of Cwl and Cw3 expressed on the same haplotype. Tissue Antigens 1978;11:350–356.
114. Lee TD, Zhao TM. Evidence for CSH1, a variant of Cw3, in an oriental population. Tissue Antigens 1987;29:71–76.

Author Affiliation

P.J. Bjorkman, Departments of Cell Biology and Medical Microbiology; *P. Parham, D.A. Lawlor, R.D. Salter, C.E. Lomen, P.D. Ennis*, Stanford University School of Medicine, Sherman Fairchild Science Building, Stanford, California 94305, USA

HLA Class I Gene Family: Characterization of Genes Encoding Non-HLA-A,B,C Proteins

Harry T. Orr

Non-HLA-A,B,C Antigens: The Rational and Initial Evidence for Their Existence

At times during the course of research on the major histocompatibility complex, the murine H-2 complex has often served as a model for the analysis of the human HLA complex. This is particularly the case in terms of the characterization of the class I MHC molecules (1). The polymorphic nature of class I molecules and their role in providing the major immunologic barrier to transplantation were first established in the mouse. The isolation and initial molecular characterization of the murine class I major transplantation antigens preceded that of the human major transplantation antigens, HLA-A, -B, and -C. Thus, because the murine class I "isoantigen," the thymus-leukemia (TL) antigen was defined as a molecule separate from the H-2 major histocompatibility antigens (2), attempts have been made to identify the human equivalent of TL. More recently, the extensive characterization at the DNA level of the murine Qa/Tla region genes (3,4) made the apparent absence of human class I nonmajor transplantation genes a notable distinction between the H-2 and HLA complexes.

Immunobiology of HLA
Volume II

Initial attempts to determine if the human MHC includes class I genes in addition to those encoding the major transplantation antigens relied on two features of the murine Qa/Tla antigens: a molecular structure consisting of an approximately 40 kDa protein associated with the 12 kd protein beta-2-microglobulin and their expression on subpopulations of lymphoid cells. Alloantisera and monoclonal antibodies have been identified that recognize non-HLA-A,B,C class I-like antigens on the surface of certain human lymphoid cells (5–13). Studies using rabbit anti-HLA and anti-beta-2-microglobulin allosera found an HLA-like protein associated with beta-2-microglobulin on the surface of the human T-lymphoid cell line MOLT-4 (5,6). Subsequent studies extended this observation to other T-cell lines and thymocytes and to the use of monoclonal antibodies (7–13). One alloantigenic series, designated HT, was reported to segregate with the HLA-A locus (8).

The most extensively characterized human thymocyte antigen, often considered to be the homologue of the mouse Tla antigens, is the antigen recognized by the CD1 monoclonal antibodies (14). The CD1 antigens have been divided into three subgroups: CD1a (originally designated T6 or HTA1), CD1b, and CD1c. These molecules consist of a glycosylated heavy chain of 49, 45, and 43 kDa, respectively, in association with beta-2-microglobulin. The corresponding cDNA clones and genomic clones have been obtained and identified through DNA transfection assays. These genes encode proteins with equal similarity to MHC class I and class II molecules. In addition, they do not share any specific sequence similarity with the murine Tla genes. Finally, the CD1 genes do not map to the HLA complex on the short arm of chromosome 6 (15). Thus, the CD1 genes, although considered to be members of the immunoglobulin supergene family, are clearly not members of the HLA class I gene family.

Southern blot analysis of human genomic DNA with HLA class I cDNA probes supports the idea that the human class I gene family consists of genes in addition to those encoding the HLA-A, -B, and -C antigens (16). Depending on the interpretation of the hybridization patterns generated with HLA class I probes, the HLA class I gene family includes between 17 and 20 different sequences. Up until only very recently, it remained uncertain whether any of these non-HLA-A,B,C DNA sequences encodes an HLA class I heavy chain.

Isolation of Non-HLA-A,B,C Class I Genes

HLA-E (17) is the first non-HLA-A,B,C class I gene to be recognized by the WHO Nomenclature Committee. Genomic clones of HLA-E have been identified by two groups (17,18) and a cDNA clone corresponding to HLA-E has been obtained from a T-cell line mRNA (19). At this conference, the isolation and characterization of three additional non-HLA-A,B,C class I genes was also reported: HLA-6.0 (20), HLA-5.4 (21), and HA2 (22). Sequence analyses of the four non-A,B,C class I genes show that each is intact and has the potential of encoding a class I heavy chain. We shall briefly describe the general features of these four genes.

HLA-E. Figure 1 (upper panel) presents the exon/intron structure of the HLA-E gene (17). Overall, the introns and exons within the HLA-E gene are of similar arrangement and size to those of other class I genes. The most notable variation in the coding region of HLA-E is the deletion of five nucleotides from exon 7 that results in an in-frame termination codon and a class I protein one amino acid shorter than most HLA-B chains, and four amino acids shorter than most HLA-A chains. Within the noncoding regions of the HLA-E gene, three Alu elements are found that are not present in other HLA class I genes. The first Alu sequence is located in the 5′ flanking region of HLA-E, Alu-2 is within intron 5, and the third Alu element is in the 3′ untranslated region. The 5′ flanking region of the HLA-E gene contains a CAAT and a TATAA box in positions similar to other class I genes; however, the variant TCTAA found within HLA-A and HLA-B genes is replaced by TATAA similar to that found in the murine H-2K, D, and L genes.

HLA-6.0. The exon/intron organization of the HLA-6.0 gene is identical to that of the HLA-A,B,C genes (Fig. 1, middle panel) (20). However, the HLA-6.0 gene differs from other HLA class I genes in that an in-frame termination codon is located at the second codon of exon 6. Thus, most of the sixth exon and all of the seventh and eighth exons of HLA-6.0 would not be translated into protein. The 5′ flanking region of the HLA-6.0 gene contains a CAAT box and variant TATAAA sequence, TCTAAA. Thus, the HLA-6.0 gene contains promoter elements conserved among the expressed HLA-A and HLA-B genes.

HLA-5.4. Depicted in the lower panel of Figure 1 are the intron and exon positions within the HLA-5.4 gene (21). Their organization and size are similar to that of the HLA-A and -B genes. HLA-5.4 differs from other class I genes in that the 3′ splice recognition site of intron 6 has a variant dinucleotide sequence: AA instead of the conserved AG sequence. This alteration in the 3′ splice site of intron 6 would mean that the nucleotides within exon 7 of the HLA-5.4 gene would not be present in HLA-5.4 mRNA. The promoter region of the HLA-5.4 gene contains a CAAT box and TCTAAA conserved among the HLA-A and HLA-B genes.

HA2. The sequence of the HA2 gene was presented at this conference. The HA2 gene is similar to the HLA-A and HLA-B genes in its exon/intron organization. It also seems that the HA2 gene is distinct from HLA-E, HLA-6.0, and HLA-5.4. The HA2 gene was isolated from a genomic library using a probe from the flanking region of an HLA-A gene (22). This probe detects a limited number of restriction fragments on a Southern blot, suggesting that the HA2 and HLA-A genes belong to a subfamily of HLA class I genes.

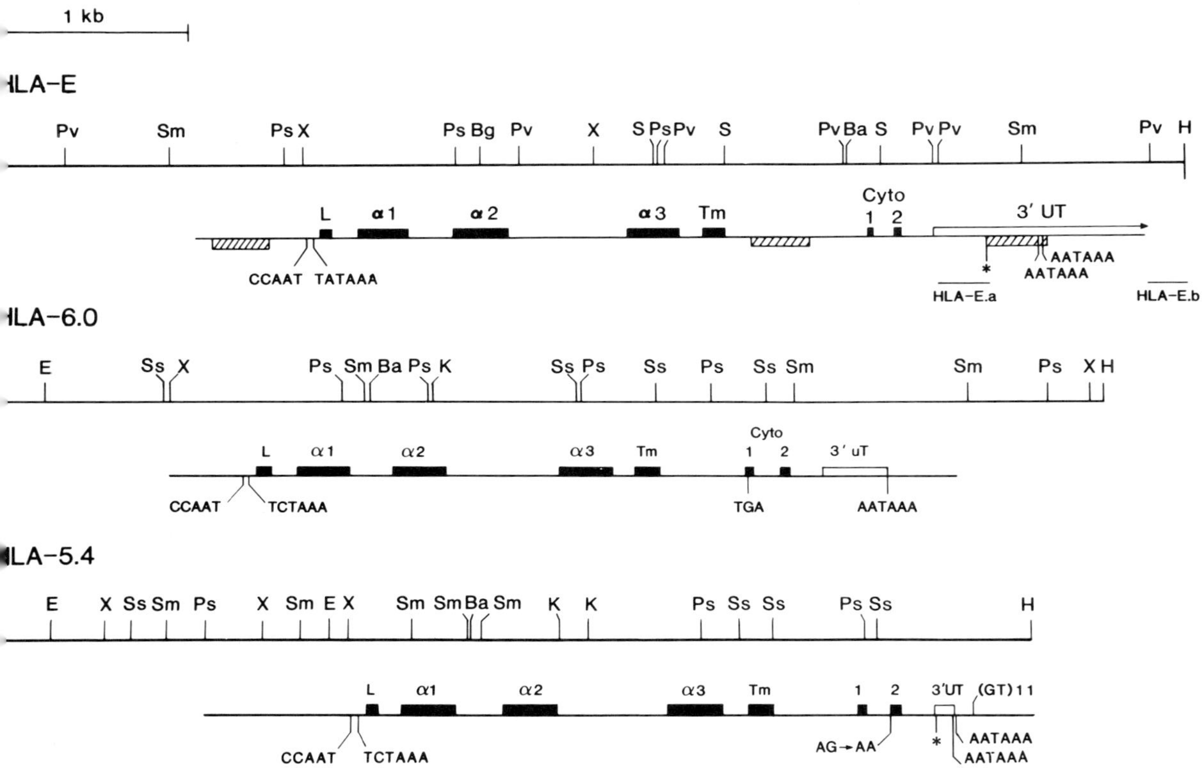

Figure 1. The exon/intron organization of the HLA-E (upper panel), HLA-6.0 (middle panel), and HLA-5.4 (lower panel) genes as determined by DNA sequencing. Exons, indicated by raised bars and labeled according to protein region, encode the following: L = leader peptide; α1, α2, and α3 = the three extracellular regions; Tm = transmembrane region; cyto 1 and 2 = the cytoplasmic regions; 3′ uT = 3′ untranslated region. The three Alu elements in HLA-E are indicated by the lower striped bars. Termination codons in HLA-E and HLA-5.4 are indicated by (*). A restriction map of each gene is presented. Restriction sites are indicated: Ba = BamHI; Bg = BglII; E = EcoRI; H = HindIII; K = KpnI; Ps = PstI; Pv = PvuII; Ss or S = SstI; Sm = SmaI; X = XbaI.

Expression of the Non-HLA-A,B,C Class I Genes

Each of the non-HLA-A,B,C class I genes listed above have been shown to encode an HLA class I heavy chain after DNA-mediated gene transfer studies. The HA2 gene has been transferred into mouse fibroblasts expressing human beta-2-microglobulin (22). Expression of the transferred class I gene at the cell surface of the mouse fibroblasts was assayed by absorption and cytofluorometry using antibodies against HLA major transplantation antigens and alloantisera that recognize class I antigenic determinants expressed on activated T cells. The mouse fibroblasts expressing HA2 DNA did not react in either assay with the monoclonal antibodies, W6/32 or B1-23-2, against monomorphic HLA class I determinants, and they did not react with any of the HLA-A, -B, or -C allele-specific monoclonal antibodies or alloantisera tested. These cells also failed to react with the HA10 alloantiserum against T-cell class I determinants; however, the cells did react with the anti-T cell class I alloantiserum HA2. HA2 is an alloantiserum that recognizes a class I antigenic determinant on HLA-A2 positive, PHA-activated T cells, and Epstein-Barr virus transformed lymphoblastoid cell lines. The HA2 determinant is also detected on peripheral blood lymphocytes and enriched B lymphocytes. Thus, this study indicated that the HLA complex contains a class I gene that encodes a class I antigen specific for activated lymphocytes, a possible analogue of a murine Qa gene.

Expression of the HLA-E, HLA-6.0, and HLA-5.4 genes has been studied following their transfer into an HLA-A,B,C-null B-lymphoblastoid cell line (23). B-lymphoblastoid, HLA-null cells containing HLA-6.0 DNA became positive at their cell surface for W6/32 binding and binding of an anti-beta-2-microglobulin monoclonal antibody, BBM.1. Increased binding of these monoclonal antibodies was not detected on cells containing either the HLA-E gene or the HLA-5.4 gene. However, sodium dodecyl sulfate-polyacrylamide gel electrophoresis (SDS-PAGE) analysis of immunoprecipitates with W6/32 and BBM.1 showed that lysates from cells into which an exogenous HLA-E gene or an exogenous HLA-5.4 gene had been transferred, contained an HLA heavy chain that coprecipitated with beta-2-microglobulin. Thus, each of the non-HLA-A,B,C class I genes, HLA-E, HLA-6.0, and HLA-5.4, encodes an HLA class I heavy chain that associates with beta-2-microglobulin. However, only the HLA-6.0 gene seems to encode a class I protein that, in association with beta-2-microglobulin, gets to the cell surface of a B-LCL after DNA-mediated gene transfer.

HLA-E gene expression has also been examined following its transfer into a mouse fibroblast cell line (18). When HLA-E DNA was transferred into a mouse L-cell line expressing only mouse beta-2-microglobulin, no surface product could be detected with either the W6/32 monoclonal antibody or with a xenoantiserum raised against a mixture of human B- or T-cell lines. Human class I protein was detectable in the cytoplasm by immunofluorescence. However, when HLA-E DNA was transferred into a mouse L-cell line expressing the human beta-2-microglobulin gene, cell-surface expression of the human class I molecule was reported to be detectable by enzyme-linked immunosorbent assay (ELISA). This result is in contrast to lack of cell surface expression in HLA-E when the gene was transferred into a human B-LCL. The difference in the results of these two studies might be explained by a difference between a fibroblast versus a B-LCL in the cell surface expression of the HLA-E encoded molecule.

The cell-specificity of HLA-E expression has been studied to a limited degree by Northern blot analysis (17,19). A gene-specific probe, subloned from the 3′ untranslated region of the gene, detects two RNA species, a 3.3 kb and a 1.95 kb band. The 1.95 kb band might contain transcripts from the HLA-E gene along with transcripts from other HLA class I genes. In contrast, the 3.3 kb band contains transcripts from only the HLA-E gene. Therefore, the 3.3 kb band was used as an indicator of HLA-E expression in several different cell types. The 3.3 kb transcript was detectable in all tissues examined, although at varying levels. Small amounts of the 3.3 kb band were found in RNA from a B-LCL, a melanoma cell line, a renal carcinoma cell line, the T-cell line MOLT-4, and a lymph node (17). A higher level of the 3.3 kb transcript was found in RNA from resting peripheral T cells. Interestingly, in peripheral T cells activated by PHA, the amount of the HLA-E 3.3 kb transcript decreased substantially (17).

Sequence Variation Among HLA Class I Proteins

The deduced amino acid sequences from the HLA-E, HLA-6.0, and HLA-5.4 genes are compared in Figure 2 to an HLA-A,B,C consensus amino acid sequence generated by Parham and colleagues (24). While the availability of the sequence of only one allele from each non-HLA-A,B,C gene does not permit a discussion of allelic variation, several points relating these genes to the HLA-A,B,C sequences are worth noting. HLA-E, HLA-6.0, and HLA-5.4 each encodes a protein that has a high percentage of sequence similarity to the consensus HLA-A,B,C protein sequence (80%, 86%, and 82%, respectively). The amino acid sequence of the alpha-3 region of the HLA-E, HLA-6.0, and HLA-5.4 proteins is the region most similar to the HLA-A,B,C consensus sequence (92%, 91%, and 93% similarity, respectively). Considerable divergence from the HLA-A,B,C sequence is evident in the amino acid sequence of the alpha-1 and -2 regions of the non-HLA-A,B,C proteins. This is particularly evident in the alpha-1 region of the HLA-5.4 protein, which is 72% similar to the HLA-A,B,C sequence, and in the alpha-2 region of the HLA-E protein, which is 70% similar to the HLA-A,B,C sequence. Of the three non-HLA-A,B,C proteins, the HLA-6.0 sequence is more closely related to the HLA-A,B,C sequence in both the alpha-1 and alpha-2 regions (81% and 85% similarity, respectively).

The recent determination of the three-dimensional structure of the HLA-A2 antigen has permitted a correlation of polymorphic and conserved residues with position in the class I molecular structure (25,26). There are 10 conserved residues within HLA-A,B,C sequences that point into the foreign antigen recognition site (i.e., M5, Y7, F22, G26, Y59, Y84, T143, K146, Y159, Y171). Eight of these residues are conserved in the HLA-E sequence, nine are conserved in the HLA-6.0 sequence, and five are conserved in the HLA-5.4 sequence. The HLA-A,B,C-conserved tyrosine residues, speculated to play a role in the recognition of a constant feature of processed antigen by a class I molecule, are conserved in the HLA-E, HLA-6.0, and HLA-5.4 sequences. Alternating patterns of conserved residues every third or fourth residue in the long helix in alpha-1 and alpha-2 noted in HLA-A,B,C are also found in HLA-E, HLA-6.0, and HLA-5.4. Thus, the high sequence similarity with HLA-A,B,C and the presence of conserved residues at specific positions in the alpha-1 and alpha-2 regions of HLA-E, HLA-6.0, and HLA-5.4 suggest that these molecules could have a three-dimensional structure that would enable them to present a processed antigen.

Chromosomal Organization of HLA Class I Genes

The position of the HLA-A, HLA-B, and HLA-C loci on the short arm of human chromosome 6 has been known for several years. Analysis of somatic cell hybrids and the segregation of chromosomal rearrangements within the MHC in families have localized the HLA-A, -B, and -C loci to the chromosomal region 6p21 (27,28). Chromosomal in situ hybridization studies using an HLA-A,B,C genomic probe indicated that all HLA class I cross-hybridizing sequences are located on chromosome 6 at band 6p21.3 (29). Recombination frequencies within families between the HLA class I major transplantation loci indicate that their genetic order and separation is: proximal to distal, HLA-B – 0.2cM – HLA-C – 0.8cM – HLA-A (30,31). Analysis of large DNA restriction fragments separated by pulsed-field gel electropho-

ALPHA 1 REGION (positions 1–90)

	Sequence	Similarity to A,B,C	# of Variable Residues: Total	# of Variable Residues: Helical
HLA-A,B,C consensus	GSHSMRYFYTSVSRPGRGEPRFIAVGYVDDTQFVRFDSDAASPRMEPRAPWIEQEGPEYWDRETQIVKAQSQTDRESLRTLRGYYNQSEA			
HLA-E	----LK--H--------------S-------------N-------V-----M----S-------RSARDTA-IF-VN-----R-------	78%	20	14
HLA-6.0	--------SAA-------------M--------------S-C---------V--------EE--RNT--HA----MN-Q-----------	81%	17	11
HLA-5.4	----L---S-A-----P----Y---E-------L-------I------E--V-----Q--EWT-GYA--NA----VA--N-LRR------	72%	25	16

ALPHA 2 REGION (positions 91–180)

	Sequence	Similarity to A,B,C	# of Variable Residues: Total	# of Variable Residues: Helical
HLA-A,B,C consensus	GSHTLQRMYGCDVGPDGRLLRGYHQYAYDGKDYIALNEDLRSWTAADTAAQITQRKWEAARVAEQLRAYLEGTCVEWLRRYLENGKETLQRA			
HLA-E	------W-H--EL---R-F----E-F-------LT----------V------SEQ-SND-SE--HQ-----D------HK---K-----LHL	70%	28	17
HLA-6.0	S-----W-I---L-S--------E---------L-----------------SK--C---N----R------------H--------M----	85%	14	7
HLA-5.4	------G-N---M------------H--------S-------------V------FY--EEY--EF-T----E-LEL------------LQ-	78%	20	15

ALPHA 3 REGION (positions 182–274)

	Sequence	Similarity to A,B,C	# of Variable Residues: Total
HLA-A,B,C consensus	DPPKTHVTHHPISDHEATLRCWALGFYPAEITLTWQRDGEDQTQDTELVETRPAGDRTFQKWAAVVVPSGEEQRYTCHVQHEGLPKPLTLRW		
HLA-E	E---------------------------------Q---GH--------------G----------------------------E-V----	92%	7
HLA-6.0	-----------VF-Y----------------I-------------V----------G-----------------------------E--M---	91%	8
HLA-5.4	----A--A--------------------------------E---------------G----------------------------Q--I---	93%	6

Figure 2. Comparison of the HLA-E, HLA-6.0, and HLA-5.4 amino acid sequences with the HLA-A,B,C consensus amino acid sequence (24). The consensus sequence is presented in the standard single-letter amino acid code. In the other sequences, positions with residues homologous to the consensus sequence are indicated by a horizontal line. Replacements are shown by the single-letter code.

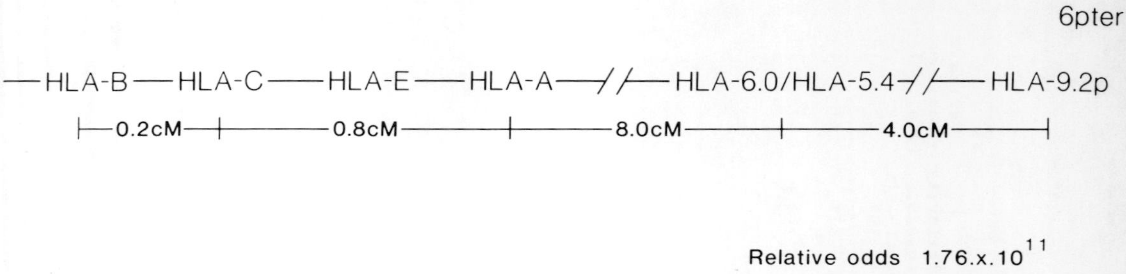

Figure 3. A genetic map of the HLA class I gene family. Distances between loci are indicated in centiMorgans (cM). The HLA-9.2p locus contains a class I pseudogene located within a 9.2 kb HindIII restriction fragment (36).

resis with a HLA-A,B,C probe suggests that the class I gene family spans a region of DNA from 1600 to 2000 kb (32,33).

Using an approach that combined the analysis of gamma-ray-induced HLA deletion mutant cell lines with an analysis of the inheritance of restriction fragment length polymorphisms in a large family, we recently placed the HLA-E, HLA-6.0, and HLA-5.4 genes relative to HLA-A, HLA-B, and HLA-C on a chromosome 6 map (Fig. 3). HLA-E is located between HLA-A and HLA-C since DNA from the locus is absent in some HLA-A deletion mutants that have not deleted HLA-C and is also absent in some HLA-B,-C deletion mutants that have not lost HLA-A DNA (17). The only region of overlap between the deletions in these two types of mutant cell lines having lost HLA-E (i.e., the location of the HLA-E locus) is between HLA-A and HLA-C. The HLA-6.0 and HLA-5.4 genes were positioned within a locus, 8.0 cM (centiMorgans) distal of HLA-A by screening individuals within large family and by analyzing the absence of HLA-6.0 and HLA-5.4 DNA in a panel of HLA deletion mutant cell lines (34). Multipoint linkage analysis gave the following gene orders and relative odds:

1. HLA-B. . . . HLA-A. . . . HLA-6.0/5.4	1.8×10^{10} odds
2. HLA-6.0/5.4. . . . HLA-B. . . . HLA-A	1.8×10^{9} odds
3. HLA-B. . . . HLA-6.0/5.4. . . . HLA-A	1 odds

A 10-fold difference in relative odds between order 1 and order 2 is not significant. However, analysis of DNA from the HLA deletion mutant cell lines with locus-specific probes showed that HLA-A and HLA-6.0/5.4 could be deleted together while the HLA-B locus remained, eliminating gene order 2. Thus, gene order 1 is the only situation that meets the results of the pedigree analysis and analysis of the HLA deletion mutants. Analysis of the family gave a genetic separation between HLA-A and HLA-6.0/5.4 of 8 cM.

Conclusions

With the isolation and expression analysis of the four non-HLA-A,B,C genes, HLA-E, HLA-6.0, HLA-5.4, and HA2, it is clear that the class I gene family of the human major histocompatibility complex does include genes other than those encoding the major transplantation antigens. While the general question concerning the presence of non-HLA-A,B,C class I genes has been answered, some very important points clearly remain to be resolved.

One point concerns the relationship between the identified non-HLA-A,B,C genes and the previously characterized human class I-like antigenic determinants. Only the HA2 gene has been shown to encode a class I molecule that is recognized by a non-HLA-A,B,C antisera. Thus, an interesting series of experiments will be to determine if either the HLA-E, HLA-6.0, or the HLA-5.4 genes encode molecules recognized by non-HLA-A,B,C antisera. It should be relatively straightforward to determine if the HLA-6.0 gene encodes a previously recognized class I-like antigenic specificity since the product of this gene has been expressed on the surface of class I-null B-LCL. However, a similar analysis of the HLA-E and HLA-5.4 products will first require obtaining transferent cells expressing these molecules on their cell surface. One point about the type of class I antigen encoded by either of these three genes is that each of them encodes a class I molecule recognized by the monomorphic monoclonal antibody W6/32. Thus, those antigenic determinants that remain after preclearing with W6/32 cannot be encoded by HLA-E, HLA-6.0, or HLA-5.4.

It is also unclear whether any of the non-HLA-A,B,C genes are the analogues of the murine Qa/Tla genes. The human non-HLA-A,B,C class I genes are no more similar in sequence to the Qa/Tla genes than they are to the mouse major transplantation genes H2-K, D, and L. Specific structural features of the non-HLA,B,C genes (e.g., the shortened cytoplasmic segment encoded by the HLA-6.0) are reminiscent of some murine Qa/Tla genes (20). In addition, the general chromosomal organization of the human class I gene family is similar to that of the murine class I genes. The non-HLA-A,B,C class I genes are located distal to HLA-C, with the HLA-A locus located in their midst. This arrangement is consistent with a suggested relationship between H-2K and HLA-A (35). In the mouse, there apparently has been a translocation of the H2-K locus from the Qa region (3). Thus, the position of the non-HLA,A,B,C genes within the MHC is similar to the location of the murine Qa/Tla genes. However, such features provide a tenuous correlation between non-HLA-A,B,C genes and the murine Qa/Tla genes. A precise correlation of the human non-HLA-A,B,C genes with murine Qa/Tla genes must await the functional definition of the respective gene products in both human and mouse.

References

1. Ploegh HL, Orr HT, Strominger JL. Major histocompatibility antigens: the human (HLA-A,B,C) and murine (H-2K, H-2D) class I molecules. Cell 1981;42:287–299.
2. Flaherty L. The TLA region antigens. In: Dorf ED (ed). The role of the Major Histocompatibility Complex in Immunobiology. New York, London: Garland STPM Press, 1981:33–57.
3. Weiss EH, Golden L, Fahrner K, et al. Organization and evolution of the class I gene family in the major histocompatibility complex of the C57BL/10 mouse. Nature 1984;310:650–655.
4. Steinzmetz M, Stephan D, Lindahl KF. Gene organization and recombinant hotspots in the mouse major histocompatibility complex. Cell 1986;44:895–904.
5. Tada N, Tanigaki N, Pressman D. Human cell membrane components bound to β2-microglobulin in T cell-type cell lines. J Immunol 1978;120:513–519.
6. Yoshizaki K, Seon BK, Barcos MP, Minowada J, Pressman D. Detection and characterization of human leukemia-associated antigens on leukemic cell lines and thymocytes I. Characterization of baboon and rabbit antisera to MOLT-4. 1979. J Immunol 1979;122:1943–1950.
7. Brodsky FM, Bodmer WF, Parham P. Characterization of a monoclonal anti-β2-microglobulin antibody and its use in the genetic and biochemical analysis of major histocompatibility antigens. Eur J Immunol 1979;9:536–545.
8. Gazit E, Terhorst C, Mahoney RJ, Yunis EJ. Alloantigens of the human T (HT) genetic region of the HLA linkage group. Human Immunol 1980;1:97–109.
9. Cotner T, Mashimo H, Kung PC, Goldstein G, Strominger JL. Human T cell surface antigens bearing a structural relationship to HLA antigens. Proc Natl Acad Sci USA 1981;78:3858–3862.
10. Knowles RW, Bodmer WF. A monoclonal antibody recognizing a human thymus leukemia-like antigen associated with β2-microglobulin. Eur J Immunol 1982;12:676–681.
11. Gazit E, Gothelf Y, Gil R, et al. Alloantibodies to PHA-activated lymphocytes detects human Qa-like antigens. J Immunol 1984;132:165–169.
12. Kahn-Perles B, Wietzerbin J, Caillol DH, Lemonnier F. Delineation of three subsets of class I human I antigens (HTA) on MOLT-4 cells: serologic and regulatory relationship to HLA class I antigens. J Immunol 1985;34: 1759–1765.
13. Fauchet R, Boscher M, Bouhallier O, et al. New class I in man: serological and molecular characterization. Human Immunol 1986;17:3–20.
14. Martin LH, Calabi F, Lefebvre F-A, Bilsland CAG, Milstein C. Structure and expression of the human thymocyte antigens CDla, CDlb, and CDlc. Proc Natl Acad Sci USA 1987;84:9189–9193.
15. Calabi F, Milstein C. A novel family of human major histocompatibility complex-related genes not mapping to chromosome 6. Nature 1986;323:540–543.
16. Orr HT, DeMars R. Class I-like HLA genes map telomeric to the HLA-A2 locus in human cells. Nature 1983;302: 534–536.
17. Koller BH, Geraghty DE, Shimizu Y, DeMars R, Orr HT. HLA-E: a novel HLA class I gene expressed in resting T lymphocytes. J Immunol 1988;141:897–904.
18. Srivastava R, Chonery MJ, Lawrence SK. Structure, expression, and molecular mapping of a divergent member of the class I HLA gene family. Proc Natl Acad Sci USA 1987;84:4224–4228.
19. Mizuno S, Trapani JA, Koller BH, Dupont B, Yang SY. Isolation and nucleotide sequence of a cDNA clone encoding a novel HLA class I gene. J Immunol 1988;140:4024–4030.
20. Geraghty DE, Koller BH, Orr HT. A human histocompatibility complex class I gene that encodes a protein with a shortened cytoplasmic segment. Proc Natl Acad Sci USA 1987;84:9145–9149.
21. Geragthy DE, Koller BH, Orr HT. A human histocompatibility complex gene composed of a class I sequence linked to a novel transcribed repetitive element. Submitted for publication.
22. Paul P, Fauchet R, Boscher MY, et al. Isolation of a human major histocompatibility complex class I gene encoding a nonubiquitous molecule expressed on activated lymphocytes. Prod Natl Acad Sci USA 1987;84:2872–2876.
23. Shimizu Y, Geraghty DE, Koller BH, Orr HT, DeMars R. Transfer and expression of three cloned human non-HLA-A,B,C class I major histocompatibility complex genes in mutant lymphoblastoid cells. Proc Natl Acad Sci USA 1988;85:227–231.
24. Parham P, Lomen CE, Lawlor DA, et al. The nature of polymorphism in HLA-A,B,C molecules. Proc Natl Acad Sci USA 1988;85:4005–4009.
25. Bjorkman PJ, Saper MA, Samraoui B, et al. Structure of the human class I histocompatibility antigen, HLA-A2. Nature 1987;329:506–512.
26. Bjorkman PJ, Saper MA, Samraoui B, et al. The foreign antigen binding site and T cell recognition regions of class I histocompatibilty antigens. Nature 1987;329:512–518.
27. Francke U, Pellegrino MA. Assignment of the major histocompatibility complex to a region of the short arm of human chromosome 6. Proc Natl Acad Sci USA 1977; 74:1147–1151.
28. Goodfellow PH, Banting G, Trowsdale J, Chambers S, Solomon E. Introduction of a human X-6 translocation chromosome into a mouse teratocarcinoma: investigation of control of HLA-A,B,C expression. Proc Natl Acad Sci USA 1982;79:1190–1194.
29. Morton CC, Kirsch IR, Nance WE, et al. Orientation of loci within the human major histocompatibility complex by chromosomal in situ hybridization. Proc Natl Acad Sci USA 1984;81:2816–2820.
30. Weitkamp LR, Lamm LU. Report of the Committee on the Genetic Constitution of Chromosome 6. Cytogenetic Cell Genet 1982;32:130–143.
31. Robson EB, Lamm LU. Report of the Committee on the Genetic Constitution of chromosome 6. Cytogenetic Cell Genet 1984;37:47–70.
32. Lawrence SK, Smith CL, Srivastava R, Cantor CR, Weisman SM. Megabase-scale mapping of the HLA gene complex by pulse field gene electrophoresis. Science 1987;235:1387–1390.
33. Carroll MC, Katzman P, Alicot EM, et al. A linkage map of the human major histocompatibility complex including the tumor necrosis factor genes. Proc Natl Acad Sci USA 1987;84:8538–8539.

34. Koller BH, Rich S, Geraghty DE, DeMars R, Orr HT. A chromosomal map of the HLA class I genes: correlation between physical and genetic separation. Submitted for publication.
35. Klein J. Evolution and function of the major histocompatibility complex. In: Parham P, Strominger J (eds.). Histocompatibility Antigens Structure and Function. London, New York: Chapman and Hall, 1982:223–239.
36. Koller BH, Geraghty D, Orr HT, Shimizu Y, DeMars R. Organization of the human class I major histocompatibility complex genes. Immunol Res 1987;6:1–10.

Author Affiliation

Department of Laboratory Medicine and Pathology, and Institute of Human Genetics, University of Minnesota, Minneapolis, MN 55455.

Molecular Structure of Human Class II Antigens

John I. Bell, John A. Todd, and Hugh O. McDevitt

There has been remarkable recent progress in characterizing the molecular structure of human class II molecules. This progress has resulted from the rapid accumulation of sequence data from both representatives of each of the major class II loci and from multiple alleles at each locus (1–4). This has permitted an understanding of both the primary structure of these molecules and an appreciation of the polymorphism at each locus. These sequence data can be best understood when analyzed in the context of the three-dimensional structure of the molecule. This can now be achieved by modeling class II sequences to the HLA-A2 crystal structure defined recently by Bjorkman and her colleagues (5,6). This provides not only basic structural information about the antigen binding site, but also allows speculation on the role of individual substitutions in the biological functions of these molecules.

Background

The loci encoded by the class II region of the human major histocompatibility complex (MHC) extend across 1.1 megabases of DNA on the short arm of chromosome 6. The order and orientation of these loci have been defined (7), and together they are responsible for the original serologic and cellular typing that led to the definition of distinct Dw or DR specificities. These serologic specificities were generated from the cluster of DR and DQ loci held in tight linkage disequilibrium at the telomeric end of each class II haplotype. Definition of the individual loci within the haplotype evolved with the identification of supertypic sera and monoclonal reagents (8,9). Two-dimensional gel electrophoresis proved to be a highly sensitive technique for identifying the products of individual loci and for confirming and extending the range of polymorphic alleles at any given locus (10). Similarly, RFLPs proved to be an effective means of further extending the number of alleles on given haplotypes and confirming the previous serologic data (11,12). This work prepared the way for the definitive characterization of polymorphism by allelic sequencing. This approach has now established both the extent and the nature of the polymorphism of many of the expressed class II loci in man (13).

Initial cloning and protein studies of class II antigens were particularly informative (1). Purified with antisera, these glycoproteins were recognized as having two chains, a 34 kd heavy (α) chain and a 29 kd light (β) chain. Each has four domains—two extracellular domains, a transmembrane region, and a cytoplasmic domain. The membrane distal domains of the β chains (β1) possess an intrachain disulphide loop between residues 15 and 79. Alpha chain first domains (α1) do not have cysteines, but have a common glycosylation site at residue 78. The second domains (α2 and β2) are recognized as having strong sequence homology to immunoglobulin. The transmembrane region is connected to the extracellular domain by a hydrophilic-connecting peptide, while the transmembrane region itself is characteristically hydrophobic, ending with a group of positively charged amino acids. A cytoplasmic tail is present in all cases, but varies in length depending on the isotype.

Within the DR subregion, DRα has been established as being nonpolymorphic across a large number of haplotypes (14). Two expressed polymorphic DRβ loci have been identified. The DRβI locus, mapping as the most centomeric of the DR loci in the haplotypes DR3, 4, and 6, is the most polymorphic. It seems to be the most important determinant of serologic reactivity, and sera reacting to it define the major serologic DR specificities (3). Expression at the RNA level of this locus may be as much as 10-fold greater than the other expressed DR locus, DRβ3 or DRβ4 (3). The DRβ3 and 4 loci are more conserved between haplotypes and are in part responsible for the supertypic cross-reactivity seen between families of haplotypes (15). These families most probably reflect a common evolutionary history as suggested by both a similarity among DRβ3 or DRβ4 sequences as well as conserved adjacent restriction fragment polymorphisms, some conservation of the DRβI sequence itself, and similarity in the pattern and distribution of loci and restriction enzyme sites at a genomic level

Immunobiology of HLA
Volume II

across the whole haplotype (16). The families defined by these criteria are the DRw52-associated family (DR3, 5, 6, 8), the DRw53-associated family (DR4, 7, 9), and the Dw21-associated family (DR2 Dw2, DR2 Dw12, DR2) (17). DRw52 and DRw53 specificities are encoded by the HLA DRβ3 and DRβ4 genes. The lack of genomic organization data for the DR2 haplotype limits the use of this terminology. On the basis of RFLPs, it has been argued that DRl haplotypes might be more closely related to the DRw52 haplotypes. However, the absence of a second expressed DRβ allele prevents confirmation of this evolutionary association (3).

The DRβ chains have a membrane distal domain of 94 amino acids and a membrane proximal domain of 84 amino acids. The cytoplasmic tail is 18 amino acids long, the hydrophobic transmembrane region is 21 residues in length, and the connecting peptide is 10 amino acids long. In all, the molecule is 237 amino acids long.

The DQ subregion has one set of expressed α- and β-chains, both of which are polymorphic. In addition, there are a pair of DX loci (DQα2, DQβ2) that are non-polymorphic, but not apparently expressed. The conservation of sequence in these DX loci despite apparent lack of expression is unusual, as all other class II pseudogenes have acquired multiple structural defects. The lack of polymorphism of both α- and β-chains in this potential DX molecule implies that if DX were once expressed, the function was not enhanced by active selection for polymorphism.

The structure of DQ products differs slightly from those at the DR locus. The DQβ locus has eight amino acids in its cytoplasmic domain due to a failure of a splice acceptor making it shorter than DRβ (13). DQα has alternative splicing, but this does not affect coding sequence (13).

The DP subregion has an αβ pseudogene pair and a pair of expressed α- and β-chains. The DPα chain is similar to the other class II α products, whereas the β-chain shows some cytoplasmic differential splicing. Polymorphism in these molecules was first defined by primed lymphocyte typing, and has been confirmed by RFLPs (19) and nucleic acid sequencing (20). No polymorphism has been recognized at either the DNA (21) or DOβ (22) locus, although both are recognized as being expressed.

Allelic Variation. The sequencing of cDNA clones from defined haplotypes now forms the basis of our understanding of the molecular basis of class II polymorphism. An impression of the nature of this polymorphism was first obtained from the allelic-sequencing studies performed in the mouse. This revealed that Eα, the homologue of DRα, had little polymorphism (23), but that Eβ, Aβ, and Aα (24–26) all contained substantial numbers of polymorphic residues and that these were concentrated in the first domain of these molecules. Particular regions of extensive polymorphism were defined and termed allelic hypervariable regions (AHVR) (24). This work primarily involved standard inbred strains and has provided one important example (the bm12 mutation) (27) of polymorphism that could have arisen from a gene conversion-like mechanism.

Table 1. Amino Acid First Domain Substitutions Between HLA DRβ Alleles

	1	2a	2b	3a	3b	4a	5a	6a	7a	7b	9a	9b
1	x	10	10	17	14	10	16	17	16	19	12	19
2a		x	13	20	17	12	15	16	15	20	10	20
2b			x	13	16	9	9	10	18	22	16	21
3a				x	12	13	9	9	20	24	20	24
3b					x	14	16	17	19	21	21	21
4a						x	11	13	18	19	16	19
5a							x	4	20	22	20	22
6a								x	19	24	20	24
7a									x	25	15	25
7b										x	19	0
9a											x	19
9b												x

The a allele is that expressed by the βI locus; the b allele is derived from the βIII locus.

Studies in man have now confirmed and extended these observations. The polymorphism in DRβ alleles shows a pattern sufficiently similar in DRβI, DRβ3 and 4 alleles that these could not be distinguished from each other on the basis of sequence. They have thus been aptly termed pseudoalleles and, for the purposes of this comparison, are considered as a single set of polymorphic alleles. In DRβ alleles, the majority of polymorphism is present in the first domain, predominantly in three major regions of allelic hypervariability (Table 1) (3). These are seen between residues 9 and 13, 26 and 33, and 67 and 74 (Fig. 1). Additional variation is seen at positions 37 and 38, 47, and 56 to 60. This pattern is also seen in DQα and DQβ alleles, although the number of allelic hypervariable regions (AHVR) does differ (Fig. 2), with there being only one prominent region in DQα (45 to 56) and only two prominent AHVRs in DQβ (4) (53 to 57 and 66 to 77). The pattern of polymorphism in DPα and DPβ is less consistent (28). The four recognized DPβ alleles showed very limited polymorphism, present at only 14 positions in the first domain. Two clusters exist, one around residues 55 to 58 and another at positions 83 to 87 in DP3β only. DPα alleles show more polymorphism than DRα, but less than DQα. The three polymorphic residues in the first domain are at positions 30, 49, and 82, and five additional polymorphic residues are present in the second domain.

The amount of sequence polymorphism relative to that predicted serologically differs among these loci. At DRβ, most of the polymorphism at a nucleic acid sequence level was predicted either serologically or with cellular typing. At DQα and DQβ, however, much polymorphism has been uncovered that was previously unrecognized serologically (4). For example, seven DQw1 β chains can be identified (DQw5, w6, 1.12, 1.9, 1.19, 1.18, and 1.AZH), and three sequences exist for DQw3 β-chain subtypes (DQw7, w8 and w9). Many of these variants were predicted on the basis of two-dimensional gel electrophoresis or RFLP analysis. The DP sequence variation identified to date correlates with that defined with primed lymphocyte typing.

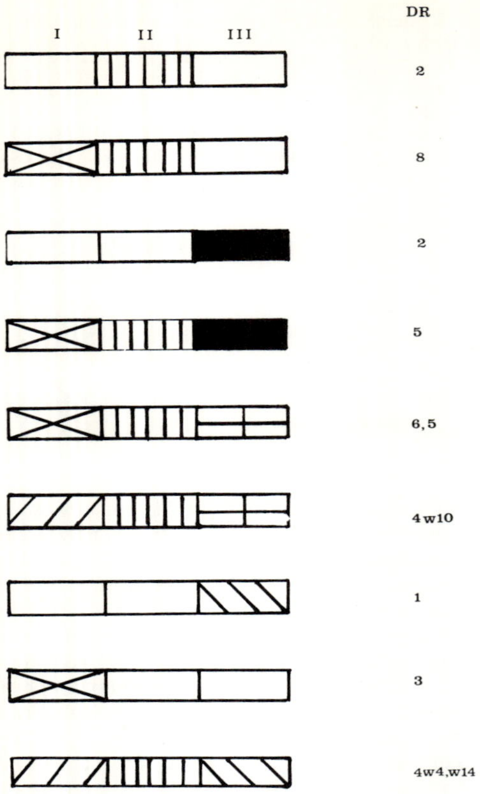

Figure 1. First domain similarities between HLA DRβ alleles. Each allele is divided into three allelic hypervariable regions, defined as I (9 to 13), II (26 to 33), and III (67 to 74). Similarities between alleles are indicated by patterns in individual AHVR. Blank squares are not homologous. This indicates the high frequency of "modular" exchange of AHVR between haplotypes.

Although referred to as allelic hypervariable regions, the islands of variability in the first domains of these molecules are not analogous to the hypervariable regions of immunoglobulins or T-cell receptors. In particular, individual residues of either DRβ, DQα, or DQβ show a distinctly limited range of substitutions. In DRβ, only residues 11, 13, and 30 show more than five different amino acid substitutions, while in DQβ no residue has more than four different amino acids. The variation at each residue is, therefore, distinctly limited. Similarly, the variation within each allelic hypervariable region is also remarkably limited. In DRβ, this limited range of polymorphic "modules" is readily apparent when the standard haplotypes are analyzed together. DRβ 3, 5, 6, and 8 all share the first allelic hypervariable region, while the second AHVR is shared between DR2, DR4, DR5, and DR6. In the third AHVR, DR1 and DR4 Dw14 are identical and differ by only a single Arg-Lys substitution (at position 71) from DR4 Dw4. Similarly the Dw10 third AHVR is shared on some DR4, DR5, and DR6 haplotypes. The similarities between haplotypes at both a single amino acid level and at the level of whole AHVRs indicate that there are stringent restrictions on the variation of these sites. The most likely explanation for this observation is that the structure of this region can be permitted to vary, but unlike immunoglobulin, cannot vary randomly because of the constraints placed on it by being a peptide binding site and by being recognized by T-cell receptors (TCR).

The shifting of individual modules between haplotypes to create new specificities was not seen in the sequencing of mouse alleles, but is readily evident in the analysis of outbred human sequences. One of the most dramatic examples of such "modular" variation is with certain DR5 alleles where the first, second, and third AHVRs are all shared by other different DRβ alleles (3). This modular variation may in part explain the difficulties observed in generating truly polymorphic monoclonal antibody reagents to class II alleles. On the basis of the known sequence data, these antibodies would be expected to cross-react between several haplotypes, and this has frequently been shown to be the case.

Little comment can be made regarding the polymorphism seen in the second domain of these class II molecules. Although less prominent than the polymorphism in the first domain, certain substitutions in this part of the molecule might have important effects on its function. For example, substitutions in the α3 domain of class I molecules may have significant effects on antigen recognition despite being physically separated from the antigen binding site. Therefore, it would be unwise to ignore these second domain polymorphisms.

Mechanism of Generation of Class II Polymorphism. The features of class II polymorphism described above give some important indications as to how this variation was generated. The variation in the first domain, particularly in the AHVR, has clearly been strongly selected for and fixed in the population. This is contrary to the view that polymorphism in the first domain is apparent only because the second domains remain relatively conserved (29). The ratio of productive to nonproductive variations in the first domain of DRβ is substantially greater than 2:3, the ratio resulting from random mutations. It is usually less than 1 in the second domain (3). This is consistent with similar findings in the mouse (24). These findings indicate that the observed first domain polymorphism has been selected for along with conservation of the second domain sequence. This contention is strongly supported by the structural data on the position of these residues in the binding site (see below).

An interesting feature of the polymorphism in these molecules is the high frequency of dinucleotide and trinucleotide changes within any one codon. The explanation for this is not clear, but suggests that single base pair mutation followed by selection is not the only mechanism for the observed diversification of these molecules.

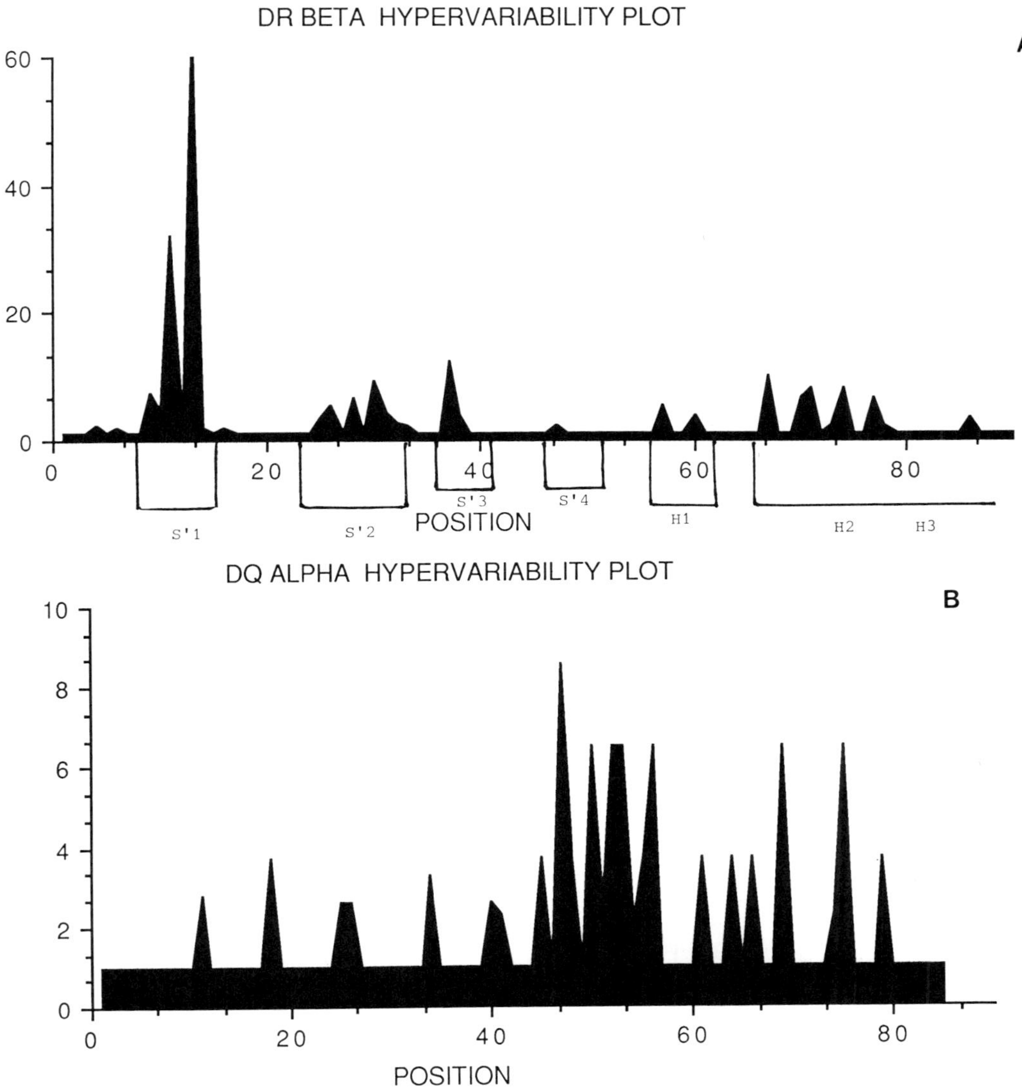

Figure 2. Line graphs of hypervariability from major polymorphic class II loci based on the formula:

$$\frac{\#\text{ of residues}}{\text{Frequency of most common residue}}$$

(see reference 37)

Only first domains are displayed. The evidence of poymorphic and conserved residues is readily observed in the strands and α-helical regions. The scale of DRβ plot is 0 to 60, whereas that for DQα and DQβ is 0 to 10.

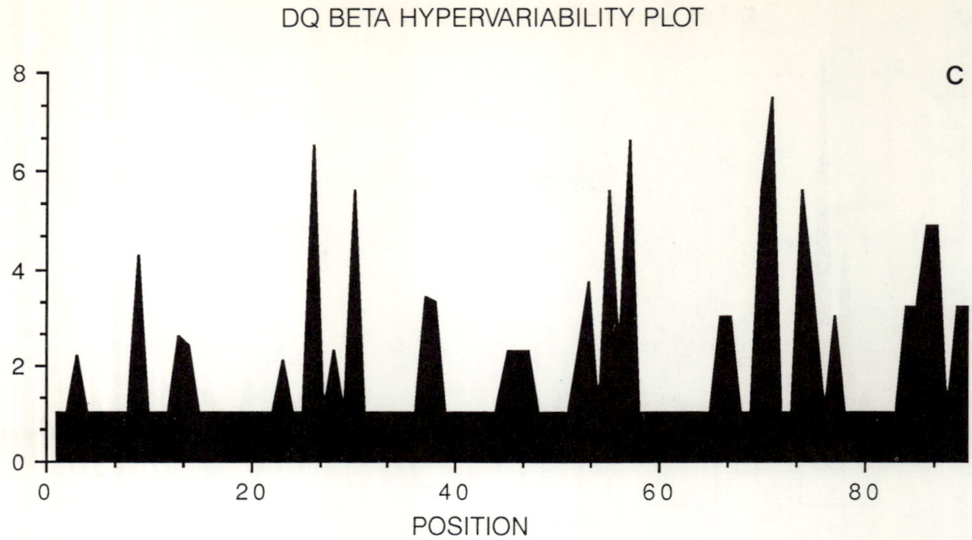

Figure 2. *(Continued)*

The observation that much of the AHVR polymorphism between alleles is shared between a limited number of alleles suggests that homologous recombination, either reciprocal or nonreciprocal, is responsible for some of the variation seen. The strongest evidence for nonreciprocal recombination or gene conversion in the class II region comes from studies of the bm12 mutation in the mouse (27). In this spontaneous mutant, the $A\beta_{bm12}$ allele received sequence from the $E\beta_b$ locus, introducing three amino acid changes and an alteration in determinants recognized by alloreactive T cells (30). This precedent was followed by the observation by Gorski and Mach (31) that similar events could explain some of the polymorphism within the human class II region. Their observation, applicable to the DR3 haplotype, has now been greatly extended to explain some polymorphism on most of the DR and DQ alleles (3). The majority of these recombination events, transferring modules of sequence information, occur within a single locus or subregion. Some examples of genetic transfer of information between subregions do exist. For example, position 78 in DRβ7 and 9 has all three nucleic acids (GTG) variant from a conserved TAC in other DRβ alleles. This GTG is found in position 78 of some DQβ alleles, suggesting a donor origin in this locus. Regardless of the appeal of gene conversion as a mechanism for the generation of this polymorphism, no direct evidence for this has been available in human populations. Equally plausible is the mechanism of multiple intralocus recombination. Such a mechanism could explain the modular variation within these alleles and, in some circumstances, remains by far the most likely possibility. The DR2 Dw21 DRβ chains, for example, show strong homology to the DR2 Dw12 β chains, except for the region between residues 30 and 60 (17). Over this region, the DRβI and DRβ3 sequences have been exchanged between the two alleles. The easiest (but untestable) hypothesis for this variation is a double homologous recombination event between these two loci on a Dw12 haplotype (17, 32).

Three-Dimensional Structural Correlations. An exciting advance in our understanding of the role of individual polymorphisms within class II molecules have come from the three-dimensional analysis of crystals for the class I molecule HLA-A2 (Fig. 4) (5). Bjorkman and her colleagues (6) derived the three-dimensional structure from these data and have identified a prominent groove on the top of this molecule formed by the α1 and α2 domains, which is thought to represent the peptide binding site. It has an appropriate location, size, and shape to bind short peptides. Further, the distributions of both polymorphic and conserved residues in and adjacent to this site are consistent with its putative role. Alterations in the residues produce the expected change in T-cell recognition, either due to alteration in peptide binding or a change in T-cell recognition.

The structure of this region in class I molecules includes two antiparallel α-helices sitting above a row of β-strands that form the floor of a groove bounded by the helices. The α-helices are encoded by α1 and α2 domain sequences, respectively, and each domain also contributes several β-strands for the peptide binding site. This binding site contains 15 of the 17 residues with a variability index greater than 6 (6).

The significance of these data to class II structure is obvious. These observations have permitted the alignment of class II and class I sequences to establish whether these molecules share similar tertiary structure (33). Such an analysis has shown that despite a relative

```
             S1                S2            S3          S4     HI                 H2

                 10      20          30          40      50          60       70        80       90

A2α1    GSHSMRYFFTSVSRPGRGEPRFIAVGYVDDTQFVRFDSDAASQRMEPRAPWIEQEGPEYWDGETRKVKAHSQTHRVDLGTLRGYYNQSEA
                          + -                                                                  +++

DQw1α   ADHVSCGVNLYQF-YGP-SG--QYTHEFDDEEFYVDL-------ERKETAWRWPEFSKFGGFDPQGALRNMAVAKHNLNIMIKRYNSTAA
        * *10        20   *        *** * **        40       50        60  *   *  *  *    *
                                                                 H1               70       80
                                                                                  H2

            S1'        S2'            S3'     S4'        H1'          H2'              H3'

                100      110      120     130      140       150     160      170
A2α2    GSHTVQRMYGCDVGSDWRFLRGYHQYAYDGKDYIALKEDLRSWTAADMAAQTTKHKWEAAHVAEQLRAYLEGTCVEWLRRYLENGKETLQRT
                                          S–S

                               + -                                   + -

DQw1β   EDFVYQFKGLCYFTNGTERVRGVTRHIYNREEYVRFDSDVGVYRAVTPQGRPVAEYWNSQEVLEGARASVDRVCRHNYEVATRGILQRR
DR1β    PRFLWQLKFECHFFNGTERVRLLERCIYNQEESVRFDSDVGEYRAVTELGRPDAEYWNSQDLLEQRRRAVDTYCRHNYGVGESFTVQRR
            10     *    20   **    *        *  *  ***    **     *    60**      **70**     **  *          90
                                   30          40          50                            80
```

Figure 3. Sequence of HLA-A2 α1 and α2 domains aligned with DQα and DRβ/DQβ sequences. Disulphide and salt bridges conserved between class I and class II are shown. +++ = Conserved glycosylation site in DQα and α1 HLA-A2. * = Residues conserved in all class I and class II molecules, although the residue may be different in each class. S1-S4 = The four β-strands associated with HLA-A2 α1 domain and class II α. S′-S4′ refers to the β-strands of the HLA-A2 α2 domains and class II β-chains. H1 and H2, H1′, H2′, and H3′ refer to the α-helices in class I α1/class II α or class I α2/class II β, respectively. Alignment based on reports by Brown et al. (33).

lack of homology between class I and class II molecules at a sequence level, the α1 and α2 domains of class I can be aligned to the α1 and β1 domains of class II, respectively (33). The alignment of these molecules can be achieved using the glycosylation site at residue 86 (class I) or residue 81 (class II DQα-chain) and the cysteines at residues 101 and 164 (class I) and 15 and 79 (class II β-chain). Advantage has been taken of polymorphic or conserved residues in the β-strands or α-helices to identify their counterparts in class II sequences. The β-strand residues show an alternating pattern of polymorphism, with every other residue pointing into the site. In particular, the class I α2 polymorphic residues in β-strands S1′ and S2′ (95, 97, and 99 in S1′, 114 and 116 in S2′) can be aligned with a similar periodicity in polymorphic residues of class II β (9, 11, and 13 or 26, 28, and 30). A similar alignment can be readily made for the two α-helices (33). In HLA-A2, conserved residues occur along a single face of each the α1 and α2 helices. These occur every third or fourth residue in class I (α1 residues 68, 72, 75, 78, and 84; α2 residues 148, 153, 154, 157, 158, 161, 164, 165, and 168). In comparison to class II α- and β-chain sequences, these can be readily aligned to residues α63, 67, 70, 73, and 79 or β62, 68, 69, 72, 76, 80, and 83. These residues must represent the conserved side of each α-helix. Between these residues, in both class I and class II lie the most polymorphic residues, facing either up or into the binding site and affecting T-cell recognition or peptide binding. In class I, these residues are α1 69, 74, 79, 80, 81 or α2 151, 156, and 163. In class II, these residues correlate with DQα-chain residues 69, 75, 76, and 80 or β-chain residues 67, 70, 71, and 74.

The DQ α-chains have a prominent region of polymorphism (45 to 56), which may lie in the first helical region, but does not correlate with the presence of class I α1 polymorphism over this region. The class I polymorphism at the beginning of the second helix (41 to 47) is not seen in class II α-chains (33).

In this model of class II structure (Fig. 3), intramolecular interactions between residues are conserved between class I and class II. The residues in HLA-A2 associated with salt bridges across the open loops between strands S1 and S2 (α-1) and strands S2′ and S3′ are conserved in class II molecules. These are a histidine at position 7 (DQα) or position 5 (DRα and DPα) to the aspartic acid at position 30 of DQα or 28 of DRα/DPα and the arginine at position 25 in DQβ or DRβ to the aspartic acid at position 43. The arginine is missing in DPβ. A disulphide bridge is also present in class II molecules between position 15 in DR or DQβ and position 79. The conservation of these potential interactions

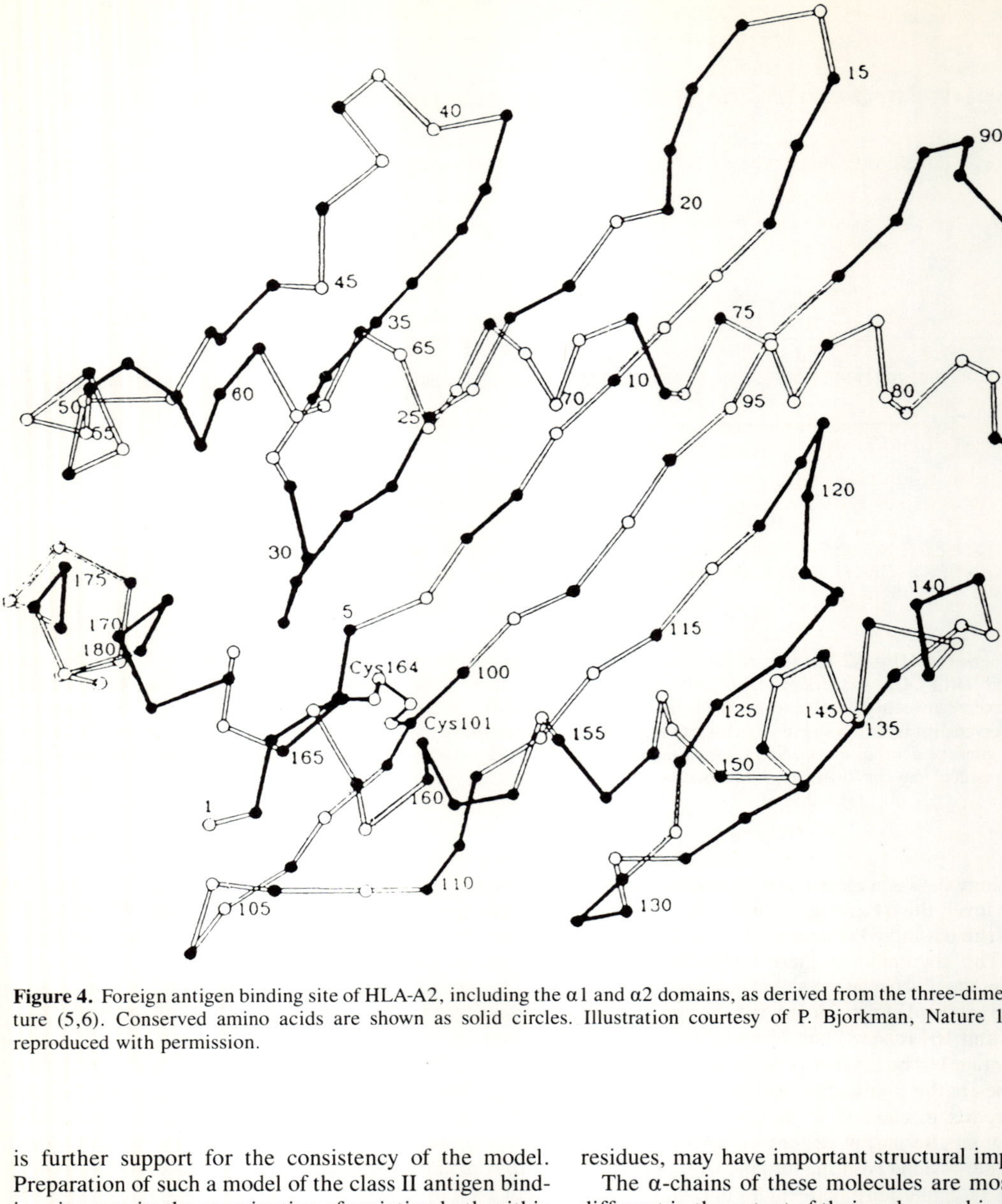

Figure 4. Foreign antigen binding site of HLA-A2, including the α1 and α2 domains, as derived from the three-dimensional structure (5,6). Conserved amino acids are shown as solid circles. Illustration courtesy of P. Bjorkman, Nature 1987;329:516, reproduced with permission.

is further support for the consistency of the model. Preparation of such a model of the class II antigen binding site permits the examination of variation both within and between class II loci and enables one to predict the significance of prominent variation particularly in the binding site (33).

Isotypic Structural Variation. The products of the class I loci show remarkable similarity in their structure, implying a common configuration with polymorphism in individual residues. It is more difficult to infer a common fine structure among class II isotypes, although the general configuration may be conserved at the binding site. The differences between DR, DQ, and DP products, both in terms of their conserved and their polymorphic residues, may have important structural implications.

The α-chains of these molecules are most obviously different in the extent of their polymorphism. The DRα chain possesses no polymorphism, while the DPα chain has only three polymorphic residues adjacent to the binding site. DQα, however, has extensive polymorphism, particularly along the α-helix on one side of the binding site. This relative lack of polymorphism in DPα and DRα is distinct from the pattern seen in class I molecules, and implies that selective forces have prevented variation from a particular sequence in this region. This is further supported by the conservation of Eα in the mouse. The lack of variation in DRα is particularly difficult to explain in the presence of extensive selected polymorphism in the binding site of DQα.

Of interest is the ability of these α-chains to associate with other β-chains. In the mouse, the N-terminal portions of the α1 and β1 domains are thought to be largely responsible for chain pairing. In man, Long and his colleagues (38) have shown that DRα and DPα will pair with DR, DQ, and DP β-chains, but DQα will not pair with other isotypes. This has been demonstrated in both transient and stable expression systems. DRα and DPα are very similar to each other across the N-terminal domain and are both relatively nonpolymorphic. DQα is much different across this domain and displays substantial polymorphism. The ability of DRα and DPβ to form, interisotypic pairs substantially extends the number of potential heterodimers that can be generated and the lack of polymorphism may be necessary to achieve such promiscuity in pairing.

When considering the variation within and between isotypes along the proposed first β-strand (S1), there are very substantial differences between class II isotypes but little polymorphism within each isotype. The loop region between S1 and S2 is clearly different from the equivalent region in class I molecules and differs even between class II isotypes. The second β-strand (S2) has polymorphic residues in DQα at DQα25 and 26 and is quite distinct in sequence over this region (DRα 20–23) in DRα and DPα compared with the DQα residues 24–27.

The major polymorphic residue on strand 3 (S3) in HLA-A2 is at residue 32, and this equates with position 34 in DQα, also polymorphic. The equivalent position in DPα is 30, and this is one of three polymorphic residues in the first domain. Over this β-stand region, the three class II molecules vary at only a few residues, but more variation is again obvious in the loop region between S3 and S4. This loop is also different from that seen in the HLA-A2.

The first α-helix is polymorphic in DQα, but differs little between or within the DRα and DPα isotypes. There is a marked difference in DQα alleles, which in some instances lack position 56. The second α-helix begins in DQα immediately after a proline that would break any previous helical structure. This helix contains further polymorphism in DQα. Again, DRα and DPα are conserved across much of the helical region, but these two molecules do diverge substantially at the end of the helix.

The second half of the peptide binding site is contributed by the β-chain of the class II molecule. These sequences can be matched more easily to the α2 domain of HLA-A2 than can the α1 domain be matched to the class II α-chain (33). The S1′ region of these molecules is highly polymorphic in DR, DQ, and DP. The DPβ chains have an S2′, which is also unique because it is missing two residues as compared with the other isotypes. One of these residues is part of the conserved molecular salt bridge between S2′ and S3′, and its absence may have significant structural implications. Position 57 at the beginning of the first α-helix is a conserved aspartic acid in most DPβ and DRβ alleles, but is variant in DQβ. Since this residue is probably also associated with a salt bridge to the α-chain (33), aspartic acid-negative variants of DQβ may have unique structures.

The majority of the helical polymorphism in DRβ and DQβ is found in the same region, namely positions 70 to 77. Across this α-helix, there is striking conservation of the residues lying between the polymorphic residues, even between all the isotypes. Class II positions 68, 69, 72, and 76 are all highly conserved between isotypes, even among DPβ chains, implying that this basic portion of the binding site is conserved in structure, but varies only in the residues facing into or above the site.

Functional Considerations

Class II MHC molecules are specific ligands for peptides. This has been clearly demonstrated in experiments utilizing purified class II molecules and peptides in equilibrium dialysis (34) or alternatively in a gel filtration system (35). The recognition that these molecules might conform to the structural pattern of HLA-A2 (33) and that the location of the key polymorphic residues can be identified within the binding site permits an analysis of peptide-MHC interactions that may provide an explanation of how peptide fits into and interacts with residues in the binding site. Such analyses have already begun. Two DR1- restricted responses have been used to study the residues in two peptides, identifying those that interact with DR1 and those that interact with their respective T-cell receptors (36). This analysis has suggested that if the peptides have a helical conformation, DR1 contact residues can be defined and exchanged between peptides. Such analyses could be extended to explain the interaction between any other class II molecule and antigen. An understanding of this interaction is one of the final steps toward a molecular understanding of antigen presentation.

The functional role of individual residues in class II molecules can be partly deduced from mutants, either spontaneous (bm12) or generated by in vitro mutagenesis (39). Such studies confirm the prominent role of α1 and β1 domains in T-cell recognition. In addition, they have revealed that variation of single polymorphic residues in the first domain can alter responsiveness of some but not necessarily all T-cell clones recognizing antigen in the context of that particular MHC molecule. For example, mutagenesis of positions 9 and 13 in the $A\beta_b$ allele produced no change in reactivity with a panel of monoclonal antibodies, but T-cell clones (both alloreactive- and antigen-specific) were affected in widely variable fashions by individual mutations (40). The mutations in this study were in the first β-strand (S1′) or in the H2′ α-helix (positions 65 to 67). These latter changes are in the same region as the bm12 mutation substitutions. The variable effects of these mutations on individual T-cell clones indicates that within the tertiary conformation of the class II molcule, multiple T-cell recognition sites may exist and these interactions are likely to be complex.

Conclusions

The linear sequence of the products of various HLA class II loci has provided substantial insights into the nature and mechanism of generation of the polymorphism in these molecules. This detailed analysis has provided an understanding of the reactivity of the antisera used to define them. Within the framework of the three-dimensional structure of these molecules, particularly of their peptide binding site, these data become even more valuable. They provide an appreciation of isotypic variation in this region and permit the analysis of peptide MHC interactions at a molecular level.

References

1. Kaufman JE, Auffray C, Korman AJ, Shackelford DA, Strominger S. The class II molecules of the human and murine major histocompatibility complex. Cell 1984;36:1.
2. Long EO, Wake CT, Gorski J, Mach B. Complete sequence of an HLA-DRβ chain deduced from a cDNA clone and identification of multiple non-allelic DRβ chains. EMBO J 1983;2:389–394.
3. Bell JI, Denney D, Foster L, Belt T, Todd JA, McDevitt HO. Allelic variation in the DR subregion of the human major histocompatibility complex. P.N.A.S. 1987;84: 6234–6238.
4. Todd JA, Bell JI, McDevitt HO. HLA-DQβ gene contributes to susceptibility and resistance to insulin-dependent diabetes mellitus. Nature 1987;329:599.
5. Bjorkman PJ, Saper MA, Samraoui B, Bennett WS, Strominger JL, Wiley DC. Structure of the human class I histocompatibility antigen HLA-A2. Nature 1987;329: 506–512.
6. Bjorkman PJ, Saper MA, Samraoui B, Bennett WS, Strominger JL, Wiley DC. The foreign antigen binding site and T cell recognition regions of class I histocompatibility antigens. Nature 1987;329:512.
7. Hardy DA, Bell JI, Long EO, Lindsten T, McDevitt HO. Nature 1986;323:453–455.
8. Park MS, Terasaki PI, Nakata S, Aoki D. Supertypic DR groups: MT1, MT2, and MT3. In: Terasaki P (ed). Histocompatibility Testing 1980. Berlin, Springer-Verlag, 1980.
9. Fuggle SV, Errasti P, Daar AS, Fabre JW, Ting A, Morris PJ. The localization of major histocompatibility complex (HLA-ABC and DR) antigens in 46 kidneys: differences in HLA-DR staining of tubules among kidneys. Transplantation 1983;35:385.
10. Charron DJ, McDevitt H. Characterization of HLA-D region antigens by two-dimensional gel electrophoresis. J Exp Med 1980;152:18s–36s.
11. Bell JI, Denney D, MacMurray A, Foster L, Watling D, McDevitt HO. Molecular mapping of class II polymorphisms in the human major histocompatibility complex I. DRβ'. J Immunol 1987;139:562–573.
12. Cohen-Haguenauer O, Robbins E, Massart C, et al. A systematic study of HLA class II-β DNA restriction fragments in insulin-dependent diabetes mellitus. P.N.A.S. 1985;82:3335.
13. Bill P, Wiman K, Jonsson A-K, et al. Both α and β chains of HLA DC class II histocompatibility antigens display extensive polymorphism in their amino terminal domain. EMBO J 1984;3:447.
14. Lee J, Trowsdale J, Travers PJ, et al. Sequence of an HLA-DR α-chain cDNA clone and intron-exon organization of the corresponding gene. Nature 1982;299:750.
15. Gorski J, Tosi R, Strubin M, Rabourdin-Combe C, Mach B. Serological and immunochemical analysis of the products of a single HLA DRα and DRβ chain expressed in a mouse cell line after DNA mediated cotranformation. J Exp Med 1985;162:105–116.
16. Rollini P, Mach B, Gorski J. Linkage map of three DRβ chain genes: evidence for a recent duplication event. P.N.A.S. 1985;82:7197–7201.
17. Lee BSM, Rust NA, McMichael AJ, McDevitt HO. HLA-DR2 subtypes form an additional supertypic family of DRβ alleles. P.N.A.S. 1987;84:4591–4595.
18. Trowsdale J, Kelly A, Lee J, Carson S, Austin P, Travers P. Linkage map of two HLA-SBβ and two HLA-SBα-related genes: an intron in one of the SBβ genes contains a processed pseudogene. Cell 1984;38:241.
19. Heyldig-Nielsen JJ, Morling N, Odum N, et al. Restriction fragment length polymorphism of the HLA DP sub-region and correlations with HLA DP phenotypes. P.N.A.S. 1987;84:1644.
20. Kappes DJ, Arnot D, Okada K, Strominger JL. Structure and polymorphism of the HLA class II SB light chain genes. EMBO J 1984;3:2985.
21. Trowdale J, Kelly A. DZα, a new human class II chain gene. EMBO J 1985.
22. Tonnelle C, DeMars R, Long EO. DOβ: a new β chain in HLA-D with a distinct regulation of expression. EMBO J 1985;4:2839–2847.
23. Mathis D, Benoist C, Williams V, Kanter M, McDevitt H. The murine Eα immune response gene. Cell 1983;32: 745.
24. Benoist CO, Mathis DJ, Kanter MR, Williams VE, McDevitt HO. Regions of allelic hypervariability in the murine Aα immune response gene. Cell 1983;34:169.
25. Estess P, Begovich AB, Koo M, Jones PP, McDevitt HO. P.N.A.S. 1986;83:3594–3598.
26. Mengle-Gaw L, McDevitt HO. Isolation and characterization of a cDNA clone for the murine I-Eβ polypeptide chain. P.N.A.S. 1983;80:7621.
27. McIntyre K, Seidman J. Nucleotide sequence of mutant I-$A\beta_{bm12}$ gene is evidence for genetic exchange between mouse immune response genes. Nature 1984;308:551.
28. Trowsdale J, Young JAT, Kelly AP, et al. Structure, sequence and polymorphism in the HLA-D region. Immunol Rev 1985;85:1–43.
29. Gustafsson K, Wiman L, Emmoth E, et al. Mutations and selection in the generation of class II histocompatibility antigen polymorphism. EMBO J 1984;3:1655.
30. Mengle-Gaw L, Connor S, McDevitt HO, Fathman CG. Gene conversion between murine class II major histocompatibility complex loci. Functional and molecular evidence from the BM12 mutant. J Exp Med 1984;160:1184.
31. Gorski J, Mach B. Polymorphism of human Ia antigens: gene conversion between two DRβ loci results in a new HLA-D/DR specificity. Nature 1986;322:67.
32. Wu S, Saunders TL, Bach FH. Polymorphism of human Ia antigens generated by reciprocal intergenic exchange. Nature 1986;324:676.
33. Brown JH, Jardetzky T, Saper MA, Samraoui B, Bjorkman P, Wiley DC. A hypothetical model of the foreign antigen binding site of class II histocompatibility antigens. Nature 1988;332:845.
34. Allen P, Matsueda G, Evans R, Dunbar J, Marshall G, Unanue E. Identification of the T cell and Ia contact residues of a T cell antigenic epitope. Nature. 1987;327: 713–715.

35. Buus S, Sette A, Colon S, Jenis D, Grey H. Isolation and characterization of antigen-Ia complexes involved in T cell recognition. Cell 1986;47:1071–1077.
36. Rothbard JB, Lechler RI, Howland K, et al. Structural model of HLA-DR1 restricted T cell antigen recognition. Cell 1988;52:515–523.
37. Kabat EA, Wu TT, Bilofsky H. Sequences of immunoglobulin chains. U.S. Department of Health Education and Welfare, N.I.H., Publication No 80:2008, 1979.
38. Long EO, et al. Isotypic Diversity and Function of HLA Class I Antigen. In: Silver J (ed). The Molecular Biology of Class II Antigens. Boca Raton, Florida: CRC Press, 1988.
39. Glimcher LH, Griffith IJ. Mutations of class II MHC molecules. Immunology Today 8:274.
40. Cohn L, Glimcher L, Waldmann R, et al. Identification of functional regions on I-A^G by site directed mutagenesis. PNAS 1986;83:747–751.

Author Affiliations

John I. Bell and John A. Todd, Nuffield Departments of Clinical Medicine and Surgery, John Radcliffe Hospital, Headington, Oxford OX3 9DU England; *Hugh O. McDevitt*, Departments of Medical Microbiology and Medicine, Stanford University School of Medicine, Stanford, California 94053, USA

Regulation of HLA Class II Gene Expression

Janet S. Lee

Introduction

Most developmentally regulated genes that have been studied intensively by molecular techniques are expressed at limited times during differentiation. HLA class II genes, in contrast, are expressed constitutively in a wide variety of cell types, and are inducible in other cells by soluble factors such as gamma interferon and interleukin-4 (1–4). It is probable that cells of different tissues regulate HLA class II gene expression in distinct ways. This makes the HLA class II system an ideal model for the study of gene regulation. Moreover, because class II proteins are necessary for presentation of foreign antigens to T cells, their presence or absence, or even their relative quantitative levels at the cell surface, probably control the type or magnitude of local immune cell interactions (5,6). Thus, not only are studies in this area important in general molecular genetics, they are fundamental to understanding the mechanisms that regulate the immune response in man.

HLA class II antigens were first described as cell surface heterodimers recognized by human alloantisera (1). Serologic specificities correlated with differences detected by mixed lymphocyte culture typing. Early studies of the homologous antigens in the mouse had already indicated that these class II antigens were the probable products of the immune response genes (7). Subsequent serologic, biochemical, and molecular genetic analyses provided a more complete but complicated view of the HLA region. During the last several years, the genomic organization of class II genes in both man and mouse has been elucidated (8–11). The results indicate that the human class II region is larger and contains several more genes that the murine equivalent. These differences in complexity may be reflected by variations in expression of products of individual loci. Many general findings regarding murine class II gene expression can be applied to the human system. Nevertheless, some disparities in tissue distribution between the two species remain, and both *cis*- and *trans*-acting regulatory elements may exhibit variations that are responsible for these differences.

Expression of HLA Class II Genes

Tissue Specificity. Early studies indicated that determinants recognized by alloantisera distinct from HLA A and B antigens, were found primarily or exclusively on B cells and macrophages (1). This finding contrasted with the apparently ubiquitous distribution of the class I antigens and provided further evidence for an additional HLA locus at the time. Refinements in serology and biochemical definition of the HLA class II antigens using monoclonal antibodies have allowed a more thorough analysis of tissue distribution [reviewed in reference 2]. Table 1 lists the tissues or cells in which class II antigens are found constitutively in man and mouse. These are primarily hematopoietic cells and a few related cell types (e.g., Langerhans cells and dendritic cells in the skin) (2,12). Many cells that lack constitutive expression of the class II antigens, however, can be induced to express the antigens with gamma-interferon (3). A partial list of these tissues is also shown in Table 1. In addition, many tumor types can be induced to express class II antigens. Tumor necrosis factor and lymphotoxin may act synergistically with interferons to amplify the expression of the class II molecules at the cell surface (13). Like gamma-interferon induction of class II antigens in macrophages and epithelial tissues, interleukin-4 can induce class II antigens in pre-B cells (4). Other factors may have a repressive effect on class II expression. For instance, prostaglandin E appears to repress class II synthesis in hematopoietic cells that express the antigens (14). Other factors may affect levels

Immunobiology of HLA
Volume II

Table 1. Tissue Specificity of Class II Expression

	Human	Mouse
Constitutive expression		
B cells	+	+
Monocytes	+	+/−
Erythroid precursors	+	−
Langerhans cells	+	+
Kupffer cells	+	?
Inducible expression		
T cells (MLC, PHA, ConA)	+	−?
NK cells (MLC)	+	−
Monocytes (gamma-IF)	+	+
Vascular endothelium (gamma-IF)	+	+
Thymic epithelium (gamma-IF)	+	+
Keratinocytes (gamma-IF)	+	−
Melanocytes (gamma-IF)	+	−
Astrocytes/glial cells (gamma-IF)	+	+
Fibroblasts (gamma-IF)	+	+

of class II expression, presumably via their effects of growth and differentiation.

Perhaps the most striking difference between the tissue distribution of class II antigens between man and mouse is the presence of class II on activated T cells in man and their probable absence on most activated T cells in the mouse (15,16). Early studies using alloantisera suggested that Ia molecules could be found on murine concanavalin A-activated T cells, but this has not been confirmed using monoclonal antibodies or molecular probes that detect mRNA. It is possible that class II proteins are involved in antigen-specific HLA-restricted interactions among T cells in man, adding another level of complexity to regulation of T-cell proliferation in an immune response. Another apparent difference in class II expression between man and mouse is the presence of the class II surface proteins on virtually all peritoneal macrophages and monocytes in man, and only a subset of the equivalent cells in the mouse (17). The only other species in which the distribution of class II proteins have been well studied are rat and rabbit (18,19). In rat, expression of class II genes is similar to that in the mouse, while rabbit class II genes are organized and expressed in a manner similar to those in man.

Many nonlymphoid tissues in which HLA class II antigens are expressed are not easily studied. Only cell lines derived from malignant tumors have been amenable to thorough biochemical and molecular genetic analysis. In the hematopoietic system, the cell lines, HL60 and U937, derived from tumors of the myeloid lineage, have been used as examples of cells inducible with gamma-interferon to express class II antigens (20,21). However, variations in the phenotypes of sublines of these cells have made some studies difficult to interpret. A number of normal epithelial tissues can also be induced with gamma-interferon to express the class II antigens, as well as some endothelial tissues and fibroblasts (22–25). In addition, astrocytes and probably other cells in the brain can be induced (26–28). Only cells derived from germ cell or early embryonic tissues, such as choriocarcinomas, teratocarcinomas, and a few others, have not been stimulated to express the class II antigens (3).

Another nonlymphoid tissue that has been used to begin to elucidate control mechanisms for class II expression is the skin (12). Analysis of tissue sections of skin shows that keratinocytes and Langerhans cells can, under certain circumstances, express HLA class II antigens. However, permanent cell lines representing these cell types are not available for study. In contrast, normal melanocytes have been shown to be inducible with gamma-interferon (3) and, after introduction of activated *ras* oncogenes, to express class II antigens (29); in addition, the transformed counterparts of melanocytes —melanomas—show an intriguing variety of phenotypes with respect to class II expression (3,27,30). Some melanoma lines express class II antigens constitutively, others are inducible with gamma-interferon, and a few cannot be induced to express class II. Where variables like allelic polymorphism are controlled, for instance in studies of melanoma cell lines derived from metastatic tumors in the same patient, this system provides a useful model for studying class II regulation.

HLA Class II Antigens in Differentiating Lymphoid Cells. Although the primary recognized function of the class II proteins is participation in presentation by macrophages or B cells of foreign antigens to T cells, class II proteins may have additional functions. For example, class II products are present on the surfaces of erythroid, myeloid, and granulocytic precursors in the bone marrow (31,32). In this environment, they are unlikely to encounter either foreign antigen to present to the immune system or the mature effector T cells that generate or amplify a response. It is possible that the class II proteins in this context participate in cellular interactions regulating stem cell development. Although specific antigen is not present in the thymus, the presence of class II proteins on thymic epithelium is important in the education and maturation of antigen-specific T cells (33). Again, class II proteins must be important in interactions with the developing T cells, but in an antigen-independent manner.

Aberrant Expression of HLA Class II Antigens in Autoimmune Diseases. A number of diseases are associated with aberrant HLA class II gene expression. It is clear that during transplant rejection, class II expression is elevated in grafted and surrounding tissue (34,35). This is presumably a result of local production of gamma-interferon. The class II proteins are also found in unusually high levels in disseminated host tissues in graft-versus-host disease (36). In some diseases (for instance, systemic lupus erythematosus [SLE]), class II expression in tissues is abnormally low (37). In the combined immunodeficiency bare lymphocyte syndrome (see below), HLA class I and/or class II antigens are absent on the surfaces. However, in several autoimmune diseases, such as rheumatoid arthritis and autoimmune thyroiditis, class II expression may occur in tissues in which these genes are not normally expressed (38–40). Recently, it was shown that aberrant constitutive expression of class II genes in pancreatic cells of transgenic mice was sufficient to cause the development of insulin-dependent diabetes in these mice (41,42). The I-A α and β genes of the host haplotype were introduced separately into the mice under the specific control of the human insulin gene promoter (41). Only when mice carrying the α and β transgenes were bred did the progeny that carried both introduced genes express class II antigens on their pancreatic cells and develop the disease. Furthermore, when the gamma-interferon gene was expressed in a similar transgenic mouse under the control of the insulin promoter, local induction of class II antigens in the pancreas occurred, followed by inflammation, although diabetes did not always develop in these mice. Whether a viral infection is necessary to trigger the onset of diabetes and whether the response is similar to that in natural diabetics can now be studied with these animals. In an independent but similar study, however, transgenic mice expressing an I-E^b antigen in the pancreas consistently developed diabetes, but without evidence for an autoimmune mechanism (42). Moreover, though the mice do not normally express the I-E^b antigens due to a mutation in the I-E α gene, the transgenic mice are tolerant to the I-E^b antigens instead of recognizing them as alloantigens. Again, these mice can now be used to study both the development of insulin-dependent diabetes and of tolerance to nonlymphoid tissue antigens.

Expression of HLA Class II Antigens of Tumors. As mentioned earlier, many types of human tumors, hematopoietic and epithelial in origin, express or can be induced with gamma-interferon to express class II antigens (3). To date, melanomas are the best studied examples. Melanomas can be classified according to certain morphologic and antigenic characteristics which may reflect distinct stages in melanocyte differentiation (43). This classification led to the idea that melanoma cells originate from class II-positive precursors of normal melanocytes. However, it is also possible that class II expression is a consequence of malignant transformation. Support for this latter alternative comes from recent experiments in which normal melanocytes in culture were infected with murine retroviruses carrying viral *ras* oncogenes (29). Expression of the *ras* oncogene products in the melanocytes after spread of the virus through the culture caused a dramatic induction of DP, DQ, and DR antigens, even though, in the absence of gamma-interferon, melanocytes never express class II. Because culture supernatants did not contain any gamma-interferon, *ras* seems to induce class II via a distinct mechanism in these cells. The infected melanocytes showed some other characteristics displayed by melanomas (e.g., augmented expression of G_{D3} ganglioside and the ability to form colonies in soft agar). Nevertheless, they were not completely transformed, and thus class II expression is thought to be an early marker in melanocyte transformation.

Although *ras* oncogene products can induce class II proteins in melanocytes, similar experiments with class II-negative melanoma cell lines of both human and murine origin did not show an induction (29, and unpublished observations). Moreover, studies of cell lines derived from other tissues, such as fibroblasts and renal epithelial cells, showed no change in class II expression on infection with retroviruses containing *ras* genes (A. Albino, personal communication). However, when recombinant retroviruses containing H-*ras* and N-*ras* genes were used to infect B-cell variants carrying mutations in regulatory genes that code for positive *trans*-acting factors, the class II antigens were reexpressed (44). The introduced *ras* gene products had no effect on the expression of other genes, including the class I genes, β_2 microglobulin, and invariant chain. Interestingly, the expression of DR and DQ, but not DP, was specifically increased in the RJ2.2.5 cell line, while only an increase in DQ expression was observed in the 6.1.6 cell line (44,45). In both cases, the reactivation of class II was only partial. Complementation tests by somatic cell fusion experiments show that the mutations are in different genes. Together these findings suggest that the mutations do not involve *ras* proto-oncogenes. It is probable that activation of class II expression by *ras* represents a physiologic pathway in the modulation of gene expression distinct from regulation by gamma-interferon and interleukin-4 (29,44,45). Since treatment of both mutant cells with phorbol myristic acid (a diacylglycerol analog and activator of protein kinase C) can also partially restore class II expression, it is likely that the induction involves the turnover of phosphatidyl inositol, also probably regulated by *ras* gene products (45). Two conflicting interpretations regarding the significance of class II expression on malignant tissues can be construed. First, it is possible that induction of class II antigens at the surfaces of cells in response to oncogene activation may increase the susceptibility of cells to recognition by the immune system. Second, inappropriate or altered expression of class II antigens on transformed cells may enable them to evade recognition by T cells.

Differential Regulation of DP, DQ, and DR Locus Expression. Evidence for differential regulation of class II expression has only recently accumulated. Because the DP and DQ loci were defined much later than DR, and are often expressed at lower levels than is DR, thorough analysis of their expression was not possible until development of monoclonal antibodies that are monomorphic with respect to polymorphisms, but locus-specific. In most cells or tissues that express class II antigens constitutively, DP, DQ, and DR are coexpressed (2). However, some cell lines derived from lymphomas and leukemias have been shown to express DP or DQ in the absence of DR (46,47). Whether such cells reflect rare cells found normally or an aberrant characteristic of the malignant phenotype is not known. In some cell

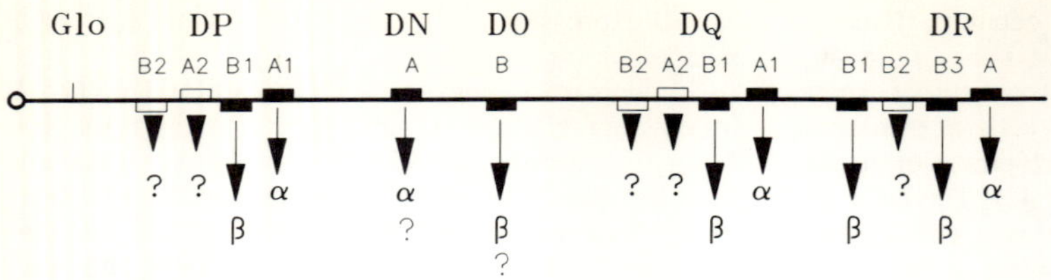

Figure 1. Map of the HLA class II region.

lines, DP and DR, but not DQ, are inducible with gamma-interferon (48). Perhaps, this reflects an unknown functional difference between DQ and the other class II antigens. Many of the results have now been confirmed by analyzing specific RNAs from the respective cell lines.

Other HLA Class II Genes. At least 15 class II genes have been identified in the HLA class II region; a map of this region is shown in Figure 1. The functions of many of the genes (DP2, DN, DO, and DQ2) have not been amenable to study because no protein products for these genes have yet been identified (49–53). The DN α and DO β genes are clearly expressed as RNAs, but at levels much lower than the DP, DQ, and DR genes (50,52). The DO β gene is regulated in a pattern distinct from the other class II genes, while DN α is expressed with a similar pattern, but at much lower levels. The DP2 genes are probably pseudogenes in at least some haplotypes, and thus cannot be expressed. The DQ2 genes, on the other hand, have been completely sequenced and no crippling mutations can be found (51,54,55). Nevertheless, RNA products for these genes have never been detected, albeit in limited surveys of cell lines. It is certainly possible that they are expressed under restricted circumstances.

Trans-Acting Factors Governing HLA Class II Expression. Several lines of evidence indicate that the class II genes are regulated positively by *trans*-acting factors. To date, the most illuminating work in this area has come from the analysis of regulatory mutants. The first such mutant was identified a decade ago by Gladstone and Pious (56). After ethyl-methane sulfonic acid mutagenesis of a B-lymphoblastoid cell line and negative selection with alloantibodies and complement, a cell line, called 6.1.6, no longer expressing HLA-DR was derived. Further analysis showed that DP and DQ antigens were absent as well, yet somatic cell fusion experiments showed that though expression was absent, the genes were still present (57–59). One peculiar, but still unexplained, feature of this mutant was its high reversion rate (60). Moreover, the reversion appeared to affect only DR antigens of one haplotype, not DP and DQ antigens.

A second mutant was isolated after mutagenesis by gamma-irradiation of the Burkitt's lymphoma cell line Raji (61). Phenotypically, this mutant seemed to be identical to the first one, such that although all of the recognized class II antigens were missing from the cell surface, the class II region was unaltered. This mutant line was fused with murine B cells, and its HLA class II genes were reexpressed (62). Thus the mutation also appears to affect a regulatory factor rather than the structural genes. The two regulatory mutants isolated independently by the two laboratories were fused, and class II antigens were reexpressed (C. Hume and J.S.L., unpublished observations). Thus, the cells appear to bear complementary, hence different, mutations. Accolla and co-workers have gone on to demonstrate that a possible gene encoding the *trans*-acting factor is localized to murine chromosome 16 (63). Furthermore, the ability to complement the mutation appears to be transferrable by DNA through transfection (64). Two other human class II regulatory mutants have been isolated (65). Both mutants, like the one derived by Accolla (61), were derived from the cell line Raji; thus somatic cell fusion experiments should be performed to determine whether the mutations are in fact distinguishable. A low level of expression of HLA-DR and DQ occurs in one mutant, but transcription is reduced in both.

Mutants lacking murine class II antigen expression have also been isolated (66). When I-A–negative mutants were characterized, the defect appeared to be at the transcriptional level for the A_β, but not for the other class II α or β genes. Like the human B cell class II-negative mutants, deletion or rearrangement in the class II region did not seem to account for the lack of expression of I-A and I-E antigens in two of the mutants. However, because the only transcriptional defect appeared to be in A_β RNA expression, the nature of these mutations is puzzling. Moreover, when exogeneous A_α and A_β genes were introduced into one mutant by DNA-mediated gene transfer, these exogenous genes were expressed. Thus, a mutation of a gene for a *trans*-acting factor similar to the defective genes in either human B-cell mutant is unlikely to explain the phenotypes of two of the mutants. Another enigma is the finding that the I-E proteins were synthesized but not transported to the cell surface in these mutants. Thus both transcriptional and posttranscriptional defects were found, suggesting that two independent mutations probably produced the phenotype in these mutants. A third mutant lacked I-A expression and yet I-E antigens could still be expressed. This suggests that I-A and I-E expression may well be governed by distinct control mechanisms. Table 2 summarizes phenotypic and genetic characteristics of the class II regulatory mutants identified so far.

Bare Lymphocyte Syndrome. Congenital immunodeficiencies that are associated with a lack of HLA class I or

Table 2. Tumors Expressing HLA Class II Antigens

Constitutive/Inducible	Negative
Lymphomas	Teratocarcinomas
Leukemias	Choriocarcinomas
Melanomas	Neuroblastomas
Renal carcinomas	
Ovarian carcinomas	
Colon carcinomas	
Lung carcinomas	
Bladder carcinomas	
Breast carcinomas	

Table 3. Regulatory Mutants Defective for HLA Class II Expression Derived by In Vitro Mutagenesis

Cell Line	Reference	Characteristics
6.1.6	56–60	No surface antigen
		Reduced transcription
		High reversion rate
		Complemented by fusion
	45	Complemented by *ras*, TPA
RJ2.2.5	61–64	No surface antigen
		Reduced transcription
		Complemented by fusion with lymphoid, but not myeloid lines
	45	Complemented be *ras*
P1	65	No surface antigen
P2		Reduced transcription
G1	66	No surface antigen
G2		No I-A mRNAs
		Cytoplasmic Eβ
		Transfected class II genes expressed
		Two mutations?
G3	66	No surface I-A
		I-E expressed on surface

class II gene expression have been observed in the population, although at very low frequency (67). Initially, these deficiencies were termed bare lymphocyte syndrome (BLS), and are now classified as combined immunodeficiencies associated with defective expression of HLA antigens (68–70). For the limited number of families in which the data allow genetic analysis, these syndromes behave as recessive defects (68,71). Because BLS can affect class I or class II regulation and result in a wide variety of symptoms, it is assumed that BLSs are a collection of immunodeficiencies arising from mutations of different regulatory genes. Three categories have been established: Type I in which only class I antigen expression is affected, Type II where only class II is affected, and Type III in which class II antigens are negative and class I antigens are decreased (68).

Consequences of the lack of surface class II antigen expression in those patients with the type II or III BLS range from mild to severe (Table 4). They result from loss of lymphocyte function, absence of specific antibody production and delayed hypersensitivity, and increased susceptibility to infection (67–74). All of these emphasize the critical role that class II antigens play in the immune response.

Study of BLS has been greatly facilitated by the ability to culture B-cell lines from some of the patients after Epstein Barr virus (EBV) transformation, but this has been possible in by no means all of the patients (75). Some reports have indicated that the cells of some patients revert to a class II-positive phenotype after EBV transformation, although in other patient cell lines, the class II-negative phenotype is stable for months to years in culture (76,77). Analysis of genomic DNA either from peripheral blood or cultured B cells from patients with BLS has indicated that the HLA class II genes are intact even though expression is not observed (78). Thus, BLS defects correspond to mutations in *trans*-acting regulatory genes, natural counterparts of the mutants derived in vitro.

Bare lymphocyte syndrome cells lack cell surface class II expression because the genes are not expressed as mRNA (76,77). One exception to this rule is a BLS cell line that expresses a very small quantity of DQ α mRNA, but none for any other known class II gene (77). The significance of this expression is not known, but again suggests the possibility that the DQ genes may be regulated differently from the DR and DP genes. Because no mRNA for class II genes is usually found in BLS cell lines, transcription of the genes was assessed by nuclear run-on assays. Regulation of most genes that have been studied takes place primarily at the transcriptional level; thus it was not surprising that the three BLS cell lines analyzed so far do not transcribe the genes (79–81).

Another study of a patient with BLS has shown in addition to a lack of HLA class II expression, a defect in the expression of the CD 21 B-cell differentiation antigen (82). Inability to transform B cells with EBV is a probable consequence of this phenotype because CD 21 acts as a receptor for EBV (83). Culturing B cells in the presence of gamma-interferon or interleukin-4 had no affect on the expression of class II or CD 21. However, monocytes derived from the same patient were induced to express class II proteins by treatment with gamma-interferon (75). The authors concluded, therefore, that the mechanisms regulating class II expression in the B cells are distinct from those regulating expression in the

Table 4. Characteristics of Bare Lymphocyte Syndrome

Bare Lymphocyte Syndrome Type	II	III
Hypogammaglobulinemia	Yes	Yes
Altered proliferative response to antigens	Yes	Yes
Impaired homing of lymphocytes to peripheral organs	Some	Yes
Low CD 4 (helper) cells	Some	Yes
Helper cell function absent	Some	Yes
Altered delayed type hypersensitivity	No	Yes

See also references 67–82.

monocytic lineage. These findings reinforce previous conclusions based on somatic cell fusion experiments with a regulatory B-cell mutant and of a mouse monocyte line that regulation in the two lineages is distinct (84).

Molecular Mechanisms for Class II Gene Regulation

Methylation of Class II Genes. Methylation of cytosine residues is a major postsynthetic modification of DNA in eukaryotes (85). The altered nucleotides in a sequence probably affect chromatin structure and influence interactions between *trans*-acting protein factors and *cis*-acting regulatory elements within and flanking a gene. Results from many studies suggest that genes are methylated when cells have terminally differentiated and no longer express them (86). In contrast, the murine class I H-2K gene is hypomethylated in embryonal carcinoma cells when the gene is not expressed, and activation of H-2K transcription is correlated with an increase in methylation (87).

In studies on the methylation of HLA class II genes, even more conflicting results have been obtained. Initial studies comparing the methylation of HLA DR α genes in B cells and T cells from the same individuals showed that the DNA was at least partially unmethylated in the B cells that expressed DR, and heavily methylated in the nonexpressing T cell lines (88). Nevertheless, during short-term culture, DR expression is increased on peripheral blood T cells from normal or HTLV-I infected individuals, while there is no decrease in methylation of the DR α genes. When a T-cell line, which lacks DR expression and is highly methylated in the DR α gene is fused with the class II deletion mutant 721.174, a decrease in methylation is associated with induction of expression (89). Another study, using the monocytic line U937, showed that in normal conditions when the DR α gene is not expressed, the gene is methylated. Yet, demethylation with 5-azacytidine does not induce DR expression. Only treatment with gamma-interferon in addition to 5-azacytidine caused DR induction (90). On the other hand, the class II negative regulatory mutant 6.1.6 was less methylated at DR α than was its parent B-cell line (91). More confusing results were obtained when methylation of DR α in melanoma and erythroid lines were studied. The DR-expressing melanoma line was highly methylated, while no methylated sequences in DR α were detectable in the nonexpressing erythroid line (92). Finally, to clarify the role of methylation, melanoma cell lines that were all derived from the same primary tumor were studied. Results indicated that methylation was correlated positively with DR expression, although treatment with 5-azacytidine on lines expressing DR had no effect or actually increased expression (93). Thus, in some systems where polymorphic methylation substrates can be ruled out, as in the melanoma lines, a positive correlation was found. While in cells of different lineages, B cells and T cells, even though from the same individual, the nonexpressed DR α genes from the T cells were more highly methylated. It seems that if methylation can control gene expression at all, it may only be operating within a lineage and in conjunction with other mechanisms of gene control.

Cis-Acting Regulatory Elements in Class II Genes. The ability to transfer endogenous genes into different cell types and measure transient expression of a protein product or specific RNA has allowed the identification of controlling elements within promoter regions or even within the bodies of several genes (94). Viral- and subsequently cellular tissue-specific enhancer elements were initially identified in this manner (94,95). An enhancer was identified in a region upstream of the I-E β gene (96), and similar elements have been detected in the DR α promoter and first intron (97). Examination of the sensitivity to DNase I of DR α sequences within isolated nuclei indicated that proteins were probably bound at the locations of the putative enhancer elements because the configuration of the chromatin in those regions was altered, making them hypersensitive to the DNase I (98). Not surprisingly, another DNase I hypersensitive site was located in the promoter of the DR α gene.

Comparisons of the regions immediately upstream from the mRNA start sites within several class II genes and the gene for the class II-associated invariant chain have shown that at least two sequence elements, termed X and Y boxes, are conserved in all the promoters (99,100). These consensus elements have now been shown to be substrates for the binding of nuclear protein factors using gel retardation and DNase I footprinting assays (101–103). Recently, the presence of both positive and negative regulatory sequence elements in the region flanking the DQ β genes has been suggested on the basis of activity of progressive deletion constructs of the promoter for the gene in stably transfected human fibroblasts (104). Deletion of sequences from −2500 to −159 nucleotides (upstream from the mRNA cap site) indicated the presence of a negative regulatory element within this region. The 30 nucleotides adjacent (to −128) appeared to contain another negative element, the W box, because constructs missing this sequence directed a high level of transcription that was unaffected by treatment with gamma-interferon. A further 20 nucleotide deletion (to −107) within the X box decreased transcription, and deletion to −82 at the 5− boundary of the Y box again caused an increase in transcription efficiency. For the DR α gene, 300 nucleotides of upstream sequence were shown to provide the *cis*-acting sequences required for induction of the gene with gamma-interferon in a human glioblastoma multiforme line (105). For expression in transfected B cells, a fragment containing an upstream sequence of the DR α gene (including both X and Y consensus elements) was sufficient to direct transcription of the introduced construct. Together these results indicate a complex combination of negative and positive controlling elements in the class II promoters. A schematic representation of these control elements in class II gene promoters is shown in Figure 2. Some caution, however, is needed in interpreting experi-

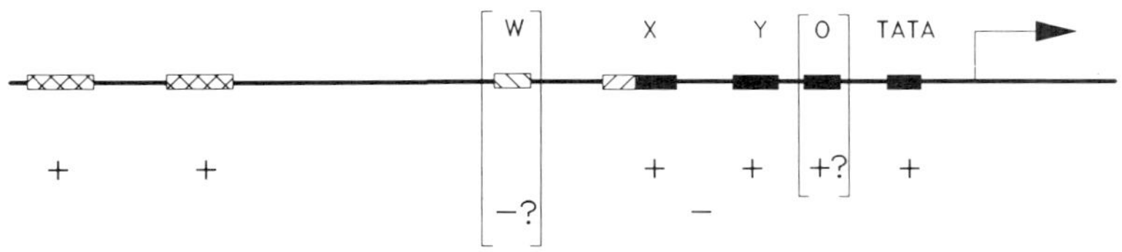

Figure 2. Schematic map of *cis*-acting elements in class II gene promoters. Consensus boxes X and Y as well as the TATA motif exist generally in class II promoters. The W box and O (octamer) are found in only one or a few class II genes. Upstream elements (unnamed) have been identified in constructs of the E_α gene introduced into transgenic mice (33). + and − indicate the probable effects on class II transcription. The arrow indicates the transcription start site.

ments using deletion constructs because the relative position of promoter elements can certainly affect their function (106); random sequence replacement as well as point mutation constructs should be analyzed to confirm these findings. Another sequence that appears to be involved in the control of DR α gene transcription is an octamer sequence found upstream of all immunoglobulin variable regions (107). Intriguingly, the octamer is present downstream from the X and Y boxes and upstream from the mRNA start site in the DR α gene (81). In the immunoglobulin system, the octamer and a TATA box alone are sufficient to direct tissue-specific transcription (108). However, in the DR α gene, removal of the X and Y sequences completely abolishes transcription directed by transfected constructs in vivo (105) and by templates in in vitro transcription systems (C. Hume, unpublished observations). Thus, the octamer itself behaves differently under control of sequences within the first 40 nucleotides upstream from the cap site or possibly under the control of other elements within the class II and immunoglobulin genes.

The nature of the control of tissue-specific transcription of the I-E α gene was addressed by analysis of expression in mice carrying I-E α transgenes with varying promoter constructs (103). This gene carries the X and Y boxes, but unlike its human homolog DR α, I-E α lacks the immunoglobulin octamer sequence (109). Deletion of either the X or Y boxes drastically reduces the efficiency of transcription of the gene. Nevertheless, tissue specificity of transcription of the gene is retained in some of the transgenic lines carrying the deleted constructs (103). Inducible expression of the I-E α gene is achieved only when a functional Y box is present in the promoter. Deletion of this Y box sequence within the construct completely abolished the induction of I-E expression in macrophages, while deletion of the X box generally reduced expression, but gamma-interferon treatment of the cells still led to a small increase in expression. A peculiar finding in some of these experiments was the demonstration that transcription of the gene was directed by some of the promoter element deletion constructs, but the transcripts were initiated in regions upstream of the recognized promoter sequences. Thus, interpretations based on the expression of reporter genes, like chloramphenicol acetyltransferase (CAT), should be made with caution if the sites for initiation of transcription are not known.

In an independent study of the expression of various constructs of E_α transgenes introduced into the mouse germline, at least 2 kb of 5′-flanking sequences were required to achieve appropriate tissue-specific expression (33). These findings are in agreement with those of the previous study. A construct with 1.4 kb of upstream sequences was expressed only in adherent spleen cells and thymus, and not in B cells. However, peritoneal exudate cells were still inducible for E_α expression with gamma-interferon. Thus, sequences within a region between −2 and 1.4 kb appear to contain elements that promote expression in B cells. In fact, six enhancer core-like sequences were found in this region. Comparisons of the E_α gene with the DR α indicated that DR α, but not any of the other class II genes available for analysis, also contains these enhancer elements (33). These studies suggest that promoters of genes expressed in a complex tissue-specific manner may be much larger than originally anticipated.

Trans-Acting Protein Factors Controlling Transcription. DNA binding protein factors that are required for the transcription of several genes are beginning to be identified and purified. For example, several distinct proteins bind to the octamer that is present upstream of the immunoglobulin variable regions (110,111). Two of these octamer binding proteins have been studied in detail. One of the factors is ubiquitous, and another is present only in B cells. The octamer also found in histone gene promoters is thought to bind the ubiquitous protein. The octamer element in the DR α promoter clearly binds one or more of these protein factors found in nuclear extracts from B cells (81). Presumably, this is identical to the B cell-specific factor directing transcription of the immunoglobulin genes in B cells. On the other hand, since the octamer and the TATA box in the DR α are not sufficient to direct transcription either in vivo or in vitro, a distinct protein factor could be binding to that element. It will be of interest to determine whether other octamer binding factors are involved in DR α transcription in nonlymphoid tissues. Of even greater interest, however, will be clarification of how the presence of the octamer in the DR α gene determines the differences of expression between the DR versus DQ and DP loci, and more particularly, between DR and I-E.

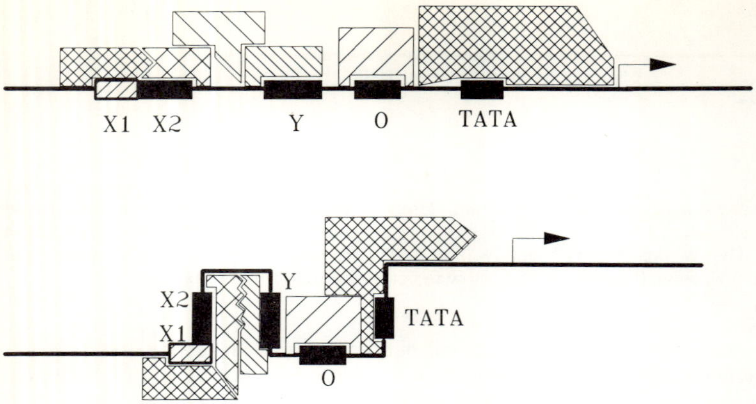

Figure 3. Models for the assembly of class II transcription complexes. X1, X2, Y and O *cis*-acting elements of the promoter are indicated by rectangles. Irregular polygons indicate nuclear binding proteins. The upper panel indicates a linear scheme; the lower panel shows a scheme including secondary structure in the DNA.

The class II consensus Y box contains within its core the sequence ATTGG, which is actually an inverted CCAAT box found in many promoters (112). This sequence has been shown to bind several nuclear proteins in in vitro assays and presumably binds a specific protein in vivo to mediate transcription. Detailed analysis of binding to the Y or inverted CCAAT box has now indicated that a protein (NF-Y) binds with a high affinity to the I-E α Y box, and to CCAAT boxes in other promoters (α-globin, herpes simplex virus thymidine kinase, and β-globin) with much lower affinities (112). NF-Y is apparently distinct from several other proteins that have been shown to bind CCAAT boxes. Another report examined the binding of nuclear factors to the HLA-DQ β promoter (101). Again, binding was detected within the large region of the promoter, between −142 and −60 nucleotides upstream of the DQβ mRNA start site. Oddly, the DQ β gene contains a normal CCAAT box at −50 in addition to the Y box at approximately −70 to −80, but like the I-E α gene and the other class II genes, DQ β lacks the immunoglobulin octamer found at the equivalent position in DR α. Thus, it is possible that two distinct CCAAT binding proteins are associated with the transcription complex for the DQ β promoter.

Demonstrations of transcriptional defects for the mutants derived in vitro and cell lines established from patients with the class II-negative bare lymphocyte syndrome have prompted analysis of binding factors from nuclear extracts of these cell lines. Gel retardation assays using a fragment from the DR α promoter did not show reproducible differences of those complexes formed with proteins from normal B cells and from the mutants (81). Fragments from the DR α promoter that include the X, Y, and octamer consensus sequences clearly bind nuclear proteins in all extracts tested. DNase I footprint analysis, however, clearly demonstrates the lack of protection by protein nuclear extracts from BLS cell lines over a region upstream of and including the 5′ half of the X consensus element of DR α (81). Thus, the binding of multiple proteins to the class II promoter elements and the alteration of this binding in the BLS mutants suggest that a large protein complex is needed to initiate the transcription of these genes. Since even cell lines that do not express class II also contain nuclear factors that appear to bind the promoter elements, it is possible that different sequence-specific binding factors may be found in different tissues (112). A model for the positive action of multiple binding proteins on the DR α promoter in B lymphocytes is shown in Figure 3. In the first scheme, a linear array of proteins binds to activate RNA polymerase II. More likely, however, the DNA in the promoter bends, and nuclear binding proteins interact as in the second scheme. Tissues that do not transcribe the genes clearly contain nuclear factors that bind the DNA at the X and Y boxes. It is possible that they are distinct proteins though they bind the same sequence, or that modifications such as glycosylation or acetylation affect their functions in different tissues. Additional proteins not requiring contact with the DNA may be superimposed on the bound complex, and thereby control transcription.

Posttranscriptional Processing. Although most eukaryotic genes are regulated at the level of transcription, many processing steps are required before mRNA is translated into protein and the protein transported to the appropriate site of action. After the initiation of transcription, the RNA must be elongated, terminated, spliced, and transported to the cytoplasm. Thus, the quantitative level of mature mRNA can be regulated at any of these steps. Furthermore, stability of the mRNA species may also regulate the eventual quantity of protein produced. A suggestion that control of RNA elongation may play a role in regulating class II expression comes from recent findings that the first exon of the DR α gene is transcribed with much greater efficiency in nuclear run-on assays than is the body of the gene (81). It is possible that sequences in the first intron of the gene are involved in binding proteins that either facilitate elongation or prematurely terminate transcription.

Relative levels of class II antigen expression may control myriad interactions between cells during an immune

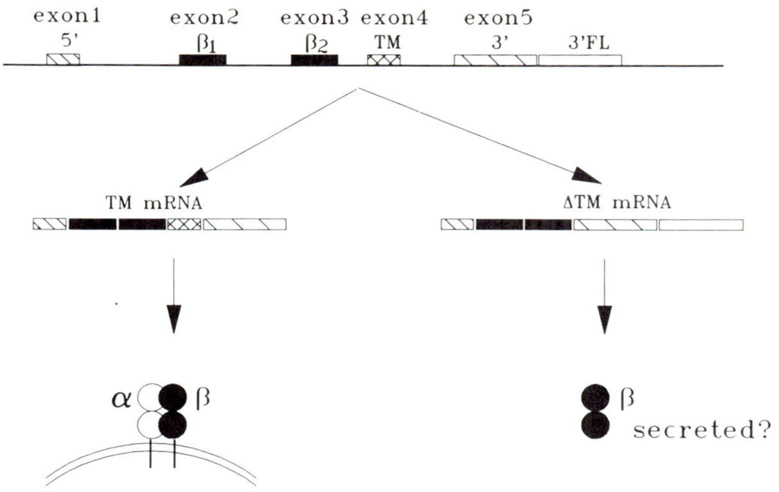

Figure 4. Alternative splicing and polyadenylation of a class II β gene. Exons of the DQ β gene are depicted. The first RNA product represents the "classic" form of DQ β mRNA, which is subsequently translated into a membrane-bound protein. The second RNA represents the alternatively spliced RNA which contains the 3′ flanking region as a result of polyadenylation readthrough. A potential protein product would be secreted.

response (6). Recently, it was shown that ATTTA sequences in the 3′ untranslated regions of mRNAs for some growth factors and oncogenes can affect the stability of their corresponding mRNAs under certain conditions (113). Some DQ α alleles contain these sequences (C. Hume and J.S.L., unpublished observations). It is feasible that such polymorphic controlling elements or those as yet unidentified could play a role in the regulation of gene expression for different HLA class II alleles. This is a particularly interesting possibility in light of recent results showing that readthrough of polyadenylation signals occurs in the transcription of some DQ β alleles (114). In addition, comparisons of sequences of different DQ α alleles previously showed that alternative splice sites dependent on sequence polymorphisms within the untranslated regions are used (115,116). However, the splicing differences do not affect the coding regions and, thus, do not affect the structure of the protein products. Analysis of DP and DR β genes has indicated that alternative splicing can create molecules with different cytoplasmic residues beyond the transmembrane sequence (117,118). Sequence analysis of a novel HLA DQ beta cDNA clone and corresponding amino acids of the putative protein showed that alternative splicing can also produce potentially secreted molecules (114). In some, but not all, DQ β alleles, exon 3 can be fused to exon 5. A diagram of the alternative splicing for DQ β transcripts is shown in Figure 4. Though exon 4 was not included, the novel splicing event does not shift the reading frame, and a putative product should terminate with the same four amino acids as found in the common DQ beta chain. Because the 5′ sequences are unaltered with respect to the more common mRNAs for DQ β, the alternatively spliced mRNA should be translated. This certainly provides an additional mechanism that may create polymorphism in products of class II alleles. Cells that produced alternatively spliced mRNAs also showed evidence for polyadenylation readthrough, but the two processing events were not in complete association. These findings suggest that controlling elements for splicing patterns and polyadenylation for DQ mRNAs are within the gene or closely linked.

Alpha and beta chains of class II antigens normally associate soon after their synthesis in the endoplasmic reticulum and appear at the plasma membrane as heterodimers (119,120). In the absence of its partner, a single β or α chain is not usually transported to the cell surface (121). Thus, it is possible, but seems intuitively unlikely, that a DQ β chain lacking a transmembrane domain would be secreted as a single chain. It could be secreted if there is an equivalent transmembrane lacking DQ α chain. Another potential structure is an association of the transmembrane lacking DQβ with a normal DQ α chain. Such structural alterations found to be associated with certain class II alleles could profoundly affect the functions of the antigens within the immune response. It is possible that certain associations of HLA class II alleles and diseases may be explained by alternative splicing affecting the membrane structure of corresponding antigens.

No evidence for control of class II expression via translational or posttranslational control has yet been found, apart from mutants identified for the murine class II gene expression (66). Nevertheless, the finding that the glycosylation of class II antigens was different in melanoma cells versus B cells (122) suggests the possibility that carboyhydrate residues at the cell surface could affect the function or stability of the class II proteins at the cell surface.

Conclusions and Speculations

Analysis of the tissue distribution of HLA class II genes and the modulation of their expression by soluble factors or by the expression of regulatory genes has led to the view that complex and elaborate controls govern the expression of the class II genes. Both naturally occurring mutations (patients bare lymphocyte syndrome and cell lines derived from them) and mutations generated in cell lines in vitro have been important tools in efforts to define regulatory genes encoding positive *trans*-acting factors. Together these data indicate that at least four complementing mutations in such genes exist, and it is probable that more may be found.

The molecular mechanisms involved in class II gene regulation are now beginning to be elucidated. Transfer of class II genes or hybrids of the class II promoters and reporter genes into different cell types have begun to identify *cis*-acting sequence elements that quantitatively affect gene expression. Both positive and negative control elements have been postulated from these studies. This approach has been extended to the introduction of class II transgenes into the germlines of mice. Even more complexity is observed in these experiments because expression in different tissue types is clearly governed by sequences quite distant from the mRNA start site. Detection of nuclear proteins binding to class II promoter sequences has also been refined recently. Several nuclear factors from B-cell lines and cell lines of other tissues have been shown to bind specifically to particular sequences in class II promoters; the mutations responsible for defective class II expression in bare lymphocyte syndrome B-cell lines are manifested in the lack of binding of protein factors to sequences upstream and including part of the X consensus element. Because more than one mutation produces this phenotype, it is unlikely that the absence of one binding protein can explain the defect. More probably, modifications of protein interactions are affected by the mutations.

Structural polymorphism of HLA class II genes has been a focus of major interest over that last several years. It is probable that regulatory polymorphisms govern the function of class II alleles as well. Alternative splicing of some alleles produces different 3′ untranslated regions containing sequences involved in the regulation of mRNA stability. Readthrough of polyadenylation signals may also be responsible for quantitative regulation of mRNA levels. As yet, polymorphisms within promoters and introns that may affect transcription levels or RNA processing are unexplored, but almost certainly exist. Strikingly, polymorphism in some class II β alleles appears to be responsible for the generation of alternatively spliced mRNAs missing transmembrane exons. The products of such mRNAs are likely to be secreted and profoundly affect interactions with antigen-specific T cells in the immune response.

Altered or imbalanced expression of class II antigens in general or of individual subloci may affect interactions between antigen presenting cells and T cells. There are already indications that distinct factors may act to regulate the transcription of individual loci, particularly distinguishing DQ from DP and DR. Identification of these genes that regulate class II gene expression quantitatively is a goal of intensive research. The genes and their products may be the targets of mutations or epigenetic alterations in diseases of autoimmune etiology or malignancies in which aberrant expression of HLA class II genes is found. Furthermore, it is possible that polymorphisms in regulatory genes may play a role in some HLA-associated diseases, contributing to disease susceptibility. Continued efforts to elucidate mechanisms for class II gene regulation should eventually allow fine manipulation of class II gene expression in vitro and in vivo. This will make it possible to evaluate more precisely the functions of class II antigens in different tissues and develop new approaches for allogeneic transplantation and treatment of many diseases.

References

1. Winchester RJ, Kunkel HG. The human Ia system. Adv Immunol 1980;28:221–285.
2. Radka SF, Charron DJ, Brodsky FM. Class II molecules of the major histocompatibility complex considered as differentiation markers. Human Immunol 1986;16:390–400.
3. Houghton AN, Thomson TM, Gross D, Oettgen HF, Old LJ. Surface antigens of melanoma and melanocytes. Specificity of induction of Ia antigens by human gamma-interferon. J Exp Med 1984;160:255–269.
4. Noelle R, Krammer PH, Ohara J, Uhr JW, Vitetta ES. Increased expression of Ia antigens on resting B cells: an additional role for B-cell growth factor. Proc Natl Acad Sci USA 1984;81:6149–6153.
5. Unanue ER. Antigen-presenting function of the macrophage. Ann Rev Immunol 1984;2:395–428.
6. Janeway CA, Bottomly K, Babich J, et al. Quantitative variation in Ia antigen expression plays a central role in immune regulation. Immunol Today 1984;5:99–105.
7. Jones P, Murphy DB, McDevitt HO. Two gene control of the expression of a murine Ia antigen. J Exp Med 1978; 148:925–939.
8. Hardy DA, Bell JI, Long EO, Lindsten T, McDevitt HO. Pulsed-field gel electrophoresis mapping of the class II region of the human major histocompatibility complex. Nature 1986;323:453–456.
9. Lawrance SK, Smith CL, Srivastava R, Cantor CR, Weissman S. Megabase-scale mapping of the HLA gene complex by pulsed field gel electrophoresis. Science 1987;235:1387–1390.
10. Carroll MC, Katzman P, Alicot EM, et al. Linkage map of the human major histocompatibility complex including the tumor necrosis factor genes. Proc Natl Acad Sci USA 1987;84:8535–8539.
11. Hood L, Steinmetz M, Malissen B. Genes of the major histocompatibility complex of the mouse. Ann Rev Immunol 1983;1:529–568.
12. Edelson RL, Fink JM. The immunologic function of skin. Sci Am 1985;2:46–53.
13. Chang RJ, Lee SH. Effects of interferon-gamma and tumor necrosis factor-α on the expression of an Ia antigen on a murine macrophage cell line. J Immunol 1986;137: 2853–2856.
14. Snyder DS, Beller DI, Unanue ER. Prostaglandins modulate macrophage Ia expression. Nature 1982;299:163–165.
15. Mengle-Gaw L, McDevitt HO. Genetics and expression of mouse Ia antigens. Ann Rev Immunol 1985;3:367–396.
16. Lorber MC, Loken MR, Stall AM, Fitch FW. I-A antigens on cloned alloreactive murine T lymphocytes are acquired passively. J Immunol 1982;128:2798–2803.

17. Gonwa TA, Picker LJ, Raff HV, Goyert SM, Silver J, Stobo JD. Antigen-presenting capabilities of human monocytes correlates with their expression of HLA-DS, an Ia determinant distinct from HLA-DR. J Immunol 1983;130:706–711.
18. Fukumoto T, McMaster WR, Williams AF. Mouse monoclonal antibodies against rat major histocompatibility antigens. Two Ia antigens and expression of Ia and class I antigens in rat thymus. Eur J Immunol 1982;12:237–243.
19. Kulaga H, Sogn JA, Weissman JD, et al. Expression patterns of MHC class II genes in rabbit tissues indicate class homology to human counterparts. J Immunol 1987;139: 587–592.
20. Koeffler PH, Ranyard J, Yelton L, Billing R, Bohman R. Gamma-interferon induces expression of the HLA-D antigens on normal and leukemic human myeloid cells. Proc Natl Acad Sci USA 1984;81:4080–4084.
21. Peterlin MB, Gonwa TA, Stobo JD. Expression of HLA-DR by a human monocyte cell line is under transcriptional control. J Mol Cell Immunol 1984;1:191–200.
22. Collins T, Korman AJ, Wake CT, et al. Immune interferon activates multiple class II major histocompatibility complex genes and the associated invariant chain gene in human endothelial cells and dermal fibroblasts. Proc Natl Acad Sci USA 1984;81:4917–4921.
23. Rosa F, Hatat D, Abadie A, Wallach D, Revel M, Fellous M. Differential regulation of HLA-DR mRNAs and cell surface antigens. EMBO J 1983;2:1585–1589.
24. Steeg PS, Moore RN, Johnson HM, Oppenheim JJ. Regulation of murine macrophage Ia antigen expression by a lymphokine with immune interferon activity. J Exp Med 1982;156:1780–1793.
25. King DP, Jones PP. Induction of Ia and H-2 antigens on a macrophage cell line by immune interferon. J Immunol 1983;131:315–318.
26. Ting JP-Y, Shigekawa BL, Linthicom DS, Weiner LP, Frelinger JA. Expression and synthesis of murine immune response-associated (Ia) antigens by brain cells. Proc Natl Acad Sci USA 1981;78:3170–3174.
27. Natali PG, deMartine C, Quaranta V, et al. Expression of Ia-like antigens in human non-lymphoid tissues. Transplantation 1981;31:75–78.
28. Fontana A, Fierz W, Wekerle H. Astrocytes present myelin basic protein to encephalitogenic T-cell lines. Nature 1984;307:273–276.
29. Albino AP, Houghton AN, Eisinger M, et al. Class II histocompatibility antigen expression in human melanocytes transformed by Harvey murine sarcoma virus (Ha-MSV) and Kirsten MSV retroviruses. J Exp Med 1986;164:1710–1722.
30. Albino AP, Lloyd KO, Houghton AN, Oettgen HF, Old LJ. Heterogeneity in surface antigen and glycoprotein expression of cell lines derived from different melanoma metastases of the same patient. J Exp Med 1981;154:1764–1778.
31. Guy K, van Heyningen V. An ordered sequence of expression of human MHC class II antigens during B cell maturation. Immunol Today 1983;4:186–189.
32. Robinson J, Sieff C, Delia D, Edwards PAW, Greaves M. Expression of cell-surface HLA-DR, HLA-ABC and glycophorin during erythroid differentiation. Nature 1981;289:68–71.
33. Widera G, Burkly LC, Pinkert CA, et al. Transgenic mice selectively lacking MHC class II (I-E) antigen expression on B cells: an *in vivo* approach to investigate Ia gene function. Cell 1987;51:175–187.
34. deWaal RMW, Bogman MJJ, Maass CN, Cornelissen LMH, Tax WJM, Koene RAP. Variable expression of Ia antigens on the vascular endothelium of mouse skin allografts. Nature 1983;303:426–429.
35. Milton AD, Fabre JW. Massive induction of donor-type class I and class II major histocompatibility complex antigens in rejecting cardiac allografts in the rat. J Exp Med 1985;161:98–112.
36. Hickey WF, Kimura H. Graft-vs.-host disease elicits expression of class I and class II histocompatibility antigens and the presence of scattered lymphocytes in rat central nervous system. Proc Natl Acad Sci USA 1987;84: 2082–2086.
37. Sano H, Compton LJ, Shiomi N, Steinberg AD, Jackson RA, Sasaki T. Low expression of human histocompatibility leukocyte antigen-DR is associated with hypermethylation of human histocompatibility leukocyte antigen-DRα gene regions in B cells from patients with systemic lupus erythrematosus. J Clin Invest 1985;76:1314–1322.
38. Teyton L, Lotteau V, Turmel P, et al. HLA-DR, DQ, and DP antigen expression in rheumatoid synovial cells: a biochemical and quantitative study. J Immunol 1987;138:1730–1738.
39. Traugott U. Multiple sclerosis: relevance of class I and class II MHC expressing cells to lesion development. J Neuroimmunol 1987;16:283–302.
40. Bottazzo GF, Todd I, Mirakian R, Belfiore A, Pujol-Borrell R. Organ-specific autoimmunity: a 1986 overview. Immunol Rev 1986;94:137–169.
41. Sarvetnick N, Liggitt D, Pitts SL, Hansen SE, Stewart TA. Insulin-dependent diabetes mellitus induced in transgenic mice by ectopic expression of class II MHC and interferon-gamma. Cell 1988;52:773–782.
42. Lo D, Burkly LC, Widera G, et al. Diabetes and tolerance in transgenic mice expressing class II MHC molecules in pancreatic beta cells. Cell 1988;53:159–168.
43. Houghton AN, Eisinger M, Albino AP, Cairncross JG, Old LJ. Surface antigens of melanocytes and melanomas—markers of melanocyte differentiation and melanoma subsets. J Exp Med 1982;156:1755–1766.
44. Hume CR, Accolla RS, Lee JS. Defective HLA class II expression in a regulatory mutant is partially complemented by activated *ras* oncogenes. Proc Natl Acad Sci USA 1987;84:8603–8607.
45. Hume CR, Accolla RS, Pious D, Lee JA. Activated *ras* oncogenes partially complement the transcriptional defects in class II negative mutant B-cell lines. This volume.
46. Hume CR, Accolla RS, Pious D, Lee JS. Modulation of HLA class II gene expression by activated *ras* oncogenes and phorbol esters. 1988 (Submitted).
47. Symington FW, Levine F, Braun M, et al. Differential Ia antigen expression by autologous human erythroid and B lymphoblastoid cell lines. J Immunol 1985;135:1026–1032.
48. Maurer DH, Hanke JH, Mickelson E, Rich RR, Pollack MS. Differential presentation of HLA-DR, DQ, and DP restriction elements by interferon-gamma-treated dermal fibroblasts. J Immunol 1987;139:715–723.
49. Spielman RS, Lee J, Bodmer WF, Bodmer JG, Trowsdale J. Six HLA-D region α-chain genes on human chromosome 6: polymorphisms and associations of DCα-related sequences with DR types. Proc Natl Acad Sci USA 1984;81:3461–3465.

50. Tonnelle C, DeMars R, Long EO. DOβ: a new β chain gene in HLA-D with a distinct regulation of expression. EMBO J 1985;4:2839–2847.
51. Auffray C, Lillie JW, Arnot D, Grossberger D, Kappes D, Strominger JL. Isotypic and allotypic variation of human class II histocompatibility antigen α-chain genes. Nature 1984;308:327–333.
52. Trowsdale J, Kelly A. The human HLA class II α chain gene DZα is distinct from genes in the DP, DQ and DR subregions. EMBO J 1985;4:2231–2237.
53. Nomenclature for factors of the HLA system. World Health Org 1987.
54. Jonsson A-K, Hyldig-Nielsen J-J, Servenius B, et al. Class II genes of the human major histocompatibility complex. Comparisons of the DQ and DX α and β genes. J Biol Chem 1987;262:8767–8777.
55. Korman AJ, Frantz JD, Strominger JL, Mulligan RC. Expression of human class II major histocompatibility complex antigens using retrovirus vectors. Proc Natl Acad Sci USA 1987;84:2150–2154.
56. Gladstone P, Pious D. Stable variants affecting B cell alloantigens in human lymphoid cells. Nature 1978;271: 459–461.
57. Gladstone P, Pious D. Identification of a trans-acting function regulating HLA-DR expression in a DR-negative B cell variant. Somatic Cell Genetics 1980;6:285–298.
58. Levine F, Erlich HA, Mach B, Pious D. Transcriptional regulation of HLA class II and invariant chain genes. J Immunol 1985;134:637–640.
59. Salter RD, Alexander J, Levine F, Pious D, Cresswell P. Evidence for two trans-acting genes regulating HLA class II antigen expression. J Immunol 1985;135:4235–4238.
60. Levine F, Pious D. Revertants from the HLA class II regulatory mutant 6.1.6: implications for the regulation of Ia gene expression. J Immunol 1984;132:959–962.
61. Accolla RS. Human B cell variants immunoselected against a single Ia antigen subset have lost expression of several Ia antigen subsets. J Exp Med 1983;157:1053–1058.
62. Accolla RS, Scarpellino L, Carra G, Guardiola J. Trans-acting element(s) operating across species barriers positively regulate expression of major histocompatibility complex class II genes. J Exp Med 1984;162:1117–1133.
63. Accolla RS, Jotterand-Bellomo M, Scarpellino L, Maffei A, Carra G, Guardiola J. *alr*-1, a newly found locus on mouse chromosome 16 encoding a trans-acting factor for MHC class II gene expression. J Exp Med 1986;164: 369–374.
64. Guardiola J, Scarpellino L, Carra G, Accolla RS. Stable integration of mouse DNA into Ia-negative human B-lymphoma cells causes reexpression of the human Ia-positive phenotype. Proc Natl Acad Sci USA 1986;83:7415–7418.
65. Calman AF, Peterlin BM. Mutant human B cell lines deficient in class II major histocompatibility complex transcription. J Immunol 1987;139:2489–2495.
66. Glimcher LH, McKean DJ, Choi E, Seidman JG. Complex regulation of class II gene expression: analysis with class II mutant cell lines. J Immunol 1985;135:3542–3550.
67. Touraine J-L, Betuel H, Touraine F. The bare lymphocyte syndrome. In: Griscelli C, Vossen J (eds). Progress in Immunodeficiency Research and Therapy I. Amsterdam: Elsevier Science Publishers B.V. 1984:27–34.
68. Touraine J-L, Marseglia G-L, Betuel H. Thirty international cases of bare lymphocyte syndrome: biological significance of HLA antigens. Exp Hematol 1985; 13:86–87.
69. Gelfand EW, Dosch H-M. Diagnosis and classification of severe combined immunodeficiency disease. In: Birth Defects: Original Article Series, March of Dimes Birth Defects Foundation 1983;19:65–72.
70. Zegers BJM, Heijnen CJ, Roord JJ, et al. Defective expression of mononuclear cell membrane HLA antigens associated with combined immunodeficiency: impaired cellular interactions. In: Birth Defects: Original Article Series, March of Dimes Birth Defects Foundation 1983;19:93–96.
71. dePreval C, Lisowska-Grospierre B, Loche M, Griscelli C, Mach B. A trans-acting class II regulatory gene unlinked to the MHC controls expression of HLA class II genes. Nature 1985;318:291–293.
72. Hadam MR, Dopfer R, Dammer G, et al. Defective expression of HLA-D region determinants in children with congenital agammaglobulinemia and malabsorption: a new syndrome. In: Albert ED, et al. (eds). Histocompatibility Testing 1984. Berlin, Heidelberg: Springer-Verlag, 1984:645–650.
73. Zegers BJM, Heynen CJ, Roord JJ, Kuis W, Stoop JW, Ballieux RE. Combined immunodefiency with defective expression of HLA-antigens: analysis of the nature of defective monocyte-T cell interaction. In: Griscelli C, Vossen J (eds). Progress in Immunodefiency Research and Therapy I. Amsterdam: Elsevier Science Publishers B.V. 1984:35–42.
74. Rijkers GT, Roord JJ, Koning F, Kuis W, Zegers BJM. Phenotypical and functional analysis of B lymphocytes of two siblings with combined immunodeficiency and defective expression of major histocompatibility complex (MHC) class II antigens on mononuclear cells. J Clin Immunol 1987;7:98–106.
75. Plaeger-Marshall S, Haas A, Clement LT, et al. Interferon-induced expression of class II major histocompatibility antigens in the MHC class II deficiency syndrome. J Clin Immunol 1988 (In press).
76. Lisowska-Grospierre B, Charron DJ, dePreval C, et al. Defect of expression of MHC genes responsible for an abnormal HLA class I phenotype and the class II negative phenotype of lymphocytes from patients with combined immunodeficiency. In: Albert ED, et al. (eds). Histocompatibility Testing 1984. Berlin, Heidelberg: Springer-Verlag 1984:650–655.
76. Lisowska-Grospierre B, Charron DJ, dePreval C, Durandy A, Griscelli C, Mach B. A defect in the regulation of major histocompatibility complex class II gene expression in human HLA-DR negative lymphocytes from patients with combined immunodeficiency syndrome. J Clin Invest 1985;76:381–385.
77. Hume CR, Shookster LA, Lee JS. Bare lymphocyte syndrome: altered class II RNA expression in two B cell lines. 1988 (Submitted).
78. Marcadet A, Cohen D, Dausset J, Fischer A, Durandy A, Griscelli C. Genotyping with DNA probes in combined immunodeficiency syndrome with defective expression of HLA. N Engl J Med 1985;312:1287–1292.
79. Lee JS, Hume CR, Shookster L, Yang SY, Collins N. Bare lymphocyte syndrome: transcriptional regulation of HLA class II gene expression. This volume.
80. Reith W, Satola S, Amaldi I, et al. Regulation of HLA-class II genes: identification of a regulatory promoter binding protein missing in class II deficient congenital immunodeficiency. This volume.

81. Hume CR, Lee JS. Altered binding of nuclear factors to HLA class II gene promoter sequences in bare lymphocyte syndrome. 1988 (Submitted).
82. Clement LT, Plaeger-Marshall S, Haas A, Saxon A, Martin AM. The bare lymphocyte syndrome: consequences of absent class II major histocompatibility antigen expression for B lymphocyte differentiation and function. J Clin Invest 1988 (In press).
83. Fingeroth JD, Weis JJ, Tedder TF, Strominger JL, Biro PA, Fearon DT. Epstein-Barr virus receptor of human B lymphocytes is the C3d receptor (CD 21). Proc Natl Acad Sci USA 1984;81:4510–4517.
84. Maffei A, Scarpellino L, Bernard M, et al. Distinct mechanisms regulate MHC class II gene expression in B cells and macrophages. J Immunol 1987;139:942–948.
85. McGhee JD, Ginder GD. Specific DNA methylation sites in the vicinity of the chicken beta-globin genes. Nature 1979;28:419–420.
86. Bird AP. DNA methylation—how important in gene control? Nature 1984;307:503.
87. Tanaka K, Appella E, Jay G. Developmental activation of the H-2K gene is correlated with an Increase in DNA methylation. Cell 1983;35:457–465.
88. Reitz MS, Mann DL, Eiden M, Trainor CD, Clarke MF. DNA methylation and expression of HLA-DR alpha. Mol Cell Biol 1984;4:890–897.
89. Carrington MN, Salter RD, Cresswell P, Ting JP-Y. Evidence for methylation as a regulatory mechanism in HLA-DR gene expression. Immunogenetics 1985;22:219–229.
90. Wang Y, Peterlin BM. Methylation patterns of HLA-DR alpha genes in six mononuclear cell lines. Immunogenetics 1986;24:289–303.
91. Levine F, Pious D. Different roles for cytosine methylation in HLA class II gene expression. Immunogenetics 1985;22:427–440.
92. Gambari R, Del Senno L, Barbieri R, et al. Lack of correlation between hypomethylation and expression of the HLA-DR alpha gene. Eur J Immuno 1986;16:365–369.
93. Lee JS, O'Neill L. Methylation of the HLA-DR alpha gene is positively correlated with expression. Immunogenetics 1987;26:92–98.
94. Benoist C, Chambon P. *In vivo* sequence requirements of the SV40 early promoter region. Nature 1981;290:304–310.
95. Ephrussi A, Church GM, Tonegawa S, Gilbert W. B lineage specific interactions of an immunoglobulin enhancer with cellular factors *in vivo.* Science 1985;227:134–140.
96. Gilles SD, Folsom V, Tonegawa S. Cell type-specific enhancer element associated with a mouse MHC gene, E_{β}. Nature 1984;3:594–597.
97. Wang Y, Larsen AS, Peterlin BM. A tissue-specific transcriptional enhancer is found in the body of the HLA-DRα gene. J Exp Med 1987;166:625–636.
98. Peterlin BM, Hardy KJ, Larsen AS. Chromatin structure of the HLA-DRα gene in different functional states of major histocompatibility complex class II gene expression. Mol Cell Biol 1987;7:1967–1972.
99. Saito H, Maki RA, Clayton LK, Tonegawa S. Complete primary structures of the Eβ chain and gene of the mouse major histocompatibility complex. Proc Natl Acad Sci USA 1983;80:5520–5524.
100. O'Sullivan D, Larhammar D, Wilson MC, Peterson PA, Quaranta V. Structure of the human Ia-associated invariant (gamma)-chain gene: identification of the 5′ sequences shared with major histocompatibility complex class II genes. Proc Natl Acad Sci USA 1986;83:4484–4488.
101. Miwa K, Doyle C, Strominger JL. Sequence-specific interactions of human class II major histocompatibility complex genes. Proc Natl Acad Sci USA 1987;84:4939–4943.
102. Sherman PA, Basta PV, Ting JP-Y. Upstream DNA sequences required for tissue-specific expression of the HLA-DRα gene. Proc Natl Acad Sci USA 1987;84: 4254–4258.
103. Dorn A, Durand B, Marfing C, Le Meur M, Benoist C, Mathis D. Conserved major histocompatibility complex class II boxes-X and Y-are transcriptional control elements and specifically bind nuclear proteins. Proc Natl Acad Sci USA 1987;84:6249–6253.
104. Boss J, Strominger JL. Regulation of a transfected human class II major histocompatibility complex gene in human fibroblasts. Proc Natl Acad Sci USA 1986;83: 9139–9143.
105. Basta PV, Sherman PA, Ting JP-Y. Identification of an interferon-gamma response region 5′ of the human histocompatibility leukocyte antigen DRα chain gene which is active in human glioblastoma multiforme lines. J Immunol 1987;138:1275–1280.
106. Takahashi K, Vigneron M, Matthes M, Wildeman A, Zenke M, Chambon P. Requirement of stereospecific alignments for initiation from the simian virus 40 early promoter. Nature 1986;319:121–126.
107. Mason JO, Williams GT, Neuberger MS. Transcription cell type specificity is conferred by an immunoglobulin V_H gene promoter that includes a functional consensus sequence. Cell 1985;41:479–487.
108. Wirth T, Staudt L, Baltimore D. An octamer oligonucleotide upstream of a TATA motif is sufficient for lymphoid-specific promoter activity. Nature 1987;329: 174–178.
109. Mathis DJ, Benoist CO, Williams VE, Kanter MR, McDevitt HO. The murine Eα immune response gene. Cell 1983;32:745–754.
110. Landolfi NF, Capra JD, Tucker PW. Interaction of cell-type specific nuclear proteins with immunoglobulin V_H promoter region sequences. Nature 1986;323:548–551.
111. Staudt LM, Singh H, Sen R, Wirth T, Sharp PA, Baltimore D. A lymphoid specific protein binding to the octamer motif of immunoglobulin genes. Nature 1986;323:640–643.
112. Dorn A, Bollekens J, Staub A, Benoist C, Mathis D. A multiplicity of CCAAT box binding proteins. Cell 1987;50:863–872.
113. Shaw G, Kamen R. A conserved AU sequence from the 3′ untranslated region of GM-CSF mRNA mediates selective mRNA degradation. Cell 1986;46:659–667.
114. Briata P, Sartoris S, Lee JS. Alternative splicing and polyadenylation produce novel HLA-DQ beta transcripts. 1988 (Submitted).
115. Auffray C, Lillie JW, Korman AJ, et al. Structure and expression of HLA-DQα and DXα genes: interallelic alternate splicing of the HLA-DQα gene and functional splicing of the HLA-DXα gene using a retroviral vector. Immunogenetics 1987;26:63–73.
116. Schenning L, Larhammar D, Bill P, et al. Both α and β chains of HLA-DC class II histocompatibility antigens display extensive polymorphism in their amino-terminal domains. EMBO J 1984;3:447–452.
117. Kappes DJ, Arnot D, Okada K, Strominger JL. Structure and polymorphism of the HLA class II SB light chain genes. EMBO J 1984;3:2985–2993.

118. Cairns JS, Curtsinger JM, Dahl CA, Freeman S, Alter BJ, Bach FH. Sequence polymorphism of HLA DR β1 alleles relating to T-cell recognized determinants. Nature 1985;317:166–168.
119. Owen MJ, Kissonerghis A-M, Lodish HF, Crumpton MJ. Biosynthesis and maturation of HLA-DR antigens *in vivo*. J Biol Chem 1981;256:8987–8993.
120. Kvist S, Wiman K, Claesson L, Peterson PA, Dobberstein B. Membrane insertion and oligomeric assembly of HLA-DR histocompatibility antigens. Cell 1982;29:61–69.
121. Murphy DB, Jones PP, Loken MR, McDevitt HO. Interaction between I region loci influences the expression of a cell surface Ia antigen. Proc Natl Acad Sci USA 1980;77:5404–5408.
122. Alexander S, Hubbard SC, Strominger JL. HLA-DR antigens of autologous melanoma and B lymphobastoid cell lines: differences in glycosylation but not protein structure. J Immunol 1984;133:315–320.

Acknowledgments. I thank Drs. B. Dupont, C. Hume, S. Kambhu, L. Shookster, P. Briata, M. Sheffery, N. Flomenberg, N. Kernan, R. Knowles, and S.Y. Yang for helpful discussions and encouragement. The work from my laboratory is supported by the Multiple Sclerosis Society (RG 1937-A-1) and the National Institutes of Health (GM3968-01, CA20194-11, and CA22507-11).

Author Affiliation

Immunology Program, Memorial Sloan Kettering Cancer Center, 1275 York Avenue, New York, New York 10021, USA

Molecular Genetics of the Class III Region of the HLA Complex

Perrin C. White

The class III region of the HLA complex is located between the class I and class II regions. Whereas HLA class I and class II genes encode sets of structurally related, highly polymorphic transplantation antigens, the class III region consists of at least four types of genes, only two of which have obvious functions in the immune system. Unlike the highly polymorphic class I and II genes, class III genes display a low to moderate degree of genetic variability, but are nevertheless subject to evolutionary mechanisms observed elsewhere in the HLA complex, including gene conversions and changes in gene copy number due to unequal crossing-over during meiosis.

The order of genes in this region has been determined by mapping of overlapping cosmid clones and by large-scale restriction mapping of uncloned DNA by use of pulsed-field gradient electrophoresis (Fig. 1) (1–3). Toward the centromere from the class I (HLA-B) to the class II (HLA-DR) genes are the TNF β and α genes encoding tumor necrosis factor β (lymphotoxin) and α (cachectin), the C2 and Bf genes encoding the second component of serum complement and properdin factor B, a newly identified gene of unknown function tentatively named RD, and alternating pairs of genes encoding the fourth component of serum complement (C4A and C4B) and a cytochrome P450 enzyme, steroid 21-hydroxylase (CYP21A and B). This paper will briefly review the molecular biology of each of these genes.

Tumor Necrosis Factors (TNF α and β)

Tumor necrosis factors (TNF) α and β are secreted, respectively, by macrophages and T lymphocytes during the inflammatory response (4–7). Along with other lymphokines such as interleukin-1 and interferon-α, TNFα has a wide range of effects on many cells involved in inflammation; it activates T and B lymphocytes, promotes granulocyte attachment, and stimulates granulocytes to produce toxic oxygen radicals. It inhibits lipoprotein lipase, an enzyme involved in fat storage; this inhibition is apparently an important mechanism by which cachexia develops in patients with chronic infections or cancer. Thus, TNFα is also termed "cachectin." It has both angiogenic and angiotoxic effects, the latter mediating, in certain experimental systems, the tumor necrosis for which the factor is named. TNFα also has direct cytotoxic effects on some tumor cell lines. TNFα stimulates transcription of class I genes, which may constitute another pathway by which TNFα modulates the immune response.

TNFβ has many of the same effects as TNFα, although it has been characterized primarily by in vitro cytotoxicity against certain cell lines, rather than by its in vivo effects. Whereas TNFα is best produced in vitro from macrophages stimulated with bacterial lipopolysaccharide, TNFβ is secreted by mitogen-activated lymphocytes.

TNFα is a single polypeptide synthesized as a proprotein with a signal peptide of 76 amino acid residues. The mature protein contains 156 residues and has a molecular mass of 17 kD, but multimeric forms may exist. The protein is encoded by a 1.6 kb mRNA (8).

TNFβ is encoded by a 1.3 kb mRNA. The protein is synthesized with a 34 residue signal sequence; the secreted form contains 171 residues with a predicted molecular mass of 19 kD. The protein is glycosylated in vivo so that the observed size of natural TNFβ is 25 kD. Secreted TNFα and β are identical in 35% of amino acid residues (9); the signal sequences have no obvious homology.

Each of the TNF genes is about 3 kb long and contains three introns, only the last of which is in a homo-

Immunobiology of HLA
Volume II

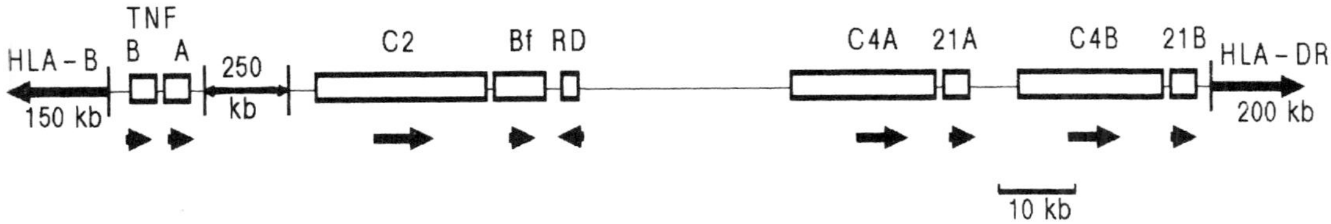

Figure 1. Molecular map of the HLA class III region on chromosome 6p. The centromere is to the right. Distances to the class I region (HLA-B) and class II region (HLA-DR) are shown in kilobases (kb). The map is drawn to scale except between the tumor necrosis factor genes (TNF A and B) and C2. Genes C2, C4A, C4B, and Bf, respectively, encode the second and fourth components of serum complement and factor B; "21A" and "21B" are the CYP21A and CYP21B genes encoding steroid 21-hydroxylase (CYP21A is a pseudogene), and gene RD is of unknown function. The arrow below each gene indicates the direction of transcription.

logous position within the coding sequences of the two genes (10).

Based on analysis of patterns of hybridization to human-mouse somatic cell hybrid lines, the genes for TNFα and β were mapped to chromosome 6 between 6p23 and 6q12 (10), a region that encompasses the HLA complex located at 6p21.3. The presence of these genes within the HLA complex was demonstrated by Southern blot hybridization to DNA samples from cell lines carrying radiation-induced deletions of various parts of the HLA complex. These experiments localized the TNF genes to the region between HLA-A and DR (11). Large-scale restriction mapping shows that the genes are located about 600 kilobasepairs (kb) centromeric to HLA-B (2,3).

Second Component of Serum Complement (C2) and Factor B (Bf)

The serum complement system consists of a series of proteins that form a proteolytic cascade in response to the presence of antigen-antibody complexes or other stimuli, resulting in the insertion of tubular complexes into the membrane of target cells, with consequent disruption of cellular integrity (12).

The second component of complement (C2) and factor B are serine proteases that take part in two pathways that activate complement (13). C2 participates in the classic (antibody–dependent) pathway as follows: the first component of complement, C1, binds to antigen-antibody complexes and a protease activity is formed, activating C4 and C2. Activated C4 and C2 then bind to the antigen-antibody complexes and become a protease that activates C3. Activated C3, when complexed with C4 and C2, forms C5 convertase, initiating formation of the actual lytic complex by the later components of complement (C5–9). The proteolytic active site of both the C3 and C5 convertase activities resides on the C2 polypeptide (14).

Factor B functions in the alternative pathway of complement activation, which may be triggered by antigen-antibody complexes or by other substances such as bacterial cell wall polysaccharides. Activated C3 binds to antibody or polysaccharides, becomes associated with factor B, and the C3B complex is split by factor D to form a C3 convertase. Upon binding of additional activated C3 molecules, the $(C3)_nB$ complex develops a C5 convertase activity that initiates formation of the C5–9 lytic complex, as occurs in the classic pathway. The proteolytic active site for both the C3 and C5 convertases of the alternative pathway is located on the factor B molecule, so that the role of factor B in the alternative pathway corresponds closely to that of C2 in the classic pathway (13).

C2 and factor B are glycoproteins with molecular masses of 102 and 90 kD, respectively. The difference in size is due mainly to differences in carbohydrate content (16% for C2 vs. 9% for factor B). They are identical in about 37% of amino acid residues. They are both activated by single proteolytic cleavages; C2 is cleaved by activated C1 to fragments of 30 and 70 kD, whereas factor B is cleaved by factor D to fragments of 30 and 60 kD. The proteolytic active site for C3 and C5 convertase is on the larger fragment in each case. These 60 to 70 kD peptides are markedly larger than most serine proteases, which are usually about 30 kD in size.

C2 and factor B are encoded by mRNAs of 2.9 and 2.6 kb, respectively (15). The chromosomal genes encoding C2 and factor B (C2 and Bf) are, respectively, 18 and 6 kb long, separated by less than 500 bp, and transcribed in the same direction. The Bf gene contains 18 exons, of which the last 8 encode the serine protease region (16).

C2 and factor B exhibit a relatively low degree of genetic variation compared with class I or class II antigens. Complete deficiency of C2 occurs in 1:10,000 individuals; about 40% of these patients develop autoimmune diseases (17,18). This deficiency is not associated with a major DNA deletion or rearrangement, but instead seems to be caused by a defect preventing synthesis of the mRNA (19). Structural variants of C2 (C2A and C2B) have been detected by differences in electrophoretic mobility, but 94% of individuals are homozygous for the common allele C2C. The C2 gene was localized to the HLA complex because both C2 deficiency and C2 structural variants are HLA-linked (20,21).

Complete deficiency of factor B has not been observed. There are two common electrophoretic variants,

F and S, and at least two rarer variants, F_1 and S_1. As is the case for C2, these variants are inherited as alleles at a single locus that is HLA-linked (22).

Relatively few DNA polymorphisms have been noted in these genes, consistent with the low degree of structural variation observed (23).

Gene RD

This gene was first identified in the mouse H-2 major histocompatibility complex during a systematic reexamination of cosmid clones to determine if additional genes were located in the murine class III (S) region (24). It was reasoned that coding regions of genes are more highly conserved than introns or intergenic regions, so that segments containing functional genes would hybridize to DNA of other species, whereas "empty" segments would not. A subclone from the region between the genes for Bf and Slp (a C4 homolog) hybridized to genomic DNA of mouse, rat, hamster, and man. This subclone was then found to hybridize at low levels to a 1.7 kb mRNA species in all tissues tested. A cDNA library was screened with the genomic subclone, and a full-length cDNA clone was isolated. Sequence analysis demonstrated that the clone encoded a 375 residue, 42 kD polypeptide with no homology to any known protein. The predicted peptide has a high proportion of charged amino acids and lacks a hydrophobic signal peptide or a mitochondrial extra peptide; thus, the protein is probably not secreted and is not located in cell membranes or mitochondria. The central region of the polypeptide consists of 52 residues of strictly alternating basic and acidic amino acids such as arginine (R) and aspartic acid (D); hence the tentative name of the gene.

The RD gene is transcribed in the opposite direction from the C2 and C4 genes. Examination of cosmid clones containing human genomic DNA from the same region reveals that a homologous gene to RD is again present between the Bf and C4A genes. The sequence of the gene appears to be highly conserved; the function of the gene product has not yet been elucidated.

Fourth Component of Serum Complement (C4A and C4B)

The fourth component of complement is synthesized as a single polypeptide of 200 kD that undergoes proteolytic cleavage to the mature protein, which contains three chains of 75, 95, and 30 kD termed beta, alpha and gamma, respectively (25). C4 participates in activation of the classic complement pathway; C1 activates C4 by cleaving a 7 kD peptide from the amino terminus of the alpha chain. Activated C2 and C4 form the C3 convertase activity (26).

C4 exists as two nonallelic electrophoretic variants termed C4A (fast) and C4B (slow). There are numerous electrophoretic variants inherited as alleles at both the C4A and C4B loci, of which at least seven alleles at each locus occur with a frequency of at least 1% (27). In addition, null alleles (absence of detectable protein) occur with a frequency of about 10% at C4A and 20% at C4B (28).

Homozygous null alleles of C4A or C4B are correlated, respectively, with absence of the red cell antigens, Rodgers and Chido (29). These antigens have been determined to represent fragments of the alpha chain of C4A and C4B, respectively.

Complete absence of C4 protein is a very rare condition, in contrast to the frequent null alleles observed for C4A and C4B. Patients with complete absence of C4 are prone to frequent infections and autoimmune disease (30).

Patients with null alleles of C4A are also prone to autoimmune disease, in particular systemic lupus erythematosus (31). Although this might be due to nonrandom associations between C4A null alleles and alleles of other linked genes (see below), it may also represent a direct biological effect of the absent C4A protein. The biological activities of C4A and C4B differ; for example, C4A binds less rapidly than C4B to antibody-coated erythrocytes. Apparently, C4B binds polysaccharides on the erythrocyte surface more actively than C4A, whereas C4A binds protein antigens more effectively than C4B (32).

Complete absence of C4 (33), electrophoretic variants, and null alleles at C4A and C4B (28) are all inherited as HLA-linked traits, thus demonstrating that the C4A and C4B genes are located within the HLA complex.

C4A and C4B are both encoded by mRNAs that are 5.5 kb long. Sequence analysis of the corresponding cDNAs has revealed only 14 nucleotide differences between C4A and C4B cDNAs from among 4600 basepairs compared. Twelve nucleotide differences resulting in nine amino acid changes are located within the region encoding the alpha chain. Six of the amino acid differences within a stretch of 90 residues are specific for C4A or C4B, of which 4 are clustered within 6 residues. Thus, C4A and C4B are extremely similar structurally, and isotypic differences are confined to a very small portion of the C4 polypeptide (34).

The genomic gene encoding C4A is 22 kb long, whereas the C4B gene is found in two polymorphic forms of 16 and 22 kb ("short" and "long" C4B) that differ by 7 kb in an intron in the 5′ portion of the gene (35,36). The C4 genes alternate with the two CYP21 genes encoding steroid 21-hydroxylase (see below) (37,38); C4A is about 30 kb away from the factor B gene.

Steroid 21-Hydroxylase (CYP21)

The steroid hormone cortisol is synthesized from cholesterol in the adrenal cortex in five enzymatic steps: the cholesterol side chain is cleaved to form pregnenolone, which is dehydrogenated to progesterone, and successively hydroxylated at the 17, 21, and 11 positions to yield cortisol. A deficiency in 21-hydroxylase activity prevents cortisol from being synthesized normally, resulting in the syndrome of congenital adrenal hyperplasia. In this disorder, accumulated cortisol precursors are shunted into the androgen biosynthetic pathway, and so males and females have signs and symptoms of androgen excess including ambiguous genitalia in females, rapid somatic growth with accelerated skeletal maturation and short adult stature, and abnormal pubertal development. Most patients with the severe classic form of this disorder have a parallel impairment in their ability to synthesize the mineralocorticoid hormone, aldosterone, and, if untreated, may die shortly after birth from renal salt loss (39).

The classic form of 21-hydroxylase deficiency occurs in about 1/10,000 births, and a mild, frequently asymptomatic, "nonclassic" form of the enzyme deficiency is found in about 1% of the genearl Caucasian population.

Steroid 21-hydroxylase deficiency is inherited as an autosomal-recessive trait linked to the HLA complex (40). The severe and mild forms of the disorder are inherited as alleles at a single locus (41).

The 21-hydroxylase enzyme is a specific cytochrome P450 termed P450c21 or P450XXIA, a heme-containing monooxygenase with a molecular mass of 52 kD (494 amino acid residues) that is located in the membrane of the endoplasmic reticulum. It accepts electrons from a flavoprotein cytochrome reductase that is not substrate-specific (42)

The mRNA encoding P450c21 is 2.0 kb long. The corresponding genomic gene, CYP21B, is quite compact, containing 10 exons within 3.1 kb. CYP21B is located immediately 3′ to C4B and is transcribed in the same direction as C4B. The CYP21A gene is found in the analogous position 3′ to C4A (37,38). Although CYP21A and CYP21B are 98% identical in exons, CYP21A is a pseudogene (i.e., it does not yield a detectable mRNA or a protein product) due to several deleterious mutations (43,44).

Mutations causing 21-hydroxylase deficiency have been extensively studied. Because C4 and CYP21 genes develop mutations by similar mechanisms, these mutations will be discussed together.

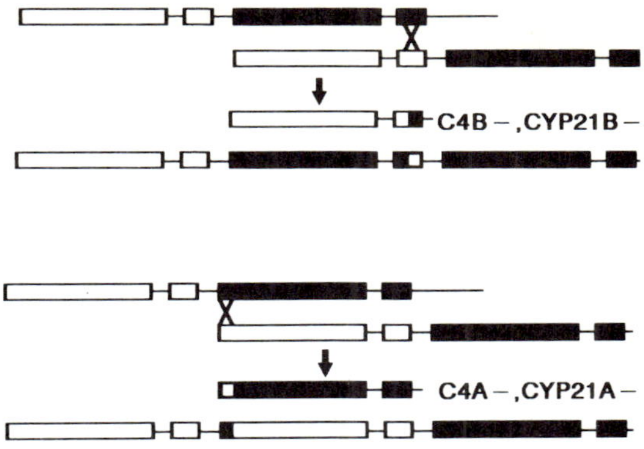

Figure 2. Unequal cross-overs in the C4-CYP21 region. These cross-overs occur after misalignment of the tandemly duplicated pairs of genes, and produce one chromosome with one pair of C4-CYP21 genes, and one chromosome with three pairs of these genes. If the cross-over occurs between the CYP21A pseudogene (white) and CYP21B (black), one chromosome will contain only the C4A gene and a single CYP21 gene that will have the 5′ end of CYP21A and the 3′ end of CYP21B. Because deleterious mutations occur near the 5′ end of CYP21A, this single remaining CYP21 gene is nonfunctional. In effect, this cross-over deletes C4B and CYP21B. Such a deletion, which is an allele for steroid 21-hydroxylase deficiency, is typically associated with the HLA haplotype, A3;Bw47;DR7. The other chromosome will have one C4A gene (white) and two C4B genes (black), similar to the arrangement associated with HLA-B14;DR1. If the cross-over occurs between the 5′ ends of the C4 genes, one chromosome will have a single C4B-like gene (the differences distinguishing C4A and C4B occur in the middle of the gene) and a single CYP21B gene. This arrangement is typical of the HLA haplotype, A1;B8;DR3. There is no detecable abnormality of steroid metabolism, but this haplotype is associated with autoimmune disorders.

Mechanisms of Allelism in C4 and CYP21 Genes

The class III region contains four sets of duplicated genes: TNFα and TNFβ, C2 and Bf, C4A and C4B, and CYP21A and CYP21B. The latter two pairs alternate and were presumably generated by a duplication of a 30 kb segment carrying a single C4 and CYP21 gene.

TNFα and TNFβ are only about 28% identical in amino acid sequence, while C2 and Bf are 37% identical. Thus, it is remarkable that C4A and C4B are more than 99% identical, and the amino acid sequence derived from the sequence of the CYP21A pseudogene is 98% identical to the sequence derived from CYP21B. This high degree of similarity even extends to nucleotide sequences in introns, which are presumably not subject to the same conservative forces as coding regions.

Either the C4 and CYP21 genes were duplicated much more recently than the TNF and C2-Bf genes, or one or more mechanisms are causing the pairs of C4 and CYP21 genes to evolve in parallel. The mouse has a tandem duplication of the C4 and CYP21 genes in an arrangement similar (but not identical) to that seen in man (45), and cattle apparently have two CYP21 genes (46). This suggests that the C4 and CYP21 genes were originally duplicated prior to mammalian speciation, making parallel evolution a more likely explanation for the observed degree of similarity between the members of each pair.

The existence of this tandem duplication suggests two primary mechanisms for parallel evolution. If the tandem pairs become misaligned during meiosis, unequal crossing-over could occur, resulting in chromosomes with one or three pairs of genes (Fig. 2). Repeated cycles of deletion and duplication would obviously tend to homogenize the sequences of each pair of genes. Alternatively, smaller segments could be transferred from one member of a pair to the other, a process termed gene conversion.

In fact, both of these processes apparently contribute to the evolution of the C4 and CYP21 genes. More than 5% of all chromosomes carry deletions of C4A and CYP21A, so that about half of all C4A null alleles result

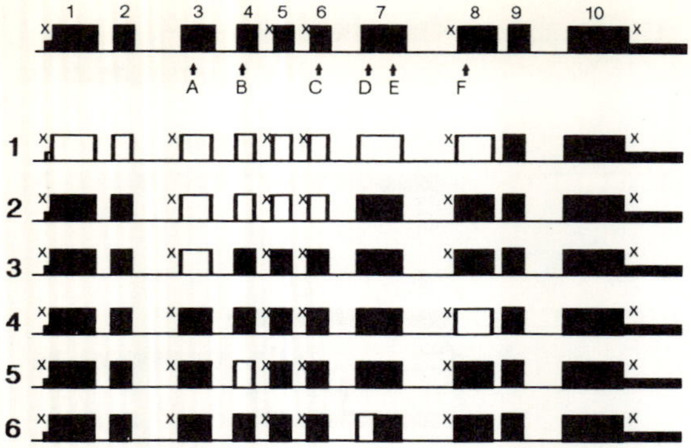

Figure 3. Gene conversions causing steroid 21-hydroxylase deficiency. A diagram of a CYP21 gene is displayed at the top. Exons are symbolized by the black rectangles and are numbered. The "x"s mark the location of *chi*-like sequences that may promote recombination. Some deleterious mutations in the CYP21A gene are marked by letters: A = eight base-pair deletion in the coding sequence leading to a frameshift; B = point mutation changing Ile-172 to Asn; C = cluster of four point mutations affecting codons 235 to 238; D = point mutation changing Val-281 to Leu; E = single nucleotide insertion causing a frameshift; and F = nonsense mutation in codon 318. The CYP21A-like region transferred into each mutant CYP21B gene is indicated by white exons. Mutant genes 1 to 4 include frameshifts or nonsense mutations and completely prevent synthesis of a functional enzyme (53–56). Mutant gene 5 is associated with disease of intermediate severity (56), and mutant gene 6 with mild "nonclassic" disease (57).

from deletions of C4A (47). Although there is no obvious disadvantage to the absence of the CYP21A pseudogene, C4A null alleles, as mentioned, carry an increased risk of autoimmune disease. About 25% of all classic 21-hydroxylase deficiency alleles result from a deletion of CYP21B (and C4B) (48–50). Based on a disease frequency of 1/10,000, this suggests that about 0.5% of chromosomes carry a deletion of C4B and CYP21B despite the fact that this disorder was often fatal prior to the availability of adequate steroid replacement therapy. The prevalence of chromosomes with three pairs of C4 and CYP21 genes is not precisely known, but based on other HLA associations (see below), haplotypes known to carry a second copy each of CYP21A and C4B may be carried on more than 10% of chromosomes in some populations. The unequal cross-overs causing these deletions and duplications can presumably occur anywhere along the normally duplicated segment carrying each C4 and CYP21 gene, and intragenic cross-overs will obviously produce hybrid genes (e.g., a cross-over from C4A on one chromosome to C4B on the other will produce one chromosome that has a single C4 gene containing the 5′ end of C4A and the 3′ end of C4B [51]). This will also tend to minimize differences between the C4 or the CYP21 genes.

Gene conversions of both C4 and CYP21 genes have been observed. Because the isotypic differences distinguishing C4A and C4B are tightly clustered within a small region of the alpha chain, it is apparent that gene conversions involving the area encoding this region would result in C4A gene products that behave electrophoretically and antigenically like C4B (or vice versa). This would, in effect, constitute a null allele of C4A. About half of C4A null alleles and almost all C4B null alleles are not caused by deletions of the corresponding gene; because null alleles at these loci are very frequent, this suggests either that mutations preventing protein synthesis are extraordinarily common, or that gene conversions cause many chromosomes to carry two C4A-like genes or two C4B-like genes instead of one of each. It appears that the latter explanation is correct (51,52).

Similarly, cloning and sequence analysis and oligonucleotide hybridizations have shown that many 21-hydroxylase-deficient alleles result from gene conversions that transfer deleterious mutations from the CYP21A pseudogene to CYP21B (Fig. 3). These mutations range in severity from frameshifts (53) and nonsense mutations (54,55), which completely prevent synthesis of a functional enzyme, to a nonconservative substitution (Ile-172 to Asn) that apparently results in an impaired but not completely inactive enzyme (56)), to a conservative substitution (Val-281 to Leu) associated with the mild, nonclassic form of the disease (57). Point mutations that are not gene conversions (e.g., Ser-269 to Thr) have also been reported (58).

If conversions of the C4 and CYP21 genes are so frequent, why haven't the TNF and C2-Bf pairs of genes remained more homologous? Possibly, some evolutionary pressure existed for rapid evolution of these latter genes; presumably, once the members of each pair had diverged sufficiently, unequal crossing-over and gene conversion would become less likely. Alternatively, there may be structural features of the C4 and CYP21 genes that promote recombinational events. Sequences resembling the recombination-promoting *chi* sequence of bacteriophage lambda (59) have been observed near gene conversion sites elsewhere in the major histocompatibility complex (60), and inspection of the sequences of CYP21A and B reveals at least five *chi*-like sequences (GCTGGGG) within the introns of each gene. Possibly, a high concentration of such sequences (5 in 3 kb rather than the 1 every 8 kb expected in random DNA) promotes frequent recombination of these genes.

Genetic Linkage Disequilibrium

Certain combinations of alleles at HLA loci, or haplotypes, occur at a frequency much higher than expected based on the frequency of each allele in the population (61). The existence of such combinations is referred to as genetic linkage disequilibrium. For example, the HLA antigens A1;B8;DR3 occur together on about 5% of all chromosomes in the Caucasian population (62). Such haplotypes often include characteristic alleles at class III loci; HLA-A1;B8;DR3 almost always includes a deletion of C4A and CYP21A. Another, much less frequent haplotype, A3;Bw47;DR7, is associated with a deletion of C4B and CYP21B and, thus, almost always carries a 21-hydroxylase-deficient allele (63). The frequent haplotype B14;DR1 carries a duplicated C4B gene and a duplicated CYP21A gene (64).

These nonrandom associations of alleles occur over a genetic distance of at least 2 centiMorgans. It is possible in some cases that insufficient time has elapsed for alleles at the various loci to have diverged by recombination, but the very high prevalence of certain haplotypes suggests that they have a selective advantage and/or that they suppress recombination by some mechanism.

The t haplotypes involve the H-2 major histocompatibility complex in mice. These haplotypes are usually associated with an apparent suppression of recombination due to an inversion of all or part of the H-2 complex (65). Particular t haplotypes are often present on about 25% of chromosomes in different wild mouse populations, even though they carry various recessive lethal mutations. This occurs because males that are heterozygous for t haplotypes preferentially transmit the affected chromosome to their offspring, a phenomenon termed male transmission ratio distortion.

It has been proposed that certain human HLA haplotypes such as A1;B8;DR3 are analogous to murine t haplotypes (60), but they are not associated with recessive lethal mutations, and male transmission ratio distortion has not been consistently demonstrated for any HLA haplotype. These haplotypes do not carry chromosomal inversions. However, several of the more common HLA haplotypes are associated with deletions or duplications of C4 and CYP21 genes, and it is tempting to speculate that the loss or addition of genetic material in this central region of the HLA complex is sufficient to cause misalignment of the HLA complex during meiosis, thus suppressing recombination with the other chromosome.

Conclusion

If the class III region is defined as the entire chromosomal segment between the class I (HLA-B) and class II (HLA-DR) genes, it covers 1000 kb of DNA. Thus far, four distinct types of genes have been located in the class III region. While the genes encoding factor B (Bf) and the second (C2) and fourth (C4A and B) components of complement were originally localized to this region by classic genetic approaches, molecular genetic techniques have been necessary to identify structural genes for tumor necrosis factor (TNFα) and lymphotoxin (TNFβ), steroid 21-hydroxylase (CYP21B), and the RD protein. As the Bf, C2, RD, C4A, CYP21A, C4B, and CYP21B genes are all found within a single 100 kb segment, it is apparent that many additional genes could be located within the class III region if the same high density of genes is maintained throughout. Although the complement components and tumor necrosis factors have obvious functions within the immune system, the steroid 21-hydroxylase enzyme and RD protein do not, and a teleologic explanation for their presence within the HLA complex is not obvious. The identification of additional genes within the class III region may thus provide further clues as to the overall structure, function, and evolution of the HLA complex.

References

1. Lawrence SK, Smith CL, Srivastava R, Cantor CR, Weissman SM. Megabase-scale mapping of the HLA gene complex by pulsed field gel electrophoresis. Science 1987;235:1387.
2. Dunham I, Sargent CA, Trowsdale J, Campbell RD. Molecular mapping of the human major histocompatibility complex by pulsed-field gel electrophoresis. Proc Natl Acad Sci USA 1987;84:7237.
3. Carroll MC, Katzman P, Alicot EM, et al. Linkage map of the human major histocompatibility complex including the tumor necrosis factor genes. Proc Natl Acad Sci USA 1987;84:8535.
4. Old LJ. Tumor necrosis factor (TNF). Science 1985; 230:630.
5. Beutler B, Cerami A. Cachectin and tumour necrosis factor as two sides of the same biological coin. Nature 1986; 320:584.
6. Drysdale BE, Agarwal S, Shin HA. Macrophage-mediated tumoricidal activity: mechanisms of activation and cytotoxicity. Progr Allergy 1988;40:111.
7. Ruddle NH, Homer R. The role of lymphotoxin in inflammation. Progr Allergy 1988;40:162.
8. Pennica D, Nedwin GE, Hayflick JS, et al. Human tumour necrosis factor: precursor structure, expression and homology to lymphotoxin. Nature 1984;312:724.
9. Gray PW, Aggarwal BB, Benton CV, et al. Cloning and expression of cDNA for human lymphotoxin, a lymphokine with tumour necrosis activity. Nature 1984;312:721.
10. Nedwin GE, Naylor SL, Sakaguchi AY, et al. Human lymphotoxin and tumor necrosis factor genes: structure, homology and chromosomal localization. Nucl Acids Res 1985;13:6361.
11. Spies T, Morton CC, Nedospasov SA, Fiers W, Pious D, Strominger JL. Genes for the tumor necrosis factors α and β are linked to the human major histocompatibility complex. Proc Natl Acad Sci USA 1986;83:8699.
12. Rother K, Rother U. The reactivity of the complement system. Progr Allergy 1986;39:8.
13. Porter RR. The complement components coded in the major histocompatibility complexes and their biological activities. Immunol Rev 1985;87:7.
14. Campbell RD, Bentley DR. The structure and genetics of the C2 and factor B genes. Immunol Rev 1985;87:19.
15. Morley BA, Campbell RD. Internal homologies of the Ba fragment from human complement component Factor B, a class III MHC antigen. EMBO J 1984;3:153.
16. Campbell RD, Porter RR. Molecular cloning and characterization of the gene coding for human complement protein Factor B. Proc Natl Acad Sci USA 1983;80:4464.
17. Glass D, Raum D, Gibson D, Stillman JS, Schur PH. Inherited deficiency of the second component of complement. J Clin Invest 1976;58:853.
18. Ruddy S. Complement deficiencies 3. The second component. Progr Allergy 1986;39:250.
19. Cole FS, Whitehead AS, Auerbach HS, et al. The molecular basis for the genetic deficiency of the second component of human complement. N Engl J Med 1985;313:11.
20. Meo T, Atkinson JP, Bernoco M, Bernoco D, Ceppellini R. Structural heterogeneity of C2 complement protein and its genetic variants in man: a new polymorphism of the HLA region. Proc Natl Acad Sci USA 1977;74:1672.

21. Fu SM, Stern R, Kunkel HG, et al. MLC-determinants and C2 deficiency: LD-7a associated with C2 deficiency in four families. J Exp Med 1975;142:495.
22. Allen FH. Linkage of HL-A and GBG. Vox Sang 1974;27:382.
23. Cross SJ, Edwards JH, Bentley DR, Campbell RD. DNA polymorphism of the C2 and factor B genes. Immunogenetics 1985;21:39.
24. Levi-Strauss M, Carroll MC, Steinmetz M, Meo T. A previously undetected MHC gene with an unusual periodic structure. Science 1988;240:201.
25. Goldberger G, Colton HR. Precursor complement protein (Pro-C4) is converted in vitro to native C4 by plasmin. Nature 1980;286:514.
26. Carroll MC, Belt KT, Palsdottir A, Yu Y. Molecular genetics of the fourth component of human complement and 21-hydroxylase. Immunol Rev 1985;87:39.
27. Schendel DJ, O'Neill GJ, Wank R. MHC-linked class III genes, analysis of C4 gene frequencies, complotypes and associations with distinct HLA haplotypes in German caucasians. Immunogenetics 1984;20:23.
28. O'Neill GJ, Yang SY, Dupont B. Two HLA-linked loci controlling the fourth component of human complement. Proc Natl Acad Sci USA 1978;75:5165.
29. O'Neill GJ, Yang SY, Tegoli J, Berger R, Dupont B. Chido and Rodgers blood groups are distinct antigenic components of human C4. Nature 1978;273:668.
30. Hauptmann G, Goetz I, Uring-Lambert B, Grossehaus E. Complement deficiency 2. The fourth component. Progr Allergy 1985;39:232.
31. Dawkins RL, Christiansen FT, Kay PH, et al. Disease associations with complotypes, supratypes and haplotypes. Immunol Rev 1983;70:5.
32. Law SKA, Dodds AW, Porter RR. A comparison of the properties of two classes, C4A and C4B, of the human complement component C4. EMBO J 1984;3:1819.
33. Rittner C, Hauptmann G, Grosse-Wilde H, Grosshans E, Tongio MM, Mayer S. Linkage between HL-A (major histocompatibility complex) and genes controlling synthesis of the fourth component of complement. In: Histocompatibility Testing. Copenhagen: Munksgaard, 1975:945.
34. Belt KT, Carroll MC, Porter RR. The structural basis of the multiple forms of human complement component C4. Cell 1984;36:907.
35. Carroll MC, Campbell RD, Bentley DR, Porter RR. A molecular map of the major histocompatibility complex class III region of man linking complement genes C4, C2 and factor B. Nature 1984;307:237.
36. Prentice HL, Schneider PM, Strominger JL. C4B gene polymorphism detected in a human cosmid clone. Immunogenetics 1986;23:274.
37. Carroll MC, Campbell RD, Porter RR. The mapping of 21-hydroxylase genes adjacent to complement component genes in HLA, the major histocompatibility complex in man. Proc Natl Acad Sci USA 1985;82:521.
38. White PC, Grossberger D, Onufer BJ, New MI, Dupont B, Strominger JL. Proc Natl Acad Sci USA 1985;82:1089.
39. White PC, New MI, Dupont B. Congenital adrenal hyperplasia. N Engl J Med 1987;316:1519, 1580.
40. Dupont B, Oberfield SE, Smithwick EM, Lee TD, Levine LS. Close genetic linkage between HLA and congenital adrenal hyperplasia (21-hydroxylase deficiency). Lancet 1977;2:1309.
41. Kohn B, Levine LS, Pollack MS, et al. Late-onset steroid 21-hydroxylase deficiency: a variant of classical congenital adrenal hyperplasia. J Clin Endocrinol Metab 1982;55:817.
42. Kominami S, Ochi H, Kobayashi Y, Takemori S. Studies on the steroid hydroxylation system in adrenal cortex microsomes: purification and characterization of cytochrome P-450 specific for steroid C-21 hydroxylation. J Biol Chem 1980;255:3386.
43. Higashi Y, Yoshioka H, Yamane M, Gotoh O, Fujii-Kuriyama Y. Complete nucleotide sequence of two steroid 21-hydroxylase genes tandemly arranged in human chromosome: a pseudogene and a genuine gene. Proc Natl Acad Sci USA 1986;83:2841.
44. White PC, New M, Dupont B. Structure of human steroid 21-hydroxylase genes. Proc Natl Acad Sci USA 1986; 83:5111.
45. White PC, Chaplin DD, Weis JH, et al. Two steroid 21-hydroxylase genes are located in the murine S region. Nature 1984;312:465.
46. Chung BC, Matteson KJ, Miller WL. Structure of a bovine gene for P-450c21 (steroid 21-hydroxylase) defines a novel cytochrome P-450 gene family. Proc Natl Acad Sci USA 1986;83:4243.
47. Carroll MC, Palsdottir A, Belt KT, Porter RR. Deletion of complement C4 and steroid 21-hydroxylase genes in the HLA class III region. EMBO J 1985;4:2547.
48. Rumbsy G, Carroll MC, Porter RR, Grant DB, Hjelm M. Deletion of the steroid 21-hydroxylase and complement C4 genes in congenital adrenal hyperplasia. J Med Genet 1986;23:204.
49. Jospe N, Donohoue PA, Van Dop C, McLean RH, Bias WB, Migeon CJ. Prevalence of polymorphic 21-hydroxylase gene (CA21HB) mutations in salt-losing congenital adrenal hyperplasia. Biochem Biophys Res Commun 1987;142:798.
50. White PC, Vitek A, New MI, Dupont B. Characterization of frequent deletions causing steroid 21-hydroxylase deficiency. Proc Natl Acad Sci USA 1988;85:4436.
51. Yu CY, Campbell RD. Definitive RFLPs to distinguish between the human complement C4A/C4B isotypes and the major Rodgers/Chido determinants. Immunogenetics 1987;25:383.
52. Palsdottir A, Arnason A, Fossdal R, Jenssen O. Gene organization of haplotypes expressing 2 different C4A allotypes. Hum Genet 1987;76:220.
53. Harada F, Kimur A, Iwanaga T, Shimozawa K, Yata J, Sasazuki T. Gene-conversion-like events cause steroid 21-hydroxylase deficiency in congenital adrenal hyperplasia. Proc Natl Acad Sci USA 1987;84:8091.
54. Globerman H, Amor M, Parker KL, New MI, White PC. A nonsense mutation causing steroid 21-hydroxylase deficiency. J Clin Invest 1988;82:139.
55. Higashi Y, Tanae A, Inoue H, Fujii-Kuriyama Y. Evidence for frequent gene conversion in the steroid 21-hydroxylase P-450 (C21) gene: implications for steroid 21-hydroxylase deficiency. Am J Hum Genet 1988;42:17.
56. Amor M, Parker KL, Globerman H, New MI, White PC. Mutation in the CYP21B gene (Ile-172 to Asn) causes steroid 21-hydroxylase deficiency. Proc Natl Acad Sci USA 1988;85:1600.
57. Speiser PW, New MI, White PC. Molecular genetic analysis of non-classic steroid 21-hydroxylase deficiency associated with the HLA haplotype B14;DR1. N Engl J Med 1988;319:19.
58. Rodrigues NR, Dunham I, Yu CY, Carroll MC, Porter RR, Campbell RD. Molecular characterization of the HLA-

linked steroid 21-hydroxylase B gene from an individual with congenital adrenal hyperplasia. EMBO J 1987;6:1653.
59. Smith GR, Kunes SM, Schultz DW, Taylor A, Triman KL. Structure of chi hotspots of generalized recombination. Cell 1981;24:429.
60. Wu S, Saunders TL, Bach FH. Polymorphism of human Ia antigens generated by reciprocal intergenic exchange between two DR β loci. Nature 1986;324:676.
61. Awdeh ZL, Raum D, Yunis EJ, Alper CA. Extended HLA/complement allele haplotypes: evidence for T/t-like complex in man. Proc Natl Acad Sci USA 1983;80:259.
62. Baur MP, Neugebauer M, Albert ED. Reference tables of 3 locus haplotype frequencies and delta values in Caucasians, Orientals and Negroids. In: Albert ED, Baur MP, Mayr WR (eds). Histocompatibility Testing 1984. Berlin: Springer-Verlag, 1985:756.
63. White PC, New MI, Dupont B. HLA-linked congenital adrenal hyperplasia results from a defective gene encoding a cytochrome P-450 specific for steroid 21-hydroxylation. Proc Natl Acad Sci USA 1984;81:1986.
64. Werkmeister JW, New MI, Dupont B, White PC. Frequent deletion and duplication of the steroid 21-hydroxylase genes. Am J Hum Genet 1986;39:461.
65. Silver LM, Artzt K. Recombination suppression of mouse T-haplotype due to chromatin mismatching. Nature 1981;290:68.

Author Affiliation

Department of Pediatrics, Cornell University Medical College, New York, NY 10021, USA

Part 2: Original Contributions

Section I
Genetic Organization of the HLA Region: MHC Mapping

Mapping of the Human Major Histocompatibility Complex by Pulsed Field Gel Electrophoresis

Ian Dunham, Carole A. Sargent, John Trowsdale, and R. Duncan Campbell

Genetic analysis of recombinant HLA haplotypes in family studies has established that the class I loci (HLA-A, -B, and -C) lie telomeric to the class II (HLA-D) loci. Located between HLA-DR and HLA-B are the class III genes, encoding the serum complement proteins C2, Factor B, C4A and C4B as well as two copies of the steroid 21-hydroxylase (21-OHase) gene (1,2). The orientation and exact location of these genes relative to the class I and class II genes has not been determined due to the absence of informative recombination events. By a combination of cosmid walking and genomic mapping using pulsed field gel electrophoresis (PFGE) (3,4) we have derived a map of ~3800 kb of DNA covering the HLA region. This has allowed the position and orientation of the complement genes in the class III region to be defined, the C2 gene being telomeric to the C4 gene. In addition, the positions of the TNFα and β genes within the HLA region have been determined. They lie ~390 kb telomeric of the C2 gene and ~250 kb centromeric of the HLA-B locus.

In order to minimize mapping problems due to possible haplotype-specific RFLPs, genomic DNA for both PFGE analysis and construction of cosmid libraries was prepared from a HLA-homozygous cell line (HLA type -A2, B7, Cw7, DR2, C2C, BfS, C4A 3, C4B Q0) (5). A series of overlapping cosmid clones was isolated covering ~230 kb of genomic DNA containing the genes for C2, factor B, C4A, and 21-OHase B (6). This cluster of cosmid clones was mapped for restriction enzymes that cut rarely in the genome. A single *Not*I site was defined ~25 kb from the 5′ end of the C2 gene. A single copy hybridization probe (probe J, Fig. 1) was isolated from the cosmid cluster distal to the *Not*I site and ~50 kb from the C2 gene.

For genomic restriction mapping by PFGE, high molecular weight DNA prepared in agarose blocks from the HLA homozygous cell line was digested with a range of infrequently cutting restriction enzymes (see (7) for review), alone or in double digest combinations. The resulting DNA restriction fragments were separated by PFGE and transferred by Southern blotting to nylon membranes. These filters were hybridized sequentially to a panel of ^{32}P-labelled probes from across the MHC, with the filters being stripped between each hybrization (see Fig. 1) (6). The sizes of fragments observed for single digests with a number of enzymes are summarized in Table I.

Hybridization of DQA, DRA, and 21-OHase probes to *Not*I-digested DNA identified a common 980 kb *Not*I fragment linking the DQ and DR subregions with the

Abbreviations. PFGE = pulsed field gel electrophoresis; HLA = human leukocyte antigen; MHC = major histocompatibility complex; TNF = tumour necrosis factor; kb = kilobase; 21-OHase = steroid 21-hydroxylase.

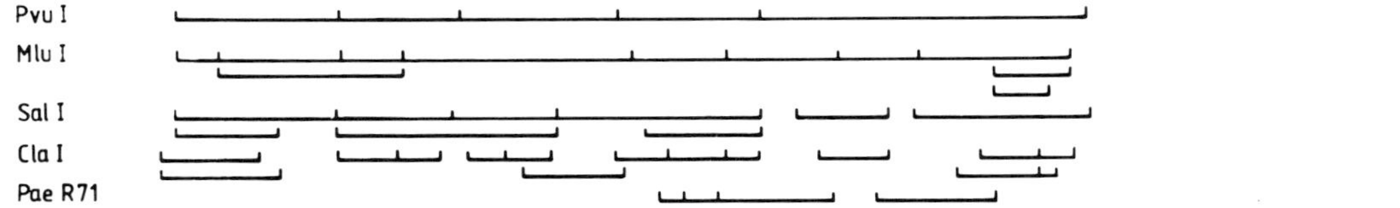

Figure 1. Molecular map of the human MHC. A genomic restriction map obtained by Southern blot analysis of genomic DNA from a HLA-homozygous cell line (A2, B7, Cw7, DR2, C2C, BfS, C4A 3, C4B Q0) separated by PFGE after single and double digestion with the enzymes shown (see ref (6) for methods). The centromere is to the left. Closed boxes indicate the positions of genes whereas open boxes indicate the limits of the region hybridising to the probes used. The positions of probes used in the analysis are shown by closed circles. Where partial digestion was obtained with some enzymes two or more sets of data are shown. The results reported here confirm the limits of the class II subregions, and to indicate the approximate positions of the class II genes the map of Hardy et al. (8) has been superimposed on the restriction fragment map.

Table 1. Size of Restriction Fragments (kb) Observed After Hybridization of a Panel of MHC-Region Probes to Southern Blots of Single Restriction Enzyme Digests of the DNA*,†

Probe	Enzyme						
	*Not*I	*Nru*I	*Pvu*I	*Mlu*I	*Sal*I	*Cla*I	*Pae*R7I
Class II							
DP	425	>1200	390	B100 A440 A290	380 240	280 230	ND
DQA	980	>1200	290	550 440 150	530 280	140 110	ND
DR	980	>1200	380	550	530 250	B80 A240 A110	ND
Class III							
21OHase	980	640	340	225	490 280	140	60
J	210	640	340	270	490 280	85	280
L	270	640	780	190	220	ND	–
TNFα	270	100	780	190	220	165	290
Class I							
HLA-B	950	375	780	360 180 130	420	195 140	290
HLA-C	950	375	780	360 180 130	420	80 40	ND
HLA-ABC	675 400	–	–	ND	ND	ND	ND

*See (6) for details of probes.

†For the DP and DR subregions, where different fragments were observed with A and B gene probes, these are indicated by A and B. Multiple sets of data indicate that partial digestion was obtained and the major mapped fragments are given. – indicates that no fragment was observed using PFGE; ND indicates not determined.

complement loci (Fig. 1). This fragment can be split with *Nru*I to give a 700 kb fragment with the DRA probe and a fragment of 280 kb identified with the 21-OHase probe. However, probe J hybridizes to a 210 kb *Not*I fragment on the same blot. Since the exact position of the *Not*I site between the 21-OHase B gene and probe J is known from the cosmid map, these results establish the orientation of the complement genes relative to DRA, the C2 gene being telomeric to the 21-OHase gene (Fig. 1).

Hybridization of class II and class III probes to Southern blots of the DNA digested with the enzymes shown in Table 1 in single and double digest combinations, allowed the map in Figure 1 to be constructed. In the *Sal*I digests the 21-OHase probe hybridizes to a partial fragment of 490 kb which does not contain DRA. Double digestion with *Not*I/*Sal*I maps a *Sal*I site 85 kb telomeric to the *Not*I site. Hence the minimum distance between the 21-OHase B and DRA genes must be 300 kb. Our map of the class II subregions shows striking similarities with that published by Hardy et al. (8) and confirms the HLA-D subregion organisation. Comparison of the maps and our mapping of partial *Cla*I digests (results not shown) is consistent with the DRA gene lying no more than 360 kb from the 21-OHase B gene.

The genes for TNFα and β have been linked to the human MHC (9) and in the mouse have been mapped about 70 kb from the H-2D region (10). To establish the position of these genes in the human MHC, a TNFα probe was hybridized to *Pvu*I digested DNA and a 780 kb fragment was identified that also contained the HLA-B and -C loci. To confirm this linkage with the MHC we isolated 82 kb of genomic DNA in overlapping cosmid clones containing the TNFα gene and mapped them for infrequently cutting restriction enzyme sites (6). On the basis of these data a single copy hybridization probe (probe L, Fig. 1) was isolated distal to the TNFα gene and the single *Nru*I site in the cosmids. Hybridization of probe L to *Nru*I digested DNA detected the same 640 kb *Nru*I fragment that contains the complement genes. Hence, the complement genes and TNFα are linked, and C2 lies 390 kb centromeric of the TNFα gene. Comparison of the published map of Nedospasov et al (11) with our cosmid map infers that TNFβ is telomeric of TNFα.

The distance between the TNFα and HLA-B loci was estimated by hybridization of these probes to blots of *Pae*R7I digested DNA. A common 290 kb *Pae*R7I fragment was observed. The *Pae*R7I site defining the 5′ end of this fragment lies within the TNFα gene. Double digestion data allowed the map as shown in Figure 1 to be constructed and established that the TNFα and β genes lie ~250 kb centromeric of the HLA-B locus.

The general class I probe hybridizes to *Not*I fragments that include 1100 kb of DNA. The B and C loci are separated by at least 80 kb on the basis of hybridization of B- and C-locus specific probes to *Mlu*I, *Cla*I and *Mlu*I/*Cla*I double digests. Additional data suggest that the HLA-A locus is separated from the HLA-C locus by at least 600 kb (Fig. 1).

Based on the long range genomic restriction site mapping presented here we have established a map of the human MHC (Fig. 1) spanning up to ~3800 kb. We have orientated the complement and 21-OHase genes relative to the class I and class II loci, C2 being telomeric to 21-OHase B. The distance between the 21-OHase B gene and the DRA gene is 300–360 kb and that between the C2 gene and the HLA-B locus is ~650 kb. We have located the genes for TNFα and β within the MHC, the TNFα gene lying 390 kb telomeric of C2. The location of the genes for the lymphokines TNFα and β within the MHC is of particular interest because of their possible role in the pathogenesis of MHC-linked autoimmune disorders and because they raise the possibility that other, as yet unidentified, genes may reside within the large regions of DNA separating the class I, II, and III loci.

References

1. Carroll MC, Campbell RD, Bentley DR, Porter RR. A molecular map of the human major histocompatibility complex class III region linking complement genes C4, C2, and factor B. Nature 1984;307:237–241.
2. Carroll MC, Campbell RD, Porter RR. Mapping of steroid 21-hydroxylase genes adjacent to complement component C4 in HLA, the major histocompatibility complex in man. Proc Natl Acad Sci USA 1985;82:521–525.
3. Schwartz DC, Cantor CR. Separation of yeast chromosome-sized DNAs by pulsed field gradient gel electrophoresis. Cell 1984;37:67–75.
4. Carle GF, Olson MV. Separation of chromosomal DNA molecules from yeast by orthogonal-field-alteration gel electrophoresis. Nucleic Acids Res 1984;12:5647–5664.
5. Yu CY, Campbell RD. Definitive RFLPs to distinguish between the human complement C4A/C4B isotypes and the major Rodgers/Chido determinants: Application to the study of C4 null alleles. Immunogenetics 1987;25:383–390.
6. Dunham I, Sargent CA, Trowsdale J, Campbell RD. Molecular map of the human major histocompatibility complex by pulsed-field gel electrophoresis. Proc Natl Acad Sci USA 1987;84:7237–7241.
7. van Ommen GJB, Verkerk JMH. Restriction analysis of chromosomal DNA in a size range up to two million base pairs by pulsed field gradient electrophoresis. In: Davies KE (ed): Human genetic disease. A practical approach. Oxford, UK: IRL Press Ltd, 1986, pp. 113–133.
8. Hardy DA, Bell JI, Long EO, Lindsten T, McDevitt HO. Mapping of the class II region of the human major histocompatibility complex by pulsed-field gel electrophoresis. Nature 1986;323:453–455.
9. Spies T, Morton CC, Nedospasov SA, Fiers W, Pious D, Strominger JL. Genes for the tumor necrosis factors α and β are linked to the human major histocompatibility complex. Proc Natl Acad Sci USA 1986;83:8699–8702.
10. Müller U, Jongeneel CV, Nedospasov SA, Lindahl KF, Steinmetz M. Tumour necrosis factor and lymphotoxin genes map close to H-2D in the mouse major histocompatibility complex. Nature 1987;325:265–267.
11. Nedospasov SA, Shakov AN, Turetskaya RL, Mett VA, Georgiev GP, Dobrynin VN, et al. Molecular cloning of human genes coding for tumour necrosis factor: Tandem arrangement of the alpha- and beta-genes in a short segment (6 kilobase pairs) of the human genome. Proc Acad Sci USSR (Dokl Acad Nauk SSSR) 1985;285:1487–1490.

Author Affiliations

Ian Dunham, Carole A. Sargent, R. Duncan Campbell, MRC Immunochemistry Unit, Department of Biochemistry, University of Oxford, Oxford; *John Trowsdale*, Imperial Cancer Research Fund, London, UK

Towards a Physical Map of the HLA Complex

Andreas Ziegler, Katharina Bloemer, Heike Pohla, Elisabeth Weiss, Peter Schneider, and Jiannis Ragoussis

The human major histocompatibility (HLA) complex is located on the short arm of chromosome 6 in the 6p21.31→6p21.33 region (1,2). There are three clusters of genes, the HLA class I, II, and III regions. Whereas the class III loci are only moderately polymorphic (see (3) for review), the class I and II genes which encode cell surface glycoproteins show an extreme degree of polymorphism. There are a minimum of 17 class I loci (4) and at least 15 genes for class II alpha and beta chains (5). In addition, we (6) and others (7) have recently demonstrated that also the loci for tumor necrosis factor (TNFA) and lymphotoxin (TNFB) are part of the HLA complex [see also Ragoussis et al., this volume]. A precise knowledge of the exact arrangement of HLA genes and the physical distances involved is essential for an understanding of several important properties of the HLA system, among them the association of haplotypes with diseases (8) and the phenomenon of linkage disequilibrium between alleles of different loci (9). Towards this end, we have assigned individual HLA genes to large DNA restriction fragments separated by pulsed field gel electrophoresis or related techniques. This allowed us to construct a molecular map of the HLA complex spanning at least 4000 kilobasepairs (kb).

Mutant cell lines with monosomy 6 or HLA-hemizygosity were employed since possible interpretative problems with haplotype-specific restriction fragment length polymorphisms could thus be avoided. All mutants were derived from BJAB-B95.8.6 lymphoma cells with the HLA haplotypes Al, Cw4, B35, Bw6, DR5, DRw52, DQw3, DPw4, and A2, C-, B13, Bw4, DRw6, DRw52, DQwl, DPw2 (1,2,10). The data presented here were mainly derived from the monosomy 6 mutant cell line BM19.7 which retains the A2 haplotype including two loci each of C4 and 21-steroid hydroxylase.

Large genomic DNA fragments were generated with various restriction enzymes as described (6) and separated predominantly by field inversion gel electrophoresis [FIGE, (11)] or the newly developed technique of rotaphoresis (12). Sequential hybridization of DNA probes [Table 1] to Southern blots was carried out under

Table 1. DNA Probes Employed in this Study

Designation	Specificity
pMF 28	Common Class I
HLA-B27-3′	HLA-B (-C)
pCD2.6H6.2	HLA-B-spec. (5′)
pCD2.6T2.2	HLA-B-spec. (3′)
EX1.35	HLA-A-spec. (3′)
pCAH5SH2.9	TNFB (5′)
pCAH5KB5	TNFB (5′)
pCAH5P6	TNFA (3′)
pCAH5P11	TNFA (3′)

The origin of further DNA probes for the class II and III regions has been described (6,13).

stringent conditions (6). The relative position of a gene-containing restriction fragment generated with a particular enzyme was usually proven by employing multiple double digests with other restriction enzymes in addition. The sizes of restriction fragments were estimated by reference to yeast chromosomes and phage λ concatemers.

Two enzymes were found to be particularly useful in obtaining a large scale physical map of the HLA complex (Fig. 1), namely Not I and Nru I. We were able to confirm our initial observations (13) that the complement region genes as well as the DRα, DRβ, DQα, DQβ, DXα, and DXβ genes were all present on a single Not I fragment with a length of approximately 1000 kb. This fragment also included the DVβ gene (not shown). However, the four DP region loci, the DZα, and the DOβ genes were located on a smaller Not I fragment with a size of about 600 kb. Since the DP region was known to be centromeric to the DQ and DR loci (14), these results prove that the DZα and DOβ genes are located centromeric to the DR and DQ regions. Digests with the restriction enzyme Mlu I demonstrate also that the DRβI gene must be located between the DRβIII and the DQ/DX loci. Data obtained with the mutant BM3.1 (not shown) indicate that the maximal size of the chromosomal fragment carrying the DRα,β as well as the DQ and DX genes is not longer than 500 kb.

The maximal length of the class II region, 1000 kb, could be estimated from the size of an Nru I fragment containing all known class II genes. This more detailed analysis extends the results presented previously (13), and is fully in line with the careful analysis carried out by Hardy and coworkers (15).

Only one enzyme, Not I, was found which generated an approximately 1000-kb long fragment containing not only class II genes, but all class III loci in addition (7,13,16). Double digests employing Not I and Nru I proved that the complement region genes were located

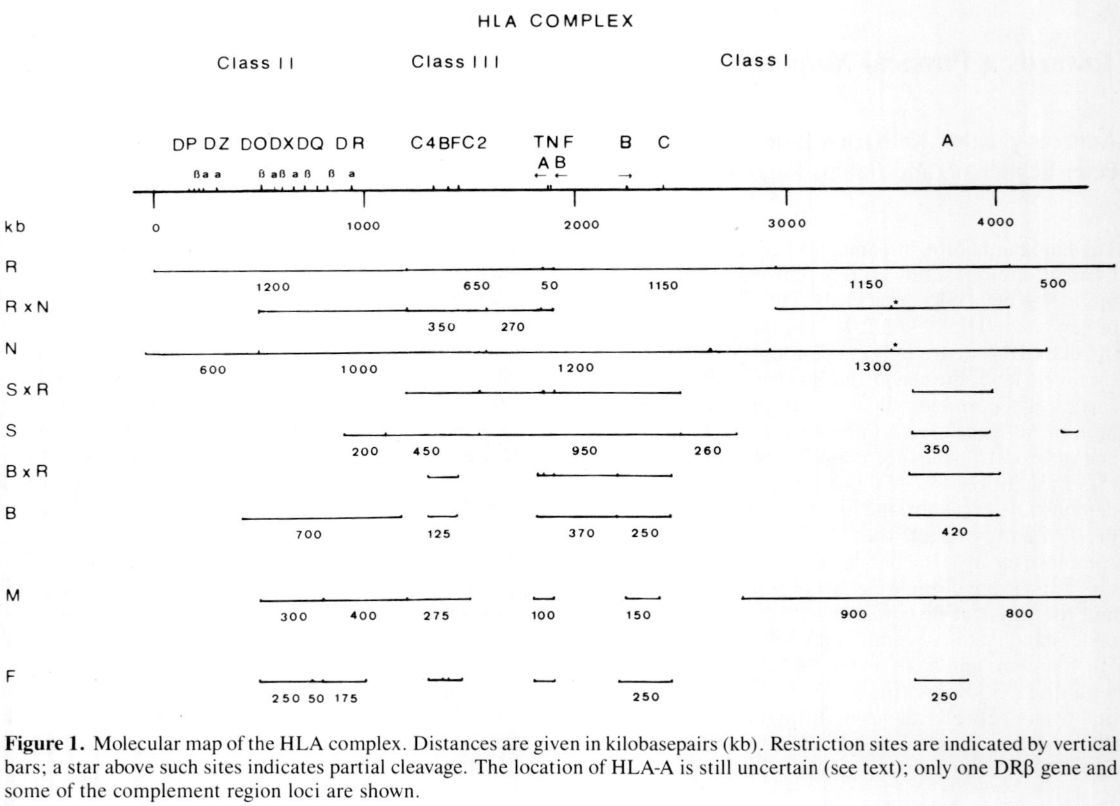

Figure 1. Molecular map of the HLA complex. Distances are given in kilobasepairs (kb). Restriction sites are indicated by vertical bars; a star above such sites indicates partial cleavage. The location of HLA-A is still uncertain (see text); only one DRβ gene and some of the complement region loci are shown.

in the telomeric 300 kb of this Not I fragment, 250 to 370 kb away from DRα. However, we were unable to define the orientation of the class III loci on chromosome 6, because no infrequently cutting enzyme was found which generated a fragment containing only one end of this region in physical linkage with a class I or a class II gene. Only Sfi I generated several small class III region fragments, but these were uninformative for the purpose of orienting the genes relative to other HLA loci. Dunham and coworkers (7) combined chromosomal walking and pulsed field gel analysis and found that C2 is the most distal class III locus. They estimate the DRα to 21-OH distance as 300 to 360 kb, in line with our data.

We and others (13,16) were initially unable to find a restriction fragment containing both class III and class I genes. This has led to a serious misinterpretation of mapping data generated with pulsed field gels since the distance between these two regions was estimated as only approximately 50 kb (16). There is in fact a distance of at least 650 kb between C2 and HLA-B (6,7, Ragoussis et al, this volume). It was possible to bridge this large gap since we found that the TNFA and TNFB loci were located between the class III genes and the HLA-B locus. The restriction endonuclease *Nru*I generated a 700-kb long fragment which hybridized with probes for C2, C4, and the region 3′ of TNFA. Often another band with a length of approximately 650 kb was observed, and probes form the TNFA3-′ region detected a fragment of ~50 kb. The most likely explanation for these results is the presence of the C2, C4, and TNF genes on a fragment which can be partially cleaved at its distal end to generate a 50-kb fragment carrying the TNFA locus. The transcriptional orientation of the TNF loci could be determined as well, because the long Nru I-fragment did not hybridize to probes from the region 5′ of TNFB.

Linkage of the TNF genes with the class I region was obtained using several enzymes: Not I, Sal I, and BssH II. The latter was particularly informative, because a 370-kb fragment was found which contained the TNF loci and hybridized to a genomic probe specific for the 5′ end of the HLA-B gene. However, the HLA-B gene itself and genomic probes derived from its 3′ end detected another 250-kb long BssH II fragment. This result allowed to orient the HLA-B gene relative to the proximally located TNF loci. Therefore, the TNF genes are transcribed in the same direction (5′-3′-centromere) as the complement region genes, whereas the reverse is true for the HLA-B locus (5′-3′-6pter).

We encountered considerable difficulties in assigning additional class I sequences to defined chromosomal fragments, mainly for two reasons: there were only very few locus-specific probes at our disposition, and the number of restriction enzymes giving fragments informative for the construction of a large scale physical map was found to be small. However, the map depicted in Figure 1 is fully compatible not only with our results, but also with those of other groups (7,16), even taking the analysis of different HLA haplotypes into account. Hybridizations with an HLA-B/HLA-C-3′ crossreactive probe revealed that the HLA-C gene was present on the same 250-kb BssH II fragment carrying the HLA-B locus. Since the 950 kb long Sal I fragment (Fig. 1) could be arranged accurately in the HLA complex, an additional enzyme proves that the HLA-B-HLA-C distance cannot be larger than 250 to 300 kb. In fact, it is probably not longer than 150 kb (data using Mlu I). The application of Not I-digested BM 19.7 DNA in combination with various class I probes (Table 1) made it possible to assign the HLA-A locus to a 1300-kb long fragment which was clearly distinct from that hybridizing to HLA-B locus probes. In some blots, a smaller Not I fragment (~700 kb) instead of the 1300 kb piece was found to hybridize with the A-probe. Again, this is probably due to cleavage of a normally methylated Not I restriction site. Unfortunately, double digests with Not I and Nru I were not very informative with regard to the localization of HLA-A, because the data were only compatible with an 1150-kb Nru I fragment being part of the large Not I piece. Double digests with Not I and Mlu I proved to be of more value, and gave a 450-kb long HLA-A containing fragment (not shown in Fig. 1). These results are supported by digestions with other enzymes and show that HLA-A must be located in the proximal or distal 450 kb DNA of the 1300 kb Not I fragment.

Hybridization with the "common" class I-probe proved that a 500-kb long Nru I fragment existed containing additional HLA class I sequences (not identical with HLA-A, B, or C). The most plausible localization of this fragment is telomeric, adjacent to the 1150-kb Nru I fragment carrying HLA-A. Since the "common" class I-probe detects also the RS5 gene (E. Weiss, unpublished results), it is possible that this locus (17) is part of that presumably most telomeric HLA gene-containing Nru I fragment.

The physical map of the HLA complex derived from our results (Fig. 1) is supported by evidence from other studies (5,6,7,13,15,16,18) using a variety of experimental approaches. The following conclusions can be drawn from the work presented here:

1) The total length of the HLA complex must be at least 4000 kb, and probably does not exceed 4500 kb. The latter figure implies that there are no internal Not I or Nru I restriction fragments in the class I region which do not hybridize with the "common" class I probe employed here.

2) The approximate sizes of the HLA regions are: ~2000 kb for class I, ~1000 kb for class II, and 1200 kb for the chromosomal region between HLA-B and HLA-DR, which contains not only the "classical" class III genes but also TNFA and TNFB.

3) The transcriptional orientations of the HLA-B, TNFA, TNFB loci could be determined by combining the data of Nedospasov and coworkers (19) with our own.

4) It will be misleading to rely only on one (13) or two (16) infrequently cutting restriction enzymes for large scale physical mapping. Our study proves that the application of even 6 different enzymes may not be sufficient to obtain a completely unambiguous map. Furthermore,

preliminary data using the partial monosomy 6 mutant BM 28.7 (HLA-Al, CW4, B35) and the HLA-homozygous Epstein Barr Virus-transformed B cell line H2LCL (HLA-A3, B7) indicate that there will be considerable differences in the lengths of HLA chromosomal DNA telomeric to the complement genes. This highlights the importance of mutant cell lines (or somatic hybrids) retaining only one human chromosome 6 for this type of approach towards gene mapping.

Finally, we would like to point out the relatively good agreement between our physical map and that derived from recombinational events in families giving a distance of 3.2 cM between HLA-A and HLA-DP. If one assumes that 1000 kb roughly equals 1 cM (9), the overall fit is quite convincing. Further studies in this area will undoubtedly be devoted at obtaining an exact map of the entire HLA class I region and at the identification of further genes belonging to the HLA complex (see (20) for a novel TNFA-linked gene in the mouse).

Acknowledgments. This work was supported by the Deutsche Forschungsgemeinschaft through SFB 120 (project A4) and the Genzentrum Munchen.

References

1. Spring B, Fonatsch C, Muller C, Pawelec G, Kömpf J, Wernet P, et al. Refinement of HLA gene mapping with induced B-cell line mutants. Immunogenetics 1985;21:227.
2. Ziegler A, Zimmer F-J, Fonatsch C, So A, Trowsdale J. Mapping of HLA class I and II genes using HLA deletion mutant cell lines. HGM8. Cytogenet Cell Genet 1985;40: 788.
3. Rittner C, Schneider PM. Genetics and polymorphism of complement components. In: Rother K, Till GO (eds): The complement system. Berlin: Springer, 1987, in press.
4. Strachan T. Genetics and polymorphism: class I antigens. Br Med Bull 1987;43:1.
5. Trowsdale J. Genetics and polymorphism: class II antigens. Br Med Bull 1987;43:15.
6. Ragoussis J, Bloemer K, Weiss E, Ziegler A. Localization of the genes for tumor necrosis factor and lymphotoxin between the HLA class I and III regions by field inversion gel electrophoresis. Immunogenetics, 1988;27:66.
7. Dunham I, Sargent CA, Trowsdale J, Campbell D. Molecular map of the human major histocompatibility complex by pulsed field gel electrophoresis. Proc Natl Acad Sci USA, 1987; in press.
8. Tiwari LJ, Terasaki PI. HLA and disease associations. Berlin: Springer Verlag, 1985.
9. Olaisen B, Sakaguchi AY, Naylor SL. Report of the committee on the genetic constitution of chromosomes 5 and 6. HGM9. Cytogenet Cell Genet, 1987; in press.
10. Ziegler A, Muller C, Heinig J, Radka SF, Kömpf J, Fonatsch C. Monosomy 6 in a human lymphoma cell line induced by selection with a monoclonal antibody. Immunobiol 1985; 169:455.
11. Carle GF, Frank M, Olson MV. Electrophoretic separations of large DNA molecules in one-dimensional electric fields that are periodically inverted. Science 1986;232:65.
12. Ziegler A, Geiger K-H, Ragoussis J, Szalay G. A new electrophoresis apparatus for separating very large DNA molecules. J Clin Chem Clin Biochem 1987;25:578.
13. Ragoussis J, van der Bliek A, Trowsdale J, Ziegler A. Mapping of HLA genes using pulsed field gradient electrophoresis. FEBS Lett 1986;204:1.
14. Kavathas P, DeMars R, Bach FH, Shaw S. SB: A new HLA-linked human histocompatibility gene defined using HLA-mutant cell lines. Nature 1981; 293:747.
15. Hardy DA, Bell JI, Long EO, Lindsten T, McDevitt HO. Mapping of the class II region of the human major histocompatibility complex by pulsed-field gel electrophoresis. Nature 1986;323:453.
16. Lawrance SK, Smith CL, Srivastava R, Cantor CR, Weissman SM. Megabase-scale mapping of the HLA gene complex by pulsed field gel electrophoresis. Science 1987; 235:1387.
17. Srivastava R, Chorney MJ, Lawrance SK, Pan J, Smith Z, Smith CL, et al. Structure, expression and molecular mapping of a divergent member of the class I HLA gene family. Proc Natl Acad Sci USA 1987;84:4224.
18. Erlich H, Lee JS, Petersen JW, Bugawan T, DeMars R. Molecular analysis of HLA class I and class II antigen loss mutants reveals a homozygous deletion of the DR, DQ, and part of the DP region: implications for class II gene order. Hum Immunol 1986;16:205.
19. Nedospasov SA, Shakov AN, Turetskaya RL, Mett VA, Azizov MM, Georgiev GP, et al. Tandem arrangement of genes coding for tumor necrosis factor (TNFα) and lymphotoxin (TNFβ) in the human genome. Cold Spring Harbor Symp Quant Biol 1986;51:611.
20. Tsuge I, Shen F-W, Steinmetz M, Boyse EA. A gene in the H-2S:H-2D interval of the major histocompatibility complex which is transcribed in B cells and macrophages. Immunogenetics 1987;26:378.

Author Affiliations

Andreas Ziegler,* *Jiannis Ragoussis*, Medizinische Klinik und Medizinisch-Naturwissenschaftliches Forschungszentrum, University of Tübingen, Tübingen; *Katharina Bloemer*, *Heike Pohla*, *Elisabeth Weiss*, Institute für Immunologie, University of München, München; *Peter Schneider*, Institut für Rechtsmedizin, University of Mainz, Mainz, Federal Republic of Germany

---

*Present address: Institut für Experimentelle Immunologie, Philipps-Universität, Deutschhausstraβe 1, D-3550 Marburg/Germany.

Organization of the HLA Class I Region*

Pierre Pontarotti, Giovanna Chimini, Catherine Nguyen, Joëlle Boretto, and Bertrand R. Jordan

HLA class I genes including HLA-A, -B, and -C genes plus 20 to 30 other related sequences are probably all located in the same chromosome region, band p 21.3 of chromosome 6 (1). The genetic map of this region indicates that HLA-A, -B, and -C genes are contained within a segment of 1 centimorgan; the position of the other related genes is not known with certainty, although it appears that most of them are clustered close by and telomeric to the HLA-A gene (2,3).

This region has already been investigated by pulsed field gel (PFG) electrophoresis (4), but so far no complete and continuous physical map has been established independently of the genetic map. This is not surprising since analysis of this region with this technology is made difficult by the high degree of similarity of the class I sequences present in the HLA complex, which requires the use of locus-specific probes to identify unambiguously fragments containing HLA-A, -B, or -C genes and by the striking CpG richness in 5′ regions shared by all HLA class I sequences available so far, which may constitute "putative HTF" (Hpa II Tiny Fragments) islands, where clusters of rare cutting enzyme sites are present (5).

*This work was supported by institutional grants from Inserm and CNRS to the Centre d'Immunologie, and by grants from ARC and FRM to B.R.J. C.N. was supported by FRM postdoctoral fellowship.

In this paper, we present results from a series of PFG experiments in which we examined the pattern of bands obtained by complete digestion with "rare cutter" enzymes of DNA from the HHK cell line (homozygous for HLA, A3, B7, Cw7). The fragments obtained are relatively small (100 to 400 kilobasepairs) and the HLA-A and -B genes are contained in different fragments (i.e., not physically linked). They are not as small, however, as they should be if *each* HLA sequence contained an HTF island, thus either many of the other unsequenced HLA class I-related genes do not contain CpG-rich regions, or many of these CpG-rich regions are methylated, which raises the question of how their "CpG richness" has been preserved since C methylation in CpG dinucleotides leads to C−T transitions (6). Finally using 5′ and 3′ flanking HTF island probes (these HTF islands are present in the 5′ region of B7, Cw7, and A3 genes), we were able to link B and C genes.

As every human and murine class I gene sequenced so far displays a CpG-rich region containing a cluster of rare cutting enzyme sites, as detailed in Figure 1 for four human genes (7–10), it is conceivable that all the other yet unsequenced, class I genes display similar features. If all these HFT-like regions were really unmethylated, the enzymes used in this study, which are mostly methylation sensitive, should cut the 1,000 kb region containing 30 genes into fragments of 30 kb.

PGF experiments with DNA of HHK, digested in this case with Mlu I (Fig. 2), show that we obtained with

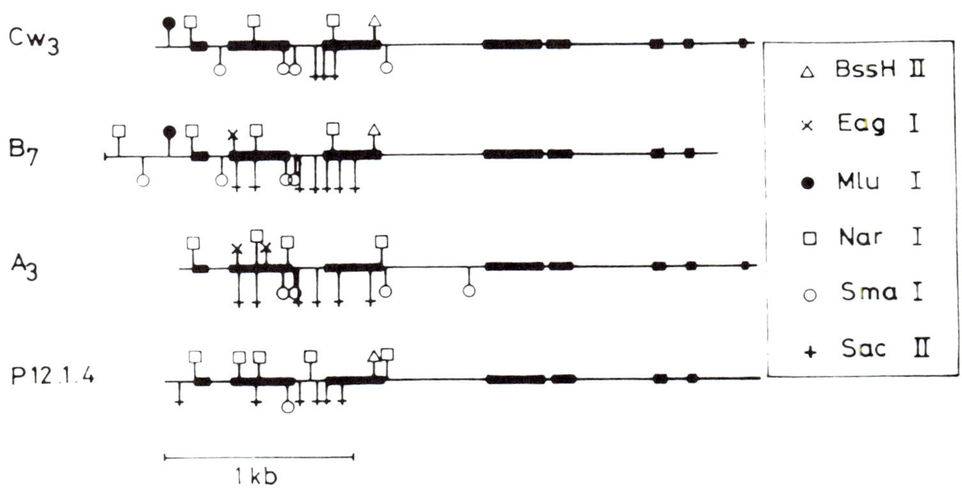

Figure 1. Rare cutter restriction enzyme maps of four representative HLA class I genes. Enzyme symbols are given on the right panel. Distribution of sites is strikingly skewed towards the 5′ region in all cases.

Immunobiology of HLA
Volume II

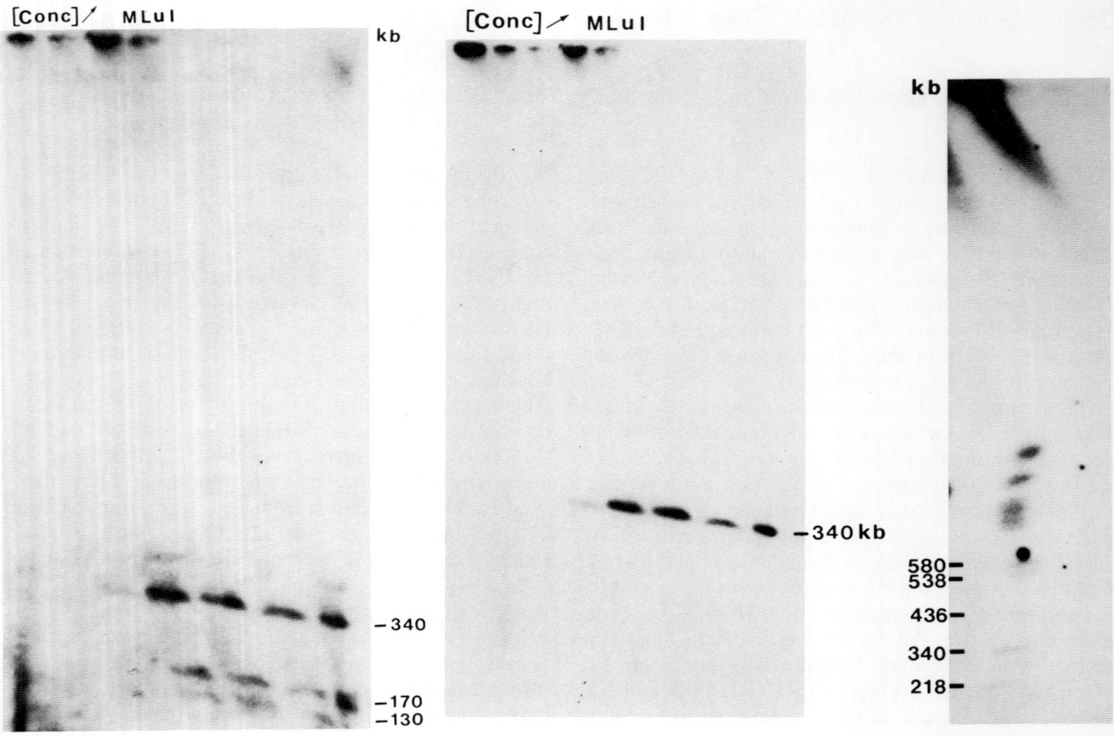

Figure 2. PFG pattern of Mlu I-digested DNAs of HHK cell line. The same blot was hybridized successively with A (a), B and C locus-specific probes (not shown) and with the general HLA probe (b). Yeast chromosomes (Y) prepared from GY 22 strain and used as size markers, were hybridized in parallel with ^{32}P-labeled yeast DNA (c). Sizes as calibrated with respect to a ladder of linear DNA concotamers are: 220, 340, 440, 530, 580, 680, 720, 780, 820, 900, 1,150 kb. Relevant sizes are given in kilobases. A dehybridization step was performed between each hybridization. Techniques used for these figures are described elsewhere (11).

general and locus-specific probes fragments which are larger than expected fragments. Similar results were obtained with other restriction digests (BssH II, Nae I, Not I, etc., data not shown).

The paradox elicited by our results can be explained either by the disappearance of CpG-rich region from most of human class I genes, which is unlikely as we know that they are present in the K, D, and Qa region of murine genome, or by the hypothesis that CpG-rich regions while maintained are nonetheless methylated and therefore not accessible to digestion in contrast to the "HTF islands" 5′ to A, B, C, and p12.1.4 genes (data not shown). Yet such methylation would lead the CpG sequences to change to TpG. The two possible explanations are that even though these sequences *are* methylated: 1) C−T transitions did not occur, or 2) C−T transition did occur but frequent sequence exchanges (reciprocal or unidirectional) restore or rather conserve the CpG richness of these regions. The latter hypothesis ties in well with the demonstration (particularly compelling in the mouse system) that gene-conversion-like events *do* occur between class I genes and pseudogenes, and that these occur essentially in the polymorphic (and CpG rich) 5′ end of the genes (12).

PFG gels of HHK DNA cut by Mlu I (Fig. 2) show, when hybridized with a HLA general probe, a strongly hybridizing fragment and two weakly hydridizing fragments. The strongly hybridizing fragment corresponds to the one detected by the A locus-specific probe and the weak ones to C and B (cf. hybridization with 3′ of C and 3′ of B probes shown in Fig. 3). As the general probe should hybridize to a similar extent, under the conditions used, with most class I genes and pseudogenes (data not shown) the strongly hybridizing fragment can be expected to contain most of the class I sequences. The same result (i.e., the presence of a single strongly hybridizing fragment) has been obtained following digestion with several other "rare cutter" enzymes (BssH II, Nae I, Not I, results not shown) and therefore strengthens the hypothesis of a clustering of most HLA class I genes around A gene.

We thus believe that cleavable sites (HTF islands) are present at least within the *expressed* HLA genes (HLA-A, -B, -C) and that the fragments seen by PFG

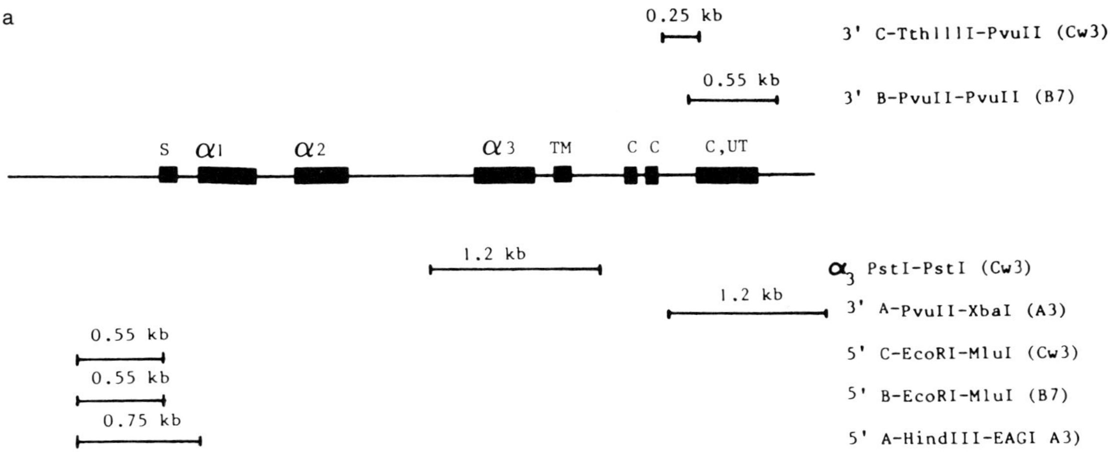

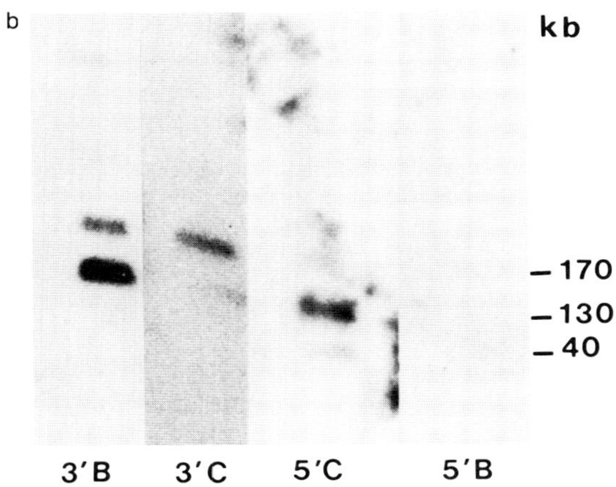

Figure 3. *a*, 5′ and 3′ locus-specific probe used in this work (13–15). *b*, Because the 5′ Mlu I site is present on B7 and in Cw7 (data not shown), we constructed specific probes 5′ from these sites in order to determine if the 5′ end of B7 could be on the same Mlu I fragment as the 3′ end of Cw7 or the 3′ end of B7 on the same fragment as the 5′ end of Cw7. We show in this figure the PGF pattern of the Mlu I-digested DNAs of the HHK cell line. The same blot was hybridized successively with 3′ of B-, 3′ of C-, 5′ of C-, and 5′ of B-specific probes. The 3′ of B-specific probe which cross-hybridizes weakly with C sequences detects a 130-kb fragment. The 3′ of C probe hybridizes with a 170-kb fragment. The 5′ of the C-specific probe hybridizes with the 130-kb fragment. Thus the 5′ of C gene and 3′ of B gene are present in the same Mlu I fragment. The 5′ of B probe hybridizes with a 40-kb fragment.

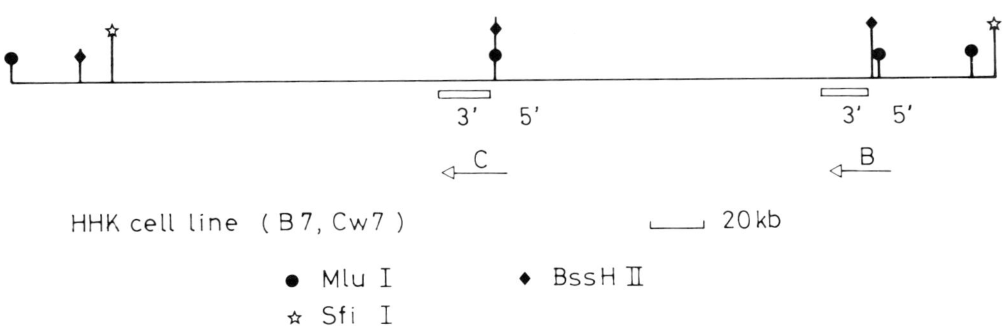

Figure 4. Large-scale genomic organization of HLA-B and -C regions. This was done with data obtained by PFG analysis using the different enzymes put into this map.

analysis are bounded by HTF islands which generally coincide with the 5′ region of expressed genes (5).

As we noted before by the hybridization of 3′ probe from A, B, C genes on PFG blot of HHK DNA cut with different rare cutter enzymes, we were not able to link those genes; this is due to the HTF islands present in 5′ part of the A, B, C genes. We therefore derived 5′-specific probes for A, B, and C genes, flanking these HTF islands in order to overcome the linkage gaps.

Indeed, with the 5′ probe of B and C sequence we were able to link B and C genes (Figs. 3 and 4) and we concluded that: 1) B and C genes were really separated by an HTF island represented by the 5′ part of the C gene; 2) the distance between B and C genes is approximately 130 kb, in agreement with the genetic distance of 0.1 cm found by linkage analysis; 3) B and C genes are in head-to-tail orientation.

On the other hand we were unable to link A to B or C genes even with the help of the 5′ probes. This may be due to the presence of other HTF islands around HLA-A, as suggested by the hybridization to the same Mlu I and Sfi I fragment of a 3′-specific probe for p12.1.4 and the A-specific probe (data not shown). Subcloning of a new 5′ HTF flanking probe for p12.1.4, which should allow us to go on in our linkage investigation, is currently in progress.

Acknowledgments. We are very grateful to G.B. Ferrara for providing LCLS, to Joséphine Sire, Juan Antonio Rey, and many colleagues at CIML for help and discussions, and to Mrs. Chantal Vaffides for typing this manuscript.

References

1. Morton CC, Kirsh IR, Nance WE, Evans GA, Korman AJ, Strominger JL. Orientation of loci within the human major histocompatibility complex by chromosomal in situ hybridization. Proc Natl Acad Sci USA 1984;81:2816–2820.
2. Orr HT, Bach FH, Ploegh HL, Strominger JL, Kavathas P, De Mars R. Use of HLA loss mutants to analyse the structure of human major histocompatibility complex. Nature 1982;296:454–456.
3. Srivastava R, Chorney MJ, Lawrance SK, Pan J, Smith Z, Smith CL, et al. Structure, expression, and molecular mapping of a divergent member of the class I HLA gene family. Proc Natl Acad Sci USA 1987;84:4224–4228.
4. Lawrance SR, Smith CL, Srivastava R, Cantor CR, Weissman SM. Megabase-scale mapping of the HLA gene complex by pulsed field gel electrophoresis. Science 1987; 235:1387–1390.
5. Lindsay S, Bird A. Use of restriction enzymes to detect potention gene sequences in mammalian DNA. Nature 1987;327:336–338.
6. Bird AP. CpG rich islands and the function of DNA methylation. Nature 1986;321:200–213.
7. Strachan T, Sodoyer R, Damotte M, Jordan BR. Complete nucleotide sequence of a functional class I HLA gene, HLA A3. Implication for the evolution of HLA genes. Embo J 1984;3(4):879–885.
8. Srivastava R, Duceman BW, Biro PA, Ashwani KS, Weissman SH. Molecular organization of the class I genes of human major histocompatibility complex. Immunol Rev 1985;84:93–121.
9. Sodoyer R, Damotte M, Delovitch TL, Trucy J, Jordan BR, Strachan T. Complete nucleotide sequence of a gene encoding a functional human class I histocompatibility antigen (HLA CW3). Embo J 1984;3:879–885.
10. Malissen M, Malissen B, Jordan BR. Exon-Intron organization and complete nucleotide sequence of an HLA gene. Proc Natl Acad Sci USA 1982;79:893–897.
11. Nguyen C, Pontarotti P, Birnbaum D, Chimini G, Rey JA, Mattei JF, et al. Large scale physical mapping in the q27 region of the human X chromosome: The coagulation factor IX gene and the mcf.2 transforming sequence are separated by at most 270 kilobase pairs and are surrounded by several "HTF islands." Embo J 1987;11:3285–3289.
12. Pease LR, Schulze DH, Pfaffenbach GM, Nathenson SG. Spontaneous H-2 mutants provide evidence that a copy mechanism analogous to gene conversion generates polymorphism in the major histocompatibility complex. Natl Acad Sci USA 1983;80:242–246.
13. Koller BH, Sidwell B, de Mars R, Orr HT. Isolation of HLA locus specific DNA probes from the 3′ untranslated region. Proc Natl Acad Sci USA 1984;81:5175–5178.
14. Coppin HL, Denny DW, Weissman SM Jr, McDevitt HO. HLA-B locus polymorphism: Studies with a specific hybridization probe. Proc Natl Acad Sci USA 1985;82: 8614–8618.
15. Strachan T, Dodge AB, Smillie D, Dyer PA, Sodoyer R, Jordan BR, et al. An HLA C specific DNA probe. Immunogenetics 1986;23:115–120.

Author Affiliations

Pierre Pontarotti, *Giovanna Chimini*, *Joëlle Boretto*, and *Bertrand R. Jordan*, Centre d'Immunologie Inserm-CNRS de Marseille-Luming Case 906, 13283 Marseille Cedex 9, France; *Catherine Nguyen*, Inserm U242, Hopital d Enfants de la Timone, Marseille, France.

Mapping of HLA Class II Genes by Pulsed Field Gel Electrophoresis and Size Polymorphism

Hidetoshi Inoko, Kimiyoshi Tsuji, Vicky Groves, and John Trowsdale

The HLA-D histocompatibility region encodes for polymorphic cell-surface class II antigens composed of a 34 kd α chain and a 28 kd β chain. This region is divided into several subregions:, DP, DO/DZ, DQ and DR, although each contains a complex array of genes and pseudogenes. In this report, the organization of the genes in the HLA-D region of serologically different HLA-homozygous B cell lines was examined by pulsed field gel electrophoresis (1,2) (PFGE) using α and β chain probes from all of the subregions including DVβ (3) (a new HLA class II pseudogene) and a DX region-specific probe. The map was determined to be: DPβ-DPα-DO(DZ)α-DOβ-DXβ-DXα-DVβ-DQβ-DQα-DRβ-DRα. Interestingly, the distances not only between DQα and DRβ but also between DVβ and DQβ were found to vary markedly among different HLA haplotypes.

High MW genomic DNA was prepared in agarose blocks (4) from the B cell line MANN(DR7) homozygous for the HLA region and cleaved with ten infrequently cutting restriction enzymes, namely: *Sfi*I, *Cla*I, *Sal*I, *Sac*II, *BssH*II, *Nar*I, *Mlu*I, *Not*I, *Nae*I, and *Nru*I, singly and in pairwise combinations to construct a molecular map. Digested DNA were separated on PFGE gels, transferred to nylon membranes, and hybridized sequentially with 11 probes from the HLA class II region: the six cDNA probes of DRα, DRβ, DQα, DOβ, DPα, and DPβ (which were distributed for the Southern blot studies in the Tenth International Histocompatibility Workshop), the DO(DZ) α probe (1.1 kb cDNA fragment) (J. Young et al, unpublished), the DVβ probe (2 kb *Pst*I genomic fragment from λHMCβ19) (3), the DXβ upstream probe (0.4 kb *EcoR*I genomic fragment 20 kb away from the 5′ end of the DXβ gene) (H. Inoko et al, unpublished), and the β1 exon probe (0.95 kb *Hind*III-*Pst*I genomic fragment 15 kb downstream from the DRα gene) (5).

Figure 1 shows the results of a typical experiment using the same filter with the DRα (a), DRβ (b), and DQα (c) probes. *Sfi*I digestion was informative. Two bands of 260 kb and 180 kb were obtained with the DQα probe when this enzyme was used (Fig. 1). This probe detects both DQα and DXα due to their close homology (6). Since the smaller, 180 kb, *Sfi*I fragment hybridized to the DXβ upstream probe, the 260 kb and 180 kb fragments represent DQα and DXα, respectively. Similarly, the DOβ probe hybridized with the 180 kb fragment (data not shown) and the 260 kb DQα band was shared with one the DRβ hybridizing bands (Fig. 1C), establishing the order of the four subregions as: DO-DX-DQ-DR. *Nar*I and *Sac*II digestion produced 150 kb bands that hybridized with both the DOβ and DXβ upstream probes but not with DQα or DQβ, permitting the orientation of the DXβ and DXα loci to be determined, DXβ being

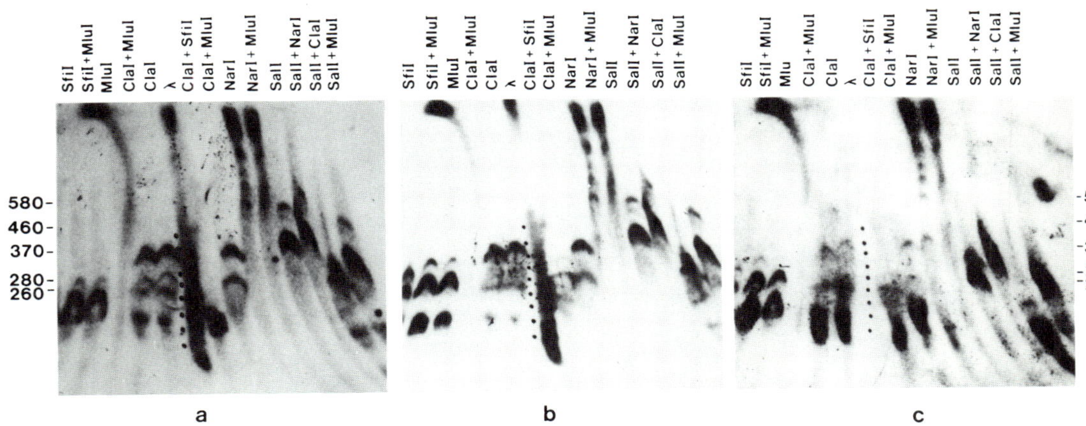

Figure 1. PFGE analysis of genomic DNA from the human B cell line MANN(DR7). The probe used were a, DRα b, DRβ c, DQα. The black dots in a-c show the positions of multimers of λCI857S7 which has a unit size of 48.5 kb. In a and b, visualization of the markers was hybridization to ^{32}P-λ DNA. Yeast chromosomes separated in the tracks at both ends of the gel provided additional markers (sizes in kb are shown). *Cla*I and *Nar*I sites around the DR and DQ subregions that are sensitive to partial cleavage were not taken into account for restriction mapping (Fig. 2). *Sal*I digestion also produced some faint bands due to partial cleavage.

Immunobiology of HLA
Volume II

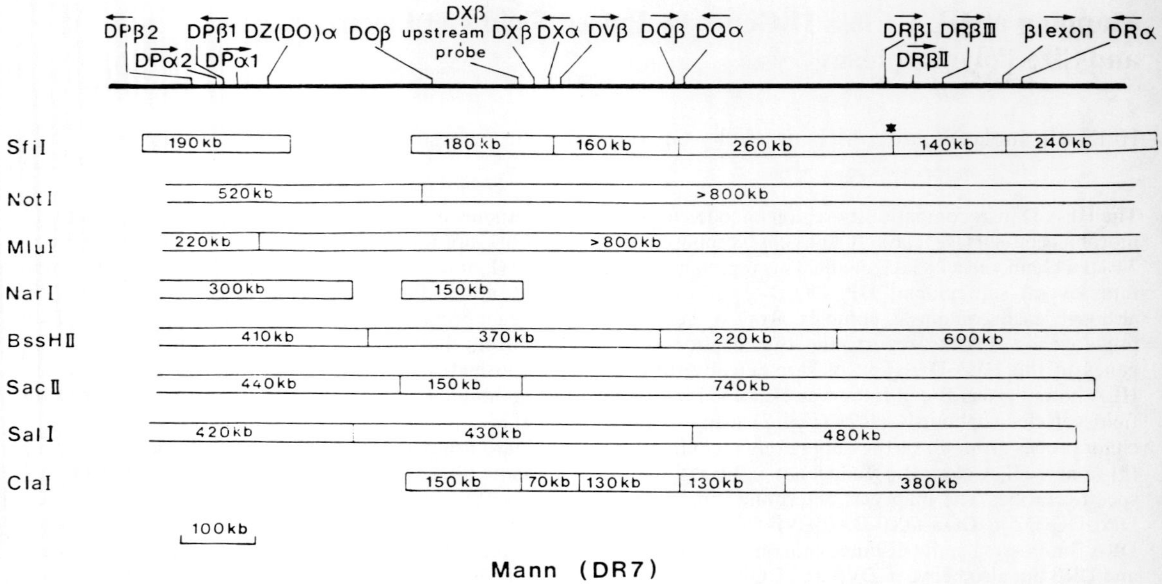

Figure 2. Long range restriction map of the HLA class II region of a DR7(MANN) haplotype. Fragments obtained with each enzyme are illustrated to scale below the class II genes with which hybridization has been demonstrated. Open spaces mean that no fragment could be identified due to lack of probes available or complex band patterns as a result from partial digestion possibly caused by methylation of the cytosine residue. Restriction sites were mapped in relation to each other by double digestion. Unambiguous restriction sites due to partial digestion are not included (e.g., *Cla*I sites within the DR subregion as shown in Fig. 1). The location of the DPα2 and DPβ2 genes and the gene order of the DRβI, DRβII, DRβIII, and β1 exon locus are assigned on the basis of gene cloning studies (8,9). In addition, the fourth gene, DRβIV, has recently been identified in DR4 haplotype (10), probably located within the DR subregion. Arrows indicate gene orientation (5′ → 3′). *restriction sites sensitive to partial cleavage.

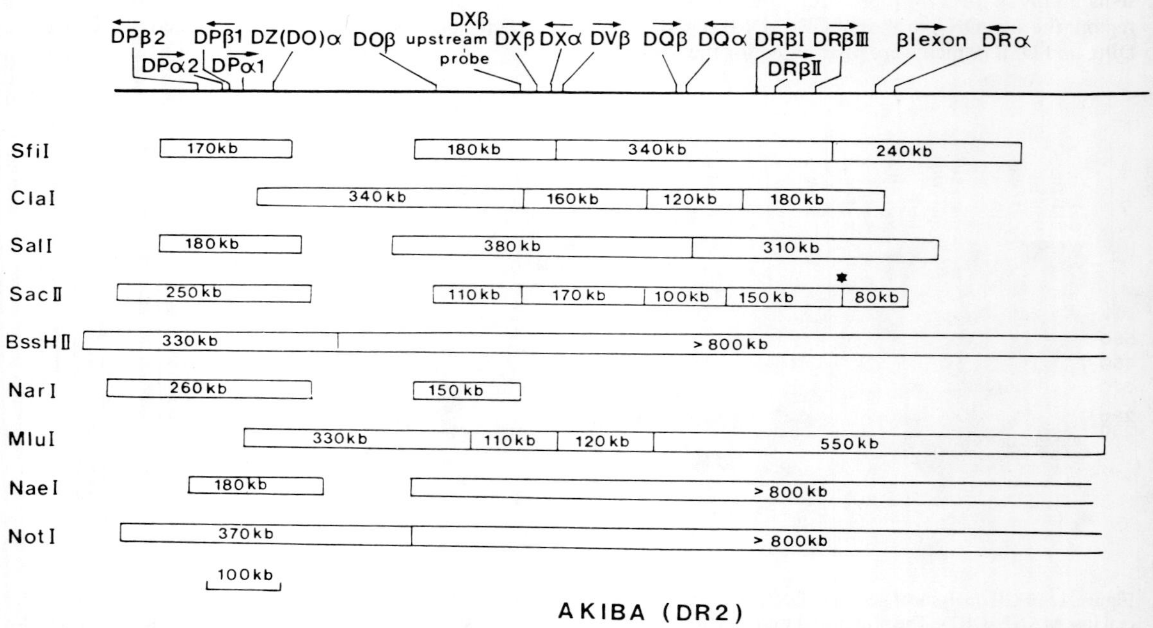

Figure 3. Long-range restriction maps of the HLA class II region of a DR2(AKIBA) haplotype. Also in DR2 haplotype, three DRβ genes (DRβI, DRβII, and DRβIII) have recently been defined by cosmid gene cloning studies (H. Inoko et al, unpublished). See the legend to Figure 2 for details.

proximal to DXα (Fig. 2) because DXα is located 15 kb upstream of DXβ, established by gene cloning (6,7).

The DQβ probe detected another distinct *Sfi*I band at 160 kb which by elimination of the fragments above must represent the DQβ gene. Since DQα and DQβ are 15 kb apart, established by gene cloning (6,7), the gene order is clearly determined to be DXβ-DXα-DQβ-DQα, and the distance between DXα and DQβ is more than 140 kb. This 160 kb *Sfi*I fragment positive with DQβ was shared only with the DVβ hybridizing band, although the DVβ probe detected the similar bands to DXα and DXβ in other cases suggesting close linkage of DVβ and the DX subregion. Based on these results, the gene order is established as: DOβ-DXβ-DXα-DVβ-DQβ-DQα. The linkage of DVβ and DXα was confirmed by cosmid cloning and the distance between them was only 15 kb (H. Inoko et al, manuscript in preparation).

In a similar manner, detailed restriction maps of MANN and AKIBA(DR2Dw12) were consistently constructed as summarized in Figures 2 and 3, respectively. The DO(DZ)α probe detected the similar bands to DPα, which is consistent with the placement of DO(DZ)α close to DPα. However, when DNA from AKIBA was used, the 340 kb *Cla*I and 330 kb *Mlu*I fragments hybridizing to DOβ also encompassed DOα, but not DPα, determining the gene order: centromere-DPβ-DPα-DO(DZ)α-DOβ. The distance between adjacent subregions are fairly large, namely about 200 kb between DO(DZ)α and DOβ, 150 kb between DOβ and DXβ, and 130 kb between DVβ and DQβ. These distances were consistent for both MANN(DR7) and AKIBA(DR2).

In contrast, it was quite surprising that the distance between DQα and DRβ was found to differ considerably in the two haplotypes, being about 80 kb in DR2 and 240 kb in DR7. Then, in order to measure the distance between the DQα and DRβ genes in other haplotypes, similar PFGE analysis was extended to samples from other B cell lines with different HLA specificities from DR1 to DRw8: IBW4(DR1), PGF(DR2Dw2), WT49(DR3), WT51(DR4), JVM(DR5), WT46(DR6), and OLL(DR8). As summarized in Table 1, DR1, DR2Dw2, DR3, DR5, DR6, and DR8 haplotypes possess the shorter genomic sequence between DQα and DRβ with the same size (80 kb) as AKIBA(DR2Dw12), whereas the DR4 haplotype contains the longer one with the same size (240 kb) as Mann(DR7). Furthermore, the distance between the DVβ and DQβ genes was also found to vary among different HLA haplotypes (Table 1). DR5 and DR6 haplotypes possess the longer genomic sequence between DVβ and DQβ with the size of 260 kb and 170 kb, respectively, whereas DR2 and DR7 haplotypes contain the shorter one with the size of 130 kb as described above.

The PFGE analysis presented here established the gene order in the HLA class II gene region as: centromere-DPβ-DPα-DO(DZ)α-DOβ-DXβ-DXα-DVβ-DQβ-DQα-(DRβI-DRβII-DRβIII)-DRα-telomere. There are big gaps or open spaces between adjacent subregions that are large enough to encode new HLA class II genes or other expressed non-HLA genes which may be responsible for the susceptibility to HLA-associated diseases. The DVβ gene is mapped very close to the 5′ end of DXα, which is consistent with the fact that genetic polymorphisms in DVβ perfectly correlate with variations at the DXα locus, but not at the DQ and DR loci by RFLP analysis (H. Inoko et al, manuscript in preparation). The DQ and DR loci are well known to be in strong linkage disequilibrium (11,12), indicating that a hot spot for recombination is present between DVβ and DQα.

Our results suggesting the size difference in the distance between the DQ-DR and DV-DQ regions are quite striking. It remains unknown whether this size difference has any biologic significance, but it is tempting to propose that the size difference in this area among HLA haplotypes is correlated with the susceptibility to various DR- and/or DQ-associated diseases. Another possibility, which may be more likely, is that these regions between DQα and DRβ and between DVβ and DQβ only represent junk or selfish-DNAs that have been accumulated as a result of gene duplication or gene conversion mechanism, which is known to operate frequently around this area (10,12,13). Anyway, this interesting observation raises questions about the genetic mechanisms that may account for such differences, and elucidation of this molecular mechanisms should provide a better understanding of the dynamics of the major histocompatibility complex.

Table 1. Distances (kb) Between the DQ-DR and DV-DQ Regions in Different HLA Haplotypes

	Distance (kb)	
HLA Haplotype	DQα–DRβ	DVβ–DQα
IBW4(DR1)	80 kb	ND
PGF(DR2Dw2)	80 kb	ND
AKIBA(DR2Dw12)	80 kb	130 kb
WT49(DR3)	80 kb	ND
WT51(DR4)	240 kb	ND
JVM(DR5)	80 kb	260 kb
WT46(DRw6)	80 kb	170 kb
MANN(DR7)	240 kb	130 kb
OLL(DRw8)	80 kb	ND

ND = Not determined.

References

1. Schwartz DC, Cantor CR. Separation of yeast chromosome-sized DNAs by pulsed field gel electrophoresis. Cell 1984;37:67.
2. Carle GF, Olsen MV. Separation of chromosomal DNA molecules from yeast by orthogonal-field-alteration gel electrophoresis. Nucl Acids Res 1984;12:5647.
3. Ando A, Inoko H, Awataguchi S, Sato T, Tsuji K. Isolation and characterization of genomic clones for new HLA class II antigen light chains. In: Aizawa M (ed): HLA in Asia-Oceania. Sapporo, Hokkaido University Press, 1986; p. 859.
4. Bernards A, Kooter JM, Michels PAM, Moberts RMP, Borst P. Pulsed field gradient electrophoresis of DNA digested in agarose allows the sizing of the large duplication

unit of a surface antigen gene in trypanosomes. Gene 1986; 42:313.
5. Meunier HF, Carson S, Bodmer WF, Trowsdale J. An isolated β1 exon next to the DRα gene in the HLA-D region. Immunogenetics 1986;23:172.
6. Jonsson AK, Hyldig-Nielsen JJ, Servenius B, Larhammar D, Anderson G, Jorgensen F, et al. Class II genes of the human major histocompatibility complex. Comparisons of the DQ and DXα and β genes. J Biol Chem 1987;262:8778.
7. Okada K, Boss JM, Prentice H, Mangler R, Auffray C, Lillie J, et al. Gene organization of DC and DX subregion of the human major histocompatibility complex. Proc Natl Acad Sci USA 1985;82:3410.
8. Spies T, Sorrentino R, Boss JM, Okada K, Strominger JL. Structural organization of the DR subregion of the human major histocompatibility complex. Proc Natl Acad Sci USA 1985;82:5165.
9. Ando A, Inoko H, Kimura M, Ogata S, Tsuji K. Isolation and allelic polymorphism of cDNA clones and genomic clones of HLA-DP heavy and light chains. Human Immunol 1986;17:355.
10. Anderson G, Larhammer D, Widmark E, Servenius B, Peterson PA, Rask L. Class II genes of the human major histocompatibility complex. Organization and evolution of the DRβ gene. J Biol Chem 1987;262:8748.
11. Bodmer WF. The HLA system 1984. In: Albert ED et al (eds): Histocompatibility Testing 1984. Heidelberg, Springer-Verlag, 1984; p. 11.
12. Trowsdale Y, Young JAT, Kelly AP, Austin PJ, Carson S, Meunier H, et al. Structure, sequence and polymorphism in the HLA-D region. Immunol Rev 1985;85:4.
13. Figueroa F, Klein J. The evolution of MHC class II genes. Immunol Today 1986; 7:78.

Author Affiliations

Hidetoshi Inoko, Department of Transplantation, School of Medicine, Tokai University, Isehara, Japan and (present address) Human Immunogenetics Laboratory, Imperial Cancer Research Fund, London, England; *Kimiyoshi Tsuji*, Department of Transplantation, School of Medicine, Tokai University, Isehara, Japan; *Vicky Groves, John Trowsdale*, Human Immunogenetics Laboratory, Imperial Cancer Research Fund, London, England

The Tumor Necrosis Factor (TNFA) and Lymphotoxin (TNFB) Genes Map Between the HLA Class III and Class I Regions

J. Ragoussis, K. Bloemer, E.H. Weiss, and A. Ziegler

The aim of this study is to map the genes for tumor necrosis factor (TNFA) and lymphotoxin (TNFB) to a distinct location in the HLA complex. TNFA and TNFB have been mapped between HLA-DR and HLA-A or centromeric to HLA-DP by analyzing HLA deletion mutant cell lines (1). Pulsed field gel electrophoresis (2) and field inversion gel electrophoresis (FIGE) (3) allowed an even more precise localization between the class III and class I regions.

We improved the validity of the physical map and determined the transcriptional orientation of the TNFA and TNFB loci as well as the HLA-B gene.

Large DNA fragments from the monosomy 6 mutant cell line BM 19.7 (3) were prepared and separated as

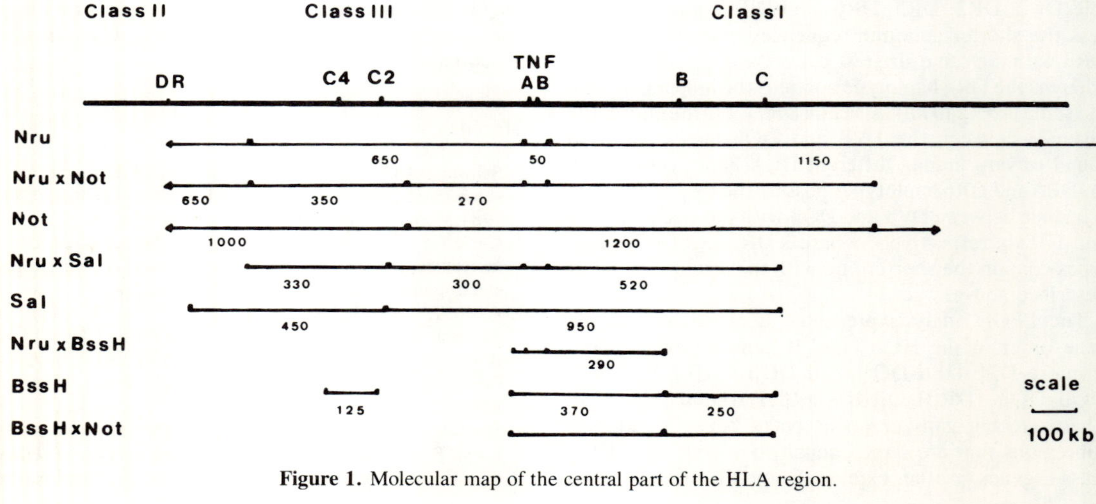

Figure 1. Molecular map of the central part of the HLA region.

Immunobiology of HLA
Volume II

described (3). After FIGE, the DNA was transferred onto nylon membranes by Southern blotting and hybridized to DNA probes from the TNFA, TNFB, Class-I, and Class-III loci. We were able to link the Class III genes to the region flanking the 3′ end of TNFA (excluding the genomic region 5′ of TNFB) in a 650 kb Nru I fragment. Both TNF genes were linked to the HLA-B gene in a 1200 kb Not I and a 950 kb Sal I fragment. A genomic probe derived 35 kb from the 5′ end of the HLA-B gene hybridized to the same 370 kb BssH II fragment as both TNF genes. The HLA-B gene and a probe derived 30 kb 3′ of it hybridized to a different *BssH*II fragment of about 270 kb length.

Double digestions established that the TNF genes are situated about 350 kb centromeric to the HLA-B gene in the A2-B13 haplotype. According to our results, the transcriptional orientations of the TNFA and TNFB genes on one hand and the HLA-B locus on the other hand are "head to head." Our results make it likely that no additional class I genes are located between HLA-B and the complement region (Fig. 1).

Acknowledgments. This work was supported by the Deutsche Forschungsgemeinschaft through SFB 120 (Project A4) and the Genzentrum München.

References

1. Spies T, Morton CC, Nedospasov SA, Fiers W, Pious D, Strominger JL. Genes for the tumor necrosis factor α and β are linked to the major histocompatibility complex. Proc Natl Acad Sci USA 1986;83:8699–8702.
2. Dunham I, Sargent CA, Trowsdale J, Campbell DR. Molecular map of the human histocompatibility complex by pulsed-field gel electrophoresis. Proc Natl Acad Sci USA 1987; 84:7237–7241.
3. Ragoussis J, Bloemer K, Weiss EH, Ziegler A. Localization of the genes for tumor necrosis factor and lymphotoxin between the HLA class I and III regions by field inversion gel electrophoresis. Immunogenetics 1988;27:66–69.

Author Affiliations

J. Ragoussis, *A. Ziegler*, Medizinische Klinik and Medizinisch-Naturwissenschaftliches Forschungszentrum, University of Tübingen, Tübingen; *K. Bloemer*, *E.H. Weiss*, Institut für Immunologie, University of München, München, FRG

A Human t-Complex-Related Gene (TCP1) Is Not Closely Linked to the HLA Complex

Andreas Ziegler, Georg Gradl, Jiannis Ragoussis, and Christa Fonatsch

The t-complex of the mouse constitutes a 20 to 30 megabasepairs long genetic region on the centromeric part of chromosome 17 in linkage with the major histocompatibility complex (MHC), the H-2 system (1,2). Due to some bizarre genetic properties, this region has attracted considerable interest, and the possible existence of a human homologue of the t-complex has been controversial (3,4,5). A mouse chromosome 17 can either carry a t haplotype, which often occurs among wild mice, or the "wild type" (+) form. Since t haplotypes are transmitted at a much higher than the expected 50% ratio from male t/+ mice to their offspring, recessive male sterility or embryonic lethality genes which are part of these haplotypes, remain in the population (6). All complete t haplotypes are distinguished from standard chromosome 17 DNA by two nonoverlapping large inversions (7–10), as a consequence of which recombination between t haplotypes and wild type DNA is suppressed. If t haplotypes existed in humans at an analogous chromosomal position, this property could contribute to linkage disequilibrium in the HLA complex. To clarify this question, we have determined the chromosomal localization of the human homologue (TCPl) (11) of the Tcp-l locus which is part of the proximal t-complex and codes for an abundant testicular germ cell protein in the mouse.

The human TCPl cDNA probe pB1.4^{hum} (12) was radiolabeled by nick-translation with all four tritiated nucleotides to a specific activity of 1.2×10^8 dpm/μg of DNA. In situ hybridization to metaphase spreads of phytohemagglutinin-stimulated human peripheral blood cells was done as described (11). Q-banded chromosomes were analyzed and the distribution of grains over chromosome bands was marked on a human chromosome idiogram at the 400-band stage.

Chromosomes could clearly be identified in 153 metaphases in which a total of 545 grains (3.56 grains per metaphase) were detected. 61 grains (11.2% of total) were found to be located on chromosome 6 (5.75% expected) with 41 grains (67.2%) in the 6q23→qter region. 26.8% of all labeled metaphases revealed grains in the region 6q23→qter with a maximum over bands 6q25→q27. Chi^2-values were highly significant with $p<0.001$. A slight overrepresentation of grains on chromosome arms 1q, 3p, 10q, 12p, 15q and 20p was observed as well (Table 1), which probably indicates sequence relationships between these chromosome arms and the pB1.4^{hum} probe. However, no overrepresenta-

Table 1. Silver Grains Observed Over Chromosome Arms and χ^2 Values

Chr. Arm	Obs.	Exp.	χ^2	Chr. Arm	Obs.	Exp.	χ^2
1p	13	14.652	0.186	13p	3	2.198	0.293
1q	31	15.568	15.298	13q	17	11.099	3.138
2p	16	10.916	2.368	14p	4	2.198	1.478
2q	18	17.766	0.003	14q	16	10.623	2.722
3p	22	11.905	8.561	15p	4	2.198	1.478
3q	19	13.407	2.334	15q	21	9.707	13.138
4p	4	6.410	0.906	16p	5	4.396	0.083
4q	21	16.117	1.479	16q	11	6.593	2.945
5p	8	5.861	0.781	17p	3	3.590	0.097
5q	22	15.201	3.041	17q	10	7.326	0.976
6p	10	7.509	0.826	18p	1	2.491	0.892
6q	51	13.370	105.911	18q	11	7.326	1.843
7p	11	6.777	2.632	19p	6	4.029	0.964
7q	17	11.172	3.040	19q	5	4.762	0.012
8p	5	5.495	0.045	20p	10	3.846	9.846
8q	16	11.538	1.725	20q	2	4.689	1.542
9p	8	5.678	0.950	21p	2	1.905	0.005
9q	15	11.172	1.312	21q	7	4.212	1.845
10p	10	5.311	4.139	22p	4	2.198	1.478
10q	19	11.355	5.147	22q	5	4.835	0.006
11p	7	6.227	0.096	Xp	2	5.750	2.446
11q	15	9.707	2.886	Xq	9	9.141	0.002
12p	10	4.762	5.762	Yp	0	0.300	0.300
12q	18	12.088	2.892	Yq	1	1.000	0.000

Sum observed: 545 grains Sum expected: 545.000 grains
Sum χ^2: 219.844 Sum of metaphases: 153

tion of grains was found on 7p, contrary to the report by Willison and coworkers (13), who otherwise confirm our observations, and show furthermore, that the only functional human TCPl gene is that mapped by us to 6q25→q27.

These results prove that TCPl is not closely linked to the human MHC, which has been mapped previously using HLA deletion mutant cell lines to 6p21.31, distal →p21.33, distal (14). Studies carried out by Blanche and coworkers (15) confirm our conclusion, because they demonstrate that TCPl cannot be part of a linkage map covering 45 cM on 6p including HLA, glyoxalase 1 (GLOl), phosphoglyceratekinase 2 (PGK2) as well as other 6p loci.

Obviously, it will be very interesting to determine the precise localization of further human homologues of t-complex genes. It seems quite possible that some of these loci might be TCPl-linked and localized on 6q, whereas others could be detected in a chromosomal location analogous to that found in the mouse on 6p near the HLA complex. There are at least two types of DNA probes which could contribute to a solution of this question: sequences from T66 elements which are part of the t-complex (16) cross-hybridize with human DNA, and it is likely that a human homologue of the Tcp-1-linked functional locus Rp17 (17) also exists. The possibility to assign genes to very large DNA fragments (see e.g., Ziegler et al, and Ragoussis et al, this volume) will undoubtedly contribute to a rapid analysis of linkage relationships between these genetic regions.

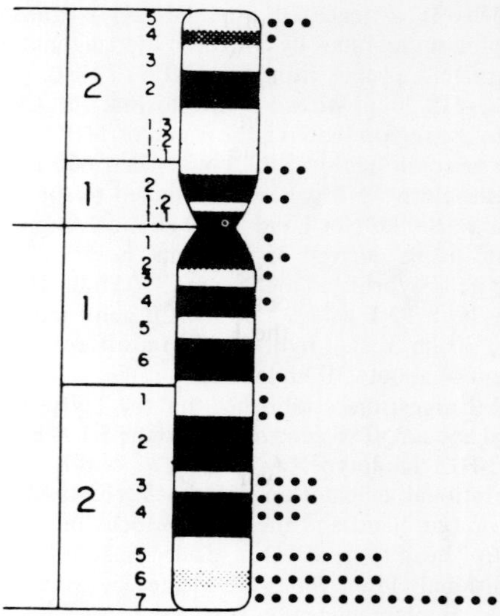

Figure 1. Idiogram of chromosome 6, showing the distribution of silver grains in 153 metaphases.

So far, at least three other genes have been located near TCPl: the locus for the estrogen receptor (ESR) maps to 6q24→q27, the glyceraldehyde-3-phosphate dehydrogenase like 8 gene (GAPDL8) to 6q25→q27, and plasminogen (PLG) to 6q26→q27 (see (18) for details).

Because of the effects which t haplotypes exert on embryonic lethality and male fertility, it may be of interest to search for comparable effects in humans as well. Chase and coworkers (19) determined that the chromosomal region 6q26→qter appears to be always involved in the trisomy 6q syndrome which leads to multiple malformations in affected individuals. Furthermore, in one case of a child suffering from this anomaly (trisomic for 6q23→qter), it was reported (20) that the father who carried a balanced reciprocal translocation involving chromosomes 6q and 3p, was oligozoospermic and subfertile. The parental history also involved one spontaneous abortion at three months gestation. Although it is possible that there is no relationship to TCPl or another 6q locus with homology to a t-complex gene in the mouse, it is intriguing that chromosomal breaks near TCPl or Tcp-l occur in both species.

Acknowledgments. This work was supported in part by the Deutsche Forschungsgemeinschaft through SFB120 (project A4) and Fo91/11-1. We thank Dr. Keith Willison for donating the pB1.4^{hum} probe. We are grateful to Ms. Ulrike Kolbus for her excellent technical assistance.

References

1. Silver L. Mouse t haplotypes. Ann Rev Genet 1985; 19: 179.
2. Rogers JH. The mouse t complex is composed of two separate inversions. Trends Genet 1986; 18:145.
3. Bobrow M, Bodmer JG, Bodmer WF, McDevitt HO, Lorber J, Swift P. The search for a human equivalent of the mouse T-locus—Negative results from a study of HLA types in spina bifida. Tissue Antigens 1975; 5:234.
4. Fellous M, Boue J, Malbrunot C, Wollman E, Sasportes M, Van Cong N, et al. A five generation family with sacral agenesis and spina bifida: Possible similarities with the mouse T-locus. Am J Med Genet 1982; 12:465.
5. Awdeh ZL, Raum D, Yunis EJ, Alper CA. Extended HLA complement allele haplotypes: Evidence for T/t-like complex in man. Proc Natl Acad Sci USA 1983; 80:259.
6. Lewontin RC. Interdeme selection controlling a polymorphism in the house mouse. Am Nat 1962; 887:65.
7. Artzt K, McCormick P, Bennett D. Gene mapping within the T/t complex of the mouse. I. T-lethal genes are nonallelic. Cell 1982; 28:463.
8. Artzt K, Shin HS, Bennett D. Gene mapping within the T/t-complex of the mouse. II. Anomalous position of the H-2 complex in t-haplotypes. Cell 1982; 28:471.
9. Herrmann B, Bucan M, Mains PE, Frischauf A-M, Silver L, Lehrach H. Genetic analysis of the proximal portion of the mouse t complex: Evidence for a second inversion within t haplotypes. Cell 1986; 44:469.
10. Sarvetnick N, Fox HS, Mann E, Mains P, Elliott RW, Silver LM. Nonhomologous pairing in mouse t haplotype heterozygotes can produce recombinant chromosomes with adjacent duplications and deletions. Genetics 1986; 113:723.
11. Fonatsch C, Gradl G, Ragoussis J, Ziegler A. Assignment of the TCP1 locus to the long arm of human chromosome 6 by in situ hybridization. Cytogenet Cell Genet 1987; 45:109.
12. Willison K, Dudley K, Spurr N, Goodfellow P. Chromosomal assignment of TCP-1 the human homologue of a mouse t-complex locus. HGM8. Cytogenet Cell Genet 1985; 40:779.
13. Willison K, Kelly A, Dudley K, Goodfellow P, Spurr N, Groves V, et al. The human homologue of the mouse t-complex gene, TCP1, is located on chromosome 6 but is not near the HLA region. Embo J 1987; 6:1967.
14. Spring B, Fonatsch C, Muller C, Pawelec G, Kompf J, Wernet P, et al. Refinement of HLA gene mapping with induced B-cell line mutants. Immunogenetics 1985; 21:227.
15. Blanche H, Massart C, Dausset J, Cann H. TCP1 is not linked to HLA, GLO1, PGK1P2 and other markers in a 45 cM map of the short arm of chromosome 6. HGM9. Cytogenet Cell Genet 1987; in press.
16. Schimenti J, Vold L, Socolow D, Silver LM. An unstable family of large DNA elements in the center of the mouse t complex. J Mol Biol 1987; 194:583.
17. Mann EA, Silver LM, Elliott RW. Genetic analysis of a mouse t complex locus that is homologous to a kidney cDNA clone. Genetics 1986; 114:993.
18. Olaisen B, Sakaguchi AY, Naylor SL. Report of the committee on the genetic constitution of chromosomes 5 and 6. HGM9. Cytogenet Cell Genet 1987; in press.
19. Chase C, Jalal SM, Martsolf JT, Wasdahl WA. Duplication 6q24–6qter in an infant from a balanced paternal translocation. Am J Med Genet 1983; 14:347.
20. Franchino CJ, Beneck D, Greco MA, Wolman SR. Partial trisomy 6q: Case report with necropsy findings. J Med Genet 1987; 24:300.

Author Affiliations

Andreas Ziegler,* *Jiannis Ragoussis*, Medizinische Klinik und Medizinisch-Naturwissenschaftliches Forschungszentrum, University of Tübingen, Tübingen; *Georg Gradl*, *Christa Fonatsch*, Institut für Humangenetik, Medizinische Universität, Lübeck, Federal Republic of Germany

---

*Present address: Institut für Experimentelle Immunologie, Philipps-Universität, Deutschhausstraße 1, D-3550 Marburg/Germany.

Section II
Class I Histocompatibility Antigens: Genes, Structure, and Allotypes

Molecular Characterization of Serologic Recognition Sites in the HLA-A2 Molecule Using Oligonucleotide-Generated Site-Specific Mutants

Jesús Santos-Aguado, Juan J. Yunis, E. Milford, Edmond J. Yunis, and Jack L. Strominger

Abstract: The specific amino acid residues involved in the serologic specificity of the HLA-A2 molecule have been investigated using a combination of oligonucleotide site-directed mutagenesis, DNA-mediated gene transfer, and immunologic techniques (IIF followed by flow cytometry and a complement-dependent cytotoxicity assay).

Human major histocompatibility class I antigens (HLA) constitute one of the most polymorphic families of molecules known to date. They serve as the major barrier to tissue transplantation by eliciting allospecific antibodies and by acting as target antigens for allospecific cytotoxic T lymphocytes (CTLs) (1). Also, they participate in the recognition and lysis of virus infected cells (2) or chemically modified target cells (3). The characterization of the antigenic determinants of HLA molecules recognized by antibodies or CTLs are important in our understanding of the molecular basis of the immune response. The aim of this study was the localization and characterization of the specific residues of the HLA-A2 molecule that are recognized by B or T cells. This paper refers to the definition of serologic epitopes. The accompanying paper in this volume (pp. 27–30) focuses on the antigenic determinants that anti-HLA-A2 allospecific CTLs recognize in the HLA-A2 molecule.

Our approach to study structure-function relationships in HLA class I MHC molecules involved the manipulation of the genomic clone that codes the HLA-A2.1 molecule using synthetic oligonucleotides (4) to introduce specific substitutions in the HLA-A2.1 gene. Once the mutant molecules were generated and characterized at the DNA level (e.g., sequencing), the effect of a particular substitution in the serologic definition of the molecule was analyzed at the surface of transfected human (rhabdomyosarcoma, RD: HLA-A1, -Bw51, -B14) or mouse fibroblast cell lines (Ltk- or LJ26: Ltk+ expressing human β2m) using a large panel of mouse HLA-specific monoclonal antibodies or human alloantisera obtained from the Tenth Histocompatibility Workshop (Antigen Society #4, HLA-A2). The transfection into mouse cells facilitated: a) the structural analysis of the mutated heavy chain in association with mouse or human β2m; and b) the analysis of the mutated molecules with some MAbs that recognize the endogenous class I genes expressed by the RD cell line. The use of this approach offered the advantage over previous exon-shuffling experiments because the only change introduced in the molecule was the amino acid(s) targeted by the mutagenesis protocol in such a way that we could study the role of a particular residue in the structure and/or reactivity of the molecule.

HLA-A2 is one of the best characterized human MHC class I antigens. The sequences of the major subtype (HLA-A2.1) (5) and several members of a related family have been obtained (6); structural analysis of immunoselected mutants (7) and CTL natural variants (A2.2(M7) (8), A2.2(WT49) (9), A2.3(DK1) (8), A2.4a(CLA) (10) and A2.4b(KNE) (11) has been used to determine residues that may be directly involved in the serologic specificity and CTL recognition of the molecule. Synthetic peptides (12) and monoclonal antibodies (13) have been used to delineate the antigenic map of the molecule and the three-dimensional structure has recently been established (14,15). Interestingly, none of the known mutants or variants displayed any alteration in the largest region of diversity in MHC class I molecules spanning residues 62 to 83 of the heavy chain. We studied the importance of this region in the serologic specificity of the molecule using several mutagenic oligonucleotides to replace systematically the HLA-A2 residues in this segment with those present at the same position in the HLA-B7 antigen. Thus, not only did we identify the loss of a given specificity but also the gain of a different one. By the sequential use of the in vitro mutagenesis protocol (4) we have been able to create an HLA-A2 molecule carrying most of the amino acid residues of HLA-B7 origin in this polymorphic segment. Additional point mutations were introduced in positions that have been previously shown of importance for the recognition of the molecule (aa 9 (10), 43 (8,9), 107 (7), 152, 156 (8,9)). The molecules created were designed according to the specific change introduced (see Table 1 and ref. 16).

Table 1. Amino Acid Substitutions in Oligonucleotide-Generated Site-Specific HLA-A2 Mutants

HLA-A2M_{9}	(F to Y)
-A2M_{43}	(Q to R)
-A2M_{6263}	(GE to RN)
-A2M_{6566}	(RK to QI)
-A2M_{65}	(R to Q)
-A2M_{66}	(K to I)
-A2M_{70}	(H to Q)
-A2M_{74}	(H to D)
-A2M_{7074}	(HH to QD)
-A2M_{6570}	(65R 66K 70H to 65Q 66I 70Q)
-A2M_{6574}	(65R 66K 70H 74H to 65Q 66I 70Q 74D)
-A2M_{7080}	(70H 74H 76V 77D 79G 80T to 70Q 74D 76E 77S 79R 80N)
-A2M_{6580}	(65R 70H 74H 76V 77D 79G 80T to 65Q 70Q 74D 76E 77S 79R 80N)
-A2M_{107}	(W to G)
-A2M_{152}	(V to E)
-A2M_{156}	(L to S)
-A2$M_{43/152}$	(QV to RE)
-A2$M_{43/156}$	(QL to RR)

Immunobiology of HLA
Volume II

The importance of the amino acid substitutions in the serologic specificity of the molecule was analyzed at the surface of transfected human or mouse cell lines by a) a large panel of mouse and rat monoclonal antibodies that recognize monomorphic and polymorphic determinants in HLA molecules (Table 2) and b) human alloantisera currently used in tissue typing obtained from the Tenth Histocompatibility Workshop.

Monoclonal antibodies were obtained from ascites fluids and were used at saturating conditions. IIF and FACS analysis was carried out as previously described (4). The results obtained in both human and mouse cells are listed in Table 3. Briefly, the results suggest that the polymorphic region is not responsible for the majority of serologic determinants (although we cannot exclude the interactions with other regions of the molecule) of HLA-A2 and HLA-B7 molecules, since amino acid substitutions in that region do not have a major effect in the recognition of the molecule by HLA-A2 or HLA-B7 specific MAbs (with the exception of MA2.1 and SFR8-B6). As expected the recognition of monomorphic MAbs (W6/32, BB7.7) or HLA-A locus (A131) MAbs is not affected by any of the mutations introduced.

At least three independent determinants have been identified in the HLA-A2 molecule [Table 3 and ref. 16]: the first one is localized in the α_1 domain between residues 62 to 66 and is defined by the -A2/-B17 crossreactive MAb MA2.1. The specific involvement of these residues in MA2.1 binding is exemplified by the mutant -A2M$_{7080}$ which shows a normal pattern of reactivity with all the MAbs including MA2.1. After the introduction of the substitution at position 65 (A2M$_{6580}$), the reactivity with MA2.1 is abolished. The second determinant (-A2/Aw69) is localized in the second domain and is defined by the MAbs BB7.2, PA2.1, and MA2.2, although they define slightly different epitopes according to their pattern of reactivity with the -A2M$_{107}$ mutant. Previous results have also suggested the importance of residue 161 in the formation of this determinant (17). A third determinant (-A2/-Aw68/-Aw69) is defined by MAbs 4B3, A2A28M1, and CR351-11 and is not affected by any of the mutations introduced. The differential reactivity of these MAbs with HLA-A2 CTL variants (18,19) and natural hybrids (6) suggest that this determinant may be in fact a mixture. None of the HLA-B7-specific MAbs recognize the new molecules with the exception of SFRB8-B6, a rat MAb that recognizes the -Bw6 supertypic determinant present in -B (and some -C) molecules. This MAb has considerable reactivity with the -A2M$_{7080}$ and -A2M$_{6580}$ molecules in mouse or human transfected cells. Thus, the mutant -A2M$_{7080}$ carries all the HLA-A2 serologic determinants plus a -Bw6 epitope.

To determine if the epitopes recognized by mouse MAbs are the same as those recognized by alloantibodies, we have analyzed the molecules transfected into the human RD cell line with alloantisera of the antigen society #4 (HLA-A2) provided by the Tenth Histocompatibility Workshop. The analysis was carried out by a modification of the standard complement-dependent cytotoxicity assay (20). The adherent RD cells were first incubated with alloantisera followed by rabbit complement (A,B,C typing). The reading was performed in an inverted fluorescent microscope with a rhodamine filter after staining the killed cells with ethidium bromide. The results are listed in Table 4. Only 14 alloantisera of the original 70 have been fully characterized with the panel of transfected cells because the majority of these antibodies (54) gave positive reactions with the RD cell line or with the HLA-B7 RD transfectant or inconsistency in their pattern of reactivity in repeated experiments. The reactivity of the alloantisera with cells expressing the natural hybrids -Aw68 and -Aw69 was also determined.

These results confirm previous data (Table 3) suggesting that residues from the highly polymorphic region *do not* have a major role in the formation of the serologic determinants of the molecule and the specific relevance of residues 65 to 66 and 107 (see reactivity of #4234, #4262, #4264 and #4223 with mutations involving residues 65 or 66 and #4234 with -A2m$_{107}$). Alloantisera 4262 and 4264 are also -A2/-B17 supertypic. One alloantisera (#4248: A2/Aw68/Aw69) had an interesting pattern of reaction since its reactivity was mostly affected by combined mutations at positions 43, 152, and 156, although these substitutions did not have a major efect in the reactivity with mouse MAbs. Interestingly, a similar pattern was obtained with some allospecific CTLs, since combined (but not individual) changes at positions 43/156 had also an important effect in the cellular recognition of the HLA-A2 molecule.

Additional work is currently in progress to analyze the mutant molecules transfected into mouse cells to identify additional informative sera among those that were toxic to untransfected RD cells. Moreover, antigen societies #15 (HLA-Bw6), #10 (HLA-B17), and HLA-B7 will be used to determine whether or not the mutants gained some epitopes detected by these alloantisera.

The results obtained with mouse MAbs and human alloantisera confirmed the presence of multiple serologic epitopes in what we know as HLA-A2 serologic specificity. The serologic epitopes identified are located on the α helix in the α_1 domain (residues 62–66 for MA2.1 and probably 79–80 for the -Bw6 determinant) and on an exposed loop (residue 107) respectively, thus appearing accessible to direct recognition by an antibody (14–15).

Based on our results and the knowledge of the tertiary structure of the HLA-A2 molecule it will be possible to design future experiments of mutagenesis that will contribute to our understanding of how B cells recognize serologic epitopes in the human HLA-A2 molecule.

Acknowledgments. We thank V. Bertolino and P. Lopez for their excellent assistance with the EPICS V analyzer. We specially thank P. Parham, A. McMichael, S.Y. Yang, S. Ferrone, and F.C. Grumet for generously providing the MAbs. The LJ26 cell line was obtained from L. Herzenberg.

This work was supported by grants A1-15669 (to J.L. Strominger; CA20531, HL29583 and the American Red Cross (to E.J. Yunis).

Table 2. Specificity, Characteristics and Reference of the Monoclonal Antibodies Used

Name	Class	Specificity	Reference
W6/32	IgG_{2a}	Monomorphic	Barnstable, Cell 14, 9 (1978)
MHM.5	IgG_1	B, C locus some A	Ellis, Hum. Immunol. 13, 13 (1985)
BB7.7	IgG_{2b}	Monomorphic	Brodsky, Immunol. Rev. 47, 3 (1979)
		Combinatorial, human	
4B3	IgG_{2a}	A2, Aw68, Aw69	Yang, Immunogen. 19, 217 (1984)
A2A28M1	IgM	A2, Aw68, Aw69	Grumet. Hum. Immunol. 5, 61 (1982)
CR11-351	IgM	A2, Aw68, Aw69	Russo, Immunogen. 18, 23 (1983)
PA2.1	IgG_1	A2, Aw69	Parham, Nature 276, 397 (1978)
MA2.2		A2, Aw69	Parham (1981)
BB7.2	IgG_{2b}	A2, Aw69	Parham, Hum. Immunol. 3, 277 (1981)
MA2.1	IgG_1	A2, B17	McMichael, Hum. Immunol. 1, 121 (1980)
A131	IgG_1	A locus	Kornbluth, J. Immunol. 134, 728 (1985)
4E	IgG_{2a}	B locus	Yang, Immunogen. 19, 217 (1984)
ME1	IgG_1	B7, B27, Bw22	Ellis, Hum. Immunol. 5, 49 (1982)
BB7.1	IgG_1	B7, Bw42	Brodsky, Immunol. Rev. 47, 3 (1979)
B27M1	IgM	B7, B27	Grumet, Lancet 174 (1981)
B27M2	IgM	B27, B47	Grumet, Hum. Immunol. 5, 61 (1982)
MB40.1	IgG_1	Broad polymorphic, B40, A28, B7 > A2	Parham, Hum. Immunol. 1, 131 (1980)
MB40.2	IgG_1	B7, Bw60, Bw41	Parham, Immunogen. 13, 509 (1981)
MB40.3	IgG_1	B7, Bw60, Bw61	Parham, Immunogen. 13, 509 (1981)
MB40.4	IgG_1	Broad polymorphic B7 not A2	Brodsky, J. Immunol. 128, 129 (1982)
MB40.7	IgG_1	Bw6 related	Parham, Hum. Immunol. 15, 44 (1986)
BB7.6	IgG_1	Bw6 related	Brodsky, Immunol. Rev. 47, 3 (1979)
SFR8-B6	Ig$G_{2b\ rat}$	Bw6	Radka, J. Immunol. 128, 2804 (1982)
MCLB.2	Ig	Broad polymorphic B7 not A2	Brodsky, J. Immunol. 128, 129 (1982)
KA28.5	IgG_1	Broad polymorphic no A2 or B7	Brodsky, J. Immunol. 128, 129 (1982)
PA2FA3	IgG_1	Broad polymorphic A2 not B7	Brodsky, J. Immunol. 128, 129 (1982)
BBM.1	IgG_{2b}	B2-microglobulin	Brodsky, Eur. J. Immunol. 9, 536 (1979)

Table 3. Analysis of Transfected Human and Mouse Cells with the Panel of HLA-Reactive Monoclonal Antibodies

	W6/32	NHM.5	BB7.7	4B3	A2A28M1	CR351-11	PA2.1	MA2.2	BB7.2	MA2.1	A131	4E	ME1	BB7.1	B27M1	B27M2	MB40.1	MB40.2	MB40.3	MB40.4	MB40.7	BB7.6	SFR8-B6	MC2	KA28.5	PA2FA3	BBM.1
HLA-A2.1	+	−	+	+	+	+	+	+	+	+	+	−	−	−	−	−	+	−	−	−	−	−	−	−	−	+	+
-B7	+	+	+	−	−	−	−	−	−	−	−	+	+	+	+	−	+	+	+	+	+	+	+	+	−	−	+
-A2M(9)	+	−	+	+	+	+	+	+	+	+	+	−	−	−	−	−	+	−	−	−	−	−	−	−	−	+	+
-A2M(43)	+	−	+	+	+	+	+	+	+	+	+	−	−	−	−	−	+	−	−	−	−	−	−	−	−	+	+
-A2M(6263)	+	−	+	+	+	+	+	+	+	−	+	−	−	−	−	−	+	−	−	−	−	−	−	−	−	+	+
-A2M(6566)	+	−	+	+	+	+	+	+	+	−	+	−	−	−	−	−	+	−	−	−	−	−	−	−	−	+	+
-A2M(65)	+	−	+	+	+	+	+	+	+	−	+	−	−	−	−	−	+	−	−	−	−	−	−	−	−	+	+
-A2M(66)	+	−	+	+	+	+	+	+	+	−	+	−	−	−	−	−	+	−	−	−	−	−	−	−	−	+	+
-A2M(70)	+	−	+	+	+	+	+	+	+	+	+	−	−	−	−	−	+	−	−	−	−	−	−	−	−	+	+
-A2M(74)	+	−	+	+	+	+	+	+	+	+	+	−	−	−	−	−	+	−	−	−	−	−	−	−	−	+	+
-A2M(7074)	+	−	+	+	+	+	+	+	+	+	+	−	−	−	−	−	+	−	−	−	−	−	−	−	−	+	+
-A2M(6570)	+	−	+	+	+	+	+	+	+	−	+	−	−	−	−	−	+	−	−	−	−	−	−	−	−	+	+
-A2M(6574)	+	−	+	+	+	+	+	+	+	−	+	−	−	−	−	−	+	−	−	−	−	−	−	−	−	+	+
-A2M(7080)	+	−	+	+	+	+	+	+	+	+	+	−	−	−	−	−	+	−	−	−	−	−	+	−	−	+	+
-A2M(6580)	+	−	+	+	+	+	+	+	+	−	+	−	−	−	−	−	+	−	−	−	−	−	+	−	−	+	+
-A2M(107)	+	−	+	+	+	+	−	−	−	+	+	−	−	−	−	−	+	−	−	−	−	−	−	−	−	+	+
-A2M(152)	+	−	+	+	+	+	+	+	+	+	+	−	−	−	−	−	+	−	−	−	−	−	−	−	−	+	+
-A2M(156)	+	−	+	+	+	+	+	+	+	+	+	−	−	−	−	−	+	−	−	−	−	−	−	−	−	+	+
-A2M(43152)	+	−	+	+	+	+	+	+	+	+	+	−	−	−	−	−	+	−	−	−	−	−	−	−	−	+	+
-A2M(43156)	+	−	+	+	+	+	+	+	+	+	+	−	−	−	−	−	+	−	−	−	−	−	−	−	−	+	+

Table 4. Analysis of Human Transfected Cell Line With Alloantisera (Antigen Society #4, HLA-A2)

70 Sera
14 Informative Sera
56 Endogenous Cross-reactivity

	NEG	4205	4206	4225	4234	4235	4236	4240	4247	4248	4256	4262	4264	4223	4265	POS
RD	1	1	1	1	1	1	1	1	1	1	1	1	1	1	1	8
A2	1	8	8	8	8	8	8	8	8	8	8	8	8	8	8	8
B7	1	1	1	1	1	1	1	1	1	1	1	1	1	1	1	8
9	1	8	8	8	8	8	8	8	8	8	8	8	8	8	8	8
43	1	8	8	8	8	8	8	8	8	4	8	8	8	8	8	8
43/152	1	8	8	8	8	8	8	8	8	1	8	8	8	8	8	8
43/156	1	8	8	8	8	8	8	8	8	1	8	8	8	8	8	8
152	1	8	8	8	8	8	8	8	8	4	8	8	8	8	8	8
156	1	8	8	8	8	8	8	8	8	8	8	8	8	8	8	8
107	1	8	8	8	1	8	8	8	8	8	8	8	8	8	8	8
65-80	1	8	8	8	1	8	8	8	8	8	8	8	8	1	8	8
70-80	1	8	8	8	8	8	8	8	8	8	8	8	8	8	8	8
65-74	1	8	8	8	8	8	8	8	8	8	8	8	4	8	8	8
70	1	8	8	8	8	8	8	8	8	8	8	8	8	8	8	8
74	1	8	8	8	8	8	8	8	8	8	8	8	8	8	8	8
70-74	1	8	8	8	8	8	8	8	8	8	8	8	8	8	8	8
65-66	1	8	8	8	8	8	8	8	8	8	8	8	2/4	1	8	8
65	1	8	8	8	4	8	8	8	8	8	8	4	1	1	8	8
66	1	8	8	8	1	8	8	8	8	*	8	4	2/4	1	8	8
62-63	1	1	1	1	8	1	8	8	0	8	1	1	1	1	8	8

*Artifact.

SCORE:			
0 = Not Valid		4 = 40–60% dead cells	Intermediate reaction
1 = 0–20% dead cells.	Negative reaction	6 = 60–80% dead cells	Positive reaction
2 = 20–40% dead cells	Negative reaction	8 = 80–100% dead cells	Positive reaction

References

1. Zinkernagel R, Doherty PC. MHC-restricted cytotoxic T cells: Studies on the biological role of polymorphic major transplantation antigens determining T-cell restriction, specificity, function and responsiveness. Adv Immunol 1979;27:52.
2. McMichael AJ, Ting A, Zweerink HJ, Askonas BA. HLA-restriction of cell mediated lysis of influenza virus infected human cells. Nature 1977;270:524.
3. Shearer GM. Cell-mediated cytotoxicity to trinitrophenyl-modified syngeneic lymphocytes. Eur J Immunol 1974;4:527.
4. Santos-Aguado J, Biro PA, Furhmann U, Strominger JL, Barbosa JA. Amino acid sequences in the α_1 domain and not glycosylation are important for HLA-A2/β2m association and cell surface expression. Mol Cell Biol 1987;7:982.
5. Koller BH, Orr HT. Cloning and complete sequence of an HLA-A2 gene: Analysis of two HLA-A alleles at the nucleotide level. J Immunol 1985;134:2727.
6. Holmes N, Parham P. Exon shuffling "in vivo" can generate novel HLA class I molecules. Embo J 1985;4:2849.
7. Krangel MS, Taketani S, Pious D, Strominger JL. HLA-A2 mutants immunoselected in vitro: definition of residues contributing to an HLA-A2-specific serological determinant. J Exp Med 1983;157:324.
8. Mattson DH, Handy DE, Bradley DA, Coligan JE, Cowan EP, Biddison WE. DNA sequences of the genes that encode the CTL-define HLA-A2 variants M7 and DKl. Immunogenetics 1987;26:190.
9. Holmes N, Ennis P, Wan AM, Denney DW, Parham P. Multiple genetic mechanisms have contributed to the generation of the HLA-A2/A28 family of class I MHC molecules. J Immunol 1987;139:936.
10. Ezquerra A, Domenech N, van der Poel JJ, Strominger JL, Vega MA, Lopez de Castro JL. Molecular analysis of an HLA-A2 functional variant CLA defined by cytolytic T lymphocytes. J Immunol 1986;137:1642.
11. Domenech N, Ezquerra A, Castaño R, Lopez de Castro JA. Structural analysis of an HLA-A2 functional variant KNE defined by cytolytic T lymphocytes. Implications for the mapping of HLA-A2-specific T-cell epitopes. Immunogenetics 1988;27:196.
12. Ways JP, Rothbard JB, Parham P. Amino acid residues 56 to 69 of HLA-A2 specify an antigenic determinant shared by HLA-A2 and HLA-B17. J Immunol 1986;137:218.
13. Ways JP, Parham P. The antigenic structure of HLA-A2: An analysis with competitive binding assays and monoclonal antibodies. J Immunol 1983;131:856.
14. Borkjman PJ, Saper MA, Samraoui B, Bennett WS, Strominger JL, Willey DC. Structure of the human class I histocompatibility antigen HLA-A2. Nature 1987;329:506.
15. Borkjman PJ, Saper MA, Samraoui B, Bennett WS, Strominger JL, Willey DC. Structural identification of the foreign antigen binding site and T cell recognition regions of class I histocompatibility antigens. Nature 1987;329:512.
16. Santos-Aguado J, Barbosa JA, Biro PA, Strominger JL. Molecular characterization of serologic recognition sites in the human HLA-A2 molecule. J. Immunol. (in press)
17. Taketani S, Krangel MS, Pious D, Strominger JL. Structural analysis of HLA-A2 antigen from immunoselected mutant 8.6.1.: Further definition of an HLA-A2-specific serological determinant. J Immunol 1983;131:2935.

18. Russo C, Ng A-K, Pellegrino MA, Ferrone S. The monoclonal antibody CR11-351 discriminate HLA-A2 variants identified by T cells. Immunogenetics 1983;18:23.
19. Russo C, Pellegrino MA, Ferrone S. Analysis of the repertoire of HLA antibodies with anti-idiotypes to a murine anti-HLA-A2-A28 monoclonal antibody. Hum Immunol 1984;10:57.
20. Yunis JJ, Santos-Aguado J, Yunis EJ. Unpublished.

Author Affiliations

Jesús Santos-Aguado, Jack L. Strominger, Divisions of Tumor Virology; *Juan J. Yunis, Edmond J. Yunis*, Division of Immunogenetics, Dana-Farber Cancer Institute; *E. Milford*, Tissue Typing Laboratory, Brigham & Women's Hospital, Harvard Medical School, Boston, MA, USA

Molecular Characterization of Allospecific Cytotoxic T-Lymphocyte Recognition Sites in the HLA-A2 Molecule Using Oligonucleotide-Generated Site-Specific Mutants

Jesús Santos-Aguado, Mary A.V. Crimmins, Steven J. Mentzer, Jack L. Strominger, and Steven J. Burakoff

Abstract: The complexity and fine specificity of the allospecific T-cell response generated against the human HLA-A2.1 molecule and the characterization of the antigenic determinants that anti-HLA-A2 alloreactive CTLs recognize in the molecule has been analyzed using as targets, cell lines expressing HLA-A2 CTL-variants and the human rhabdomyosarcoma cell line (RD) transfected and expressing HLA-A2 mutants obtained by site-directed mutagenesis.

T lymphocytes mediate the cellular response of the immune system. Cytotoxic T lymphocytes recognize and lyse cells expressing foreign antigens in the context of self major histocompatibility (MHC) molecules (MHC restriction) (1). Additionally, allospecific CTLs efficiently lyse cells expressing foreign histocompatibility antigens (allorecognition). In order to completely understand how T cells perform their immune function in the organism, first the regions in contact between the T-cell receptor and the antigen must be characterized precisely and second, the biophysical characteristics of this interaction should be determined (2). In this report we investigate the molecular characteristics of the antigenic determinants that anti-HLA-A2 allospecific CTLs recognize in the human HLA-A2 molecule.

The study of the HLA-A2 specificity has the potential advantage of having the sequence of the major subtype and several natural hybrids already determined (3,4). Furthermore, cells carrying those molecules have been used as estimulators and targets in the generation of human CTL clones (5). Cells expressing serologically defined HLA-A2 molecules have been distinguished in viral (influenza (6) or Epstein-Barr (7)) and minor histocompatibility (H-Y and HA (8)) systems by HLA-A2 restricted CTLs. Additionally, the use of anti-HLA-A2 allorestricted CTLs (9), biochemical criteria (IEF) (10), and finally, protein (11–14) or DNA sequencing (15,16) analysis of their isolated heavy chains permits definition of the polymorphism within the serologically defined HLA-A2 molecule. Since some of these CTL variants carry multiple amino acid differences with respect to the HLA-A2.1 molecule, the use of mutants generated by site-directed mutagenesis permits to directly address the contribution that individual positions may have to the cellular recognition of the molecule (e.g., 9, 43, 152, and 156).

Peripheral blood mononuclear cells from donor MC (HLA-A3, -Aw30; -B7, -B13; Cw6; DR4) purified on a Ficoll-Hypaque gradient and the irradiated (10.000R) Epstein-Barr transformed lymphoblastoid cell line JY (HLA-A2.1, -B7; DR4,6) were used as responder and stimulator cells respectively for primary and secondary MLC as previously described (17). CTLs cultures were maintained in RPMI 1640 medium (M.A. Bioproducts, Bethesda, MD) supplemented with 10% heat-inactivated foetal calf serum (M.A. Bioproducts), 2mM glutamine, and 10% HYCLONE. The primary culture was maintained with weekly stimulation of JY irradiated cells for five to six weeks and the cells cloned by limiting dilution at less than one cell per well in flat-bottom 96-well microtiter plates after being separated on a Ficoll-Hypaque gradient. CTL clones were maintained in human-conditioned medium containing 10% interleukin 2 (IL-2). Clones were expanded in 16-mm wells and their specificity and cytolytic activity was initially characterized with the cell lines: JY, MICH (HLA-A2.1, 32;B27,15;C2;DR5) and PFG (HLA-A3;B7;DR3) (Table 1-A). Of the original 220 clones obtained, 174 were anti-HLA-A2, 7 anti-HLA-B7, and 32 showed no cytolytic activity. Lytic activity was assessed in standard 4-hour ^{51}Cr-release assay (18). Fifty-six of these anti-HLA-A2 clones were tested against several cell lines expressing previously characterized HLA-A2 CTL variants: M7(HLA-A2.2) (15), DK1(HLA-A2.3) (15), CLA(HLA-A2.4a) (13), and KNE(HLA-A2.4b) (14). Four different reaction patterns were obtained (Table 1-B): 1) CTL clones that recognize all the lymphoblastoid cells lines (9); 2) CTL clones that *do not* recognize the CLA cell line expressing the A2.4a variant (8); 3) CTL clones that *do*

Immunobiology of HLA
Volume II

Table 1. Specificity of Human CTL Clones on Cells Expressing HLA-A2, -B7, and HLA-A2 CTL Variants

A. Reactivity of CTL Clones With -A2 or -B7 Typed Cell Lines

Clones Characterized	220
Target Cell Line	Number of + Clones
. JY (A2.1, B7) +	181
. MICH (A2.1,32, B27,15) +	174
. PGF (A3, B7) +	7
. NON CYTOTOXIC	39

B. Characterization With HLA-A2 CTL Variants

Clones Characterized	56

C. Fine Specificity of Human CTL Clones on Lymphoblastoid and Transfected RD Cell Lines

No. of Clones	JY (A2.1)	M7 (A2.2)	DK1 (A2.3)	KNE (A2.4)	CLA (A2.4)
10	+	+	−	+	+
29	+	+	−	+	−
8	+	+	+	+	−
9	+	+	+	+	+

Specific cytotoxicity was calculated as percent cytotoxicity = 100 × (cpm experimental release − cpm spontaneous release)/ (total cpm − cpm spontaneous release). + = same percentage of specific ^{51}Cr release as RD-HLA-A2 transfectant; ± = half; − = same as mock transfectant. * Indicates that they were induced with gamma interferon due to problems in surface expression.

not recognize the cell lines DK1 (A2.3) and CLA (29) and 4) clones that *do not* recognize the A2.3 molecule (DK1) but have a normal pattern of reactivity with the rest of the CTL-variant cell lines (10). The HLA specificity of these clones was further confirmed by cytotoxicity assays with the RD cell line transfected with the HLA-A2.1 and -B7 genes; percentage of nonspecific lysis was also determined in a mock transfectant. Depending upon the specific CTL clone used this percentage was 5 to 10%. Their reactivity with the natural hybrids -Aw68

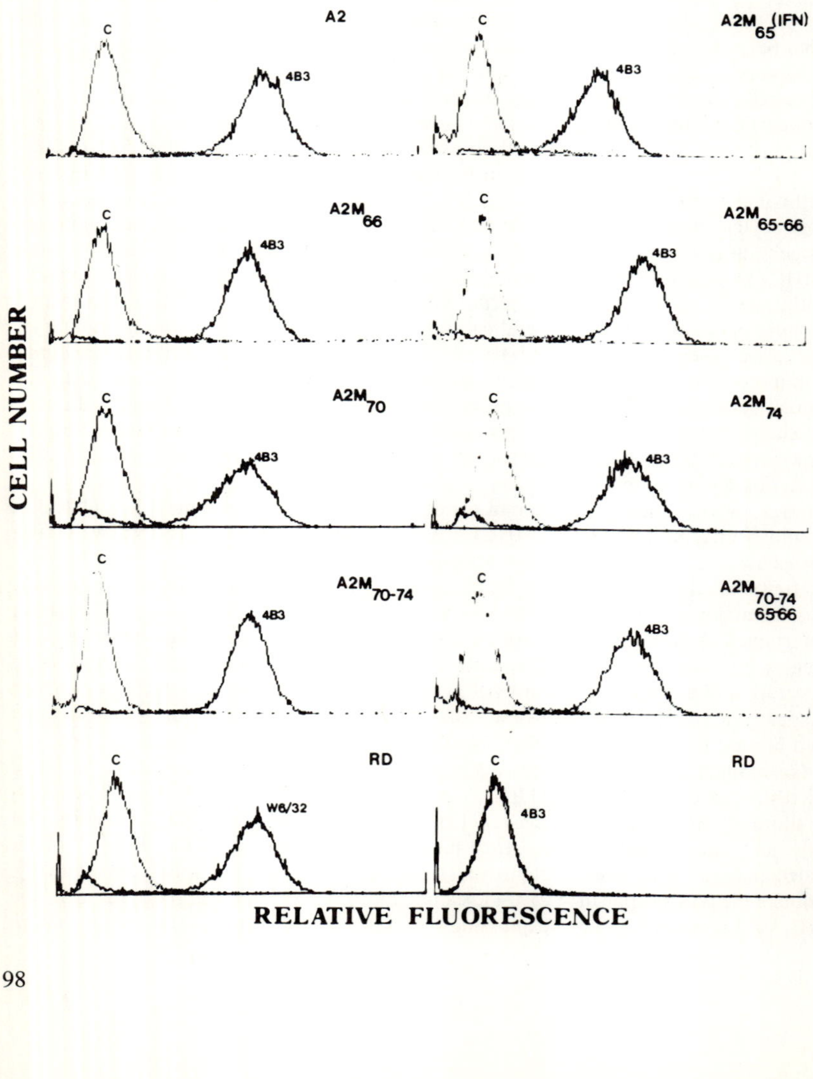

Figure 1. The histograms represent the similar levels of surface expression of the oligonucleotide-generated site-specific HLA-A2 mutants after sterile cell-sorter analysis. Their levels are comparable to the endogeneous MHC class I molecules as determined by binding with W6/32.

Table 2.

	JY(A2,B7)	MICH(A2)	PGF(B7)	M7(A2.2)	DKI(A2.3)	KNE(A2.4)	CLA(A2.4)	LB(Aw68)	IDF(Aw69)	RD control	RD-A2	-B7	-M9	-M43	-M6263*	-M65*	-M66	-M6566	-M70	-M74	-M7074	-M6574	-M7080	-M6580	-M107	-M152	-M43/152	-M156	-M43/156
w1	+	+	-	+	+	+	-	+	-	-	+	-		+		-		-	+/-	+/-	-	-	-	-	+		-		
w2	+	+	-	+	+	+	-	-	-	-	+	-	-	+	-	+/-	-	+/-	+/-	+	-	-	-	-	+	+	+	+	-
w41	+	+	-	+	+	+	+	+	-	-	+	-	+	+	-	-	-	-	-	-	-	-	-	-	+	+	+	+	-
w74	+	+	-	+	+	+	+	+	-	-	+	-	+	+	-	-	-	-	+	-+	+	-	+	-	+	+	+	+	-
w42	+	+	-	+	-	+	+	-	-	-	+	-	+	+	-	-	-	-	+	+	+	-	+/-	+/-	+/-	-	-	-	-
w44	+	+	-	+	-	+	+	+	+	-	+	-	+	+	-	-	+	+/-	+	+	+	+/-	+/-	-	+	+	+	+	+
w46	+	+	-	+	-	+	-	-	-	-	+	-	-	+			+/-	+/-	+	+	+	-	-	-	+	-	-	+	-
w128	+	+	-	+	-	+	-	-	-	-	+	-	-	+	-	+	-	-	+	+	+	-	+	+	+	+	+	-	-
w169	+	+	-	+	-	+	-	+	-	-	+	-	-	+	-	-	+/-	+/-	+	+	-	-	-	-	+	-	-	+	+
w193	+	+	-	+	-	+			-	-	+	-	+	+	-	-		-			+	-	-		+	+/-	+/-	+/-	+/-
w195	+	+	-	+	-	+			-	-	+	-	+	+	-	+		+			+	-	-		+	-	-	+	+/-
w219	+	+	-	+	-	+			-	-	+	-	+	+	-	+/-		-			+	-	+/-		+	-	-	+	+
w164	+	+	-	+	-	+			-	-	+	-	+	+				+			+	+	+		+	-	-	+	+

and -Aw69 (4) was also determined. Three different effector to target ratios (6, 2, 0,7: 1) were normally used in every experiment for all clones. All the effector cells have homogeneous CD3+, CD8+, and CD4− phenotype as determined by IIF and FACS analysis (data not shown).

The fine specificity of some of these clones was further determined using the HLA-A2 site-specific mutants generated with synthetic oligonucleotides in double-stranded vectors (19) and transfected into the human RD cell line. The mutants were designed according to the specific amino acid(s) changed (ref. 20 and table I on page 23). Once the presence of the particular mutation was confirmed, the mutant plasmid DNA was then used as a template for additional mutagenesis experiments in order to create the different combinations (20). The transfected cell lines were then selected for similar levels of expression of the mutated molecules using the sort capabilities of an Epics V analyzer under sterile conditions (Fig. 1).

The results shown in Table 2 indicate the complexity of the clonal T-cell response. None of the clones analyzed in detail so far display the same pattern of reactivity with cells expressing CTL variants, natural hybrids (Aw68, Aw69), and the oligonucleotide site-generated mutants. This suggests that multiple and different epitopes may be seen by allospecific CTLs in the human HLA-A2 molecule. Two of the changes introduced (positions 43 and 107) did not have a major effect on the recognition of the molecule by the CTL clones analyzed (28 and 15 clones respectively). The introduction of the change at position 156 in -A2M$_{43}$ (-A2M$_{43/156}$) had nevertheless a major effect in the recognition of the molecule (see Table 2, clones w2, w41, w74, w44, w46). Interestingly, the majority of the substitutions in the highly polymorphic region spanning residues 62 to 83 had a drastic effect in the recognition of the molecule by anti-HLA-A2 allospecific CTLs. Different results have been obtained when these mutants were used to analyze the effect of the substitutions in the serologic specificity of the HLA-A2 molecule (20).

CTL clones that display the same pattern of reactivity with CTL variants or natural hybrids (e.g., clones w41 and w74 or w46 and w128) can be readily distinguished for their reactivity with the site-specific mutants created (see Table 2). Thus, although the reactivity of clones w41 and w74 is affected by substitutions in the α_1 domain (not α_2) of the HLA-A2.1 molecule, changes in every position from residue 62 to 83 have a drastic effect in the reactivity of w41, while w74 is only affected by substitutions at positions 6566 (even -A2M$_{7080}$ is recognized). A similar phenomenon occurred with clones w46 and w128; while residues 70 to 80 and 152 are important in the recognition of w46, the reactivity of the w128 clone is only affected by puntual changes at positions 66 and 156 (even the mutant -A2M$_{6580}$ in which residue 66 is unchanged is recognized as the original -A2.1 molecule). In addition, inhibition of cytolysis of some of the CTL clones using MAbs against monomorphic determinants of class I (W6/32) or class II (LB3.1) molecules, HLA-A2 (4B3, MA2.1, PA2.1), -B7 (ME1), CD3 (OKT3), CD4 (OKT4), CD8 (OKT8) and lymphocyte function-associated antigens LFA-1 (TS1/18, TS1/22), LFA-2(TS2/18) and LFA-3(TS2/9) was also performed in the JY cell line. The killing of the clones was inhibited by HLA class I monomorphic or HLA-A2 specific MAbs but not by HLA class II monomorphic or HLA-B7 specific MAbs. Additionally, MAb to LFA-1, LFA-2, LFA-3, CD3, CD8 but not CD4 were able to inhibit the CTL clones (data not shown). The percentage of inhibition varied according with the clone being tested. This, as has been previously suggested (21), may reflect the high/low avidity of the particular CTL clone.

Since the X-ray crystallographic structure of the HLA-A2 molecule has been recently determined (22,23), we can locate the particular substitution in the structure and interpret the results obtained when we introduce a particular change. According to their location, the mutated residues can be classified in five categories: 1) out of the binding site (43 and 107), 2) facing into the site (eg63, 66, 70, 74, 77, 80, 152, and 156), 3) facing top of the binding pocket (62, 65, and 76), 4) in the floor (9) and 5) pointing away from the site (79). In summary, the reactivity of all the clones is affected by substitutions in the α helix localized in the α_1 domain and residues 65, 66 and 152 seem to be critical for the cellular recognition of the molecule (24). CTL clones w2, w41, w74 and w44 recognize determinants mainly in the α_1 domain since mutations at positions 152, 156, and 43/152 do not alter their lytic activity. A different result was obtained when residues 43 and 156 combined are substituted—this may reflect the effect of a conformational change created in the molecule (24).

The results presented in this study together with the analysis of the transfected cell lines in virus-restricted systems (see A. McMichael, et al., this volume) will contribute to elucidate the cellular and molecular mechanisms implicated in the recognition and elimination of foreign antigens by T lymphocytes.

Acknowledgments. We thank V. Bertolino and P. Lopez for their excellent assistance with the EPICS V analyzer. This work was supported by NIH grants to J.L. Strominger and S.J. Burakoff.

References

1. Zinkernagel R, Doherty PC. MCH-restricted cytotoxic T cells: Studies on the biological role of polymorphic major transplantation antigens determining T-cell restriction, specificity, function and responsiveness. Adv Immunol 1979;27:52.
2. Sung K-LP, Sung LA, Crimmins MAV, Burakoff SJ, Chien S. Determination of junction avidity of cytolytic T cell and target cell. Science 1986;234:1405.
3. Koller BH, Orr HT. Cloning and complete sequence of an HLA-A2 gene: Analysis of two HLA-A alleles at the nucleotide level. J Immunol 1985;134:2727.
4. Holmes N, Parham P. Exon shuffling "in vivo" can generate novel HLA class I molecules. EMBO J 1985;4:2849.
5. Clayberger C, Holmes N, Wang PL, Koller TD, Parham P, Krensky AM. Determinants recognized by human cytotoxic T cells on a natural hybrid class I HLA molecule. J Exp Med 1985;162:1709.
6. Biddison WE, Ward FE, Shearer GM, Shaw S. The self determinants recognized by human virus-immune T cells can be distinguished from the serologically defined HLA antigens. J Immunol 1980;124:548.
7. Gaston JSH, Rickinson AB, Epstein MA. Epstein-Barr virus specific cytotoxic T lymphocytes as probes for HLA polymorphisms. Heterogeneity of T cell restricting determinants associated with the serologically defined HLA-A2 antigens. J Exp Med 1983;158:280.
8. Goulmy E, van der Poel JJ, Giphart M, van Rood JJ. Analysis of the functional epitopes on different HLA-A2 molecules. Immunogenetics 1984;20:13.
9. van der Poel JJ, Pool J, Goulmy E, van Rood JJ. Differential recognition of the serologically defined HLA-A2 antigen by allogeneic cytotoxic T cells. Immunogenetics 1983;17:599.
10. van der Poel JJ, Molders H, Thompson A, Ploegh HL. Definition of four HLA-A2 subtypes by CML typing and biochemical analysis. Immunogenetics 1983;17:609.
11. Krangel MS, Taketani S, Biddison WE, Strong DM, Strominger JL. Comparative structural analysis of HLA-A2 antigens distinguishable by cytotoxic T lymphocytes: Variants M7 and DR1. Biochemistry 1982;21:6313.
12. Krangel MS, Biddison WE, Strominger JL. Comparative structural analysis of HLA-A2 antigens distinguishable by cytotoxic T lymphocytes. II. Variant DK1: Evidence for a discrete CTL recognition region. J Immunol 1983;130: 1856.
13. Ezquerra A, Domenech N, van der Poel JJ, Strominger JL, Vega MA, Lopez de Castro JA. Molecular analysis of an HLA-A2 functional variant CLA defined by cytolytic T lymphocytes. J Immunol 1986;137:1642.
14. Domenech N, Ezquerra A, Castaño R, Lopez de Castro JA. Structural analysis of an HLA-A2 functional variant KNE defined by cytolytic T lymphocytes. Implications for the mapping of HLA-A2-specific T-cell epitopes. Immunogenetics 1988;27:196.
15. Mattson DH, Handy DE, Bradley DA, Coligan JE, Cowan EP, Biddison WE. DNA sequences of the genes that encode the CTL-define HLA-A2 variants M7 and DK1. Immunogenetics 1987;26:190.
16. Holmes N, Ennis P, Wan AM, Denney DW, Parham P. Multiple genetic mechanisms have contributed to the generation of the HLA-A2/A28 family of class I MHC molecules. J Immunol 1987;139:936.
17. Krensky AM, Reiss CS, Mier JW, Strominger JL, Burakoff SJ. Long-term human cytolytic T cell lines allospecific for HLA-DR6 antigen are OKT4+. Proc Natl Acad Sci USA 1982;79:2365.
18. Krensky AM, Clayberger C, Reiss CS, Strominger JL, Burakoff SJ. Specificity of OKT4+ cytotoxic T lymphocyte clones. J Immunol 1982;129:2001.
19. Santos-Aguado J, Biro PA, Fuhrmann U, Strominger JL, Barbosa JA. Amino acid sequences in the α_1 domain and not glycosylation are important for HLA-A2/β2m association and cell surface expression. Mol Cell Biol 1987;7:982.
20. Santos-Aguado J, Barbosa JA, Biro PA, Strominger JL. Molecular characterization of serologic recognition sites in the human HLA-A2 molecule. J Immunol 1988. (in press).
21. Mentzer SJ, Barbosa JA, Strominger JL, Biro PA, Burakoff SJ. Species-restricted recognition of transfected HLA-A2 and HLA-B7 by human CTL clones. J Immunol 1986;137:408.
22. Borkjman PJ, Saper MA, Samraoui B, Bennett WS, Strominger JL, Willey DC. Structure of the human class I histocompatibility antigen HLA-A2. Nature 1987;329:506.
23. Borkjman PJ, Saper MA, Samraoui B, Bennett WS, Strominger JL, Willey DC. Structural identification of the foreign antigen binding site and T cell recognition regions of class I histocompatibility antigens. Nature 1987;329: 512.
24. Santos-Aguado J, Crimmins MAV, Mentzer SJ, Strominger JL, Burakoff SJ. Molecular characterization and fine specificity of anti-HLA-A2 allospecific CTL response. Manuscript in preparation.

Author Affiliations

Jesús Santos-Aguado, Jack L. Strominger, Division of Tumor Virology; *Mary A.V. Crimmins, Steven J. Mentzer, Steven J. Burakoff*, Division of Pediatric Oncology, Dana-Farber Cancer Institute, Harvard Medical School, Boston, MA, USA

Mapping of Serologic and CTL-Defined Epitopes on HLA-A2 by Site-Directed Mutagenesis

Kevin T. Hogan, John P. Ridge, Scott F. Walk, Peter Parham, and Victor H. Engelhard

Site-directed mutagenesis of HLA-A2.1 has been used to identify the amino acid substitutions in HLA-A2.3 that are responsible for the lack of recognition of the latter molecule by the A2/A28 specific antibody, CR11-351 (1) and by HLA-A2.1 specific CTL (2–4). Analysis of the mutated molecules revealed that the epitope recognized by the CR11-351 antibody was completely lost when Thr was substituted for Ala_{149}. In contrast, substitutions of Glu for Val_{152} or Trp for Leu_{156} had no effect on the reactivity of this antibody, either singly or in combination. These findings implicate Ala_{149} as being critical in the formation of the epitope recognized by CR11-351. In contrast, murine HLA-A2.1 specific CTL clones were found to be sensitive to changes at amino acids 152 and 156, but were unaffected by substitution at amino acid 149. Through the use of hemi-exon shuffling between HLA-A2.1 and HLA-A2.2, the role of substitutions at amino acids 9, 43, and 95 in the formation of CTL-defined epitopes was also studied. The results indicate that the substitution of Trp for Leu_{156} can account for the loss of all CTL reactivity against HLA-A2.2 whereas only some CTL clones are affected by amino acid changes at 9, 43, and 95.

The amino acid differences between HLA-A2.1 and HLA-A2.3 (5–9) indicate that residues 149, 152, and 156, either alone or in combination affect the epitopes that are recognized by the monoclonal antibody, CR11-351 and by HLA-A2.1 specific CTL clones. To evaluate the role of each of these positions on CR11-351 reactivity, we have used site-directed mutagenesis to replace the amino acids that occur at these positions in HLA-A2.1 with those that occur in HLA-A2.3. The genes thus constructed, as well as the genes encoding HLA-A2.1 and HLA-A2.2, were subsequently transfected into the class I negative line Hmy2.C1R (10) and are listed in Table 1. These include the three single amino acid changes (LTXX-10, LXEX-8, and LXXW-9) and the complete triple change (LTEW-13) corresponding to HLA-A2.3.

All of the transfectants used in this study had similar levels of cell surface expression as judged by binding with the polymorphic antibody MA2.1 (Table 2) and the monomorphic antibody W6/32 (data not shown). As previously shown by Russo et al. (1) these antibodies exhibited high levels of binding to both JY and HSB (HLA-A2.1) and WT49 (HLA-A2.2), whereas only MA2.1 bound well to DK1B (HLA-A2.3). Similar results were obtained with the transfectants LA2.1-5 (HLA-A2.1), LA2.2-6 (HLA-A2.2), and LTEW-13 that expresses an A2 gene constructed to contain the amino acid substitutions found in HLA-A2.3. The substitution of Thr for Ala_{149} (LTXX-10) resulted in a complete loss of CR11-351 reactivity whereas the substitution of either Glu for Val_{152} (LXEX-8) or Trp for Leu_{156} (LXXW-9) had no effect (Table 2). The substitution of both amino acids 152 and 156 in the same molecule (LXEW-3) also had no effect on the epitope recognized by CR11-351. These results indicate that the substitution of Thr for Ala_{149} in HLA-A2.3 is entirely responsible for the loss of the CR11-351 epitope. Ala_{149} occurs on an exposed face of an α-helix, adjacent to residues 142 and 145, that have previously been implicated in formation of the CR11-351 epitope (9). It is therefore likely that all three of these residues are critical in the formation of this epitope.

The role of amino acid substitutions in HLA-A2.2 and A2.3 in the formation of epitopes recognized by HLA-A2.1 specific CTL was also evaluated. In addition to the mutants described above, hemi-exon shuffles were done between HLA-A2.1 and HLA-A2.2 to construct a mole-

Table 1. Amino Acid Sequences of Mutant Class I Molecules

	Amino Acid Residue					
Transfectant	9	43	95	149	152	156
LA2.1-5	Phe	Gln	Val	Ala	Val	Leu
LA2.2-6	Tyr	Arg	Leu	–*	–	Trp
LA2.2/1-18	Tyr	Arg	Leu	–	–	–
LTEW-13	–	–	–	Thr	Glu	Trp
LTXX-10	–	–	–	Thr	–	–
LXEX-8	–	–	–	–	Glu	–
LXXW-9	–	–	–	–	–	Trp
LXEW-3	–	–	–	–	Glu	Trp

*Signifies that the amino acid is identical to that occurring in HLA-A2.1.

Table 2. Recognition of HLA-A2.1 Prototypic and Mutant Molecules by the Polymorphic Antibodies MA2.1 and CR11-351*

Cell	MA2.1	CR11-351
JY	71,074	81,190
HSB	46,715	33,011
WT49	70,179	54,685
DK1B	45,013	1,227
LA2.1-5	45,394	45,461
LTEW-13	54,749	1,480
LTXX-10	49,360	2,514
LXEX-8	46,732	45,810
LXXW-9	56,637	53,285
LXEW-3	60,863	44,964

*The data represent the mean of triplicate samples after subtraction of background counts using second antibody alone.

Immunobiology of HLA
Volume II

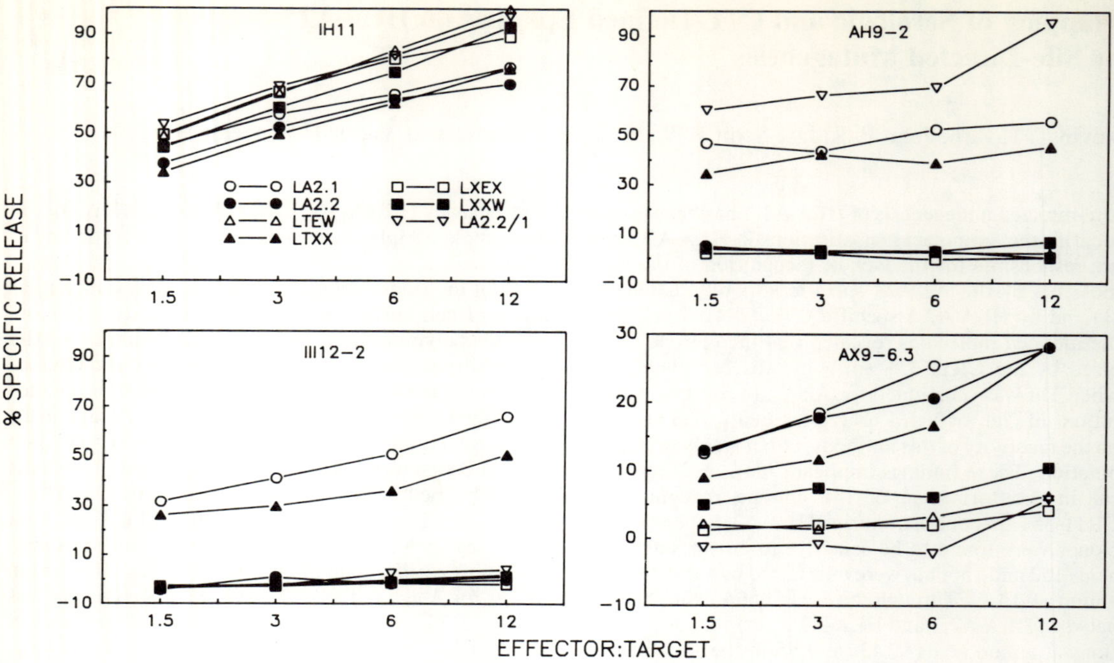

Figure 1. Recognition of transfectant cell lines expressing HLA-A2.1 prototypic and mutant molecules by xenogeneic CTL.

cule that only had HLA-A2.2 substitutions at amino acids 9, 43, and 95 (LA2.2/1-18) (Table 1). Analysis of these mutants with a panel of 9 xenogeneic CTL revealed 4 basic patterns of recognition (Fig. 1). Clone IH11 recognizes HLA-A2.1, A2.2, A2.3, and all of the mutants equally well. This result indicates that all of these molecules are expressed at levels high enough to be recognized by CTL. The second pattern of recognition is characterized by clone AH9-2, which recognizes A2.1 but not A2.2 or A2.3. The substitution of Thr for Ala_{149} has no effect on the epitope recognized by AH9-2, whereas either the substitution of Glu for Val_{152} or Trp for Leu_{156} completely destroys it. Interestingly, LA2.2/1 is recognized by AH9-2, indicating that the epitope recognized is not affected by amino acid changes at 9, 43, and 95 that occur in HLA-A2.2. Clone III12-2 has the same pattern of recognition as AH9-2, except that it does not recognize LA2.2/1. This indicates that the epitope recognized by this clone can be destroyed by substitutions at amino acids 9, 43, and 95 or 152 or 156. A final recognition pattern was exhibited by clone AX9-6.3, which recognizes both HLA-A2.1 and -A2.2. Although the epitope recognized by this clone was destroyed by a substitution at residue 152, it was also destroyed by substitution of HLA-A2.2 residues at either 9, 43, and 95, or 156. Thus, the epitope recognized by this clone is formed by a complex interaction among these residues.

These results indicate that Ala_{149} in HLA-A2 is unimportant in the formation of the epitopes recognized by a large group of HLA-A2 specific CTL. However, the epitopes recognized by all but one CTL clone were sensitive to substitutions at either amino acids 152 or 156 alone, suggesting that both of these residues are important in epitope formation. In addition, the epitopes recognized by many of these clones were also affected by substitutions at amino acids 9, 43, and 95. However, no evidence was obtained by epitopes that are altered by amino acids 9, 43, and 95 and insensitive to changes at amino acids 152 or 156. This indicates that amino acids 152 and 156 play a dominant role in epitope formation for the CTL clones used in this study. All of these residues are found in the "antigen-binding pocket" of the HLA-A2 molecule (11,12). The alteration of individual CTL epitopes by substitutions at amino acids 9, 43, and 95, or 152 or 156 is consistent with the hypothesis that these epitopes are formed by peptides whose binding or conformation is influenced by these amino acids.

Acknowledgments. This work was supported by National Institutes of Health grants AI20963 and AI21393. V.H.E. is the recipient of National Institutes of Health Research Career Development Award CA00835 and K.T.H. is a postdoctoral trainee supported by National Institutes of Health grant CA09109.

References

1. Russo C, Ng A-K, Pellegrino MA, Ferrone S. The monoclonal antibody CR11-351 discriminates HLA-A2 variants identified by T cells. Immunogenetics 1983;18:23.
2. Biddison WE, Ward FE, Shearer GM, Shaw S. The self determinants recognized by human virus-immune T cells can be distinguished from the serologically defined HLA antigens. J Immunol 1980;124:548.
3. Biddison WE, Kostyu DD, Strominger JL, Krangel MS. Delineation of immunologically and biochemically distinct HLA-A2 antigens. J Immunol 1982;129:730.
4. van der Poel JJ, Pool J, Goulmy E, Giphart MJ, van Rood JJ. Recognition of distinct epitopes on the HLA-A2 antigen by cytotoxic T lymphocytes. Hum Immunol 1986;16:247.
5. Krangel MS, Taketani S, Biddison WE, Strong DM, Strominger JL. Comparative structural analysis of HLA-A2 antigens distinguishable by cytotoxic T lymphocytes: variants M7 and DR1. Biochemistry 1982;21:6313.
6. Krangel MS, Taketani S, Pious D, Strominger JL. HLA-A2 mutants immunoselected in vitro. Definition of residues contributing to an HLA-A2-specific serological determinant. J Exp Med 1983;157:324.
7. Krangel MS, Biddison WE, Strominger JL. Comparative structural analysis of HLA-A2 antigens distinguishable by cytotoxic T lymphocytes. II. Variant DK1: evidence for a discrete CTL recognition region. J Immunol 1983;130:1856.
8. Ezquerr A, Domenech N, van der Poel JJ, Strominger JL, Vega MA, Lopez de Castro JA. Molecular analysis of an HLA-A2 functional variant CLA defined by cytolytic T lymphocytes. J Immunol 1986;137:1642.
9. Holmes N, Ennis P, Wan AM, Denney DW, Parham P. Multiple genetic mechanisms have contributed to the generation of the HLA-A2/A28 family of class I MHC molecules. J Immunol 1987;139:936.
10. Storkus W, Howell DN, Salter RD, Dawson JR, Cresswell P. NK susceptibility varies inversely with target cell class I HLA antigen expression. J Immunol 1987;138:1657.
11. Bjorkman PJ, Saper MA, Samraouri B, Bennet WW, Strominger JL. Structure of the human class I histocompatibility antigen, HLA-A2. Nature 1987;329:506.
12. Bjorkman PJ, Saper MA, Samraouri B, Bennet WW, Strominger JL. The foreign antigen binding site and T cell recognition regions of class I histocompatibility antigens. Nature 1987;329:512.

Author Affiliations

Kevin T. Hogan, John P. Ridge, Scott F. Walk, Victor H. Engelhard, Department of Microbiology, University of Virginia School of Medicine, Charlottesville, VA; *Peter Parham*, Department of Structural Biology, Stanford University School of Medicine, Stanford, CA, USA

Cytotoxic T Cell and Antibody Recognition of HLA-A2,A28 Molecules Analyzed by Site Directed Mutagenesis

Russell D. Salter, Carol Clayberger, Catherine E. Lomen, Alan M. Krensky, and Peter Parham

Abstract: We have previously identified tryptophan at position 107 of HLA-A2 and Aw69 as a critical residue for recognition by allospecific cytotoxic T cells (CTL) and monoclonal antibodies (MAb) (1). When tryptophan was substituted for glycine at position 107 of HLA-Aw68.1 by site-directed mutagenesis, the HLA-A2, AW69 specific MAb BB7.2 and two allospecific CTL clones were able to recognize the mutated molecule. Substituting the A2 residues valine and arginine at positions 95 and 97 respectively of HLA-Aw68.1 had no augmenting effect on BB7.2 binding, but was recognized by two HLA-A2 allospecific clones not reactive with wild type HLA-Aw68.1. We have now mutated HLA-Aw69 to the sequence found in HLA-Aw68 at these three positions and confirmed that tryptophan 107 is crucial for forming the BB7.2 determinant. In these experiments, no effect on CTL reactivity was seen after changing tryptophan to glycine at position 107. Mutations at positions 95 and 97 however did affect reactivity with several HLA-A2, Aw69 specific clones. We have also mutated a number of conserved residues of HLA-A2 and observed that substitutions at two of them, positions 5 and 93, have some effect on allospecific CTL recognition. In the three-dimensional structure of HLA-A2, three of these five critical residues are located in the putative peptide binding site (2,3).

Site directed mutants were made in the HLA-Aw68.1 gene as described (1). More recently, for mutagenesis of HLA-A2.1, an Amersham mutagenesis kit was used. Fragments of the mutant genes containing exons corresponding to the α 1 and 2 domains were isolated and reinserted into a wild type HLA-A2 gene in the vector pHEBO (4). After electroporation into a class I negative

Table 1. ^{125}I Rabbit Anti-Mouse Bound Monoclonal Antibody

Cell	MB40.5	CR11	MA2.1	BB7.2	SP2
A2.1 CIR	27120*	27788	20314	31889	342
Aw69 CIR	20872	19049	927	19733	234
69m.107	19786	18446	958	786	240
69m.95,97	13455	12020	1060	12208	270
Aw68 CIR	27917	21578	1220	870	343
CIR	9006	1258	1096	969	333

*Results are the mean of duplicates.

Table 2. % Lysis (E:T=5)

Cell	AJY	ALI1.1
A2.1 CIR	90*	71
Aw69 CIR	51	38
69m.107	39	35
69m.95,97	32	4
CIR	8	0

*Results are the mean of triplicates.

B cell line called CIR (5), the expressed mutants were characterized by antibody binding and CTL assays. Table 1 shows the results of a cell binding radioimmunoassay using the indicated primary antibodies. The mutant genes are expressed on CIR as shown by the binding of MB40.5 and CR11 351. CR11 351 is specific for HLA-A2,Aw69, and Aw68 molecules and reacts with each of the transfectants. BB7.2 reacts with A2.1,Aw69, and the Aw69 position 95,97 mutant (69m.95,97) only. The Aw69 molecule lacking tryptophan at position 107 has lost reactivity with BB7.2. Table 2 shows analysis of these mutants with two CTL; one an A2 specific clone called ALI1.1, the second a line from which ALI1.1 was derived called AJY. The line AJY is well characterized (6,7) and has been used to derive clones reactive with HLA-A2, Aw69, and Aw68. Whereas AJY lyses all of the transfectants, ALI1.1 lyses all but 69m95,97. Other CTL clones have been found with this same pattern of lysis, but it should be noted that these do not constitute the majority of clones we have tested. No HLA-A2, Aw69 specific clones were found which lost reactivity with 69m.107.

Table 3 shows the positions we have mutated in HLA-A2.1 and analyzed with antibodies and CTL clones. Only mutations at positions 5 and 93 had detectable effects on CTL reactivity. Most of the clones still reacted with these mutants. None of the mutations had any measurable effect on reactivity with the panel of antibodies shown in Table 1.

Table 3. HLA-A2 Conserved Residues

Mutated Residue		Expressed in CIR Cells	CTL Recognition
5	M>F	+	lost ALI1.1
5	M>L	ND	ND
35	R>V	+	+
37	D>H	ND	ND
37	D>N	+	+
37	D>Y	+	+
38	S>N	ND	ND
38	S>I	+	+
42	S>T	+	+
58	E>V	+	+
74	H>D	+	+
93	H>P	ND	ND
93	H>L	+	lost ALI1.1
121	K>R	+	+
157	R>T	+	+
158	A>H	+	+
169	R>H	ND	ND
177	E>K	+	+

Our results show that single point mutations of HLA-A2.1 have little effect on allospecific CTL recognition. There are several possible interpretations of these results. The simplest explanation is that single amino acid substitutions at conserved positions make very little difference in the overall avidity of the interaction measured by CTL lysis. However, it may be that the HLA-A2-specific clones we have used are of quite high avidity due to the process used to select them, i.e., multiple rounds of stimulation before cloning. Alternatively, allorecognition itself may be relatively insensitive to single changes in these molecules, although this does not seem to be the case with single changes at polymorphic positions in the human or mouse class I variants, which show striking alterations in both allospecific and restricted recognition (8,9). It would clearly be desirable to analyze these mutants with other types of T cells, specifically antigen-specific, HLA-A2-restricted CTL. This is especially true for mutants in the putative peptide binding site, such as positions 5 and 74, both of which point into the site.

References

1. Salter RD, Clayberger C, Lomen CE, Krensky AM, Parham P. In vitro mutagenesis at a single residue introduces B and T cell epitopes into a class I HLA molecule. J Exp Med 1987; 166:283.
2. Bjorkman PJ, Saper MA, Samraoui B, Bennett WS, Strominger JL, Wiley DC. Structure of the human class I histocompatibility antigen, HLA-A2. Nature 1987; 329:506.
3. Bjorkman PJ, Saper MA, Samraoui B, Bennett WS, Strominger JL, Wiley DC. The foreign antigen binding site and T cell recognition regions of class I histocompatibility antigens. Nature 1987; 329:512.
4. Sugden B, Marsh K, Yates J. A vector that replicates as a plasmid and can be efficiently selected in B lymphoblasts transformed by Epstein-Barr virus. Mol Cell Biol 1985; 5:410.
5. Storkus W, Howell DN, Salter RD, Dawson JR, Cresswell P. NK susceptibility varies inversely with target cell class I HLA antigen expression. J Immunol 1987; 138:1657.
6. Clayberger C, Holmes N, Wang PL, Koller TD, Parham P, Krensky AM. Determinants recognized by human cytotoxic T cells on a natural hybrid class I HLA molecule. J Exp Med 1985; 162:1709.
7. Parham P, Clayberger C, Zorn SL, Ludwig DS, Schoolnik GK, Krensky AM. Inhibition of alloreactive cytotoxic T lymphocytes by peptides from the alpha 2 domain of HLA-A2. Nature 1987; 325:625.
8. Nairn R, Yamaga K, Nathenson SG. Biochemistry of the gene products from MHC mutants. Ann Rev Genet 1980; 14:241.
9. Lopez de Castro JA, Barbosa JA, Krangel MS, Biro PA, Strominger JL. Structural analysis of functional sites of class I HLA antigens. Immunol Rev 1985; 85:149.

Author Affiliations

Russell D. Salter, Catherine E. Lomen, Peter Parham, Department of Cell Biology; *Carol Clayberger, Alan M. Krensky*, Department of Pediatrics, Stanford University Medical Center, Stanford, CA, USA

Peptides Derived From HLA-A2 Modulate Lysis by HLA-A2-Specific Cytotoxic T Lymphocytes

Alan M. Krensky, Peter Parham, Jonathan Rothbard, David S. Ludwig, Gary K. Schoolnik, and Carol Clayberger

Understanding the molecular basis of T-cell recognition of major histocompatibility complex (MHC) antigens may allow new diagnostic and therapeutic approaches to transplantation. Recently it has been shown that in general T cells do not recognize antigens in their native form but rather as short linear protein fragments in the context of self MHC (1–3). These findings prompted us to investigate what role peptides derived from HLA molecules might play in human allorecognition. Two peptides corresponding to regions containing sequences unique to HLA-A2 were selected. These peptides specifically inhibit cytolysis by HLA-A2-specific cytotoxic T lymphocytes (CTL). Moreover, one of these peptides can sensitize a related HLA molecule to lysis by HLA-A2-specific CTL. The recently published crystal structure of HLA-A2 now makes it possible to draw more precise structure/function correlations in understanding the role of HLA molecules as peptide binding proteins.

HLA-A2 was selected for these studies for the following reasons: HLA-A2, as defined by histocompatibility typing, is expressed by over 50% of the population; HLA-A2 and several related molecules (HLA-Aw68.1, -Aw68.2, -Aw69 and -B17) have been cloned and sequenced (4,5); a number of monoclonal antibodies that distinguish individual members of the HLA-A2 family are available (6); and the three-dimensional crystal structure of HLA-A2 has been solved (7,8).

We elected to synthesize peptides surrounding the tryptophan at residue 107 and the glycine at position 62 (Fig. 1). Of the more than 30 HLA class I molecules that have been sequenced to date only HLA-A2 has a tryptophan at 107; all others have a glycine at this position (9). Studies with monoclonal antibodies, T-cell clones, and site-directed mutagenesis of HLA genes have demonstrated that the tryptophan at position 107 is important for both B and T cell recognition (9,10).

The glycine at position 62 comprises an epitope that is shared by HLA-A2 and HLA-B17 and can be identified with antibodies (11). By using stimulator cells that expressed HLA-A2 and/or HLA-B17, we were able to derive CTL clones which lysed all targets expressing HLA-A2 and HLA-B17 but no others. Thus, these CTL are specific for the same epitope recognized by B cells (12).

Peptides corresponding to residues 56-69 or 98-113 of HLA-A2 and the closely related HLA-Aw68 molecule were tested for effects on cytolysis. Peptides corresponding to residues 98-113 of HLA-A2 but not HLA-Aw68 specifically inhibited HLA-A2 specific T-cell lines and most HLA-A2-specific CTL clones, whereas peptides corresponding to other sequences (including A2.56-69) did not (Table 1). The peptide corresponding to residues 56-69 of HLA-A2 specifically inhibited the HLA-A2/B17 specific CTL. Peptide A2.98-113 had only a small effect while the corresponding peptides from HLA-Aw68 had no effect. Using a larger panel of CTL of defined specificities, we found that there was a strict correlation between CTL specificity and inhibition by a particular peptide. The A2.52-69 peptide inhibited only HLA-A2/B17-specific CTL while the A2.98-113 peptide inhibited only HLA-A2 specific CTL.

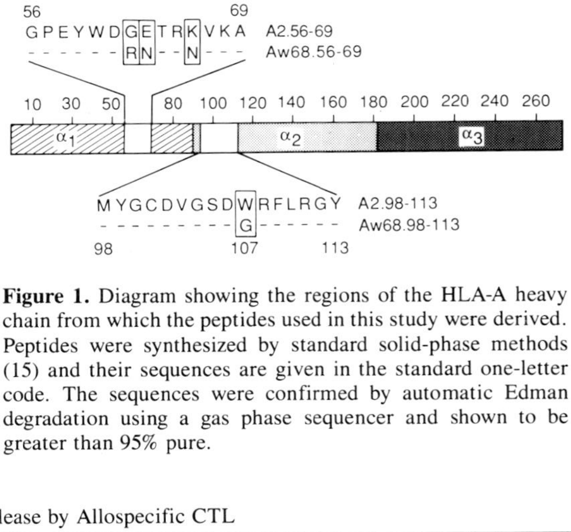

Figure 1. Diagram showing the regions of the HLA-A heavy chain from which the peptides used in this study were derived. Peptides were synthesized by standard solid-phase methods (15) and their sequences are given in the standard one-letter code. The sequences were confirmed by automatic Edman degradation using a gas phase sequencer and shown to be greater than 95% pure.

Table 1. Effects of Peptides on Cytolysis and Serine Esterase Release by Allospecific CTL

CTL	Specificity of CTL	Inhibition of Cytolysis/Esterase Release by Peptide			
		A2.98-113	Aw68.98-113	A2.56-69	Aw68.56-61
CTL A2	HLA-A2	+	−	−	−
Clone AL8.1	HLA-Aw68/Aw69	−	−	−	−
Clone AI5.1	HLA-Aw69	−	−	−	−
Clone A2/B17	HLA-A2/B17	weak	−	+	−
Clone B17	HLA-B17	−	−	−	−
CTL DR6	DR6	−	−	−	−

Immunobiology of HLA
Volume II

Table 2. Specificity of Sensitization to Lysis by Peptide

	Target Cell Expressing HLA-				
	Aw69	Aw68.1	Aw68.2	A3	A2.1
	% Specific Cytolysis				
Clone A2/B17*					
+Medium	0	0	0	0	70
+A2.56-69 (150μM)	52	14	8	0	55
CTL A2					
+Medium	0	0	0	0	85
+A2.98-113 (150μM)	0	0	0	0	37

*CTL:target=5:1. Peptide was present throughout the assay.

Table 3. Locus of Peptide Effect

Pretreatment		% Effect	
Effector	Target	Inhibition	Induction
peptide	medium	51	13
medium	peptide	5	100
peptide	peptide	57	37

10^6 CTL or ^{51}Cr-labeled targets were treated for 30 min with medium or 150μM peptide at 37°, washed, and used in a CTL assay. For inhibition, the CTL were CTL-A2, target: JY (HLA-A2,2;B7,7,DR4,6), E:T=1:1, peptide: A2.98-113. Specific release was 55% for medium alone and 24% for maximal inhibition. For induction, the CTL were clone A2/B17, target: IDF (HLA-Aw69,26;B18,38;DR5), E:T = 5:1, peptide: A2.56-69. Specific release was 10% for medium alone and 47% for maximal induction.

We also investigated the ability of these peptides to sensitize targets not normally recognized by HLA-A2-specific CTL to lysis (Table 2). No effects were observed using a wide combination of CTL and HLA-A2 negative targets. We were however able to show sensitization of HLA-Aw69 expressing targets to lysis by the HLA-A2/B17-specific CTL when peptide A2.56-69 was included in the cytotoxicity assay (13). HLA-A2 and HLA-Aw69 differ by only six amino acids and all of these differences lie in the α_1 domain (14). Furthermore, three of these differences are present in the 56-69 region. It should be noted that in no instance could peptides corresponding to sequences around position 107 induce cytolysis (Table 2).

Studies were next undertaken to define the locus of induction/inhibition (Table 3). CTL and target cells were pretreated with appropriate peptides or medium, washed, and tested in a standard cytotoxicity assay (13). Sensitization occurred only when the target cells were pretreated with peptide whereas inhibition occurred only when the CTL were pretreated with peptide. The high correlation between CTL specificity and ability of peptides to inhibit lysis suggests that peptide interacts with the T-cell receptor. The ability of peptide A2.56-69 to sensitize only HLA-Aw69 expressing targets suggests that, in this case, peptide interacts with the HLA molecule.

Although these peptides were originally selected based on sequence analyses, we now know their precise location within the three-dimensional structure of HLA-A2 (7,8). The α_1 and α_2 domains of HLA-A2 interact to form a "peptide binding region" which is composed of a series of β strands overlaid by two opposing α helices. Residues 62-65 lie on the α helix of this peptide binding site while residue 107 lies on a turn at a distance from the peptide binding site. Our data suggest that there may be two separate modes of T cell recognition of HLA-A2: one involving the peptide binding region and another involving a turn away from this region.

Acknowledgments. We thank K. Drickamer for amino acid sequence analysis and Linda Hopper for preparation of the manuscript.

This work was supported by grants from the American Cancer Society (RD256), Weingart Foundation, and NIH (DK35008 and AI22039). P.P. is a scholar of the Leukemia Society of America; A.M.K. is the recipient of an American Heart Association Clinician-Scientist Award.

References

1. Ziegler K, Unanue ER. Identification of a macrophage antigen processing event required for I-region-restricted antigen presentation to T lymphocytes. J Immunol 1982;127:1869.
2. Townsend ARM, Gotch FM, Davey J. Cytotoxic T cells recognize fragments of influenza protein. Cell 1985;42:457.
3. Buss S, Colon S, Smith C, Freed JH, Miles C, Grey HM. Interaction between a "processed" ovalbumin peptide and Ia molecules. Proc Natl Acad Sci USA 1986;83:3968.
4. Holmes N, Ennis P, Wan A, Denney D, Parham P. Multiple genetic mechanisms have contributed to the generation of the HLA-A2/A28 family of Class I MHC molecules. J Immunol 1987;139:936.
5. Ways JP, Coppin H, Parham P. The complete primary structure of HLA-Bw58. J Biol Chem 1985;260:11924.
6. Ways JP, Coppin H, Parham P. The binding of monoclonal antibodies to cell surface molecules: A quantitative analysis with IgG against two alloantigenic determinants of HLA-A2. Biochem J 1983;216:423.
7. Bjorkman PJ, Saper MA, Samraoui B, Bennett WA, Strominger JL, Wiley DC. Structure of the human class I

histocompatibility antigen, HLA-A2. Nature 1987;329: 506.
8. Bjorkman PJ, Saper MA, Samraoui B, Bennett WS, Strominger JL, Wiley DC. The foreign antigen-binding site and T cell recognition regions of class I histocompatibility antigens. Nature 1987;329:512.
9. Parham P, Clayberger C, Zorn SL, Ludwig DS, Schoolnik GK, Krensky AM. Inhibition of alloreactive cytotoxic T lymphocytes by peptides from the alpha$_2$ domain of HLA-A2. Nature 1987;325:625.
10. Salter RD, Clayberger C, Lomen CE, Krensky AM, Parham P. In vitro mutagenesis at a single residue introduces B and T cell epitopes into a Class I HLA molecule. J Exp Med 1987;166:283.
11. Ways J, Rothbard JB, Parham P. Amino acid residues of 56-69 of HLA-A2 specify an antigenic determinant shared by HLA-A2 and HLA-B17. J Immunol 1986;137:217.
12. Seong R, Clayberger C, Krensky AM, Parnes J. Rescue of Daudi cell HLA expression by transfection of the mouse β_2-microglobulin gene. J Exp Med 1988;167:288.
13. Clayberger C, Parham P, Rothbard J, Ludwig DS, Schoolnik GK, Krensky AM. HLA-A2 peptides can inhibit or enhance cytoloysis by human allogeneic T lymphocytes. Nature 1987;330:763.
14. Holmes N, Parham P. Exon shuffling in vivo can generate novel HLA class I molecules. Eur Mol Biol Organ J 1986;4:2849.
15. Erickson BW, Merrifield RB. Solid phase peptide synthesis. In: Neurath H, Hill RL (eds): The Proteins, Vol. 2, 3rd ed. New York: Academic Press, 1970, p 255.

Author Affiliations

Alan M. Krensky, Carol Clayberger, Department of Pediatrics; *Peter Parham*, Department of Cell Biology; *David S. Ludwig, Gary K. Schoolnik*, Department of Medical Microbiology, Stanford University School of Medicine, Stanford, CA, USA; *Jonathan Rothbard*, Imperial Cancer Research Fund, London, England

Variant and Mutant HLA-A2 Molecules Affect Recognition of Influenza Matrix Peptide 56–68 by Cytotoxic T Lymphocytes*

Andrew J. McMichael, Frances M. Gotch, Jonathan B. Rothbard, Jesús Santos-Aguado, and Jack L. Strominger

We have previously shown that HLA-A2-restricted influenza A virus-specific cytotoxic T lymphocytes (CTL) recognize the influenza matrix protein, expressed in target cells by recombinant vaccinia virus (1). We were able to identify a synthetic peptide comprising amino acids 56–68 which substituted for virus on target cells (2). The CTL response to this peptide is determined in humans by the presence of HLA-A2 and the most likely explanation is that peptide binds to the HLA-A2 molecule. The three-dimensional structure of HLA-A2 was recently reported by Bjorkman et al. (3,4). A putative peptide binding site was identified in the molecule and many of the polymorphic residues in HLA-A2 are close to this site. To assess the nature of the A2-peptide interaction we have examined cytotoxic T-cell recognition of peptide in association with variant HLA-A2 molecules, both naturally derived and generated by site-directed mutagenesis.

Natural variants of HLA-A2 were first recognized by Biddison et al. (5) who found that some influenza virus-infected target cells were not lysed by virus-specific CTL that shared the serological specificity of HLA-A2. They showed that these were electrophoretic variants of the HLA-A2 molecules which were subsequently named A2.1 (normal), A2.2 (basic on isoelectric focusing), and A2.3 (acidic on isoelectric focusing) (5,6). More recently, variant A2.4 (which has a normal position on isoelectric focusing) and two splits of A2.2 (A2.2F and A2.2Y) have been identified (6,7). The amino acid sequence of several of these molecules has now been determined by Krangel et al. (8), Mattson et al. (9), Holmes et al. (10), and Ezquerra et al. (11). The variants differ in 1, 3, or 4 amino acid residues from HLA A2.1 at one or more of positions 9, 43, 95, 149, 152, and 156. All of these except 43 point in to the putative peptide binding groove (4).

In this study we determined the ability of influenza A virus-specific CTL from donors with HLA-A2 variant molecules to recognize the matrix peptide 56–68. The results shown in Table 1 indicate that CTL from donors with A2.2F, A2.2Y, A2.3, all of which respond to influenza virus, failed to recognize this peptide. Similarly, CTL from donors with the closely related HLA types A68 and 69 failed to react with the peptide. In contrast, CTL from all A2.1 donors so far tested lysed HLA-A2.1 target cells that had been treated with the matrix peptide. These CTL failed to lyse peptide-treated target cells that carried A2.2F, A2.2Y, A2.3, A68, and AW69.

The results with the A2 variants therefore implicated one of more of positions 9, 43, 95, 149, 152, and 156 in the peptide interaction. Furthermore, the findings with A69, which has the $\alpha 1$ domain of A68 and the $\alpha 2$ domain

*This work was supported by grants from the MRC and NIH.

Table 1. Recognition of Matrix Peptide 55–73 by Human Influenza A Virus-Specific CTL

Donor	HLA Type*		E:T Ratio	Percent Specific Lysis of Autologous Target Cells†		
				Ax31	55–73	Nil
KW	A2, 2;	B7, 51	40:1	48	44	10
IH	A2, 3;	B7, 62	18:1	31	21	5
JM	A2, 2;	B51, 62	40:1	29	22	2
GP	A2, 24;	B7, 44	50:1	37	26	3
AT	A2, 11	B54, 60	50:1	32	31	9
C212	A2, 2;	B8	3:1	59	68	3
FM	A2, 2.2Y;	B15, 49	50:1	53	36	19
FM‡	A2, 2.2Y;	B15, 49	50:1	49	10	10
AM	A2.2Y, 32;	B49, 60	50:1	75	8	8
RT	A2.2F, 3;	B7, 57	34:1	33	6	5
JJ	A2.2F, 11;	B35, 50	40:1	25	0	0
DEW	A2.3, 24;	B13, 27	40:1	33	7	8
WI	A69;	B51, 22	40:1	42	2	2
DC	A1, 68;	B8, 18	6:1	28	10	10
NC	Al, 24;	B7, 44	60:1	28	10	10
LG	Al, 29;	B37, 44	25:1	25	2	3
NR	Al, 3;	B8, 60	40:1	23	5	9
CB	Al, 30;	B7, 44	40:1	28	10	12

*The common variant of HLA-A2 (A2.1) is shown as A2; other variants are referred to as A2.2F, A2.2Y, and A2.3.
†Target cells were either infected with influenza Ax31 virus (which was also used to stimulate the CTL cultures) or uninfected but pulsed for 1 h in matrix peptide 55–73 at 30 μm, or uninfected (Nil).
‡Tested on target cells that share only A2.2Y.

Table 2. CTL Recognition of Matrix Peptide 55–68 in Association with HLA-A2 Mutants

Transfectant*		JM-6 CTL (6 μm)†	JM-8 CTL (0.08 μm)†
A2		25‡	33
mock		9	nt.
9	F-Y	13	9
43	Q-R	18	29
62,63	GR-RN	11	7
66	K-I	7	2
70	H-Q	14	29
74	H-D	30	24
107	W-G	26	50
152	V-E	9	10
156	L-S	15	11

*Transfections were made into the RD rhabdomyosarcoma cell line (ref. 12). Mutations are shown with the residue number (e.g., 9), the amino acid in A2.1 (e.g., F), and then in the mutant (e.g., Y).
†Peptide concentration in the 4-h chromium release assay.
‡Results shown as percent specific lysis for two CTL lines, JM-6 and JM-8.

of A2 (10), suggested that some of residues 62, 65, 66, 70, and 74 could also be critical. To explore this further we tested target cells that had been transfected with HLA-A2 cDNA with single-point mutations (to the B7 sequence) generated by site-directed mutagenesis (12). As effector cells we used CTL from donor JM (A2.1) that had been restimulated with peptide and interleukin-2 for 3 to 8 weeks. These CD8 T cells were oligoclonal rather than monoclonal so that impairment of recognition of mutant target cells was more likely to reflect peptide binding to A2 than interaction with a T-cell receptor. The results shown in Table 2 indicate that position 43 had no effect whereas changes at 9, 152, and 156 greatly reduced peptide recognition. In addition, mutations at positions 62, 63, and 66 had some effect. In one CTL line (JM-6), but not another (JM-8) (both derived from the same donor), a change at position 70 abolished recognition. There was also some variability in the degree of impairment seen with the 9, 62–63, and 66 mutants by different T-cell lines. With the other mutants the results with the two CTL lines were consistent. All of the mutants were shown to express HLA-A2 by binding monoclonal antibodies and were lysed by one or more alloreactive CTL clones (ref. 12; J.L. Strominger et al., unpublished results).

When the residues that affected CTL recognition of the matrix peptide 56–68 are plotted on the A2 structure derived by Bjorkman et al. (3,4) (Fig. 1), it is clear that sites on both α helices are involved, and so is the adjacent position 9, on the floor of the groove. Additional data from a set of A2-A3 hybrid molecules implicates residues 114 and/or 116, both of which are sited in the floor of the groove and point upwards (F. Healy, J.P. Levy, B. Jordan, and F. Gotch, unpublished data).

In addition to modifying the HLA-A2 molecule we have also varied the peptide. These data will be published in detail elsewhere, but briefly the essential findings are as follows. By cutting into the peptide from both ends it was shown that the minimum effective peptide was the 57–68 12-mer KGILGFVFTLTV. Recognition was impaired by moving the frame one position toward

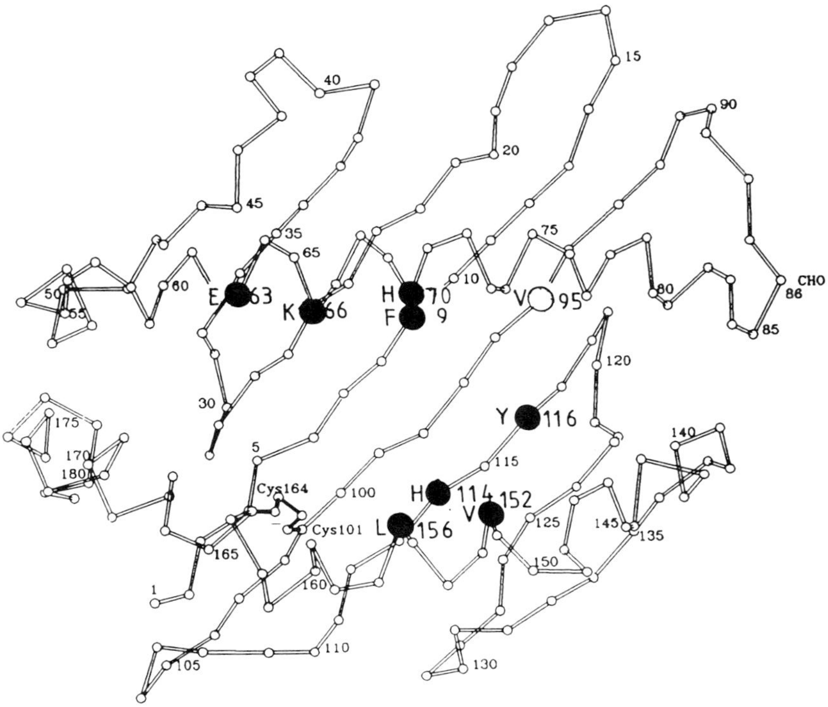

Figure 1. The location of the residues involved in presentation of the influenza A virus matrix protein peptide 56–68 as shown by solid dots. The open dot, position 95, indicates a residue that may be important because it varies in the A2-2 variants. The peptide could make contact with all of these residues if it adopts an α helical conformation. The figure is based on the HLA-A2 structure as described by Bjorkman et al. in references 3 and 4.

the NH_2 terminus or one position toward the COOH terminus. Thus the initial lysine (K) and terminal valine (V) would appear to be involved in peptide-HLA-A2 interaction. We then tested a set of peptides each of which had an alanine substituted at one position between 59 and 67. This had the most dramatic effect when alanine was substituted for glycine at position 61. No recognition was observed at any concentration and other amino acid substitutions in this position gave identical negative results. Alanine substitutions at other positions in the peptide had partial effects but these could be overcome to an extent by raising the peptide concentration. Thus the glycine at position 61 is uniquely crucial. One possible explanation for this is that the glycine, the smallest amino acid, fits opposite the lysine at position 66 in HLA-A2. This lysine protrudes into the putative peptide binding groove (see ref. 4). If the peptide adopts an α helical conformation, residues 62(F) and 65(T) of the peptide could then point up toward the T-cell receptor. Substitution at these positions with lysine completely abolished recognition. The peptide with lysine at position 65 competed effectively with the natural peptide, which implied that it binds to the A2 molecule in a similar, if not identical, position. In contrast, the peptide with the alanine substitution at position 61 did not compete, which implied that this does not fit in to the A2 structure. The data are, therefore, consistent with the peptide adopting an α helix with the glycine-61 residue opposite lysine-66 of HLA-A2. In this conformation the peptide could interact with all of the residues identified as important in the variant and mutant A2 molecules. In addition there would be hydrophobic face of the peptide interacting with the hydrophobic face of the α2 domain α helix and a possible salt bridge between the glutamic acid at position 62 on A2 and the lysine at position 57 of the peptide.

Although such a model would fit all of the data that we have available at present, it is far from proven and other models are possible. For instance, the differences in recognition of the matrix peptide with the α1 helix mutants (62–63, 66, 70) may mean that the peptide can fit into the groove in more than one way. Also substitutions at one position could alter configuration and therefore binding at other distant positions. The solution will probably have to await crystallization of a peptide-A2 complex and such work is currently in progress. In the mean time, the model presented here may facilitate attempts to predict peptides from other virus proteins which will be presented to CTL by HLA-A2.

Acknowledgments. We are grateful to Drs. A. Townsend and P. Robbins for helpful discussions and Clare Crew for preparing the manuscript.

References

1. Gotch FM, McMichael AJ, Smith GL, Moss B. Identification of the viral molecules recognized by influenza specific human cytotoxic T lymphocytes. J Exp Med 1986;165: 408–416.

2. Gotch F., Rothbard J, Howland K, Townsend A, McMichael A. Cytotoxic T lymphocytes recognise a fragment of influenza virus matrix protein in association with HLA-A2. Nature 1987;326:881–882.
3. Bjorkman PJ, Saper MA, Samraoui B, Bennett WS, Strominger JL, Wiley DC. Structure of the human class I histocompatibility antigen, HLA-A2. Nature 1987;329: 506–511.
4. Bjorkman PJ, Saper MA, Samraoui B, Bennett WS, Strominger JL, Wiley DC. The foreign antigen binding site and T cell recognition regions of class I histocompatibility antigens. Nature 1987;329:512–516.
5. Biddison WE, Krangel MS, Strominger JL, Ward FE, Shearer GM, Shaw S. Virus-immune cytotoxic T cells recognise structural differences between serologically indistinguishable HLA-A2 molecules. Hum Immunol 1980;3:225–232.
6. Van der Poel JJ, Molders M, Thompson A, Ploegh HL. Definition of four HLA-A2 subtypes by CML typing and biochemical analysis. Immunogenetics 1983;17:609–621.
7. Gotch FM, Kelly C, Ellis SA, Wallace L, Rickinson AB, Van Der Poel JJ, et al. Characterisation of the HLA-A2.2 subtype: T cell evidence for further heterogeneity. Immunogenetics 1985;21:11–23.
8. Krangel MS, Biddison WE, Strominger JL. Comparative structural analysis of the HLA-A2 antigens distinguished by cytotoxic T lymphocytes: II. Variant DK1. Evidence for a discrete CTL recognition region. J Immunol 1983;130: 1856–1862.
9. Mattson DH, Hardy DE, Bradley DA, Coligan JE, Cowan EP, Biddison WE. DNA sequences of the genes that encode the CTL-defined HLA-AR variants M7 and DK1. Immunogenetics, in press.
10. Holmes N, Ennis P, Wan AM, Denney DW, Parham P. Multiple genetic mechanisms have contributed to the generation of the HLA-A2/A28 family of class I MHC molecules. J Immunol, in press.
11. Ezquerra A, Domenech N, Van der Poel J, Strominger JL, Vega MA, Lopez de Castro JA. Molecular analysis of an HLA-A2 functional variant CLA defined by cytolytic T lymphocytes. J Immunol 1986;137:1642–1649.
12. Santos-Aguado J, Barbosa JA, Biro A, Strominger JL. Molecular characterization of serological recognition sites in the human HLA-A2 molecule. J Immunol (in press).

Author Affiliations

Andrew J. McMichael and *Frances M. Gotch*, Nuffield Department of Medicine, John Radcliffe Hospital, Oxford, England; *Jonathan B. Rothbard*, Imperial Cancer Research Fund Laboratories, London, England; *Jesús Santos-Aguado* and *Jack L. Strominger*, Dana-Farber Centre, Boston, MA, USA

Class I-Presented Influenza Peptides Predicted by an Algorithm that Selects Class II-Presented Peptides*

Victor E. Reyes, L. Thomas Chin, and Robert E. Humphreys

Class II MHC-restricted, $CD4^+$ T cells recognize processed foreign peptides that form amphipathic helices. It also has been shown that class I MHC-restricted, $CD8^+$ T cells recognize foreign peptide fragments rather than entire proteins. In light of this knowledge and the facts that 1) $CD4^+$ and $CD8^+$ T cells use the same gene families to generate their receptors, and 2) class I and class II MHC products hold many structural similarities, we hypothesized that peptides recognized by $CD4^+$ and $CD8^+$ T cells might have some structural homologies. To test this hypothesis, we screened a series of proteins, which contain $CD8^+$ cytotoxic T lymphocyte (CTL)-recognized epitopes, using a strip-of-helix algorithm that has been shown to predict well class II MHC-restricted peptides (1). This algorithm searched for amphipathic α helices by calculating the average hydrophobicity of amino acids at positions n, n+4, n+7, n+11, n+14, and n+18 in a linear sequence, and could also be adapted to analyze 3_{10} helices or β-pleated sheets.

We screened the following influenza virus proteins: matrix, nucleoprotein (NP), and hemagglutinin HA_2 for putative amphipathic α helices or 3_{10} helices and then compared the predictions of the strip-of-helix hydrophobicity algorithm with the actual peptides found experimentally to be $CD8^+$ T cell presented. Our screen revealed that the known CTL epitopes were well predicted by our algorithm. For instance, influenza virus matrix protein Leu_{55}-Gly_{73}, which is presented to $CD8^+$ T cells in the context of HLA-A2 (2), was the first peptide selected by our algorithm (Leu_{55}-Leu_{66}). Two other influenza virus peptides, NP Ser_{335}-Gly_{349} (3) and HA_2 Val_{183}-Gly_{205} (4), were predicted by peptides Ser_{335}-Ile_{347} and Ser_{180}-Ile_{189}, respectively, in the form of 3_{10} helices. NP peptide Ser_{50}-Arg_{65}, which has been reported to be H-$2K^k$ restricted, did not rank among the first five peptides (prior to exclusions due to the presence of proline), but was identified as the sixth ranked peptide (Leu_{56}-Ala_{70}, index 2.70) for a four-cycle α helix. It is possible that the other peptides selected by our algorithm might be presented by some other class I

* This work was supported in part by NIH grant CA 37645. L.T. Chin was a Betty Lee Stone Fellow of the American Cancer Society, Massachusetts Division and was supported by training grant R25CA36762.

Table 1. Strip-of-Helix Hydrophobicity Ranking of CTL-Recognized Peptides

Protein	Reported Peptide (Restriction Element)	Helix	Predicted Peptide	Index
Matrix	Leu_{55}-Gly_{73} (HLA-A2)	α	55–66	3.72
			132–143	3.70
			125–136	3.02
		3_{10}	22–31	2.90
NP	Ser_{50}-Arg_{65} (H-$2K^k$)	α	190–201	4.42
			186–197	4.17
			25–36	3.77
	Ser_{335}-Gly_{349}		56–70	2.70
	(HLA-B37)	3_{10}	338–347	3.72
			30–39	2.85
HA_2	Val_{183}-Gly_{205} (H-$2K^d$)	α	48–59	3.70
			73–84	3.42
			52–63	3.35
		3_{10}	180–189	2.40

or class II MHC alleles. Studies are in progress to investigate the allelic specificity of binding of these peptides to immunopurified class I and class II molecules.

Since our strip-of-helix algorithm predicted class I as well as class II MHC-restricted, T cell-presented peptides, one can suggest that peptides recognized by both $CD4^+$ and $CD8^+$ T cells are structurally homologous in the sense of being amphipathic α or 3_{10} helices. If both sets of peptides have the same tertiary structure, then one could assume that the desetope on class I and class II MHC antigens is also structurally homologous.

References

1. Stille CJ, Thomas LJ, Reyes VE, Humphreys RE. Hydrophobic strip-of-helix algorithm for selection of T cell-presented peptides. Mol Immunol 1987;24:1021.
2. Gotch F, Rothbard J, Howland K, Townsend A, McMichael A. Cytotoxic T lymphocytes recognize a fragment of influenza virus matrix protein in association with HLA-A2. Nature 1987;326:881
3. McMichael AJ, Gotch FM, Rothbard J. HLA-B37 determines an influenza A virus nucleoprotein epitope recognized by cytotoxic T lymphocytes. J Exp Med 1986;164:1397.
4. Braciale TJ, Braciale VL, Winkler M, Stroynowski I, Hood L, Sambrook J, et al. On the role of the transmembrane anchor sequence of influenza hemagglutinin in target cell recognition by class I MHC-restricted, hemagglutinin-specific cytolytic T lymphocytes. J Exp Med 1987;166:678.

Author Affiliation

Victor E. Reyes, *L. Thomas Chin*, and *Robert E. Humphreys*, Department of Pharmacology, University of Massachusetts Medical School, Worcester, MA 01655, USA

Structure of HLA-B27 Subtypes: Evolutionary Implications

Susana Rojo and José A. López de Castro

The HLA-B27 antigen is a well-defined serologic specificity, but at least six structurally different subtypes can be distinguished by biochemical analyses and by cytolytic T lymphocytes. These subtypes are designated as B*2701 to B*2706. The available population studies show an uneven distribution of the various B27 subtypes among the major ethnic groups, B*2705 being predominant in the overall human population (1,2). The structure of these six subtypes has been established by comparative peptide mapping and protein sequence analysis (3, and references therein). Their structural differences are summarized in Table 1. Three structural patterns are readily apparent from this comparison: 1) B*2705, B*2702 and B*2701 differ from each other by two or three substitutions within residues 74 to 81; 2) B*2704 and B*2706 both differ from B27.1 by two amino acid changes at position 77 and 152. B*2706 has two additional substitutions at residues 114 and 116; and 3) B*2703 differs from B*2705 by a single amino acid change. The origin of these structural patterns could be accounted for by three different genetic mechanisms. First, both B*2702 and B*2701 could have arisen by single gene conversion-like events from B*2705. Potential donor sequences for both subtypes are likely to exist in the HLA gene repertoire. Second, the nature of the two non-clustered amino acid differences between B*2705 and B*2704, involving at least two and one base changes respectively, would require a minimum of two genetic steps—either sequential recombination or recombination and point mutation. A further gene conversion-like event would probably be necessary to explain the two additional differences of B*2706. Third, the structure of B*2703 is compatible with its origin from B*2705 by a single point mutation.

When the structural and genetic relationships among the six B27 subtypes are correlated with their ethnic distribution, an evolutionary pattern emerges that is consistent with the diversification of HLA-B27 taking place *after* the differentiation of the major ethnic groups. Thus, those subtypes found almost exclusively

Table 1. Amino Acid Differences Among HLA-B27 Subtypes*

Position	B27.1	B27.2	B27.f	B27.3	B27.4	B27.d
59	Tyr	–	–	–	–	His
74	Asp	–	Tyr	–	–	–
77	Asp	Asn	Asn	Ser	Ser	–
80	Thr	Ile	–	–	–	–
81	Leu	Ala	Ala	–	–	–
114	His	–	–	–	Asp	–
116	Asp	–	–	–	Tyr	–
152	Val	–	–	Glu	Glu	–

*Dashes indicate identity with B27.1 at that position.
The new official designation for HLA-B27 subtypes is the following: B*2701(B27f), B*2702(B27.2), B*2703(B27d), B*2704(B27.3), B*2705(B27.1), and B*2706(B27.4).

in Caucasians, B*2702 and B*2701 (1,2), display an analogous pattern of multiple clustered substitutions within the same segment of the molecule that could have arisen in each case by a single genetic event. Similarly B*2704, which is the predominant subtype in Orientals, and B*2706, which has only been found in a few individuals from this ethnic group (1), both have the same two changes at positions 77 and 152 respective to B*2705 and would have required a minimum of two (B*2704) or three (B*2706) genetic steps in their evolution from B*2705. Presumably, B*2706 would have evolved from B*2704 by a more recent event that would have introduced the two changes at positions 114 and 116. The B*2703 antigen has only been confirmed in an American black person, but a likely second example has been found within the same ethnic group (2,3). African blacks virtually lack HLA-B27 and the presence of this antigen in the American black population is considered to be the result of penetration of the Caucasian allele. Thus, it is conceivable that *B2703 is the product of a very recent mutation in the B*2705 gene within the black population. Alternatively, this mutation could have occurred in the Caucasian population and would have been introduced into American blacks through racial admixture, being then amplified within this ethnic group.

Finally, all six HLA-B27 subtypes are distinguishable by cytolytic T lymphocytes. Thus, the structural polymorphism of this antigen is likely to be functionally relevant, contributing to increase the repertoire of restriction elements in the human population for a changing variety of pathogens.

Acknowledgments. Supported in part by grants from the U.S.-Spain Joint Committee for Scientific and Technological Cooperation (83/054) and the Fondo de Investigaciones Sanitarias de la Seguridad Social of Spain (84/793 and 85/1149).

References

1. Breur-Vriesendorp BS, Neefjes JC, Huis B, van Seventer GA, Ploegh HL, Iványi P. Identification of new B27 subtypes (B27C and 27D) prevalent in Oriental populations. Hum Immunol 1986;16:163.
2. Choo SY, Antonelli P, Nisperos B, Nepom GT, Hansen JA. Six variants of HLA-B27 identified by isoelectric focusing. Immunogenetics 1986;23:23.
3. Rojo S, Aparicio P, Hansen JA, Choo SY, López de Castro JA. Structural analysis of an HLA-B27 functional variant, B27d, detected in American Blacks. J Immunol 1987; 139:3396.

Author Affiliation

Susana Rojo, José A. Lopez de Castro, Department of Immunology, Fundación Jiménez Díaz (C.S.I.C.), Madrid, Spain

Cellular Allorecognition of HLA-B27: Clonal Heterogeneity and Identification of Immunodominant Sites

Pedro Aparicio, Dolores Jaraquemada, and José A. López de Castro

Eighteen CD3$^+$CD4$^-$CD8$^+$ HLA-B27 allospecific cytolytic T-lymphocyte (CTL) clones were isolated by limiting dilution of HLA-B27-negative responder cells stimulated with HLA-B27.1-positive lymphoblastoid cells. The specificity and avidity of these clones was determined from panel analyses, cold-target competition studies, inhibition with anti-HLA monoclonal antibodies, and blocking studies with anti-CD3 and anti-CD8 antibodies. Target HLA antigens of known structure, including all the structurally characterized HLA-B27 subtypes, were used in order to examine how the polymorphism at particular amino acid sequence positions modulates the fine specificity of each CTL clone. As shown in Table 1, seven reaction patterns were distinguished on the basis of their differential reactivaties with the various B27 subtypes and of crossreactions with HLA-B40* or HLA-DR2. The crossreactivity of CTL clones 23, 27, and 47 with B27.1 and HLA-DR2 has been analyzed in detail elsewhere (1). Otherwise, these reaction patterns were characterized by: lack of recognition of those B27 subtypes (B27.3 and B27.4) with a change at position 152; and altered reactivity towards a variant with changes in the segment 77-81 (B27.2) and sometimes with those differing at residues 74-81 (B27f) or at position 59 (B27d). No clones were found that reacted identically with all three B27.1, B27.2, and

Table 1. Reaction Patterns of B27-Specific Alloreactive CTL Clones

Reaction Pattern	Clone	B27.1	B27.2 (77,80,81)	B27.d (59)	B27.f (74,77,81)	B27.3 (77,152)	B27.4 (77,114,116,152)	B40*	DR2
I	39,40,53	+++	−	−	−	−	−	−	−
II	23,27,47	+++	−	−	−	−	−	−	+++
III	28	+++	++	−	−	−	−	−	−
IV	1,3,6,7,26,29	+++	++	+++	−	−	−	−	−
V	46,52	+++	−	+++	−	−	−	+++	−
VI	2,15	+++	−	−	+++	−	−	+	−
VII	48	+++	+	+	+++	−	−	+++	−

The amino acid sequence positions in which each B27 subtype differs from B27.1 are given in parenthesis. The new official designation for HLA-B27 subtypes is the following: B*2701(B27f), B*2702(B27.2), B*2703(B27d), B*2704(B27.3), B*2705(B27.1), and B*2706(B27.4).

B27f o which recognized B27.3 or B27.4. The fact that all eighteen analyzed clones recognized epitopes involving residues 77-81 and 152 suggests that these positions are involved in an immunodominant structure of the HLA-B27.1 molecule. Because all the clones were obtained from only one responder individual, the assignment of putative immunodominant regions is relative to this individual response.

Several clones were cross-reactive with HLA-B40* (Table 1), an antigen that differs from B27.1 at 16 positions in the amino acid sequence (2). This is in contrast with the general lack of recognition of B27.3 and B27.4, differing from B27.1 by two and four amino acid changes respectively. Thus it appears that, in terms of CTL recognition, there is no straight correlation between structural and functional relatedness. Very limited amino acid changes may, in certain cases, give rise to a new antigen that is more dissimilar in CTL recognition than other, structurally more diverse, antigens. The molecular bases for the cross-reaction between B27.1 and B40* are not readily apparent, but it is possible that the concomitant presence of certain amino acid changes may have mutually compensatory effects. For instance, CTL clones 46 and 52 do not recognize those B27 subtypes with changes in the segment 74-81 (B27.2 and B27f, Table 1), but do recognize HLA-B40* which also has three amino acid differences with B27.1 within residues 74 to 81 (2). Thus, some of the additional changes outside this segment must have a compensatory effect on the structure of the relevant epitope of HLA-B40*, so that it becomes indistinguishable from that of B27.1 for the corresponding CTL clone. Presumably, concomitant changes in a number of positions could result in a new structure with equivalent antigenic properties in its interaction with the clonotypic T-cell receptor. This sort of effect might also be invoked to explain the differential recognition of B27.2 and B27f, both with changes in the same segment, by some CTL clones (Table 1).

Acknowledgments. Supported in part by grants from the U.S.-Spain Joint Committee for Scientific and Technological Cooperation (83/054) and the Fondo de Investigaciones Sanitarias de la Seguridad Social of Spain (85/1149).

References

1. Aparicio P, Jaraquemada D, López de Castro JA. Alloreactive cytolytic T cell clones with dual recognition of HLA-B27 and HLA-DR2 antigens. Selective involvement of CD8 in their class I-directed cytotoxicity. J Exp Med 1987;165:428.
2. Aparicio P, Rojo S, Jaraquemada D, López de Castro JA. Fine specificity of HLA-B27 cellular allorecognition. HLA-B27f is a functional variant distinguishable by cytolytic T cell clones. J Immunol 1987;139:837.

Author Affiliation

Pedro Aparicio, Dolores Jaraquemada, and *José A. López de Castro,* Department of Immunology, Fundación Jiménez Diaz (C.S.I.C.), Madrid, Spain

Molecular Analysis of the Variant Alloantigen HLA-B27d Identifies a Unique Single Amino Acid Substitution

Sung Yoon Choo, Thomas St. John, Harry T. Orr, and John A. Hansen

HLA-B27 is a human major histocompatibility complex class I product defined by its antigenic specificity with conventional alloantisera. Detailed studies using monoclonal antibodies (MAb), cytotoxic T lymphocytes (CTL), and isoelectric focusing (IEF) gel electrophoresis have demonstrated heterogeneity in the B27 antigen. We have previously identified a unique variant molecule of B27 designated B27d which is distinguished from other B27 variants by isoelectric point, serologic reactivity, and by a cloned CTL recognition (1). In this

Table 1. Amino Acid Substitutions at Variable Positions in the α 1 and α 2 Domains of the Six HLA-B27 Variants

Variant Allele*	Alpha 1 Domain					Alpha 2 Domain		
	59	74	77	80	81	114	116	152
B27a,B27.1	Tyr	Asp	Asp	Thr	Leu	His	Asp	Val
B27b,B27.3	–	–	Ser	–	–	–	–	Glu
B27d	His	–	–	–	–	–	–	–
B27e,B27.2	–	–	Asn	Ile	Ala	–	–	–
B27f	–	Tyr	Asn	–	Ala	–	–	–
B27D,B27.4	–	–	Ser	–	–	Asp	Tyr	Glu

*Different designations of B27 variants were correlated by comparing IEF and serological reactivity patterns as has been previously described (1).

study, we report the cloning and sequencing of the gene encoding B27d.

A partial genomic library from the B27d variant cell line "CHI" (HLA-A31,32;B27d,18;Cw2,w7) was constructed in bacteriophage λ 607 with *Eco*RI-cleaved size-selected DNA fragments of approximately 6–8 kb in length. This size-selection was performed to increase the probability of cloning HLA-B locus genes which were assumed to be contained in either 6.5 or 6.8 kb *Eco*RI fragments. Three clones of 6.8 kb inserts were isolated and each hybridized with a nearly full-length HLA-B7 cDNA probe and with two genomic probes derived from the 5′- and 3′- ends of a HLA-B7 gene, suggesting that they might contain full-length B locus genes. Two of the clones, Y317 and Y442, showed identical restriction endonuclease maps with that reported for a gene encoding B27a (2). Each of the three clones was transfected into the mouse fibroblast cell line J27 which expresses human β_2-microglobulin. Transfected cells were examined for the surface expression of HLA molecules by direct immunofluorescence using FITC-conjugated MAb. The J27 cells transfected with clones Y317 and Y442 reacted with the anti-class I MAb W6/32, the B27-reactive MAb P7.1, and the B27-specific MAb GS 145.2. In contrast, the J27 cells transfected with the third clone reacted with MAb W6/32 but not with MAb P7.1 or GS 145.2. These results indicated that clones Y317 and Y442 both encode for the B27d molecule. A nucleotide sequence of exons, introns, and 3′-untranslated region of clone Y442 was determined. In the coding region, the B27d gene was found to differ from the B27a gene by only a single base at codon 59 (B27a: *T*AT – B27d: *C*AT) in exon 2. This nucleotide change results in Tyr to His substitution at position 59 in the α 1 domain. His residue at position 59 in B27d is unique among the known class I sequences. This single amino acid change is apparently sufficient to alter the epitope(s) recognized by one B27-specific antibody and a B27a-specific CTL clone. Previous primary structural analysis of the other five B27 variants has revealed differences of 2–4 amino acids (3). The combined structural data on the B27 variants indicate that HLA-B27 represents a family of closely related B locus alleles that share the B27 allospecificity and differ by a limited number (one to four) of amino acid substitutions (see Table 1).

Acknowledgments. This work has been supported by CA 18029 awarded by the National Cancer Institute, National Institutes of Health and by a grant from Genetic Systems Corporation, Seattle, WA, USA.

References

1. Choo SY, Antonelli P, Nisperos B, Nepom GT, Hansen JA. Six variants of HLA-B27 identified by isoelectric focusing. Immunogenetics 1986;23:24–29.
2. Weiss EH, Kuon W, Dorner C, Lang M, Riethmuller G. Organization, sequence and expression of the HLA-B27 gene: A molecular approach to analyze HLA and disease associations. Immunobiol 1985;170:367–380.
3. Rojo S, Aparicio P, Choo SY, Hansen JA, Lopez de Castro JA. Structural analysis of an HLA-B27 population variant, B27f: Multiple patterns of amino acid changes within a single polypeptide segment generate polymorphism in HLA-B27. J Immunol 1987;139:831–836.

Author Affiliations

Sung Yoon Choo, *Thomas St. John*, *Harry T. Orr*, and *John A. Hansen*, The Puget Sound Blood Center, Fred Hutchinson Cancer Research Center, and University of Washington School of Medicine, Seattle, WA; and the Department of Laboratory Medicine, University of Minnesota, Minneapolis, MN, USA

CTL Recognize Certain Bacterial Antigens in Association with HLA-B27*

Carmel M. Edwards, John S. Sullivan, Linda I. Upfold, and Andrew F. Geczy

Abstract: Cytotoxic T lymphocytes (CTL) can be raised by stimulating in vitro the peripheral blood mononuclear cells (PBMC) of HLA-B27-positive normal individuals (B27$^+$AS$^-$) with autologous cells "modified" by cross-reactive bacterial antigens or modifying factor. These CTL specifically lyse PBMC from HLA-B27-positive patients with ankylosing spondylitis (B27$^+$AS$^+$) but not PBMC from HLA-B27-negative AS patients (B27$^-$AS$^+$) or from HLA-B27-positive (B27$^+$AS$^-$) or -negative (B27$^-$AS$^-$) normal controls. Studies with HLA-loss variants of a B27-positive lymphoblastoid cell line have shown that mutants that no longer express B27 cannot be "modified" with factor and they fail to function as targets in the CTL assay. These results suggest that CTL can recognize certain bacterial antigens in the context of HLA-B27.

The importance of the association between the tissue type HLA-B27 and ankylosing spondylitis remains undisputed, but its biological significance is still unresolved. Our group has extensively studied the cross-reactivity between certain enteric bacteria and HLA-B27 and/or associated structure on the surface of cells from HLA-B27 positive patients with ankylosing spondylitis (B27$^+$AS$^+$). We have demonstrated that antisera raised against a broad spectrum of enteric bacteria are capable of lysing the cells of B27$^+$AS$^+$ individuals, but fail to lyse those of B27$^+$ or B27$^-$ normal individuals (B27$^+$AS$^-$, B27$^-$AS$^-$) (1,2). In addition, these bacteria shed into their culture media a factor that is capable of "modifying" B27$^+$AS$^-$ cells such that they behave serologically as B27$^+$AS$^+$ cells (3,4). We speculate that in B27$^+$ individuals this "modification" may lead to the development in vivo of cytotoxic T lymphocytes (CTL) against modified B27. Indeed, we have shown that CTL can be raised in vitro by stimulating B27$^+$AS$^+$ cells with autologous cells modified by the cross-reactive bacterial antigen/s (factor) (5). The importance of the antigen HLA-B27 to the expression of this cell surface modification has been supported by studies with HLA loss variants of a B27$^+$ lymphoblastoid cell line (LCL). We report here that, even after "modification," those mutants that no longer express HLA-B27 are unable to function as targets for the in vitro CTL.

Although it is 14 years since the original discovery of the strong association between HLA-B27 and AS (6,7), it still remains to be elucidated whether the antigen HLA-B27 itself plays a key role in the pathogenesis of the disease or whether it is in close linkage disequilibrium with a disease susceptibility gene (8). Failure to detect such a disease susceptibility gene, together with a lack of evidence in family studies for independent segregation of HLA-B27 from AS, have cast doubts on the latter hypothesis. The work presented in this paper, with HLA loss variants of a B27$^+$ LCL, support the concept of a direct role for HLA-B27 in disease pathogenesis.

The method used for the in vitro production of CTL to the HLA-B27-associated determinant(s) was that of Geczy et al. (5), with minor modifications. Briefly, CTL were raised by incubating 10^7 peripheral blood mononuclear cells (PBMC) (responder cells) with 10^7 in vitro "modified" mitomycin-treated autologous PBMC (stimulator cells) in 20 ml RPMI-1640 medium supplemented with 20% human AB serum. After six days at 37°C in 5% CO_2, a standard 4-hour CTL assay was performed. In order to establish their specificity, a large number (n = 15) of CTL were raised in this manner and subsequently tested against a range of targets (all ≥ 5) that included cells from the following categories: B27$^+$AS$^+$, B27$^-$AS$^+$, B27$^+$AS$^-$, B27$^-$AS$^-$, as well as "modified" B27$^+$AS$^-$ and B27$^-$AS$^-$ cells. The results clearly indicate that these CTL specifically recognize B27$^+$AS$^+$ cells and B27$^+$AS$^-$ cells modified with cross-reactive culture filtrate. To confirm the requirement for HLA-B27 in the specificity of these CTL, modified and unmodified HLA loss variants of an HLA-B27$^+$ LCL (MM) have been used as targets. The mutant lines of LCL MM were originally derived by Nicklas and Bach (9); the HLA phenotypes expressed by each of the mutants are listed in Table 1.

Table 2 summarizes the results of experiments where these loss mutants have been used as targets for the in vitro generated CTL. Clearly, the only cells capable of being lysed by these CTL are "modified" MM and "modified" 4F4. Therefore, it appears that both HLA-B27 and an associated structure conferred on the cell line by modification with "cross-reactive" culture filtrate, are necessary for specific target cell lysis.

As we have previously shown that antisera raised in rabbits to "cross-reactive" organisms specifically recognize only B27$^+$AS$^+$ cells and "modified" B27$^+$AS$^-$ cells in an antibody-complement dependent ^{51}Cr release assay (3,4), we decided to test the loss mutants in this assay

Table 1. HLA Typings of B27 Loss Variants of a Lymphoblastoid Cell Line (MM)

Cell Line	HLA Type			
MM	A2	Cw5	B44	DR1
	A24	Cw2	B27	DR4
4F4	–	–	–	–
	A24	Cw2	B27	DR4
1D4	A2	Cw5	B44	DR1
	–	–	–	–
1B10	A2	Cw5	B44	DR1
	A24	Cw2	–	DR4

Immunobiology of HLA
Volume II

Table 2. CTL to B27-Associated Determinants Recognize Modifying Factor in the Presence of HLA-B27

Cell Line	Modifying Factor*	% Specific ^{51}Cr Release†
MM	Absent	5 ± 8
	Present	67 ± 12
4F4	Absent	3 ± 3
	Present	74 ± 8
1D4	Absent	9 ± 10
	Present	12 ± 11
1B10	Absent	5 ± 5
	Present	5 ± 6

*Cell lines were "modified" by incubating 10^7 cells, overnight at 37°C in 5% CO_2 in 1.9 ml RPMI-1640 medium supplemented with 20% human AB serum, with 100 µl of a 20-fold concentrate of culture filtrate from a "cross-reactive" bacterium.

†% specific ^{51}Cr release $= \frac{\text{experimental-spontaneous release}}{\text{total-spontaneous release}}$

system. The results given in Table 3 indicate that for successful recognition by "cross-reactive" antisera, both HLA-B27 and the "disease-associated" determinant must be present on the cell, thus lending support to our CTL findings.

To further address the role of HLA-B27 in disease expression by B27$^+$AS$^+$ patients, we have used immunochemical techniques to confirm the HLA antigen status of LCL MM and its HLA loss variants. Briefly, immunoprecipitations were performed on radioactively labeled detergent extracts of the various cell lines using an anti-B27 monoclonal antibody. The immunoprecipitated material was then electrophoresed on a one-dimensional isoelectric focusing gel, the results of which are shown in Figure 1. This method is described in detail by Upfold et al. (Upfold L.I., J.S. Sullivan, L.E. McGuigan, and A.F. Geczy, manuscript in preparation). Figure 1 shows that MM and 4F4 have an identifiable band that corresponds with HLA-B27. Immunoprecipitation of these cell lines was also performed with an anti-class I monoclonal antibody – W6/32 (results not shown)

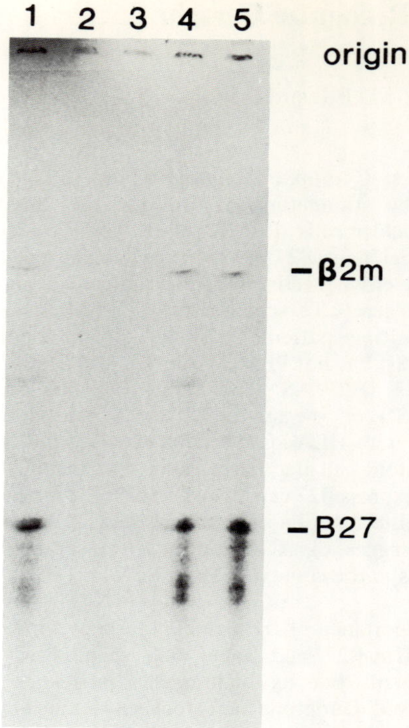

Figure 1. HLA-B27 loss variants of a lymphoblastoid cell line (MM) were immunoprecipitated with an anti-B27 monoclonal antibody, then electrophoresed on a one-dimensional isoelectric focusing gel. Lane 1 = MM; 2 = 1B10; 3 = 1D4; 4 = 4F4; 5 = B27$^+$ control (LCL jes thom).

and again, only MM and 4F4 displayed evidence of HLA-B27.

This work has demonstrated, both by cellular techniques (CTL assay) and serologically (antibody-complement dependent ^{51}Cr release asay), that the "disease associated" determinant (conferred on the cells by incubation with "cross-reactive" culture filtrate) is not detectable on the HLA-B27 loss variants which have been "modified." It appears, therefore, that either HLA-B27 is necessary for the expression of the disease-associated determinant, or that HLA-B27 forms part of the cross-reactive determinant or complex.

It could be argued that using this single series of HLA loss mutants to determine the involvement of HLA-B27 in AS may not be conclusive as these cell lines may have lost other gene(s) in addition to HLA-B27. However, in view of the large number of PBMC of various HLA specificities that have also been used as targets in our assays, it is reasonable to assume that HLA-B27 is a necessary element in the recognition by CTL. Evidence from loss mutants from other sources would clarify this matter (10). Preliminary studies in the in vitro assay have demonstrated that prior incubation of the "modified" parental line MM with a monoclonal anti-B27 antibody completely blocks CTL-induced lysis of this cell line (results not shown). This provides additional evi-

Table 3. Antisera to "Cross-Reactive" Bacteria Recognize Modifying Factor in the Presence of HLA-B27

Cell Line	Modifying Factor*	% Specific ^{51}Cr Release†
MM	Absent	1 ± 2
	Present	57 ± 11
4F4	Absent	2 ± 2
	Present	72 ± 7
1D4	Absent	3 ± 6
	Present	3 ± 4
1B10	Absent	1 ± 2
	Present	1 ± 1

*Cell lines were "modified" by incubating 10^7 cells, overnight at 37°C in 5% CO_2 in 1.9 ml RPMI-1640 medium supplemented with 20% human AB serum, with 100 µl of a 20-fold concentrate of culture filtrate from a "cross-reactive" bacterium.

†% specific ^{51}Cr release $= \frac{\text{experimental-spontaneous release}}{\text{total-spontaneous release}}$

dence that in the in vitro CTL assay, HLA-B27 is necessary for the detection of the "disease-associated" determinant or complex.

In the present study we have shown that the in vitro CTL only recognize bacterial antigens in association with HLA–B27. The implication of this for disease pathogenesis is that if CTL specific for "modified" B27⁺ cells were generated in vivo, they might kill target cells bearing HLA-B27 in association with certain factors derived from "cross-reactive" bacteria. Such an effector mechanism could trigger an inflammatory reaction, a characteristic lesion in the early stages of AS, and in some way initiate the disease process in susceptible individuals.

Acknowledgments. This work was supported by a grant (870131) from the National Health and Medical Research Council of Australia.

References

1. Geczy AF, Alexander K, Bashir HV, Edmonds JP, Upfold L, Sullivan J. HLA-B27, *Klebsiella*, and ankylosing spondylitis: Biological and chemical studies. Immunol Rev 1983;70:23.
2. Prendergast JK, Sullivan JS, Geczy AF, Upfold LI, Edmonds JP, Bashir HV, et al. Possible role of enteric organisms in the pathogenesis of ankylosing spondylitis and other seronegative arthropathies. Infec Immun 1983;41:935.
3. Geczy AF, Alexander K, Bashir HV, Edmonds J. A factor(s) in *Klebsiella* culture filtrates specifically modifies an HLA-B27-associated cell-surface component. Nature 1980;283:782.
4. Sullivan J, Upfold L, Geczy AF, Bashir HV, Edmonds JP. Immunochemical characterization of *Klebsiella* antigens which specifically modify an HLA-B27- associated cell-surface component. Hum Immunol 1982;5:295.
5. Geczy AF, McGuigan LE, Sullivan JS, Edmonds JP. Cytotoxic T lymphocytes against disease-associated determinant(s) in ankylosing spondylitis. J Exp Med 1986;164:932.
6. Schlosstein L, Terasaki PI, Bluestone R, Pearson CM. High association of HLA antigen B27 with ankylosing spondylitis. N Engl J Med 1973;288:704.
7. Brewerton DA, Caffrey M, Hart FD, James DCO, Nicholls A, Sturrock RD. Ankylosing spondylitis and H2-A27. Lancet 1983;i:904.
8. Upfold LI, Sullivan JS, Prendergast JK, Geczy AF. HLA-B27: Speculations on the nature of its involvement in ankylosing spondylitis. Prog Allergy 1985;36:177.
9. Nicklas JA, Miyachi Y, Taurog JD, Wee S-L, Chen L-K, Grumet FC, Bach FH. HLA loss variants of a B27⁺ lymphoblastoid cell line: Genetic and cellular characterization. Hum Immunol 1984;11:19.
10. Levine F, Erlich H, Mach B, Leach R, White R, Pious D. Deletion mapping of HLA and chromosome 6p genes. Proc Natl Acad Sci USA 1985;82:3741.

Author Affiliation

Carmel M. Edwards, John S. Sullivan, Linda I. Upfold, Andrew F. Geczy, N.S.W. Red Cross Blood Transfusion Service, Sydney, Australia

HLA-Bw62 Variants Identified by a Cytotoxic T-Lymphocyte Clone and Isoelectric Focusing Gel Electrophoresis

Li-an Fan, Sung Yoon Choo, Patrick G. Beatty, and John Hansen

Abstract: We have used a cytotoxic T-lymphocyte clone (CTLc) and isoelectric focusing (IEF) gel electrophoresis to test for heterogeneity among HLA-Bw62 antigens. One CTLc, specific for Epstein-Barr virus (EBV) transformed B-lymphoblastoid cell lines (LCL) and restricted by an HLA class I determinant, was found to react with 13 of 15 Bw62⁺ target cells tested. IEF gel electrophoesis showed these CTLc-defined variants differed by net charge and therefore represent Bw62 structural variants. Altogether, four distinct Bw62 molecules were identified. These results indicate that Bw62 is a public specificity shared by four or more unique B-locus alleles.

CTL specific for virus-infected cells recognize both virus-associated antigens and certain restriction elements usually associated with HLA class I molecules (1–3). We have prepared antigen-specific HLA-restricted CTLc by culturing peripheral blood lymphocytes with cells from an autologous EBV-transformed LCL.

Preliminary data from a specificity screening experiment suggested that one CTLc, No. 6, might be specific for LCL expressing HLA-Bw62. LCL from 101 unrelated donors, including 75 cell lines from the 10WS, were tested in two or more independent experiments. Thirteen LCL were lysed by CTLc No. 6 (Table 1). A significant correlation was found with HLA-Bw62 ($\chi^2 = 84.7$, $p < 0.0001$). All thirteen of the LCL lysed by clone No. 6 were Bw62⁺; however, two additional Bw62⁺ LCL, "WEWAK I" and "BB", were not lysed. The donor for LCL "BB" was the propositus for a family study that included two parents and three siblings (Table 2). The mother of this family, "DB," was homozygous for Bw62. The propositus "BB" and sibling 1 and sibling 2 shared the maternal "d" haplotype while sibling 3 inherited the

Table 1. Reactivity of CTLc No. 6 Tested with a Target Cell Panel of EBV-Transformed B-Lymphoblastoid Cell Lines*

Target Cells†	HLA A	HLA B	HLA DR	Specific Lysis‡
9024	2,11	w62,35	4	25%
9031	2	w62	4	41%
9032	2	w62	4	90%
9060	1	w62	w6	95%
9064	2	w62	w6	45%
9071	31	w62	w8	66%
9072	31	w62	w8	95%
HA	2	w62	4	74%
DWI	2,24	w62,35	8	57%
MJ4	2	w62,35	4	77%
CRI	2,31	w62	6(14)	61%
WWI	2,24	w62,w60	6(14)	48%
HAS15	2	w62,w54	4	32%
WEWAK I	11,24	w62,27	2	0%
BB	31,24	w62,51	4,8	0%

*CTLc No. 6 was primed in mixed lymphocyte culture with irradiated cells from an autologous EBV-transformed LCL, and cloned by limiting dilution. The donor "HA" was included in the 9WS (#9w0403) and the 10WS (#10w0151), and is known to be homozygous for the HLA genotype: A2;Cw3;Bw62;Bw6; DR4;DRw53;DQw3 (TA10⁻);Dw4. LCL from donor "HA" were irradiated with 9000 R and 5×10^4 cells were co-cultured with 2×10^6 autologous peripheral blood lymphocytes in 3 ml nickel wells (Costar, MA) for 10 days. These primed cells were then cloned by limiting dilution in the presence of EBV-LCL and recombinant IL-2 (Amgen, 25 U/ml) at day 4 after restimulation. Clones cytolytic for the priming EBV-LCL but not for autologous PBL-PHA blasts were isolated and expanded for further analysis.

†Only $Bw62^+$ target cells are shown. All 86 $Bw62^-$ target cells tested gave specific release values of $\leq$ 15% and thus were not lysed (data not shown).

‡CTL cytolytic activity was assayed by incubating 2×10^4 CTL effector cells with 1×10^3 ^{51}Cr-labeled target cells for 4 h in a 0.2 ml U-bottomed microtest plate. Total ^{51}Cr-release (TR) was measured following detergent lysis of target cells, and spontaneous release (SR) was measured by culturing ^{51}Cr-labeled target cells alone. Specific release was calculated as follows: % specific release = (experimental − SR)/(TR-SR) × 100.

maternal "c" haplotype. LCL from the propositus and siblings 1 and 2 were not lysed by CTLc No. 6; however, LCL from the mother and sibling 3 were lysed, suggesting that the Bw62 molecule on "c" haplotype but not the Bw62 molecule on "d" haplotype was recognized by clone No. 6.

Table 2. Segregation Analysis of HLA-Bw62 Variants*

Target Cell	Relationship	HLA Haplotypes	Specific Lysis
"RB"	Father	a,b	0%
"DB"	Mother	c,d	55%
"BB"	Propositus	a,d	0%
"AB"	Sibling 1	a/b,d	0%
"VB"	Sibling 2	b,d	−1%
"SB"	Sibling 3	b,c	32%

*HLA haplotypes: paternal, "a" − A31;B51;Cw3;DR4/ "b" − A24;B27;C−;DR4; maternal, "c" − A24;Bw62;Cw1;DR4/ "d" − A24;Bw62;Cw1;DRw8. Individual "AB" has inherited a paternal recombinant haplotype ("a/b").

To determine if the recognition by CTLc No. 6 involved an HLA class I molecule, monoclonal antibody (MAb) blocking studies were performed using the anti-class I MAb, W6/32 (4), and the anti-class II MAb, 7.2 (5). Significant blocking was observed only with the MAb W6/32, and not with the MAb 7.2 (Table 3), suggesting that the restriction element for this anti-EBV CTLc was associated with an HLA class I molecule.

These data suggested that CTLc No. 6 was specific for a nominal antigen present on EBV-transformed B cell lines and restricted by an epitope present on certain Bw62 antigens. To determine if this variable reactivity of $Bw62^+$ cells with CTLc No. 6 was indicative of Bw62 structural differences, Bw62 molecules from selected cells were immunoprecipitated with MAb 4E (6) and analyzed by IEF gel (7). Four distinct variants of Bw62 (Bw62a, Bw62b, Bw62c, and Bw62d) were identified on IEF (Fig. 1). The family study of "BB" was informative for demonstrating the segregation of two of the Bw62 variants, Bw62a and Bw62d, with haplotypes "c" and "d", respectively (Fig. 1, lanes 2 and 1). "WEWAK I," which was not lysed by CTLc No. 6, expresses a Bw62b (Fig. 1, lane 4).

From the combined data of panel and family studies, it was apparent that CTLc No. 6 was restricted by an epitope shared by Bw62a and Bw62c. The Bw62 variants lacking the restriction element recognized by CTLc No. 6, Bw62b and Bw62d, were each present in only a single LCL from our original cell panel. However, the failure of the Bw62d variant to elicit specific recognition by CTLc No. 6 was confirmed by family study. The two Bw62 variants nonreactive with CTLc No. 6 originated from non-Caucasian donors. LCL "WEWAK I" (Bw62b) was derived from an individual who was a native of New

Table 3. Competitive Blocking of CTL No. 6 Cytolytic Activity with Anti-HLA Class I and Class II Monoclonal Antibodies*

MAb W6/32				MAb 7.2				
22.5	5.6	0.56	0.056	Dilutions				
μg/ml	μg/ml	μg/ml	μg/ml	1:40	1:160	1:1600	1:16000	Medium Control
41%	62%	74%	75%	74%	87%	80%	80%	78%

*CTLc No. 6 was assayed for cytotoxic activity against EBV-LCL "HA" at an effector to target cell ratio of 10:1. MAb W6/32 (anti-HLA class I, nonpolymorphic determinant) and MAb 7.2 (anti-HLA class II, DR+DQ, nonpolymorphic determinant) were added to the cytotoxic assay at increasing concentrations. Data represent mean % specific release for duplicate or triplicate cultures.

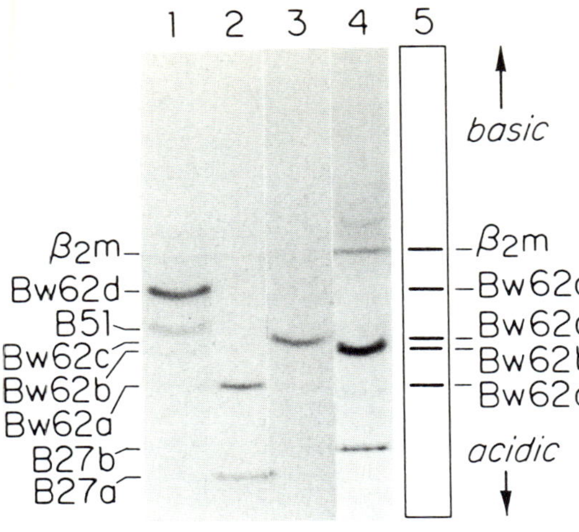

Figure 1. Bw62 molecules representing four distinct Bw62 variants are compared by IEF gel electrophoresis. HLA-B locus antigens were specifically immunoprecipitated with MAb 4E from ^{35}S-methionine labeled cells and analyzed on the IEF gel. Lane 1 – LCL "BB"; lane 2 – LCL "SB"; lane 3 – LCL "HA"; lane 4 – LCL "WEWAK I"; and lane 5 – a schematic composite representation of the relative IEF positions noted for the four Bw62 variants (Bw62a, Bw62b, Bw62c, and Bw62d, from most acidic to most basic).

Guinea (8) and LCL "BB" (Bw62d) from an individual who was a native American from the Pacific Northwest. These results are similar to those observed by others which have demonstrated that the Bw62 antigens of Orientals are different from the Bw62 antigens commonly found in Caucasians (9,10).

In conclusion, the serologically defined HLA-Bw62 allospecificity does not represent one B locus product, but appears to be either a crossreactive complex of determinants or a single distinct public determinant shared by four or more unique B locus alleles.

Acknowledgments. This work has been supported by grant number CA 18029 awarded from the National Cancer Institute and by grant number HL 33775 from the National Heart, Lung and Blood Institute, National Institutes of Health. Dr. Fan is a Research Associate from Shanghai Institute of Immunology, Shanghai Second Medical University, Shanghai, China.

References

1. Gaston JSH, Richinson AB, Epstein MA. Epstein-Barr virus-specific cytotoxic T lymphocytes as probes of HLA polymorphism (heterogeneity of T cell-restricting determinants associated with the serologically defined HLA-A2 antigen). J Exp Med 1983;158:280.
2. Kress HG, Kreth HW. HLA restriction of secondary mumps-specific cytotoxic T lymphocyte. J Immunol 1982;129:844.
3. Toubert A, Gomard E, Grumet FC, Amor B, Muller JY, Levy J-P. Identification of several functional subgroups of HLA-B27 by restriction of the activity of antiviral T killer lymphocytes. Immunogenetics 1984;20:513.
4. Parham P, Barnstable CJ, Bodmer W. Use of a monoclonal antibody (W6/32) in structural studies of HLA-A,B,C antigens. J Immunol 1979;123:342.
5. Hansen JA, Martin PJ, Nowinski RC. Monoclonal antibodies identify a novel T-cell antigen and Ia antigens of human lymphocytes. Immunogenetics 1980;10:247–260.
6. Yang SY, Morishima Y, Collins NH, Alton T, Pollack MS, Yunis EJ, et al. Comparison of one-dimensional IEF patterns for serologically detectable HLA-A and B allotypes. Immunogenetics 1984;19:217.
7. Choo SY, Antonelli P, Nisperos B, Nepom GT, Hansen JA. Six variants of HLA-B27 identified by isoelectric focusing. Immunogenetics 1986;23:24.
8. Rooney CM, Rickinson AB, Moss DJ, Lenoir GM, Epstein MA. Paired Epstein-Barr virus-carrying lymphoma and lymphoblastoid cell lines from Burkitt's lymphoma patients: Comparative sensitivity to non-specific and to allo-specific cytotoxic responses in vitro. Int J Cancer 1984;34:339.
9. Neefjes JJ, Breur-Vriesendorp BS, van Seventer GA, Ivanyi P, Ploegh HL. An improved biochemical method for the analysis of HLA class I antigens. Definition of new HLA class I subtypes. Human Immunol 1986;16:169.
10. Clark BD, Park MS, Tonai R, Terasaki PI. Serologic and biochemical analysis of HLA-B15 and B5 complexes. Human Immunol 1987;20:47.

Author Affiliations

Li-an Fan, Fred Hutchinson Cancer Research Center and Department of Medicine, University of Washington; *Sung Yoon Choo, Patrick G. Beatty, John A. Hansen*, The Puget Sound Blood Center, Fred Hutchinson Cancer Research Center and Department of Medicine, University of Washington, Seattle, WA, USA

Comparison of the Amino Acid Sequences Encoded by the HLA-Bw47 and HLA-B13 Genes

Jacqueline Zemmour, Peter D. Ennis, Peter Parham, and Bo Dupont

Congenital adrenal hyperplasia due to 21-hydroxylase deficiency (21-OHD) is a disorder of cortisol and aldosterone biosynthesis that is closely linked to the HLA-B locus (1,2). HLA-Bw47, a very rare antigen, is strongly associated with a deletion of 21-OHB gene among patients carrying 21-OHD. This allele frequently occurs on the *HLA-A3;Cw6;DR7* haplotype which always includes a null allele at the locus encoding the fourth component of complement C4B. Moreover, the more common allele HLA-B13, which is not associated with the deletion of 21-OHB and C4B genes, shares the same pattern of genetic linkage disequilibrium. In addition, these HLA-B13 and HLA-Bw47 alleles show serological cross reactivity with alloantisera and have identical isoelectric points (3). On the basis of these structural similarities, we have postulated that Bw47 was the product of the B13 gene which became mutated during the genetic deletion of the 21-hydroxylase locus (4).

Here, we report the derived amino acid sequences of the HLA-Bw47 and HLA-B13 genes. These alleles differ by 14 amino acid substitutions. The complex pattern of substitutions indicates that HLA-Bw47 and HLA-B13 alleles are not related by a simple mutational event.

HLA-B13 and HLA-Bw47 genes were isolated from partial genomic libraries constructed from the homozygous LBF (B13) and PLH (Bw47) cell lines as described elsewhere (5). These libraries were screened with HLA locus-specific probes (6). Four putative B locus genes were obtained from the PLH library and two from the LBF library. These clones contained 6.5 Kb inserts that were subcloned into pUC18 (B13) and SP65 (Bw47) plasmids. These cloned genes were identified by transfection into mouse L cells (7). The expressed product of clone 21B codes for B13, which reacted with the monospecific anti-B13 monoclonal antibody, SY1 (8). The clone J3Z codes for Bw47 as assessed by isoelectric focusing, reactivity with antibody 4E, which reacts with all HLA-B molecules, and its lack of reactivity with SY1.

The comparison of the nucleotide sequences between HLA-B13 and HLA-Bw47 showed 28 nucleotide differences resulting in 14 amino acid substitutions: five in the α1 domain, eight in the α2 domain, and one in the transmembrane region. These substitutions are clustered into regions previously described as polymorphic in HLA-A,B,C molecules. In opposition to HLA-B13, which shows five novel substitutions found throughout the α1 and α2 domains, HLA-Bw47 has no residues that are not shared with other HLA molecules, which probably explains the lack of specific Bw47 epitopes. The amino acid comparison of the external domains of HLA-Bw47 with other HLA-B molecules reveals regions of particular homology (Fig. 1, Table 1). In the α1 domain HLA-B13 and HLA-Bw47 only share the first 32 residues, whereas HLA-B44 and HLA-Bw47 are more closely related, only differing by three residues clustered into a region from residue 77 to residue 82. Similar substitution patterns have been described in H-2Kb mutants and are hypothesized to result from gene conversion events (9). Our results indicate that the exons coding for the α1 domain of HLA-B44 and HLA-Bw47 may have recently diverged from a common ancestor, while the relationship between HLA-B13 and HLA-Bw47 probably involves a more ancient divergence. It is striking that the region encompassing residues 62-76, which is in the more polymorphic part of the class I molecule, has an identical sequence in Bw47, B13, B44, Bw60, and B40. This structural similarity could explain the serological cross reactivity observed between molecules in the HLA-B12 and HLA-B40 groups (10).

Table 1. Pairwise Differences Between HLA-B13 and HLA-Bw47 with Other HLA-B Amino Acid Sequences in the α1 and α2 Domains

	B44	B27.1	B27.2	Bw58	B7.1	Bw60	B40*	B13	Bw47
α1 Domain									
B13	4	14	13	12	17	9	11	–	5
Bw47	3	10	13	14	16	5	7	5	–
α2 Domain									
B13	11	7	7	7	12	11	6	–	8
Bw47	7	2	2	8	11	10	6	8	–
α1 + α2 Domains									
B13	15	21	20	19	29	20	17	–	13
Bw47	10	12	15	22	27	15	13	13	–

Immunobiology of HLA
Volume II

α1 domain

```
           1       10        20        30        40        50        60        70        80        90
consensus  GSHSMRYFYTSVSRPGRGEPRFIAVGYVDDTQFVRFDSDAASPRMEPRAPWIEQEGPEYWDRETQIVKAQSQTDREDLRTLRGYYNQSEA
      B44  ----------AM-----------T-------L--------T---K---------------------S-TNT--Y--N---AAR-------
      B13  ----------AM-----------T----------------T----A--------------------S-TNT--Y--N---ALR-------
     Bw47  ----------AM-----------T-------L--------T---K---------------------S-TNT--Y-------LR-------
    B27.1  --------H--------------T-------L------------E---------------------C--KA----------LR-------
    B27.2  --------H--------------T-------L------------E---------------------C--KA-----N--IALR-------
    B27.3  --------H--------------T-------L------------E---------------------C--KA-----S----LR-------
    B27.4  --------H--------------T-------L------------E---------------------C--KA-----S----LR-------
     Bw58  ----------AM--------------------------------T----------------G--RNM--SA--Y--N--IALR-------
     B7.1  -----------------------S--------------------E-----------------N---Y---A-----S--N----------
     B7.2  -----------------------S--------------------E-----------------N---Y---A-----S--N----------
     Bw60  --------H-AM-----------T-------L--------T---K---------------------S-TNT--Y--S--N----------
     B40*  --------H--------------T-------L--------T---K---------------------S-TNT--Y--S--N----------
      A24  --------S---------------------------------Q------------------E--GK---H------N--IALR-------
```

α2 domain

```
                    100       110       120       130       140       150       160       170       180
Consensus  GSHTLQRMYGCDVGPDGRLLRGYHQYAYDGKDYIALNEDLRSWTAADTAAQITQRKWEAARVAEQLRAYLEGTCVEWLRRYLENGKETLQRA
      B44  ---II------------------D-D--------------S------------------------D------L---S---------------
      B13  ----W-T-----L---------HN-L--------------S-------------L-----------------E-------------------
     Bw47  --------F----------------D--------------S-------------------------------E-------------------
    B27.1  ------N------------------D--------------S-------------------------------E-------------------
    B27.2  ------N------------------D--------------S-------------------------------E-------------------
    B27.3  ------N------------------D--------------S--------------------E----------E-------------------
    B27.4  ------N----------------D----------------S--------------------E----------E-------------------
     Bw58  ---II-------L---------HD-S--------------S-------------------------------L-------------------
     B7.1  ------S---------------HD-------------------------------------E---R------E-------------DK-E--
     B7.2  ------S---------------HD-X-----------------------------------E---R------E-------------DK-E--
     Bw60  ----------------------HN----------------------------S---L---------------E-------------DK-E--
     B40*  ------S---------------HN------------------------------------------------E-------------------
      A24  ------M-F-----S---F-----------------K----------M-----K------H----Q---------DG---------------
```

α3 domain

```
                  190       200       210       220       230       240       250       260       270
Consensus  DPPKTHVTHHPISDHEATLRCWALGFYPAEITLTWQRDGEDQTQDTELVETRPAGDGTFQKWAAVVVPSGEEQRYTCHVQHEGLPKPLTLRW
      B44  ----------------------S---------------------------------R-----------------------------------
      B13  --------------------------------------------------------R-----------------------------------
     Bw47  --------------------------------------------------------R-----------------------------------
    B27.1  --------------------------------------------------------R-----------------------------------
    B27.2  --------------------------------------------------------R-----------------------------------
    B27.3  --------------------------------------------------------R-----------------------------------
    B27.4  --------------------------------------------------------R-----------------------------------
     Bw58  -----------V--------------------------------------------R-----------------------------------
     B7.1  --------------------------------------------------------R-----------------------------------
     B7.2  --------------------------------------------------------R-----------------------------------
     Bw60  --------------------------------------------------------R-----------------------------------
     B40*  --------------------------------------------------------R-----------------------------------
      A24  ------M-------------------------------------------------------------------------------------
```

Figure 1. Amino acid comparison of HLA-B13 and HLA-Bw47 in the three external domains with other HLA-B genes. The dashes correspond to identities with a consensus sequence deduced from 23 HLA-A, -B and -C amino acid sequences (12).

In the α2 domain HLA-Bw47 diverges from HLA-B13 and HLA-B44, but is closely related to HLA-B27. The differences between Bw47 and B27 comprise a pair of substitutions at residues 97 and 99. The serological cross reactivity demonstrated between HLA-Bw47 and B27 (11) probably involve residues of the α2 domain.

The structural comparison of the external domains of HLA-B13 and HLA-Bw47 shows that they differ by a large number of substitutions. It is likely that the cross reactivity between B13 and Bw47 arises from epitopes that involve residues of the α1 domain. HLA-Bw47 is more closely related to HLA-B44 in the α1 domain and to HLA-B27 in the α2 domain. This raises the possibility that a reciprocal recombination between these or related genes may generate an intermediate designated as HLA-Bx. Then, a non-reciprocal recombination between the HLA-Bx and HLA-B13 haplotypes might generate HLA-Bw47 with concomitant deletion of 21-OHB and C4B genes. This hypothesis is highly speculative but is in agreement with the patterns of genetic linkage disequilibrium and the disease association.

Acknowledgments. This work was supported by NIH grants PO 1 CA22507, PO 30 CA08784 (B. Dupont) and AI24258 (P. Parham). P.P. is a scholar of the Leukemia Society of America.

References

1. White PC, New MI, Dupont B. Congenital adrenal hyperplasia (Part 1). N Engl J Med 1987;316:1519.

2. White PC, New MI, Dupont B. Congenital adrenal hyperplasia (Part 2). N Engl J Med 1987;316:1580.
3. Yang SY, Morishima Y, Collins NH, Alton T, Pollack MS, Yunis EJ, et al. Comparison of one-dimensional IEF patterns for serologically detectable HLA-A and HLA-B allotypes. Immunogenetics 1984;19:217.
4. O'Neill GJ, Dupont B, Pollack MS, Levine LS, New MI. Complement C4 allotypes in congenital adrenal hyperplasia due to 21-hydroxylase deficiency: Further evidence for different allelic variants at the 21-hydroxylase locus. Clin Immunol Immunopath 1982;23:312.
5. Holmes N, Parham P. Exon shuffling in vitro can generate novel HLA class I molecules. EMBO J 1985;4:2849.
6. Koller BH, Sidwell B, DeMars R, Orr HT. Isolation of HLA locus-specific DNA probes from the 3′ untranslated region. Proc Natl Acad Sci USA 1984;81:5175.
7. Wigler M, Pellicer A, Silverstein S, Axel R. Biochemical transfer of single-copy eukaryotic genes using total cellular DNA as donor. Cell 1978;14:725.
8. Yang SY, Chang A, Olivero R, Relias V, Yunis EJ. IEF patterns of HLA-B13 antigens from Orientals and Caucasians. Immunogenetics 1985;21:125.
9. Nathenson SG, Geliebter J, Pfaffenbach GM, Zeff RA. Murine major histocompatibility complex class-I mutants: Molecular analysis and structure-function implications. Ann Rev Immunol 1986;4:471.
10. Darke C. The HLA-B12 and -B40 cross-reactive groups and their serological relationships. Tissue Antigens 1983; 21:281.
11. Grumet FC, Fendly BM, Fish L, Foung S, Engleman EG. Monoclonal antibody (B27M2) subdivising HLA-B27. Human Immunol 1982;5:61.
12. Parham P, Clayberger C, Zorn SL, Ludwig DS, Schoolnick GK, Krensky AM. Inhibition of alloreactive cytotoxic T lymphocytes by peptides from the α2 domain of HLA-A2. Nature 1987;325:625.

Author Affiliations

Jacqueline Zemmour, Bo Dupont, Human Immunogenetics Laboratory, Sloan Kettering Institute for Cancer Research, New York, NY; *Peter D. Ennis, Peter Parham*, Department of Cell Biology, Stanford University, Stanford, CA, USA

Cloning and Analysis of the cDNA Encoding New HLA-C Antigen Cx52

Hajime Takata, Hidetoshi Inoko, Asako Ando, Bunji Watanabe, Kimiyoshi Tsuji, and Hisami Iri

HLA-C locus, a class I antigen locus, frequently has a serologically unclassifiable "blank" specificity (CwBL). It is unclear whether HLA-C genes linked to the characteristic haplotypes of A24-Bw52-CwBL-DR2-DQw1 and Aw33-B44-CwBL-DRw13-DQw1 in Japanese (tentatively named Cx52 and Cx44, respectively) really exist and if so, are actually expressed. To address these questions, we have isolated and characterized some class I antigens cDNA derived from the former haplotype.

DNA probes were prepared from HLA-Cw3 genomic clone p42 (1), which were PvuII-digested 1.0kb fragment (PvuII-1.0kb) including 3′UT region, PvuII-digested 5′-end 1.7kb fragment (PvuII-1.7kb), and 248bp digested with PvuII and Tth111I (C250) (2), including exon 7 and intron 7. HLA-B7 cDNA (pDP001, 1.5kb) (isolated by Biro et al.) which were also employed for DNA probe. They were used after labeling with radioactive ^{32}P.

Southern hybridization experiments revealed that EcoRI-cleaved genomic DNAs from AKIBA, consanguineous HLA-homozygote of the former haplotype, and L-KT11 and KIA1566 with the homozygous latter haplotype were strongly hybridized on 7.9kb and 7.6kb to C250, respectively.

The cDNA library constructed from AKIBA was employed to isolate cDNA clones encoding putative Cx52 antigen. Using colony hybridization with the probes of PvuII-1.0kb, PvuII-1.7kb, and pDP001, 29 cDNA clones, with inserts longer than 1.0kb, were picked up through the screening of 100,000 clones as candidates for class I antigens. By restriction enzyme-mapping analysis of selected clones, three isotypes of cDNA clones were identified as typically represented by pC68, pB55, and pA01. They appeared to encode Cx52, Bw52 and A24, respectively. That is, the restriction enzyme map of pB55 was quite similar to that of B27 cDNA, pBl reported by Szöts et al. On the other hand, that of pA01 was slightly different from that of exons of A24 gene, cos27 sequenced by N'Guyen et al. The difference exists within the second domain, showing a possible difference of the A24 genes between Caucasians and Japanese. The cDNA pC68 appeared to encode a new class I protein, probably Cx52, because the restriction enzyme map (Fig. 1) was not identical to that of Cw3 exons (1), but strongly hybridized to the HLA-C probes, PvuII-1.0kb and C250 from Cw3 genomic clone.

Complete nucleotide sequence alignment of pC68 cDNA indicates higher homology among pC68, Cwl cDNA, Cw2 exons and Cw3 exons. Especially in exon 6 to 8, which is HLA-C locus specific region (2), they show complete identity as demonstrated in Table 1.

Immunobiology of HLA
Volume II

Figure 1. Restriction enzyme map of pC68 cDNA clone in comparison with Cw3 (p42) exon. The arrows indicate the sequencing strategy. P; Pst I, V; PvuII, S; Sac I, R; Rsa I.

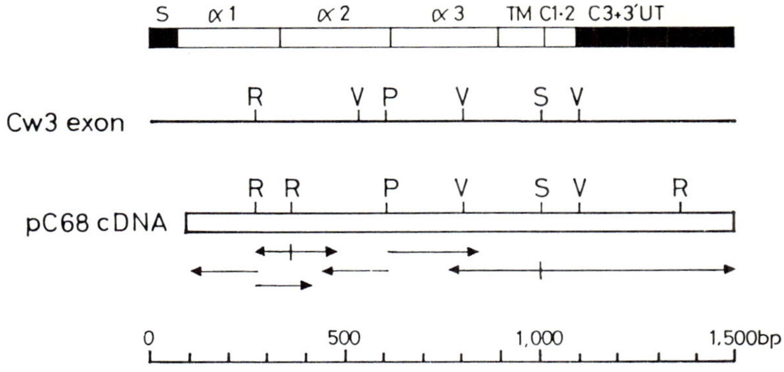

Amino acid sequences are compared in every domain with already reported class I proteins. The results show higher homology (86.3 to 100%) among Cwl, Cw2, Cw3, and new Cx52 protein coded by pC68.

These results support the idea that the inability to define C antigens serologically is not due to HLA-C gene deletion.

Four genomic clones were picked up through the screening of 400,000 clones of AKIBA genomic library as the candidates of the gene encoding Cx52. Studies are in progress to identify the genomic clones encoding Cx52.

Acknowledgments. We thank Dr. B.R. Jordan and Dr. S.M. Weissman for providing Cw3 genomic clone and B7 cDNA clone, respectively.

References

1. Sodyer R, Damotte M, Delovitch TL, Trucy J, Jordan BR, Strachan T. Complete nucleotide sequence of a gene encoding a functional human class I histocompatibility antigen (HLA-CW3). Embo J 1984;3:879.
2. Strachan T, Dodge AB, Smillie D, Dyer PA, Sodyer R, Jordan BR et al. An HLA-C-specific DNA probe. Immunogenetics 1986;23:115.

Author Affiliations

Hajime Takata, *Hisami Iri*, Department of Clinical Laboratories, Keio University School of Medicine, Tokyo; *Hidetoshi Inoko*, *Asako Ando*, *Kimiyoshi Tsuji*, Department of Transplantation II, Tokai University, Kanagawa; *Bunji Watanabe*, Jikei Medical College, Tokyo, Japan

Table 1. Nucleotide Sequence Homology(%) in Exon 6-8 of New HLA-C cDNA, pC68 in Comparison with Already Reported Class I cDNAs and Genes(Exon)

		Cwl	Cw2	Cw3	A24	B27	Reported by
Cx52	(pC68)	100.0	100.0	100.0	84.9	91.9	
Cwl	(pC1)	–	100.0	100.0	84.9	91.9	Szöts et al.
Cw2	(I7)		–	100.0	84.9	91.9	Güssow et al.
Cw3	(p42)			–	84.9	91.9	Sodyer et al.
A24	(cos 27)				–	90.7	N'Guyen et al.
B27	(pB1)					–	Szöts et al.

A C Locus-Related HLA Class I Molecule on Human Trophoblast

Shirley A. Ellis, Tom Strachan, and Andrew J. McMichael

Abstract: We have previously demonstrated the presence of an HLA class I-like molecule on human extravillous trophoblast and on a choriocarcinoma cell line, BeWo, neither of which appear to express HLA class I A, B, or C (1). In this study, we describe the characterisation of an HLA class I molecule at the nucleotide level, in the BeWo cell line. Three class I DNA probes were used in a Northern blot analysis to detect mRNA from class I or related genes. We found hybridization with all three probes, the strongest being seen with the pC800 probe (2), which corresponds to an 800bp section of the Cw3 gene, containing exons 7 and 8. We have made a cDNA library and screened it for positive clones using the p44/2 probe (3). The sequences which we have obtained are clearly derived from a single class I gene, since they are identical to each other and show a high level of homology to published class I sequences. The sequence is C-locus related, but is not identical to Cwl, Cw2, or Cw3.

The genetic complexity of the class I MHC loci on human chromosome 6 suggests that there may be functional class I loci in addition to HLA A, B, and C, although some have now been identified as pseudogenes (4) and gene fragments. In the rat, a unique class I placental antigen has been identified which only carries shared class I antigenic determinants (5) and there is evidence for expression of a similar molecule in baboon placenta. There are examples of human cells which express tissue restricted, nonclassical ABC class I molecules (6,7). We are carrying out investigations to determine whether the molecule which we have described on human trophoblast (1) also falls into this category. We have used the adherent human choriocarcinoma cell line BeWo, which is trophoblast derived, as a source of RNA for our experiments, since we have shown previously that it expresses a molecule which shares all of the characteristics of that isolated from human placental amniochorion.

mRNA, prepared by a standard technique (8), from the BeWo cell line, and from a control EBV transformed B cell line, was investigated for the presence of HLA class I or related sequences, by Northern blot analysis. The three class I DNA probes used in the study hybridized strongly to the control RNA, but much more weakly to BeWo RNA. It was clear, however, that the strongest hybridisation was seen with the pC800 probe, which will hybridize most efficiently to C locus sequences, but is not locus specific. A cDNA library was made from BeWo mRNA, using the method described by Efstratidis and Villa-Komaroff (9) and Wickens et al. (10), and was screened using the p44/2 probe (which includes a full length HLA A3 gene). Three positive clones were selected for further study. Sequencing by the dideoxy method (11) has enabled us to map the position of the clones, both in relation to each other and to known class I gene sequences. In the regions where the three clones overlap, the nucleotide sequence is identical, which indicates that they are probably all derived from the same gene. Exon 5 in class I genes, the transmembrane region, has a diagnostic locus-specific region, spanning amino acids 292 to 300, where there is a large degree of interlocus divergence, but absolute identity within the C locus sequences. In Exon 5, the BeWo class I nucleotide sequence is identical to that of Cwl, Cw2, and Cw3 (the only published C locus sequences (12,13) (Fig. 1)). However, in other regions of the sequence there are nucleotide differences between Cwl, Cw2, Cw3, and BeWo, but no changes unique to BeWo. We are therefore confident in assigning the BeWo sequence to the HLA C locus family—it is clearly not, however, Cwl, Cw2, or Cw3.

The sequence data that we have obtained from 3 clones derived from a BeWo cDNA library leads us to draw several conclusions concerning the HLA class I expression of this cell line. Since screening of the library was carried out using the P44/2 probe, which could not produce a bias towards C locus sequences, it is surprising that all 3 positive clones sequenced are C-locus related, and identical. In other cell lines described, message for normal C locus products is significantly less abundant than for A and B locus products, and so this result confirms our view that the cell line is not expressing any A or B locus products (or that they are extremely rare); it may however express a single C-locus allele. Without a full-length sequence we cannot yet rule out the possibility that this gene is coding for a serologically definable C-locus product (other than Cwl, 2, or 3), or a modified C-locus product which may correspond to the 40kd surface expressed molecule which we have described.

There has previously been no evidence for normal C-locus products acting as restriction elements for cytotoxic T cells, and no other explanation as to their function has been suggested, indeed, it has been postulated that they may have lost any function they may once have had (13). Our result therefore raises interesting questions regarding a novel function for HLA-C (or a C-related molecule). It has been suggested that the role of class I-like molecules on trophoblast may be to induce specific suppression of the maternal immune response, although the mechanism by which this could occur has not been satisfactorily explained.

Until we have further information, and a greater understanding of the factors which regulate expression of HLA class I in amniochorion, the function of this class I-related molecule will remain unknown. However, the presence of significant amounts of the molecule

Immunobiology of HLA
Volume II

```
EXON 5   (TM)
                                   280                                     290
Cw1  -AG CCG TCT TCC CAG CCC ACC ATC CCC ATC GTG GGC ATC GTT GCT GGC CTG GCT
Cw2  --- --A --- --- --- --- --- --- --- --- --- --- --- --- --- --- --- ---
Cw3  --- --- --- --- --- --- --- --- --- --- --- --- --- --- --- --- --- ---
BeWo --- --- --- --- --- --- --- --- --- --- --- --- --- --- --- --- --- ---

                                               300
Cw1  GTC CTG GCT GTC CTA GCT GTC CTA GGA GCT GTG GTG GCT GTT GTG ATG TGT AGG
Cw2  --- --- --- --- --- --- --- --- --- --- --- --- --- --- --- --- --- ---
Cw3  --- --- --- --- --- --- --- --- --- --- --- --- --- --- --- --- --- ---
BeWo --- --- --- --- --- --- --- --- --- --- --- --- --- --- --- --- --- ---

         310
Cw1  AGG AAG AGC TCA G
Cw2  --- --- --- --- -
Cw3  --- --- --- --- -
BeWo --- --- --- --- -
```

Figure 1. The nucleotide sequence in Exon 5 (transmembrane region) of Cw1, Cw2, Cw3, and BeWo (BeWo sequence derived from three separate, identical cDNA clones).

in this tissue suggests a role in the maintenance of pregnancy.

References

1. Ellis SA, Sargent IL, Redman CWG, McMichael AJ. Evidence for a novel HLA antigen found on human extravillous trophoblast and a choriocarcinoma cell line. Immunology 1986;59:595–601.
2. Strachan T, Dodge AB, Smillie D, Dyer PA, Sodoyer R, Jordan BR, et al. An HLA-C-specific DNA probe. Immunogenetics 1986;23:115–120.
3. Strachan T, Sodoyer R, Damotte M, Jordan BR. Complete nucleotide sequence of a functional class I HLA gene, HLA-A3: Implication for the evolution of HLA genes. EMBO J 1984;3(4):887–894.
4. Biro PA, Pan J, Sood AK, Kole R, Reddy VB, Weissman SM. Sequences of human repetitive DNA, non globulin genes, and major histocompatibility locus genes. III. The MHC complex. Cold Spring Harbor Symp. Quant. Biol. 1983;47:1079–1086.
5. Macpherson TA, Ho H, Kunz HW, Gill TJ. Localisation of the Pa antigen on the placenta of the rat. Transplantation 1986;41:392.
6. Srivastava R, Chorney MJ, Lawrence BK, Pan J, Smith Z, Smith CL et al. Structure, expression and molecular mapping of a divergent member of the class I HLA gene family. PNAS 1987;84:4224–4228.
7. Orgad S, Young S, Gazit E, Relias V, Zaizor R, Lyson S et al. Expression of extra class I MHC antigens on T cell ALL lymphoblasts. Human Immunology 1985;12:133.
8. Aviv H, Leder P. Purification of biologically active globin messenger RNA by chromatography on oligothymidylic acid-cellulose. PNAS 1972;69:1408.
9. Efstratidis A, Villa-Komaroff L. Cloning of double-stranded DNA. In: Stelon JK, Hollaender A (eds): Genetic engineering. New York: Plenum Press, 1979, Vol 1:15.
10. Wickens MP, Buell GN, Schimke RT. Synthesis of double-stranded DNA complementary to lysozyme, ovumucord, and ovalbumin mRNAs. J Biol Chem 1978;253:5462–5467.
11. Sanger F, Nicklen S, Coulson AR. DNA sequencing with chain-terminating inhibitors. PNAS 1977;74:5463–5467.
12. Sodoyer R, Damotte M, Delovitch TL, Trucy J, Jordan BR, Strachan T. The complete nucleotide sequence of a gene encoding a functional human class I histocompatibility antigen (HLA-Cw3). EMBO J 1984;3:879–885.
13. Güssow D, Rein RS, Meijer I, de Hoog W, Seeman GUA, Hochstenbach FM, Ploegh HL. Isolation, expression, and the primary structure of HLA-Cwl and HLA-Cw2 genes: Evolutionary aspects. Immunogenetics 1987;25:313–322.

Author Affiliations

Shirley A. Ellis, Andrew J. McMichael, Nuffield Department of Clinical Medicine, John Radcliffe Hospital, Oxford; *Tom Strachan*, Department Medical Genetics, St. Mary's Hospital, Manchester, United Kingdom

How Similar Are Chimpanzee and Human Class I Antigens?

David A. Lawlor, Frances E. Ward, and Peter Parham

Although the HLA and H-2 systems have provided considerable information of the structure and function of the class I MHC molecules, little is known concerning the evolution of the genes during the process of mammalian radiation. This is largely due to the fact that the two best studied species (mouse and man) diverged 75 to 90 million years ago. Comparison of nucleotide sequences has suggested that some of the current HLA-A,B,C alleles are of relatively ancient origin; however, sequence information with species closely related to man has been lacking. To investigate the possibility that HLA alleles predate primate speciation, our efforts are directed towards analyzing the class I molecules of the chimpanzee, a species with which we shared an ancestor some 3.5 to 5 million years ago.

Our initial results, reported here, are with a cDNA library obtained from the EBV-transformed chimp B-cell line, Tank. Clones containing sequences corresponding to β-2-microglobulin and class I molecules were isolated and subcloned into M13 for DNA sequencing. A panel of HLA sequences, obtained in our lab, has been utilized to assess the similarity between chimp and human class I antigens.

The chimp β-2-microglobulin gene encodes a peptide that is identical to the human (1). At the nucleotide level, there are only four bases in the coding region that distinguish chimp from human and none of these involve a coding change. This rules out the possibility that the light chain could contribute inter-species polymorphisms. A 1350 bp cDNA clone (not full-length) hybridized to the human class I probe. The analysis of the nucleotide distinctions between the chimp and human class I genes revealed that the chimp gene is most similar to the human B-locus alleles. In both exon 2 and 3, the chimp gene has the highest degree of similarity with Bw47. The percent similarity of the chimp gene is less with the C-locus alleles and lowest with the A-locus alleles.

The deduced peptide sequences for the external domains of the chimp and HLA molecules are compared in Table 1. The chimp molecule has greatest similarity with the HLA-B, less with the HLA-C and the least similarity with HLA-A molecules. The α3 domain of the chimp molecule is identical with the majority of the HLA-B molecules, which supports the contention that the gene is the chimp homologue for a B-locus allele. In spite of the high degree of similarity, the chimp antigen contains several residues not observed in the corresponding position of the HLA-B molecules.

Table 1. Comparison of Human and Chimp Class I Heavy Chains

HLA Molecule	% Similarity*
Bw47	93.4
B40*	92.1
B44.1	90.5
Bw41	91.7
Cw1	86.7
A24	85.5
A32	85.1
Cw3	84.6
A2.1	82.2
A1	82.2

*241 residues of the predicted chimp and human peptide sequences were compared, starting at residue 34 in α1 and extending to the end of the α3 domain.

References

1. Suggs SV, Wallace RB, Hirose T, Kawashima EH, Itakura K. Use of synthetic oligonucleotides as hybridization probes: Isolation of cloned cDNA sequences for human beta-2-microglobulin. Proc Natl Acad Sci USA 1981;78:6613.

Author Affiliations

David A. Lawlor, Peter Parham, Stanford University Medical Center, Stanford, CA; *Frances E. Ward*, Duke University, Durham, NC, USA

Immunobiology of HLA
Volume II

Serological and Biochemical Studies of an HLA-A Locus Public Epitope*

Dolores J. Schendel and Andreas Ziegler

Several alloantisera have been described that react with public specificities of human class I molecules that are shared by several loci or by several alleles (1,2). The exact characterization of these epitopes has remained problematic because of the polyclonal nature of the alloantisera and since their titers are generally low, hindering extensive absorption analyses and making them unsuitable for biochemical studies. Mouse monoclonal antibodies (MAb) offer new means to examine HLA public specificities. We have isolated and characterized such a MAb that reacts with a public epitope shared by several HLA-A alleles (3). Following immunization with 3-day phytohemagglutinin-stimulated human lymphocytes, the spleen cells of a (C57BL/6 × BALB/c) F1 female mouse were fused with the myeloma P3/X63Ag 8.6.5.3. The MüA80 hybridoma was selected for further characterization since it revealed a limited polymorphism when tested on a random panel (75% positive; 25% negative). The epitope, A80, that is defined by this reagent was localized to the MHC class I region since the MAb showed differential binding to a set of MHC class I deletion mutants. The segregation pattern in an HLA-A:C recombinant family suggested that the antigen expressing the A80 epitope is either encoded by the HLA-A locus or is associated with another gene mapping telomeric to HLA-C. To distinguish between these possibilities the molecules precipitated by MüA80 were analyzed in a sodium dodecyl sulfate polyacrylamide gel, yielding bands with molecular weights of 45 kD and 12 kD that are characteristic for class I molecules. The isoelectrifocusing patterns of five different cells showed conclusively that MüA80 precipitates HLA-A molecules; no bands corresponding to HLA-B or HLA-C antigens were detected. Immunofluorescence binding studies of cells of 86 HLA-typed donors, 44 of which were HLA-A homozygous, allowed the A80 epitope to be assigned to seven HLA-A alleles (Fig. 1).

The public specificity defined by MüA80 is confined to HLA-A alleles and, therefore, presumably represents a sequence difference that arose after the separation of the classical HLA-A, -B, and -C loci. Whether the A80 epitope will be expressed by any novel class I molecules encoded by genes mapping telomeric to HLA-A, remains to be determined. As with the Bw4 and Bw6 specificities, definition of the A80 public epitope may be useful in clarifying some splits of HLA-A antigens. Precise definitions of public epitopes with monoclonal antibodies will enable evaluation of their roles in transplantation and immune responses. Reagents like MüA80 are also useful for analyzing locus-specific expression of class I molecules in DNA-transformed cells and in tumors.

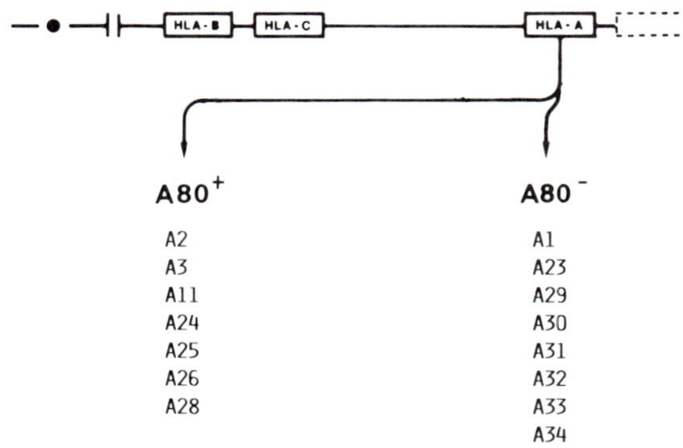

Figure 1. The A80 epitope defined by monoclonal antibody MüA80 was found to be present on seven alleles of the HLA-A locus.

*This work was supported by the Deutsch Forschungsgemeinschaft through SFB 120 (project A4) and SFB 217 (project B2). We thank Dr. Rudolf Wank for providing the HLA typed cells.

References

1. Bodmer JG. Joint report from the Sixth International Histocompatibility Workshop. In: Kissmeyer-Nielsen F (ed): Histocompatibility Testing 1975. Copenhagen: Munksgaard, 1973, p 21.
2. Konoeda Y, Terasaki PI, Wakisaka A, Park MS, Mickey MR. Public determinants of HLA indicated by pregnancy antibodies. Transplantation 1986;41:253.
3. Schendel DJ, Ziegler A. Serological and biochemical characterization of HLA-A molecules bearing a public epitope. Immunogenetics 1988;27:61.

Author Affiliations

Dolores J. Schendel, Institute of Immunology, University of Munich, Goethestrasse 31, D-8000 Munich 2, Federal Republic of Germany; *Andreas Ziegler*, Medical Clinic and Medical-Natural Science Research Center, University of Tübingen, D-7400 Tübingen, Federal Republic of Germany

Characteristics of HLA Class I Antigens in Human Sera

Lia M.E. Dobbe, Nico J. Stam, Jacques J. Neefjes, and Marius J. Giphart

Class I antigens deriving from classical or nonclassical major histocompatibility complex class I genes have been detected in sera of many different species (1). Those present in human serum – as inferred from microcytotoxicity-inhibition assays (2) – have not been analyzed biochemically in detail yet.

On the assumption that human serum class I antigens, as those found in sera of other species, would be associated with β2m, we set out to isolate these molecules by means of a Sepharose 4B coupled rabbit anti-human β2m antiserum. After separation by SDS-PAGE and transfer to nitrocellulose, the isolated molecules were detected by means of a rabbit antiserum directed towards denatured HLA class I heavy chains (RaHC) (3).

Molecules thus isolated display a complex pattern consisting of various molecules in the MW range of 44kD to 35kD. Major bands usually are observed at 44kD, 40kD, 37kD, and 35kD.

Only the 44kD molecules were found to possess a hydrophobic transmembrane region (TM) as they partitioned into the detergent phase upon TX114 phase separation.

An antiserum raised against a conserved cytoplasmic tail peptide (RaCT) was used to investigate the presence of a regular cytoplasmic tail in the serum class I molecules. Whereas the 44kD molecules generally reacted with this antiserum, 40kD molecules reactive with this antiserum were present in only a minority (2 of 13) of samples tested. None of the 37kD or 35kD molecules were found to react with RaCT. These results are summarized in Table 1.

From these data we conclude that the 44kD molecules most likely are intact class I antigens that have been shed from the cell membrane.

Table 1. Characteristics of Serum HLA Class I Antigens

MW Species	RaHC Reactivity	RaCT Reactivity	Detergent Binding
44KD	+	+	+
40KD	+	+/−	−
37-35KD	+	−	−

Alternative splicing events, resulting in exact deletion of the TM exon as described for a leucemic T-cell line (4), do not seem to play an important role in vivo, as only in two of 13 individuals 40kD molecules reactive with the RaCT serum were found. However, alternative splicing events resulting in a shift in reading frame would not be picked up by this method and thus cannot be ruled out.

The 37 and 35kD molecules most likely are generated by proteolysis, a process that also might generate some of the 40kD molecules. The idea that some of the non-44kD molecules derive from Q10-like genes provides an interesting alternative for these explanations.

Various ideas about a function of serum class I antigens have been put forward. However, no general evidence for a role in the immune response has been obtained so far. Our results, which add to the great diversity of these molecules in and between species as known so far, make it even more difficult to perform studies on the possible function of these molecules.

Acknowledgments. Financial support for this study was obtained from the Dutch Organisation for the Advancement of Pure Research (ZWO), the J.A. Cohen Institute for Radiation Protection (IRS), and the Maurits and Anna de Kock Foundation.

References

1. Gussow D, Ploegh H. Soluble Class I antigens: A conundrum with no solution? Immunol Today 1987;8:220–221.
2. van Rood JJ, van Leeuwen A, van Santen MCT. Anti HL-A2 inhibitor in normal serum. Nature 1970;226:366–367.
3. Dobbe CME, Stãm NJ, Neefjes JJ, Giphart MJ. Biochemical complexity of serum HLA Class I antigens. Immunogen. in press.
4. Krangel MS. Secretion of HLA-A and -B antigens via an alternative RNA splicing pathway. J Exp Med 1986;173: 1173–1190.

Author Affiliations

Lia M.E. Dobbe, Marius J. Giphart, Department of Immunohaematology and Bloodbank, University Hospital, Leiden; *Nico J. Stam, Jacques J. Neefjes*, The Netherlands Cancer Institute, Amsterdam, The Netherlands

The *H-2K*^{kml} Mutation: A Single Nucleotide Substitution Causes Multiple Functional Differences in a Murine Class I MHC Molecule

J.M. Martinko, J.C. Solheim, and J. Geliebter

The polymorphic MHC class I glycoproteins are implicated in a number of immunologic reactions involving T-cell recognition. Among these polymorphic molecules in the murine H-2 complex, a number of mutants have been identified. One of these mutants was mapped to the *H-2K*^{k} allele in the CBA inbred mouse strain and resulted in graft rejection and graft versus host reactions. This mutant was designated *H-2K*^{kml} and the mouse strain carrying it was designated CBA.M523 (1). Analysis of the structural variation in the *H-2K*^{kml} molecule identified a single nucleotide substitution at codon 152, resulting in a change from Asp (GAT) in K^k to Ala (GCT) in K^{kml} (2). This report documents the completion of the nucleotide sequence coding for the extracellular domains of the altered K^{kml} and compares the known K^k coding sequence from the AKR strain (3) to that of the CBA strain which is ancestral to the CBA.M523 mutant-bearing strain.

Nucleotide sequence information was obtained by chain termination sequencing methods using sequence-specific deoxyoligonucleotide primers and either mRNA or uncloned cDNA as template material (4). Template mRNA was isolated from mouse livers by standard methods using oligo(dT)-cellulose chromatography. A K^k sequence-specific ^{32}P-labeled deoxyoligonucleotide and the mRNA were denatured and annealed. The annealed oligomer was then extended with reverse transcriptase and terminated using dideoxyoligonucleotides to yield direct sequence information. Alternatively, unlabeled K^k sequence-specific deoxyoligonucleotides were used to prime mRNA and were extended with reverse transcriptase. The resulting cDNA copies were then sequenced using complementary, sequence-specific, ^{32}P-labeled deoxyoligonucleotides, DNA polymerase I (Klenow fragment) or reverse transcriptase and appropriate dideoxyoligonucleotides, yielding sequence in the opposite orientation as compared to the mRNA analysis. Using a number of K^k-sequence specific deoxyoligonucleotide primers in this system, we defined the sequence of the K^{kml} transcript from codons 1-292, which includes the coding sequence for the N, C1, and C2 protein domains, as well as a portion of the transmembrane protein domain. From this information, the primary structure of the immunologically recognizable extracellular protein domains of the K^{kml} glycoprotein was deduced. As compared to the known K^k sequence from the AKR mouse strain (3), the K^{kml} transcript differed from K^k only at codon 152, resulting in the GAT (Asp) → GCT (Ala) alteration already reported. Because the AKR mouse strain and the CBA strain are derived from disparate genetic stocks, the possibility remained that the codon 152 alteration was a strain-specific variation. However, Figure 1 shows that the sequence derived from the AKR strain, the CBA strain, and the CBA.M523 strain for the *K* allele in the area of the mutation is identical except for the A → C alteration in K^{kml}. Both K^k sources possess the identical information at codon 152. Since the sequence of the mutant allele is identical to that of AKR-derived K^k at every other position in the coding sequence and since the CBA.M523 strain was derived directly from a CBA progenitor, we conclude that the immunologically recognizable K^k glycoproteins derived from two different sources are identical.

The existence of a single base pair alteration in K^{kml} that results in a single amino acid alteration in the expressed class I molecule shows that the altered biological activity seen between K^k and K^{kml} is due to the single amino acid/nucleotide substitution at codon 152. This amino acid is now known to be involved in the antigen-binding site and/or the T-cell recognition site of the class I molecule (5). This mutation is the only one of several class I variants that shows a single amino acid change; all other 152 variants have other amino acid alterations. Thus, this study proves that amino acid 152 is an important part of the T-cell receptor-foreign antigen-MHC class I complex involved in antigen presentation.

Comparison with the other class I substitutions at 152 shows that all substitutions involve transversion events. Although we cannot rule out the possibility of a point mutation giving rise to K^{kml}, we favor a recombination

```
                 L  I  T  K  H  K  W  E  Q  A  G  A/D A  E  R  D  R  A  Y  L
141-160 M523   CTGATCACCAAACACAAGTGGGAGCAGGCTGGTGCTGCAGAGAGAGACCGGGCCTACCTG
        AKR    -----------------------------------A------------------------
        CBA                      -----------------A-----------
```

Figure 1. Nucleotide and deduced amino acid sequence ofK^{kml} from CBA.M523 mice and comparison to K^k from the RDM-4 cell line of AKR mouse origin (Arnold et al., 1984) from codons 141-160. The sequences are identical at all positions except for the A → C substitution at codon 152 which results in an Asp → Ala (D → A) substitution at the amino acid level. The sequence of K^k from the CBA parent of M523 is identical to the AKR-derived sequence in the region of the mutation as shown. A total of 876 nucleotides, spanning the entire extracellular protein domains of K^{kml}, have been compared.

Immunobiology of HLA
Volume II

mechanism involving segmental exchange (gene conversion model) because of the complex alterations involving amino acid 152 in other class I variants and because of the exclusive occurrence of the transversion substitutions in the alterations at the 152 codon.

Acknowledgment. This work was supported by a grant to J.M. from the National Institutes of Health (AI-21738).

References

1. Blandova Z, Mnatsakanyan YA, Egorov I. Study of H-2 mutations in mice. VI. M523, a new K end mutant. Immunogenetics 1975;2:291.
2. Martinko JM, Solheim JC, Geliebter J. The *H-2K*^{kml} mutation: Nucleotide sequence and comparative analysis. Molec Immunol 1987;24:197.
3. Arnold B, Burgert HG, Archibald AL, Kvist S. Complete nucleotide sequence of the H-2K^{k} gene. Comparison of three H-2K locus alleles. Nucleic Acids Res 1984;24:9473.
4. Geliebter J, Zeff RA, Melvold RW, Nathenson SG. Mitotic recombination in germ cells generates two major histocompatibility complex mutant genes shown to be identical by RNA sequence analysis: K^{bm9} and K^{bm6}. Proc Natl Acad Sci USA 1986;83:3371.
5. Bjorkman PJ, Saper MA, Samroui B, Bennett WS, Strominger JC, Wiley DC. The foreign antigen binding site and T cell recognition regions of class I histocompatibility antigens. Nature 1987;329:512.

Author Affiliations

J.M. Martinko, J.C. Solheim, Department of Microbiology, Southern Illinois University, Carbondale, IL; *J. Geliebter*, The Rockefeller University, New York, NY, USA

Molecular Mapping of Antigenic and Cytotoxic T Lymphocyte Recognition Determinants on the Major Histocompatibility Complex Class I Molecule K^{b} Using In Vitro Mutant Cell Lines

Parthasarathi Ajitkumar, Steven Geier, Francescopaolo Borriello, Mayuma Nakagawa, Jeffrey Bluestone, and Stanley Nathenson

MHC class I molecule plays a crucial role in the cell-mediated immune response by functioning as the recognition element for cytotoxic T lymphocyte (CTL) to discriminate self-antigens from foreign (nonself) antigens (1). Based on the previous extensive studies using a series of spontaneous in vivo mouse mutants expressing altered class I K^{b} molecules with multiple clustered amino acid substitutions, we had postulated that CTL interacts with the "recognition regions" formed by amino acid residues in the polypeptide stretches from residues 70 to 90 in the α1 domain and from 150 to 180 in the α2 domain (2). In the present study, using allogeneic anti-K^{b} CTL clones, K^{b}-specific monoclonal antibodies (MAbs) and a series of mutant cell lines expressing K^{b} molecules with single point mutations, we have identified some of the amino acid residues on the two polypeptide stretches that are important for the recognition of the K^{b} molecule by CTL and MAbs (3). While the MAbs recognized specific regions in either the α1 or α2 domain of the molecule, CTL appeared to simultaneously interact with specific amino acid residues on the recognition regions of the two domains.

The in vitro mutant cell lines, generated by chemical mutagenesis using ethylmethonyl sulfate (EMS) or ethylnitrosourea (ENU), were selected and serologically characterized using K^{b}-specific MAbs, EH-144, B8-24-3, K9-136, 28-13-3, K9-178, 5F1-2-14, and Y-3 (4). The point mutations were determined by direct dideoxy sequencing of K^{b} mRNA using K^{b} gene-specific oligomers as primers (5). The alloreactive CTL recognition pattern of the mutant K^{b} molecules was studied using anti-K^{b} CTL clones raised in naturally occurring mutant mice (bm mutants) expressing K^{b} molecules with clustered multiple amino acid substitutions (6). The CTL and MAb recognition pattern of the mutants were then correlated with the position and nature of the amino acid substitutions in order to determine the amino acid residues important for the recognition process (3).

Single point mutations were found at positions corresponding to the amino acids 80, 82, and 90 in the α1 domain; 138, 141, 150, 162, 166 and 174 in the α2 domain; and 192 in the α3 domain. The amino acids threonine, leucine, and glycine at the positions 80, 82, and 90, respectively, were found to be important for the recognition of the K^{b} molecule by the MAb EH-144. Interestingly, of the three residues, only leucine at the position 82, but neither threonine at 80 nor glycine at 90, appeared to be important for the recognition by the B8-24-3 MAb. The change at the amino acid residue 90 disrupts the recognition by both the MAbs EH-144 and K9-136. Thus the glycine residue at the amino acid position 90 is important for the recognition of the K^{b} molecule by the K9-136 MAb. An interesting feature of the recognition requirements noted for the MAb K9-136 was the participation of the amino acid residue leucine at position 82. The mutation of this residue to phenylalanine in the mutant R8.313 abolished recognition of the K^b molecule by this antibody. But the change of leucine to proline in R8.208 did not affect the recognition by K9-136. Thus, the nature of the amino acid residue at position 82 was found to be a deciding factor in the recognition of the K^b molecule by this MAb. Mutations of the amino acids at positions 80, 82, and 90 did not affect the binding of other MAbs 28-13-3, K9-178, 5F1.2.14, and Y-3. The determinants recognized by the MAb 5F1.2.14 was distinctly different from that recognized by others. The 5F1.2.14 MAb was found to recognize a determinant formed by the participation of at least the three amino acid residues at 162, 166, and 174. The Y-3 MAb recognized a different region on the α2 domain of the K^b molecule. The recognition determinants for the MAb Y-3 seemed to be defined by at least the residues methionine, leucine, and alanine at positions 138, 141, and 150, respectively. The mutations of the amino acids at 138, 141, 150, 162, 166, and 174 positions did not affect the recognition of the K^b molecule by the MAbs EH-144, B8-24-3, K9-136, and 28-13-3.

The allogeneic CTL recognition of the somatic cell variants showed a different pattern as compared to that by MAbs. The pattern of recognition of the mutant K^b molecules by CTL clones was dependent upon two factors. One was the site of point mutation on the K^b molecule and the second factor was the bm mutant mouse (probably the features of the alloreactive CTL repertoire in the mouse) in which the anti-K^b CTL clones were generated. The recognition of the K^b molecule by the anti-K^b CTL clones from bm8 (9 clones), bm10 (6 clones), and bm11 (4 clones) mice was abolished by mutations of two amino acid residues, leucine and glutamic acid at positions 82 and 166, respectively. The CTL clones showed a variation with respect to the importance of other residues for recognition. The amino acid residues glycine and leucine at positions 90 and 141, respectively, were not found to be important for the recognition of the K^b molecule by almost all the CTL clones tested. While for the majority of the bm8 anti-K^b CTL clones the residue methionine at position 138 was found to be important, it was dispensible for the CTL clones of the bm10 mouse origin. A similar variation was noticed with respect to the relevance of the amino acid residues threonine, glycine, and histidine at positions 80, 162, and 192, respectively, for the CTL clones raised in the bm8 mouse. On the contrary, for all the bm10 anti-K^b CTL clones tested, the amino acid residues threonine and glycine, but not histidine, were found to be important for the recognition. Thus the data clearly identify some of the amino acid residues that are important for the recognition of the K^b molecule by CTL. However, these are definitely not the only residues recognized by CTL, since the MAbs available for the selection of the mutants were limited in number. The data also show that the recognition of the K^b molecule by CTL requires the participation of amino acid residues in the α1 and α2 domains and that CTL of different origin recognize different residues on the molecule.

Thus, in the present study we have mapped some of the amino acid residues important for the recognition of the K^b molecule by MAbs and CTL. The data clearly imply that while MAbs recognize amino acid residues at discrete regions on either the α1 or the α2 domain of the class I molecule, CTL interacts with the residues on both the domains simultaneously.

Acknowledgements. We thank J. Trojnacki, R. Raval, D. McGovern, and E. Palmieri-Lehner for excellent technical assistance, and Catherine Whelan for the preparation of the manuscript. P. Ajitkumar was supported by a postdoctoral fellowship from Cancer Research Institute, New York.

References

1. Zinkernagel RM, Doherty PC. MHC-Restricted cytotoxic T cells: Studies on the biological role of polymorphic major transplantation antigens determining T-cell restriction-specificity, function, and responsiveness. Adv Immunol 1979;27:51.
2. Nathenson SG, Geliebter J, Pfaffenbach GM, Zeff RA. Murine Major histocompatibility complex class I mutants: Molecular analysis and structure-function implications. Annu Rev Immunol 1986;4:471.
3. Ajitkumar P, Geier SS, Kesari KV, Borriello F, Nakagawa M, Bluestone JA, et al. Evidence that multiple residues on both the α-helices of the class I MHC molecule are simultaneously recognized by the T cell receptor. Cell 1988, in press.
4. Geier SS, Zeff RA, McGovern DM, Rajan TV, Nathenson SG. An approach to the study of structure-function relationships of MHC class I molecules: Isolation and serologic characterization of H-2K^b somatic cell variants. J Immunol 1986;137:1239.
5. Geliebter J. Dideoxy sequencing of RNA and uncloned cDNA. Focus 1987;9(1):5.
6. Bluestone JA, Langlet C, Geier SS, Nathenson SG, Foo M, Schmitt-Verhulst A-M. Somatic cell variants express altered H-2K^b allodeterminants recognized by cytolytic T cell clones. J Immunol 1986;137:1244.

Author Affiliations

Parthasarathi Ajitkumar, Steven Geier, Francescopaolo Borriello, Mayuma Nakagawa, and *Stanley Nathenson*, Department of Cell Biology, Albert Einstein College of Medicine, Bronx, NY 10461, USA; *Jeffrey Bluestone*, Transplantation Biology Section, National Institutes of Health, Bethesda, MD 20892, USA. Current address of S. Geier: Transplantation Laboratory, VA Medical Center, 508 Fulton St., Durham, NC 27705, USA

Section III
Class I Histocompatibility Antigens: Transfection and Gene Expression

Isolation of an HLA-A:,B:,C Null Human Cell Line Capable of Expressing Transferred Class I Genes*

Yoji Shimizu and Robert DeMars

The functional and molecular analysis of the human leukocyte antigen (HLA) complex has benefited from the generation of γ-ray-induced mutant cells that have lost expression of various HLA gene products. The technique of γ-ray mutagenesis and negative immunoselection (1) has been used to isolate a human B-lymphoblastoid cell line (B-LCL) that does not express the serologically defined HLA-A,B,C class I antigens but can serve as a recipient cell for cloned HLA class I genes introduced by DNA-mediated gene transfer. This HLA-A,B,C null human cell has enabled us to produce B-LCL transferents that express only one HLA-A, -B, or -C antigen—that encoded by a transferred class I gene.

γ-Irradiation and immunoselection with the anti-class I monoclonal antibody (MAb) W6/32 were used to isolate the HLA-A,B,C null mutant .221 from mutant .184, which already had a homozygous deletion of the HLA-A and -B loci but still expresses HLA-C antigen. Antibody binding studies show that .221 and the β2-microglobulin-deficient LCL Daudi bind similar amounts of W6/32 and the β2-microglobulin-specific MAb BBM.1 (Table 1). HLA-A,B,C mRNA is not detectable in .221 but a normal amount of β2-microglobulin mRNA is present. Very small amounts of two non-A,B,C class I antigens are present, but not on the surfaces of .221 cells (2).

Genomic genes for the HLA-A2, -A1, -B5, and -B8 class I antigens were cloned from the B-LCL 721. Each gene was transferred into mutant .221 using the plasmid vectors pHeBo, pHPT32, and pHPTe (1). Antibody binding studies showed that transferents expressed class I antigen at a level at least equal to expression of the same endogenous gene in a comparable LCL, as suggested in Table 1. The transfer of a cloned HLA-C gene into .221 resulted in class I antigen expression at a level comparable to the HLA-A,B null, HLA-C positive B-LCL .184 (not shown).

As an illustration of the usefulness of this mutant cell-gene transfer system, a panel of class I gene transferents of .221 was used to demonstrate an inverse correlation between susceptibility to natural killer (NK) cell lysis and class I expression (Table 1). Specifically, while the HLA-A,B,C positive LCL .45 was resistent to NK attack, mutant .221 showed an increased susceptibility to NK lysis. The transfer and expression of HLA-A or HLA-C genes resulted in a slight reduction in NK susceptibility while the expression of an HLA-B antigen resulted in reduced NK susceptibility similar to that of mutant .45. These results with gene transferents show more directly than could be determined with antigen-loss mutants alone that class I expression on target cells influences NK susceptibility, in agreement with other studies that have suggested such a relationship (3).

Table 1. Lysability by NK Cells in Relation to Expression of Transferred Cloned Class I HLA Genes by HLA-A,B,C Null B-LCL Mutant .221

Cell	BBM.1 Binding	% Specific Lysis	
		Expt. 1	Expt. 2
K562	2.2	37	65
.45	100	1.0	13
.221	1.8	27	34
pHeBo→.221	1.7	28	NT
pHeBo(A2)→.221	50.6	20	NT
pHeBo(A1)→.221	78.2	19	NT
pHPTe→.221	2.5	NT	43
pHPTe(C)→.221	17	NT	29
pHPTe(B8)→.221	146	NT	11

K562 is a human erythroleukemia cell line commonly used as a positive control target for NK lysis. Human B-LCL mutant .45 has one copy of the MHC and expresses HLA-A2, -B5, and -C, while mutant .221 completely lacks these antigens. Class I antigen expression was assayed with an EPICS C flow cytometer as binding of the anti-β2-microglobulin MAb BBM.1; linear values of binding were normalized to binding to .45, which was given an arbitrary value of 100. Cell lines were tested for NK susceptibility in a 6-hr Cr release assay using freshly isolated peripheral blood lymphocytes from normal, healthy donors. Experiments 1 and 2 were performed on different occasions. NT = not tested.

Other potential uses for class I gene transferents produced with mutant .221 include the production and analysis of antibody and cellular reagents, analysis of class I structure-function relationships, isolation of class I protein for crystallographic analysis, and the identification of non-A,B,C class I genes (2).

References

1. Shimizu Y, Koller B, Geraghty D, Orr H, Shaw S, Kavathas P, DeMars R. Transfer of cloned human class I major histocompatibility complex genes into HLA mutant human lymphoblastoid cells. Mol Cell Biol 1986;6;:1074–1087.
2. Shimizu Y, Geraghty DE, Koller BH, Orr HT, DeMars R. Transfer and expression of three cloned human non-HLA-A,

*This is paper No. 2956 from the Laboratory of Genetics of the University of Wisconsin (Madison). This work was supported by NIH grant AI 15486.

-B, -C class I histocompatibility complex genes in mutant lymphoblastoid cells. Proc Natl Acad Sci USA, 1988;85: 277–231.

3. Harel-Bellan A, Quillet A, Marchiol C, DeMars R, Tursz T, Fradelizi D. Natural killer susceptibility of human cells may be regulated by genes in the HLA region on chromosome 6. Proc Natl Acad Sci USA 1986;83:5688–5692.

Author Affiliations

Laboratory of Genetics and Department of Human Oncology, University of Wisconsin, Madison, WI 53706, USA.
Current address of *Y. Shimizu*: Immunology Branch, National Cancer Institute, National Institutes of Health, Bethesda, MD 20892, USA

Selective Increase of Expression of HLA Class I Molecules by HLA-A3, HLA-B7, and HLA-Cw3 Transfected P815 Mastocytoma Cells After Subsequent Transfection by the Human β2-Microglobulin Gene

B.M. Perarnau, A. Gillet, M. Barad, R. Hakem, and F.A. Lemonnier

Association with murine β2-microglobulin (β2-m) enables cell surface expression of HLA class I heavy chains following transfection of murine cells with HLA-A3, HLA-B7, and HLA-Cw3 class I genes (1,2). However, detailed serologic analysis of the resulting HLA molecules suggests that their structure is significantly altered (3,4). This heterospecific association was, furthermore, reported to result in markedly reduced translocation of HLA-B27 molecules from the endoplasmic reticulum to the cell surface (5). The following experiments indicate that this later observation could be extended to the HLA-A3, HLA-B7, and HLA-Cw3 heavy chains.

Clones of P815 (H-2^d) mastocytoma cells (6), transfected with either HLA-A3, HLA-B7, or HLA-Cw3 class I genes were further transfected with the human β2-m gene (which was isolated from a genomic library) and the expression of HLA class I molecules by parental and daughter surtransfected cells were then compared. The cell surface expression of HLA-A3, HLA-B7, and HLA-Cw3 molecules was evaluated by indirect immunofluorescence. We selected anti-HLA class I monoclonal antibodies (MAb) directed at antigenic determinants that are expressed by HLA molecules independently of the human or murine nature of the β2-m associated to the heavy chain (3). A 40-fold increase of the expression of HLA-B7 molecules was observed in the presence of human β2-m. An 11- and 6-fold increase was also observed for HLA-A3 and HLA-Cw3 molecules respectively as shown in Table 1.

These values were constant, as illustrated in Table 1, no matter which clone of daughter surtransfected cells was studied. This enhancement was specific since under the same experimental conditions only a 1- to 3-

Table 1. Increased Expression of HLA Class I Molecules by HLA-A3, HLA-B7, and HLA-Cw3 P815 Cells After Surtransfection with Human β2-m

	Mean Fluorescence After Incubation with MAbs			
Target Cell	B1.23.2 B9.12.1*	20-8-4S	28-14-8S	23A-5-21S 27-11-13S⁺
P815 HLA-A3	6.7*	63.5	28.9	77.4
P815 HLA-A3 × Hβ2-m C1	72.9*	172.6	86.6	107.2
P815 HLA-A3 × Hβ2-m C8	81.1*	137.6	65.8	55.9
P815 HLA-B7	.9	96.3	42.1	22.0⁺
P815 HLA-B7 × Hβ2-m C6	42.0	130.2	73.0	45.9⁺
P815 HLA-B7 × Hβ2-m C19	46.5	ND	ND	ND
P815 HLA-Cw3	22.0	80.6	22.6	15.4
P815 HLA-Cw3 × Hβ2-m C1	134.9	91.0	30.8	18.2
P815 HLA-Cw3 × Hβ2-m C12	121.8	106.4	50.2	50.3

HLA-A3, HLA-B7 or HLA-Cw3 P815 parental single transfected cells and clones (C) of human β2-m (Hβ2-m) daughter double-transfected cells were first incubated with saturating concentrations of B1.23.2 (anti HLA-B-C) (7), B9.12.1 (anti HLA-A, -B, -C) (8), 20-8-4S (anti H-2 K^d) (9), 28-14-8S (anti H-2 D^d) (9), 23A-5-21S (anti H-2 D^d) (9) or 27-11-13S (anti H-2 D^d) (9) MAbs. The bound antibodies were subsequently revealed with saturating amount of FITC-conjugating sheep anti-mouse immunoglobulin antibodies. After washing and fixation, the cells were analyzed with a fluorescence activating cell sorter. The mean fluorescence was evaluated by integration of the curve derived for each cell population.

Immunobiology of HLA
Volume II

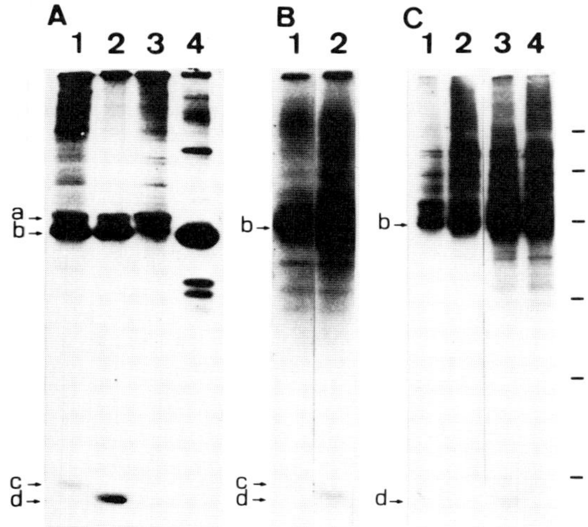

Figure 1. Preferential pairing of human heavy and light chains in HLA × human β2-m P815 double transfected cells. Daudi (A, lane 4), P815 HTR (A, lane 3), P815 HLA-A3 (A, lane 1), HLA-Cw3 (B, lane 1), and HLA-B7 (C, lanes 1–2); parental single transfected cells, P815 HLA-A3 × Hβ2-m (A, lane 2), HLA-Cw3 × Hβ2-m (B lane 2), HLA-B7 × Hβ2-m (C, lanes 3–4) were internally labeled with ^{35}S methionine. Immunoprecipitations studies were carried out with either a rabbit anti-HLA serum (A, B, C) or HC-10 MAb (C lanes 2–4) (10). Unlabeled protein markers of 94 000, 66 000, 45 000, 29 000, 24 000, 12 000 daltons were included in the gel. The following bands can be identified: a, murine H-2K^d heavy chains; b, human HLA class I heavy chains; c, murine β2-m; and d, human β2-m.

fold increase of the expression of endogenous H-2^d molecules occurred. Biochemical studies were undertaken to determine which step of the biosynthesis of the HLA class I heavy chains was altered in the absence of human β2-m. Dot-blot analysis of total cellular RNA, using HLA class I specific DNA probes, indicated that the transcriptional activity of the heavy chain class I genes was not modified following introduction of the human β2-m gene (data not shown).

Immunoprecipitation studies were performed following internal labeling with ^{32}S methionine using either a rabbit antiserum which recognizes HLA class I heavy chains whether or not associated with β2-m, or HC-10 MAb, which reacts only with isolated HLA class I heavy chains (10). Equal amounts of 44 kilodalton molecules were precipitated by the rabbit antiserum from HLA-A3, HLA-B7, HLA-Cw3 transfected parental and corresponding human β2-m daughter surtransfected cell lysates, as shown in Figure 1.

These molecules correspond to HLA class I heavy chains and could also be precipitated by the same serum from a Daudi cell lysate, however, some cross reactivity with murine H-2 class I heavy chains was observed. These results suggest that the traductional step of the biosynthesis of HLA class I heavy chains was not modified in human β2-m surtransfected daughter cells. It was also noticed that the human β2-m migrates slightly further than the murine light chain on a single one-dimensional electrophoresis gel (Figure 1). The murine β2-m contains five methionine residues. No detectable amount of this light chain could be precipitated by the rabbit anti-HLA serum from lysates of HLA-A3, HLA-B7, and HLA-Cw3 P815 cells surtransfected by the human β2-m gene. On the contrary, the human β2-m, which contains only one methionine residue, was always coprecipitated from these lysates by the same antiserum. It became evident therefore that human HLA-A3, HLA-B7, and HLA-Cw3 class I heavy chains preferentially (or exclusively ?) associate with human β2-m in P815 cells surtransfected by the human β2-m gene. In the case of P815 HLA-B7 single transfected cells, it was even impossible, under our experimental conditions, to coprecipitate the murine light chain with the human heavy one suggesting a very weak association. A comparison was made of the amount of HLA class I heavy chains precipitated from lysates of HLA-B7 parental and human β2-m surtransfected daughter cells, by either the rabbit antiserum (reacting with free and β2-m associated heavy chains) or HC-10 MAb (reacting with free heavy chains exclusively).

The results shown in Figure 1 suggest that, in the absence of human β2-m, there is a relative accumulation of free HLA-class I heavy chains (band b in lane 2 is larger than in lane 1, whereas it tends to be smaller in lane 4 than in lane 3). This observation was further substantiated by pulse and chase experiments (data not shown) which indicate that, in the absence of human β2-m, HC-10 reactive molecules could be detected 2½ hours after the chase, whereas in the presence of the human light chain, they had disappeared after 30 minutes. The susceptibility to endoglycosidase H of this HC-10 reacting material was verified and led to the conclusion that, in the absence of human β2-m, free HLA-B7 heavy chains accumulate in the endoplasmic reticulum.

Altogether, these data indicate that in murine transfected cells, association of human HLA-A3, HLA-B7, and HLA-Cw3 heavy chains with human instead of murine β2-m results in more abundant cell surface expression of HLA class I molecules, extending to three other allelic products the observation already made for HLA-B27 (5).

Acknowledgments. We would like to thank Drs. B.R. Jordan and C. N'Guyen for providing us the opportunity to screen their human genomic library; Drs. H.L. Ploegh and D. Sachs for their generous gifts of MAbs; and Dr. K. Harper for careful reading of the manuscript.

This work was supported by grants from Institut National de la Santé et de la Recherche Médicale, Centre National de la Recherche Scientifique, and Association pour la Recherche sur le Cancer.

References

1. Lemonnier FA, Dubreuil PL, Layet C, Malissen M, Bourrel D, Mercier P, et al. Transformation of LMTK⁻ cells with purified HLA class I genes. II serological characterization of HLA-A3 and -Cw3 molecules. Immunogenetics 1983;18:65.
2. Barbosa JA, Kamarck ME, Biro PA, Weismann SM, Ruddle FH. Identification of human genomic clones coding the major histocompatibility antigens HLA-A2 and HLA-B7 by DNA mediated gene transfer. Proc Natl Acad Sci USA 1982;79:6327.
3. Ferrier P, Layet C, Caillol DH, Jordan BR, Lemonnier FA. The association between murine β2-microglobulin and HLA class I heavy chains results in serologically detectable conformational changes of both chains. J Immunol 1985;135:1281.
4. Kahn-Perlès B, Boyer C, Arnold B, Sanderson AR, Ferrier P, Lemonnier FA. Acquisition of HLA class I W6/32 defined antigenic determinant by heavy chains from different species following association with bovine β2-microglobulin. J Immunol 1987;138:2190.
5. Rein RS, Seeman GHA, Neffjes JJ, Hochstenbach FMH, Stam NJ, Ploegh HL. Association with β2-microglobulin controls the expression of transfected human class I genes. J Immunol 1987;138:1178.
6. Maryansky JL, Moretta A, Jordan B, De Plaen E, Van Pel A, Boon T, et al. Human T cell recognition of cloned HLA class I gene products expressed on DNA transfectants of mouse mastocytoma P815. Eur J Immunol 1985;15:1111.
7. Rebaï N, Malissen B. Structural and genetic analysis of HLA class I molecules using monoclonal xeno antibodies. Tissue Antigens 1983;22:107.
8. Malissen B, Rebai N, Liabeuf A, Mawas C. Human cytotoxic T cell structures associated with expression of cytolysis. I. Analysis at the clonal level of the cytolysis inhibiting effect of 7 monoclonal antibodies. Eur J Immunol 1982;9:739.
9. Ozato K, Sachs DH. Monoclonal antibodies to mouse MHC antigens. III. Hybridoma antibodies reacting to antigens of the H-2^b haplotype reveal genetic control of isotype expression. J Immunol 1980;126:317.
10. Stam NJ, Spits H, Ploegh HL. Monoclonal antibodies raised against denatured HLA-B locus heavy chains permit biochemical characterization of certain HLA-C locus products. J Immunol 1986;137:2299.

Author Affiliations

B.M. Perarnau, A. Gillet, M. Barad, R. Hakem, and *F.A. Lemonnier,* Centre d'Immunologie INSERM-CNRS de Marseille-Luminy, Marseille, France

The Cloning, Exon Shuffling, and Expression of the HLA-Cw1 Gene

Shinichi Mizuno, Jennifer Ng, Bo Dupont, and Soo Young Yang

Abstract: A genomic clone that encodes the HLA-CW1 gene was isolated from the B-lymphoblastoid cell line HOM2 (10th WS ID #9005). The gene was successfully expressed on murine L cells (J27) that had already been transfected with human beta-2 microglobulin. However, the level of expression was much lower than the HLA-A2 transfectant. The isoelectric focusing pattern for the HLA-Cw1 antigen precipitated with the monoclonal antibody W6/32 from the metabolically labeled transfectant showed three distinct protein bands of equal intensity. Two hybrid genes were constructed from the HLA-Cw1 and HLA-A2 genes. These hybrid genes were transfected into J27 and expressed on the cell surface. Isoelectric focusing analysis of these hybrid genes showed that a multiplicity of the HLA-Cw1 gene products are controlled by the 5′ portion of the HLA-Cw1 gene.

The EMBL 3 phage lambda genomic library was derived from consanguineous homozygous human lymphoblastoid cell line HOM2 (typing: HLA-A3, B27, Cw1:10th WS ID #9005). One clone designated W5, was isolated and hybridized strongly with the C locus-specific probe pHLA-1d (1). The W5 contained a 19.5kb insert. Southern blot analysis of W5 showed that a single band of 3.9kb hybridized with the pan-HLA probe pHLA-B7 (2) after XbaI digestion. The 3.9kb XbaI fragment was subcloned into a plasmid pUC18 (3) and was designated pHLA-Cw1. The W5 as well as pHLA-Cw1 were transfected into mouse L cell J27 (4) by standard calcium phosphate precipitation. The J27 had been transfected with human beta-2 microglobulin. Since both W5 and pHLA-Cw1 were expressed on the cell surface of J27 at the same level, the transfectants carrying the pHLA-Cw1 were used for further characterization.

Absorption studies showed that the transfectants expressing the pHLA-Cw1 gene products were able to absorb the HLA-Cw1-specific alloantisera, but were unable to absorb the HLA-A3-specific alloantisera, indicating that pHLA-Cw1 is indeed carrying a gene coding for the HLA-Cw1 allele.

Biochemical analysis of the transfectants showed that pHLA-Cw1 encoded 43 KD protein, which was immunoprecipitated using a monoclonal antibody W6/32 on sodium dodecyl sulfate-polyacrylamide gel electrophoresis (SDS-PAGE). Furthermore, isoelectric focusing (IEF) gel analysis showed three protein bands with identical intensity that are also found on the IEF gel from the parental lymphoblastoid cell line HOM2. This finding shows that HLA-Cw1 antigens encoded by pHLA-Cw1

Immunobiology of HLA
Volume II

are biochemically identical to those from HOM2. Since such multiple bands have not been clearly assigned on IEF gel for HLA-A and -B locus products, the multiplicity of the bands on IEF gel may be unique to the HLA-C locus products (five bands have recently assigned for HLA-Cw2 by Gussow et al. [5]).

To investigate the multiplicity of the HLA-Cwl gene products and their weak expression on the cell surface, we constructed two exon-shuffled hybrid genes using the pHLA-2a (6) and pHLA-Cwl. The pHLA-2a contained a 5.2kb HindIII genomic sequence of HLA-A2. The hybrid pHLA-A2/Cwl contained the alpha-1 and alpha-2 domains of HLA-A2 and the alpha-3, transmembrane region and cytoplasmic domains of HLA-Cwl, whereas the HLA-Cwl/A2 construct contained the opposite exons of the pHLA-A2/Cwl. These hybrid genes as well as pHLA-2a were transfected into J27 and the levels of cell surface expression were compared by fluorescence-activated cell sorter (FACS) analysis. The expression of the pHLA-Cwl/A2 antigens was as low as that of the HLA-Cwl antigens. On the other hand, the expression of the pHLA-A2/Cwl was significantly higher than that of HLA-Cwl, but still much lower than that of HLA-A2. The molecular weight of the hybrid molecules was confirmed to be 43 KD by SDS-PAGE.

Our sequence analysis of the 5′ untranslated region of pHLA-Cwl was shown to be identical to that of the reported HLA-Cwl sequence (5). The 5′ untranslated region of pHLA-Cwl was quite similar to the corresponding sequence of HLA-A2 (6) and did have CAT and TATA boxes around the same positions. These results indicate that different levels of surface expression of the HLA-2 and HLA-Cwl antigens are not solely attributable to the difference in promotor activity. This is particularly supported by the fact that pHLA-A2/Cwl, which has a 5′ portion of the A2 gene, had much weaker expression compared with that of the native HLA-A2 gene. It is conceivable, however, that the alpha-3 domain from the HLA-Cwl gene in pHLA-A2/Cwl has a weak association with beta-2 microglobulin, thereby making the expression of this hybrid molecule low. If indeed the association of class I heavy chain with beta-2 microglobulin is playing a crucial role in the expression of the antigens, the alpha-1 and/or alpha-2 domain also seem to mediate interaction of the heavy chain with beta-2 microglobulin. This is indicated by the low expression of the hybrid molecule pHLA-Cwl/A2.

Isoelectric focusing analysis of the hybrid gene products showed that the pHLA-A2/Cwl produced a single peptide, whereas the pHLA-Cwl/A2 gave rise to multiple bands on IEF (Fig. 1). All of these hybrid gene products had an acidic shift in isoelectric points (pI) when compared with that of HLA-A2, indicating that HLA-Cwl polypeptides carried a greater number of acidic amino acid residues throughout the antigen domains that did those of HLA-A2. The multiple protein IEF bands produced by the pHLA-Cwl/A2, but not by pHLA-A2/Cwl, indicate that the 5′ portion of HLA-Cwl gene segment (from the 5′ untranslated region to the beginning of the intron 3) is responsible for the multiplicity of the HLA-Cwl antigen. Since the IEF gel separates the polypeptides according to their charge differences, branches of sugar chain, phosphorylation of amino acid residues, and the polypeptide backbone itself can cause multiple bands from a single gene. The first two possibilities seem unlikely because sialic acids that contribute to most of the charge difference in the sugar chain are removed by neuraminidase during IEF. Phosphorylation and other amino acid modification is also unlikely because our results show that the cytoplasmic region that is the primary target for amino acid residue modification is not responsible for the multiple bands in IEF (pHLA-Cwl/A2 encoded a single protein). Alternative splicing of the 5′ portion of the gene could provide an explanation, as has been previously demonstrated for the H-2K gene (7,8). Although this remains to be formally proven, these findings make HLA-C antigens unique among HLA antigen.

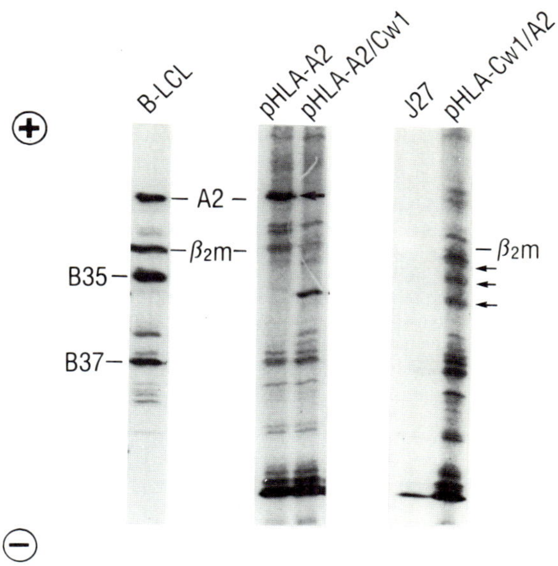

Figure 1. IEF gel analysis of hybrid gene products. A homozygous B-cell line (A2, B35, B57) was used for HLA-A2 control. *Arrows* indicate the products of transfected genes.

Acknowledgment. We thank Dr. Harry T. Orr for kindly providing us with pHLA-B7, pHLA-ld, and pHLA-2a.

References

1. Strachan T, Dodge AB, Smillie D, Dyer PA, Sodoyer R, Jordan BR, Harris R. An HLA-C specific DNA probe. Immunogenetics 1986;23:115.

This work was supported by grants P01 CA 22507 (BD) and P30 CA 08748 (BD).

2. Sood AD, Pereira D, Weissman SM. Isolation and partial nucleotide sequence of a cDNA clone for human histocompatibility antigen HLA-B by use of an oligodeoxynucleotide primer. Proc Natl Acad Sci USA 1981;78:616.
3. Norrander J, Dempe T, Messing J. Construction of improved M13 vectors using oligodeoxynucleotide-directed mutagenesis. Gene 1983;26:101.
4. Kavathas P, Herzenberg LA. Stable transformation of mouse L cells for human membrane T-cell differentiation antigens, HLA and beta2-microglobulin: selection by fluorescence-activated cell sorting. Proc Natl Acad Sci USA 1983;80:524.
5. Gussow D, Rein RS, Meijer I, de Hoog W, Seeman GHA, Hochstenbach FM, Ploegh HL. Isolation, expression, and the primary structure of HLA-Cw1 and HLA-Cw2 genes: evolutionary aspects. Immunogenetics 1987;25:313.
6. Koller BH, Sidwell B, DeMars R, Orr HT. Isolation of HLA locus-specific DNA probes from the 3′-untranslated region. Proc Natl Acad Sci USA 1984;81:5175.
7. Lalanne J-L, Cochet M, Kummer A-M, Gachelin G, Kourilsky P. Different exon-intron organization at the 5′ part of a mouse class I gene is used to generate a novel H-2K^d-related mRNA. Proc Natl Acad Sci USA 1983;80:7561.
8. Transy C, Lalanne J-L, Kourilsky P. Alternative splicing in the 5′ moiety of the H-2 K^d gene transcript. EMBO J 1984;3:2283.

Author Affiliations

Shinichi Mizuno, *Jennifer Ng*, *Bo Dupont*, and *Soo Young Yang*, Laboratories of Biochemical Immunogenetics and Human Immunogenetics, Sloan-Kettering Institute for Cancer Research, 1275 York Avenue, New York, NY 10021, USA

Weak Stimulation of Murine Cytotoxic T Cells by Human Class I Molecules Maps to the α1 + α2 Domains

Victor H. Engelhard, Mark Holterman, and Ai-Xuan T. Le

Class I antigens function as recognition elements in the interaction of cytotoxic T lymphocytes (CTL[3]) with virally infected cells (1), hapten-modified cells (2), cells displaying minor histocompatibility antigens (3), or tumor cells (4). These molecules are also the primary structures recognized by CTL elicited against allogeneic cells (5). Thus, histocompatibility antigens are essential in the formation of structures that lead to activation of CTL precursors (CTLp) or effector cell function. Additional support for this concept comes from the observation that the response of T cells of one species to cells of another is restricted entirely to recognition of the MHC products on the stimulator cells (6–10), and that the T cells elicited are specific for unique, rather than species-specific epitopes. Surprisingly, however, in view of the lower homology between HLA and H-2 molecules than between H-2 alleles, such xenogeneic responses are much weaker than allogeneic responses (6,7,11,12). In this study, the basis for this low level of response has been investigated.

Using limiting dilution analyses in naive C57BL/6 (H-2^b) mice, it was found that the frequency of CTLp

Table 1. Frequency of CTL Precursor in Naive C57BL/6 Spleens Capable of Responding to Allo- and Xenoantigens Expressed on Human or Murine Cells

Stimulator	Target	Antigen	Mean	Range	N
EL4-L^d	EL4-L^d	H-2L^d	1 in 2.2×10^4	$1.2–3.1 \times 10^4$	4
JY	MST	HLA-B7	1 in 1.1×10^6	1.1×10^6	2
JY	HSB	HLA-A2	1 in 1.8×10^6	$0.88–2.8 \times 10^6$	4
JM-L^d	JM-L^d	H-2L^d	1 in 4.9×10^4	$2.4–9.6 \times 10^4$	3
	EL4-L^d	H-2L^d	1 in 2.8×10^4	$2.3–3.3 \times 10^4$	2
EA6	EA6	HLA-A2	1 in 3.0×10^5	$1.5–4.4 \times 10^5$	2
	HSB	HLA-A2	1 in 5.4×10^5	$4.7–6.1 \times 10^5$	2
EB1	EB1	HLA-B7	1 in 4.7×10^5	$3.2–5.9 \times 10^5$	3
	MST/JM	HLA-B7	1 in 3.9×10^5	$3.4–4.4 \times 10^5$	2

Precursor frequencies were measured as previously described (11). The cells used include: EL4-L^d, an EL4 transfectant expressing H-2L^d; JY, a human B-cell line (HLA-A2,2;-B7,7;-DR4,6); HSB, a human T-cell line (HLA-A1,2;-B12,17;-DR neg); MST, a human B-cell line (HLA-A3,3;-B7,7;-DR2,2); JM-L^d, a human T-cell line (HLA-A3,25;-B7,37;-DR neg) expressing H-2L^d; EA6, an EL4 transfectant expressing HLA-A2; and EB1, an EL4 transfectant expressing HLA-B7.

Immunobiology of HLA
Volume II

Table 2. Frequency of CTL Precursors in Primed C57BL/6 Spleens Capable of Responding to HLA Antigens

Stimulator Cell Used				
In Vivo	In Vitro	Target	Precursor Frequency	N
JY	MST	JM	1 in 3.5×10^5	2
JY	JY/HSB	HSB	1 in 4.4×10^5	3

Details are given in the footnotes to Table 1.

capable of responding to a transfectant of the syngeneic EL4 cell line that expresses the allogeneic H-2L^d antigen (EL4-L^d) was in the range of 1 in 2.2×10^4 (Table 1). In contrast, the frequency of CTLp specific for HLA-A2 or -B7 on the human lymphoblastoid cell line JY was less than 1 in 10^6 spleen cells. Spleen cells from an animal primed with JY and restimulated in vitro with HSB, which shares only HLA-A2 with JY, or with MST, which shares only HLA-B7 with JY, gave CTLp frequencies of 1 in 4.4×10^5 and 1 in 3.5×10^5, respectively (Table 2). The ability to increase the frequency of these CTLp by in vivo priming argues against suppression as an explanation for the low frequency in naive animals. However, the secondary response to HLA antigens was still one to two orders of magnitude weaker than a primary allogeneic response.

One possible explanation for these results is that only a small fraction of murine T cells are capable of recognizing epitopes on HLA antigens. Another possibility is that stimulation requires a species-specific interaction between accessory molecules on the T cells and ligands on the stimulator cells. To investigate this latter possibility the cytotoxic response to H-2L^d and HLA-A2 expressed on either human (JM) or syngeneic murine (EL4) cells was compared. H-2L^d positive JM or EL4 transfectants stimulated approximately equal responses in naive animals (Table 1). The fine specificity of short-term clones indicated that CTL stimulated by either cell cross-reactively recognized the other (not shown). These results indicate that the cellular background plays little, if any, role in the ability of CTLp to respond to H-2 alloantigens.

When EL4 transfectants expressing HLA-A2 (EA6) or HLA-B7 (EB1) were used to stimulate a response in naive animals, CTLp responses of approximately 1 in 3-5 $\times 10^5$ were seen (Table 1). These values suggest that the murine cell background enhanced the response of CTLp to HLA antigens. Interestingly, stimulation of CTLp with murine transfectants also enhanced the frequency of CTLp measured using human cells as targets. Thus, this background difference appears most important during the stimulation of CTLp, rather than recognition by CTL effectors. The participation of the cellular background, presumably due to accessory molecule interactions, may be more important during CTLp stimulation by human class I molecules than during stimulation by murine alloantigens. Nonetheless, the precursor frequencies seen are still at least an order of magnitude lower than those to H-2 alloantigens. Thus, there is still a strong disparity in the response of murine CTL to murine and human class I molecules that is independent of cellular background. These results are in accord with the hypothesis that there have been evolutionary changes between HLA and H-2 antigens that render xenogeneic antigens less recognizable as transplantation antigens.

Table 3. Frequency of CTL Precursors in Naive C57BL/6 Mice Responding to Murine Transfectants Expressing H-2L^d, HLA-B7, or HLA/H-2 Chimeric Molecules

Stimulator	Target	Precursor Frequency
EL4-L^d	EL4-L^d	1 in 4.4×10^4
	EL4-L^d/B7	1 in 3.0×10^4
EL4-L^d/B7	EL4-L^d/B7	1 in 1.4×10^4
	EL4-L^d	1 in 4.0×10^4
EL4-B7	EL4-B7	1 in 3.2×10^5
	ELA-B7/L^d	1 in 5.1×10^5
	MST	1 in 4.4×10^5
EL4-B7/L^d	EL4-B7/L^d	1 in 5.4×10^5
	EL4-B7	1 in 1.1×10^6
	MST	1 in 1.0×10^6

The cells used include: EL4-L^d, EL4 cells expressing H-2L^d; EL4-L^d/B7, EL4 cells expressing the H-2L^d/HLA-B7 chimeric molecule; EL4-B7, EL4 cells expressing HLA-B7; EL4-B7/L^d; EL4 cells expressing the HLA-B7/H-2L^d chimeric molecule; and MST, a human cell line expressing HLA-B7.

The regions of the HLA molecule determining the strength of this response were mapped using chimeric H-2/HLA antigens (13). Molecules expressing either the $\alpha1$ and $\alpha2$ domains of H-2L^d and the $\alpha3$, TM, and CT domains of HLA-B7 (L^d/B7), or the $\alpha1$ and $\alpha2$ domains of HLA-B7 and the $\alpha3$, TM, and CT domains of H-2L^d (B7/L^d) were expressed in syngeneic EL4 cells. CTLp frequencies in naive mice stimulated with transfectants expressing either H-2L^d or L^d/B7 were similar (Table 3), and the majority of clones stimulated with either antigen exhibited cross-reactive recognition of the other (not shown). CTLp frequencies in naive mice stimulated with transfectants expressing either HLA-B7 or B7/L^d were also similar (Table 3). The majority of the positive wells from clonal dilutions contained CTL that recognized only the murine cells, and not human cells expressing HLA-B7. Of this group, approximately half recognized both the parental molecule and the chimeric construct, whereas the remainder recognized only the stimulating antigen (not shown). These results indicate that the frequency of CTLp is controlled by the $\alpha1 + \alpha2$ domain while the species of origin of the $\alpha3$, TM, and CT domains is largely irrelevant. However, these latter domains do influence the fine specificity of HLA-B7-specific clones. A possible explanation for this is that this region directly alters epitopes on the HLA-B7 molecule. Alternatively, this region may contain epitopes that are processed and become associated with H-2 molecules for recognition by H-2 restricted CTL (11,14–16).

The weak T-cell response to xenogeneic class I antigens remains a paradox. One possible explanation is that during T-cell ontogeny, the T-cell repertoire develops to recognize epitopes that closely resemble syngeneic H-2 molecules. Therefore, CTL recognize epitopes on murine alloantigens more easily than those on human class I molecules. However, HLA-specific murine CTL distinguish between different HLA alloantigens, indicating that allele-specific and not species-specific determinants are the principal polymorphic determinants recognized by such T cells (11,17,18). This observation is difficult to reconcile with a model in which the repertoire develops to preferentially recognize polymorphic structures that most closely resemble syngeneic H-2 molecules. Another possible explanation is that MHC antigens express conserved structures that mark them as ligands for T cells. T cells might not interact as well with xenogeneic MHC antigens if such determinants had undergone evolutionary modification. Recognition of these determinants could occur either through the T-cell receptor (19), or through accessory molecules such as CD8 (reviewed in 20). Further experimentation will be required to distinguish among these possibilities.

Acknowledgments. Supported by PHS research grants Al20963 and Al21393.

References

1. Doherty PC, Gotze D, Trinchieri G, Zinkernagel RM. Models for recognition of virally modified cells by immune thymus derived lymphocytes. Immunogenetics 1976;3:517.
2. Shearer GM, Rhen TG, Garbarino CA. Cell mediated lympholysis of trinitrophenyl-modified autologous lymphocytes. Effector cell specificity to modified cell surface components controlled by the H-2K and H-2D serological regions of the murine major histocompatibility complex. J Exp Med 1975;141:1348.
3. Bevan MJ. In a radiation chimaera, host H-2 antigens determine immune responsiveness of donor cytotoxic cells. Nature 1977;269:417.
4. Gooding L. Specificity of killing by cytotoxic lymphocytes generated in vivo and in vitro to syngeneic SV40 transformed cells. J Immunol 1977;118:920.
5. Cerottini J-C, Brunner KT. Cell mediated cytotoxicity, allograft rejection, and tumor immunity. Adv Immunol 1974;18:67.
6. Burakoff SJ, Ratnofsky SE, Benacerraf B. Mouse cytolytic T-lymphocytes induces by xenogeneic rat stimulator cells exhibit specificity for H-2 complex alloantigens. Proc Natl Acad Sci USA 1977;74:4572.
7. Asantila T, Vahala J, Toivanen P. Immunogenetics 1974;3:272.
8. Lindahl KF, Bach FG. Human lymphocytes recognise mouse alloantigens. Nature 1975;254:607.
9. Carnaud C, Fadai Ghotbi M, Lesavre P, Bach J-F. Education of human lymphocytes against mouse cells: Specific recognition of H-2 antigens. Eur J Immunol 1977;7:81.
10. Engelhard VH, Kaufman JF, Strominger JL, Burakoff SJ. Specificity of mouse cytotoxic T lymphocytes stimulated with either HLA-A and -B or HLA-DR antigens reconstituted into phospholipid vesicles. J Exp Med 1980;152:54s.
11. Holterman MJ, Engelhard VH. HLA antigens expressed on murine cells are preferentially recognized by murine cytotoxic T cells in the context of the H-2 major histocompatibility complex. Proc Natl Acad Sci USA 1986;83:9699.
12. Payne LN, Jaffe WP. Graft against host reactions produced in chicken embryos with blood from heterologous avian donors. Transpl Bull 1962;30:20.
13. Engelhard VH, Yannelli JR, Evans GA, Walk SF, Holterman MJ. Construction of novel class I histocompatibility antigens by interspecies exon shuffling. J Immunol 1985;134:4218.
14. Maryanski JL, Pala P, Corradin G, Jordan BR, Cerottini J-C. H-2-restricted cytolytic T cells specific for HLA can recognize a synthetic HLA peptide. Nature 1986;324:578.
15. Maryanski JL, Accolla RS, Jordan B. H2-restricted recognition of cloned HLA class I gene products expressed in mouse cells. J Immunol 1986;136:4340.
16. Gomard E, Begue B, Sodoyer S, Maryanski JL, Jordan BR, Levy JP. Murine cells expressing an HLA molecule are specifically lysed by HLA-restricted antiviral human T cells. Nature 1986;319:153.
17. Engelhard VH, Benjamin C. Isolation and characterization of monoclonal mouse cytotoxic T lymphocytes with specificity for HLA-A, B or -DR alloantigens. J Immunol 1982;129:2621.
18. Yannelli JR, Moore LC, Engelhard VH. Multiple epitopes on human and murine cells expressing HLA-B7 as defined by specific murine cytotoxic T cell clones. J Immunol 1985;135:900.
19. Engelhard VH, Benjamin C. Xenogeneic cytotoxic T-cell clones recognize alloantigenic determinants on HLA-A2. Immunogenetics 1983;18:461.
20. Goverman J, Hunkapiller T, Hood LE. A speculative view of the multicomponent nature of T cell antigen recognition. Cell 1986;45:475.

Author Affiliations

Victor H. Engelhard, *Mark Holterman*, Department of Microbiology; *Ai-Xuan T. Le*, Department of Surgery, University of Virginia School of Medicine, Charlottesville, VA, USA

Section IV
Class I Histocompatibility Antigens: Regulation

Phorbol Ester (PMA)-Activated Protein Kinase C Phosphorylates the Class I HLA Heavy Chain*

Jennifer Ng, Shinichi Mizuno, Bo Dupont, and Soo Young Yang

Abstract: The intracellular domains of the class I HLA heavy chain contain a series of amino acid residues that can be phosphorylated by various protein kinases. Among them, two serine residues are phosphorylated by cAMP-dependent protein kinase (1–3), and a tyrosine residue can be phosphorylated in vitro by the pp60 src kinase of the rous sarcoma virus (4). We now report that the class I HLA heavy chains are phosphorylated on treatment with PMA in a dose-dependent manner in B-lymphoblastoid cell lines and purified T lymphocytes. According to IEF gel analysis, only one amino acid residue is phosphorylated by the phorbol ester-activated protein kinase C. We also observed that all HLA-A and HLA-B allelic products tested are phosphorylated by protein kinase C. This indicates that the substrate residue for protein kinase C in class I HLA antigens is well conserved.

The HLA class I molecules are known to be involved in graft rejection and cytotoxic T-lymphocyte (CTL) recognition of virus-infected cells (5). Class I antigens consist of a polymorphic integral membrane glycoprotein and an invariant light chain (beta-2 microglobulin) (6). The carboxyterminus of the heavy chain contains both a hydrophilic stretch of 32 amino acid residues and a penultimate sequence of 25 amino acid residues that is markedly hydrophobic (1,7). The catalytic subunit of the cAMP-dependent protein kinase has been demonstrated to recognize sites of phosphorylation on substrates that contain at least two positively charged residues, followed by one or two intervening residues leading to a serine (8) or threonine (9) residue. A tyrosine residue has been found to be phosphorylated in vitro by the pp60 src kinase of the rous sarcoma virus (4). Protein kinase C can also phosphorylate a variety of receptors. The treatment of intact cells with phorbol esters, which directly activate protein kinase C, also leads to receptor phosphorylation (10,11).

In our studies, class I HLA heavy chains were phosphorylated by treatment of B-lymphoblastoid cell lines (B-LCL) or purified T cells with phorol 12 myristate 13-acetate (PMA) in a dose-dependent manner. The dose response of B-LCL to PMA shows an increasing amount of phosphorylation of the class I heavy chain from 1 ng/ml to 1 μg/ml PMA (the optimal phosphorylation). The amount of phosphorylation did not increase with PMA concentrations over 1 μg/ml (100 μg/ml was the maximal concentrations of PMA used). The results we obtained using purified peripheral T lymphocytes were similar, with an optimal phosphorylation at 1 μg/ml PMA. We tested the phosphorylation of class I products in the presence of PMA on five consanguinous homozygous human lymphoblastoid cell lines – BH (WS ID# 9046: A2, B13), DBB (WS ID# 9052; A2, B57), YAR (WS ID# 9026: A26, B38), MGAR (WS ID# 9014: A26, B8), and LWAGS (WS ID# 9079: A33, B14). For each homozygous cell line, there were two distinct bands that could be assigned to their respective HLA-A and HLA-B haplotypes when the cells were metabolically labeled with in the absence of PMA (Fig. 1A). When the cells were labeled in the presence of 10 μg/ml PMA, four strong distinct bands were observed for each cell line (Fig. 1B). Each of the new bands observed after the addition of PMA was shifted by 1 unit in the acidic direction of the isoelectric focusing (IEF) gel. To show that these new bands were a result of phosphorylation, we labeled the cell lines with ^{32}P-orthophosphate in the presence of PMA. The acidic shift of the phosphorylated bands on the gel was identical for all alleles tested (Fig. 1A and D). These data indicate that only one amino acid residue among the protein kinase substrates in the class I heavy chain is phosphorylated by the phorbol ester-activated protein kinase C, and the substrate residue for protein kinase C is well conserved. Basal levels of phosphorylation were observed in these cell lines in the absence of PMA (Fig. 1C).

T cells can be activated by a variety of stimuli including plant lectins and monoclonal antibodies directed against certain cell surface molecules (12). We have tested phytohemaglutinin (PHA), PMA, monoclonal antibody T3 (CD3), and a mitogenic combination of CD2 antibodies (9.1 and 9.6) to examine whether HLA class I antigens are phosphorylated when T cells are activated by these stimuli. Interestingly, while all of these stimuli activated T cells, only PMA phosphorylated class I antigens (not shown). PHA had a negative effect on phosphorylation. When T cells were activated by PHA, the basal level of class I antigen phosphorylation was significantly reduced (not shown). PHA exhibited no effect on the phosphorylation of heavy chains on B-LCLs. These data suggest that phosphorylation of HLA class I antigens is regulated by a series of mechanisms that have an important immunological function. It is conceivable that different kinds of protein kinases are involved and/or regulatory elements exist for phosphorylation of the HLA class I antigens. Receptor phosphorylation affects a variety of functions. In many cases, phosphorylation negatively affects the function of

This work was supported by grants P01 CA 22507 and P30 CA 08748.

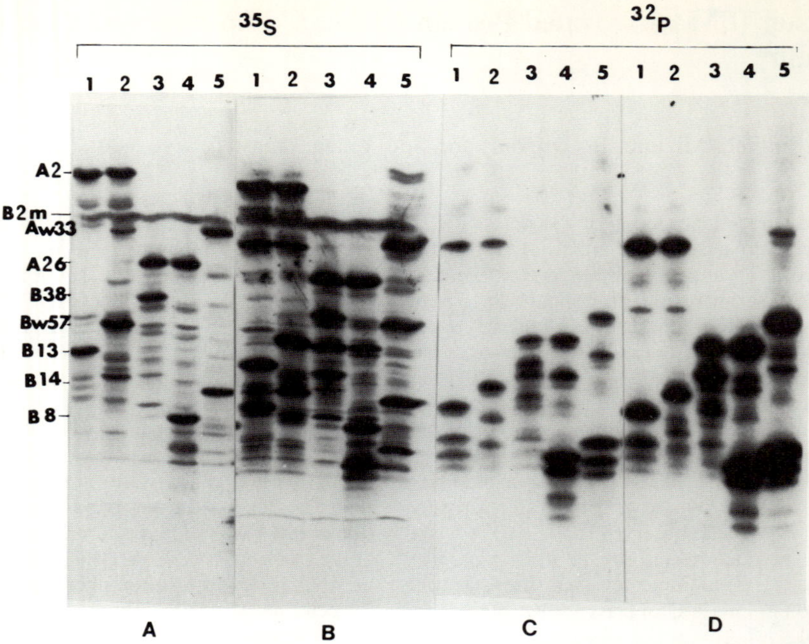

Figure 1. Here, 5×10^6 B cells were labeled with ^{35}S-methionine (^{35}S) for 12 hours (*panels A and B*) or with 0.4 mCi of ^{32}P-orthophosphate (^{32}P) for 2 hours (*panels C and D*) in the presence (*panels B and D*) or absence (*panels A and C*) of 10 μg/ml PMA. The class I heavy chain products were then analyzed by IEF gel analysis. Cell lines – 1: BH (WS ID# 9046: A2, B13), 2: DBB (WS ID# 9052: A2, B57); 3: YAR (WS ID# 9026: A26, B38), 4: MGAR (WS ID# 9014: A26, B8), and 5: LWAGS (WS ID# 9079: A33, B14).

receptors. In most instances, negative modulations are associated with phosphorylation of serine and/or threonine residues on the receptor proteins. In contrast, the tyrosine kinase receptors appear to be susceptible to positive modulation by phosphorylation. The conservation of so many potential sites of phosphorylation in class I genes is unlikely to have been maintained by chance. The phosphorylation of various sites may affect different regulatory processes such as endocytosis, interaction with acto-myosin filaments, or other unknown functions. The analysis of sites of phosphorylation within the HLA molecule should reveal whether or not, and if so, how such modifications regulate these important cell surface molecules.

References

1. Robb RJ, Terhorst C, Strominger JL. Sequence of the COOH-terminal hydrophilic region of histocompatibility antigens HLA-A2 and HLA-B7. J Biol Chem 1978;253:5319.
2. Pober JS, Guild BC, Strominger JL. Phosphorylation in vivo and in vitro of human histocompatibility antigens (HLA-A and HLA-B) in the carboxy-terminal intracellular domain. Proc Natl Acad Sci USA 1978;75:6002.
3. Guild BC, Strominger JL. HLA-A2 antigen phosphorylation in vitro by cyclic AMP-dependent protein kinase. J Biol Chem 1984;259:13504.
4. Guild BC, Erikson RL, Strominger JL. HLA-A2 and HLA-B7 antigens are phosphorylated in vitro by Rous sarcoma virus kinase ($pp60^{v\text{-}src}$) at a tyrosine residue encoded in a highly conserved exon of the intracellular domain. Proc Natl Acad Sci USA 1983;80:2894.
5. Zinkernagel RM, Doherty PC. MHC-restricted cytotoxic T cells: studies on the biological role of polymorphic major transplantation antigens determining T-cell restriction-specificity, function, and responsiveness. Adv Immunol 1979;27:51.
6. Ploegh HL, Orr HT, Strominger JL. Major histocompatibility antigens: The human (HLA-A, -B, -C) and murine (H-2K, H-2D) class I molecules. Cell 1981;24:287.
7. Springer TA, Strominger JL. Detergent-soluble HLA antigens contain a hydrophilic region at the COOH-terminus and a penultimate hydrophobic region. Proc Natl Acad Sci USA 1976;73:2481.
8. Yokoyama K, Nathenson S. Intramolecular organization of class I H-2 MHC antigens; localization of the alloantigenic determinants and the beta-2m binding site to different regions of the H-2 K^b glycoprotein. J Immunol 1983;130:1419.
9. Ways JP, Lawlor DA, Wan AM, Parham P. Immunogenetics 1987 (in press).
10. Krangle MS, Biddison WE, Strominger JL. Comparative structural analysis of HLA-A2 antigens distinguishable by cytotoxic T lymphocytes. J Immunol 1983;130:1856.
11. Russo C, Ng AK, Pellegrino MA, Ferrone S. The mono-

clonal antibody CR11-351 discriminates HLA-A2 variants identified by T cells. Immunogenetics 1983;18:23.
12. Yang SY, Chouaib S, Dupont B. A common pathway for T lymphocyte activation involving both the CD3-Ti complex and CD2 sheep erythrocyte receptor determinants. J Immunol 1986;137:1097.

Author Affiliations

Jennifer Ng, *Shinichi Mizuno*, *Bo Dupont*, and *Soo Young Yang*, Laboratories of Biochemical Immunogenetics and Human Immunogenetics, Sloan-Kettering Institute for Cancer Research, New York, NY, USA

Differential Modulation by Anti-HLA Class I Monoclonal Antibodies of T-Cell Proliferation Induced via CD2 and CD3 Pathways*

Maria Caterina Turco, Mario De Felice, Soo Young Yang, Soldano Ferrone, and Salvatore Venuta

Activation of human T cells is known to occur through two different pathways involving CD2 and CD3 molecules (for review, see Ref. 1). HLA class I antigens have been found unexpectedly to regulate T-cell proliferation induced through these two pathways (2–10). In this paper we compare the effect of monoclonal antibodies (MAbs) to monomorphic and polymorphic determinants of HLA class I antigens on the proliferation of T cell stimulated with the anti-CD2 MAb 9-1 + 9.6 or with the anti-CD3 MAb OKT3.

The anti-HLA-A2,A28 MAb CR11-351 and the MAb Q6/64, recognizing a determinant restricted to the gene products of the HLA-B locus, enhance the activation of T cells stimulated with the anti-CD2 MAb 9-1 + 9.6, but inhibit the activation of T cells stimulated with the anti-CD3 MAb OKT3 or with the anti-Ti MAb BMA 031. The MAb W6/32 to a monomorphic determinant of HLA class I antigens has a marked inhibitory effect on the MAb OKT3- or BMA 031-induced T-cell proliferation from all the donors tested, but enhances the proliferation of 9-1 + 9.6 activated T cells from less than 30% of the donors tested; the degree of enhancement is low. The anti-β_2-microglobulin (β_2-μ) MAb NAMB-1 does not affect the proliferation of T cells activated via the CD2 or CD3 pathway (Table 1).

The dose response and the kinetics of the effect of anti-HLA class I MAbs on the two pathways are similar. As shown in Figure 1, the MAb CR11-351 exerts its maximal effect on the two pathways at the concentration of 1 μg/ml, 50% of the maximal effect at the concentration of about 0.3 μg/ml, and less than 25% at the concentration of about 0.20 μg/ml. The modulatory effect of the MAb Q6/64 on the proliferation of T cells activated via the two pathways is markedly reduced when it is added after 20 hours of incubation with the mitogenic MAbs. Anti-HLA class I MAbs modulate the CD2 and CD3 pathways by interacting with the corresponding determinants, since $F(ab')_2$ fragments are as effective as the whole Ig. Furthermore, MAb to polymorphic determinants of HLA class I antigen display a modulatory effect only when tested with peripheral blood mononuclear cells (PBMC) with the appropriate HLA phenotype.

This work was supported in part by Consiglio Nazionale delle Ricerche projects "Oncologia" and "Ingegneria genetica e basi molecolari delle malattie ereditarie," by grants from Associazione Italiana per la Ricerca sul Cancro, by NIH grants AI21384 and CA39559, and by a grant from the International Juvenile Diabetes Foundation.

Table 1. Effect of Anti-HLA Class I Monoclonal Antibodies on T-Cell Proliferation Induced via CD2 or CD3-Ti Pathways*

MAb	HLA Specificity	MAb 9-1 + 9.6		MAb OKT3		MAb BMA 031	
		cpm × 10^{-3}	% Variation	cpm × 10^{-3}	% Variation	cpm × 10^{-3}	% Variation
		67.8 ± 4.5		86.2 ± 9.7		48.8 ± 3.9	
W6/32	Framework	65.5 ± 5.6	(−4)	2.7 ± .1	(−97)	2.0 ± 1.3	(−96)
Q6/64	B locus restricted	112.0 ± 1.8	(+65)	1.8 ± .2	(−98)	19.6 ± 2.5	(−60)
CR11-351	A2,A28	138.1 ± 19.7	(+103)	16.4 ± 1.5	(−81)	12.9 ± .8	(−73)
NAMB-1	β_2-μ	60.2 ± 6.2	(−11)	82.9 ± 7.5	(−4)	N.T.†	

*Peripheral blood mononuclear cells (PBMC) (5 × 10^5/ml) were incubated with the MAb 9-1 + 9.6 (2 μg/ml), MAb OKT3 (5 ng/ml), or BMA 031 (10 μg/ml) in the presence of anti-HLA class I MAbs (10 μg/ml). After a 3-day incubation at 37°C, [^{3}H]dT was added to each well; after an additional 15-hour incubation, cells were harvested on glass fiber strips and [^{3}H]dT incorporation was measured in a β counter.
†Not tested.

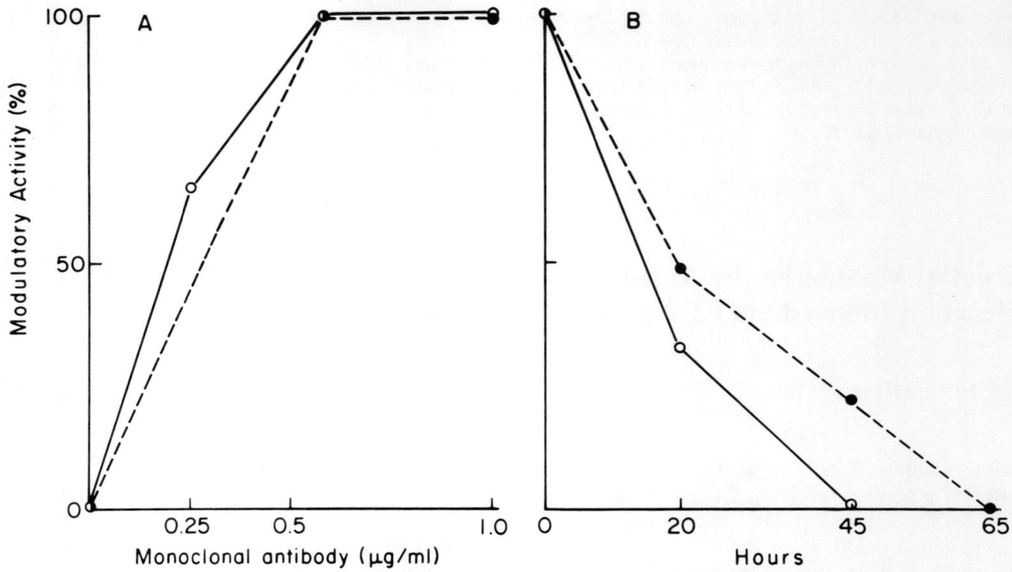

Figure 1. Dose-dependent modulation by anti-HLA-A2,A28 MAb CR11-351 (A) and kinetics of the modulation by anti-HLA Class I MAb Q6/64 (B) of the proliferation of HLA-A2-positive T cells stimulated with the anti-CD2 MoAb 9–1 + 9.6 (○ – ○) and with the anti-CD3 MAb OKT3 (● – ●). In A, peripheral blood mononuclear cells were incubated at 37°C with MAb 9–1 + 9.6 (2 μg/ml) or with MAb OKT3 (5 ng/ml) in the presence of MAb CR11-351. In B, peripheral blood mononuclear cells were incubated with MAb 9–1 + 9.6 (2 μg/ml) or with MAb OKT3 (5 ng/ml); at the indicated times, the MAb Q6/64 (10 μg/ml) was added. In both experiments following a total 72-hour incubation at 37°C [^{3}H]thymidine incorporation was measured. Results are expressed as percentage of maximal effect.

Both IL-2 secretion, measured by a CTLL assay, and expression of IL-2 and transferrin receptors, measured by immunofluorescence staining with MAbs, are inhibited in PBMC stimulated with MAb OKT3 in the presence of anti-HLA MAb CR11-351, Q6/64, and W6/32. Furthermore, inhibition of IL-2 receptor expression, as well as of tritiated thymidine ([^{3}H]dT) incorporation, is not overcome by the addition of exogenous IL-2 to the cultures. In contrast, in PBMC stimulated with the MAb 9–1 + 9.6 in the presence of anti-HLA MAb Q6/64, the percentage of cells expressing Tac antigen and transferrin receptor is enhanced; no change in IL-2 secretion is detected.

Table 2. Effect of Anti-HLA Class I MAb Q6/64 on Tac Antigen and Transferrin Receptor Expression by T Cells Activated via CD2 or CD3 Pathway*

MAb 9–1 + 9.6	MAb OKT3	MAb Q6/64	Tac Antigen	Transferrin Receptor
–	–	–	2† (41)‡	1 (56)
+	–	–	25 (85)	23 (89)
+	–	+	57 (83)	42 (79)
–	+	–	34 (98)	30 (96)
–	+	+	4 (56)	1 (96)

*PBMC (5 × 10^5/ml) were incubated at 37°C for 3 days with anti-CD2 MAb 9–1 + 9.6 (2 μg/ml) and with anti-CD3 MAb OKT3 (5 ng/ml) in the presence of anti-HLA class I MAb Q6/64 (10 μg/ml). Then, cells were washed, incubated at 5 × 10^6/ml with fluorescein isothyocianate-conjugated MAb anti-Tac and antitransferrin receptor for 30 minutes at 4°C, washed again and analyzed with a cytofluorograph Ortho Spectrum III.
†Percent of positive cells.
‡Mean intensity of fluorescence.

Experiments with MAbs to polymorphic determinants of HLA antigens and cell populations reconstituted with T cells and accessory cells with the appropriate HLA phenotype have shown that the modulatory effect of anti-HLA class I MAbs on the CD2 and CD3 pathways reflects their interaction with T lymphocytes and not with accessory cells. Specifically, the anti-HLA-A2,A28 MAb CR11-351 modulates the proliferation of HLA-A2-positive T cells mixed with HLA-A2,A28-negative accessory cells and stimulated with either MAb OKT3 or MAb 9–1 + 9.6. On the other hand, no effect is detected on HLA-A2,A28-negative T cells cocultured with HLA-A2-positive accessory cells and activated via either the CD2 or CD3 pathway.

The data that we have summarized indicate that MAbs to both polymorphic and monomorphic determinants of HLA class I antigens display a differential effect on the proliferation of T cells activated via the CD2 or CD3 pathway. This difference is not only in terms of enhancement and inhibition of T-cell proliferation induced via

Table 3. Cell Target in the Modulation by Anti-HLA-A2,A28 MAb CR11-351 of the Proliferation of T Cells Activated via the CD2 or CD3 Pathway*

HLA Phenotype		Mitogenic	MAb	[³H]dT Incorporation	
T Cells	Monocytes	MAb	CR11-351	cpm × 10^{-3}	% Variation
A2,A26	A24,A24	9–1 + 9.6	−	22.8 ± 3.7	
A2,A26	A24,A24	9–1 + 9.6	+	53.6 ± 3.2	+135
A24,A24	A2,A26	9–1 + 9.6	−	31.6 ± 2.8	
A24,A24	A2,A26	9–1 + 9.6	+	35.6 ± 1.2	+13
A2,A26	A24,A24	OKT3	−	35.7 ± 1.6	
A2,A26	A24,A24	OKT3	+	6.8 ± 0.9	−81
A24,A24	A2,A26	OKT3	−	31.0 ± 2.0	
A24,A24	A2,A26	OKT3	+	28.5 ± 1.8	−8

*T cells were purified from peripheral blood mononuclear cells (PBMC) by sequential centrifugation on a Percoll density gradient and incubation with anti-HLA class II MAb OKIaI and rabbit complement. Monocytes were purified from PBMC by centifugation through a 46% Percoll density gradient. T cells (5 × 10^5/ml) were mixed with monocytes (5 × 10^4/ml) and incubated for 3 days at 37°C with MAb 9.1 + 9–6 (2 μg/ml) or MAb OKT3 (5 ng/ml) in the presence of anti-HLA class I MAbs (10 μg/ml). Then [³H]dT incorporation was measured.

the former and the latter pathways, respectively, but also in terms of degree of involvement of a given determinant of HLA class I antigens in the two pathways. The differential role of a determinant in the two pathways does not correlate with its polymorphic or monomorphic nature and with the IgG subclass of the corresponding MAb. Whether the affinity of MAbs plays a role in this phenomenon remains to be determined. The effect of anti-HLA class I MAbs on the two pathways is similar in terms of dose response, kinetics, lack of involvement of Fc portion, interaction with target cells, and modulated steps of the series of events leading to T-cell proliferation.

The mechanism(s) underlying the modulatory effect of anti-HLA class I MAbs on T-cell proliferation is only speculative at this time. MAbs may induce interactions of HLA class I antigens with molecules involved in T-cell activation and/or may trigger signals affecting the level of molecules that control T-cell proliferation. The latter possibility is suggested by changes in the level of mRNA for IL-2 and IL-2 receptor associated with the stimulation of T cells via the CD2 and CD3 pathways in the presence of anti-HLA class I MAbs.

Our study has shown that in addition to their role as restricting elements in T-cell killing, HLA class I antigens regulate T-cell proliferation and that anti-HLA class I MAbs may represent useful reagents to modulate T-cell proliferation. The recent reports (11,12) that HLA class I antigens regulate also the proliferation of B cells suggest that this may be a general phenomenon. Whether abnormalities in the expression of HLA class I antigens like those associated with malignant transformation of cells play a role in their uncontrolled proliferation remains to be determined.

Acknowledgments. The authors wish to thank Dr. P.J. Martin (University of Washington, Seattle, WA, USA) and Drs. A. Afting and R. Kurrle (Behringwerke AG, Marburg, West Germany) for providing monoclonal antibodies and to acknowledge the excellent secretarial assistance of Mrs. Edwina L. Jones and Ms. Deborah O. Kimbo.

References

1. Alcover A, Ramarli D, Richardson NE, Chang H-C, Reinherz EL. Functional and molecular aspects of human T lymphocyte activation via T3-Ti and T11 pathways. Immunol Rev 1987;95:5.
2. Akiyama Y, Zicht R, Ferrone S, Bonnard GD, Herberman RB. Effect of monoclonal antibodies (MoAb) to Class I and Class II HLA antigens on lectin- and MoAb OKT3-induced lymphocyte proliferation. Cell Immunol 1985;91:477.
3. Bonagura VR, Nordli DR, Pernis B. Antibodies against monomorphic determinants of the alfa chain of Class I human leukocyte antigens inhibit the reaction of human lymphocytes to mitogens and tetanus toxoid. Immunol Invest 1985;14:233.
4. Turco MC, De Felice M, Corbo L, Morrone G, Mertelsmann R, Ferrone S, et al. Regulatory role of a monomorphic determinant of HLA Class I antigens in T cell proliferation. J Immunol 1985;135:2268.
5. Taylor DS, Nowell PC, Kornbluth J. Functional role of HLA Class I-cell surface molecules in human T lymphocyte activation and proliferation. Proc Natl Sci USA 1986; 83:4446.
6. Dasgupta JD, Cemach K, Dubey DP, Yunis EJ, Amos DB. The role of class I histocompatibility antigens in the regulation of T-cell activation. Proc Natl Acad Sci USA 1987; 84:1094.
7. Dasgupta JD, Yunis EJ. Receptor-like role of HLA-Class I antigens: Regulation of T cell activation. J Immunol 1987;139:672.
8. De Felice M, Turco MC, Carandente Giarrusso P, Corbo L, Pizzano R. Martinelli V, et al. Differential regulatory role of monomorphic and polymorphic determinants of HLA Class I antigens in monoclonal antibody (MoAb) OKT3-induced T cell proliferation. J Immunol 1987;139:2683.
9. Huet S, Boumsell L, Raynal B, Degos L, Dausset J, Bernard A. Role in T-cell activation for HLA Class I molecules from accessory cells: Further distinction between activation signals delivered to T cells via CD2 and CD3 molecules. Proc Natl Acad Sci USA 1987;84:7222.

10. Turco MC, De Felice M, Corbo L, Carandente Giarrusso P, Yang SY, Ferrone S, et al. Enhancement by anti HLA Class I monoclonal antibodies of T cell proliferation induced via CD2 molecule. J Immunol 1988;7:2275.
11. Giudizi MG, Biagiotti R, Almerigogna F, Alessi A, Tiri A, Del Prete GF, et al. Role of HLA Class I and Class II antigens in activation and differentiation of B cells. Cell Immunol 1987;108:97.
12. Taylor DS, Nowell PC, Kornbluth J. Anti-HLA Class I antibodies inhibit the T cell-independent proliferation of human B lymphocytes. J Immunol 1987;139:1792.

Author Affiliations

Maria Caterina Turco and *Mario De Felice*, Istituto di Scienze Biochimiche, II Facoltá di Medicina e Chirurgia, Napoli, Italy; *Soo Young Yang*, Immunology Program, Memorial Sloan-Kettering Cancer Research Center, New York, NY, USA; *Soldano Ferrone*, Department of Microbiology and Immunology, New York Medical College, Valhalla, NY, USA; *Salvatore Venuta*, Istituto di Oncologia Clinica e Sperimentale, Facoltá di Medicina e Chirurgia di Catanzaro, Universitá di Reggio Calabria, Catanzaro, Italy

Anti-HLA Class I Antibodies Alter Gene Expression in Human Natural Killer Cells

Jacki Kornbluth and Richard G. Hoover

Abstract: Natural killer (NK) function can be augmented by a variety of agents, including interleukin 2 (IL2) and interferon (IFN) (1–3). Anti-HLA class I and anti-LFA-1 monoclonal antibodies (MAbs) also affect NK lysis at the level of the effector cell (4–8). These modulators were used to assess potential changes in gene expression in a human NK clone (NK3.3) associated with activation or inhibition of lytic function. We have found that augmentation of NK3.3 lysis by IL2, IFN-β, and the anti-HLA A-locus-specific MAb 131 (9) is associated with increased levels of message encoding the proto-oncogene c-myb. Conversely, inhibition of lysis by the anti-HLA A,B,C-specific MAb W6/32 (10) results in decreased c-myb expression. The anti-LFA-1 MAb ICII (8), which inhibits the lysis of K562 but not of MOLT-4 target cells by NK3.3, has no effect on c-myb expression. These results indicate that anti-HLA class I MAbs alter gene expression in NK cells and that c-myb expression may be associated with NK lytic function.

Table 1. Effect of IL2, IFN, Anti-HLA Class I, and Anti-LFA-1 MAbs on NK3.3 Lytic Function

Pretreatment of NK3.3		Lysis of Target Cells (LU)	
Condition	Pretreatment (h)	K562	MOLT-4
A. –	–	<0.1	17
IFN	2	22	525
IL2	2	27	267
B. IL2	2	25	240
IL2 + W6/32	2	9	90
IL2 + 131	2	71	333
IL2 + ICII	2	<0.1	318
C. –	–	0.3	104
IFN	6	88	1111
IL2	6	123	1709
D. IL2	6	118	1821
IL2 + W6/32	6	56	556
IL2 + 131	6	167	7326
IL2 + ICII	6	55	7448

In A and C, NK3.3 cells were "rested" overnight in medium lacking IL2. Cells were then treated for 2 or 6 h with IL2 (2 U/ml) or IFN-β (5000 U/ml). In B and D, NK3.3 cells were used without resting. IL2 (2 U/ml) was present in these cultures. All MAbs were used at 1 μg/ml. LU = number of effectors mediating 20% lysis of target cells per 10^6 NK3.3 cells in a 4-h ^{51}Cr-release assay.

The major histocompatibility complex (MHC) plays a central role in regulating the immune response. We recently defined two additional functions of MHC molecules: regulation of the cytolytic activity of human NK cells, and of the activation and proliferation of human T and B lymphocytes (4,5,11,12). To determine how MHC molecules are involved in NK lysis, we have examined changes in gene expression that may occur as a result of anti-HLA class I MAbs binding to the NK cell surface. Other modulators of NK function, IL2, IFN-β, and anti-LFA-1 MAbs, were used for comparison.

The source of NK cells for this study is the IL2-dependent NK clone 3.3, which has been extensively characterized (13,14). Functionally and phenotypically these cells represent the majority of NK cells in peripheral blood (CD2+, CD3−, CD4−, CD8−, CD16+, NKH-1+, T-cell receptor α, β, γ-) and express the activation antigens CD25, and HLA-DR, DP, and DQ. In addition, their lytic function can be modulated by the same agents that affect peripheral blood NK activity. To assess the effects of the cytokines IL2 and IFN-β on NK3.3 lysis, cells are incubated ("rested") for 18 to 24 hours in medium lacking IL2 to stop proliferation and decrease their killing activity. Adding recombinant IL2 (Cetus Corp.) or recombinant IFN-β (Triton Bio-

Immunobiology of HLA
Volume II

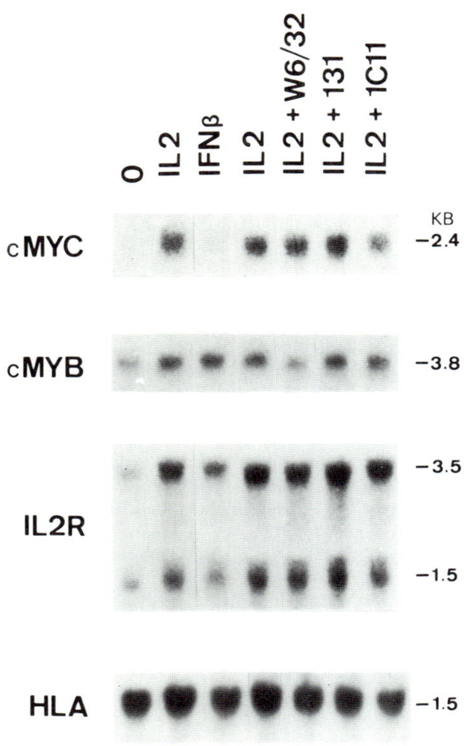

Figure 1. Changes in gene expression in NK3.3 cells after 6-hour treatment with IL2, IFN, anti-HLA MAbs W6/32 and 131, and anti-LFA-1 MAb ICII were assessed by RNA blot analysis using 10 μg of total cellular RNA per lane. Treatments correspond with those described in Table 1 (C and D).

sciences) at this time substantially augments lysis. To study the effects of MAbs on NK function, cells maintained in IL2 are used without resting. Table 1 shows how the lytic activity of NK3.3 increases dramatically after two-hour treatment with IL2 or IFN. The augmentation of lysis is even greater after six hours. Results are expressed in lytic units (LU), where 1 LU represents 20% lysis of target cells per 10^6 effector cells. Two-hour or six-hour treatment with the anti-HLA A,B,C-specific MAb W6/32 substantially inhibits lytic function. In contrast, the anti-HLA A locus-specific MAb 131 markedly augments lysis of both K562 and MOLT-4 targets. The anti-LFA-1 MAb ICII inhibits the lysis of K562 by NK3.3 but either has no effect or, in some cases, augments the lysis of MOLT-4 by NK3.3.

In parallel with the cytolytic studies, total cellular RNA was isolated from "rested" NK3.3 cells and from those that had been treated for six hours with either IL2, IFN, or MAbs. RNA blot analysis was performed to determine the level of expression of various genes that might be important in NK3.3 function, such as the proto-oncogenes c-myc and c-myb, and those coding for IL2 receptors (IL2R) and HLA class I molecules (15,16).

Of the messages analyzed, c-myb levels correlate well with the effect of the anti-HLA class I MAbs on lysis (Fig. 1). Rested NK3.3 cells have very low levels of c-myb mRNA (lane 1). Expression is markedly increased by both IL2 and IFN treatment (lanes 2, 3, and 4). Moreover, six-hour treatment with the inhibitory anti-HLA class I MAb W6/32 inhibits the induction of c-myb mRNA by IL2 (lane 5), whereas the anti-HLA A locus-specific MAb 131, which augments lysis, also augments c-myb expression.

The anti-LFA-1 MAb ICII, which has target cell specificity in its effect on NK3.3 lysis, does not significantly affect the induction of c-myb expression by IL2. These changes in c-myb levels are specific in that HLA class I mRNA levels are unaffected by any of these treatments (Fig. 1).

Expression of c-myc correlates with exposure of NK3.3 to IL2, regardless of whether the anti-HLA MAbs W6/32 and 131 are present. However, a slight decrease in c-myc expression is sometimes seen in the W6/32-treated and ICII-treated NK3.3 cells. IFN does not induce c-myc expression, indicating that c-myc levels do not correlate with the extent of NK lysis.

Changes in IL2R expression were also assessed. IL2R mRNA levels are very low in rested NK3.3 cells. Six-hour incubation with either IL2 or IFN increases IL2R mRNA levels, with IL2 more stimulatory than IFN. None of the antibody treatments significantly alters IL2R mRNA levels in IL2-stimulated NK3.3 cells. Increased levels of IL2R mRNA are therefore induced by both IL2 and IFN, but do not correlate with the degree of lytic activity mediated by NK3.3.

These results demonstrate that alterations in gene expression by NK3.3 occur when its lytic activity is modulated by a variety of agents. IFN and IL2, which rapidly augment lysis, induce increased levels of c-myb and IL2R message. However, IL2R expression is not affected by the inhibitory antibody W6/32 or the lysis-augmenting antibody 131, indicating that it does not correlate with the level of lytic activity. Increased c-myb message does correlate with level of lysis, in that the anti-HLA class I MAbs W6/32 and 131 influence c-myb expression in NK3.3 in a manner analogous to their effects on NK lysis. Perturbation of MHC class I molecules on NK3.3 by antibody binding therefore has a direct effect on gene expression in these cells. In contrast, the binding of anti-LFA-1 MAb ICII to the NK cell surface does not alter expression of any of the genes examined. Preliminary studies using peripheral blood NK cells have yielded similar results to date. Work is in progress to further elucidate the potential role of c-myb in NK lysis and to determine the mechanism by which MHC class I molecules affect gene expression in NK cells.

Acknowledgments. This work is supported by grants CA-37827 and CA-41165 from the National Institutes of Health.

References

1. Trinchieri G, Santoli D. Enhancement of human natural killer cell activity by interferon. J Immunol 1978;120: 1845.
2. Kuribayashi K, Gillis S, Kern DE, Henney CS. Murine NK cell cultures: Effects of interleukin 2 and interferon on cell growth and cytotoxic reactivity. J Immunol 1981;126: 2321.
3. Ortaldo J, Herberman R, Pestka S. Augmentation of human natural killer cells with human leukocyte and human recombinant leukocyte interferon. In: Herberman RB (ed): NK Cells and Other Natural Effector Cells. New York, Academic Press, 1982, p 349.
4. Kornbluth J, Wilson DB. Monoclonal antibodies directed against HLA molecules affect the lytic and proliferative behavior of a cloned line of human natural killer cells. Hum Immunol 1984;11:239.
5. Kornbluth J, Spear B, Raab SS, Wilson DB. Evidence for the role of class I and class II HLA antigens in the lytic function of a cloned line of human natural killer cells. J Immunol 1985;134:728.
6. Krensky AM, Sanchez-Madrid F, Robbins E, Nagy JA, Springer TA, Burakoff SJ. The functional significance, distribution, and structure of LFA-1, LFA-2, and LFA-3: Cell surface antigens associated with CTL-target interactions. J Immunol 1983;131:611.
7. Schmidt RE, Bartley G, Levine H, Schlossman SF, Ritz J. Functional characterization of LFA-1 antigens in the interaction of human NK clones and target cells. J Immunol 1985;135:1020.
8. Hart MK, Kornbluth J, Main EK, Spear BT, Taylor J, Wilson DB. Lymphocyte function-associated antigen One (LFA-1) and natural killer (NK) cell activity: LFA-1 is not necessary for all killer:target cell interactions. Cell Immunol 1987;109:306.
9. Spear BT, Kornbluth J, Strominger JL, Wilson DB. Evidence for a shared HLA-A intra-locus determinant defined by monoclonal antibody 131. J Exp Med 1985;162: 1802.
10. Parham P, Barnstable CJ, Bodmer WF. Use of a monoclonal antibody (W6/32) in structural studies of HLA-A, B, and C antigens. J Immunol 1979;123:342.
11. Taylor DS, Nowell PC, Kornbluth J. Functional role of HLA class I cell surface molecules in human T lymphocyte activation and proliferation. Proc Natl Acad Sci USA 1986;83:4446.
12. Taylor DS, Nowell PC, Kornbluth J. Anti-HLA class I antibodies inhibit the T cell-independent proliferation of human B lymphocytes. J Immunol 1987;139:1792.
13. Kornbluth J, Flomenberg N, Dupont B. Cell surface phenotype of a cloned line of human natural killer cells. J Immunol 1982;129:2831.
14. Leiden JM, Gottesdiener KM, Quertermous T, Coury L, Bray RA, Gottschalk L, et al. T cell receptor gene rearrangement in human natural killer cells: Natural killer activity is not dependent upon the rearrangement and expression of T cell receptor alpha, beta, or gamma genes. Immunogenet 1988;27:231.
15. Reed JC, Alpers JD, Nowell PC, Hoover RG. Sequential expression of proto-oncogenes during lectin-stimulated mitogenesis of normal human lymphocytes. Proc Natl Acad Sci USA 1986;83:3982.
16. Bishop JM. Cellular oncogenes and retroviruses. Ann Rev Biochem 1983;52:301.

Author Affiliations

Jacki Kornbluth and *Richard G. Hoover*, Departments of Immunology and Hematopathology, University of Texas M.D. Anderson Cancer Center, Houston, TX 77030, USA

Adenovirus Inversely Modulates Target Cell Class I Antigen Expression and Sensitivity to Natural Killing*

Walter J. Storkus, Peter Cresswell, Eric B. Patterson, and Jeffrey R. Dawson

Natural killer (NK) cell-mediated cytolysis occurs without apparent major histocompatibility complex (MHC) restriction (1). However, recent evidence suggests that the level of MHC class I antigens expressed by the target cell may influence its sensitivity to natural killing in vitro (2–4) as well as in vivo (4,5). Here we have examined the known adenovirus suppression of class I antigen expression as it relates to target cell sensitivity to NK. Adenovirus infection of NK sensitive, but not resistant targets results in increased susceptibility to NK. Enhanced target sensitivity is specific for NK as virus-infected and noninfected targets produce comparable results in antibody-dependent cellular cytotoxicity (ADCC) and complement-mediated cytolysis (CMC) assays. NK effector cells appear to recognize common structures on both virally infected and noninfected targets and do not recognize novel viral specificities. We conclude that class I antigens may modulate the recognition of resident target structures by NK cells.

NK cells spontaneously lyse autologous and allogeneic target cells with comparable efficacy (1). The recognition molecules on both the effector and target cell required for such interaction have neither been identified nor well characterized to date. Cell-surface

*This research was supported by National Institutes of Health (NIH) grants CA 42890 and AI 15775.

Address correspondence to: Dr. Jeffrey R. Dawson, Box 3010, Duke University Medical Center, Durham, NC 27710.

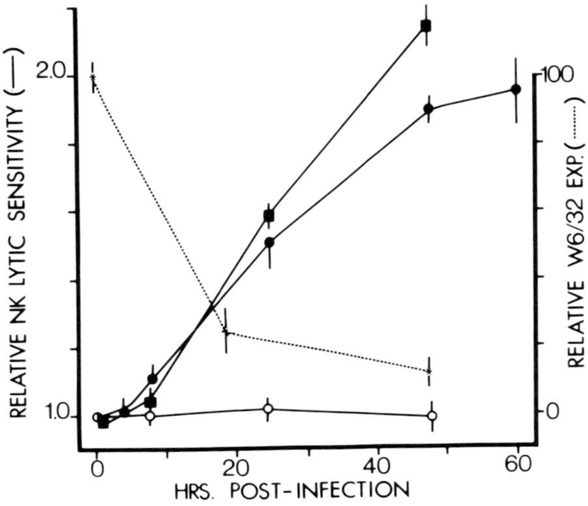

◄ **Figure 1.** Kinetics of adenovirus modulation of target cell sensitivity to natural killing and class I expression. HSB-A3 cells were infected with adenovirus serotype 2 (closed circles) or serotype 12 (closed squares) as described by Andersson et al. (8). At varying times postinfection, cells were harvested and tested for lytic sensitivity to NK in 4-hour chromium release assays and phenotyped for class I expression using the W6/32 monoclonal antibody in indirect immunofluorescence assays. These techniques are described elsewhere in detail (3). Lytic results are based on lytic unit (LU) determinations and are reported relative to the uninfected HSB values (open circles). One LU is defined as the number of effectors required to generate 20% specific lysis (3). Relative W6/32 expression is calculated from mean fluorescence channel data and is reported relative to expression on uninfected HSB cells (i.e., 100%). Since the adenovirus 2 and 12 gave class I values within standard error of each other, we are reporting only the adenovirus 2 data (asterisks) to indicate a typical reduction in W6/32 expression.

Table 1. Effects of Adenovirus Infection on NK Sensitivity

Cell Line	Ad Inf.	NK Rel. (LU)	NK Rel. Conj. (%)	Rel. W6/32 (%)	Rel. T9 (%)
HSB-A3	–	1.0 (70)	1.0 (20)	100	100
	2	1.8 ± 0.2	1.4 ± 0.1	17	97
	12	1.9 ± 0.3	1.5 ± 0.1	21	93
HSB-F10	–	1.0 (120)	1.0 (42)	100	100
	2	1.2 ± 0.1	1.2 ± 0.1	20	98
	12	1.2 ± 0.1	1.2 ± 0.1	11	97
Molt-4	–	1.0 (74)	1.0 (26)	100	100
	2	3.0 ± 0.4	2.3 ± 0.8	12	95
	12	2.3 ± 0.2	2.0 ± 0.5	20	90
CEM	–	1.0 (195)	1.0 (33)	100	100
	2	1.8 ± 0.1	1.4 ± 0.2	15	105
	12	1.7 ± 0.2	1.2 ± 0.1	22	99
CEM.NK	–	1.0 (2)	1.0 (4)	100	100
	2	0.9 ± 0.1	1.0 ± 0.1	20	92
	12	1.0 ± 0.1	1.1 ± 0.2	25	91
OC2008	–	1.0 (5)	1.0 (3)	100	100
	2	0.9 ± 0.2	1.1 ± 0.3	37	96
	12	1.1 ± 0.2	0.9 ± 0.2	22	106
174 × CEM	–	1.0 (42)	1.0 (17)	100	100
	2	1.9 ± 0.1	1.7 ± 0.2	11	89
	12	1.7 ± 0.1	1.5 ± 0.3	24	93

Adenovirus was obtained from the American Type Culture Collection and infections were performed as described by Andersson et al. (8) and verified by specific adenovirus probe in situ hybridization or by indirect immunofluorescence using a monoclonal antibody which recognizes all adenovirus serotypes (data not shown). The T-cell lymphoblastoid lines (HSB-A3, HSB-F10, CEM, CEM.NK), the B × T cell hybrid 174 × CEM, and the ovarian carcinoma OC2008 used as target cells have been previously characterized (3,9,10). Effector cells are 60 to 90% NK cells based on the expression of the Leu-7, Leu-11b, OKM1, and NKH-1 markers (3) (data not shown). NK conjugate and cytolytic assays were performed as previously described (3). One lytic unit (LU) is defined as the number of effector cells required to generate 20% specific chromium release and is expressed as LU/10^7 effector cells. Numbers in parentheses represent control LU or control conjugate frequency actual values. Relative W6/32 and T9 (OKT9, antitransferrin receptor) are based on mean fluorescence channel values determined by indirect immunofluorescence assays as previously described (3).

molecules that may modulate NK recognition of target cells have been identified however, namely class I antigens (2,3).

To examine class I-mediated modulation further, we chose to investigate the effect of adenovirus infection on the NK susceptibility of a series of well-defined NK-sensitive and -resistant targets. Target cells were infected with either the nononcogenic serotype 2 or oncogenic serotype 12 adenovirus at 1,000 PFU/cell. Adenovirus 2 encodes an early region 3 glycoprotein, termed E3/19k (Da), that blocks class I biosynthesis (at a posttranslational step) and its subsequent cell surface expression (6–8). Adenovirus 12 inhibits class I biosynthesis at a posttranscriptional, pretranslational step by diminishing the quantity of translatable class I mRNA (6–8).

The kinetics of increased susceptibility to NK following adenovirus infection are indicated in Figure 1. Enhanced NK-mediated killing is evident by 8 hours and continues to rise throughout the incubation period assayed. In subsequent experiments, 48 hour fixed incubations were used. The kinetics with which sensitivity increases correlate well with class I antigen disappearance on the virally infected cells.

Adenovirus serotype 2 or 12 infection of the NK-sensitive targets HSB-A3, HSB-F10, MOLT-4, CEM, and 174 × CEM increased their susceptibility to NK by 20 to 200%, and conjugate frequency by 20 to 100% (Table 1). The NK-resistant targets CEM.NK and OC2008 exhibited no modulation of NK conjugate frequency or cytolytic sensitivity when infected with either serotype of adenovirus. Regardless of its initial NK sensitivity, each virally infected cell line exhibited significant reduction in class I antigen expression (60 to 95%) as detected by the W6/32 monoclonal antibody in indirect immunofluorescence assays monitored by flow cytometry.

Of particular note are the HSB subclones. The HSB-A3 and F10 subclones differ significantly from each other in relative class I antigen expression and NK sensitivity (3). However, following adenovirus infection each of these targets approaches a maximal sensitivity of 140 lytic units (LU). This suggests that each subclone may express similar levels of NK target structure(s), the sensitivity to NK being modulated by differential class I antigen expression. It also suggests that the class I-mediated effects are occurring posttranslationally since both adenovirus 2 and 12 produce comparable results. This is consistent with prior findings in a panel of B × T cell hybrids (3).

Equally striking is the CEM and CEM.NK pair of cell lines in which CEM.NK represents an immunoselected, NK-resistant subclone of CEM (9). Although these cell lines express comparable levels of class I antigens (9), which are markedly reduced upon adenovirus infection, only the CEM cell line shows increased susceptibility to NK. These data suggest that CEM.NK does not express NK target structure(s), or at the very minimum, few or no NK binding structure(s) (since binding is required for subsequent cytolysis [1]). The OC2008 cell line appears similarly devoid of NK target structure(s). These data imply that the simple loss of class I antigen expression is not sufficient to confer sensitivity to natural killing in a resistant target, but may affect susceptibility quantitatively in a constitutively sensitive target.

Table 2. NK Effector Cells Do Not Recognize Novel Viral Determinants

Target (T)	Ad	% Cold Target Inhibition at I:T Ratio of: 0.3/1	1/1	10/1	50/1
HSB-A3	–	18	55	100	100
	2	13	48	96	98
	12	11	43	91	99
HSB-F10	–	9	40	84	99
	2	8	43	86	96
	12	12	43	85	95

Chromium release assays (3) were modified to examine cold target (I, inhibitor) inhibition of NK cytolysis of chromium-labeled targets (T). The indicated I:T ratios were used at varying NK:T ratios, the resulting lytic units (LU) calculated, and percent inhibition of control (I:T = 0) LU determined. The above data represent one representative experiment of three performed.

Table 3. Adenovirus-Induced Augmented Lytic Sensitivity Is Specific for NK

Cell Line	Adenovirus	NK Sensitivity: LU/10 EC	Conj. (%)	ADCC (%)	CMC (%)	Rel. % W6/32
HSB-A3	–	144 ± 18	20 ± 1	72 ± 5	87 ± 3	100
HSB-A3	2	235 ± 10	33 ± 2	65 ± 7	85 ± 3	36
HSB-A3	12	341 ± 17	38 ± 2	66 ± 9	84 ± 6	19

The indicated HSB-A3 cell lines were used as targets in NK, ADCC, and CMC assays. NK cytolytic and conjugate assays were performed as described earlier (3). ADCC assays were performed as described by Howell et al. (9) using plastic adherent macrophages effectors (by-products of NK purification (3)) and the 4F2 monoclonal antibody (11), which reacts comparably with both virus-infected and noninfected targets (data not shown). ADCC results are reported for E:T = 50:1. Some of the ADCC-prepared targets were used in CMC assays. Antibody-coated targets were incubated with 1/4 dilution of baby rabbit complement for 60 min at 37°C. The resulting specific chromium release is reported. Relative W6/32 expression was determined as described previously. Reported is one representative experiment of four run.

The CEM.NK and OC2008 results also indicate that novel viral specificities are not being recognized by NK effector cells. This observation is further supported by cold-target inhibition data (Table 2). Uninfected HSB-F10 was used as a cold-target inhibitor of control and adenovirus infected HSB-A3 and HSB-F10. Within each panel, comparable inhibition frequencies are found, and at sufficiently high cold/hot target ratios (i.e., 50:1) inhibition is complete. This is consistent with NK recognition of control, uninfected target specificities alone and supports a similar finding in the rat NK system (7).

Adenovirus-induced augmentation of target cell lytic sensitivity is specific for natural killing (Table 3). HSB-A3 and Molt-4 targets were infected with either serotype 2 or 12 adenovirus and subsequently assayed for lytic susceptibility to NK, ADCC, or CMC. As indicated, significant sensitivity differences were apparent only in NK assays.

In summary, these data suggest that adenovirus infection of human NK-sensitive target cells leads to diminished class I antigen expression and elevated susceptibility to natural killing. This augmented lytic sensitivity is specific for NK and does not involve recognition of novel viral determinants. We conclude that class I antigens may modulate NK target susceptibility at some posttranslational step if the NK target structure(s) and class I antigens are expressed concurrently on the target cell membrane. Further experiments are now in progress to better define this intriguing relationship.

Acknowledgment. The authors would like to thank Dr. Jack Keene for advice and assistance.

References

1. Ortaldo JR, Herberman RB. Heterogeneity of natural killer cells. Annu Rev Immunol 1984;2:359.
2. Harel-Bellan A, Quillet A, Marchiol C, DeMars R, Tursz T, Fradelizi D. Natural killer susceptibility of human cells may be regulated by genes in the HLA region of chromosome 6. Proc Natl Acad Sci USA 1986;83:5688.
3. Storkus WJ, Howell DN, Salter RD, Dawson JR, Cresswell P. NK susceptibility varies inversely with target cell class I HLA antigen expression. J Immunol 1987;138:1657.
4. Karre K, Ljunggren HG, Piontek G, Kiessling R. Selective rejection of H-2 deficient lymphoma variants suggests alternative immune defense strategy. Nature 1986;319: 675.
5. Kawano Y-I, Taniguchi K, Toshitani A, Nomoto K. Synergistic defense system by cooperative natural effectors against metastasis of B16 melanoma cells in H-2-associated control: Different behavior of H-2+ and H-2− cells in metastatic processes. J Immunol 1986;136:4729.
6. Severinsson L, Martens I, Peterson PA. Differential association between two human MHC class I antigens and an adenoviral glycoprotein. J Immunol 1986;137:1003.
7. Raska K Jr, Gallimore PH. An inverse relation of the oncogenic potential of adenovirus-transformed cells and their sensitivity to killing by syngeneic natural killer cells. Virology 1982;123:8.
8. Andersson M, Paabo S, Nilsson T, Peterson PA. Impaired intracellular transport of class I MHC antigens as a possible means for adenovirus to evade immune surveillance. Cell 1985;43:215.
9. Howell DN, Andreotti PE, Dawson JR, Cresswell P. Natural killer target antigens as inducers of interferon: Studies with an immunoselected, natural killing resistant human T-lymphoblastoid cell line. J Immunol 1985;134: 971.
10. Shau H, Koren HS, Dawson JR. Human natural killing against ovarian carcinoma. Br J Cancer 1983;47:687.
11. Haynes BF. Human T lymphocyte antigens as defined by monoclonal antibodies. Immunol Rev 1981;57:127.

Author Affiliations

Walter J. Storkus, *Peter Cresswell*, *Eric B. Patterson*, and *Jeffrey R. Dawson*, Department of Microbiology and Immunology, Duke University Medical Center, Durham, NC 27710, USA

Differential Regulation of HLA-B Antigen Expression by Interferon

Helmuth Schmidt, Elisabeth Weiss, Volker Gekeler, Hans-Jörg Bühring, Gabriele Engler-Blum, Ulrike Reichmann, and Claudia Müller

Quantitative differences in the expression of HLA antigens have been shown to influence T-cell mediated immune responses. On murine lymphocytes and thymocytes (1), as well as on human cell lines (2), differential expression of MHC class I antigens has been described. For further insight into regulation of HLA class I antigen expression, DNA-mediated gene transfer has been applied. L(tk⁻) cells transfected with the genes coding for the heavy chain of the HLA-A2 and -B7 antigen revealed a two to three times more increased expression of the HLA-B compared to the HLA-A locus molecule after stimulation with mouse interferon (3).

To further investigate a possibly stronger inducibility of all HLA-B locus antigens, genomic DNA clones coding for the heavy chains of the HLA-B27 and HLA-B51 antigens were also integrated into L(tk⁻) cells by DNA-mediated gene transfer. After cotransfection with the thymidine kinase gene of herpes simplex virus and selec-

Table 1. Inducibility of HLA-B Antigens by Interferon

Used Antibody		W6/32.HL		4E	
HLA-gene	IFN	cpm . .	-fold ind.	cpm . .	-fold ind.
none	–	255 .		204 .	
	+	259 .		192 .	
HLA-B7	–	1220 .		1450 .	
	+	4471 .	3.7	6062 .	4.2
HLA-B27	–	3976 .		4602 .	
	+	7471 .	1.9	8226 .	1.8
HLA-B51	–	1337 .		1670 .	
	+	1826 .	1.4	2908 .	1.7

Cpm bound by 10^5 transfected cells tested. Results are average values from three tests. Monoclonal antibodies W6/32.HL recognized a common epitope of all HLA class I antigens and 4E recognized HLA-B and -C antigens.

tion in HAT medium (DMEM containing 15 μg/ml hypoxanthin, 5 μg/ml thymidine and 1 μg/ml aminopterin), single colonies of transfected cells were grown up to mass culture. Expression of HLA class I antigens on the cell surface of transfected cells was established by an indirect radioactive binding assay using W6/32.HL as specific antibody and ^{125}J labeled protein A or ^{125}J labeled gout anti-mouse IgG as detection systems. Bound radioactivity was determined after cell lysis. Using further mono- and polymorphic monoclonal antibodies, the expressed HLA-molecules were shown to carry appropriate epitopes of the HLA-B27 or -B51 alleles. Southern blot analysis revealed that 5 to 500 copies of the transfected HLA gene were integrated into the high molecular weight DNA of different transfected mouse L cell lines. Maximal enhancement of HLA class I expression on cell lines transfected with HLA genes were achieved with 1000 U mouse interferon per ml medium after 36h stimulation. For comparative analysis of the inducibility of the different HLA-B locus antigens, 10^4 U/ml mouse interferon were applied for 48 h. Irrespective of the monomorphic or polymorphic antibody used in binding assays, HLA-B7 was found to be the strongest inducible gene of the three HLA-B alleles tested (Table 1). At least five independent transfected cell lines were tested for each used gene. Because of limitation of space only representative results obtained with one such cell line tested with two monoclonal antibodies are shown. A comparative simultaneous increase of H2 antigens in all investigated cell lines suggested that this phenomenon was not related to alterations of cellular sensitivity to interferon after transfection affecting regulation of HLA-antigen expression. Furthermore, because cell lines with integration of comparative gene copies were investigated, this study suggests transcriptional or translational differences in the regulation of HLA-B genes by interferon.

Acknowledgments. Supported by the Deutsche Forschungsgemeinschaft, SFB 120.

References

1. Tartakovsky B, De Baetselier P, Segal S. Serological detection of H-2K- and H-2D-gene products: Principal differences between T and B lymphocytes in expression of H-2D-encoded alloantigens. Immunogenetics 1980;11:585.
2. Burrone OR, Kefford RF, Gilmore D, Milstein C. Stimulation of HLA-A,B,C by IFN-alpha. The derivation of Molt 4 variants and the differential expression of HLA-A,B,C subsets. EMBO J 1985;4:2855.
3. Schmidt H, Forsthuber T, Bühring H-J, Müller CA. Differential expression of the HLA-B7 and the HLA-A2 gene in transfected mouse L(tk⁻) cells after stimulation by mouse interferon. Immunobiol 1987;174:51.

Author Affiliations

Helmuth Schmidt, Hans-Jörg Bühring, Gabriele Engler Blum, Ulrike Reichmann, Claudia Müller, Medizinische Klinik, Tübingen; *Elisabeth Weiss,* Institut für Immunologie, München; *Volker Gekeler,* Physiologisch-Chemisches Institut, Tübingen, FRG

Section V
Class I Histocompatibility Antigens: New Class I, TL/Qa-Like, and Others

Transfer and Expression of Human Non-A,B,C Class I Genes in Human HLA-A,B,C Null Lymphoblastoid Cells*

Yoji Shimizu, Beverly H. Koller, Daniel E. Geraghty, Harry T. Orr, and Robert DeMars

Southern blotting analysis indicates that there are more class I genes in the major histocompatibility complex (MHC) than antigens that have been defined serologically (1,2). In the mouse, some of these class I genes have been shown to encode the Qa and TL antigens (3,4). The identification of similar class I antigens in the human is much less defined. We report here the identification of three expressible human leukocyte antigen (HLA) non-A,B,C class I genes on Hind III-generated 5.4-kb (HLA-5.4), 6.0-kb (HLA-6.0), and 6.2-kb (HLA-6.2) DNA fragments derived from lymphoblastoid cell line (LCL) 721. Each gene was transferred into LCL. 221, which does not express HLA-A,B,C antigens because of γ-ray-induced mutations in the HLA complex. Although cell surface expression of only the HLA-6.0 antigen was observed, each gene produced a distinct gene class I α chain that coprecipitated with β2-microglobulin. These results demonstrate that at least three HLA non-A,B,C class I genes encode class I α chains that can be expressed in human cells.

The three non-A,B,C class I genes were isolated from a cosmid library of LCL 721 DNA. Nucleotide sequence analysis showed the HLA-5.4 (D.E. Geraghty, B.H. Koller, H.T. Orr, manuscript in preparation), HLA-6.0 (5), and HLA-6.2 (6) genes to be intact class I genes. Each gene was subcloned into the nonintegrating vector pHeBo (7), transferred by electroporation into the mutant LCL .221 (8,9), and hygromycin-resistant transferents were isolated. Mutant .221 was derived from the HLA-A,B null mutant .184 (1) after γ-irradiation and immunoselection with the anti-class I monoclonal antibody (MAb) W6/32 (10). HLA-A,B,C antigens are not expressed in .221 as judged by antibody binding and Northern blotting mRNA analysis but genomic HLA-A,B,C genes transferred into .221 are expressed at levels seen in LCLs containing the same endogenous gene (12,13). Southern blotting DNA analysis indicates that intact HLA-5.4, HLA-6.0, and HLA-6.2 genes are also present in .221 (13).

The HLA-A,B,C null phenotype of .221 made it possible to monitor the expression of transferred non-A,B,C genes by assaying the binding of the anti-β2-microglobulin MAb BBM.1 to transferents. While pHeBo(5.4)→.221 and pHeBo(6.2)→.221 transferents did not show increased BBM.1 binding relative to pHeBo→.221 control transferents, the mean amounts of BBM.1 unmistakably bound by pHeBo(6.0)→.221 transferents were 60% of that bound to HLA-C on mutant .184 (1). The anti-class I MAb W6/32 also bound to pHeBo(6.0)→.221 transferents, indicating that the HLA-6.0 antigen shares an epitope common to HLA-A,B,C antigens.

Northern blotting mRNA analysis shows that the HLA-5.4, HLA-6.0, and HLA-6.2 genes were transcribed after transfer into .221 (13). While the sizes of the HLA-6.0 and HLA-6.2 transcripts were not distinguishable from those of HLA-A,B,C transcripts, HLA-5.4-specific mRNA was distinctly shorter than HLA-A,B,C mRNA (12). The amount of transcript in each type of transferent was far less than HLA-A,B,C mRNA detectable in LCL 721, even when Southern blotting analysis indicated that there were 5 to 10 extrachromosomal copies of the transferred gene in some pHeBo (5.4)→.221, pHeBo(6.0)→.221, and pHeBo(6.2)→.221 transferents (13).

One-dimensional sodium dodecyl sulfate gel electrophoresis of BBM.1 immunoprecipitates of [^{35}S]methionine-labeled cell extracts showed that HLA-5.4, HLA-6.2, and HLA-6.0 transferents contained α chains that coprecipitated with β2-microglobulin (12). Each α chain was smaller than 43 to 44-kD HLA-A,B,C α chains: 40 kD (HLA-5.4), 41 kD (HLA-6.2), and 38 kD (HLA-6.0). The abundance of each non-A,B,C α chain was ≤1% of the total amount of HLA-A,B,C antigens in the HLA-A,B,C positive LCL .114. Each non-A,B,C α chain was also immunoprecipitated by W6/32, indicating that all three α chains share determinants with HLA-A,B,C antigens (12).

The absence of HLA-A,B,C antigen expression in mutant .221 has also revealed the presence of two β2-microglobulin-associated class I-like α chains that are specifically immunoprecipitated by BBM.1 and W6/32 (12). The sizes of these two α chains correspond to those found in pHeBo(5.4)→.221 and pHeBo(6.2)→.221 transferents, suggesting very low expression of two endogenous non-A,B,C class I genes in this LCL.

These results (summarized in Table 1) demonstrate that three HLA non-A,B,C class I genes can be expressed after transfer into human cells. Other investigators have recently demonstrated the expression of non-A,B,C class I genes after transfer into mouse L cells (14,15). The relationship of these genes to HLA-5.4, HLA-6.0, and HLA-6.2 is not clear at this time, although the class I gene described by Paul et al. (15) is not likely to be HLA-5.4, HLA-6.0, or HLA-6.2 since its gene

*This is paper No. 2964 from the Laboratory of Genetics of the University of Wisconsin (Madison). This work was supported by NIH grants AI 15486 (R.D.) and AI 18124 (H.T.O.) and American Cancer Society grant IM 379 (H.T.O.). Address correspondence to: Dr. Yoji Shimizu, Immunology Branch, National Cancer Institute, National Institutes of Health, Bethesda, Maryland 20892.

Table 1. Comparison of Cell Surface Antibody Binding, mRNA Transcript Size, and Molecular Weight of Class I α Chains Produced by Transferents of Mutant .221 Containing the Recombinant Plasmid Constructs pHeBo(A1), pHeBO(5.4), pHeBo(6.0), or pHeBo(6.2)

Transferent	Cell Surface Binding*		Transcript Size†	α Chain Size‡
	BBM.1	W6/32		
pHeBo(A1)→.221TG	+	+	Normal	43–44 kD
pHeBo(5.4)→.221TG	−	−	Shorter than normal	40 kD
pHeBo(6.0)→.221TG	+	+	Normal	38 kD
pHeBo(6.2)→.221TG	−	−	Normal	41 kD

*Antibody binding was measured at saturating antibody concentrations with an EPICS C flow cytometer as described (8).
†Cytoplasmic RNA samples were subjected to electrophoresis and Northern blotting analysis with an HLA-B7 cDNA clone as a hybridization probe as described (12).
‡Extracts of cells labeled in medium containing [^{35}S]methionine were subjected to immunoprecipitation with MAbs BBM.1 or W6/32 and subjected to one-dimensional gel electrophoresis in a sodium dodecyl sulfate (SDS) gel followed by autoradiography (12).

product is not recognized by W6/32 in L-cell transferents. Serological information about other non-A,B,C antigens does not permit their identification with the HLA-5.4, HLA-6.0, and HLA-6.2 genes, although TCA antigens (16) appear to correspond most closely to the HLA-6.0 antigen. The lack of cell surface expression of the HLA-5.4 and HLA-6.2 antigens may be significant in light of evidence regarding the secretion of Qa antigens by certain cell types (17,18) and the anchoring of the Q7 antigen by a phospholipid tail in hepatoma cells (19). Since the transferents described here predominantly express just one class I α chain, they should be useful as immunogens for the generation of specific reagents that can be used to analyze the expression of these non-A,B,C antigens in various cell types.

Acknowledgments. We thank Sandra Guthrie and Carol Kass for flow cytometry and Dr. Donna Paulnock for assistance with the immunoprecipitations.

References

1. Flavell RA, Allen H, Burkly LC, Sherman DH, Waneck GL, Widera G. Molecular biology of the H-2 histocompatibility complex. Science 1986;233:437.
2. Orr HT, DeMars R. Mapping of class I DNA sequences within the human major histocompatibility complex. Immunogenetics 1983;18:489.
3. Flaherty L. The Tla region antigens. In Dorf M (ed): The role of the major histocompatibility complex in immunobiology. New York, Garland STPM Press, 1981, p 33.
4. Robinson PJ. Structure and expression of polypeptides encoded in the mouse Qa region. Immunol Res 1987; 6:46.
5. Geraghty DE, Koller BH, Orr HT. A human major histocompatibility class I gene that encodes a protein with a shortened cytoplasmic segment. Proc Natl Acad Sci USA 1987;84:9145–9149.
6. Koller BH, Geraghty DE, Shimizu Y, DeMars R, Orr HT. HLA-E: A novel class I gene expressed in resting T lymphocytes. J Immunol 1988;141:897–904.
7. Sugden B, Marsh K, Yates J. A vector that replicates as a plasmid in B-lymphoblasts transformed by Epstein-Barr virus. Mol Cell Biol 1985;5:410.
8. Shimizu Y, Koller B, Geraghty D, Orr H, Shaw S, Kavathas P, et al. Transfer of cloned human class I major histocompatibility complex genes into HLA mutant lymphoblastoid cells. Mol Cell Biol 1986;6:1074.
9. Knutson JC, Yee D. Electroporation: Parameters affecting transfer of DNA into mammalian cells. Anal Biochem 1987;164:44.
10. DeMars R, Rudersdorf R, Chang C, Petersen J, Strandtmann J, Korn N, et al. Mutations that impair a posttranscriptional step in expression of HLA-A and -B antigens. Proc Natl Acad Sci USA 1985;82:8183.
11. Kavathas P, Bach FH, DeMars R. Gamma ray-induced loss of expression of HLA and glyoxylase I alleles in lymphoblastoid cells. Proc Natl Acad Sci USA 1980;77:4251.
12. Shimizu Y, Geraghty DE, Koller BH, Orr HT, DeMars R. Transfer and expression of three cloned human non-HLA-A, -B, -C class I histocompatibility complex genes in mutant lymphoblastoid cells. Proc Natl Acad Sci USA 1988;85:227–231.
13. Shimizu Y. Use of DNA-mediated gene transfer and HLA mutant B-lymphoblastoid cells to study the expression of human class I major histocompatibility complex genes. Ph.D. thesis, University of Wisconsin-Madison, 1987.
14. Srivastava R, Chorney MJ, Lawrance SK, Pan J, Smith Z, Smith CL, et al. Structure, expression, and molecular mapping of a divergent member of the class I HLA gene family. Proc Natl Acad Sci USA 1987;84:4224.
15. Paul P, Fauchet R, Boscher MY, Sayagh B, Masset M, Medrignac G, et al. Isolation of a human major histocompatibility complex class I gene encoding a nonubiquitous molecule expressed on activated lymphocytes. Proc Natl Acad Sci USA 1987;84:2872.
16. van Leeuwen A, Giphart MJ, de Groot G, Morolli B, Festenstein H, Nijenhuis LE, et al. Two different T-cell systems in humans, one of which is probably equivalent to Qa or Tla in mice. Hum Immunol 1985;12:235.
17. Soloski MJ, Vernachio J, Einhorn G, Lattimore A. Qa gene expression: Biosynthesis and secretion of Qa-2 molecules in activated T cells. Proc Natl Acad Sci USA 1986; 83:2949.

18. Maloy WL, Coligan JE, Barra Y, Jay G. Detection of a secreted form of the murine H-2 class I antigen with an antibody against its predicted carboxyl terminus. Proc Natl Acad Sci USA 1984;81:1216.
19. Stroynowski I, Soloski M, Low MG, Hood L. A single gene encodes soluble and membrane-bound forms of the major histocompatibility Qa-2 antigen: Anchoring of the product by a phospholipid tail. Cell 1987;50:759.

Author Affiliations

Yoji Shimizu and *Robert DeMars*, Laboratory of Genetics and Department of Human Oncology, University of Wisconsin, Madison, WI 53706, USA; *Beverly H. Koller*, *Daniel E. Geraghty*, and *Harry T. Orr*, Department of Laboratory Medicine and Pathology and the Institute of Human Genetics, University of Minnesota, Minneapolis, MN 55455, USA

Activated T-Cells Express a Non-HLA-ABC Class I Gene that Is Inducible with Gamma-Interferon*

Shinichi Mizuno, Joseph A. Trapani, Bo Dupont, and Soo Young Yang

Abstract: A cDNA clone, designated JTW15, has been isolated from an activated T-cell cDNA library and found to encode a novel, divergent class I HLA gene transcript. This transcript, JTW15, is 1665 nucleotides in length and represents a full-length copy of a 1.7kb mRNA detected in Northern blot analyses. Comparison of the derived amino acid sequence with those of other HLA class I proteins showed JTW15 to be the most divergent HLA class I gene isolated thus far, the level of sequence homology being lowest for the alpha-1 and alpha-2 domains. The JTW15 gene transcript was found in a broad variety of cell types and was inducible in cells treated with gamma-interferon. This non-HLA-ABC class I gene is a candidate to encode a human Qa-like gene transcript.

A cDNA library was constructed in lambda-gtll using RNA derived from a bulk culture of a healthy donor's T lymphocytes (HLA-A29, w33, -B14, w62, -Cw6, w8), which were stimulated by co-culture with allogeneic, lethally irradiated target cells. A full length HLA class I probe, pHLA-B7 (1), was used to screen the library, and 40 reactive clones were further analyzed using the HLA-A, -B, and -C locus-specific probes pHLA-2a.1 (2), pHLA-1.1 (2), and pHLA-ld (3). Only one clone, designated JTW15, failed to hybridize with any of these probes and was, thus, a candidate to encode a non-ABC class I HLA gene transcript.

Nucleotide sequence analysis confirmed that JTW15 encoded a full-length class I cDNA clone. Comparison of the predicted amino acid sequence of JTW15 with those of classical HLA class I alleles showed that the average homology was greatest in the alpha-3 domain (91%) (Table 1). However, homologies for the alpha-1 and alpha-2 domains were far lower and averaged only 76 and 66%, respectively. By comparison, the average homology among alleles at different HLA-ABC loci ranged from 76 to 88% (alpha-1 domain), from 78 to 88% (alpha-2 domain), and from 91 to 98% (alpha-3 domain). Although this degree of divergence of JTW15 with the HLA-ABC genes suggested that it may represent a human Qa/Tla-like gene, it was notable that the degree of similarity between JTW15 and the cloned murine Qa genes was not greater than that observed with classic H-2 gene sequences. Furthermore, JTW15 had a markedly reduced homology with a cloned murine Tla gene (Table 1).

Comparison of the nucleotide sequence of JTW15 with the exons of recently cloned atypical class I gene, designated HLA-E (17) showed a high degree of homol-

Table 1. Percentage of Amino Acid Sequence Homologies Between JTW15 and Previously Cloned Human and Murine Class I Genes

	First External Domain	Second External Domain	Third External Domain	Reference
HLA-A2	78	65	87	4
HLA-A3	78	67	92	5
HLA-B7	78	65	90	6
HLA-B27	76	66	92	7
HLA-Cw3	72	65	93	8
H-2K^b	69	61	74	9
H-2D^b	67	66	78	10
27.1	71	63	80	11
Q7	71	63	79	12
Q10	76	63	77	13
Tla	54	49	82	14
RS5	83	95	98	15

*This work was supported by grants P01 CA 22507 (BD) and P30 CA 08748 (BD). J.A.T. has been supported as a Clinical Scholar from the Norman and Rosita Winston Foundation and is currently supported by a C.J. Martin Travelling Fellowship from the National Health and Medical Research Council of Australia.

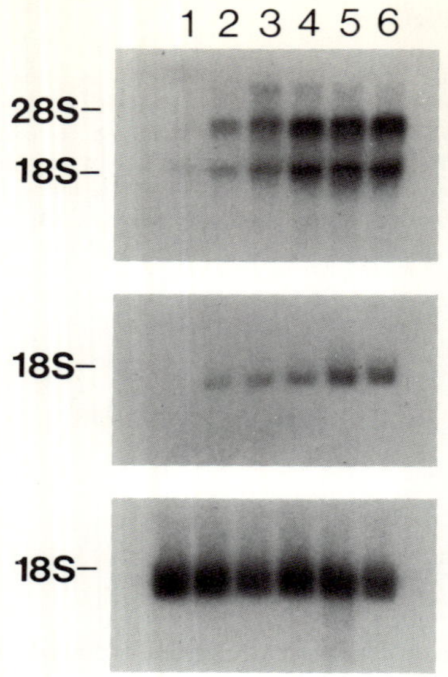

Figure 1. Northern blot analysis of RNA derived from K562 (*lane 1*), U937 cells cultured in the absence of gamma-interferon (*lane 2*) and from U937 cells cultured with 100U/mL gamma-interferon for 1 hour (*lane 3*), 6 hours (*lane 4*), 12 hours (*lane 5*), or 24 hours (*lane 6*). The FNA was probed with 3' untranslated region probe from JTW15 (*panel A*), HLA-DR beta (*panel B*), and a probe detecting gamma-actin (*panel C*). A faint additional signal of approximately 4.5 kb is indicated by the arrow in *panel A*.

ogy, with only a single amino acid change at residue 83 in the alpha-1 domain. Furthermore, when JTW15 was aligned with the exon sequences of another nonclassic class I gene, RS5 (15), the level of homology was similarly high except for two short regions in exon 1 and exon 2. In each case, the loss of homology could be restored either wholely (exon 1) or in part (exon 2) by the insertion or deletion of a single nucleotide in RS5.

A 240-base nucleotide probe corresponding to bases 1124 to 1364 of the 3' untranslated region of JTW15 was used to determine the tissue distribution of JTW15 gene transcripts. To confirm the specificity of this probe (i.e., to rule out binding to HLA-ABC transcripts), dot blot analysis was performed using DNA clones from HLA-A, -B, and -C alleles and JTW15. With highly stringent washing of the blots, the JTW15 3' untranslated probe showed no cross-activity with HLA-A, -B, and -C alleles even after extended radiographic exposure.

Two transcripts of ~1.7kb and ~3.3kb were found to be present in Northern blot analysis using a variety of cell lines and tissues. Although the levels of expression varied, both transcripts were found in all cell types tested. Expression was generally low in T-ALL cell lines such a MOLT-4, CEM, and HPB-ALL, but was higher in B-lymphoblastoid cell lines and in the Burkitt's lymphoma cell line, RAJI. Expression was also observed in tissues such as adult thymus, adrenal, and testis. The tissue distribution for JTW15 transcripts suggested that JTW15 is not the human homologue of a murine Tla gene, not was it consistent with the restricted distribution of Qa proteins reported thus far. Interestingly, however, Qa gene transcripts that recently detected in a variety of murine tissues which have previously were thought not to express detectable Qa protein (16). The true tissue distribution for murine Qa is, therefore, not entirely clear.

The level of expression of JTW15 could be markedly augmented in U937 cells by culture in medium containing gamma-interferon (Fig. 1A). Treatment of U937 cells with gamma-interferon also resulted in the appearance of a previously undetected mRNA species of approximately 4.5kb. This form of transcript could not be detected in cell cultures grown in the absence of gamma-interferon, even with extended autoradiography.

The finding that the mRNA for JTW15 exists in at least three different forms, with one form (4.5kb) apparent only after stimulation with gamma-interferon, suggests either that the transcript may be spliced in alternative ways or that at least three different polyadenylation sites exist. Alternatively, it is conceivable that treatment of cells with interferon is responsible for inducing the expression of other class I gene(s) that have larger transcripts, but which possess a 3' untranslated region that is closely homologous with or identical to that of JTW15.

The variability in the expression of JTW15 among different types of immunologic cells and the change in its level of expression on stimulation with gamma-interferon constitute strong evidence that the protein encoded by JTW15 has an immunologic function or a role in differentiation. It remains unclear whether this divergent gene represents isolation of a gene equivalent to murine Qa.

Acknowledgment. We thank Dr. Harry T. Orr for kindly providing the DNA probes pHLA-B7, pHLA-2a.1, pHLA-1.1, and pHLA-1d and for sharing the nucleotide sequence of "HLA-E" prior to publication.

References

1. Sood AK, Pereira D, Weissman SM. Isolation and partial nucleotide sequence of a cDNA clone for human histocompatibility antigen HLA-B7 by use of an oligodeoxynucleotide primer. Proc Natl Acad Sci USA 1981;78:616.
2. Koller BH, Sidwell B, DeMars R, Orr HT. Isolation of HLA locus-specific DNA probes from the 3'-untranslated region. Proc Natl Acad Sci USA 1984;81:5175.
3. Strachan T, Dodge AB, Smillie D, Dyer PA, Sodoyer R, Jordan BR, Harris R. An HLA-C-specific DNA probe. Immunogenetics 1986;23:115.
4. Koller BH, Orr HT, Cloning and complete sequence of an HLA-A2 gene: analysis of two HLA-A alleles at the nucleotide level. J Immunol 1985;134:2727.
5. Strachan T, Sodoyer R, Damotte M, Jordan BR. Complete nucleotide sequence of a functional class I HLA gene, HLA-A3: implication for the evolution of HLA genes. EMBO J 1984;3:887.

6. Biro PA, Pan J, Sood AK, Kole R, Reddy VB, Weissman SM. Sequence of human repetitive DNA, non-alpha-globin genes, and major histocompatibility locus genes. III. The major histocompatibility complex. Cold Spring Harbor Symp Quant Biol 1983;47:1079.
7. Seemann GHA, Rein RS, Brown CS, Ploegh HL. Gene conversion-like mechanisms may generate polymorphism in human class I genes. EMBO J 1986;5:547.
8. Sodoyer R, Damotte M, Delovitch TL, Trucy J, Jordan BR, Strachan T. Complete nucleotide sequence of a gene encoding a functional human class I histocompatibility antigen (HLA-Cw3). EMBO J 1984;3:879.
9. Weiss E, Golden L, Zakut R, Mellor A, Fahrner K, Kvist S, Flavell RA. The DNA sequence of the H $2K^b$ gene: evidence for gene conversion as a mechanism for the generation of polymorphism in histocompatibility antigens. EMBO J 1983;2:453.
10. Maloy L, Coligan JE. Primary Structure of the $H\text{-}2D^b$ Alloantigen. II. Additional amino acid sequence information, localization of a third site of glycosylation and evidence for K and D region specific sequences. Immunogenetics 1982;16:11.
11. Steinmetz M, Moore KW, Frelinger JG, Sher BT, Shen F-W, Boyse EA, Hood L. A pseudogene homologous to mouse transplantation antigens: transplantation antigens are encoded by eight exons that correlate with protein domains. Cell 1981;25:683.
12. Devlin JJ, Weiss EH, Paulson M, Flavell RA. Duplicated gene pairs and alleles of class I genes in the Qa2 region of the murine major histocompatibility complex: a comparison. EMBO J 1985;4:3203 .
13. Mellor AL, Weiss EH, Kress M, Jay G, Flavell RA. A non-polymorphic class I gene in the murine major histocompatibility complex. Cell 1984;36:139.
14. Fisher DA, Hunt SW III, Hood L. Structure of a gene encoding a murine thymus leukemia antigen, and organization of Tla genes in the BALB/c mouse. J Exp Med 1985;162:528.
15. Srivastava R, Chorney MJ, Lawrance SK, Pan J, Smith Z, Smith CL, Wiessman SM. Structure, expression, and molecular mapping of a divergent member of the class I HLA gene family. Proc Natl Acad Sci USA 1987;84:4224.
16. Palmer MJ, Frelinger JA, Widespread transcription of a Qa region gene in adult mice. J Exp Med 1987;166:95.
17. Koller BH, Geraghty DE, DeMars R, Orr HT. HLA-E: a novel class I gene expressed in resting T lymphocytes. J Immunol 1988;141:897.

Author Affiliations

Laboratories of Biochemical Immunogenetics and Human Immunogenetics, Memorial Sloan-Kettering Insitute for Cancer Research New York, NY, USA

A Novel Class I HLA Gene

P. Andrew Biro, Barbara Herbut, and Hilliard Festenstein

Abstract: We present the sequence of a human T cell-derived cDNA clone, p6, which represents the transcript of an expressed MHC class I-related, non-HLA-A, -B, or -C gene. The sequence of p6 shows strong homology with that of the classic class I genes but has sufficient unique features, especially in its 3′ untranslated region, to establish it as a novel gene product. The gene may be the human equivalent of the well-studied murine Qa or, possibly, T1 antigens.

There is serologic, biochemical and genetic evidence for the existence of human functional MHC class I-related non-HLA -A, -B, or -C genes that are linked to the MHC and whose products are β2-microglobulin-associated (1–3). A number of such genes have recently been identified (4–6) although, to date, none have been fully characterized. Their gene products may be the human equivalents of the better characterized murine Qa or T1 antigens and may have similar functions.

The sequence of a partial cDNA clone that codes for one such antigen is presented here. The clone was isolated from a cDNA library (a kind gift of D. Dialynas, Dept. of Biochemistry, Harvard University) constructed from a human T-cell line, HPB-ALL, using oligonucleotide linkers. T cells from a variety of species have been shown to express class I related genes (1,2,6–8). A panel of clones was established by probing the library with HLA class I cDNA fragments. Under very high stringency conditions (0.6M NaCl, 0.06M sodium citrate, 50% formamide at 50 to 55°C), clones derived form the HLA-A or HLA-B loci will hybridize preferentially to their allelic equivalents and show only limited cross-hybridization to other class I genes (P.A. Biro, unpublished). The clone described here, designated p6, was identified from the panel of class I cDNA clones by its relatively weak hybridization with both HLA-A (9) and HLA-B (10) locus probes. The p6 insert was isolated as an Eco RI fragment, subcloned in both orientations into M13 and sequenced by the dideoxy chain termination method using synthetic oligonucleotide primers. The sequence of p6 is sufficiently different form that of the HLA-A, -B, and -C genes, especially in its 3′ untranslated region, to confirm its identity as a novel class I gene.

Figure 1 compares the sequence of p6 with that of other class I genes (11,12). Unlike the products of the murine T1 locus (13), the p6 gene shares a very similar structural organization with the other HLA genes, especially in its 3′ end that codes for the cytoplasmic protein domains. Although it is clearly a class I gene product, it is the most highly diverged member of the

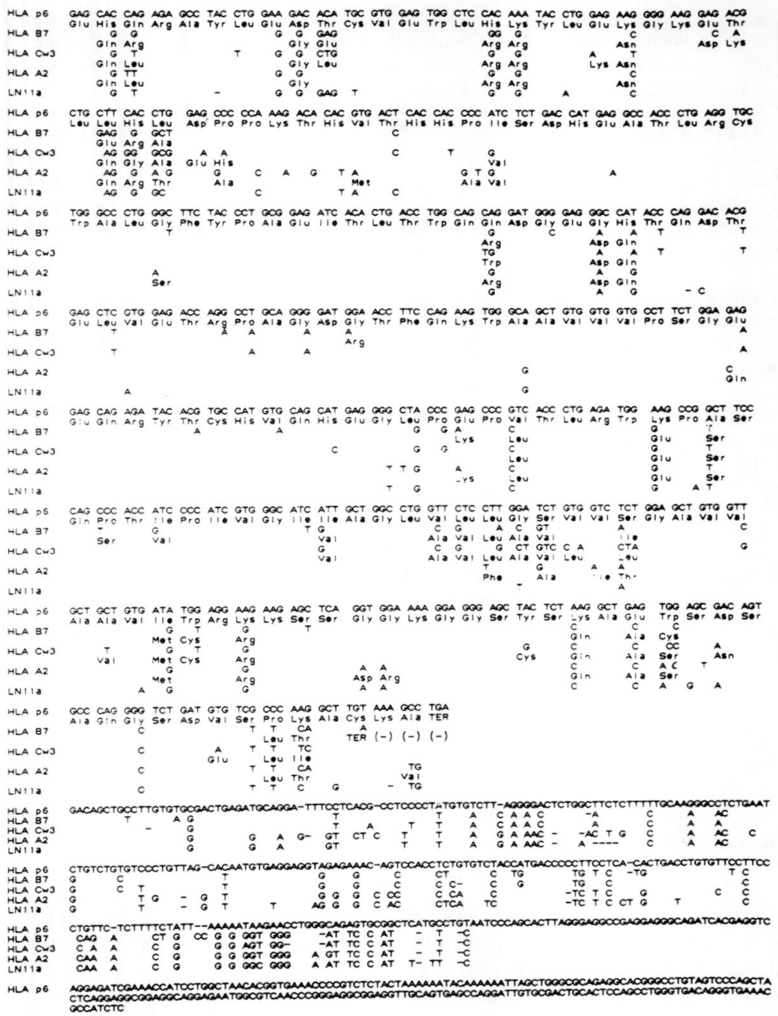

Figure 1. Sequence comparison of p6 with other class I HLA genes.

class I series (Table 1), showing 16% overall divergence in its nucleotide sequence and 20% divergence in its amino acid sequence. The nucleotide sequence divergence among the other genes is only 5 to 10%.

The most conserved region is the α3 domain, indicating that the protein product may associate with β2-microglobulin. The α2 and transmembrane regions share slightly more homology with the HLA-A-like genes (A2 and LN11a) than with the HLA-B-like genes (B7 and Cw3), suggesting that this gene may map closer to the HLA-A locus than to the -B or -C loci. The amino acid sequence of the transmembrane region is the most highly diverged and lacks a translation termination codon that is found in some murine Qa genes (14). The expressed antigen is probably membrane-bound and not secreted. The first 260 bases of the 3′ untranslated region are highly diverged (25%), after which point there is no homology with the classic class I genes. The sequence is interrupted by an *Alu* (15) repeat element. Although such repeated sequences are found in murine class I genes (16), they have not been reported in HLA-A, -B, or -C genes.

The sequence of this clone is very similar to, but not identical with, RS5, recently reported by Srivastava et al. (5). For example, p6 has a tyrosine residue at amino acid position 27 of the α3 domain, whereas RS5 has leucine and p6 has proline in the second position of the transmembrane domain where RS5 has alanine. The frameshift mutation described in the second cytoplasmic exon of RS5 appears to be absent in p6. However, in view of the extremely high sequence homology (~99%) between the two clones, it remains entirely possible that they are allelic variants of the same locus. Although p6 lacks a canonical AATAAA polyadenylation sequence, it terminates in a run of poly (A), demonstrating that it is a faithful copy of the 3′ terminus of its template mRNA. Instead, the sequence GGGTGAAA resides 10 bases upstream of the poly (A) tract.

Table 1. Percentage Divergence Among Class I HLA Genes

	α2		α3		TM		Cytoplasmic		3' Untranslated	Overall	
Region	aa	Bases	aa	Bases	aa	Bases	aa	Bases	Bases	aa	Bases
p6											
B7	41	23	7	7	36	15	33	11	25	22	16
Cw3	38	23	8	7	36	18	30	15	21	21	15
A2	31	15	12	8	21	7	30	16	30	19	16
LN11a		21		5		7		19	31		16
B7											
Cw3	24	14	4	3	13	10	22	9	9	12	5
A2	21	14	8	11	28	12	15	6	16	15	7
LN11a		10		8		13		15	15		7
Cw3											
A2	17	11	7	10	28	15	26	15	16	16	8
LN11a		11		7		16		17	16		8
A2											
LN11a		11		5		5		7	10		4

Only the first 262 bases of the 3' untranslated region have been included in the comparison. The base figures for overall divergence include the 3' untranslated region. aa = amino acids.

The presence of the *Alu* sequence makes the present form of the clone unsuitable for use as a probe in Southern and Northern blotting experiments. However, preliminary experiments have indicated that it detects a single band of about 5.0 to 6.0 kb in human DNA digested with a variety of restriction enzymes including Hind III and Bam HI. The data suggest that the gene is present in single copy in the haploid genome.

We conclude that p6 represents a clone for a transcribed gene that can code for a novel cell-surface HLA class I-like antigen, expressed in at least one T-cell line (HPB-ALL). It is distinct from the HLA-A,-B, or -C genes but is highly homologous to RS5 (5) and may be allelic to it or the product of a second closely related gene.

Note. The sequence of p6 is very similar to that of the HLA-E gene (17), suggesting it is an allele of the HLA-E locus.

Acknowledgments. We thank Shirley Cutbush for assistance with the analysis of the sequence data.

This work was supported by the Cancer Research Campaign and, in the initial stages, by the NIH.

References

1. van Leeuwen A, Giphart MJ, de Groot G, Morolli B, Festenstein H, Nijenhuis LE, van Rood JJ. Two different T-cell systems in humans, one of which is probably equivalent to Qa or Tla in mice. Hum Immunol 1985;12:235.
2. Orgad S, Yang SY, Gazit E, Relias V, Zaizov R, Lyson S, et al. Expression of extra class I major histocompatibility antigens on T-cell acute lymphoblastic leukemia (ALL) lymphoblasts. Hum Immunol 1985;12:133.
3. Mitsuishi Y, Falkenrodt A, Urlacher A, Tongio MM, Mayer S. New human MHC class I antigens segregating with HLA-A antigens detected on some lymphocyte subpopulations. Hum Immunol 1986;15:175.
4. Duceman BW, Ness D, Rende R, Chorney MJ, Srivastava R, Greenspan DS, et al. HLA JY328: Mapping studies and expression of a polymorphic HLA class I gene. Immunogenetics 1987;23:90.
5. Srivastava R, Chorney MJ, Lawrance SK, Pan J, Smith Z, Smith CL, et al. Structure, expression, and molecular mapping of a divergent member of the class I HLA gene family, Proc Natl Acad Sci USA 1987;84:4224.
6. Paul P, Fauchet R, Boscher MY, Sayagh B, Masset M, Medrignac G, et al. Isolation of a human major histocompatibility complex class I gene encoding a nonubiquitous molecule expressed on activated lymphocytes. Proc Natl Acad Sci USA 1987;84:2872.
7. Ehrlich R, Lifshitz R, Pescovitz MD, Rudikoff S, Singer D. Tissue-specific expression of a divergent member of a class I MHC gene family. J Immunol 1987;139:593.
8. Rebiere M-C, Marche PN, Kindt TJ. A rabbit class I major histocompatibility complex gene with a T cell-specific expression pattern. J Immunol 1987;139:2066.
9. Arnot D, Lillie JW, Auffray C, Kappes D, Strominger JL. Inter-locus and intra-allelic polymorphisms of HLA class I antigen gene mRNA. Immunogenetics 1984;20:237.
10. Sood AK, Pan J, Biro PA, Pereira D, Srivastava R, Reddy VB, et al. Structure and polymorphism of class I MHC antigen mRNA. Immunogenetics 1985;22:101.
11. Srivastava R, Duceman BW, Biro PA, Sood AK, Weissman SM. Molecular organization of the class I genes of the human MHC. Immunol Rev 1985;84:93.
12. Sodoyer R, Damotte M, Delovitch TL, Trucy J, Jordan BR, Strachan T. Complete nucleotide sequence of a gene encoding a functional human class I histocompatibility antigen (HLA-Cw3). EMBO 1984;3:879.
13. Pontarotti PA, Mashimo H, Zeff RA, Fisher DA, Hood L, Mellor A, et al. Conservation and diversity in the class I genes of the major histocompatibility complex: Sequence analysis of a *Tlab* gene and comparison with a *Tlac* gene. Proc Natl Acad Sci USA 1986;83:1782.
14. Kress MD, Cosman D, Khoury G, Jay G. Secretion of a transplantation-related antigen. Cell 1983;36:189.
15. Jelinek WR, Toomey TP, Leinwand L, Duncan C, Biro PA, Choudary PV, et al. Ubiquitous, interspersed repeated

sequences in mammalian genomes. Proc Natl Acad Sci USA 1980;77:1398.

16. Hood L, Steinmetz M, Malissen B. Genes of the major histocompatibility complex of the mouse. Ann Rev Immunol 1983;1:529.
17. Koller BH, Geraghty DE, Shimizu Y, DeMars R, Orr HT. HLA-E. A novel HLA class I gene expressed in resting lymphocytes. J Immunol 1988;141:897.

Author Affiliations

P. Andrew Biro, Department of Immunology, London Hospital Medical College, London, England and Division of Tumor Virology, Dana-Farber Cancer Institute, Boston, MA, USA; *Barbara Herbut; Hilliard Festenstein*, Department of Immunology, London Hospital Medical College, London, England

Isolation from the Human MHC of a Class I Gene Encoding a Nonubiquitous Antigen Expressed on Activated Cells

P. Paul, R. Fauchet, M.Y. Boscher, P. Villedieu, M. Masset, G. Merdrignac, J. Dausset, L. Degos, and D. Cohen

The class I gene family of the human major histocompatibility complex (MHC) consists of 20 or more homologous genes. Among these genes and located in the distal portion of the MHC are the HLA-A, -B, and -C loci encoding highly polymorphic, ubiquitous cell surface antigens; HLA-A is the most distal of these loci. Human analogues of the murine class I Qa and Tla loci also probably exist in the MHC. In the mouse MHC, the Qa and Tla gene 1) encode class I antigens that are restricted in tissue distribution to some lymphoid and early hematopoeitic cells (Qa) or only thymocytes (Tla), 2) show little polymorphism, and 3) are mapped more distally in the MHC, as compared to H2-K,D,L genes encoding class I murine transplantation antigens which are equivalent to HLA-A, -B and -C. In humans, several serological reagents define novel class I antigens, whose distribution is restricted to specific subsets of lymphocytes (1–7). HA (human activated) alloantisera specific for phytohemagglutinin-activated T lymphocytes (TPHA) detect possible

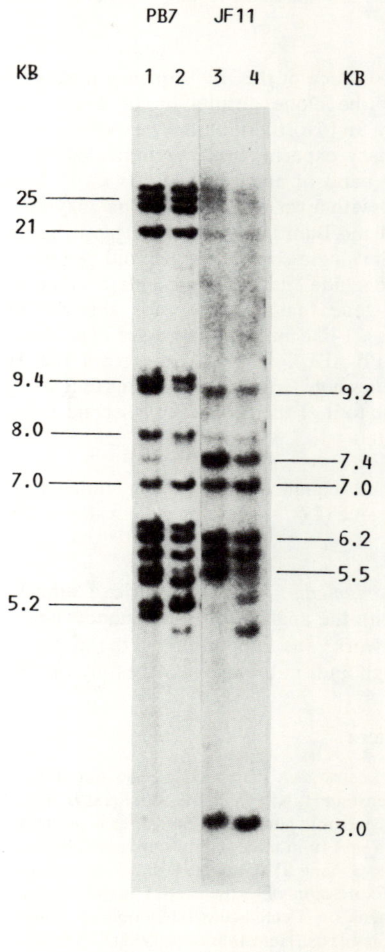

Figure 1. Southern blot comparison of Hind III-digested DNAs from unrelated individuals using PB7 and JF11 probes. High molecular weight DNA was prepared from blood samples of unrelated individuals. Approximately 9 μg of each DNA were digested with restriction endonuclease Hind III, subjected to electrophoresis for 40 hours in a 0.7% agarose gel, at 30 volts, and transferred onto a nylon membrane (MSI). Successive hybridizations were carried out on the same membrane with PB7 (lanes 1 and 2), then JF11 (lanes 3 and 4). JF11 is a 0.8-kb Hind III fragment flanking the 4.8-kb Hind III fragment containing the A11 gene. PB7 is a quasi-full-length class I cDNA clone (11). Both probes were nick-translated to a specific activity of 3.10^8 cpm/g, and the filters were hybridized for 40 hours at 42 °C in 50% formamide, 0.1% Denhart s solution, 6 × SSC (standard saline citrate), 5% dextran sulfate, 0.5% sodium dodecyl sulfate, and 200 g/ml of salmon sperm DNA. The filters were washed 15′ in 2 × SSC and 0.1% sodium dodecyl sulfate at room temperature, then 15′ at 65%, and revealed by autoradiography.

human Qa-like antigens, showing polymorphism associated with that of HLA-A specificities (6,7), probably reflecting linkage disequilibrium of the genes encoding these two groups of antigens. In an attempt to isolate the genes coding for these novel human class I alloantigens, we constructed an intergenic probe form the 5′ flanking region of an HLA-A gene. This probe detects only a limited number of class I restriction fragments on genomic blots. When polymorphic, these restriction fragments exhibit linkage disequilibrium with HLA-A alleles; some of these fragments have previously been mapped telomeric to the HLA-A locus (8) where, by analogy with the murine MHC, human Qa-like genes might be located. A clone from one of these transformed cells specifically reacts with an alloantiserum defining a set of nonubiquitous class I molecules present in particular, on HLA-A2-positive PHA-activated T cells. Data presented here provide evidence for the isolation of a gene encoding a non-HLA-A,B,C class I molecule which could be the human equivalent of a Qa murine antigen.

As opposed to the mouse system where use of congenic mice has allowed a very extensive description of the

Table 1. Analysis of Transfected Cells by Microlymphocytotoxicity

	Lysis						
	PBLs		Activated T Cells		EBV-Transformed B-Cell Lines		
Reagents	$A2^+,A26^-$	$A2^-,A26^+$	$A2^+,A26^-$	$A2^-,A26^+$	$A2^+,A26^-$	$A2^-,A26^+$	$A2^+,A26^+$ (O2BN5)
mAb W6.32	+	+	+	+	+	+	+
Absorbed with PBLs	−	−	−	−	−	−	−
Absorbed with J27	+	+	+	+	+	+	+
Absorbed with J27, then J19X14 (A26)	−	−	−	−	−	−	−
Absorbed with J27, then JP2B2	+	+	+	+	+	+	+
Anti-A26 alloserum	−	+	−	+	−	+	+
Absorbed with HLA-A26 PBLs	−	−	−	−	−	−	−
Absorbed with J27	−	+	−	+	−	+	+
Absorbed with J27, then J19X14 (A26)	−	−	−	−	−	−	−
Absorbed with J27, then JP2B2	−	+	−	+	−	+	+
mAb anti-A2, A28	+	−	+	−	+	−	+
Absorbed with HLA-A2 PBLs	−	−	−	−	−	−	−
Absorbed with J27	+	−	+	−	+	−	+
Absorbed with J27, then J19X14 (A26)	NT	−	+	−	+	−	+
Absorbed with J27, then JP2B2	+	−	+	−	+	−	+
HA2 alloantiserum	−	−	+	−	+	−	+
Absorbed with HLA-A2 PHA-activated T cells	−	−	−	−	−	−	−
Absorbed with J27	−	−	+	−	+	−	+
Absorbed with J27, then J19X14 (A26)	−	−	+	−	+	−	+
Absorbed with J27, then JP2B2	−	−	−	−	−	−	−
HA10 alloantiserum	−	−	−	+	−	+	+
Absorbed with HLA-A26 PHA-activated T cells	−	−	−	−	−	−	−
Absorbed with J27	−	−	−	+	−	+	+
Absorbed with J27, then J19X14 (A26)	−	−	−	+	−	+	+
Absorbed with J27, then JP2B2	−	−	−	+	−	+	+

Reagents used were : Monoclonals: W6.32 monomorphic monoclonal antibody (MAb) (18), MAb anti-A2 (XV17) (16); alloantisera: HLA-A26 allo-antiserum, HA10 and HA2 that react respectively with HLA-A10 and HLA-A2 PHA-activated T cells and EBV-transformed B cells, and not with resting T or B cells. Target cells for absorption were: cryopreserved human blood lymphocytes (PBL) obtained from a ficoll hypaque gradient. PHA-activated T cells: PBL were resuspended in RPMI 1640 medium (Gibco), 10% fetal calf serum (FCS) at a concentration of 1.10^6 cells/ml and phytohemaglutinin (PHA, Difco Lab.) at a final concentration of 1:2000. On day 3, 10% of interleukin 2 (IL2, Boehringer) was added to a fresh culture medium; on day 5 activated cells were harvested and cryopreserved. J27: L cells expressing human β2-microglobin (14). J19X14: J27 cells transfected by an HLA-A26 gene from individual 02BN5 typed as CA,Cw^-, Bw51/A26, Cw7, B8. JP2B2: J27 cells transfected by a P2B2 cosmid isolated from the O2BN5 cosmid library. Transfection and selection of transfected clones were performed as described (10). Absorptions were performed twice: 100 μl of reagent for 10 10^6 cells, 30 min, at 37°C and 30 min, at 4°C on PBL, PHA-activated cells, J27, J19X14 or JP2B2 transfected cells. Each reagent was also absorbed with J27 cells before absorption with J19X14 or JP2B2. Absorbed and nonabsorbed reagents were tested on a panel of cryopreserved PBL or PHA-activated T cells from 10 HLA-A2- positive, A26-negative individuals, 6 HLA-A26-positive and A2-negative individuals; EBV-transformed B-cell lines from 3 HLA-A2-positive, A26-negative individuals; 3 HLA-A26-positive, A2-negative individuals, and from 2BN5 (HLA-A2, A26). A standard complement dependent microlymphocytotoxicity method was performed for 1 hr at 22°C before and after complement addition.

organization of the class I gene family, the lack of well-characterized recombinants has made localization of non-HLA-A,B,C class I genes in humans very difficult. Sequence analysis of noncoding sequences of class I genes in the mouse show that Qa and H-2K, D, and L genes are closely related and that the Tla cluster probably diverged very early from the H-2/Qa 2-3 ancestral gene (9). By analogy, human Qa like genes could also be related to HLA-A, -B, -C genes, diverging relatively late in evolution from a common ancestral gene. If this is the case, one can assume that some intergenic sequences involved in the duplication event generating HLA-A,B,C and Qa-like subfamilies of genes are still shared by members of these subfamilies, while other class I Tla equivalent genes have different flanking sequences due to their earlier divergence.

To test this hypothesis and to try to identify Qa-like genes, we constructed an intergenic probe from sequences flanking the HLA-A11 gene, assuming that this probe could discriminate between A,B,C and Qa genes and other class I sequences. This JF11 probe is a 0.8-kb Hind III fragment, free of repetitive sequences, located in DNA regions flanking the 4.8-kb Hind III fragment containing the complete HLA-All gene (10).

The JF11 probe was tested for hybridization with Hind III-digested genomic DNA from 44 unrelated individuals. This probe detects 6 to 12 Hind III fragments on Southern blots. Hybridization of the same genomic blots with the almost full-length HLA-B7 cDNA clone (11) (PB7) shows that most of the bands detected by JF11 comigrate with PB7 hybridizing fragments (Fig. 1). Both probes detect polymorphic and nonpolymorphic fragments, but JF11 clearly detects less class I restriction fragments (6 to 12 Hind III fragments) as compared with PB7 (12 to 25 Hind III fragments). The polymorphic restriction endonuclease fragments detected by JF11 were all found to be in linkage disequilibrium with HLA-A specificities (A9, A33, A31, A11), suggesting that these restriction fragments are closely linked to the HLA-A locus. Moreover, some of the Hind III restriction fragments detected by JF11 (5.5 kb, 8 kb, 9.2 kb) have been mapped telomeric to the HLA-A locus using human mutant cell lines (11).

It is interesting to note that only HLA-A-related polymorphic restriction fragments are revealed by JF11.

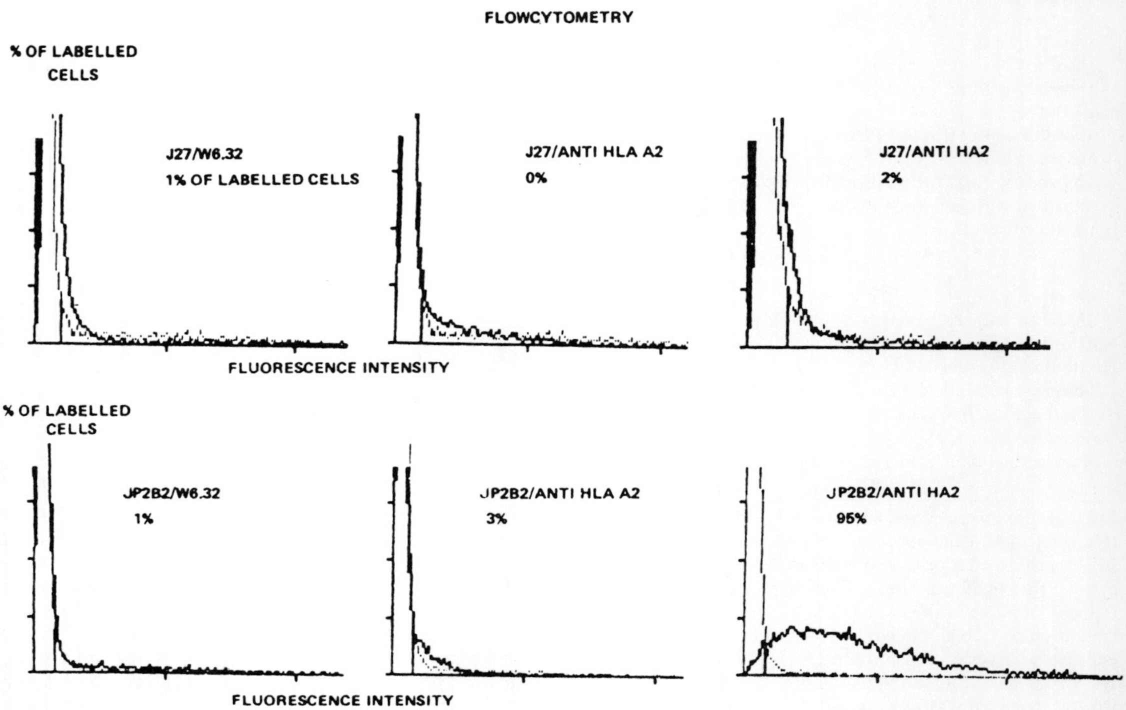

Figure 2. Analysis by flow cytometry of 1.10^4 J27 or JP2B2 transfected cells was based on analysis on the cell sorter Ortho 50H (West Wood, MA). Two monoclonal antibodies (MAb) were used : W6.32 or anti-HLA-A2 supernatants after incubation for 30 minutes at 4°C with 1.10^5 cells. For HA2 studies, an alloantiserum was first absorbed 3 times on J27 cells to eliminate nonspecific binding, 1.10^5 cells were then incubated for 1 hour at 37°C and for 30 minutes at 4°C with 90 µl of absorbed antiserum. Cells were then labeled with FITC (fluorescein isothiocyanate) conjugated antibody and antibody binding was assessed by measuring cell fluorescence at 525 nm. The upper histograms show J27 labeling with W6.32 anti-HLA-A2 MAb and HA2 alloantiserum. The JP2B2 labeling is shown underneath.

JF11 does not hybridize with restriction fragments that have been shown to be specific for the HLA-B locus (12,13) under stringent washing conditions. However, under relaxed conditions of rinsing, the HLA-B-specific Hind III restriction fragments are faintly hybridized by JF11 (Fig. 1).

These data, together with the evidence that JF11 does not detect bands known to hybridize with HLA-B-specific probes, suggest that JF11 detects a subfamily of class I genes, some of which are in linkage disequilibrium with HLA-A or already mapped telomeric to HLA-A. The analogy with the mouse MHC suggests that the genes belonging to this subfamily of class I genes would be good candidates for encoding human Qa-like antigens. In order to isolate the genes corresponding to the JF11 hybridizing fragments, the JF11 probe was used to screen a human cosmid library of 250,000 clones, constructed from DNA of an individual typed as HLA-A2Cw-B51/A26 Cw7 B8. One cosmid clone, P2B2, which contained a 5.5-kb Hind III fragment hybridizing to both PB7 and JF11, was introduced into mouse L cells expressing human β2-microglobulin (14) by use of the $CA_3(PO4_2/DNA$-mediated gene transfer method (15). Cell surface expression of new class I gene determinants was assessed by absorption and cytofluorometry using alloantisera that recognize novel class I antigens. We used two such human-activated (HA) sera, HA2 and HA10, alloantisera that recognize antigenic determinants present respectively on HLA-A2 or HLA-A10 (A25 + A26) TPHA blasts and Epstein-Barr virus (EBV) lymphoblastoid cell lines, and absent from peripheral blood lymphocytes and enriched peripheral blood B lymphocytes. HA10 alloantisera failed to react with JP2B2, a cell clone transfected with P2B2 cosmid DNA, while HA2 alloantisera showed strong specific reactivity with the transfected cells as shown by both absorption (Table 1) and cytofluorometry (Fig. 2). The 5.5-kb Hind III fragment of clone P2B2 has been mapped outside of the HLA-A locus, and we confirmed that HLA-A molecules were not expressed at the surface of JP2B2 transfected cells using, as controls, anti HLA-A26 alloantisera and an anti-HLA-A2,A28 monoclonal antibody (16) (Table and Fig.2). An anti-HLA-B8 monoclonal antibody (17) and anti-HLA-B51 and Cw7 alloantisera also did not react with the transfectants. W6.32 (18) (Table 1 and Fig. 2) and B1-23-2 (19) (data not shown), two monomorphic anti-HLA class I heavy chain monoclonal antibodies, failed to react with JP2B2 transfected cells. This work provides evidence that the expression of HA antigens, which are in strong disequilibrium with HLA-A specificities, is not the result of alternate splicing of an HLA-A gene. The HLA-A2 and HLA-A26 coding gene are each contained in a 5.2-kb PB7 hybridizing Hind III fragment which is found to be associated with all HLA-A alleles except HLA-A1, -A11, and -A9 (20). The HLA-A2, A26 individual from whom the P2B2 clone was isolated (O2BN5) is homozygous for the 5.2-kb, PB7 hybridizing Hind III fragment. This fragment is clearly distinct from the 5.5-kb Hind III fragment encoding an HA2 determinant.

In these studies, we used a cloned genomic fragment detecting a subfamily of human class I genes, to isolate HLA non-A,B,C class I genes. We show that: 1) a functional human gene, encoding a novel class I antigen specifically expressed on activated lymphocytes, is present in the MHC and could be the equivalent of Qa; 2) this gene can be expressed in a mouse fibroblasts bearing human β2-microglobulin; and 3) this gene is part of a cluster of HLA-A-related class I genes defined through homology of the flanking sequences.

References

1. Ferrara GB, Longo A, Tosi R, Biagini A, Menucci M, Carminati G. HLA segregation on human T cell subpopulation markers. Hum Immunol 1981;2:213.
2. Kim SJ, Christiansen FT, Gosar J, Silver DM, Pollack MS, Dupont B. Frequency of alloantibodies reacting with PHA activated T lymphocytes, unexplainable by known HLA activities. Hum Immunol 1980;4:347.
3. Gazit E, Terhost C, Yunis EJ. The human "T" genetic region of the HLA-linkage group is a polymorphism detected on lectin-actovated lymphocytes. Nature 1980;284:275.
4. Van Leeuwen A, Giphart MJ, de Groot G, Morolli B, Festenstein H, Nijenhuis LE, et al. Two different T cell systems in human, one of which is probably equivalent to Qa ot Tla in mice. Hum Immunol 1985;12:235.
5. Lima G, Wollman E, Lepage V, Degos L, Dausset J. Alloantigens expressed on activated human T cells different from HLA-A,-B,-C and -DR antigens. Immunogenetics 1981;13:529.
6. Fauchet R, Charron D, Bouhallier O, Genetet N, Merdrignac G, Turmel P. Class I MHC linked determinants expressed on activated T cells. In: Albert ED, Baur MP, Mayr WR (eds): Histocompatibility Testing 1984. Berlin, Springer-Verlag, 1985.
7. Mitsuishi Y, Falkenrodt A, Urlacher A, Tongio MM, Mayer S. New human MHC class I antigens segregating with HLA-A antigens detected on some lymphocyte subpopulations. Hum Immunol 1986;15:175–197.
8. Orr HT, De Mars R. Class I like HLA genes map telomeric to the HLA-A2 locus in human cells. Nature, 1983;303: 534.
9. Ronne H, Widmark E, Rask L, Peterson PA. Intron sequences reveal evolutionary relationships among major histocompatibility complex class I genes. Proc Natl Acad Sci USA 1985;82:5860–5864.
10. Paul P, Lepage V, Sayagh B, Metzger JJ, Pla M, Boumsell L, et al. Serological expression after sequential double transfection with purified HLA-A11 gene of mouse fibroblasts carrying human β2m. Immunogenetics 1985;22:1–8.
11. Sood AK, Peira D, Weisman SM. Isolation and partial nucleotide sequence of a cDNA clone for human histocompatibility antigen HLA-B by sue of an oligodeoxynucleotide. Proc Natl Acad Sci USA 1981;78:616–620.
12. Coppin HL, Denny DW, Weissman SM, McDevitt HD, HLA-B locus polymorphism: studies with a specific hybridization probe. Proc Natl Acad Sci USA 1985;82:8614–8618.
13. Steere K, Siudwell B, Leach R, Ward FE, Taurog JD, Orr H. Use of DNA probes from the 5′ flanking region of the HLA-B gene to examine polymorphism at the HLA-B locus. Hum Immunol 1986;16:137–147.

14. Kavathas P, Herzenberg LA. Stable transformation of mouse L cells for human T-cell differentiation antigens, HLA and B-2-microglobulin: selection by fluorescence-activated cell sorting. Proc Natl Acad Sci USA 1983; 80:524–528.
15. Gorman C. High efficiency gene transfer into mammalian cells. In: Glover DM (ed): DNA Cloning, Vol. II. IRL Press 1985 p. 143.
16. Bourel D, Fauchet R, Chevrinais AM, Dejour G, Merdrignac G, Genetet B, Anti HLA-A2 and -A28 monoclonal antibody: Production and study of the cross reaction. Tissue Antigens 1984;23,n°5.
17. Fauchet R, Bodmer JG, Kennedy LJ, Mazilli MC, Muller C, Raffoux C, et al. In: Albert ED, Baur MP, Mayr WR (eds): Histocompatibility Testing 1984. Berlin-Heidelberg, Springer-Verlag, 1985, pp. 211–217.
18. Parham P, Barnstahle CJ, Bodmer WF. Use of a monoclonal antibody (W6/32) in structural studies of HLA A,B,C antigens. J Immunol 1979;123:342.
19. Rebai N, Malissen B. Structural and genetic analyses of HLA-class I molecules using monoclonal xenoantibodies. Tissue Antigens 1983;22:107.
20. Cohen D, Paul P, Le Gall I, Marcadet A, Font MP, Cohen-Haguenauer O, et al. Molecular genetics of class I and II MHC antigens II. Immunol rev 1985;87–105.

Author Affiliations

P. Paul, P. Villedieu, and *L. Degos*, Unité de Recherches sur l'immuonogénétique de la transplantation humaine (INSERM U.93), hôpital Saint-Louis, 75010 Paris, France; *R. Fauchet, M.Y. Boscher*, and *G. Merdrignac*, Laboratoire d'histocompatibilité, Centre Régional de Transfusion Sanguine, 35000 Rennes, France; *M. Masset, J. Dausset*, and *D. Cohen*, Centre d'Etude du polymorphisme Humain, 3 rue d'Ulm, 75005 Paris, France

Cellular Distribution and Expression Regulation of HA Molecules (New Class I-Like Antigens)

R. Fauchet, M. Boscher, N. Genetet, O. Bouhallier, P. Paul, and B. Genetet

Abstract: HA antigen distribution, as assessed with 11 HA alloantisera on a group of activated and nonactivated cells, confirm their non-ubiquitous distribution as well as their epitopic regulation by IL2, IFNα, and γ, which is evidence of their modulation from the existing gene. From their expressivity on a group of pathologic cells, such as chronic leukemia (CLL), acute leukemias (ANLL and T or NT ALL), we can assess their importance at an early stage of cellular differentiation, as confirmed by the observed marker correlation CD9+/HA+. Also, analysis of the transfectant cells, hybridized with a subfamily probe of a class I gene recognized by an HA2 alloserum, brings evidence of the existence of a functional gene encoding non-HLA molecules independent from HLA molecules.

The HLA-A region of chromosome 6 appears more and more as a complex multiloci region. Indeed, several studies have shown that the number of ADN fragments of the human genome hybridized with a class I probe is greater than could have been implied by the three HLA-A,B,C loci, and is now as high as 16 or 17 (1). From a serologic definition (2), several systems are now proposed that characterize these new class I genes: Gazit's HT system (3), (4), Van Leeuwen's TCA (5), and Tongio's PHA-L (6) in disequilibrium with the HLA-A genes. In previous studies, we have demonstrated the existence on PHA T cells (7) of new epitopic markers in close disequilibrium with HLA antigens, such epitopes being implicated in the cell activation process. Analyzing the structure of the identified molecule revealed that it matched that of a 41K–12K dimer, the independence of which from the HLA-A molecule was achieved through depletion and sequential immunoprecipitation for HLA-A3(8). These "class I-like" antigens were denominated HA antigens (Human Activated antigens).

The aim of the present study is to establish the cellular distribution of the new markers, to approach their epitopic regulation, to clarify their genetic control, and to venture an answer to the question of their equivalence with the mouse Qa-T1 genes. This approach is essentially serologic (micro-lymphocytotoxicity) and cytoflueometric, thanks to 11 highly selected local allosera gathered in five clusters: HA1, HA2, HA3, HA9, and HA10. The numeral indicates the specificity of the disequilibrium with the HLA-A genes (e.g., HA1 = disequilibrium with HLA-A1).

The target cells under study compose four major groups: normal cells (activated, nonactivated); two cell

Table 1. Cellular Distribution

Nonactivated Cells		Activated Cells		Cell Lines
T lymphocytes	95	T PHA	126	MOLT 4
B lymphocytes	95	T Allo	50	
Monocytes	20	ZBV	47	
No expression		Expression		Expression

Numerical figures indicate the number of cells treated in each cellular variety.

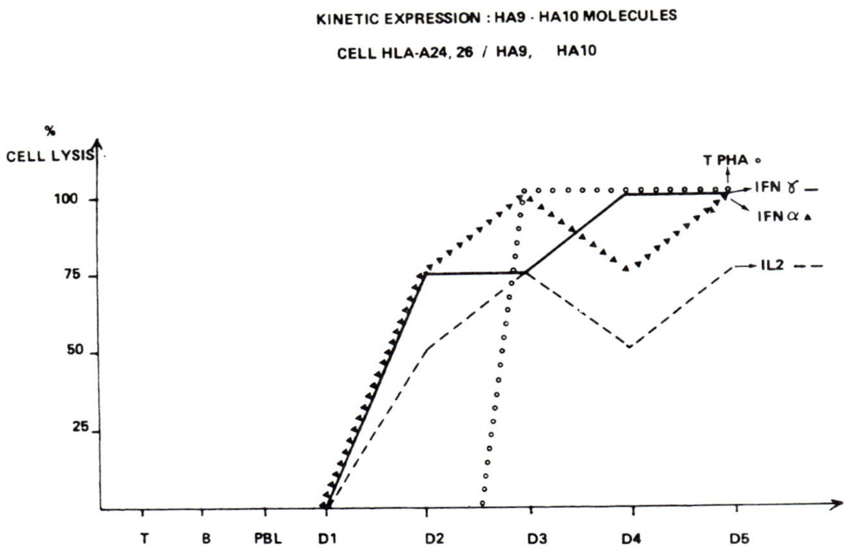

Figure 1. Kinetic Expression: HA9·HA10 Molecules Cell HLA·A24,26 / HA9, HA10.

lines (MOLT 4, K562); transfected cells; and pathological cells including Chronic Lymphoid leukemia (CLL), Acute Non-Lymphoblastic Leukemia (ANLL) and Lymphoblastic Leukemia ALL (T ALL and NT ALL).

NO expression was traceable on nonactivated cells. The number of cases studied is shown in brackets: T lymphocytes (95), B Lymphocytes (95), monocytes (20), thymocytes (2). When the cells are activated in vitro, the HA antigens are expressed: PHA (126), allogenic T (50), EBV cell lines (47) (Table 1). Treatment of PBL by IFNα and IFNγ (500 IU), IL2 (10 IU) induces the expressivity of the HA molecules. Expression kinetics on the treated PBL is shown in Figure 1. On D5, expression with IFNα and γ is identical to that obtained through the PHA mitogen, and remains lower with IL2.

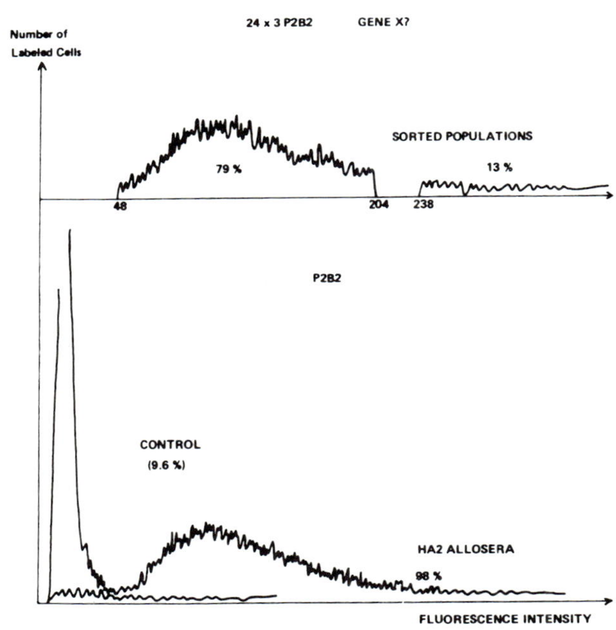

Figure 2. Flow cytometry of P_2B_2—cells labeled with HA_2 antiserum. In the first step, 75% of the cells were labeled by HA_2 antibody; among the sorted cells, 98% were labeled by HA_2 antibody.

Table 2. Pathologic Cells: 54 Cases

Chronic Leukemia		Acute Leukemia		
	CLL	ANLL	T ALL	NT ALL
Total (n=)	18	11	10	15
HA + (N=)	2*	1	5*	6*
HA − (N=)	16	10	5	9
	*=One CD9		*=50% One CD9	*=40% All CD9

CLL: Expression was weak in two cases.
TALL: Positive cells were found in pre-T leukemia.

The MOLT 4 line (9) HLA-A1/HLA-A10 expresses the HA1, HA10 antigens only when the culture has been performed in the presence of IFNα or γ. As to K562 (10), HA molecules are undetectable with or without IFN.

In order to confirm the hypothesis of an existing human equivalent of the mouse Qa, a JF11 intergenic probe was elaborated (P. Paul, D. Cohen) with sequences flanking the HLA-A11 gene. JF11 detects a limited number of 6-12 HIND III restriction fragments which, when polymorphic, are in disequilibrium with the HLA-A specificities (11).

A cosmid library was constructed from the DNA of a healthy individual 02BN5 (HLA A2/A26, Cw7, Cw.), and screened by JF 11. One cosmid 24 × 3 carrying 5.5 HIND III fragment was transfected in a mouse cell line J 27. The transfectant clone P2B2 was analyzed by cytotoxicity (absorption-elution) and cytofluometry against HLA monoclonal antibodies—w6/32 (12), A2+A28 (12), HA allo-antisera HA2, HA10 for the identification of human activated antigens on the transfectant surface. W6/32, A2, A28 monoclonal antibodies, HA10 failed to react with P2B2, whereas HA2 alloserum showed a specific reactivity with the transfectant cells, both by absorption and flow cytometry (Fig. 2).

Fifty-four pathologic cells were studied, which were phenotyped as to the T and B line differentiation markers. Among the pathologic cells, 18 originate from chronic lymphoid leukemia, two of which express HA molecules, and one is CD9 positive; 36 come from acute leukemia. HA expressivity is retrieved in only one case out of the 11 ANLL.

In the TALL group (10 cases), 50% express HA antigens. Taking into consideration the state of maturation when analyzing this expression, the HA antigens are expressed in three of the four pre-T leukemia cases, in one out of four immature T, and in one of the two mature T cases. This cell also expresses the CD9 marker.

Expressivity is most significant in the NT acute leukemia cells (15). In this group, HA antigens are expressed in six cases. Correlating expressivity to the differentiation markers reveals the correlation between HA and CD9 markers: all HA+ cells are CD9+. In other words, there is no HA expression without CD9 (Table 2).

Up to now, the linkage disequilibrium between HLA antigens and HA markers had appeared to be rather close. This exhaustive distribution study has revealed the existence of unexpected associations under two forms of expressivity: either through dissociation of HLA-HA, in five cases—four normal cells (HLA-A2, A29/HA1, HA2; HLA-A3, A28/HA3, HA2; HLA-A2, A29/HA2, HA1; HLA A2,A2/HA2,HA1, and in one T ALL leukemia: HLA-A2-A2/HA2, HA1—or in triplets—HLA A2,A23/HA2,HA9,HA1; HLA A3,A26/HA3, HA10,HA1. Two family surveys have allowed to follow the segregation of the unexpected HLA marker (Fig. 3). HA1 is transmitted through haplotype b. In this situation family, the father is homozygous HLA-A2,A2 and the possible occurrence of a triplet HA2(haplotype a) HA2,HA1 (haplotype b) cannot be excluded nor demonstrated.

To conclude, the expressivity of nonubiquitous HA antigens can be modulated by IL2 (14), IFNα, and γ (13) in PBL as well as in the MOLT 4 line, as has been

Family 3

Father	a/b	HA2 HLA-A2 / HA1 HLA-A29	Siblings	b/c	HA1 HLA-A29 / HA. HLA-A29
Mother	c/d	HA. HLA-A29 / HA3 HLA-A3		a/c	(HA2) HLA A2 / HA. HLA-A29

Family 4

Father	a/b	HA2 HLA-A2 / HA1 HLA-A2	Siblings	b/d	HA1 HLA-A2 / HA2 HLA-A2
Mother	c/d	HA2 HLA-A2 / HA2 HLA-A2		a/d	HA2 HLA-A2 / HA2 HLA-A2
				b/c	HA1 HLA-A2 / HA2 HLA-A2
				a/c	HA2 HLA-A2 / HA2 HLA-A2

Figure 3. Family studies: HLA-HA marker dissociation.

done in previous studies. These structures then clearly appear as molecules implicated in the activation process. Their presence on pathologic cells (ALL) (14,15) is due to the early differentiation states of T and NT lines, where expression, correlated to the presence of the CD9 marker, confirms the position of HA molecules among the molecules present at the early states of cell differentiation.

The existence of unexpected HLA-HA association brings further evidence of the system complexity, and a multi-loci hypothesis cannot be disregarded. Biochemistry could only demonstrate the independence of the HLA-A3-HA3 molecules. Transfectant data from a probe that recognizes a class I human gene subfamily proves the existence of a functional gene encoding non-HLA molecules which could be the human equivalent of murine Qa (16,17).

Acknowledgments. Work supported by A.R.C. (Association pour la Recherche sur le Cancer) and I.N.S.E.R.M. (Institut National de la Santé et de la Recherche Médicale) (grant no. 86305)

References

1. Koller BM, Geraghty O, Orr HT, Shimizu Y, Demars R. Organization of the human class I major Histocompatibility complex genes. Immunol Research 1987;6:1-10.
2. Ferrara GB, Longo A, Tosi R, Biagini A, Menucci M, Carminati G. HLA segregation on human T cell sub-population markers. Human Immunology 1981;2-2:213-223.
3. Gazit E, Terhorst C, Mahoney RJ, Yunis EJ. Alloantigens of human T(HT) genetic region of the HLA linkage group. Human Immunology 1980;1:97–109.
4. Gazit E, Gothelf Y, Gil R, Orgad S, Pitman TB, Watson LM, et al. Alloantibodies to PHA-activated lymphocytes detect human Qa-like antigens. J Immunol 1984;132:165.
5. Leeuwen AV, Giphart MD, de Groot G, Morolli B, Festenstein H, Nidenhuis LE, et al. Two different T cell systems in humans, one of which is probably equivalent to Qa or Tla in mice. Human Immunology 1985;12:235-2346.
6. Mitsuishi Y, Falkenroot A, Tongio MM, Mayers N. New human MHC class I antigens segregating with HLA-A antigens detected on some lymphocyte subpopulation. Human Immunology 1986;15:175-187.
7. Fauchet R, Charron D, Bouhallier O, Genetet N, Merdrignac G, Turmel P, et al. Class I MHC-linked determinants expressed on activated T cells. In: Albert ED, et al. (eds): Histocompatibility testing. Berlin, Springer-Verlag, 1984.
8. Fauchet R, Boscher M, Bouhallier O, Merdrignac G, Genetet B, Turmel P, et al. New class I in man: Serological and molecular characterization. Human Immunology 1986;17:3-20.
9. Tada N, Tanigaki N, Pressman D. Human cell membrane components bound to be microglobulin in T cell lines. J Immunol 1987;2:120, 513-519.
10. Sutherland J, Nannoni P, Rosa F, Huyat D, Turner R, Fellous M, Induction of the expression of HLA class I antigens on K562 by interferon and sodium butyrate. Human Immunology 1985;12:65-72.
11. Paul P, Fauchet R, Boscher M, Sayagh R, Masset M, Merdrignac G, et al. Isolation of a human MHC class I gene encoding a nonubiquitous molecule expressed on activated lymphocytes. Proc Nat Acad Science USA 1987;2872.
12. Fauchet R, Bodmer JG, Kennedy JL, Mazzilli MC, Muller C, Raffoux C, et al. HLA-A,B,C monoclonal antibodies. In: Albert EA, et al (eds): Histocompatibility testing. Berlin, Springer-Verlag, 1984.
13. Gazit E, Gil R, Altshaler R. Lymphocytes modulated by β interferon express non-HLA-A,B,C class I antigens. Human Immunology 1987;18:53-67.
14. Van Leeuwen A, Schrier PI, Giphart MJ, Nordemeer IA, Ruiter DJ, Rubinstein P, et al. TCA: A polymorphic genetic marker in leukemias and melanoma cell lines. Blood 1986;67:1139.
15. Peuccio D, Crepaldi T, Lovisone E, Paolino F, Foa R, Castagnoli A, et al. HLA class I like antigen expression on human leukemic cells. Tissue Antigens 1987;30:76-83.
16. Fahrner K, Hogan BLM, Flavell RA. Transcription of H2 and Qa genes in embryonic and adult mice. EMBO J 1987;5:1265-1271.
17. Lew A, Maloy L, Cougan JF. Characteristics of the expression of the immunosoluble class I molecule Σ 10. J Immunol 1986;254-258.

Author Affiliations

R. Fauchet, M Boscher, N. Genetet, O. Bouhallier, B Genetet, Laboratoire d'histocompatibilité, C.R.T.S. de Rennes, Rennes France; *P. Paul,* Inserm 93, Paris, France

A Human T-Cell Clone Recognizes a Qa- or Tla-Like Antigen in Man

Rudolf Wank

From molecular and biochemical studies it is known that the murine Tla and Qa antigenic systems structurally resemble MHC class I antigens. In man alloantisera have been found that define antigens, now called HT, which show restricted expression on phytohemagglutinin (PHA)-activated peripheral blood lymphocytes (PBL), B lymphocytes, and some leukemic cells (1). Antigens of the HT system are HLA associated (1,2) and as with Qa antigens, are not expressed on thymocytes (3). Another very similar system, called TCA (T-cell associated) has been identified which is also HLA linked, but may be distinct from the HT system since no HLA-A antigen association has been found (4). These genetic systems have been defined serologically and it is not known whether the antigens can be recognized by human T cells. The human T cell clone 234, raised between HLA-D identical siblings differing for the HLA-A,B segment and for part of the complement region, recognizes an HLA-A24 associated antigen. This is expressed on peripheral blood lymphocytes (PBL), Epstein-Barr-Virus transformed lymphoblastoid cell lines (LCL), PHA-blasts and, to a lower degree, on T-cell lines but is not expressed on resting T-cells and minimally, if at all, on resting B cells. Clone 234 also responded to HLA-A24 positive thymocytes from one individual and to HLA-A24 negative preleukemic cells from another individual. The HLA-A24 negative preleukemic cells stimulated the clone as strongly as HLA-A24 positive PHA blasts or T-Cell lines, but they lost their stimulatory capacity upon PHA stimulation or IL-2 expansion. The specificity pattern of the determinant stimulating clone 234 resembles both HT (Qa-like) and Tla. Further studies are required to clarify the exact nature of the stimulating antigen. These results indicate that some human T-cells may recognize novel class I-like antigens.

Table 1. Proliferative and Cytotoxic Responses Against Resting and Activated Lymphocyte Subpopulations of T-Cell Clone 234*

	Test system	
Stimulating Cells	PLT†	CML‡
Resting		
B-lymphocytes§	((+))	–
T-lymphocytes§	–	–
PBL	++	–
Activated		
PHA-blasts‖	+	+
T-cell lines#	(+)	–
EBV-LCL**	++	++

*Clone phenotype is CD4 positive.
†Standard PLT testing using 10^5 irradiated stimulating cells and 10^3 cloned T cells.
‡Standard cell mediated lympholysis assay using chromium-51 labeled target cells.
§T and B lymphocytes were separated by nylon wool purification. T-lymphocytes consist of 99% T cells; B-lymphocytes of 90% B cells.
‖Peripheral blood lymphocytes (PBL) were stimulated with 1% phytohemagglutinin for 72 hours.

Acknowledgments. This work was supported by the Deutsche Forschungsgemeinschaft through SFB 217 (project A4).

References

1. Gazit E, Terhorst C, Yunis E. The human "T" genetic region of the HLA linkage group is a polymorphism detected on lectin-activated lymphocytes. Nature 1980;284:275.
2. Ferrara GB, Longo A, Tosi R, Biagini A, Mennucci M, Carminati G. HLA segregation of human T-cell subpopulation markers. Human Immunology 1981;2:213.
3. Gazit E, Gothelf Y, Gil R, Orgad S, Pitman TB, Watson ALM, et al. Alloantibodies to PHA-activated lymphocytes detect human Qa-like antigens In: Albert ED, et al. (eds): Histocompatibility Testing 1984. Berlin, Springer Verlag, 1984; p. 455.
4. Van Leeuwen A, Festenstein H, van Rood JJ. Diallelic alloantigenic systems on subsets of T-cells. Human Immunology 1982;4:109.

Author Affiliations

Institute of Immunology, University of Munich, Munich, Federal Republic of Germany

Comparative Analysis of TL^b and TL^c Gene Organization

G. Dalon Brown, Gregory Egan, Yechin Choi, and Daniel Meruelo

Abstract: Cosmids representing $T9^b$ through $T13^b$ were isolated from a radiation leukemia virus (RadLV)-induced C57BL/6 thymoma-derived cell line. These cosmids were co-linear with cosmids isolated from C57BL/10 mice (1). Also isolated were two previously unreported *H-2* genes, $T14^b$ and $T15^b$ (2), which have been shown to be present and linked to $T13^b$ in both C57BL/6 and C57BL/10 mice. A collection of four low-copy probes (2; C. Pampeno and D. Meruelo, manuscript in preparation) were prepared from the $T13^b$-$T15^b$ region and used to compare the organization of TL^b and TL^c both in genomic and cosmid DNAs. The region of DNA encoding $T11^b$ through $T15^b$ does not have a counterpart in the BALB/c genome. All isolated TL^b genes are located on a contiguous stretch of DNA. In the BALB/c mouse, it appears that two regions of DNA from cluster A have been duplicated, thereby generating the *TL* cluster B, $T11^c$-$T17^c$ (3). Limited reorganization has led to similar, yet unique, gene organization in these two as yet unlinked TL^c regions.

The TL^b region of C57BL/10 (B10) mice has been shown to contain two retroviral-like sequences, *TLev1* and *TLev2* (4). As part of a study aimed at determining whether these elements are involved in *TL* gene expression, this region was reisolated from a RadLV-induced C57BL/6 (B6) thymoma-derived cell line (2). As a result, the TL^b gene map has been extended by approximately 30 kb to include *T14* and *T15*, two previously unreported genes (see Fig. 1A). The B6-derived cosmids isolated are referred to as c3, c15, c116, and c153 and span *T9* through *T15*. The proximity of *TLev1* to the genes *T13, T14*, and *T15* prompted the examination of these genes in greater detail (2). Exons 4, 5, and 6 as well as parts of intervening sequences (IVS) 4, 5, and 6 have been sequenced (2). The DNA flanking the site of *TLev1* integration has been sequenced (4) as has the first four exons of *T14* (2,4). Several observations resulting from this effort are: *T13* either encodes a secreted protein or is a pseudogene by virtue of a termination codon in exon 4; *T13* is highly conserved with other *H-2* sequences but shows highest homology with genes from outside the *TL* region, i.e., *H-2K, H-2L*, and *H-2D*; *T14* is physically disrupted by *TLev1*; *T14* has very low homology as compared to other previously sequenced *H-2* genes, $<60\%$ in exons 1, 2, and 3; while overall *H-2* homology in *T14* is low, the sequences immediately flanking *TLev1* (90-bp to the left and 75-bp to the right) is conserved with the 3' IVS 1 and 5' exon 2 of other *H-2* genes; and *T15* appears to be a complete *H-2* gene but has not yet been sequenced.

The cloning and analysis of this stretch of DNA generated a collection of DNAs that proved useful as low-copy probes. These probes were used to compare the organization of the *TL* regions of B6 mouse (TL^b, *TL* non-expresser) with that of BALB/c (TL^c, low *TL* expresser). Genomic DNAs as well as a collection of cosmid clones representing $T1^b$-$T15^b$ (1,2) and $T1^c$-$T18^c$ (3,5) were probed and the hybridization patterns are shown in Figure 1. Also shown are hybridization patterns of two probes, pTLA4 and pTLA5, described by Fisher et al. (3). Based on previously published data (3), the alignments of the TL^b and TL^c maps appeared relatively simple. The data in Figure 1 demonstrates that while homologous regions exist in the two haplotypes, considerable polymorphism does exist. Hybridization with any of the low-copy probes shows that the *T13-T15* region of B6 mice is not represented in BALB/c; in addition, genes homologous to $T11$-$T12^b$ have not been linked to either of the BALB/c clusters.

The probes used in this study were *TL200*, a 200-bp *Bam*HI-*Hae*III fragment isolated from the 0.54-kb *Bam*HI fragment located 5' of gene *T7* (1,2). This probe hybridizes 5' of *T5, T7*, and *T13* as well as in the *Qa* region. *TLev1* 5' flanking probe is an 800-bp *Bgl*II-*Hpa*II fragment isolated from pTLev1; the *TLev1* 3' flanking probe is a 150-bp *Bgl*II-*Pst*I fragment also from pTLev1 (C. Pampeno and D. Meruelo, manuscript in preparation). The *T15*-region probe is a *Hind*III-*Hae*III 600-bp fragment isolated from a 1.5-kb *Hind*III fragment located 3' to *T15* (2). Also shown in Figure 1 are the *Bam*HI restriction sites and the placement of individual cosmids used in this study. Shown below the hybridization locations are the sizes of the *Bam*HI fragments to which hybridization occurs.

It would appear that $T1$-$T8^b$ is homologous to $T1$-$T7^c$ as judged by the hybridization profiles. Close examination of the *T9-T10* region of each haplotype indicates that while homology exists, reorganization has occurred following divergence of the two strains. Comparison of the two BALB/c clusters indicates that $T1$-$T3^c$ is closely related to $T11$-$T13^c$ and likewise that $T6$-$T10^c$ is related to, yet divergent from $T14$-$T17^c$. Interestingly, even though these two clusters appear to have evolved by duplication, $T4$-$T5^c$ does not have a homolog in the second TL^c cluster. We have also examined $T18^c$ (6), another unlinked *TL* gene from BALB/c, and no hybridization with the low-copy probes was observed. This entire region has been examined using other enzymes as well as by double digestion. These results are more fully discussed elsewhere (2). The $T7$-$T8^b$ is highly conserved with $T6$-$T7^c$ as is $T10^b$, $T9^c$, and $T17^c$. In fact, more diversity is observed between the two BALB/c clusters (i.e., $T14$-$T15^c$ vs $T6$-$T7^c$ and $T16$-$T17^c$ vs $T9$-$T10^c$) than is seen between cluster A of BALB/c and B6 ($T5$-$T8^b$ vs $T4$-$T7^c$).

Immunobiology of HLA
Volume II

The endogenous retroviral elements found in B6 are both absent from BALB/c. Furthermore the entire region encompassing these elements is missing in the *TL*^{c} haplotype, i.e., *T11-T15*^{b}. Data is available that *TLev1* has been deleted from BALB/c or its progenitors during evolution and strain divergence (C. Pampeno and D. Meruelo, manuscript in preparation). Thus, by comparing the *TL* regions of various inbred and wild mouse strains with these and other *TL* specific low-copy probes, one can look at expansions and contractions that have occurred in the mammalian genome. The significance, if any, of the *TLev* sequences in *TL* non-expressor B6 mice and absence in the *TL* expressing BALB/c strain is not clear at this time.

Acknowledgments. We would like to thank Christine Pampeno for supplying the *TLev1* flanking probes, the 5' flanking sequences, and for critical comments of the manuscript.

D.M. is an established Investigator of the American Heart Association. This work was supported by NIH grants CA22247, CA31346, CA35482, and CA41420. G.D.B. is supported by NIH Research Fellowship grant CA07860.

References

1. Weiss,EH, Golden L, Fahrner K, Mellor AL, Devlin JJ, Bullman H, et al. Organization and evolution of the class I

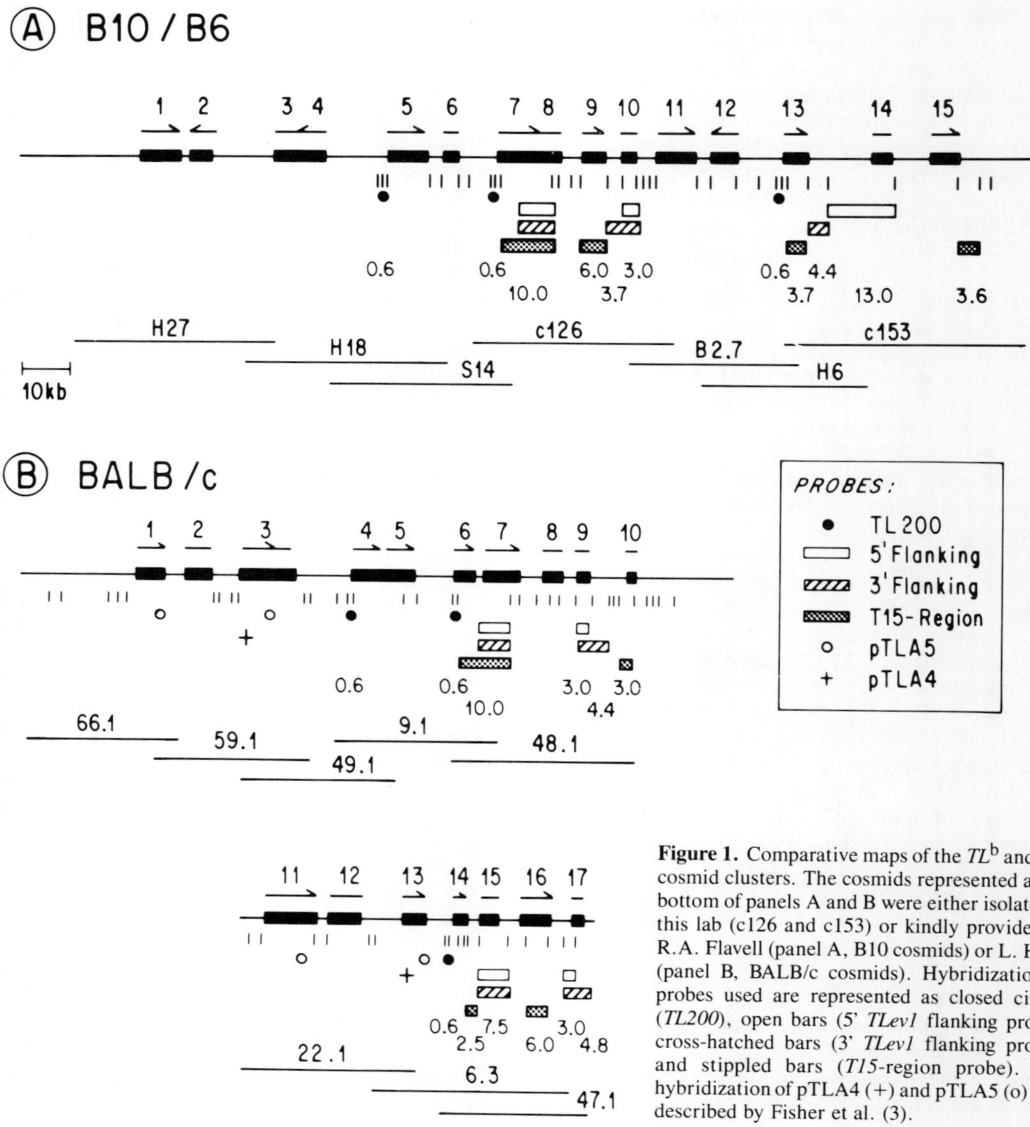

Figure 1. Comparative maps of the *TL*^{b} and *TL*^{c} cosmid clusters. The cosmids represented at the bottom of panels A and B were either isolated in this lab (c126 and c153) or kindly provided by R.A. Flavell (panel A, B10 cosmids) or L. Hood (panel B, BALB/c cosmids). Hybridization of probes used are represented as closed circles (*TL200*), open bars (5' *TLev1* flanking probe), cross-hatched bars (3' *TLev1* flanking probe), and stippled bars (*T15*-region probe). The hybridization of pTLA4 (+) and pTLA5 (o) is as described by Fisher et al. (3).

gene family in the major histocompatibility complex of the C57BL/10 mouse. Nature 1984;310:650.
2. Brown GD, Choi Y, Egan G, Meruelo D. Extension of the *H-2 TL*[b] molecular map: Isolation and characterization of *T13, T14* and *T15* from the C57BL/6 mouse. Immunogenetics 1988;27:239.
3. Fisher DA, Hunt SW, Hood L. Structure of a gene encoding a murine thymus leukemia antigen, and organization of *Tla* genes in the BALB/c mouse. J Exp Med 1985;162: 528.
4. Pampeno CL, Meruelo D. Isolation of a retroviruslike sequence from the *TL* locus of the C57BL/10 murine major histocompatibility complex. J Virol 1986;58:296.
5. Steinmetz M, Winoto A, Minard K, Hood L. Clusters of genes encoding mouse transplantation antigens. Cell 1982; 28:489.

Author Affiliation

Department of Pathology and Kaplan Cancer Center, New York University Medical Center, New York, NY 10016, USA

Study of an Endogenous Retrovirus-Like Locus Reveals Genetic Polymorphisms Related to Mouse TLA Haplotypes*

C. Pampeno and D. Meruelo

A retrovirus-like element (*TLev1*) is located within the *Tla* locus of the C57BL/10 major histocompatibility complex (1). Low copy probes have been isolated from sequences flanking the *TLev1* integration site to examine the distribution of *TLev1* among inbred mouse strains having genotypically determined variations in TL antigen expression. Table 1 shows that the *TLev1* genome was present in 20 out of 20 mouse strains of the *Tla*[b], nonexpressing (TL−), haplotype. Inbred strains of *Tla*[a,c,e], expressing (TL+), haplotypes were not found to contain the *TLev*1 locus. Certain TL⁺ strains, including 3 *Tla*[f] inbred strains and an *H-2* congenic strain, A.CA (*Tla*[d]), were found to contain *TLev1* genomes indicating that the retrovirus does not exclusively correlate with the TL⁻ phenotype.

The *TLev1* flanking probes were also found to cross-hybridize with other sequences within the *TL* locus. The organization of these sequences appear characteristic for different haplotypes.

Genomic DNA from mice of different species and subspecies was analyzed for the presence of the *TLev1* locus. The *TLev1* locus was found to be present in populations of several *Mus Musculus* subspecies including, *M.m. domesticus*, *M.m. musculus*, *M.m. castaneus*, and *M.m. molossinus*. These results indicate that *TLev1* integrated into a common ancestor of the species *Mus musculus; TLev1* integration can therefore be dated to at least 1 to 2.5 million years ago, the time estimated for the divergence of the *Mus musculus* subspecies (2). The absence of the *TLev1* locus from certain inbred and wild mouse strains consequently reflects the occurrence of excision or deletion events within the *TL* locus following the separation of these mouse populations.

Several observations argue that *TLev1* was lost from the murine genome as a result of genetic rearrangements in the *TL* locus: 1) Fluctuations in the number of sequences cross-hybridizing with the *TLev1* flanking probes occur in inbred mice as well as different wild populations of *M.m. domesticus*. 2) No traces of viral excision recombination events could be found in TL⁻ mouse strains. 3) The region of C57BL/10 containing the *TLev1* locus, genes *T11-T15*, is absent from the BALB/c genome (3).

Table 1. Strain Distribution of *TLevl*

Tla Haplotype*	*TLevl* Present	Mouse Strains
b (0–0.5%)	+	C57BL/10, C57BL/6, AKR/J, C3H/HeJ, C3H/Disn, CBA/J, CE/J, DBA/1J, HRS/J, MA/MyJ, RF/J, SM/J, ST/bJ, B6.K2, B10.A(4R), B10.GD, B10.Q, B10.S, A.SW, A.Tla[b]
d (100%)	+	A.CA
f (5%)	+	LP/J, C57/L, 129/J, B10.129(6M)
c (5%)	−	BALB/cJ, DBA/2J, LG/J, B10.D2, B10.TL, B10.HTT, A.AL
a (100%)	−	A/J, BUB/BnJ, SJL/J, SWR/J, B10.A(3R), B10.A(5R), B10.AKM, B10.BR, B10.G, B10.S(7R), B10.T(6R),NZB
e (100%)	−	BDP/J, B10.P

**Tla* haplotypes taken from reference 4. Numbers in parentheses refer to levels of TL antigen expression relative to *Tla*[a] haplotype (100%) (4).

*This work was supported by NIH grants CA22247, CA31346, CA35482, and CA41420. D.M. is an established investigator of the American Heart Association.

Immunobiology of HLA
Volume II

It is suggested that the loss of the *TLev1* locus from certain murine genomes reflects evolutionary rearrangements in the TL region; the haplotype-related organization of cross-hybridizing *TLev1* flanking sequences suggests that the resultant diversity may relate to the differential expression of TL antigens among mouse strains.

References

1. Pampeno CL, Meruelo D. Isolation of a retroviruslike sequence from the *TL* locus of the C57BL/10 murine major histocompatibility complex. J Virol 1986;58:296.
2. Yonekawa H, Moriwaki K, Goloh O, Hayashi JI, Wantanabe J, Miyashita N, et al. Evolutionary relationships among 5 subspecies of *Mus Musculus* based on restriction enzyme cleavage patterns of mitochondrial DNA. Genetics 1981;98:801.
3. Brown GD, Choi Y, Egan G, Meruelo D, Extension of the *H-2* TL^b molecular map: Isolation and characterization of *T13, T14,* and *T15* from the C57BL/6 mouse. Immunogenetics 1988;27:239.
4. Klein J, Figueroa F, Klein D. *H-2* haplotypes, genes, and antigens: Second listing. I. Non-*H-2* loci on chromosome 17. Immunogenetics 1982;16:285.
5. Michealson J, Boyse E, Chorney M, Flaherty L, Fleissner E, Hammerling U, et al. The biochemical genetics of the *Qa-TL* region. Transplant Proc 1983;15:2033.

Author Affiliation

Department of Pathology and Kaplan Cancer Center, New York University Medical Center, 550 1st Avenue, New York, NY, 10016 USA

Section VI
Class I Histocompatibility Antigens: HLA-DRB1

Analysis of Isotypic and Allotypic Sequence Variation in the HLA-DRβ Region Using the In Vitro Enzymatic Amplification of Specific DNA Segments

Henry A. Erlich, Stephen J. Scharf, Christopher M. Long, and Glenn T. Horn

Abstract: DRβ sequence polymorphism has been analyzed using the polymerase chain reaction in vitro amplification method. The amplified DRβ segments have been cloned and sequenced to determine the extent of allotypic (allele) and isotypic (locus) variation. The sequences have been assigned to either the DRβ1 or the second DRβ chain locus on specific haplotypes. This sequence information can serve as the basis for a rapid DNA typing method using a dot-blot format with sequence-specific oligonucleotide probes. The patchwork pattern of polymorphism revealed by this analysis may reflect gene conversion-like mechanisms between different alleles and different loci. A previously unreported sequence, termed DRβX, is described and helps define one of the three major evolutionary groups of DR haplotypes.

Polymorphism in the HLA-DR antigens has been determined by serologic typing (the DR specificities) and by mixed-lymphocyte culture (MLC) typing (the Dw specificities). Some of the serologically defined class II haplotypes (e.g., DR2, DR4, and DRw6) are heterogeneous by Dw typing. Recently, further genetic heterogeneity within the DR types has been revealed by the 2D gel analysis of DRβ-chains, by specific T-cell clones, and by restriction fragment length polymorphism (RFLP) analysis using class II α and β probes (reviewed in 1–3). Molecular genetic analysis of the HLA-DR region using overlapping cosmid clones (4,5) has shown that most haplotypes contain two polymorphic β loci and one relatively invariant α locus. Some haplotypes also contain DRβ pseudogenes (6). The map order is centromere, DP, DQ, DRβ-1 which encodes the DR specificity and a second β-chain locus, β-III or β-IV which encodes the public DRw52 or DRw53 specificity. Until very recently, the analysis of DRβ sequence polymorphism required the construction of cDNA or genomic libraries and the isolation of clones for subsequent sequencing. Here, we describe the application of the polymerase chain reaction (PCR) method (7,8) to amplify and clone (9) the polymorphic second exon of the various DRβ loci. This

```
      |-----------GH46------------>
  1   CCGGATCCTTCGTGTCCCCAGACCACGTTTCTTGGAGCAGGTTAAACATGAGTGTCATTT
      ------                           |----> EXON 2
      Bam H-I

 61   CTTCAACGGGACGGAGCGGGTGCGGTTCCTGGACAGATACTTCTATCACCAAGAGGAGTA

121   CGTGCGCTTCGACAGCGACGTGGGGGAGTACCGGGCGGTGACGGAGCTGGGGCGGCCTGA

                                       |-------GH78-------|
181   TGCCGAGTACTGGAACAGCCAGAAGGACATCCTGGAAGACGAGCGGGCCGCGGTGGACAC

        Pst I
        ------
241   CTACTGCAGACACAACTACGGGGTTGTGGAGAGCTTCACAGTGCAGCGGCGA
          <-----------GH50-----------|

Sequence of the PCR primers

GH46: CCGGATCCTTCGTGTCCCCACAGCACG   27-mer

GH50: CTCCCCAACCCCGTAGTTGTGTCTGCA   27-mer

Sequence of the Dw10 sequence specific oligomer

GH78: GACATCCTGGAAGACGAGC   19-mer
```

Figure 1. Sequence of the PCR-amplified DRβ segment.

Immunobiology of HLA
Volume II

rapid procedure has allowed the analysis of isotypic (locus) and allotypic (allele) sequence variation from many different haplotypes from normal individuals as well as from insulin-dependent diabetes mellitus (IDDM) patients. In addition, the extensive sequence variation revealed by this analysis can be detected in a simple and rapid dot-blot format using PCR-amplified DNA and sequence-specific oligonucleotide probes (10).

The oligonucleotide primers used to amplify a 272-bp fragment from the DRβ loci and the restriction sites used for cloning are shown in Figure 1. The sequence of the primers contains a region highly conserved among all DRβ alleles to allow the amplification of this fragment from any DNA or cell sample. As an illustrative example, the sequence of a "subtyping" oligonucleotide probe specific for an epitope found in DR4,Dw10 alleles as well as in some other DRβ-1 alleles is also shown. We have determined the sequence of PCR-amplified DRβ clones from over 100 haplotypes. The nucleotide sequence and the protein translation derived from many

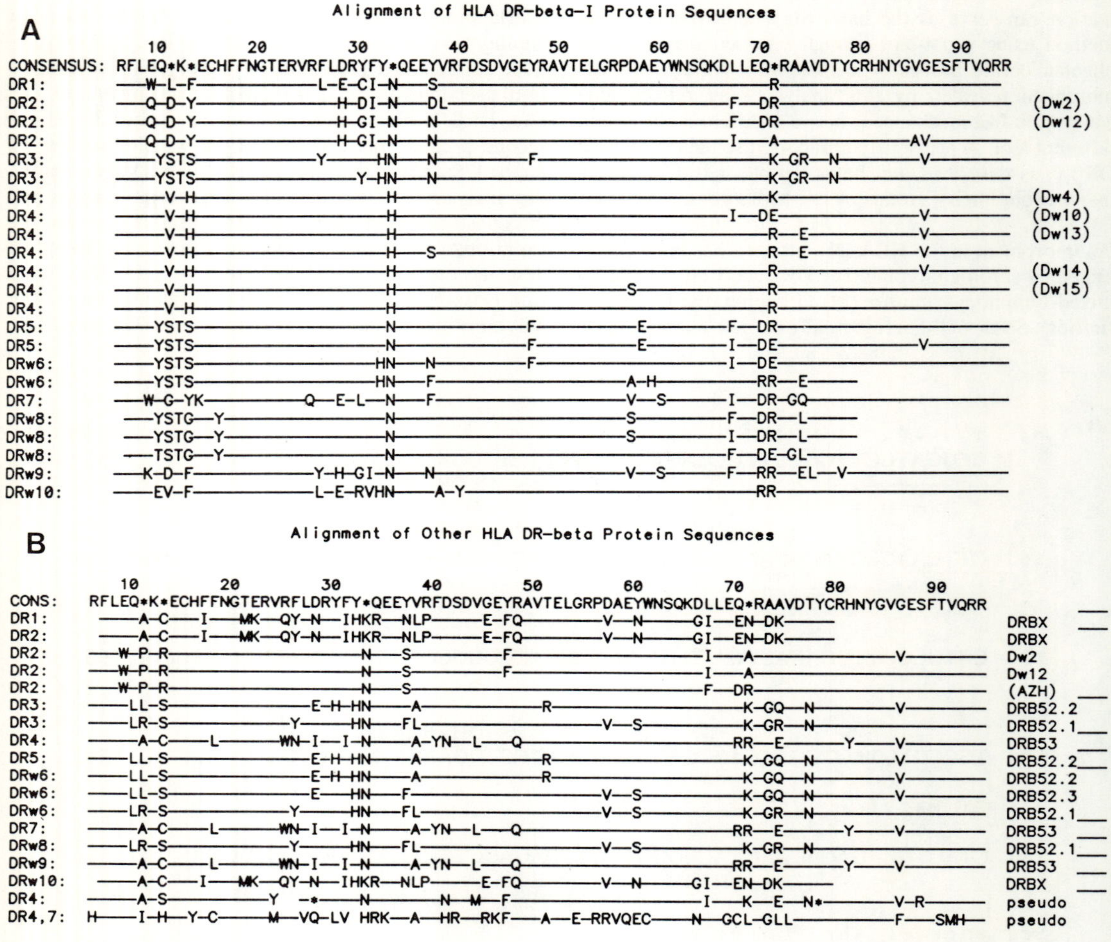

Figure 2. The inferred protein sequence from PCR, cDNA, or genomic clones associated with the DR serotypes DR1-DRw10 is shown in the one-letter code for DRβ-1 (**A**) and for the other DRβ loci (**B**). Sequences have been obtained by the chain termination method of Sanger from M13 clones derived from over 100 individual haplotypes from cell lines and individual blood samples. In **A**, the DR4 sequences are from seven different haplotypes and from Gregerson et al (22). The second DR5 DRβ-I sequence is from a published sequence from the cell line JVM (15). The first DRw6 DRβ-I sequence is associated with the DRw13 subtype and the second with the DRw14 subtype. The DRw10 sequence is from the SHY HTC. In **B**, DR2 sequences are from (14). DRBX designates a sequence unique to DR1, DR2, and DRw10 haplotypes, which may be a pseudogene (see text). The β-III (52.1, 52.2, 52.3) sequences are derived from 15 different haplotypes. For any given allele, the longest available sequence is shown. A more complete listing of sequences and citations will be published elsewhere.

of these clones will be published elsewhere, given the space constraints of this report. In general, all the sequences could be assigned either to the β-1 locus, to the β-III locus (for DRw52 haplotypes), or to the β-IV locus (for DRw53 haplotypes), based on our previous analysis of cDNA and genomic sequences from HTC's and from family members (11) and from published work of other labs.

A summary alignment of the inferred protein sequence from our clones and from published reports is shown in Figure 2. The DR1,2 and DRw10 haplotypes contain a previously unreported sequence, which we have designated DRBX. It is invariant and has not been found as a cDNA clone so it may represent a pseudogene unique to these three haplotypes. The DR1, DR2, and DRw10 haplotypes also lack both the β-III locus and the β-IV locus and type negatively for DRw52 and DRw53 public specificities. This observation suggests a common evolutionary origin for these haplotypes. Thus, analysis of DRβ isotypic variation suggests three major evolutionary groups of DR haplotypes: 1) DRw52 defined by the β-III locus variants on DR3,5, and 6 haplotypes; 2) DRw53 defined by the invariant β-IV on DR4,7, and 9 haplotypes; and 3) a group defined by the DRBX sequence on DR1,2, and DRw10 haplotypes. The second functional β-chain locus for the DR2 haplotype is not closely related to the β-III locus variants (52.1, 52.2, 52.3) or to the invariant β-IV sequence (53) and the DR1 and DRw10 haplotypes do not appear to contain a second β-chain locus. These observations are consistent with the notion that these haplotypes constitute a distinctive evolutionary group.

Given the available sequence information, it is now possible to identify specific allelic sequence variants associated with each DR serotype (DR1-DRw10). These data can serve as the basis for a simple HLA DNA typing system using oligonucleotide probes to "type" PCR-amplified samples. At the DNA level, however, many of the DR haplotypes are heterogeneous, raising the possibility of specific oligonucleotide based "subtyping." An example of a DR-typing oligonucleotide specific for DRw8 is shown in Figure 3A and of a DR4 subtyping oligonucleotide panel is shown in Figure 3B. Although these data employed a ^{32}P-labeled probe, the enzyme-labeled probes reported for analysis of β-globin and HLA-DQα variation (12) can also be used for DR typing. Given the patchwork pattern of polymorphic sequences shared by several different alleles or loci, it is sometimes difficult to assign an oligonucleotide-defined sequence to a specific haplotype in a heterozygous individual. This problem can be addressed by either digesting the PCR-amplified DNA with a restriction endonuclease that cleaves one of the two alleles and analyzing the digest by Southern blot analysis or by using a panel of probes to define each allele (13).

Variation at the DRβ-I locus subdivides the DR2, DR4, DR5, DRw6, and DRw8 haplotypes. (A rare DRβ-I allele from a DR1/DR1 IDDM patient (BG) is not included in this summary and will be discussed elsewhere.) Variation at the β-III locus (52.1, 52.2, and 52.3) subdivides the DR3 and DRw6 haplotypes. The

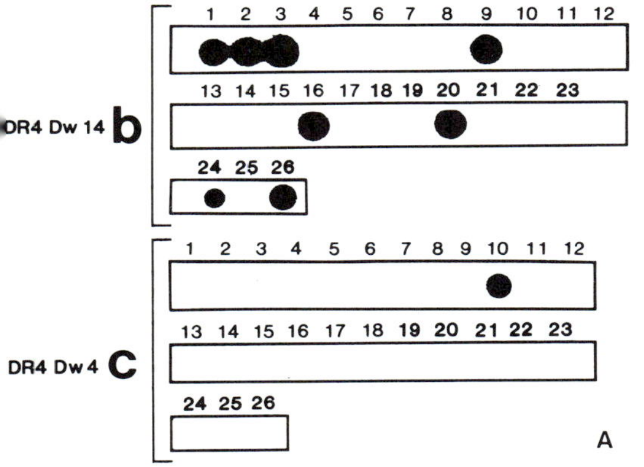

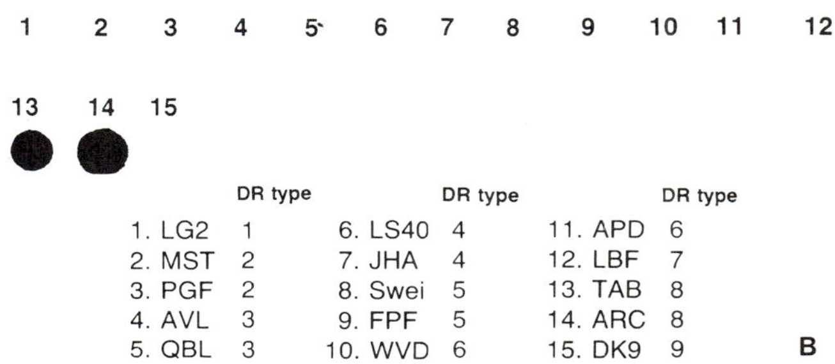

Figure 3. Dot-blot oligonucleotide probe typing. DRβ sequences were amplified from a panel of cells, spotted on nylon membranes, and hybridized to ^{32}P-labeled oligonucleotides as described (10). **A** shows results obtained with the DR4 subtyping probes, GH 101 (GGAGCAGAAGCGGGCCGCG), specific for Dw4 (**panel a**), GH 78 (see Fig. 1 for sequence) specific for Dw10 (**panel b**), and GH 100 (GGAGCAGAGGCGGGCCGCG), specific for Dw14 (**panel c**). The GH 100 probe hybridizes to the DR1 DRβ-I allele as well as to the DR4,Dw14 allele due to the shared sequences in the third hypervariable region (see Fig. 2 and text). **B** shows results obtained with the GH 102 probe (GAATAACACTCACCCGTAG) specific for DRw8 using PCR-amplified DNA from an HTC panel.

52.1 β-III variant of DR3 corresponds to the HLA-B8, DR3 haplotype while most nonB8, DR3 haplotypes contain the 52.2 β-III variant, confirming our earlier suggestion based on DRα and DRβ RFLP, that the DR region of B8,DR3 haplotypes would differ from that of other DR3 haplotypes (13). The DR2 haplotypes are heterogeneous at both the β-I and the second β-chain locus (14). It is likely that as more sequences become available, new β-I and β-III haplotypes will be identified.

One of the striking aspects of DRβ sequence variation is the patchwork pattern of polymorphic residues, consistent with a gene-conversion-like mechanism for generating allotypic and isotypic diversity. Short sequence segments are shared among several different alleles or loci, giving rise to a kind of "natural epitope shuffling." For example, a DRβ-I sequence from residue 68–72 (the "IDE" epitope) is present on DR4,Dw10 alleles, most DRw6 alleles, a rare DR5 allele (Fig. 2A) (15), and a DR1 allele from the IDDM patient BG (Scharf et al, manuscript in preparation). Similarly, a DRβ-I sequence in the same area, termed the third hypervariable region (HV3), is shared by the DR4,Dw14 allele and most DR1 alleles. This epitope correlates with the reactivity pattern of some T-cell clones (16) and with genetic susceptibility to rheumatoid arthritis (RA), a disease associated with DR4 and DR1 (17). Another such DRβ-I epitope in HV3 is shared between the DR4,DW4 allele and the DR3 alleles. This shared epitope is homologous with a hexapeptide encoded by EBV open-reading frame (BALF4).) As pointed out previously, other sequences are shared between DRβ-I and DRβ-III alleles (18).

The availability of DRβ sequence information from normal individuals and from patients with specific autoimmune disease makes possible the analysis of HLA disease associations at the molecular level. The DR4 susceptibility to *Pemphigus vulgaris* is strongly associated with the DRβ-I Dw10 allele, discussed in more detail in Scharf et al (this volume). IDDM is associated with DR3 and DR4 and with DQβ allelic variation (Horn et al, this volume). DRβ allelic variation may also be associated with IDDM. For the DR4 susceptibility, the frequency of the DRβ-I Dw10 or Dw4 subtypes defined by oligonucleotides is increased in DR4+ patients relative to DR4+ controls (69% (18/26) vs. 25% (6/20)). The mobile "IDE" epitope associated with Dw10 was also observed in a DR1 DRβ-I allele from a DR1/DR1 IDDM patient. For DR1/4 patients, most (8/9) DR4 haplotypes are Dw4 but the distribution of DQβ alleles is similar in patients and controls. The analysis of a larger panel of DR4 patients and controls is currently underway. For the DR3 susceptibility, the frequency of the DRβ-III allele 52.1 associated with the B8,DR3 haplotype is higher in DR3+ patients than in DR3+ controls (unpublished). A study of a larger panel of DR3 and DRw6 haplotypes in patients and controls is required to assess the significance of the β-III association. The charge of the polymorphic residue at position 57 of the DQβ chain has been shown to correlate with IDDM susceptibility (20,21). Similarly, the DR3 β-III variation involves a Val-57 (52.1) and Asp-57 (52.2) polymorphism. In general, disease associated haplotypes may contain several class II sequences capable of conferring susceptibility. These putative susceptibility alleles and haplotypes are also present in normal individuals. An environmental "triggering" agent, such as a viral infection, may be required to elicit the disease in genetically predisposed individuals.

Acknowledgments. We are grateful to Jerry Nepom, Eric Mickelson, and Brian Tait for providing HLA-typed cells or DNA samples and to Malcolm Simons and Michael Sheehy for helpful discussions.

References

1. Giles RC, Capra JD. Structure, function, and genetics of human class II molecules. Adv in Immunol 1985;37:1.
2. Erlich HA, Horn GT, Bugawan TL, Scharf S. The molecular analysis of HLA class II gene polymorphism. In: Schacter B et al (eds): Molecular and cellular biology of histocompatibility antigens. New York, 1986, ASHI, p 93.
3. Strominger J. Biology of the human histocompatibility leukocyte antigen (HLA) system and a hypothesis regarding the generation of autoimmune disease. J Clin Invest 1985; 77:1411.
4. Rollini P, Mach B, Gorski J. Linkage map of three HLA-DR β-chain genes: Evidence for a recent duplication event. Proc Natl Acad Sci USA 1985;82:7197.
5. Spies T, Sorrentino R, Boss J, Okada K, Strominger J. Structural organization of the DR subregion of the human major histocompatibility complex. Proc Natl Acad Sci USA 1985;82:5165.
6. Jonsson A-K, Hyldig-Nielsen J-J, Servenius B, Larhammar D, Andersson G, Jorgensen F, et al. Class II genes of the human major histocompatibility complex. J Biol Chem 1987;262:8767.
7. Mullis KB, Faloona F. Specific synthesis of DNA in vitro via a polymerase catalyzed chain reaction. Methods in Enzymology 1987;155:335.
8. Saiki RK, Scharf S, Faloona F, Mullis KB, Horn GT, Erlich HA, et al. Enzymatic amplification of β-globin genomic sequences and restriction site analysis for diagnosis of sickle cell anemia. Science 1985;230:1350.
9. Scharf S, Horn GT, Erlich HE. Direct cloning and sequence analysis of enzymatically amplified genomic sequences. Science 1986;233:1076.
10. Saiki RK, Bugawan TL, Horn GT, Mullis KB, Erlich HA. Analysis of enzymatically amplified β-globin and HLA-DQα DNA with allele-specific oligonucleotide probes. Nature 1986;324:163.
11. Horn GT, Bugawan TL, Long CM, Manos MM, Erlich HA. Sequence analysis of HLA class II genes from insulin-dependent diabetic individuals. Hum Immunol 1988;21:249.
12. Saiki R, Chang C-A, Levenson CH, Boehm CD, Kazazian HH, Erlich HA. Rapid genetic analysis of enzymatically-amplified DNA with non-radioactive allele-specific oligonucleotide probes. Am J Human Genetics 1987;41:A237, Abstract.
13. Stetler D, Grumet FC, Erlich HA. Polymorphic restriction endonuclease sites linked to the HLA-DRα gene: Localization and use as genetic markers of insulin-dependent diabetes. Proc Natl Acad Sci USA 1985;82:8100–8104.

14. Curtsinger J, Hilden JM, Cairns JS, Bach FH. Evolutionary and genetic implications of sequence variation in two nonallelic HLA-DR β-chain cDNA sequences. Proc Natl Acad Sci USA 1987;84:209.
15. Bell JI, Denney Jr D, Foster L, Belt T, Todd JA, McDevitt HO. Allelic variation in the DR subregion of the human major histocompatibility complex. Proc Natl Acad Sci USA 1987;84:6234.
16. Weyand CM, Goronzy J, Fathman CG. Human T-cell clones used to define functional epitopes on HLA class II molecules. Proc Natl Acad Sci USA 1986;83:762.
17. Nepom GT, Hansen J, Nepom B. The molecular basis for HLA class II associations with rheumatoid arthritis. J Clin Immunol 1987;7(1):1.
18. Gorski J, Mach B. Polymorphism of human la antigens: Gene conversion between two DRβ loci results in a new HLA-D/DR specificity. Nature 1986;322:67.
19. Scharf S, et al. HLA Class II allelic variations and susceptibility to *Pemphigus Vulgaris*. Proc Natl Acad Sci USA 1988;85:3504–3508.
20. Horn GT, Bugawan TL, Long CM, Erlich HA. Allelic sequence variation of HLA-DQ genes: Relationship to serology and disease susceptibility. Proc Natl Acad Sci USA 1988;85:6012–6016.
21. Todd JA, Bell JI, McDevitt HO. HLA-DQβ gene contributes to susceptibility and resistance to insulin-dependent diabetes mellitus. Nature 1987;329:599.
22. Gregerson PK, Shen M, Song Q-L, Merryman P, Degar S, Seki T, et al. Molecular diversity of HLA-DR4 haplotypes. Proc Natl Acad Sci USA 1986;83:2642.

Author Affiliation

Department of Human Genetics, Cetus Corporation Emeryville, CA, 94608 USA

Two Different T-Cell Perspectives on the DR1 Alloantigen

David D. Eckels, Thomas W. Sell, Jay B. Hunter, and Susan A. Hackbarth

Exposure to thymic MHC molecules may influence the T-cell allorecognition repertoire. Therefore, since we have observed such a plethora of distinct fine specificities in alloreactive T-cell clones[1,2], we postulated that T cells derived from allogeneic donors might recognize alloantigens with different fine specificities. To test this hypothesis, three series of alloreactive T-cell clones were derived by limiting dilution as previously described (1): series 61 (DR2;DRw13 anti-DR1); series 62 (DR2;DRw14 anti-DR1;DR2s); series 63 (DR2;DRw13 anti-DR1;DR2s). DR1-associated allodeterminants should have been primarily recognized because DQ and DP antigens were shared. However, T cells from series 62 would have differentiated in the presence of DR2 and DRw14, whereas those from series 61 and 63 would have developed in the presence of DR2 and DRw13.

Fifty-three alloreactive clones from the three priming series were analyzed for proliferative responses on the Workshop LCL panel (Fig. 1). In addition to workshop protocols, including testing in three separate assays, questionable specificity assignments were confirmed by titration experiments using LCLs at increasing concentrations (10 to 80 $\times$ 10^3/well) in order to determine that a missed specificity was not due to low cell-surface expression of the allodeterminant. Statistical analyses were performed on triplicate medians using the T_{max} analysis described by Mendell and colleagues[3] after first subtracting LCL medium backgrounds and square root transformation. This approach compares favorably with a 100% DNV transformation for the statistical analysis of secondary lymphocyte proliferation assays. Anti-DR, not anti-DQ monoclonal antibodies, blocked 17 out of 18 TLCs analyzed in standard Workshop protocols. This was confirmed by more extensive testing with purified L243(DR), Genox3.53(DQ) and B7/21(DP) monoclonal antibodies.

Despite our efforts to limit priming against multiple antigens, a surprising number of distinct clonal specificities was obtained, some of which were shared among the priming series and other which were not. Each series had at least one TLC that recognized only the nine DR1$^+$ Workshop LCLs and a subset of DR1$^+$ LCLs that excluded the B14$^+$, DR1$^+$ LCL's 9002, 9078, and 9079. Taken together, these data suggest that there are at least two, and possibly three, T-cell specificity groups correlating with what may be distinct epitopes on the DR1 molecule. Interestingly, additional reactions were observed on cells bearing DR2, DR3, DR4, DR5, DRw6, DR7, DRw8, and DRw9 associated with DR1 either singly (e.g., 1+4 or 1+7) or grouped in various combinations (e.g., 1+5+6+8). Serologic patterns from local Workshop analyses also show a remarkably similar set of associations when the data were clustered without bias. Although the specificities of clones within each priming series are not exactly concordant, there was a remarkable degree of overlap on four particular cell groups with TLCs from each series recognizing DR1$^+$, DR4$^+$, DR5$^+$, and DRw6$^+$ LCLs. Thus, T cells from allogeneic donors appear to recognize potentially similar epitopes. In this regard, we suggest that the DR1 molecule may possess, minimally, four major epitopes that can be independently recognized by groups of T-cell

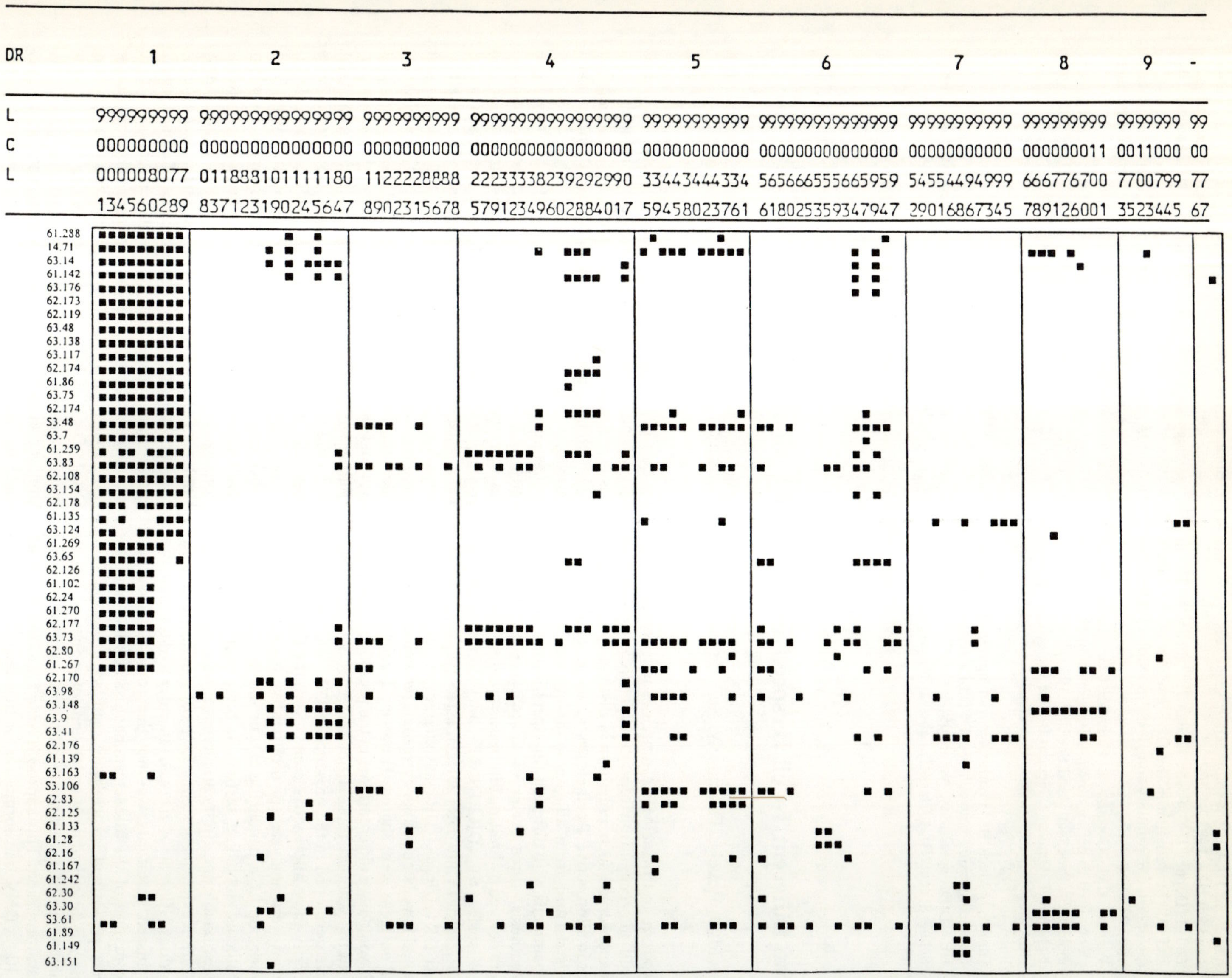

Figure 1. Proliferative responses of 53 alloreactive clones from the three priming series on the Workshop LCL panel.

clones from different genetic origins. Further analyses from our laboratory may even suggest that, somewhat analogous to antigens such as influenza hemagglutinin[4], there may exist a continuum of antigenic determinants on the surfaces of allogeneic molecules that are recognized by alloreactive T-cells (Eckels et al., manuscript submitted).

References

1. Rosen-Bronson S, Johnson AH, Hartzman RJ, Eckels DD. Immunogenetics 1986;23:368.
2. Eckels DD, Hartzman RJ. Immunogenetics 1982;16:117.
3. Mendell NR, Lee KL, Reinsmoen N, Yunis E, Amos B, Emme L. Transplant Proc 1977;9:99.
4. Lamb JR, Eckels DD, Lake P, Woody JN, Green N. Nature 1982;300:66.

Author Affiliation

Immunogenetics Research Section, Blood Center SE Wisconsin, Milwaukee, WI USA

Polymorphism of the DR1 Haplotype: Structural and Functional Analysis

Barbara L. Ziff, Carolyn Katovich Hurley, Sandra Rosen-Bronson, Ting Tang, Robert Hartzman, Jack Silver, Peter Gregersen, David Eckels, and Armead H. Johnson

HLA-DR haplotypes vary in the number of Dw subtypes defined by the mixed lymphocyte reaction (MLR). DR2 and DR4 have been shown to be quite complex (1,2) whereas DR1 is relatively homogeneous in populations of Northern European extraction (3). In Caucasians, almost all DR1 individuals express a DR1,DQw5,Dw1 haplotype. The DR1 haplotype in American blacks is always associated with DQw5, but approximately 50% of DR1 positive American blacks express a Dw allele undefined by mixed lymphocyte typing (Dw blank or Dw-) (4). Undefined Dw specificities associated with DR1 have been reported in Nigerians (5) and in HLA-B14-related haplotypes (6–8). It is not clear whether the DR1,Dw- haplotype found in American blacks, which is not associated with HLA-B14, is the same variant described by others. The structure and function of the American black DR1,DQw5,Dw- haplotype have been studied and compared to the Caucasian DR1 haplotype using alloproliferative T-cell clones and cDNA sequencing.

Six alloproliferative T-cell clones, primed against a Caucasian DR1 haplotype, have been tested with a panel of 204 stimulator cells including 21 DR1 positive cells from both blacks and Caucasians. Only one clone responded positively to all of the DR1 positive stimulators, a second clone responded positively to all of the DR1 stimulator cells except two. The other four clones showed a heterogeneity of response that was most evident when tested with blacks expressing the DR1, DQw5,Dw- haplotype. The results from two of these clones are shown in Table 1. Clone S14.32 responds to almost all of the DR1 stimulators previously typed by homozygous typing cells (HTC) as either Dw1 or Dw-, thus reflecting the serologic typing of these cells. In contrast, clone S14.50 responds positively to ten out of 13 DR1,Dw1 cells but only to two out of eight DR1,Dw- cells. Therefore, the variability observed with the MLR can be paralleled by alloproliferative T-cell clone reactivity. In addition, failure to observe a complete correlation with the Dw typing response suggests that the T-cell clones detect additional variability within the DR1 haplotypes.

The DRBl polypeptide encoded by the DR1,DQw5, Dw- haplotype has been characterized using cDNA sequence analysis. B lymphoblastoid cell line (B-LCL) 1568 (DR1,w18(3);DQw5,w4;Dw-,-) from an American black individual expressing the variant DR1 haplotype was selected for analysis. The amino acid sequence of the first domain (amino acids 1-94) of the DRBl chain encoded by this cell line is almost identical to the DRBl chain sequences from B-LCL 45.1, a DR1,Dw1 HLA deletion mutant cell line (10) and B-LCL LG2, a DR1,Dw1 HTC (11) (Table 2). The sequences differ at amino acid positions 85 and 86. Position 85 is an alanine in the DR1,Dw- sequence while a valine is present at this position in the DR1,Dw1 cell lines. The DR1,Dw- DRBl gene encodes a valine at position 86 while a glycine is encoded in the Dw1 cell lines.

In DR4, as in DR1, DNA sequence analysis of the structural basis of MLR-defined HLA-D specificities indicate that there are different DR polypeptides associated with each HLA-D type. DQw3-associated subsets of DR4 molecules differ in a small number of amino acids predominantly in the third variable region (amino acids 67-74) and at position 86 of the DRBl protein sequence (13). This variability ranges from five amino acid substitutions between DR4,Dw4 and DR4, Dw10 to one amino acid difference at position 74 between DR4,Dw13 and DR4,Dw14. Although haplotypes of DR1 and DR4 differ in extent of variability present, residue 86 is a common site of polymorphism in both.

Table 1. DR1 Associated Alloproliferative T-Cell Clones*

	Stimulators							Responders	
		HLA Type							
ID#	Ethnic Code†	A	B	Dw	DR(w)		DQw	S14.50	S14.32
144	C	1,1	8,8	1,–	1,3	52	1,2	+	+
226	C	1,1	8,8	1,–	1,4	53	1,3	+	+
388	C	11,1	35,8	1,–	1,3	52	1,2	+	+
954	C	30,29	13,60	1,–	1,4	53	1,3	+	+
1403	C	2,24	55,7	1,2	1,2	53	1,1	+	+
1129	B	11,34	55,58	1,–	1,5	52	1,3	+	+
1159	C	1,1	12,8	1,2	1,2		1	+	+
215	C	1	8	1,3	1,3	52	1,3	+	+
1068	B	2,3	35,58	1,–	1,13	52	1	+	+
1011	C	23,28	49,18	1,–	1,4		1,3	+	–
1432	B	11,33	51,53	1,7	1,7	53	1,2	–	–
500	C	2,3	35,35	1,5	1,5	52	1,3	–	+
1202	B	28	45	1,18	1,13	52	1	–	+
802	C	1	8	–,3	1,3	52	1,2	+	+
1233	B	30,–	72,53	–,18	1,13	52	1	–	+
1569	B	2,26	45,39	–,–	1,10		1	+	+
1298	C	33,2	14,55	–,–	1,2		1	–	+
1096	B	1,30	8,63	–,–	1,4	53	1,3	–	+
632	C	3,3	14,41	–,–	1,4	53	1,–	–	+
1568	B	30,34	71,42	–,–	1,3	52	1,–	–	+
1431	B	31,33	57,35	–,–	1,8	52	1,3	–	+
Number of responses to non-DR1+ stimulators:								0	2

*Methodology has been described (5).
†C=Caucasian, B=black.

Table 2. DR1 Beta

	1									10		
DR1,Dw-*								LEU	TRP	GLN	LEU	LYS
DR1,Dw1†	GLY	ASP	THR	ARG	PRO	ARG	PHE	–	–	–	–	–
								20				
DR1,Dw-	PHE	GLU	CYS	HIS	PHE	PHE	ASN	GLY	THR	GLU	ARG	VAL
DR1,Dw1	–	–	–	–	–	–	–	–	–	–	–	–
						30						
DR1,Dw-	ARG	LEU	LEU	GLU	ARG	CYS	ILE	TYR	ASN	GLN	GLU	GLU
DR1,Dw1	–	–	–	–	–	–	–	–	–	–	–	–
				40								
DR1,Dw-	SER	VAL	ARG	PHE	ASP	SER	ASP	VAL	GLY	GLU	TYR	ARG
DR1,Dw1	–	–	–	–	–	–	–	–	–	–	–	–
		50										60
DR1,Dw-	ALA	VAL	THR	GLU	LEU	GLY	ARG	PRO	ASP	ALA	GLU	TYR
DR1,Dw1	–	–	–	–	–	–	–	–	–	–	–	–
										70		
DR1,Dw-	TRP	ASN	SER	GLN	LYS	ASP	LEU	LEU	GLU	GLN	ARG	ARG
DR1,Dw1	–	–	–	–	–	–	–	–	–	–	–	–
								80				
DR1,Dw-	ALA	ALA	VAL	ASP	THR	TYR	CYS	ARG	HIS	ASN	TYR	GLY
DR1,Dw1	–	–	–	–	–	–	–	–	–	–	–	–
						90						
DR1,Dw-	ALA	VAL	GLU	SER	PHE	THR	VAL	GLN	ARG	ARG		
DR1,Dw1	VAL	GLY	–	–	–	–	–	–	–	–		

*Preparation of cDNA library, selection of clones and cDNA sequencing have been described (8).
†(6,7).

The two amino acid replacement substitutions identified in the DRBl sequence of DR1,Dw- represent a relatively minor difference in the two DR1 haplotypes. Amino acids 85 and 86 are predicted to be on the outside face of the DR molecule in a three-dimensional structural model (14) making them accessible to T-cell recognition in an MLR. These two amino acid changes may, therefore, be responsible for the observed MLR stimulation. The DRBl sequence variation observed in the MLR-defined subsets of DR1 and DR4 is an indication of the importance of the DR molecule in HLA-D typing.

The diversity of the HLA class II molecules in man is particularly evident when comparisons between different populations are made. As the vast majority of HLA-D region studies have analyzed the polymorphism in European and Asian (primarily Japanese) populations, we have begun an analysis of the HLA-D region in American blacks to measure the extent of class II diversity. The very modest change in DRBl chain sequence between the two different DR1 haplotypes found in these widely divergent (15–17) population groups may be an indication that the degree of polymorphism in the major histocompatibility complex will be limited.

Acknowledgment. This research is supported by ONR N00014-83-K-0410 (CKH), ACS IM351 (CKH) and NIH AI22005 (JS).

References

1. Reinsmoen N, Layrisse Z, Betuel H, Freidel A, Bach F. A study of HLA-DR2 associated HLA-Dw/LD specificities. Human Immunology 1984;11:105.
2. Reinsmoen N, Bach F. Five HLA-D clusters associated with HLA-DR4. Human Immunology 1982;4:249.
3. Dupont B, Braun D, Yunis E, Carpenter C. Joint report on HLA-D by cellular typing. In: Terasake P (ed): Histocompatibility Testing. Los Angeles, CA, UCLA Tissue Typing Laboratory, 1980, p 229.
4. Johnson AH Joint Report: American Blacks. In: Aizawa M (ed): HLA in Asia-Oceania. Sapporo, Japan, Hokkaido University Press, 1986, p 320.
5. Jaraquemada D, Okoye R, Ollier W, Awad J, Festenstein H. HLA-D and DR in different populations. In: Albert E, et al (eds): Histocompatibility Testing. Berlin, Springer-Verlag, 1984, p 472.
6. Cohen NA, Friedmann F, Szafer A, Amar Cohen D, Brautbar C. Polymorphism of the HLA DR1 haplotype in the Israeli population investigated at the serological, cellular, and genomic levels. Immunogenetics 1986;23:252.
7. Suciu-Foca NM, Godfrey R, Khan K, Woodward C, Rohowsky E, Reed M, et al. New HLA-D alleles associated with DR1 and DR2. Tissue Antigens 1981;17:294.
8. Davis, J, Cook R, Brown M, Pollack M, Van M, Rich R. Unusual functional and biochemical properties of DR molecules from the HLA-DR1,B14 haplotype. Trans Proc 1987;XIX:8.
9. Rosen-Bronson S, Johnson A, Hartzman R, Eckels D. Human allospecific TLCs generated against HLA antigens associated with DR1 through DR8. I. Growth and specificity analysis. Immunogenetics 1986;23:368.
10. Tonnelle C, DeMars R, Long E. DO beta: a new beta chain gene in HLA-D with a distinct regulation of expression. Embo J 1985;4:2839.
11. Todd J, Bell J, McDevitt H. HLA-DQ beta gene contributes to susceptibility and resistance to insulin-dependent diabetes mellitus. Nature 1987;329:599.
12. Hurley CK, Gregersen P, Steiner N, Bell J, Hartzman R, Nepom G, et al. Polymorphism of the HLA-D region in American Blacks: A DR3 haplotype generated by recombination. J Immunol 1988;140:885.
13. Gregersen P, Shen M, Song Q, Merryman P, Degar S, Seki T, et al. Molecular diversity of HLA-DR4 haplotypes. Proc Natl Acad Sci USA 1986;83:2642.
14. Norcross M, Kanehisa M. The predicted structure of the Ia $beta_1$ domain. Scand J Immunol 1985;21:511.
15. Reed TE. Caucasian genes in American Negroes. Science 1969;165:762.
16. Wainscoat J, Hill A, Boyce A, Flint J, Hernadez M, Thein S, et al. Evolutionary relationships of human populations from an analysis of nuclear DNA polymorphisms. Nature 1986;319:491.
17. Cann R, Stoneking M, Wilson A. Mitochondrial DNA and human evolution. Nature 1987;325:31.

Author Affiliations

Barbara L. Ziff, Carolyn Katovich Hurley, Sandra Rosen-Bronson, Ting Tang, Armead H. Johnson, Departments of Microbiology and Pediatrics, Georgetown University, Washington, DC; *Robert Hartzman*, Naval Medical Research Institute, Bethesda, MD; *Jack Silver, Peter Gregersen*, Hospital for Joint Diseases, New York, NY; *David Eckels*, The Blood Center of Southeastern Wisconsin, Milwaukee, WI, USA

The Heterogeneity of HLA-DR2

Machteld Oudshoorn, Ernette D. Du Toit, Robert W. Martell, and Derek G. Taljaard

In 1978, Richiardi and co-workers first reported the division of HLA-DR2 into two subtypic specificities, named T060 and T061 (1). Several authors have since reported splits of DR2 (2). During the Ninth International Histocompatibility Workshop we subdivided DR2 into three variants (3). The classic antigen (DRw15), a DR2 short variant (DRw16), and a new Black specificity (DR2 LUMCT) were detected in our laboratory.

Using the Tenth Workshop Core Set sera we studied 55 unrelated individuals with HLA-DR2 from three distinct South African population groups: Blacks, Cape Coloureds, and Caucasoids. As shown in Table 1, we were once again able to subdivide DR2 into the three abovementioned subtypes. The conventional DR2 specificity DRw15 was clearly defined by positive reactions with the DR2 sera 1030, 1025, 1027, 1031, 1032, 1029, 1024, 1026, and 1023. HLA-DRw16 was defined by positive reactions with the aforementioned DR2 sera, with the exception of 1031 and 1032. This split is the same as the DR2.2 variant reported by Mervart et al. during this Workshop (4). DR2 LUM(CT) could be identified by positive reactions with sera 1030, 1025, 1027, 1031, and 1032 and negative reactions with sera 1029, 1024, 1026, and 1023 and could thus be easily distinguished from the DR2.3 variant reported by Mervart et al (4). All three DR2 subgroups segregated clearly in families. HLA-DRw15 was the only split seen in the Caucasoids tested, and was also the most frequent in the other populations studied. HLA-DRw16 was seen in only two of 21 Cape Coloureds with HLA-DR2, whereas DR2 LUM(CT), virtually confined to Blacks, was seen in four unrelated individuals.

We further investigated the DR2 LUM(CT) specificity using the Tenth Workshop Southern blot protocol. Genomic DNA was obtained from one family (DUT 138) and one individual (DUT 0015) with the DR2 LUM(CT) specificity, as well as from two local DRw15(Dw2) homozygous typing cells (JDU and ZSI) and digested with the restriction endonucleases *Taq* I, *BamI* and *Eco* RI. Using the Workshop DR-β probe, different RFLPs were observed for DRw/15 long and DR2 LUM(CT) with each of these enzymes (Table 2). The RFLPs detected with the DQ-β probe were identical in these DR2 variants (data not shown).

In conclusion, we have shown that DR2 can be subdivided into three serological subgroups: DRw15, DRw16, and DR2 LUM(CT). Furthermore, DRw15 long and DR2 LUM(CT) could be distinguished by RFLPs using *Taq* I, *Bam* HI and *Eco* RI with the Tenth Workshop DR-β probe.

Table 1. Reaction Patterns with DR2 Core Set Sera

	10 WS DR2 Sera								
DR2 Specificity	1030	1025	1027	1031	1032	1029	1024	1026	1023
DRw15	+	+	+	+	+	+	+	+	+
DRw16	+	+	+	–	–	+	+	+	+
DR2 LUM(CT)	+	+	+	+	+	–	–	–	–

Table 2. DR-β RFLPs in Individuals with HLA-DRw15 and DR2 LUM(CT)

DR	w15/w15	w15/w15	2LUM/1 –(CT)	2LUM (CT)		
DQ	w6/w6	w6/w6	w6/w5	w6		
Dw	2/2	2/2	–/1	–		
Cell ID	JDU	ZSI	0015	a*	Enzyme	kb
DR-β	+	+	+	+	Taq I	2.1
Probe	+	+	+	+	Taq I	1.7
	+	+	+	+	Eco RI	6.4
	+	+	+	+	Eco RI	5.3
	+	+	+	+	Eco RI	3.8
	+	+	–	–	Taq I	11.5
	+	+	–	–	Taq I	1.45
	+	+	–	–	Eco RI	4.9
	+	+	–	–	Bam HI	4.85
	–	–	+	+	Taq I	4.6
	–	–	+	+	Bam HI	4.9

*Haplotype a of family 138. + = fragment present; – = fragment absent.

References

1. Richiardi P, Belvedere M, Borelli I, De Marchi M, Curtoni ES. Split of HLA-DRw2 into subtypic specificities closely correlated to two HLA-D products. Immunogenet 1978;7: 57.
2. Batchelor JR, Hors J, Dodi IA, Betuel H, Borelli I. Antigen report: HLA-DR2. In: Albert ED, et al (eds): Histocompatibility Testing 1984. Berlin, Springer-Verlag, 1984, 186.
3. Oudshoorn M, Campbell EM, du Toit ED. Splits of HLA-DR2. In Ninth International Histocompatibility Workshop and Conference Newsletter V:29.
4. Mervart H, Mody M, Wong KH. HLA DR2 subtypes. In Tenth International Histocompatibility Workshop Newsletter 2:6.

Author Affiliation

Provincial Laboratory for Tissue Immunology, Private Bag 4, Observatory 7935, Cape Town, South Africa

Immunobiology of HLA
Volume II

Correlation of Primary Sequence and Functional Determinants of HLA Class II Molecules

N.L. Reinsmoen and F.H. Bach

The structure-function relationship of HLA class II molecules remains one of the main areas of interest in HLA immunogenetics. While X-ray crystallographic studies are one ultimate goal, association of primary sequences of polymorphic genes with functional T lymphocyte-defined (LD) determinants represents a first approximation to this goal; our interpretations must be considered in that light. Data accumulating in two areas are of value in this regard. First, we and others (1–6) have provided DNA sequence information comparing polymorphic genes associated with Dw/LD subtype polymorphism of certain DR specificities. Second, we have used cloned T-cell reagents to define polymorphic epitopes associated with DR and DQ (7,8). We report here that sharing of amino acid (aa) sequences in the third, as well as possibly the second, hypervariable regions (HVR-aa 67–74 and aa 26–31) in several cases appears to correlate with sharing of LD epitopes.

T-cell clones were generated from cells primed against a Dw2 HTC or the Dw21 (FJO), which has been previously identified (8,9) to define the lymphocyte-defined (LD) subtype variously referred to as MN2, FJO, AZH, WT24, and is distinct from Dw2 and Dw12. The LD specificity Dw21 is associated with the "DR2 short" serologic specificity DRw16. A related specificity initially defined in Warao Indians (10), Dw22, has been shown by DNA sequencing to express one DR β chain (DR β1) identical to Dw21, and the second differing by only one amino acid (11); however, the DQ specificity of Dw22 is DQw3, in contrast to DQw1 for Dw21.

Those clones, determined to be DR- or DQ-directed based on monoclonal antibody inhibition assays, were studied by panel cell analysis utilizing local and workshop DR2-negative and DR2-positive lymphoblastoid cell line (LCL) targets. Previous studies have demonstrated the ability of these cloned reagents to define various determinants, such as those shared by most DR2 subtypes, those shared by Dw21 and Dw22 or by Dw2 and Dw12, and those unique to each subgroup (8). Representative clones in Table 1 illustrate reactivity patterns: clone 2B-8 for Dw2 and Dw12 and clone M-69 for Dw21 and Dw22. Approximately 50% (16/30) of the DR2-directed clones also detected determinants on DR2-negative cells; three common patterns of reactivity were observed as illustrated in Table 1 by representative clones 3F-30, 4F-132, and 3F-6. These DR2-directed clones appear to detect determinants predominantly shared by DR5+, DR8+, and DR5+ cells or only by DR8+ cells. Monoclonal antibody inhibition assays of target cells representative of DR5 and DR8 indicate that the anti-DR2 clones also recognize DR-associated determinants on the DR5 and DR8 cells (data not shown). The available DNA and protein sequencing results have revealed primary sequences shared by DR2, DRw8, and DR5-positive cells as illustrated in Table 2. The shared sequence polymorphism in the third hypervariable region (amino acids 67 to 74) of the first domain can explain many (most) of the reactions observed with these DR-directed T-cell clones.

For a few clones, reactivity related to sequence polymorphism of the second hypervariable region cannot be ruled out, as illustrated by clone 4F-132. Clone 4F-132 lyses DR2 cells of all four subtypes, as well as most DR5 and DRw8-positive cells. This clone may be detecting the shared sequence at positions 68 to 73 of all cells, regardless of the β chain on which the sequence is found. Alternatively, it may be detecting the shared sequence at positions 26 to 31 found on these cells. These findings are consistent with the recognition of "linear" determinants.

The patterns of reactivity for some clones can only be explained if sequence polymorphisms of the second and third hypervariable regions are considered together, as illustrated by the three clones 2B-8, M-69, and 3F-30. Clone 2B-8, which lyses Dw2 and Dw12 cells, but not cells of other specificities, may detect the β3 molecule of those specificities. Although the Dw21 and Dw22 cells share the same amino acids at positions 67 to 74 on their β1 chain, the sequences at 26 to 31 are different from Dw2/12 cells. (The converse argument can also be made.) Recognition by clone M-69 also appears to require identity at amino acid positions 26 to 31 and 67 to 74, perhaps detecting the β1 molecule of Dw21 and Dw22, since only cells of those types share both groups of sequences on the same chain. Clone 3F-30, which lyses Dw21, Dw22, and most DR5 cells, may be detecting the 68–74 sequence on the β3 molecules; position 74 appears important since DRw8 cells differ at that position and are not detected. Position 67 does not appear critical in this comparison since Dw22 cells differ but are detected. Again, all three cell types share the same sequences at amino acid positions 26 to 31. These three patterns of clonal reactivity may thus require identify at the second and third hypervariable regions. Amino acids 67 and 74 may be of importance in explaining some patterns of clonal reactivity and not others.

Some patterns of recognition cannot be explained by primary sequences and suggest the detection of con-

Table 1. Investigation of Determinants Shared by DR2-Positive and DR2-Negative Haplotypes

				T-Cell Clones				
WKSP #	Local ID	DR	Dw/LD	2B-8	M-69	3F-30	4F-132	3F-6
	FJO	(w16)	21	–	++	++	++	+
9009	KAS 011	(w16)	21	–	++	++	+	–
9012	WJR	(w16)	21	–	+	+	nt	–
9015	WT24	(w16)	21	–	++	+	+	–
9016	RML	(w16)	22	–	++	++	++	–
9008	DO	2	2	+	–	–	nt	–
9010	AMAI	2	2	++	–	–	nt	–
9013	SCHU	2	2	++	–	–	nt	–
9017	WT8	2	2	+	–	–	+	–
9014	MGAR	2	nt	+	–	–	nt	–
9011	E418	2	12	+	–	–	++	–
9035	JBUSH	5	5	–	–	++	++	–
9037	SWE1	5	5	–	–	–	++	–
9043	BM21	5	5	–	–	++	+	–
9044	BRIP	5	5	–	–	++	–	–
9045	Tubo	11	5	–		++	++	–
9041	JO	5	?	–	–	+	–	–
9039	JUM	5	5v	–	–	+	–	–
9042	TISI	11	?	–	–	–	++	–
9036	SPO	11	DB2	–	–	++	–	–
9040	BM15	5	5	–	–	–	–	–
9067	BT8	8	8.1	–	–	–	++	++
9068	BM9	8	8.1	–	–	–	++	++
9069	MAD	8	8.1	–	–	–	++	++
9066	TAB	8	8.2	–	–	–	–	+
9071	OLGA	8	8.2	–	–	–	+	nt

T-cell clones were generated utilizing T-cell cloning procedures previously described (7). The cytolytic clones were categorized as being DR-, DQ-, DP-, or class I-directed based on monoclonal antibody inhibition assays. Those clones determined to be DR- or DQ-directed were analyzed further by testing with local target cells and workshop lymphoblastoid cell lines well characterized for all HLA specificities. The results are symbolized in ranges of percent cytotoxicity: $<10\% = -$; $10–25\% = \pm$; $>25\%$ ++. The CML technique counting auger rays by means of a β scintillation counter has been previously described (7).

Table 2. Shared DNA Sequences—DR2, DR5, and DRw8 Haplotypes

		Amino Acid Position											
DR/Dw	DR β Chain	26	28	30	31	//	37	38	//	67	70	71	74
DR5	β1	F	D	Y	F		Y	V		F	D	R	A
DRw8	β1	–	–	–	–		–	–		–	–	–	L
Dw21	β3	–	–	–	–		S	–		–	–	–	–
Dw22	β3	–	–	–	–		S	–		L	–	–	–
Dw2	β3	–	–	–	–		S	–		I	Q	A	–
Dw12	β3	–	–	–	–		S	–		I	Q	A	–
Dw2	β1	–	H	D	I		D	L		–	–	–	–
Dw12	β1	–	H	G	I		N	–		–	–	–	–
Dw21	β1	–	H	G	I		N	–		I	Q	A	–
Dw22	β1	–	H	G	I		S	–		I	Q	A	–
DR1	β1	L	E	C	I		S	–		L	Q	–	–

The informative amino acids that differ in the second and third hypervariable regions as well as position 37 and 38 are listed (3,12,13). The β3 and β1 designations are used to designate the constantly migrating β chain (β3) and variably migrating (β1) chain as determined by 2D-PAGE analysis of DR2 haplotypes (14). β3 also designates the chain with less amino acid variation, i.e., DRw52 and 53 in relevant haplotypes.

Table 3. Correlation of T-Cell Responses and DQ β DNA Sequences

Specificity			T-Cell Clone Reaction Pattern				Amino Acid Positions						
DQw	DR	Dw	A	B	C	D	66	67	70	71	74	75	77
1	2	2	+	−	+	+	E	V	G	T	E	L	T
1	2	MN2	−	+	+	−	−	−	−	A	S	V	R
1	1	Dw1	−	+	+	−	−	−	−	A	S	V	R
1	2	Dw12	−	−	−	−	D	I	R	−	−	−	−
1	6	Dw18	+	−	+	−	−	−	−	−	−	−	−
1	6	Dw19	−	−	+	−	−	−	R	−	−	−	−
1	6	Dw9	−	+	+	−	−	−	−	A	S	V	R
DQw7			(−)	(−)	(+)	(−)	−	−	R	−	−	−	−
DQw8			(−)	(−)	(−)	(−)	−	−	R	−	−	−	−
DQw4		Dw15	−	−	−	−	D	I	E	D	S	V	−
DQw4		Dw8.1	−	−	−	−	D	I	E	D	S	V	−
2		Dw3	−	−	−	−	D	I	R	K	A	V	R

The informative amino acids which differ in the third hypervariable region are listed (Saunders et al (15), manuscript in preparation). The T-cell clone recognition patterns were compiled from several clones demonstrating the same patterns of reactivity (+ = >10% lysis, − = <10% lysis).

formational determinants. Clone 3F-6 lyses the sensitizing cell (FJO), but not other Dw21-positive cells, and lyses all DRw8 cells tested. The Dw21 and DRw8 cells do not share a unique polymorphic sequence. Among several possible explanations is the detection of a determinant unique to FJO (i.e., presentation of a minor histocompatibility antigen or microheterogeneity of the β chain), which is functionally shared by DRw8 positive cells. This type of reactivity cannot be explained by primary sequence and may constitute a conformational epitope.

Similar correlation was observed for DQ-directed T-cell reactivity and shared amino acid sequences of the third hypervariable region of the DQ β chain, as illustrated schematically in Table 3. Some patterns of panel cell reactivity (A and B) can be explained by sequence polymorphism involving amino acids 66 to 77. Clones of pattern A detect only Dw2+ and Dw18+, cells which have identical sequences in this area. Other clones lyse Dw21, Dw1, and Dw9 positive cells (pattern B), which have identical sequences at positions 66 to 77. The reactivity of other clones (pattern C) is associated with polymorphism of amino acids 66 and 67. The reactivity pattern C illustrated detection of cells representing several specificities. Based on limited serologic information, the reactivity did appear to correlate with the DQw7 split of DQw3. The DQw7 negative cells, which apparently are not detected by these clones, also share these amino acids but differ by four other amino acids elsewhere in the chain, which may influence T-cell recognition. For some patterns of reactivity, there is no apparent correlation with primary sequences. The Dw2 specific pattern (D) does not appear to be explained by amino acid polymorphism in this region or elsewhere on either chain. Conformational determinants related to the β chain or to the αβ dimer might be responsible for this reactivity.

In summary, our overall results indicate that sequence polymorphism of the third hypervariable region of DR β and of DQ β can explain patterns of reactivity for most of the DR-directed and DQ-directed clones. The determinants detected may represent i) linear determinants of the third or second hypervariable region; ii) linear, or more likely, conformational determinants involving both the second and third hypervariable regions; and iii) conformational determinants apparently not directly related to primary sequences. We offer these interpretations as only first approximations; other factors need further evaluation. Future investigations will focus on the possible immunodominance of these various determinants and the extent to which determinants are shared among haplotypes.

Acknowledgments. This is paper #480 from the Immunobiology Research Center, University of Minnesota, Minneapolis, MN 55455. This work was supported in part by NIH grants AI 17687, AI 18326, AI 19007, and AI 22682.

References

1. Cairns S, Curtsinger JM, Dahl CA, Freeman S, Alter BJ, Bach FH. Sequence polymorphism of HLA-DR β1 alleles relating to T cell-recognized determinants. Nature 1985; 317:166–168.
2. Curtsinger JM, Hilden JM, Cairns JS, Bach FH. Evolutionary and genetic implications of sequence variation in two non-allelic HLA-DR β cDNA sequences. Proc Natl Acad Sci USA 1987;84:209–213.
3. Wu S, Saunders T, Bach FH. Nature 1986;324:676–679.
4. Wu S, Yabe T, Madden M, Saunders TL, Bach FH. J Immunol 1987;138:2593–2599.
5. Gregersen PK, et al. Proc Natl Acad Sci USA 1986;83: 2642–2646.
6. Lee BSM, Rust NA, McMichael AJ, McDevitt HO. Proc Natl Acad Sci USA 1987;84:4591–4595.
7. Reinsmoen NL, Bach FH. Hum Immunol 1986;16:329–346.
8. Reinsmoen NL, Bach FH. Hum Immunol 1987;20:13–32.
9. Reinsmoen NL, Layrisse Z, Betuel H, Bach FH. Hum Immunol 1984;11:105–115.

10. Layrisse A, Simoney N, Park MS, Terasaki PI. Transplant Proc 1979;11:1178.
11. Liu C-P, Wu S, Bach FH. In: Dupont B (ed): Immunobiology of HLA: Immunogenetics and Histocompatibility. New York, Springer-Verlag, 1988.
12. Treber VL, Abruzzini LF, Didier DR, Schwartz BD, Rotwein P. J Biol Chem 1986;261:2738–2742.
13. Bell JI, et al. Proc Natl Acad Sci USA 1987;84:6234–6238.
14. Yabe T, Reinsmoen NL, Wu S, Bach FH. Hum Immunol 1987;19:127–137.
15. Todd JA, Bell JI, McDevitt HO. Nature 1987;329:599–604.

Author Affiliation

Cellular Histocompatibility Laboratory, University of Minnesota Hospital, Minneapolis, MN 55455, USA

Molecular Studies of a Rare HLA Haplotype: Implications for Mechanisms of Generating Class II Gene Polymorphisms*

Chih-Pin Liu†, Shaokee Wu, and Fritz H. Bach

For both HLA-DR and -DQ antigens there are serologically detectable (e.g., DR1, DR2, etc. and DQw1, DQw3, etc.) as well as Dw/LD T lymphocyte-defined (e.g., Dw2, LD-MN2 and LD-5a, etc.) polymorphisms (1). Haplotypes encoding DR2 can be subdivided into Dw2, Dw12, and DR2.3 (including FJO and MN2) subtypes based on distinct T-cell responses (2). DR components of a putative additional LD subtype of DR2, LD-5a, have been related to the DR2.3 subtype, although they are associated with DQw3 rather than with DQw1. Cytotoxic T-cell clones can differentiate FJO (DR2/MN2/DQw1) from REM (DR2/LD-5a/DQw3) (3) for DR as well as DQ molecules (4). In an attempt to study the mechanisms for generation of polymorphisms related to Dw/LD in class II antigens, we isolated and sequenced cDNA clones from REM representing the polymorphic DRB, DQB, and DQA genes and compared these with sequences of other DR and DQ genes. The results demonstrated that the DR2 β2 chains (DRβ2 in this paper shall be called DRB3) must be involved in allo-specific T-cell recognition, and that LD-5a (Dw22) should be considered independently of the DR2.3 (Dw21) subtype, even for DR. Our data suggest that the DQw3 serological specificity may be mainly defined by the DQB chain rather than the A chain. A mechanism is suggested for the generation of the DR2/Dw22/DQw3 haplotype.

The sequences of DRB1 and DRB3 chains of REM (DR2/Dw22/DQw3) are identical to those of FJO (DR2/Dw21/DQw1) (5,6) except for a single amino acid

* This is paper #481 from the Immunobiology Research Center, supported in part by NIH grants AI 17687, 18326, 19007, and 22682. Results included herein overlap with those in a paper (Liu et al.) published in the J Immunol Vol. 140, pp. 3631, 1988.

†The current address of CPL is: University of Wisconsin, Madison, WI 53706.

Table 1. Differences in Amino Acid Sequences of DRB Chains from the Four DR2 Dw Subtypes

Chain and	First Domain										Second Domain				CP
Subtype	6	30	37	38	47	67	70	71	85	86	110	135	150	157	203
DRB1															
Dw22	C	G	N	V	Y	I	Q	A	A	V	Q	G	N	I	I
Dw21	–	–	–	–	–	–	–	–	–	–	–	–	–	–	–
Dwl2	NA	NA	NA	NA	–	F	D	R	V	G	–	S	–	T	V
Dw2	R	D	D	L	–	F	D	R	V	G	E	S	NA	NA	NA
DRB3															
Dw22	R	Y	S	V	Y	L	D	R	V	G	Q	G	M	T	V
Dw21	–	–	–	–	–	F	–	–	–	–	–	–	–	–	–
Dwl2	–	–	–	–	F	I	Q	A	–	–	–	–	–	–	–
Dw2	–	–	–	–	F	I	Q	A	–	V	–	–	D	–	–

The table lists the amino acid sequence differences of DRBl and DRB3 chains from the four different DR2 Dw/LD subtypes. Positions not shown in the table are identical in all four DRBl or DRB3 chains (see Refs. 5, 6, and 17 for complete amino acid residues). The two Dw2 sequences are from protein sequencing of two DRβ chains of a DR2/Dw2 homozygous cell line (16). The other sequences are deduced from cDNA clones isolated and sequenced in our laboratory (5,6). The Dw22 sequences are from homozygous typing cell (HTC) line, REM (DR2/Dw22/DQw3) (3). For detailed molecular cloning methods and sequencing strategy, refer to Refs. 5 and 6. Residues not available for comparison are marked as NA. CP refers to the connecting peptide. The results included in this and other tables and in the figure are taken from Liu et al., J Immunol Vol. 140, pp. 3631, 1988.

Immunobiology of HLA
Volume II

Table 2. Summary of Amino Acid Homology

	DR2, Dw22, DQw3 (REM)			
	DQ		DR	
	B	A	B1	B3
	a.a.1-94	a.a.13-87*	a.a.1-94	a.a.1-94
DR4/Dw4 DQw3	100%	87.3%	86.2%	79.8%
DR5 DQw3	97.9%	100%	81.9%	84.0%
DR3(Raji) DQw2	85.1%	100%	86.2%	NA
DR2/Dw21 DQw1	80.9%	–[†]	100%	98.9%
DR2/Dw2 DQw1	88.3%	–[†]	90.4%	94.7%

The first domain of DR- and DQB chains and DQA chains from REM and from other cell lines are compared. The percent homology of amino acid sequences are listed. For DQB: DR4 from Ref. 10, DR3 from Ref. 17, DR2 from Ref. 14 and manuscript in preparation. For DQA: DR4 from Ref. 13, DR5 from Ref. 12, DR3 from Ref. 11. For DRB: see Table 1 legend; DR3 from the pII-β-3 clones of Ref. 18, DR5 from Ref. 19. NA = not available.

*cDNA sequence of REM DQA from the first 12 amino acids are not available for comparison.

[†]According to our unpublished data, percent homology of amino acids of A chains between DR2/DQw1 and DR2/Dw22/DQw3 is less than 80% in the first domain.

change at position 67 on DRB3 chains with leucine in REM and phenylalanine in FJO (Tables 1 and 2). The current data are consistent with our previous report (2) showing apparent identity of migration of B1 and B3 proteins of REM and FJO by using two-dimensional polyacrylamide electrophoresis (2D-PAGE) analysis, in that both amino acids at position 67 of DRB3 chains are uncharged ones and do not affect electrophoretic mobility. The several DR-directed cytotoxic T-lymphocyte clones that differentially recognize target cells expressing Dw21 and those expressing Dw22 (4) are presumably related to this difference at position 67 of DRB3 chains, given the fact that the DRB1 sequences of REM and the Dw21 group are identical and that the third DRB gene in DR2 is a pseudogene (7). DR2-positive cells such as REM (DQw3) (3) expressing Dw22 should be separated from the Dw21 (FJO, MN2) group of cells, both with regard to the difference for DQ (DQw1 versus DQw3) and to the difference on the DRB3 chain that is involved in allo-specific T-cell recognition.

In DR4-positive cells, the DRB3 molecules of Dw4, Dw10, and Dw15 are identical (8). Therefore, the Dw subtypes of DR4 are mainly defined by their polymorphic DRB1 molecules. We have previously reported that in DR4-positive cells, single or few amino acid differences in the third hypervariable region (HVR) and at position 86 (in the fourth HVR) within the first domain of DRB1 chains can elicit T-cell responses (9). The present data, together with our previous reports (5,6), strongly support the concept that DRB3 chains of DR2-positive cells (in the third HVR and position 86 also) are probably involved in differential T-cell responses.

One possible schema that would explain the generation of the Dw/LD subtype polymorphism of the DR2 haplotypes would be the following. Let us assume that the original haplotype was Dw2 or Dw12. The sequence homology of the DRB1 and DRB3 chains in the known polymorphic sequences of Dw2 and Dw12 is very high (Table 1). The presence of one amino acid change in DRB3 chains at position 86 distinguishes Dw2 from Dw12 and probably accounts for the differential response of T lymphocytes to them. We have proposed that DRB genes of Dw21 may then be formed by a reciprocal genetic exchange, including amino acids 67–72, in the third HVR between DRB1 and DRB3 genes of either Dw2 or Dw12 (5); Dw21 is thus separated from Dw2 and Dw12. Dw22, as discussed above, diverged from Dw21 due to an amino acid change in the third HVR on the DRB3 chain. Therefore, it seems most likely that the DRB3 chains of DR2-positive cells are involved in allo-specific T-cell recognition, and single or few amino acid differences within the third hypervariable region and at position 86 are sufficient to induce differential T-cell responses (e.g., between Dw22 and Dw21, and between Dw2 and Dw12; see Reinsmoen and Bach, this volume). Differences in the second domain, such as the two between Dw2 and Dw12, may play a much less important, or no, role in allo-specific T-cell recognition.

The DQB sequence of REM is identical to the sequences of DQB in the DR4/DQw3 haplotypes (Table 2) (10). The cDNA clone encoding DQA is a truncated one and lacks the first 12 amino acids at the NH_2 terminus (Table 2). However, the available coding sequence is identical to DQA sequences of both DR3/DQw2 (11) and DR5/DQw3 (12) haplotypes except for a single silent mutation at the third nucleotide of amino acid position 171 (data not shown). Furthermore, this DQA sequence is different from the DQA sequences of the

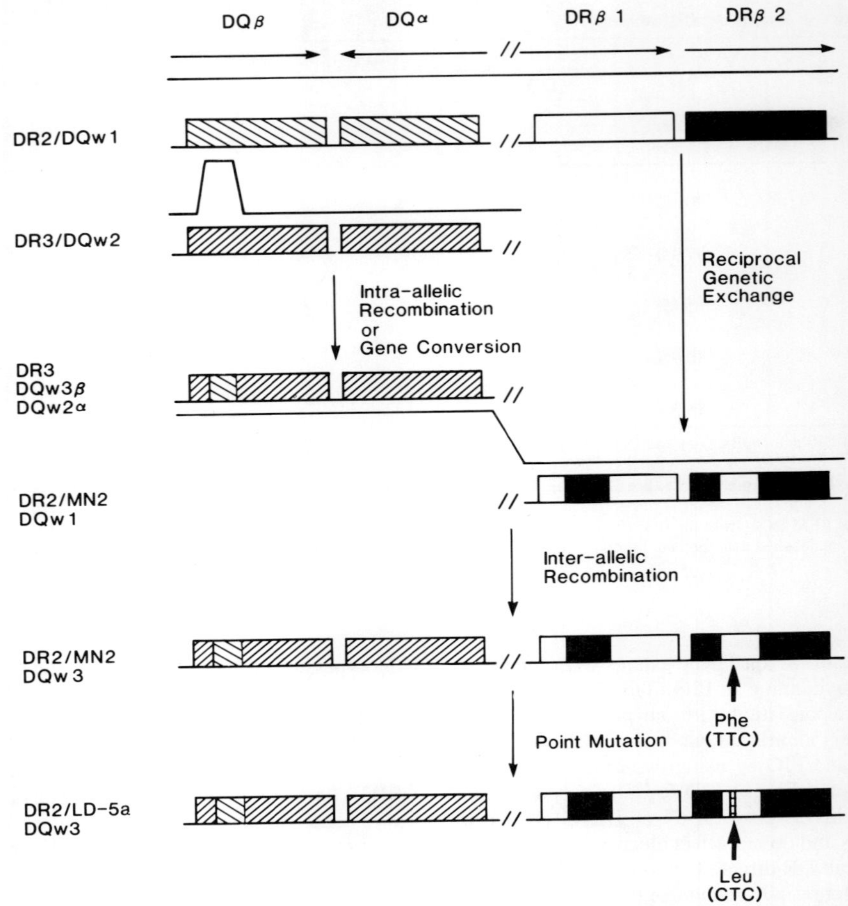

Figure 1. Proposed multiple genetic events for the generation of DR2/LD-5a/DQw3 haplotype and related haplotypes.

DR4/DQw3 haplotypes (13). Therefore, the combination in REM of a "DQw3B gene" with a "DQw2A gene" resulting in a DQw3 serological specificity suggests that DQw3 in this instance is mainly defined by the DQB gene product. This is consistent with a recent study of a DR5/DQw3 haplotype which has a similar DQ combination (12). However, the precise relationship of serologically defined specificities and the A-B dimers certainly requires further investigations.

The association of a DQw3B and a DQw2A gene, such as the one found in REM, agrees well with a genetic model that we have proposed (i.e., that intra-allelic recombinations or a gene conversion event occurred between DQw1B and DQw2B genes that resulted in the formation of the DQw3B gene (14). In this model, part of the first domain sequence of the DQw3B gene may be derived from the DQB gene of a DQw1 haplotype, and the rest of the sequence from a DQw2B gene (see Figure 1). Given the order of class II genes on chromosome 6 (15), and the proposed origin of the DQw3B gene, one would expect that the DQA and DRB genes associated with the DQw3B gene in REM would be from the same DQw2 haplotype.

There is no report of recombination between DQ and DR, although there is speculation in this regard (12). In the present study, REM (DR2/DQw3) DRB sequences are essentially identical to those of a DQw1-positive cell (FJO) (5,6), rather than to a DQw2-positive cell. This strongly suggests that a recombinational event occurred between DQA and DRB during the generation of the REM haplotype (Table 2).

We propose here a genetic mechanisms for the generation of the DR2/Dw22/DQw3 haplotype (Fig. 1). Intra-allelic recombination or gene conversion between the DQB genes of a DR2/DQw1 and a DR3/DQw2 haplotype result in the formation of the DQw3B gene which is associated with a DQw2A gene (14). A recombination between the DQA and DRB genes of the newly formed haplotype and the DR2/Dw21/DQw1 haplotype then results in the suggested DR2/Dw21/DQw3 haplotype.

Dw21 is formed through a reciprocal genetic exchange between DRB1 and -B3 genes of either Dw2 or Dw12 (5). A point mutation, which occurred at some time, results in an amino acid change at position 67 in the DRB3 gene of the Dw21 haplotype and causes an altered T-cell response, resulting in the formation of the Dw22 specificity.

The presence of only one difference between the DRB3 genes of Dw22 and Dw21 suggests a very recent divergence of these two genes. The Dw22 specificity and the association of DR2 with DQw3 are present in the Warao population with high frequency (3), but not in the Caucasian population. These facts suggest that the point mutation and the proposed recombination between DR and DQ occurred after the isolation of the Warao Indian tribe. Further molecular and genetic studies of this long-isolated Warao population may be an unusual opportunity to gain insight into the genetic mechanisms generating the isotypic and allotypic diversity of HLA class II genes.

References

1. Bach FH, van Rood JJ. The major histocompatibility complex–genetics and biology. N Engl J Med 1976;295:806.
2. Yabe T, Reinsmoen NL, Wu S, Bach FH. Protein analysis of HLA-DR2.3 haplotype. Hum Immunol, 1987;19:127.
3. Layrisse Z, Simoney N, Park MS, Terasaki PI. HLA-D and DRw determinants in an American indigenous isolate. Transplant Proc 1979;XI:1788.
4. Reinsmoen NL, Bach FH. T cell clonal analysis of HLA-DR2 haplotypes: Panel cell analysis. Hum Immunol 1987;20:13–32.
5. Wu S, Saunders TL, Bach FH. Polymorphism of human Ia antigens generated by reciprocal intergenic exchange between two DR β loci. Nature 1986;324:676.
6. Wu S, Yabe T, Madden M, Saunders TL, Bach FH. cDNA cloning and sequencing reveals that the electrophoretically constant DR β2 molecules, as well as the variable DR β1 molecules, from HLA-DR2 subtypes have different amino acid sequences including a hypervariable region for a functionally important epitope. J Immunol 1987;138:2953.
7. Larhammar D, Servenius B, Rask L, Peterson PA. Characterization of an HLA-DR β pseudogene. Proc Natl Acad Sci USA 1985;82:1475.
8. Gregersen PK, Shen M, Song Q-L, Merryman P, Degar S, Seki T, et al. Molecular diversity of HLA-DR4 haplotypes. Proc Natl Acad Sci USA 1986;83:2642.
9. Cairns JS, Curtsinger JM, Dahl CA, Freeman S, Alter BJ, Bach FH. Sequence polymorphism of HLA DR β1 alleles relating to T-cell-recognized determinants. Nature 1985;317:166.
10. Hilden JM, Curtsinger JM, Cairns JS, Bach FH. DQ β sequences in HLA-DR4 haplotypes. Hum Immunol 1987;18:261.
11. Schenning L, Larhammar D, Bill P, Wiman K, Jonsson A-K, Rask L et al. Both α and β chains of HLA-DC class II histocompatibility antigens display extensive polymorphism in their amino-termino domains. EMBO J 1984;3:447.
12. Schiffenbauer J, Didier DK, Klearman M, Rice K, Shuman S, Tieber VL, et al. Complete sequence of the HLA DQ α and DQ β cDNA from a DR5/DQw3 cell line. J Immunol 1987;139:228.
13. Auffray C, Lillie JW, Arnot D, Grossberger D, Kappes D, Strominger JL, et al. Isotypic and allotypic variation of human class II histocompatibility antigen α-chain genes. Nature 1984;308:327.
14. Wu S, Saunders T, Bach FH. 1987. Polymorphism of HLA-DQ β cDNAs from individuals of different DR2 subtypes: Evolution of DQ β genes. Fed Proc, 1987;46:943.
15. Auffray C, Strominger JL. Molecular genetics of the human major histocompatibility complex. Adv Hum Genet 1985;15:197.
16. Kratzin H, Thinnes FP, Kruse T, Barnikal HU, Wernet P, Hilschmann N. Structure of human class II antigens expressed by a homozygous lymphoblastoid B cell line. In: Solheim BG, Moler E, Ferrone S (eds): HLA Class II Antigens. New York, Springer-Verlag, 1986.
17. Larhammar D, Schenning L, Gustafsson K, Wiman K, Claesson L, et al. 1982. Complete amino acid sequence of an HLA-DR antigen-like β chain as predicted from the nucleotide sequence: Similarities with immunoglobulin and HLA-A, -B, and -C antigens. Proc Natl Acad Sci USA 1982;79:3687.
18. Gustafsson K, Wiman K, Emmoth E, Larhammar D, Bohme J, Hyldig-Nielsen JJ, et al. Mutations and selection in the generation of class II histocompatibility antigen polymorphism. EMBO J 1984;3:1655.
19. Dider DK, Schiffenbauer J, Shuman S, Abruzzini LF, Gorski J, Watling DL, et al. Characterization of two distinct DRβ-chain alleles at the βIII locus of the DR5 haplotype: βIII alleles are highly conserved. J Immunol 1986;137:2627.

Author Affiliations

Chih-Pin Liu: University of Wisconsin, Madison; *Shaokee Wu* and *Fritz H. Bach:* Immunobiology Research Center, Departments of Laboratory Medicine/Pathology and Surgery, University of Minnesota, Minneapolis, MN 55455, USA

Identification of Dw2, Dw12, and "Short" DR2 Splits with Sequential Exon-Specific DRβ, DQβ, and DQα cDNA Probes

Jeffrey L. Bidwell, Elizabeth A. Bidwell, David A. Savage, Derek Middleton, Celine Cullen, Peter T. Klouda, and Ben A. Bradley

The HLA-DR2 specificity has been divided by alloantisera (DR2 and "short" DR2) and by T cells (Dw2, Dw12, DB9 (LD-5a), FJO, and LD-MN2). The FJO and LD-MN2 splits appear to correlate with the short DR2 serotypes. Here, we report the correlation between the phenotypically recognized splits of DR2, and genotypes as defined by restriction fragment length polymorphism (RFLP) analysis. Genomic DNAs from 84 DR2-positive individuals and 10 homozygous typing cells were hybridized sequentially with short (exon-specific) cDNA probes isolated from the following clones: pRTV1 (DRβ), 512bp *Pst*I fragment; pII-β-1 (DQβ), 627bp *Ava*I fragment; pDCH1 (DQα), 797bp *Pst*I fragment. Methods and interpretation of DR and DQ alleles were detailed previously (1). Figure 1 shows the DRβ, DQβ, and DQα RFLP patterns observed in DR2-DQw1 haplotypes, either shown to segregate in family studies, or predicted from association due to linkage disequilibrium. Each split of DR2 defined by phenotypic criteria is associated with a particular combination of RFLPs for DRβ (*DRβ2* or *DRβ2sh*), DQβ (*DQβ1a*, *DQβ1b* or *DQβx*), and DQα (*DQα1b* or *DQα1c*). Thus, of 121 DR2-DQw1 RFLP combinations examined, 105 were *DRβ2-DQβ1b-DQα1b*. This combination has been previously reported in DR2-Dw2 splits; four were *DRβ2-DQβx-DQα1c*, previously reported in DR2-Dw12 splits; nine were *DRβ2sh-DQβ1a-DQα1b*, which has been described in the 'short' DR2 splits FJO and LD-MN2, and was shown in this study to segregate within one family. FJO and LD-MN2 are reported to possess indistinguishable RFLPs (2) and identical nucleotide sequences for the β1 exon of the $DR\beta_1$ gene (3). The remaining novel RFLP combination, *DRβ2-DQβx-DQα1b*, designated "DR2-Akhtar," was uncovered in three Asian individuals from a single family and may be related to DB9 (LD-5a) based on similarities with published DQβ RFLP data, but shows the "long" DR2 DRβ RFLP (*DRβ2*), whereas the DB9 (LD-5a) DRβ RFLP as reported does not correspond to either *DRβ2* or *DRβ2sh* (2,4). Investigation of DR2 haplotypes from other non-Caucasoid populations is likely to reveal further novel combinations of DRβ, DQβ, and DQα RFLPs.

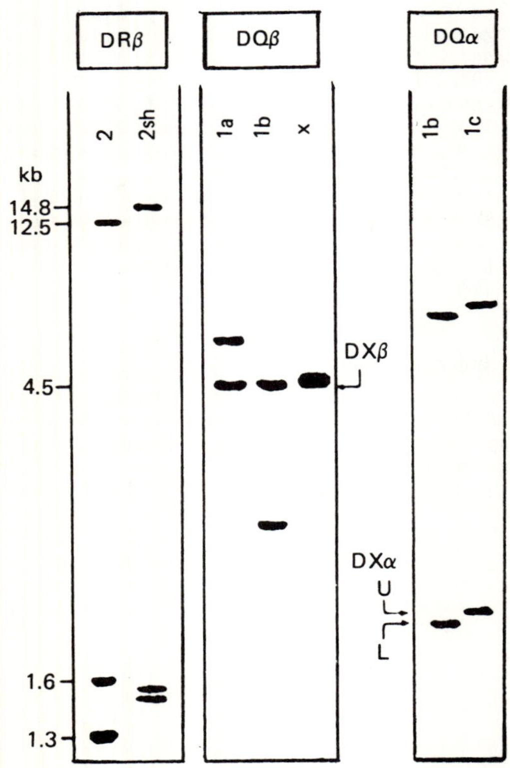

Figure 1. *Taq*I-specific DR2-DQw1 RFLP patterns revealed with exon-specific DRβ, DQβ, and DQα cDNA probes

References

1. Bidwell JL, Bidwell EA, Savage DA, Middleton D, Klouda PT, Bradley BA. A DNA-RFLP typing system which positively identifies serologically well-defined and ill-defined HLA-DR and DQ alleles, including DRw10. Transplantation 1988;45:640–646.
2. Segall M, Noreen H, Schluender L, Swenson M, Barbosa J, Bach FH. DR2[+] haplotypes in insulin-dependent diabetes: Analysis of DNA restriction fragment length polymorphisms. Hum Immunol 1986;17:61–68.
3. Freeman SM, Saunders TL, Madden M, Segall M, Bach FH, Wu S. Comparison of $DR\beta_1$ alleles from diabetic and normal individuals. Hum Immunol 1987;19:1–6.
4. Segall M, Noreen H, Schluender L, Bach FH. DNA restriction fragment length polymorphisms characteristic for Dw subtypes of DR2. Hum Immunol 1986;15:336–343.

Author Affiliations

Jeffrey L. Bidwell, Elizabeth A. Bidwell, Peter T. Klouda, Ben A. Bradley, United Kingdom Transplant Service, Bristol, England; *David A. Savage, Derek Middleton, Celine Cullen*, Belfast City Hospital, Lisburn Road, Belfast BT9 7AD, Northern Ireland

Immunobiology of HLA
Volume II

Subdivision of Haplotype HLA-DR3 by Biochemical and Functional Analysis

S.H. Sacks, A. Bushell, K.J. Wood, and A.J. McMichael

We and others (1,2) have investigated molecular HLA class II polymorphism in the haplotype HLA-DR3 in an attempt to correlate structural heterogeneity at the protein and DNA levels with T-cell specificities. HLA class II antigens were immunoprecipitated from HLA-DR3 homozygous B-lymphoblastoid cell lines with SG171 (antimonomorphic DR) or TAL 1B5 (anti-α chain) monoclonal antibodies, and examined by two-dimensional nonequilibrium pH gradient- sodium dodecyl sulfate (SDS)-polyacrylamide gel electrophoresis (PAGE). Genomic DNA was digested with TaqI, BamHI, or RsaI restriction enzymes and hybridized with DRβ or DQβ gene probes. Mixed lymphocyte reaction was carried out by incubation of 5.10^4 fresh peripheral blood lymphocytes (PBL) (responders) with 5.10^4 irradiated PBL (stimulators) for 5 days, followed by addition of $[^3H]$thymidine and determination of incorporated ^3H.

Table 1. Proliferation Induced in Mixed Lymphocyte Culture, Given as Percent Stimulation (Double Normalized Values)

	Stimulators							
	JG	MB	MSTR	JS	CB	ATO	FG	DH
HLA-DR:	3	3	3	3	2,4	1	1,7	7,11
HLA-B:	18	8	8	8				
DR3 DRβ-2:	–*	–†	–†	–†				
Responders								
JG	1	21	28	25	98	96	90	81
MB	59	1	1	27	100	96	60	81
MSTR	94	1	0	1	90	108	102	78
JS	58	21	2	0	100	98	93	79
CB	59	108	120	103	0	88	140	108
ATO	106	93	127	97	92	1	75	130
FG	80	67	71	59	107	81	1	92
DH	106	107	80	103	96	102	98	0

†Possession of β-2a chain,
*Possession of β-2b chain, where β-2a is on the basic side and β-2b is on the acidic side of the invariant β-1 DRβ chain on 2D gels of HLA-DR3 antigens.

In a panel of 17 HLA-DR3 homozygous B-cell lines, we identified two subtypes of HLA-DR3 that were distinguishable on the basis of polymorphism in one of the two DRβ polypeptide chains we detected, corresponding with differences in DRβ restriction fragment patterns. These polymorphisms exhibited significant cosegregation with the B8 DR3 (p = .0007) and B18 DR3 (p = .0087) haplotypes, respectively. In proliferative experiments, cells displaying the B18 DR3-type polymorphism strongly stimulated cells of donor B8 DR3 type, at a level comparable with allogeneic stimulation (Table 1). Stimulation in the reverse responder-stimulator combinations was weak.

We conclude that charge differences in the DRβ chains resolved on focusing gels correlate with differences in proliferative responses within the HLA-DR3 haplotype, and together these identify T-cell recognition sites on the B18 DR3-associated polymorphic DRβ chain that induce proliferation in B8 DR3 cells. This T-cell epitope on B18 DR3 cells may correspond to the DRw52-linked specificity that has been referred to as LB-Q1 (2).

References

1. Sacks S, Bushell A, Rust N, Karagiannis J, Jewell D, Ledingham J, et al. Functional and biochemical subtypes of the haplotype HLA-DR3 in patients with celiac disease or idiopathic membranous nephropathy. Hum Immunol, 1987; 20:175.
2. Bontrop R, Tilanus M, Mikulski M, von Eggermond M, Termijtelen A, Giphart M. Polymorphisms within the HLA-DR3 haplotypes. Immunogenetics 1986;23:401.

Author Affiliations

S.H. Sacks and *A.J. McMichael*, Nuffield Department of Clinical Medicine, John Radcliffe Hospital, Headington, Oxford, Ox3 9DU, England; *A. Bushell* and *K.J. Wood*, Nuffield Department of Surgery, John Radcliffe Hospital, Headington, Oxford, OX3 9DU, England

Characterization of Multiple HLA-DR3 Haplotypes*

Tom Cotner, Elizabeth Mellins, and Donald A. Pious

The HLA-DR allospecificities, DR1-DRw14, are defined by using selected alloantisera. Analysis at the biochemical and molecular level has demonstrated DRβ structural heterogeneity within certain haplotypes such as DR2, DR4, and DRw8 (1–4). Although DR3 is well defined (5), a recent variant has been described at the biochemical and molecular level (6,7). Here we describe three DR3 subtypes that can be distinguished by a combination of two-dimensional gel electrophoresis, monoclonal antibody (MAb) reactivity, quantitative variation in expression, and reactivity with DRw52-specific T-cell clones.

Results and Discussion

The biochemical characterization of DR molecules was performed on metabolically labeled homozygous typing cells (HTC) representative of each of the three DR3 subtypes. SC-JB is representative of the common Caucasian or DR3a subtype; WT49, QBL, and BCR are HTC representatives of the second DR3 (or DR3b subtype); and the third subtype, represented by the HTC, LEM, is a common Chinese DR3. Two-dimensional (2D) gels of immunoprecipitated DR molecules that have been treated with endoglycosidase F to remove carbohydrate charge heterogeneity are shown in Figure 1. All three patterns contain two relatively basic DRβ chains, which are expressed at markedly different levels. A possible fourth variant apparently expresses both DRβ chains at comparable levels (Fig. 1, fourth panel). The more abundantly expressed chain in DR3a cells is the β1 gene product, whereas the less abundant, slightly more basic chain is a β3 chain. This was determined by eluting the radiolabeled chains from a 2D gel, obtaining their amino-terminal sequence, and comparing these sequences with those amino acid sequences predicted from published DR3B1 and B3 gene sequences (9). There are two further indications that the abundant, slightly more acidic chain is the β1 chain. The first is that the β1 and β3 mRNA levels correlate with expression at the protein level. Use of β1- and β3-specific oligonucleotide probes has shown that β1 mRNA is about 4- to 5-fold more abundant than the β3 mRNA in DR3a cell lines (date not shown). A critical link between the β1 gene, β1 mRNA and the abundant β chain is provided by a DR3B1 mutant 9.4.3. Due to a transcriptional defect (8,9), it fails to express β1 mRNA, as determined by using gene-specific oligonucleotide probes. It also fails to synthesize the abundant β chain, whereas expression of β3 mRNA and the β3 chain are essentially normal.

The second subtype, DR3b, was found in a number of cell lines in which the flanking HLA-B allele was not HLA-B8. This subtype expresses an abundant β chain that cofocuses with the β1 chain of the DR3a subtype. It is characterized by its less abundant β chain, which occupies a position about two charges more acidic than βI. Two DRβ cDNA clones have been sequenced from one of these HTC, WT49 (10). The DR3B1 sequence is

*This work was supported by grants GM15883 and AI16689 from the National Institutes of Health.

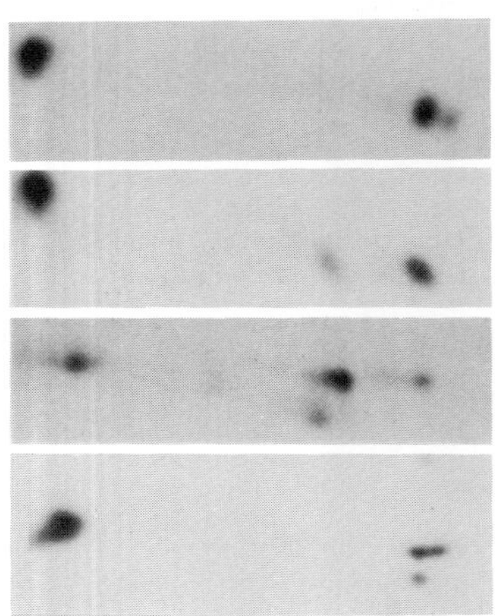

Figure 1. Two-dimensional gels of immunoprecipitated DR molecules representative of the three DR3 subtypes. A fourth variant in which the two DRβ chains are expressed at comparable levels is shown in the fourth panel.

Table 1. Relative Abundance of DR3β Chains

DR Type	Number of Determinations	β Chain Abundance Relative to DRα: Major β	Minor β	Ratio of Major Chain to Minor Chain
DR3a	15	53 ± 7	9 ± 1	5.9 ± 0.2
DR3b	9	63 ± 10	19 ± 4	3.3 ± 0.6
DR3c	8	60 ± 8*	19 ± 3*	3.2 ± 0.5

Densitometric scans were performed on the DR region of 2D gels of endoglycosidase F-treated DR molecules from cells from each of the three subtypes. The intensities of each DRβ spot were normalized to that of DRα for each film .

*Note that the major or abundant chain in the DR3c cell lines corresponds in charge to the minor β chain of DR3b cells and the minor chain in DR3c to the major chain in DR3b.

Immunobiology of HLA
Volume II

identical to the B1 sequence obtained from DR3a cells, whereas the β3 cDNA should encode a chain that differs in 11 positions from the β3 chain of DR3a and should be about two charges more acidic than the DR3a β3 chain (10). For this reason we assume that the chain that cofocuses with the β1 chain in DR3a is the β1 chain and the more acidic chain is β3. This subtype almost certainly corresponds to one previously described (6). We have found that it is associated with DR3 cells containing HLA-B17, B18, B21, Bw22, Bw50, and B57.

The 2D gel pattern of DR3 from Chinese subjects is similar to the DR3b pattern but can be distinguished by several criteria. First, even though the DR3β chains from this cell correspond in charge to those of DR3b, the relative abundance of the chains is reversed. The basic and most abundant β chain (probably βI) of DR3b is 3- to 4-fold more abundant than the acidic chain of DR3b, whereas in the third DR3 subtype it is the acidic chain that is 3- to 4-fold more abundant (see Table 1 for summary of quantitative expression). Preliminary results from Northern blots with β1 oligonucleotide probes indicate that the Chinese DR3 has β1 mRNA levels that are comparable to those in DR3a or DR3b cells. The hybridizations were performed under conditions of stringency such that as little as a single mismatch would result in no signal. This is consistent with the interpretation that the abundant, more acidic β chain in the third subtype is a B1 gene product, and as such, represents a new DRB variant. The basic and presumptive β3 chain of the third DR3 variant may also be new alloreactive T-cell clones which can distinguish three or more distinct DRw52 specificities have demonstrated that the third DR3 variant may have an additional DRβ3 determinant not present on any of the DR3a or DR3b cell lines (11). The fact that the DR3c HTC, LEM, appears to differ from the other DR3 haplotypes in both its β1 chain (isoelectric point) and its β3 chain (both antibody and T-cell determinants) calls into question whether LEM is truly a DR3 cell or represents a new DR3-like allospecificity which is not distinguishable by current allosera. Sequencing of the DRβ genes from this cell line should permit gene assignments of the expressed DRβ chains, and should explain the relationship to other DR3 haplotypes.

References

1. Nepom GT, Nepom BS, Wilson ME, Mickelson E, Antonelli P, Hansen JA. Multiple Ia-like molecules characterize HLA-DR2-associated haplotypes which differ in HLA-D. Hum Immunol 1984;10:143.
2. Nepom BS, Nepom GT, Mickelson E, Antonelli P, Hansen JA. Electrophoretic analysis of human HLA-DR antigens from HLA-DR4 homozygous cell lines: Correlation between β chain diversity and HLA-D. Proc Natl Acad Sci USA 1983;80:6962.
3. Suzuki M, Yabe T, Satake M, Juji T, Hamaguchi H. Two-dimensional gel electrophoretic analysis of the MT3 and DR4 molecules from the different D-typed cells. J Exp Med 1984;160:751.
4. Baldwin GC, Mickelson EM, Hansen JA, Nisperos B, Antonelli P, et al. Electrophoretic variation between class II molecules expressed on HLA DRw8 homozygous typing cells reveals multiple distinct haplotypes. Immunogenetics 1985;21:49.
5. Fuller TC, Park MS, Thorston E. In: Albert ED, Baur MP, Mayr WR (eds): Histocompatibility Testing. Berlin, Springer-Verlag, 1984, p 724.
6. Bontrop R, Tilanus M, Mikulski M, van Eggermond M, Termijtelen A, Giphart M. Polymorphisms within the HLA-DR3 haplotypes. Immunogenetics, 1986;23:401.
7. Gorski J, Tilanus M, Giphart M, Mach B. Oligonucleotide genotyping shows that alleles at the HLA-DRIII locus of the DRw52 supertypic group segregate independently of known DR or Dw specificities. Immunogenetics, 1987; 25:79.
8. Pious D, Dixon L, Levine F, Cotner T, Johnson R. HLA class II regulation and structure. Analysis with HLA-DR3 and HLA-DP point mutants. J Exp Med 1985;162:1193.
9. Cotner T, Charbonneau H, Mellins E, Pious D. mRNA abundance, rather than differences in subunit assembly, determine differential expression of HLA-DRβ1 and DRβ3 molecules. J Biol Chem, in press.
10. Bell JI, Denney D, Foster L, Belt T, Todd JA, McDevitt HO. Allelic variation in the DR subregion of the human major histocompatibility complex. Proc Natl Acad Sci USA 1987;84:6234.
11. Mickelson EM, Masewicz SA, Cotner T, Hansen JA. Variants of HLA-DRw52 defined by T lymphocyte clones. Hum Immunol 1988;22:263.

Author Affiliation

The Departments of Pediatrics and Genetics, University of Washington, Seattle, WA 98195, USA

DR3 Heterogeneity in American Blacks

Armead H. Johnson, Ting Fang-Tang, Sandra Rosen-Bronson, Fu-Meei Robbins, Noriko Steiner, Robert J. Hartzman, Peter K. Gregersen, Jack Silver, Barbara Ziff and Carolyn K. Hurley

The heterogeneity of DR3-associated haplotypes in American Blacks is striking when compared with that in the Caucasoid populations. In American Blacks, only 7.7% of DR3-positive individuals are Dw3. Three DR3-associated haplotypes are generally observed in the American Black population. Two appear to be unique to Blacks and are 1) A30, Bw42, Dw-, DRw18, DRw52a, DQw4, and 2) Bw53 or Bw71, Dw3v, DRw17, DRw52b, DQw2. HLA-D typing responses for these haplotypes are Dw- (DNV > 70) and Dw3v (DNV 35–55), respectively. The third DR3 haplotype is identical to the Caucasian A1, B8, Dw3, DRw17, DRw52a, DQw2.

Functional differences between these haplotypes were investigated by generating alloreactive T-cell clones in several priming combinations. The priming combinations are given in Table 1. Differences between stimulator and responder are underlined. Alloreactive T-cell clones (TLC), which discriminate between the DRw18, DQw4 and the DRw17,DQw2 haplotypes, were detected in each of the four primings. Representative examples of TLC are given in Table 2. All DR3-positive stimulator cells were recognized by one of the two groups of TLCs that defined the DR3 subtype. These TLC are blocked by DR-specific (L243) but not by DQ-specific (SFR PI.2) or by DP-specific (B7/21) monoclonal antibodies.

The cDNA sequence of the DRBl gene from the DRw18, DQw4 Black haplotype indicates that this gene encodes a unique polypeptide chain, whereas the DRBl gene from the Bw71/Bw53,DRw17,DQw2 haplotype encodes a polypeptide chain identical to the DRBl gene product of the B8,DRw17,DQw2 homozygous typing cell, AVL. These two polypeptide chains vary at amino acid residues 26, 28, 47, and 86. Inspection of the published amino acid sequences for DR1-DRw9 reveals that the combination of amino acid residues located within the second variable region on the DR3 haplotypes represent unique DRBl sequences. Residues 47 and 86 are both located outside variable regions and are variable in all DR molecules, having one of two alternative amino acids. Therefore, we hypothesize that these DR3 subtype-specific T-cell clones are recognizing an epitope determined by amino acid residues 26 and 28. If the three-dimensional structure of class II molecules can be related to that of class I, amino acid residues 26 and 28 would be positioned in the antigen-binding site. Using this model, one can hypothesize how these structural changes might cause specific functional differences between DR3 subtypes in antigen/T-cell receptor interactions and in allorecognition.

Acknowledgment. This work was supported in part by National Institutes of Health Grants AI23371 (A.J.) and AI22005 (J.S.) and Office of Naval Research Contract N0014-83-K-0410 (C.K.H.).

Author Affiliations

Armead H. Johnson, Ting Fang-Tang, Sandra Rosen-Bronson, Fu-Meei Robbins, Noriko Steiner, Barbara Ziff, and Carolyn K. Hurley, Georgetown University School of Medicine, Departments of Pediatrics and Microbiology, Washington, DC; *Peter K. Gregersen and Jack Silver*, Hospital for Joint Diseases, New York, NY; *Robert J. Hartzman*, Naval Medical Research Institute, Immunobiology and Transplantation Department, Bethesda, MD

Table 1. Priming Combinations for Alloproliferative T-Cell Clones

Priming		HLA Type				Race
		D	DR	DRw	DQ	
S3	R	Dw2,w2	DR2,2		DQw6,w6	C
	S	Dw3,w5	DRw17,w11	DRw52a,w52b	DQw2,w7	C
TC1	R	Dw-,w8	DR1,w8	(DRw52)	DQw1,w4	C
	S	Dw-,w8	DRw18,w8	DRw52a,(w52)	DQw4,w4	B
TC2	R	Dw3,w5	DRw17,w11	DRw52a,w52b	DQw2,w7	C
	S	Dw-,w-	DRw18,w11	DRw52a,w52b	DQw4,w7	B
TC4	R	Dw-,w-	DRw18,w11	DRw52a,w52b	DQw4,w7	B
	S	Dw3,w5	DRw17,w11	DRw52a,w52b	DQw2,w7	C

B = Black; C = Caucasian; R = responser; S = stimulator.

Table 2. DR3 Subtypes Detected by Alloproliferative T-Cell Clones

Stimulators							Clones							
		HLA Type												
ID #	Race	B	D	DR		DQ	S3.11	S3.81	TC4.52	TC1.18	TC1.6	TC1.15	TC2.40	TC2.31
1563	B	42,45	0,0	18,11	52a,52b	4,7	−	+	−	+	+	+	+	+
2698	B	42,44	0.2	18,2	52a,−	4,6	−	−	−	+	+	+	+	+
2041	B	50,44	0,0	18,2	52a,−	4,6	−	−	−	+	+	+	+	+
2710	B	58,58	0,0	18,13v	52a,52b	4,5		−	−	+	+	+	+	+
1087	B	42,60	0,4	18,4	52a,53	4,8	−	−	−	−	+	+	+	+
1559	B	35,27	0,8	18,8	52a	4,7	−	−	−	+	+	+	+	+
373	B	71,53	3v,0	17,17	52b,52b	2,2	+	+	+	+	−	−	−	−
1064	B	71	3v,0	17,8	52b	2,7	+	+	+	+	−	−	−	−
1560	B	42,49	3v,0	17,13	52b,52c	2,6	+	+	−	+	−	−	−	−
1053	B	71,58	3v,7	17,7	52b,53	2,2	+	+	+	+	−	−	−	−
1061	L	27,7	3v,2	17,2	52b	2,6	+	+	+	+	−	−	−	−
1060	L	8,44	3,4	17,4	52a,53	2,8	+	+	+	+	−	−	−	−
1063	B	8,7	3,0	17,13v	52a	2,5	+	+	+	+	−	−	−	−
958	C	41,60	3,14	17,4	52a,53	2,8	+	+	+	+	−	−	−	−
144	C	8	3,0	17,1	52a,−	2,5	+	+	+	+	−	−	−	−
127	C	8,64	3,7	17,7	52a,53	2,2	+	+	NT	+	−	−	−	−
533	C	8,8	3,3	17,17	52a,52a	2,2	+	+	+	+	−	−	−	−
6	C	8	3	17,17	52a	2,2	+	+	+	+	−	−	−	−
12	C	8	3	17,17	52a	2,2	+	+	+	+	−	−	−	−
381	C	8,35	0,0	17,8	52a,−	2,4	+	+	+	+	−	−	−	−
1077	C	8,13	3,5	17,11	52a,52b	2,7	+	+	+	+	−	−	−	−
938	C	27,44	5,4	11,4	52b,53	7,8	−	−	−	−	−	−	−	−
930	C	8,7	11,2	7,2	−,−	3,6	−	−	NT	−	−	−	−	−
1145	C	7,18	0,2	13,2	52b,−	6,6	−	−	NT	−	−	−	−	−
946	C	13,35	1,8	1,8	−,−	5,4	−	−	−	−	−	−	−	−
101	C	58,41	5,19	11,13	52b,52c	7,6	−	−	−	−	−	−	−	−
1579	C	62,65	8,0	8,1	−,−	4,5	−	−	−	−	−	−	−	−
162	C	27,37	0,9	10,14	−,52b	5,5	−	−	−	−	−	−	−	−
1426	C	62,55	18,9	13,14	52*	6,5	−	−	−	−	−	−	−	−
2015	B	45,27	0,0	11,11	52a,52b	6,7	−	−	−	−	−	−	−	−
1129	B	58,55	0,1	11,1	52b,−	7,5	−	−	−	−	−	−	−	−
258	B	71,35	19,0	13,11	52,52*	6,6	−	−	−	+	−	−	−	−
260	B	14,35	0,0	2,5v	52,52*	6,7	−	−	−	−	−	−	−	−
2006	B	58,53	5,0	11,14	52b	6,5	−	+	−	−	−	−	−	−
1066	B	58,45	0,0	13v,8	52b,−	5,7	−	−	−	−	−	−	−	−
2040	B	42,71	0,0	8,1	−,−	7,5	−	−	−	−	−	−	−	−
1096	B	8,63	0,0	1,4	−,53	5,8	−	−	−	−	−	−	−	−
1116	B	57,49	NT	7,4	53,53	2,8	−	−	−	−	−	−	−	−
1099	B	39,58	0,2	10,2	−,−	5,6	−	−	−	−	−	−	−	−
1406	B	45,58	NT	2,9	−,53	6,2	−	−	−	−	−	−	−	−

*DRw52 subtype unknown.
Race: C=Caucasian, B=Black, L=Lumbee Indian. NT= Not tested.

Correlation Between Function and Expression of the Two DR Molecules in DR3 Haplotypes

Dominique Jaques, José Berdoz, Eduardo Martinez, Michel Jeannet, Bernard Mach, and Claude Irlé

Abstract: We have reported elsewhere in these proceedings that T-cell clones restricted by distinct DR molecules expressed in DR3 haplotypes, namely the α-βI and the α-βIII heterodimers, contain the DR3 and DRw52 epitopes, respectively. We have now correlated the quantitative responsiveness of these Tetanus toxoid (TT)-specific, DR3 and DRw52-restricted T-cell clones, respectively, with the quantitative expression of these two distinct DR molecules. T-cell responsiveness was compared after normalizing proliferation. Expression was determined by mRNA measurements and microfluorometry using DR-specific monoclonal antibodies. Our results show that responsiveness of clones to TT presented by DR3, which is expressed at a higher level, exceeds that of TT restricted by DRw52.

In mice, quantitative variations in the constitutive expression of MHC class II molecules have been shown to influence T-cell responsiveness to antigen (1). Similarly, a correlation between class II immunofluorescence staining intensity and primary MLR stimulation, respectively antigen presentation, has been found (2,3). Previous reports have indicated differences between the expression of individual class II molecules (4,5), and suggested that these could explain the relative immunodominance of DR in terms of T-lymphocyte activation (6). The potential importance of MHC expression during autoimmune reactions has been stressed repeatedly (7–9). Hyperinducibility of MHC class II expression by astrocytes was recently correlated with the susceptibility for experimental autoimmune encephalomyelitis observed in certain mouse strains (10). Using TT-specific T-cell clones restricted by either DR3 or by DRw52, we have recently been able to map these restriction determinants to two distinct DR molecules (11). Together with the use of lymphoblastoid cell lines as antigen presenting cells, the T-cell clones reported here provide a homogeneous and sensitive assay for the correlation between the level of expression of these distinct DR molecules, and their ability to stimulate antigen specific T-cell responses.

Responses of several clones restricted by either DR βI or DRβIII gene products were compared. Because proliferative T-cell clones differ in their reactivity not only between each other, but also between experiments, a common reference response was determined to "normalize" ^{3}H-thymidine incorporations. We defined the 100% reference response as the proliferation of a given clone to a standard dosis of exogenous Il-2 (100 U rIL-2/ml). When the proliferation of T-cell clones to a panel of antigen-presenting cells were compared, we noticed that the reference response of two DRw52-restricted clones (clone 16, 52"a"; clone 10, 52"b") was always quite comparable, but lower than that of two DR3-restricted clones (clone 49 and clone 5). The lower response to DRw52 APC was not DR3 haplotype-dependent, because DRw52 APC-typed DR3, DR5 or DRw6 were equally effective in stimulating proliferation. In addition, the level of response between clone 16 (DRw52a) and clone 10 (DRw52b) to 52a/52b APC did not differ. Furthermore, this finding was not dependent on the nature of APC tested: both PBMC and B LCL presented TT less effectively via the DRw52 (βIII) molecule. The DR3/DRw52 relative response ratios between PBMC and LCL were comparable (typical results are shown in Table 1). Interestingly, the relative responsiveness of the T-cell clones to TT-LCL was two- to fourfold higher than to TT-PBMC (Table 1). This "immunodominance" of DR βI over DR βIII gene products could relate to differences in their structure or their expression. Sequence comparisons have shown that differences between DR βI genes and between DR βIII genes occur in strikingly distinct patterns in the first domain. While in DR βI they are clustered in the hypervariable regions; in DR βIII they are widely scattered throughout the first domain. Both molecules could present antigen. Therefore we wanted to know if a relation between function and expression could be found.

This was investigated by two independent methods. First, the fluorescence intensity of APC that were stained

Table 1. Il-2-Normalized Response Index of TT-Specific, DR3- or DRw52-Restricted T-Cell Clones to TT-Incubated PBMC and B-LCL

Responders		Stimulators		
Clones		PBMC		LCL
Tested	Restriction	DR3,w6	DR3,-	DR3,3
49	DR3	1.1	1.15	3.04
5	DR3	1.08	1.1	–
10	DRw52	.56	.01	1.56
16	DRw52	.01	.43	–
Ratio DR3: DRw52		1.96	2.67	1.95

Results represent the response indexes calculated using mean CPM following 56h of culture, with ^{3}H-TDR incorporation during the last 8h. Standard deviations were always less than 15%. The responsiveness was calculated as follows:

$$\text{Response index} = \frac{\text{CPM (clone+APC+TT)} - \text{CPM (clone+APC}-\text{TT)}}{\text{CPM (clone+rIl-2)} - \text{CPM (clone}-\text{rIl-2)}}$$

rIl-2 (gift from Biogen SA, Geneva, Switzerland) was used at 100U/ml. Experiments with PBMC were performed using 50,000 cells/well incubated during 4h with TT (1:1250) before irradiation (3500R). Experiments with LCL were performed using 25,000 cells/well incubated overnight with TT (1:1000) before irradiation (3500R).

Immunobiology of HLA
Volume II

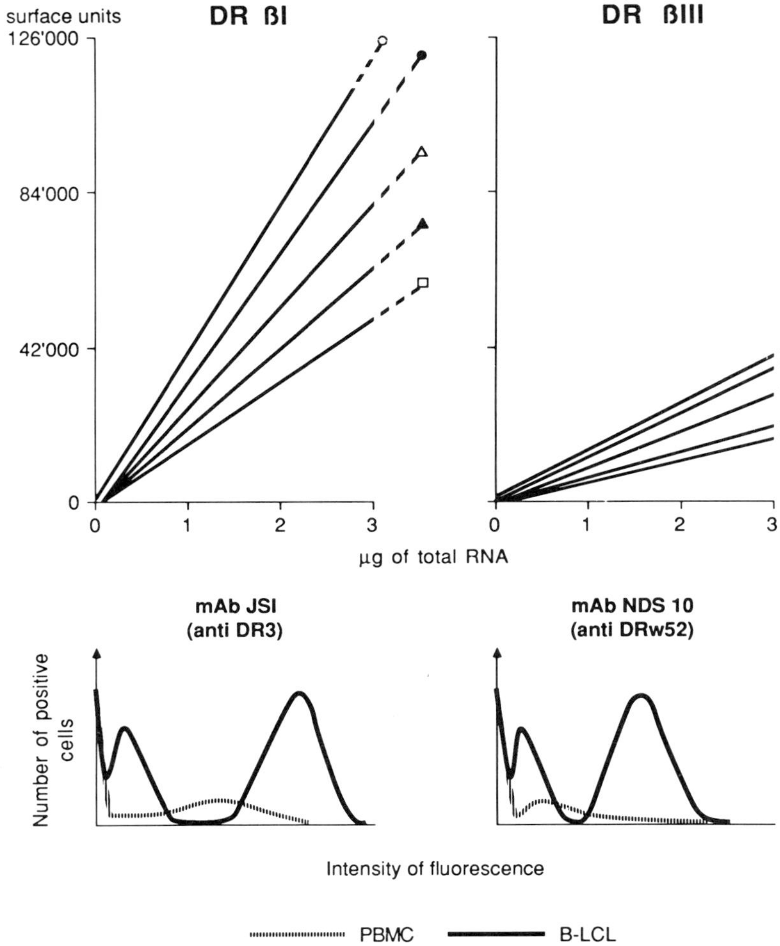

Figure 1. The upper panels show the intensity of autoradiographic signals determined by densitometry as a function of total mRNA isolated from four B-LCL cell preparations: ○ = VAVY; ● = LO081785; △ = DUCAF; ▲ = QBL, and □ = L0081785, after incubation with 500U/ml of recombinant gamma-interferon for 24 h. The left upper panel shows hybridizations of mRNA using the DR βI specific, [32]-P-labeled oligonucleotide probe 6/1, as described in ref. 12. The right panel shows the same mRNA preparation after hybridization using the DR βIII specific probe (6.2 for 52a haplotypes, 6.3 for 52b haplotypes). The slopes of the curves shown were corrected according to differences in hybrid stability and specific activity between the oligonucleotide probes used, and represent linear regressions calculated by the method of least squares from three different concentrations of mRNA (1ug, 2ug, and 3ug).

The lower panels show PBMC 2778, ····, (DR3,w6;DRw52,52; oligotyping 52b/b) and B-LCL QBL, — (DR3,3;DRw52,w52; oligotyping 52b,b) stained by indirect immunofluorescence using either (left panel) monoclonal antibody JSI (to DR3), or (right panel) monoclonal antibody NDS 10 (to DRw52) and a fluorescein-conjugated goat anti-mouse IgG. Cells were studied by fluorescent flow cytometry (D. Wohlwend, Centre de Fluorometrie, CMU, Geneva)

using MAb known to inhibit DR3- and DRw52-restricted T-cell clones was determined by fluorescence flow cytometry (FACS analysis). The saturating concentration of MAb was determined in each instance by maximal staining according to FACS analysis. A representative result is shown in Figure 1, illustrating that the intensity of surface expression of the DRw52 determinant recognized by the monoclonal antibody NDS10 is lower than the expression of the DR3 epitope recognized by the monoclonal antibody JSI. Second, the DR βI and DR βIII mRNA levels contained in APC (B LCL cells) was determined. In these experiments, a lower β III level was always found. Representative experiments are shown in Figure 1.

The best interpretation of our results is that differences in the level of expression, rather than differences

Table 2. Response Index of TT-Specific, DR3 or DRw52- Restricted T-Cell Clones to TT-Incubated PBMC and B-LCL Treated or Not With Gamma-Interferon

Responders		Stimulators			
		PBMC DR3,3		LCL DR3,3	
Clones Tested	Restriction	+INF	−INF	+INF	−INF
49	DR3	1.67	.78	2.6	2.52
10	DRw52	.73	.29	1.61	1.59
Ratio					
DR3: DRw52		2.28	2.68	1.61	1.58

The response index was calculated as described in legend of Table 1. PBMC or LCL were incubated during 24h with 250u of gamma-interferon or culture medium only. TT was added 4h prior to the end of incubations for PBMC, and together with gamma-interferon for LCL.

in the intrinsic function of these distinct DR molecules, determine their different ability to present antigen to T cells. This possibility was further investigated in experiments testing the ability of these distinct DR molecules to present TT to T cells after the treatment of APC (peripheral blood mononuclear cells, and B-LCL cells) with gamma interferon, to up-regulate class II expression. This increased the responsiveness of DR3- and DRw52-restricted clones to TT-PBMC, but not to TT-B-LCL cells (Table 2). Similarly, immunofluorescence analysis showed that gamma-interferon was found to up-regulate DR expression only in PBMC. This is also paralleled by the increased levels of DR βI and βIII mRNA after gamma-interferon treatment of monocytes/macrophage preparations, which are the main APC in PBMC (12), and the expected lack of increment of DR mRNA in B-LCL cells following gamma-interferon (Figure 1).

Together, our results show that both DR molecules are functional and expressed at different levels. They indicate that this quantitative difference of expression can contribute to the immunodominance of the DR βI over the DR βIII gene product.

Acknowledgment. This work was supported by a grant from the Swiss National Research Foundation (3.912.086).

References

1. Matis LA, Glimcher LH, Paul WE, Schwarz RH. Magnitude of response of histocompatibility-restricted T-cell clones is a function of the product of the concentrations of antigen and Ia molecules. Proc Natl Acad Sci USA 1983; 80:6019.
2. Nunez G, Ball EJ, Myers LK, Stastny P. Allostimulating cells in man. Quantitative variation in the expression of HLA-DR and HLA-DQ molecules influences T-cell activation. Immunogenetics 1985;22:85.
3. Triebel F, De Roquefeuil S, Blanc C, Charron DJ, Debre P. Expression of MHC class II and Tac antigens on IL2-activated human T cell clones that can stimulate in MLR, AMLR, PLT and can present antigen. Human Immunology 1986;15:302.
4. Bontrop R, Ottenhoff T, Van Miltenburg R, Elferink D, de Vries R, Giphart M. Quantitative and qualitative differences in HLA-DR molecules correlated with antigen-presentation capacity. Eur J Immunol 1986;16:133.
5. Ottenhoff THM, Elferink DG, Hermans J, de Vries RP. HLA class II restriction repertoire of antigen-specific T cells. I. The main restriction determinants for antigen presentation are associated with HLA-D/DR and not with DP and DQ. Human Immunology 1985;13:105.
6. Bach FH, Reinsmoen NL. The role of HLA-DR and HLA-DQ products in T lymphocyte activation and in contributing to "Dw specificities." Human Immunology 1986;16:271.
7. Ballardini G, Bianchi FB, Doniach D, Mirakian R, Pisi E, Bottazzo GF. Aberrant expression of HLA-DR antigens on bileduct epithelium in primary biliary cirrhosis: Relevance to pathogenesis. Lancet 1984;ii:1009.
8. Londei M, Lamb JR, Bottazzo GF, Feldmann M. Epithelial cells expressing aberrant MHC class II determinants can present antigen to cloned human T cells. Nature 1984; 312:639.
9. Mason DW, Dallman M, Barclay AN. Graft-versus-host disease induces expression of Ia antigen in rat epidermal cells and gut epithelium. Nature 1981;293:150.
10. Massa PT, Ter Meulen V, Fontana A. Hyperinducibility of Ia antigen on astrocytes correlates with strain-specific susceptibility to experimental autoimmune encephalomyelitis. Proc Natl Acad Sci USA 1987;84:4219.
11. Irlé C, Jaques D, Berdoz J, Martinez E, Jeannet M, Mach B. Correlation between function and expression of the two DR molecules in DR3 haplotypes. Histocompatibility Testing (in press).
12. Berdoz J, Gorski J, Termijtelen A-M, Dayer J-M, Irlé C, Schendel D, et al. Constitutive and induced expression of the individual HLA-DR βI and βIII chain loci in different cell types. J Immunol 1987;139:1336.

Author Affiliations

*Dominique Jaques, Eduardo Martinez, Michel Jeannet, Claude Irlé,** Transplantation Immunology Unit, Division of Immunology and Allergology, Department of Medicine, Hôpital Cantonal Universitaire, Geneva; *Jose Berdoz, Bernard Mach*, Department of Microbiology, University of Geneva Medical School, Geneva, Switzerland

*For correspondence.

Single Amino Acid Substitutions Within the HLA-DR4β_1 Chain Create Multiple T-Cell Recognition Sites

J. Goronzy and C. M. Weyand

The ability of major histocompatibility complex (MHC) class II molecules to specifically bind antigenic peptides defines their role as immune response genes (1,2). To study the relationship between polymorphic specificities on class II molecules and T-cell function we have used alloreactive human T-cell clones. Utilizing a panel of T-cell clones we have dissected T-cell epitopes expressed by the five different subtypes of the HLA-DR4 family. HLA-DR4 alleles are characterized by a limited sequence polymorphism residing within the third hypervariable region of the HLA-DRβ_1 chain (3,4). Correlation of T-cell reactivities to allele-specific amino acid substitutions allowed the molecular mapping of a number of T-cell determinants encoded within the third hypervariable region of the β_1 chain. A segment of 15 amino acids spanning positions 71 to 86 appears to be crucially involved in the formation of multiple T-cell epitopes. An extraordinary diversity for T-cell responses is created by substitutions at residues 71, 74, and 86, each of which contributes equally to a cluster of overlapping T-cell recognition sites. These data provide evidence that single amino acids within the third hypervariable region play a critical role for the three-dimensional structures recognized as T-cell receptor ligands and that a limited sequence polymorphism can induce multiple sites that differentially stimulate individual T lymphocytes.

Cloning of alloreactive human T cells has provided the means for analyzing structure-function relationships of HLA class II molecules (5). To map allospecific determinants expressed on DR molecules of the HLA-DR4 haplotype we have made use of the fact that the DR4 family is comprised of five different subtypes that are characterized by a limited sequence polymorphism. Amino acid substitutions in positions 57, 67, 70, 71, 74, and 86 of the first extracellular domain provide the basis for the allelic variation within the DR4 family (Fig. 1). We have generated a panel of allospecific T-cell clones to identify T-cell recognition sites and to analyze the contribution of different amino acid exchanges within the third hypervariable region to the formation of these T-cell determinants. Peripheral blood lymphocytes of normal donors were stimulated with DR4$^+$ stimulator cells, cloned and expanded (6). Individual clones were characterized on a panel of homozygous typing lines. As shown in Figure 2, a wide variety of reactivity patterns emerged when the clones were tested on the DR4 subtypes. Ten independently derived T-cell clones defined at least seven different determinants. A single subtype, Dw14, was recognized by seven distinct T-cell clones. However, most of the clones did not discriminate between the different subtypes. T-cell interaction sites were shared among different members of the DR4 family.

Localization of the multiple T-cell epitopes was accomplished by blocking studies with monoclonal antibodies. Antibodies reactive to monomorphic determinants of HLA-DP, -DQ, and -DR molecules were used to inhibit the proliferative response of the different T-cell clones. All of the clones shown in Figure 2 were inhibited by four different HLA-DR reactive monoclonal antibodies. However, we found pronounced differences in inhibition with an individual blocking pattern for each T-cell clone. Two exemplary inhibition patterns are shown in Figure 3. The response of clone G1-2 was strongly inhibited by monoclonal antibody P4.1 and to a lesser extent by the antibodies L227, L243, and L203. In contrast, antibody L227 did not affect proliferation of clone A1-6, which was poorly inhibited by antibody L203 and slightly better by antibodies L243 and P4.1. The finding that monoclonal antibodies reactive to monomorphic structures on HLA molecules inhibited

AA	67	--	--	70	71	--	--	74	-- -- -- -- -- -- -- -- -- -- --	86
Dw 4	Leu	—	—	Gln	Lys	—	—	Ala	— — — — — — — — — — —	Gly
Dw 13	=	—	—	=	Arg	—	—	Glu	— — — — — — — — — — —	Val
Dw 14	=	—	—	=	Arg	—	—	=	— — — — — — — — — — —	Val
Dw 15	=	—	—	=	Arg	—	—	=	— — — — — — — — — — —	=
Dw 10	Ile	—	—	Asp	Glu	—	—	=	— — — — — — — — — — —	Val

Figure 1. Amino acid sequence polymorphism of the five HLA-DR4 subtypes residing within the third hypervariable region of the HLA-DRβ_1 chain (3).

% relative response

	HLA-Dw	G1-2	GD2-30	C3-4	JA4-3	JD2-40	G1-1	F3-29	A1-6	E2-3	E1-10
BOD	4,4	0	0	3	2	0	0	0	76	36	80
BAR	4,4	1	1	n.t.	2	0	0	0	20	45	100
BSM	4,4	1	0	3	1	1	2	0	99	48	28
JAH	4,4	5	2	0	2	0	0	1	69	n.t.	70
AL10	10,10	1	3	3	2	0	82	2	97	1	0
TS10	10,10	0	1	1	2	0	95	0	82	0	1
SST	13,13	0	0	n.t.	11	0	3	95	11	2	0
JHA13	13,13	1	0	0	15	0	2	70	13	2	2
KT2	13,13	1	0	2	92	4	4	67	7	2	3
LS40	14,14	100	24	67	38	100	100	100	84	3	7
THO	14,14	98	42	78	40	80	77	35	100	0	4
BIN40	14,14	84	100	100	47	82	67	65	94	0	2
ISA	14,14	86	46	62	100	89	88	95	63	0	9
HAS15	15,15	0	1	52	0	80	1	0	55	100	3

Figure 2. Proliferative responses of 10 alloreactive T-cell clones on HLA-DR4⁻ homozygous typing lines (HTC) representing the five HLA-DR4 subtypes. Human alloreactive T-cell clones were cloned by limiting dilution technique and maintained in tissue culture in the presence of recombinant interleukin-2 as described (6). To define the fine specificity, triplicates of 1×10^4 established T-cell clones were stimulated with 5×10^4 irradiated (10,000 rads) HTC in 96-well microplates. Microcultures were pulsed with 1 μCi per well of [^{3}H] thymidine after 48 hours and harvested after an additional 12 hours of culture. Results are expressed as percent maximal response defined by the following formula: The proliferative response of a clone in the presence of stimulator cells minus counts per minute (cpm) incorporated by stimulator cells alone was divided by the thymidine uptake of the highest responder-stimulator combination in each assay × 100. A relative response of less than 20% was considered negative. Maximal responses were in the range between 12,487 cpm (clone A1-6) to 38,856 cpm (clone JD2-40).

the proliferation of T-cell clones which recognize polymorphic sites strongly suggests that allosteric changes are induced by the binding of the antibody molecule. Different monoclonal anti-DR antibodies differentially affected recognition of the individual T-cell clones, indicating the existence of distinct conformations functionally important in T-cell interaction.

It was surprising to find that most T-cell determinants defined by the clones were not subtype specific. Provided that allospecific T-cell determinants reflect T-cell restriction sites used for the recognition of antigenic peptides, this finding suggests that individuals distinguished by the expression of different HLA-DR4 alleles share multiple lines of T-cell responses. Correlation of reactivity pattern with allelic sequence differences allowed us to analyze the contribution of single amino acid substitutions to the formation of T-cell epitopes. As shown in Figure 4, seven distinct T-cell epitopes were defined. All different T-cell epitopes are included in a 15-amino acid segment. Residues 71, 74, and 86 play a crucial role in the formation of these multiple T-cell epitopes. All three crucial amino acid positions are involved in the creation of more than one T-cell epitope, suggesting that the amino acid sequence 71–86 correlates to a cluster of overlapping T-cell interaction sites. The T-cell determinant characteristic for Dw14 covers the full length of the 15-amino acid stretch. Thus, the sequence 71–86 defines the minimal size of a T-cell recognition site. It is conceivable to speculate that the length of the epitope is not unique for the Dw14-specific site, but rather represents the size of T-cell determinants which are recognized as nonlinear structures.

Analysis of three-dimensional protein structures from primary sequences have predicted an α-helical forma-

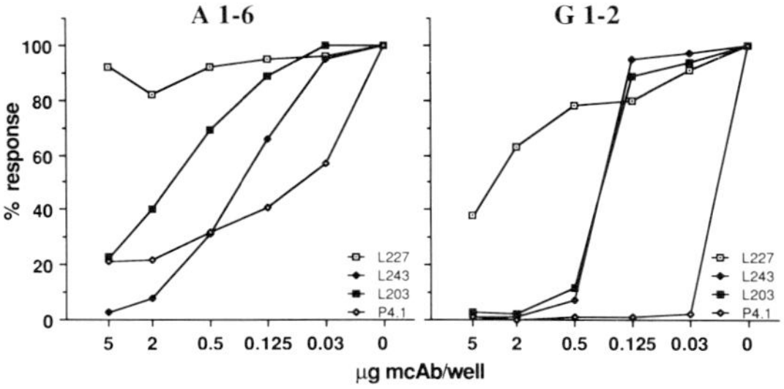

Figure 3. Inhibition of T-cell proliferation by monoclonal anti-HLA-DR antibodies. Increasing concentrations of four different monoclonal antibodies (MAb) specific for monomorphic determinants on HLA-DR molecules (L243, L203, L227, P4.1) were included at the beginning of the coculture experiment. T-cell clones were stimulated with HLA-Dw14⁺ stimulator cells as described in Figure 1.

tion of the β_1 chain around position 71–74 (7). α-helical structures are preferentially stimulatory for T lymphocytes, suggesting that the third hypervariable region is presented to T lymphocytes in an α-helical formation (8). It is not quite clear how position 86 contributes to the formation of the antigenic determinants, however, most defined T-cell epitopes were dramatically affected by position 86. The clonal analysis presented here cannot determine the precise structural features of the T-cell epitopes recognized by T lymphocytes. Positions 71, 74, and 86 might be directly involved in the physical contact with the T-cell receptor molecule and thus represent the histotope (9). However, it cannot be excluded that single substitutions indirectly influence the three-dimensional configuration recognized by T cells. Exon shuffling experiments as well as mapping of T-cell stimulatory sites on site-directed mutant MHC molecules have provided evidence that T cells react with three-dimensional structures accessible on the surface of the folded α-β MHC dimers (10). Data presented here indicate that multiple overlapping T-cell epitopes are formed by a 15-amino acid segment within the third hypervariable region of HLA-DRβ_1 chain. One possible explanation is that the multiple T-cell epitopes are arranged on the surface of the HLA molecule as physically distinct regions. The important contribution of residues 71, 74, and 86, however, raises the possibility that T lymphocytes recognize a major determinant, the fine structure of which is modulated by subtype-specific amino acid substitutions. Quantitative variations in the avidity of different T-cell receptor molecules for that single site would then be the basis for the observed complexity in T-cell responses. In line with this model is the recent report that MHC molecules possess a single binding site for antigenic peptides (1,2). If antigenic fragments associated with single binding sites on HLA molecules are directly recognized by the T-cell receptor molecule, the region of direct physical contact between the T-cell receptor molecule and the HLA epitope has to be limited. One possible model is that amino acid positions arranged around the embedded antigenic fragment are corecognized by the T-cell receptor. We have generated antigen-specific T-cell clones which recognize the protein PPD in restriction to different alleles of the DR4 family. So far, we were able to generate T-cell clones that appear to be restricted to some of the T-cell epitopes defined by the alloreactive T-cell clones. Correlation of reactivity pattern with sequence polymorphism again suggested that the 15-amino acid segment spanning positions 71 to 86 is crucial for the restricted recognition of these T-cell clones. We are now planning to extend our sample of antigen-specific T-cell clones to answer the question whether all T-cell specificities defined by allospecific

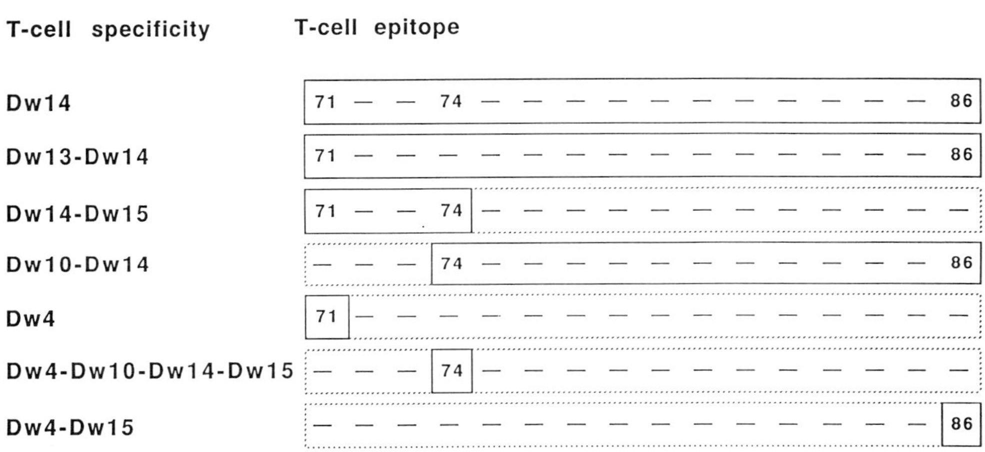

Figure 4. Molecular mapping of seven different epitopes within the third hypervariable region of the HLA-DR4β_1 chain. The cross-reactivity pattern of individual T-cell clones was used to define amino acid positions crucial for T-cell recognition.

clones can be repeated with antigen-restricted T-cell clones.

Acknowledgments. We are grateful to Ms. K.R. Jurutka and Ms. B. Saur for their assistance in the preparation of this manuscript. We thank the German Cancer Research Center for technical support. We appreciate the encouragement and the support of Dr. C.G. Fathman.

References

1. Babbit DP, Allen P, Matsueda G, Haber E, Unanue E. Binding of immunogenic peptides to Ia histocompatibility molecules. Nature 1985;317:359.
2. Guillet JG, Lai M-Z, Briner TJ, Buus S, Sette A, Grey HM, et al. Immunological self, non-self discrimination. Science 1987;235:865.
3. Gregerson PK, Shen M, Song Q, Merryman P, Degar S, Sehi T, et al. Molecular diversity of HLA-DR4 haplotypes. Proc Natl Acad Sci USA 1986;83:2642.
4. Cairns JS, Curtsinger JM, Dahl CA, Freeman S, Alter BJ, Bach FH. Sequence polymorphism of HLA-DR β_1 alleles relating to T-cell-recognized determinants. Nature 1985; 317:166.
5. Weyand CM, Goronzy J, Fathman CG. Human T-cell clones used to define functional epitopes on HLA Class II molecules. Proc Natl Acad Sci USA 1986;83:762.
6. Goronzy J, Weyand CM, Fathman CG. Cloning of human alloreactive T-cells. In: Sabato G, (ed): Methods in Enzymology. Immunochemical Techniques. New York, Academic Press, in press.
7. Norcross MA, Kanehisa M. The predicted structure of the IA β_1 domain. Scand J Immunol 1985;21:511.
8. De Lisi C, Berzofsky JA. T cell antigenic sites tend to be amphipathic structures. Proc Natl Acad Sci USA 1985; 82:7048.
9. Schwartz RH. Immune response (Ir) genes of the murine major histocompatibility complex. Adv Immunol 1986; 38:31.
10. Germain RW, Ashwell JD, Lechler RI, Margulies DH, Nickerson RM, Suzuki G, et al. "Exon shuffling" maps control of antibody- and T-cell-recognition sites to the NH_2-terminal domain of the class II major histocompatibility polypeptide A_β. Proc Natl Acad Sci, USA 1985; 82:2940.

Author Affiliation

Medizinische Klinik und Poliklinik V (Chairman Prof. Hunstein), University of Heidelberg, 6900 Heidelberg, Federal Republic of Germany

DR4 Recognized by PLT Clones

T. Kaneko, M. Ohkubo, K. Itoh, and N. Kashiwagi

Primed lymphocyte typing (PLT) cells raised in a MLR generally recognize HLA-D region products (1). In this study we examined the serologically defined DR4 by our four PLT clones.

The cloning was carried out using the single-cell manipulation method. By this method three clones were isolated from the responder OM (Dw1,w19;DR1,w13; DRw52;DQw1; DPw4) primed by a HLA-DR4 homozygous cell KT3 (Dw15; DR4; DRw53; DQw4; DPw5) and one clone was isolated from the responder IK (Dw7; DR5,7; DRw52, w53; DQw2, w3) primed by another DR4 homozygous cell KT2 (DKT2; DR4; DRw53; DQw3; DPw5) (2). To determine specificities of the PLT clones, a panel study was carried out (Table 1). The stimulator panel used consisted of 13 HLA-D homozygous cells which included nine DR4 homozygous cells (Table 1). When a positive response was defined as 50% or more relative to the priming cell, as shown in Table 1, it was found that all clones reacted shortly against DR4 on the stimulator panel. Next, we examined their relation to the DR4 subtypes that had been split by Dw specificities. As a result, one clone 5D1 could be identified as detecting DKT2 specifically, while two clones M1E8 and 2PD7, which showed a common reaction pat-

Table 1. Panel Study of Four PLT Clones

Panel Cells	HLA			Precent Response of PLT Clones			
	D	DR	DQ	M1E8	2PD7	5G2	5D1
KT3[†]	w15	4,w53	w4	<u>100</u>	<u>100</u>	<u>100</u>	14
KT3B	w15	4,w53	w4	<u>96</u>	<u>80</u>	<u>58</u>	10
KT9	w15	4,w53	w4	<u>52</u>	<u>60</u>	32	22
KT2[※]	DKT2	4,w53	w3	7	2	<u>74</u>	<u>100</u>
KT13	DKT2	4,w53	w3	17	17	<u>91</u>	<u>153</u>
HBH	w4	4,w53	w3	16	38	46	44
LEO10	w10	4,w53	w3	1	18	18	22
SSTO	w13	4,w53	w3	1	3	30	27
LS40	w14	4,w53	w3	20	2	18	5
KT14	Dw23	9,w53	w3	1	2	7	15
PIT	w17	7,w53	w3	5	11	7	27
KT7	w12	2,–	w1	3	9	7	10
KT11	w19	w13,w52	w1	2	1	7	10

*Results were expressed as relative responses to a priming cell stimulation. The control responses were: 16,055 cpm for M1E8; 4,223 cpm for 2PD7; 11,283 cpm for 5G2; and 13,369 cpm for 5D1. Underlined numbers are positive relative responses defined as that of 50% or more.
[†]Priming cell for M1E8, 2PD7, and 5G2.
[※]Priming cell for 5D1.

Immunobiology of HLA
Volume II

Table 2. Effect of Monoclonal Antibodies on Responses of PLT Clones Stimulated with KT3

Monoclonal Antibodies*	PLT Clones			
	Specificities	M1E8	2PD7	5G2
109d6 (ref. 3)	DRw53	28†	23	15
L243 (ref. 4)	DR	77	81	37
I2 (ref. 5)	DR	61	79	78
Hu46 (ref. 6)	DQw4	19	4	12

*All monoclonal antibodies were used at a final dilution of 1:80.
†Number indicates percent inhibition calculated as:

$$\frac{(\text{cpm without antibody} - \text{cpm with antibody})}{(\text{cpm without antibody})} \times 100$$

Responses without antibody were: 19,174 cpm for M1E8; 6,467 cpm for 2PD7; and 13,702 cpm for 5G2.

Table 3. Effect of Workshop Monoclonal Antibodies on PLT Clone 5D1

Monoclonal Antibodies	Specificities	Percent Inhibition of 5D1 Responses*
L243	DR	100
PL8	DR	100
GSP4.1	DR	96
TU35	DR + DP	99
TU39	DR + DQ + DP	92
2D6	DR + DQ + DP	96
1A3	DQ	6
TU22	DQ	13
B7/21	DP	25
PL15	DP	−5
NFCD.M1	DR4	−2
CCCL20	DR4 + DR1	10
PL3	DRs53	16
LEU10	DQw1 + DQw3	10
w6/32	Class I	15

*5D1 was stimulated with an Epstein-Barr virus-transformed B-cell line derived from KT2. All monoclonal antibodies were used at a final dilution of 1:40. The response in the absence of antibodies was 36,552 cpm. Underlined numbers are percent inhibition greater than 30%.

tern on the panel, appeared to detect a unique determinant on Dw15. As the remainder clone 5G2 showed a reactivity to cover these two subtypes, this clone seemed to recognize a common determinant shared by the two subtypes.

To confirm if the determinants detected by these clones are indeed located on the DR4 molecule carried by the priming cell, an inhibition assay of the PLT response by anticlass II monoclonal antibodies was carried out (Tables 2 and 3). The results for the three clones M1E8, 2PD7, and 5G2 are shown in Table 2 and the results for 5D1 are shown in Table 3. By these results we were able to locate all the PLT determinants on the DR4 molecules carried by their priming cells.

In conclusion, we could split DR4 into at least three subtypes by our two PLT specificities, one of which was defined by two clones, M1E8 and 2PD7, and another one by a clone, 5D1. Interestingly, the former specificity seems to be identical to Dw15 and the latter to DKT2.

References

1. Zeevi A, et al. Immunogenetics 1982;16:209.
2. Kaneko T, et al. In: Aizawa M (ed): HLA in Asia-Oceania. 1986, p 425.
3. Maeda H, Hirata R. In: Albert ED (ed): Histocompatibility Testing. 1984, p 511.
4. Bono MR, Strominger JL. Nature 1982;299:836–838.
5. Igarashi M. Personal communication.
6. Kunikame H, et al. In: Aizawa M (ed): HLA in Asia-Oceania. 1986, p 830.

Author Affiliation

Department of Transplantation Immunology, Kitasato University, Kanagawa 228 Japan

A Novel DR5 B1 (DRw11) Chain with a Distinctive Third Diversity Region*

Richard M. Crapper, Peter K. Gregersen, Sicy H. Lee, Antonio Nunez-Roldan, Parvin Merryman, and Robert J. Winchester

Abstract: We have utilized the polymerase chain reaction technique using primers specific for the conserved region flanking the first domain of DRB genes to amplify specific cDNA from the lymphoblastoid cell line OPBU. Two distinct DRB chain sequences were obtained from 43 of 45 clones. One has a nucleotide sequence that has been characterized in our laboratory as that of the DRw10B chain. A second DRB chain appears to be a novel DR5 allele. This variant DR5B1 chain (OP5) is identical in its first and second diversity regions to previously sequenced DR5B1 chains in the DR5 cell lines SWEI and FPA. In contrast, the third diversity region of OP5 differed in sequence from previously defined DR5B1 chains and was identical to the sequences of DRw6a and DR4, Dw10 DRB1 chains in this region.

In the course of analyzing disease associations of DR5 haplotypes (1,2), a DR5/DRw10 heterozygous hispanic individual (OPBU) was selected for detailed study. The serological characterization was A33, B37, DRw10, DQw1 and A32, B35, DR5 (DRw11), DRw52, DQw3. The DR5-containing haplotype of OPBU was reactive with the 7.3.19.1 monoclonal antibody but was negative for reaction with the monoclonal 17-3-3S, differentiating this DR5 haplotype from the DR5 variant DRw12 in GM3160 (3), but emphasizing its similarity to the prototype DR5 cell line SWEI.

The polymerase chain reaction technique (4) was utilized to amplify OPBU cDNA DRB genes and 45 clones were sequenced. Twenty-two were of the DRw10 B gene (5) and 21 were DR5 B1(BI) chain genes of which the prototype was designated OP5. No DRB3 sequences were isolated.

The nucleotide sequence of the OP5, and the sequences of the SWEI-DR5, DRw6a, and DR4, Dw10 are shown in Figure 1. The nucleotide sequence to amino acid position 66, including the first and second diversity regions, was identical to the previously sequenced DR5 B1 chains in the DR5 cell lines SWEI (6) and FPA (7).

```
                                                     10                                      20
SWEI-DR5        GGG GAC ACC AGA CCA CGT TTC TTG GAG TAC TCT ACG TCT GAG TGT CAT TTC TTC AAT GGG
OP5-DR5         --- --- --- --- --- --- --- --- --- --- --- --- --- --- --- --- --- --- --- ---
DRw6a(HHK)      --- --- --- --- --- --- --- --- --- --- --- --- --- --- --- --- --- --- --- ---
DR4,Dw10(FS)    --- --- --- C-- --- --- --- --- --- C-G GT- -AA CA- --- --- --- --- --- --C ---

                                                     30                                      40
DR5             ACG GAG CGG GTG CGG TTC CTG GAC AGA TAC TTC TAT AAC CAA GAG GAG TAC GTG CGC TTC
OP5             --- --- --- --- --- --- --- --- --- --- --- --- --- --- --- --- --- --- --- ---
DRw6a           --- --- --- --- --- --- --- --- --- --- --- C-- --- --G --- --- A-- --- --- ---
DR4,Dw10        --- --- --- --- --- --- --- --- --- --- --- --- C-- --- --- --- --- --- --- ---

                                                     50                                      60
DR5             GAC AGC GAC GTG GGG GAG TTC CGG GCG GTG ACG GAG CTG GGG CGG CCT GAT GAG GAG TAC
OP5             --- --- --- --- --- --- --- --- --- --- --- --- --- --- --- --- --- --- --- ---
DRw6a           --- --- --- --- --- --- --- --- --- --- --- --- --- --- --- --- --- -CC --- ---
DR4,Dw10        --- --- --- --- --- --- -A- --- --- --- --- --- --- --- --- --- --- -CC --- ---

                                                     70
DR5             TGG AAC AGC CAG AAG GAC TTC CTG GAA GAC AGG CGG GCC GCG GTG GAC ACC TAC
OP5             --- --- --- --- --- --- A-- --- --- --- GA- --- --- --- --- --- --- ---
DRw6a           --- --- --- --- --- --- A-- --- --- --- GA- --- --- --- --- --- --- ---
DR4,Dw10        --- --- --- --- --- --- A-- --- --- --- GA- --- --- --- --- --- --- ---
```

Figure 1. DRB1 first-domain nucleotide sequence of the DR5 haplotype OP5. The published first-domain nucleotide sequences of the other DRB1 alleles are given for comparison. cDNA was synthesized from 10 μg of total cellular RNA derived from the lymphoblastoid cell line OPBU. cDNA was subjected to 30 cycles of DNA polymerase 1 Klenow fragment amplification. The DRB first-domain oligonucleotide primers were: 1436 (5'GTGACACTGATGGTGCT-3') annealing in the leader sequence and 1437 (5'-CAGACCAGGAGGTTGTG-3') annealing at the beginning of the second domain.

Immunobiology of HLA
Volume II

From position 59 to position 78 the sequence was identical to that previously determined for DRw6a (8) and DR4, Dw10 (9,10) in this region. A summary of the nucleotide and amino acid sequences of OP5, SWEI-DR5, DRw6a, and DR4, Dw10 from positions 67 to 72 is shown in Table 1. Due to the method used for ligating the amplified gene into the Bluescript vector, the final 46 nucleotides of the first domain could not be determined.

Restriction fragment length polymorphism (RFLP) analysis was performed comparing OPBU with SWEI and a number of other heterozygous HLA-DR5-positive individuals. The restriction enzymes *TaqI* and *BamHI* were used and hybridization was performed with a full-length DRB cDNA probe. Similar restriction fragment patterns were found in OPBU and each of the other DR5 cells with these two enzymes, despite their nucleotide differences, as has been reported for certain other haplotypes (11).

Exchange of DNA involving both class I and class II MHC genes was first postulated on the basis of amino acid replacement data and has been described in several systems (12,13). The mechanism of such events may involve recombinatorial exchanges or gene conversion-like events, as have been described in yeast. Gene conversion (or the closely related phenomenon of double recombination), both intra- and interchromosomal, has been invoked in the generation of allelic polymorphisms. In particular, gene conversion involving the third diversity region, as in the present case, has been implicated in the generation of the DR3 haplotype from DRw6a (8) and in the production of diversity among the Dw subtypes of DR4 (10).

It is not possible from our data to tell whether the novel DRB1 gene OP5 was generated by gene recombination or gene conversion. As DQw3 is strongly linked with DR4 and 5, the DQw3 serological type is compatible with the 5′ end of OP5 being identical with DR5. The fact that DRw52 is supertypic to DR3,5 and -w6 is consistent with the OP5 gene being generated by a single recombination with a DRw6a gene or by a gene conversion-like event with either a DRw6a or a DR4, Dw10 gene.

Recently two allelic forms (a and b) of the DRw52 gene have been described (14), distinguishable by sequencing or by oligonucleotide probes. DRw52a is associated with DRw6a and DRw52b with DR5. Concerning OP5, a determination of DRw52a would favor a single recombination with a DRw6a haplotype and make a discrete gene conversion or double recombination unlikely, whereas DRw52b would favor a local gene conversion-like event and make a single recombination less likely. However, out of the 45 clones sequenced, a DRB3 gene was not obtained. The OP5-containing haplotype is reactive with the 7.3.19.1 monoclonal antibody and studies have shown that this antibody recognizes DRB3 chains in DR5 and DRw6 haplotypes (15). Thus, it would appear that there is a DRB3 gene expressed in OPBU, although probably at a low level.

After this work was completed, the sequence of the first domain of the DRB1 chain of another DR5 cell line, JVM, was reported and this was identical to OP5 (16). JVM has been oligotyped as DRw52b and it is most likely that this will also be the subtype occurring with OP5. As DRw52b is associated with DR5 and not DRw6a, this would favor a local gene conversion-like event as the mechanism responsible for the generation of the OP5-containing haplotype. This local gene conversion mechanism, as distinct from a major recombination, is also supported by the RFLP data showing no difference in the restriction fragment pattern between OPBU, SWEI, and other DR5-positive cell lines.

Despite the difference between the structure of OP5 and the prototypical DR5-SWEI, OPBU typed unequivocally with all available specific DR5 alloantisera. This is consistent with the finding that certain serologically determined HLA-DR specificities depend primarily on the first and second diversity region structure rather than on the third. For instance, in the HLA-DR4 supertypic group there are 5 Dw subtypes, all with different third diversity region sequences but identical first and second diversity regions, and while HLA-DR1 is identical to HLA-DR4, Dw14 in the third diversity region, the first and second regions differ.

Table 1. A Summary of Nucleotide and Amino-Acid Differences in the Third Diversity Region of the First Domain of DRBl Chains

a-a position	67	68	69	70	71	72
SWEI-DR5	TTC Phe	CTG Leu	GAA Glu	GAC Asp	AGG Arg	CGG Arg
OP5-DR5 DRw6a DR4,Dw10	ATC Ile	CTG Leu	GAA Glu	GAC Asp	GAG Glu	CGG Arg

Acknowledgments. The authors wish to thank Dr. J. Silver for excellent advice and helpful discussions, Ms. K. Barr for her skillful technical assistance, and Ms. B. Reynaud for her assistance in preparing the manuscript.

References

1. Nunez-Roldan A, Szer I, Toguchi T, Cuttner J, Winchester R. Association of certain Ia allotypes with the occurrence of chronic lymphocytic leukemia, recognition by a monoclonal anti-Ia reagent of a susceptibility determinant not in the DR series. J Exp Med 1982;156:1872.
2. Lo Galbo P, Nunez-Roldan A, Cuttner J, Winchester R. Association of HLA-DR5 with the occurrences of Hodgkin's Disease in a New York City population. Disease Markers 1985;3:39.
3. Toguchi T, Burmester GR, Nunez-Roldan A, Gregersen P, Seremetis S, Lee S, et al. Evidence for the separate molecular expression of four distinct polymorphic Ia epitopes on cells of DR4 homozygous individuals. Hum Immunol 1984;10:69
4. Saiki RK, Scharf S, Faloona F, Mullis KB, Horn GT, Erlich HA. Enzymatic amplification of B-Globin genomic sequences and restriction site analysis for diagnosis of sickle cell anemia. Science 1985;230:1350.

5. Merryman P, Gregersen P, Lee S, Silver J, Nunez-Roldan A, Crapper R, et al. Nucleotide sequence of a DRw10β chain gene; Identity of the third diversity region with that of the DRw53 allele of the β2 locus and as the probable site encoding a polymorphic MHC class II epitope. Submitted for publication.
6. Tieber VL, Abruzzini LF, Didier DK, Schwartz BD, Rotwein P. Complete characterization and sequence of an HLA class II DRβ chain cDNA from the DR5 haplotype. J Biol Chem 1986;261:2738.
7. Nunez-Roldan A, Gregersen PK, Winchester RJ, Silver J. Analysis of class II genes from a DR5, DQw1 haplotype: Implications for haplotype evolution. Hum Immunol 1986;17:178.
8. Gorski J, Mach B. Polymorphism of human Ia antigens: Gene conversion between two DRβ loci results in a new HLA-D/DR specificity. Nature 1986;322:67.
9. Cairns JS, Curtsinger JM, Dahl CA, Freeman S, Alter BJ, Bach FH. Sequence polymorphism of HLA DRβ1 alleles relating to T-cell-recognized determinants. Nature 1985;317:166.
10. Gregersen PK, Shen M, Song Q, Merryman P, Degar S, Maccari J, et al. Molecular diversity of HLA-DR4 haplotypes. Proc Natl Acad Sci USA, 1986;83:2642.
11. Coppin HL, Villedieu P, Thomsen M, Giraud P, Cambon-Thomsen A, Charron D. HLA-D/DR-BON: A new biochemically and cellularly defined polymorphism indistinguishable by RFLP from DR1. Immunogenetics 1987; 25:130.
12. Plunkett ML, Coligan JE, David CS, Freed JH. Intragenic recombination in an Eβ gene for a murine Ia antigen. J Exp Med 1982;155:937.
13. Lopez de Castro JA, Strominger JL, Strong DM, Orr HT. Structure of cross reactive human histocompatibility antigens HLA-A28 and GKA-A2: Possible implication for generation of HLA polymorphism. Proc Natl Acad Sci USA 1982;79:3813.
14. Gorski J, Tilanis M, Giphart M, Mach M. Oligonucleotide genotyping shows that alleles at the *HLA-DRβ111* locus of the DRw52 supertypic group segregate independently of known DR or Dw specificities. Immunogenetics 1987; 25:79.
15. Karr RW, Lipman LK, Alber C, Koning F, Goeken NE. Ia molecular localization of the DRw52 allodeterminant and the DRw52-like determinants defined by monoclonal antibodies. J Immunol 1985;135:2642.
16. Bell JI, Denney D, Foster L, Belt T, Todd JA, McDevitt HO. Allelic variation in the DR subregion of the human major histocompatibility complex. Proc Natl Acad Sci USA 1987;84:6234.

Author Affiliation

Department of Rheumatic Diseases, Hospital for Joint Diseases, New York University School of Medicine, New York, NY, USA. Current address of *Antonio Nunez-Roldan*: Auda Manual Siurot 3, Bloque 8, 41013 Seville, Spain

Restriction Fragment Length Polymorphism in Serologically Defined DR5: The Difference Between DRw11 and DRw12

R. Tonai, T. Takenouchi, A. Roumanas, M. Kanatani, and P.I. Terasaki

The development of oligonucleotide probes has improved investigations into multiple gene family-like HLA genes by eliminating the cross hybridization bands observed with cDNA probes. We therefore examined the serologically ill-defined antigen DR5 using Bam HI restriction endonuclease with an oligonucleotide probe specific for DR-β (37mer, probe site: amino acid residues 81-93) (1). The Ninth International Histocompatibility Workshop split DR5 into DRw11 and DRw12 and recently, a monoclonal antibody HU-39, which recognizes DN1, has been reported as useful in typing DRw12. Therefore, we also investigated RFLP in families and donors that possess DN1 in order to understand the relationship between DN1 and DR5.

Figure 1 shows the RFLP results of nine well-characterized DR5 homozygous cell lines. Two patterns are evident. Of the seven cells possessing a single 4.5kb band band, two cells (IDF and SWEIG) are designated DRw11. Of the two cells (SH,MIKAGAMI) displaying a 4.2kb and 4.5kb band pattern, SH is DRw12. According to the banding patterns obtained with DR homozygous cell lines (Table 1) (1), we can determine that cells typed as DR5 possess the same 4.2kb and 4.5kb bands seen in Figure 1. Twenty-two of the DR5 cells serologically typed as DRw11 have the single 4.5 kb band, whereas three cells (#1, #5, and R73), serologically DR5 but not DRw11 have the same 4.2kb and 4.5kb band pattern as SH. This data strongly suggests that within DR5, cells that display only the 4.5kb band belong to DRw11 whereas the 4.2kb and 4.5kb band pattern belongs to DRw12.

The banding pattern of DRw11 and DRw12 clearly segregates in family members. In the mother, DR4 and DRw12 banding patterns can be determined because pattern overlap can be deduced from the distinct double intensity of the 4.2kb band. Also, DR4 lacks a 4.5kb band. In the father (#2), DR2 and DRw11 banding patterns overlap at the 4.5kb band. Segregation of DRw11 as one band is demonstrated in children (#3,#4) where the 4.5kb band is present and no other overlap appears. In child #5, typed as DRw11,DRw12, the 4.5kb band of DRw11 overlaps with the 4.2kb and 4.5kb band pattern

Immunobiology of HLA
Volume II

of DRw12 resulting in a double intensity 4.5kb band and normal intensity 4.2kb band.

All five cells typed as DRw12 and displaying the two-band pattern also possessed DN1. However, nine cells (P338, P340, P342, P334, P333, p337, p209, 931, J.C., 894) positive to DN1 but not DRw12 did not show the double-band pattern of DRw12 but displayed a 4.5kb band. This observation was confirmed in Family B. In this family, DN1 appears to segregate with DRw8. Also, study of HU-39 has shown that this antibody binds to the DR molecule and reacts to subpopulations of DR5 and DRw8 (2). These observations show DN1 is not DRw12 and suggest DN1 most probably recognizes a determinant associated with DR5, DRw6, or DRw8. Thus, in the case of the DR5 split, DRw12 is distinguished from DRw11 by the presence of the DN1 determinant. In conclusion, we have shown a molecular basis for the difference between DRw11 and DRw12 and propose DN1 is not DRw12 but a determinant on the DR molecule.

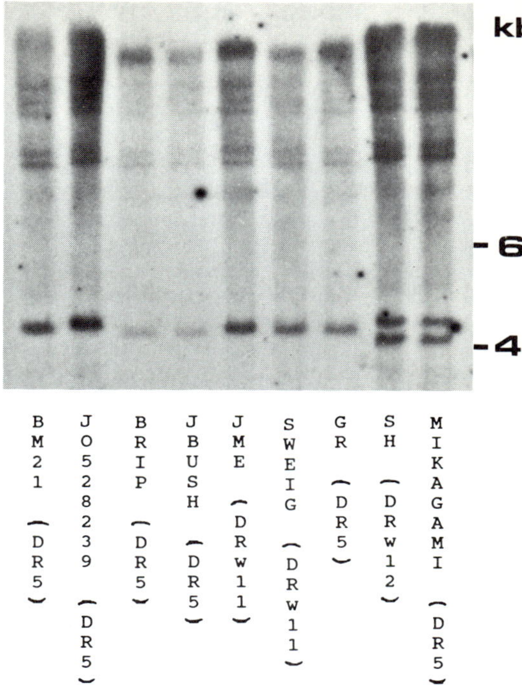

Figure 1. RFLP of DR homozygous cell lines digested by Bam HI with synthetic oligonucleotide DRβ probe.

References

1. Takenouchi T, Wakisaka A, Toyoda H, Clark WS, Terasaki PI. The use of synthetic oligonucleotides probe for HLA DNA typing. Proceedings of the 3rd Asia-Oceania Histocompatibility Workshop Conference, M Aizawa, ed, 1986, pp 855–858.
2. Nakayama T, Kazumasa O, et al. A cytotoxic monoclonal antibody (HU-39) that detects DRw8 + DRw12. Hum Immunol 1987;19:117–126.

Author Affiliations

UCLA Tissue Typing Laboratory, Los Angeles, CA, USA

Table 1. Summary of RFLP Patterns From Bam HI Digestion—DR β Oligonucleotide Probe

	L	H	E	G	E	E	E	W	G	8	G	E	W	K	J	G	M	E	W	P	M	G	K	E
	G	O	B	M	B	B	B	T	M	1QR		A	T	M	R	I	B	T	I	A	I		Y	B
	2	M	V	3	V	V	V	4	3	6			2	E		K	V	4	T	N				V
DR		2	S	1	C	K	A	9	0							A	S	6	O	N				I
Homozygous			A	6	M		K		9							G	H							S
Cell Lines			1	G			I	8								A								K
							B									M								
							A									I								
DR	1	1	1	2	2	2	2	3	3	3	3	4	4	4	5	5	5	12	13	w7	7	ww8	w9	w9
KB																								
6.8	+				+																			
6.0													+	+	+									
5.2	+	+	+	+	+	+	+		+	+	+	+	+	+			+	+	+	+	+		+	+
4.5			+	+	+	+	+	+	+	+	+				+	+	+	+	+	+	+	+		
4.2	+	+	+	+	+	+	+		+	+	+	+	+	+			+	+		+	+		+	+

HLA-DRw6 and Its Complexity

Ernette D. du Toit, Machteld Oudshoorn, Robert W. Martell, and Ken J. MacGregor

HLA-DRw6 was first defined during the Seventh Workshop, but its definition remained difficult after the Eighth Workshop (1). During the Ninth Workshop the existence of subgroups, i.e., DRw13 and DRw14 was accepted, but there were problems in the distinction of DRw12 from DRw6 positive, DQw1 negative specificities (2). This study has reevaluated the definition of the HLA-DRw6 complex in South African populations. The majority of individuals were Negroes and Cape Coloureds. This sample included 57 unrelated DRw6 haplotypes consisting of 50 DRw13 and 7 DRw14 haplotypes. The group included five families with DRw13 and two with DRw14. Using the reaction patterns indicated in Table 1(a), we assigned the DRw8, DRw12, DRw13, and DRw14 specificities without difficulty in most cases. HLA-DRw12 was always seen with DQw7, whereas both HLA-DRw13 and DRw14 occurred with DQw1.

Table 1. Reaction Patterns with DRw8, DRw12, DRw13, DRw14 and the New Specificities DRw12x6 and DRw8x14

Specificity		8	12	11+ 13	11+ 13+ 14	3+ 13+ 14	14	52*
		1 1	9 9	1 1 1	1 1	1 1	1 9	3
		0 0	9 0	1 1 1	1 1	1 1	1 0	0
		8 9	9 5	3 3 2	3 2	5 3	3 6	6
		7 1	9 0	4 3 6	2 3	1 5	6 0	3
DR	DQ							
(a) Classical Specificities								
8	4/7/1	+ +	− −	− − −	− −	− −	− −	−
12	7	− −	+ +	− − −	− −	− −	− −	−
13	1	− −	− −	w + w	+ +	+ +	− −	+
14	1	− −	− −	− − −	+ +	+ +	+ +	+
(b) New Variants								
12x6	7	− −	+ +	− − −	+ +	w +	− −	−
12x6	1	− −	+ +	− − −	w +	w −	+ +	−
8x14	1	+ +	− −	− − −	+ +	+ +	+ +	−

*Negative with DRw8 and DRw12.

In several families the observed segregation patterns could not be applied without resulting in the assignment of "triplets" or even "quadruplets." In order to explain this, we have postulated the existence of two new DR specificities, i.e., *DRw12x6 and DRw8x14* (Table 1(b)). DRw12x6 was positive with all DRw12 sera as well as some DRw6 sera, and was seen in association with DQw1 and DQw7. The DRw12x6,DQw1 haplotype was seen exclusively with the locally described Dw specificity BME in South African Negroes. The extended Negro haplotype A30, Cw6, Bw58, BfS, C4A3, C4B2, 92, DRw12x6, DRw52, DQw1, DwBME was seen in six of the seven individuals with DRw12x6,DQw1. The DRw12x6, DQw7 haplotype was seen only in the mixed descent group and occurred in association with the locally described Dw specificity JOH. Several DQw1 positive cells, which reacted equally well with DRw8 and DRw14 sera, were assigned the specificity DRw8x14 and occurred only in those of mixed descent.

Southern blot hybridization of *Taq* I digested DNA, using the Tenth Workshop Protocol and cDNA probes, was performed on cells from families, including individuals who were homozygous for the new variants (Table 2). Using the DR-β probe, we have shown that the DRw12x6, DQw7 specificity differed from DRw12 (HERLUF) by the presence of a 9.7 Kb fragment and the absence of the 11.3 Kb fragment. It also differed from

Table 2. *Taq* I RFLPs

				DR-β					DQ-α					
ID	DRw	DQw	Dw	11.3†	9.7	7.0	4.3	4.0	7.1	6.5	6.3	4.9	4.7	2.7
HLF*	12	7	HLF	+	−	−	+	+	−	−	−	+	−	−
028*	12x6	7	JOH	−	+	−	+	+	−	−	+	−	−	−
263	12x6	7	JOH	−	+	−	+	+	−	−	+	−	−	−
221	12x6	7	−	−	+	−	+	+	−	−	−	−	+	−
039*	12x6	1	BME	+	−	−	+	+	−	−	−	−	−	+
NSI*	13	1	19	−	+	+	+	−	−	+	−	−	−	−
HHK*	13	1	18	−	+	+	−	+	+	−	−	−	−	−
221	8x14	1	−	−	+	+	−	+	−	−	−	−	−	+
020*	14	1	9	+	−	+	−	+	−	−	−	−	−	+
287	8x14	1	−	+	−	+	−	+	−	−	−	−	−	+
221	8x14	1	−	+	−	+	−	+	−	−	−	−	−	+
SPL	8	4	8.2	−	−	−	−	−	−	−	−	−	−	−

*Homozygous typing cell; †fragment size in Kb.

the DRw13 and DRw14 haplotypes by the absence of a 7.0 Kb fragment. The DRw12x6,DQw1 haplotype had features of both the classic DRw12, DQw7 and DRw14,DQw1 haplotypes when hybridized with DR-β and DQ-α probes. The various DRw8x14 haplotypes revealed different RFLPs when hybridized with the DR-β probe, suggesting this group to be heterogeneous. At least one of these haplotypes had unique DNA bands, with features of both the DRw13, Dw18 and DRw14 specificities.

On the basis of the serologic and Southern blot hybridization patterns observed in several families with DRw6 related specificities, we have postulated the existence of at least two new DR variants. It is conceivable that these new DR variants, DRw12x6 and DRw8x14, are, in fact, hybrid haplotypes.

References

1. Bodmer JG, Pickbourne P, Richards S. Joint Report:Ia serology. In Bodmer WF, et al. (eds): Histocompatibility Testing 1977. Copenhagen, Munksgaard, 1977, 35.
2. Schreuder GMT, Kennedy LJ, Gebuhrer L, Awad J, Betuel H, Degos L, et al. Antigen Report: HLA-DRw6 and its subgroups HLA-DRw13 and DRw14. In: Albert ED, et al. (eds): Histocompatibility Testing 1984. Berlin, Springer-Verlag, 1984, 192.

Author Affiliations

Ernette D. du Toit, Machteld Oudshoorn, Robert W. Martell, Provincial Laboratory for Tissue Immunology, Cape Town; *Ken J. MacGregor,* Department of Computer Science, University of Cape Town, Cape Town, South Africa

DRw6 Polymorphism in American Blacks*

Kyung-Wha Lee, Armead H. Johnson, Ting Tang, Noriko Steiner, Fu Meei Robbins, Robert Hartzman, and Carolyn Katovich Hurley

The DRw6 "specificity" is found on a complex family of at least 10 related haplotypes. DRw6 haplotypes are expressed at a phenotype frequency of 45% in American Blacks. We have studied the polymorphism of HLA-DRw6-related haplotypes in the American Black population using serology, cellular typing, and DNA restriction fragment length polymorphism (RFLP). Several unique DRw6-related haplotypes have been identified.

Using serology, five different DRw6 haplotypes have been identified. Using lysostripping experiments, we postulate a minimum of four DQw1 subtypes associated with DRw13 (DQw6.6a, DQw6.6b, DQw5.6a, and DQw5.6b) and a minimum of two DRw13 subtypes (DRw13 and DRw13v). The HLA-D association with the various DQw1 subtypes was analyzed. Most DRw13, DQw6.6a cells HLA-D type as Dw18, whereas most DRw13, DQw6.6b cells HLA-D type as Dw19. The DRw13, DQw5.6a cells and DRw13v, DQw5.6b cells do not type for any known HLA-D specificity (Dw-).

American Black families expressing different HLA-DRw6 haplotypes were analyzed using RFLP to further define the molecular basis of the DRw6 subgroups. Following *Taq*I restriction enzyme digestion, DRB gene fragments can be associated with particular DRw6-related haplotypes. Two different DRw6-related haplotypes from American Blacks exhibit unique fragments. A 4.6-kb fragment is associated with the DRw13v,

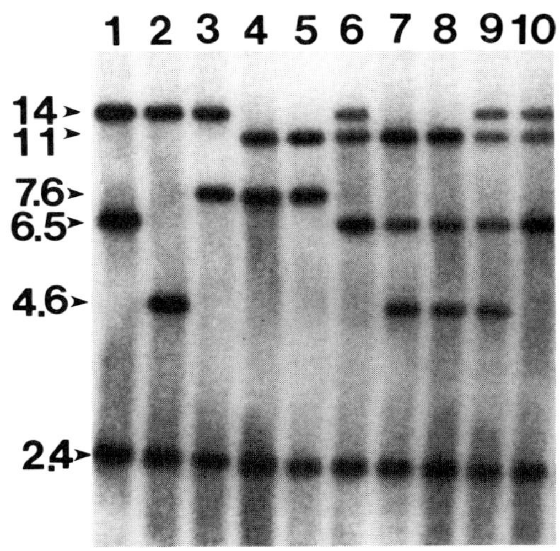

Figure 1. Southern blot analysis of DNA isolated from Epstein-Barr virus-transformed B-cell lines. Genomic DNAs were digested with *Taq*I and hybridized with a ^{32}P-labeled full-length DRB probe as previously described (1). Reference cells: 1 = GM3105A (DR5,DQw3,Dw5); 2 = BM16 (DRw12,DQw7, DB6; WS ID#9038); 3 = APD (DRw13,DQw6.6a,Dw18); 4 = HHK (DRw1,DQw6.6a,Dw18); 5 = ARENT (DRw13, DQw6.6a,Dw18). Family members: 6 = 2015 (a:DRw11, DQw1,Dw0/b:DRw11,DQw7,Dw0); 7 = 1183 (c:DRw13v1, DQw5.6b,Dw0/d:DRw13v2,DQw3,Dw0); 8 = 1180 (a/c); 9 = 1179 (b/c); 10 = 1181 (b/d).

*This research was supported by ONR grant N00014-83-K-0410 (to C.K.H.) and NIH grant AI23371 (to A.H.J.).

DQw5.6a and a 6.5-kb *Taq*I fragment is associated with a variant of DRw6 which could not be serologically discriminated. Figure 1 shows the segregation of the two unique fragments in a Black family expressing both variant DRw6 haplotypes. In contrast, DRw13/Dw18 (APD, HHK, and ARENT) cells contain a 7.6-kb *Taq*I fragment. Homozygous typing cells for DRw13/Dw19 (SLE), DRw14/Dw9 (EK/OH), and DRw14/Dw16 (AMALA) also have the 7.6-kb *Taq*I fragment (data not shown). A DRB gene fragment at 14 kb associated with Dw25, a DRw52 subtype defined by primed lymphocyte typing and related to the DRw52b allele (2,3), could also be identified in the *Taq*I digest. Dw24 and Dw26 cells were found to carry a 11-kb *Taq*I fragment (see also Ref. 3). Using *Taq*I and a DQA probe, five groups of DQw1A gene fragments can be associated with different DRw6 haplotypes. Also, three groups of DQB gene fragments can be detected using the *BamH*I enzyme in DRw6 haplotypes. Ten different DRw6-related haplotypes can be differentiated based on RFLP patterns and correlate with subdivisions obtained using alloantisera and cellular typing. Based on DNA fragment sizes, the DRw6 variants may arise from differences in DRBl, DRB3, DQA, or DQB gene regions.

References

1. Hurley CK, Gregersen P, Steiner N, Bell J, Hartzman R, Nepom G, et al. Polymorphism of the HLA-D region in American Blacks: A DR3 haplotype generated by recombination. J Immunol, 1988;140:885.
2. Gorski J, Tilanus M, Giphart M, Mach B. Oligonucleotide genotyping shows that alleles at the HLA-DR beta III locus of the DRw52 supertypic group segregate independently of known DR or Dw specificities. Immunogenetics 1987; 25:79.
3. Termijtelen A, Tilanus M, Engelen I, Koning F, van Rood J. Molecular localization of LB-Q1, a DRw52-like T-cell recognition epitope and identification at the genomic level of associated shared hybridizing fragments. Hum Immunol 1987;19:255.

Author Affiliations

Kyung-Wha Lee, Armead H. Johnson, Ting Tang, Noriko Steiner, Fu Meei Robbins, and *Carolyn Katovich Hurley*, Departments of Microbiology and Pediatrics, Georgetown University, Washington, DC, USA; *Robert Hartzman*, Naval Medical Research Institute, Bethesda, MD, USA

Molecular Diversity of HLA-DRw6 Haplotypes

T.C. Takahashi, H.T. Kao, J.C. Tang, P. Gregersen, J. Silver, and C.Y. Wang

Two subdivisions of the HLA-DRw6 haplotype have recently been defined serologically as DRw13 and DRw14. We sought to investigate the molecular basis for this serological difference by sequencing the DRβ1 chains from a DRw13-typed and a DRw14-typed cell line. Two cell lines, CBG (DRw13, DQw1, DRw52) and AMALA (DRw14, DQw3, DRw52) were used for this study.

cDNA libraries were constructed from both cell lines using Okayama-Berg cloning vectors (1), and full-length cDNA clones corresponding to the DRβ1 chains were isolated using an oligonucleotide probe corresponding to the 5′ end of the mRNA. Two independently isolated clones were isolated from each library, and each was sequenced bidirectionally using the dideoxy chain termination method (2,3) with oligonucleotide primers corresponding to constant regions of the DRβ1 chain. Ten nucleotide differences giving rise to six amino acid changes were observed between the DRβ1 chains of CBG and AMALA. All observed differences occurred in the first domain of the molecule, with four of the six amino acid changes occurring in the third hypervariable region. We propose that these changes give rise to the serological differences observed between the two cell lines.

Further comparison of the nucleotide sequence was made with that of known sequences. The nucleotide sequence of the DRβ1 chain from CBG is identical to a previously described DR6a, DRβ1 chain sequence described by Gorski and Mach (4). The sequence of the DRβ1 chain from the AMALA cell line has not been described before. This sequence, in fact, bears striking resemblance to the DRβ1 sequence from DR1, with complete identity from amino acid positions 38 to 94 (Fig. 1). At 9 of the 10 nucleotides where AMALA and CBG differ, AMALA is identical to DR1, DRβ1 (5,6). To explain this apparent identity to the DR1 sequences, we suggest that an interchromosomal gene conversion event between a DR1, DRβ1 chain and a DR6, DRβ1 chain may have given rise to a new DR6, DRβ1 chain as seen in the AMALA cell line.

References

1. Okayama H, Berg P. High-efficiency cloning of full-length cDNA. Mol Cell Biol 1983;2:161–170.

```
                                                   10                                                20
       Gly Asp Thr Arg Pro Arg Phe Leu Glu Tyr Ser Thr Ser Glu Cys His Phe Phe Asn Gly
CBG    GGG GAC ACC AGA CCA CGT TTC TTG GAG TAC TCT ACG TCT GAG TGT CAT TTC TTC AAT GGG
AMALA  --- --- --- --- --- --- --- --- --- --- --- --- --- --- --- --- --- --- --- ---
DR1    --- --- --- C-- --- --- --- --- TG- C-G CT- -A- -T- --A --- --- --- --- --- ---

                                                   30                                                40
       Thr Glu Arg Val Arg Phe Leu Asp Arg Tyr Phe His Asn Gln Glu Glu Asn Val Arg Phe
CBG    ACG GAG CGG GTG CGG TTC CTG GAC AGA TAC TTC CAT AAC CAG GAG GAG AAC GTG CGC TTC
AMALA  --- --- --- --- --- --- --- --G --- --- --- --- --- --- --- --- --- --- --- ---
DR1    --- --- --- --- --- --G --- --A --- -G- A-- T-- --- --A --- --- TC- --- --- ---

                                                   50                                                60
       Asp Ser Asp Val Gly Glu Phe Arg Ala Val Thr Glu Leu Gly Arg Pro Asp Ala Glu Tyr
CBG    GAC AGC GAC GTG GGG GAG TTC CGG GCG GTG ACG GAG CTG GGG CGG CCT GAT GCC GAG TAC
AMALA  --- --- --- --- --- --- -A- --- --- --- --- --- --- --- --- --- --- --- --- ---
DR1    --- --- --- --- --- --- -A- --- --- --- --- --- --- --- --- --- --- --- --- ---

                                                   70                                                80
       Trp Asn Ser Gln Lys Asp Ile Leu Glu Asp Glu Arg Ala Ala Val Asp Thr Tyr Cys Arg
CBG    TGG AAC AGC CAG AAG GAC ATC CTG GAA GAC GAG CGG GCC GCG GTG GAC ACC TAC TGC AGA
AMALA  --- --- --- --- --- --- C-- --- --G C-G AG- --- --- --- --- --- --- --- --- ---
DR1    --- --- --- --- --- --- C-- --- --G C-G AG- --- --- --- --- --- --- --- --- ---

                                                   90                                               100
       His Asn Tyr Gly Val Val Glu Ser Phe Thr Val Gln Arg Arg Val His Pro Lys Val Thr
CBG    CAC AAC TAC GGG GTT GTG GAG AGC TTC ACA GTG CAG CGG CGA GTC CAT CCT AAG GTG ACT
AMALA  --- --- --- --- --- -GT --- --- --- --- --- --- --- --- --- --- --- --- --- ---
DR1    --- --- --- --- --- -GT --- --- --- --- --- --- --- --- --G G-G --- --- --- ---
```

Figure 1. DNA sequence and predicted protein sequence of the first domain of the DRβ1 chains from the CBG (DRw13) and AMALA (DRw14) cell lines. Also shown is a comparison to the nucleotide sequence from the same region in a DR1, DRβ1 chain (5,6).

2. Sanger F, Nicklen S, Coulson AR. DNA sequencing with chain-terminating inhibitors. Proc Natl Acad Sci USA 1977; 74:5463–5467.
3. Chen EY, Seeburg P. Supercoil sequencing: A fast and simple method for sequencing plasmid DNA. DNA 1985;4: 165–170.
4. Gorski J, Mach B. Polymorphism of human Ia antigens: Gene conversion between two DRβ loci results in a new HLA/DR specificity. Nature 1986;322:67–70.
5. Tonnelle C, DeMars R, Long EO. DOB: A new β chain in HLA-D with a distinct regulation of expression. EMBO J 1985;4:2839–2847.
6. Gregersen P, et al. (1987). Unpublished observations.

Author Affiliations

T.C. Takahashi, H.T. Kao, J.C. Tang, and *C.Y. Wang*, Molecular Biology Division, United Biomedical, Inc., 2 Nevada Drive, Lake Success, NY 11042, USA; *P. Gregersen* and *J. Silver*, Molecular Biology Unit, Department of Rheumatic Diseases, The Hospital for Joint Disease, New York, NY 10003, USA

Sequence Analysis of the DRw13 β Chain Genes:: The Dw19 Specificity May Be Encoded by the DRB3 Locus

Jack Gorski, David D. Eckels, Jean-Marie Tiercy, Catherine Ucla, and Bernard Mach

Abstract: The first domain sequences of the two expressed DRB chain genes from a DRw6/Dw19 cell line, WT46, were determined. The B1 sequence varied by only a single amino acid from that reported earlier from a DRw6/Dw18 cell line, HHK. The B3 sequence, however, defines a new DRw52 allele referred to as DRw52c. In an accompanying paper (Tiercy et al., these proceedings) we show, using allele-specific oligonucleotide probes, that DRw52c is only associated with DRw6. Here, analysis of DRw6 cells, whose Dw specificities are known, shows that the presence of the DRw52c allele correlates with the presence of Dw19. In view of the almost identical B1 sequence, we propose that the T-cell recognition specificity, Dw19, when encoded by DR molecules, is encoded by the B3 locus. Thus this form of Dw19 is an allele of LBQ1, LBQ4, BO, other T-cell specificities encoded by the B3 locus. These sequences, together with the recent crystal structure of HLA, permit a modeling of the DRw52 chains that reveals polymorphic antigenic surfaces implicated in alloreactive T-cell recognition.

The HLA-DR haplotypes constituting the DRw52 supertypic group are characterized by the presence of two expressed DR β chain loci which are the product of a relatively recent gene duplication (1). One of these loci, B1, encodes the private DR specificities and a public specificity; the other, B3, encodes only the supertypic specificity. We have been analyzing the sequence polymorphism of this particular group of haplotypes (2). Alleles at the B1 locus so far identified are DR3, DR5, DRw6/Dw18, and DRw6/Dw9. Alleles at the B3 locus are DRw52a and DRw52b. DR3 and DRw6/Dw18 can be associated with either DRw52a or DRw52b (3).

With the identification of two alleles at the B3 locus, allele-specific oligonucleotide hybridization was used to correlate the presence of these alleles with known phenotypes. A strong correlation was found with a number of allo T-cell and serological specificities. These include the T-cell specificities LBQ1, LBQ4 (4, and Termijtelen et al. these proceedings), BO (5, and Rowe et al. these proceedings). The correlation would break down in the case of a subset of DRw6 cells, some of which were known to be DRw13/Dw19. To further investigate this we cloned the DRB1 and B3 cDNAs from a DRw13/Dw19 cell line, WT46, and sequenced the polymorphic first domains. The amino acid sequence of the first domain of the B1 chain was almost identical to that of the B1 chain from a DRw13/Dw18 cell line HHK (Table 1a). The B3 chain, however, represents a new allele that differs from both of the previously identified DRw52 alleles (Table 1b). It is of interest to note that this allele is a pastiche of the DRw52a and DRw52b allelic sequences and may represent the result of a gene conversion.

The difference between the Dw18 and Dw19 associated DRB1 chain is a Val to Gly change at position 86. This polymorphism is frequently encountered in other specificities. For example, a DRw52b chain from a DR5 cell line (Swei) differs from that of a DRw52b chain from a DRw6 line in only this position, and B3 sequence from a DR2/Dw2 differs from that of DR2/Dw12 only in this position.

The small difference found in the B1 chains appears to be insufficient for determining the Dw18/Dw19 specificity associated with the cells from which the sequences were derived. The B3 differences are much more likely to be responsible for generating different T-cell recognition epitopes. To test this we first determined that DRw52c was indeed only associated with DRw6 haplotypes (6, and Tiercy et al., this volume). We then analyzed a panel of DRw6 cells corresponding to the DRw6 LCL of the Tenth Workshop using RNA oligonucleotide typing (7) to determine their DRw52 type (Table 2). The Tenth Workshop number designation is given for each cell as well as the preliminary Dw type assigned to that cell and its DRw52 oligo typing. As can be seen, the DRw52c assignment correlated with the Dw19 T cell typing. This result indicates that the Dw19, or that sub-

Table 1.

Locus B1	
a.a. position	86
WT46 (DRw13/Dw19)	V
HHK (DRw13/Dw18)	G

B3											
a.a. position	11	26	28	30	37	38	51	57	60	74	86
WT46 (DRw52c)	L	F	E	Y	F	V	T	V	S	Q	C
HHK (DRw52a)	R	Y	D	Y	F	L	T	V	S	R	Q
4/6 (DRw52b)	L	F	E	H	Y	A	R	D	Y	Q	V

Immunobiology of HLA
Volume II

Table 2. DRw52 Typing of DRw6 Cell Lines Using Oligonucleotide Probes

Cell ID		DW Type	DRw52 Type
9053	HOR	19	c
9054	EK	9	b
9055	HO	19 var	c
9056	KOSE	19 & 9	b & c
9057	TEM	9	b
9058	OMW	18	N.D.
9059	SLE	19	c
9060	CB	18	b
9061	ABO	9	b
9062	WDV	18	a
9063	WT47	19	c
9064	AMAL	16	a
9065	HHK	18	a
9097	EMJ	19	c

set of the Dw19 specificity encoded by DR, may be encoded by the same chain that encodes DRw52c.

DRw52c positive cells (Dw19) are not recognized by many of the T-cell reagents that recognize DRw52a (LBQ4) and DRw52b (LBQ1, BO) positive cells and have been shown by monoclonal antibody blocking experiments to be encoded on the DRB3 chain (4,5). A Dw19-restricted T-cell reagent (reported as part of the Tenth Workshop) was also blocked by monoclonal antibodies specific for the DRB3 chain in the DRw52 haplotypes (Mickelson et al, these proceedings).

The sequence of these three DRw52 alleles along with the recent elucidation of HLA structure (8,9) can allow us to model possible recognition elements that stimulate the various alloreactive T-cell reagents (Fig. 1). An alloreactive DRw52a T cell would recognize a spatial cluster of differences on the first and second β sheets associated with the DRw52c β chain as well as a polymorphic difference on the α helix pointing sideways into the pocket (Fig 1A). A DRw52b T cell would recognize the DRw52c chain at a different location, closer to the edge of the 52c molecule (Fig 1B). Therefore, the priming combinations determine what antigenic surface the T cell is allowed to "see." Sometimes these will be unique to a DR molecule, at other times shared by two DR molecules resulting in a "public" T-cell specificity. It is of great interest to note that the polymorphisms of these DRw52 alleles is restricted in such a manner as to form a cluster with just the right geometry to be recognized by an Ig-like receptor molecule (circles in Fig. 2), which in this case is the T-cell receptor. The role of an endogenous peptide in the DR binding pocket, if such exists, is not clear for the moment. By definition, in those situations where allorecognition is exhibited, either the peptide is not interfering with recognition (for example it binds elsewhere in the pocket) or it may even contribute to the binding via allosteric influences of the polymorphic DR residues. These forms of analysis, where primary structural data are combined with functional data, will prove to be extremely useful in our efforts to understand the restriction of T-cell recognition.

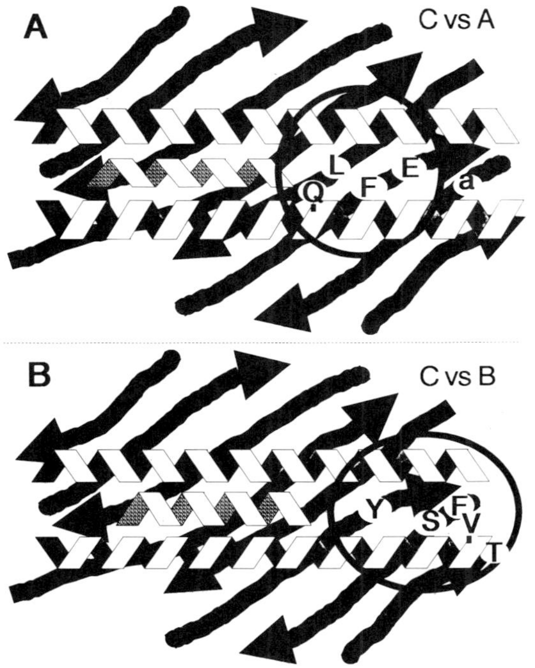

Figure 1. Schematic representation of the two external domains of DRw52. The molecule is modeled after the class I crystal structure (6,7) with the β1 domain of DR being equivalent to the α2 domain of HLA-A2. The arrows represent β sheets, the leftmost four from the DR α chain, the rightmost four from the β chain. The top α helix is from the DRα chain, the bottom from the β chain. Amino acid polymorphism is shown using the one-letter code. Upper case: extending up out of the plane of paper. Lower case: extending down. The circle represents the size of an immunoglobulin-like combining region. A peptide is drawn occupying the DR binding pocket but not interfering with the polymorphic region (see text for discussion).

Acknowledgments. This work was supported by The Blood Center Research Foundation and the Swiss National Science Foundation. We would like to thank our many colleagues for discussions.

References

1. Rollini P, Mach B, Gorski J. Linkage map of three HLA-DR β-chain genes: Evidence for a recent duplication event. Proc Natl Acad Sci USA 1985;82:7197
2. Gorski J, Mach B. Polymorphism of human Ia antigens: Gene conversion between two DR β loci results in a HLA-D/DR specificity. Nature 1986;322:67.
3. Gorski J, Tilanus M, Giphart M, Mach B. Oligonucleotide genotyping shows that alleles at the HLA-DR βIII locus of the DRw52 supertypic group segregate independently of known DR or Dw Specificities. Immunogenetics 1987; 25:79.
4. Tiercy J-M, Gorski J, Jeannet M, Mach B. Identification and distribution of three serologically undetected alleles

of HLA-DR by oligonucleotide typing. Proc Natl Acad Sci USA (in press).
5. Termijtelen A, Tilanus M, Engelen I, Koning F, van Rood JJ. Molecular localization of LB-Q1, a a DRw52-like T-cell recognition epitope and identification at the genomic level of associated shared hybridizing fragments. Hum Immunol 1987;19:255–267.
6. Sheehy MJ, Rowe JR, Koning F, Jorgensen L. Functional Polymorphism of the HLA-DR beta-III chain. Hum Immunol (in press).
7. Ucla C, van Rood JJ, Gorski J, Mach B. Analysis of HLA-D micropolymorphism by a simple procedure: RNA oligonucleotide hybridization. J Clin Invest 1987;80:1155.
8. Bjorkman PJ, Saper MA, Samraoui B, Bennett, WS, Strominger JL, Wiley DC. Structure of the human class I histocompatibility antigen, HLA-A2. Nature 1987;329: 506.
9. Bjorkman PJ, Saper MA, Samraoui B, Bennett WS, Strominger JL, Wiley DC. The foreign antigen binding site and T cell recognition regions of class I histocompatibility antigens. Nature 1987;329:512.

Author Affiliations

Jack Gorski, David D. Eckels, Immunogenetics Research Section, The Blood Center of Southeastern Wisconsin, Milwaukee, WI, USA; *Jean-Marie Tiercy, Catherine Ucla, Bernard Mach*, Department of Microbiology, University of Geneva Medical School, Geneva, Switzerland

DR and DQ α and β Polymorphism in DRw13- and DRw52+-Related Haplotypes

Alain Haziot, Catherine Freidel, Virginia Lepage, Lucette Gebuhrer, Marilyn Sasportes, Laurent Degos, Herve Betuel, and Dominique Charron

We reported previously the diversity of DR molecules found in association with DRw13 haplotypes (1,2). In order to dissect the complexity of DRw13 haplotypes at the molecular level, we performed two-dimensional polyacrylamide gel electrophoresis (2D-PAGE) analysis of DQα and -β chains from DRw13 cells. Molecular relationships between DRw13 and other DRw52+ haplotypes were established by comparison of DRβ, DQα, or DQβ chains.

We have previously shown (1) that DRw13-DQw1-Dw18 cells and DRw13-DQw1-Dw19 cells have a common DRβ1 chain (DRβ1a) and two different DRβ2 chains (DRβ2a and DRβ2b), as summarized in Table 1. In this study, we identified a molecular variant form of DRw13-DQw1-DW18 haplotypes with a more acidic DRβ2 chain, DRβ2c, while the DRβ1a is conserved. The DQα and -β chains associated with these haplotypes were analyzed in 2D gel with, in the first dimension, a pH 3.5 to 10 NEPHGE gel for the β chains, and a pH 5 to 7 isoelectric focusing (IEF) gel for the α chains. The α chains as well as the β chains of all these cells migrated to identical positions on the gels. Comigration experiments confirmed the electrophoretic identity of DQ molecules in Dw18 and Dw19 cells. Our previous results showed that DRw13-DQw3-9w1802 cells expressed a DRβ2 identical to the DRβ2a of most Dw18 cells in association with a more acidic DRβ1, DRβ1b. The DQα chain and the DQβ chain associated with this haplotype were distinct from the DQα and -β chains of DRw13-DQw1 cells. Within other DRw52+ haplotypes, the DR or DQ molecules of DRw12-DQw3-DB6 and HLA B8-DR3-DQw2 haplotypes were compared with the class II molecules found in the DRw13 cells and described above. The DRw12-DQw3-DB6 cell expressed a DRβ2c identical to the one found in the Dw18 variant, and expressed DRβ1b, DQα and DQβ chains similar to the ones detected in the DRw13-DQw3-9w1802 cells. In HLA B8-DR3, two DRβ chains were detected, one of them being identical to the DRβ2a detected in most Dw18 cells and in the DRw13-DQw3 cells.

In conclusion, we showed the complex molecular relationships within DRw52 haplotypes. Particularly, considering the polymorphism of the DRβ2 chain, DRw52 haplotypes are subdivided in most Dw18, DR3, and DRw13-DQw3 cells, variant Dw18 and DRw12-DB6 cells, and Dw19 cells, while three distinct DRβ1 chains are found in Dw19 and all Dw18 cells, DRw13-DQw3 and DRw12-DB6 cells, and DR3 cells.

Table 1. DRβ, DQα, and DQα in DRw52 Haplotypes

	DRβ1	DRβ2	DQ(α+β)
Dw18	a	a	DQw1
Dw18var	a	c	DQw1
Dw19	a	b	DQw1
DRw13-9w1802	b	a	DQw3
DRw12-DB6	b	c	DQw3
B8-DR3	c	a	DQw2

Immunoprecipitation of [^{35}S]methionine-labeled molecules as done with MAbs D1.12 for DR molecules and G2a5, IIb3, 33.1, and PLM2 for DQ molecules.

References

1. Haziot A, Lepage V, Freidel AC, Betuel H, Degos L, Charron DJ. Biochemistry of HLA-DRw6:Evidence for seven distinct haplotypes. J Immunol 1986;136:3767.
2. Haziot A, Michon J, Sterkers G, Degos L, Levy JP, Charron DJ. Molecular localization to a unique DR β-chain of the restriction element of an anti-influenza T cell clone. J Immunol 1987;138:1834.

Author Affiliations

Alain Haziot and *Dominique Charron*, Laboratoire d'Immunogenetique Moleculaire des Cordeliers, 15 Rue de l'Ecole de Medecine, 75006 Paris, France; *Catherine Freidel, Lucette Gebuhrer*, and *Herve Betuel*, CTS, Lyon, France; *Virginia Lepage, Marilyn Sasportes* and *Laurent Degos*, Inserm U93, Paris, France

Structural Mapping of T-Cell Clones Restriction Elements Suggests That a DQα-DRβ Hybrid Molecule Could Be Functional in DRw13 Cells

Alain Haziot, Jean Michon, Ghislaine Sterkers, Laurent Degos, Jean-Paul Levy, and Dominique Charron

Abstract: The activation of competent T cells requires the recognition of both an antigen and a major histocompatibility complex (MHC) molecule. The strong polymorphism of HLA-D genes, their homologies, and their complex associations in various haplotypes hindered the localization of functional determinants at the polypeptidic level. Here, using a selected panel of DRw13-DQw1 cells we looked for correlations between the two-dimensional gel profiles of class II antigens and the reactivity of antigen-specific T-cell clones. The polymorphism of DRβ, DQα, and DQβ chains encoded by DRw13-DQw1 haplotypes (Dw18 and Dw19) was studied by two-dimensional polyacrylamide gel electrophoresis (2D-PAGE), revealing one DRβ1 and two DRβ2 chains and no electrophoretic polymorphism of DQα and DQβ chains. Eight T-cell clones specific for influenza A-infected DRw13-DQw1 antigen presenting cell (APC) were generated from a Dw19 homozygous individual. Three patterns of reactivity were observed: 1) three clones proliferated with Dw18 and Dw19 APC and were blocked by anti-DR and anti-DRβ1 monoclonal antibody (MAb), showing a restriction by the DRαβ1 molecule; (2) three clones proliferated with Dw19 APC and were blocked by anti-DR MAb, (The restriction by the DRαβ2 unique to Dw19 is very likely); 3) two clones proliferated with Dw19 APC. Since these clones were blocked by both anti-DQ and anti-DR MAb and since we did not observe any structural difference by 2D gel analysis in DQ products from Dw18 and Dw19 cells, these results suggest a restriction by a hybrid DQα-DRβ2 molecule.

We have previously investigated the polymorphism of DR and DQ α and β chains in DRw13-DQw1 cells using 2D gel electrophoresis (1,2, A. Haziot et al., this volume). We have shown that all DRw13-DQw1 cells possessed an identical DRβ1 chain. In contrast, the DRβ2a chain detected in DRw13-DQw1-Dw18 cells was distinct from the DRβ2b chain present in DRw13-DQw1-Dw19 cells (Table 1). As shown in Figure 1 (A,B, and D,E) the DQα and -β chains detected in Dw18 and Dw19 cells migrated in identical position in 2D gels. When, in comigration experiments, DQ molecules from Dw18 and Dw19 cells were run on the same gel, no additional spot was detected compared to the DQ patterns obtained when DQ molecules from each cell type were separately run in a gel (Fig. 1, C and F). Thus, Dw18 and Dw19 cells express electrophoretically identical DRα, DRβ1, DQα, and DQβ chains and a different second DRβ chain, DRβ2a and DRβ2b, respectively.

Eight influenza A-specific proliferative T-cell clones were obtained from a DRw13-DQw1-Dw19 individual. As previously reported (3) no DRw13-negative cell could present the viral antigens to these clones. Moreover, as summarized in Table 2, three T-cell clones (C11, C16, and C17) proliferated in the presence of Dw18 and Dw19 infected cells, while only the Dw19 cells, matching the autologous cells, could present the viral antigens to the five other T-cell clones (C14, C21, C23, C12, and C20).

Blocking experiments with anti-HLA-class II-specific monoclonal antibodies (MAbs) showed (Table 3) that only anti-DR MAbs inhibited the proliferation of the Dw18- and Dw19-restricted clones C11, C16, and C17. These three clones appeared similar to a previously

Table 1. DRβ Chains Detected by Two-Dimensional Polyacrylamide Gel Electrophoresis in Dw18 and Dw19 Cells

	DRβ1	DRβ2a	DRβ2b
Dw18	+	+	−
Dw19	+	−	+

Immunobiology of HLA
Volume II

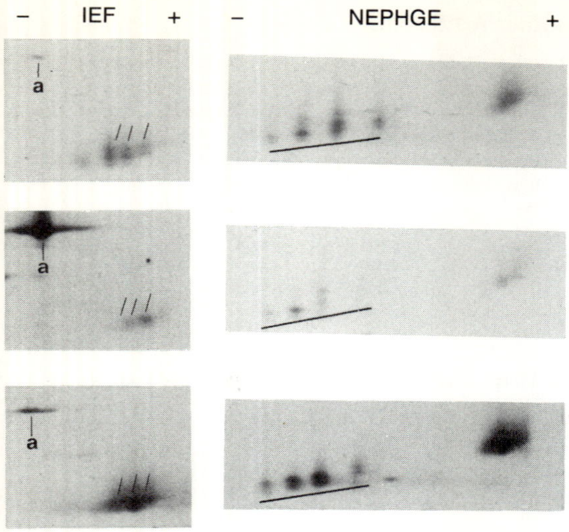

Figure 1. Two-dimensional gel electrophoresis identity of DQ molecules in Dw18 and Dw19 cells: DQβ and DQα spots from Dw18 and Dw19 cells migrate in identical positions and superpose when comigrated in one gel. Cells were labeled with [^{35}S]methionine and the DQ molecules were precipitated with the anti-DQw1 MAb G2a5. In the first dimension, β chains were separated on a pH 3.5 to 10 NEPHGE gel and α chains on a pH 5 to 7 isoelectric focusing (IEF) gel, as described (1,5). In comigration experiments, samples of the two cells were mixed and run as described above. The acidic ends of the gels are on the right, the basic ones on the left. a = Actin. A and B, DQ chains from HBS4 (Dw18) and M3P (Dw19), respectively; C, Comigration; D and E, DQ chains from HBS4 and M3P; F, comigration.

described T-cell clone C2 obtained from the same individual in a previous cloning experiment (4) and for which the restriction element was localized to the DRβ1 chain of DRw13-DQw1 cells (2). Similarly, considering the inhibition of the clones C11, C16, and C17 by the DRβ1-specific MAb ILR2 and the absence of proliferation of these clones in the presence of infected DR3 or

Table 2. Patterns of Restriction of Eight Influenza Virus-Specific T-Cell Clones by Dw18 and Dw19 Cells

	C11	C16	C17	C14	C21	C23	C12	C20
Dw18	+	+	+	–	–	–	–	–
Dw19	+	+	+	+	+	+	+	+

Cloned cells were cultured 3 days with irradiated and virus-treated stimulator cells, pulsed the last 18 hr with [^{3}H]TdR, harvested and counted, and described (3).

DRw13-DQw3-9w 1802 cells which express the DRβ2a also present in Dw18 cells (A. Haziot et al., this volume), we concluded that the clones C11, C16, and C17 are restricted by the DRα-β1 molecule which is common to Dw18 and Dw19 cells.

Out of five Dw19-restricted T-cell clones, three clones (C14, C21, and C23) were strongly blocked by anti-DR MAbs and not by anti-DQ MAb and thus appeared to be restricted by the DRα-β2b molecule which is absent from Dw18 cells and unique to Dw19 cells. Accordingly, the DRβ1-specific MAb ILR2 did not block the clone C14.

The two other Dw19-restricted clones, C12 and C20, were strongly blocked by anti-DQ MAb and, although to a lower extent, also by anti-DR MAb. Since no blocking of the clones C12 and C20 was observed with the control MAb (anti-DP B7/21 and anti-HLA class I W6/32), the inhibition induced by the anti-DR MAb appeared specific and not due to a background activity. Since the DQ molecule is electrophoretically identical by 2D gel analysis in the Dw18 and Dw19 cells studied, and since the clones C12 and C20 were restricted only by Dw19 cells, the existence of an epitope common to DR and DQ molecules which would restrict these clones is unlikely. It is as well very unlikely that clones C12 and C20 each include two T-cell populations since these clones were obtained independently and exhibit very similar patterns of reactivity. Alternatively, DQw1 α and β chains could differ only by a very limited number of conservative substitutions that would not affect the electrophoretic mobility of the molecule. Whether those would affect T-cell recognition is questionable. In any case, this alternative would not explain the anti-DR monoclonal anti-

Table 3. Inhibition of the Proliferation of T-Cell Clones by Monoclonal Antibodies

MAb	Anti-DR				Anti-DRβ1	Anti-DQ			Controls	
Clones	LKT111	NG2	L243	D1.12	ILR2	BT3/4	33.1	Leu-10	B7/21	W6/32
C11	+++	+++	+++	++	+++	–	+	–	–	–
C16	+++	+++	++	+	ND	–	–	–	–	–
C17	+++	+++	++	+	++	–	–	–	–	–
C14	+++	+	++	+	–	–	–	–	–	–
C21	+++	+++	+++	++	ND	–	–	–	–	–
C23	+++	+++	+++	++	ND	–	+	–	–	–
C12	++	+	++	++	–	+++	+++	++	–	–
C20	++	+	++	+	–	+++	+++	++	–	–

Ten dilutions of ascites fluids were added at the initiation of the culture, as described (3). The results are expressed as the percentage of inhibition: +++, 60 to 100%; ++, 30 to 60%; +, 10 to 30%; –, less than 10%; ND, not done.

body blocking effect observed for these two clones. Therefore, these results suggest that a hybrid HLA class II molecule composed of one DR chain and one DQ chain is implicated in the restriction of the clones C12 and C20. A single mixed-isotype DR-DQ dimer is unique to Dw19 cells. It is formed by the association of the DQα chain and the DRβ2b chain. The restriction by this DQα-DRβ2b molecule can explain the patterns of reactivity of the clones C12 and C20. Accordingly, the anti-DRβ1 chain MAb ILR2 did not inhibit the proliferation of C12 and C20. Recently, the existence of a mismatched isotype DRα-DQβ molecule in a nontransfected DR2 B-cell line was reported (5). Here, these results suggest that a hybrid DR-DQ molecule could be functional in antigen presentation to T cells.

In conclusion, we show that antigen presentation can occur in association either with DRα-β1 or with DRα-β2 molecules. In addition, the ability of a DQα-DRβ2 molecule to be efficient in immune responses, as suggested here, would greatly increase the repertoire of functional HLA class II determinants.

References

1. Haziot A, Lepage V, Freidel AC, Betuel H, Degos L, Charron DJ. Biochemistry of HLA-DRw6:Evidence for seven distinct haplotypes. J Immunol, 1986;136:3767.
2. Haziot A, Michon J, Sterkers G, Degos L, Levy JP, Charron DJ. Molecular localization to a unique DR β-chain of the restriction element of an anti-influenza T cell clone. J Immunol 1987;138:1834.
3. Michon J, Henin Y, Sterkers G, Freidel C, Gebuhrer L, Betuel H, et al. Highly polymorphic products of both HLA-DR and HLA-DQ genes contribute to the polymorphism of the HLA-DRw13 haplotype. Immunogenetics 1986; 24:177.
4. Michon J, Sterkers G, Haziot A, Freidel AC, Gebuhrer L, Betuel H, et al. Despite the lack of monospecific anti-DRw13 anti-sera, a corresponding class II determinant can be defined by a DRw13 restricted anti-influenza virus T cell clone. Hum Immunol 1986;15:272.
5. Lotteau V, Teyton L, Burroughs D, Charron D. A novel HLA class II molecule (DRα-DQβ) created by mismatched isotype pairing. Nature 1987;329:339.

Author Affiliations

Alain Haziot and *Dominique Charron*, Laboratoire d'Immunogenetique Moleculaire, 15 rue de l'Ecole de Medecine, 75006 Paris, France; *Jean Michon, Ghislaine Sterkers*, and *Jean-Paul Levy*, Inserm U152, Paris, France

Restriction Fragment Length Polymorphism of HLA-DR7 Alleles and Association with HLA-B Antigens

Elizabeth A. Bidwell, Jeffrey L. Bidwell, David A. Savage, Derek Middleton, Trevor J. Jones, Peter T. Klouda, and Ben A. Bradley

Although HLA-DR7 appears to be serologically homogeneous (1), it may be split according to epitopes recognized by T cells (Dw7, Dw11) and is normally found in association with either DQw2 or DQw3. This report describes the correlation between restriction fragment length polymorphism (RFLP) and phenotypic splits of DR7. Genomic DNAs from 74 DR7-positive individuals and from eight DR7-homozygous typing cells were analyzed by sequential hybridization with short (exon-specific) DRβ, DQβ, and DQα cDNA probes (J. L. Bidwell et al, this volume). Figure 1 shows the RFLP patterns revealed by each of these probes. Four combinations of RFLP patterns were established by analysis of segregation within families, or by predicted association due to linkage disequilibrium:

1. *DRβ7*[1]-DQβ*2b*-*DQα3* (n=44), previously shown to correlate with DR7-Dw7-DQw2 phenotypes, and associated in this study predominantly with HLA-B12 (n=22) or B14 (n=4);
2. *DRβ7*[2]-DQβ*2b*-*DQα3* (n=15), previously shown also to correlate with DR7-Dw7-DQw2 phenotypes, but associated in this study predominantly with B13 (n=4) or B50 (n=2);
3. *DRβ7*[2]-DQβ*3a*-*DQα3* (n=32), previously shown to correlate with DR7-Dw11-DQw3 phenotypes and associated in this study predominantly with B57 (n=12). We have reported elsewhere that this RFLP combination is indistinguishable from that revealed by DRw9-DQw3 phenotypes (2);
4. *DRβ7*[2]-DQβ*2c*-*DQα3:* this combination was found in one Nigerian individual and is characterized by a previously unreported DQβ *Taq*I RFLP, *DQβ2c*, differing from *DQβ2b* in that the former lacks a 7.2kb signal and possesses a 4.0kb signal (Fig. 1).

The correlation between RFLP and DQ/Dw splits of DR7 has been reported previously (2,3). This study confirms these correlations and in addition, describes further associations between HLA-B antigens and observed HLA-DR and DQ RFLPs.

Immunobiology of HLA
Volume II

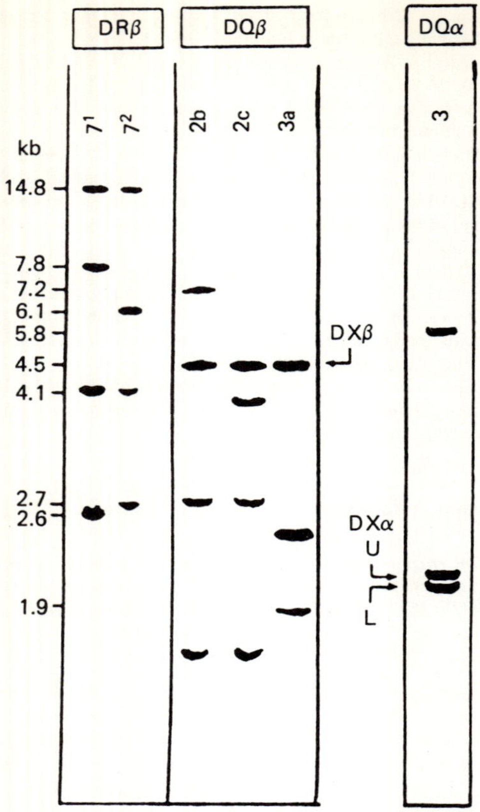

Figure 1. TaqI-specific DR7 RFLP patterns revealed with exon-specific DRβ, DQβ, and DQα cDNA probes. ◄

References

1. Jakobsen BK, Terasaki PI, Svejgaard A. In: Albert ED, et al (eds): Histocompatibility Testing 1984. Berlin, Springer-Verlag, 1984, p 198.
2. Bidwell JL, Bidwell EA, Savage DA, Middleton D, Klouda PT, Bradley BA. A DNA-RFLP typing system which positively identifies serologically well-defined and ill-defined HLA-DR and DQ alleles, including DRw10. Transplantation 1988;45:640–646.
3. Trucco M, Duquesnoy RJ. Polymorphism of the HLA-DQ subregion. Immunology Today 1986;7:297–300.

Author Affiliations

Elizabeth A. Bidwell, Jeffrey L. Bidwell, Trevor J. Jones, Peter T. Klouda, Ben A. Bradley, United Kingdom Transplant Service, Bristol, England; *David A. Savage, Derek Middleton*, Belfast City Hospital, Belfast, Northern Ireland, U.K.

Two-Dimensional IEF Analysis of DR, DQ, and DP Molecules from DRW8-Carrying Haplotypes

Hiroo Maeda, Ranko Hirata, and Takeo Juji

HLA-DR, DQ, and DP molecules were isolated from eight HLA-DRw8 homozygous B-cell lines using the monoclonal antibodies (MAb) submitted to the 10th International Histocompatibility Workshop. These molecules were analyzed by two-dimensional IEF gel electrophoresis as previously described (1).

A total of three α and three to four β chains were isolated from each cell line; one DR, one DQ, and one or two DP molecules. Among these B-cell lines, two distinct DR β chains (DRβ1 and DRβ2) and three DQ β chains (DQβ1, DQβ2, DQβ3) were identified making four different DR/DQ haplotypes; DRβ1/DQβ1 (#9066 TAB089), DRβ1/DQβ2 (#9067 BTB, #9068, #9069), DRβ1/DQβ3 (#9070 LUY), and DRβ2/DQβ2 (#9071 OLGA, #9072, #9100). Both DR molecules were isolated by PL8; DRβ2 was one charge unit acidic to DRβ1. DQβ1 and DQβ2 were isolated by PLM13, while DQβ3 was precipitated by PLM12; DQβ1 corresponded to DQw1 β chain, DQβ2 to DQw4 β chain, and DQβ3 to DQw7(TA10+) β chain from the DRw8-carrying haplotypes, respectively.

Complex DR/DQ haplotype combinations were generated; the first three groups ("TAB", "BTB", and "LUY") shared the DRβ1 chain but carried different DQ β chains from each other, while "OLGA" shared the DQβ2 chain with "BTB" but carried a distinct DRβ chain (DRβ2). The data strongly suggest that the DRw8 prototype would be the "BTB" type (DRβ1/DQβ2), which would have generated new haplotypes by making recombinations between DR and DQ subregions with DQw1

Immunobiology of HLA
Volume II

Table 1. DR, DQ, and DP β Chains from a Panel of DRw8 B-Cell Lines

WS ID	Cell Name	DR/Dw	DQ*	DP[†]	DRβ	DQβ	DPβ
#9066	TAB089	w8/DB7	w1	7	β1	β1	β4
#9067	BTB	w8/w8	w4	4	β1	β2	β5
#9068	BM9	w8/w8	w4	2B?	β1	β2	β2
#9069	MADURA	w8/w8	w4	4	β1	β2	β5
#9070	LUY	w8/8.3	w7	1,4	β1	β3	β3,β5
#9071	OLGA	w8/8.2	w4	3,4?	β2	β2	β2,β4
#9072	SPACH	w8/8.2	w4	4B?	β2	β2	β4
#9100	OLL	w8/–	w4	3,4B?	β2	β2	β4,β?

*DQ types locally determined; [†]DP types by Eckels.

("TAB", DRβ1/DQβ1) or DQw3 ("LUY", DRβ1/DQβ3). The "OLGA" type (DRβ2/DQβ2) might be generated by mutations of DRβ1 into DRβ2 from "BTB" since both DR β chains carry the DRw8 epitope and are most likely evolutionally related.

Another interesting finding is that four different DR/DQ haplotypes perfectly correlated with HLA-D types; "TAB" correlated with DB7, "BTB" with Dw8, "LUY" with Dw8.3, and "OLGA" with Dw8.2. In the primary MLR in one combination of these HTC cells, DR molecules are different and may be stimulatory to each other, while in another combination DQ molecules are different and may be stimulatory. Thus, DQ molecules as well as DR molecules appear to be responsible for the HLA-D specificity on the DRw8-carrying haplotype.

These DR/DQ haplotypes were found differently with one or two DP molecules isolated by B7/21. The tentative nomenclature was given to these DP molecules as previously published (2), where five distinct DP β chains (DPβ1-DPβ5) were described. Four DPβ chains (DPβ2-DPβ5) were identified (Table 1). A correlation between DPβ5 and the cellularly defined DPw4 antigen was confirmed. However, the other three DP β chains were not correlated with any DP antigens in this study.

References

1. Maeda H, Hirata R, Okuyama M, Thompson A, Tohyama H. Two-dimensional gel analysis of a second family of class II molecules by polymorphic HLA-DR4, 5, and w9 monoclonal antibodies. J Immunol 1984;132:2478.
2. Maeda H, Hirata R, Juji T, Katagiri M. Isolation and characterization of 9C4-reactive class II molecules. In: Aizawa M (ed): HLA in Asia-Oceania 1986. Sapporo, Hokkaido University Press, 1986, 444.

Author Affiliations

Hiroo Maeda, Ranko Hirata, Blood Transfusion Service, Saitama Medical Center, Saitama Medical School, Kawagoe, Saitama; *Takeo Juji*, Blood Transfusion Service, Tokyo University Hospital, Tokyo, Japan

Identification of a Cellularly Defined DRW8 Subtype

M. Bonneville, J.F. Moreau, M.L. Cheneau, F. Bonneville, E. Blokland, J. Pool, E. Goulmy, J.D. Bignon, J.Y. Muller, and J.P. Soulillou

Cellular mechanisms involved in the allograft rejection process remain highly controversial. Using monoclonal antibodies, several studies have demonstrated convincingly that most of the infiltrating mononuclear cells are T lymphocytes. Moreover, several investigators have developed techniques for culturing T cells out of various human allografts in order to delineate functions of intragraft T-cell populations.

From a rejected human kidney graft, a limiting dilution technique (1) was used to clone a large number of graft-invading cells with clonal efficiency of 1/13. Out of 55 clones, 20 were tested for 1) proliferative activity in primed lymphocyte typing (PLT); 2) cytotoxic activity in cell-mediated lympholysis (CML); and 3) cytotoxic activity inhibited by monoclonal antibodies (MAb). A large panel of well HLA-defined cells (PBL, PHA-blasts, and B lymphoblastoid cell lines–LCL) were used as stimulator or target cells. This panel included all DRW8 positive cells from workshop homozygous B.LCL (BM9, TAB 089, MADURA, BTB, OLGA, LUY, S. PACHEO, OLL, SPL), seven heterozygous DRW8 positive cells from Blood Bank of Leiden (Netherlands), and cells from recipient BER (DR2-DRW6) and donor DAB (DR5-DRW8). The following MoAbs were used to inhibit cytotoxicity experiments against donor-BLCL: W6.32, LEU 10, NDS 10, 1A3, B7/21, 2D6, GSP 4.1, NDS 13, UK 8.1, CHE 249.2, MAD 88 (all workshop reagents), and BT 2.9 (anti-class II), D1.12 (anti DR), VI.15 (anti DR), and BM.50 (anti DR).

Clone "1D9" was selected for its ability to proliferate with and to kill only but not all cells bearing DRW8 phenotype (Table 1). This clone was significantly

Table 1. Proliferative and Cytotoxic Activities of Clone 1D9

Stimulators (PTL) or Targets (CML)	HLA DR/DQ	Clone 1D9 Responder/Effector in PLT (cpm)*	in CML (% lysis)†
Le. #15 (BLCL)	7-W8/W2	21040	25/22
Le. #16 (BLCL)‡	1-W8/W1	3285	-/-
Le. #24 (PHA blasts)	W8–W13/W1	10383	-/-
Le. #26 (PHA blasts)	1-W8/W1	11010	-/-
Le. #28 (PHA blasts)	W8–W14/W1	10121	-/-
Le. #29 (PHA blasts)	W8–W13/W1	13327	-/-
Le. #31 (BLCL)‡	3–W8/W2	5055	-/-
LUY (workshop) (BLCL)	W8/W3	18573	24/12
BER (recipient BLCL)	2–W14/W1	2100	-/-
DAB (donor BLCL)	5–W8/W3	20600	16/17

*Results expressed as mean (triplicate) cpm of 3H–TdR uptake of clone cultured 72 hours with stimulating cells; †% specific 51 Cr release calculated at two effector/target ratios (20:1/4:1); ‡Leiden #16 and #31 were not recognized by 1D9.

inhibited by broad anti-class II and anti-monomorphic DR MoAb, but not by anti-DQ or anti-DP MoAb. Monoclonal antibody MAD 88 (anti-DRW8) elicited against DRW8 positive cell MADURA, which was lysed by clone "1D9," did not block "1D9" cytotoxicity directed against donor cell DAB (DRW8 positive). But this antibody labeled in immunofluorescence the Leiden #16 cell (DRW8 positive), which in its turn was not lysed by "1D9," clearly suggesting that "1D9" and MoAb MAD 88 recognized two different epitopes on the β chain of the DRW8 molecule.

In order to identify at the DNA level this cellularly defined DRW8 subtype, a restriction fragment length polymorphism (RFLP) study was carried out to distinguish a DNA polymorphism of these DRW8 cells according to their sensitivity to clone "1D9." Genomic DNAs from nine DRW8 positive homozygous workshop cells, from two recognized heterozygous DRW8 positive cells (Leiden #15 and donor DAB), and from a non-recognized heterozygous DRW8 positive cell (Leiden #16) were digested with Eco RI, Bam HI, Taq I, and Hind III and then hybridized with workshop DR β, DQ α, and DQ β probes. With DR β probe and all enzymes tested, DRW8 haplotype revealed a characteristic pattern distinct from those of others DR specificities. Moreover a specific fragment of 8.9 Kb was noted with Taq 1. But no polymorphism was found at the DNA level because the same RFLP pattern was observed for all DRW8 positive cells including the non-recognized Leiden #16 cell. All the restriction enzyme used to hybridize the DQ α and DQ β probes revealed three different DQ patterns. These results were concordant with those previously reported by serology (2).

In summary, clone 1D9 specific for the kidney donor cells is described. It killed neither K562 nor autologous BLCL. On a large panel of well HLA-characterized cells it recognized a majority of DRW8 positive cells (15/17 cells). On the other hand, all DRW8 negative cells (N = 22) were not recognized. Proliferative and cytotoxic activities were concordant. RFLP analysis of DRW8 positive cells either recognized or not by 1D9 did not revealed differences with DR β probe. Hybridization with DQ α and DQ β probes exhibited three different patterns without any relationship to 1D9 reactivity. 1D9 T-cell clone might be directed against a DRW8 subtype which would need further investigations (other restriction enzymes and probes) at the DNA level. Alternatively, the DRW8 molecule might be the restriction element for some yet unknown minor alloantigens.

Acknowledgments. We thank D. Gauvin for her excellent technical assistance and Ms. V. Gallee for preparation of the manuscript.

References

1. Moreau JF, Bonneville M, Peyrat MA, Godard A, Jacques Y, Desgranges C et al. T lymphocyte cloning from rejected human kidney allografts. J Clin Invest 1986;78:874.
2. Betuel H, Gebuhrer L, Schreuder GMT, Layrisse Z, Arnaiz Villena A, Goldmann SF. In: Albert ED, et al (eds): Antigen report: HLA-DRW8 in histocompatibility testing 1984. Berlin, Springer-Verlag, p 198.

Author Affiliations

M. Bonneville, J.F. Moreau, J.P. Soulillou, INSERM U211, Nantes; *M.L. Cheneau, F. Bonneville, J.D. Bignon, J.Y. Muller*, Centre Transfusion, Nantes, France; *E. Blockland, J. Pool, E. Goulmy*, University Hospital, Leiden, The Netherlands

Molecular Polymorphism of HLA-DRw8

Angelica Olivo, Santiago Rodríguez de Cordoba, Pablo Rubinstein, and Clara Gorodezky

The phenotypic hallmark of the HLA complex is a high degree of polymorphism and class II polymorphic differences, especially in the B chain genes, have been associated with immune response and with disease susceptibilities in humans. The RFLPs are now widely used to demonstrate subtle variations in single antigens and to characterize new alleles, serologically defined as blanks. As examples: two splits of DRw52 (DRw52a and DRw52b) have been found recently using oligonucleotides (1); BglI digests have shown a different restriction pattern in DRw8 haplotypes when analyzing the DR3, DRw6, DRw8, 8w13 group (2).

We report here the DNA restriction pattern of three families who have a DRw8-DQwx haplotype. Two families (no. 1 and no. 2) have a mestizo ethnic background and the third is of Caucasian origin. The analysis was done using the Southern blot technique according to the 10th International Workshop protocol. DNA was digested with six restriction endonucleases (EcoRI, TaqI, BamHI, EcoRV, PvuII, and HindIII). Hybridization was performed with DRβ,DQβ,DQα,DPβ, and class I probes, provided as part of the workshop reagents, and labeled with ^{32}P.

Hybridization of genomic BamHI digested DNA with a DRβ probe and of EcoRV digested with a DQβ probe identified genetic polymorphism of DRw8 with different restriction sites. In family no. 2, a particular DRw8 DRβ BamHI, 11.8 Kb fragment was present in the DRw8 members, which does not exist in the DRw8 individuals of the other families. Interestingly, this fragment is also present in another sibling of the same family who has DR1 in one haplotype and DRw6 in the other one. Moreover, the band has a double intensity in the mother who is DRw6/DRw8 (Fig. 1, p. 230). These results are consistent with the hypothesis of ancestral evolutionary relationship between DRw8 and DRw6 genes. It seems also that DRw8 is in linkage disequilibrium with two different serologically undefined DQ products (DQwx), because in family no. 2 where the 11.8 Kb DRβ BamHI fragment is present, an EcoRV DQβ 14.0 Kb band is absent, but exists in the DRw8 individuals of the other two families (no. 1 and no. 3). The DRw8 characteristic band, DRβ Taq1 of 8.5 Kb, previously described (3), is present in every DRw8 subject in this study. We are thus suggesting a new split of DRw8 that may be in linkage disequilibrium with a particular undefined DQ specificity, in agreement with protein biochemistry findings currently in press (4). The results are encouraging since we reported in Mexicans that DRw8-DQw1 and DRw8-DQwx are the most common combinations of DRw8. Studies with a larger panel of DRw8 cells are currently in progress to obtain more information on these possible new DR and DQ gene products.

Acknowledgments. This paper was supported in part by NIH contract NO1-A1-62-523.

References

1. Angelini G, Preval C, Gorski J, Mach, B. High-resolution analysis of the human HLA-DR polymorphism by hybridization with sequence-specific oligonucleotide probes. Proc Natl Acad Sci USA 1986;83:4489.
2. Angelini G, Preval C, Ferrara GB, Mach, B. Analysis and segregation of HLA-DR by DNA typing: Correlation with supertypic specificities. In: Albert ED et al (eds): Histocompatibility Testing, 1984. Berlin, Springer-Verlag, p 582.
3. Kohonen-Corish MJK, Serjeantson SW. HLA-DRB gene polymorphisms revealed by Taq1 correlate with HLA-DR specificities Human Immunol 1986;15:263.
4. Rodríguez de Cordova S, Núñez-Roldan A, Winchester R, Mollen P, Walker M, Geensberg-Fellner F, et al. Molecular characterization by high resolution by isoelectric focusing of the products encoded by the class II loci of the MHC in humans. Hum Immunol (in press).

Author Affiliations

Angelica Olivo, Clara Gorodezky, Departamento de Inmunogenética, Instituto de Salubridad y Enfermedades Tropicales, SSA, Mexico City, Mexico; *Santiago Rodríguez de Cordoba, Pablo Rubinstein*, Department of Immunogenetics, The Lindsley F. Kimball Research Institute of the New York Blood Center, New York, NY, USA

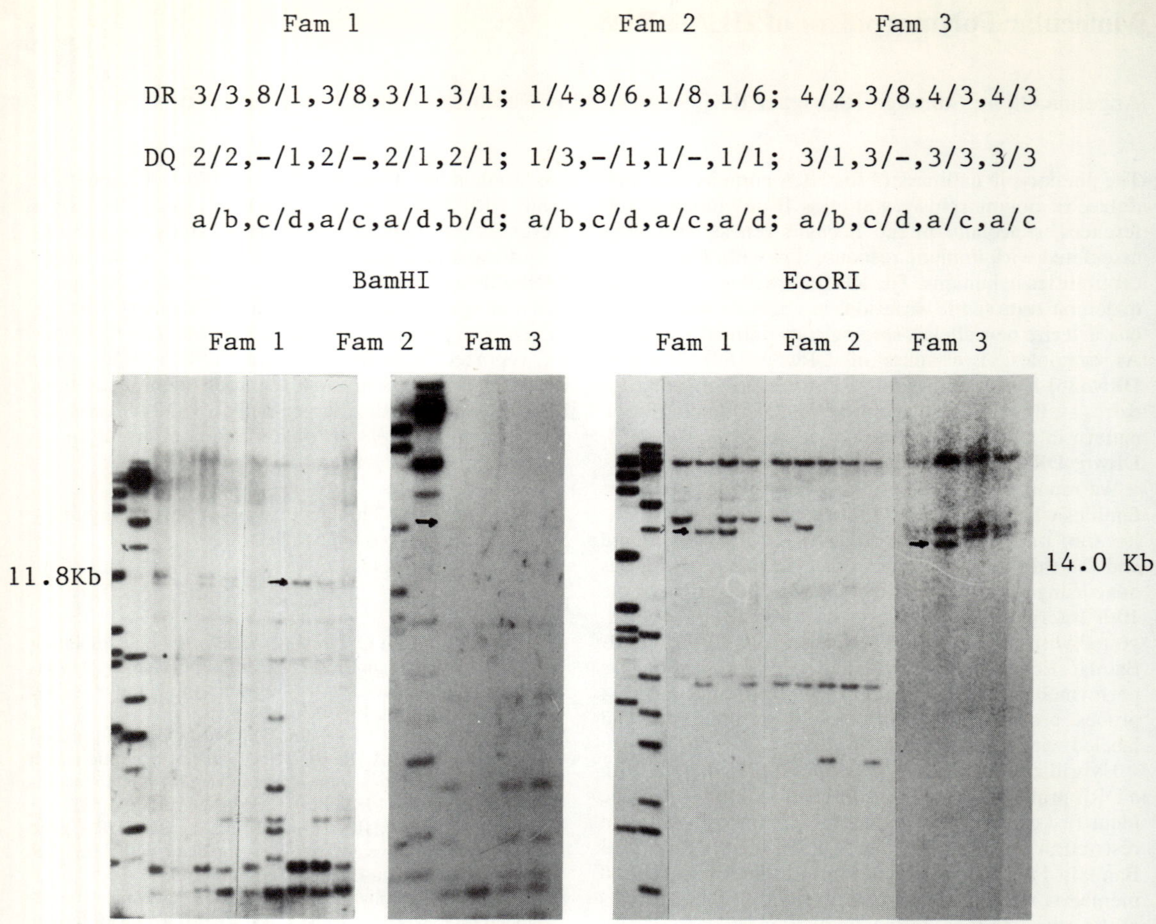

Figure 1. Autoradiograms of blots from experiments of three families where DNA samples were digested by BamHI and hybridized with a DRB probe (left) and EcoRV hybridized with a DQB probe (right). Arrow shows the BamHI 11.8 Kb DRB fragment present in DRw8 individuals of family 2 and absent in families 1 and 3. (left). DQB EcoRV 14.0 Kb fragment is shown by an arrow in families 1 and 3, which is absent in family 2 (right).

Biochemical Studies of the DRw9 Specificity in Caucasians

R. Fauchet, O. Bouhallier, A. Gaulier, G. Semana, V. Lotteau, and D.J. Charron

DRw9 specificity has been definitively considered at the Eighth Workshop as different from DR4-DR7 specificities. Today, DRw9 is serologically well defined by monomorphic and plurispecific allosera. DRw9 is commonly linked to DQw3.

The frequency in the Caucasian population is low (0.20%), but reaches 1.5% in our local Caucasian population against 23% in Japanese population. DRw9 molecule has been studied by two-dimensional gel electrophoresis according to P. Jones' technique, from DRw9 haplotypes of nine EBV cell lines (five unrelated and four from a family with DRw9-HTC). DR1, DR2, DR7 homozygous cell lines were used for complementary tests. Four monoclonal antibodies: D1 12 (anti-DR), TAL1 B5 (anti-DRα), BT2 9 (anti-B Class II), LEU10 (anti-DQw1,anti-DQw3) made possible the identification of DRw9 haplotype. DR-DQ typing was assessed by local and Ninth Workshop sera. A, B, DR, DRw, DQ typing cytotoxicity and RFLP typing are given in Table 1.

Table 1. Material

Subject		DR	DRw	DQw Serology	DQw RFLP
Family 1					
Father	A	2–9	53	1–3 Ta10–	Ta10–
Mother	B	4–9	53	3–3 Ta10+	Ta10+ Ta10–
Sib	1	9–9	53	3–3 Ta10–	Ta10–
Sib	2	2–9	53	1–3 Ta10–	Ta10–
Unrelated					
1		2–9	53	1–3 Ta10–	Ta10–
2		5–9	53	3 Ta10–	Ta10+-Ta10–
3		7–9	53	2–3 Ta10–	Ta10–
4		4–9	53	3 Ta10+	ND
5		9	52–53	1–3 Ta10–	Ta10–
6		1–1		1–1	
7		2–2		1–1	

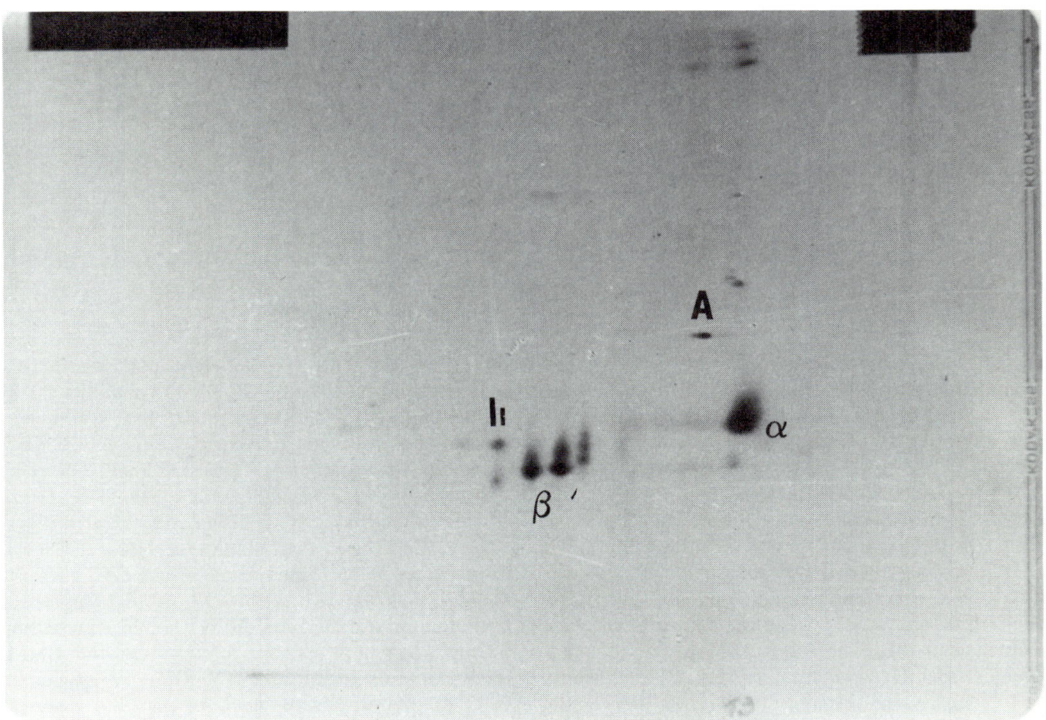

Figure 1. The common DRw9 B1 chain pattern found in all DRw9 cell studied.

Different molecules can be immunoprecipitated from DRw9 haplotype:

DRw9 β*1 chain (D1 12, TAL 1 B5, monoclonal antibodies). A common DRw9* β1 chain pattern was found in all the DRw9 cells studied, formed by four spots, the most basic of which is located vertically below the Ii invariant spot (Fig. 1, p. 231). In heterozygous haplotypes (family members or unrelated), two DR β chains were found, one specific to the DRw9 β1 chain, and the other one associated either to DR2 or DR4 or DR5 or DR7 β chain. Moreover, a common precursor DR1-DR2-DRw9 β1 chain can be postulated. Indeed, comigration tests of DR1-DR1, DR2-DR2, DRw9-DRw9 HTC showed that the most basic spot migrates identically for the three determinants and corresponds to the precursor spot by pulse test. However, the other more acidic spots of DRw9 β1 chain 2D pattern were not superposable with the DR1 β or DR2 β chain patterns and define specifically one DRw9 β1 chain (1).

DRw9 β2 chain was faintly immunoprecipitated by BT2-9 and could correspond to the DRw53 chain. In Japanese data, an associated β2 chain DRw9 has been described using DRw53 alloantisera (2).

DQw3 molecule (Leu10 MAb). The DRw9 cells typed DQw3,Ta10−possessed a DQw3 β chain formed by three spots different from the DR spots by their electrophoretic mobility and their molecular weight. The DQw3 β chain pattern of DRw9 cells, all typed DQw3, Ta10−appeared different from the DQw3, Ta10+ pattern. In a DR4,DRw9,DQw3, DQw3, Ta 10+ Ta10−, two DQw3 β chain existed: one basic DQw3 (Ta10+) and the other one DQw3 (Ta10−) more acidic. Data were confirmed by RFLP.

In conclusion, in DRw9 haplotype, different β chains were immuno-precipitated: a DRw9-specific β1 chain presenting the same precursor with DR1, DR2 β1 chains; a weak DR β2 chain, which may correspond to the DRw53 β chain; a DQ β chain defining the DQW3 Ta10− associated variant. No DRw9 variants were found in the studied material.

References

1. Yabe T, Suzuki M, Mukai R, Satake M, Juji T, Hamaguchi H. Analysis of the DRw9 antigen using two-dimensional gel electrophoresis and alloantisera. Jpn J Human Genet 1984;29:17.
2. Maeda H, Hirata R, Thompson A, Mukai R. Molecular characterization of three HLA class II molecules on DR4 and DRw9 haplotypes: Serologic and structural relationship at the polypeptides level. Human Immunology 1986;15:1.

Author Affiliations

R. Fauchet, O. Bouhallier, A. Gauthier, G. Semana, Laboratoire d'Histocompatibilité—Centre Régional de Transfusion Sanguine, Rennes; *V. Lotteau, D.J. Charron*, Laboratoire Immunogénétique Moléculaire, Institut des Cordelier, Paris, France

Molecular Definition of the DRW10 Haplotype

Sicy H. Lee, Antonio Nunez-Roldan, Parvin Merryman, Peter K. Gregersen, Richard M. Crapper, and Robert J. Winchester

Seven individuals with rheumatoid arthritis who gave a typing reaction with DRw10 and partial reactions with DR1 alloantisera were studied with the aim of better defining the DRw10 haplotype. Family studies of these individuals by 2-D gel analysis demonstrated that the DRw10 haplotype encodes reactivity with Ab109d6 that is expressed on a β chain with a characteristic basic mobility. Ab 109d6 otherwise reacts with an epitope on an acidic $DR\beta_4$ molecule that bears the DRw53 specificity (1,2), but this $DR\beta_4$ chain was not identified. Southern blot analysis of genomic DNA from these individuals and the panel of workshop cells digested with TaqI or BamHI were done using a full length DRβ probe to identify RFLPs that characterize the DRw10 haplotype. This haplotype exhibited a pattern readily distinguishable from DR1, 4 or 7. Using TaqI digestion (Table 1A), we identified a distinctive 12.2 Kb fragment unique to the DRw10 haplotype and readily distinguished from the 13.8 Kb DRw53 and the 11.3 Kb DRw52a fragments. The 6.0 Kb TaqI fragment was absent in DRw10 but present in DR1. A 4.6 Kb TaqI fragment was common to both DR1 and DRw10, but absent in DR4 or 7. Further evidence that DRw10 can be distinguished from DR1 is provided by BamHI digestion (Table 1B). The DRw10 haplotype had a 4.3 Kb fragment absent in DR1, but present also in DR2 and DRw52 (DR3, 5, 13 and 14). A 5.07 Kb BamHI fragment was common to both DR1 and DRw10 but distinct from the 5.26 Kb fragment specific for DR4 and the 4.80 fragment common to DRw53. The DR1 haplotypes had a 3.92 fragment absent in DRw10. The delineation between DR1 and DRw10 was best demonstrated in the RFLP profile of the SC family (Fig. 1). The 6.0 Kb TaqI and 3.92 Kb BamHI fragments specific for DR1 were

Immunobiology of HLA
Volume II

Table 1. Summary of RFLP Profile

A. TAQI-DRβ Probe

Fragment Size in Kb	Proposed Specificity	LISC 1,7	SASC NASC 1,W10	DOSC MSSC 1,12	MASC 12,W10	OPBU 5,W10	RUBU 4,W10	REGA 1,W10	ASP-GAR 5,W10	TULA 3,W10	IRCA 4,W10	MAXKAH 5,W10	WT 100 1	DEB-PAG 3	3103 4(Dw4)	LS40 4(Dw14)	TS10 5(Dw10	SWEI 5	JES-THOM 1	HHK 6
13.8	DRw53	+					+				+			+	+	+				
12.2	DRw10		+		+	+	+	+	+	+	+	+								
11.3	DRw52a			+	+	+			+	+		+		+				+		
10.9	DR3.6													+						+
7.00	DR3,6,7									+				+		+				
6.30	DRw11,7					+			+									+		
6.00	DR1	+	+	+				+					+							+
5.60	DR4						+				+			+	+	+				
5.30	DR4						+				+			+	+	+				
4.60	DR1+w10	+	+	+	+	+	+	+	+	+	+	+	+							
4.50	DR12			+	+															
4.12	DRw53+DQ2	+				+			+	+		+		+				+		+
3.15	DRw53	+		+	+		+				+					+	+	+		
2.84	DRw53	nd	nd	nd	nd		+				+			+	+	+				

B. BamHI-DRB Probe

Fragment Size in Kb	Proposed Specificity	LISC 1,7	SASC NASC 1,W10	MSSC 1,12	MASC 12,W10	RUBU ,4,W10	REGA 1,W10	PAGA 1,4	TULA 3,W10	ASP-GAR W11,W10	WT-100 1	HHK W6	JES-THOM 1	3103 4(W4)	LS40 5(W14)	TS10 4(W10)	SWEI 5
11.3	DRw53			+	+				+	+		+					+
6.80	DRw53	+				+		+						+	+	+	
5.26	DR4					+		+						+	+	+	
5.07	DR1+w10	+	+	+	+	+	+	+	+	+	+	+	+				
4.80	DRw53	+				+		+						+	+	+	
4.40	DRw53	+				+		+						+	+	+	
4.30	DRw10+w52		+	+	+	+	+		+	+		+					+
4.25	DR7	+															
3.92	DR1	+	+	+			+	+	+		+		+				

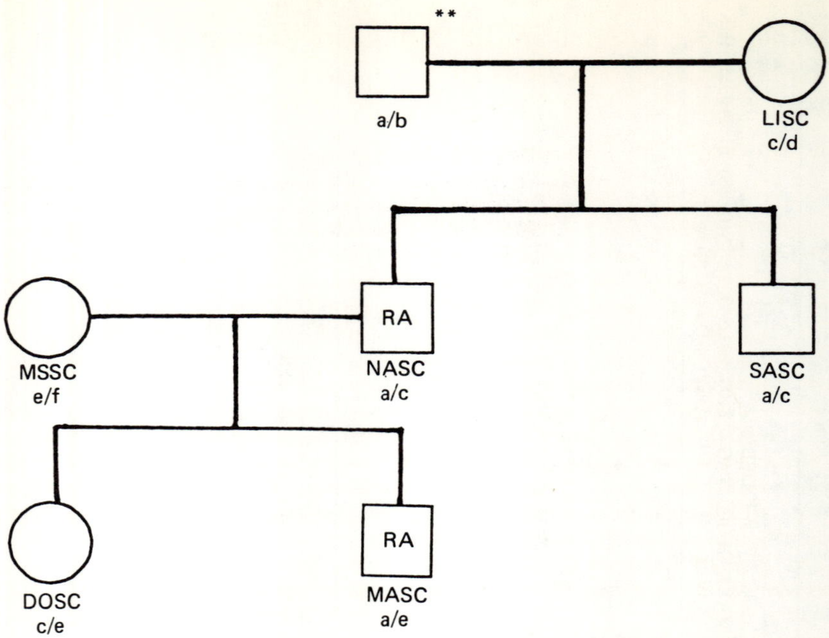

a.	A30	Cw4	B35	DRw10		DQw1	109d6 +	17 – 3 – 3S –	7.3.19.1 –
c.	A32	Cw4	B44	DR1		DQw1	109d6 –	17 – 3 – 3S –	7.3.19.1 –
d.	A3	Cw__	B14	DR7	DRw53	DQw2	109d6 +	17 – 3 – 3S +	7.3.19.1 –
e.	A28	Cw4	B35	DR12	DRw52	DQw3	109d6 –	17 – 3 – 3S +	7.3.19.1 +
f.	A2	Cw4	B14	DR1		DQw1	109d6 –	17 – 3 – 3S –	7.3.19.1 –

**Typed by interence

Figure 1. RFLP profile of the SC family, showing the delineation between DR1 and DRw10.

DR1 family members and absent in the only individual who was DRw10. Similarly, the 12.2 TaqI fragment unique to DRw10 was present only in the DRw10 and absent in all other individuals. The DRβ_1 chain genes of two DRw10 individuals, NASC and OPBU, yielded unique nucleotide sequences (3,4). No evidence for the presence of DRβ_2 genes in the DRw10 haplotype was obtained.

References

1. Lee SH, Gregersen PK, Shen H, Nunez-Roldan A, Silver J, Winchester RJ. Strong association of rheumatoid arthritis with the presence of a polymorphic Ia epitope defined by a monoclonal antibody: Comparison with the allodeterminant DR4. Rheum Int 1984;4:17–23.
2. Matsuyama T, Winchester RJ, Lee S, Shookster L, Nunez-Roldan A. Identification of the DRw10 DRβ_1 chain allele as encoding a polymorphic class II MHC epitope otherwise restricted to DRβ_2 molecules of the DRw53 type. J Immunol (in press).
3. Merryman P, Gregersen PK, Lee S, Silver J, Nunez-Roldan A, Crapper RM, et al. Nucleotide sequence of a DRw10β chain gene; identity of the third diversity region with that of the DRw53 allele of the β_2 locus and as the probable site encoding a polymorphic MHC class II epitope. J Immunol (in press).
4. Crapper RM, Gregersen PK, Lee SH, Nunez-Roldan A, Barr KM, Silver J, et al. A novel DR5 β chain with a variant third hypervariable region. Submitted for publication.

Author Affiliation

Hospital for Joint Diseases, New York University School of Medicine, New York, New York

Cloning and Sequence Analysis of the DRβ Gene Coding for the Specificity HLA-D/DR-BON

H. Coppin,[1] A. Huchenq,[1] J.M. Garnier,[2] and C. de Preval[1]

We described a new HLA-DR specificity, called provisionally DR-BON, which was defined by cellular typing (1). At the present time no specific anti-BON serologic reagent has been found. This specificity is strongly associated with DQw1 and frequently with B35 and C4A4,C4B2. After immunoprecipitation with monomorphic anti-DR monoclonal antibodies, biochemical analysis by two-dimensional gel has shown that the BON haplotype directs the synthesis of a single DRβ chain, which is the more acidic known β chain, whereas the expressed α chain remains the classical DRα polypeptide (2). In contrast, RFLP studies with seven enzymes revealed identical patterns between DRβ and DRα genes present in DR1 and DR-BON haplotypes (2). DNA analysis also demonstrated the presence of two DRβ genomic genes in this haplotype.

We have constructed a size-selected cDNA library in λ gt10 using messenger RNA isolated from an DR-BON/BON homozygous lymphoblastoid cell line. The library was screened both with a full length DRβ cDNA and with a short 5'DNA fragment. Among the 50 positive clones a single DRβ sequence was found, in agreement with the biochemical analysis. The full-length clone pBON5 was selected for sequencing. It contains a 1217 base pairs (bp) insert that includes 97 bp at the 5'end, 801 bp coding for the protein sequence (the leader peptide, the two external domains, the transmembrane and intacytoplasmic segments), and 319 bp of 3′ untranslated sequence preceding the polyadenylation site. This cDNA was subcloned in an eukaryotic expression vector, pKCR3 (gift from R. Breathnach, LGME-CNRS). After DNA transfection, mouse L cells ewpress DR-Bon molecules on their surface. The cellular and serologic epitopes of the BON antigen are under study using L cells sorted for high fluorescence staining with the monomorphic anti DR antibody, B8 13.2.

When compared with DR1 β sequence (3), 7 nucleotide differences are observed. One corresponds to a silent mutation in the leader sequence, the others are clustered in the third variable region between position 288 and 301 after the first coding base. The six clustered differences lead to three amino acid replacements between DR BON and DR1 β chains. These variations in the third variable region correspond to differences in antigenicity: at the serologic level, DR BON does not cross react with DR1 and remains a "DR-blank" specificity; at the functional level, T-cell clones specific for DR-BON do not recognize the Dw1 specificity (M. Thomsen, personal communication). Nucleotide comparisons of pBON5 with DR1 to DR9 β chain sequences (3) indicate that the BON sequence in the third variable region is identical to sequences of the βI gene of a subset of DR5 and DRw13 (Fig. 1).

The origin of the polymorphism within this region can be explained neither by similar point mutations in different haplotypes, nor by repeated double crossing-over within 15 base pairs. When DR3 and DRw6 β sequences were compared (4), it was concluded that the DR3 specificity is the consequence of an unidirectional sequence exchange from the DR βIII gene to the DR βI gene in the DRw6 haplotype. Such interpretations have been proposed to explain other examples of polymorphism of limited DNA segments among MHC genes. We propose that a similar mechanism of gene conversion occurred to create the BON sequence from a DR1 gene. Beta I DRw6 gene, or DR5, could have been the donor gene, however this cannot be ascertained solely on the basis of our cDNA sequence.

As a consequence of the structural differences between DR β genes, we had to synthesize two different oligonucleotide probes to perform a DR BON-specific DNA typing assay on Southern blots: the first oligonucleotide corresponded to the first variable region (codons for amino acids 9 to 14) and discriminates DR BON and DR1 from DR5 and DRw6, whereas the other, corresponding to the third variable region (codons for residues 67 to 72) discriminates DR BON, DR5, DRw6 from DR1.

```
                     285                        305
                      .                          .
                      .                          .
DR 1 β     ------GAC—CTC—CTG—GAG—AGG—CGG—CGG------
DR BON β   ----------A---------A—G—C—GA--------
DR w6 β          -----A---------A—G—C—GA-----
```

Figure 1. Sequence comparison inthe third variable region of DR β chain genes (3, 4).

References

1. Cambon-Thomsen A, Thomsen M, Abbal M, Sommer E, Calot M, Arnaud J, Ohayon E. A new HLA-D specificity associated with DR-blank: D-BON. Tissue Antigens 1986;27:256–261.
2. Coppin H, Villedieu P, Thomsen M, Giraud P, Cambon-Thomsen A, Charron D. HLA-D/DR BON: A new biochemically and cellularly defined polymorphism indistinguishable by RFLP from DR1. Immunogenetics 1987;25: 130–132.
3. Bell JI, Denney D, Foster L, Belt T, Todd JA, McDevitt HO. Allelic variations in the DR subregion of the human major histocompatibility complex. Proc Natl Acad Sci USA 1987;84:6234–6238.
4. Gorsky J, Mach B. Polymorphism of human Ia antigens: Gene conversion between two DRβ loci results in a new HLA-D/DR specificity. Nature 1986;322:67–70.

Author Affiliations

[1]CHU Inserm U.100, Chu Purpan, 31052 Toulouse, Cedex, France; CRPG CNRS, CHU Purpan, 31052 Toulouse, Cedex, [2]Laboratoire de Genetique Moleculaire des Eucaryotes, 67085 Strasbourg, France

Structure of HLA-D/DR-BON Antigen

J. Tkaczuk, M. Thomsen, M. Calot, A. Cambon-Thomsen, and E. Ohayon

The D-DR BON antigen was individualized among the DR blank specificities in 1986. Cells of this specificity were characterized by serologic unreactivity with all the allo-antisera of the 9th Histocompatibility Workshop and as being DQw1-positive. Studies in the French family BON by mixed lymphocyte culture allowed us to isolate a homozygous typing cell (HTC) BON/BON which allowed further definition of the antigen at the membrane (two-dimensional gel electrophoresis) and at the DNA level (2). Here, we describe more recent results obtained by serology, T-cell clones, and two-dimensional gel electrophoresis.

Either peripheral blood mononuclear cells (PMBC) or Epstein-Barr virus- (EBV)-transformed cells were used for serologic study of the HTC D/DR-BON and other cells with the monoclonal antibody (MoAb) HU-30 (a gift from M. Aizawa [3]) and the core MoAbs of the 10th Workshop. The HLA-DR reactivity was performed by a two-color fluorescence technique according to Van Rood et al. (1976), and with minor modifications for EBV-transformed cells.

Alloreactive T-cell clones were generated toward D/DR-BON homozygous cells by limiting dilution. Clones were fed every 2 days with IL2. They were restimulated weekly.

For two-dimensional gel analysis, cells were labeled with ^{35}S-methionine, before classical lysis. First and second dimensions of the two-dimensional gel analysis were performed according to the 10th Workshop protocol, for alpha and beta runs. When NEPHGE (nonequilibrium pH gradient electrophoresis) was realized, pH 3.5 to 10 ampholytes, 5^{30} h migration, and polarity inversion were applied.

In the serologic study, PBMC-D/DR-BON was found to give a marked reactivity with the anti-DR2 HU-30 MoAb. The epitope recognized by HU-30 on D/DR-BON cells was also found on PBMC of other specificities, such as DR1 and (DRw6) cells, when using an 1:500 diluted ascites. For greater dilutions ($>$1:100 000), only DR2 and BON EBV-transformed cells were still reactive. Thus, HU-30 seems to be a good tool to identify the BON specificity in the absence of DR1,DR2 or DRW6 antigens on the same cell. These results must be related to the reactivity of some of the produced T-cell clones, which recognized BON HTC, but also other DR cell specificities, i.e. DR2 and DRw 13.

From a biochemical point of view, we confirm here the constitution of the DRβ BON pattern (formed by four spots) being more acidic than the DRβ chain of DR1 HTC. Among them, the third spot from the left seems to be a major DRβ spot (Fig. 1a).

The DQβ chain as well as the DRβ chain are characterized by a very similar acidic pattern of four spots. Here, however, the first of the four spots looks more intense and seems to be a major DQβ spot (Fig. 1b). The NEPHGE DQ pattern shows similar basic spots on the left of invariant spot both in DR1 and DR-BON cells.

The DRα chain of DR-BON HTC does not seem to be different when compared with DR1, except that the 6th more acidic spot of the upper DR line is absent (Fig. 1c).

Like in DR1 cells, the DQα chain has a major spot on the lowest DQ/DP line of spots located just under the second DRα spot.

The DPα chain of DR-BON HTC precipitate with B7-21 monoclonal antibody shows a polymorphic spot, which is also seen on Fig. 1c with the antibody Tu39. Unlike what was found in the DR1 cells tested, this DP spot was located under the third DRα spot. It could be related to the DPw4 specificity of D/DR-BON HTC.

Thus, we showed that the D/DR-BON specificity could be defined by serology with HU-30 monoclonal antibody. Two-dimensional gel analysis confirmed that DRβ chain of D/DR-BON is very acidic and that the DQw1β chain of D/DR-BON cells seems polymorphic when compared to the DQw1 β chain of DR1. Further biochemical and T-cell clone studies will be performed to determine D/DR-BON specificity.

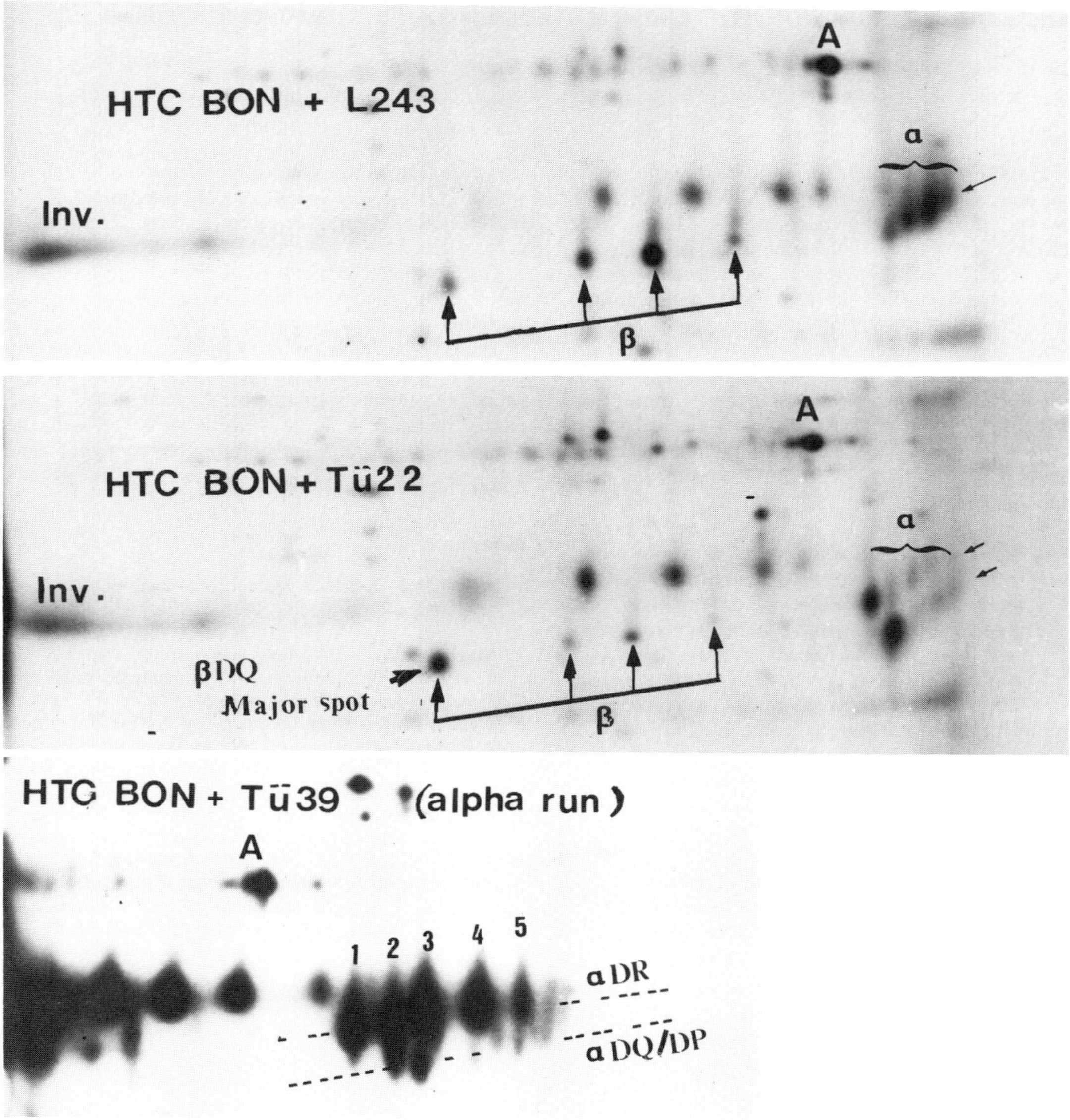

Figure 1. β and α two-dimensional gel patterns given by HTC D/DR-BON with different monoclonal antibodies: L243, Tü22, and Tü39.

Acknowledgments. We are greatly indebted to Dr. R. Knowles for providing Workshop monoclonal antibodies. We thank Nathalie Woloszyn, Daniele Clement, Claudie Offer, and Monique Sorbara for their technical assistance.

References

1. Cambon-Thomsen A, Thomsen M, Abbal M, Sommer E, Calot M, Ohayon E. A new HLA-D specificity associated with DR blank: D-BON. Tissue Antigen 1986;27:256–261.
2. Coppin H, Villadieu P, Thomsen M, Giraud P, Cambon-Thomsen A, Charron D. HLA-D/DR-BON: new biochemically and cellularly defined polymorphism indistinguishable by RFLP from DR1. Immunogenetics 1987;25:130–132.
3. Kashara M, Ogasawara K, Ikeda H, Okuyama T, Ishikawa N, Takenouchi T, Wakisaka A, Kikuchi Y, Aizawa M. A monoclonal antibody that detects a polymorphic determinant common to HLA-DR1 and HLA-DR2. Tissue Antigen 1983;21:105.

Author Affiliations

Chu Purpan, 31052 Toulouse Cedex, France

Biochemical Analysis of HLA Class II Antigens Using Transfected Cells

J. Alejandro Madrigal, David Wilkinson, Hitoshi Ikeda, John Trowsdale, and Julia G. Bodmer

The class II products are expressed from a minimum of three sets of loci: HLA-DR, HLA-DQ, and HLA-DP. However, these loci are complex; for example, each haploid genome may express: one α and two or more β genes in the HLA-DR subregion, at least one α and one β gene in the HLA-DQ subregion and another α and β pair in the HLA-DP region (1,2).

The analysis of the heterogeneity of the HLA-D gene products has been facilitated greatly by the production of monoclonal antibodies (MAb), together with the use of 2D-IEF gels. However, in many cases the MAb cross-react with other class II products, and often the analysis of class II 2D-IEF patterns from cells expressing multiple loci is complex (3).

Recently the production, by transfection, of mouse L cells expressing a single human HLA-D region product has provided the basis for a novel approach to analyzing the structure-function relationships of MHC gene products (4,5). In J. Trowsdale's laboratory a variety of HLA class II transfectant cell lines expressing a single locus product have been produced, and these products have been good reagents for the study of the biochemical complexity of the MHC. For instance: The LDR7 cell line is a fibroblastoid mouse cell line transfected with the HLA-DR7 α and β genes obtained from a cDNA library from the HLA-DR7 homozygous lymphoblastoid cell line MANN. This LDR7 cell line expresses high levels of cell surface HLA-DR7 as evidenced by flow microfluorimetic analysis. Biochemical studies of these LDR7 cell line using different MAbs, clearly show one α and one β chain that are similar to the α and β chain of the lymphoblastoid cell line MANN (Fig. 1A). Interestingly, three different spots were observed at the acidic end of the gel on the 2D-IEF from the LDR7 (Fig. 1B and 1C). These spots had a different isoelectric point and molecular weight. However, when appropriate conditions were set up to define the α region, only one family of spots representing different stages of glycosylation from the same α chain were observed (Fig. 1D).

Conclusions: The use of HLA class II transfected cells has been of great help in dissecting HLA complexity, allowing the unambiguous assignment of protein products to particular genes. 2D-IEF analysis of the class II products obtained from transfected cell lines did not show any obvious significant difference in the α and β chains structure when they were compared with their homologous product on homozygous lymphoblastoid cell lines.

References

1. Bodmer WF. The HLA system. In: Albert ED, et al (eds): Histocompatibility Testing. Berlin, Springer-Verlag, 1984, p 11.
2. Trowsdale J, Young JAT, Kelly AP, Austin PJ, Carson S, Meunier H, et al. Structure, sequence and polymorphism in the HLA-D region. Immunol Rev 85:5
3. Crumpton MJ, Bodmer JG, Bodmer WF, Heyes JM, Lindsay JJ, Rudd CE. Biochemistry of class II antigens. Workshop report. In: Albert ED, et al (eds): Histocompatibility Testing. Berlin, Springer-Verlag, 1984, p 29.
4. Austin P, Trowsdale J, Rudd C, Bodmer W, Feldmann M, Lamb J. Functional expression of HLA-DP genes transfected into mouse fibroblasts. Nature 1985;313:61.
5. Heyes J, Austin P, Bodmer J, Bodmer W, Madrigal A, Mazzilli MC, et al. Monoclonal antibodies to HLA-DP-transfected mouse L cells. Proc Natl Acad Sci USA 83:3417.

Author Affiliation

Imperial Cancer Research Fund, London WC2A 3PX, UK

►

Figure 1. 2D-IEF gels profiles of the HLA-DR7 homozygous lymphoblastoid cell line MANN and of the HLA-DR7 transfected fibroblast mouse cell line LDR7 using different MAbs: 17.3.3, SG465 and TAL-1B5. The α chains are represented at the left side of the figure and β chains are at the basic end (right-hand side). **A**. Short labeling time (30 min) of the MANN cell line, using the MAb 17.3.3, the β chain occupied the extreme basic end of the gel at the right-hand side and it is represented as a single spot, and the α chain is at the left side. **B**. using the MAb SG465, and **C**. the MAb 17.3.3, the 2D-IEF pattern of the LDR7 mouse cell line shows an α and a β chain that have similar positions to the MANN lymphoblastoid cell lines. **D**. Immunoprecipitates from the LDR7 transfected cell lines and the MAb TAL-1B5 were analyzed in 2D-IEF with conditions chosen to separate the α chains.

Immunobiology of HLA
Volume II

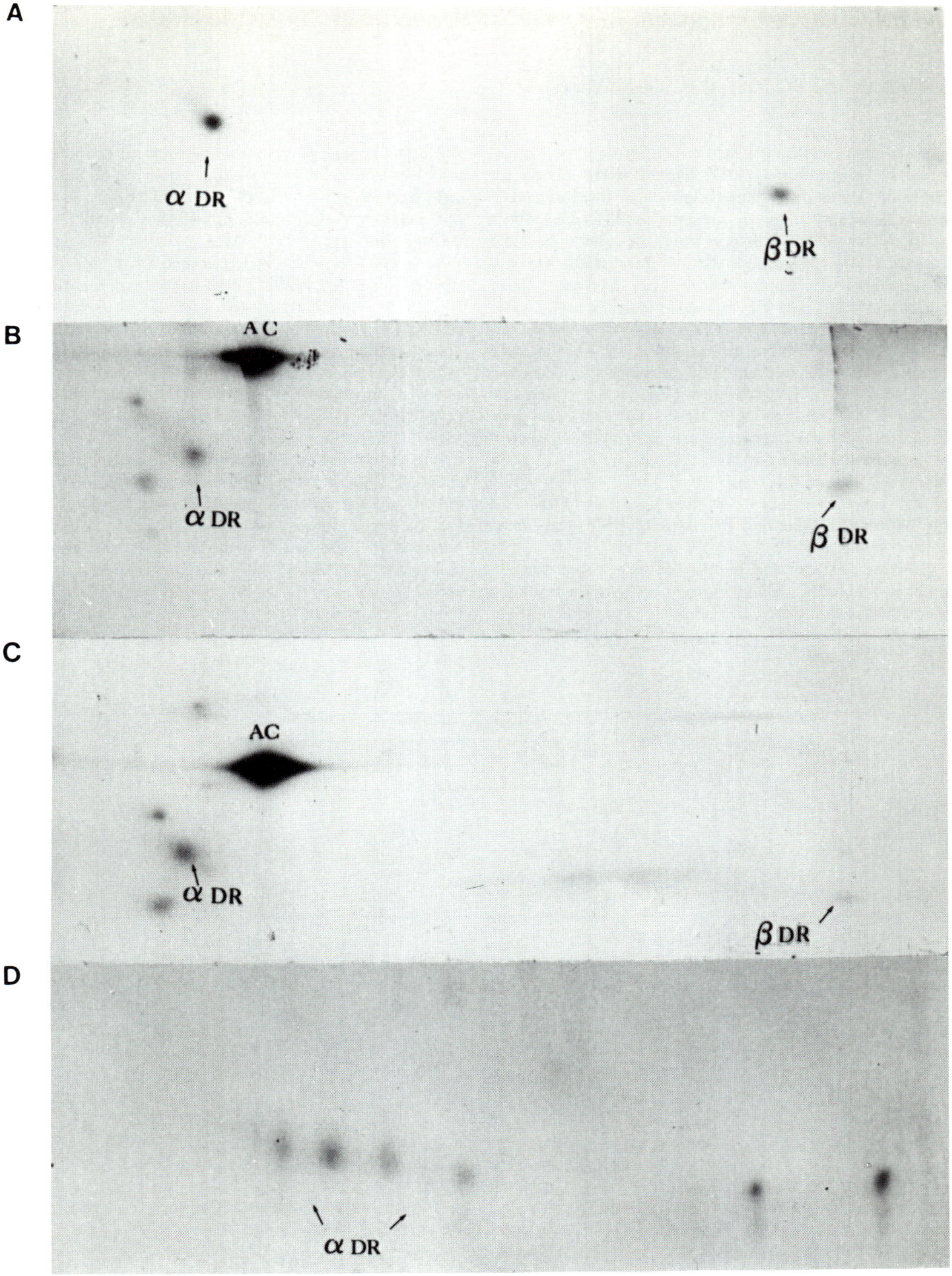
A
α DR
β DR
B
AC
α DR
β DR
C
AC
α DR
β DR
D
α DR

Quantitation of Component Molecules of Precipitated Class II Complexes

J. Lindsay, J.F. Maddox, and J.G. Bodmer

The use of a scanning instrument that detects β emissions from dried gels has enabled us to quantitate emission levels from individual class II products. If radiolabeled molecules are presented as discrete spots or bands on an SDS/PAGE gel, they can be detected and analyzed. The instrument used is an AMBIS Mark II β scanner (Automated Microbiology Systems, Inc.), which detects β particles passed through a mylar filter into an argon-methane mix. Localization of each ionization event is determined, and a two-dimensional representation is built up. The scanner is operated from an IBM PC AT-X, which allows data to be stored and retrieved for subsequent analysis. We examined profiles from cells expressing class II products to determine whether the location of the epitope on the dimer had an effect on the ratio of α and β chains bound by monoclonal antibody. We then used this to examine the expression of α and β chains in P815-transfected cells.

Both human DR and mouse I-E have three methionines in the mature α chain sequence, with an extra two in the signal sequence. The β chains of mice and humans have two methionines in the second domain plus three in the signal sequence. Overall, the total ratio of α to β in terms of counts could be expected to be almost equal with bias toward more α than β chain counts, assuming more mature than immature molecules would be bound by an antibody.

We used TAL 1B5, which is anti-DRα, and M1.18, which is against the β1 domain to precipitate from PRIESS (DR4) lymphoblastoid cells biosynthetically radiolabeled with ^{35}S-methionine. Complexes were separated on an isoelectric focusing first dimension and SDS/PAGE second dimension. The dried gels were scanned, and by dividing the class II profiles using a grid (Fig. 1), the counts from each product could be totalled and background counts subtracted. Levels of actual counts depended on the length of time the gels were scanned. Ratios of α to β counts were calculated for both anti-α and anti-β chain precipitations (Table 1). With the α chain antibody, we observed a ratio of 2:1 α:β, whereas with the anti-β chain antibody, the ratio of β chains had increased (i.e., 5:4 α:β). These results would seem to indicate that the antibodies were precipitating from a free pool of polypeptides, as well as from complexes.

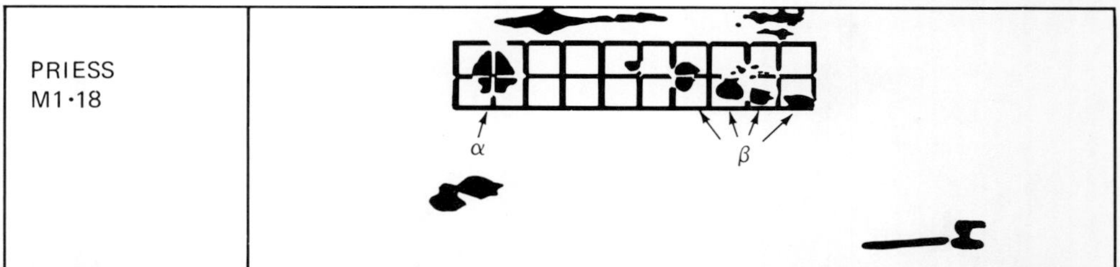

QUANTITATION OF GRID FOR SCAN 6nov-1

grid total counts: 81005 grid total CPM: 184.3 Results Range: full

BOX #	COUNTS	CPM	S.D. CPM	% OF GRID	S.D. %	BOX #	COUNTS	CPM	S.D. CPM	% OF GRID	S.D. %
1	6407	8.2	0.4	7.9	0.37	2	8061	10.4	0.4	10.0	0.42
3	3257	4.2	0.3	4.0	0.26	4	3249	4.2	0.3	4.0	0.26
5	2783	3.6	0.2	3.4	0.24	6	3556	4.6	0.3	4.4	0.27
7	4764	6.1	0.3	5.9	0.32	8	4047	5.2	0.3	5.0	0.29
9	4170	5.4	0.3	5.1	0.29	10	3844	4.9	0.3	4.7	0.28
11	3679	4.7	0.3	4.5	0.28	12	4525	5.8	0.3	5.6	0.31
13	2816	3.6	0.2	3.5	0.24	14	2041	2.6	0.2	2.5	0.20
15	1764	2.3	0.2	2.2	0.19	16	2529	3.3	0.2	3.1	0.23
17	3609	4.6	0.3	4.5	0.27	18	5930	7.6	0.4	7.3	0.35
19	4799	6.2	0.3	5.9	0.32	20	5175	6.7	0.3	6.4	0.33

*not significantly different than 0/+ not significantly different than 100 percent

Figure 1. Grid for quantitation of component molecules of precipitated class II complexes.

Immunobiology of HLA
Volume II

Table 1. Ratios of α to β Counts for Anti-α and Anti-β Chain Precipitations

Ab	Cell	α cpm	β cpm	Ratio α:β
IB5	PRIESS	2800	1877	4:2·6
14.4S	P815 #5	508	605	5:6
M1.18	PRIESS	1530	1253	5:4
M1.18	P815 #5	–	250	–

This has been checked on other lymphoblastoid cell lines, and so far has been consistent. We used data from the lymphoblastoid cell lines to look at the construction and expression of class II chains on transfected P815 mastocytoma cells. A βI domain from a DR4 sequence was introduced into a mouse I-E β gene. This was cotransfected with mouse I-E gene into P815 cells. Low expression of class II was observed on the surface by FACS analysis. We used the anti-β chain antibody M1.18, which is known to bind the first domain of DR4β and the mouse anti-α chain antibody 14.4.4S (1) to precipitate class II complexes from these cells. After separation on IEF SDS/PAGE, the dried gels were scanned. Results obtained from these scans with the mouse anti-α chain antibody gave a ratio of 5:6 α:β (i.e., slightly more β than α counts) while the anti-β chain antibody precipitated β with no detectable α. This would seem to indicate that antibodies do precipitate from a free pool of polypeptides, and that the ratio of α to β chains in a precipitated complex is affected by the position of the antibody binding site on the dimer. The less than equal counts of α chains in the anti-α chain P815 precipitation could be a result of little or no free immature α chains or an excess of immature β chains. The conclusion is that the level of α chain is the limiting factor in the level of surface expression of the dimer on the transfected cells. Studies are ongoing using this scanning system to investigate the relative affinity of antibodies towards class II and other complexes.

References

1. Ozato K, Mayer N, Sachs DH. Hybridoma cell lines secreting monoclonal antibodies to mouse H-2 and Ia antigens. J Immunol 1980;124:533–540.

Author Affiliations

Imperial Cancer Research Fund, P.O. Box 123, Lincoln's Inn Fields, London WC2A 3PX, UK

Section VII
Class II Histocompatibility Antigens: HLA-DRB3 and HLA-DRB4

Epitope Mapping of HLA-DR Antigens with the Use of DNA-Transfected Cells

Christine C. Berte, Jack Gorski, Walter Reith, and Bernard Mach

Abstract: This study reports on a direct correlation between individual polymorphic markers (serological determinants) and the products of individual class II genes and alleles. The DR product of the different alleles of loci DRB1 and DRB3 (formally βI and βIII) from three DRw6 and two DR3 haplotypes, all with known DNA sequences, have been expressed in DNA-transfected Ia-negative mouse cells. Four major patterns of reactivity of anti-DRw52 monoclonal antibodies (MAbs) with these transfectant lines have been identified: 1) The product of both the 52a allele of locus DRB3 (certain DR3 and DRw6 haplotypes) and the DRB1 locus of DR3 react with certain MAbs. This epitope is the phenotypic counterpart of the gene conversion event which generated the DR3 gene. 2) The product of the 52b allele of locus DRB3 reacts with one MAb. 3) Certain anti-DRw52 MAbs react with the 52a, 52b,and 52c alleles of DRB3 as well as with DRB1 of DR3. 4) Other less specific anti-DRw52 MAbs react with the gene products of both DRB1 and -B3 of DR3 and DRw6. This study has allowed us to map specific DRw52-like epitopes to individual genes and to individual amino acids in the sequence of the various DRβ chains and to propose a localization of these epitopes in a putative three-dimensional model of class II molecules, based on the class I structure.

The multiplicity and diversity of HLA class II antigens, including DR, DQ, and DP molecules, at the surface of positive cells render very difficult the understanding of the genetic basis of serological and functional allopolymorphism. In the case of HLA-DR molecules, which are quantitatively and functionally the most important class II antigens, two distinct DR molecules are expressed and the polymorphism is determined by the β chains components encoded by locus B1 and B3 (1). At least 14 serological specificities have been described so far (2). In addition, several haplotypes have been grouped in supertypic families on the basis of shared reactivity with the same supertypic serum: the DRw52 group including the DR3, -5, -w6, -w8 haplotypes and the DRw53 group including the DR4 and DR7.

The exact contribution of a given gene product to the HLA class II reactivity of specific antisera and of restricted T cells has been dramatically simplified by the use of cells transfected with a single DRB chain gene (3), since all results can now be directly referred to the polypeptide encoded by the particular DR gene used for transfection (4). We have chosen to study the product of the DRB1 and -B3 loci from three DRw6 and two DR3 haplotypes. These genes include the three recently identified alleles of locus DRB3, which encodes the supertypic specificity DRw52 (52a, 52b, and 52c) (5,6).

Eight B-chain genes were isolated from genomic and cDNA libraries and characterized (5,7). Sequencing of these genes has allowed assignment of each gene to locus B1 and B3, based on the molecular mapping reported earlier (7). From the analysis of the sequences of these genes, we have shown earlier that the variability is clustered in three hypervariable segments in the β1 domain of the alleles of locus B1 and that the DR3 sequence results from the introduction of two short DNA segments from a B3 locus (allele 52a) by gene conversion into a DRB1 locus of DRw6 haplotypes (5). For the less polymorphic DRB3 locus, three alleles have been defined: 52a present in some of the DR3 and DRw6 haplotypes; 52b present in other DR3 and DRw6 haplotypes and in all DR5 haplotype; 52c present in some DRw6 haplotypes encoding Dw19 (6,8).

Genomic DRB genes and cDNA genes, inserted in a plasmid expression vector carrying the neomycin resistance, were introduced in a mouse L cell by cotransfection with the DRA gene. A panel of mouse cell clones expressing individual DR genes was constructed. The different genes used and the transfectants generated are described in Table 1. We have tested the reactivity of

Table 1. Panel of DR-Expressing Ltk Cell Lines and Source of DRB Chain Genes Transfected

Source of the Genes		Ltk Cell Lines Transfected with the DRB Genes			
		Locus B1		Locus B3	
Cell Line	Typing by Serology	Allele	Transfectant	Allele	Transfectant
WT46	DRw6,Dw19,DRw52	DR6	LR6.3/6*	DRw52c	LR6.3/52c
HHK	DRw6,Dw18,DRw52	DR6	ND	DRw52a	LR6.2/52a
AVL	DR3,DRw52	DR3	LR3.1/3	DRw52a	LR3.1/52a
QBL	DR3,DRw52	DR3	LR3.2/3	DRw52b	LR3.2/52b
DR4/6	DR4/DRw6,DRw52	DR6	ND	DRw52b	LR6.1/52b

Transfections were performed as described previously (10).
*Our present denomination of the transfectants indicates the cell transfected (L for L cells), the HLA-D subregion (R for DR), the haplotype (DR3 or DR6) of the cell line used as source of the gene, and the number of the transfectant and the allele used (3,6,52a,52b, and 52c).

Immunobiology of HLA
Volume II

Table 2. Reactivity of Anti-DRw52 MAbs Tested with Transfected Mouse Cells

MAb	Reactivity with Transfectants DR6	DR3	52a	52b	52c	Amino Acids Recognized (Class I Homologous)	Putative Position of Exposed Epitope
7.3.19.1	+	+	+	+	+	13 (99) or 32 (118)	H at position 32, β strand *
NDS9	−	+	+	−	−	26 (112) or 74 (159)	R at position 74, α helix *
NDS10	ND	−	−	+	−	30 (116) or 51 (137)	H at position 30, β strand *
SJ1.6B1	−	+	+	+	+	71 (156), 73 (158) or 77 (162)	K at position 71, α helix * G at position 73, α helix ↑ N at position 77, α helix ↑

Individual transfectants are described in Table 1. Amino acids unique to the alleles recognized are indicated, together with the homologous position in the α 2 domain of HLA Class I. The Class I position in bracket that are underlined are accessible residues as listed in Ref. 15. By analogy with the structural model of HLA class I (15), the putative location of the exposed epitopes is given for each MAb. * = Residue pointing towards the site; ↑ = residue on an α helix pointing up.

these transfectants with several anti-DRw52 and one "anti-DR3" MAbs. Four patterns of reactivity, presented in Table 2, were established: 1) NDS9 (S.Fuggle et al., this volume), like TR81 antiserum (10), reacted with the products of the 52a allele of locus DRB3 (found in DR3 or DRw6 haplotypes) and the DRB1 locus of DR3. The epitope is therefore the counterpart of the gene conversion event and the amino acids involved in the recognition must be coded by the segment of DNA carried over from the 52a gene to the DRw6 B1 locus to generate the DRB1 of DR3. The only amino acids common only to the products of these DR3 (B1 locus) and DRw52a (B3 locus) genes are tyrosine on position 26 and arginine on position 74. 2) a MAb (NDS10) (S. Fuggle et al., this volume) and TR22 antiserum (11) recognize only the 52b alleles of locus DRB3 (found in some DR3 and DRw6 haplotypes). The only amino acids characteristic of this 52b allele are histidine on position 30 and arginine on position 51. 3) The third pattern of reactivity is represented by SJ1.6B1 (12), which reacts with the product of DRB1 of DR3 and also with the products of all three alleles of the DRB3 locus. It is negative with the product of DRw6 B1 locus. This MAb has been previously described as an anti-DR3 MAb (12). It is in fact specific for both DRw52 and DR3, and its reactivity as DR3 MAb can be explained by the higher level of expression of locus B1 compared to locus B3 (13). In this case, reactivity with the DRB3 products may have been interpreted as tail reactions. The epitope of this MAb must involve the amino acids in position 71 (lysine), 73 (glycine), and /or 77 (threonine). 4) The fourth pattern of reactivity, exemplified by MAb 7.3.19.1 (14), is characterized by reactivity with all the genes products of the first and the third locus of DR3 and DRw6. MAb 7.3.19.1 recognized therefore a fairly public epitope on the products of DRB1 and -B3 and on several alleles, which involves either serine on position 12 or histidine on position 32.

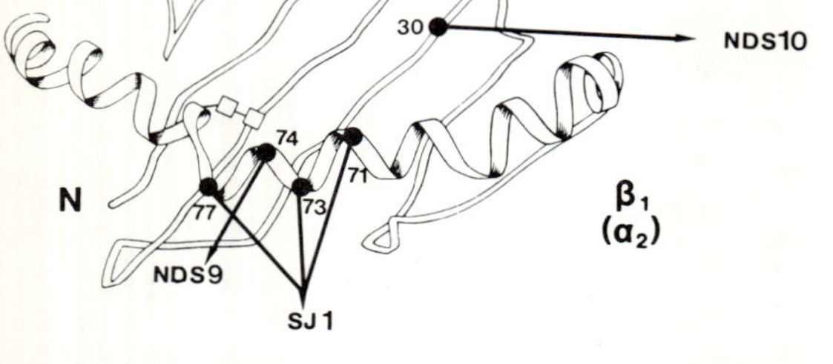

Figure 1. Tentative homologous model for HLA class II structure based on the published three-dimensional model for HLA class I (15) with the position of the exposed amino acids recognized by each MAb.

A knowledge of the amino acids epitopes recognized by MAb is important for further studies of the functional and structural aspects of DR molecules. By testing reactivities of different DRw6 cells with MAbs NDS9 and NDS10, it is now possible to type them as 52a or 52b. Assignment of MAb reactivities to one or the other DRB locus, is useful for the analysis of inhibition of T-cell clones proliferation by these MAbs. And finally, the identification of the specific amino acid that reacts with a given MAb will help in the interpretation of the structure and the conformation of these DR molecules. On the basis of the published three-dimensional model for HLA class I (15) and of the remarkable analogy of conserved and variable residues in the sequence of the α2 domain of class I and the β1 domain of class II, it has been possible to position the amino acids recognized by each MAb on a tentative homologous model for HLA class II structure (manuscript in preparation) (see Table 2 and Fig. 1). Interestingly, some MAb epitopes are located in the bottom of the groove, which was not expected from the class I data (15).

Acknowledgments. Technical assistance of Ms. M. Zufferey is deeply appreciated. Flow cytometric analysis were performed at the Fluorocytometry Unit with the excellent technical assistance of D. Wholwend.

References

1. Lee J, Trowsdale J, Bodmer W. cDNA clones coding for the heavy chain of human HLA-DR antigen. Proc Natl Acad Sci USA 1982;79:545.
2. Albert ED, Baur MP, Mayr WR. Histocompatibility Testing 1984. Berlin, Springer-Verlag, 1984.
3. Rabourdin-Combe C, Mach B. Expression of HLA-DR antigens at the surface of mouse L cells cotransfected with human genes. Nature 1983;303:670.
4. Gorski J, Tosi R, Strubin M, Rabourdin-Combe C, Mach B. Serological and immunochemical analysis of the products of a single HLA DR-α and DR-β chain gene expressed in a mouse cell line after DNA-mediated co-transformation reveals that the β chain carries a known supertypic specificity. J Exp Med 1985;162:105.
5. Gorski J, Mach B. Polymorphism of human Ia antigens: Gene conversion between two DR β loci results in a new HLA-D/DR specificity. Nature 1986;322:67.
6. Tiercy JM, Gorski, Jeannet M, Mach B. Identification and distribution of three serologically undetected alleles of HLA-DR by oligonucleotide typing. Proc Natl Sci USA, in press.
7. Rollini P, Mach B, Gorski J. Linkage map of three HLA-DR β-chain genes: Evidence for a recent duplication event. Proc Natl Acad Sci USA 1985;82:7197.
8. Gorski J, Tilanus M, Giphart M, Mach B. Oligonucleotide genotyping shows that alleles at the HLA-DR β3 locus of the DRw52 supertypic group segregate independently of known DR or Dw specificities. Immunogenetics 1987; 25:79.
9. Nadler LM, Stashenko P, Russel H, Pesando JM, Yunis EJ, Schlossman SF. Monoclonal antibodies defining serologically distinct HLA-D/DR related Ia-like antigens in man. Hum Immunol. 1981;1:77.
10. Berte CC, Tanigaki N, Tosi R, Gorski J, Mach B. Serological recognition of HLA-DR allodeterminant corresponding to DNA sequence involved in gene conversion. Immunogenetics, in press.
11. Tosi R, Tanigaki N, de Préval C, Gorski J, Mach B. Immunochemical analysis of a cell transfected with an HLA-DR gene reveals a new alloantigenic specificity within HLA-DRw52. Eur J Immunol 16:1603.
12. Sachs JA, Fernandez N, Kurpisz M, Okoye R, Ogilvie J, Awad J, Labeta M, Festenstein H. Serological biochemical and functional characterization of three different HLA-DR monoclonal antibodies derived from C57BL6 mice. Tissue Antigens 1986;28:199.
13. Berdoz J, Gorski J, Termijelen AM, Dayer JM, Irle C, Schendel D, Mach B. Constitutive and induced expression of the individual HLA-DR β and α chain loci in different cell types. J Immunol 1987;139:1336.
14. Koning F, Schreuder I, Giphart M, Bruning H. A mouse monoclonal antibody detecting a DR-related MT2-like specificity: Serology and biochemistry. Hum Immunol 1984;9:221.
15. Bjorkman PJ, Saper MA, Samraoui B, Bennett WS, Strominger JL, Wiley DC. Structure of the human class I histocompatibility antigen, HLA-A2. Nature 1987;329:506.

Author Affiliation

Department of Microbiology, University of Geneva, Geneva, Switzerland

Monoclonal Antibody Definition of Polymorphism Within DRw52

S.V. Fuggle, C. Carter, J.I. Bell, and P.J. Morris

Recent sequence data of HLA-DRβ chains has revealed two allelic forms of the DR3, 5, w6, w8 associated antigen DRw52, the Dw24 and Dw25 alleles (1,2). We have studied the serologic and biochemical nature of a DRw52-related antigen defined by the monoclonal antibody (MAb) NDS10. This antibody recognizes a subpopulation of DRw52-typed cells and probably reacts with an epitope on Dw25.

NDS10 was characterized together with a DR3 reactive antibody NDS9 and three other MAbs having different, but DRw52-related specificities (NDS10 anti-DR5, <3, <w6; NDS11 anti-DR3, 5, w6, NDS12 anti-DR3, 5, w6, w8; NDS13 anti-DR3, 5, w6, w8+ (3).

In competitive binding assays using the homozygous DR3-typed cell WT49 as a target, NDS10 binding was strongly inhibited by the other DRw52-like antibodies suggesting a linkage between these epitopes. NDS9 only slightly inhibited NDS10 binding suggesting that these epitopes may be spatially distinct. Two-dimensional gel analyses of immunoprecipitates from WT49 cell lysate confirmed these results. Co-electrophoresis of individual immunoprecipitates and sequential depletion studies demonstrated that, at least in the DR3 haplotype studied, NDS10 was present on the same molecule as the other DRw52 determinants. In contrast, NDS9 was present on a molecule with different electrophoretic mobility. The 2-D gel profile of the mixture of NDS9 + 10 immunoprecipitates was similar to the profiles obtained with monomorphic anti-DR reagents, thus on WT49 NDS9 probably reacts with the DRβ1 and NDS10 with the DRβ111 gene products.

The serologic specificity of NDS10 was defined on a

Table 1. Population Distribution of Dw25 Defined by NDS10

HLA-Type	NDS10 Reactivity +	NDS10 Reactivity −
DR3	13	26
A1, B8, DR3	0	18
B18, DR3	9	0
B8 (non A1), DR 3	2	0
DR3 linked to other B locus antigens	2	8
DR5		
w11	26	0
w12	2	1
DRw6		
w13	3	14
w14	9	1
DRw8	0	9

panel of 104 DRw52 typed cells in indirect binding assays (Table 1). NDS10 reacted with a subpopulation of DR3 typed cells. A detailed analysis revealed that whereas all 9 B18, DR3 typed cells were NDS10 positive, the 18 cells expressing the A1, B8, DR3 haplotype were NDS10 negative. However, two cells where B8, DR3 was linked to A locus alleles other than A1 were reactive with NDS10. There was no distinct reactivity pattern when DR3 was associated with other B locus antigens; two such cells were positive and eight negative. All 26 DR5, wll-typed cells reacted with NDS10 whereas one of three DR5,w12-typed cells was negative. Although NDS10 reacted with some DRw13 and most DRw14-typed cells, the reactivity could not be assigned to any particular Dw type. All nine DRw8-typed cells were NDS10 negative.

The distribution of NDS10 reactivity within DR3 typed cells, being negative on A1, B8, DR3 cells, is consistent with the epitope recognized by NDS10 being present on the DRβ111 gene product, Dw25. In particular, the epitope is likely to include one or more of the 11 variable amino acids that comprise the only structural differences between Dw24 and Dw25. The distribution of Dw25 as defined by NDS10 is in accordance with that described by oligonucleotide genotyping (4).

Acknowledgments. This work was supported by grants from the National Kidney Research Fund and the Medical Research Council of the UK.

References

1. Gorski J, Mach B. Polymorphism of human Ia antigens: Gene conversion between two DRβ loci results in a new HLA-D/DR specificity. Nature 1986;322:67–70.
2. Todd JA, Bell JI, McDevitt HO. HLA-DQβ gene contributes to susceptibility and resistance to insulin dependent diabetes mellitus. Nature 1987;329:599–604.
3. Fuggle SV, Carter C, Watts FW, Kirkley J, Morris PJ. Monoclonal antibody definition of multiple polymorphic epitopes on HLA-DRw52. Hum Immunol 1987;20:249–264.
4. Gorski J, Tilanus M, Giphart M, Mach B. Oligonucleotide genotyping shows that alleles at the HLA-DRβ111 locus of the HLA-DRw52 supertypic group segregate independently of known DR or Dw specificities. Immunogenetics 1987; 25:79-83.

Author Affiliation

Nuffield Department of Surgery, University of Oxford, Oxford, U.K.

Oligonucleotide Typing Analysis of the Polymorphism of DRB1 and DRB3 Genes Within DRw52 Haplotypes

Jean-Marie Tiercy, Jack Gorski, Michel Jeannet, and Bernard Mach

Abstract: The HLA class II antigens are present as heterodimers of α and β chains on the surface of B cells, macrophages, and activated T cells (1). The polymorphism encoded in the class II DP, DQ, and DR subregions provide the structural basis for the immune responsiveness to foreign antigens. In particular the class II determinants play a key role in the MHC-restricted activation of T helper cells (2). The requirement for class II matching in transplantation and the HLA class II disease associations underline the medical importance of class II antigens. Cloning and sequencing data revealed that the genetic complexity of class II genes is greater than was expected from the serologic analyses (3,4). Using locus- and allele-specific oligonucleotide probes, we have been able to make true allelic comparisons of the two expressed DRβ chain loci, DRBl and DRB3, formerly called DRβI and DRβIII (5,6), within a large panel of DRw52 individuals, including DR3, DR5, DRw6 and DRw8 haplotypes. Three alleles of DRw52 have been detected (7) and extended haplotypes were identified: A1,B8,DR3,52a; DR5(wll), 52b; DRw13,Dw19,52c and DRw14,Dw9,52b. Four

Immunobiology of HLA
Volume II

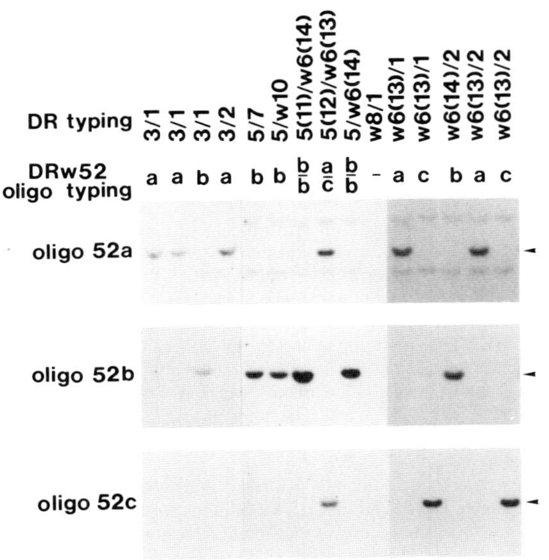

Figure 1. Oligo typing of a representative panel of DR3, 5, w6, and w8 cells obtained from randomly selected Caucasoid blood donors. EcoRI-digested DNA was hybridized (7,9) successively with the allele-specific oligonucleotide probes 52a, 52b, and 52c, respectively. DR typing and DRw52 oligotyping are indicated above each lane. Arrows point to the 3.3 Kb EcoRI fragment corresponding to the DRB3 first domain exon. Samples in lanes 1, 2, and 4 are from individuals with the A1B8DR3 haplotype. The DRw52c allele is exclusively found in linkage disequilibrium with DRw13 haplotypes.

splits of DRw6 corresponding to the Dw9, Dw16, Dw18 and Dw19 cellular specificities can now be discriminated by oligonucleotide typing.

Three allele-specific oligonucleotide probes, referred to as oligos 52a, 52b and 52c (7), were synthesized on the basis of DRB3 sequences derived from HHK (6), DR4/w6 (8) and WT46 (Gorski et al., this volume), respectively. These oligos were used to analyze the distribution of these three serologically undetected alleles within a panel of 82 DRw52 haplotypes from healthy Caucasian blood donors (7). An example of this analysis is shown in Figure 1. The oligo typing results, summarized in Table 1, show that:

1. all A1,B8,DR3 individuals are 52a;
2. Ax,By,DR3 individuals are either 52a or b;
3. DR5 individuals are almost all 52b (see below);
4. The 52c allele is never found in linkage disequilibrium with DR3 or DR5 haplotypes;
5. DRw13 haplotypes can be either 52a, 52b or 52c;
6. DRw14 haplotypes are almost all 52b;
7. none of the three oligos did hybridize to the DRw8 haplotypes. DRw8 haplotypes were, however, all positive with a DRw8 oligo probe derived from a DRw8 βI sequence (10).

Table 1. Extended Haplotypes Within Caucasian Individuals

A1 B8 DR3 DRw52a
Ax By DR3 DRw52a/52b
Ax B18 DR3 DRw52b
DR5 (DRw11) DRw52b
DR5 (DRw12) DRw52a/52b
DRw13 Dw18 DRw52a/52b
DRw13 Dw19 DRw52c
DRw14 Dw9 DRw52b

We also found that the DRw52 oligo typing results correlate with the TaqI/DRβ (WS probe: DRβ) restriction fragment length polymorphism (RFLP) data: a TaqI 12kb-fragment is observed in all 52a individuals regardless of their serological typing. DRw52c allele is also associated with the presence of the 12kb-fragment, whereas the DRw52b allele is associated with a TaqI 15kb-fragment.

DR5 individuals from our panel were characterized for DRw52 alleles as well as for the DRB1 locus, using an oligonucleotide derived from the sequence of DR5/Dw5 Svei homozygous typing cell (HTC) (11), referred to as oligo DR5I. Almost all the DR5 cells tested were oligo DR5I-positive and typed as DRw52b, corresponding to the extended haplotype described in Table 1. This pattern was also found in BM21 (WS ID # 9043), ATH, and BOB HTCs which are all serologically typed as DRwll. Although few of the DR5 individuals of our panel were typed for the serologic splits DRw11 and DRw12, none of the DRw12 cells tested so far hybridize with oligo DR5I. This shows that the DRBl sequence of DRw12 cells differs from the sequence of Svei HTC. Furthermore, we have evidence that the DRw12 specificity can be subdivided on the basis of its linkage disequilibrium with DRw52 alleles into two groups, namely DRw12-DRw52a and DRw12-DRw52b. An example of this rare split is shown in Figure 1 (lane 8). These two splits of DRw12 also correlate with different TaqI/DRβ RFLP patterns. We now find that the DRw12-DRw52a haplotype corresponds to the Dw HAG (DRw13/DQw7) specificity. A fourth DR5 haplotype consists of the cell line JVM (WS ID # 9039) (DR wll/LDVII), which is positive with oligo DR5I, is oligo typed as DRw52b, but differs from the DR5βI sequence of Svei at the third hypervariable region (10).

The Dw6/DRw6 complex can be defined by cellular techniques and by serology. DRw6 specificity is divided into two serologic subtypes DRw13 and DRw14 (12). Furthermore, MLC reactivity subdivides DRw13 into Dw18 and Dw19 and DRw14 into Dw9 and Dw16 (13,14). However, the lack of monospecific alloantisera and the multiple splits make it difficult to provide a reliable and routine typing for this specificity. Using allele-specific oligonucleotide probes we can now identify the Dw9, Dw16, Dw18, and Dw19 splits of DRw6 by analysis of both DRBl and DRB3 loci (15). The distribution of the DRw52 alleles within the DRw13 and DRw14 haplotypes (Table 1) (7) were confirmed and the following extended haplotypes were observed: DRw13,Dw18, DRw52a/52b; DRw13,Dw19,DRw52c; and DRw14,Dw9, DRw52b. Using well-characterized DRw6 cells we have observed the exclusive distribution of the DRw52c allele within Dw19 cells, thus confirming our previous hypothesis that DRw52c contributes to the MLC reactivity defining Dw19 (7). The correlation of the cellular specificity Dw19 with the DRw52c allele is analyzed in more detail in the accompanying paper by Gorski et al. (this volume). In fact, due to the very strong linkage disequilibrium of Dw19 with DRw52c, oligo typing for Dw19 can be achieved by using a DRw52c-specific oligonucleotide.

The use of HLA genotyping by hybridization with locus- and allele-specific oligonucleotide probes represents a more direct and sensitive procedure than the RFLP technique. This new approach allows i) the replacement of cellular typing in certain cases such as the Dw splits within DR2, DR4 and DRw6 and ii) the analysis of hidden alleles identified by DNA sequencing but impossible or difficult to type by current techniques (7). This is important for all three loci DR, DQ, and DP polymorphism. For example we can identify within DQβw3 the DQ3.1, DQ3.2 and DQblank alleles.

The oligo typing procedure is especially important as a complement to serologic typing in the search for HLA-identical unrelated donors for bone marrow transplantation. Furthermore, the analysis of extended haplotypes, as presented in this paper, might also improve the correlation between HLA class II gene polymorphism and susceptibility to autoimmune diseases.

Acknowledgments. The expert technical assistance of P. Chabbey and A. Morrison is acknowledged.

This work was supported by grants 3.911.086 (to M.J. and J.-M.T.) and 3.468.83 and 3.468.86 (to B.M.) from the Swiss National Science Foundation, and by a grant from the Carlos and Elisa De Reuter Foundation.

References

1. Kaufman J, Auffray C, Korman A, Shackelford D, Strominger J. The class II molecules of the human and murine major histocompatibility complex. Cell 1984;36:1.
2. Schwartz RH. T-lymphocyte recognition of antigens in association with gene products of the major histocompatibility complex. Annu Rev Immunol 1985;3:237.
3. Mach B, Gorski J, Rollini P, Berte C, Amaldi I, Berdoz J, et al. Polymorphism and regulation of HLA class II genes of the major histocompatibility complex. Cold Spring Harbor Symp Quant Biol 1986;51:67.
4. Saiki RK, Bugawan TL, Horn GT, Mullis KB, Erlich HA. Analysis of enzymatically amplified β-globin and HLA-DQα with allele-specific oligonucleotide probes. Nature 1986;324:163.
5. Rollini P, Mach B, Gorski J. Linkage map of three HLA-DR β-chain genes: Evidence for a recent duplication event. Proc Natl Acad Sci USA 1985;82:7197.
6. Gorski J, Mach B. Polymorphism of human Ia antigens: Gene conversion between two DRβ loci results in a new HLA-D/DR specificity. Nature 1986;322:67.
7. Tiercy J-M, Gorski J, Jeannet M, Mach B, Identification and distribution of three serologically undetected alleles of HLA-DR by oligonucleotide. DNA typing analysis. Proc Natl Acad Sci USA 1988;85:198.
8. Long EO, Wake CT, Gorski J, Mach B. Complete sequence of an HLA-DRβ chain deduced from a cDNA clone and identification of multiple non-allelic DRβ chain genes. EMBO J 1983;2:389.
9. Angelini G, de Préval C, Gorski J, Mach B. High resolution analysis of the human HLA-DR polymorphism by hybridization with sequence-specific oligonucleotide probes. Proc Natl Acad Sci USA 1986;83:4489.
10. Bell JI, Denney Jr D, Foster L, Belt T, Todd JA, McDevitt H. Allelic variation in the DR subregion of the human major histocompatibility complex. Proc Natl Acad Sci USA 1987;84:6234.
11. Tieber VL, Abruzzini LF, Didier DK, Schwartz BD, Rotwein P. Complete characterization and sequence of an HLA class II DRβ chain cDNA from the DR5 haplotype. J Biol Chem 1986;261:2738.
12. Schreuder GMT, Kennedy LJ, Gebuhrer L, Awad J, Bétuel H, Degos L, et al. Antigen report: HLA-DRw6 and its subgroups HLA-DRw13 and DRw14. In: Albert ED, et al. (eds): Histocompatibility Testing 1984. Berlin, Springer Verlag, 1984, p 192.
13. Termijtelen A, Schreuder GMT, Mickelson EM, van Rood JJ. Ninth International Histocompatibility Workshop pretesting of Dw6 HTCs: Two subtypes of Dw6. Tissue Antigens 1984;24:73.
14. WHO-Nomenclature Report: Nomenclature for factors of the HLA system 1984. Tissue Antigens 1984;24:73.
15. Tiercy J-M, Gorski J, Bétuel H, Freidel AC, Gebuhrer L, Jeannet M, et al. DNA typing of DRw6 subtypes: correlation with DRB1 and DRB3 allelic sequences by hybridization with oligonucleotide probes. Human Immunol 1988 (in press).

Author Affiliations

Jean-Marie Tiercy, Michel Jeannet, Transplantation Immunology Unit, Hôpital Cantonal Universitaire, Geneva; *Jack Gorski, Bernard Mach*, Department of Microbiology, University of Geneva, Switzerland

Subtypes of DRw52: Different Typing Techniques Reveal Very Similar but Distinct Typing Results

Annemarie Termijtelen, Fu Meei Robbins, Willeke E.M. Schroeijers, Vera P.G. Jacobs, Jack Gorski, Roberto Tosi, Nobuyuki Tanigaki, Marcel G.J. Tilanus, and Jon J. van Rood

Abstract: LB-Q1* is a cellularly defined subtype of DRw52. The presence of LB-Q1 can be determined with proliferative alloantigen specific T-cell lines or clones (1,2). We have identified the T-cell recognition epitope LB-Q4, a postulated allele of LB-Q1. Association studies between LB-Q and established HLA-DR antigens revealed that both antigens LB-Q1 and LB-Q4, could be present in combination with DR3, DR5, and DRw6, but not with DRw8.

LB-Q1 was shown to map to the DRβIII gene (2). Other polymorphisms for DRβIII within DRw52 have previously been described. Gene sequence analyses revealed three alleles, DRw52a, b and c (3,4; J.M. Tiercy personal communication). Alloantisera were defined recognizing the DRw52-subtypes Tr22 and Tr81 (5,6). RFLPs were found, correlating with the presence or absence of LB-Q1 (2).

In an extended collaborative study, 22 well-defined homozygous typing cells were compared in four different techniques. Although all techniques showed very similar reaction patterns, none of these patterns were completely identical.

To estimate the gene frequencies of LB-Q1 and LB-Q4, a random panel of 93 Dutch, Caucasoid individuals was typed with 1 LB-Q1-specific T-cell clone prepared in Washington (FRL4C40), and 1 LB-Q1-specific (c21) and 1 LB-Q4-specific (c1437) T-cell clone prepared in Leiden. The typing results obtained with the 2 LB-Q1-specific T-cell clones were 100% concordant. The relative ease with which LB-Q1-specific T-cell clones are found stresses its relevance as T-cell recognition epitope. Gene frequencies for LB-Q are given in Table 1. The gene frequencies for LB-Q1 and LB-Q4 were similar (.19 ± 0.03 and .23 ±0.03, respectively). Only seven out of 70 DRw52 positive individuals were LB-Q blank. These individuals were 2× DRw8 positive, 4× DRw13 positive and 1× DRw8, DRw13 heterozygous.

Table 1. Gene Frequencies of LB-Q in a Random Dutch Population of 93 Individuals

	Pf	Gf ± S.E.
DRw52	70/93 = 0.75	0.50 ± 0.04
DRw52/LB-Q1	32/93 = 0.34	0.19 ± 0.03
DRw52/LB-Q4	38/93 = 0.41	0.23 ± 0.03
DRw52/LB-Q blank*	7/93 = 0.08	0.04 ± 0.01

*The DRw52/LB-Q blank individuals were 2x DRw8 positive, 4x DRw13 positive, and 1x DRw8, DRw13 heterozygous.

To study the associations of of LB-Q1 and LB-Q4 with the established HLA-DR antigens, a total of 125 DR homozygous B cell lines, including the full workshop set, were typed. The results are depicted in Table 2. LB-Q1 and LB-Q4 were only found in association with DRw52. Both can be associated with DR3, DR5 and DRw6, but not with DRw8. Also the HLA-DR subtypes DRw12, DRw13, and DRw14 can be associated with either LB-Q1 or LB-Q4. Only DRw11 seems to be exclusively associated with LB-Q1. In this panel, five out of 22 DRw13-positive individuals were LB-Q blank. The findings depicted in Table 2 were concordant with the analysis of a panel of heterozygous individuals (n=221) (data not shown).

Proliferation inhibition studies with well-characterized monoclonal antibodies previously demonstrated that LB-Q1 is located on DRβIII (2). In the scope of the present workshop, the LB-Q4 specific clone c1437 was also used in inhibition experiments. The monoclonal antibody 7.3.19.1 (DRw52-like, reactive with DRβIII in DR5 and DRw6 haplotypes, reactive with DRβI + DRβIII in DR3 haplotypes (7)) did inhibit proliferation of c1437 in all instances. Therefore LB-Q4 is most likely also located on DRβIII. Neither of the monoclonal antibodies UK8.1 (DRw52) or JS.1 (DR3) inhibited,† even when a DR3 stimulator cell was used. This finding suggests that these monoclonal antibodies recognize DRβI defined structures.

Additional polymorphisms of DRβIII within DRw52 have been described. Sequence analysis of the encoding exons of the first domain revealed three alleles, 52a, b, and c. Oligonucleotides could be prepared to distinguish

*LB-Q1 now designated Dw25; LB-Q4 now designated Dw24.
†The blocking capacity of UK8.1 and JS.1 was demonstrated when using a DR3(βI)-specific T-cell clone (data not shown).

Table 2. Associations of LB-Q with Established HLA-DR Specificities in 125 DR Homogzygous B Cell Lines

	N	LB-Q1	LB-Q4	LB-Q blank
DRw52	66	0.47*	0.30	0.23
DR3	15	0.33	0.67	0
DRw11(5)	12	1.00	0	0
DRw12(5)	2	0.50	0.50	0
DRw13(w6)	22	0.45	0.32	0.23
DRw14(w6)	5	0.80	0.20	0
DRw8	10	0	0	1.00

*31/66 = 0.47 of DRw52 positive individuals typed LB-Q1, etc.

Immunobiology of HLA
Volume II

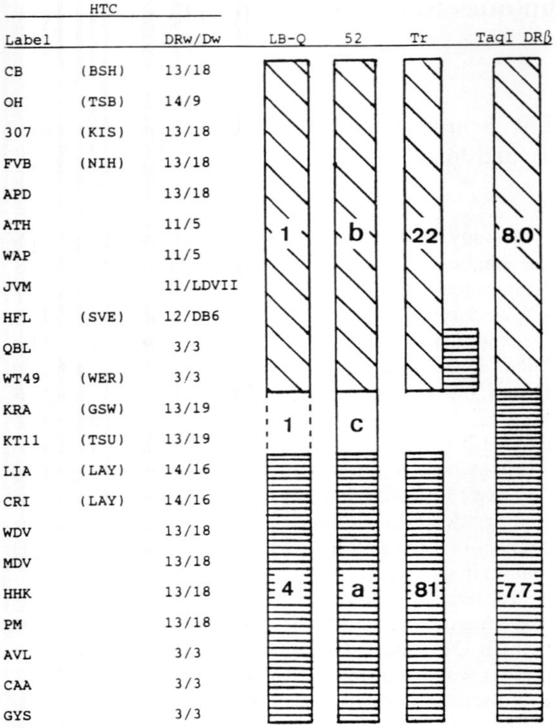

Figure 1. Comparative typing of 22 DRw52-positive, DR-homozygous cells. LB-Q1 and 4 are defined by T-cell clones; 52a, b, and c, with oligonucleotide typing; Tr22 and 81 with serology; TaqI DR 8.0 and 7.7 are the fragment sizes in kb of hybridizing fragments obtained after digestion of DNA with TaqI and hybridization with a full-length DR cDNA probe.

between these alleles (3,4). Using alloantisera and cells, transfected with DRβIII genes, two serological subtypes of DRw52, Tr22 (5), and Tr81 (6) could be identified. DNA analysis of a panel of DR homozygous B cell lines, digested with TaqI, EcoRI, or PvUII, and hybridized with a full-length DRβ cDNA probe, revealed RFLP's compatible with the presence or absence of LB-Q1 (2).

A comparative study, exploring all techniques on the same set of B cell lines has been performed (8). The results depicted in Figure 1 were obtained: LB-Q1 was associated with 52b, Tr22, and TaqI DRβ8.0; LB-Q4 was associated with 52a, Tr81, and TaqI DRβ7.7. Anomalous reactions were observed in the Dw19 HTCs KRA and KT11 which typed LB-Q1 weak, 52c, Tr blank, and TaqI DRβ7.7. Moreover, the LB-Q1-positive DR3 HTCs QBL and WT49 were heterozygous for Tr81 and Tr22. It is believed that in these cells, the Tr81 determinant is present on the DRβI chain (6).

In conclusion, all typing results were very similar but small differences were observed. This finding points to a complex of epitopes on class II molecules, which can so far only be studied by combining the different typing techniques.

Acknowledgments. The authors wish to thank Tiny van Westerop for typing the manuscript.

Supported by the J.A. Cohen Institute for Radiopathology and Radiation Protection (IRS) (Grant nr. 3.3.1.).

References

1. Termijtelen A, Van den Berge SJ, van Rood JJ. LB-Q1 and LB-Q2: Two determinants defined in the primed lymphocyte test and independent of HLA-D/DR, MB/LB-E, or SB. Hum Immunol 1983;8:11.
2. Termijtelen A, Tilanus MGJ, Engelen I, Koning F, van Rood JJ. Molecular localization of LB-Q1, a DRw52-like T-cell recognition epitope and identification at the genomic level of associated shared hybridizing fragments. Hum Immunol 1987;19:255.
3. Gorski J, Mach B. Polymorphism of human Ia antigens: Gene conversion between two DRβ loci results in a new HLA-D/DR specificity. Nature 1986;322:67.
4. Gorski J, Tilanus MGJ, Giphart MJ, Mach B. Oligonucleotide genotyping shows that alleles at the HLA-DR βIII locus of the DRw52 supertypic group segregate independently of known DR or Dw specificities. Immunogenetics 1987; 25:79.
5. Tosi R, Tanigaki N, De Préval C, Gorski J, Mach B. Immunochemical analysis of a cell transfected with an HLA-DR gene reveals a new aloantigenic specificity within HLA-DRw52. Eur J Immunol 1986;16:1603.
6. Berte C, Tanigaki N, Tosi R, Gorski J, Mach B. Serological recognition of HLA-DR allodeterminant corresponding to DNA sequence involved in gene conversion. Immunogenetics 1988;27:167.
7. Koning F, Schreuder I, Giphart M, Bruning H. A mouse monoclonal antibody detecting a DR-related MT2-like specificity: Serology and biochemistry. Hum Immonol 1984; 9:221.
8. Termijtelen A, Gorski J, Robbins FM, Tanigaki N, Tosi R, Tilanus MGJ, Schroeijers WEM, van Rood JJ. Correlations between polymorphisms at the DNA and at the protein level of DRw52 haplotypes, revealed with a variety of techniques. Hum Immunol 1988;22:171.

Author Affiliations

Annemarie Termijtelen, Willeke E.M. Schroeijers, Vera P.G. Jacobs, Marcel G.J. Tilanus, Jon J. van Rood, Department of Immunohaematology & Blood Bank, University Hospital, Leiden, The Netherlands; *Fu M. Robbins*, Georgetown University Medical Center, Immunology Oncology Division, Washington DC; *Jack Gorski*, The Blood Center of Southeastern Wisconsin, Milwaukee, WI, USA; *Roberto Tosi*, Laboratory of Cell Biology, Consiglio Nazionale delle Ricerche, Rome, Italy; *Nobuyuki Tanigaki*, Department of Molecular Immunology, Roswell Park Memorial Institute, Buffalo, NY, USA

Variants of HLA-DRw52 Defined by Proliferative T-Lymphocyte Clones

Eric M. Mickelson, Susan A. Masewicz, and John A. Hansen

Abstract: HLA-DRw52 is a serologically defined class II allospecificity that is found in strong positive linkage disequilibrium with DR3, DR5, DRw6 and DRw8. DRw52 and its allelic form DRw53 are the products of a second DRβ locus, previously designated DRβ2 (1) or DRβ3 (2), that is distinct from the DRβ1 locus controlling the expression of the DR1 to DRw14 series of alloantigens. Polymorphism within the DRw52 gene has been demonstrated by DNA sequencing (3-5) and variation among DRw52 antigens has been demonstrated by serological analysis with monoclonal antibodies (MAb) (6) and by 2-D gel electrophoresis (7). To determine whether these DRw52 structural variants might be functionally significant with respect to T-cell recognition, we have attempted to generate DRw52-reactive proliferative T-lymphocyte clones (PTLc) by selective priming with DRw52+ cells (8). PTLc derived from these primings were tested for specific proliferative activity against cells from the Tenth Workshop (10WS) cell-line panel and a local panel. Three different DRw52-specific PTLc were isolated. The reactivity patterns of these PTLc suggested that there are at least five DRw52 variants.

PTLc C6 and E3 were generated by haplotype priming against DRw13,DRw52,Dw18 and PTLc ZUK16 by haplotype priming against DRw13,DRw52,Dw19. Table 1 shows the results of testing these clones against a DRw52+ cell panel. PTLc C6 reacted only with certain DRw52+ cells (B8, DR3+; DRw13,Dw18+; DRw13, LD"HAG"+; and DRw14,Dw16+) and did not react with any DRw8+ cells. PTLc E3 reacted with a second set of DRw52+ cells, including B18,DR3+, DRw11,Dw5+, DRw11,DB2+, DRw12,DB6+, DRw13,Dw18+, DRw14,Dw9+ cells, and a local DRw8+ cell (KGU) of East Indian origin. KGU is a new HLA-D (HTC-defined) specificity that is distinct from the provisional Dw"8.1", "8.2" and "8.3" clusters previously described (9). PTLc ZUK16 reacted with a third set of DRw52+ cells, including DRw13,Dw19+ and DRw13,Dw"blank"+ cells. Both PTLc E3 and ZUK16 reacted with two of the HCL in the panel, TEM (10WS 9057) and LEM (B58,DR3, DRw52,Dw"blank"), generating a fourth pattern of reactivity that suggested the existence of a fourth unique DRw52 variant comprised of 52b- and 52c-like epitopes. A fifth reactivity pattern was apparent by the failure of the three PTLc to react with one of the DRw52+ HCL, OMW (10W9058). The five DRw52 variant reactivity patterns are illustrated in Table 2.

In addition to the PTLc described above, two additional PTLc, C5 and C3, originally primed against DRw6+ stimulator cells, were isolated that appeared to detect distinct determinants associated with DRw52b. Table 3 compares the reactivity of DRw52+ cells with PTLc E3 (discussed above) and PTLc C5 and C3. PTLc E3 was reactive with all six of the DRw52+ cells, but PTLc C5 and C3 each failed to respond to two of the six cell lines, suggesting that there might be additional polymorphism among the genes encoding DRw52b antigens that are associated with DRw11(DR5)+ haplotypes.

Table 1. Reactivity of HLA-DRw52-Specific Proliferative T-Lymphocyte Clones with 10WS and Local Cell Panels

Cell Panel				T-Lymphocyte Clone		
Name	WS#	DRβI	Dw	C6	E3	ZUK16
COX	9022	3	3	+	−	−
VAVY	9023	3	3	+	−	−
RSH	9021	3	*	+	−	−
LOO8	9018	3	3	−	+	−
DUCAF	9019	3	3	−	+	−
LEM	−	3	*	−	+	+
JBUSH	9035	w11	5	−	+	−
SWEI	9037	w11	5	−	+	−
BM15	9040	w11	5	−	+	−
J0528	9041	w11	5	−	+	−
BM21	9043	w11	5	−	+	−
BRIP	9044	w11	5	−	+	−
TUBO	9045	w11	5	−	+	−
JVM	9039	w11	*	−	+	−
TISI	9042	w11	*	−	+	−
SPO	9036	w11	DB2	−	+	−
BM16	9038	w12	DB6	−	+	−
WDV	9062	w13	18	+	−	−
HHK	9065	w13	18	+	−	−
CB	9060	w13	18	−	+	−
OMW	9058	w13	18	−	−	−
HOR	9053	w13	19	−	−	+
HO301	9055	w13	*	−	−	+
SLE	9059	w13	19	−	−	+
WT47	9063	w13	19	−	−	+
EMJ	9097	w13	19	−	−	+
EK	9054	w14	9	−	+	−
ABO	9061	w14	9	−	+	−
TEM	9057	w14	*	−	+	+
AMAL	9064	w14	16	+	−	−
LZL	9099	w14	16	+	−	−
BTB	9067	w8	"8.1"	−	−	−
BM9	9068	w8	"8.1"	−	−	−
MAD	9069	w8	"8.1"	−	−	−
OLG	9071	w8	"8.2"	−	−	−
SPACH	9072	w8	"8.2"	−	−	−
TAB	9066	w8	"8.3"	−	−	−
LUY	9070	w8	"8.3"	−	−	−
KGU	−	w8	*	−	+	−

*Cells RSH, LEM, JVM, TISI, and HO301 type as Dw"blank"; cell TEM types as Dw9-like; cell KGU is of East Indian origin and appears to represent a fourth DRw8-associated HLA-D (HTC-defined) cluster.

Table 2. DRw52 Variants Defined by Cloned Proliferative T Lymphocytes

Cell Panel*			PTLc-Defined DRw52 Variants				
Name	DRβI	Dw	52a	52b	52c	52d	52e
VAVY	3	3	+	−	−	−	−
DUCAF	3	3	−	+	−	−	−
LEM	3	†	−	−	−	+	−
SWEI	w11	5	−	+	−	−	−
SPO	w11	DB2	−	+	−	−	−
HERL	w12	DB6	−	+	−	−	−
HHK	w13	18	+	−	−	−	−
HAG	w13	†	+	−	−	−	−
CB	w13	18	−	+	−	−	−
OMW	w13	†	−	−	−	−	+
SLE	w13	19	−	−	+	−	−
EK	w14	9	−	+	−	−	−
TEM	w14	†	−	−	−	+	−
AMAL	w14	16	+	−	−	−	−
MAD	w8	8	−	−	−	−	−
OLGA	w8	8	−	−	−	−	−
TAB	w8	8	−	−	−	−	−
KGU	w8	†	−	+	−	−	−

*A representative cell from each DR/Dw haplotype is listed.
†Cell LEM types as Dw"blank"; cell HAG (9W1802) is DRw13,DQw3+ and types as LD"HAG"; cell OMW types as Dw18-like; cell TEM types as Dw9-like; cell KGU does not correspond to previously defined Dw8 clusters.

To further establish that these PTLc were reacting with products of the second DRβ locus, blocking studies were carried out using 10WS MAb (Table 4). MAbs P4.1, PL8, and L243, directed against nonpolymorphic DR-associated determinants, completely blocked the reactivity of PTLc C6 and E3, and partially but significantly blocked PTLc ZUK16. MAb 7.3.19.1, directed against a polymorphic determinant previously shown to be associated with the DRβ3 chain of DR5,DRw52+ and DRw6,DRw52+ haplotypes (10), completely blocked the reactivity of the three PTLc. These results confirmed that the determinants detected by PTLc C6, E3 and ZUK16 were associated with the second DRβ locus product and not with DRβI or DQ chains.

The PTLc analyzed in this study were originally primed against DRw13(w6)+ haplotypes, and as is apparent from Tables 1 and 2, did not detect determinants on the majority of DRw8,w52+ cells. These results suggest that the DRβ3 molecules found on DRw8 haplotypes are relatively divergent from the DRβ3 molecules associated with DR3, DR5, and DRw6 haplotypes. Only the DRw8+ HCL KGU, derived from an East Indian donor and expressing the DRw8"I" allospecificity (11), was able to activate a DRw52-specific T-cell clone (clone E3, specific for DRw52b).

Table 3. Possible variants of DRw52b

	Tenth Workshop Cell Line*					
	9037	9041	9045	9042	9039	9043
PTLc	SWEI	J052	TUBO	TISI	JVM	BM21
E3	+†	+	+	+	+	+
C5	+	+	+	+	−	−
C3	+	+	−	−	+	+

*Six DR5+ HCL from the 10WS tested as stimulator cells against DRw52-reactive PTLc E3, C5, and C3.
†Positive responses (+) determined by centroid cluster analysis.

Table 4. Inhibition of DRw52-Specific PTLc with 10WS Monoclonal Antibodies

Monoclonal	PTLc		
Antibody*	C6	E3	ZUK16
P4.1	95†	99	70
PL8	98	100	80
L243	101	100	45
UK8.1	4	0	0
7.3.19.1	98	107	96

*P4.1, PL8, and L243 are nonpolymorphic, anti-DR mAb; UK8.1 and 7.3.19.1 are polymorphic MAb specific for DRw52-like determinants.
†Percent inhibition of reference proliferative response against DRw6+ stimulator cells.

The 52a, 52b, and 52c variants identified here by PTLc correspond to DRβ3 genomic variants previously defined by DNA sequencing (3; J. Gorski, personal communication). These three T-cell-defined variants also correspond to the structural variants of DRw52 molecules from DRw6+ haplotypes (Dw9, Dw16, Dw18, and Dw19) that have been defined by 2-D gel electrophoresis (7). The 52a and 52b variants are similar to the T-cell-defined LB-Q4 and LB-Q1 specificities described by Termijtelen et al. (12,13). In addition, the reactivity patterns of PTLc E3, C3, and C5 suggest the presence of further variation within the DRw52b allele that is associated with DR5+ haplotypes. Recent data from DNA sequencing studies, demonstrating polymorphism of the DRβ3 chains of a DR5(DRw11)+ cell line (SWEI; 10WS 9037), would tend to support the existence of additional polymorphism within DR5-associated DRw52 molecules (4).

The use of allospecific PTLc has the potential for the definitive discrimination of functionally relevant class II structural polymorphisms. Selected PTLc should be particularly useful as typing reagents for transplantation and disease association studies.

Acknowledgments. Supported by Grant CA 18029 from the National Cancer Institute, National Institutes of Health.

References

1. Bodmer WF. The HLA system, 1984. In: Albert ED et al. (eds): Histocompatibility testing 1984. Springer-Verlag, Berlin, 1986, p 11.
2. Gorski J, Tilanus M, Giphart M, Mach B. Oligonucleotide genotyping shows that alleles at the HLA-DRβIII locus of

the DRw52 supertypic group segregate independently of known DR or Dw specificities. Immunogenetics 1987; 25:79.
3. Gorski J, Mach B. Polymorphism of human Ia antigens: Gene conversion between two DRβ loci results in a new HLA-D/DR specificity. Nature 1986;322:67.
4. Didier DK, Schiffenbauer J, Shuman S, Abruzzini LF, Gorski J, Watling DL, et al. Characterization of two distinct DRβ chain alleles at the βIII locus of the DR5 haplotype: βIII alleles are highly conserved. J Immunol 1986; 137:2627.
5. Erlich HA, Horn GT, Bugawan TL, Scharf S. The molecular analysis of HLA class II gene polymorphism. In: Molecular and Cellular Biology of Histocompatibility Antigens. ASHI, New York, 1987, p 93.
6. Tosi R, Tanigaki N, de Preval C, Gorski J, Mach B. Immunochemical analysis of a cell transfected with an HLA-DR gene reveals a new alloantigenic specificity within HLA-DRw52. Eur J Immunol, 1986;16:1603.
7. Bosch ML, Termijtelen A, Gerrets R, Schreuder GMT, Bontrop RE, Giphart MJ. Polymorphisms within the HLA-DRw6 haplotype. II. Protein charge heterogeneity reflects MLC subtyping of HLA-DRw6 homozygous cells. Immunogenetics, 1985;22:23.
8. Mickelson EM, Beatty PG, Storb R, Hansen JA. Immune responses in an untransfused patient with aplastic anemia: Analysis of cytolytic and proliferative T cell clones. Human Immunol 1984;10:189.
9. Mickelson EM, Nisperos B, Layrisse Z, Kim SJ, Thomas ED, Hansen JA. Analysis of the HLA-DRw8/Dw8 haplotype: Recognition of three distinct antigen complexes in Caucasians, Native Americans and Orientals. Immunogenet 1985;17:399
10. Koning R, Schreuder I, Giphart M, Bruning H. A mouse monoclonal antibody detecting a DR-related MT2-like specificity: Serology and biochemistry. Human Immunol 1984;9:221.
11. Awad J, Festenstein H. Preliminary local analysis of selected 10th International Histocompatibility Workshop class II coreset antisera. Tenth International Histocompatibility Workshop Newsletter, 1987;1, p 12.
12. Termijtelen AW, Schroeyers W, Jacobs V. LB-Q1 and LB-Q4 in DRw52 positive B-LCL. Tenth International Histocompatibility Workshop Newsletter, 1987;1(6), p 3.
13. Termijtelen A, van den Berge SJ, van Rood JJ. LB-Q1 and LB-Q2: Two determinants defined in the primed lymphocyte test and independent of HLA-D/DR, MB/LB-E, or SB. Human Immunol 1983;8:1.

Author Affiliations

Eric M. Mickelson, Fred Hutchinson Cancer Research Center and Department of Medicine, University of Washington; *Susan A. Masewicz*, The Puget Sound Blood Center; *John A. Hansen*, The Puget Sound Blood Center, Fred Hutchinson Cancer Research Center, and Department of Medicine, University of Washington, Seattle, WA, USA

Functional and Structural Analysis of HLA-DRw52

Claude Irlé, Dominique Jaques, Jean-Marie Tiercy, Michel Jeannet, and Bernard Mach

Abstract: Antigen-specific T-cell responses are triggered following recognition by the T-cell receptor of antigenic determinants presented by major histocompatibility complex (MHC) molecules (1). Recently, the multigenic organization of the MHC class II region has been unravelled (2). In DRw52 haplotypes, the DR region was shown to contain one α gene and three β genes (3). One of these latter is a pseudogene. Two different DR molecules, corresponding to distinct α-β heterodimers (β I and β III), are expressed. The β I gene product contains the epitopes defined by DR1-w14 alloantisera, whereas the β III gene product contains the supertypic DRw52 epitope(s)(4). The functional role of these two molecules remains, however, a matter of debate. This question was investigated using TT-specific and DR-restricted cloned T cells, We now present evidence that these two molecules are functional. We also show that T-cell clones can functionally discriminate between the different alleles of the DRw52 locus which can be identified by allele-specific oligonucleotide probes.

Antigen (tetanus toxoid, TT) specific T-cell clones were obtained after cloning of peripheral blood mononuclear cells (PBMC) obtained from immune volunteer donors (# 1417:DR3,X and # 2278:DR3,w6) after in vitro restimulation with TT (dilution 1:5000). Blasts were isolated by density gradient centrifugation using Percoll gradients and cloned in culture medium (RPMI1640, supplemented with 10% heat-inactivated human AB serum, antibiotics, and IL-2) exactly as described elsewhere (5), at a concentration of .3 blasts/well, in the presence of autologous irradiated (3500 Rad) feeders. The latter were previously incubated for 4 hours with TT. Growing cultures were tested for TT specificity using the autologous PBMC as antigen presenting cell (APC), and different antigens (diphtheria toxoid, PPD, streptokinase, proteus antigen, trichophyton antigen, candida antigen) in addition to TT. Selected antigen-specific clones were than expanded by periodic restimulation with autologous TT-APC and IL-2. These clones were then retested using panels of TT-APC representing a large spectrum of DR specificities, including DR1 to DRw14 lymphoblastoid B cell lines (LCL) available for these studies in the present workshop.

Several DR3-restricted and TT-specific clones were found, including clone 49 used in the present study. Two

APC #	1417	QBL	AVL	1249	891	1165	ME	2778	3135	1213	2134
DR-type	3/x	3/3	3/3	2/3	3/w5	2/w5	1/w5	3/w6	2/w6	w6/7	2/w8

clone 49c	+	+	+	+	+	-	-	+	-	-	-
clone 16c	+	-	+	+	+	-	-	-	+	+	-
clone 10c	-	+	-	-	+	+	+	+	+	-	-

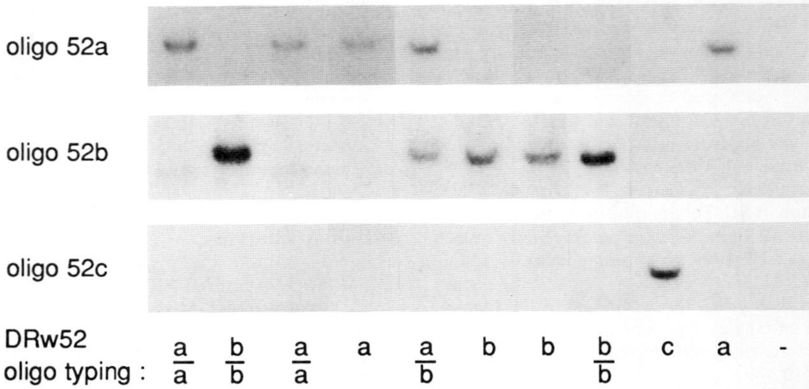

Figure 1. A selection of panel cells used to characterize clones 49, 16, and 10 are shown. The reactivities of the clones are indicated using + or − symbols. These correspond to specific proliferation of these clones to TT-APC. "+" represent stimulation of 10x or more over background proliferation (which was always less than 1000 CPM). Oligonucleotide DNA typing using 52a, 52b, and 52c locus and allele specific probes was performed exactly as described elsewhere (9), and results are indicated under each lane.

other clones recognized subsets of DR3 including, respectively, the homozygous typing cell (HTC) AVL (clone 16), or the HTC QBL (clone 10). The panel studies showed that clone 16 reacted to a subset DR3 and DRw6 TT-APC, while cone 10 was stimulated by other DR3 and DRw6, as well as by all DR5 TT-APC. These results are summarized schematically on Figure 1. This distribution corresponds exactly to that of the three alleles of DRβIII recently described, namely 52a, 52b, and 52c(6).

The supertypic specificity DRw52 includes DR3, DR5, DRw6, and DRw8. Our results suggested that TCC 10 and 16 reacted to TT presented by distinct epitopes included in DRw52. This possibility was further investigated in family studies. A summary of a typical family is shown in Table 1. These experiments indicate independent allelic segregation of the restriction elements of clone 10 and clone 16 among DRw52 APC.

The restriction of T-cell clones can be defined by MAb blocking studies. Therefore, clones 49, 16, and 10 were

Table 1. Segregation Study of DRw52 Epitopes Restricting TT-Responses of Clone 10 and Clone 16 in a Family, Compared to 52A and B Oligonucleotide Typing

Responders		Stimulation Indices by TT-Antigen Presenting Cells			
Clones Tested	Restriction	Father (DR1,3)	Mother (DR1,5)	Sib 1 (DR1,5)	Sib 2 (DR3,5)
Clone 49	DR3	184	0.8	1.0	150
Clone 10	<DRw52	3.4	140	27	102
Clone 16	<DRw52	49	1.1	0.6	16
Oligo 52A		+	−	−	+
Oligo 52B		−	+	+	+

Results represent means of ^{3}H-TdR-incorporation (last 8h of culture) in 56h triplicate cultures, between 20,000 responders/and 50,000 irradiated (3500R) stimulators/per 200 ul culture. These were used to determine stimulation indices. Standard deviations were always less than 20%.

Oligonucleotide typing was performed as described in Figure 1, and showed agreement between results of oligo 52a and clone 16, and between oligo 52b and clone 10.

Table 2. Inhibition of TT-Specific Proliferation of Clones 49, 10, and 16 by Anti-Class II (mAb)

MAb	Specificity	Clones Tested (% Inhibition) 49[1]	10[1]	16[2]
DR,DQ,DP	Tu39	95	99	–
DR	D1/12	85	95	99
	GSP4.1	93	99	–
	PL8	99	99	–
	L243	99	99	–
DR,DP	2D6	99	99	–
DR,DP	Tu35	93	99	–
DR3	UL65	0	14	–
DR3	16.23	88	5	–
DRw52	NDS10	0	99	–
DRw52	NDS13	0	98	–
DRw52	7.3.19.1	99	99	99
DRw52	UK8.1	0	0	–
DQw	Tu22	35	40	50
DQw	1A3	43	22	–
DQw2	CL	50	20	
DP	B7/21	2	18	0
DP	PL15	50	38	–

Results represent means of triplicate cultures with MAb-containing ascites at a final dilution of 1:200. % inhibition was calculated as follows:

$$\% \text{ inh.} = \frac{(CPM_{REF,Cell+TT}+MAb)-(CPM_{REF,Cell-TT}+MAb)}{(CPM_{REF,Cell+TT}) \qquad -(CPM_{REF,Cell-TT})} \times 100.$$

Reference cells were 1) B-LCL Ducaf (DR3.3) for clone 49 and 10; 2) B-LCL 1417 (DR3,-) for clone 16.

tested in blocking experiments using a panel of MAb specific for DR, DQ, and DP framework epitopes, most of which were available to Tenth Workshop participants. Among the MAb recognizing DR framework epitopes that were tested, all but one (MAb IIIE3) inhibited completely the TT-specific proliferation of these clones at high concentrations (less than 1:200 dil. of ascites, Table 2). However, at higher dilutions, some DR framework antibodies were more effectively inhibiting clone 10 (DR52-restricted) than clone 49 (DR3-restricted). Similar differential blocking has been observed before (7). This could relate to functional or structural differences between the two DR molecules, in addition to differences in their expression, or to steric hindrance, or a combination of these possibilities. Antibodies to DQ and some antibodies to DP inhibited significantly at very high concentration only (1:40 dilution of ascites). This was probably due to either weak cross-reactivities with DR, or to nonspecific effects of proteins contained in ascites fluids.

Further evidence for a DRw52 associated restriction was obtained in blocking experiments performed with a panel of Tenth Workshop MAb binding to DR3 and DRw52 associated epitopes. Interestingly, MAb 16.23 (DR3 and some DRw6) and JSI (DR3) inhibited clone 49, but not clone 10. In contrast, MAb NDS 10 (DR3,5, and DRw6) and NDS13 (DR3,5,w6,w8) blocked only clone 10, but not clone 49. The MAb 7.3.19.1 (DRw52 -like) inhibited all three clones. The MAb UK8.1 (DRw52) and UL65(DR3) inhibited the DR3 and the DRw52 restricted clones only at 1:40, but not any more at 1:200. The functional distinction noticed between the DR3 MAb could therefore relate to different DR3 epitopes, determined by either region. Indeed, DR3 has resulted from a gene conversion of two DRβIII segments of 52a into different hypervariable regions of a DRβI locus of DRw6 (8). The functional distinctions noticed between DRw52 MAb could also relate to structural differences between βIII genes. Indeed, the allelic differences between DRβIII genes are scattered widely throughout the first domain (2,8). These could result in functionally different DR52-like epitopes, recognized by different DRw52-like MAb.

A selection of the panel cells studied with clones 49,16, and 10 were analyzed by oligonucleotide typing with the recently described 52a, b, and c locus and allele-specific oligonucleotide probes (6). All the APC that stimulated only clone 16 were found to type 52a, and all the APC stimulating only clone 10 were typed 5.2b. One APC with only one DRw52 haplotype (DR2,w6,DRw52) stimulated unexpectedly both clone 10 and clone 16. This cell was found to be 52c by oligo typing, indicating that the 52c gene product contains (R) cross-reactive restriction site(s). These results show that clones 10 and 16 define indeed the three distinct alleles of DRw52 recognized by oligonucleotide probes, More importantly, they demonstrate the relevance of the products of locus DRβIII in restriction and presentation.

Acknowledgment. This research was supported by grant No. 3.9120.86 from the Swiss Natl. Research Foundation, and by grant No. 38 from the Elisa DeReuter Foundation.

References

1. Schwartz RH. Fugue in T-lymphocyte recognition. Nature 1987;326:738.
2. Mach B, Gorski J, Rollini P, Berte C, Amaldi I, Berdoz J, et al. Polymorphism and regulation of HLA class II genes of the major histocompatibility complex. Cold Spring Harbor Quant Biol 1986;51:67.
3. Rollini P, Mach B, Gorski J. Linkage map of three HLA-DR β-chain genes: Evidence for a recent duplication event. Proc Natl Acad Sci USA 1985;82:7197.
4. Gorski J, Tosi R, Strubin M, Rabourdin-Combe C, Mach B. Serological and immunochemical analysis of the products of a single HLA DR-a and DR-β chain gene expressed in a mouse cell line after DNA-mediated cotransformation reveals that the b carries a known supertypic specificity. J Exp Med 1985;162:105.
5. Irlé C, Beatty PG, Mickelson E, Thomas ED, Hansen JA. alloreactive T cell responses between HLA-identical siblings. Transplantation 1985;40:329.
6. Tiercy J-M, Gorski J, Jeannet M, Mach B. Identification and distribution of three serologically undetected alleles of HLA-

DR by oligonucleotide DNA typing analysis. Proc Natl Acad Sci USA (in press).

7. Triebel F, Missenard-Leblond M, Couty M-C, Charron DJ, Debré P. Differential inhibition of human antigen-specific T cell clone proliferative responses by distinct monoclonal anti-DR antibodies. J Immunol 1984;132:1773.
8. Gorski J, Mach B. Polymorphims of human Ia antigens: Gene conversion between two DR β loci results in a new HLA-D/DR specificity. Nature 1986;322:67.
9. Angelini G, de Preval C, Gorski J, Mach B. High-resolution analysis of the human HLA-DR polymorphism by hybridization with sequence-specific oligonucleotide probes. Proc Natl Acad Sci USA 1986;83:4489.

Author Affiliations

Claude Irlé, Dominique Jaques, Michel Jeannet, Transplantation Immunology Unit, Division of Immunology and Allergology, Department of Medicine, Hôpital Cantonal Universitaire, Geneva; *Bernard Mach*, Department of Microbiology, University of Geneva Medical School, Geneva; *Jean-Marie Tiercy*, Transplantation Immunology Unit, Division of Immunology and Allergology, Department Medicine, Hôpital Cantonal Universitaire, Geneva; Department of Microbiology, University of Geneva Medical School, Geneva; Switzerland

Immunochemical Assessment of DRw52 Specificity by the Use of HLA Class II Transfectant Molecules

Nobuyuki Tanigaki, Christine C. Berte, Roberto Tosi, Jack Gorski, Bernard Mach, and Giovanni B. Ferrara

Abstract: DRw52, previously MT2, is a supertypic specificity that is tightly associated with DR3, DR5, (DRw11, DRw12), DRw6 (DRw13, DRw14), and DRw8. There is no doubt that this specificity is confined to DR molecules, but it is not clear whether this complex specificity, defined by the cytotoxicity reaction and representing probably several associated and cross-reactive determinants, is localized either on the DRBl or the DRB3 gene product, or on the both gene products. Utilizing mouse L cell transfectants, each expressing a single DRB gene product, we have analyzed "DRw52" at the molecular level rather than at the cellular level, and have sorted out three different specificities, provisionally called 52/R1, 52/R2, and 52/R3, that show the haplotype distribution conforming to DRw52 or DRw52-like specificity and yet differ in the gene localization.

Table 1. Immunochemically Definable Specificities of the "DRw52" Family

a) DR Haplotype Distribution

Specificity	The distribution in DRw52 haplotype:							
	DR3	DR3	DR5 w11	DR5 w12	DRw6	DRw6	DRw6	DRw8
	52a	52b	52b	52b	52a	52b	52c	
TR22	−	+	+	+	−	+	−	−
TR81	+	+	−	−	+	−	−	−
52/R1	+	+	+	+	+	+	+	+
52/R2	+	+	+	−	+	+	+	+w
52/R3	+	+	+	−	+	+	+	−

b) Gene Localization

Specificity	The localization in DR genes:									
	DR3	DR3	DR3	DR5	DR5	DR5	DRw6	DRw6	DRw6	DRw6
	3	52a	52b	w11	w12	52b	w6	52a	52b	52c
	B1	B3	B3	B1	B1	B3	B1	B3	B3	B3
TR22	−	−	+	?	?	+	−	−	+	−
TR81	+	+	−	−	−	−	−	+	−	−
52/R1	+	+	+	+	+	+	+	+	+	?
52/R2	+	−	−	+	−	−	+	−	−	?
52/R3	+	+	−	+	−	−	+	+	−	+w

Table 2. Expression of the 52/R1 and 52/R2 Determinants on the DRB1 and DRB3 Loci Products

Specificity Antiserum Labeled Antigen			52/R1 Fe88/37 LR3.1/52a (DRB3 52a)	52/R2 Fe117/30 AVL (DR3,3)
Inhibitor			Relative Inhibition* (Means)	
Human B cell line				
DR haplotype		Number tested		
DR3	52a	6	0.690	0.589
DR3	52b	3	1.690	0.524
DRw11	52b	4	0.975	0.825
DRw12	52b	2	0.815	0.003
DRw6	52a	4	0.692	0.432
DRw6	52b	4	1.050	0.455
DRw6	52c	5	1.196	0.394
DRw8		4	0.360	0.087
Mouse transfectant				
Name		Transfected gene		
LR3.1/52a		DRB3 52a	1.160	0.007
LR6.1/52b		DRB3 52b	0.967	0.008
LR3.1/3		DRB1 DR3	0.625	0.555

*The extent of expression was quantitated by the binding inhibition assay (7). Graded amounts of the cell lysate of each B cell line or mouse transfectant were incubated with a fixed amount of antiserum for 4 h and then with a fixed amount (~20,000 cpm) of ^{125}I-labeled preparation overnight, both at room temperature. Then an optimal amount of goat anti-human Ig antiserum was added and the resultant immunoprecipitate was washed and the radioactivity was determined. The inhibitory activity thus determined was expressed as the activity per unit of the class II activity determined by the use of rabbit anti-class II antiserum R7147 (7).

Table 3. Distribution of the 52/R3 Determinant Recognized by the Mouse Monoclonal Antibody MCS7

^{125}I-transfectant class II preparation			% binding* with		
Cell source	Locus	Allele	Anti-class II R7147	Anti-DRα U115	Anti-DRw52-like MCS7
LR3.1/3	DRB1	DR3	9.3	8.2	4.0 (42.0)
LR3.1/52a	DRB3	52a	52.2	46.6	36.0 (68.9)
LR6.1/52b	DRB3	52b	32.3	28.5	0.0 (0.0)
LR6.3/52c	DRB3	52c	59.1	29.2	12.0 (20.3)

*The binding activity was determined by the direct binding assay (7). A fixed amount (20 ul, ~20,000 cmp) of ^{125}I-labeled preparation was incubated with MCS7 ascitic fluid (2 ul) for 1 h at 37 C and then with an optimal amount of goat anti-mouse Ig antiserum for 1 h at room temperature. The resultant precipitate was washed and the radioactivity was determined. A control with normal mouse serum was included in each assay. The values in parentheses indicate the percent class II binding which is the proportion of the maximum binding obtained with rabbit anti-class II antiserum R7147.

Two distinct functional DRB chain loci, DRB1 and DRB3 have been identified in DRw52 haplotypes (1). DNA sequencing of their genes from several DRw52 haplotypes have revealed the presence of three different alleles, provisionally called 52a, 52b, and 52c, in the DRB3 locus (2). These alleles show different distribution in the DRw52 haplotypes: DRw6 can be either 52a, 52b, or 52c; DR3 is either 52a or 52b; and DR5 is solely 52b (3). The cloned genes have been transfected into mouse L cells (4,5) and the expressed class II molecules have been isolated and characterized by the radioimmunochemical methods (4,6).

Table 1 shows the haplotype distribution (Table 1a) and the gene localization (Table 1b) of the 52/R1, 52/R2, and 52/R3 specificities that were determined by the analysis of the binding reactions between anti-DRw52 antisera and HLA class II molecules isolated from mouse L cell transfectants and from human B cell lines. The loci assignments were made based on direct information when the corresponding transfectants were available. The other conclusions were deduced either from the sequence identity or from the subset analysis of class II molecules derived from human B cell lines.

Table 1 also shows data for two DRB3 specificities, TR81 (4), and TR22 (6). They are included in DRw52 and are allelic at the DRB3 locus. In addition, TR81 is also present in the DRB1 locus product in the DR3 haplotype. This specificity has been shown to correspond to the gene conversion event that led to the generation of the DRB1 gene of DR3 haplotype (4). The other specificities, 52/R1, 52/R2, and 52/R3 show haplotype distribution more closely related to the classical DRw52 pattern. Their detailed data are given in Tables 2 and 3.

52/R1: a DRw52 determinant localized on both DR loci products (Table 2). Antiserum Fe88/37 was found to give strong binding of the class II molecules isolated from transfectants LR3.1/52a (expressing DRB3 52a gene product) and LR6.1/52b (expressing DRB3 52b gene product). To test the distribution of the determinant recognized on these transfectant molecules, an inhibition assay (7) was set up using the Fe88/37-^{125}I-LR3.1/52a binding reaction. This assay defines the 52/R1 specificity. Lysates from either human B cell lines or mouse transfectants were used as inhibitors. Their inhibitory activity was determined using graded amounts of cell lysate and expressed as the activity unit per one unit of the HLA class II activity determined by the use of rabbit antiserum R7147 mostly directed to the framework structure of DR molecules (7). This compensates for the wide differences in class II concentration in the different cell lysates. The binding involves a DRB3 determinant since the radiolabeled antigen is a DRB3 molecule. That this determinant is also located on DRB1 products is shown directly by the inhibition given by transfectant LR3.1/3 expressing a DRB1 product and indirectly, on a quantitative basis, by the inhibition with B cells. In all haplotypes, the relative inhibition was clearly higher than that expected if the determinant was expressed on DRB3 molecules only. If the inhibition is due to DRB3 molecules only, it would give a value corresponding to 1/4 to 1/5 of the inhibition by the homologous antigen

(LR3.1/52a), since DRB3 expression is that much lower than DRBl expression (4). The data also show that the determinant responsible to this 52/R1 specificity is not exactly the same in the two loci. In fact, the inhibition by LR3.1/3 and by B cell lines of some haplotypes is significantly lower than the value expected in the case of identity, indicating involvement of cross-reactive determinants.

52/R2: a "DRw52-short" determinant localized on DRBl product only. The distribution of this determinant is similar to that of 52/R1 with the difference that DRw12 cells are negative and DRw8 cells are weakly positive (Table 2). The specificity is defined based on the binding of the AVL (DR3,3) class II preparation by alloantiserum Fe117/30. All B cell lysates gave a relative inhibition close to that of the homologous antigen (AVL) except DRw8 and DRw12. However, of the three transfectants tested, only DRBl transfectant LR3,1/3 inhibited, whereas two DRB3 transfectants, LR3.1/52a and LR6.1/52b, did not. This indicates that the determinant recognized in this reaction is a DRBl determinant.

52/R3: a DRw52-like epitope recognized by mouse monoclonal antibody MCS7. This epitope has a distribution very similar to that of 52/R2, being absent from DRw8 and DRw12 haplotypes (Table 1). However, MCS7 binding of transfectant class II preparations shows a lack of reactivity with the DRB3 52a gene product (Table 3). Therefore this specificity is localized on the DRBl product only in the haplotypes carrying the 52b allele (such as DR5, as demonstrated in a previous work (8)), and on both DRBl and DRB3 loci products in the haplotypes carrying the 52a or 52c alleles.

Acknowledgments. This investigation was supported by the USPHS grant AI20251, by the Swiss National Science Foundation grant 3.468-0.83, by the C.N.R. Progetto Finalizzato Ingegneria Genetica-Basi Molecolarei delle Malattie Ereditarie and Progetto Finalizzato oncologia N.86.00659.44 and N.86.00487.04.

References

1. Rollini P, Mach B, Gorski J. Characterization of an HLA-DR beta pseudogene in the DRw52 supertypic group. Immunogenetics 1987;25:336.
2. Gorski J, Mach B. Polymorphism of human Ia antigens: Gene conversion between two DR beta loci results in a new HLA-D/DR specificity. Nature 1986;322:67.
3. Gorski J, Tilanus M, Giphart M, Mach B. Oligonucleotide genotyping shows that alleles at the HLA-DRβIII locus of the DRw52 supertypic group segregate independently of known DR or Dw specificities. Immunogenetics 1987;25:70.
4. Berte CC, Tanigaki N, Tosi R, Gorski J, Mach B. Serological recognition of HLA-DR allodeterminant corresponding to DNA sequence involved in gene conversion. Immunogenetics 1988;27:167.
5. Rabourdin-Combe C, Mach B. Expression of HLA-DR antigens at the surface of mouse L cells co-transfected with cloned human genes. Nature 1983;303:670.
6. Tosi R, Tanigaki N, de Preval C, Gorski J, Mach B. Immunochemical analysis of a cell transfected with an HLA-DR gene reveals a new alloantigenic specificity within HLA-DRw52. Eur J Immunol 1987;16:1603.
7. Tanigaki N, Tosi R. Assessment of the specificity of human Ia alloantisera and alloantigens by the use of radioiodinated human Ia antigens. Tissue Antigens 1982;20:1.
8. Tanigaki N, Tosi R, Sagawa K, Minowada J, Ferrara GB. The distribution of DR5, MT2, and MB3 specificities on human Ia subsets. Immunogenetics 1983;17:371.

Author Affiliations

Nobuyuki Tanigaki, Department of Molecular Immunology, Roswell Park Memorial Institute, Buffalo, NY, USA; *Christine C. Berte, Jack Gorski, Bernard Mach,* Department de Microbiologie, Universite de Geneve, Faculte de Medicine, Geneva, Switzerland; *Roberto Tosi*, Istituto di Biologia Cellulare, CNR, Rome, Italy; *Giovanni B. Ferrara,* Istituto Nationale per la Ricerca sul Cancro, Genova, Italy

Three Subtypes of DRw52 Molecules Among DR3, DR5, and DRw6 Haplotypes

Ranko Hirata, Hiroo Maeda, Hiroshi Tohyama, and Tetsuo Ishikawa

In the HLA-DR subregion, two allelic series of DR molecules, DR1-DRw14 and DRw52-DRw53, have been recognized. DRw52 is supertypic to DR and is found in association with DR3, DR5, DRw6, and DRw8. We recently produced a monoclonal antibody PLM16 that strongly reacted with DR3, DR5, and DRw6 and thus showed the DRw52-like specificity. The antibody bound to a single DR β chain from DR5 and DRw6 haplotypes but two DR β chains from DR3 haplotype. These DR β chains carrying the DRw52-like determinant, i.e., DRw52 molecules. from cells of DR3, DR5, and DRw6 types were directly compared by one-dimensional IEF gel electrophoresis and immunoblotting methods. Three subtypes of DRw52 β chains became evident; the most acidic type, DRw52a, was found in association with DR3, DRw11, DRw13, and DRw14; an intermediate one, DRw52b, with DRw13; and the most basic one, DRw52c, with DRw12 (1).

In this brief report, we describe the DRw52 subtypes from B-cell lines of DR3, DR5, and DRw6 types in the 10th Workshop (Table 1). The DR3 cells associated with two different DRw52 β chains, DRw52a and DRw52c,

Immunobiology of HLA
Volume II

Table 1. DRw52 β Chain Subtypes Among DR3, DR5, and DRw6 Haplotypes

WS ID	Cell Name	DR/Dw	DQ*	DR β	DRw52β
#9018	LO081785	3/w3	NT†	3	a
#9019	DUCAF	3	NT	3	a
#9020	QBL	3/w3	NT	3	a
#9085	EJ32B	3	NT	3	a
	WT49	3	w2	3	a
#9021	RSH	3	WA	3	c
#9022	COX	3	NT	3	c
#9023	VAVY	3/w3	NT	3	c
#9071	STEINLIN	3	NT	3	c
#9086	L0541265	3	NT	3	c
#9053	HOR	w13,w6/w19	w1	13	b
#9055	HO301	w6/	w1	13	b
#9059	SLE005	w13,w6/w19	NT	13	b
#9063	MT47	w6/	w1	13	b
	WT46	w6/w19	w1	13	b
#9058	OMW	w13,w6/	w1	13	c
#9065	HHKB	w13,w6/w18	w1	13	c
#9062	WDV	w13,w6/w18	w1	13	c
#9060	CB6B	w13,w6/w18	NT	13	a
	APD	w6/w18	w1	13	a
#9065	KOSE	w14,w6/	NT	13/14	a b
#9054	EK	w6/w9	w1	14	a
#9061	ABO	w6/w9	w1	14	a
#9057	TEM	w6/	w1	14	a
#9064	AMALA	w14,w6/w16	w3(TA10)	HR8‡	c
#9099	LZL	w6	w3(TA10)	HR8	c
	SH	w12	w3(TA10)	HR8	c
#9037	SWEIG007	w11, 5/w5	w3(TA10)	11	a
	THR	w11, 5/w5	w3(TA10)	11	a

*HLA-DQ type was defined by local monoclonal antibody.
†NT,not tested.
‡HR8, a new DR subtype of DRw12.

although the DR3 β chain (DRβ3) was the same and was shared among these cells. All the DR3 cells from the HLA-A1-B8 haplotype carried DRw52c. The DRw13(w6) cells carried the same DR β chain (DRβ13) but three different DRw52 β chains. All the cells typed for Dw19 carried DRw52b, whereas the cells of Dw18 type carried either DRw52a or DRw52c just as the DR3 cells did. This phenomenon is of interest; both groups of cells are typed identically (Dw18 or Dw3), indicating that neither DRw52a not DRw52c molecule is stimulatory within Dw18 or Dw3 homozygous typing (HTC) cells in the primary MLR. In contrast, either Dw18 or Dw19 HTC cells are stimulatory to each other, where DRw52 molecules are discrepant and may be responsible for the stimulation. The DRw6/Dw9 cells carried their own individual DR β chain (DRβ14) but all had DRw52a. One cell of DRw6/Dw16 type also carried its specific DR β chain (DRβ HR8) but its DRw52 molecule was DRw52c. This DR/DRw52 β chain type was identical to that of a B-cell line SH, that was typed for a new subtype of DRw12, HR8. Two cells of DRw11(5)/Dw5 carried their own DR β chain (DRβ11) but shared one of DRw52 molecules, DRw52a.

These data indicate that: 1. Three DRw52 β chain subtypes were confirmed among DR3, DR5, and DRw6 cells. 2. Five different DR β chains were recognized which correlated with the officially recognized DR types: DR3, DRw11, DRw12, DRw13, and DRw14. 3. A total of eight DR/DRw52 haplotypes are recognized biochemically: DR3/DRw52a, DRw52c; DRw11/DRw52a; DRw12(HR8)/DRw52c; DRw13/DRw52a, DRw52b, DRw52c; DRw14/DRw52a. 4. Distinct DRw52 molecules are not always stimulatory in the primary MLR, for example, Dw3 or Dw18.

References

1. Hirata R, Ishikawa T, Maeda H. Three subtypes of DRw52 β chains detected by one-dimensional isoelectric focusing gel electrophoresis and immunoblotting.

Author Affiliations

Ranko Hirata, Hiroo Maeda, Hiroshi Tohyama, Blood Transfusion Service, Saitama Medical Center, Saitama Medical School, Kawagoe, Saitama; *Tetsuo Ishikawa*, The 2nd Department of Surgery, Hiroshima University School of Medicine, Hiroshima, Japan

Three Functionally Distinct DRw52 Subtypes Are Alleles of the DRβIII Locus*

James R. Rowe, Frits Koning, Lars Jorgensen, and Michael J. Sheehy

Several clusters of class II HLA genes contribute to the variation in human antigen-presenting capacity. In the HLA-DR cluster, most of the variation is due to the highly polymorphic DRβI gene. Recent work by others (1) has shown some nucleotide and implied amino acid sequence variation in DRβIII chains, but this variation was not known to be functionally significant. We previously described a subset of HLA-DRw52 called "BO1," defined by T-cell clones B33-25 and B33-30, that is

*This work was supported in part by grant DR-38209 from the National Institutes of Health, grant 5-538 from the March of Dimes Birth Defects Foundation, and an Established Investigator Award from the American Red Cross.

Immunobiology of HLA
Volume II

Table 1. Blocking by Monoclonal Antibodies Shows DRβIII Specificity of Clones 25 and 30

A. Blocking of Responses to DR5 Cells

	MAb (Chain Specificity)	
Resp. cells	PdV5.2 (broad)	7.3.19.1 (DRβIII)*
Clones 25 and 30	o	+
Bulk-primed	+	o

B. Blocking of Responses to DR3 Cells

	MAb (Chain Specificity)	
Resp. cells	I-LR2 (DRβI)	7.3.19.1 (DRβI+βIII)*
Clones 25 and 30	o	++
Anti-DR3	++	+/o

+ indicates blocking.

*Chain specificity of this antibody differs on DR3 and DR5 cells.

found on all cells with DR5 and a subset of DR3 and DRw6 cells. Testing 58 lymphoblastoid cell lines (LCLs) each having a single DRw52 haplotype, showed that clone 30 has somewhat broader reactivity than clone 25 and that, together, the two clones define three DRw52 subtypes, DRw52.1 (25−30−), DRw52.2 (25+30+), and DRw52.3 (25−30+). DR3 cells were either DRw52.1 (strongly correlating with HLA-B8) or DRw52.2 (strongly correlating with HLA-B18). The DRw6 cells fell into all three groups: 26% DRw52.1, 26% DRw52.2, and 48% DRw52.3. Both DR5 and DRw8 cells gave uniform results; all DR5 cells were DRw52.2 and all DRw8 cells fell into the double-negative (DRw52.1) category. The DRw8 cells probably should not be considered as DRw52.1, however, as explained below.

A series of experiments, using monoclonal antibodies to block T-cell proliferation, was undertaken to determine which HLA glycoprotein chain(s) determine these DRw52 subtypes (Table 1). Responses of both clones to a DR5-positive cell line are strongly blocked by monoclonal antibody (MAb) 7.3.19.1, which binds only to DRβIII chains on DR5 cells. Reciprocal controls, for both clones and antibody, showed this blocking to be specific. Thus, on this DR5-positive cell line, the DRw52 subtypes must be expressed on DRβIII.

On DR3 cells, two DRw52-like MAbs show reciprocal blocking specificity for clones 25 and 30 (anti-DRw52 subtypes) versus two other T-cell clones (anti-DR3). MAb 7.3.19.1, which binds to both DRβI and DRβIII chains of DR3 cells, strongly blocks the anti-DRw52 subtype responses but less strongly blocks the anti-DR3 responses. MAb I-LR2, which binds only to DRβI chains on DR3 cells, strongly blocks the anti-DR3 responses but has no effect on the anti-DRw52 subtype responses. Thus we conclude that the DRw52 subtypes are expressed on DRβIII chains of both DR3 and DR5 cells. Unpublished studies with Drs. B. Mach, J. Gorski, and C. Ucla indicate that this is true for DRw6 cells as well.

Table 2. T-Cell Clones Divide DRw52-Associated DRβI Alleles into Three or Four Categories

	52.1 (Dw24)	52.2 (Dw25)	52.3 (Dw26)	w52-Neg. or 52.4
DR3	+	+	o	o
DR5	o	+	o	o
DRw6	+	+	+	o
DRw8	o	o	o	+

Fifty-eight samples tested; + indicates combination was observed.

Although DRw8 falls into the DRw52.1 category (negative for both clones), other studies indicate that the DRw52 on DRw8 haplotypes differs from that on DR3, −5, and −w6 haplotypes (2,3). Thus DRw8 haplotypes must either have a fourth DRw52 subtype or be DRw52-negative. They may even be DRβIII-null since, as noted elsewhere, they seem to have only one DRβ gene. For these reasons, we propose dividing DRw52 into three or four subtypes, 52.1-52.3 or 52.1-52.4 (Table 2), depending on whether DRw8 cells are considered DRw52-positive.

These studies are described elsewhere in greater detail (4).

References

1. Gorski J, Rollini P, Kawashima E, Long E, Mach B. Multiplicity of human class II β chain genes. UCLA Symposia on Molecular and Cellular Biology (New Series). 1984;18:47.
2. Hurley CK, Giles RC, Nunez G, DeMars R, Nadler L, Winchester R, et al. Molecular localization of the human class II MT2 and MT3 determinants. J Exp Med 1984;160:472.
3. Koning FI, Schreuder M, Giphart M, Bruning H. A mouse monoclonal antibody detecting a DR-related MT2-like specificity. Hum Immunol 1984;9:221.
4. Sheehy MJ, Rowe JR, Koning F, Jorgensen L. Functional polymorphism of the HLA-DR beta-III chain. Hum Immunol, in press.

Author Affiliations

The American Red Cross Blood Services, Madison WI 53705, USA; the Department of Medicine, University of Wisconsin, Madison, WI 53792, USA; and the National Institute of Allergy and Infectious Disease, Bethesda, MD 20892, USA

The Allelic Determinants HLA-DR3 and HLA-DR5 and the Supertypic Determinant HLA-DRw52 May Be Located on the Same DRβ Chain

M. Ballas, T.H. Eiermann, A. Wölpl, and S.F. Goldmann

The molecular nature of the supertypic specificity HLA-DRw52 is still a matter of controversy. There is general agreement that the DRw52 epitope resides on DRβ chains, but previous attempts could not clearly assign this determinant to one or both DRβ chains expressed by one allotype (1). This was mainly due to the fact that only "DRw52-like" antibodies were available that are negative on DRw8 . We recently produced a new monoclonal antibody, UL-52/124, that recognizes HLA-DRw52 in association with HLA-DR3, DR5, DRw6, and DRw8. We used this antibody in comparative immunoprecipitation experiments together with L243* (DR monomorphic), SFR-DR5* (HLA-DR5 specific) and MAD 88* (HLA-DRw8 specific). The cell lines examined were homozygous on the DR locus.

In a preliminary study, immunoprecipitation with UL-52/124 and subsequent SDS-gel electrophoresis was performed on a panel of radiolabeled cell lines representing the haplotypes DR1-DRw8. As expected, UL-52/124 was only reactive with DR3, DR5, DRw6, and DRw8 cells, giving an SDS-gel pattern typical for Ia-molecules. No molecules were precipitated from DR1, DR2, DR4, and DR7 cells.

In order to detect the subclass of Ia-molecules recognized by UL-52/124, 2-D gel electrophoresis with the immunoprecipitates of UL-52/124 was performed and compared to those of L243 and SFR-DR5.

Two cell lines representing the allotypes DRw6 (KLE) and DRw8 (FRE) expressed two DRβ chains that were precipitated with L243. UL-52/124 bound to only one β chain. This β chain was identical to one of the DRβ chains recognized by L243. So, for both of these cell lines the DRw52 determinant was found to reside exclusively on a subset of DRβ chains. Because immunoprecipitation with MAD 88 was negative, we could not decide whether DRw8 and DRw52 reside on the same or on different DRβ chains.

Only one DRβ chain was detectable in the L243 precipitated material from the cell lines RUP and BAE (DR3). This was the only β chain that was precipitated with UL-52/124. Thus, these cell lines obviously express both DR3 and DRw52 on one and the same DRβ chain.

TUBO, a lymphoblastoid cell line homozygous for DR5, was found to express three DRβ chains when immunoprecipitated with L243. One of these DRβ chains, conventionally named $DR\beta_1$, was also recognized by SFR-DR5, so this is the β chain that carries the polymorphic determinant HLA-DR5. Two DRβ chains bound to UL-52/124, one of them being identical to $DR\beta_1$. This means that $DR\beta_1$ bears both the polymorphic epitope HLA-DR5 and the supertypic specificity HLA-DRw52. The second β chain recognized by UL-52/124 ($DR\beta_2$) expresses solely, the DRw52 epitope.

When compared with other DR5 cells tested, this $DR\beta_2$ chain was only detectable in DRw11 subtypes, whereas the third DRβ chain that was not recognized by UL-52/124 but was precipitated with L243, was specific for DRw12 cells. These data suggest that TUBO is heterozygous DRw11/DRw12 and that UL-52/124 binds to both DRβ chains expressed by DRw11 cells but not to $DR\beta_1$ from DRw12 cells.

To briefly summarize the results we obtained so far, DRw52 was found:

on a subset of DRβ chains from DRw6 and DRw8 homozygous cell lines;
on the only DRβ chain detectable in DR3 cell lines; and
on both DRβ chains expressed by DRw11 allotypes, but not on $DR\beta_1$ from DRw12 cells.

So, there is considerable complexity within the subpopulation of DRw52-bearing molecules and the general consideration that the polymorphic determinants DR1-DRw14 are located on $DR\beta_1$, whereas the supertypic specificities DRw52,53 reside on $DR\beta_2$ should be revised. As most of the DRw52 associated allospecificities exhibit extensive polymorphism on the product level, further investigations have to be carried out.

Reference

1. Karr RW, Lipman LK, Alber C, Koning F, Goeken NE. Ia molecular localization of the DRw52 allodeterminant and the DRw52-like determinants defined by monoclonal antibodies. J Immunol 1985;135:2642.

Author Affiliations

M. Ballas, T.H. Eiermann, A. Wölpl, S.F. Goldmann, Department of Transplantation Immunology, Red Cross Blood Bank, Ulm, and Department of Transfusion Medicine, University of Ulm, D-7900 Ulm, Federal Republic of Germany

---

*Included in the Tenth Workshop panel.

Immunobiology of HLA
Volume II

An Abnormal DRB4 (DRw53) mRNA Is Produced by DR7 Individuals Who Fail to Express the Gene Product

Vivien R. Sutton, Jack Gorski, and Robert W. Knowles

Abstract: All DR4 and most DR7 (DR7,DQw2) individuals express the products of two DRβ genes, DRBl and DRB4 (DRw53). However, expression of the DRB4 gene is defective in some DR7 individuals (DR7,DQw3), as biochemical studies have failed to detect the product of the DRB4 gene (1). This lack of expression correlates with a Hind III RFLP, detected with a DRβ cDNA probe (2,3). However, a detailed Southern blot analysis has now demonstrated that this altered Hind III fragment maps to the DRBl gene, which is expressed normally, and no RFLP has been detected in the DRB4 gene. In a Northern blot analysis using a DRB4 gene-specific oligonucleotide probe, individuals who lack expression of the DRβ-IV protein product were found to express an abnormally large and heterogeneous mRNA. Furthermore, RNA from such individuals fails to completely protect RNA probes complementary to 5′ sequences of a normal DRB4 gene (5′ UT and first two exons). Thus, it appears likely that the abnormal gene transcript results from a defect in the 5′ region of the DRB4 gene and that absence of a normal transcript results in lack of DRβ-IV protein expression.

In order to determine the defect responsible for lack of expression of the protein product of the DRB4 (DRw53) gene in some DR7 individuals, Southern and Northern blot analyses and RNase protection studies were performed. Comparisons were made of the DRB4 gene from individuals who had either normal or defective expression of this gene [Table 1]. Southern blots were used to identify RFLPs which might demonstrate structural alterations in the gene. Previously, an RFLP has been described which can differentiate DR4 or DR7 individuals who express the DRB4 gene product from DR7 individuals who do not (2,3). An 8.5-kb Hind III fragment was detected, using a DRβ cDNA probe, in individuals with normal expression of the gene, whereas in individuals whose expression of the DRB4 gene was defective, a 9.4-kb Hind III fragment was found instead. Complete correlation has been found between the altered Hind III fragment and the lack of expression of the DRB4 gene. On this basis, it had been suggested that this polymorphism mapped to the DRB4 gene and that the alteration leading to the larger Hind III fragment may also be responsible for the defective expression of the DRB4 gene. However, the normal 8.5-kb Hind III fragment is not present in the restriction enzyme map of two overlapping genomic clones encoding the DRB4 gene from a DR4 individual. Rather, this fragment maps to the 3′ region of the DRBl gene, located 100 kb from the DRB4 gene (data not shown).

Other RFLPs were therefore sought by comparing the restriction enzyme fragments present in the cloned DRB4 gene (DR4) with those present in the DRB4 gene from a DR7,DQw2 individual (MANN) and a DR7,DQw3 individual (DBB). DNA was restricted with Eco RI, Hind III, Kpn I, Sac I, or Xba I and analyzed with a series of DRβ cDNA and genomic probes which hybridized to different coding and noncoding sequences of the DRB4 gene. Each of the restriction enzyme fragments which could be identified in the cloned DRB4 gene was also present in both DR7 haplotypes examined. For example, a 7.0kb Xba I fragment located in the 3′ region of the genomic clone was detected in genomic DNA from both DBB and MANN. This analysis demonstrated that the DRB4 gene was structurally intact, since no RFLP was found.

In order to determine whether the DRB4 gene is transcribed, DRB4-specific anti-sense RNA probes were used in RNase protection assays and oligonucleotide probes were used in Northern blot analysis. Three probes were constructed by subcloning fragments from a

Table 1. Characteristics of Cell Lines Used in this Study

Cell Lines: (WS ID#)	DR typing*	PL3† mAb	109d6† mAb	DRB4‡ chain	Dw* typing	DQ* typing
COX (9022)	DR3	–	–	–§	Dw3	DQw2
BOLETH (9031)	DR4	+	+	+	Dw4	DQw3
SAVC (9034)	DR4	+	+	+	DR4	DQw3
MANN (MOU) (9050)	DR7	+	+	+	Dw7	DQw2
DBB (9052)	DR7	+	–	–	Dw11	DQw3
CFKB	DR7	+	+‖	+‖	Dw11#	DQw3
KT12 (9073)	DR9	+	+	+	DB5	DQw3

*The HLA-DR, -Dw, and -DQ typing results are from previous reports: MANN (1), DBB (1), COX (4), CFKB (1), or were supplied when submitted to the 10th Workshop.
†Binding of the monoclonal antibodies was determined by indirect immunofluorescence.
‡Presence of the DRB4 chain was determined by the characteristic 2-D gel pattern after immunoprecipitation with 109d6.
§COX has a DRw52 DRβ3 chain, which is not precipitated by PL3 or 109d6.
‖CFKB is heterozygous at the A locus, heterozygous for the RFLP in the DRβ-I and assumed to be heterozygous at the DRB4 gene.
#Results from the 8th Workshop.

Immunobiology of HLA
Volume II

DRB4 cDNA clone (5) into the pGEM3 vector (Promega). Anti-sense RNA transcripts were produced *in vitro* using SP6 polymerase, according to the protocol of Melton et al. (6). The probes, as illustrated in Figure 1, were complementary to : (a) the 5′ UT region and signal sequence, (b) the exon encoding the first external domain, and (c) the exons encoding the second external domain, transmembrane, cytoplasmic and 3′ UT region.

Only RNA from individuals who express the DRB4 gene, BOLETH (DR4), MANN (DR7), and KT12 (DR9) were able to completely protect the RNA transcript complementary to the exon encoding the first external domain (Fig. 1). RNA from the negative control, a DR3 individual (COX) was unable to protect the probe, although some small fragments (less than 70 bp) were maintained. In contrast, RNA from DBB did, however, provide considerable protection to the probe. A unique band, approximately 40 bp smaller than the fully protected probe, was evident, indicating that DRB4-specific RNA was transcribed in this individual. Similarly, using the probe complementary to the 5′ UT region and the signal sequence, RNA from DBB was able to only partially protect the probe when compared to the protection achieved with RNA from MANN (Fig. 1). In contrast, the 3′ region probe was protected equally by RNA from both DBB and MANN. Thus it is clear that RNA was transcribed from the DRB4 gene in individuals who are defective in the expression of that gene. However, as the probes complementary to the 5′ region sequences were only partially protected, these sequences have been altered compared to the normal DRB4 gene while the integrity of the 3′ region of the DRB4 gene has been maintained.

The alteration in the 5′ region of the gene was further investigated by Northern blot analysis of the DRB4 gene transcripts using a gene-specific oligonucleotide probe. The use of a DRβ cDNA probe in Northern blot analysis was not informative due to cross-hybridization with the DRBl and DQB gene transcripts. Therefore, a 20-base sequence from the second exon of the normal DRB4 gene was chosen to construct an anti-sense oligonucleotide probe. Specificity of the probe was demonstrated by its failure to hybridize to RNA from a DR3 individual (COX) and to RNA from a DR7 BI gene-transfectant. The probe hybridized to a 1.4-kb transcript in DR7 (MANN) and DR4 (BOLETH,SAVC) individuals whose expression of the DRB4 gene product is normal. A DRB4 gene transcript was detected in RNA from DBB, but it was larger and quite heterogeneous compared to normal. A broad band at about 1.5 to 1.6 kb was evident, which extended into a smear up to about 4 kb.

It would thus appear that the failure of DR7,DQw3 individuals to express the DRβ-IV chain is related to their inability to produce a normal DRB4 gene transcript. Alteration in the 5′ sequences of the DRB4 gene apparently causes the gene to be incorrectly processed, resulting in an abnormal transcript. If a protein is translated from this transcript, it would appear to be unstable or sufficiently altered such that it was not detected in the previous biochemical studies (1).

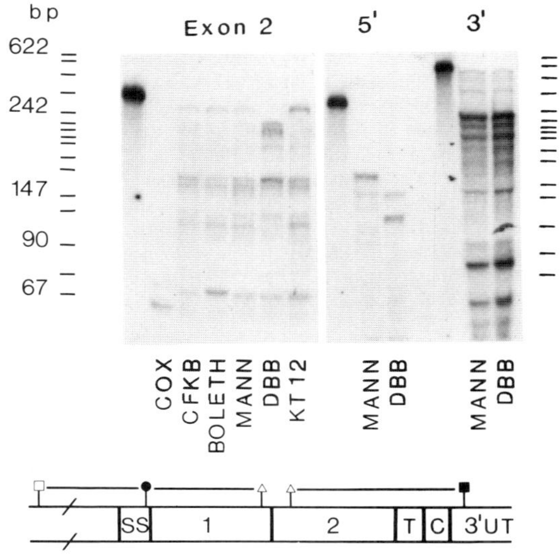

Figure 1. RNase protection assays demonstrate that the 5′ region of the DRB4 gene is altered in individuals lacking expression of that gene. Anti-sense RNA probes were constructed to (a) the 5′ UT and signal sequence (Bam HI-Sac I) [5′], (b) the exon for the first external domain (Sac I-Pst I) [Exon 2], and (c) the exons for the second external domain, transmembrane, cytoplasmic and 3′ UT region (Pst I-Hind III) [3′]. Total RNA (20 ug) was used for each individual, MANN (DR7), BOLETH (DR4), KT12 (DR9), DBB (DR7), COX (DR3) and CFKB (DR7) [see Table1]. Following hybridization of the ^{32}P-labeled probes, the samples were treated with RNase A and RNase T_1, and the protected RNA hybrids were analyzed on 8% acrylamide-urea gels. ⚲ = Bam HI; ● = Sac I; ⇧ = Pst I; ■ = Hind III.

Acknowledgments. The DRB4 cDNA clone was kindly provided by Dr. Peter Gregersen, the DR7B1-transfectant by David Wilkinson and Dr. John Trowsdale, and the mAb 109d6 by Dr. Robert Winchester. This work was supported, in part, by the Society of Memorial Sloan-Kettering Cancer Center, by an Institutional Grant from the American Cancer Society, and by NIH grant AI22916 from the US Public Health Science. VRS was supported by a fellowship from the Damon Runyon-Walter Winchell Cancer Fund and an Alan and Christine Curtis Fellowship from the Cancer Research Institute.

References

1. Knowles RW, Flomenberg N, Horibe K, Winchester R, Radka S, Dupont B. Complexity of the supertypic HLA-DRw53 specificity: Two distinct epitopes differentially expressed on one or all of the DRβ-chains depending on the HLA-DR allotype. J. Immunol. 1986;137:2618.
2. Paulsen G, Qvigstad E, Guadernack G, Rask L, Winchester

R, Thorsby E. Identification, at the genomic level, of an HLA-DR restriction element for cloned antigen-specific T4 cells J. Exp. Med. 1985;161:1569.
3. Angelini G, Tanigaki N, Tosi R, Ferrara GB. Southern blot and microfingerprinting analysis of two DR7 haplotypes. Immunogenetics. 1986;24:63.
4. Schreuder GMT, Doxiadis I, Parlevliet J, Grosse-Wilde H, HLA-DR, DQ, LB and TA10 Specificities of Ninth Workshop Homozygous Typing Cells. In: E.D. Albert, M.P. Baur, and W.R. Mayr, (eds.): Histocompatibility Testing, 1984; Springer-Verlag, Berlin.
5. Karr RW, Gregersen PK, Obata F, Goldberg D, Maccari J, Alber C, et al. Analysis of DRβ and DQB chain cDNA clones from a DR7 haplotype. J. Immunol. 1986; 137:2886.
6. Melton DA, Krieg PA, Rebagliati MR, Maniatis T, Zinn K, Green MR. In vitro biosynthesis of biologically active RNA and RNA hybridization probes from plasmids containing a bacteriophage SP6 promoter. Nucleic Acid Res. 1984;12: 7035.

Author Affiliations

Vivien R. Sutton and Robert W. Knowles, Sloan-Kettering Institute, New York, NY; *Jack Gorski*, The Blood Center of Southeast Wisconsin, Milwaukee, WI, USA

Evidence for a Hot Spot for Generalized Gene Conversion in the Second Exon of MHC Genes

Marnix L. Bosch, Annemarie Termijtelen, Jon J. van Rood, and Marius J. Giphart

Abstract: In the past, several mechanisms including recombination and gene conversion have been proposed to account for the high degree of polymorphism observed in MHC genes. We have compared the 22 available HLA class II second exon sequences, as well as numerous class I murine H-2 I-E sequences. Our analysis revealed the existence of regions of unexpected sequence identity in immunogenetically unrelated genes. These regions are preceded by strongly conserved, tandemly repeated, sequence blocks that show significant homology to sequences that have been related to high recombination frequencies in other biological systems. We conclude that the observed sequence organization indicates the existence of a hot spot for gene conversion in the second exon of MHC genes.

The polymorphic regions of MHC antigens are implicated as the antigen-presenting entities of the molecules (1). In view of the broad antigenic challenge presented by an ever-changing environment, the maintenance of a high degree of polymorphism is of prime importance to species that rely on an immune system for defense against invading microorganisms, etc. Therefore, mechanisms must exist that continuously generate new polymorphisms in MHC genes. Several authors have provided evidence for gene conversion events between genes of one haplotype (2) or of different haplotypes (3), based on analyses of limited numbers of sequences. We have tried to identify general requirements for DNA sequence exchange by comparing all known DR β gene sequences.

The number of DR β genes in the DR subregion in the different HLA haplotypes is thought to have arisen from one primordial gene through gene duplication. It has been postulated that the two well-defined DR supertypic specificities, DRw52 and DRw53, reflect characteristics of the two original DR β genes that originate from the first gene duplication, and have diverged since, giving rise to these two gene families (2). We have tried to divide the known DR β second exon sequences shown in Figure 1 into gene families on the basis of sequence similarity. To do this we compared sequences at the 5′ and 3′ ends. The division we arrived at is depicted in Figure 1, the two gene families are arbitrarily named "L" and "M." The sequence characteristics upon which this division is based are for family "L" AT at position 21 and the sequences G.C and AT at positions 210-212 and 222-223, respectively. These nucleotides are indicated by small rectangles at the indicated positions. For family "M" those positions read A, C.G and CC, respectively (indicated with circles). Whereas family "L" only contains sequences from DRw52 expressing haplotypes, family "M" contains sequences from both DRw52- (#6-10) and DRw53- (#11-18) expressing haplotypes, as well as sequences derived from the DR1 and DR2 haplotypes, which express neither DRw52 nor DRw53. A striking observation is that the "L" family contains only DR β III gene sequences of DRw52 expressing haplotypes, whereas all DRw52-haplotype-derived sequences in the "M" family are all from the DR β I genes (2,4). In Figure 2 we have drawn a dendrogram in which the evolution of the now existing DRw52 and DRw53 expressing haplotypes is represented based on these data.

At positions 249–250 we find either GT or TG in each sequence and occurring in both "L" and "M" families. This is indicative of DNA sequence exchange between different genes after the formation of the "L" and "M" gene families. A closer look at the sequences preceding position 250 reveals that some genes that share either of the GT or TG sequences actually show sequence identity

Immunobiology of HLA
Volume II

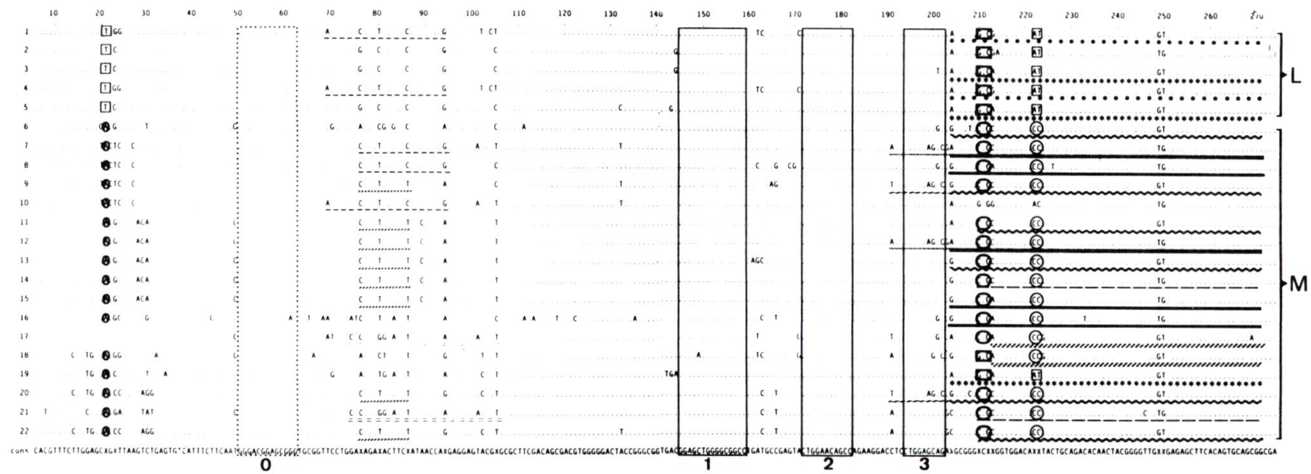

Figure 1. Representation of DR β second exon sequences. Dots indicate identity to the consensus sequence (bottom line). See text for details. The sequences were derived from the following references: #1, 2, 4, 5, 7, 8 and 10: ref. 2; #3 and 9:ref. 6; #6: ref. 11 and M.L. Bosch, unpublished observations; #11-16: ref. 7 and 9; #17 and 18: ref. 8; #19: ref. 10 and #20-22: ref. 3.

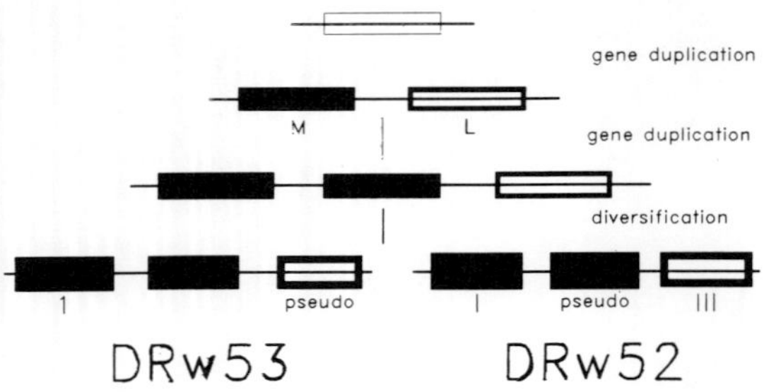

Figure 2. Evolutionary relationship between the DRw52 and DRw53 haplotypes. A primordial DR β gene duplicated originally to yield the ancestors of the "L" and "M" gene families. Another duplication of the "M" gene ensued, giving two "M" gene copies per haplotype. In the following process of diversification pseudogenes arose: haplotypes with two remaining functional "M" family copies ended up expressing the DRw53 determinant whereas the functional "L" gene on the other haplotype gave rise to the DR beta III gene of DRw52 haplotypes and carries the DRw52 determinant. It has been noted previously by others that gene conversion-like events may have carried the DRw52 determinant also to the DR beta I ("M" family) gene of these haplotypes (2).

over a longer stretch of nucleotides, up to around position 200. These sequence identities are marked in Figure 1 with underlinings (identical lines for identical sequences and different lines for different sequences). Most genes that show sequence identity in this region (pos. 200-250) are very dissimilar preceding position 200 (see e.g., genes #9 and #20, and also #3, #5 and #19). A similar phenomenon, although less striking, is found at the 3′ end of position 70: genes that are dissimilar in sequence preceding position 70 show sequence identity from positions 70-105 (also marked in Figure 1). Most likely this and the above-mentioned sequence identities have been caused by DNA exchange (either through recombination or through gene conversion-like events) between genes that had diverged through mutations after the duplication of the postulated primordial DR β gene (see also Figure 2). This exchange between different genes also explains the occurrence of "L" family sequences in "M" family genes (as in genes #18 and #19, pos. 210-212). The single base differences in genes #8 and #16 at positions 226 and 232 probably are the result of recent point mutations.

The blocks of sequence indicated Figure 1 and marked 0, 1, 2, and 3 stand out because of their relative conservation. It has been noted by others that part of the sequence included in block 1 has significant homology to the phage λ-χ sequence, and that this sequence also occurred in other MHC class II genes (3). Part of block 1 is repeated in block 2 and 3, and a preceding homologous sequence is designated block 0. In Figure 3 we show the homology of this tandemly repeated block to the χ sequence and to the human minisatellite repeat sequence, which is also associated with high recombination frequencies (5). Additionally we have added the

CHI		. C . . . TGG
MINISATELLITE		GGAGGTGGGCAGGARC
		|||| |||| ||
DR BETA	1	GGAGCTGGGGCGGCC
	2	. . . . AA . . . . .
	3	. . . . A . . A .
I-E BETA	1	C . . . .
	2	. . . . AA . A . . .
	3	. . . . A . . AA
CLASS I	1	. . . . . C . C . . GC . . .
	2	T . . . . A .

Figure 3. Comparison of the conserved sequence blocks observed in DR β genes to the minisatellite repeat sequence (5) and the λ-χ sequence, and also to homologous sequences observed in HLA class I and murine I-E β genes.

homologous sequences from other MHC genes: the murine I-E β genes and the human class I genes. Analysis of the published sequences of those genes reveals similar regions of unexpected sequence identity as noted here for the HLA class II genes (data not shown).

In this paper we have presented evidence for frequent DNA sequence exchange between homologous MHC genes, resulting in regions of unexpected sequence identity between otherwise (immunogenetically) unrelated genes. Preceding these regions we observed the occurrence of sequence blocks with high homology to sequences known to be related to high recombination frequency. We would like to postulate that these sequence blocks serve as initiation sites and/or breakpoints in homologous recombination or gene conversion, and are therefore responsible for the observed DNA exchange. Currently we are investigating whether these sites serve as binding sites for DNA-binding proteins.

References

1. Ronchese F, Schwartz RH, Germain RN. Functionally distinct subsites on a class II major histocompatibility complex molecule. Nature 1987;329:254.
2. Gorski J, Mach B. Polymorphisms of human Ia antigens: Gene conversion between two DR beta loci results in a new HLA-D/DR specificity. Nature 1986;322:67.
3. Wu S, Saunders TL, Bach FH. Polymorphism of human Ia antigens: reciprocal intergenic exchange between two DR beta loci. Nature 1986; 324:676.
4. Rollini P, Mach B, Gorski J. Linkage map of three HLA-DR beta-chain genes: Evidence for a recent duplication event. Proc Natl Acad Sci USA 1985;82:7197.
5. Jeffreys AJ, Wilson V, Thein SL. Hypervariable 'minisatellite' regions in human DNA. Nature 1985;314:67.
6. Didier DK, Schiffenbauer J, Shuman S, Abruzzini LF, Gorski J, Watling DL, et al. Characterization of two distinct DR beta chain alleles at the beta III locus of the DR5 haplotype: Beta III alleles are highly conserved. J Immunol 1886; 137:2627.
7. Gregersen PK, Shen M, Song QL, Merryman P, Degar S, Seki T, et al. Molecular diversity of HLA-DR4 haplotypes. Proc Natl Acad Sci USA 1986;83:2642.

8. Gregersen PK, Moriuchi T, Karr RW, Obata F, Moriuchi J, Maccari J, et al. Polymorphism of HLA-DR beta chains in DR4, -7 and -9 haplotypes: Implications for the mechanisms of allelic variation. Proc Natl Acad Sci USA 1986; 83:9149.
9. Cairns JS, Curtsinger JM, Dahl CA, Freeman S, Alter BJ, Bach FH. Sequence polymorphism of HLA-DR beta I alleles relating to T-cell-recognized determinants. Nature 1985;317:166.
10. Bell JI, Estess P, St. John T, Saiki R, Watling DL, Ehrlich HA, et al. DNA sequence and characterization of human class II major histocompatibility complex beta chains from the DR1 haplotype. Proc Natl Acad Sci USA 1985;82:3405.

Author Affiliations

Marnix L. Bosch, Department of Immunobiology, RIVM, Bilthoven, and Department of Immunohaematology and Bloodbank, University Hospital, Leiden, The Netherlands; *Annemarie Termijtelen, Jon J. van Rood, Marius J. Giphart*, Department of Immunohaematology and Bloodbank, University Hospital, Leiden, The Netherlands

Section VIII
Class II Histocompatibility Antigens: HLA-DQ

DRα-DQβ: A Novel Class II Molecule Whose Expression is Influenced by the Ratio of α β Chains Transcripts

Vincent Lotteau, Luc Teyton, Pascale Turmel, and Dominique Charron

Human major histocompatibility complex (MHC) class II molecules are heterodimeric integral membrane glycoproteins composed of non covalently associated α and β chains. At least three class II isotypes are formed by α - β pairing (DR, DQ, DP)(1). Haplotype mismatched α - β associations have been described resulting in the expression of hybrid HLA-DQ molecules (2). We have recently reported the existence of a different type of hybrid molecule (DRα-DQβ) arising by mixed isotype pairing in human B cell lines (3). Here we provide evidence for the influence of the relative amount of mRNA on this interisotypic pairing phenomenom.

Several EBV transformed B cell lines were selected for DQw1 serologic specificity. Two cell lines (FJO, CA) are homozygous for the three isotypes (4,5). Isa and Mic are from the same family and are heterozygous for DR but express DQw1 molecules from both haplotypes. In order to detect the interisotypic heterodimers DRα-DQβ in these cells, we performed the following experiments. DR molecules were removed from unlabeled total protein extract by successive immunoprecipitation with an anti-DR complex monoclonal antibody (MAb) D1.12 before immunoprecipitation with an anti DQw1 β MAb G2a5(6,7). The first and sixth D1.12 precipitates and G2a5 precipitate were dissociated and reprecipitated with the MAb HC2-1 (specific for DRα chain) (8) The immunoprecipitates were run in ID-SDS-PAGE, transferred to nitrocellulose, and probed with HC2-1.As shown in Figure 1, after total removal of DR complex, a DRα chain can be detected in G2a5 immunoprecipitate from Isa and Mic but not FJO and CA. We concluded that DRα-DQβ interisotypic heterodimers were present in Isa and Mic but not in FJO and CA. The CA-DQw1 β chain is also present in Isa cells and in another heterozygous cell line from the same family which expresses DRα-DQβ molecule (data not shown). Therefore this permissive haplotype does not result automatically in interisotypic heterodimer. For this reason we investigated the influence of quantitative variation of mRNA on the interisotypic pairing. Total RNA was extracted by successive phenolchloroform and LiCl precipitation. RNA was run in agarose gel, transferred to gene screen membrane, and successively hybridized with the four probes specific for DRα, DRβ, DQα, DQβ (9). The resulting signal on autoradiographies was scanned. Since this system is not calibrated with a reference curve for

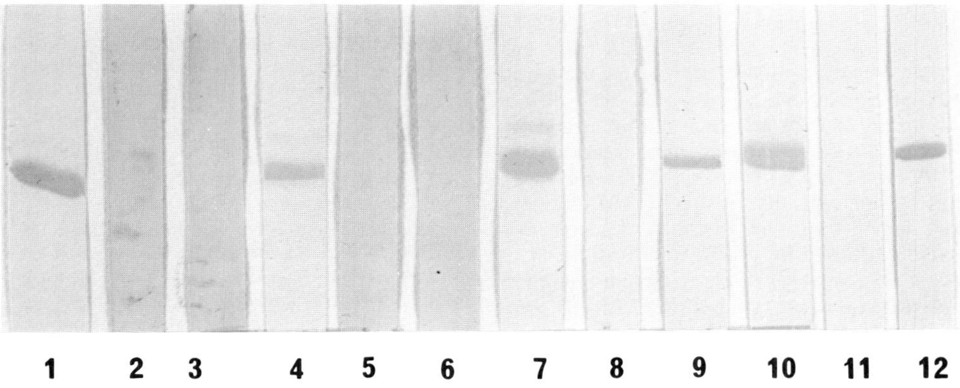

Figure 1. Detection of DRα chain in G2a5 immunoprecipitate. Unlabeled cells were extracted in 0.5% Nonidet P-40 buffer. The immunoprecipitations were performed as described (14). The first and sixth D1.12 precipitates and G2a5 precipitate were resuspended in 30% of 1% SDS, boiled for 3 minutes and diluted with 170 ml of NET buffer. Each sample was then immunoprecipitated with HC2-1, run in ID-SDS-PAGE and transferred to nitrocellulose. The membrane was incubated in a 200 mM NaCL, 50 mM Tris, 3% BSA, pH 7.4 blocking solution for 1 hour. This blocking solution supplemented with 0.10% Triton-X-100 was used in further incubations and washes. After overnight incubation with HC2-1, the specific binding to DRα chain was revealed by an enzymatic technique using a biotinylated anti-mouse antibody and a peroxidase-labeled streptavidine complex (Amersham). Lanes 1–3 = FJO; Lanes 4–6 = CA; Lanes 7–9 = Isa; Lanes 10–12 = Mic.; Lanes 1, 4, 7, 10 = DRα chain detected in the first D1.12 immunoprecipitate. Lanes 2, 5, 8, 11 = No DRα chain can be detected in the sixth D1.12 immunoprecipitate. Lanes 3, 6, 9, 12 = Supernatant of sixth D1.12 precipitate was precipitated by G2a5 before reprecipitation with HC2-1. No DRα chain can be detected in FJO and CA (Lanes 3 and 6). In contrast, a DRα chain can be detected in Isa and Mic (Lanes 9 and 12).

Immunobiology of HLA
Volume II

Table 1. DRα DQβ Mismatched Isotypic Pairing: Comparison with Ratio of Chain Transcripts

Cell Ratio	FJO	CA	Isa	Mic
DRα/DRβ	1.25	1.45	4.66	6.80
DQα/DQβ	2.53	0.63	0.87	0.84
DRα/DQα	0.82	2.12	1.81	1.43
DQβ/DRβ	0.61	1.09	2.94	5.88
DRα-DQβ interisotipic heterodimer	−	−	+	+

Pellets of 20.10^6 cells were extracted in Tris 0.2 M pH 8.8 EDTA 25 mM, SDS 1%, LiCl 0.1 M, dextran T40, phenol pH 7.5, chloroform isoamyalcohol (24:1). After centrifugation, supernatants were treated with successive phenolchloroform isoamyalcohol extractions. Supernatants were precipitated with ethanol +0.2 M ammonium acetate and resuspended in 5 mM Tris, pH 7.5, LiCP 2 M for a second precipitation. RNA was resuspended in H_2O, and 20 μg were run in 1% agarose gel before transfer to gene screen membrane. Sheets were successively hybridized with the probes specific for DRα, DQα, DRβ, and DQβ. The resulting signals were scanned to give a number proportional to the surface and intensity of the band.

each probe, the ratios represent arbitrary numbers that do not correspond to a given amount of mRNA. Therefore the interpretation of the results must be restricted to the comparison of the same ratio between different cells. The ratios are reported in Table 1. The interisotypic pairing DRα-DQβ can occur in Isa and Mic which DRα/β and DQβ/DRα ratios are high relative to FJO and CA, in which no hybrid molecules could be detected. This result can reflect an excess of DRα and DQβ mRNA and/or a lower amount of DRβ mRNA in these two cells compared to FJO and CA. Although the data gives important information about the variation of RNA ratios among several cells, they do not indicate an excess of DRα and DQβ mRNA over DRβ mRNA. Berdoz et al. recently reported an excess of DRβ mRNA over DRα mRNA in T and B cells with more α mRNA in T cells than in B cells (10). Previous studies have demonstrated the possibility of interspecies association of DRα with I-Aβ (11). Transfection of I-E and I-A subregion genes have shown that the mixed isotyped pairing can occur in the form of Eα-Aβ complex in L cells and is highly influenced by the allelic polymorphism of the Aβ chain (12). These data support the idea that the DRα (I-Eβ) has a high capacity of association with a β chain from another isotype. The same group also demonstrated that the haplotype mismatched associations need high levels of Aα and Aβ mRNA (13). The mixed isotype pairing in man could be influenced by at least two factors. First, the DQβ chain polymorphism could influence the pairing with DRα and define permissive haplotypes. Second, in permissive haplotypes the mixed isotype heterodimers could be the result of nonfavored pairing when all the favored pairing (intraisotypic) has occurred. In the case of DRα-DQβ heterodimers one could expect that DRα or DQβ or both, in excess relative to their regular pattern, will compete with DQα and DRβ respectively to create the interisotypic molecule. The variation of the normal balance of chains of the three isotypes could lead to new class II phenotypes. We are currently performing transfection experiments with sense and antisense cDNA to change the balance of α and β mRNA in order to manipulate the formation of new heterodimers. However since the synthesis rate of protein can be influenced by many factors it would be ultimately useful to quantify the amount of proteins rather than of mRNA. The presence of such interisotypic α-β chain pairing particularly should be considered for future investigation of the HLA class II expression in autoimmunity.

Acknowledgments. We thank F. Brodsky, P. Parham, J. Silver, S. Goyert, J.L. Strominger, R. Accola, V. van Heyningen, and M. Steel for providing monoclonal antibodies; C. Auffray for helpful discussions and comments; and M. Brandel for typing the manuscript. This work was supported by Inserm, ARC, ARSEP, and LNCC grants.

References

1. Kaufmann JF, Auffray C, Korman AJ, Shackelford DA, Strominger JL. The class II molecules of the human and murine major histocompatibility complex. Cell 1984;36:1.
2. Charron DJ, Lotteau V, Turmel P. Hybrid HLA-DC antigens provide molecular evidence for gene trans-complementation. Nature 1984;312:157.
3. Lotteau V, Teyton L, Burroughs D, Charron D. DRα-DQβ: A novel HLA class II molecule created by mismatched isotype pairing. Nature 1987;329-339.
4. Friedel AC, Betuel H, Gebuhrer L, Farre A, Lambert J. Distinct subtypes of HLA-D associated with DR2.9th Histocompatibility W.S. News Letter n°7, 1984 p 24.
5. Charron D, McDevitt HO. Characterization of HLA-D region antigens by two-dimensional gel electrophoresis. Molecular genotyping. J Exp Med 1980;152:18s.
6. Carrel S, Tosi R, Gross N, Tanigaki N, Carmagnola AL, Accola RS. Subsets of human Ia-like molecules defined by monoclonal antibodies. Mol Immunol 1981;18:403.
7. Brodsky FM, Parham P, Barnstable CJ, Crumpton MJ, Bodmer WF. Monoclonal antibodies for analysis of the HLA system Immunol Rev 1979;47:3.
8. Knudsen PJ, Stominger JL, A monoclonal antibody that recognizes the chain of HLA-DR antigens. Hum Immunol 1986;15:150.
9. Piatier-Tonneau D, Turmel P, Auffray C, Charron D. Construction of chain and locus specific HLA class II DNA probes. Study of HLA class II transcripts in leukemias. J. Immunol 1986;137:2050.
10. Berdoz J, Gorski J, Termijtelen AM, Dayer JM, Irle C, Schendel D, et al. Constitutive and induced expression of the individual HLA-DRβ and α chain loci in different cell types. J Immunol. 1987;139:1336.
11. Norcross MA, Raghupathy R, Strominger JL, Germain RD. Transfected human B lymphoblastoid cells express the mouse IA β ^{d}-chain in association with DRα. J Immunol 1986;137:1714.
12. Germain RN, Quill H. Unexpected expression of a unique

mixed-isotype class II MHC molecule by transfected L-cells. Nature 1986;320:72.
13. Germain RN, Bentley DM, Quill H. Influence of allelic polymorphism on the assembly and surface expression of class II MHC (Ia) molecules. Cell 1985;43:233.
14. Jones PP. Analysis of radiolabeled lymphocyte proteins by one- and two-dimensional polyacrylamide gel electrophoresis. In: Mishell BB, Shiigi SM (eds): Selected Method in Cellular Immunology. San Francisco, W.H. Freeman and Company.

Author Affiliations

Laboratoire d'Immunogénétique Moléculaire, Institut Biomédical des Cordeliers, Paris, France

Biochemical Analysis of DQ α Polymorphisms Defined by a Monoclonal Antibody

Susan F. Radka, Janet V. Johnston, David Schwarz, and Barbara S. Nepom

Abstract: Most of the presently identified DQ serologic specificities appear to be polymorphisms of DQ β genes, although polymorphism clearly exists in the α genes as well. We previously described a DQ α polymorphism detected by monoclonal antibody SFR20-DQα5, produced against Sweig, a DR5, DQw3 homozygous BLCL. The epitope detected by this antibody is present on cells carrying the DR3, 5, and w8 specificities (1). Biochemical analysis of the DQ α molecules isolated by a DQ monomorphic monoclonal antibody from a series of DQα5⁺ and DQα5⁻ BLCL reveal at least two different pI patterns. In general, all DQα5⁺ cell lines associated with DR3 and DR5 display DQ α chains of indistinguishable pI, which are more basic than those of DQα5⁻ BLCL. In contrast, DRw8 BLCL, all of which are DQα5⁺, have DQ α chains of considerably more acidic electrophoretic mobility than other DQα5⁺ cell lines. A comparison of first domain amino acid sequences of DQ α chains from DQα5⁺ vs. DQα5⁻ haplotypes show virtual identity among DQα5⁺ α chains, thus eliminating first domain substitutions as being responsible for pI differences between DRw8 and other DQα5⁺ chains. Gene conversion or intragenic recombination might be a possible explanation for the observed results.

HLA-DR homozygous BLCL were radiolabeled with ^{3}H-leucine. DQ molecules were isolated by the monoclonal antibody SFR20-DQM, which detects a monomorphic determinant on DQ beta chains, then treated with neuraminidase (C. perfringens). Electrophoresis in two dimensions was performed by the method of Shackelford et al. (2) with modifications employing 3.5-10 and 4-6 ampholytes at a 1:2 ratio. A comparison of the electrophoretic mobilities of DQ α chains isolated from

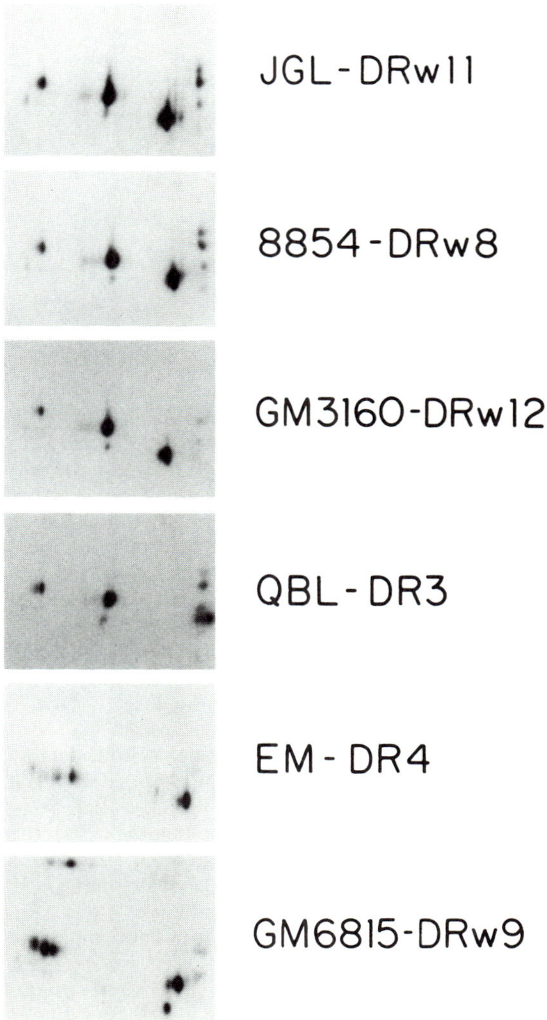

►

Figure 1. Comparison of the electrophoretic mobilities of DQ α chains isolated from various β lymphoblastoid cell lines that are either DQα5⁺ or DQα5⁻.

Table 1. Comparison of Relative pI with DQα5 Reactivity in a Series of HLA Homozygous Cell Lines

Cell	DR	DQ	DQα BASIC pI	DQα5 Reactivity
JMIT	1	1	−	−
CMcG	2	1	−	−
ELD	w6	1	−	−
JB	7	2	−	−
JK	7	w3,w9	−	−
HA	4	w3,w8	−	−
ER	4	w3,w7	−	−
LS40	4	w3,w8	−	−
DKB	w9	w3,w9	−	−
KOZ	9	w3,w9	−	−
SWEIG	5	w3,w7	+	+
JVM	5	w3,w7	+	+
THR	5	w3,w7	+	+
JGL	5	w3,w7	+	+
HERLUFF	5	w3,w7	+	+
QBL	3	2	+	+
COX	3	2	+	+
VAVY	3	2	+	+
DUCAF	3	2	+	+
RSH	3	w4	−	+
8854	w8?	w3,w7	+	+
LUY	w8	w3,w7	−	+
ARC	w8	w4	−	+
OLN	w8	w4	−	+
OLG	w8	w4	−	+
MAD	w8	w4	−	+
MMRA	w8	w4	−	+
HAS-15	4	w4	−	−
KT3	4	w4	−	−

various B lymphoblastoid cell lines (BLCL) that are either DQα5⁺ or DQα5⁻, is presented in Figure 1. QBL (DR3, DQw2), Sweig (DR5, DQw3), and 8854 (DRw8, DQw3) are DQα5⁺, and all possess DQ α chains of indistinguishable electrophoretic mobility that are considerably more basic than those of EM (DR4, DQw3) and DKB (DRw9, DQw3), which are DQα5⁻. Further analysis is underway to determine whether there is microheterogeneity in the pI's of lDQ α chains of DQα5⁻ BLCL.

Table 1 lists the homozygous BLCL whose DQ α chains we have examined, and their reactivity with the DQα5 monoclonal antibody. All the DR3, DQw2, and DR5, DQw3 BLCL examined have DQ α chains with the characteristic basic electrophoretic mobility. No DQα5⁻ BLCL have DQ α's of this basic pI. Surprisingly six of seven DRw8 BLCL have DQ α chains of considerably more acidic pI than the DQα5⁺ prototype pattern, even though all are bound strongly by the DQα5 antibody. The typing of the 8854 BLCL is uncertain and in fact, the cell may not be DRw8 positive. The other exception is the RSH line, a DR3, DQw4 BLCL, which is DQα5⁺, but has an acidic electrophoretic mobility.

An examination of the published first domain amino acid sequences for the DQ α chains of several lines, some of which we have examined in these studies. is presented in Table 2 (3-5). The first domain sequences of the DR3, DQw2 haplotypes are identical to those of all the DR5, DQw3 haplotypes so far published, and are concordant with our findings of identical electrophoretic mobility. The DRw8 sequence shown here (MAD, DRw8, DQw4), is identical to that of the DR5 and DR3-associated DQ α's, except for conservative changes at positions 69 and 75. Amino acid substitutions unique to these three haplotypes are found at positions 40, 47, 50, 51, and 53 (amino acid 76 is also found in DQα's

Table 2. Comparison of DQα First Domain Amino Acid Sequences from Several Haplotypes

Ref #											10
3	DR1 DQw1,DQw5	E	D	I	V	A	D	H	V	A	S
4	DR4 DQw3,w7	*	*	*	*	*	*	*	*	*	*
5	DR3 DQw2	*	*	*	*	*	*	*	*	*	*
4	DR5 DQw3,w7	*	*	*	*	*	*	*	*	*	*
4	DRw8 DQw4	*	*	*	*	*	*	*	*	*	*
Ref #											20
3	DR1 DQw1,DQw5	C	G	V	N	L	Y	Q	F	Y	G
4	DR4 DQw3,w7	Y	*	*	*	*	*	*	S	*	*
5	DR3 DQw2	Y	*	*	*	*	*	*	S	*	*
4	DR5 DQw3,w7	Y	*	*	*	*	*	*	S	*	*
4	DRw8 DQw4	Y	*	*	*	*	*	*	S	*	*
Ref #											30
3	DR1 DQw1,DQw5	P	S	G	Q	Y	T	H	E	F	D
4	DR4 DQw3,w7	*	*	*	*	*	S	*	*	*	*
5	DR3 DQw2	*	*	*	*	*	*	*	*	*	*
4	DR5 DQw3,w7	*	*	*	*	*	*	*	*	*	*
4	DRw8 DQw4	*	*	*	*	*	*	*	*	*	*
Ref #											40
3	DR1 DQw1.1	G	D	E	E	F	Y	V	D	L	E
4	DR4 DQw3,w7	*	*	*	*	*	*	*	*	*	*
5	DR3 DQw2	*	*	*	Q	*	*	*	*	*	G
4	DR5 DQw3,w7	*	*	*	Q	*	*	*	*	*	G
4	DRw8 DQw4	*	*	*	Q	*	*	*	*	*	G
Ref #											50
3	DR1 DQw1.1	R	K	E	T	A	W	R	W	P	E
4	DR4 DQw3.1	*	*	*	*	V	*	Q	L	*	L
5	DR3 DQw2	*	*	*	*	*	V	*	C	L	V
4	DR5 DQw3.1	*	*	*	V	*	C	L	*	*	V
4	DRw8 BLANK	*	*	*	*	V	*	C	L	*	V
Ref #											60
3	DR1 DQw1.1	F	S	K	F	G	G	F	D	P	Q
4	DR4 DQw3.1	*	R	R	*	R	R	*	*	*	*
5	DR3 DQw2	L	R	Q	*	R	( )	*	*	*	*
4	DR5 DQw3.1	L	R	Q	*	R	( )	*	*	*	*
4	DRw8 BLANK	L	R	Q	*	R	( )	*	*	*	*
Ref #											70
3	DR1 DQw1.1	G	A	L	R	H	M	A	V	A	K
4	DR4 DQw3.1	F	*	*	T	*	I	*	*	L	*
5	DR3 DQw2	F	*	*	T	*	I	*	*	L	*
4	DR5 DQw3.1	F	*	*	T	*	I	*	*	L	*
4	DRw8 BLANK	F	*	*	T	*	I	*	*	T	*
Ref #											80
3	DR1 DQw1.1	H	H	L	N	I	M	I	K	R	Y
4	DR4 DQw3.1	*	*	*	*	*	V	*	*	*	S
5	DR3 DQw2	*	*	*	*	S	L	*	*	*	S
4	DR5 DQw3.1	*	*	*	*	S	L	*	*	*	S
4	DRw8 BLANK	*	*	*	*	*	L	*	*	*	S

of DR7 haplotypes) and any or all of these might be responsible for the DQα5 epitope, if it is a first domain epitope. These results would further suggest that amino acid changes in the second domain of the DQ α chain of DRw8 haplotypes are responsible for the electrophoretic mobility differences between these and other DQα5+ haplotypes. It is intriguing to speculate that either gene conversion events or intragenic recombination might be responsible for transferring several amino acids that must have occurred to result in such large pI differences between DRw8 and other DQα5+ haplotypes. Resolution of this dilemma must await sequencing and other molecular analyses.

References

1. Amar A, Radka SF, Holbeck SL, Kim SJ, Nepom BS, Nelson K, et al. Characterization of specific HLA-DQα allospecificities by genomic, biochemical, and serologic analysis. J Immunol 1987;138:3986.
2. Shackelford DA, Lampson LA, Strominger JL. Analysis of HLA-DR antigens by monoclonal antibodies: recognition of conformational differences in biosynthetic intermediates. J Immunol 1981;127:1403.
3. Saiki RK, Bugawan TL, Horn GT, Mullis KB, Erlich HA. Analysis of enzymatically amplified β-globin and HLA-DQα DNA with allele-specific oligonucleotide probes. Nature 1986;324:163.
4. Todd JA, Bell JI, McDevitt HO. HLA-DQβ gene contributes to susceptibility and resistance to insulin-dependent diabetes mellitus. Nature 1987;329:599.
5. Shenning L, Larhammer D, Bill P, Wiman K, Jonsson AK, Rask L, et al. Both α and β chains of HLA-DC class II histocompatibility antigens display extensive polymorphism in their amino terminal domains. Embo J 1984;3:447.

Author Affiliation

Genetic Systems Corporation, Seattle, WA, USA

At Least Three Allelic Forms of the α- and β-chain Genes Encode DQw1-Positive Molecules

Massimo Trucco, Alessandra Care, Patricia Compagnone-Post, Craig Robinson, Isabella Cascino, and Emilia Turco

To date, three HLA-DQ allelic serologic specificities have been officially recognized: DQw1, 2, and 3. These serologic DQ markers are strongly associated with groups of DR antigens, namely DQw1 with DR1, 2, and w6; DQw2 with DR3 and 7; and DQw3 with DR4, 5, and w9. Although biochemical polymorphism had been recognized in both α and β chains, it was believed that the DQw1, 2, and 3 specificities were expressed by DQ-β chains. We studied the relationship existing between biochemical polymorphism as determined serologically, and restriction fragment length polymorphism (RFLP) at the DNA level. We observed that unique DQ-α RFLP patterns could be assigned to groups of DR antigens. Likewise, DQ-β RFLP patterns were found to be associated with different DR groups. The restriction enzymes PstI and TaqI were particularly useful in this analysis and, using these enzymes, five α and seven β allelic patterns were characterized.

We devised terminology in which the two restriction enzymes PstI and TaqI used to define the DQ-RFLP patterns are represented as "p" and "t," respectively, and the DQ-α and DQ-β probes as "a" and "b," respectively. Characteristic band patterns are differentiated from each other with a number. Because the DQap1 form, but none of the DQ-β forms, correlated with the serologically defined DQwl specificity, then DQwl may be an α-chain allelic marker and not a β-chain marker. We therefore called it AQ1 (1). DQap1 has always been found to be associated with one of three different TaqI α band patterns named ap1.t1, ap1.t2, and ap1.t6, because they are generally found in DR1, DR2, and DRw6 positive cells, respectively (2). We also found three DQ-β patterns associated with those α patterns in DQw1 positive cells. This is summarized in Table 1.

In order to test whether these patterns correlated with the phenotype, we attempted to generate alloreactive T-cell clones from mixed lymphocyte reactions (MLR) in which the responder and stimulator population were identical serologically (DR2, DQw1) and differed only in their DQ-β RFLP pattern (bp1.t2 versus bp4.t1). Such

Table 1. DR-Associated DNA Polymorphisms of DQ-α and DQ-β and Corresponding Cell Surface Determinants

Serology		RFLP		
DR	DQw	DQ-α	DQ-β	Phenotype
1	1	ap1.t1	bp1.t1	AQ1.1,BQ1.1
2	1	ap1.t2	bp1.t2	AQ1.2,BQ1.2
			bp4.t1	AQ1.2,BQ4.1
6	1	ap1.t6	bp4.t1	AQ1.6,BQ4.1
		ap1.t1	bp1.t1	AQ1.1,BQ1.1
		ap1.t2	bp1.t2	AQ1.2,BQ1.2

T-cell clones were generated and were specific for stimulator cells carrying specific pairs of α/β DQ patterns (ap1.t2,bp1.t2 and ap1.t2,bp4.t1, respectively), probably recognizing different, corresponding, α- and β-chain determinants (AQ1.2,BQ1.2 versus AQ1.2,BQ4.1). Although these results were consistent with the working hypothesis, they did not unequivocally demonstrate that the phenotypic differences recognized by our T-cell clones reflected coding sequence differences between the two DQ-β chains.

To obtain conclusive evidence, we isolated the two DQβ-chain genes corresponding to the bp1.t2 and bp4.t1 patterns, and obtained complete nucleotide sequences for these genes. Likewise, complementary DNA corresponding to DQap1.t2, ap1.t6 fragment patterns were also isolated and sequenced (3). The comparison of our sequences with each other and with others found in the literature confirmed that the deduced amino acid sequences were correlated precisely with the various DQ allelic forms determined by RFLP analysis.

Based on this evidence, we concluded that:

1) DQw1-positive DQ molecules comprise different allelic forms of DQ-α and DQ-β chains that are strongly associated with different DR specificities.
2) These α and β forms can be consistently identified both by RFLP analysis and using T-cell clones.
3) Our complementary T-cell clones are able to discriminate between two DQw1-positive molecules of stimulator cells carrying the same α chain, but a different β chain.
4) Two allelic genes encode the two DR2-associated DQ-β chains that carry the determinants recognized by T-cell clones.
5) These two DR2-associated β genes encode for chains that differ in six amino acid residues of the first domain and vary in nine and 11 residues, respectively, when compared with a DR1-associated DQ-β gene (bp1.t1) product.
6) These amino acid residues must be located on those portions of the beta chain that participate in the shaping of the recognition site; the corresponding determinants are defined here as BQ1.1, BQ1.2, and BQ4.1.
7) The α genes show a similar correlation between RFLP pattern and nucleotide sequence.
8) At least three different allelic forms of DQ-alpha chains, carrying the AQ1.1, AQ1.2, and AQ1.6 determinant, respectively, are present in DQw1-positive molecules.

Acknowledgment. The work was supported in part by grants Ca-44977 and DK-24021 from the National Institutes of Health, Bethesda, MD USA.

References

1. Trucco M, Duquesnoy RJ. Polymorphisms of the HLA-DQ subregion. Immunol Today 1986;7:297.
2. Trucco M, Rosenshine S, Cascino I, Duquesnoy RJ. Analysis of HLA-DQ subregion polymorphisms using sequential restriction endonuclease digestion of genomic DNA. Immunogenetics 1986;24:184.
3. Turco E, Care' A, Compagnone-Post P, Robinson C, Cascino I, Trucco M. Allelic forms of the alpha and beta chain genes encoding DQw1-positive heterodimers. Immunogenetics 1987;26:282.

Author Affiliations

The Pittsburgh Cancer Institute and Department of Pathology, University of Pittsburgh School of Medicine, Pittsburgh, PA USA

Definition of Polymorphism Within DR2-Associated DQw1 Molecules

S.V. Fuggle, N.A. Rust, C. Carter, A.J. McMichael, and P.J. Morris

The anti-HLA class II monoclonal antibody NDSl5.38, produced and characterized in our laboratory, was assigned a DR2 specificity on the basis that on the panel of cell lines tested, it reacted exclusively with DR2-typed cells (1). Additional data from competitive inhibition experiments and saturating binding assays demonstrated that NDSl5.38 reacted with a subpopulation of the class II antigens detected by a monomorphic anti-DR, DQ, DP reagent NFK2. Two-dimensional gel profiles confirmed these findings, but in a previous analysis the NDS15.38 profile was thought to be distinct from the profile obtained with the anti-DQw1 antibody SDR1.2. (2). More recent studies, including the workshop analysis have shown that the antibody reacts with the DQw1 molecule on a subpopulation of DR2-typed cells.

Two-dimensional gel analysis of cells representing different Dw subtypes of DR2 revealed structural polymorphism of the DQβ chain. Sequence data confirmed that this variability resided at the genetic level (3,4). In this report, we demonstrate that while NDS15.38 is reactive with the DQw1 molecule associated with Dw2 and Dw12 typed cells, the, DR2, non-Dw2, non-Dw12-typed cells UCA and AZH, which express different polymorphic forms of DQw1, are unreactive with the antibody (Table 1). NDS15.38 was negative on all DR1 typed cells, and thus does not detect the DR1-

Table 1. NDS 15.38 Reactivity with DR2-Typed Cell Lines

Cell line	HLA class II type	NDS 15, 38 Reactivity
PGF	DR2, DQwl, Dw2	+
EA181324	DR2, DQwl, Dw12	+
AZH	DR2, DQwl, Dw21	−
UCA	DR2, DQwl, non Dw2, non Dw12	−
REM	DR2, DQw3, Dw22	−

associated DQw1 molecule. These results are consistent with amino acid sequence data, which reveal that the DQβ sequence of the DR2-associated DQw1, "AZH" molecule is almost identical to the DR1, DQw1 molecule, differing only in one residue at position 57.

Acknowledgment. This work was supported by grants from the National Kidney Research Fund and Medical Research Council of the United Kingdom.

References

1. Fuggle SV, Kirkley J, Ting A, Morris PJ. Immunogenetics 1983;18:415.
2. Makgoba MW, Fuggle SV, McMichael AJ, Morris PJ. Nature 1983;301:531.
3. Lee BSM, Bell JI, Rust NA, McDevitt HO. Immunogenetics 1987;26:85.
4. Todd J, Bell JI, McDevitt HO. Nature 1987;329:599.

Author Affiliations

S.V. Fuggle, C. Carter, and P.J. Morris, Nuffield Department of Surgery; *N.A. Rust, and A.J. McMichael*, Nuffield Department of Medicine, University of Oxford, Oxford, U.K.

Division of HLA.DQw1 by DNA Typing and by an Alloserum: DR1, Subtypes of DR2 and DRw6; DRw10 and Rare Haplotypes DRw8-DQw1, DRw11-DQw1 Fit in Two Different Clusters

H. Betuel, L. Gebuhrer, M.P. Font, and A.C. Freidel

A tight linkage disequilibrium exists between antigens coded by HLA.DR and DQ. So far three alleles, DQw1, DQw2, DQw3, are recognised. Each codes for a product composed of two polymorphic glycoprotein molecules, α and β. DQw1 includes DR1, 2, w6, and w10. It has been evident in the last years that: 1) other DQ alleles exist, and 2) that the groups of linked DR antigens are not homogeneous and can assemble in separate clusters. These divisions, at least for DQw1 and DQw3, are based on serologic evidence, cellular reactivity, and DNA analysis. We shall comment on the division of DQw1.

Methodology

Serologic typing HLA was performed in a standard NIH test using Ninth Histocompatibility Workshop (HW) and local antisera (1). Cellular typing HLA.D region was performed in standard MLC using previous HW as well as local homozygous typing cells (HTCs) (2). DNA was digested with six restriction enzymes: Bam HI (B), Eco RI (E.I), Eco RV (E.V.), Hind III (H), Pvu II (P), Taq (T). After Southern blotting the fragments were hybridized successively with DRA, DQB and DQA probes (3).

We investigated an alloantiserum LY 1327 which reacted with 63.5% of DQw1 cells (1). This apparent division of DQw1 was further examined in 11 families and 33 genotyped individuals of Caucasoid origin. Because anti-DR antibodies may coexist in certain sera with anti-DQ specificity, we have included unusual but helpful DR-DQ haplotypes such as DRw8-DQwl, DRw11-DQw1, DRw13-DQw3, DRw14-DQw3. We also added a split of DR1 , called DR1x or DR"BR", recognized cellularly by HTC DwBON distinct from Dw1, (see ref. 1). A variant of DR2, different from DRw16, which did not type as Dw2, Dw12, Dw22, Dw21 (DR2sh), provisionally called DwGEN, was also included (2).

Results and Discussion

Serum LY 1327 reacted positively with DR1, DR1x, DRw16, DR2 (DwGEN), DRw10, DRw14 (Dw9) cells. The following DQw1-linked antigens remained negative, namely DR2 (Dw2 and Dw12), DRw13 (Dw18 and Dw19), as well as DRw8-DQw1, DRw11-DQw1. Segregation of haplotypes was evident in informative families. It must be noted that the rare associations DRw14-DQw3 and DRw13-DQw3 were unreactive. Alloserum LY 1327, in addition to its anti-DQw1 activity, is also anti-DR1 and anti-DRw10. As regards hybridization with *DRB probe*, two points are emphasized: five fragments correlated only with DRw10 cells, and DR1x cannot be distinguished from DR1 on the DNA level. We have previously commented on the fragments associating with DR2 and DRw6 subtypes (3). The results obtained with *DQB probe* were more complex and yielded various clusters. Three bands correlated with all DR antigens linked to DQw1 and two bands (H3.3 kb, P3.65 kb) with DQw1 and DQw8. Cluster DQw5, previously described (3), consisting of three bands (B 3.02, B 5.9, RV 2.7),is associated with DR1, DRw16, DR2 (DwGEN), DRw14. In the present study can be added DR1x, DRw10 and DRw8X14, a specificity observed by Du Toit in Cape Coloured. These DR specificities match perfectly with the division of DQw1 recognized serologically by alloserum LY 1327. We have found a fragment (P 3.05 kb) correlating with these DQw5 specificities and DQw8 as well. The second division of DQw1, cluster DQw8 composed of three fragments RI 9.5, RI 2.3, RV 5.5), also described (3) as correlating with DR2 (Dw2), DRw11-DQw1, DRw13 (Dw18 and Dw19), is also associated in the present study with DRw8-DQw1. Two allelic fragments contribute to the distinction between Dw18 and Dw19, namely, T 5.4 associated with DQw5, as well as Dw19 and T 2.9 correlating with DQw6 with the exception of Dw19.

The use of *DQA probe* has shown that the fragments obtained can be classified in two groups. A group of three distinct clusters associated with DQw1, with DRw52, and with DRw53. A second group of allelic bands without any correlation with DR or DQ specificities corresponds probably to DX. Futhermore, the different DRw6, subtypes, both serological (DRw13, DRw14) and cellular (Dw18, Dw19), can be distinguished by distinct DQα fragments. In the DRw52 cluster three fragments (EI 11.8, H 5.8, T 4.7) correlate as expected with DR3, DRw8, DRw11 and only with DRw6-DQw3. The relevant point is that these specificities do not fit in this cluster when they are linked to DQw1 whereas they do correlate when they are (unusually) linked to DQw3, as for example DRw6, DRw8.

Conclusion

Serologic data and genomic fragments are concordant and support the existence of two DQw1 splits. Each of these divisions contain different subtypes of DR2 and of DRw6 that are individualized by serology, cellular typing and RFLP.

References

1. Gebuhrer L, Betuel H, Lambert J, Freidel AC, Farre A. A division of HLA-DQw1 associated with DR1, DR1x-DQw1, DR2 short, DRw10 and DRw14 - I- Definition by an alloantiserum LY 1327. Hum Immunol 1987;18:235.
2. Freidel AC, Betuel H, Gebuhrer L, Farre A, Lambert J. Cellular and serological subtypes of HLA.DR2. Tissue Antigens 1987;29:129.
3. Font MP, Gebuhrer L, Betuel H, Freidel AC, Dausset J, Cohen D. HLA.DR2, .DR5 and DRw6 subtypes correlate with HLA.DR and DQ restriction fragments length polymorphism. Proc Natl Acad Sci USA 1986;76:3687.

Author Affiliations

H. Betuel, L. Gebuhrer, A.C. Freidel, Histocompatibility Laboratory, Blood Transfusion Centre, Lyon; *M.P. Font*, INSERM, U.33, Paris, France

The DQβ Chain Dictates DQ Type for DQw2 and DQw3

J. Schiffenbauer, D.K. Didier, S. Shuman, V. Tieber, M. Klearman, K. Rice, and B.D. Schwartz

The association of a number of diseases with the HLA-DR5 haplotype led us to sequence the expressed class II molecules of this haplotype to establish structure-function relationships, and to possibly elucidate a structural basis for predisposition to disease. We have made a cDNA library from the DR5 cell line, Swei (DR5, DRw52, Dw25, DQw7(w3), DPw4) and report here the sequence of the cDNA's corresponding to the DQw7α and DQw7β chains from this cell line (1). From our analysis and that of Saiki et al. (2), DQα appears to have one region of major variability. In contrast, DQβ appears to have four regions of variability. A comparison of DQα and DQβ chain sequences from DQw7 with those from other DQ types indicates that DQw7α derived from DR5 (denoted DQw7α (DR5)) is distinct from DQw7α (DR4), and DQw2α (DR3) is different from DQw2α (DR7) (3). Moreover, DQw7α (DR5) is identical to DQw2α (DR3). In contrast, DQw7β (DR5) is almost identical to DQw7β (DR4), and DQw2β (DR3) is identical to DQw2β (DR7). Thus, at least for DQw2 and w7, the DQβ chain dictates the DQ specificity. In addition, these data indicate that recombination occurs between DQA1 and DQB1. In contrast to data from DQw2 and DQw7, sequence data has indicated that DQw1α from DR1, 2, and w6 differ by only one amino acid and that the first domains of DQw1β derived from DR1 and DRw6 are identical. Thus, it is possible that both DQα and β chains contribute to the formation of the DQw1 epitope.

DQ sequence comparisons suggest most differences are attributable to multiple point mutations. However,

Immunobiology of HLA
Volume II

Figure 1. Comparison of amino acid sequence of DQα and DQβ. Tick marks indicate differences in sequence.

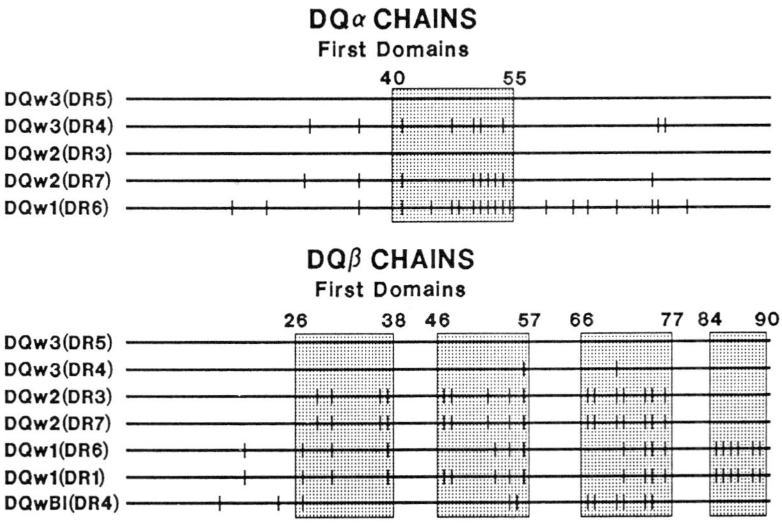

for DQα, comparison of DQA1 and DQA2 indicates that variability has also been generated by codon additions and deletions as well as point mutations. Furthermore, for DQβ, substitutions such as Asp-Ile in DQw2 and DQw blank for Glu-Val in DQw7 and DQw1 at positions 66 and 67 suggest that gene conversion events are also responsible for this variability. Of note, a χ-like sequence (5′-GCTGGGG-3′) located near amino acid 53, is 36 base pairs upstream from the third variable region of DQB1. A similar χ-like sequence has been suggested as a possible signal for gene conversion or recombination in DRB1 (4), and our data suggest a similar mechanism may be operating in DQB1.

The associations of DR and DQ antigens suggest that recombination can occur at a number of sites within the HLA-D region. DR5, DRw6, and DRw8 can be associated with both DQw1 or DQw3. Furthermore, one allele of DQA1 is in linkage disequilibrium with DRw53 and DR4, 7 and 9; whereas a second allele of DQA1 is in linkage disequilibrium with DRw52 and DR3, 5, w6 and w8. In contrast alleles of DQB1 are in linkage disequilibrium with other groups of DR molecules. Taken together these data suggest that recombination can occur between DRB1 and DQA1 as well as between DQA1 and DQB1. In addition, other work (5) suggests that recombination can occur between DRB1 and DRB3, as well as between DQA2 and DQB1. Thus there appear to be several potential sites of recombination and not a single hot spot of recombination in this region.

The polymorphism of both DQα and DQβ chains allows for the potential formation of DQ hybrid molecules and thus, epitopes unique to these molecules. Further structural analysis of DQ molecules should allow further understanding of the relationship between these molecules and disease.

Acknowledgments. The authors would like to thank Lorraine Whiteley and Pat Stewart for preparation of the manuscript.

Supported in part by grants AI 18925 and AI 15322.

References

1. Schiffenbauer J, Didier DK, Klearman M, Rice K, Shuman S, Tieber VL, et al. Complete sequence of the HLADQα and DQβ cDNA from a DR5/DQw3 cell line. J Immunol 1987;139:228-233.
2. Saiki RK, Bugaman TL, Horn GT, Mullis KB, Erlich HA. Analysis of enzymatically amplified β globin and HLADQα DNA with allele specific oligonucleotide probes. Nature 1986;324:163.
3. Song QL, Gregersen PK, Karr RW, Silver J. Recombination between DQα and DQβ genes generates histocompatibility leukocyte antigen class II haplotype diversity. J Immunol 1987;139:2993.
4. Wu S, Saunders TK, Bach FH. Polymorphism of human Ia antigens generated by reciprocal intergenic exchange between two DRβ loci. Nature 1986;324:676.
5. Gorski J, Tilanus M, Giphardt M, Mach B. Oligonucleotide genotyping shows that alleles at the HLA-$DR\beta_{III}$ locus of the DRw52 supertypic group segregate independently of known DR or Dw specificities. Immunogenetics 1987;25:79.

Author Affiliations

Division of Rheumatology, Howard Hughes Medical Institute and Washington University School of Medicine, St. Louis, MO, USA

Different DQw2 Are Associated with DR3 and DR7 as Revealed by RFLP Analysis

P. Avoustin, H.L. Coppin, A. Zerbib, A. Cambon-Thomsen, M. Thomsen, J. Tkaczuk, E. Ohayon, and C. de Preval

HLA-DQ antigens were first characterized by serologic assays. Three types of HLA-DQ alloantigens have been defined: HLA-DQw1, -DQw2, -DQw3. At the genomic level, two genes encoding the α chain (αDQ, αDX) and the β chain (βDQ, βDX) have been identified within the HLA DQ region of the major histocompatibility complex. The DQw2 specificity is in strong linkage disequilibrium with HLA-DR3 and -DR7 alleles. Recently, biochemical studies and RFLP analysis allowed the split of HLA-DQw1 and DQw3 types in several subgroups (1). Immunoprecipitation and two-dimensional gel analysis revealed a βDQ chain polymorphism between DQw2 molecules isolated either from cells bearing DR3 or DR7, DQw2 haplotypes.

In the present RFLP study, we analyzed genomic DNA of cells bearing B8 DR3 DQw2, B18 DR3 DQw2, DR7 DQw2, and DR9 DQw2 haplotypes. Analysis was performed with four restriction enzymes, Bam HI, Hinc II, Hind III, and RsaI, on 16 homozygous typing cells obtained from local donors and from the IX histocompatibility workshop (Table I). Seven cells presented the HLA-DR3, DQw2 haplotype, and nine the HLA-DR7, DQw2 haplotype. Two families were also studied with these enzymes. In family CA, the father bears a HLA-DR3, -DQw2 haplotype and the mother a HLA-DR7, -DQw2 haplotype. In the family MO, the father possesses a HLA-DRw9, -DQw2 haplotype and the mother a HLA-DRw9, -DQw3 haplotype.

After hybridization with a full length β DQ cDNA probe, the HLA-DQw2 patterns could be split in two subgroups with three restriction enzymes (Bam HI, Hind III, RSA1). These subgroups correlated with HLA-DR3 or HLA-DR7. Within each subgroup specific fragments could be observed (Table 2). In the CA family, specific DQ fragment segregated according to DR3 or DR7 inheritance; for example, after digestion with Rsa 1, a 2.8 Kb fragment segregates with DR3 and a 1.75 Kb fragment with DR7. With Hinc II restriction enzyme four patterns could be defined, two linked to HLA-DR7 and two linked to HLA-DR3. Among the HLA-DR3 DNA patterns one correlates with the B8 DR3 DQw2 haplotype or a B17 DR3 DQw2 haplotype, and the other with the B18 DR3 DQw2 haplotype. This result is in agreement with those previously described by Wescott and colleagues (2).

The family MO was analyzed to study the DQ linked to DRw9. In this family, the father was typed by serology a: A24, B35, DR9, DQw2; b: A11, Bw62, DR6, DQw1. And the mother, c: A1, B8, DR3, DQw2; d: A1, B8, DRw9, DQw3. One of the children bears the a/c haplotypes and thus was serologically type DRw9 DQw2.3. After digestion of his DNA with Rsa 1 restriction enzyme, none of the specific DQw2 fragment linked to DR3 or DR7 could be found. This result indicates that at least at the genomic level DQw2 linked to DRw9 seems to be different from DQw2 linked to DR3 or DR7.

The same filter was dehybridized and rehybridized with a full length αDQ cDNA probe. Again with Bam HI, Hind III and Rsa I DQw2 could be split in two pat-

Table 1.

Name	HLA-A	HLA-B	DR	Dw	DQ
AVL	1	8	3	3?	2
DUC	30	18	3	3	2
LAI			3	3	
QBL	26	18	3	3?	2
SOM	30	18	3	3	
STE	1	8	3	3	
WT 49	2	17	3	3?	2
BH 13	2	13	7	17	2
BOR	30	13	7	17	2
CAS	29,30	13,44	7	7	
DEL	24,29	8,44	7	7	
IBW9	33	14	7		2
LAR	2,28	44	7	7	
LBF	30	13	7		2
MANN	29	44	7	7	
ROB	9,28	14,44	7	7	

Table 2. Specific Bands of DQw2

	Bam HI		Hind III		Rsa I	
αDQ	DR3	DR7	DR3	DR7	DR3	DR7
9Kb		+++				
7,8			+++			
5,9				+++		
1,05					+	
0,85						+

	Hinc II		Hind III		Bam HI		Rsa I	
βDQ	DR3	DR7	DR3	DR7	DR3	DR7	DR3	DR7
16Kb				+++				
14,5			+++					
12	+++							
10		+						
6,6					++			
3,1						++		
2,8							+++	
1,75								+++

+++ represents strong hybridization, ++ medium hybridization, and + light hybridization.

Immunobiology of HLA
Volume II

terns that correlate either with HLA-DR3 DQw2 or HLA-DR7 DQw2 haplotypes (Table 2).

A more complex picture was observed with Hinc II. Seven DR7 DQw2 cells out of nine present a DNA pattern characterized by a 2.6 Kb fragment; this pattern is also shared by the DNA of the cells bearing either the B18 DR3 DQw2 or B17 DR3 DQw2 haplotypes. The DNA pattern of one of the other DR7 DQw2 cells (IBW9) was identical to the pattern associated to the B8 DR3 DQw2 haplotype which is characterized by a fragment of 2.2 Kb. The last DR7 DQw2 cell line (BOR) exhibits both fragments (2.6 Kb and 2.2 Kb). Thus the splits of the DQw2 β chain correlate with HLA-DR3 and HLA-DR7 allelism and within DR3 specificity, βDQw2 patterns were found to correlate with B8 DR3 DQw2 and B18 DR3 DQw2 haplotypes. In contrast, splits obtained with specific Hinc II αDQ fragment genes could not be completely associated with any haplotypes. Further studies on this subject will be performed using four specific probes.

References

1. Bontrop RE, Baas EJ, Otting NO, Schreuder GMT, Giphart MJ. DQ alpha and beta chain isoelectric point differences and their relations to serologically defined HLA-DQ allospecificities. Immunogenetics 1987;25:305–312.
2. Wescott MZ, Awdeh ZL, Yunis EJ, Alper CA. Molecular analysis distinguishes two HLA-DR3 bearing major histocompatibility complex extended haplotypes. Immunogenetics 1987;26:370–374.

Author Affiliation

Inserm U100, Cedex, France

Evolution of DQw2-Related Haplotypes

Carolyn Katovich Hurley, Noriko Steiner, Peter Gregersen, Jack Silver, Robert Hartzman, Gerald Nepom, and Armead H. Johnson

Abstract: The majority of HLA-DR alleles have been grouped into two families on the basis of supertypic serologic specificities, DRw52 and DRw53. Although members of each family share a common evolutionary pathway, there is evidence that the DRw52 and DRw53 pathways have crossed producing haplotypes that express genes found in both families. Haplotypes that share DQw2-like α genes are an example of this convergence. The haplotypes in this group include DRw52-associated DR3 (DQw2 and DQw4), DR5 (DQw3), DRw8 (DQw4), and DRw- (DQw2); and DRw53-associated DR7(DQw2 and DQw3) and DRw9 (DQw2). Some of these haplotypes share the DQw2 serologic specificity whereas others share only a DQw2-like α gene. A unique DQB2 *Hinc* II DNA restriction fragment is found associated with many of these haplotypes. Reciprocal recombination, mutation, and gene conversion play a major role in the generation of diversity in these haplotypes.

Several DR3 haplotypes are found in the DQw2-related haplotype family. Two of these haplotypes, DRw17(3), DQw2,Dw3v and DRw18(3),DQw4,Dw-, are commonly found in the American black population. Only half of the DR3-positive American blacks possess the DRw17(3),DQw2 haplotype commonly observed in Caucasians. The remainder of DR3 individuals express a DQw4 allele (DQ blank or DQw-) (1). The DRw18(3), DQw4 individuals fail to type for any established HLA-D specificity using mixed lymphocyte typing. (This haplotype is also found in RSH, WS ID#9021.) The majority of the DRw17(3),DQw2 black individuals give a low level of stimulation (Dw3v) with Dw3 homozygous typing cells (HTC). The usual Caucasian haplotype is DRw17(3),DQw2,Dw3.

We have recently described the structural basis of the DRw18(3)-associated DQw4 molecule (2). The amino acid sequence of the first domain of the DQw2-like (DQw2v) α chain encoded by this haplotype is presented in Table 1. The sequence is identical to a DR3- and DR5-associated DQw2 sequence (3–5) with two exceptions, positions 69 (threonine/leucine) and 75 (isoleucine/serine). Based on a comparison with DQw1 and DQw3 α sequences (4), the original DQw2 (DQw2°) α sequence probably contained a leucine at position 69 and an isoleucine at position 75. Thus, both DQw2 and DQw2v α polypeptide sequences appear to have diverged from DQw2° by one replacement substitution.

DR7,DQw2 and DR7,DQw3 haplotypes also express a DQw2-like alpha polypeptide (6,7). Although the DR7-associated polypeptide bears similarity to both DQw2 and DQw3 alpha sequences, it also exhibits a one amino acid deletion at position 55, which is a unique feature of DQw2 α polypeptides. The amino acids at positions 69 and 75 in the DR7-associated DQ α chain are shared with the DQw2° sequence. In addition, nine replacement substitutions are observed. The majority of these are shared with DQw3 alpha chains suggesting a gene conversion-like event in the generation of this diversity.

The sequence of the DQB1 gene associated with the DRw18(3),DQw4 haplotype is distinct from the

Table 1. DQA1

	1									10			
DR18(3),DQw4	GLU	ASP	ILE	VAL	ALA	ASP	HIS	VAL	ALA	SER	TYR	GLY	VAL
DRw17(3),DQw2*	–	–	–	–	–	–	–	–	–	–	–	–	–
DR7,DQw2†	–	–	–	–	–	–	–	–	–	–	–	–	–
							20						
DRw18(3),DQw4	ASN	LEU	TYR	GLN	SER	TYR	GLY	PRO	SER	GLY	GLN	TYR	THR
DRw17(3),DQw2	–	–	–	–	–	–	–	–	–	–	–	–	–
DR7,DQw2	–	–	–	–	–	–	–	–	–	–	–	PHE	–
				30									
DRw18(3),DQw4	HIS	GLU	PHE	ASP	GLY	ASP	GLU	GLN	PHE	TYR	VAL	ASP	LEU
DRw17(3),DQw2	–	–	–	–	–	–	–	–	–	–	–	–	–
DR7,DQw2	–	–	–	–	–	–	–	GLU	–	–	–	–	–
	40										50		
DRw18(3),DQw4	GLY	ARG	LYS	GLU	THR	VAL	TRP	CYS	LEU	PRO	VAL	LEU	ARG
DRw17(3),DQw2	–	–	–	–	–	–	–	–	–	–	–	–	–
DR7,DQw2	GLU	–	–	–	–	–	–	LYS	–	–	LEU	PHE	HIS
								60					
DRw18(3),DQw4	GLN	PHE		ARG	PHE	ASP	PRO	GLN	PHE	ALA	LEU	THR	ASN
DRw17(3),DQw2	–	–		–	–	–	–	–	–	–	–	–	–
DR7,DQw2	ARG	LEU		–	–	–	–	–	–	–	–	–	–
					70								
DRw18(3),DQw4	ILE	ALA	VAL	THR	LYS	HIS	ASN	LEU	ASN	ILE	LEU	ILE	LYS
DRw17(3),DQw2	–	–	–	LEU	–	–	–	–	–	SER	–	–	–
DR7,DQw2	–	–	–	LEU	–	–	–	–	–	–	–	–	–
		80											
DRw18(3),DQw4	ARG	SER	ASN	SER	THR	ALA	ALA	THR	ASN				
DRw17(3),DQw2	–	–	–	–	–	–	–	–	–				
DR7,DQw2	–	–	–	–	–	–	–	–	–				

*(3,4); †(6).

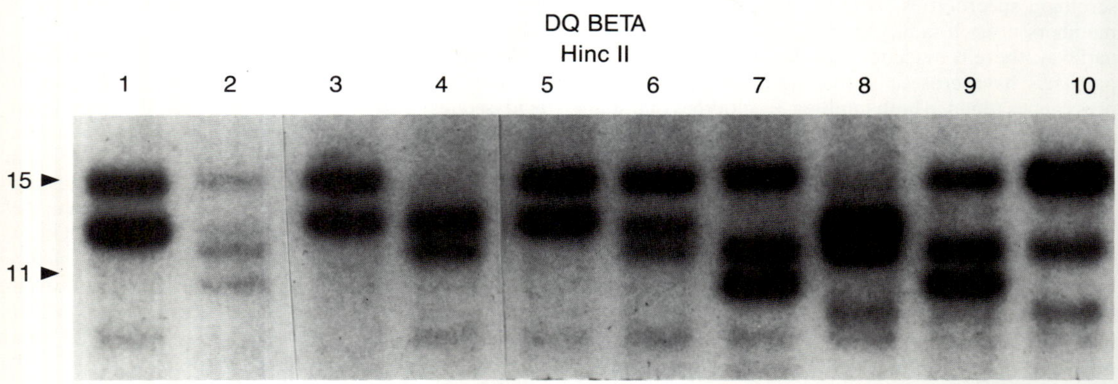

Figure 1. Southern blot analysis of DNA isolated from EBV-transformed B-cell lines. Genomic DNAs were digested with *Hinc* II and hybridized with a ^{32}P-labelled full-length DQ β probe as previously described (2). 1 = RSH (WS ID# 9021) (DRw18(3);DQw4); 2 = 1014 (DRw18(3),2;DQw4,1); 3 = QBL (WS ID# 9020) (DRw17(3);DQw2); 4 = AVL (DRw17(3);DQw2); 5 = 373 (DRw17(3);DQw2;Dw3v); 6 = 2704 (DRw-,8;DQw2,3); 7 = 1083 (DRw9,13;DQw2,1); 8 = Fisher (DR7;DQw3); 9 = 1249 (DR7,5;DQw2,1); 10 = WET (DR7;DQw2).

DQw2 β sequence from WT49 (DRw17(3),DQw2) (8) and corresponds instead to the DQBl sequence from KT3 DR4, DQw4) (9). The sequences differ only at position 23 where the DRw18(3)-associated sequence contains an arginine compared to a leucine in KT3. All other DQ molecules sequenced from cells expressing DQw1 and DQw3 contain an arginine at this position. Therefore, the expressed DRw18(3)-associated DQw4 heterodimer is composed of a DQw2-like α chain combined with a DQw4 β chain. An identical DQ α and β heterodimer is found in a DR8,DQw4 cell (10). The association of the DR3 haplotype with a new DQBl gene represents a recombination event between the DQw2 DQA1 gene and the DQw2 DQBl gene commonly associated with DR3 to generate a DQw2 α/DQw4 β association. Recombination between DQAl and DQBl genes has also generated other members of the DQw2 family, DR5, DQw3 (5) and DR7,DQw3 (7).

Restriction fragment length polymorphism was used to examine the relationships among the DQw2-related haplotypes. A unique 15 kb *Hinc* II DQ β restriction fragment was observed in unrelated individuals expressing DRw18(3),DQw4, some DRw17(3),DQw2, DR7,DQw2, and DR9,DQw2 haplotypes (Figure 1) (11,12) . Segregation analysis in an informative family (#014) shows the 15 kb *Hinc* II fragment associated with the DRw18(3),DQw4 and DRw-,DQw2 haplotypes expressed by the family (2). This restriction fragment is also present in the HLA-B18-associated DRw17(3), DQw2 HTC QBL (WS ID#9020), but is absent in the HLA-B8-associated DRw17(3),DQw2 homozygous typing cell (HTC) AVL. Instead, AVL has an 11 kb fragment. In the DR3 haplotypes, the *Hinc* II fragments at 15 kb and 11 kb have been identified as DQB2 genes using a DQB2 oligonucleotide probe (2). Thus, it appears that the usually invariant 11 kb *Hinc* II DQB2 gene fragment is found as a 15 kb fragment in some DQw2-related haplotypes.

Diversity in the DQw2-related haplotypes has been generated by reciprocal recombination, point mutation, and gene conversion. Recombination has been particularly important in the evolution of these haplotypes generating new combinations of class II genes that can be used to mediate immune responsiveness.

Acknowledgments. This research is supported by ONR N00014-83-K-0410 (CKH), NIH AI23371 (AJ), and NIH AI22005 (JS).

References

1. Johnson, A. Joint report: American blacks. In: Aizawa M (ed): HLA in Asia-Oceania, Sapporo, Japan, Hokkaido University Press, 1986, p 320.
2. Hurley CK, Gregersen P, Steiner N, Bell J, Hartzman R, Nepom G, et al. Polymorphism of the HLA-D region in American blacks: A DR3 haplotype generated by recombination. J Immunol 1988;140:885.
3. Schenning L, Larhammar D, Bill P, Wiman K, Jonsson A, Rask L, et al. Both alpha and beta chains of HLA-DC class II histocompatibility antigens display extensive polymorphism in their amino-terminal domains. EMBO J 1984;3:447.
4. Saiki R, Bugawan T, Horn G, Mullis K, Erlich H. Analysis of enzymatically amplified beta-globin and HLA-DQ alpha DNA with allele-specific oligonucleotide probes. Nature 1986;324:163.
5. Schiffenbauer J, Didier D, Klearman M, Rice K, Shuman S, Tieber V, et al. Complete sequence of the HLA DQ alpha and DQ beta cDNA from a DR5/DQw3 cell line. J Immunol 1987;139:228.
6. Chang H, Moriuchi T, Silver J. The heavy chain of human B-cell alloantigen HLA-DS has a variable N-terminal region and a constant immunoglobulin-like region. Nature 1983; 305:813.
7. Song Q, Gregersen P, Karr R, Silver J. Recombination between DQ alpha and DQ beta genes generates human HLA class II haplotype diversity. J Immunol 1987;139:2993.
8. Boss J, Strominger J. Cloning and sequence analysis of the human major histocompatibility complex gene DC-3 beta. Proc Natl Acad Sci USA 1984;81:5199.
9. Gregersen P, Shen M, Song Q, Merryman P, Degar S, Seki T, et al. Molecular diversity of HLA-DR4 haplotypes. Proc Natl Acad Sci USA 1986;83:2642.
10. Todd J, Bell J, McDevitt H. HLA-DQbeta gene contributes to susceptibility and resistance to insulin-dependent diabetes mellitus. Nature 1987;329:599.
11. MacMurray A, Bell J, Denney D, Watling D, Foster L, McDevitt H. Molecular mapping class II polymorphisms in the human major histocompatibility complex. II. DQ beta. J Immunol 1987;139:574.
12. Bell J, Smoot S, Newby C, Toyka K, Rassenti L, Smith K, et al. HLA-DQ beta-chain polymorphism linked to a myasthenia gravis. Lancet 1986;1:1058.

Author Affiliations

Carolyn Katovich Hurley, Noriko Steiner, Armead H. Johnson, Departments of Microbiology and Pediatrics, Georgetown University, Washington, DC; *Peter Gregersen, Jack Silver*, Hospital for Joint Diseases, New York, NY; *Robert Hartzman*, Naval Medical Research Institute, Bethesda, MD; *Gerald Nepom*, Virginia Mason Research Center, Seattle, WA, USA

Subunit and Spatial Relationships of HLA-DQ Polymorphisms That Subdivide the DQw3 Specificity

S.F. Radka, K. Nelson, and J.V. Johnston

Polymorphism of the HLA-DQ subregion is more extensive than classical DQ serologic specificities presently define. Recently, a DQ polymorphism termed TA10 DQw7 was defined (1) which has 100% correlation with a restriction length polymorphism of the DQ β gene (2). The TA10 epitope was found to exist on a subset of β chains carrying the DQw3 epitope, which have an electrophoretic mobility distinct from that of TA10⁻ DQw3⁺ β chains. To determine more precisely the spatial relationships between the TA10 and DQw3 epitopes, we have analyzed several monoclonal antibodies detecting these specificities by Western blotting and competitive inhibition of antibody binding.

Sweig (DR5, DQw3 TA10⁺ DQw7) membranes were prepared as described previously and treated to separate the class II α and β chains by a procedure involving acid treatment (4). This protocol has been shown to preserve some class II epitopes that are destroyed by the more conventional methods of boiling and electrophoresis in 1% or greater SDS concentrations. Purified monoclonal antibodies were obtained for labelling with FITC and as unlabelled blocking reagents by affinity chromatography on a monoclonal rat anti-mouse kappa column.

The summary of isotypes, serological specificities, and subunit reactivities of the antibodies is presented in Table 1. Only the 17.15 antibody consistently reacts with a separated β chain. GS159.1 immunoblots a β chain inconsistently, whereas the other antibodies so far have not reacted with separated chains under the mildest chain separation conditions we have used (acid treatment of membranes, electrophoresis in 0.1% SDS under nonreducing conditions).

The ability of each of these polymorphic anti-DQ antibodies to inhibit the binding of FITC labelled antibodies is presented in Table 2. Each antibody was tested for its ability to block (blocking antibody) the binding of each of the other antibodies and reciprocally, to be blocked (FITC-labelled antibody) by each of the other antibodies. In summary, each of the four anti-DQw3 monoclonal antibodies was able to block binding of the other in a reciprocal fashion. However, of the three DQw3 antibodies generated against a TA10⁻ haplotype DQw8 (17.15, 285-8B67, and 215-3E5), only the IgM 17.15 antibody was able to inhibit the binding of the anti-TA10 antibody 159.1, and the inhibition was incomplete. Conversely, 159.1 was unable to inhibit the binding of the two IgG antibodies generated against DQw8 haplotypes (285-8B6.7 and 295-3E5). It seems likely that the reciprocal partial blocking between 159.1 and 17.15 is due to hindrance of or by the IgM antibody. In contrast, the anti-DQw3 antibody generated against the TA10⁺ DQw7

Table 2. Inhibition of FITC-Labelled Antibody Binding by Unlabelled Antibodies

Blocking Antibody	FITC-Labelled Antibody	% Inhibition*
GS 295-3E5 (DQw3)	GS 159.1 (TA10)	4
GS 285-8B6.7 (DQw3)	GS 159.1 (TA10)	2
17.15 (DQw3)	GS 159.1 (TA10)	68
SFR 20-DQw3 (DQw3)	GS 159.1 (TA10)	93
GS 159.1 (TA10)	GS 295-3E5 (DQw3)	7
GS 285-8B6.7 (DQw3)	GS 295-3E5 (DQw3)	81
17.15 (DQw3)	GS 295-3E5 (DQw3)	99
SFR 20-DQw3 (DQw3)	GS 295-3E5 (DQw3)	100
GS 295-3E5	GS 285-8B6.7 (DQw3)	88
GS 159.1	GS 285-8B6.7 (DQw3)	1
17.15	GS 285-8B6.7 (DQw3)	96
SFR 20-DQw3	GS 285-8B6.7 (DQw3)	95
GS 285 8B6.7	17.15 (DQw3)	69
GS 295-3E5	17.15 (DQw3)	84
GS 159.1	17.15 (DQw3)	56
SFR 20-DQw3	17.15 (DQw3)	100
17.15	SFR 20-DQw3 (DQw3)	99
GS 285-8B6.7	SFR 20-DQw3 (DQw3)	80
GS 295-3E5	SFR 20-DQw3 (DQw3)	86
GS 159.1	SFR 20-DQw3 (DQw3)	87

*% Inhibition of peak channel fluorescence =

$$\frac{\text{blocked} - \text{background fluorescence}}{\text{unblocked} - \text{background fluorescence}}$$

Table 1. Reactivity of Anti-DQ Monoclonal Antibodies on Sweig (HLA-DR5, DQw3, TA10⁺)

Antibody Name	Immunogen	Antibody Isotype	Specificity	Subunit Reactivity in Western Blotting
17.15	RHu (DR3,4)	Mouse IgM	DQw3	αβ, β
GS 159.1	ER (DR4,Ta10⁺)	Mouse IgG	TA10 (DQw7)	αβ, β
SFR20-DQw3	Sweig (DR5,TA10⁺)	Rat IgG	DQw3	αβ
GS 285-8B6	HA (DR4,TA10⁻)	Mouse IgG	DQw3	αβ
GS 295-3E5	KT2 (DR4,TA10⁻)	Mouse IgG	DQw3	αβ
SFR20-DQM	Sweig (DR5,TA10⁺)	Rat IgG	DQ monomorphic	αβ, β

Immunobiology of HLA
Volume II

haplotype (SFR20-DQw3) is able to block completely anti-TA10 binding and vice versa. These data would suggest that the TA10 epitope is spatially distinct from the DQw3 epitope defined by 285-8B6.7 and 2953-E5, whereas the epitope identified by SFR20-DQw3 lies between the other two.

In conclusion, these data support the premise that there are multiple DQw3-like epitopes, some spatially distinct from the TA10 epitope and others spatially close. All of these epitopes appear to be β determinants, but have tertiary structural requirements for their formation.

References

1. Schreuder GMT, Doxiadis T, Parleviet J, Grosse-Wilde H. DR DQ LB and TA10 specificities of Ninth Workshop homozygous typing cells. In: Albert ED, Baur MP, Mayr WR (Eds.): Histocompatibility Testing. Berlin, Springer-Verlag, 1984, p 243.
2. Kim SJ, Holbeck SL, Nisperos B, Hansen JA, Maeda H, Nepom GT. Identification of a polymorphic variant associated with HLA-DQw3 and characterized by specific restriction sites within the DQ β gene. Proc Natl Acad Sci USA 1985;82:8139.
3. Radka SF, Scott RG, Stewart SA. Molecular complexity of HLA-DQw3: The TA10 determinant is located on a subset of DQw3 β chains. Hum Immunol 1987;18:287
4. Cohen BB, Moxley M, Crichton D, Deane DL, Steele CM. A mild procedure for separating polypeptide chains prior to immunoprecipitation and Western blotting analysis. J Immunol Meth 1984;75:99.

Author Affiliation

Genetic Systems Corp, Seattle, WA 98121, USA

Serology and RFLP Analysis of the Splits of DQw3

Jodee Sedalnick, Edward J. Ball, Shari Clark, Lori Dombrausky, Carolina Salvador, and Peter Stastny

The DQ specificity DQw7 was described during the Ninth Workshop by alloantisera (1), a monoclonal antibody (2), and TA10-restricted T-cell line (3). Subsequently, this specificity was shown to be associated with a unique DQwβ restriction fragment length polymorphism (RFLP) pattern with several different enzymes.

In the Tenth Workshop, DQw7 was described by a cluster of monoclonal antibodies and antisera (10W3119, 3120, 3121, 9051, 9052, 1107) with r values of 0.594–0.872 in our tes t ing. Another cluster identified by antibodies 10W3088, 1217, 3086, 3090, 3122, 9055, and 9056 was negatively correlated with DQw7 and was given the name IIB3 for the lead antibody of the group. This cluster was shown to have r values of 0.519–0.780. The specificity defined by the IIB3 cluster included most, if not all, cells bearing DQw1, DQw4, and DQw3 negative for DQw7. With the addition of the WA and IIB3 DQ antigens one now has the opportunity to distinguish the following phenotypes serologically: DQw7/DQw8 and DQw7/DQw7, as shown in family studies in Table 1 (p. 288).

The DQw7 coding in DQw3 haplotypes was as follows: DR2 (2/2), DR4 (13/36), DR5 (38/38), DRw6 (6/6), DR7 (0/10), DRw8 (8/12), and DRw9 (0/3). All other DQw3 haplotypes were IIB3 positive as were DQw1 and WA haplotypes. With the exception of one cell (DR2,3; DQw1,DQw4,DQw7), DQw7 was found only in DQw3-positive cells.

The 10WS B-cell lines and selected panel members were studied for DQβ RFLP patterns with the enzyme BgI II. The DQw7-positive cells were exclusively associated with a 4.3-kb restriction fragment. DQw7-negative DQw3 cells all had a 6.0-kb band, which was also found in DQw4-positive cells.

References

1. Nunez G, Latimer MJ, Ball EJ, Myers LK, Stastny P. Serologic detection of new polymorphisms on DQ molecules distinct from the DQw1,2, and 3 specificities. In: Albert I, Baur MP, Mayr WR, (ed): Histocompatibility testing. Heidelberg, Springer-Verlag, 1984, pp 421–422.
2. Maeda H. Mouse monoclonal antibody detects a new polymorphic Ia determinant other than HLA-DR antigen; A possible allele of DC-1. Tissue Antigens 1984;23:163–170.
3. Ball EJ, Stastny P. Antigen-specific HLA-restricted human T-cell lines. II. A GAT-specific T-cell line restricted by a determinant carried by an HLA-DQ molecule. Immunogenetics 1984;20:547–564.

Author Affiliation

Department of Internal Medicine, The University of Texas Health Science Center at Dallas, Southwestern Medical School, 5323 Harry Hines Blvd., Dallas, TX 75235, USA

Table 1. Segregation of TA10 and 11B3 in Family Studies

FAM Memb	Hap	HLA DR	HLA DQw	Cell ID	DQw1 111133 111100 556699 342512	DQw2 1333 1111 7000 0567	DQw3 11113333399 11110111100 77786111155 67986124578	DQw3TP 333991 111001 212550 190127	IIB3 1333399 2000100 1889255 7860256
Family 2000									
	a)	DR7		DQw3	IIB3	c)	DR4 DQw3	TA10	
	b)	DR3		DQw2	–	d)	DR4 DQw3	IIB3	
F-PA	a,b	3,7	2,3	2001	111311	8468	88882488888	111112	6118668
F-MA	c,d	4,4	3,3	2002	181111	1111	88866888866	886888	4468666
SIB1	a,c	4,7	3,3	2003	161411	6141	88888688866	886488	4666144
SIB2	b,c	3,4	2,3	2004	164110	8486	82818884824	884888	2111112
SIB3	b,d	3,4	2,3	2005	111111	8888	88888888888	111114	8888888
SIB4	a,d	4,7	3,3	2006	111111	1211	88888888888	111114	8888886
Family 1000									
	a)	DR3		DQw2	–	c)	DR4 DQw3	TA10	
	b)	DRw13		DQw1	IIB3	d)	DR2 DQw1	IIB3	
F-PA	a,b	3,13	1,2	1001	888888	8888	11121111111	111116	8888888
F-MA	c,d	2,4	1,3	1002	886866	4111	16884888888	861686	6288688
SIB1	a,d	2,3	1,2	1003	481882	8662	11111111111	111111	4188222
SIB2	b,c	4,13	1,3	1004	486686	2111	24666846824	666866	4661621
SIB3	b,d	2,13	1,1	1005	888686	2111	11111111111	111111	8148211
SIB4	a,c	3,4	2,3	1006	111214	4884	66446862814	888882	4111111

The HLA B17,BfS,C4A6,B1,DR7 and DQw3.2 Supratype Marks an Extensive Chromosomal Segment Shared Between Different Ethnic Groups

P.H. Kay, J. Williamson, Katsushi Tokunaga, R.L. Dawkins, F.T. Christiansen, P. Charoenwong, and E. Martin

Previously, we have shown that the supratype defined by HLA B17,BfS,C4A6,B1,DR7, as well as a specific 12kb BglII/Alu7 fragment (17,6,1,7), occurs in Caucasoids and Orientals with population frequencies of approximately 1 to 6% an 6% respectively (1). It is often assumed that marked polymorphism of MHC loci reflects increased susceptibility to mutational and recombinant events important in generation of functional diversity. As a consequence, conservation of MHC supratypes between different racial groups would not be expected. Since HLA B17,DR7 may be distinguished serologically and at the genomic level from other DR7-bearing supratypes (2,3), we have examined genomic DNA and DQ subtypes in a further attempt to distinguish between the 17,6,1,7 supratype in different ethnic groups.

Following digestion with Hind III, four of four examples of genomic DNA from unrelated Thai/Chinese subjects heterozygous for 17,6,1,7 had a fragment of approximately 10kb hybridizing with a DRβ probe as did genomic DNA from the homozygous 17,6,1,7 Tenth Histocompatibility Workshop Caucasoid cell line 9052. By contrast, four of four examples of the Oriental supratype HLA B44,BfS,C4AQO,B1,DR7 (1) and all examples of Caucasoid HLA B44,BfF,C4A3,B1,DR7 tested lacked the 10kb DR β/Hind III fragment. Further, genomic DNA from one Oriental and two Caucasoid examples of 17,6,1,7 (including cell line 9052) was subject to PFGE Southern blotting. Following digestion with MluI and PvuI all examples of 17,6,1,7 had fragments of sizes approximately 180kb and 730kb respectively hybridizing with a TNF probe. Futhermore, all examples of 17,6,1,7 showed an insertion of some 70kb within the class II region (Tokunaga et al, this volume). DQ typing of EBV-transformed cell lines or B-cell enriched peripheral blood lymphocytes revealed that four of the four Thai/Chinese subjects with 17,6,1,7 were DQw3 positive and DQw2 negative, as was the cell line 9052. Further, all examples of 17,6,1,7 from the different ethnic groups were DQw3.2 positive (3).

Immunobiology of HLA
Volume II

Tenth International Histocompatibility Workshop core sera 1176, 1177, 1179, 1188, 3066, 3111, 3112, 3114, 3115, 9057, and 9058 were used to assign DQw3, whereas core sera 1217, 3086, 3088, 3090, 3122, 9055, and 9056 were used to assign DQw3.2. Thus far, it is not possible to distinguish between 17,6,1,7 in Caucasoids and Orientals.

From these results it seems likely that 17,6,1,7 is an ancient supratype that has been maintained in fixed form throughout the process of raciation. Even if present in Orientals because of recent racial admixture, 17,6,1,7 must be extremely stable to achieve such high population frequencies. We are not aware of any other supratype that is similar in these respects.

Portions of the HLA A1,B8,BfS,C4AQO,B1,DR3 (1,8,3,) supratype are found in Oriental patients with autoimmune disease (4) but the population frequency is much lower than for 17,6,1,7. Further, we frequently find evidence of fragmentation of 1,8,3 in Europeans suggesting fragility in relatively recent times (Gin et al, this volume). Similarly, HLA Bw46,BfS,C4A4,B2,DRw9 is frequent in the Chinese and is rare (with DRw9) in Japanese. We conclude that 17,6,1,7 is possibly both ancient and stable. Further analysis may provide insight into possible mechanisms such as *cis* interaction and recombination suppression. We predict that different supratypes will differ in these respects.

References

1. Kay PH, Martin E, Dawkins RL, Charoenwong P. Class III gene rearrangements in Thai/Chinese supratypes containing null or defective C4 alleles. Immunogenetics 1988;27: 46–50.
2. Sutton VR, Knowles RW. Defective expression of the DR beta-2 chain in HLA Bw57,DR7,DQ3 homozygous individuals is associated with polymorphic restriction fragment mapping 3′ to the exon encoding of the first external domain of a DR beta chain. Hum Immunol 1986;17:123–124.
3. Williamson J, Horton B, Christiansen FT, Dawkins RL. The relationships between DQW3 splits and supratypes containing DR4,DR7 and DRw9. Tenth International Histocompatibility Workshop Newsletter 1987;6:13–14.
4. Cobain TJ, Kay PH, Martin E, Dawkins RL, Hawkins BR. Immunogenetics of serum IgA deficiency and related autoimmune disease. In Aisawa M (ed): HLA in Asia Oceania, Sapporo, Japan, Hokkaido University Press, 1986, pp 750–754.

Author Affiliations

P.H. Kay, J. Williamson, Katsushi Tokunanga, R.L. Dawkins, F.T. Christiansen, E. Martin, Departments of Clinical Immunology, Royal Perth Hospital, The Queen Elizabeth II Medical Centre and University of Western Australia, Perth, Western Australia; *P. Charoenwong*, Department of Microbiology, Faculty of Medicine, Chulalongkorn Hospital, Bangkok, Thailand

Epitopes on HLA-DQw3 Molecules Recognized by T-Lymphocyte Clones

Knut E.A. Lundin, Gustav Gaudernack, Gunnar Paulsen, Ludvig M. Sollid, and Erik Thorsby

Abstract: The DQ region encodes HLA molecules where three different allo-specificities, DQw1, DQw2, and DQw3, may be recognized by antisera. The DQw3 antigen, found on DR4, -5, -7, -w8, and -w9 haplotypes, can be divided into DQw7 and DQw8 and w9 by restriction fragment length polymorphism (RFLP) analysis and certain monoclonal antibodies (MAb) (1). To study epitopes on DQw8 molecules that may be recognized by T cells, T-lymphocyte clones (TLCs) were generated in a mixed lymphocyte culture (MLC) in a combination identical at the DR loci, but differing for DQw8. Four CD4+ TLCs define epitopes on DQ molecules exclusively expressed by DR4,DQw8 haplotypes, but not by DR7,DQw9 or DRw9,DQw9 haplotypes. When the reactivity of these TLCs is compared with published HLA sequence data, the results indicate that residue 57 of the DQ β chain most probably is involved in the fine specificity of these T-lymphocyte clones.

The HLA glycoprotein cell-membrane molecules are instrumental in antigen activation of T cells because T cells can only recognize antigenic fragments together with "self" HLA molecules. The HLA molecules are extensively polymorphic and the polymorphic parts determine their ability to bind different antigenic fragments as well as interaction sites with the T-cell receptor (2,3). Recent structural studies of HLA class I molecules indicate that polymorphic residues, which are important either in their antigen recognition sites or in the sites interacting with the T-cell receptor, also are important residues in the epitopes being recognized by alloreactive T cells (3). The situation for class II molecules may be similar. Thus, T-cells recognizing allodeterminants on HLA class II molecules may be a powerful tool to investigate functionally important HLA class II polymorphisms.

The HLA class II region encodes at least three series of functional molecules, DR, DQ, and DP. Three different DQ molecules were recognized by alloantisera at the Ninth International Histocompatibility Workshop; DQw1, DQw2, and DQw3 (4). Evidence exists, however, that the DQ series is far more polymorphic (5).

Table 1. Proliferative Responses of T-Cell Clones BL22, BL25, and BL33 Against Original Stimulator (BE) and Some 10th IHWS "Core Cell Lines"

Stimulator Cells				Proliferative Response (%RR)		
WS#	Name	DR	B-LCL alone	BL22	BL25	BL33
	None			796	313	130
	BE	3/4	591	8229	2405	4380
9025	DEU	4	738	659(<0)	407(<0)	499(<0)
9030	JHAF	4	282	703(<0)	725(7)	683(10)
9027	PF97387	4	719	753(<0)	715(<0)	601(<0)
9026	YAR	4	340	6051(72)	2745(126)	2809(87)
9034	SAVC	4	1322	8129(88)	2415(47)	1101(<0)
9031	BOLETH	4	954	9889(119)	4140(173)	4269(118)
9029	WT 51	4	1037	6141(63)	4369(182)	3744(96)
9033	BM 14	4	1645	9529(104)	2691(44)	5742(147)
8024	KT 17	4	1011	8026(91)	2438(67)	3082(72)
9050	MOU	7	345	681(<0)	783(4)	617(5)
9052	DBB	7	345	333(<0)	421(<0)	137(<0)
9071	OLGA	w8	405	380(<0)	350(<0)	214(<0)
9070	LUY	w8	409	410(<0)	448(<0)	414(<0)
9075	DKB	w9	479	817(<0)	908(<0)	660(2)
9074	HID	w9	910	635(<0)	727(<0)	674(<0)
9073	KT 12	w9	558	375(<0)	375(<0)	748(<0)

Values are given as mean cpm of triplicate cultures and % relative response; i.e., relative to the response towards original stimulator (following 10th IHWS protocol procedure). Positive restimulation is underlined.

Alloreactive TLCs were established from MLC. 8×10^6 peripheral blood mononuclear cells (PBM) from responder (BL) were mixed with 2×10^6 3.000 rads gamma-irradiated PBM from the stimulator (BE) in complete medium (CM:RPMI 1640 supplemented with 20% pooled, inactivated human serum, and antibiotics). The cell culture was restimulated after 10 days with 10.000 rad irradiated EBV transformed B-LCLs from the original stimulator. On the third day of restimulation, IL-2 receptor (Tac) positive cells were extracted using monodisperse, paramagnetic particles M-450 (Dynal, Norway) coated with the α-Tac monoclonal antibody (A gift from T. Waldmann). The Tac positive cells were cloned by limiting dilution, using CM supplemented with IL-2 (Biotest, FRG), and B-LCLs from the original stimulating donor as feeder cells. Clones were propagated by weekly restimulation with stimulating donor B-LCLs in IL-2 containing medium, and stored frozen in liquid nitrogen for later expansions. In proliferative assays, 1×10^4 cloned T cells were restimulated with $2{,}5 \times 10^4$ irradiated B-LCLs in CM in triplicate cultures. ^{3}[H]-thymidine was added after 48 hours, and cultures harvested 18 hours thereafter.

Two allelic variants of DQw3 have been defined, DQw7 and DQw8 and w9 (1). DQw7 is characterized by a DQBl specific 3.5 kb Bam HI fragment in RFLP analysis, but can also be defined by a monoclonal antibody (MAb) (6) and by TLCs (7). DQw8 and w9 is defined by an 11.8 kb Bam HI fragment in RFLP studies, and DQw8 is found among 90 to 95% of the DR4 positive insulin dependent diabetes mellitus (IDDM) patients (8–12). No TLCs have previously been reported defining epitopes on DQw8.

We previously reported preliminary MLC data indicating that allostimulatory epitopes were associated with DQw8 (12). To further study T-cell activating epitopes on DQw8, TLCs were established from an MLC between a healthy responder (BL=HLA-A2;B12,15;Cw3; DR3,4; Dw3,4; DQw2,w7) and an IDDM patient as stimulator (BE=HLA-A1; B8,27;Cw2;DR3,4;Dw3,4; DQw2,w8). Three of these TLCs–BL22, 25 and 33, all derived from the same MLC and with the phenotype CD2+, CD3+, CD4+, and CD8−, were expanded for the Tenth International Histocompatibility Workshop (IHWS) T-cell clone class II studies.

They were first tested against 70 10th IHWS homozygous typing cells (HTC). Some of the results obtained in our laboratory are given in Table 1. The results clearly indicate that these three different TLCs recognize epitopes present exclusively on some DR4 haplotypes.

The TLCs were then tested against the original stimulating cells (BE) and other cells giving a positive restimulation response in the presence of different dilutions of various monoclonal antibodies (MAbs) directed against the three different HLA class II series. Data from studies of TLC BL22 are shown in Figure 1, but similar data were obtained for BL25 and BL33. The results demonstrate that only MAbs directed at DQ molecules were able to inhibit the response.

The TCL BL33 and a fourth clone obtained from the same MLC, BL23, originally gave high responses towards original stimulator cells (data not shown). However, following further expansion, freezing and thawing, they required exogenous Interleukin-2 (IL-2) or phorbol myristate acetate (PMA) (13,14) to be able to proliferate in the presence of stimulating cells (Table 2). IL-2 or

Figure 1. Various anti HLA monoclonal antibodies were tested for their ability to inhibit the proliferative response of TLC BL 22. In this experiment, HTC #9031 BOLETH was used as stimulator, and MAbs used at 1:400 dilution of ascites. ^{3}[H]-thymidine incorporation for TLC alone in CM was 354±31 cpm, for B-LCL alone 954±60 cpm. *Methods*: 2,5×10^{4} irradiated B-LCLs were incubated in CM containing the MAb in question for 30 minutes prior to the addition of 1×10^{4} cloned T cells, and the MAb was present throughout the proliferative assay. MAbs shown here are 10th IHWS MAbs L243(DR), Leu10(DQw1,w3), IA3(DQ), Tu 22(DQ), Tu 35(DR,DP), and 2D6(DR,DP).

►

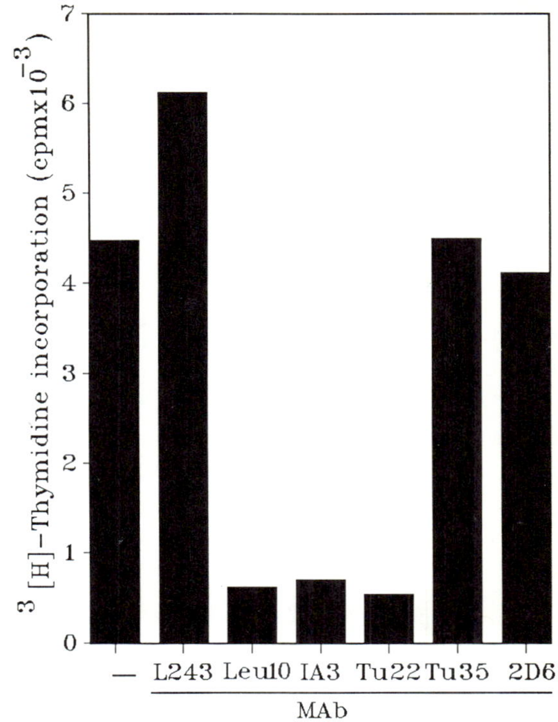

PMA alone or in the presence of autologous stimulating cells (BL) gave only a weak proliferative response.

To be better able to define the DQ epitope being recognized, the four TLCs were then tested against a selected panel of 46 HTCs from the 10th IHWS "core cell panel." BL22 and BL25 were tested with the standard 10th IHWS protocol; BL23 and BL33 also using IL-2 or PMA as costimulating agent. The HTCs used as stimulating cells were typed with some DQ-specific MAbs in simple, rosette forming assay. DNA from the same HTCs were also investigated in RFLP analysis using a DQB1 specific cDNA probe and the restriction enzyme Bam HI. DQw7 was said to be present when the HTC reacted with MAbs 17.15 (anti DQw3) and PLM 2 (anti TA 10), and a 3,5 kb fragment was seen in RFLP analysis. DQw8 or w9 was said to be present when MAbs 17.15 and IIB3 gave positive results, and a 11.8 kb fragment was seen in RFLP analysis.

The overall results are given in Table 3. It can be seen that all four TLCs tested only responded to HTCs carrying the DR4,DQw8 haplotype. None of the TLCs were able to be restimulated with HTCs carrying the DQ9 specificity (defined as given above) on DR7 or DRw9 haplotypes. TLC BL22, BL23, and BL25 seem to have an identical specificity, whereas BL 33 apparently has a slightly more narrow specificity because it was not restimulated (in repeated experiments) with # 9034 SAVC.

Sequences in the first domain of the HLA class II α and β chains have been shown to be essential for T-cell recognition (15). Sequence data from DR4,DQw7 and

Table 2. Activation Requirements of T-Cell Clones BL23 and BL33

	DNA Synthesis (incremental cpm)*					
	BL23			BL33		
Stimulating B-LCL	Alone	+IL-2	+PMA	Alone	+IL-2	+PMA
None	158	373	143	75	2185	83
Orig. stim. (BE)	316	9518	12203	1090	12400	3746
Autologous (BL)	0	817	244	0	1549	0
#9031 BOLETH	1798	17800	16036	4380	14879	7007
#9026 YAR	3250	17642	16919	2010	16331	7205
#9034 SAVC	1466	14256	17940	82	2783	0
#9030 JHAF	0	539	0	0	2356	46
#9075 DKB	26	565	0	4	3399	0

*Mean cpm of triplicates − cpm of B-LCL cultured alone.
TLCs BL23 and BL33 were propagated with weekly restimulation with original stimulator B-LCLs in CM enriched with 5U/ml recombinant IL-2 (Amersham, UK). Proliferative assays were performed seven days after the last addition of feeder cells. 2×10^{4} cloned T cells were restimulated with 2×10^{4} irradiated B-LCLs in CM alone, or in CM with the addition of 5U/ml rec IL-2, or PMA (Sigma, USA) to a final concentration of 10^{-8} M. ^{3}[H]-thymidine was added after 48 hours, and cultures harvested 18 hours thereafter.

Table 3. Reactivity of T-Cell Clones BL22, BL23, BL25, and BL33 Against Various HLA Homozygous Stimulating Cells

WS#	Cell ID	DR	17.15	TA10	2B3	Bam HI	DQ	BL22	BL23	BL25	BL33
9003	KAS 116	1	−	−	+	nt	w1	−	−	−	−
9004	JES THOM	1	−	−	−	nt	w1	−	−	−	−
9017	WT 8	2	−	−	+	nt	w1	−	−	−	−
9011	E 4181324	2	−	−	+	nt	w1	−	−	−	−
9023	VAVY	3	−	−	−	nt	w2	−	−	−	−
9019	DUCAF	3	−	−	−	nt	w2	−	−	−	−
9025	DEU	4	+	+	−	3,5	w7	−	−	−	−
9030	JHAF	4	+	+	−	3,5	w7	−	−	−	−
9027	PF 97387	4	+	+	−	3,5	w7	−	−	−	−
9031	BOLETH	4	+	−	+	11,8	w8	+	+	+	+
9026	YAR	4	+	−	+	11,8	w8	+	+	+	+
9034	SAVC	4	+	−	+	11,8	w8	+	+	+	−
9029	WT 51	4	+	−	+	11,8	w8	+	+	+	+
9032	BSM	4	+	−	+	11,8	w8	/	+	/	+
9033	BM 14	4	+	−	+	11,8	w8	+	+	+	+
9024	KT17	4	+	−	+	11,8	w8	+	+	+	+
9038	BM 16	5	+	+	−	3,5	w7	−	−	−	−
9039	JVM	5	+	+	−	3,5	w7	−	−	−	−
9060	CBG-IB	6	−	−	+	nt	w1	−	−	−	−
9059	SLE-005	6	−	−	+	nt	w1	−	−	−	−
9050	MOU	7	−	−	−	nt	w2	−	−	−	−
9052	DBB	7	+	−	+	11,8	w9	−	−	−	−
9071	OLGA	w8	−	−	+	11,8	−	−	−	−	−
9070	LUY	w8	+	+	−	3,5	w7	−	−	−	−
9075	DKB	w9	+	−	+	11,8	w9	−	−	−	−
9074	HID	w9	+	−	+	11,8	w9	−	−	−	−
9073	KT 12	w9	+	−	+	11,8	w9	−	−	−	−

/ = Technical failure; nt = not tested.

A synopsis of HTC panel investigations with TLCs BL22, BL23, BL25, and BL33. For BL22 and BL25, experiments were performed as in Table 1. For BL23 and BL33, the experiments were performed as in Table 2. Positive (+) and negative (−) restimulation responses were of a magnitude similar to that shown in Tables 1 and 2. DQ types were determined in a simple rosette forming assay (G. Gaudernack, manuscript in preparation); monodisperse particles M-450 (Dynal, Oslo, Norway) were coated with various monoclonal antibodies and allowed to react with B-LCLs in the wells of a 96-well round-bottomed plate. A small aliquot was observed under the microscope, and the test said to be positive when more than 70% of the cells formed rosettes. MAbs used here were 10th IHWS MAbs 17.15 (anti DQw3), PLM2 (anti TA10 (DQw7)), and IIB3 (anti-DQ, not TA10). The Bam HI kb fragments were determined in RFLP tests, performed as previously described (12).

DR4,DQw8 haplotypes show that there are four aminoacid differences in the first domain of their DQβ chains, whereas the α chains are identical in the first domain (10,16,17). When comparing the DQβ sequence of the DR4, DQw8 cell line BOLETH (16) with the published sequence for the DRw9,DQw9 cell line DKB (17) (Table 3), the only difference in the α and β chain first domain is a β chain Ala to Asp substitution at position 57. Because our TLCs do not respond to cells expressing the DR4, DQw7, or the DRw9, DQw9 haplotypes (Table 3), the data strongly suggest that residue 57 of the DQβ chain is critical for the epitope being recognized by these T-cell clones. Ala at this position seems, however, not to be sufficient alone. Cells expressing DQw2, which has Ala at position 57 of its β chain, but differs at several nearby positions as well as in the α chain, are nonstimulatory. The explanation that TLC 33 is not restimulated by HTC # 9034 SAVC, which induce restimulation responses of the other three TLCs, is unknown and must await DQ sequence data of this and the other DR4, DQw8 HTCs.

We conclude that all four TLCs investigated recognize an epitope on DQ molecules, where Ala at position 57 on the β chain plays a critical role. Ala at this position also may contribute to susceptibility to IDDM (17), thus the T-cell clones may recognize an epitope of functional importance for IDDM.

Acknowledgments. This work was supported by The Norwegian Cancer Society. We thank Anne Brit Thoresen for technical assistance. Participants of the 10th International Histocompatibility Workshop are acknowledged for supplying HTCs an monoclonal antibodies.

Knut E.A. Lundin, Gustav Gaudernack, and Gunnar Paulsen are research fellows of the Norwegian Cancer Society

References

1. Kim SJ, Holbeck SL, Nisperos B, Hansen JA, Maeda H, Nepom GT. Identification of a polymorphic variant associated with HLA-DQw3 and characterized by specific

restriction sites within the DQ β-chain gene. Proc Natl Acad Sci USA 1985;82:8139.
2. Buus S, Sette A, Grey HM. The interaction between protein-derived immunogenic peptides and Ia. Immunol Reviews 1987;98:115.
3. Bjorkman PJ, Saper MA, Samraoui B, Bennett WS, Strominger JL, Wiley DC. The foreign antigen binding site and T cell recognition regions of class I histocompatibility antigens. Nature 1987;329:512.
4. Albert ED, Baur MP, Mayr WR, (eds): Histocompatibility Testing. Heidelberg, Springer-Verlag, 1984.
5. Trucco M, Duquesnoy RJ. Polymorphism of the HLA-DQ subregion. Immunology Today 1986;7:297.
6. Maeda H. Mouse monoclonal antibody detects a new polymorphic Ia determinant other than HLA-DR antigen: A possible allele of DC-1. Tissue Antigens 1984;23:163.
7. Pawelec G, Wernet P. HLA-DQw3 dissected by cloned alloreactive T cells. Immunogenetics 1987;25:356.
8. Nepom BS, Palmer J, Kim SJ, Hansen JA, Holbeck SL, Nepom GT. Specific genomic markers for the HLA-DQ subregion discriminate between DR4+ insulin-dependent diabetes mellitus and DR4+ seropositive juvenile rheumatoid arthritis. J Exp Med 1986;164:345.
9. Cohen-Haguenauer O, Robbins E, Massart C, Busson M, Deschamps I, Hors J, et al. A systematic study of HLA class II-β DNA restriction fragments in insulin-dependent diabetes mellitus. Proc Natl Acad Sci USA 1985;82:3335.
10. Michelsen B, Lernmark Å. Molecular cloning of a polymorphic DNA endonuclease fragment associates insulin-dependent diabetes mellitus with HLA-DQ. J Clin Invest 1987;79:1144.
11. Böhme J, Carlsson B, Wallin J, Möller E, Persson B, Peterson PA, et al. Only one DQ-β restriction fragment pattern of each DR specificity is associated with insulin-dependent diabetes. J Immunol 1986;137:941.
12. Bruserud Ø, Paulsen G, Markussen G, Lundin K, Thoresen AB, Thorsby E. Genomic HLA-DQβ polymorphism associated with insulin-dependent diabetes mellitus. Analysis of possible functional significance. Scand J Immunol 1987; 25:235.
13. Minami M, Ebner SA, Stadecker MJ, Dorf ME. The effects of phorbol ester on alloantigen presentation. J Immunol 1987;138:393.
14. Goronzy J, Weyand C, Imboden J, Manager B, Fathman CG. Heterogeneity of signal requirements in T cell activation within a panel of human proliferative T cell clones. J Immunol 1987;138:3087.
15. Cairns JS, Curtsinger JM, Dahl CA, Freeman S, Alter BJ, Bach FH. Sequence polymorphism of HLA DRβ1 alleles relating to T-cell-recognized determinants. Nature 1985;317:166.
16. Larhammar D, Hylding-Jensen JJ, Servenius B, Andersson G, Rask L, Peterson PA. Exon-intron organization and complete nucleotide sequence of a human major histocompatibility antigen DCβ gene. Proc Natl Acad Sci USA 1983;80:7313.
17. Todd JA, Bell JI, McDevitt HO. HLA-DQβ gene contributes to susceptibility and resistance to insulin-dependent diabetes mellitus. Nature 1987;329:599.

Author Affiliation

Institute of Transplantation Immunology, The National Hospital, Rikshospitalet. Oslo, Norway

DQw3 Subtypes Linked to DRw9 Specificity: Biochemistry and RFLP Definition

G. Semana, F. Quillivic, O. Bouhallier, A. Gaulier, D.J. Charron, and R. Fauchet

Class II genes of the HLA D region of chromosome 6 include three families: DR, DQ and DP. Within each locus, a polymorphism has been described for most of those genes, especially β genes but also DQα genes. This generates an extensive polymorphism of HLA class II molecules.

The aim of this study is to define through serologic, biochemical, and RFLP techniques, the DRw9 haplotype and particularly the DR DQ linkage disequilibrium. The DRw9 specificity, rare in Caucasian (0.20%) and more frequent in the Japanese population (23%), is associated with the DQw3 allele. We studied a family of four subjects with a DRw9 homozygous member and seven unrelated DRw9 subjects. All cells were HLA typed for A,B, DR, DQ loci including the DQw3 Ta10 subdivisions (two Ta10 MoAbs and six Ta10 allosera were used). The DQw3 molecule was isolated by two MoAbs (Leu 10 and II B 3) and then analysed on two-dimensional electrophoresis (NEPHGE-SDS). DNA was digested by six restriction enzymes and hybridized with DQα and DQβ probes. RFLP were compared to a panel of 15 homozygous cells (DR1 to DRw9). All the DRw9 cells are serologically defined as DQw3 Ta10−.

Immunoprecipitation with LEU 10 MoAb (DQw1, DQw3) revealed two DQw3 β chain patterns: a family of three basic spots associated with DRw9 and DR4 (Ta10−) and another family of more acidic spots associated with DR5 and DR4 (Ta10+). II B 3 MoAb, considered as recognizing the DQw3, Ta10− variant, only immunoprecipitated the DQ β chain associated with DRw9 and DR4 (Ta10−) with the same pattern than LEU 10.

This serologically and biochemically defined DQw3 polymorphism (1) was confirmed by RFLP studies. Indeed with four enzymes (Hind III,Bgl II, Eco RI, and Bam HI) hybridization with the DQ β probe separated DQw3 into two variants (Table 1) perfectly correlated

Table 1. DQw3 Hybridization Revealed Two Ta10 Correlated DQw3 Variants

	DQw3 (DR5) Ta10+	DQw3 (DR4) Ta10+	DQw3 (DRw9) Ta10−	DQw3 (DR4) Ta10−
Hind III	3.4 Kb+	3.4 Kb+	3.4 Kb−	3.4 Kb−
Bgl II	4.1 Kb+	4.1 Kb+	5.7 Kb+	5.7 Kb+
	5.7 Kb−	5.7 Kb−	4.1 Kb−	4.1 Kb−
Eco R I	11.3 Kb+	11.3 Kb+	11.3 Kb−	11.3 Kb−
Bam H I	12 Kb−	12 Kb−	12 Kb+	12 Kb+
	3.3 Kb+	3.3 Kb+	3.3 Kb−	3.3 Kb−

with Ta10 subdivisions (2). In all cases, DRw9 cells yielded the DQw3 Ta10− pattern.

Otherwise, hybridization with DQα probe of the DNA digested by Bgl II revealed a polymorphism in the studied family (DRw9 13.1 Kb+, DRw9 13.1 Kb −). This polymorphism could correspond to a cross-hybridization with DXα locus genes. As a conclusion, DRw9 is associated with the DQw3 Ta10− variant, which is generated by the DQ β gene polymorphism (3). The DQ α probe hybridization demonstrated a polymorphism within this DRw9 specificity. Assuming that this polymorphism involves the encoding segments of the DQα genes, its presence would suggest that there are two DQ heterodimers associated with DRw9.

References

1. Radka SF, Scott RG, Stewart SJ. Molecular complexity of HLA DQw3: The Ta10 determinant is located on a subset of DQw3 β chains. Human Immunology 1987;18:287–300.
2. Kim SE, Holbeck SL, Nisperos B, Hansen JA, Maeda H, Nepom GT. Identification of a polymorphic variant associated with HLA-DQw3 and characterized by specific restriction sites within the DQ β chain gene. Proc Natl Acad Sci USA 1985;82:8139.
3. Maeda H, Hirata R, Thompson A, Mukai R. Molecular characterization of three HLA class II molecules DR4 and DRw9 haplotypes. Human Immunology 1986;15:1–15.

Author Affiliations

G. Semana, F. Quillivic, O. Bouhallier, A. Gauthier, R. Fauchet, Laboratoire d'histocompatibilité, Centre Régional de Transfusion Sanguine, Rennes; *D.J. Charron*, Laboratoire d'immuno-génétique—Institut des Cordeliers, Paris, France

A New DR Antigen, Associated with DQw7, Defined Serologically and by Restriction Fragment Length Polymorphism

Lorna J. Kennedy, Susan Tonks, Steven G.E. Marsh, J. Alejandro Madrigal, and Julia G. Bodmer

The inheritance of a new DR antigen, COR, was studied in a two-generation family. The COR antigen has been inherited by two of four children clearly segregating with Al Cw7 B7 and DQw7. This family has now been typed in the last three International Histocompatibility Workshops, and only in the recent Tenth Workshop have any sera or monoclonal antibodies with COR reactivity been identified. These include 10W1008 (DR1++), 10W3010 (DR1+DR7+DR9+), and 10W3011 (DR1,2,7,9,w10,+), which all also react with DR"BON" (A. Cambon-Thomsen, personal communication).

To determine whether the COR antigen is expressed on the cell surface, two-dimensional gel electrophoresis was performed for each member of the family plus appropriate homozygous controls (data not shown). Two sets of DRβ spots were found for each family member, indicating that the COR gene is expressed. The COR pattern of DR β spots closely resembles those of a homozygous DR"BR" B cell line.

Restriction fragment length polymorphism studies were then undertaken to look for a unique banding pattern corresponding to COR. Using an EcoR I digest hybridized with a DR β probe, a 9.8 kb band was seen to segregate in the family on the same haplotype as COR (see Fig. 1).

These data indicate that COR may be similar to DR"BON" (1) or DR"BR" (2,3) (which may also be equivalents). The RFLP data are showing a unique band for COR which has to be confirmed.

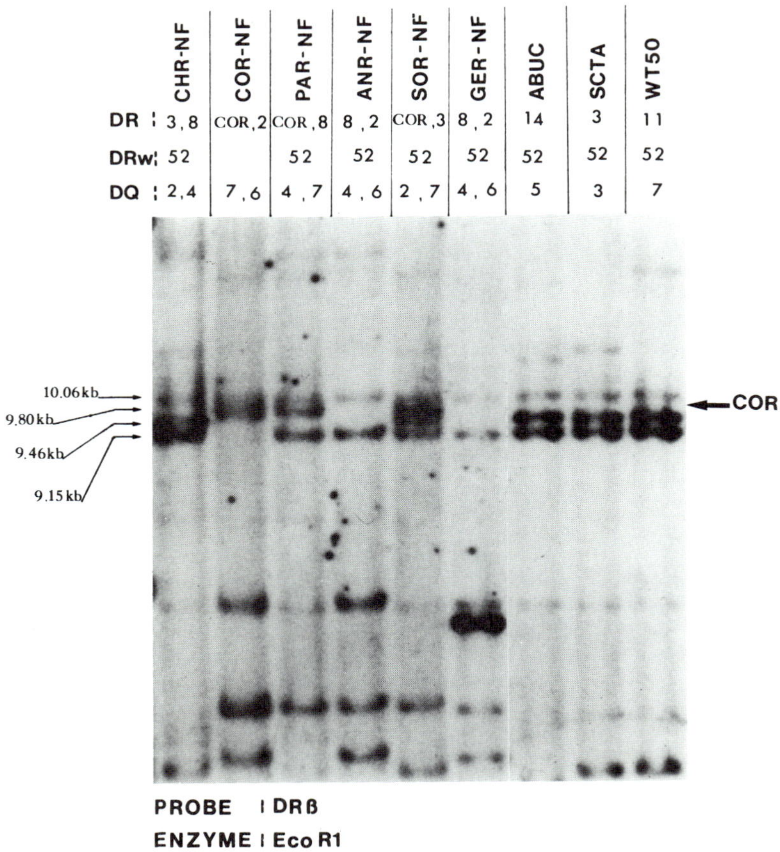

Figure 1. An EcoR I digest of the family RED, hybridized with a DR β probe, showing a specific 9.8 kb band segregating with COR. The first six lanes contain the father (a: A1,Cw7,B8,DR3, DQw2; b: A2,Cw2,B27,DRw8, DQw4), mother (c: A1,Cw7,B7, COR, DQw7; d: A2,Cw7,B8, DR2,DQw1) and four children with haplotypes bc, bd, ac, and bd, respectively. The final three lanes show three control cell lines that distinguish a 10.06 kb band from the 9.80 kb COR band.

Immunobiology of HLA
Volume II

References

1. Coppin HL, Villedieu P, Thomsen M, Giraud P, Cambon-Thomsen A, Charron D. HLA-D/DR-BON: a New Biochemically and Cellularly Defined Polymorphism Indistinguishable by RFLP from DR1. Immunogenetics 1987;25:130–132.
2. Bidwell JL, Jarrold EA, Laundy GJ, Klouda PT, Bradley BA. Molecular genetics of HLA-DRβ2: allogenotypes of DR1 and DR Br are indistinguishable. Tissue Antigens 1986;27:99–101.
3. Middleton D, Savage DA, Bidwell JL, Cullen C. Association of HLA-DRBr with HLA-DQw3. Tissue Antigens 1988;31:79–82.

Author Affiliations

Tissue Antigen Laboratory, Imperial Cancer Research Fund, London, England

Biochemical and Genomic Characterization of HLA-DQ Gene Products Associated with DR3, DR4, and DR5 Haplotypes

Dimitri S. Monos, Massimo Trucco, Richard S. Spielman, Susan F. Radka, Chester M. Zmijewski, and Malek Kamoun

The polymorphism of HLA antigens detected by various methods (i.e., conventional serology, monoclonal antibodies, T-cell clones, electrophoresis, or restriction fragment analysis) reflects differences among cells at the DNA level, at the protein level, or at the assembly/post-translational level. Correlation of the polymorphisms identified by these individual procedures provides insight into the structural/functional relationship of these antigens.

In this report, the polymorphism of the DQ α and β chains associated with DR3, DR4, and DR5 haplotypes of HLA-D homozygous cell lines was examined electrophoretically and related to the restriction fragment polymorphism identified after digestion with Pst I and Southern blotting. As shown in Table 1, the DQ α chains from DR3 and DR5 haplotypes were identical (DQ α 1), but could be distinguished from that of the DR4 (DQ α 2) by two-dimensional electrophoresis (2D-PAGE). Correspondingly, hybridization with a DQ α probe to genomic DNA from DR3 and DR5 revealed identical RFLP patterns; in particular, a 4.1-kb fragment common to both, but not present in the DR4 cell lines. The RFLP pattern among the DR4 cell lines was characterized by a 2.2-kb fragment. Recent reports demonstrate that the DQ α alleles described at the DNA level correspond to the polymorphic determinants defined serologically and by T-cell clones (1–3). Additional heterogeneity of DQw3 α chains of DR4 haplotypes has been recently defined by 2D-PAGE, but not by RFLP (4).

Table 1. Correlation Between DQ α or β Pst I Restriction Fragments and DQ α or β Chain Variations Among DR3, DR4, and DR5 Haplotypes

	DR3	DR5	DR4
DQα probe	4.1 Kb	4.1 Kb	2.2 Kb
DQα chain	DQα1	DQα1	DQα2
DQβ probe	5.5 Kb	5.0 Kb	5.0 Kb
DQβ chain	DQw2β	DQw3.1	DQw3.1 or DQw3.2
SFR20-DQβ5	−	+	+(DQw3.1) or −(DQw3.2)

For the biochemical determination of DQ α chains, the following cell lines were used: DR3 (QBL, WT20), DR4 (HA,108-4), DR5 (JVM,RM8).
For the biochemical determination of DQ β chains, the following cell lines were used: DR3 (QBL,WT20), DR4 (HA,108-4, ER,KT2,TAS,FS), DR5 (JVM,RM8).
DNA was extracted, digested with Pst I, and hybridized with DQ α or DQ β probes from the following cell lines: DR3 (WT20,RM2,RM3,RM5,RM15,RM7,WT49), DR4 (WT51, EM,KT2,KT3,HA,THO,TAS), DR5 (JVM,WT50,RM13, RM8).
All cell lines were tested for reactivity with SFR20-DQβ5.

Concerning the identification of DQ β chains. three patterns of DQ β chains were identified. The most basic DQ β chain was found with the DR3 haplotype. A second pattern, which was the most acidic, was found with DR5 a and subgroup of DR4 haplotypes and was designated DQw3.1. The third pattern, designated DQw3.2, was found with another subgroup of the DR4 haplotype. Furthermore, from the restriction fragment analysis with a DQ β probe, it appears that the 5.5-kb Pst I fragment of the DR3 haplotype is associated with the DQw2 β chain, while the 5.0-kb fragment of the DR4 and DR5 haplotypes is associated with the DQ β chain of the DR5 and the DQw3.1 or DQw3.2 β chain of the DR4 (Table 1). The reactivity of the monoclonal antibody SFR20-DQ β 5 (DQ-TA10 specific) with all the DR5 and the DQw3.1 subtype of the DR4 cell lines further suggests identity between the β chains of DR5 and the DQw3.1 of the DR4. The heterogeneity of DQw3 β chains associated with the DR4 haplotype is undetectable with Pst I, but has been shown with other restriction enzymes such as Bam HI and Taq I (5,6).

These results from 2D-PAGE support the conclusion from RFLPs that the DQ α chain of DR5 is indistinguishable from the DQ α of DR3, and the DQ β chain of DR5 chain is indistinguishable from the DQw3.1 β

Immunobiology of HLA
Volume II

chain of DR4. Thus, the DQ β of DR5 and some DR4 is the same, but it differs from that of DR3. Therefore, it is most likely that the DQ serologic determinants that characterize DR4/DR5 (DQw3) and DR3 (DQw2) reside on the β chains.

Acknowledgments. The authors wish to thank Peter Lo for technical assistance and Nancie-Marie Beck for excellent secretarial work. This work was supported by grants DK 38011, DK 19525 and DK 35047 from the National Institutes of Health and by grant JDF-185321 from the Juvenile Diabetes Foundation.

References

1. Spielman RS, Lee J, Bodmer WF, Bodmer JG, Trowsdale J. Six HLA-D region α-chain genes on human chromosome 6: Polymorphisms and associations of DC α-related sequences with DR types. Proc Natl Acad Sci USA 1984;81:3461.
2. Rosenshine S, Cascino I, Zeevi A, Duquesnoy RJ, Trucco M. DQ α and β RFLP reveals the composition of the DQ molecule recognized by T-cell clones. Immunogenetics 1986;23:187
3. Cascino I, Rosenshine S, Turco E, Marrari M, Duquesnoy RJ, Trucco M. Relationship between DQ α and DQ β RFLP and cell surface polymorphisms of class II HLA antigens J Immunogenetics 1986;13:387.
4. Bontrop RE, Baas EJ, Otting N, Schrender MT, Griphart M. Molecular diversity of HLA-DQ. Immunogenetics 1987; 25:305.
5. Kim SJ, Holbeck SL, Nisperos B, Hansen JA, Maeda H, Nepom GT. Identification of a polymorphic variant associated with HLA-DQw3 and characterized by simple restriction sites within the DQ β-chain gene. Proc Natl Acad Sci USA 1985;82:8139
6. Monos DS, Spielman RS, Gogolin KJ, Radka SF, Baker L, Zmijewski CM, et al. HLA-DQw3.2 allele of the DR4 haplotype is associated with insulin-dependent diabetes; Correlation between DQ β restriction fragments and DQ β chain variation. Immunogenetics 1987;26:299.

Author Affiliations

Dimitri S. Monos, Richard S. Spielman, Chester M. Zmijewski, Malek Kamoun, Department of Pathology and Laboratory Medicine, University of Pennsylvania, Philadelphia, PA; *Massimo Trucco*, Pittsburgh Cancer Center, Pittsburgh, PA; *Susan F. Radka*, Genetic Systems, Seattle, WA

Cellular Analyses of DQ Antigens

Y. Tajima, H. Kojima, N. Ishikawa, S. Hawkin, Y. Fukasawa, A. Takada, A. Wakisaka, and M. Aizawa

A large number of structurally polymorphic class II antigens have been demonstrated serologically and biochemically. Recent advances in techniques to generate and propagate clonal population of T lymphocytes have provided a new basis to identify the molecular properties of HLA antigens. Indeed, there have been many reports of cellular analyses of HLA antigens by using allospecific T lymphocytes. However, HLA-DQ antigens associated with DR4 and DRw8 are not always defined as any of the known specificities. By using monoclonal antibody HU-46(10W3101), we proposed a novel DQ specificity, DQWa now designated as DQw4, which filled their blanks.

To investigate the DQ antigen DR4 and DRw8 cells more precisely, alloreactive CTL clones were produced by using the DR4 or DRw8 cells as a stimulator. Five informative CTL clones were obtained. The OW-B6 and OW-B7 were generated against DW15/DR4/DQw4-typed LCL. Panel studies revealed that OW-B7 reacted with only Dw15/DR4-typed LCLs and OW-B6 with Dw15/DR4- and Dw8/DRw8-typed ones (Table 1). The reactivity of OW-B7 was inhibited by anti-DR MoAb (HU-20) and OW-B6 by anti-DQ MoAb HU-46 in a dose-dependent manner (Table 2). These results revealed that OW-B7 recognized DR4 molecules specific to Dw15 and that OW-B6 recognized DQw4 determinant, which is common to DR4- and DRw8-typed LCLs. Reactivity of OW-B6 was in accordance with that of HU-46(10W3101).

On the other hand, TGAF-3 and TERD-2, which were generated against DW8/DRw8- and Dw4/DR4-typed LCLs respectively, reacted with DW8/DRw8/DQw4-typed LCLs without any cross reaction with DW15/DR4/DQw4-typed LCLs. Because their reactivity was inhibited by anti-DQ MoAbs, they were thought to discriminate DQw4 of DRw8 LCLs from that of DR4 LCLs. This result was confirmed by the fact that 2-D profiles of HU-46 exhibited different electrophoretic patterns of DQw4 between DR4 and DRw8. Moreover, further splits of DQw4 were suggested because the reactivity of TGAF-3 and that of TERD-2 were mutually exclusive against DRw8/DQw4 panels. By using these two CTL clones, DW8/DRw8 could be subdivided into three groups, Dw8.1, Dw8.2 and Dw8.3, corresponding to D-typing. In other words, Dw8.1 was defined by TGAF-3, Dw8.2 by TERD-2, and Dw8.3 was negative with both CTL clones. These facts indicated that molecular heterogeneity of DQw4 antigen plays a crucial role for D-Typing among Dw8-typed LCLs. The TERD-2 also reacted with DR4/DQw3-typed cells, indicating a shared DQ determinant among DR4/DQw3

Table 1. Cytotoxicity of CTL Clones

				% Cytotoxicity		
Cell Lines	D	DR	DQ	OW-B6	OW-B7	5-44
EBV-Sa	w1,w1	1,1	w1,w1	–	–	2.9
EBH-Hi-4	w1,w1	1,1	w1,w1	–	–	1.1
EBV-Wa	w15,w15	4,4	w4,w4	67	71	7.8
L-KT9	w15,w15	4,4	w4,w4	34	62	6.5
EBV-Fuk	w1,w15	1,4	w1,w4	56	64	37.1
EBV-Wak	w1,w15	1,4	w1,w4	72	68	24.1
EBV-OS-1	w1,w4	1,4	w1,w3	–	–	6.1
KT2	KT2,KT2	4,4	w3,w3	7	22	–
L-KT13	KT2,KT2	4,4	w3,w3	11	–	–
ER	w4,w4	4,4	w3,w3	9	22	–
WALK	w4,w4	4,4	w3,w3	11	9	–
LEO-172	w4,w4	4,4	w3,w3	−5	−1	–
DA	w10,w10	4,4	w3,w3	–	24	–
GI	w8,w8	w8,w8	w4,w4	46	18	–
K562	–	–	–	–	–	7.5

and DRw8/DQw4 typed LCL which have been undefined previously.

The 5-44 stimulated with DR1-DQw1/DR4-DQw4 heterozygous LCL lysed only DQw1/DQw4 heterozygotes, but none of the DQw1 or DQw4 homozygotes. Its cytotoxicity was inhibited by HU-46. If DR1/DQw1 and DR4/DQw4 homozygotes are thought to be parents, DR1-DQw1/DR4-DQw4 heterozygote is considered as F1. Therefore, 5-44 was supposed to recognize a novel determinant that appeared on the hybrid molecules composed of DQw1 and DQw4. The results that HU-46 precipitated two α chains and one β chain of the DQ molecule from DQw1/DQw4 heterozygotes indicated HU-46 recognized the β chain of DQw4. Because the reactivity of 5-44 was inhibited by HU-46, 5-44 was shown to recognize the hybrid molecule composed of both α chain from DQw1 and β chain from DQw4.

Table 2. Reactivity of DRw8-LCLs

Cell ID	A	B	D	DR	DQ	HU-46	TGAF-3	TERD-2
MADURA	2	40	w8.1	w8	w4	+	*106.9%	39.6
							78.5	34.4
BTB	2	35	w8.1	w8	w4	+	115.6	41.0
							84.7	73.3
BM9	2	35	w8	w8	w4	+	40.0	74.5
							63.8	106.4
OLG	31	62	w8.2	w8	w4	+	23.6	98.5
							29.0	111.7
OLL	31	62	w8	w8	w4	+	32.2	87.8
							38.4	94.0
SPACH	31	62	w8.2	w8	w4	+	75.6	102.3
							37.8	125.2
SPL	31	62	w8	w8	w4	+	32.2	106.8
							16.8	85.3
TAB	2	46	w8.3	w8	w1	–	3.1	2.3
							10.4	21.4
LUY	2	51	w8.3	w8	w3	–	7.8	18.6
							10.3	34.9

*Relative cytotoxicity which is calculated by the formula of

$$\frac{\text{\% cytotoxicity against experimental cell}}{\text{\% cytotoxicity against immunizing cell}} \times 100$$

References

1. Ishikawa N, Kojima H, Nakayama T, Kunikane H, Hawkin S, Fukasawa Y, et al. Detection of a novel HLA-DQ specificity: Serological and immunohistochemical analyses by a monoclonal antibody. Immunogenetics 1987;26:143–149.
2. Kojima H, Fukasawa Y, Ishikawa N, Tajima Y, Wakisaka A, Aizawa M. Detection of a novel HLA-DQ specificity II, Cellular analysis by cytotoxic T-cell clones. Immunogenetics 1988;27:145–147.

Author Affiliation

Department of Pathology, Hokkaido University School of Medicine, Sapporo, Japan

DQw4: A New Allele of HLA-DQ

Lori Dombrausky, Edward J. Ball, Carol Long, Laura Gentry, Jodee Sedalnick, and Peter Stastny

With the definition of the alleles DQw1, 2, and 3, there remains a high proportion of DQ-"blank" haplotypes (24% Caucasian, (9th Workshop central analysis). This was particularly so in haplotypes such as DRw8 (79%) and DR3 (42%). The antibody HU46, (9W921, 10W3101) contributed by Aizawa et al. (1,2), was found to define a new DQ specificity associated with DR4, Dw15, and DRw8. Recently an alloantiserum recognizing this antigen has also been described (3).

Ten panel cells were DQw4-positive. Four workshop B-cell lines were DQw4+ with 10W3101 in the 10WS ELISA protocol. All DQw4+ cells were also positive for the new DQ-IIB3 cluster (10W1217, 3088, 3086, 3090, 3122, 9055, 9056). Thus, this cluster of reagents is useful for the definition of DQw4 in some cases.

All of our DQw4+ cells could be accounted for by DR3 (5 cells) and DRw8 (9 cells), all of which had formerly been considered DQ-"blank." Considering DQw1,

Immunobiology of HLA
Volume II

Table 1.

HLA Antigens					DQ-β Bgl II (kb)		DQ-α Bgl II (kb)		
DR	DQw4	DQw3	DQw7	IIB3	5.6	4.1	12.2	7.9	3.4
8	+	−	−	+	+	−	+	−	+
3	+	−	−	+	+	−	NT	+	−
4	−	+	−	+	+	−	−	−	−
4	−	+	+	−	−	+	−	−	−

Table 2. Correlation of DQB/BGL II RFLP with WA and DQw3

RFLP Band (kb) vs Antigen	++	+−	−+	−−	X^2Yates	r
4.1/DQw7	7	1	0	20	18.9	.82
5.6/DQw3+TN	12	7	1	9	5.5	.44
5.6/TN-DQw4+	6	10	1	7	.6	.16
5.6/DQw3+ DQw7-or DQw4+	18	0	2	6	13.6	.72

2, 3 and Wa as alleles, no "triplet" were observed (95 observations.

Southern blot analysis was done on DNA from 10WS B-cell lines and local cells using the enzyme BgI II and DQ-α (DQA) or DQ-β (DQB) probes (Tables 1 and 2). All DQw4+ and DQw3TN (TA10-negative) cells had a 5.6 kb fragment hybridizing to the DQB probe. A 4.1 kb band was allelic to the 5.6 kb band and appeared only in DQw7 (TA10-positive) cells. Similar distinctions were also made with Taq I enzyme. The DQA probe and BgI II showed two different patterns associated with, but not unique to DQw4+ cells. The same pattern was seen in DQw4 and DQw3 DR8 cells. These data suggest that DQw4 is encoded by DQB rather than DQA.

DQw4 appears to be allelic to DQw1, w2, and w3. It is completely included in the new specificity IIB3, which also includes DQw1 and DQw3TN.

References

1. Bodmer J, Kennedy L, Aizawa M, Dawkins R, Lepage V, Mazzilli M, et al. HLA-D region monoclonal antibodies. In: Albert ED, Bowr MP, Mayr WR (eds): Histocompatibility Testing 1984. Berlin, Heidelberg: Springer-Verlag 1984:217.
2. Kunikane H, Ishikawa N, Nakayama T, Fukasawa Y, Kojima H, Hawkin S, et al. HLA-DQWA: A new DQ specificity recognized by monoclonal antibody. Third Asia-Oceania Histocompatibility Workshop Conference Abstract, 1986.
3. Matsuki K, Juji T, Satake M. A novel HLA class II alloantigen detected by an alloantiserum TK2. Human Immunol 1987;18:247.

Author Affiliation

Department of Internal Medicine, University of Texas Health Science Center at Dallas, Southwestern Medical School, Dallas, TX, USA

Comparative Analysis of the Two-Dimensional Gel Profiles of the Monoclonal Antibodies Tu22, IA3, and SPV-L3

Ronald E. Bontrop, Nel Otting, and Marius J. Giphart

HLA class II molecules were isolated from a large panel of Epstein Barr virus-transformed B-cell lines by immunoprecipitation with the monoclonal antibodies Tu22, IA3, and SPV-L3. The radiolabeled immunocomplexes were analyzed by two-D gel electrophoresis. Such assays demonstrated that SPV-L3 (not used in this Workshop; therefore, see reference 1) is truly locus-specific and isolates only one type of HLA-DQ molecules from HLA-DR-DQ homozygous B-cell lines that originate from consanguineous origin.

In contrast to SPV-L3 reactivity, IA3 and Tu22 do not have a DQ locus-specific reaction profile. Mixing experiments showed that IA3 and especially Tu22 are reactive with HLA-DQ and HLA-DR molecules. On most IA3 and Tu22 two-D profiles, DQ-alpha, DQ-beta and DR-alpha and -beta spots are detected. In some examples Tu22 does even precipitate HLA-DP structures (Fig. 1). The amount of HLA-DR molecules isolated appears to be dependent on the HLA-DR-DQ serotype of the cell lines studied. For example, from HLA-DR5 and HLA-DRw6 positive B-cell lines, tiny amounts of HLA-DQ molecules are isolated with Tu22 (Fig. 1), whereas in the example of HLA-DR1-DQw5 positive cell lines, more HLA-DQ than HLA-DR molecules are isolated (Fig. 2). This unexpected binding reactivity may explain why Tu22 is able to inhibit the proliferative response of HLA-DR-restricted T-cell lines or clones (unpublished observations). For that reason, it is speculated that Tu22, and to a lesser extent IA3, may recognize epitopes that are shared between HLA-DQ and HLA-DR molecules. In contrast, SPV-L3 appears to recognize a backbone structure that is exclusively located on HLA-DQ molecules (Fig. 2). Furthermore, this study did not demonstrate the presence of DX region products. This

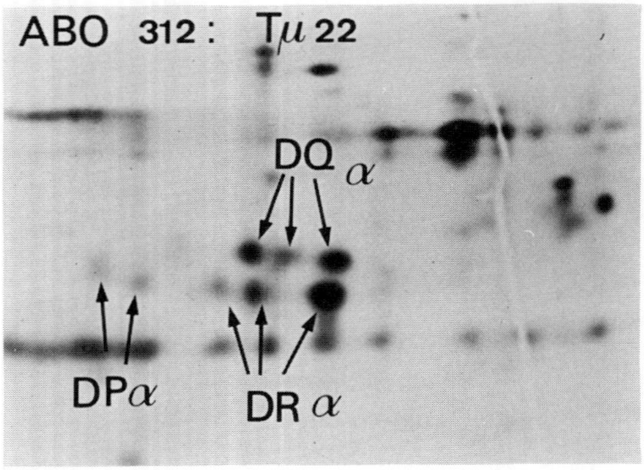

Figure 1. Two-dimensional gel profile of the alpha-chains that can be isolated from the cell line ABO 312 with monoclonal antibody Tu22. This gel represents a typical alpha-chain gradient as it was developed for the biochemistry section of this Workshop. As can be seen, Tu22 isolates high amounts of DR-alpha chains in addition to DQ-alpha, and in this example even some DP-alpha chains.

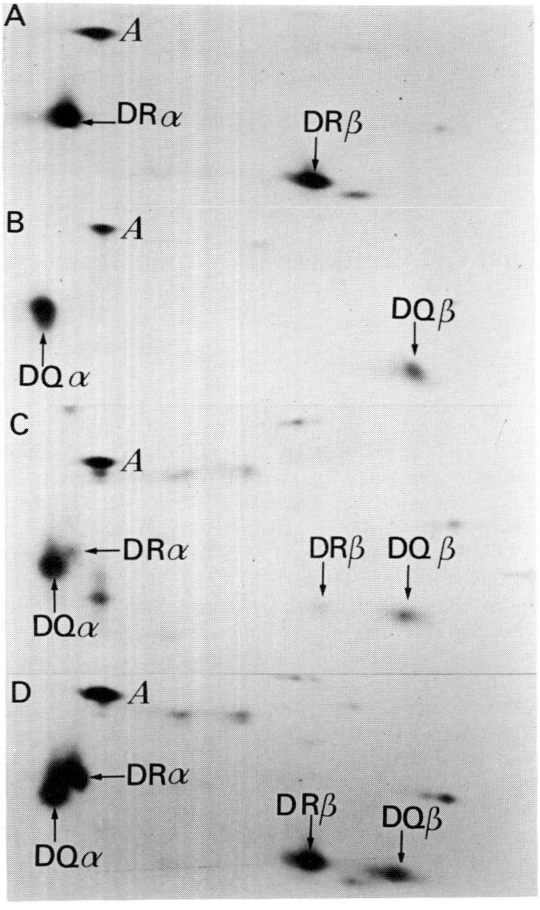

indicates that none of the three monoclonal antibodies tested is reactive with DX region products or that EBV-transformed B-cell lines have a low HLA-DX expression profile or that EBV B-cell lines simply do not express DX region products.

Apart from this, on SPV-L3 two-D profiles, DR-alpha chains are sometimes detected whereas absolutely no DR-beta chains are observed (data not shown). The amount of DR-alpha chains detected in this way does not seem to be constant, but appears to be dependent on the serotype of the cell line studied. These data suggest that SPV-L3 possibly isolates DR-alpha–DQ-beta dimers from EBV-BLCL extracts. The existence of such mismatched isotype pairing has recently been described (2).

References

1. Bontrop RE, Baas EJ, Otting N, Schreuder GMTh, Giphart MJ. Molecular diversity of HLA-DQ: DQ alpha chain isoelectric point differences and their relation to serologically defined HLA-DQ allospecificities. Immunogenetics 1987;25:305.
2. Lotteau V, Teyton L, Burroughs D, Charron D. A novel HLA class II molecule (DR -DQ) created by mismatched isotype pairing. Nature 1987;329:339.

Author Affiliation

Department of Immunohematology and Blood Bank, University Hospital Leiden, P.O. Box 9600, 2300 RC Leiden. Ronald E. Bontrop's present address: Primate Center TNO, Lange Kleiweg 151, Rijswijk, The Netherlands

◄
Figure 2. Two-dimensional gel profiles of the HLA class II molecules isolated from the nonworkshop cell line OOS:HLA-DR1-DQw5. Immunoprecipitates have been treated with neuraminidase to reduce sialic acid heterogeneity. A: HLA-DR molecules isolated with the DR locus-specific monoclonal antibody B8.11.2. B: HLA-DQ molecules isolated with the DQ-specific monoclonal antibody SPV-L3. C: The HLA-class II molecules (HLA-DR and HLA-DQ) isolated with Tu22. D: A mixture of preparations A, B, and C.

Evidence for Intra HLA-DR and -DQ Subregion Crossing Over Events

R.E. Bontrop, N. Otting, B.G. Elferink, M.G.J. Tilanus, R.R.P. de Vries, and M.J. Giphart

The HLA-DR and -DQ region of 47 homozygous typing cell lines were analyzed by a variety of techniques such as restriction fragment length polymorphism analysis, gel electrophoretic screening of the gene products, and in some cases, by testing MHC class II restriction in antigen presentation assays (1,2). Among these 47 cell lines, 16 different HLA-DR-DQ haplotypes were observed. Their genetic make up has been summarized in Table 1. It is concluded that some cell lines that express the same type of DRβI allele may express different kinds of DRβIII alleles. For example, cell lines positive for the AVL and QBL DRw17-DQw2 haplotypes do share the same $DR\beta I_3$ gene product but AVL or QBL-like cell lines express $DR\beta III_3$ and $DR\beta III_5$ gene products, respectively. This $DR\beta III_5$ gene product is often expressed by HLA-DR5 positive cell lines and is involved in the expression of the T-cell recognition epitope LB-Q1 (3,4).

Another example of such a situation where different DRβI and DRβIII locus combinations are observed is provided by cell lines with the HLF and LIA like haplotypes (Table 1). In this case the difference for an HLA-DRβIII gene product seems to affect both the HLA-D and DR specificity (1, Table 1).

Within the panel studied, also different HLA-DQ α-DQ β chain gene associations are noticed. For instance, the cell lines AVL, QBL, SAR, ATH, JVM, HLF, LIA, and LUY all appear to express the DQ α 2.3 allele (Table 1). However, cell lines with the AVL and QBL (DRw17-DQw2)-like haplotypes do express the DQβ 2.3 gene product; the SAR cell line (DRw17-DQw2) expresses the DQβ 2.7 gene product which is also expressed by DR7-DQw2 positive cell lines, whereas all the other cell lines mentioned above express the DQβ 3.4 (TA10) allele. The fact that both the polymorphic DQ α and β chain may bear determinants that contribute to the HLA-DQ allospecificity has been discussed in detail elsewhere (2).

Such different HLA-DRβI-DRβIII and DQα-DQβ associations cannot have been generated by gene conversion or by point mutations. We would like to propose that such different class II α and β associations may have been arisen by crossover events. Given the fact that, at least in some examples, the DRβIII gene product and both the polymorphic DQ α and β chain gene may bear determinants contributing to the HLA-DR or HLA-DQ allodefinition, such intra HlA-DR and -DQ crossing over events may explain some of the "unusual" HLA-DR and -DQ associations that do exist. Furthermore, exchange of functional DR β chain genes between related and unrelated class II haplotypes increases the complexity and number of different HLA class II immune response gene sets within the population.

Table 1. HLA-DR-DQ Haplotypes of HTC Studied

Reference Cell	HLA-type	N*	DQ β	DQ α	DR βI	DR βIII
LUY	DRw8/Dw8/DQw7	1	3.4	2.3	w8a	–†
LIA	DRw14/Dw16/DQw7	2	3.4	2.3	w12	3
HLF	DRw12/DB6/DQw7	1	3.4	2.3	w12	5
ATH	DRw11/Dw5/DQw7	4	3.4	2.3	w11a	5
QBL	DRw17/Dw3/DQw2	3	2.3	2.3	3	5
AVL	DRw17/Dw3/DQw2	7	2.3	2.3	3	3
SAR	DRw17/Dw3/DQw2	2	2.7	2.3	3	3
MWM	DR4/Dw4/DQw7	3	3.4	2.3	4a	4
BSM	DR4/Dw4/DQw8	3	1.2	2.7	4a	4
EKR	DR7/Dw7/DQw2	2	2.7	2.7	7a	4
IWB	DRw15/Dw2/DQw6	5	1.2	1.2	2a	2
SAS	DRw15/Dw12/DQw6	2	1.12	1.12	2b	2
HHK	DRw13/Dw18/DQw6	3	1.2	1.12	w13	3
KRA	DRw13/Dw19/DQw6	2	1.19	1.2	w13	19
BRU	DRw14/Dw9/DQw5	2	1.1	1.1	w14	5
OOS	DR1/Dw1/DQw5	5	1.1	1.1	1	–

*N = number of cells tested with identical haplotype as reference cell; † = gene product absent.

Acknowledgments. This work was supported in part by the Dutch Foundation for Medical Research (MEDIGON), which is subsidized by the Dutch Organization for the Advancement of Pure research (ZWO) and the J.A. Cohen Institute for Radiopathology and Radiation Protection (IRS).

References

1. Bontrop RE, Tilanus MGJ, Mikulski MMA, Elferink DG, Termijtelen A, de Vries RRP, et al. Polymorphism and complexity of HLA-DR: Evidence for intra-HLA-DR region crossing over events. Immunogenetics 1988;27:40.
2. Bontrop RE, Baas EJ, Otting N, Schreuder GMTh, Giphart MJ. Molecular diversity of HLA-DQ: DP alpha and beta chain isoelectric point differences and their relation to serologically defined HLA-DQ allospecificities. Immunogenetics 1987;25:305.
3. Termijtelen A, Tilanus MGJ, Engelen I, Koning F, van Rood JJ. Molecular localization of LB-Q1, a DRw52-like T-cell recognition epitope and identification at the genomic level of associated shared hybridizing fragments. Hum Immunol 1987;19:255.
4. Bontrop R, Tilanus M, Mikulski M, van Eggermond M, Termijtelen A, Giphart MJ. Polymorphisms within the HLA-DR3 haplotypes 1. HLA-DR polymorphisms detected at the protein and DNA levels are reflected by T-cell recognition. Immunogenetics 1986;23:401.

Author Affiliation

Department of Immunohaematology and Blood Bank, University Hospital Leiden, Leiden, The Netherlands

Some Supratypes Extend to DQ

G.C. Saueracker, E. Martin, and R.L. Dawkins

Supratypes are particular combinations of alleles at multiple MHC loci which occur in a population. For many diseases, supratypes provide the best marker for the disease gene within the MHC (1). Supratypes are predictable, they identify alleles at intervening loci, and identify the intervening DNA sequence. Determining the relative risk for each of the alleles defining the disease-associated supratype does not provide a measure for the distance between the disease and marker genes within the supratype. Gene(s) predisposing to disease may be located anywhere along the supratype. We have attempted to define the extent of a number of disease-associated supratypes and thus define the limits of the region within which the disease predisposition gene(s) lies. In particular we examined the supratypes A1,B8, BfS,C4AQ0,C4B1,DR3 (1,8,3) and B18,BfF1,C4A3, C4BQ0, DR3 (18,F1,3) as they have similar DR and DQ serology.

Briefly, DNA from five EBV-transformed cell lines homozygous for 18,F1,3(9018,9019,9020,9085, and one local) and five cell lines homozygous for 1,8,3 (9022,9023,9086,9088, and one local) were tested with enzyme/probe combinations that split DR3-bearing supratypes at class II genes. RFLP's between genes were detected by pulse field gel electrophoresis (PFGE) of large fragments bearing class II genes.

At DR, both of the enzyme/probe combinations Taq I/DR β and Pvu I/DR β show RFLP's that distinguish 1,8,3 from 18,F1,3 and that are entirely supratype specific. The Hinc II/DQ β combination also distinguishes between the 18,F1,3 and 1,8,3 supratypes. However, one 18,F1,3 homozygous cell line (9018) has a composite pattern implying heterozygosity at DQ.

In the class II region, Mlu I generates large fragments that are detected using DQ, DP, and DR probes. The 18,F1,3 gives a fragment of approximately 1000 kb, which includes all three genes. However, in 1,8,3, this fragment is cut at a site between DQ and DP to generate two smaller fragments (of approximately 310 and 680 kb). Cell line 9018 again shows a composite pattern. Two other 18,F1,3 cell lines (9019 and local) show a composite pattern of the typical 1000 kb band and two smaller bands (850 and 150 kb). These results imply heterozygosity in three of the five 18,F1,3 cell lines tested. In contrast, all five 1,8,3 cell lines tested show the typical pattern. However, some show an additional weak band at about 1000 kb.

Using the Hinc II/DP α combination, three patterns could be demonstrated. These patterns appear to correlate with known DP types. Despite the limited polymorphism of DP, all three patterns are evident among the 1,8,3 cell lines. In contrast, seven of ten 18,F1,3 haplotypes demonstrate the same RFLP, which is uncommon in the other lines we examined ($\chi^2 = 17$ $p \geq 0.001$).

These results show that the 1,8,3 supratype is stable and predictable through DR and DQ to within at least 310 kb of DP, but does not extend to DP. Thus 1,8,3-associated disease predisposition genes must lie telomeric of DP.

In contrast, some 18,F1,3s show variation within the DQ region. However, haplotypes that show the typical 18,F1,3 fragments in this region are also consistent at DP, possibly as a reflection of the limited polymorphism of DP. If a consistent DP association is maintained after further splitting of DP as a result of Tenth Workshop activities, this may prove important in characterizing 18,F1,3. However, two cells known to originate from individuals with IDDM (9018 submitted by this lab and local) are apparently heterozygous within the DQ region. As 18,F1,3 homozygosity has the highest relative risk for IDDM (2), it seems likely that the disease gene is located telomeric of the DQ variable region (8757).

References

1. Dawkins RL, Christiansen FT, Kay PH, Garlepp M, McCluskey J, Hollingsworth P, et al. Disease associations with complotypes, supratypes and haplotypes. Immunological Rev 1983;70:5–22.
2. Kelly H, McCann VJ, Kay PH, Dawkins RL. Susceptibility to IDDM is marked by MHC supratypes rather than individual alleles. Immunogenetics 1985;643–651.

Author Affiliations

Departments of Clinical Immunology, Royal Perth Hospital, Queen Elizabeth II Medical Centre, and University of Western Australia, Perth, Australia

Immunobiology of HLA
Volume II

HLA Class II Restriction Fragment Length Polymorphism (RFLP) with Regard to the Class II Antigens

Marcel G.J. Tilanus and Marius J. Giphart

Abstract: Mutational events in the genome may result in the creation of new restriction sites. The restriction fragment length variation obtained in Southern blot analyses, however, includes polymorphism caused by nonprotein coding sequences. The question is to what extent RFLP determinations contribute to the identification of polymorphisms related to class II antigens will be discussed.

Introduction

The HLA class II multigenic family has been described by serologic and cellular methods, by protein analyses, and by RFLP determinations. The RFLP relies on the variation of the localization of restriction sites, either in the exons or introns. If, for instance, a point mutation occurs in a restriction site, a new RFLP will be the result. In many cases, correlations have been found between RFLP and serologic cellular and protein polymorphism. Besides such correlations, RFLP reveals an additional polymorphism, defining DNA subtypes of the class II specificities. The main question thus seems to be to what extent this additional polymorphism is significant in the description of relevant polymorphism in the HLA class II region.

Results and Discussion

In a genomic sequence (DX α), we determined the presence of the restriction sites PvuII, BglII, KpnI, HaeIII, TaqI, and AluI by computer analyses, and wondered how often new restriction sites could be created by a single nucleotide substitution. We therefore tested the presence of the sequence of the restriction enzyme sites with an "N" at one of the positions of the nucleotides in their recognizing sequence: the position with "N" matches with G, A, T, and C and, for instance, in the genomic DX α sequence, it was determined how often the mutated PvuII sequences NAGCTG, CNGCTG, CANCTG, CAGNTG, CAGCNG, and CAGCTN were present. Besides one real PvuII site (CAGCTG), nine other such sites with a single nucleotide difference were found. For the other restriction enzymes mentioned above, similar analyses were performed (Table 1). If point mutations alter the DX α sequence at the "N" positions to create new restriction sites then these alterations are candidates for the generation of additional RFLP.

We then analyzed whether the thus created restriction enzyme sites resulted in a change of an amino acid, affected a splice site, or formed a new stop codon. These changes may indeed lead to a potential relevant RFLP. The substitutions that did not lead to such changes are considered as "silent" mutations as well as changes in the introns and the 3′ untranslated sequence. Table 2 shows the number of potential "relevant" and "silent" restriction sites created by simulated nucleotide substitutions.

The percentages of the substitutions leading to a potentially relevant RFLP vary from 33% (for PvuII) to 66% (for KpnI); the numerous silent replacements reach 66% (PvuII). Some of the silent nucleotide substitutions in introns might affect regulatory sequences, and thus might have functional implications in the (rate of) expression of molecules.

Although RFLP is useful in the description of the gene organization of the HLA region, these data justify the question of how far one should go on to describe additional polymorphisms. Since only relatively small percentages of the created restriction sites are potentially relevant, one should not only describe additional RFLP in DNA analyses, but also correlate them to antigen characteristics as defined by serologic methods, cellular reactivity, antigen presentation, protein analysis, and disease associations.

Table 1. Number of Restriction Sites Present in the DX α Genomic Sequence and the Number of Restriction Sites that Could Be Created by Nucleotide Substitutions

Restriction Enzyme	Sequence	No. of Sites in DX α	No. of Sites After Simulated Point Mutations
PvuII	CAGCTG	1	9
BglII	AGATCT	1	6
KpnI	GGTACC	0	3
HaeIII	GGCC	6	45
TaqI	TCGA	0	38
AluI	AGCT	5	60

Table 2. "Relevant" and "Silent" Restriction Sites

Restriction Enzyme	"Relevant" Sites	"Silent" Sites
PvuII	3 (33%)	6 (66%)
BglII	3 (50%)	3 (50%)
KpnI	2 (66%)	1 (33%)
HaeIII	27 (60%)	18 (40%)
TaqI	14 (37%)	24 (63%)
AluI	32 (53%)	28 (47%)

In a number of cases, RFLP analyses are of particular importance. It has been shown that RFLP obtained with selected subprobes correlates with the expected number of genes in a haplotype. Whereas serologic and cellular techniques characterize products of one DR β gene, RFLP analyses allow the characterization of all DR β genes, including pseudogenes. Besides, RFLP can be used for typing purposes, as shown in the Tenth Histocompatibility Workshop. As an additional method to conventional DR typing, RFLP analyses is the choice of typing technique in cells in which no expression of class II molecules can be detected; furthermore, in analyzing DR "blank" haplotypes, in the identification of real homozygosity in, for example, DR homozygous individuals and in cases where serologic typing is difficult due to the scarcity of monospecific alloantisera. In these cases, RFLP analysis is a valuable contribution to class II typing.

Conclusion

As a result of point mutations, additional polymorphism is easily understandable. A significant part of these additional polymorphisms, however, is the consequence of silent mutations, bearing no significant relevance (i.e., do not lead to amino acid or signal site variation). Analyzing additional polymorphisms should therefore always include correlations with serologic cellular, protein, antigen presentation, or disease association characteristics.

Author Affiliation

Marcel G.J. Tilanus, Department of Cell Biology, Marijkeweg 40, 6709 PG Wageningen, The Netherlands; *Marius J. Giphart*, Department of Immunohematology and Bloodbank, Rijnsburgerweg 10, 2333 AA Leiden, The Netherlands

Section IX
Class II Histocompatibility Antigens: HLA-DP

Class II RFLP Analysis Using β Domain Specific Probes

Massimo Trucco, Arnold Fritsch, and Emilia Turco

The human class II antigens are encoded by a series of linked α and β genes at the DP, DQ, and DR loci, present in the HLA-D region, which is located on the short arm of chromosome 6. Although sequence analysis of these highly polymorphic genes is the most detailed way to determine the genetic basis of these polymorphisms, restriction fragment length polymorphism (RFLP) analysis is an easier but still very informative approach. In particular, this technique has been very useful for studying DQ-α and -β gene polymorphisms, leading to the elucidation of DQ patterns that had been inadequately explained by serology (1).

RFLP analysis of endonuclease-digested genomic DNA hybridized with different class II β chain-specific probes has been frequently hampered by cross-hybridization. Although a certain degree of cross-hybridization is expected to be present based on the extent of homology existing between the DP, DQ, and DR β gene sequences, it was somewhat surprising that the cross-hybridization persisted despite high stringency conditions. In particular, it was not clear if a particular stretch of the gene sequence was responsible for this cross-hybridization, and, if this were the case, which gene stretch had to be removed in order to eliminate, or at least drastically reduce, the cross-hybridization.

To answer to these questions we subcloned "domain-specific" segments of full-length cDNA cloned genes encoding class II β chains of DP (2), DQ (3), and DR (4) molecules, and used each of the subcloned fragments as a labeled probe against the blotted fragments of the other genes to recognize the cross-reactive element(s).

Six domain specific probes, DPb1 (i.e., DP first β domain), DPb2 (i.e., DP second β domain), DQb1, DQb2, DRb1, and DRb2 were used in turn against the same fragments, which were separated on a 1.3% agarose gel and blotted on a Nytran membrane. At the conditions of stringency we used, all the b1 fragments showed a cross-hybridization pattern when challenged as probes

```
DP -   GCAGGTTTCT GGCGCCCCCC GGACAGTGGC TCTGACGGCG TTACTGATGG TGCTGCTCAC
DQ -   GCGGATCCCC GGAGACCTTC GGGTAGCAAC TGTCACCTTG ATGCTGGCGA TGCTGAGCTC
DR -   GAAGCTCCCT GGAGG-CTCC AGCTTGGCAG CGTTGACAGT GACACTGATG GTGCTGAGCT

DP -   ATCTGTGGTC C-AGGGCAGG GCCACTCCAG AGAA--TTAC CTTTTCCAGG --GACGGCAG
DQ -   -CCTACTGGC TGAGGGCAGA GACTCTCCCG AGGA--TTTC GTGTACCAGT TTAAGGGCAT
DR -   CCCGACTGGC T-TTCGCTGG GG-ACACCCG ACCACGTTTC TTGGAGCTGC TTAAGTCTGA

                                                     ▼
DP -   GAA-TGCTAC GCGTT--TAA TGGGACACAG CGCTTCCTGG AG--AGATAC ATCTACAACC
DQ -   GTGCTACTTC ACCAACGGGA CGGAGCGC-G TGCGTCTTGT AACCAGACAC ATCTATAACC
DR -   GTGTCATTTC TTCAATGGGA CGGAGCGG-G TGCGGTTCCT GGAGAGACAC TTCCATAACC

DP -   GGGAGGAGTT CGCGCGCTTC GACAGCGACG TGGGGGAGTT CCGGGCGGTG ACGGAGCTGG
DQ -   GAGAGGAGTA CGCGCGCTTC GACAGCGACG TGGGTGTGTA CCGGGCGGTG ACGCCGCAGG
DR -   AGGAGGAGTA CGCGCGCTTC GACAGCGACG TGGGGGAGTA CCGGGCGGTG AGGGAGCTGG

DP -   GGCGGCCTGC TGCGGAGTAC TGGAACAGCC AGAAGGACAT CCTGGAGGAG AAGCGGGCAG
DQ -   GGCGGCCTGT TGCCGAGTAC TGGAACAGCC AGAAGGAAGT CCTGGAGAGG ACCCGGGCGG
DR -   GGCGGCCTGA TGCCGAGTAC TGGAACAGCC AGAAGGACCT CCTGGAGCAG AAGCGGG---

DP -   TGCCGGACAG GATGTGCAGA CACAACTACG AGCTGGGCGG GCCCATGACC CTGCA
DQ -   AGTTGGACAC GGTGTGCAGA CACAACTACG AGGTGGGGTA CCGCGGGATC CTGCA
DR -   ------GCCA GGTGGACAAT TAC
             ▲
```

Figure 1. Comparison of b1 fragment nucleotide sequences from DP-, DQ-, and DR-β genes. Triangles limit the homology region rich in C-G residues. Hyphens were introduced to maximize base pair matching between sequences.

with the b1 fragments of the other two genes. They did not hybridize, however, with either their own b2 fragment or the b2 fragments from the other two genes. When we used instead the b2 fragments as labeled probes, they did not cross-hybridize with the b2 fragments from the other two genes, and did not hybridize with any of the b1 fragments.

This suggested that the cross-hybridization was due to a stretch of nucleotides present in the first domain of these β chains. When we compared the sequence of the three β genes, we found an highly conserved region between the second and third hypervariable regions of the first domain. Furthermore, this region of high homology between the sequences of DP, DQ, and DR genes was found to be particularly rich in C and G residues, from 63.9% in the case of DR region to the 67.4% in the DQ and the 69.5% in the DP involved regions. In contrast, the percentages of C and G in the homology stretches of the second domain were considerably lower (i.e., DP: 51.9%, DQ: 59.5%, and DR: 54.2%) (Fig. 1). The removal of the DNA segment containing the C-G rich region from the fragments used as labeled probes completely eliminated or drastically reduced the cross-hybridization.

Therefore, the appropriate use of probes specific for the class II second domain allows a more precise and reliable RFLP analysis than does the use of full-length gene probes. Also, this technique is simpler than nucleotide sequencing.

Acknowledgments. The work was supported in part by grants CA-44977 and DK-24021 from the National Institutes of Health, Bethesda, MD, USA.

References

1. Trucco M, Duquesnoy RJ. Polymorphisms of the HLA-DQ subregion. Immunol Today 1986;7:297.
2. Compagnone-Post P, Turco E, Robinson C, Trucco M. The beta chains of DP4 molecules from different haplotypes are encoded by the same gene. Genomics 1988;2:8.
3. Turco E, Care A, Compagnone-Post P, Robinson C, Cascino I, Trucco M. Allelic forms of the alpha and beta-chain genes encoding DQw1-positive heterodimers. Immunogenetics 1987;26:282.
4. Long E, Wake C, Gorski J, Mach B. Complete sequence of an HLA-DR beta chain deduced from a cDNA clone and identification of multiple non allelic DR beta chain genes, EMBO J 1983;2:289.

Author Affiliation

The Pittsburgh Cancer Institute and Department of Pathology, University of Pittsburgh School of Medicine, Scaife Hall, Pittsburgh, PA 15261, USA

DNA Polymorphism of HLA-DP: Further Studies of the Restriction Fragment Length Polymorphism of DPα and DPβ

Jens J. Hyldig-Nielsen, Niels Morling, Niels Ødum, Lars P. Ryder, Per Platz, Bodil Jakobsen, and Arne Svejgaard

Abstract: The restriction fragment length polymorphism (RFLP) of HLA-DP was investigated in 88 individuals who had been typed for DPw1-w6 and the local specificity CDP-HEI with the primed lymphocyte typing (PLT) technique. Chromosomal DNA was digested with ten restriction enzymes and hybridized with DPα and DPβ probes. We found nine different DPα markers and 14 different DPβ markers. Each marker was defined by from one to six different DNA fragments. Seven of the markers were absolutely or very strongly associated with a single DP antigen. Other markers were associated with two or more DP antigens while a few markers showed no associations to known DP antigens. On the basis of these investigations, it seems possible by means of three enzymes, MspI, BglII, and TaqI, and DPα and DPβ probes to type for the following HLA-DP determinants: DPw1, w2, w3 + w6, w4, w5, and CDP-HEI.

The human HLA system is located on the short arm of chromosome no. 6. It is divided into three regions: Class I, II, and III. By serologic and cellular techniques, the Class II region has been further split into the DR, DQ, and DP subregions. Biochemical studies have shown that the HLA Class II molecules consist of α and β chains each of which are encoded by separate α and β genes. The polymorphism of the HLA-DP antigens is defined with the PLT technique (1). By means of the RFLP technique, it has become possible to study the polymorphism of HLA genes in a large number of individuals. Previously, we have reported on our initial RFLP studies of the HLA-DP region (2) and here we present the data of a larger panel including a number of selected individuals with rare HLA-DP types.

Chromosomal DNA was prepared from a panel of 88 unrelated individuals who had been typed for DPw1-w6 and the locally defined "new" specificity, CDP-HEI,

Immunobiology of HLA
Volume II

Table 1. Correlations Between RFLP with a DP_β Probe* and HLA-DP Antigens

Marker	Fragment†	HLA-DPw	Fragment Versus DP ++	+−	−+	−−	N	P‡
1	MspI 4.9	1	13	1	0	71	85	1.9×10^{-14}
2	RsaI 1.7	2+8	17	11	1	56	85	1.0×10^{-9}
3	MspI 2.1	2	11	2	1	71	85	4.3×10^{-11}
4	MspI 1.1	4	58	5	0	22	85	6.4×10^{-16}
5	MspI 1.8	8	5	0	0	80	85	3.0×10^{-8}
6	MspI 3.0	3+5+6	29	8	0	48	85	8.6×10^{-16}
7	PstI 2.9	1+3+5+6+8	44	8	0	34	86	1.2×10^{-16}
8	BglII 20.0	1+3+5+6	40	9	0	37	86	3.8×10^{-16}
9	RsaI 2.1	1+2+4+5	62	1	0	5	68	5.8×10^{-7}
10	MspI 1.9	2+4+5	72	3	0	10	85	9.1×10^{-11}
11	BglII 17.3	2+4+8	72	2	0	12	86	6.0×10^{-13}
12	PstI 2.4	2+4	70	3	0	13	86	6.4×10^{-13}
13	HindIII 5.0	2+3+6+8	33	23	5	25	86	1.4×10^{-4}
14	HindIII 5.3	1+4+5	60	9	5	12	86	6.5×10^{-6}

*A cDNA clone complementary to the $DP\beta_1$-gene (7) was used as probe.
†For each marker, only data on one representative fragment of the group of fragments is shown.
‡Fisher's exact test. All p values were significant ($p < 0.05$) after correction for the multiple comparisons.

with the PLT technique (3). The DNA was digested with one of ten different restriction enzymes, separated in agarose gels, blotted onto nitrocellulose filters, and hybridized with α and β probes for DR, DQ, and DP, all by standard techniques (2).

Hybridization with the DPβ probe gave rise to 34 polymorphic fragments. Four fragments (Table 1 and Fig. 1, marker nos. 1, 3, 4, and 5) correlated strongly with a single HLA-DP antigen: DPw1, w2, w4, and CDP-HEI (in the tables and figure called "DPw8"). A large number of bands were associated with two or more HLA-DP antigens (Table 1 and Fig. 1, marker nos. 2 and 6–14). Three fragments did not correlate to any DP-antigen. Most of these "new" markers were found in individuals with only one recognized DP antigen and the markers might belong to the same series as do markers nos. 1–12. It remains to be seen whether there are unknown DP antigens associated with these DNA markers.

The two groups of DPβ fragments represented by HindIII 5.0 kb and 5.3 kb (Table 1, marker nos. 13 and 14) seemed to belong to another series of allotypes than did the fragments 1–12. On the basis of restriction fragment mapping of genomic clones, the DPβ fragments HindIII 5.0 kb and 5.3 kb have been allocated to the DPβ2 pseudogene (4). Thus, the markers nos. 13 and 14 most probably reflect allotypes of the DPβ2 pseudogene.

Hybridization with the DPα probe gave rise to 24 polymorphic fragments. Three fragments correlated with DPw1, DPw2, and DPw5 (Table 2 and Fig. 1, marker nos. 15, 21, and 16, respectively). The other six DNA markers (Table 2 and Fig. 1) divided the known DP antigens into two groups, one including DPw1, w5, and w8 and another group including DPw2, w3, w4, and

Table 2. Correlations Between RFLP with a DPα Probe* and HLA-DP Antigens

Marker	Fragment†	HLA-DPw	Fragment Versus DP ++	+−	−+	−−	N	P‡
15	BglII 3.5	1	12	14	1	60	87	5.9×10^{-7}
16	TaqI 6.5	5	7	1	1	75	84	1.4×10^{-8}
17	BglII 3.8	5+8	12	2	1	72	87	6.5×10^{-12}
18	MspI 6.5	1+5	14	1	1	51	67	2.3×10^{-12}
19	TaqI 6.3	1+8	18	13	1	52	84	3.4×10^{-9}
20	PvuII 3.9	1+5+8	26	12	0	50	88	1.9×10^{-13}
21	TaqI 13.5	2	11	1	2	70	84	4.9×10^{-11}
22	TaqI 14.0	3+4+6	71	3	0	10	84	1.0×10^{-10}
23	BglII 2.3	2+3+4+6	82	1	0	3	86	4.0×10^{-5}

*A genomic *Hind*III subclone complementary to the $DP\alpha_1$-gene (5) was used as probe.
†For each marker, only data on one representative fragment of the group of fragments is shown.
‡Fisher's exact test. All p values except that of BglII 2.3 kb (marker no. 23) were significant ($p < 0.05$) after correction for the multiple comparisons.

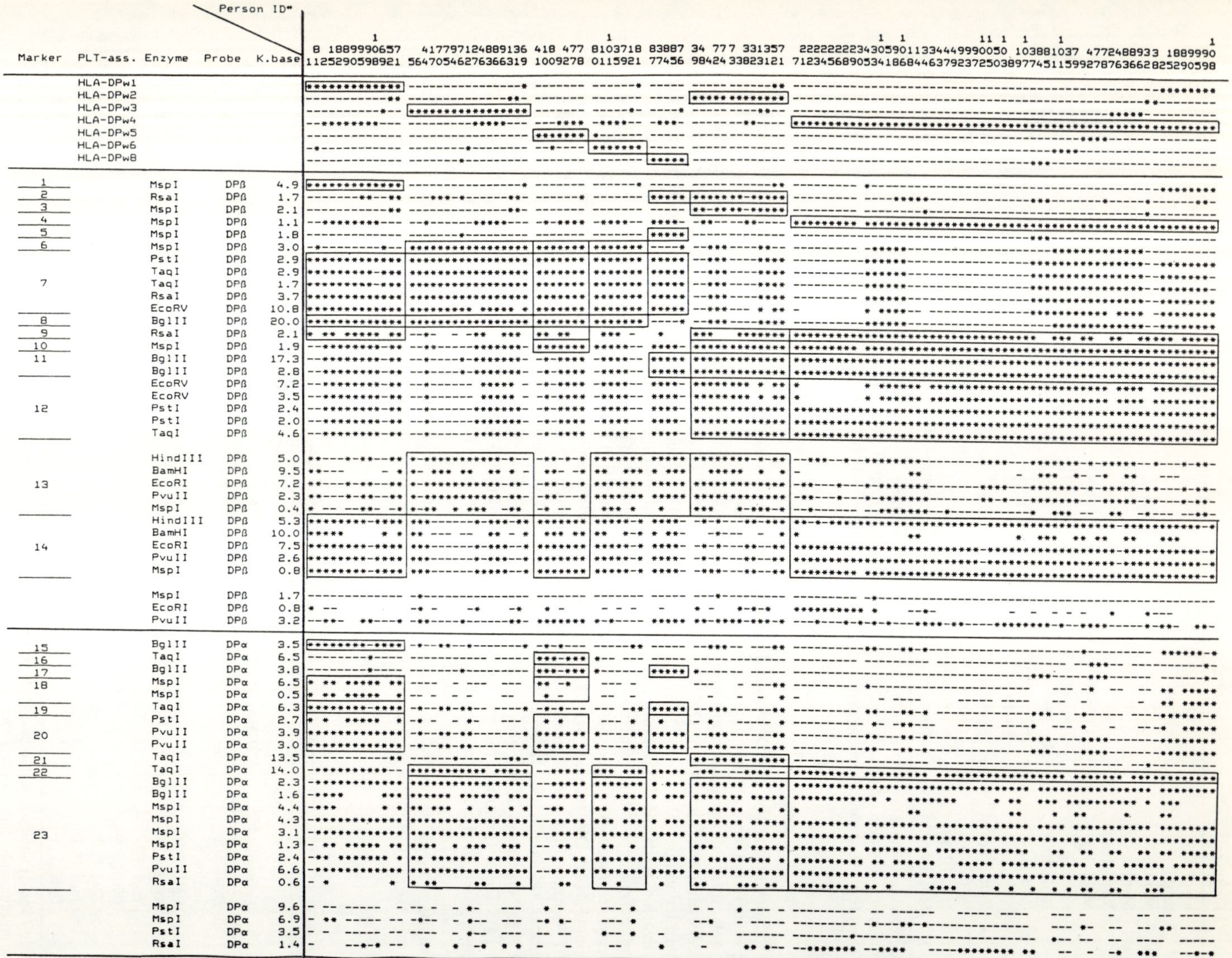

Figure 1. DP fragmentogram. HLA-DP types, DPβ and DPα polymorphic fragments in 88 unrelated individuals. Columns correspond to panel members ordered according to DP-assignments. Individuals with two DP antigens are included twice. The presence of a fragment is indicated with *, absence with -, and no symbol indicates not done. Note that subject ID numbers are printed vertically—e.g., $^{8}_{1}$ is ID no. 81.

w6. At least four other fragments were found, but it is not known whether corresponding allomorfic determinants at the cell surface exist.

The homology between the DPα1 and DPα2 genes is only approximately 60% (5) and our DPα1 probe did not cross hybridize to DPα2 or DZα (2,6). Thus, the DPα data presented here reflect the polymorphism of DPα1.

Using the RFLP technique with three enzymes—MspI, BglII, and TaqI—and a DPα and a DPβ probe, it is possible to "DNA type" for HLA-DPw1, w2, w3+w6, w4, w5, and the local specificity CDP-HEI. In general, the DNA markers were broader than HLA-DP indicating that further refinement of the RFLP technique, e.g., probing with oligonucleotides, may be of value for future studies of the HLA-DP gene polymorphism. The RFLP technique is well suited for studies of the polymorphism of HLA genes in larger populations and it seems as if the RFLP typing technique, with minor modifications, may substitute for HLA-DP typing with the PLT technique.

Acknowledgments. Supported by grants from the Danish Medical Research Council, The Danish Rheumatism Association, and The Danish Multiple Sclerosis Association.

References

1. Shaw SM, Pollack S, Payne SM, Johnson AH. HLA-linked B-cell alloantigens of a new segregant series: Population and family studies of the SB antigens. Hum Immunol 1980;1:177.
2. Hyldig-Nielsen JJ, Morling N, Ødum N, Ryder LP, Platz P, Jakobsen B, et al. Restriction fragment length polymorphism of the HLA-DP subregion and correlations to HLA-DP phenotypes. Proc Natl Acad Sci USA 1987;84: 1644.
3. Ødum N, Hartzman R, Jakobsen BK, Morling N, Platz P, Robbins F-M, et al. The HLA-DP polymorphism of Denmark investigated by local and international PLT reagents. Definition of two "new" DP antigens. Tissue Antigens 1986; 28:105.
4. Kappes DJ, Arnot D, Okada K, Strominger JL. Structure and polymorphism of the HLA class II SB light chain genes. EMBO J 1984;3:2985.
5. Servenius B, Gustafsson K, Widmark E, Emmoth E, Andersson G, Larhammer D, et al. Molecular map of the human HLA-SB (HLA-DP) region, and sequences of an SB-alfa (DP-alfa) pseudogene. EMBO J 1984;3:3209.
6. Trowsdale J, Young JOAT, Kelly AP, Austin PJ, Carson S, Meunier H, et al. Structure, sequence, and polymorphism in the HLA-D region. Immunol Rev 1985;85:5.
7. Gustafsson K, Emmoth E, Widmark E, Böhme J, Peterson PA, Rask L. Isolation of a cDNA clone coding for an SB β-chain. Nature 1984;309:76.

Author Affiliation

Tissue Typing Laboratory of the Department of Clinical Immunology, State University Hospital (Rigshospitalet), Copenhagen, Denmark

HLA-DP Region Complexity by CDC, RFLP, and Cellular Assays

Min Sik Park, Toshinao Takenouchi, Paul I. Terasaki, Richard Tonai, and Aida Barbetti

HLA-DP, one of the class II complexes, had been originally typed only by the PLT method using cellular reagents (since 1980). Now, it can be defined by RFLP, but not by serologic methods. Strong secondary and weak primary proliferation responses were identified by Mawas (1), Wank (2), and Shaw (3), who classified them as SB antigens (SB1–5), and were officially designated by the 9th International Histocompatibility Workshop in 1984.

We have identified a few monoclonals as well as alloantibodies for DP gene products or the gene products associated with the DP locus, using PLT defined panel cells as a reference for serologic typing and RFLP band pattern analysis. In this study, more than 5000 sera were absorbed with platelets to remove class I activities. Those DP-like antibodies were selected and reacted to the PLT-typed reference cell panels that were obtained from many different labs and 10th Workshop reference cells. The characteristics of the probes, enzymes, and bands have been described in a simple format (probe-enzyme-kb).

Table 1 shows the summary of a battery of SP3–6 and SP4 antibodies to 128 cell lines. SP3–6 was positive to

Table 1. SP3-6/SP4 Antibody Reaction to Panel

ab/ag	++	+−	−+	−−	Inclusion %
SP3-6/DPw3	31	7	1	89	96
SP3-6/DPw6	4	3	0	89	100
SP3-6/DPw3,w6	35	6	1	86	97
SP4/DPw4	62				92
SP4/DPW2	26				87
SP4/DPw1	20				80
SP4/DPw5	9				70

Positive cells for: SP3-6; 9003,9004,9076 and local lines (DUGE,AKIBA,TOK) SP4; 9003,9010,9018,9030,9043,9048, 9076 (all DP-unknown).

Table 2. Phenotypes Defined by Cellular (DP), Serologic (CDC), and RFLP Using MspI Enzyme with DP β Probe

			w1	w2	w2	w3 w5 w6 11	w4	w2 w4 w5 w1	New 11	???	???
Cellid	DP	CDC	4.8	4.3	2.1	3.1	1.1	1.9	1.7	4.1	1.8
LAZ -028	w1	4	+								
LUY -227	w1,w4	4	+				+	+			
H9B -259	w1,w4	4	+				+	+			
GLS -032	w1,w4	4	+				+	+			
PFO -322	w1,w4	4	+				+	+			
H9 -174	w1,w4	4	+				+	+			
LAY -446	w1,w3	3,2	+			+					
JNY -417	w1,w3	4,3	+			+					
MCRA-154	w1,??	4	+			+			+		PB11
DUGE-150	w1,??	3,4	+			+			+		PB11
PPA -111	NT	4				+			+		PB11
7526 -291	??,	3,4				+			+		PB11
KOSE -340	w2,L1	3,4,2		+	+	+		+	+		PB11
MAN -006	w2	2,4		+	+			+			
DKY -016	w2	2,5		+	+			+			
WT51 -326	w2,??	4		+	+			+			
TAB -484	w7	NT		+	+			+			
QBL -462	w2,L1	NT		+	+			+			
BM92 -349	2b,	4,2		+	+			+			
DAAL-146	w2,w3	3,4		+	+	+		+			
S11 -177	w2,w5	4,2		+	+	+		+			
BERR -235	w2,w6	4,3		+		+		+			
DAU -109	w2,4?	2		+	+		+	+			
KT -116	w2,w5	4,2,5		+	+			+			
BR1 -331	w2,4B,L1	2,5,4		+	+		+	+			
IO5 -426	w4	4,5		+			+	+			
ABO -352	w4	4,5		+			+	+			
NIE -084	w3,w6	3,5				+					
THO -196	w3,w6	3,5				+					
SLE -013	w3	3				+					
AZH -271	w3	NT				+					
COX -139	w3	3				+					
BM15 -327	w3	3				+					
LOO -315	w3,L1	3,4				+					
E3b -263	w6	3,4				+					
HAS -015	w5	5				+		+		+	+
HAS -200	w5	5				+		+		+	+
KT2 -247	w5	5,4				+		+			
KT2 -115	w5	5,4				+		+			
KT17 -117	w5,??	5				+		+			
DEM -312	w3,w4	3,4				+	+	+			
MM -057	w3,w4	3,4				+	+	+			
DUNE-035	w5	2,4				+	+	+			
KT12 -119	w5,w4	5,4				+	+	+			
LL -424	??	5,4?				+	+	+			
AKI -186	w5,CP63	4,5,3				+	+	+			
TOK -187	CP63	3				+	+	+			
GEV -005	NT	5				+	+	+			
MS7 -415	w5,w4	5,4				+	+	+			
BOD -085	w5,w4	4,2				+	+	+			
TIS -332	4b,L1	5,2				+	+	+			
GOD -031	w4,1??	4				+	+	+			
PE11 -466	w4	NT					+	+			
GW -042	w4	4					+	+			
MAY -204	w4	4					+	+			
TS1 -185	w4	4					+	+			
BOR -216	w4	4					+	+			
EM -036	w4	4					+	+			
BTB -351	w4	4,2					+	+			

Table 2. *Continued.*

			w1	w2	w2	w3 w5 w6 11	w4	w2 w4 w5 w1	New 11	???	???
Cellid	DP	CDC	4.8	4.3	2.1	3.1	1.1	1.9	1.7	4.1	1.8
DEU -464	w4	NT					+	+			
YAR -465	w4	NT					+	+			
K40 -380	NT	4					+	+			
AMAL-345	L1	2					+	+			
MLF -489	4b	NT					+	+			
BOB -488	4b	NT					+	+			
LBUF -333	w6,??	3,4					+	+			
MM -058	w4	5					+	+			

DPw3,w6 and six other reactions. The false-positive cells are listed. There were three different allosera used to produce this table, one human monoclonal from Dr. Ferrara for SP3–6, three allosera for SP3–6 and DR5, one alloserum for SP3–6 and DR7. It is interesting to find that SP3–6 antibodies are often found with DR5 specificities. A similar monoclonal antibody ILR1 (DPw3, DR5, DPw2, unknown) has been found (4), but it also contains additional DPw2 and an unknown specificity.

The SP4 antibody shows the inclusion value of DPw4,w2,w1 and DPw5. The majority of the cells positive to SP4 but not having any of the other antigens carry an unknown specificity and are listed below. The SP4 antibody may have detected a common epitope shared by DPw4,w2 (SP4–2) and DPw1,w5 (SP1–5). The primitive split of SP4–2, and SP1–5 is based on SP4 antibody at a higher dilution reacting very weakly to DPw1,w5 positive cells, but strongly positive to DPw4,w2.

Another monoclonal antibody: DP11.1 (5) (not shown) was positive to DPw2 and DPw4. Lotteau et al. (6) indicated that two allelic patterns of DPa chains are associated with different DP antigens: one chain for DPw2,w4, and DPw3,w6 and the other chain for DPw1,w5 and Cp63. These independent studies provide information that may suggest that there are at least three different supertypic antigens in the DPb region such as DP2x4,DP1x5, and DP3x6. This SP4 antibody then is presumably positive to two different supertypes, DP4–2 and DP1–5.

Table 2 shows the phenotypes derived from the cellular, serologic, and DNA method of cell lines. Sixty-seven cells were digested with MspI enzyme and hybridized with a DP β probe obtained from Dr. Benhard Mach. The b-Msp-4.8 band was seen only in DPw1-positive cells. The DPw2-positive cells are highly correlated to b-Msp-4.3 and b-Msp-2.1. Thirty-five of the cell lines produced a b-Msp-3.1 band. Of these all 12 are Dpw3, 11 out of 12 are DPw5, and 4 out of 5 are DPw6 (exception: 10w9048, LBUF). The b-Msp-1.1 band was very specific for DPW4 and all these cells were included into a b-Msp-1.9 band which is designated to a corresponding SP4 typing.

Those cells having a b-Msp-1.7 band are all DP unknown and the b-Msp-3.1 band that is present in the same cells indicates a new DP antigen and is tentatively designated as PB11. This indicates the SP3–6 antibody may have a new specificity in addition to DPw3, w6, w5, Cp63. This new PB11 is not same as Cp63.

Table 3 shows the relation between SP2 alloantibody and DPw2 antigen. The heterogenicity within DPw2 was also observed by others (6,7). There are 21 false-positive and 10 false-negative cells but only those cell lines used for the 10th workshop are listed because the lines obtained from other labs were not typed against the complete set of DP reagent.

Most false-positive cells are related to an unknown DP or DPw4 antigen and can be due to very high frequencies or a variant of DPw2,w4 epitope. A new DP-like antigen (Jet) described by Odum et al. (8) was associated with DPw2, w6 and DP blank. Six positive cells have well-known DP antigens. This antigen is segregated with DPw6 into two informative recombinant families. Three possibilities were suggested: DP-like, associated with DPw2,w6 and blank, or a new locus product in linkage disequilibrium with DP. The 10th core cell lines: 10w9034-SAV and 10w9050-MOU, which have been identified as Jet, are positive to SP7 antibody. Whether SP7 is detecting the new DP-"Jet" or not will

Table 3. SP2 and SP5 Reactions to Panel

	SP2				
ab/ag	++	+−	−+	−−	Inclusion
DPw2	17	20	10	80	63%

Positive cells (only 10w lines are listed) 9064,9034,9092,9004,9001,9051,9080,9067.
Negative cells 9093,9029,9085,9008.

	SP5				
ab/ag	++	+−	−+	−−	Inclusion
DPw5	9	32	4	83	69%

Positive cells, 9061, 9045, 9083, 9015, 9071, 9079, 9004, 9005, 9044, 9067, 9069, 9080

Table 4. HLA-DP Supertypic Model

a	b				
DPA1	DPw4	SP4-2			DP11.1
DPA1	DPw2	SP4-2		ILR1	DP11.1
DPA1	DPw3		SP3-6	ILR1	
DPA1	DPw6		SP3	ILR1	
DPA2	DPw1	SP1-5			
DPA2	DPw5	SP1-5			
DPA2	*Cp63	SP4b?	SP3-6?		

*Not enough sample to be sure.

remain to be clarified. Below these results are listed an interesting monoclonal antibody's (SP5) reaction to the panels. Nine out of 13 DPw5 cells are positive, leaving four as false-negative, and 20 out of 32 false-positive cells are DPw4. These extra positives have to remain until the heterogenicity DPw4 antigen is clearly defined. Only the 10th core cell lines are listed.

Table 4 shows the simplification of DP supertypic model based on: 1. our monoclonal which is positive to DPw2,w4 and DPw1,w5 but DPW1,w5 were weakly positive at higher dilution; 2. DP11.1 (DPw2,w4 specific) by Bodmer et al.; 3. a protein study done by Lotteau where two molecules in DP α chain were described – one for DPw2,w4,w3,w6 and the other one for DPw1,w5; and 4. the cluster of RFLP band are also similar combinations shown by (5,9) and our local data: b-MspI-3.1 and b-Msp-1.9.

The serologic analysis shows rather supertypic reactions to DP antigens. Those cells positive to SP3–6 antisera have a RFLP band pattern b-MspI-3.10, which is shared by DPw3, w6, Cp63 and DPw5. A supertypic monoclonal SP4 reactive for DPw4,w2 and DPw1,w5 reacts with those cells that shared the band pattern of b-Msp-1.9. These serologic data patterns are a supertype which can be analogous to DR b-III(drw52), and DR b-IV(drw53) which are not allelic gene products demonstrated recently by Mach group (10). However, more than one β gene expressed on the cell surface has not yet been discovered.

The SP2 antibody does not exactly correlate to DPw2 but the false-positive cells are mainly DPw4 or of unknown antigens indicating a split of DPw2, or DPw2 related a new specificity like "Jet." The SP5 monoclonal antibody is related to DPw5 and a small fraction of DPw4 positive cells. The fact that some of these antigens detected by SP2 or SP5 are encoded by another gene whose alleles are in a strong linkage disequilibrium with alleles of DP cannot be completely excluded.

We also found that DP typing by RFLP pattern analysis can be done using Msp I enzyme with DP b probe but DPw5 and w6 requires a-BglII since DP3, DPw5 and w6 are in the same band in MspI with DP b probe.

References

1. Mawas C, Charmot D, Mercier P. Split of HLA-D into two regions alpha and beta by a recombination between HLA-d and GLO. 1.Study in a family and primed lymphocyte typing for determinants coded by the beta region. Tissue Antigens 1980;15:458–466.
2. Wank R, Schendel D, Hansen JA, Dupont B. The lymphocyte restimulation system: Evaluation by intra-HLA-D group priming. Immunogenetics 1978;6:107–115.
3. Shaw S, Johnson AH, Shearer GM. Evidence for a new segregant series of B cell antigens that are encoded in the HLA-D region and that stimulate secondary allogenic proliferative and cytotoxic responses. Exp Med 1980;152: 565–580.
4. Shaw S, Demars R, Schlossman S, Smith PL, Lampson LA, Nadler LM. Serologic identification of the human secondary b cell antigens. J Exp Med 1982;156:731–743.
5. Bodmer J, Bodmer W, Heyes J, So A, Tonks S, Trowsdale J, Young J. Identification of HLA-DP polymorphism with DPa and DPb probes and monoclonal antibodies: Correlation with primed lymphocyte typing. Proc Natl Acad Sci USA 1987;84:4596–4600.
6. Lotteau V, Teyton L, Tongio MM, Soulier A, Thomsen M, Sasportes M, Charron D. Biochemical polymorphism of the HLA-DP heavy chain. Immunogenetics 1987;25:403–407.
7. Beall SS, Lawrence J, Bradley D, Mattson DH, Singer D, Biddson W. B-chain gene rearrangements and Vb gene usage in DPw2-specific T cells. J Immunol 1987;139: 1320–1325.
8. Odum N, Hofmann B, Hyldig-Nielsen JJ, Jakobsen BK, Morling N, Platz P, et al. A "new" supertypic HLA-DP related determinant detected by PLT. Tissue Antigens 1987;29:101–109.
9. Hyldig-Nielsen JJ, Morling N, Odum N, Ryder LP, Plata P, Jakobsen B, et al. Restriction fragment length polymorphism of the HLA-DP subregion and correlations to HLA-DP phenotypes. Proc Natl Acad Sci USA 1987;84:1644–1648.
10. Gorshi J, Rollini P, Mach B. Structural comparison of the genes of two HLA-DR supertypic groups: The loci encoding DRW52 and DRW53 are not truly allelic. Immunogenetics 1987;25:397–402.

Author Affiliation

UCLA Tissue Typing Laboratory, Department of Surgery, School of Medicine, University of California, Los Angeles, CA, USA

Biochemical Polymorphism of the DPα and DPβ Chains

Santiago Rodriguez de Cordoba, Patricia Marshall, Fredda Ginsberg-Fellner, and Pablo Rubinstein

The human class II HLA-DP molecules are heterodimers of a 34–36,000 d polypeptide (DPα-chain) non-covalently associated to a 28–29,000 d polypeptide (DPβ-chain) with both the DPα and DPβ chains being encoded by genes located in the most centromeric region of the human major histocompatibility complex (MHC) (1–3). DP molecules, as the other class II molecules DR and DQ, can act as restriction elements for T-cell recognition of foreign antigens (4) and represent important histocompatibility antigens. Although the genetic polymorphism of the DP molecules can be shown by the use of serologic and cellular reagents (5,6), the precise biochemical characterization of the different polymorphic variants has been made difficult by the low levels of expression of these molecules and by their extensive microheterogeneity, particularly that of the DPβ chain. We report here on results of the biochemical analysis of the DPα and DPβ products in 137 individuals from 25 multiplex insulin-dependent diabetes mellitus (IDDM) families which allowed for the direct observation by segregation of the DP molecules encoded by 94 haplotypes. Two DPα and 7 DPβ isoelectric focusing (IEF) variants were characterized. Haplotypic associations between the DP IEF-variants and those of the DR and DQ loci were encountered which suggest a way to discriminate between high and low IDDM risk among otherwise indistinguishable HLA haplotypes.

DP molecules were specifically immunoprecipitated using the monomorphic anti-DP monoclonal reagent B7/21 (7) from each of the 137 ^{35}S-methionine-labelled EBV-transformed lymphoblastoid cell lines (LCLs). Following desialylation the DPα and DPβ chains were separated in 12.5% SDS-polyacrylamide gels, electroeluted and analyzed in IEF gels as previously described (8). Neither DRα, DQα nor DRβ, DQβ chains were observed in the DPα and DPβ preparations from the B7/21 immunoprecipitates.

Both DPα and DPβ chains were found to be polymorphic among these LCLs and the different DPα and DPβ IEF-variants segregated in families with the serologically defined HLA haplotypes.

Two DPα alleles, DPαa and DPαb, were characterized. Both DPαa and DPαb IEF-banding patterns are composed of three bands of which the middle band is the most intense (Fig. 1). The nature of this IEF microheterogeneity is unknown, but it is extremely reproducible and most likely related to the prolonged labeling times (14h) used in these experiments. The acidic DPαa variant was found to be the most frequent (84.7%) in this population and the basic DPαb variant (15.3%) was found in linkage disequilibrium with two DPβ IEF variants, DPβb and DPβd (see below). These two DPα IEF variants reported here appear to correspond to those recently described by Lotteau et al (9).

The DPβ chain is far more polymorphic than the DPα chain. Figure 2a shows the IEF banding patterns of the DPβ chains immunoprecipitated by MAb B7/21 from 11 selected LCLs (most of which are heterozygous) to illustrate six of the seven DPβ IEF variants characterized thus far. Like the DPα chain, each DPβ IEF variant is characterized by a distinct banding pattern composed of three (or four) bands, again the middle one being the most intense. Figure 2b shows a diagrammatic representation of the bands that compose each DPβ pattern and the relative focusing position of the different DPβ IEF-banding patterns. Also shown in Figure 2b is a provisional nomenclature for the DPβ IEF variants based on their focusing positions, DPβa being the most acidic one. Gene frequencies for the DPβ IEF variants in our population sample were as follows: DPβa: 0.022; DPβb: 0.045; DPβc: 0.250; DPβd: 0.045; DPβe: 0.181; DPβf: 0.397; and DPβg: 0.011. Four haplotypes could not be DPβ typed. They presented IEF-banding patterns that might represent additional DPβ variants.

Preliminary data, not shown here, on the relationships between the DPα/DPβ IEF variants and the DPw1–6 specificities indicate correlations between DPw3 and DPβc, DPαa; DPw4 and DPβf, DPαa; DPw6 and DPβa, DPαa; and DPw5 and DPβd, DPαb. DPw1 and DPw2 specificities were heterogeneous with regard to their associated DPα and DPβ IEF variants.

Haplotypic associations between the DPα and DPβ variants and other class II loci (DR and DQ), as well as the existence of recombination between DQ and DP were evaluated by the analysis of the segregation of the DR, DQ and DP IEF variants in the 25 families. In agreement with the estimated 1 to 3% recombination frequency between DQ and DP (10,11), one DQ-DP recombinant haplotype was observed among a total of 86 informative meioses for these two loci. Somewhat unexpectedly, because of this high recombination frequency, we found strong linkage disequilibrium between DPβb,

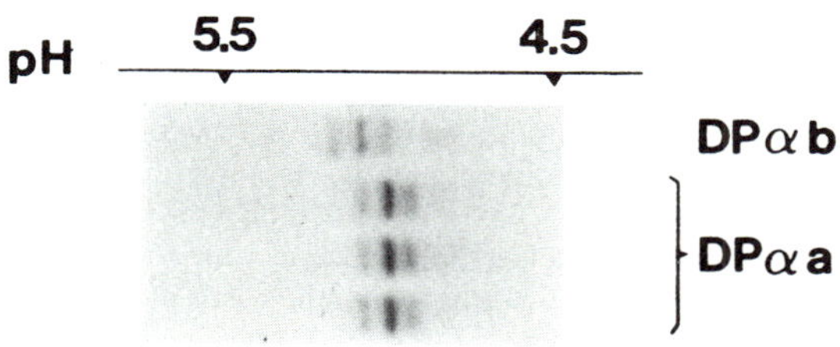

Figure 1. IEF-banding patterns of the two DPα variants. The figure shows the IEF-banding patterns of the desialylated DPα chains precipitated from three homozygotes DPαa and one homozygote DPαb.

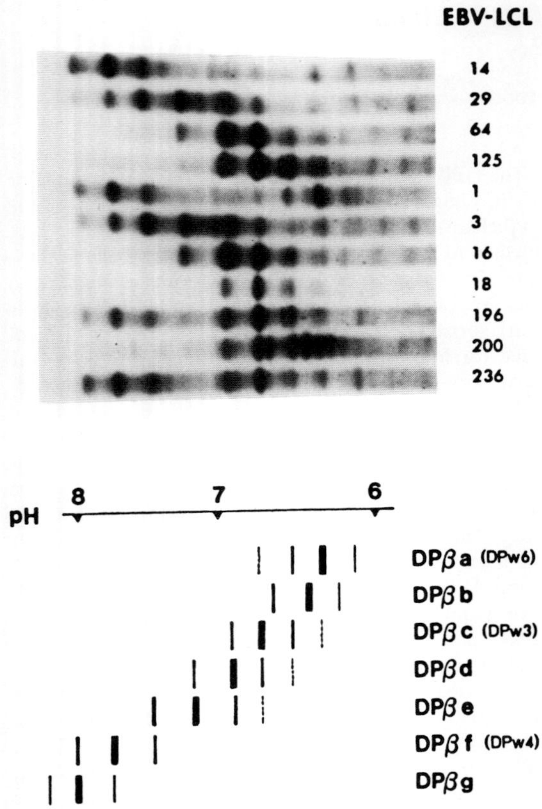

Figure 2. DPβ chain polymorphism. Top) IEF-banding patterns of the desialylated DPβ chains precipitated by MAb B7/21 from 11 local EBV LCLs illustrating the following phenotypes: 14, DPβf; 29, DPβf, DPβe; 64, DPβd, DPβc; 125, DPβc; 1, DPβf, DPβa; 3, DPβf, DPβe; 16, DPβd, DPβc; 18, DPβc; 196, DPβf, DPβc; 200, DPβc, DPβb and 236, DPβf, DPβc. Inheritance with the HLA haplotypes was demonstrated in family studies for all the DPβ variants shown. Bottom) Diagrammatic representation of the IEF-banding patterns that identify each of the seven DPβ variants observed thus far. In parenthesis are indicated the DPw1–6 specificities which best correlate with the IEF DPβ variants.

DPαb and DR7, DQw2 ($p = 0.00017$) and a significantly different association of DPβc and DPβf with DR3 and DR4 haplotypes where DPβc occurred more frequently than DPβf among DR3 haplotypes and the opposite among DR4 haplotypes.

The population sample analyzed here consisted of 25 multiplex IDDM families, thus the haplotypic associations reported could either reflect those occurring in the general population or be secondary to the selection of HLA haplotypes determined by the association with IDDM. In a preliminary attempt to solve this question, we have selected those HLA haplotypes that are inherited by the offspring and separated them in "IDDM affected" and "control" haplotypes based on the criteria that "control" haplotypes were those exclusively present in healthy sibs of IDDM patients. All three DR7-DQw2-DPβb-DPαb haplotypes were among the four "IDDM affected" DR7 haplotypes whereas the only DR7 "control" haplotype in our sample was associated with DPβc, DPαa. Similarly, DPβc was encoded by seven out of 12 "IDDM affected" DR3 haplotypes but by none of the four "control" DR3 haplotypes. Clearly, the number of DR3 and DR7 haplotypes included here is too small to draw definitive conclusions but the results suggest that the DP IEF polymorphism might be helpful as a genetic marker in the discrimination between high and low IDDM-risks among otherwise indistinguishable HLA haplotypes.

Acknowledgments. We thank Dr. F. H. Bach for his generous gift of MAb B7/21; Ms. C. Carrier and Mr. N. Mollen for their help with the LCLs; Ms. Tellervo Huima for photographic work; and Ms. A. Varipapa for preparation of the manuscript. This work was supported in part by grant 187601 from the Juvenile Diabetes Foundation and a grant from the American Diabetes Association.

References

1. Shaw S, Kavathas P, Pollack MS, Charmot D, Mawas C. Family studies define a new histocompatibility locus, SB, between HLA-DR and GLO. Nature 1981;293:745–747.
2. Kavathas P, DeMars R, Bach FH, Shaw S. SB: a new HLA-linked human histocompatibility gene defined using HLA-mutant cell lines. Nature 1981;293:747–749.
3. Hardy DA, Bell JI, Long EO, Lindsten T, McDevitt HO. Mapping of the class II region of the human major histocompatibility complex by pulsed-field gel electrophoresis. Nature 1986;323:453–455.
4. Eckels DD, Lake P, Lamb JR, Johnson AH, Shaw S, Woody JN, et al. SB-restricted presentation of influenza and herpes simplex virus antigens to human T-lymphocyte clones. Nature 1983;301:716–718.
5. Shaw S, Johnson AH, Shearer GM. Evidence for a new segregant series of B cell antigens that are encoded in the HLA-D region and that stimulate secondary allogeneic proliferative and cytotoxic responses. J Exp Med 1980; 152(3):565–580.
6. Bodmer J, Bodmer W, Heyes J, So A, Tonks S, Trowsdale J, et al. Identification of HLA-DP polymorphism with DPα and DPβ probes and monoclonal antibodies: Correlation with primed lymphocyte typing. Proc Natl Acad Sci USA 1987;84:4596–4600.
7. Watson AJ, DeMars R, Trowbridge IS, Bach FH. Detection of a novel human class II HLA antigen. Nature 1983; 304:358–361.
8. Rodriguez de Cordoba S, Nunez-Roldan A, Winchester R, Marshall P, Carrier C, Mollen N, et al. Molecular characterization by high-resolution isoelectric focusing of the products encoded by the class II region loci of the major histocompatibility complex in humans. I. DR and DQ gene variants. Hum Immunol 1987;20:71–93.
9. Lotteau V, Teyton L, Tongio M, Soulier A, Thomsen M, Sasportes M, et al. Biochemical polymorphism of the HLA-DP heavy chain. Immunogenetics 1987;25:403–407.
10. Termijtelen A, Meera Khan P, Shaw S, van Rood JJ. Mapping SB in relation to HLA and GLO1 using cells from first

cousin marriages offspring. Immunogenetics 1983;18: 503–512.
11. Shaw S, Duquesnoy RJ, Smith PL. Population studies of the HLA-linked SB antigens. Immunogenetics 1981;14:153–162.

Author Affiliations

Santiago Rodriguez de Cordoba, Department of Immunogenetics, The Lindsley F. Kimball Research Institute of The New York Blood Center, New York, NY 10021, USA and Centro de Investigaciones Biologicas, Consejo Superior de Investigaciones Cientificas (CSIC), 28006 Madrid, Spain; *Patricia Marshall, Pablo Rubinstein*, Department of Immunogenetics, The Lindsley F. Kimball Research Institute of the New York Blood Center, New York, NY; *Fredda Ginsberg Fellner*, Department of Pediatrics, Mount Sinai School of Medicine of the City University of New York, New York, NY 10029, USA

Proliferative Response by Stimulation of DP-Transfectants

Takako Nakatsuji, Tadayuki Sato, Kanako Ito, Fumiko Tsuchida, Asako Ando, Nobuhiko Sano, Naoto Matsuno, Hidetoshi Inoko, Yukio Koide, Takato O. Yoshida, and Kimiyoshi Tsuji

Abstract: DP antigens are products of the HLA-D region (class II antigens) along with DR and DQ. They were first defined by the primed lymphocyte typing (PLT) system (1–3). Whereas alloantisera exist against DR and DQ, no single DP specific alloantiserum has yet been discovered (4). The role of HLA-DP in transplantation immunology has not been defined. By introducing DP gene clones into cell lines, the resultant transfectant cells will allow the expression of DP antigens and hence enable their study. Using such transfectant cells as stimulator cells, the T-cell proliferative response was measured. In this study, it was observed that DP transfectant cells could generate a T-cell response.

The DP region contains two A and two B genes. One set of DPA genes (λHMC3) and B genes (λHMC351) (5) (Fig. 1) were introduced into murine L cells (6), human premonocytoid cells U937 (7), and human T-lymphoma cells CEM (8), using a liposome-mediated transfer technique (9). L cell, U937, and CEM cells do not express HLA class II antigens. Co-transfection was done with pSV-2 Neo gene and the selection medium used was G418(400 μg/ml geneticine, Gibco). The transfectant cells grown were analyzed by flow cytometry using anti-class II monoclonal antibody, Sa3 (10) (Fig. 2) and SG520 (11). Positive cells expressing HLA-DP antigens were selected by a fluorescent-activated cell sorter (FACS) and then kept frozen. To identify qualitative and quantitative antigen expression, two-dimensional polyacrylamide gel electrophoresis was performed using DP transfectant cell lysates immunoprecipitated with monoclonal antibody SG520. Patterns were obtained using a Bio-image analyzer (14).

As shown in Figure 3, patterns determined from DP-transfected L cells were compared with those made by L cells, lymphoblastoid cell line RML (haplotype DRw16, Dw22, DQw7), DQ-transfected L cells, and DR-transfected L cells. The pattern of DP-transfected L cells had one A and one B spot; these were absent in the non-transfected cells. These spots were located at a lower molecular weight position than the corresponding spots from lymphoblastoid cell line RML. The pattern of L cells transfected with DR, DQ, DP appeared at a lower molecular weight position than lymphoblastoid cell line RML. Also, some extra spots were observed on the pattern from transfectant cells. They corresponded with invariant (Ii) spots of the lymphoblastoid cell line RML (Fig. 3C). If these were mouse invariant gene products, DR, DQ, and DP transfected L cells seem to use mouse invariant genes for their expression.

The secondary proliferative response (PLT assay) was measured using DP-transfectants as stimulator cells. Stimulator cells used were DP-transfected L cells, DP-transfected U937, and DP-transfected CEM. Eight kinds of DP typing reagents (DPw1, w2, w3, w4, w5, w6, w7, and Cp63) were generated by priming T cells of allogeneic donors who were matched for all the other known HLA types (12). As shown in Figure 4, a positive stimulation was caused by DP-transfected cells in the PLT assay. With DP-transfected cells of human origin, DP-transfected CEM and DP-transfected U937, a DP-allospecific reaction could be measured. DPw3 specific clones were activated by DP-transfected CEM. This "extra" reaction could be due to the origin of CEM cells. The cell line was derived from a T-cell lymphoma which does not express class II antigens. However the cells may originally have had a suppressed DPw3 gene which may have become desuppressed, i.e., expressed after transfection. DP-transfected L cells were found to strongly stimulate T-cell activity in PLT assays. This result was not DP-specific. Moreover DP-transfected L cells could stimulate primary MLR response (data not shown). The cause could be quantitative, the amount of DP antigen expressed on transfectant cells being larger, although we

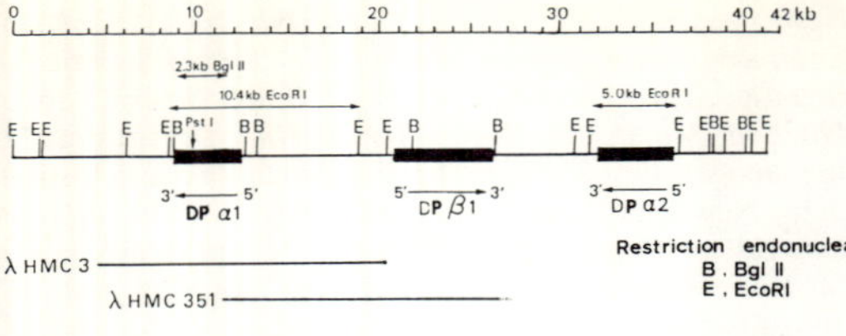

Figure 1. Schema shows the restriction map of clone λHMC3 and λHMC351 present on the HLA-DPα chains gene and DPβ chain gene respectively (5). Genomic libraries were screened with radioactively labeled DNA (3–6 × 10^5 cpm per filter) by colony and plaque hybridization as described by Maniatis and co-workers (15). The human gene library was constructed in the phage vector Charon 4A (16).

The liposome-mediated transfer technique was used (9). A solution of 2 μM phosphatidylserine was dried under reduced nitrogen pressure with an evaporator for 2 h. Five hundred microliters of 10 mM Tris buffer with 0.145 M sodium chloride (Tris-NaCl, pH 7.4) was then added and mixed. The aqueous suspension was sonicated for 5 min in a bath-type sonicator (Branson Sonic Power Co.) with 50% duty cycle. Five hundred microliters of Tris-NaCl with 20 mM calcium chloride was then added. The suspension was incubated for 1 h at 37°C and centrifuged at 4000 rpm for 10 min. One hundred microliters of phosphate-buffered saline (PBS) containing 10 μg each of DPα and DPβ DNA and 1 μg of pSV-2 Neo DNA (selection marker) were added to the pellet. Finally, the DPα, DPβ, and pSV-2 Neo DNA were coencapsulated into the large unilamellar liposome by the addition of 100 μl of Tris-NaCl with 30 mM ethylenediamine tetraacetate followed by brief mixing. The DNA-loaded liposomes were then incubated for 30 min at 37°C, centrifuged at 15000 rpm for 20 min, and resuspended in 500 μl PBS with 2 mM magnesium chloride (2 mM $MgCl_2$-PBS). Approximately 5 × 10^6 parental cells(Ltk$^-$, U937, or CEM, previously described) were washed twice with PBS to remove fetal calf serum proteins and were resuspended in 500 μl of 2 mM $MgCl_2$-PBS. Five hundred microliters of the DNA-loaded liposome suspension was mixed with the cells, and this mixture was then shaken for 30 min at 37°C. The cells were then seeded at a density of 2.5 × 10^5 cells per well into a 24-well tissue culture plate (Corning). The culture medium was changed, 24 h later, to Neo selection medium containing 400 μg/ml Geneticine (Gibco, Gand Island, New York). After 2 to 3 weeks, Neo-positive colonies from each well were roughly scored, dispersed into six wells of a 24-well plate, and grown as a heterogeneous population of cotransfectants.

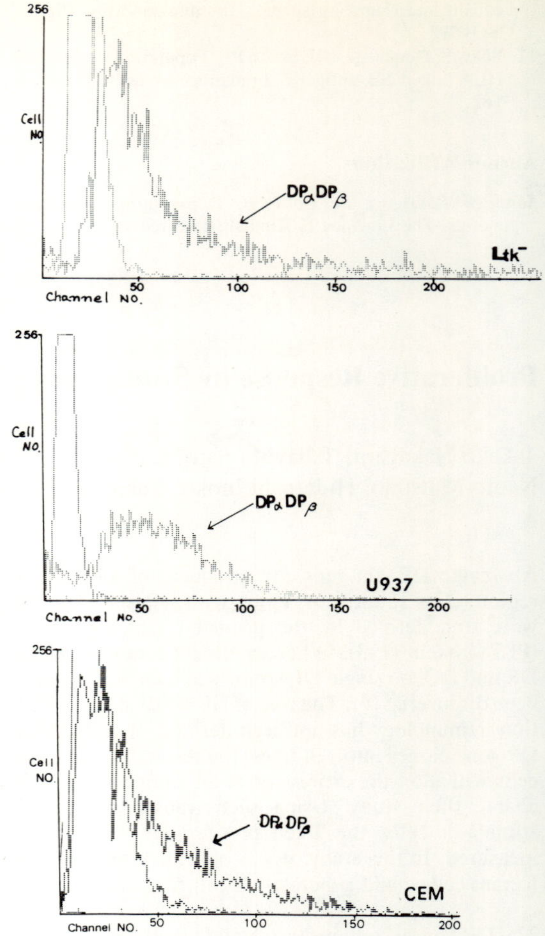

Figure 2. Surface expression of DP transfectants (DP transfected L cell, DP transfected U937, and DP transfected CEM) was analyzed by FACS using HLA class II monoclonal antibody Sa3 (10). Fluorescence intensity was plotted as a linear fraction of cell number. Ten thousand cells were analyzed for each histogram.

Cell sorting was performed by the following method. Cells (5 × 10^6) were washed twice in cold RPMI 1640 containing 5% fetal calf serum (5% FCS-RPMI) and incubated at 4°C with 500 μl of antibody (ascites) at a 1:100 dilution in 5% FCS-RPMI for 30 min. After being washed twice at 4°C with 5% FCS-RPMI, cells were incubated for 15 min at room temperature at a 1:15 dilution of FITC-conjugated $F(ab')_2$ goat anti-mouse immunoglobulin antiserum (Cappel). After two washes, cells were suspended in cold 5% FCS-RPMI and sorted with an Epics-V.

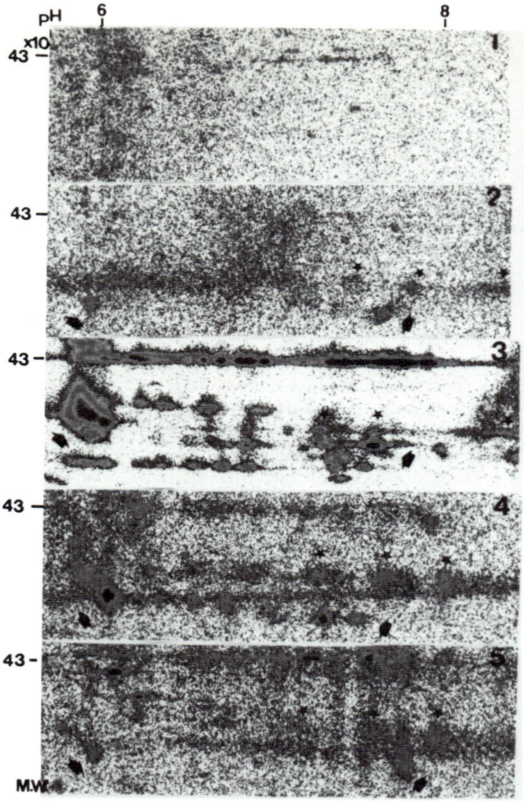

Figure 3. Two-dimensional gel patterns were analyzed using a Bio-image analyzer (14) (Fuji photo film). The cell lysates: nontransfected L cell (A), DP-transfected L cell (B), B-lymphoblastoid cell line RML (C), DR-transfected L cell (D), DQ transfected L cell (E) were immunoprecipitated with a monoclonal antibody SG520 (11). (SG520 was a gift from Dr. S. Goyert. B-lymphoblastoid cell line RML was distributed for the 10th International HLA Workshop Biochemistry Study. Two-dimensional electrophoresis was performed by the method described by Dr. R. Knowles (17) in the Tenth International HLA Workshop Biochemistry Study.) Each transfectant (1×10^7) was biosynthetically labeled with 0.1 m Ci ^{35} S-methionine (NEN), whereas RML cells (2×10^6) were labeled with 25 μ Ci ^{35}S-methionine. After electrophoresis, gels were dried and the autoradiograms were analyzed by a Bio-Image analyzer. The radioactive dried gels were exposed to an imaging plate coated with minute crystals of photostimulable phosphorus in an organic binder for 168 hours in a lead box. Then the imaging plate was scanned with a fine laser beam. The fluorescent light emitted by the laser was detected by a photomultiplier tube, and the resulting electric signal was converted into digital data. A computer then analyzed radiographic pictures and relative radioactivity. The DP-transfected L cell (B) precipitated at 30–31 K molecular weight (A chain spots like) and 24–25 K molecular weight (B chain spots like) the same as the DR-transfected L cells (D) and DQ transfected L cells (E). Moreover, all these transfectants precipitated at 31–34 K molecular weight. These spots were corresponded to invariant spots of RML (C). The relative radioactivity among these patterns (ratio) was as follow; (A) 0.5: (B) 1.0: (C) 8.0: (D) 2.0: (E) 1.5.

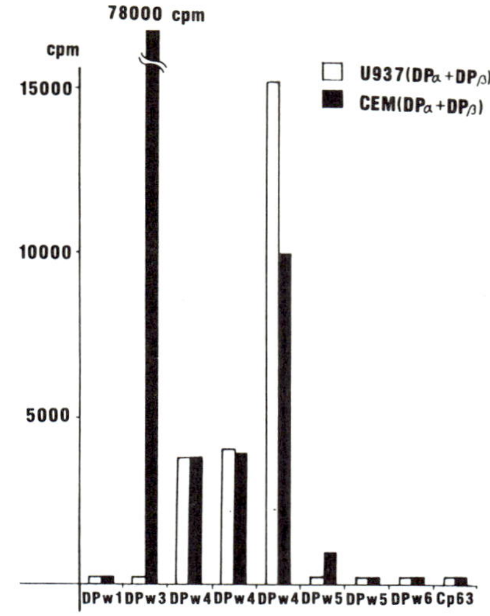

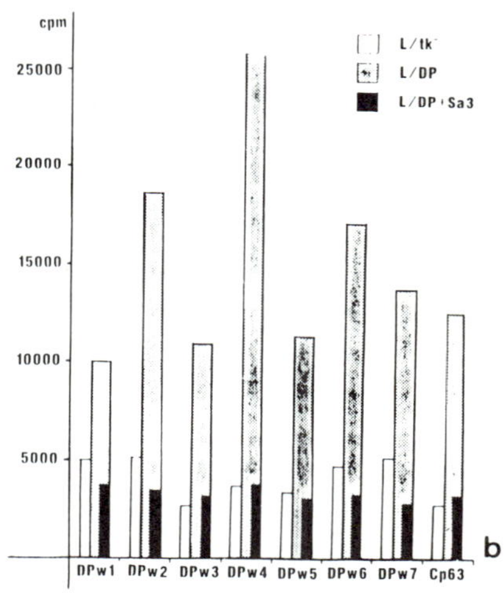

Figure 4. PLT assay using DP transfectants as stimulator cells: (A) DP-transfected CEM and DP transfected U937, (B) DP-transfected L cells. Nontransfected L cell, U937 and CEM were used as control cells. (The PLT reagents DPw1, DPw2, DPw3, DPw4, DPw5, DPw6 were a gift from Dr. R. Hartzman. DPw7 clones were kindly given by Dr. N. Reinsmoen.) Cp63 clones were developed in our Laboratory (18). These were used at a concentration of 1×10^4 cells per well. The stimulator cell irradiated with 10000 rad of ^{60}Co were used at 5×10^4 cells per well. They were cultured together in a 96-well U-bottom tray (Corning), supplemented with 10% AB$^+$ human serum in RPMI-1640, for 48 hours at 37°C in a 5% CO_2 incubator. The cultures were pulsed for 16 hours with 1.0 μ Ci ^{3}H-thymidine (specific activity 20 Ci/mM : NEN). The inhibition assay (B) was performed by adding monoclonal antibody (Sa3 hybridoma supernatant) at the beginning and keeping them in culture throughout. Non-transfected CEM and U937 cell showed low background activity (data not shown).

have to consider a high background activity by mouse cell antigen. A functional study of DP-transfectant cells showed that influenza virus induce a response in cloned DP-restricted human T lymphocytes (13). The choice of the recipient L cells may be very important. A poor choice of cells can lead to difficulties in MLR or PLT studies with transfectant cells. In conclusion it can be said that DP-transfectant cells can stimulate human T-cell proliferation, although the antigen expressed has been found to be of lower molecular weight than normal stimulating antigens.

References

1. Sheehey MJ, Sondel PM, Bach ML, Wank R, Bach FH. HL-A LD (lymphocyte defined) typing: a rapid assay with primed lymphocytes. Science 1975;188:1308.
2. Termijtelen A, Bradley BA, van Rood JJ. A new determinant, defined by PLT, coded for in the HLA region and apparently independent of the HLA-D and DR loci. Tissue Antigen 1980;15:267.
3. Shaw S, Pollack MS, Payne SM, Johnson AH. HLA-linked B cell alloantigens of a new segregant series: population and family studies of the SB antigens. Hum Immunol 1980;1:177.
4. Johnson AH, Thorsby E, Nakatsuji T, Fang T, Moen T, Hartzman RJ. Recognition of an HLA-DPw1 specific alloantiserum raised by planned immunization. Hum Immunol 1986;17:21.
5. Ando A, Inoko H, Kimura M, Ogata S, Tsuji K. Isolation and allelic polymorphism of cDNA clones and genomic clones of HLA-DP heavy and light chains. Hum Immunol 1986;17:355.
6. Goodenow RS, McMillan M, Nicolson M, Sher BT, Eakle K, Davidson N, et al. Identification of the class I genes of the mouse major histocompatibility complex by DNA-mediated gene transfer. Nature 1982;300:237.
7. Sundstrom C, Nilsson K. Establishment and characterization of a human histocytic lymphoma cell line (U-937). Int J Cancer 1976;17:565.
8. Foley GE, Lazarus H, Sidney F, Uzman BG, Boone BA, McCarthy RE. Continuous culture of human lymphoblasts from peripheral blood of a child with acute leukemia. Cancer 1965;18:522.
9. Nakatsuji T, Inoko H, Ando A, Sato T, Koide Y, Tadakuma T, et al. The role of transfected HLA-DQ genes in the mixed lymphocyte reaction-like condition. Immunogenetics 1987;25:1.
10. Ikewaki N, Nakatsuji T, Inoko H, Ando A, Sato K, Bao YH, et al. Monoclonal antibody against HLA-DP antigen on transfected cells. In: Aizawa M (ed): HLA in Asia-Oceania 1986, The Proceeding of the 3rd Asia-Oceania Histocompatibility Workshop Conference. Sapporo, Hokkaido University Press, 1986.
11. Goyert SM, Silver J. Further characterization of HLA-DS molecules: implications for studies assessing the role of human Ia molecules in cell interactions and disease susceptibility. Proc Natl Acad Sci USA 1983;80:5719.
12. Hartzman RJ, Shaw S, Robbins FM. Origin and expansion of SB(DP)-specific reagents characterized in the Ninth International Workshop: Workshop report. In: Albert ED, Baur MP, Mayr WR (eds). Histocompatibility Testing 1984. Berlin, Heidelberg, Springer-Verlag, 1984.
13. Austin P, Trowsdale J, Rudd C, Bodmer W, Feldmann M, Lamb J. Functional expression of HLA-DP genes transfected into mouse fibroblasts. Nature 1985;313:61.
14. Amemiya Y, Wakabayashi K, Tanaka H, Ueno Y, Miyahara J. Laser-stimulated luminescence used to measure X-ray diffraction of a contracting striated muscle. Science 1987; 237:164.
15. Maniatis T, Fritsch EF, Sanbrook J. In situ hybridization of bacterial colonies of bacteriophage plaques. In: Maniatis T (ed): Molecular Cloning. Cold Spring Harbor, NY, Cold Spring Harbor Laboratory, 1982.
16. Lawn RM, Fritsch EF, Parker RC, Blake G, Maniatis T. The isolation and characterization of linked δ- and B-β globulin genes from a cloned library of human DNA. Cell 15:1157.
17. Knowles R. Detailed procedures. In 10th International Histocompatibility Workshop Biochemistry Study of class II antigens. Circular #4, 1986.
18. Nose Y, Nakatsuji T, Tsuji K, Reinsmoen NL. A new HLA-DP(SB) specificity (Cp63) defined in Japanese and Caucasian cell populations by cloned PLT cells. Hum Immunol 1985;13:145.

Author Affiliations

Takako Nakatsuji, Kanako Ito, Asako Ando, Nobuhiko Sano, Naoto Matsuno, Hidetoshi Inoko, Kimiyoshi Tsuji, Department of Transplantation, Tokai University School of Medicine, Isehara; Kanagawa 259–11, Japan; *Tadayuki Sato, Fumiko Tsuchida*, Transplantation Immunology Center, Tokai University Hospital, Isehara, Kanagawa 259–11, Japan; *Yukio Koide, Takato O. Yoshida*, Department of Microbiology and Immunology, Hamamatsu University School of Medicine, Hamamatsu, Shizuoka 430–31, Japan

Analysis of HLA-DP Allelic Sequence Polymorphism Using the In Vitro Enzymatic Amplification of DPα and DPβ Loci

T.L. Bugawan, G.T. Horn, J.A. Hansen, E.M. Mickelson, G. Angelini, G.B. Ferrara, C.M. Long, and H.A. Erlich

Accurate typing of the HLA-DP antigens can provide useful information when considering allogeneic bone marrow transplantation. As with the HLA-DR and DQ antigens, the DP molecules may also play a role in susceptibility to certain immunological disorders. DP typing is currently performed using the primed lymphocyte typing technique (1). This assay, biochemical methods (2), and RFLP analysis (3) have all suggested additional polymorphism beyond the major DPw1–6 types. We are determining the sequence of DPα and DPβ alleles using the PCR DNA amplification technique (4), examining a number of cell lines and individuals who show unusual DP typing reactions. Figure 1 shows a summary of our ongoing analysis. Thus far, only two alleles for the DPα locus have been identified, with the DPA1 allele being more frequent. We have seen no correlation between these DPα alleles and the DPw phenotypes. The DPβ locus, however, exhibits extensive polymorphism. Splits of the haplotypes encoding the DPw4 and DPw2 (and possibly DPw3) types have been seen. Additionally, five new alleles from cells typed as DP (blank) were observed. Thus far, 15 different DPβ alleles have been sequenced. Oligonucleotide probes (5) to several regions of DPβ have been synthesized and are currently under evaluation. The dispersed nature of the DPβ polymorphism, as well as the pattern by which identical "epitopes" appear in different alleles, will complicate the application of these probes to a DP typing assay.

References

1. Mickelson, EM, Hansen JA. The Mixed Lymphocyte Culture Reaction and Typing for HLA-D Determinants. In "AACHT Laboratory Manual, Vol. 4", p. 1, Zachary and Braun, eds. 1981.
2. Lotteau V, Teyton L, Tongio M, Thomsen M, Sasportes M, Charron D. Biochemical Polymorphism of the HLA-DP Heavy Chain. Immunogenetics 1987;25:403.

A. DP-alpha

```
                10        20        30        40        50        60        70        80
DPA1:     DHVSTYAAFVQTHRPTGEFMFEFDEDEMFYVDLDKKETVWHLEEFGQAFSFEAQGGLANIAILNNNLNTLIQRSNHTQATN   (LB)
DPA2:     ---------------------------Q------------------R--------------------------------A-   (Daudi)
(SXA):    -------E-------S--Y------E-Q---N--E--M--P-P--IHT-D-G--R-I-G-VMARKH---R-NG KQ-W--D
          ------->                                                                <------
```

B. DP-beta

```
              10        20        30        40        50        60        70        80        90
DPB4.1:   NYLFQGRQECYAFNGTQRFLERYIYNREEFARFDSDVGEFRAVTELGRPAAEYWNSQKDILEEKRAVPDRMCRHNYELGGPMTLQRR   (HHK)
DPB4.2:     ----------------------------V------------------DE----------------------------------   (APD)
DPB2.1:   ------------------------------V------------------DE------------E-----------------------   (WJR)
DPB2.2:     ---------------------------LV------------------E-------------E-----------------------   (QBL)
DPB8:       ----------------------------V------------------DE------------E------V-------DEAV-----   (Piaz)
DPB5:       ---------------------------LV------------------E----------------------------DEAV-----   (HAS)
DPB7:     ----------------------------------------------------------------------V-------DEAV-----   (Raji)
DPB3:     --VY-L------------------------V------------------DED-------L----------V-------DEAV-----   (SLE)
DPB6:       VY-L------------------------V------------------DED-------L---E--------------DEAV---     (JMOS)
DPB11:      VY-L---------------------Q-Y-----------------------------L---R--------------DEAV---     (CRK)
DPB1:       VY-------------------------Y----------------------------------------V-------DEAV---     (LUY)
DPB10:      VH-L------------------------V------------------DE------------E------V-------DEAV---     (BM21)
DPB9:       VH-L------------------------V------------------DED-----------E------V-------DEAV---     (TOK)
DPB12:    --VH-L-----G---------S--------V------------------DED-----------E------V--R----DEAV-----   (AKIBA)
(SXB):    -SVY-E------------VVDGL------YVH---A----L--M-----IG--F-----FM-R---EV-KV---K---ME-LIR---   (Priess)
      ----->                                                                                   <-----
           |-A-|                    |-B-|              |-C-|    |-D-|    |-E-|   |-F-|
```

Figure 1. Comparison of HLA-DP allelic variation. DNA sequences were translated to the standard one-letter amino acid codes and then aligned to show variable residues. A dash indicates identity with the prototypic sequence listed above each group of alleles. Positions within the mature protein are listed on top, our working designation for the allele is shown at the left, and one source of the allele is shown at right. Most of the alleles have been isolated from three or more sources. The sequence of the adjacent DP (SX) pseudogenes are included for comparison. (A) Sequences of the HLA-DPα alleles. In our analysis, the DPA1 allele was isolated from cell lines COX, DKY, LUY, and QBL, while DPA2 allele was seen in cell lines LUY and TOK. (B) Sequences of the HLA-DPβ alleles. The designation of these DPβ alleles has been chosen to coincide when possible to the DPw type of the source(s). Alleles isolated from DP (blank) haplotypes are designated as DPB7–12.

3. Hyldig-Nielsen JJ, Morling N, Odum N, Ryder LP, Platz P, Jakobsen B, et al. Restriction Fragment Length Polymorphism of the HLA-DP Subregion and Correlations to HLA-DP Phenotypes. Proc. Natl. Acad. Sci. (USA) 1987;84:1644.
4. Scharf SJ, Horn GT, Erlich HA. Direct Cloning and Sequence Analysis of Enzymatically Amplified Genomic Sequences. Science 1986;233:1076.
5. Bugawan TL, Horn GT, Long CM, Michelson E, Hansen JA, Ferrara GB, Angelini G, Erlich HA. Analysis of HLA-DP Allelic Sequence Polymorphism Using the In Vitro Enzymatic DNA Amplification of DPα and DPβ Loci. J Immunol 1988;141:4024.

Author Affiliations

G.T. Horn, T.L. Bugawan, C.M. Long, H.A. Erlich, Department of Human Genetics, Cetus Corporation, Emeryville, CA; *J.A. Hansen, E.M. Mickelson*, Fred Hutchinson Cancer Research Center, Puget Sound Blood Center, Seattle, WA, USA; *G. Angelini, G.B. Ferrara*, Istituto Tumori, Genova, Italy

Molecular Analysis of HLA-DP: DP β Chain Charge Heterogeneity Correlates with PLT Subtyping

R.E. Bontrop, N. Otting, E.J. Baas, and M.J. Giphart

HLA-DP molecules were isolated from Epstein-Barr virus-transformed B cell lines by immunoprecipitation with monoclonal antibody (MAb) B7/21.2 and analyzed by two-dimensional gel electrophoresis. By mixing HLA-DP molecules with HLA-DR or HLA-DQ molecules, which were separately isolated with the locus-specific MAbs B8.11.2 (HLA-DR) and SPV-L3 (HLA-DQ), and subsequent analysis on the same gel, DP α and β chains could be distinguished from DR and DQ polypeptides. DP β chains show (unlike DR β and DQ β chains) under normal conditions a poor penetration into the first-dimension IEF gels. Part of these β chains do not reach equilibrium and even tend to comigrate with the DQ α chain. It has been suggested that DP β chains are either markedly less soluble or perhaps more prone to aggregate than DR and DQ β chains (1). This latter possibility seems to be true and may be due to the HLA-DP β chain-specific cysteine residue found at position 214 (2), which may be involved in aggregate formation caused by disulfide bonds. Increasing the amount of β-mercaptoethanol in the IEF sample buffer indeed did improve the penetration of DP β chains in the first-dimension gels.

Table 1. HLA Class II Specificities of the Cell Lines Used in this Study

Cell	HLA				DP β Isoelectric Point Type		
Line	-D	-DR	-DQ	-DP	DPβw1	DPβw2	DPβw4
OOS	w1	1	w1	w2		+	
IWB	w2	2	w1	w2		+	
WIB	w2	2	w1	w2		+	
NOL	w2	2	w1	w4			+
QBL	w3	3	w2	w2		+	
CAA	w3	3	w2	w1,w4	+		+
HAR	w3	3	w2	w4			+
BSM	w4	4	w3	w2		+	
ATH	w5	w11	w3	w2		+	
LUY	w8	w8	w3	w1,w4	+		+

Within the panel tested, three allelic charge varieties of the DP β chain could be distinguished corresponding to the PLT defined specificities HLA-DPw1, -DPw2 and -DPw4 (Table 1). Resuming, DPw1 β chains appeared to have a more basic isoelectric point than the DPw2 β chains and the DPw4 β chains are more negatively charged than the DPw1 β polypeptides. Because all cell lines tested appear to express DP α chains with identical electrophoretic mobility and a polymorphic DP β chain, it may be concluded that these results suggest that DP β polypeptides bear the epitopes detected by PLT subtyping for DP specificities. However, this may not be true for all HLA-DP specificities since it has recently been demonstrated that DPα chains may exhibit some polymorphism (3) and besides this, cell lines with the HLA-DPw3, -DPw5, or -DPw6 specificities are lacking within our panel.

Acknowledgments. This work was supported in part by the Dutch Foundation for Medical Research (MEDIGON), which is subsidized by the Dutch Organization for the Advancement of Pure research (ZWO) and the J.A. Cohen Institute for Radiopathology and Radiation Protection (IRS).

References

1. Watson AJ, de Mars R, Trowbridge JS, Bach FH. Detection of a novel human class II HLA antigen. Nature 1983; 304:358.
2. Figueroa F, Klein J. The evolution of MHC class II antigens. Immunol. Today 1986;7:78.
3. Lotteau V, Teyton L, Tongio MM, Soulier A, Thomsen M, Sasportes M, Charron D. Biochemical polymorphism of the HLA-DP heavy chain. Immunogenetics 1987;25:403.

Author Affiliation

Department of Immunohaematology and Blood Bank, University Hospital Leiden, 2300 RC Leiden, The Netherlands

Immunobiology of HLA
Volume II

Sequence of a DPβ cDNA from the DR5, DQw7, DPw4b Cell Line Swei

Karen Rice, Micki Klearman, Daniel K. Didier, Joel Schiffenbauer, Susan L. Woulfe, Sherry Shuman, and Benjamin D. Schwartz

The HLA-DP molecules were the last of the HLA class II molecules to be identified. These molecules were originally detected by the primed lymphocyte test. Their low representation on cells compared to DR and DQ molecules, and the difficulty in generating DP-specific antibodies have resulted in these molecules being the least well studied with respect to both disease associations and biochemical and molecular biologic characterization. Indeed, very few DP genes have been sequenced to date.

As part of our ongoing effort to sequence all the expressed class II genes from the cell line Swei (DR5, $DRw52b_1$, $DRw52b_2$, DQw3, DPw4b), we have cloned and sequenced a cDNA clone corresponding to the DPw4b β chain. A λgt11 library of Swei was screened using a biotinylated DPβ probe. The clone was sequenced by dideoxy chain termination using the M13 universal primer and DPβ specific oligonucleotides.

The isolated cDNA is ~1100 bp, and encodes 27 of the 28 amino acids of the leader sequence and the 229 amino acids of the mature DPw4b β chain. The mature protein contains two extracellular domains encompassing 186 amino acids, a transition region of ten amino acids, a transmembrane region of 22 amino acids, and an intracytoplasmic region of 11 amino acids. As in other class II β chains, there is a potential N-linked glycosylation site at position 19. In contrast to other class II β chains, there is an additional potential N-linked glycosylation site at position 98 in all DPβ sequences. Cysteine residues potentially involved in disulfide bridges are located at positions 15, 77, 115, and 171.

Comparison with the DRβ (1) and DQβ (2) sequences from Swei reveals that DPβ has the shortest extracellular sequence (196 amino acids) of the three class II β chains. $DR5\beta_I$ has an extracellular sequence of 198 amino acids, a transmembrane region of 21 amino acids, and an intracytoplasmic region of 18 amino acids. DQw3β has an extracellular sequence of 198 amino acids, a transmembrane region of 21 amino acids and a ten amino acid intracytoplasmic tail. The decreased length of the extracellular domains of the DPβ chain results from a two amino acid deletion in DPβ corresponding to positions 24 and 25 of DRβ and DQβ. This two amino acid loss accounts for the shift in position of the cysteine residues found at positions 79, 117, and 173 in DRβ and DQβ to 77, 115, and 171 in DPβ.

Comparison with the limited number of other DPβ sequences available (3–7) indicates that the majority of polymorphism lies within the first domain (Fig. 1). Although the precise delineation of "variable" regions awaits the sequencing of other DPβ genes, it appears as if variable regions occur at positions 8–11, 55–57, and 84–87. Furthermore the occurrence of clusters of amino acid substitutions such as either Asp-Glu or Ala-Ala at positions 55–56, and either Gly-Gly-Pro-Met or Asp-Glu-Ala-Val at positions 84–87 in all DPβ chains sequenced to date suggest that gene conversion is at least partially responsible for the polymorphism among the DPβ genes. Finally, the single amino acid difference of Lys in DPw4bβ for Glu in DPw2β (6) at position 69 suggests this residue may be responsible for the DP specificity. However, the DPα chain may also play a role.

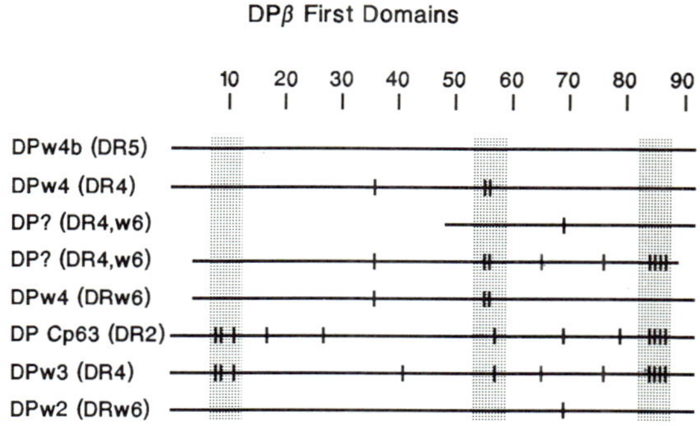

Figure 1. DPβ first domains. Horizontal lines indicate first domain sequence from several DPβ sequences. Vertical tick marks indicate amino acid differences from first sequence.

Acknowledgments. The authors wish to thank Ms. Pat Stewart and Mrs. Lorraine Whiteley for expert preparation of the manuscript.

Supported in part by NIH grants AI-18925 and AI-15322.

References

1. Tieber VL, Abruzzini LF, Didier DK, Schwartz BD, Rotwein P. Complete characterization and sequence of an HLA class II DRβ chain cDNA from the DR5 haplotype. J Biol Chem 1986;261:2738.
2. Schiffenbauer J, Didier DK, Klearman M, Rice K, Shuman S, Tieber VL, et al. Complete sequence of the HLA-DQα and DQβ cDNA from a DR5/DQw3 cell line. J Immunol 1987;139:228.
3. Gustafsson K, Widmark E, Jonssdon AK, Servenius B, Sachs DH, Larhammar D, et al. Class II genes of the human major histocompatibility complex. Evolution of the DP region as deduced from nucleotide sequences of the four genes. J Biol Chem 1987;262:8778.
4. Ando A, Inoko H, Kimura M, Ogata S, Tsuji K. Isolation and allelic polymorphism of cDNA clones and genomic clones of

HLA-DP heavy and light chains. Human Immunol 1986; 17:355.
5. Long E, Gorski J, Mach B. Structural relationship of the SB β-chain gene to the HLA-D-region genes and murine I region genes. Nature 1984;310:233.
6. Kappes DJ, Arnot D, Okada K, Strominger JL. Structure and polymorphism of the HLA class II SB light chain genes. EMBO J 1984;3:2985.
7. Roux-Dosseto M, Auffray C, Lillie JW, Boss JM, Cohen D, DeMars R, et al. Genetic mapping of a human class II antigen β-chain cDNA clone to the SB region of the HLA complex. Proc Natl Acad Sci USA 1983;80:6036.

Author Affiliations

Division of Rheumatology, Departments of Medicine and Microbiology and Immunology, and the Howard Hughes Medical Institute, Washington University School of Medicine, St. Louis, MO 63110, USA

T Cell Clones Show Polymorphism Within DPw1, DPw2, and DPw3 Specificities

Arlette Urlacher, Yoko Mitsuishi, and Marie M. Tongio

Anti-DPw1, DPw2, and DPw3 T cell clones were grown after cloning PLT reagents obtained in local intrafamilial combinations. These were selected so that stimulating and responding cells carried one haplotype in common and a second haplotype identical for DR and DQ alleles (DR and DQ typing was confirmed using the Tenth Workshop antisera), but different for DP alleles. The DP typing of the families had previously been performed with the PLT reagents GNN1 to GNN6 (kindly provided by Dr. Hartzman).

For the Workshop studies, only proliferative tests were performed, according to the Tenth Workshop protocol. Scores from 1 to 8 were attributed by the Workshop statistical data analysis (N. Flomenberg) to the reactivity of each clone tested against each stimulating lymphoblastoid cell line (WS LBL). The results were as follows.

1) Two anti-DPw1 clones were obtained, which stemmed from two intrafamilial cell combinations where the mother's responding cells were primed against the paternal DPw1 antigen carried, respectively, by one and the other of two HLA identical siblings. The HLA types of the mother and the two siblings were the following:

Mother	Sibling 1 (Re) and sibling 2 (Bea)
A24 B7 DR3 DRw52 DQw2 DP3	A1 B8 DR3 DRw52 DQw2 DPw1
A24 B35 DR11 DRw52 DQw3 DP-	A24 B7 DR3 DRw52 DQw2 DPw3

Clone no. 6067 was obtained by using the cells of child 1 (Re) as stimulators and the mother's lymphocytes as responders. Clone no. 6068 was obtained by using the cells of child 2 (Bea) as stimulators and the mother's cells as responders.

Tested on local DPw1 families, the two clones showed an identical pattern of reactivity and recognized clearly segregating epitopes within these families (data not shown).

Tested on the Tenth Workshop LBL panel (Table 1), clone no. 6067 recognized all the cell lines typed as DPw1 by the official typing labs (Dr. Hartzman) except two: LBLs no. 9058 OMW and no. 9070 LUY. No cross-reactions with antigens other than DPw1 were observed.

Clone no. 6068 recognized all the DPw1 LBLs except one: cell line no. 9021 RSH. The RFLP studies could show (Y. Mitsuishi, this volume) that cell line no. 9021 had a DP α RFLP pattern different from the other DPw1 WS LBLs. Since cell line no. 9088, which shows a different DPβ RFLP pattern within the DPw1 specificity, stimulates the proliferation of clone no. 6068, the correlation with RFLP studies suggests that the epitope recognized by clone no. 6068 is located on the DP α chain.

Comparing the results obtained with clone no. 6067 and no. 6068, it can be concluded that clone no. 6068 recognizes an epitope located on DPw1 that is different from that found with clone no. 6067. The reactivity of these two clones obtained with the same genetic combinations points out that different epitopes can be recog-

Table 1. Pattern of Reactivity of Anti-DPw1 Clones No. 6067 and No. 6068

Workshop LBL ID	Clone No. 6067	Clone No. 6068
9006	8*	8*
9021	8	1
9023	8	8
9049	8	8
9058	1	8
9070	1	8
9086	NT	+
9088	NT	+
9102	NT	+

*Results are given either in scores 1 to 8 attributed to the Tenth Workshop data analysis, or by +/− when only local data analysis was performed. NT = not tested.

Table 2. Pattern of Reactivity of Anti-DPw2 Clone No. 6064

Workshop LBL ID		Clone No. 6064
9008	DPw2	8
9011	DPw2	8
9012	DPw2	8
9017	DPw2	8
9020	DPw2	8
9029	DPw2	8
9032	DPw2	4
9036	DPw2	8
9038	DPw2	8
9039	DPw2	8
9044	DPw2	8
9045	DPw2	6
9050	DPw2	8
9056	DPw2	8
9068	DPw2	8
9085	DPw2	+
9093	DPw2	+
9103	DPw2	+
9074	DPw2	4
9063	DPw2	1
9003	DPw CDPHEI	8

nized within a single DPw1 antigen (same paternal DPw1 antigen presented to the mother's cells by lymphocytes of two HLA identical siblings.

2) One anti-DPw2 clone (no. 6064) was studied. The intrafamilial cell combination from which it was raised was the following:

Stimulating cell (child):	MA.CA :	
		A32 B44 DR11 DQw3 DPw2
		A24 B7 DR3 DQw2 DPw3
Responding cell (mother):	MA.NI.:	
		A24 B7 DR3 DQw2 DPw3
		A24 B35 DR11 DQw3 DP-

This clone recognized all the local DPw2 families and showed clear intrafamilial segregation (data not shown).

Tested on the WS LBL panel (Table 2) it recognized, without any cross-reaction, all the cells typed as DPw2, except cell line no. 9063 (WT47). The RFLP studies (Y. Mitsuishi) have shown that cell line no. 9063 has an RFLP DPβ pattern different from that of the other DPw2 LBLs. Since cell lines no. 9008 (DO 208915) and 9032 (BSM), which show a different DPw2 DPα pattern, are recognized by clone no. 6064, it can be suggested that clone no. 6064 recognizes a DPw2 epitope located on the DPβ chain.

An unexpected and unexplained result was that clone no. 6064 proliferated when stimulated by the LBL no. 9003 (KAS116), which carries the CDP HEI DPw specificity, whereas it did not recognize LBL no. 9034 SAVC carrying this same specificity (N. Odum, Newsletter no. 3, vol. 1).

Table 3. Pattern of Reactivity of Anti-DPw3 Clone No. 6066 and No. 6069

Workshop LBL ID	Clone No. 6066	Clone No. 6069
9007	8	8
9009	1	1
9015	8	8
9018	8	NT
9022	8	8
9030	8	8
9040	8	8
9059	8	8
9071	8	8
9079	+	+
9087	NT	NT
9090	+	NT
9097	+	NT
9100	+	NT

3) Two anti-DPw3 clones (no. 6066 and no. 6069) were grown from a single PLT reagent. The HLA types of the stimulating and responding cells were the following:

Stimulating cell (child):	MA.A. :	
		A1 B8 DR3 DQw2 DPw1
		A24 B7 DR11 DQw3 DPw3
Responding cell (father):	MA. Re.:	
		A1 B8 DR3 DQw2 DPw1
		A32 B44 DR11 DQw3 DPw2

Clone no. 6066 and no. 6069 showed exactly the same pattern of reactivity. All the local DPw3 families were recognized by these two clones and clear intrafamilial segregation was observed (data not shown). Tested on the Workshop LBL panel (Table 3), all the DPw3 LBLs were recognized except no. 9009 (KAS011). The RFLP studies showed that this cell line has a DPα RFLP pattern different from the other DPw3 WS LBLs. This suggests that clone no. 6066 recognizes an epitope located on the DP α gene. No cross-reactions with DPw specificities other than DPw3 were observed.

It can be concluded from the DPw epitope analysis, thanks to alloreactive PLT clones, that: 1) T cell clones raised against DPw specificities show that different epitopes within a single antigen can be recognized; and 2) correlation with RFLP studies lead us suggest that the recognized epitopes can be located on both the α and the β chains.

Author Affiliation

Centre de Transfusion Sanguine de Strasbourg, France

Section X
Class II Histocompatibility Antigens: Other HLA Class Products

HLA-DNA May Encode Two Unique Protein Products

Sandra Rosen-Bronson and Eric O. Long

DNA is a human HLA class II α-chain gene. In contrast to other class II MHC genes (HLA-DR, -DQ, and -DP), which are transcribed and processed into 1.3 kb mRNAs, the DNA gene produces two transcripts of 1.2 kb and 3.5 kb. By sequence analysis of cDNA clones, we have shown that the 3.5-kb transcript is not an unspliced precursor of the shorter transcript. The 3.5-kb transcript is fully spliced and identical to the short transcript except for an extended 3′ untranslated region. In addition, from DNA sequence analysis of several cDNA clones, we found that alternative splicing occurs. The DNA transcript is spliced either as predicted by the genomic sequence (1) or uses an alternative splice site (145 bp 5′ of the α1-exon). This alternative splice results in a change of reading frame and introduces two stop codons just 5′ of the α1-exon. Thus, the DNA gene may encode an entirely different (non-class II) secreted protein when the alternative splice is utilized.

Seven DNA cDNA clones were isolated from a λgt10 Raji cDNA library; two of these cDNA clones utilized the alternative splice acceptor. After DNA sequencing to confirm the reading frame shift, in vitro transcription and translation studies were done to test whether the alternatively spliced gene could be translated into the predicted protein product. For these studies, cDNA clones were subcloned into the Sp64 vector (2), which allowed transcription of RNA from cDNA using Sp6 RNA polymerase. In vitro transcribed RNA was subsequently translated in vitro using rabbit reticulocyte lysates. These studies showed that cDNA clones, which are conventionally spliced, are translated into a protein of the appropriate size for a typical class II α-chain. RNA from an alternatively spliced cDNA clone, on the other hand, was translated into a small peptide protein (M.W. ≅ 5000), as predicted.

Northern blot analysis using an alternative splice-specific probe showed that alternatively spliced mRNA was present in several normal class II-positive cells and tissues. Approximately 10% of the total DNA transcripts are alternatively spliced.

Currently, we are generating antibodies specific for the potential alternative protein to search for its production in various cells, as well as cells transfected with expressible DNA clones.

References

1. Trowsdale J, Kelly A. The human HLA class II α chain gene DZ α is distinct from genes in the DP, DQ, and DR subregions EMBO J 1985;4:2231.
2. Melton DA, Krieg PA, Rebagliati MR, Maniatis T, Zinn K, Green MR. Nucleic Acid Res 1984;12:7035.

Author Affiliation

Laboratory of Immunogenetics, National Institute of Allergy and Infectious Diseases, National Institutes of Health, Bethesda, MD 20892,USA

A Novel HLA-Class II Molecule Distinct from HLA-DP/DQ/DR

Nelson Fernandez, Mario O. Labeta, Marcej Kurpisz, and Hilliard Festenstein

In addition to the genes coding for the HLA-DP/-DQ/-DR products, additional loci-designated DX α and DX β (1), DO β (1,2), and DZ α (3,4) have been identified and located between the DQ and DP subregions. Several hundred kb of DNA has been found between them that may include as yet additional undiscovered class II genes. Although messenger RNAs for the DO and DZ genes have been found, it is not yet known whether these loci code for cell surface expressed products (4). Thus, the human MHC class II molecules consist of structurally well-defined heterodimers [e.g., DR,DQ, and DP] and undefined potential heterodimers [e.g., DX,DZ,DO]. Conceivably mechanisms such as antigen-driven and/or allostimulation, interisotypic chain associations and transassociations may greatly increase the repertaoire of HLA-class II heterodimers.

We provide here molecular evidence for a novel monomorphic HLA-class II molecule [provisionally designated DNN] distinct from HLA-DP/DQ/DR using a panel of lymphoblastoid cells that include HLA-loss mutants (5,6). The expression of this molecule was investigated using a set of monomorphic mouse monoclonal antibodies (MAb) including one of the IgG2a subclass designated EDU-1 (7). This antibody reacts strongly in a cell binding radioimmunoassay (CB-RIA) with HLA-DR/DQ loss mutants derived from the lymphoblastoid parental cell line 721. One of the loss mutants, 45.1, is hemizygous having lost the paternal haplotype. The

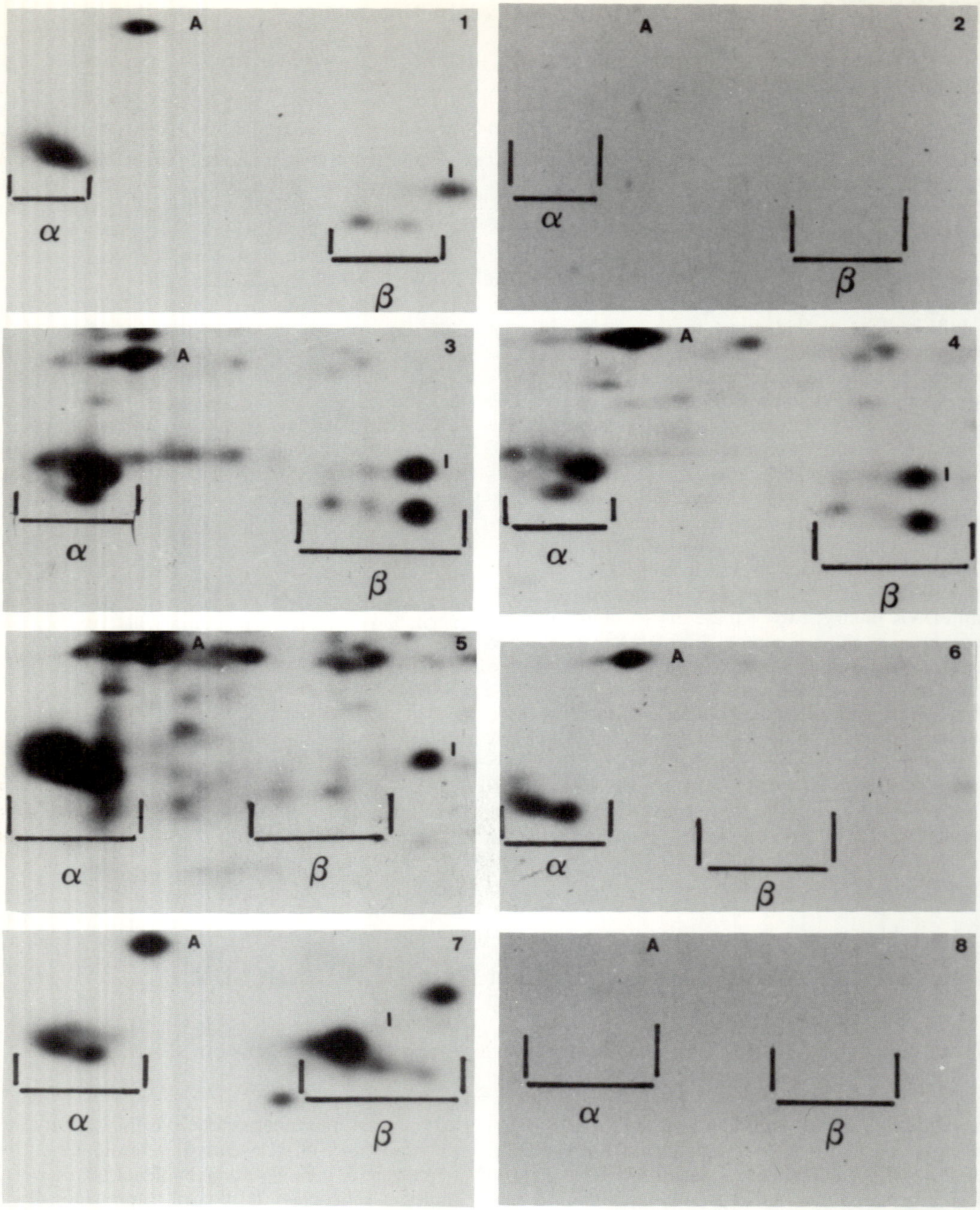

Figure 1. Non-equilibrium pH gradient gel electrophoresis (NEPHGE) analysis of the molecules immunoprecipitated by HLA-class II specific monomorphic MAbs. Panels 1 and 2 show HLA-DR antigens immunoprecipitated with L243 from cells 45.1 and 84.5 (HLA-DR/-DQ loss mutant). Panels 3 and 4 correspond to 45.1 and 84.5 immuonprecipitated with EDU-1. Panel 5 and 6 shows 45.1 and 84.5 immunoprecipitated with B7-21. Panels 7 and 8 indicate 45.1 and 84.5 immunoprecipitated with TU22. The α and β chains are assigned by overlapping acidic and basic regions of the gels. A = actin; I = invariant chain. Approximately three times as much material was used to precipitate HLA-DQ or DP as DR and six times as much for EDU-1 precipitations. All pellets containing the HLA-class II molecules were washed in lysis buffer supplemented with 0.5 m NaCl and 0.1% SDS to remove loosely associated polypeptides. The antigens were subjected to one cycle neuraminidase treatment followed by incubation in NEPHGE buffer and separated in a reverse polarity NEPHGE gel (for methods see references 17 and 18). (From: Immunogenetics 1988;27:363–369)

mutant 84.5 has in addition a deletion of the HLA-DR and -DQ loci of the maternal haplotype. In addition, the EDU-1 MAb also reacted with a local panel of homozygous EBV-transformed cell lines which includes all the HLA-class II serologically and cellularly defined specificities. The reactive molecules were further detected on allostimulated CD4-positive T-cell clones (unpublished observations) and various leukemic cells including those of myeloid origin, which lack surface expression of HLA-DQ molecules (8). These data suggest that DNN corresponds to a backbone HLA-class II determinant expressed on leukemic cells of different origin and EBV-transformed cells of different HLA phenotypes. Moreover, this antigen structure is distinct from that of HLA-DP/DQ/DR as shown by direct immunoprecipitation, and following stepwise serial immunosaturation using immunopurified DR,DQ and DP-specific monomorphic MAbs (data not shown). Comparative two-dimensional gel analysis of ^{35}S-Met labeled class II molecules isolated from the cell lines 45.1 and 84.5 (Fig. 1, panels 3 and 4) revealed that the molecule immunoprecipitated with the MAb EDU-1 has an α subunit with a unique pattern which is resolved into three distinguishable spots (pI 5.0-5.5; Mr 31 kd). These spots are differentiated by minor changes in isoelectric point and molecular weight. A further distinguishing feature of this class II molecule is that it has a strongly reacting β chain (pI 6.5-7.5; Mr 29 kd) that resolves along a short pH gradient into three regularly spaced spots. The HLA-DR molecules (Fig. 1, panel 1) precipitated with the MAb L243 from the cell line 45.1 revealed the presence of a single α chain spot (pI 5.0-5.5; Mr 33 kd) and two closely related β chain spots (pI 6.5-7.0; Mr 30 kd). A similar molecular pattern was obtained from the parental cell 721 (data not shown). These antigens were not immunoprecipitated from the HLA deletion mutant 84.5 (Fig. 1, panel 2). The HLA-DP molecules precipitated with the antibody B7-21 from the cell 45.1 and 84.5 (Fig. 1, panels 5 and 6) showed that the α chain resolves into two closely associated spots (pI 4.5-5.0; Mr 30 kd). Similar to previous reports, the HLA-DP β region spots were not clearly resolved. The HLA-DP pattern observed in the HLA-loss mutant 84.5 was clearly different from the antigen detected by EDU-1. The HLA-DQ specific monomorphic MAb TU22 (Fig. 1, panel 7) immunoprecipitated two similar and closely migrating spots in the acidic side of the gel (pI 4.5; Mr 3.1) and two related spots in the basic side of the gel (pI 6.5-7.0; Mr 28 kd). The latter pattern of α and β DQ patterns were not obtained in the HLA-DQ deleted mutant cell line.

The DX α, DX β, DZ α, DP α, DP β, but not the DR and DQ genes are present in the 84.5 deletion mutant (9). Although it is known from mRNA studies that the DO β and DZ α genes are transcribed, no corresponding surface product has so far been identified (3,4). It could be that the molecule described in this study is specified by these genes. Restriction fragment length polymorphism analysis have indicated that the DNA surrounding the DO and DZ genes is polymorphic, but this polymorphism may not reside in the coding sequences of the genes. Alternatively, the HLA class II monomorphic product described here may be specific by new as yet undescribed genes in the region between DP and DQ.

One possibility to explain the existence of DNN is that novel structures formed between non-DR/DQ α and β chains such as DP,DX,DZ and DO generate a family of novel heterodimers as a result of mixed isotypes. The existence of mixed isotype class-II molecules was first demonstrated in mouse L cells transfected with murine class-II genes (10). Moreover, a variety of HLA-DR,DQ,DP and DO containing mixed isotypes were found expressed after transfection of genes encoding different chain combinations into mouse L cells (11). These gene products were not detectable by DR,DQ or DP-specific MAb. The presence of mixed isotypes between DR α and DQ β has recently been shown on lymphoblastoid EBV-transformed cells (12).

In addition to the above, novel polymorphic class-II heterodimers designated the "fourth Ia subset" have recently been described (13). The nature of such antigen structure is unknown. There are also data from Pawelec and co-workers (14,15) demonstrating the existence of a monomorphic class II molecule termed DY which is different from DP,DQ and DR. Their conclusions are derived from functional assays in which self-reactive CD4 positive T-cell suppressor clones are activated by a putative novel monomorphic class II determinant. The suppressor effect is abrogated by the high affinity, multi-locus MAb termed TU39 which also detects class II heterodimers in addition to DR,DQ and DP (16). The relationship between DN and DY and other class II molecules is currently being investigated by immunomagnetic separation of activated T-cell subpopulations with monomorphic MAbs and immunopurification of class II molecules for peptide mapping and partial aminoacid sequencing.

Acknowledgments. We thank R. DeMars for the mutant cells, I. Trowbridge for the B7.21 monoclonal antibody, A. Ziegler for the TU22 monoclonal antibody, and J. Vives for the EDU-1 monoclonal antibody. We are grateful to D. Williams for secretarial assistance.

This work was supported by the Medical Research Cancer, the Cancer Research Campaign and the Leukaemia Research Fund of Great Britain.

References

1. Okada K, Boss J, Prentice H, Spies I, Mengler R, Auffray C, et al. Gene organisation of DC and DX subregions of the human major histocompatibility complex. Proc Natl Acad Sci USA 1985;82:3410–3414.
2. Tonnelle C, DeMars R, Long EO. DO: A new beta chain gene in HLA-D with distinct regulation of expression. EMBO J 1985;4:2847.
3. Speilman R, Lee JS, Bodmer WF, Bodmer J, Trowsdale J. Six HLA-D region alpha chain genes of DC-alpha-related sequences with DR types. Proc Natl Acad Sci USA 1984; 81:3461–3465.

4. Trowsdale J, Kelly A. The human HLA-class II alpha gene DZ is distinct from genes in the DP, DQ and DR subregion. EMBO J 1985;4:2231–2237.
5. Kavathas P, DeMars R, Bach FH, Shaw S. SB: A new HLA-linked human histocompatibility gene defined using HLA-mutant cell lines. Nature 1981;293:747–749.
6. DeMars R. Mutations that dissect the D-region of a human B-lymphoblastoid cell line. Disease Markers 1984;2:175–182.
7. Vilella R, Yaque J, Gallant MT, Vives J. The efficacy of hibridoma technology in studying surface antigens in human T and B lymphocytes. Immunologia 1983;2:58–62.
8. Navarrete C, Fernandez N, Alonso C, Festenstein H. Ontogenic and functional implications of the differential expressions of HLA-DQ antigens on leukaemic cells. Hum Immunol 1986;16:52–68.
9. Erlich H, Lee JS, Petersen JW, Bugawan T, DeMars R. Molecular analysis of HLA class I and class II antigen loss mutants reveals a homozygous deletion of the DR,DQ and part of the DP region: Implications for class II gene order. Hum Immunol 1986;16:205–219.
10. Germain RN, Quill H. Unexpected expression of a unique mixed isotype class II molecule by transfected L cells. Nature 1986;320:72.
11. Long EO, Rosen-Bronson S, Jacobson S, Sekaly EP. Molecular analysis of the diversity and function of MHC class II antigens (Abstract). Immunobiology Supplement 3:204.
12. Lotteau V, Teyton L, Burroughs D, Charron D. A novel HLA class II molecule (DR-alpha-DQ-beta) created by mismatched isotype pairing. Nature 1987;329:339–341.
13. Carra G, Acolla R. Structural analysis of human-Ia antigens reveals the existence of a fourth molecular subset distinct from DP,DQ and DR molecules. J Exp Med 1987;165: 4763–4773.
14. Pawelec G, Schneider ME, Rehbein A, Balko I, Wernet P. Dissection of suppressor cell generation in vitro. Hum Immunol 1986;17:343–354.
15. Wernet P, Pawelec CP, Schneider EM. Cellular detection of human class II antigens: Delineation of a novel HLA-DP-like suppressor restriction system, DY, the sequential expression of class II antigens and a pronounced functional flexibility of class II alloproliferative T cell clones. In: Solheim BG, Moller E, Ferrone S (eds): Human Class II Antigens. A Comprehensive Review of Structure and Function. Berlin/Heidelberg, Springer-Verlag, 1986.
16. Pawelec G, Fernandez N, Schneider M, Festenstein H, Wernet P. "DY" determinants, possibly associated with novel class II molecules stimulate autoreactive CD4+ T cells with suppressive activity. J Exp Med 1988;167: 243–261.
17. Rudd CE, Bodmer JG, Bodmer WF, Crumpton MJ. HLA-D region antigen associated invariant polypeptides as revealed by two-dimensional gel analysis. Biol Chem 1985; 260:1927–1933.
18. Fernandez N, Labeta M, Kurpisz M, Solana R, Festenstein H. HLA-DQ molecular heterogeneity in HLA-DR4-Dw4 consanguineous cell lines. Disease Markers 1986;4:191–198.

Author Affiliations

Nelson Fernandez, Mario O. Labeta, Hilliard Festenstein, Department of Immunology, The London Hospital Medical College, England; *Marcej Kurpisz,* Institute of Human Genetics, Polish Academy of Sciences, Poznan Poland

Comparative Tissue Reactivity of the Monoclonal Antibody TU39 Which Detects HLA-Class II Molecules Including a Novel Determinant Designated DY

Claudia Müller, Graham Pawelec, Nelson Fernandez, and Gerhard Müller

The monoclonal antibody (MAb) TU39 (1–3), which was used to precipitate HLA-DR/-DP/-DQ and a putative novel class II determinant, was able to block the stimulation of T-cell clones activated by "DY" a non-DR/-DP/-DQ molecule (4).

The expression of "DY" as tested by immunoperoxidase staining on frozen tissue sections, was compared with that of HLA-DP as defined by the MAb B7/21 (5). The expression of "DY" was also compared with -DQ and -DR as detected by the MAbs TU22 and TU34, respectively. In normal lymphoid organs, HLA-DP antigens are normally expressed on follicular B lymphocytes, activated T cells, dendritic, and interdigitating cells. They are also variably expressed on thymic epithelium and macrophages. They are normally absent from HLA-DQ-negative, -DR-negative and TU39-negative bile duct epithelium. They are also absent from HLA-DQ negative, -DR positive, and TU39-positive duodenal epithelium (Table 1). Selective expression of "DY" antigens on bile duct epithelium was shown to be present by local staining of TU39 in the absence of anti-DR, anti-DQ, and anti-DP labeling on one of three frozen livers. Fifteen normal kidneys did not express HLA-DP and -DQ antigens on the proximal tubular epithelium, although nine of them reacted with HLA-DR-specific antibodies. However, all renal sections revealed binding of TU39 to proximal tubular cells, as well as in three cases to the epithelium of Henle's loop. HLA-DP antigens were sometimes expressed on the capillary endothelium of normal organs, but rarely on the epithelium of larger vessels. They are absent from HLA-DQ-negative, -DR-positive sinus endothe-

Table 1. Comparative Study of the Expression of HLA-DP/-DQ/-DR and the Novel TU39 Reactive Determinants Designated "DY"

Moab	Specificity	Gut Duodenum (N=3)	Liver Bile Duct (N=3)	Kidney Prox. Tubules (N=15)	Placenta Amnionepithel. (N=2)
B7/21	DP	–	–	–	+++ ○
Tü22	DQ	–	–	–	–
Tü34	DR	++++ ●	–	++++ ○*	++++ ●
Tü39	DR/DP/DY	++++ ●	++ ○†	++++ ●	++++ ●

Gut duodenum, liver bile duct, kidney proximal tubules, and placental amnioepithelium were analyzed according to the method described in reference 3.
– absence of binding; ○ positive reaction on subpopulation of cells; ● positive reaction on all cells; ++ weak reaction; ++++ very strong reactivity.
*9 cases
†1 case.

lium of liver and spleen. In the 20th gestational week, HLA-DP and -DQ antigens are absent from the vascular endothelium of fetal tissue. However, TU39 stained the renal capillaries, arterioles, and venules of liver and kidney despite being negative for HLA-DR, -DQ and -DP antigens.

During acute and chronic GvHR of the skin after human allogeneic bone marrow transplantation, TU39-positive antigens were detected on epithelium in sequential biopsies. These molecules preceded the expression of HLA-DR and -DP. HLA-DQ antigens were only rarely expressed. A similar kinetic study of the expression of HLA-class II antigens on the epithelium of fallopian tubes was made and the expression appeared to depend on the hormonal activity of the ovary. Thus, HLA-DP and -DQ antigens seem to have a more restricted tissue distribution than HLA-DR antigens, whereas the "DY" molecule has the broadest tissue distribution and ontogenically the earliest tissue appearance. Since the "DY" determinants seem to be predominantly involved in inducing cellular suppression (4), such tissue distribution could have important implications for local organ immunity.

Acknowledgments. Supported by DFG Sonderforschungsbereich 120, Project A1 and A3 and the Medical Research Council of Great Britain

References

1. Ziegler A, Heinig J, Müller C, Götz H, Thinnes FP, Uchanska-Ziegler B, et al. Analysis by sequential immunoprecipitations of the specificities of the monoclonal antibodies TÜ22, 34, 36, 37, 39, 43, 58, and YD1/63.HLK directed against human class II antigens. Immunobiol 1986; 171:77.
2. Crumpton MJ, Bodmer JG, Bodmer WF, Heyes JM, Lindsay J, Rudd CE. Biochemistry of class II antigens: Workshop Report. In: Albert ED, et al (eds): Histocompatibility Testing 1984. Heidelberg, Springer Verlag, 1984.
3. Müller C, Ziegler A, Müller C, Hadam M, Waller HD, Wernet P. Divergent expression of HLA-DC/MB, -DR and -SB region products on normal and pathological tissues as detected by monoclonal antibodies. Immunobiol 1985;169:228.
4. Pawelec G, Fernandez N, Schneider EM, Festenstein H, Wernet P. "DY" determinants, possibly associated with novel class II molecules stimulate autoreactive CD4+ T cells with suppressive activity. J Exp Med 1988;167:243.
5. Watson AJ, DeMars R, Trowbridge IS, Bach FH. Detection of a novel human class II HLA-antigen. Nature 1983; 304:358.

Author Affiliations

Claudia Müller, Graham Pawelec, Gerhard Müller, Medical University Clinic, Departments II and III, Tübingen, FRG; *Nelson Fernandez*, Department of Immunology, The London Hospital Medical College, London, England

Section XI
Class II Histocompatibility Antigens: Human Anti-HLA Class II Monoclonal Antibodies

Human Anti-HLA Monoclonal Antibodies

Maria Pia Pistillo, Nobuyuki Tanigaki, Osvaldo Mazzoleni, Anna Longo, Guido Frumento, and Giovanni Battista Ferrara

Abstract: Since many murine anti-HLA monoclonal antibodies (MAbs) produced so far react with monomorphic determinants, human HLA-specific MAbs can be more useful as reagents for precisely defining HLA polymorphism. Here we show that polymorphic human anti-HLA MAbs can be obtained by Epstein-Barr virus (EBV)-transformation of immune B cells followed by selection and stabilization of the transformed antibody-producing cells. By this technique, so far seven cell lines, MP1, 2, 3, 4, 5, 6 and 7 secreting cytotoxic human alloantibodies of restricted HLA specificity, have been established. All the cell lines were found to produce antibodies of the IgM class and one cell line was found to produce also IgG in small amount. Four out of the six lines were only one light chain producers and this finding, together with stability in antibody production, was suggestive for monoclonality. Direct evidence for this was provided on MP1 antibody by immunochemical data.

Specific B cells obtained from peripheral blood of cell donors under planned HLA-immunization (1) have been transformed into permanent antibody-secreting lines by exposure to EBV as previously described (2,3). After few weeks of culture, the culture supernatant was tested for presence of complement-dependent cytotoxic antibodies reacting with the target cells specifically used for immunizing the cell donor.

By this procedure, seven B-lymphoblastoid cell lines, MP1, 2, 3, 4, 5, 6, and 7 that secrete cytotoxic alloantibodies with polymorphic HLA specificity have been established. The antibody-producing cells were selected by serial propagation of cell clusters (at least 3 passages) as already described (2–5) and, where possible, by limiting dilution method. All MP cell lines showed long-term stability in specific antibody production for periods ranging from one to three years of continuous culture.

The HLA-specificity pattern of MP alloantibodies has been defined by testing their cytotoxicity on an HLA reference-cell panel consisting of peripheral B lymphocytes from unrelated individuals and HLA homozygous typing cell lines. Segregation analysis within informative families indicates that these antibodies are directed against HLA specificities (data not shown). The results of the population study are represented in the histograms of Figure 1 in which the local HLA specificity assigned to the antibodies is shown.

In order to determine the immunoglobulin isotype of MP antibodies, we used cytoplasmic double staining of MP cell lines. In this technique the cells were treated with combinations of flourescein isothiocyanate (FITC)-conjugated antibody of one heavy or light chain specificity and a tetramethylrhodamine isothiocyanate (TRITC)-conjugated antibody of another heavy or light chain specificity.

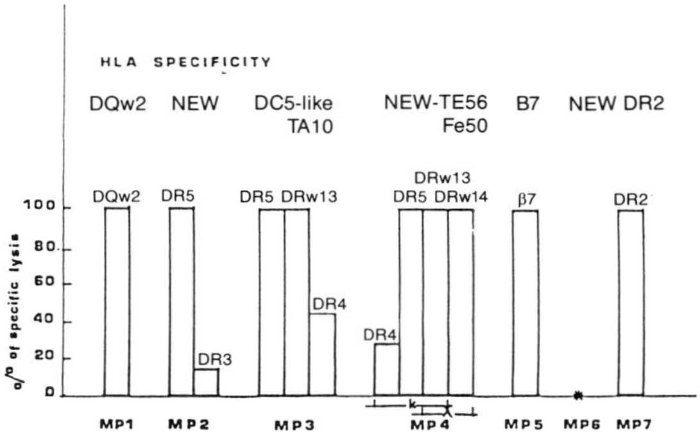

Figure 1. HLA Typing was performed by standard NIH microtoxicity assay. (*) indicates positive reaction only with B lymphocytes from the donor and from two family members.

To evaluate the monospecificity of our fluorochrome-conjugated antibodies we used control cell lines that had been characterized previously for cytoplasmic single class of heavy chain and single type of light chain. The MP1 line produces only IgM (μ,λ), the GM 1500 line (6) produces IgG (γ,k), and the MP6.1 line produces IgM (μ,k). Staining of these control cell lines confirmed the monospecificity of our anti-Ig reagents and allowed us to define their optimal dilutions.

Specific FITC-conjugated antibodies gave approximately the same percentage of stained cells as the corresponding TRITC-conjugated.

The results of Table 1 indicate that all MP cell lines except MP3 produced one single heavy chain, the μ chain. MP1, 5, 6, and 7 cell lines were only one light chain producers and MP2, 3, and 4 lines produced both κ and λ light chains. No individual cells were found to be double producers. Although MP6 parental cell line was cloned by limiting dilution at 0.5 cells/well (MP6.1),

Table 1. Cytoplasmic Heavy and Light Chain Determinations*

	Percentage of cells stained with			
Cell lines	anti-μ	anti-γ	anti-κ	anti-λ
MP1	84	0	0	88
MP2	84	0	2	90
MP3	88	4	14	66
MP4	84	0	34	54
MP5	87	0	56	0
MP6.1	73	0	97	0
MP7	52	0	0	70
GM 1500	0	88	82	0

*Cells were treated with combinations of FITC- and TRITC-conjugated antisera to human μ and γ, and κ and λ chains (Cappel, Cochranville, PA, USA).

Immunobiology of HLA
Volume II

repeated attempts to clone the other cell lines at less than 10 cells/well were unsuccessful. For this reason we separated the κ- and λ-positive cell populations of MP3 and MP4 by using panning and FACS techniques. We are now investigating the Ig isotype and the HLA specificity of the antibodies secreted by the new cell populations.

The presence of one single isotype reflects the monoclonality of MP6.1 clone and supports the hypothesis that MP1, 5, and 7 cell lines could have been cloned spontaneously at some stage through successive selection by "cluster picking" passages.

Recent immunochemical data formally confirm the monoclonality of λ-type IgM secreted by MP1 cells. Purification by affinity chromatography using anti-Fab antibodies coupled to Sepharose 4B has been carried out on the culture supernatant MP1 cell line of 162, a hybridoma between MP1 and D33 (7) that was cloned twice by the limiting dilution method. The two purified IgM preparations were labeled with ^{125}I and then subjected to isoelectrofocusing along with ^{125}I-labeled preparation of IgM obtained from a patient with Waldenstrom syndrome. The banding pattern of MP1 IgM was identical to that of 162 IgM and also was restricted in a narrow pH range comparable to that of the IgM from Waldenstrom patient (data not shown).

These results give solid evidence for the monoclonality of IgM antibody secreted by MP1 cells.

Biochemical investigations need to be performed in a similar way on the other alloantibodies to assess their monoclonality.

Acknowledgments. This work was supported in part by Progetto Finalizzato Ingegneria Genetica, CNR, in part by Progetto Finalizzato Oncologia CNR N.86.00659.44 and CNR N.86.00487.04, and also by the United States Public Health Grant AI-20251.

References

1. Ferrara GB, Tosi R, Longo A, Castellani A, Viviani C, Carminati G. A safe blood transfusion procedure for immunization against major histocompatibility complex determinants in man. Transplantation 1978;26:150.
2. Pistillo MP, Hammerling U, Dupont B, Ferrara GB. In vitro production of a human HLA alloantibody of restricted specificity (DQw2) via Epstein-Barr virus transformation. Hum Immunol 1986;15:109.
3. Pistillo MP, Tanigaki N, Mazzoleni O, Ciccone E, Hammerling U, Park M, Terasaki PI, Ferrara GB. Human lymphoblastoid cell lines secreting antibodies with restricted HLA specificity. Immunogenetics 1987;25:145.
4. Hammerling U, Kosinski S, Chua R, Livingstone P. Production of human monoclonal alloantibodies: considerations of an optimal strategy. Perspect Immunogenet Histocompatibility 1984;6:61.
5. Kosinski S, Hammerling U, Yang SY. A human monoclonal antibody to an HLA-DRw53 (MT3)-like epitope on class II antigens. Tissue Antigens 1986;28:150.
6. Croce CM, Linnenbach A, Hall W, Steplewski Z, Koprowski H. Production of human hybridomas secreting antibodies to measles virus. Nature 1980;288:488.
7. Teng NNH, Lam KS, Riera FC, Kaplan HS. Construction and testing of mouse-human heteromyelomas for human monoclonal antibody production. Proc Natl Acad Sci USA 1983;80:7308.

Author Affiliations

Maria Pia Pistillo, Osvaldo Mazzoleni, Anna Longo, Guido Frumento, Nobuyki Tanigaki, Istituto Nazionale per la Ricerca sul Cancro, Genova, Italy; Roswell Park Memorial Institute, Buffalo, NY, USA; *Giovanni Battista Ferrara*, Instituto Nazionale per la Ricerca sul Cancro, Genova, and Cattedra di Genetica Umana, I° Facoltà di Medicina, Università di Napoli and Centro Ricerche Immunoematologiche, Bergamo, Italy

A Human Monoclonal Antibody Binds to a Subset of HLA-DR Molecules on DR2/DQW1 Cell Line Cells

W.H. Marshall, Y. Wasi, and S. Drover

A human monoclonal antibody, NFLD.H1, studied serologically at the Tenth Workshop has also been examined by the two-dimensional gel methods used in the Workshop. The antibody was produced from B cells from a multiparous female, which were EBV-transformed and then fused with the KR4 fusion partner (1). It is an IgM, cytotoxic antibody, and in our hands only reacts with B cell lines and not with peripheral blood B cells. Its pattern of reaction is highly dependent on concentration and shows specificity for HLA DR2/DQw1-positive cell lines when in low concentration; however, as the concentration increases, it reacts with a broader range of cells including, at high concentrations, an autologous cell line.

To define the molecular species to which it binds, we did two types of experiment. 1) blocking with monomorphic mouse monoclonal antibodies, and 2) two-dimensional gel electrophoresis of immunoprecipitates. Its binding was effectively blocked by NFLD.M6, a monoclonal binding to a class II monomorphic determinant (data to be published). NFLD.M6 was shown to precipitate polypeptides with the same P.I. and molecular weight as did L243 (a monoclonal antibody predominantly binding to a DR monomorphic determinant) and NFLD.M1 (an antibody primarily against HLA-DR4 polypeptide chains), both of which were evaluated in the Workshop class II biochemistry section. Thus, as NFLD.M6 blocked the binding of NFLD.H1, the latter presumably also binds to HLA-DR.

NFLD.H1 was effective in immunoprecipitation with a homozygous DR2/DQw1-positive cell line, GM3161 (from The Human Genetic Mutant Cell Repository, Institute for Medical Research, Camden, NJ 08103). It precipitated only one set of spots; these were of low molecular weight (around 28 K) and lie below the lower set of standards and, curiously, there is no change of molecular weight in this series despite changes in P.I. The DR-specific monoclonal antibody L243 precipitated the same set of low molecular weight chains, but in those gels there were traces of other DR β spots as well as some putative α chains, neither of which were seen with NFLD.H1. The predominantly DQ-specific monoclonal antibody, Tu22, precipitated two prominent α chains, and the gel showed some fainter spots in the β region, including a small amount of the low molecular weight material. It is curious that NFLD.H1 only precipitated the low molecular weight polypeptides.

Of four DR2 homozygous cell lines characterized in the Workshop as having DRβ III chains of "wX" type, three showed signs of low molecular weight material migrating to the same region as that precipitated by NFLD.H1, but the spots were small compared with other DR β chain spots. This was provisionally named DR β "X" product, and was not recorded in any other cell lines (see class II biochemistry report).

The importance of NFLD.HI appears to be its ability to preferentially precipitate this low molecular weight material, which is so abundantly expressed on the cell line GM-3161. Thus a reagent is available for studying the proposed DR β gene and its product. Samples of NFLD.H1 are available to interested researchers.

Acknowledgments. The authors are grateful to Dr. R.W. Knowles and Ms. B. Kienzle for their excellent organization of the class II biochemistry and for provision of the method and nearly all the reagents.

Reference

1. Kozbor D, Lagarde AE, Roder JC. Human hybridomas constructed with antigen specific Epstein-Barr Virus-transformed cell lines. Proc Natl Acad Sci USA 1982;79:6651.

Author Affiliation

Immunology Laboratories, Faculty of Medicine, Memorial University of Newfoundland, St. John's, Newfoundland A1B 3V6

Five Human IgG Monoclonal Antibodies Obtained from a Single Polytransfused Patient Specific for Different Class I and Class II Major Histocompatibility Antigens

V. Dessi, H. Lazarus, E. Raniezi, F. Celada, R.J. Winchester, and A. Nuñez-Roldan

Five human monoclonal antibodies (MAbs) were derived from a single polytransfused patient awaiting renal transplantation. In microcytotoxicity assays, the patient's serum displayed strong positive reactions against >90% of a panel of cells representing the known human leukocyte antigen (HLA) specificities. The donor's (RR, HLA phenotype: A1, A9; B35, B37; DR3, DR4; DRw52, DRw53; DQw2, DQw3) peripheral blood lymphocytes (PBLs) were infected with Epstein-Barr (EBV), cloned, and supernatants of the virus-transformed cultures were screened for the presence of IgG antilymphocyte reactivity utilizing an ELISA method. Positive cultures were recloned and fused with the human/mouse heteromyeloma SHM.

Supernatants from five clones were selected for alloreactivity characterized by indirect immunofluorescent staining and fluorescent-activated cell sorter (FACS) analysis on homozygous typing cell lines, including those from the workshop core panel and on cell lines derived from selected families. MAb 91c2 was a perfect anti-HLA-A2+Aw68+Aw69 reagent (correlation coefficient, $r = 1$). MAb 34F11 (IgG1, λ), RRA7, and RRF9 (IgG1, κ) were reactive against all cells bearing the DQw1 phenotype ($r = 1$). MAb 2A2 (IgG1, κ) showed a high correlation with DQw1, although it was also reactive against core cells LUY and BM16 (both DQw3).

Based on light chain differences and reactivity patterns, we can conclude that at least three clones derived from the same individual recognize epitopes found on the DQw1 molecule. One of them (2A2) may also cross-react with an epitope present on a subtype of DQw3.

Author Affiliations

V. Dessi, E. Raniezi, and F. Celada, Immunology Service, University of Genova, Genova, Italy; *H. Lazarus, R.J. Winchester,* and *A. Nuñez-Roldan*, Department of Rheumatic Diseases, Hospital for Joint Diseases, New York University School of Medicine, New York, NY, USA. Address correspondence to *A. Nuñez-Roldan*, Servicio de Inmunologia, Hospital Universitario Virgen del Rocio, 41013, Sevilla, Spain.

Section XII
Class II Histocompatibility Antigens: Regulation of HLA Class II Expression

Cloning of a cDNA Encoding Class II Specific DNA Binding Protein

Daniel K. Didier, Joel Schiffenbauer, Susan L. Woulfe, and Benjamin D. Schwartz

Abstract: The class II molecules of the HLA complex are expressed primarily on immunocompetent cells including B-cells, activated T cells, and macrophages. A coordinate level of expression is seen between the class II α and β subunits. Recent studies have focused on the mechanisms underlying this coordinate level of expression, and have localized most regulatory control to the 5′ upstream region of the class II genes specifically to sequences that have been termed the Z (or W), X, and Y boxes (1–3). We designed our studies to isolate and characterize the regulatory proteins that interact with the X and Y boxes of the DRα gene. We found a number of nuclear proteins that bind to the class II box. Based on the cells from which the nuclear proteins were isolated, these proteins appear to be both negative and positive regulatory factors. We have directly cloned a cDNA corresponding to a gene that encodes a protein which appears relatively specific for the Y box, and the transcription of this gene appears inversely proportional to class II expression.

Five regions are presently thought to be involved in regulation and initiation of transcription of class II genes. Three regions designated the Z (or W), X, and Y boxes (1–3) are highly conserved and thought to be involved in tissue specific and γ-interferon (γ-IFN) inducibility of class II expression. The Y box contains an inverted CCAAT box and may be the binding site either for a ubiquitous CCAAT binding nuclear factor such as NF-1, or for a unique CCAAT binding nuclear factor designated NF-Y(4). The fourth region designated is an octamer sequence also found in the promoters of immunoglobulin genes. The fifth region designated T is the TATA box which has been implicated in accurate initiation of transcription.

We designed our studies to isolate and characterize the regulatory proteins that interact with the X and Y boxes of the DRα gene. We therefore synthesized a 48bp double stranded (ds) oligonucleotide corresponding to the 5′ region of the DRα gene which spans the X box, spacer region, and Y box: CCTAGCAACAGATG CGTCATCTCAAAATATTTTT CTGATTGGCCAAAG. This ds oligonucleotide is termed the class II box. The inverted CCAAT box within the Y box is underlined.

Nuclear proteins (5) from murine L cells and from a human B-lymphoblastoid cell line (B-LCL) binding to the class II box were assayed by Southwestern blotting (6) and gel retardation (7) using ^{32}P-end labeled class II box in the presence of poly I-C, as a noncompetitive inhibitor. Southwestern analysis revealed several proteins that bind the class II box although only two proteins showed strong binding. Some binding proteins were shared between the two cell types whereas others were specific.

Gel retardation revealed one major band of retarded mobility from L-cell nuclear extracts, and two major retarded bands from B-LCL extracts. In a cold competition experiment in which unlabeled class II box competed with the ^{32}P-labeled class II box for the nuclear proteins, the retarded bands were no longer seen, demonstrating the specificity of these reactions. The observations that the L cells do not express class II molecules, yet possess a nuclear factor that binds to the class II box suggest that this nuclear factor may have a negative regulatory effect. In contrast, the presence of nuclear factors in the class II positive B-LCL suggests that these factors have positive regulatory effects.

We next attempted to directly clone the cDNA(s) that encodes the positive regulatory protein(s) that bind to the class II box. A cDNA library from the class II positive B-LCL Swei was made using λgt11, and the amplified library of 1×10^7 clones was plated. The plates were overlaid with nitrocellulose filters saturated with isopropyl thiogalactoside (IPTG) to induce expression of the chimeric β-galactosidase fusion proteins. Filters were hybridized with ^{32}P-end labeled class II box in the presence of poly I-C. The filters were washed and autoradiographed. Upon initial screening, 34 clones were positive. After three rounds of plaque purification, only three clones remained positive. Further characterization indicated one clone was nonspecific, and the other two clones were specific, but identical. This clone was expanded and characterized.

In order to more specifically assign the specificity of the protein encoded by this cDNA, the expanded purified clone was plated, and the plates were overlaid with nitrocellulose filters saturated with IPTG to induce expression of the protein. The filters were then probed with the entire class II box, or with ds oligonucleotides corresponding to the X box, Y box, the spacer region, and a DPα coding sequence as a nonspecific control (Fig. 1).

The entire class II box bound strongly to the protein produced. The nonspecific DPα probe did not bind at all. The X box, spacer, and Y box all bound. However, the X box bound only weakly, the spacer region bound at an intermediate level, and the Y box bound most strongly. These results suggest that the cDNA we have cloned encodes a protein that has the greatest affinity for the Y box.

To study the effect of γ-IFN and phorbol myristate acetate (PMA) on the expression of this protein, the human U937 monocytoid cell line which expresses little or no class II was stimulated with PMA and γ-IFN to induce class II expression. Transcripts of RNA determining this protein were assayed by Northern analysis using the hexamer labeled cloned cDNA as a probe. The class II positive B-LCL and the class II negative murine L

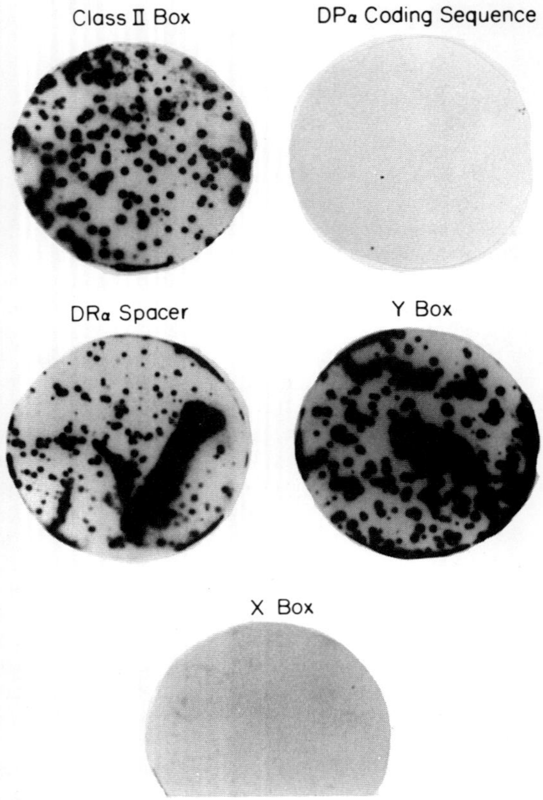

Figure 1. The clone encoding the class II box binding protein was expanded and plated. Replicate nitrocellulose filters impregnated with IPTG to induce fusion protein synthesis were hybridized with ds oligonucliotides corresponding to the class II box, the X box, Y box, spacer region, and a DPα coding sequence as a nonspecific control.

cells were included as controls. Weak hybridization was observed to RNA isolated from the B-LCL and from the U937 cells that had been stimulated to express high levels of class II molecules. A strong hybridization signal was seen with unstimulated U937 cells, and with L cells. Thus, the transcript hybridizing to the cloned cDNA inversely parallels the expression of class II molecules, suggesting that it may indeed encode a negative regulatory protein.

Footprinting is currently underway to precisely determine the region of the class II box to which this protein binds. We have also synthesized ds oligonucleotides corresponding to the CCAAT boxes for β-globin, herpes simplex virus, thymidine kinase, and murine sarcoma virus to determine if this protein is specific for the Y region CCAAT box, CCAAT boxes in general, or another portion of the Y box. These studies will more precisely determine the specificity of the isolated protein.

Note. A more complete description has been published in Didier et al. Proc Natl Acad Sci USA 1988; 85:7322.

Acknowledgments. The authors wish to thank Ms. Pat Stewart and Mrs. Lorraine Whiteley for expert preparation of the manuscript.

Research has been supported in part by NIH grants AI-18925 and AI-15322.

References

1. Sullivan KE, Calman AF, Makamishi M, Tsang S, Wang Y, Peterlin BM. A model for the transcriptional regulation of MHC class II genes. Immunology Today 1987;8:289.
2. Miwa K, Doyle C, Strominger JL. Sequence-specific interactions of nuclear factors with conserved sequences of human class II major histocompatibility complex genes. Proc Natl Acad Sci USA 1987;84:4939.
3. Dorn A, Durand B, Marfing C, Le Meur M, Benoist C, Mathis D. Conserved major histocompatibility complex class II boxes—X and Y—are transcriptional control elements and specifically bind nuclear proteins. Proc Natl Acad Sci USA 1987;84:6249.
4. Dorn A, Bollekens J, Staub A, Benoist C, Mathis D. A multiplicity of CCAAT box-binding proteins. Cell 1987;50:863.
5. Dignam JD, Lebovitz RM, Roeder RG, Accurate transcriptional initiation by RNA polymerase II in a soluble extract from isolated mammalian nuclei. Nucl Acids Res 1983;11:1475.
6. Bauer B, Steinberg J, Laemmli VK, Weintraub H. The detection of DNA-binding proteins by protein blotting. Nucl Acids Res 1980;8:1.
7. Strauss F, Varshavsky A. A protein binds to a satellite DNA repeat at three specific sites that would be brought into mutual proximity by DNA folding in the nucleosome. Cell 1984;37:889.

Author Affiliations

Howard Hughes Medical Institute and Departments of Medicine and Microbiology and Immunology, Washington University School of Medicine, St. Louis, MO 63110, USA

Regulation of HLA Class II Genes: Identification of a Regulatory Promoter Binding Protein Missing in Class II-Deficient Congenital Immunodeficiency

Walter Reith, Sarah Satola, Iris Amaldi, Carlos Herrero Sanchez, Christine Berte, Richard Ulevitch, and Bernard Mach

Abstract: Class II major histocompatibility (MHC) antigens are essential to the control of the immune response, because they present antigen to T lymphocytes and thus mediate T-cell activation (1). Moreover, the pathogenesis of autoimmune diseases may involve aberrant class II antigen expression (2–4). The study of class II gene regulation is therefore crucial for our comprehension of the immune system. We have analyzed the regulation of HLA-class II genes in two systems; induction by interferon-γ (IFN-γ) in fibroblasts, and constitutive expression in both normal B cell lines and class II-deficient B cell lines from severe combined immunodeficiency (SCID) patients (5). Induction by IFN-γ is mediated by the *de novo* synthesis of a class II inducer protein. This synthesis is an early event that precedes activation of class II gene transcription. In normal B cells, we have identified three proteins (RF-X, RF-Y, and OBP) that bind specifically to the DRA promoter. The molecular defect in SCID cells is the absence of RF-X, suggesting that this protein is a regulatory factor that is essential for normal class II gene transcription.

HLA class II antigens are encoded by a multigene family located in the D region of the MHC(6,7). Treatment of certain class II negative cells with IFN-γ leads to the coordinate induction of all the functional class II genes and associated invariant chain (In) genes (8,9). We have studied the molecular mechanisms underlying this induction in an osteosarcoma fibroblast line called 143B (18 and Reith et al, manuscript submitted).

Nuclear run-off assays show that induction of class II and In genes in IFN-γ-treated 143B cells occurs at the level of transcription. Time-course experiments demonstrate that induction is characterized by a long lag phase which precedes transcriptional activation (6 h), mRNA accumulation (9 h), and cell-surface expression of antigen (12 h). In order to determine whether this lag phase reflects a requirement for *de novo* protein synthesis, we carried out stimulations with IFN-γ in the presence of the protein synthesis inhibitor cycloheximide (CHX). CHX abolishes the induction of class II and In genes by IFN-γ, indicating that *de novo* protein synthesis is indeed essential. A detailed study, involving the addition or removal of CHX at different points in time after a 30-minute stimulation with IFN-γ, has shown that this protein synthesis is an early event that coincides with the lag phase observed for transcriptional activation of class II and In genes. We can thus conclude that the effect of IFN-γ on class II and In gene transcription is mediated by a newly synthesized class II inducer protein (Fig. 1A, top). Independent experiments suggest that an elevation of the intracellular level of cAMP can inhibit an early event during IFN-γ-mediated induction of class II genes (Ulevitch et al, manuscript in preparation), perhaps by blocking synthesis of the class II inducer protein (Fig. 1A, top).

The mechanism we propose for class II induction is different from the one involved in the induction of HLA class I genes by IFN-γ (Reith et al, submitted). In 143B cells, protein synthesis is not required for class I gene activation. In fact, CHX by itself can induce class I genes with kinetics similar to those observed for IFN-γ-mediated induction. We therefor propose that IFN-γ activates class I genes by interrupting the production of an active labile repressor that normally inhibits class I gene transcription (Fig. 1A, bottom).

Class II-deficient severe combined immunodeficiency (SCID) is characterized by an absence of constitutive and inducible class II gene expression (5,10). The mutation responsible has been shown to lie in a class II regulatory gene located outside of the MHC (10). We have made use of B lymphocyte lines from SCID patients to identify regulatory factors controlling constitutive class II gene expression. Absence of expression in SCID B cells is due to a complete lack of class II gene transcription. We have therefore identified and compared class II promoter binding proteins present in normal and SCID B cell lines (19).

The HLA-DRA promoter contains three putative cis-acting regulatory sequences: the two conserved class II boxes X and Y (11,12), and an octamer sequence that is shared by several non-class II promoters including those of SV40 and immunoglobulin genes (13,14). When DRA promoter fragments containing these three sequence elements are used in gel retardation assays, four specific protein-DNA complexes (B1-B4) are formed in nuclear extracts from normal B cells. Of these four complexes, only B2 and B4 are also formed in extracts from SCID B cells. Competition experiments done with an SV40 promoter fragment, an oligonucleotide containing the Ig octamer sequence, and an oligonucleotide covering the DRA Y box, have shown that complexes B2 and B4 are due to an octamer binding protein (OBP) [15] and a Y box binding protein (RF-Y) respectively (Fig. 1B). Methylation interference experiments suggest that RF-Y is the human homologue of the murine NF-Y factor (16,17).

Complexes B1 and B3 do not form in extracts from three different SCID B cell lines. We have therefore studied B1 and B3 in more detail by means of DNAse I protection and methylation interference experiments. The results show that B3 is due to a protein binding

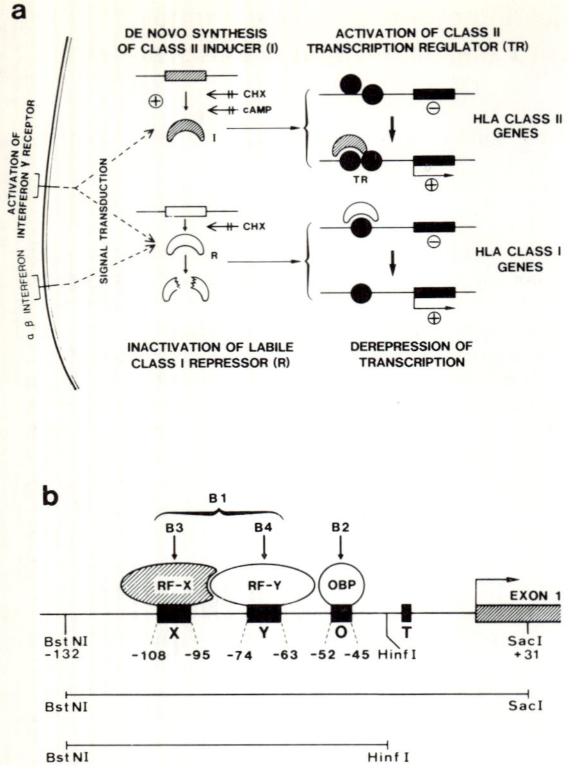

Figure 1. a) Model for the induction of HLA class II and class I genes by IFN-γ. Activation of the IFN-γ receptor leads to the de novo synthesis of a class II inducer protein (I) and the inactivation of a labile class I repressor (R). The class II inducer activates class II gene transcription by modifying or cooperating with other transcriptional regulatory factors (TR). Class I gene transcription is activated by the loss of the class I repressor. b) Model for potential regulatory proteins that bind to the DRα promoter in B cells. In normal B cells, three DNA binding proteins have been identified: an octamer-binding protein (OBP), a Y-box-binding protein (RF-Y), and an X-box-binding protein (RF-X). Protein-DNA complexes B1, B3, and B4 are due to OBP, RF-X, and RF-Y respectively. In B1 both RF-X and RF-Y are bound. In SCID B cells, RF-X is missing or inactive such that neither B1 nor B3 can be formed. Fragments used for gel retardation assays are shown below.

to the X box (RF-X), while B1 is due to the joint binding of both RF-X and RF-Y (Fig. 1B). Since neither B1 nor B3 are observed with SCID cell extracts, it can be concluded that SCID cells are deficient in RF-X. We therefore propose that the molecular defect in class II-deficient SCID is an absence of RF-X, and that RF-X is a regulatory factor that is essential for class II gene transcription in B cells.

In normal B cells, uninduced fibroblasts, and IFN-γ-induced fibroblasts, we have observed two DNAase I hypersensitive sites in the DRA promoter. These sites appear to reflect binding of RF-X because they are absent in SCID B cells (unpublished observation). RF-X may thus not be a B-cell-specific protein. Moreover, since class II genes cannot be induced by IFN-γ in SCID fibroblasts, RF-X may be required for induced as well as constitutive expression. However, RF-X must be distinct from the inducer protein that mediates activation of both class II and In genes because in SCID cells, the absence of RF-X abolishes expression of class II genes but not of the In gene.

Acknowledgments. We thank Drs. U. Schibler and P. Kourilsky for their helpful advice; C. Griscelli, B. Lisowska-Grospierre, and M. Hadams for providing the SCID B-cell lines; and E. Barras and M. Zufferey for excellent technical assistance.

This work was supported by the Swiss National Fund for Scientific Research.

References

1. Benacerraf B. Role of MHC gene products in immune regulation. Science 1981;212:1229.
2. Bottazzo GF, Todd I, Mirakian R, Belfiore A, Pujo-Borrell R. Organispecific autoimmunity: A 1986 overview. Immunol Rev 1986;94:137.
3. Massa PT, ter Meulen V, Fontana A. Hyperinducibility of Ia antigens on astrocytes correlates with strain-specific susceptibility to experimental autoimmune encephalomyelitis. Proc Natl Acad Sci USA 1987;84:4219.
4. Walker R, Cooke A, Bone AJ, Dean BM, van der Meide P, Baird JD. Induction of class II MHC antigens *in vitro* on pancreatic B cells isolated from BB/E rates. Diabetologia 1986;29:749.
5. Lisowska-Grospierre B, Charron DJ, de Preval C, Durandy A, Turmel P, Griscelli C, et al. Defect of expression of MHC genes responsible for an abnormal HLA class I phenotype and the class II negative phenotype of lymphocytes from patients with combined immunodeficiency. Histocompatibility Testing 1984;650.
6. Trowsdale J, Young JAT, Kelly AP, Austin PJ, Carson S, Meunier H, et al. Structure, sequence and polymorphism in the HLA-D region. Immunol Rev 1985;85:5.
7. Mach B, Gorski J, Rollini P, Berte C, Amaldi I, Berdoz J, et al. Polymorphism and Regulation of HLA-class II genes of the major histocompatibility complex. Cold Spring Harbor Symp Quant Biol 1986;51:67.
8. Collins T, Korman AJ, Wake CT, Boss JM, Kappes DJ, Fiers W, et al. Immune interferon activates multiple class II major histocompatibility complex genes and the associated invariant chain gene in human endothelial cells and dermal fibroblasts. Proc Natl Acad Sci USA 1984;81:4917.
9. Paulnock-King D, Sizer K, Freund Y, Jones P, Parnes J. Coordinate induction of Ia α, β, and Ii mRNA in a macrophage cell line. J Immunol 1985;135:632.
10. de Preval C, Lisowska-Grospierre B, Loch M, Griscelli C, Mach B. A trans-acting class II regulatory gene unlinked to the MHC controls expression of HLA class II genes. Nature 1985;318:291.
11. Kelly A, Trowsdale J. Complete nucleotide sequence of a functional HLA-DPβ gene and the region between the DPβ1 and DPα1 genes: Comparison of the 5′ ends of HLA-class II genes. Nucl Acids Res 1985;13:1607.
12. Okada K, Boss JM, Prentice H, Spies T, Mengler R, Auffrey C, et al. Gene organization of DC and DX sub-

regions of the human major histocompatibility complex. Proc Natl Acad Sci USA 1985;82:3410.
13. Bohmann D, Keller W, Dale T, Scholer HR, Tebb G, Mattaj IW. A transcriptional factor which binds to the enhancers of SV40, immunoglobulin heavy chain and U2 sn RNA genes. Nature 1987;325:268.
14. Rosales R, Vigneron M, Macchi M, Davidson I, Xiao JH, Chambon P. *In vitro* binding of cell-specific and ubiquitous proteins to the octamer motif of the SV40 enhancer and related motifs present in other promoters and enhancers. EMBO J 1987;6:3015.
15. Sherman PA, Basta PV, Ting JP. Upstream DNA sequences required for tissue-specific expression of the HLA-DRα gene. Proc Natl Acad Sci USA 1987;84:4254–4258.
16. Dorn A, Durand B, Marfing C, Le Meur M, Benoist C, Mathis D. The conserved MHC class II boxes-X and -Y are transcriptional control elements and specifically bind nuclear proteins. Proc Natl Acad Sci USA 1987;84:6249.
17. Dorn A, Bollekens J, Staub A, Benoist C, Mathis D. A multiplicity of CCAAT box-binding proteins. Cell 1987;50:863.
18. Amaldi I, Reith W, Berte C, Mach B. Induction of HLA class II genes by interferon γ is transcriptional and requires a trans-acting protein. J Immunol 1989; in press.
19. Reith W, Satola S, Herrero Sanchez C, Amaldi I, Lisowska-Grospierre B, Griscelli C, Hadam MR, Mach B. Congenital immunodeficiency with a regulatory defect in MHC Class II gene expression lacks a specific HLA-DR promoter binding protein, RF-X. Cell 1988;53:897–906.

Manuscripts submitted for publication

Reith W, Amaldi I, Berte C, Ulevitch R, Mach B. Regulation of MHC gene expression: Distinct mechanisms of induction of class I and class II genes by interferon gamma.

Author Affiliations

Walter Reith, Sarah Satola, Iris Amaldi, Carlos Herrero Sanchez, Christine Berte, Bernard Mach, Department of Microbiology, University of Geneva Medical School, Geneva, Switzerland; *Richard Ulevitch*, on leave from Scripps Clinic and Research Foundation, Department of Immunology, La Jolla, California, USA

Bare Lymphocyte Syndrome Transcriptional Regulation of HLA Class II Expression*

Janet S. Lee, Clifford R. Hume, Linda Shookster, Soo Young Yang, and Nancy Collins

Abstract: Coordinate regulation of HLA class II gene expression during development and induction of class II genes by factors such as γ-interferon suggests that common transacting factors control expression of the entire complex. Bare lymphocyte syndrome (BLS), a form of lethal combined immunodeficiency, characterized by the simultaneous loss of expression of class II antigens may result from mutations of such transacting factors rather than from independent mutations in several individual MHC genes (1). Furthermore, BLS has been shown by pedigree analysis in several families to be an autosomal recessive disorder not linked to the MHC(2). We have derived Epstein-Barr virus-transformed B-lymphoblastoid cell lines from the peripheral blood of two BLS patients. These cell lines express normal levels of B-cell differentiation antigens, but no HLA class II (T. Small, personal communication). No transcription of class II RNAs was found, with the exception of DQ in one patient line. Analysis of proteins binding to the DR promoters indicated that a sequence found in a complex in normal B-cell lines was not bound it the BLS lines, suggesting that at least one "transacting factor" is missing or altered in the BLS cell lines. This could be the basis for the lack of HLA class II transcription in this disease.

To verify that gross rearrangements or deletions of structural genes were not responsible for the lack of class II expression in these lines, we analyzed the HLA-DR α and β chain genes in the patient cell lines and B-cell lines derived from other family members by Southern blot analysis. All class II genes examined in the BLS patients are indistinguishable from those of their families (manuscript in preparation). Similar results showing normal hybridization patterns have also been reported by other investigators (3).

We used Northern analysis of total cytoplasmic RNA to assess the level of expression of different class II α and β chain genes. We did not detect any expression of either DR or DP α and β transcripts in the BLS cell lines. In contrast, we did detect very low levels of a transcript hybridizing to a DQ α-oligonucleotide probe in one of the BLS cell lines. This residual transcription of DQα is further support for the conclusion that the structural genes remain intact. Although an Epstein-Barr virus-transformed B-lymphoblastoid cell lines (LCL) from one BLS patient was found to reexpress HLA-DR antigens after several weeks of culture (4), the class II negative phenotype of our cell lines has been stable in culture for several months in one case and over 1 year in the other. mRNA for the invariant chain was present in normal amounts in the BLS cell lines. Thus, although the

*This work was supported by NIH grant Po1 CA20194-10, NIH grant BR SG RR-05534 , and NCI grant CA-08748.

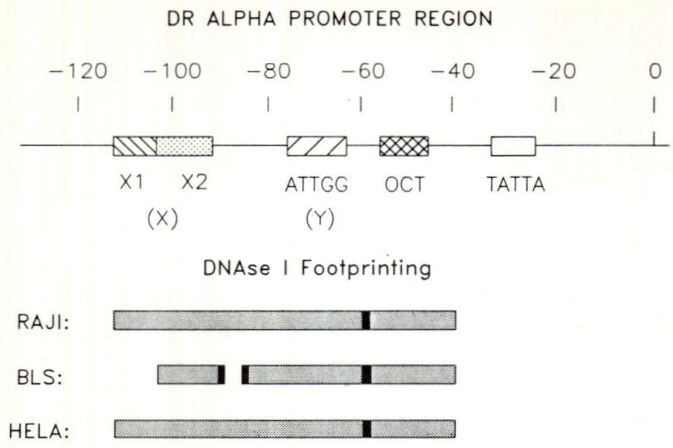

Figure 1. Schematic representation of DR promoter sequences. Stipled boxes below the promoter diagram indicate parts of the promoter bound by nuclear factors, and thus protected from DNase I digestion, extracts of Raji, BLS cell lines, and HeLa cells, respectively.

invariant chain gene is generally regulated coordinately with the class II MHC genes, there are clearly some differences in their control mechanisms. β-2-microglobulin expression and class I expression were within normal levels.

Interestingly, we detect only one DQα- specific RNA in one BLS cell line while two transcripts are seen in both parents and the sibling. In contrast, no specific hybridization to a DQβ- specific oligonucleotide was detected. RNase mapping experiments using a fragment of the DQα gene further supported our observation. DQ expression, in fact, had not previously been observed independently of DP and DR. This contrasts with expression of DR and DP in the absence of DQ in some situations (5). Positive transacting factors in addition to those required for DP and DR were thought to be necessary for DQ transcription, but this simple interpretation may not be correct. Coordinate regulation of MHC class II genes may result from the interaction of multiple factors that direct class II transcription. Lesions in regulatory genes encoding these factors may affect expression of each class II sublocus to a different extent. Nuclear run-on assays have not detected transcription of any class II sublocus genes in the BLS lines, even though DQα must be transcribed at a very low level in one of the two lines. Thus, the defect in BLS primarily affects transcription, though a low basal level of class II transcription may exist.

All class II gene promoters contain two consensus sequences, "X" and "Y," upstream from the mRNA start site (6). Although limited variation in these sequences exists, it was suggested that these sequences bind specific nuclear factors involved in gene regulation. We have used gel retardation assays to detect proteins that can form complexes with the DR and DPα promoters. Both normal B-cell lines and BLS cell lines contain nuclear proteins that bind to these sequences. However, the BLS nuclear extracts formed different complexes with the X consensus element than were found in normal B cells. DNase I footprinting also shows that proteins in BLS nuclear extracts protect smaller portions of the X element of DRα than those from normal B-cell extracts.

Figure 1 shows a model of the DRa promoter fragment depicting sequence elements that have been implicated in transcriptional regulation. In particular, a sequence just upstream from the X element is bound by proteins in extracts from normal B-cell lines but not those derived from BLS patients. It is possible that an absence or alteration of the protein that binds this region could account for the defect in the BLS patients. However, because the severity of the diseases in the two patients we have studied is drastically different, we believe that this simplistic interpretation is incorrect. Rather, intuitively we favor the idea that interactions between proteins may be involved in allowing a complex of multiple factors to form on the DNA around the X element, and that mutations in different "transacting factors" might result qualitatively in a similar pattern of binding to the class II promoters. We are now directing our efforts to the purification of the normal B-cell factor that binds to the X consensus sequence. We are also pursuing genetic approaches to the identification of the gene(s) that are defective in BLS.

Note. We have found subsequently that several class II positive EBV B-cell lines generate X box DNase I footprints similar to BLS-1. This indicates that the altered footprint we detected with BLS-1 nuclear extracts, although reproducible, may not be directly related to the transcriptional defect in BLS. We are investigating this possibility by both transfection and in vitro transcription experiments.

References

1. Touraine J-L, Betuel H, Touraine F. The bare lymphocyte syndrome. In: Griscelli C, Vossen J (eds): Progess in Immunodeficiency Research and Therapy I. Amsterdam, Elsevier 1984.
2. dePreval C, Lisowska-Grospierre B, Loch M, Griscelli C, Mach B. A transacting class II regulatory gene unlinked to the MHC controls expression of HLA class II genes. Nature 1985;318:291.
3. Marcadet A, Cohen D, Dausset J, Fischer A, Durandy A, Griscelli C. Genotyping with DNA probes in combined immunodeficiency syndrome with defective expression of HLA. N Engl J Med 1985;312:1287.
4. Lisowska-Grospierre B, Charron D, dePreval C, Durandy A, Griscelli C, Mach B. A defect in the regulation of major histocompatibility complex class II gene expression in human HLA-DR negative lymphocytes from patients with combined immunodeficiency syndrome. J Clin Invest 1985;76:381.
5. Lee JS, O'Neill L. Methylation of the HLA-DR alpha gene is positively correlated with expression. Immunogenetics 1987;26:92.
6. Korman AJ, Boss J, Spies T, Sorrentino R, Okada K, Strominger JL. Genetic complexity and expression of human class II histocompatibility antigens. Immunol Rev 1985;85:45.

Author Affiliation

Memorial Sloan-Kettering Cancer Center and Cornell University Graduate School of Medical Sciences, New York, NY, USA

DX and DQ Promoter Regions Contain Homologous Sequences That Differ in Extent of Methylation

C.E. Seyfried, W. Kwok, and G.T. Nepom

The HLA-DQ region of the MHC includes two sets of α and β genes, DQ and DX. The DQ molecule is expressed on the cell surface and participates in HLA class II-mediated immune recognition. The DX region is highly homologous to DQ by nucleotide sequencing and genomic organization, but the DXα (HLA-DQA2) and DXβ (HLA-DQB2) genes appear to be transcriptionally inactive since DX mRNA has not been detected. Recent studies of MHC class II genes have demonstrated multiple regulatory elements that may affect transcriptional control (1). One example comes from chromatin structure studies which have demonstrated that methylation of CpG islands may contribute to decreased gene expression, presumably by creating tight nucleosomal spacing and thus preventing protein-DNA interactions required for the initiation of transcription.

We have used the methylation sensitive restriction enzyme HpaII to analyze the promoter regions of the DXβ (HLA-DQB2) and DQβ (HLA-DQB1) genes. As a specific probe, we have used an oligonucleotide homologous to the region between -62 and -83 (relative to the CAP site) that is conserved among β genes within the DQ region. EcoRI or EcoRV digested DNA fragments of the DXβ and DQβ genes containing the first exon and the 5′ promoter regions were identified by characteristic restriction fragments after gel hybridization to the oligonucleotide probe described above using published techniques (2). Subsequent analysis of the DQβ fragments comparing HpaII (methylation-sensitive) and MspI (methylation-intensive) digestion demonstrated no difference in the restriction pattern for the DQβ gene (1.9 kb band), indicating that this region is not heavily methylated. In contrast, the DXβ region shows a 4.8 kb band in the HpaII pattern in comparison to a 2.6 kb band seen with MspI, indicating the presence of a methylated CpG couplet within the promoter region of the DXβ gene.

To test whether a reduction in methylation would result in an increase in DX gene expression, the cell lines (5×10^5 c/ml) were cultured for 72 hours in 10 μM 5-Azacytidine (5-AzaCR) with or without the addition of gamma interferon (IFN) at 1000 U/ml. The cells were harvested and DNA and RNA prepared. Figure 1 shows that the addition of 5-AzaCR with or without IFN resulted in a reduction in methylation as seen by hybridization of the promoter specific oligonucleotide to HpaII and MspI digested EcoRI DNA fragments. Putative DXβ mRNA was measured with an anti-sense oligonucleotide probe complementary to codons 1–13 of the DX β1 exon. The observed reduction in methylation was not sufficient to induce transcription of the DQβ gene detectable by gel hybridization even after the addition of IFN. In contrast, hybridization of the same gels to oligonucleotide probes specific for the β1 exon of the DQ gene showed the expected levels of expression.

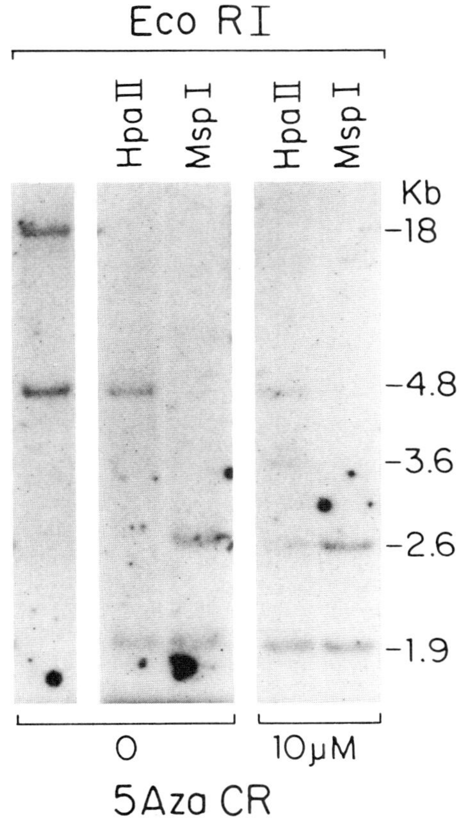

Figure 1. In situ gel hybridization of EcoRI and HpaII or MspI-digested DNA isolated from untreated or 5-AzaCR treated cell lines hybridized to the DQ locus promoter-specific oligonucleotide probe. DQ promoter fragments are indicated at 18 kb and 1.9 kb; DX promoter fragments at 4.8, 3.6, 2.6 kb.

These studies demonstrate that the DXβ promoter region is relatively more methylated than the homologous DQβ region. Nevertheless, this is not sufficient to explain the lack of expression of this gene since the reduction of methylation seen after the addition of 5-AzaCR did not result in any detectable levels of DXβ mRNA. The contribution of increased DXβ methylation to transcriptional inactivity remains to be clarified. Further investigations of the DXβ promoter region should aid in the understanding of gene regulation within the

Immunobiology of HLA
Volume II

MHC in addition to providing a model for future studies of aberrant gene expression or lack of expression in the etiology of some autoimmune disease.

Acknowledgments. This work was supported by grant IM-450 from the American Cancer Society.

References

1. Sullivan KE, Calman AF, Nakanishi M, Tsang SY, Wang Y, Peterlin BM. A model for the transcriptional regulation of MHC class II genes. Immunology Today 1987;8:289.
2. Nepom GT, Seyfried CE, Holbeck SL, Wilske KR, Nepom BS. Identification of HLA-Dw14 genes in DR4 + rheumatoid arthritis. Lancet 1986; ii (8414):1002.

Author Affiliation

C.E. Seyfried, W. Kwok, G.T. Nepom, Immunology Program, Virginia Mason Research Center, Seattle WA, USA

Activated *Ras* Oncogenes Partially Complement the Transcriptional Defects in Class II-Negative Mutant B-Cell Lines

Clifford R. Hume, Roberto S. Accolla, Donald Pious, and Janet S. Lee

Abstract: Expression of the HLA class II genes is regulated coordinately during development. This suggests that their transcription is controlled by common regulatory factors. This hypothesis is based on studies of both naturally occurring genetic defects and in vitro induced mutations that simultaneously reduce or eliminate transcription of the entire complex of genes (1–5). Others have observed that primary melanocytes can be induced to express class II genes by constitutive expression of activated *ras* oncogenes (6). We investigated the relationship between the molecular pathway by which *ras* proteins augment class II gene expression and the factors that control constitutive class II gene expression in B lymphocytes. We introduced activated *ras* oncogenes into three class II-negative human B-cell lines, two mutants derived in vitro, and a cell line (BLS-1) established from a patient with class II-deficient bare lymphocyte syndrome. Activated *ras* gene expression in the two mutants partially restored class II expression as measured by analysis of cytoplasmic RNA (7). While both DR and DQ were increased in one of the mutants, the other only demonstrated increases in DQ expression. The bare lymphocyte syndrome cell line did not respond to activated *ras* expression. In all three cell lines, levels of mRNAs for other class II and class I gene loci were unaffected. The differential responsiveness of the three mutant cell lines to constitutive *ras* expression suggests that several distinct genetic lesions may result in a class II-negative phenotype.

The class II-deficient B-cell lines RJ2.2.5 (5), 6.1.6 (4), and BLS-1 (manuscript in preparation) were infected with recombinant helper-free retroviruses containing the Tn5-neo gene either alone or in combination with the human N- or viral H- *ras* genes (7). The cell lines established after selection in G418 were tested for expression of class II genes by Northern analysis of cytoplasmic RNA. Where tested, N- and H- *ras* had comparable effects on class II expression, therefore they will be considered together.

Expression of activated *ras* genes in RJ2.2.5 led to an increased expression of both DR and DQ α and β RNA. DQα mRNA was induced in RJ-*ras* to levels of at least 50% that found in the parental B-cell line, Raji. The levels of DQ-β and DRα and β mRNAs were elevated to approximately 10% of normal lymphoblastoid cell lines. However, not all class II loci were induced in the RJ-*ras* cell lines. Of particular interest, the residual levels of DPα and DOβ mRNAs still detected in RJ2.2.5 were not altered in RJ-*ras*. Expression of DPβ remained undetectable. The levels of class I, invariant chain, and actin RNA were not induced by constitutive *ras* expression. These findings support the conclusion that the effect of *ras* is specific and not due to a change in general RNA metabolism.

The phenotype of the class II-negative mutant 6.1.6 is similar to RJ2.2.5, however, constitutive *ras* expression in 6.1.6 had a more limited effect on expression of class II region genes. DQα mRNA was induced in 6.1.6-*ras* to levels comparable to those in RJ2.2.5-*ras*, but DQα and β expression were not significantly increased. 6.1.6 expresses a low level of DRα, yet its level was unchanged in 6.1.6-*ras*. As in RJ-*ras*, expression of actin, class II, and invariant chain were unaffected in 6.1.6-*ras*. Although the BLS-*ras* cell line expressed levels of viral *ras* transcripts comparable to RJ2.2.5-*ras* and 6.1.6-*ras*, expression of class II genes in BLS-*ras* was indistinguishable from the uninfected cell line.

Several mechanisms may account for the increase in class II expression following introduction of activated *ras* genes into the mutant cell lines. Because there was no change in the basal level of cytoplasmic DPα mRNA in RJ2.2.5, or DRα mRNA in 6.1.6, it was unlikely that increases in message stability were responsible

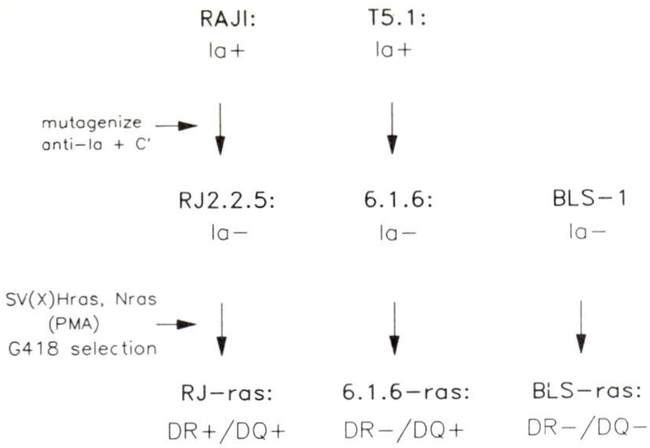

Figure 1. Outline of the effects of activated *ras* gene expression on the phenotype of class II-negative B-cell lines.

for the induction of class II genes in the *ras*-transformed cell lines. Furthermore, studies with actinomycin D indicated that class II mRNAs found in RJ2.2.5 were as stable as those in Raji.

We conducted nuclear run on experiments to determine if constitutive *ras* expression was leading to an increase in transcription rate. Transcription of DRα, DQα, and DRβ were clearly detectable in Raji; however, their apparent rates of transcription were considerably less than estimated for actin. This was surprising as cytoplasmic levels of DRα message are quite high. An explanation for this discrepancy is that because class II transcripts are very stable (half lives > 12 hr), low transcription rates generate relatively high levels of cytoplasmic mRNA. Although RJ2.2.5 transcribed levels of actin and invariant chain comparable to Raji, transcription of DRα and DRβ and DQα was undetectable. RJ2.2.5-*ras* showed a significant increase in transcription for both DR and DQ. This suggests that activated *ras* proteins enhance class II expression at least partly through an increase in transcription (manuscript in preparation).

Ras proteins have been implicated in the regulation of phosphatidylinositol turnover and protein kinase C activation (8,9). In this signaling pathway, phospholipase C cleaves phosphoinositides to yield 1,2-diacylglycerol (DAG) and inositol phosphates. These products act as second messengers in response to receptor activation, DAG as a cofactor for protein kinase C, and inositol phosphates in the mobilization of calcium (10). Because phorbol esters can act as analogues of DAG (10), we tested the ability of phorbol myristic acid (PMA) to mimic the effects of *ras* on class II expression. PMA (10 ng/ml) acted very rapidly (<30 min) to increase cytoplasmic levels of class II RNAs in RJ2.2.5 and 6.1.6. The subloci whose expression increased were the same as those affected in the respective *ras*-virus-infected cell lines. In RJ2.2.5 the levels of mRNA after 12 hours of treatment with PMA were comparable to those in RJ2.2.5-*ras*. PMA did not appear to be as efficient as *ras* in inducing DQα in 6.1.6, however, this may be due to a suboptimal concentration of PMA. As predicted, there was a negligible effect of PMA on class II expression in the BLS cell line. The comparable induction of class II genes in the mutant cell lines by *ras* and PMA suggests that both mediate their effects through alterations in PK-C activity.

References

1. Griscelli C, Durandy A, Virelizier JL, Hors J, Lepage V, Columbiani J. In: Seligman M, Hitzig H, (eds): *Primary Immunodeficiencies*. Amsterdam, Elsevier/North Holland, 1980, pp 499–503.
2. de Preval C, Lisowska-Grospierre B, Loche M, Griscelli C, Mach B. A trans-acting class II regulatory gene unlinked to the MHC controls expression of HLA class II genes. *Nature* 1985;318:291.
3. Lisowska-Grospierre B, Charron D, de Preval C, Durandy A, Griscelli C, Mach B. A defect in the regulation of major histocompatibility complex class II gene expression in human HLA-DR negative lymphocytes from patients with combined immunodeficiency syndrome. J Clin Invest 1985;76:381.
4. Gladstone P, Pious D. Stable variants affecting B-cell alloantigens in human lymphoid cells. *Nature* 1978;271:459.
5. Accolla RS. Human B-cell variants immunoselected against a single Ia antigen subset have lost expression of several Ia antigen subsets. J Exp Med 1983;157:1053.
6. Albino AP, Houghton AN, Eisinger M, Lee JS, Kantor RRS, Oliff AI, et al. Class II histocompatibility antigen expression in human melanocytes transformed by Ha-MSV and Ki-MSV retroviruses. J Exp Med 1986;164:1710.
7. Hume CR, Accolla RS, Lee JS. Defective HLA class II expression in a regulatory mutant is partially complemented by activated *ras* oncogenes. Proc Nat Acad Sci USA, in press.
8. Fleischman LF, Chahwala SB, Cantley LC. *Ras* transformed cells: Altered levels of phosphotidylinositol-4,5 bisphosphate catabolites. Science 1986;231:407.
9. Lacal JC, Moscat J, Aaronson SA. Novel source of 1,2-diacylglycerol in cells transformed by Ha-*ras* oncogene. *Nature* 1987;330:269.
10. Nishizuka Y. The role of protein kinase C in signal transduction and tumour promotion. *Nature* 1984;308:693.

Author Affiliations

Clifford R. Hume and Janet S. Lee, Memorial Sloan Kettering Cancer Center, and Cornell Graduate School of Medical Sciences, New York, NY, USA; *Roberto S. Accolla*, Ludwig Institute for Cancer Research, Lausanne Branch; *Donald Pious*, University of Washington, Seattle, WA, USA

Modulation of HLA Class II Expression Following Infection by HIV

Daniel M. Altmann, Quentin J. Sattentau, John Trowsdale, and Peter C.L. Beverley

Considerable attention has been devoted to the interactions of the human immunodeficiency virus (HIV) with the immune system, with respect both to pathogenesis of AIDS and to the generation of anti-viral immunity. In view of the fact that the cellular receptor for HIV on T_H cells is the CD4 molecule (1), a molecule whose normal role is the binding of MHC class II molecules (2), we have been interested in characterizing the interactions between HIV and class II molecules in infected T cells. Earlier observations from this laboratory indicated that the CD4/HLA class II interaction may indeed have a role in transmission of virus between T cells since some anti-HLA class II antibodies were able to block syncytium formation by infected cells (1). We now report that infection of the T-cell line H9 with HIV leads to substantial up-regulation, detectable both at the level of mRNA and surface protein expression, of some HLA class II products.

Following the finding of HIV syncytium inhibition by some MAbs against HLA-DQ (1; Sattentau et al, manuscript in preparation), a class II product expressed at a rather low level on uninfected T cells, we set out to compare HLA-DQ expression in infected and uninfected T cells. Total RNA was prepared from H9 (3), or from H9 persistently infected with the HIV isolates IIIB, MA2, or RUT, as well as from T-cell line 8166 and its persistently infected counterpart, 8166/LAV2. Northern blots of infected and uninfected T-cell RNA were initially probed with cDNA probes for DQA1 and DQB1 (Fig. 1). There were no qualitative differences seen between the size or number of transcripts in infected and uninfected T cells. However, there were dramatic quantitative differences: DQA1 and DQB1 transcripts were almost undetectable in uninfected H9 T cells, but reached (or exceeded) the level seen in the B lymphoblastoid cell line (B-LCL) MANN following infection by any of the three HIV isolates. DQ transcripts were undetectable in the T cell line 8166, both infected and uninfected.

For quantification of Northern blots, RNA loading was checked by ethidium bromide staining of formaldehyde gels and confirmed by probing filters with 7B6, which can be used as a cell-cycle independent standardization probe (4). Results for each probe were cross-checked on more than one filter, and densitometry performed on autoradiographs showing optimal signals within the linear range of the film.

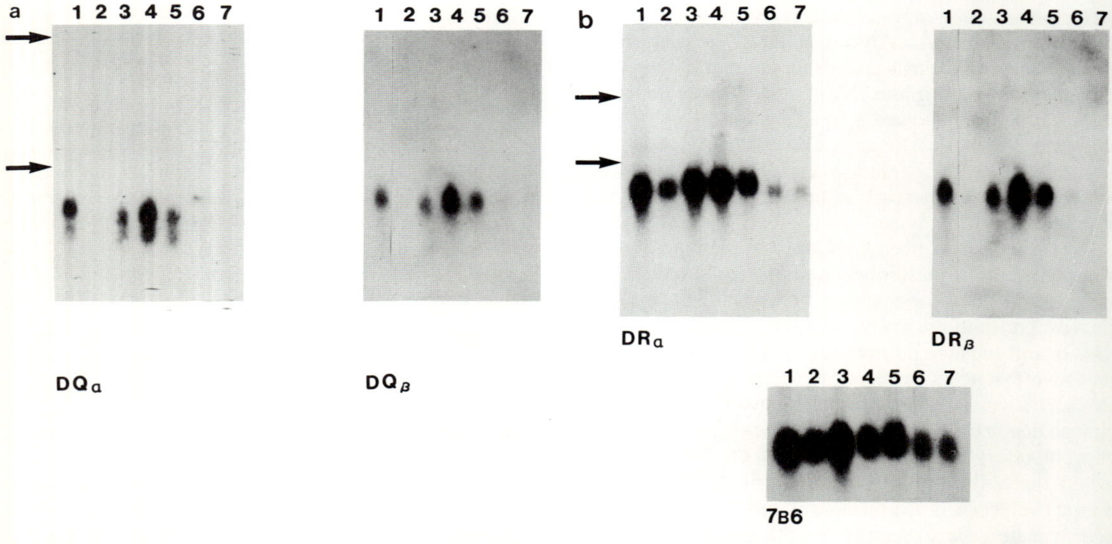

Figure 1. Northern blots of HIV-infected T cells. Total RNA was prepared by guanidinium thiocyanate denaturation followed by CsCl centrifugation from the B-lymphoblastoid cell line MANN and from T cells that were uninfected or persistently infected with HIV. RNA (20µg/track) was electrophoresed on 1% agarose formaldehyde gels, transferred to Hybond-N filters (Amersham Intl.), and hybridized with oligolabeled DNA probes. Labeled blots were then exposed to Kodak XAR-5 film at −70° for 18-72h. Arrows on the autoradiographs indicate the migration positions of the 28S and 18S ribosomal bands. *A*: Blots hybridized with DQA1 and B1 probes. The lanes are as follows: 1, MANN; 2, H9; 3, H9/IIIB; 4, H9/MA2; 5, H9/RUT; 6, 8166; 7, 8166/LAV2. *B*: A blot hybridized with DRA and DRB, and the same blot hybridized with the control probe, 7B6. The lanes are as in part A.

Table 1. Summary of HLA mRNA Induction in HIV-Infected T Cells

	mRNA Level (Arbitrary Units)*								
	cDNA Probe								
Cell Line	DQA1	DQB1	DRA	DRB	DNA	DPA1	HLA-B	Ii	TCRα
MANN	35322	49970	59219	58343	7393	75744	41987	191552	<1000 ~ †
H9	1776	<1000	26028	<1000	1620	51385	9270	18891	19545
H9/IIIB	25338	15816	56811	45501	13098	69714	24249	58520	ND
H9/MA2	71077	81853	126936	99205	72821	200844	93154	30392	16616
H9/RUT	18450	27425	59741	38974	13241	111555	59107	29764	24536

*Numbers represent peak readings detected on autoradiographs by Joyce Loebl densitometry and normalized with respect to intensity of 7B6 mRNA to eliminate errors due to loading differences.
† ~ <1000 indicates that the band came below the detection limits of the densitometer.

The finding with DQA1 and DQB1 probes are summarized in Table 1. By densitometry, the level of DQ transcripts in the infected T cells was at least 10.5-fold greater than in the uninfected cells. Analysis with a DRB probe similarly showed enhancement in the intensity of message from barely detectable levels to similar levels to those seen in the B-LCL (Fig. 1b). DRA mRNA, which was more abundant in all the T cells including the uninfected H9 and 8166 cells, showed a smaller relative change in level.

Of the other HLA class II mRNAs analyzed, DNA was also substantially up-regulated, particularly in the H9/MA2 cell line (Table 1). Although no surface expression of a DNA product has yet been identified, the gene has all the characteristics of an intact class II gene, and has been shown at the mRNA level to be γ-interferon inducible (5).

We did not find equivalent enhancement of transcripts across the entire HLA-region (as is seen, for example, following induction with γ-interferon). The level of DPA1 and HLA-B mRNAs were only moderately increased. We tested the hypothesis that the induction of class II may be the result of autocrine activation following the release of γ-interferon by the infected T cells but, in agreement with previous observations (6) there were no detectable γ-interferon transcripts in infected or uninfected cells (data not shown). Furthermore, the level of invariant chain (Ii) message which, though not on chromosome 6, is co-induced with class II genes by γ-interferon, was not significantly increased in infected cells (Table 1).

The enhanced level of class II mRNA in HIV-infected cells was reflected in surface protein expression detected by binding of labeled MAbs (Table 2). Binding of the class II framework MAb, CA2, is increased three to eight times in the infected H9 T cell. Class I and HLA-DP surface expression increased 2.4 to 6.9 times, somewhat more strikingly than apparent from Northern analysis. Intriguingly, binding of the anti-DQw1 MAb SDR4.1 (a specificity not present on the uninfected T cell) also increases in the infected cells, possibly reflecting formation of a DQw1 cross-reactive epitope by the interaction of viral antigen with surface DQ (Q. Sattentau *et al*, manuscript in preparation).

The mechanism by which HIV infection causes enhanced HLA class II expression is still not clear. Massa *et al* have reported induction of Ia on rat asrocytes via a pathway that appears to be independent of γ-interferon production, following infection with JHM virus (7). Although it is tempting to speculate that up-regulation of class II genes results from interaction of a cellular transcription factor with the HIV-tat gene, one preliminary study failed to find class II induction in T cells infected with HIV-tat III recombinant virus (8).

These enhanced levels of surface HLA class II expression are in accord with those previously reported with simian immunodeficiency virus (SIV) and with two isolates of HIV (9). Activation of class II antigens by HIV may be especially relevant to binding of HIV to CD4, when one considers that free virus produced by H9 cells has been shown to contain very substantial amounts of DR heterodimer (10). Thus the class II/CD4interaction may have a role in spread of infection both by free virus and by direct contact with infected cells. Alternatively, enhanced class II expression may be of benefit to the infected individual in presenting the virus more

Table 2. Binding of HLA Antibodies to HIV-Infected T Cells

		Antibody Binding*			
		T Cells			
Antibody	Specificity	H9/RF	H9/MA2	H9/MAc	H9/CBL20
CA2	Class II	4.10	3.33	7.47	8.11
SDR4.1	DQw1	1.37	1.81	1.40	1.85
B7/21.2	DP	2.40	3.11	4.09	4.82
W6/32	Class I	3.99	4.24	6.88	5.91

*Results are expressed as cpm ^{125}I-MAb bound to infected line/ cpm ^{125}I-MAb bound to uninfected line.

efficiently to antiviral T cells and mounting a larger response. The nature of these events should become clearer as more sophisticated models are devised for *in vitro* analysis of interactions between HIV and the immune response.

References

1. Dalgleish AG, Beverley PR, Clapham PRL, Crawford DH, Greaves MF, Weiss RA, The CD4 (T4) antigen is an essential component of the receptor for the AIDS retrovirus. Nature 1984;312:763.
2. Gay DP, Maddon R, Sekaly MA, Talle M, Godfrey E, Long G, et al. Functional interaction between human T cell protein CD4 and the major histocompatibility complex HLA-DR antigen. Nature 1987;328:626.
3. Popovic M, Sarngadharan MG, Read E, Gallo RC. Detection, isolation and continuous production of cytopathic retroviruses (HTLVIII) from patients with AIDS and pre-AIDS. Science 1984;224:497.
4. Kaczmarek L, Calabretta B, Baserga R. Expression of cell-cycle dependent genes in phytohemagglutinin-stimulated human lymphocytes. Proc Natl Acad Sci 1985;82:5375.
5. Tonnelle C, DeMars R, Long EO. DO beta: A new beta gene in HLA-D with a distinct regulation of expression. EMBO J 1985;4:2839.
6. Arya SK, Gallo RC. Human T cell growth factor (interleukin 2) and gamma interferon genes: Expression in human T lymphotropic virus type III and type I-infected cells. Proc Natl Acad Sci 1985;82:8691.
7. Massa PT Dorries R, ter Meulen V. Viral particles induce Ia antigen expression on astrocytes. Nature 1986;320:543.
8. Koka P, Dubey DP, Dasgupta JV, Yunis EJ. Modulation of MHC and L3T4 expression by HTLVI and III trans-acting factors. Abstracts of 5th HLA and H-2 Cloning Workshop 1986;112.
9. Kannagi M, Kiyotaki M, King NW, Lord CI, Letvin NL. Simian immunodeficiency virus induces class II major histocompatibility complex structures on infected target cells in vitro. J Virol 1987;61:1421.
10. Henderson LE, Sowder R, Copeland TD, Oroszlan S, Arthur LO, Robey WG, et al. Direct identification of class II histocompatibility DR proteins in preparations of human T cell lymphotropic virus type III. J Virol 1987; 61:629.

Author Affiliations

Daniel M. Altmann, John Trowsdale, Peter C.L. Beverley, Imperial Cancer Research Fund, London; *Quentin J. Sattentau*, Academic Department of Genito-Urinary Medicine, Middlesex Hospital, London, England

Differential Modulation by Cytokines of the Gene Products of the HLA-D Region in Melanoma Cells with Different Metastatic Potential

Michele Maio, Beena Gulwani, and Soldano Ferrone

The differential expression of HLA class II antigens in primary and metastatic melanoma lesions and the association between level of expression of HLA class II antigens in melanoma lesions and clinical course of the disease (for review, see Ref. 1) have stimulated interest in the characterization of the regulatory mechanisms controlling the expression of HLA class II antigens and in their role in the biology of melanoma cells. These investigations are facilitated by the availability of melanoma cell lines with different metastatic potential and of recombinant immunomodulators which may affect the expression of the gene products of the HLA-D region.

Mutagenization of the human melanoma cell line MeWo with methylmethane sulfonate followed by selection for resistance to wheat germ agglutinin has resulted in the isolation of the stable clonal lectin resistant glycosylation mutant MeM 50-10 cells. The latter are highly metastatic in nude mice and display a spreading similar to that observed in patients with melanoma (2). In this paper we compare the melanoma cell line MeWo and its autologous metastatic variant MeM 50-10 for their expression of the gene products of the HLA-D region and their susceptibility to modulation by leukocyte interferon (IFN-α), fibroblast interferon (IFN-β), immune interferon (IFN-γ) and tumor necrosis factor (TNF-α).

The melanoma cell lines MeWo and MeM 50-10 did not react in serological and immunochemical assays with the anti-HLA-DR, DQ, DP monoclonal antibody (MAb) Q5/13, with the anti-HLA-DR MAb CL413, with the anti-HLA-DQ MAb SPV-L3, and with the anti-HLA-DP MAb B7/21. The two melanoma cell lines displayed differential susceptibility to modulation of HLA class II antigens by the various types of cytokines. Following incubation with IFN-α (2000 U/ml) for 72 hours, neither MeWo nor MeM 50-10 cells acquired reactivity with anti-HLA class II monoclonal antibodies. On the other hand, HLA-DR antigens were expressed on MeM 50-10 cells following incubation with IFN-β (3,000 U/ml) or IFN-γ (1,000 U/ml) for 72 hours and on MeWo cells following incubation with TNF-α (100 U/ml) for 72 hours. Neither HLA-DQ nor HLA-DP antigens were expressed by the two cell lines following incubation with the various cytokines, since anti-HLA-DQ and DP monoclonal antibodies did nor react with the treated cell lines in serological assays. Furthermore, components with the two-dimensional electrophoretic mobility of HLA-DQ and DP antigens were not detected in immuno-

Immunobiology of HLA
Volume II

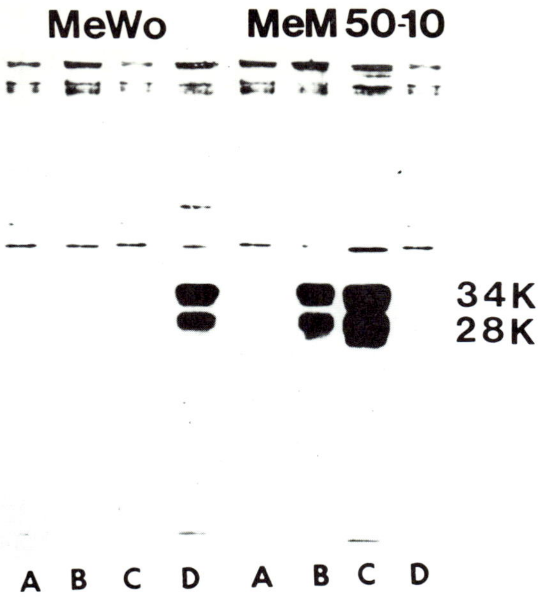

Figure 1. Sodium dodecyl sulfate-polyacrylamide gel electrophoresis of antigens immunoprecipitated by anti HLA-DR, DQ, DP MAb Q5/13 from control (lane A), IFN-β (lane B), IFN-γ (lane C), and TNF-α (lane D) treated melanoma cells MeWo and MeM 50-10.

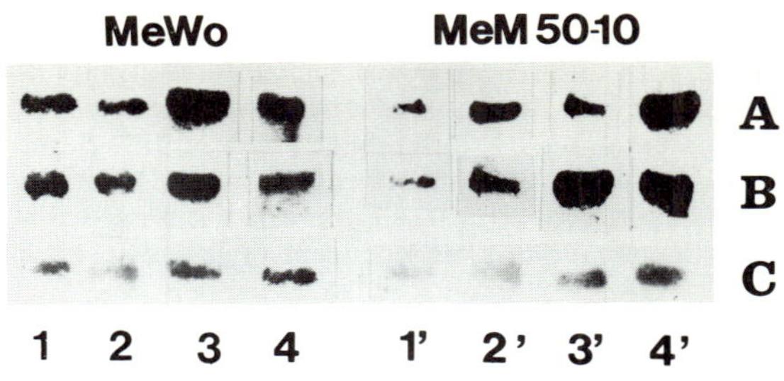

Figure 2. Northern blot analysis of HLA-DR, DQ, and DP specific mRNA in cultured melanoma cells MeWo and MeM 50-10 incubated with IFN-β, IFN-γ, or TNF-α. Control cultures (lanes 1 and 1′) and cultures added with IFN-β (final concentration 3,000 U/ml) (lanes 2 and 2′), IFN-γ (final concentration 1,000 U/ml) (lanes 3 and 3′), or TNF-α (final concentration 100 U/ml) (lanes 4 and 4′) were incubated at 37°C for 72 hours. At the end of the incubation, cells were harvested and washed twice with phosphate-buffered saline. Then cytoplasmic RNA was extracted and hybridized with probes specific for HLA-DRβ (lane A), for HLA-DQβ (lane B), and for HLA-DPβ (lane C).

precipitates with the anti-HLA-DR, DQ, DP MAb Q5/13 from cytokine-treated MeWo and MeM 50-10 cells.

To investigate the mechanism(s) underlying the differential susceptibility to induction by cytokines of HLA-DR, DQ, and DP antigens on MeWo and MeM 50-10 cells, the steady state levels of mRNA for HLA-DR, DQ, and DP antigens were analyzed by Northern blotting hybridization in the two melanoma cell lines following incubation with IFN-β, IFN-γ, or TNF-α for 72 hours. Both melanoma cell lines contain RNA species which hybridize to HLA-DRβ, DQβ, and DPβ gene probes. IFN-β and IFN-γ enhanced the level of HLA-DR and DQ mRNA in the MeM 50-10 cell line, while TNF-α induced a slight increase of HLA-DR and DQ mRNA in the MeWo cell line. No marked change was detected in the level of HLA-DP mRNA in both MeWo and MeM 50-10 cell lines following incubation with the various cytokines. These results suggest that the modulation of the gene products of the HLA-D region by IFN-β, IFN-γ, and TNF-α in the cell lines MeWo and MeM 50-10 reflects transcriptional and posttranscriptional events. Therefore TNF-α and IFN-β or IFN-γ modulate different steps of the regulatory mechanisms(s) which control the expression of HLA class II antigens.

As reported elsewhere (to be submitted), the differential susceptibility to modulation of HLA class II antigens on the two melanoma cell lines by IFN-β, IFN-γ, and TNF-α is not a general phenomenon. The MeWo and MeM 50-10 cells do not differ significantly in their modulation by these cytokines of HLA class I antigens, of the high molecular weight-melanoma associated antigen, and of the 96K melanoma-associated antigen. Furthermore, the differential susceptibility to modulation of HLA class II antigens by IFN-β and IFN-γ on

Table 1. Modulation by IFN-α, IFN-β, IFN-γ, and TNF-α of the Expression by Melanoma Cell Lines MeWo and MeM 50-10 of the Gene Products of the HLA-D Region*

	MeWo					MeM 50- 10				
MAb	Control	IFN-α	IFN-β	IFN-γ	TNF-α	Control	IFN-α	IFN-β	IFN-γ	TNF-α
CL 413	<1	<1	<1	<1	4	<1	<1	5	8	<1
SPV-L3	<1	<1	<1	<1	<1	<1	<1	<1	<1	<1
B7/21	<1	<1	<1	<1	<1	<1	<1	<1	<1	<1

*Cells were incubated with IFN-α (2,000 U/ml), IFN-β (3,000 U/ml), IFN-γ (1,000 U/ml), or TNF-α (100 U/ml) for 72 hours at 37°C. Then cells were washed and tested with the anti-HLA-DR MAb CL 413, the anti-HLA-DQ MAb SPV-L3, and the anti-HLA-DP MAb B7/21 in the binding assay with ^{125}I-labeled antimouse Ig xenoantibodies.
†cpm × 10^{-3}.

the melanoma cells MeWo and MeM 50-10 is not associated with differences in the number and affinity of receptors for the two cytokines on the two melanoma cell lines. Therefore, the differential effect of IFN-β and IFN-γ on the two melanoma cell lines may reflect an intrinsic property of each cell line which is regulated at a postmembrane level independent of ligand receptor interactions.

The biological significance of these findings remains to be determined. In view of the potential role of cytokines in the expression of HLA class II antigens by melanoma cells in vivo (3), the ability of IFN-β and TNF-α to induce these antigens may account for the expression of HLA class II antigens in melanoma lesions without an obvious source of IFN-γ. Furthermore, the appearance of HLA class II antigens in melanoma cells without a metastatic potential suggests that expression of these antigens may not be sufficient for melanoma cells to metastasize. Lastly, the differential susceptibility of the gene products of the HLA-D region to induction by cytokines may account for the heterogeneous expression of HLA-DR, DQ, and DP antigens in melanoma lesions (4).

Acknowledgments. The authors wish to acknowledge the excellent secretarial assistance of Mrs. Edwina L. Jones and Miss Deborah O. Kimbo.

This work was supported by the National Institutes of Health grants AI21384 and CA39559 and by a grant from the Juvenile Diabetes Foundation. M.M. was supported by a fellowship from the Associazione Italiana per la Ricerca Sul Cancro. Address correspondence to S.F.

References

1. Natali PG, Bigotti A, Cavaliere R, Ruiter DJ, Ferrone S. HLA Class II antigens synthesized by melanoma cells. Cancer Rev, 1988;9:34.
2. Kerbel RS, Mann MS. Single-step selection of unique human melanoma variants displaying unusually aggressive metastatic behavior in nude athymic mice. Invasion Metastasis 1984;4:31s.
3. van Vreeswink, H, Ruiter DJ, Brocker EV, Welvaart K, Ferrone S. Expression of HLA-DR, HLA-DQ and HLA-DP antigens in primary and metastatic melanoma. J Invest Dermatol, 1988;90:755.
4. Natali PG, Bigotti A, Cavalieri R, Nicotra MR, Tecce R, Manfredi D, et al. Gene products of the HLA-D region in normal and malignant tissues of nonlymphoid origin. Hum Immunol 1986;15:220.

Author Affiliation

Department of Microbiology and Immunology, New York Medical College, Valhalla, NY 10595, USA

Lymphotoxin and Class II HLA-DR Co-Precipitate?

John A. Sachs, Nelson Fernandez, Marcej Kurpisz, Mario O. Labeta, Teresa Cabrera, Khalid Razak, Marion Macey, Hilliard Festenstein, and Anthony Meager

Lymphotoxin (TNFβ), not usually present in resting peripheral blood lymphocytes, is detectable in the cytoplasm but not the cell membrane of EBV-transformed cell lines (LCLs). A 25 kD band was clearly visible from anti-DR immunoprecipitates of LCLs after SDS-PAGE analysis. A corresponding band was found after immunoprecipitates of the same LCL with anti-lymphotoxin monoclonal antibody (MAb). Immunodepletion studies with the DR and anti-lymphotoxin MAbs indicated that the two fragments were one and the same and that DR and lymphotoxin co-precipitate with the anti-DR MAb but not with the lymphotoxin MAb. This observed lymphotoxin response to virus infection and co-precipitation with class II molecules and its known effect on class II expression suggests that it may play a regulatory role in the pathogenesis of HLA-associated disorders of putative viral etiology.

The presence of lymphotoxin in supernatants of EBV-transformed cell lines was established by an ELISA assay and neutralization inhibition test using anti-lymphotoxin MAb 238/8 (1) made against recombinant human lymphotoxin (Genetech Inc) (Table 1). All cell lines tested secreted lymphotoxin and no obvious differences in the amount secreted could be related to the HLA phenotype of the cell (data not shown). Using indirect immunofluorescence with the 238/8 anti-lymphotoxin MAb, no cell surface labeling for lymphotoxin was observed, even after prior incubation of the cell lines with neuraminidase. Furthermore, immunohistochemical labeling of cross-sections of EBV-transformed cells showed staining of the cytoplasm but not the cell surface membrane and the FACS analysis patterns before and after treatment of the cells with ethanol were consistent with that of cytoplasmic staining (data not shown). Lymphotoxin appears therefore in the cytoplasm and is not an intrinsic component of the cell surface membrane.

HLA class II molecules consist of an α chain of 33,000 kD and a β chain of 29,000 kD non-covalently bound.

Immunobiology of HLA
Volume II

Table 1. Lymphotoxin and Cacechtin in Supernatants of Different Cell Cultures Assayed by ELISA

Cell Designation	Type of Cell	Days in Culture	Units/ml Lymphotoxin (TNFβ)	Units/ml Cacechtin (TNFα)
TP	PBLs	3	<1.0	13.9
TP	PBLs	6	1.5	12.5
K.W	LCL	3	1300	4.6
K.W	LCL	7	1480	4.6
K.W	LCL	10	5000	9.0
SAJ	LCL	3	120	4.1
SAJ	LCL	7	62	<3
SAJ	LCL	10	690	7.0
RDAV	LCL	3	305	3.6
RDAV	LCL	7	130	3.6
RDAV	LCL	10	1820	7.8

See ref. 1 for method.

When EBV-transformed cells are the source of class II molecules after precipitation with anti-DR monomorphic MAbs, DRα and β chains as well as actin and an invariant spot are readily discernible in 2-D SDS PAGE; an additional spot of 25 kD has been observed by us (2) and others (Class II Biochemistry Workshop data and Bontrop R, personal communication). Since lymphotoxin also has a molecular weight of about 25 kD and is controlled by the HLA chromosomal region between HLA-DR and -B (3,4), we have explored the possibility that lymphotoxin and the 25 kD spot identified in the DR precipitate are the same structure.

Monoclonal antibodies against lymphotoxin and DR were used to precipitate detergent soluble molecules from an EBV-transformed cell line labeled with ^{35}S-Methionine. In 1-D SDS PAGE gels the 25 kD molecule is clearly discernible in both the anti-DR and anti-lymphotoxin precipitates (Fig. 1, columns B1 and A1, respectively) whereas the class II α and β chains are present in the DR precipitate and not the lymphotoxin precipitate. After immunodepleting the anti-DR precipitate with anti-lymphotoxin, the 25 kD band was removed leaving the class II α and β chains (Fig. 1A, column 8). Immunodepleting the anti-lymphotoxin precipitate with anti-DR MAb also removed the 25 kD band (Fig. 1B, column 8). It is interesting to note that the bands of 60 kD are also removed following absorption with either antibody suggesting that they may be macrocomplexes of the lymphotoxin molecule as have been previously reported (3).

These studies indicate that the anti DR MAb precipitates both DR and lymphotoxin molecules, whereas the anti-lymphotoxin MAb precipitates lymphotoxin but not class II chains. We conclude that lymphotoxin and DR do co-precipitate and that lymphotoxin shares an epitope(s) with the class II α or β chains. The significance of these

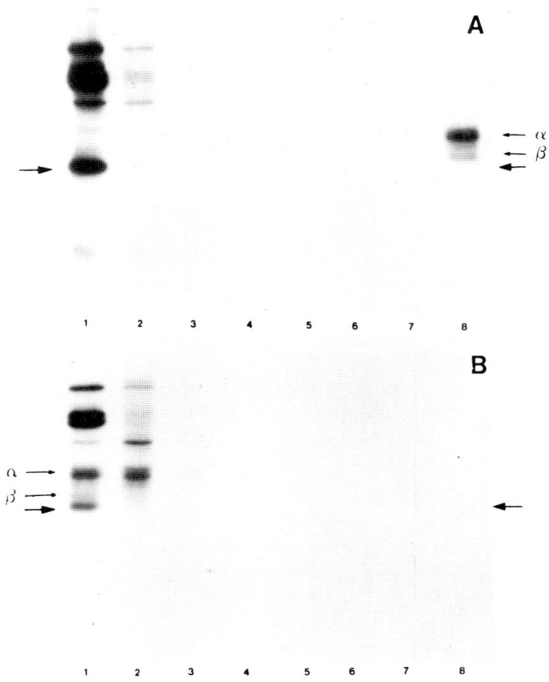

Figure 1. 1-D SDS-PAGE analysis of lymphotoxin and HLA-DR molecules extracted from the lymphoblastoid cell line JEST (HLA-DR1,1) after sequential (S) immunoprecipitation (IP) with the MAbs 238/11 (lymphotoxin) and L243 (DR). **A**: lanes 1-6, SIP with 238/11; lane 7, IP protein A alone; lane 8, IP with L243. **B**: lanes 1-6, SIP with L243; lane 7, IP with Protein A alone; lane 8, IP with 238/11. The DRα and β chains are indicated. The thick horizontal arrow shows the lymphotoxin specific band (method according to ref. 5).

findings in relation to HLA-associated disease susceptibility genes is being explored (see Fernandez et al, these proceedings (2).

Acknowledgments. This research was supported by a grant from the Wellcome Trust.

References

1. Meager A, Parti S, Leung H, Woolley J, Peil E, Sidhu s, et al. A two-site sandwich immunoradiometric assay of human lymphotoxin with monoclonal antibodies and its application. J Imm Methods 1987;104:31.
2. Fernandez N, Hitman G, Festenstein H, Garde L, Labeta MO, Stevens C, et al. The role of lymphotoxin in the pathogenesis of celiac disease. In: Dupont JA (ed): Immunobiology of HLA. New York, Springer-Verlag, 1988.
3. Gray P, Aggerwal B, Benton C, Bringman TS, Henzel WJ, Jarrett JA, et al. Cloning and expression of cDNA for human lymphotoxin, a lymphokine with tumour necrosis activity. Nature 1984;312:721.
4. Spies T, Martin C, Nedospasov S, Fiers W, Pious D, Strominger J. Genes for the two necrosis factors α and β are linked to the major histocompatibility complex. Proc Natl Acad Sci 1986;83:8702.
5. Fernandez N, Labeta MO, Kurpisz M, Solana R, Festenstein H. HLA-DQ molecular heterogeneity in HLA-DR4-Dw4 consanguineous cell lines. Dis Markers 1986;4:191.

Author Affiliations

John A. Sachs, Bone & Joint Research Unit and Department of Immunology; *Nelson Fernandez, Mario O. Labeta, Teresa Cabrera, Khalid Razak, Hilliard Festenstein*, Department of Immunology; *Marion Macey*, Department of Haematology, London Hospital Medical College, London; *Anthony Meager*, National Institute for Biological Standards & Control, South MIMMS, Enlgand; *Marcej Kurpisz*, Institute of Human Genetics, Polish Academy of Sciences, Poznan, Poland.

Expression of HLA-D Subregion Genes in Thyroid Follicular Cells Transfected with SV-40+

Klaus W.J. Badenhoop, Massimo Buscema, Antonino Belfiore, Ricardo Pujol-Borrell, John Trowsdale, and Gian Franco Bottazzo

Abstract: By transfecting HLA class II-negative thyroid cells with the plasmid px-8 containing the SV-40 early region, we obtained a continuous line of thyrocytes which maintains most of the characteristics of the parental cells. After a period of 4 months, the transfected thyrocytes showed an increased surface expression of class I molecules and a de novo expression of class II products. Different clones derived from this thyroid cell line varied in their levels of constitutive class II basal expression and also in their sensitivity to interferon-gamma (IFN-γ). This phenomenon constitutes a novel mechanism of HLA class II product induction and may be relevant to explain the in vivo inappropriate class II expression described in the autoimmune and neoplastic diseases of the thyroid.

The expression of class II genes is restricted to some cells of the immune system and of epithelia lining open body cavities, but can be induced in a variety of cells under stimulation by lectins and interferons (IFN) (1). Among the latter, IFN-γ seems to be the most important mediator in the regulation of class II expression.

HLA class II molecules are inappropriately expressed in thyrocytes from glands afflicted by autoimmune diseases (2) and also in a variety of thyroid conditions (3). Lectins and IFN-γ can induce de novo expression in thyroid monolayer cultures (4,5). HLA class I molecules are ubiquitous, their expression is also increased in autoimmune disease, and they share some of the regulatory mediators known to operate class II.

We studied HLA class II gene expression in SV-40-transfected human thyroid cells using Northern blot analysis. Briefly, human thyroid cells in monolayer culture were transfected by the calcium phosphate precipitation technique with 20 μg of DNA from the vector pSVS/px-8, a modified pBR322, containing the origin of DNA replication, the beta-lactamase gene and the SV-40 DNA, joined at its EcoRI restriction site to the pBR322 segment. After a "crisis" period at 3 months, the cells started to grow continuously. At 4 months, a subpopulation of these keratin-positive epithelial cells were found to express class II molecules spontaneously (6). Three types of continuous cultures were derived from the original plate: 1) subcultured colonies, 2) pool culture, and 3) clones derived from 2.

The transcription of class I and II genes and its regulation by IFN-γ were studied in the following cell preparations: normal thyroid cells, pre- "crisis" SV-40-transfected thyroid cells, and transformed cells from the pool cultures and from clones IF1 and IVA1. The control was provided by a lymphoblastoid cell line (FPF).

Immunofluorescence (IFL) and FACS studies had shown that cells in the pool culture (even at early stages) and in some of the clones consistently showed a more intense staining with MoAbs against class I determinants when compared with parental cells or other thyroid

Immunobiology of HLA
Volume II

cultures. Class II molecules were not detectable by IFL in the first 3 months after transfections, neither by FACS nor by Northern blotting with a DRβ probe.

Four months after transfection and just after the "crisis" period, class II molecules were detected in 4 to 8% of the cells in the pool culture. We have analyzed in detail two of the clones derived from this pool culture because they differed markedly in their basal expression of class II molecules.

In basal conditions, less that 1% of the cells from clone IF1 expressed class II, while in IVA1 more than 40% were class II-positive.

Normal thyroid cells, pre-"crisis" transfected cells, and the clone IF1 did not show any DRβ transcript in basal conditions. By contrast, the pool cell population showed a DRβ transcript and a strong hybridization signal after stimulation with 100 U/ml IFN-γ for 2 days. The DRβ transcripts were inducible in normal thyroid cells and in clone IF1 after stimulation with IFN-γ.

We then investigated the expression of the HLA-D subregion genes DP, DQ, and DO/Z. Cells from the pool culture, clone IF1, and clone IVA1 showed DPβ transcripts, which were not upregulated when treated with 100 U/ml IFN-γ. In addition, the signal was of lower intensity than DRβ signals. These data are in agreement with IFL results (data not shown).

The DQβ, DOβ, and DZα messages were weak but detectable in these cells and did not clearly increase after IFN-γ treatment as DRβ message. However, when we performed dose-response studies with the clones IF1 and IVA1 and cells were stimulated with 10 U/ml, 100 U/ml, and 1,000 U/ml and compared with the basal condition, DRβ transcripts were strongly upregulated and the same results were obtained for the DPβ messages (though at a lower level) (Fig. 1). The transcription of DQβ chains, however, was upregulated after stimulation with 100 U/ml and did not show any further enhancement after 1000 U/ml IFN-γ (data not shown).

In summary, by Northern blotting, we characterized the HLA gene expression in SV-40-transfected human thyroid cells. The de novo class II and enhancement of class I gene expression appear to occur at or even prior to the transcriptional level. All the HLA-D subregions for which probes are currently available have been examined and found to be expressed, although to a varying degree, and are increased by IFN-γ. Transcripts for DRβ and DPβ show the same pattern. On the contrary, DQβ DZα, and DOβ chains show a low level of RNA in basal conditions, and only high levels of IFN-γ can bring on a strong signal of transcription. Class I gene expression is highly elevated and is also strongly upregulated by IFN-γ. Transcripts were all of the same size as the control lymphoblastoid cell line FPF.

The mechanism by which the de novo expression of all the class II genes differentiated regulation of the subregions occurs, remains unclear. One possibility is that the SV-40 promoter insertion affects a regulatory protein that controls the expression of class I, class II, and class II invariant chain genes.* Recently it has been shown that despite the unrelatedness of the invariant chain gene to

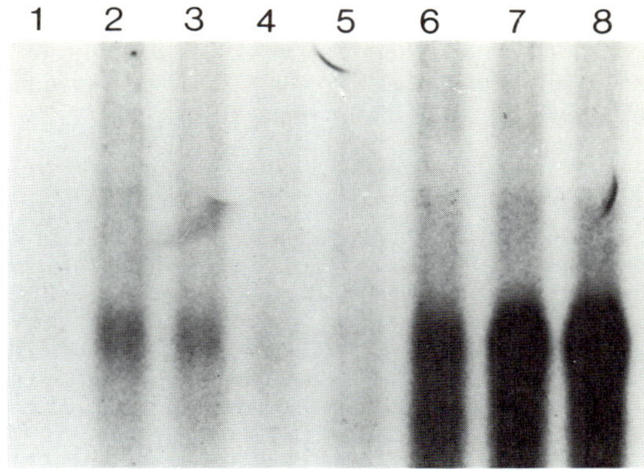

Figure 1. Autoradiograph of a Northern blot hybridized with a DPβ probe (Workshop standard). Lane 1: Clone IF1 (originated from SV-40-transfected thyroid cell pool cultures). Lane 2: IF1 stimulated with 10 U/ml IFN-γ for 2 days, with 100U/ml (Lane 3) and 1000 U/ml (Lane 4). On the right half of the blot, the same dose-response is shown for the clone IVA1. Lane 5: Clone IVA1 in basal condition. Lane 6: Clone IVA1 treated for 2 days with 10 U/ml IFN-γ, 100 U/ml (Lane 7) and 1000 U/ml (Lane 8). Briefly, total cellular RNA of 8×10^7 cells of each clone preparation was extracted in 5M guanidinium isothiocyanate over a CsCl/EDTA gradient, and 10μg were separated by denaturing formaldehyde/1% agarose gel electrophoresis and transferred onto Hybond N filters. The filters were hybridized with an oligolabeled 1 kb probe for the DPβ chain in 50% deionized formamide, 0.1% Denhardt's solution, 5 × SSPE, 1% SDS, 0.5% dextran sulphate, and 100 μg/ml salmon sperm DNA. The filters were washed down to 0.2 × SSPE, 0.1% SDS and exposed for 4 days. A strong signal of approximately 1.2 kb (calculated from marker RNA) was detected in the lymphoblastoid line FPF (not shown) and in the clone IVA1 after stimulation with 10, 100, and 1000 U/ml IFN-γ (Lanes 6–8), whereas the basal signal is weak (Lane 5). On the contrary, clone IF1 has no detectable basal level of DPβ transcript (Lane 1) and shows a moderate increase after treatment with 10 and 100 U/ml IFN-γ (Lane 2&3) and a decrease with 1000 U/ml (Lane 4), data all in agreement with IFL results.

class II genes, they share consensus sequences approximately 150 base pairs upstream from all the class II α and β chain genes (7). As this region is thought to be involved in selective gene expression in differentiated cells (8), these new thyroid cell lines are of interest to study the involvement of these sequences in regulating mechanisms that coordinate the expression of invariant, α and β class II genes in a tissue-specific fashion and in response to lymphokines. A better understanding of the tissue-specific regulation of class II should help to interpret the inappropriate class II expression observed in autoimmunity.

*There are other alternative explanations (for example, the transformation could have led to unstable phenotypes and various lines could have lost control of class II repression).

Acknowledgments. K.B. was supported by the German Academic Exchange Service (DAAD), A.B. by the Welcome Trust Foundation (UK) and R.P.B. by the Juvenile Diabetes Foundation International (USA).

References

1. Korman AJ, Boss JM, Spies T, Sorrentino R, Okada K, Strominger JL. Genetic complexity and expression of human class II histocompatibility antigens. Immunol Rev 1985; 85:45.
2. Hanafusa T, Pujol-Borrell R, Chiovato L, Russell RCG, Doniach D, Bottazzo GF. Aberrant expression of HLA-DR antigen on thyrocytes in Graves' disease: relevance for autoimmunity. Lancet 1983;ii:1111.
3. Lloyd RV, Johnson TL, Blaivas M, Sisson JC, Wilson BS. Detection of HLA-DR antigens in paraffin embedded thyroid epithelial cells with a monoclonal antibody. Am J Pathol 1985;120:106.
4. Pujol-Borrell R, Hanafusa T, Chiovato L, Bottazzo GF. Lectin-induced expression of DR antigen on human cultured follicular thyroid cells. Nature 1983;303:71.
5. Todd I, Pujol-Borrell R, Hammond LJ, Bottazzo GF, Feldman M. Interferon-gamma induces HLA-DR expression by thyroid epithelium. Clin Exp Immunol 1985;61:265.
6. Belfiore A, Pujol-Borrell R, Mauerhoff T, Mirakian R, Bottazzo GF. Effect of SV-40 transformation on HLA expression by thyroid follicular cells: arise of a population of DR positive thyrocytes. Ann Endocrinol 1986;47:17(A).
7. O'Sullivan DM, Larhammar D, Wilson MC, Peterson PA, Quaranta V. Structure of the human Ia-associated invariant (γ)-chain gene: identification of 5′ sequences shared with major histocompatibility complex class II genes. Proc Natl Acad Sci USA 1986;83:4484.
8. Edlund T, Walker MD, Barr PJ, Rutter WJ. Cell specific expression of the rat insulin gene: evidence for role of two distinct 5′ flanking elements. Science 1985;230:912.

Author Affiliations

Klaus W.J. Badenhoop, Department of Diabetes and Immunogenetics, St. Bartholomews Hospital; *Massimo Buscema, Antonino Belfiore, Ricardo Pujol-Borrell, and Gian Franco Bottazzo*, Department of Immunology, Middlesex Hospital; *John Trowsdale*, Imperial Cancer Research Fund (ICRF), London, U.K.

Structures of Class II MHC Molecules and Accessory Proteins During Trafficking in Subcellular Compartments

Quoc V. Nguyen, Lawrence J. Thomas, Patricia S. Reisert, Minzhen Xu, and Robert E. Humphreys

Synthetic rates of class II major histocompatibility complex (MHC) and associated molecules increased appreciably after 4 to 8 hours of polyclonal activation without any perceptible change in forms of those molecules. I_i was cleaved in intracellular vesicles to release a COOH-terminal p25 fragment at 20 to 40 minutes after [^{35}S] methionine labeling. The association of I_i to class II MHC antigen could protect the desetope of class II MHC molecules from ambient peptides until reaching a foreign peptide containing endosome, there catalyzing the charging of digested foreign peptides onto the desetope in a concerted mechanism of I_i release and association of the foreign peptide. I_i could also retain class II MHC antigens in an endosome until charged for transport to the cell surface.

Amphipathic, frequently helical, peptides from digests of foreign proteins bind to selected class II MHC alleles for presentation to receptors of $CD4^+$ T cells. While such peptide-class II MHC antigen complexes are very stable, their rate of formation in vitro with purified components is so slow (1.2 $M^{-1} s^{-1}$) (1) that one might propose that in vivo accessory proteins or enzymes are needed to catalyze the association process. One candidate for such a catalytic protein is I_i, the electrophoretically invariant glycoprotein noncovalently associated with class II MHC antigens, and its alternate or derived forms (2).

To characterize mechanisms in antigen processing and presentation, we examined the structures and synthetic rates of class II MHC molecules and associated proteins both during activation of resting B lymphocytes and during processing and trafficking of such complexes within polyclonally activated, human B cells.

The rate of synthesis of [^{35}S]methionine-labeled, class II MHC and I_i molecules did not increase appreciably until between 4 and 8 hours of polyclonal activation of elutriator-sized, small, resting B lymphocytes. These molecules were immunoprecipitated with anti-class II MHC or anti-I_i antibodies and examined by two-dimensional sodium dodecyl sulfate-polyacrylamide gel electrophoresis (2D SDS-PAGE). The constitutive level of I_i synthesis appeared to exceed that of class II molecules in resting cells. The changes in class II and I_i-related molecules during B-cell activation were principally in levels of synthesis rather than in structure, since no activation-specific changes in glycosylation or processing were seen. Increased synthesis of class II, I_i, Ip, γ2, γ3, and p41 proceeded together from 8 hours of activation and paralleled increased surface expression of class II molecules.

In pulse-chase studies of 72-hour formalinized *S. aureus*-stimulated B cells, separated into subcellular compartments by Percoll density gradient centrifuga-

Immunobiology of HLA
Volume II

tion, class II MHC complexes were not transported to the plasma membrane for at least 60 minutes. Since bulk flow can carry membranal components to the cell surface within a 10-minute period, one can suggest that class II complexes were sequestered in some compartment of the endoplasmic reticulum, Golgi apparatus, or endosome, pending release to bulk flow transport to the surface (upon charging with a foreign peptide). In these pulse-chase studies, class II MHC β chain and Ip were processed to more acidic forms.

A 25,000-dalton protein, shown to be derived from the COOH-terminal portion if I_i, was found at 20 to 40 minutes of pulse-chase labeling of activated cells. Cleavage if I_i to p25 was probably restricted by secondary structural features because proteolysis by trypsin, chymotrypsin, and proteinase K all led to 25-kD fragments, including the spot found in 2D electrophoresis gels from pulse-chase experiments. Monensin and chloroquine, which blocked endosomal and Golgi functions, did not reduce the appearance of p25. Since some p25 could be found in anti-class II immunoprecipitates, cleavage of I_i to p25 could occur while I_i was associated with class II molecules. The appearance of p25 was found in a subcellular fraction of Percoll density gradient-separated cells and did not appear in plasma membrane fractions even to 60 minutes after labeling.

The appearance of p25 in subcellular fractions of polyclonally activated B lymphocytes at 20 to 40 minutes after synthesis was consistent with an hypothesis that the function of I_i was either 1) to block the foreign antigen binding site of class II molecules until charging with a foreign peptide (3), or 2) to retard the class II complex from bulk transport to the surface until the II MHC complex was "completed" with a foreign peptide. Our future work will be to test whether the striking amphipathic helix in I_i, in fact, lies at the foreign antigen combining site, and what mechanisms, (e.g., flooding with T-cell-presented peptides) might release class II antigens to be transported to plasma membranes.

Acknowledgments. This work was supported by American Cancer Society institutional grant IN-129 to Q.V.N. and by NIH grant CA-37645 to R.E.H. Q.V.N. was supported by NIH fellowship AI-07272.

References

1. Buus S, Sette A, Colon SM, Jenis DM, Grey HM. Isolation and characterization of antigen-Ia complexes involved in T cell recognition. *Cell* 1986;46:1071.
2. O'Sullivan DM, Noonan D, Quaranta V. Four Ia invariant chain forms derive from a single gene by alternate splicing and alternate initiation of transcription/translation. J Exp Med 1987;166:444.
3. Elliott WL, Stille CJ, Thomas LJ, Humphreys RE. An hypothesis on the binding of an amphipathic, alpha helical sequence in I_i to the desetope of class II antigens. J Immunol 1987;138:2949.

Author Affiliations

Departments of Pediatrics, Pharmacology, and Medicine, University of Massachusetts Medical Center, 55 Lake Avenue North, Worcester, MA 01655-2397, USA.

Class II MHC Complexes with I_i, p70, and p80 (CS-I_i)

Christopher H. Sorli and Robert E. Humphreys

In order to characterize proteins that associate with class II major histocompatibility complex (MHC) molecules, perhaps to catalyze their charging with peptides from foreign antigens or to regulate antigen presentation, complexes of class II antigens and accessory proteins were cross-linked with a disulfide-containing heterobifunctional reagent and analyzed by electrophoresis in a two-dimensional, nonreduced/reduced slab gel. The following cross-linkages were identified: I_i/I_i, I_i/p70 (not CS-I_i), α/α, α/β, α chain/p80 (probably CS-I_i), and an α/β/I_i/p70/p-80 polymer(s).

[^{35}S]Methionine-labeled, detergent-solubilized Raji complexes were cross-linked with the heterobifunctional, disulfide-containing reagent sulfosuccinimidyl 2-(*p*-azidosalicylamido)ethyl-1,3'-dithiopropionate (SASD) or were mock treated. Unreduced complexes were sodium dodecyl sulfate (SDS) electrophoresed in a cylindrical gel and were then reduced and SDS electrophoresed into a slab gel. Uncross-linked complexes migrated along the diagonal of the gel, and the components of any cross-linked complex migrated in the second dimension in a vertical line which intersected the diagonal at the apparent molecular weight of the cross-linked complex. Compositions of various complexes were resolved in part because El anti-I_i (183-193) serum immunoprecipitated I_i (and γ2, γ3, p41, p25) but not CS-I_i (as demonstrated in [^{35}S]sulfate-labeled gels) and VICγ1 mAb to I_i immunoprecipitated both I_i and CS-I_i. The third antiserum used in this study, anti-p23,30 serum, recognized class II MHC α and β chains but not I_i or CS-I_i which were not associated with the α and β chain class II complex.

These experiments demonstrated two, endogenously disulfide-linked complexes present without SASD: an I_i

Immunobiology of HLA
Volume II

homodimer, and a complex of I_i with a p70 which was shown not to be CS-I_i (I_i/p70). Noncovalently associated proteins which became SASD-linked included: 1) an α chain homodimer (α/α), 2) and α, β chain heterodimer (α/β), 3) a complex of α chain with p80, which appeared to be CS-I_i, in terms of [^{35}S]sulfate labeling of VICγ1 immunoprecipitated CS-I_i and visualization after 10-hour [^{35}S]methionine labeling (α/p80), and 4) high molecular weight polymers of α chain, β chain, I_i, p70 (not CS-I_i) , and p80 (probably CS-I_i) (α/β/I_i/p70/p-80).

To analyze the time course for the association of these complexes, cells were [^{35}S]methionine labeled for 3,6,9, and 12 hours. Radiolabeled p80 (CS-I_i) and p70, which were recognized only by their associations to the class II molecules bound by the immunoprecipitating antibodies, were well represented in high molecular weight complexes only after 9 and 12 hours of labeling.

We draw several conclusions and hypotheses from these studies which confirm and extend the work of Kelner and Cresswell (1): 1) A naturally occurring, disulfide-linked homodimer of I_i is present. I_i which is dissociated from class II molecules to free the desetope might dimerize through an amphipathic helix (146-164) and be fixed by a disulfide bridge, no longer to compete for the class II antigen desetope. 2) Some complexes may either be present naturally, (e.g. the α/β heterodimer) or be identified under conditions of limiting SASD concentration because cross-links were formed only between those members, (e.g., α and p70) in some complexes containing more polymeric subunits. Thus, the α/β heterodimer, α/α homodimer, I_i/p70 (not CS-I_i) heterodimer, and α/p80 (probably CS-I_i) dimers might represent specific, prominent associations within larger complexes. 3) p70 which associates with I_i and in a large polymeric complex(es) of α/β/I_i/CS-I_i/p70, is not CS-I_i. Such late-formed complexes, including CS-I_i which is principally surface expressed, might be involved in a surface function of the class II molecules in antigen presentation to T cells (2,3). 4) p25 (which was shown separately to be a COOH-terminal proteolytic fragment of I_i) accumulates over 6 to 12 hours, and is not associated in the other identified complexes.

In future studies, we plan to: 1) identify the cellular compartments (ER-Golgi, endosome, plasma membrane, lysosome) in which these complexes are formed or to which they are trafficked; 2) relate p70 to human p67/69 or murine p72/74 which might have a role in antigen presentation; and 3) examine the structural relationship of accessory proteins to the class II desetope.

Acknowledgments. This work was supported by NIH grants CA 37645 and AM 32520.

References

1. Kelner DN, Cresswell P. Biosynthetic intermediates of HLA class II antigens from B lymphoblastoid cell lines. J Immunol 1986;137:2632.
2. Bono C, Pereira MB, O'Sullivan DM, Quaranta V, Schwartz BD. The HLA-class II-associated chondroitin sulfate proteoglycan expressed by class II positive T and monocyte-like cell lines is larger than that expressed by EBV-transformed B-lymphoblastoid cell lines. Hum Immunol 1987;18:315.
3. Sivak LE, Harris MR, Kindle C, Cullen SE. Inhibition of proteoglycan synthesis eliminates the proteoglycan form of Ia-associated invariant chain and depresses antigen presentation. J Immunol 1987;138:1319.

Author Affiliations

Departments of Pharmacology and Medicine, University of Massachusetts Medical Center, 55 Lake Avenue North, Worcester, MA 01655-2397, USA

Intracellular Signaling Molecules Responsible for the Regulation of HLA Class II Antigen Expression

Y. Koide, Y. Ina, N. Nezu, and T.O. Yoshida

The class II antigens of the major histocompatibility complex (MHC) are products of the immune response gene and play an essential role in restricting the recognition of nominal antigen by T lymphocytes. The expression of MHC class II antigens is restricted to certain cell types and is not always constitutive. The expression on certain cell types is regulated by various lymphokines (e.g., interferon (IFN)-γ) elaborated by activated T lymphocytes.

As previously reported, we observed that IFN-γ led HL-60 cells to express the HLA class II antigens, DR and DQ, in a dose-dependent manner. Furthermore, we found that cross-linking of Fc receptors (FcR) on HL-60 cells inhibited the ability of recombinant IFN-γ to induce class II antigens. This appeared to be correlated with intracellular mRNA level. Northern blot analysis revealed that IFN-γ led HL-60 to express both DRA and DRB mRNA. The cross-linking of FcR canceled the

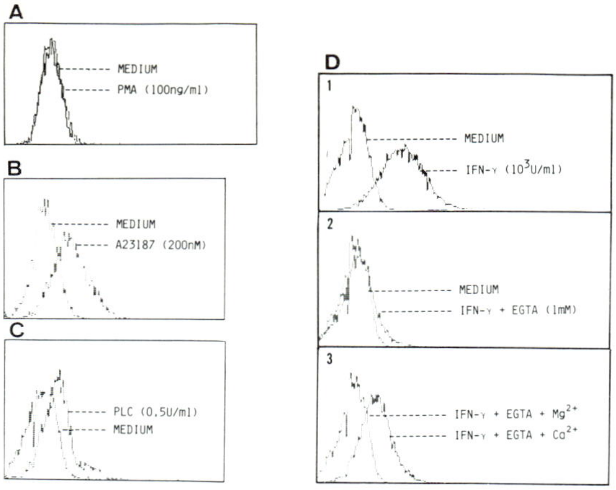

Figure 1. Effect of PMA (A), A23187 (B), PLC (C), or EGTA (D) on the induction of DR antigen expression. Cells were cultured with or without PMA (100 ng/ml) (A), A23187 (200 nM) (B), or PLC (0.5 U/ml) (C) for 24 hours and then analyzed for DR expression. D, HL-60 cells were cultured with IFN-γ (10^3 U/ml) in the presence (2) or absence (1) or EGTA (1 mM) for 24 hr. $CaCl_2$ (5mM) or $MgCl^2$ (5mM) was added to the cells suspended in medium containing IFN-γ and EGTA (1 mM) (3). DR expression was analyzed by EPICS V.

accumulation of DRA and DRB transcripts induced by IFN-γ. Therefore, experiments were designed to investigate the intracellular signaling molecules regulating the appearance of class II molecules or transcripts. The expression of DR antigen induced by IFN-γ was blocked by a calmodulin (CaM) antagonist, W-7 (1), at 20-30 μM. However, a protein kinase C (PKC) inhibitor H-7 (2) showed no effect on the class II antigen-inducing activity of IFN-γ. Furthermore, phorbol myristate acetate (PMA), a direct activator of PKC (3), was not able to induce DR antigen expression (Fig. 1A). These results indicate that IFN-γ induces class II antigen expression via the Ca^+-CaM pathway and that PKC is not involved in the antigen expression. Because a CaM branch is activated by a transient rise in the cytosolic free calcium, whether calcium influx into HL-60 is actually induced by IFN-γ was examined. Using $^{45}CaCl_2$, a remarkable and time-dependent calcium accumulation in HL-60 cells was observed in the presence of IFN-γ. We examined the direct effect of calcium influx on the induction of DR antigen expression using calcium ionophore A23187. As shown in Figure 1B, A23187 induced the expression of DR antigens. On the other hand, depletion of CA^{2+} by EGTA from culture medium resulted in failure of IFN-γ to induce DR antigen expression (Fig. 1D2). The effect of EGTA was reversed by addition of exogenous $CaCl_2$ but not $MgCl_2$. Taken together, these results suggested that IFN-γ evokes calcium influx by a so-called receptor-mediated calcium channel and activates the CaM branch of the calcium messenger system, resulting in the induction of class II antigens on the surface of HL-60 cells.

We next evaluated the second messenger generated by the cross-linking of FcR. Marked accumulation of intracellular cAMP was observed by the treatment of HL-60 with solid-phase IgG and anti-Leu 11b, both of which led to the cross-linking of FcR. Free IgG showed no effect on the level of intracellular cAMP. The preincubation of HL-60 with dibutyryl cAMP, a cAMP analogue, for 1 hour resulted in the suppression of IFN-γ-induced calcium influx, and the suppression of class II antigen-inducing activity of IFN-γ. These results suggest that the cross-linking of FcR accumulates intracellular cAMP, which results in suppression of class II antigen expression.

These intracellular events of HL-60 cells regulate the expression of HLA class II transcripts and molecules.

Acknowledgment. This work was supported by grants for research into intractable diseases from the Ministry of Health and Welfare, Japan, and a grant-in-aid for scientific research from the Ministry of Education, Science and Culture, Japan.

References

1. Hidaka H, Sasaki Y, Tanaka T, Endo T, Ohno S, Fujii Y, et al. N-(6-aminohexyl) -5-chloro-1-naphathalenesulfonamide, a calmodulin antagonist, inhibits cell proliferation. Proc Natl Acad Sci USA 1981;78:4354.
2. Hidaka H, Inagaki M, Kawamoto S, Sasaki Y. Isoquinolinesulfonamides, novel and potent inhibitors of cyclic nucleotide dependent protein kinase and protein kinase C. Biochemistry 1984;23:5036.
3. Nishizuka Y. The role of protein kinase C in cell surface signal transduction. Nature 1984;308:693.

Author Affiliation

Department of Microbiology and Immunology, Hamamatsu University School of Medicine, Hamamatsu, Japan.

HLA Class I and Class II Molecules Expressed in Monocytes, B Cells, and EBV Transformed Cells are Differentially Glycosylated

B. Merlu, D. Néel, Y. Goussault, and D.J. Charron

Comparison of asparagine-linked oligosaccharides from HLA class I and HLA class II molecules was done on normal and transformed cells excluding the influence of genetic polymorphism. B cells and monocytes were isolated from peripheral blood of a volunteer donor and an EBV transformed B lymphoblastoid cell line (C25 cell line) was obtained from the same individual. ^{3}H-Mannose radiolabeled glycoproteins were isolated with specific monoclonal antibodies (W6/32 and D1.12) and the pronase digested glycopeptides were analyzed by chromatographic methods.

Affinity chromatography on Concanavalin A Sepharose (1) (Table I) allows to discriminate between glycosylation of class I and class II molecules. Oligosaccharides substituted on HLA class I molecules were essentially biantennary structures while these on HLA class II molecules were largely distributed between biantennary, tri-, tetra-antennary, and high mannose type structures. These results could be related to the unique site of glycosylation on class I, and the 3 ones on class II molecules. Moreover, the glycosylation of HLA molecules is very much influenced by EBV transformation. The biantennary structures which are highly expressed in normal cells decrease after transformation, while an increase in more antennary complex structures is observed. These facts may suggest an overactivity of sialyltransferases in the transformed cells.

The complex structure analyses by ion exchange chromatography on DEAE SA-25 point out a difference in charge between the oligosaccharides from normal B lymphocytes and monocytes, B cells being more acidic than monocytes. These results are in agreement with the differential ability of monocytes and B cells to present antigens during the immune response as it has been shown in the mouse (2). Moreover, the EBV transformation leads to an increase of the whole charge of HLA molecules characterized by an increase of oligosaccharide structures bearing two sialic acid residues while the mono-sialylated structures were predominant in the case of normal cells. This could contribute to the greater efficiency of EBV transformed cells to present antigens compared to normal B cells.

Acknowledgment. This work was supported in part by an INSERM grant. A complete manuscript on this work is submitted for publication.

Table 1. Concanavalin A Sepharose Affinity Chromatography of HLA Molecules

	Oligosaccharide structures		
	tri-tetra-/ antennary (%)	bi-/ antennary (%)	high-/ mannose (%)
Class I			
monocytes	10	85	5
B Cells	10	85	5
C25 Cells	60	10	30
Class II			
monocytes	30	30	40
B Cells	30	30	40
C25 Cells	60	5	35

Results expressed as % of ^{3}H-Mannose incorporated in each affinity fraction.

References

1. Cummings AP, Kornfeld S. Fractionation of asparagine linked oligosaccharides by serial affinity chromatography. J Biol Chem 1982;257:11235.
2. Cowing C, Chapdelaine JM. T cells discriminate between Ia antigens expressed on allogeneic accessory cells and B cells: A potential function for carbohydrate side chains on Ia molecules. Proc Natl Acad Sci USA 1983;80:6000.

Author Affiliations

D. Néel, Y. Goussault, INSERM U. 180, Paris; *D.J. Charron*, Laboratoire d'Immunogénétique Moléculaire des Cordeliers, Paris; *B. Merlu*, INSERM U. 180 and Laboratoire d'Immunogénétique Moleculaire des Cordeliers, 15 rue de l'ecole de Medècine 75006, Paris, France

Section XIII
Antigen Presentation and HLA Restriction

Functional Analysis of HLA-DR3: A Mutational Approach*

Elizabeth Mellins, Tom Cotner, Benjamin Arp, Henry Erlich, and Donald Pious†

Introduction

Class II mutant B lymphoblastoid cell lines (B-LCL) affected in presentation of soluble antigen to HLA-restricted antigen-specific T cells can be used to characterize the molecular interactions involved in antigen presentation. We have isolated class II mutants in the DR3 haplotype, a haplotype of particular interest because of its association with several autoimmune diseases (1). Here, we describe the use of antigen loss mutants to localize restriction elements to the DRβ1, DRβ3, and non-DR dimers of the DR3 haplotype, and the use of mutants expressing altered DR3 molecules to identify regions of DR3 molecules involved in antigen presentation and in α/β chain association.

Results and Discussion

To isolate class II mutants in the DR3 haplotype, we mutagenized a DR3 expressing B-LCL with ethyl methane sulfonate and immunoselected with an allospecific monoclonal antibody, 16.23 (2). To simplify the analysis, we used as a progenitor line 8.1.6, a deletion mutant which lacks all the DR and DQ genes of one haplotype, but expresses the other haplotype, DR3, normally (3). Immunoselection produced a large group of mutants which fall into two broad classes: DR-loss mutants and DR-altered mutants.

Mutants which have lost cell surface expression of class II molecules make it possible to assess the independent contribution of individual class II loci to the restriction repertoire of T cells. The 16.23 immunoselection experiment generated two useful mutants in this regard: mutant 9.22.3, which is deleted for the single DRA gene present in 8.1.6 and has lost cell surface expression of both DRαβ1 and DRαβ3 (4); and 9.4.3, which lacks DR β1 mRNA and has lost expression of DRαβ1 dimer only (5). Both 9.22.3 and 9.4.3 express unaltered DP and DQ dimers. As an assay system for antigen presentation by the congenic antigen-loss mutants derived from 8.1.6, we generated a series of antigen-specific T cell lines from DR3 donors (6). The responses of these lines when assayed with 8.1.6 reflect T cells restricted by those allospecificities shared by 8.1.6 and the T cell donors: DR3, DRw52, DQ2, DP4.

To determine the relative use of class II molecules of the DR3 haplotype as restricting elements, we compared the ability of 8.1.6, 9.22.3, and 9.4.3 to stimulate proliferation of the antigen-specific polyclonal T cell lines. As shown in Table 1, for the tetanus toxoid specific line CT-TT and the Candida albicans specific line CT-CA, 9.22.3 stimulates 40 to 50% of the T-cell proliferation stimulated by 8.1.6, suggesting that these lines include DP and DQ restricted T cells. In assays of both the polyclonal lines (CT-TT, CT-CA) and the oligoclonal, DR-restricted lines (NJ-CA; ND-PPD; OB-TT), T-cell proliferation to 9.4.3 is higher than with 9.22.3. Because the two mutant lines differ only by cell surface expression of the DRαβ3 dimer, these data demonstrate that the DRαβ3 dimer is a functional molecule for antigen presentation. The substantial proportion of the DR-restricted T-cell response stimulated by 9.4.3 is an unexpected result, given that the β3 chain is expressed at significantly lower levels than the β1 chain (5). Further analysis

Table 1. Proliferation* to Antigen Presented by

T-Cell Donor-Antigen†	8.1.6 (Progenitor)	9.4.3 (DRβI-loss)	9.22.3 (DR-null)	6.1.6‡ (Class II-null)
CT-TT	26471	13448	10932	950
CT-CA	55927	3822	3200	0
NJ-CA	11620	8405	223	nd
ND-PPD	14272	6830	390	nd
OB-TT	7539	3809	1579	nd

*Proliferation assays were performed as described (6). Data are net specific cpm 3H-thymidine from the median of triplicate assays. No alloreactive response to 8.1.6 is observed in the absence of soluble antigen (6).
†Antigens are tetanus toxoid (TT), PPD, and C. albicans (CA).
‡6.1.6 is a class II negative mutant clone that is otherwise congenic with 8.1.6 (7). As such, it serves as a negative control for class II-restricted response.

Table 2. 16.23 Immunoselection: Mutant Phenotypes

Mutant	Cell Surface DR3 Abundance (% of 8.1.6)	2-D SDS-Page	Antigen Presentation* TT	CA	PPD
(8.1.6)	(100)	Wild Type	+	+	+
7.13.6	47	Altered βI chain	+	+	−
10.77.6	37	Altered βI chain	+	−	−
7.25.6	45	Altered βI chain	+	+	+
7.31.6	100	Altered βI chain	+	+	−
9.5.3	100	Wild Type	−	−	−
9.10.3	100	Wild Type	−	−	−

*Proliferation assays were performed as described (5).
+, Response of at least 80% of response seen with 8.1.6.
−, Response ≤10% of response seen with 8.1.6.

*Abbreviations used in this paper: B-LCL, B lymphoblastoid cell line; CA, Candida Albicans; mAb, monoclonal antibody; and TT, tetanus toxoid.

Immunobiology of HLA
Volume II

Table 3. Pattern of T-cell Responses to Mutants

Clone	Antigen† Specificity	Proliferation* to antigen presented by 8.1.6 (Progenitor)	9.22.3 (DR-null)	9.4.3 (DRβ1 loss)	9.5.3 (Mutants)	9.10.3 (Mutants)
CT_1	TT/DR3β1	+	–	–	–	–
ND_1	PPD/DR3	+	–	nd	–	–
ND_2	PPD/DR3	+	–	nd	–	–
ND_3	TT/DR3	+	–	nd	–	–
ND_4	TT/DR3	+	–	nd	–	–
62D10	Allo/DR3	+	–	–	–	–
24F7	Allo/DR3	+	–	–	–	–
CT_2	TT/DRw52	+	–	+	+	+
C6	Allo/DRw52a	+	–	+	+	+

*Proliferation assays were performed as described (6).
+, Response of at least 80% of response seen with 8.1.6.
–, Response ≤10% of response seen with 8.1.6.
†Antigens are tetanus toxoid (TT) and PPD.

using cloned T cells is required to determine whether T-cell proliferation to 9.4.3 reflects T cells which can use either DRβ1 or DRβ3 as a restricting element, or reflects T cells which are restricted solely to DRβ3.

In addition to DR-loss mutants, immunoselection with 16.23 also generated mutants which express altered DR3 molecules. Structure/function correlations in mutants of this type can identify specific residues involved in DR3-restricted antigen presentation. As shown in Table 2, we evaluated these mutants by cell surface antibody binding, two-dimensional SDS-PAGE analysis, and antigen presentation assays. Mutants 7.13.6, 10.77.6, and 7.31.6 are altered in structure and function; they express DRβ1 chains with altered isoelectric points and have lost antigen-presenting capacity for some but not all antigen-specific T cells tested. Nucleotide sequence analysis of the DR3βI gene of 7.13.6 reveals a single amino acid change in the first hypervariable region of the amino terminal domain (8). These results are consistent with the predicted functional importance of polymorphic regions of the class II β chains (9) and demonstrate the ability of a single amino acid substitution to alter activation of antigen-specific T cells. Further analysis of 7.13.6 will be necessary to determine whether the lesion in this mutant affects the binding of immunogenic peptide to the DR molecule or affects the recognition by specific T cells of the complex of DR and foreign antigen.

In addition to mapping sites involved in antigen presentation, mutants which express altered DR3 molecules may allow mapping of residues that are involved in α/β chain association. Three mutants (7.13.6, 7.25.6, 10.77.6) which express β chains with altered isoelectric points, express decreased levels of cell surface DR3 compared to the wild type line, 8.1.6 (Table 2). Northern blot analysis of these mutants reveals normal mRNA levels in each case. Thus it appears likely that the decreased level of dimer expressed by the mutants reflects abnormal chain association. In contrast, the mutant β1 chain (and dimer) of 7.31.6 is expressed at wild type levels. Data from murine exon-shuffling experiments have suggested that residues in the amino terminal portion of the β1 domain are critical for α/β chain interaction (10). Identification of the mutations in the human mutants described here will provide an independent means to locate sites of chain association.

Immunoselection experiments occasionally yield mutants which reveal hitherto unknown pathways affecting expression of cell surface molecules. For example, the isolation of the class II null mutant 6.1.6 suggested the existence of a transacting factor regulating the transcription of all class II genes (11). A group of mutants from the 16.23 immunoselection (represented by mutants 9.5.3 and 9.10.3 in Table 2) also demonstrate an unexpected phenotype. Structural alterations in 9.5.3 and 9.10.3 as detected by serology are minimal. The mutants do not bind mAb 16.23 but have unaltered reactivity with both monomorphic antibodies (4) and allospecific (DR3) antisera. Nonetheless, the DR molecules in these mutants are likely to be conformationally abnormal; the mutant dimers dissociate into monomeric chains under conditions which preserve dimer in the wild type 8.1.6 line (4). The functional abnormality in 9.5.3 and 9.10.3 is profound. The mutants are unable to stimulate most of a panel of DR3 restricted T cell clones (Table 3). Those T cells stimulated by the mutants are DR3β3 restricted, suggesting that the DRαβ1 dimer is not functional for antigen presentation. Recent preliminary evidence, however, suggests that other class II dimers in these mutants may also be unstable.

Taken together, these results raise the intriguing possibility that the lesions in this group of mutants affect a processing step common to class II gene products and of critical importance for class II molecule function.

Acknowledgments. We thank Steven Scharff at Cetus Corporation, Oakland CA for technical assistance, Barbara Ochs and Devinder Singh of our laboratory for technical assistance, and Dan Hill for preparation of the manuscript.

This work was supported by grant GM15883 and a Physician Scientist Award (to E. Mellins) from the National Institutes of Health.

References

1. Svejgaard A, Platz P, Ryder LP. HLA and disease 1982–A survey. Immun Rev 1983;70:193.
2. Johnson J, Meo T, Riethmuller G, Schendel D, Wank R. Direct demonstration of an HLA-DR allotypic determinant on the low molecular weight (bets) subunit using a mouse monoclonal antibody specific for DR3. J Exp Med 1982; 156:104.
3. Levine F, Erlich H, Mach B, Leach R, White R, Pious D. Deletion mapping of HLA and chromosome 6p genes. Proc Natl Acad Sci 1985;82:3741.
4. Pious D, Dixon L, Levine F, Cotner T, Johnson R. HLA class II regulation and structure. Analysis with HLA-DR3 and HLA-DP point mutants. J Exp Med 1985;162:1193.
5. Mellins E, Arp B, Ochs B, Erlich H, Pious D. A single amino acid substitution in the human histocompatibility leukocyte antigen DR3 β chain selectively alters antigen presentation. J Exp Med 1988;168:1531.
6. Mellins E, Woelfel M, Pious D. Importance of HLA-DQ and–DP restriction elements in T-cell responses to soluble antigens: Mutational analysis. Human Immunology 1987; 18:211.
7. Gladstone P, Pious D. Stable variants affecting B cell alloantigens in human lymphoid cells. Nature 1978;271:459.
8. Cotner T, Charbonneau H, Mellins E, Pious D. mRNA abundance rather than differences in subunit assembly determine differential expression of HLA-DRβ1 and DRβ3 molecules. J Biol Chem, in press.
9. Benoist C, Mathis D, Kanter M, Williams V, McDevitt H. Regions of allelic hypervariability in the murine Aα immune response gene. Cell 1983;34:169.
10. Baaunstein N, Germain R. Allele-specific control of Ia molecule surface expression and conformation. Proc Natl Acad Sci USA 1987;84:2921.
11. Gladstone P, Pious D. Identification of a transacting function regulating HLA-DR expression in a DR-negative B cell variant. Somat Cell Genet 1980;3:93.

Author Affiliations

Elizabeth Mellins, Tom Cotner, Benjamin Arp, Department of Pediatrics; *Donald Pious*, Department of Pediatrics and Department of Genetics, University of Washington, Seattle, WA; *Henry Erlich*, Cetus Corporation, Oakland CA, USA

Studies of MHC Class II Restricted Antigen Recognition by Human T-Cell Clones Using Transfectant Antigen-Presenting Cells (APC)

R.I. Lechler, V. Bal, K. Howland, R.P. Sekaly, E.O. Long, J.R. Lamb, and J.B. Rothbard

Abstract: Using transfected L-cell antigen-presenting cells (APC) and synthetic peptides of antigen, the recognition of antigen/MHC by two DR1-restricted T-cell clones, specific for hemagglutinin and matrix protein determinants of influenza A, was examined. Transfectants expressing DR1 functioned as efficient APC for both T-cell clones. The substitution of I-Eα for DRα led to no loss of antigen sensitivity, and the matrix protein-specific clone also gave an antigen-dependent response in the context of I-Eβ^k:I-Eα. If the amino-terminal domains of DR1 and I-E^k are modeled on the three-dimensional structure of the α_1 and α_2 domains of HLA-A2, it is striking that almost all the α_1 residues that differ between DR and I-E fall outside the putative antigen-binding groove. The β_1 domains of DR1 and I-E^k have multiple differences throughout the domain, however, the residues in the bm12 region are significantly conserved. Comparison of the sequences of antigenic peptides that are co-recognized with DR1 and I-E^k reveals four residues with conserved features, spaced 1,4,5, and 8. These four residues can be exchanged between the hemagglutinin and matrix protein peptides without loss of specific recognition. These results suggest that the antigen-binding sites of DR1 and I-E^k are similar, and that the four residues with conserved features that can be exchanged between the two peptides may form one face of an α-helix and contribute to the DR1:antigen interaction.

The activation of CD4 positive T lymphocytes by antigen requires the co-recognition by the T cell of specific antigen with a class II MHC molecule (1). It is now clear that MHC molecules can form a complex with antigen in the absence of a T-cell receptor (2,3), and it appears that most, if not all, antigens presented by a particular MHC molecule bind to a single region on that MHC molecule (4). Recognition by T cells of this binary complex is usually highly specific for both the determinant of antigen and for the allele of MHC to which it is bound. In these studies we have made use of HLA-DR1-transfected murine L cells in order to explore the specificity of two HLA-DR1-restricted human helper T-cell (T_H) clones specific for an hemagglutinin and a matrix protein determinant of influenza A (5). We have also tested the prediction that antigens which are co-recognized with the same MHC molecule have sequences in common that are physically distinct from the residues that interact with the T-cell antigen receptor.

As shown in Table 1, both T-cell clones gave an antigen-dependent, DR-1-restricted response when the DR1-expressing L cells were used as APC. Although

Table 1. DR-Restricted Recognition of Antigen Presented on DR1-Expressing Murine Fibroblasts by Human T Cell Clones

Responder	APC	Antigen	Anti-DR mAb	cpm
HA1.7	PBMC	+	−	37,624±1
HA1.7	PBMC	+	+	4,182±12
HA1.7	PBMC	−	−	433±12
HA1.7	DR1.L Cell	+	−	8,382±18
HA1.7	DR1.L Cell	+	+	74±23
HA1.7	DR1.L Cell	−	−	432±17
TLC72	PBMC	+	−	23,409±18
TLC72	PBMC	+	+	1,713±14
TLC72	PBMC	−	−	1,583±33
TLC72	DR1.L Cell	+	−	4,586±9
TLC72	DR1.L Cell	+	+	987±6
TLC72	DR1.L Cell	−	−	811±40

Cloned T cells (5×10^4/ml) were cultured with soluble antigen in the presence of irradiated histocompatible PBMC (10^5/ml) or mitomycin-C treated transfectants (5×10^4/ml) in 96 well plates. After 72 hours cultures were pulsed with ^{3}HTdR and harvested for scintillation counting 8–16 hours later. Results are expressed as mean cpm ± % error for triplicate cultures. For HA1.7 antigen was haemagglutin peptide 306–324, for TLC72 it was matrix protein peptide 18–30. Anti-DR mAb was L243 (7), used as a 25% dilution of hybridoma supernatant.

Table 2. Responses of T Cell Clones HA1.7 and TLC72 to Antigen Presented by Transfectants Expressing Human and/or Murine MHC Molecules

APC	Antigen	cpm HA1.7	cpm TLC72
DRI.L Cell	+	8,382±18	11,265±14
DRI.L Cell	−	432±18	325±12
DR1β:IEα.L Cell	+	16,692± 2	9,664±8
DR1β:IEα.L Cell	−	559±11	198±13
IEβ^{k}IEα.L Cell	+	853± 7	5,918±10 (1,100±8)
IEβ^k:IEα.L Cell	−	741±13	255±6

Culture conditions were as for Table 1. Antigen concentrations were: haemagglutin 306–324 0.3 μg/ml for HA1.7, matrix protein 18–30 30 μg/ml for TLC72. Figure in brackets represents proliferation in the presence of anti-IEα mAb, 25% culture supernatant.

PBL functioned as the most efficient APC population, it is striking that on a cell-for-cell comparison, the DR transfectant induced the same, or higher, level of proliferation by the HA1.7 T-cell clone. One of the important variables that influence the efficiency of APC function is the per cell level of expression of the MHC restriction element (6). As judged by flow cytometric analysis, the levels of DR expression on the EBV transformed cell line and the DR-1 transfectant were indistinguishable. The implications for accessory molecule function of the comparable efficiencies of human and murine APC is discussed elsewhere (8), however, these results validated the further use of L-cell transfectants as model APC for investigation of the MHC restriction specificity of these T cells.

In order to explore the restriction specificity of these T cells, two additional transfectants were used, in which one or both of the DR1 chains was substituted with products of the murine I-E locus. The results are presented in Table 2. Both the hemagglutinin, and the matrix protein-specific clones showed no loss of antigen reactivity, at any antigen dose, to the L cell expressing the human/murine hybrid MHC molecule DR1β:I-Eα at the same per cell level as the DR1β:DRα transfectants. Despite the fact that there is 20% nonidentity in the amino acid sequences of DRα_1 and I-Eα_1 domains, this result implies that α chain residues contributing to antigen and/or T-cell receptor interaction are conserved between the α chains of the two species. The HA1.7, hemagglutinin specific clone, showed no antigen response when both MHC chains on the L cell APC were substituted with murine I-E chains giving rise to an I-E^k molecule; but in striking contrast, the I-E^k transfectant was able to stimulate an antigen-dependent response by TLC 72 (Table 2). The expression of I-E^k and the absence of DR expression on this transfectant was confirmed by positive staining with mAb specific for I-Eα and β chains, and by the absence of any reactivity with anti-DR mAb. The I-E^k-restricted response was reproducible and inhibitable by anti-I-Eα mAb. Given that the level of expression of I-E on this L cell was approximately 50% of the level of DR1β:I-Eα on the parallel transfectant, there is surprisingly little loss of efficiency of antigen recognition by TLC 72 due to the substitution of I-E^k for DR1. This implies that critical residues involved in antigen and T-cell receptor (Ti) interaction are conserved between these two MHC molecules. This is discussed further below.

The second question addressed in these experiments was whether the sequences of antigenic peptides restricted by the same MHC molecule would have features in common as a consequence of binding to the same site on the restriction element. As shown in Figure 1A, when the matrix protein and hemagglutinin peptide sequences are compared and aligned on the basis of two adjacent hydrophobic residues, two additional similarities are revealed. Most strikingly lysine 21 is aligned with lysine 311. In addition, a third pair of hydrophobic residues occur at a fourth position, leucine 28 and alanine 318. As shown in Figure 1B, due to the spacing pattern 1,4,5,8, these residues would become juxtaposed if the peptides adopt a helical conformation, and form a conserved face of a helix. It is postulated that these residues contribute to the surface of the antigen that interacts with the MHC molecule, i.e., the "agretope." In view of the cross-species degeneracy of MHC restriction exhibited by TLC 72, it is also noteworthy that I-E^k-restricted T-cell epitopes also have a lysine or an arginine, a pair of adjacent hydrophobic residues, and a third hydrophobic residue with the same spacing pattern 1,4,5,8 (Fig. 1C). Conceivably, the similarity in the

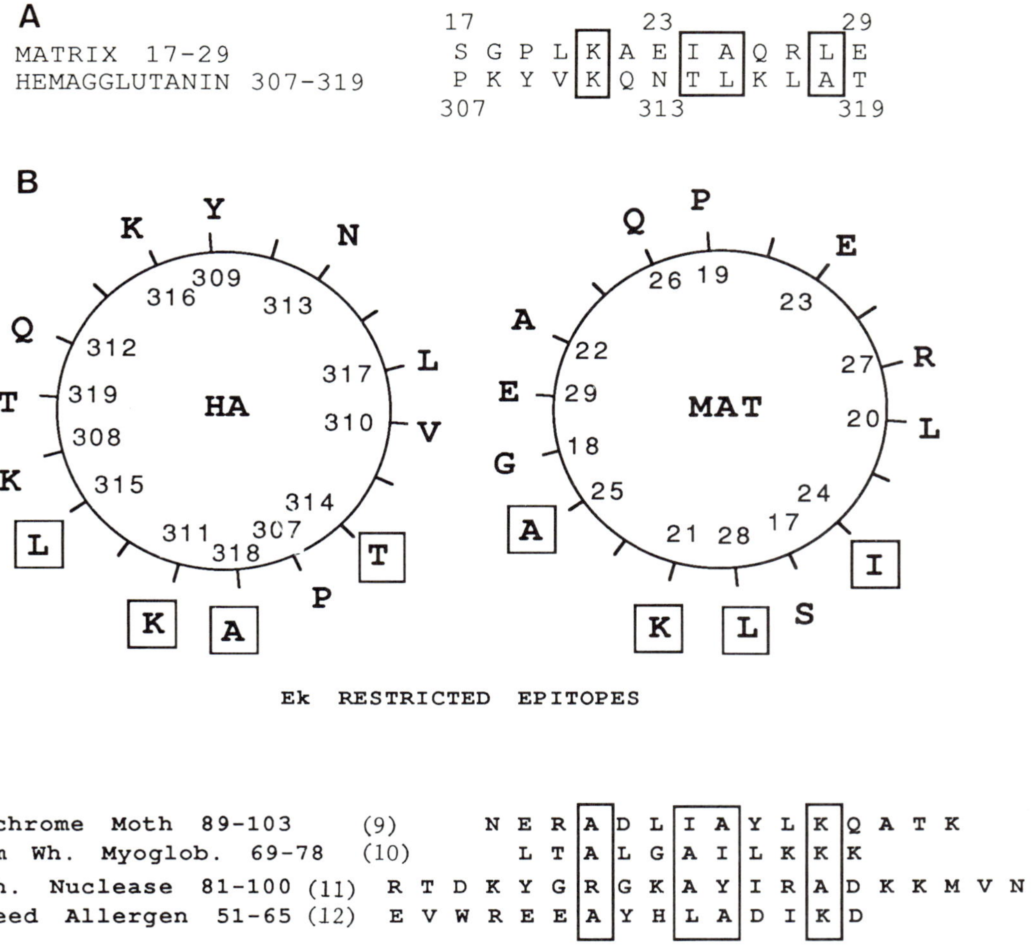

Figure 1. Comparison of matrix protein and hemagglutinin peptide sequences (**A**), helical conformation of the peptides (**B**), and I-E^k-restricted epitopes (**C**).

sequences of these peptides presented by DR1 and I-E^k may reflect common features in the antigen-binding sites of DR1 and I-E^k. However, in order for I-E^k to bind these determinants in a manner similar to the DR1 binding of the influenza peptides described above, in three cases the direction of the helix would need to be reversed. This would not necessarily violate the chiral nature of other known ligand-receptor interactions, because in a helical conformation the main chain amide nitrogens and the carbonyl oxygens are internally hydrogen bonded. Consequently they would not interact with either the MHC molecule or the T-cell receptor. Additionally, the flexibility of the amino acid side chains on both the antigen and the two receptors could be sufficient to allow similar interactions to occur regardless of the relative orientation of the helix.

The assignment of an MHC binding function to the conserved residues with the 1,4,5,8 spacing in the two flu peptides was tested by exchanging them between the two peptides while leaving the residues that are predicted to form the upper face of the helix unchanged. If the model is correct, these hybrid peptides should bind to DR1 and be recognized by the T-cell clone with specificity for the residues composing the "upper" face of the helix. As can be seen in Table 3, the hybrid peptides HAMAT 4 and MATHA 4 induce proliferative responses by HA1.7 and TLC 72, respectively. The sequence of these hybrid peptides is shown in Table 3. The details of the reactivity patterns to a large series of hybrid peptides using these two T cell clones are presented in a separate report (5).

If the two amino-terminal domains of DR1 are modelled on the recently described three-dimensional structure of HLA-A2 (13), it is very striking that all the α_1 domain residues that would be displayed within the antigen-binding groove, and those that would form the

Table 3. Responses of Clones HA1.7 and TLC72 to Haemagglutinin and Matrix Protein Hybrid Peptides

A. Antigen Concentration μg/ml	HA 307-319	cpm HAMAT IV	MP 17-29
0.03	10,117±4	265±12	656±8
0.3	33,253±2	2,712±6	478±11
3.0	35,712±5	16,167±3	712±7
30.0	22,121±2	42,403±4	736±6
B.	MP 17-29	MATHA IV.A17	HA 307-319
0.1	3,721±8	426±11	222±19
1.0	7,364±2	3,974±17	281±6
10.0	9,213±3	4,381±15	219±15
100.0	6,942±4	688±6	307±12

Peptide Sequences

HA 307-319	P	K	Y	V	**K**	Q	N	**T**	**L**	K	L	**A**	T
HAMAT IV	S	K	Y	L	**K**	Q	N	**I**	**A**	K	L	**L**	E
MATHA IV.A17	A	G	P	V	**K**	A	E	**T**	**L**	Q	R	**A**	T
MP 17-29	S	G	P	L	**K**	A	E	**I**	**A**	Q	R	**L**	E

Responses of HA1.7(A), and TLC72(B), to native and hybrid peptides of haemagglutinin and matrix proteins. Culture conditions were as for Table 1. The sequences of the four peptides are shown below. The residues shown in bold print are those with 1, 4, 5, 8 spacing that were exchanged between HA and MP in the hybrid peptides (for details see ref. 5).

α-helical strand running across the domain are conserved between DRα and I-Eα. In contrast, there is little in common between DR1β_1 and I-E$\beta_1{}^k$ domains. Perhaps the most important similarity in these β_1 domains is around the bm12 site, i.e. residues 67,70, and 71, which were altered due to the spontaneously occurring I-A bm12 mutation. Extensive functional evidence exists to illustrate the importance of this region in T-cell recognition (14). Further elucidation of the individual residues involved in physical interaction with antigen and with the T-cell receptor molecule may require the application of techniques such as site directed mutagenesis in conjunction with direct MHC/antigen binding assays.

References

1. Schwartz RH. The role of gene products of the major histocompatibility complex in T cell activation and cellular interactions. In: Paul WE (ed): Fundamentals of Immunology. New York, Raven Press, 1984, p 379.
2. Babbitt BP, Allen PM, Matsueda G, Haber E, Unanue ER. Binding of immunogenic peptides to Ia histocompatibility molecules. Nature 1985;317:359.
3. Buus S, Colon S, Smith C, Freed JH, Miles C, Grey HM. Interaction between a "processed" ovalbumin peptide and Ia molecules. Proc Natl Acad Sci USA 1986;83:3968.
4. Guillet J, Lai M, Briner J, Smith J, Gefter M. Interaction of peptide antigens and class II major histocompatibility complex antigens. Nature 1986;324:260.
5. Rothbard JR, Lechler RI, Howland K, Bal V, Eckels DD, Sekaly R, et al. Structural model of HLA DR1-restricted antigen recognition. Cell.
6. Lechler RI, Norcross MA, Germain RN. Qualitative and quantitative studies of antigen-presenting cell function by using I-A-expressing L cells. J Immunol 1985;135:2914.
7. Lampson LA, Levy R. Two populations of Ia-like molecules on a human B cell line. J Immunol 1980;125:293.
8. Lechler RI, Bal V, Rothbard JB, Sekaly R, Long EO, Lamb JR. HLA-DR-restricted antigen induced activation of human helper T lymphocyte clones using transfected murine cell lines. J Immunol, in press.
9. Heber-Katz E, Schwartz RH, Matis L, Hannum C, Fairwell T, Appella E, et al. Contribution of antigen presenting cell major histocompatibility complex gene products to the specificity of antigen induced T cell activation. J Exp Med 1982;155:1086.
10. Livingstone A, Fathman CG. The structure of T cell epitopes. Ann Rev Immunol 1987;5:477.
11. Finnegan A, Smith M, Smith J, Berzofsky J, Sachs D, Hodes R. The T cell repertoire for recognition of a phylogenetically distant protein antigen. Peptide specificity and MHC restriction of staphylococcal nuclease-specific T cell clones. J Exp Med 1986;164:897.
12. Kuisaki J, Atassi H, Atassi M. T cell recognition of ragweed allergen Ra3: Localization of the full T cell recognition profile by synthetic overlapping peptides representing the entire protein chain. Eur J Immunol 1986;16:236.
13. Bjorkman PJ, Saper MA, Samraoui B, Bennett WS, Strominger JL, Wiley DC. Structure of the human class I histocompatibility antigen, HLA-A2. Nature 1987;329:506.
14. Kanamori S, Walsh WD, Hansen TH, Tse HY. Assessment of antigen-specific restriction sites on Ia molecules as defined by the bm12 mutation. J Immunol 1984;133:2811.

Author Affiliations

R.I. Lechler, Department of Immunology, Royal Postgraduate Medical School, Hammersmith Hospital, London; *V. Bal*, *J.R. Lamb*, MRC TB and Related Infections Unit, Hammersmith Hospital, London; *K. Howland*, *J.B. Rothbard*, Laboratory of Molecular Immunology, Imperial Cancer Research Fund, London; UK; *R.P. Sekaly*, *E.O. Long*, Laboratory of Immunogenetics, National Institute of Allergy and Infectious Diseases, NIH, Bethesda, MD, USA

Hybrid Human-Mouse Class II Molecules:: Localization of Antibody Binding Sites

J.F. Maddox and J.G. Bodmer

Abstract: The technique of exon-shuffling has been used to introduce the human DR4β1 domain into the murine IE β gene. The molecule produced from this hybrid gene has the ability to pair with the murine IE α gene product as shown by the surface expression of an α-β dimer on transfected P815HTR cells. Antibodies to the β chain of HLA-DR antigens have been tested against these transfectants and it has been found that TDR31.1, M1.6, M1.18, M2.27, M2.69 and M2.72 all bind to the transfected cells whereas CA2 and DA2 fail to bind.

Assignments of chain specificity of anti-major histocompatibility complex (MHC) class II antibodies have generally in the past been made by the binding of antibodies to isolated α or β chains in Western immunoblots (1). However, because of the severe denaturing effect this method has on the class II molecule, it has been difficult to assign chain specificities to many antibodies. More recently, assignment of chain specificity to anti-MHC class II antibodies has been made by the ability of antibodies to bind to mismatched class II α and β chains on the surface of transfected cells (2). Little is known about the fine localization of epitopes within human MHC class II α and β chains as the methods for determining this, such as peptide mapping, are difficult to perform. Detailed sequence analysis of mutant murine class II molecules has enabled assignment of some antibody specificities (3). The increased utilization of genetic engineering techniques has provided other ways for localizing antibody binding epitopes within class II molecules (4,5). We have made two hybrid human-murine class II molecules that have enabled us to localize the binding of several anti-class II antibodies to the β1 domain of DR molecules.

Details of the construction of the two hybrid β genes (DR4w4/E^d and DR4Dw15/E^d) and their amino acid sequences are given in Figure 1. The hybrid genes were cotransfected with the IE α^7 gene (pJ4-Eα^7, Eα^7 was obtained from R. Germain) and the neo gene (pSV2-neo) into P815HTR cells using the calcium phosphate coprecipitation technique (6) (Table 1). The reactivity of a panel of anti-MHC class II antibodies was tested against these transfected cells in an enzyme linked immunosorbant assay (Table 2). Control transfected P815HTR cells (P815#3) were also made which expressed only the murine IE^d molecule. As most of the monoclonal antibodies that bind to human MHC class II molecules have been raised in Balb/c mice, it is unlikely that these antibodies will bind to the murine IE^d molecule, and indeed none of the antibodies tested bound to IE^d.

Six of the ten antibodies tested that bind to DR β chains bound to both the transfectants. However CA2, DA2, and M2.2, all of which bind to isolated DR β chains in Western immunoblots, failed to bind to these transfectants, suggesting that they either bind elsewhere in the β chain or that amino acid residues within the β1 domain form only part of the epitope defined by these antibodies. Localization of antibody binding sites within molecules is complicated by the fact that many epitopes are comprised of discontinuous amino acids rather than of stretches of linear amino acids. Several formulae exist for predicting the antibody binding epitopes within proteins: these involve the likelihood of occurrence of stretches of amino acids on the surface of proteins, hydrophobicity being one of the criteria used; or as part of a secondary structure such as a helix, turn, or strand (reviewed in 14). However, none of these formulae are reliable and they are of little use in predicting discontinuous epitopes. Such sites are easier to predict when the detailed crystallographic structure of the molecule is known. Recently the crystallographic structure for the HLA-A2 molecule was published (15). This in association with other data has enabled the putative assignment of several antibody epitopes (16). To date the crystallographic structure of MHC class II molecules is unknown, although it is postulated that it will be similar to that of the MHC class I molecules (15). Chimeric murine MHC class II molecules that have been created by techniques such as exon shuffling and *in vitro* mutagenesis (4,17), have been useful in defining both antibody and T-cell epitopes on IA and IE molecules, as well as the elements of the α and β chains that are important in α-β chain pairing (18).

TDR31.1, M1.6, M2.27, M2.69, and M2.72 all bind to isolated β chains in Western immunoblots, whereas M1.18 binds to the α-β dimer but not to isolated β chains. It is possible that the antibodies that bind to isolated β chains bind to linear rather than discontinuous epitopes. It was found that all of these antibodies bound to both the P815#1 and P815#5 transfected cell lines, rather than binding only to the P815#1 cell line. This indicates that these antibodies are probably binding to sites within the N-terminal 83 amino acids of the DR β chain as the DR4Dw15/E^d and IE^d sequences differ significantly from the DR4Dw4/E^d sequence for the remainder of the β1 domain. Both M2.27 and M2.69 bound more strongly to P815#5 than to P815#1 despite the lower surface expression of class II molecules on the former transfectant. The reason for this is unknown. One of the antibodies tested, X1.26, bound poorly if at all to P815#5 cells, while binding strongly to P815#1 in an ELISA using glutaraldehyde-fixed cells. It is possible

Immunobiology of HLA
Volume II

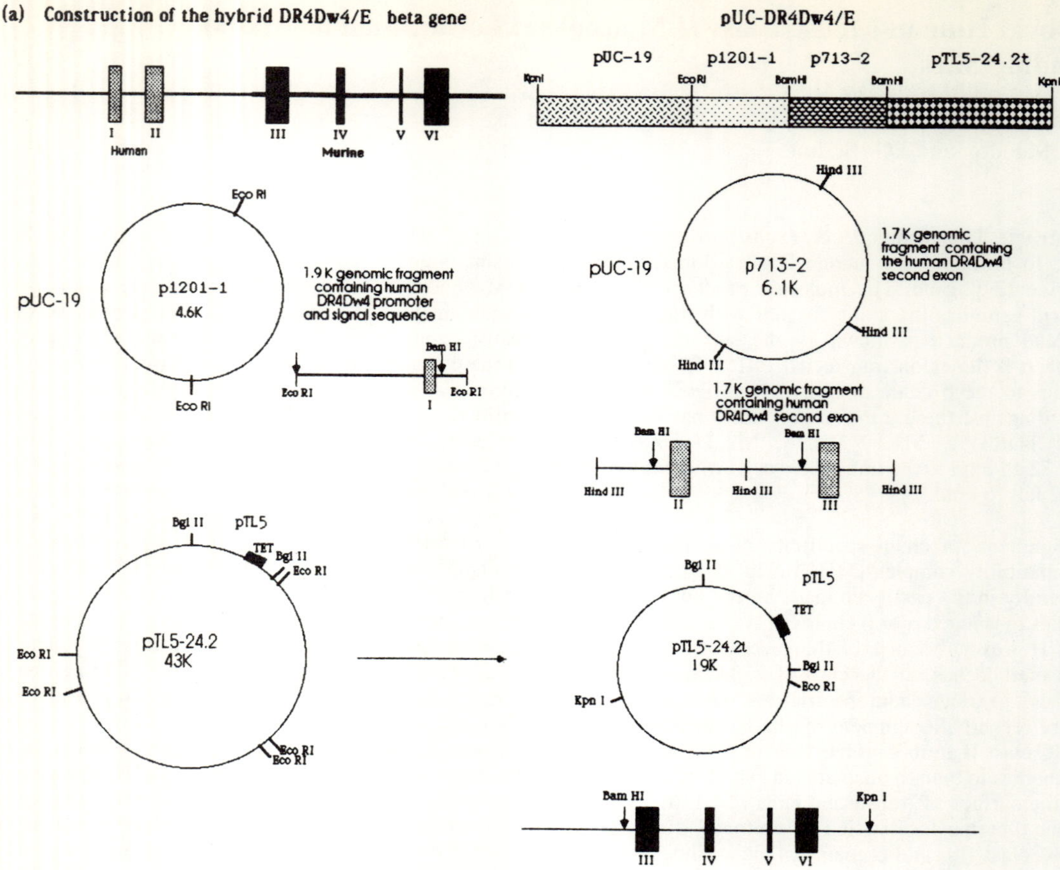

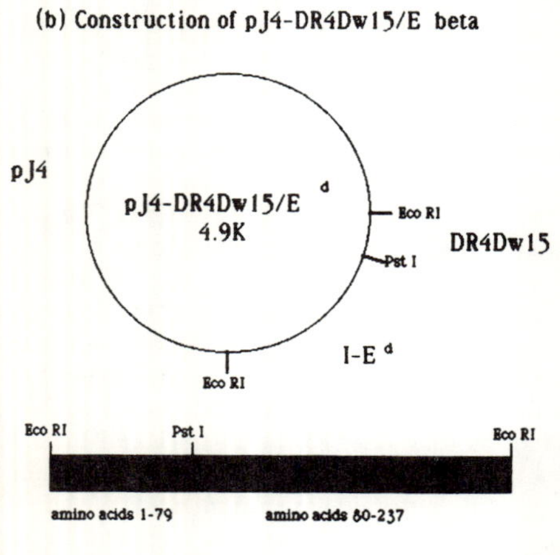

(c)

DR4Dw4/E	GDTRPRFLEQVKHECHFFNGTERVRFLDRY
DR4DW15/E	
E	VR Y TS Y QH E
DR4Dw4/E	DSDVGEYRAVTELGRPDAEYWNSQKDLLEQ
DR4Dw15/E	S
E	N PEI D
DR4Dw4/E	KRAAVDTYCRHNYGVGESFTVQRR
DR4Dw15/E	R EISDK L R
E	A S EISDK L R

Figure 1. Construction and sequences of the hybrid human-murine β chain genes. (a) pUC-DR4Dw4/E: the 1.7K p1201-1 Eco R1-Bam HI fragment containing the DR4Dw4 first exon; the 1.7K p713-2 Bam HI-Bam HI fragment containing the DR4Dw4 second exon; and the 2.9K pTL5-24.2t Bam HI-Kpn I fragment were ligated into the larger fragment of pUC-19 which had been digested with Eco RI and Kpn I. p1201-1 and p713-2 were obtained from G. Andersson. Cosmid 24.2 in pTL5 (5,6) was digested with Eco RI, the fragment containing pTL5 and exons 3-6 of the murine IE^d β gene was purified and circularised with ligase to produce pTL5-24.2t. (b) pJ4-DR4Dw15/E: the 0.4K DRβWa Eco RI-Pst I fragment containing sequence coding for the first 79 amino acids of the DR4Dw15 β gene, and the 0.8K Pst I-Eco RI fragment of the IE^d β gene were ligated into the pJ4 expression vector (J. Morgenstern) at the Eco RI site. The pJ4 expression vector uses the Moloney murine leukaemia virus long terminal repeat as the promoter for the inserted cDNA. Purified DRβ WA cDNA was obtained from H. Ikeda. (c) Sequences of DR4Dw4/E^d and DR4Dw15/E^d are compared to that of IE^d (9). The DR4 sequences were obtained from (10).

Table 1. Properties of Transfectants Expressing Hybrid Human-Murine MHC II Molecules

	P815#1	P815#3	P815#5
α chain	$E\alpha^7$	$E\alpha^7$	$E\alpha^7$
β chain	DR4Dw4/E^d	E^d	DR4Dw15^d
Resistance gene	neo	neo	neo
Selection in	G418	G418	G418
Level vs. Priess*	1/3		1/45

*The relative levels of surface expression were compared following FACS analysis of cells stained with TDR 31.1 in an indirect immunofluorescent assay (7).

that X1.26 binds to a site which comprises amino acids at the C-terminal end of the first domain, i.e., between residues 84 and 95. Using *in vitro* mutagenesis to replace small stretches of the murine IE^{d}β gene sequence with human DR β gene sequence, it is planned to further localize the binding sites of these and other anti-MHC class II antibodies.

Table 2. Reactivity of Anti-MHC Class II Antibodies vs. Transfectants

	Transfectant			
Antibody	P815#1	P815#5	P815#3	P815
TAL1B5 (8)	−	−	−	−
14.4.4S (9)	+	+	+	−
TDR31.1 (10)	+	+	−	−
B7/21 (11)	−	−	−	−
CA2 (12)	−	−	−	−
DA2 (13)	−	−	−	−
M1.6	+	+	−	−
M1.18	+	+	−	−
M2.2	−	−	−	−
M2.27	+*	+	−	−
M2.69	+*	+	−	−
M2.72	+	+	−	−
M2.86	−	−	−	−
X1.26	+	−	−	−

*M2.27 and M2.69 bound much less strongly to P815#1 than to P815#5. Monoclonal antibodies M1.6,M1.18, M2.2, M2.27, M2.69, M2.72, M2.86 and X1.26 are hitherto unpublished antibodies that bind to MHC class II molecules and were made in the Tissue Antigen Laboratory.

References

1. Cohen B, Deane D, Moxley M. Analysis of antibody specificity by Western blotting. Disease Markers 1984;2:135.
2. Landais D, Beck BN, Buerstedde J-M, Degraw S, Klein D, Koch N, et al. The assignment of chain specificities for anti-Ia monoclonal antibodies using L cell transfectants. J Immunol 1986;137:3002.
3. Beck BN, Pease LR, Bell MP, Buerstedde J-M, Nilson AE, Schlauder GC, DNA sequence analysis of I-A beta^k mutants reveals serologically immunodominant region. J Exp Med 1987;166:433.
4. Germain RN, Ashwell JD, Lechler RI, Margulies DH, Nickerson KM, Suzuki G, et al. "Exon-shuffling" maps control of antibody- and T-cell-recognition sites to the NH_2-terminal domain of the class II major histocompatibility polypeptide A beta. Proc Natl Acad Sci USA 1985;82: 2940.
5. McCluskey J, Germain RN, Margulies DH. Cell surface expression of an in vitro recombinant class II/class I major histocompatibility gene product. Cell 1985;40:247.
6. Van Pel A, De Plaen E, Boon T. Selection of highly transfectable variant from mouse mastocytoma P815. Somat Cell Mol Genet 1985;11:467.
7. Maddox JF, Mackay CR, Brandon MR. Surface antigens, SBU-T4 and SBU-T8, of sheep T lymphocytes defined by monoclonal antibodies. Immunology 1985;55:739.
8. Adams TE, Bodmer JG, Bodmer WF. Production and characterization of monoclonal antibodies recognizing the alpha chain subunits of human Ia alloantigens. Immunology 1983;50:613.
9. Ozato K, Mayer N, Sachs DH. Hybridoma cell lines secreting monoclonal antibodies to mouse H-2 and Ia antigens. J Immunol 1980;124:533.
10. de Kretser TA, Crumpton MJ, Bodmer JG, Bodmer WF. Demonstration of two distinct light chains in HLA-DR-associated antigens by two-dimensional gel electrophoresis. Eur J Immunol 1982;12:214.
11. Watson AJ, de Mars R, Trowbridge IS, Bach FH. Detection of a noval human class II antigen. Nature 1983;304:358.
12. Charron DJ, McDevitt HO. Analysis of HLA-D region-associated molecules with monoclonal antibody. Proc Natl Acad Sci USA 1979;76:6567.
13. Brodsky FM, Parham P, Barnstable CJ, Crumpton MJ, Bodmer WF. Monoclonal antibodies for analysis of the HLA system. Immunol Rev 1979;47:3
14. Hopp TP. Protein surface analysis: Methods for identifying antigenic determinants and other interaction sites. J Immunol Meth 1986;88:1.
15. Bjorkman PJ, Saper MA, Samraoui B, Bennett WS, Strominger JL, Wiley DC. Structure of the human class I histocompatibility antigen, HLA-A2. Nature 1987;329:506.
16. Bjorkman PJ, Saper MA, Samraoui B, Bennett WS, Strominger JL, Wiley DC. The foreign antigen binding site and T cell recognition regions of class I histocompatibility antigens. Nature 1987;329:512.
17. Glimcher LH, Griffith IJ. Mutations of class II MHC molecules. Immunol Today 8:274.
18. Germain RN, Quill H. Unexpected expression of a unique mixed-isotype class II MHC molecule by transfected L-cells. Nature 1987;320:72.

Author Affiliation

Tissue Antigen Laboratory, Imperial Cancer Research Fund, P.O. Box 123, Lincoln's Inn Fields, London WC2A 3PX, UK.

Antigen Mimicry
Cytotoxic T Cells Specific for Epstein-Barr Virus Recognize HLA Alloantigens*

Dolores J. Schendel, Erich Lederer, Gabriele Multhoff, and Elfriede Nößner

Immune response directed against epitopes shared by foreign antigens and normal cell surface molecules has been proposed as one mechanism of autoimmunity. Indeed T cells capable of recognizing viral antigens restricted by self major histocompatibility complex (MHC) molecules as well as certain alloantigens have been demonstrated in mouse and man (1–3). To determine the relevance of MHC-related antigen mimicry in autoimmune disease it is important to understand the structural basis of cross-reactivity that leads to an immunological response specific for both natural pathogens and MHC molecules.

We established cytotoxic T lymphocyte (CTL) lines that recognize an Epstein-Barr virus (EBV)-related determinant in association with a self-encoded MHC molecule (self-MHC + EBV) and also recognize an alloantigen expressed by some non-EBV-infected target cells (Allo). The CTL were generated from responding cells of two unrelated donors that were stimulated either with syngeneic EBV-transformed lymphoblastoid cells lines (LCL) or with normal peripheral blood lymphocytes (PBL) from selected allogeneic donors (Table 1). In all cases the CTL lines lysed syngeneic EBV-transformed LCL, but not syngeneic phytohemagglutinin-stimulated lymphoblasts. They also lysed some allogeneic target cells, as seen with the phytohemagglutinin (PHA) blasts of donors 1,2, and 3.

Preliminary analysis of the structural basis of the self-MHC + EBV target determinant suggested that the MHC component is HLA-B8, which is shared by donors A and B. Most HLA-B8 LCL were lysed by the CTL; however, several HLA-B8 LCL were not recognized as targets. This discrepancy was not explained by CTL distinction of HLA-B8 molecular variants, as defined by standard isoelectricfocusing (IEF) methods (4), since the IEF patterns were identical for all HLA-B8 LCL (data not shown). Rather, the differences may reflect variations in the EBV-related determinant. We observed fluctuations in the degree of lysis of the LCL of donor B from experiment to experiment; whereas the allospecific cytotoxicity remained uniformly high (Table 2). This suggests that the target determinant expressed by the LCL was variable and not the cytotoxic potential of the CTL. The BNLF1-MA protein of EBV has been proposed as one candidate for the lymphocyte-determined membrane antigen, LYDMA, which is a target for EBV-specific CTL, but has only been functionally defined (5). Since the levels of BNLF1-MA surface protein may vary from cell to cell, studies are now in progress to relate variations in BNLF1-MA to susceptibility of HLA-B8 LCL to CTL lysis and, using synthetic peptides, to determine whether this protein provides the EBV-related epitope of the target antigen.

As shown in Table 1, no common HLA class I or class II antigen was shared by the three allogeneic cells that could explain the allospecificity. Therefore, population and family studies were made to clarify the structural basis of this target determinant. A panel of 127 target cells was tested with CTL A.1 and A.2 (Table 3); smaller panels were analyzed with the other CTL, yielding concordant results (data not shown). Eighty-five percent of

*This work was supported by the Deutsche Forschungsgemeinschaft through SFB 217 (project B2).

Table 1. Cross-Reactive CTL Recognizing Epstein-Barr Virus-Infected Target Cells and Selected Allogeneic Target Cells

		Target Cells						
CTL	RC SC	A-PHA	A-LCL	B-PHA	B-LCL	1-PHA	2-PHA	3-PHA
A.A	A A	–	++	nt	nt	++	++	++
A.1	A 1	–	++	nt	nt	++	++	++
A.2	A 2	–	++	nt	nt	++	++	++
B.B	B B	nt	nt	–	++	++	++	++
B.3	B 3	–	++	–	++	++	++	++

CTL were generated using as responding cells (RC) peripheral blood lymphocytes from donors A and B. They were sensitized for 6 days in vitro with x-irradiated stimulating cells (SC) from donors A, B, 1, 2, or 3. SC from donors 1, 2, and 3 were normal peripheral blood lymphocytes; SC from donors A and B were EBV-transformed lymphoblastoid cell lines (LCL). They were tested in a standard chromium-release assay with PHA-lymphoblasts or LCL as target cells. The HLA antigens of the donors are as follows: donor A–HLA-A2, B8, B40, Cw-, DR2, DR3, DQw1; donor B–HLA-A1, B8, Bw57, Cw7, DR3, DR13, DQw1, DQw2; donor 1–HLA-A2, A3, B35, B37, Cw4, DR1, DR2, DQw1; donor 2–HLA-A2, B44, Cw5, DR12, DQw3; donor 3–HLA-Aw32, B44, Cw2, Cw5, DR1, DR6, DQw1.

Table 2. Variations in Lysis of B-LCL by CTL B.3

	% RCR	
Target Cells	Expt. 1	Expt. 2
B-PHA	10%	0%
B-LCL	69%	10%
3-PHA	68%	100%
3-LCL	100%	76%
1-PHA	60%	nt
1-LCL	80%	65%

CTL B.3 was tested against target cells from donors 1 and 3, using either PHA lymphoblasts or LCL, as well as against syngeneic PHA lymphoblasts or LCL. To compare the data from two experiments, the lysis is presented as a percent relative cytotoxic response (% RCR) calculated as follows: (% CML of target X) ÷ (% CML of target 3) × 100. See Figure 1 for calculation of % CML.

Table 3. HLA Association of Target Alloantigen Recognized by CTL A.1 and A.2

	HLA-B44 or HLA-B35		HLA-B5 CREG	HLA Other
Pos	40		5	2
Neg	28		15	37
	B44 I	B44 II	B35 1	B35 4
Pos	6	2	5	2
Neg	4	4	7	NT

Target cells were assigned to the positive cluster when they were lysed by both CTL with a % RCR greater than 25%. B44-I and B44-II refer to the two molecular variants of B44 (7). B35-1 and B35-4 refer to the two most common molecular variants of B35 (4).

those target cells that were lysed by both CTL were either HLA-B35 or HLA-B44, confirming earlier results from a smaller panel (6). Most of the remaining positively lysed cells expressed an antigen of the HLA-B5 serologically cross-reactive group. However, numerous HLA-B35 or HLA-B44 cells were not lysed by the CTL. We analyzed whether differential recognition of the known molecular variants of HLA-B35 (4) or HLA-B44 (7) could explain these results. This was not found to be the case (Table 3).

Family studies provided a substantial clue to the nature of allogeneic target antigen. Cells of some children were strongly lysed by the CTL, but not those of their parents and other siblings (6,8). The highly susceptible target cells were derived from HLA identical individuals, as seen with CTL B.3 (Fig. 1). Data from four families confirmed that two transcomplementing factors, both encoded by HLA-linked genes, determine the CTL allospecificity (manuscript in preparation). One component seems to be a class I molecule since lysis was blocked by a monoclonal antibody specific for class I. The second component remains unidentified. We propose that it may be a peptide, derived from an MHC-encoded protein, that associates with a class I molecule. This class I molecule may be either HLA-B35 or HLA-B44, whereby it must be postulated that the binding site (9) of these two alloantigens can accommodate the same peptide. Alternatively, the MHC molecule may be a novel class I antigen that is encoded by an allele in strong linkage disequilibrium with the alleles for HLA-B35 and HLA-B44. Recent studies suggest that an additional class I gene, mapping near the HLA-B locus, can be expressed in some haplotypes (10). Through gene cloning and transfection the role of classical versus novel class I genes should be resolved.

Several surprising points arise from these studies. It has been shown that distinct viral peptides associate with self-encoded MHC molecules forming the target antigens that are recognized by virus-specific CTL (11). Since T-cell clones recognizing both MHC-restricted nominal antigens and alloantigens have been found (1–3), it has been questioned whether a structural equivalence is required, and achieved, through association of some peptide with the allogeneic MHC molecule. The surprising finding that a peptide fragment cocrystallized with the HLA-A2 molecule (12) suggests that occupancy of the antigen binding site may be the standard state for a class I molecule. The specificities of the CTL described here imply also that a structural equivalence exists in the two target antigens seen by these CTL: self-MHC + EBV and Allo + X. In this case the component X, eventually a peptide, is encoded by an MHC-linked gene.

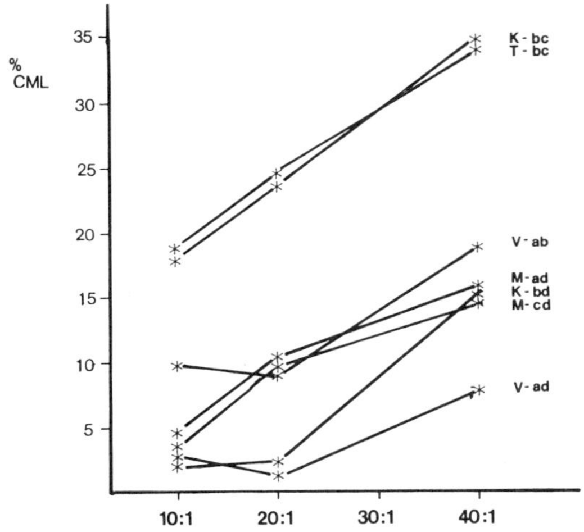

Figure 1. CTL B.3 was tested at varying effector to target cell ratios (x-axis) against phytohemagglutinin-stimulated lymphoblasts from members of an HLA-typed family. The letters designate the four HLA haplotypes which have the following antigens: a) A2, B27, Cw1, DR2, DQw-; b) A2, B13, Cw-, DR7, DQw2; c) Aw24, B35, Cw4, DR3, DQw2; d) A3, B7, Cw-, DR2, DQw-. The cell-mediated lympholysis (CML) is expressed as a percentage of the total radioactivity incorporated by the target cells and calculated as follows: (experimental release − spontaneous release) ÷ (maximum release − spontaneous release) × 100.

Cases of antigen mimicry will be dependent upon complex genetic parameters (MHC restriction elements, T-cell receptor genes) as well as complex immunological parameters (response to the pathogen and selection of the T-cell repertoire); therefore, it is surprising that identical responses were found with cells of two unrelated individuals. Interestingly, an EBV-specific CTL line described previously (3) showed partially overlapping specificities with the CTL described here. Development of similar cross-reactive immune responses in different individuals would be a precedent to be fulfilled if the mechanism of antigen mimicry plays a substantial role in HLA-associated autoimmunity.

Acknowledgment. We thank Dr. Rudolf Wank of this institute for providing the HLA typing of the cells used in these studies and Ms. Barbara Keil for excellent technical assistance.

References

1. Finberg R, Burakoff SJ, Cantor H, Benacerraf B. Biological significance of alloreactivity: T cells stimulated by Sendai virus-coated syngeneic cells specifically lyse allogeneic target cells. Proc Natl Acad Sci USA 1978;75:5145.
2. Braciale TJ, Andrew ME, Braciale VL. Simultaneous expression of H-2 restricted and alloreactive recognition by a cloned line of influenza virus-specific cytotoxic T lymphocytes. J Exp Med 1981;153:1371.
3. Gaston JSH, Rickinson AB, Epstein MA. Cross-reactivity of self-HLA-restricted Epstein-Barr virus specific cytotoxic T lymphocytes for allo-HLA determinants. J Exp Med 1983;158:1804.
4. Neefjes JJ, Breur-Vriesendorp BS, van Seventer GA, Ivanyi P, Ploegh HL. An improved biochemical method for the analysis of HLA-class I antigens. Definition of new HLA-class I subtypes. Hum Immunol 1986;16:169.
5. Modrow S, Wolf H. Characterization of two related Epstein-Barr virus-encoded proteins by synthetic oligopeptides, which are differentially expressed in Burkitts lymphoma and in vitro transformed cell lines. Proc Natl Acad Sci USA 1986;83:5703.
6. Schendel DJ, Wank R, Dupont B. Cell-mediated lympholysis: Examination of HLA genetic fine structure and complementation using cytotoxic lymphocytes. Eur J Immunol 1978;8:634.
7. Hahn AW, Schendel DJ, Wendelboe-Hansen P, Ploegh HL. Biochemical and cellular identification of HLA-Bw44 subtypes. Hum Immunol 1984;11:69.
8. Schendel DJ, Wank R. Noncodominant expression of target antigens recognized by human cytotoxic T lymphocytes. Hum Immunol 1981;2:91.
9. Bjorkman PJ, Saper MA, Samraoui B, Bennett WS, Strominger JL, Wiley DC. The foreign antigen binding site and T cell recognition of class I histocompatibility antigens. Nature 1987;329:512.
10. Duceman BW, Ness P, Rende R, Chorney MJ, Srivastava R, Greenspan DS, et al. HLA-JY328: Mapping studies and expression of a polymorphic HLA class I gene. Immunogenetics 1986;23:90.
11. Lamb JR, McMichael AJ, Rothbard JB. T-cell recognition of influenza viral antigens. Hum Immunol 1987;19:79.
12. Bjorkman PJ, Saper MA, Samraoui B, Bennett WS, Strominger JL, Wiley DC. Structure of the human class I histocompatibility antigen, HLA-A2. Nature 1987;329:506.

Author Affiliations

Institute of Immunology, University of Munich, Goethestrasse 31, D-8000 Munich 2, Federal Republic of Germany

Effect of Human Immunodeficiency Virus on Dendritic Cells Isolated from Human Peripheral Blood

Stella C. Knight and Steven Patterson

Abstract: Antigens, including viral antigens, are presented to the immune system on antigen-presenting dendritic cells (DC). For example, human DC pulsed with influenza virus in vitro subsequently initiated proliferation of autologous lymphocytes. The interaction between HIV and human DC was therefore examined. Low density mononuclear cells (5×10^5, around 25% DC plus macrophages (MO) were cultured with 10^2 -10^3 TC ID50 units of virus (H9-IIIB) and then examined by electron microscopy. After three to five days of culture with virus, DC but not MO from five out of 12 individuals showed HIV budding from the surface of cells. Immunogold labeling experiments showed that some DC expressed low levels of CD4 (T4) antigen. Occasionally the level of class II antigens appeared reduced on infected cell populations. An infection by HIV in vivo might therefore affect the initiation of immune responses by compromising the antigen-presenting function of DC.

Transient immunosuppression accompanies many virus infections (1). Some viruses produce specific suppression of the anti-viral immunity but there is also evidence of nonspecific immunosuppressive effects that are greater with more virulent strains (1). Direct infection of proliferating lymphocytes is considered important in producing the immunosuppressive effects (1,2) and the immunosuppression by HIV is also believed to operate through viral infection of T lymphocytes (3,4).

In the initiation of immune responses, antigens (including contact sensitizers (5,6) and viral antigens (18) are presented to lymphocytes by DC (7–9). However, during naturally occurring infections with Sendai virus we observed that DC from the lymph nodes of mice skin painted with contact sensitizer (6) were impaired in their capacity to initiate immune responses to the contact sensitizer (unpublished). This suggested that immunosuppression during viral infections might occur at the level of antigen-presenting cells. Our preliminary in vitro studies showed that human DC were susceptible to infection with HIV (10). In vivo changes in epidermal antigen-presenting cells, the precursors of some lymph node DC, have also been reported in HIV-infected individuals (11,12); growth of HIV in Langerhans cells (11) and reduced levels of class II antigens and ATPase (12) were reported.

In the current study we first investigated the interaction of human peripheral blood DC and influenza virus. Nonadherent, low density, mononuclear cells con-

Immunobiology of HLA
Volume II

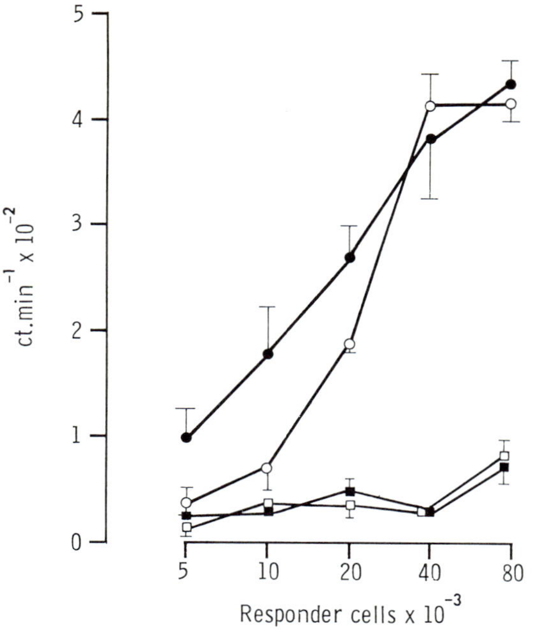

Figure 1. Stimulation of lymphocyte proliferation by syngeneic dendritic cells (DC) pulsed with influenza virus. Low density, nonadherent mononuclear cells containing 25% DC and 75% macrophages were isolated from human blood (13,16). Some were pulsed with influenza virus (Ox 47,14) and macrophages were depleted using LEU M3 + complement. Lymphocytes from the bottom of the gradient were cultured in 20 µl hanging drops (16) at different concentrations. ■ No low density cells; □ 2000 non-pulsed low density cells; ● 2000 virus-pulsed low density cells; ○ 500 virus-pulsed purified DC.

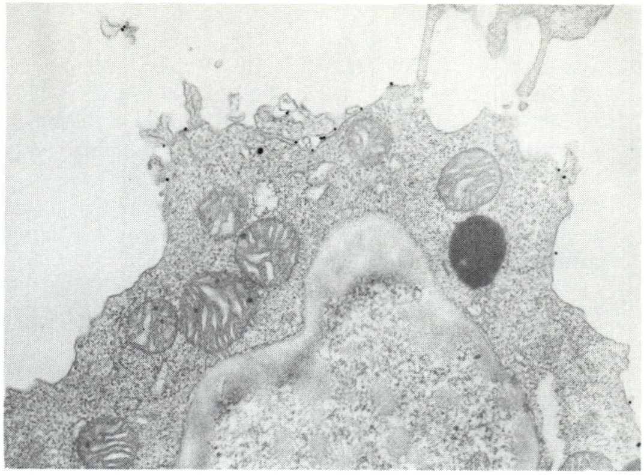

Figure 2. Dendritic cell from human peripheral blood labeled with CD4 antibody (Dako) and gold particles coated with staphylococcal protein A (13). The gold particles labeling CD4 were of 20 nm diameter.

taining around 25% DC and 75% MO were isolated from normal blood (13) and pulsed with the x47 reassortant (H_3N_2) strain of influenza virus (14). After two hours or 24 hours, viral antigen was detected on the cells using a polyclonal rabbit anti influenza virus hemagglutinin antibody and a fluorescent anti-rabbit immunoglobulin. These cells could also stimulate proliferative responses in syngeneic lymphocytes (Fig. 1). Small numbers of non-pulsed cells caused little stimulation. When MO were removed from the stimulating population with the cytolytic antibody LEU M3 and complement, only 25% of the cells remained and these still caused good stimulation (Fig. 1). Even in secondary responses to virus DC were, therefore, very potent at initiating responses. The virus-pulsed cells were examined by electron microscopy (10) one, three, or six days after exposure to antigen but no evidence of budding virus was seen. In studies of infection of mice with herpes simplex virus (8), viral antigen was seen on DC in the draining lymph nodes but there was no evidence of virus replication.

These findings contrast with the study of HIV and human DC. Low density cells ($5 \times 10^5 - 10^6$) were cultured with $10^2 - 10^3$ TC ID50 units of HIV for three to five days (10). DC from five of 11 individuals studied showed budding virus on electron microscopy but no virus was seen in the MO. A high proportion of the low density cells after five days of culture with the virus also showed staining with antibody to P24 (kindly provided by Dr. D.M. Strong) using complexes of alkaline phosphate and anti-alkaline phosphatase (17). Since the CD4 antigen has been postulated as the receptor for HIV (15), the DC were studied for expression of this antigen. After staining of the cells with a monoclonal CD4 antibody (Dako) and gold particles coated with staphylococcal protein A, some DC showed low levels of labeling (Fig. 2). The low density, nonadherent macrophages did not express CD4 and this distinguishes them from the normal adherent macrophage populations. In one experiment using anti-DR antibody and gold labeling, there appeared to be lower levels of DR on infected cells. However, measuring surface DR using the cell ELISA system developed for the Tenth Histocompatibility Workshop, only one of three experiments showed a reduced level of DR on DC infected with HIV.

DC thus express CD4 antigen and are susceptible to infection with HIV. Preliminary studies on the function of these infected cells indicate that this may severely impair their accessory function in mitogen stimulation (Macatonia, Patterson and Knight, in preparation). Our results suggest that the capacity to infect DC is a feature of HIV that may not be shared by other viruses. The growth of HIV in DC in vivo may compromise the antigen-presenting function of these cells and be instrumental in the production of the profound immunosuppressive effects of this virus.

Acknowledgments. We are grateful to P. Bedford and J. Gross for technical assistance and Mrs. Margaret Runnicles for typing the manuscript.

References

1. Denman AM, Bacon TH, Pelton BK. Viruses: Immunosuppressive effects. Phil Trans R Soc Lond B 1983;303:137.
2. Knight SC, Najera R. The effect of rubella virus on the growth of lymphoma cells and normal lymphocytes in vitro. Microbios 1969;2:185.
3. Montagnier L. Lymphadenopathy associated virus: Its role in the pathogenesis of AIDS and related diseases. Progress in Allergy 1986;37:46.
4. Gallo RC, Sarngadharan MG, Popovic M, Shaw GM, Hahn B, Wong-Staal F, et al. HTLV-III and the etiology of AIDS. Progress in Allergy 1986;37:1.
5. Knight SC, Krejci J, Malkovsky M, Colizzi V, Gautam A, Asherson G. Role of dendritic cells in the initiation of immune responses to contact sensitisers. Cell Immunol 1985;94:427.
6. Macatonia SE, Knight SC, Edwards AJ, Fryer P, Griffiths S. Localization of antigen on lymph node dendritic cells after exposure to the contact sensitizer fluorescein isothiocyanate. J Exp Med 1987;166:1654.
7. Steinman RM, Nussenzweig MC. Dendritic cells: Features and functions. Immunol Review 1980;53:127.
8. Knight SC. Veiled cells–'dendritic' cells of the peripheral lymph. Immunobiol 1984;168:349.
9. Austyn J. Lymphoid dendritic cells. Immunology 1987; 62:161.
10. Patterson S, Knight SC. Susceptibility of human peripheral blood dendritic cells to infection by human immunodeficiency virus. J Gen Virol 1987;68:1177.
11. Tschachler E, Groh V, Popovic H, Mann D, Konrad K, Safai B, et al. Epidermal Langerhans cells—a target for HTLV III/LAV infection. J Invest Dermatol 1987;88:233.
12. Belsito DV, Sanchez R, Baer MD, Ballantine F, Thorbecke GJ. Reduced Langerhans cell Ia antigen and ATPase activity in patients with the acquired immunodeficiency syndrome. N Engl J Med 1984;310:1279.
13. Knight SC, Fryer P, Griffiths S, Harding B. Class II histocompatibility antigens or human dendritic cells. Immunology 1987;61:21.
14. Patterson S, Gross J, Oxford J. The intracellular distribution of influenza matrix and nucleo-protein in infected cells and their relationship to hemagglutinin in the plasma membrane. J Gen Virol 1988;69:1859.
15. Dalgleish AG, Beverley PCL, Clapham PR, Crawford DH, Greaves MF, Weiss RA. The CD4 (T4) antigen is an essential component of the reception for the AIDS retrovirus. Nature 1984;312:763.
16. Knight S, Farrant J, Bryant A, Edwards AJ, Burman S, Lever A, et al. Non-adherent, low density cells from human peripheral blood contain dendritic cells and monocytes, both with veiled morphology. Immunology 1986;57:595.
17. Mason DY. Immunocytochemical labeling of monoclonal antibodies by the ABAPP Immunoalkaline Phosphatase Technique. In: Bullock GR, Petrusz P (eds): Techniques in Immunocytochemistry. New York, Academic Press, 1985 p 25.
18. Seid JM, Dickson HA, Liberto MC, Bonino L, Knight SC, Nash AA. Interaction of dendritic cells with herpes simplex virus: Role in the induction of an immune response (submitted).

Author Affiliation

Division of Immunological Medicine, Clinical Research Centre, Middlesex HA1 3UJ, U.K.

Alloreactive T4-Cell Clones Recognizing HLA Class II Antigen in Linkage Disequilibrium with HLA-A1*

F. Pazderka, L. Olson, and J.B. Dossetor

Cloning of alloactivated T cells has become a unique tool for immunogenetic analysis of MHC-encoded determinants that, in many cases, cannot be distinguished by serological techniques.

In this study, we have utilized cloned T-cell reagents for dissection of the complexity of class II HLA antigens. The clones were derived from bulk MLC between unrelated individuals. HLA phenotypes of responding and stimulating cells were A2, A24, B7, B51, Cw4, Cw7, DR2, DR4, DQw1, DQw3 and A1, B7, B8, Cw-, DR2, DR3, DQw1, respectively. Cloning was performed by limiting dilution using established techniques (1). Reactivity of clones was assayed by primed lymphocyte testing (PLT) against original stimulator and panel of unrelated cells. In total, nine PLT-reactive clones were derived from bulk MLC-19. Of those, six clones showed pattern of reactivity associated with DR3. Two clones gave reactivity that could not be correlated with any serologically detected specificities, but since DP phenotypes of responding and stimulating cells as well as panel cells were not known, the possibility of DP specificity of these clones cannot be excluded. This possibility is being explored presently.

Clone 19-2, upon testing in PLT with panel of 23 unrelated cells, showed a pattern of reactivity corresponding to HLA-A1 specificity ($r = 0.91$). It did not show any correlation with DR or DQ. Identical pattern of reactivity was observed upon typing members of four informative families where only haplotypes carrying A1 were reactive in PLT with clone 19-2. The results of typing families Br and Am are shown in Table 1. As can be seen from the table, all members of family Br carrying A1-positive haplotype (c) are typed positively by clone 19-2. In family Am, A1-positive haplo-

*This project was supported by Canadian Red Cross Society BTS research grant 2036.

Immunobiology of HLA
Volume II

Table 1. PLT Reactivity of Clone 19-2 with Cells from Two Families

Family Br	PLT (cpm)	Family Am	PLT (cpm)
Father ab	100	Father ab	2,430
Mother cd	1,220	Mother cd	60
Child 1 ad	90	Child 1 bd	40
Child 2 ac	2,710	Child 2 a/bc	70
Child 3 ad	50	Child 3 ad	5,680
Child 4 bc	1,780		
Child 5 bd	60		

HLA haplotypes in the families were as follows. Family Br: a-A3, B35, Cw4, DRw6; b-A32, Bw62, Cw3, DR2; c-A1, B8, Cw-, DR3; d-A2, B51, Cw-, DRw8. Family Am: a-A1, Bw57, Cw6, DR7; b-A3, B7, Cw-, DR2; c-A2, Bw62, Cw3, DRw8; d-A11, B8, Cw7, DR3. Genotype of child #2 in family Am was A3, B7, Cw7, DR7/A2, Bw62, Cw3, DRw8.

type (a) is present on cells of the father and child #3, which are also typed by clone 19-2. Child #2, however, is A1 negative due to a crossing-over and is also negative in PLT with clone 19-2.

Fluorescent-activated cell sorter (FACS) analysis of clone 19-2 revealed that phenotypically this clone is composed entirely of T4 cells. To verify clonal origin of these cells, recloning of clone 19-2 was performed. No deviations from pattern of reactivity or T-cell phenotype were observed, confirming clonal origin of alloreactive cells. To evaluate the specificity of target antigen recognized by clone 19-2 blocking experiments were performed using a panel of monoclonal antibodies (MAb) and alloantibody anti-A1, following the technique of Myers et al. (2). The results of three experiments are shown in Table 2. As can be seen from the table, anti-DQ (Leu-10) MAb did not result in any blocking of PLT reactivity. Paradoxically, anti-class I MAb (M18 and W6/32) as well as anti-A1 antibody did not result in inhibition of PLT reactivity. Reactivity was completely blocked by anti-DR and anti-DP (Becton-Dickinson) MAbs. We hypothesize that clone 19-2 is recognizing class II determinant shared by DR and DP which occurs in linkage disequilibrium with HLA-A1.

Table 2. MAb Blocking of PLT Reactivity of Clone 19-2 Against Cells of Original Stimulator

Ab	Exp. 1	Exp. 2	Exp. 3
–	100.0	100.0	100.0
M18	97.9	ND	86.5
W6/32	ND	135.1	74.0
A1	77.1	ND	ND
DR	3.7	13.8	6.5
DQ	ND	88.3	78.4
DP	ND	3.2	10.3

Results expressed as percentage of reactivity against MAb-treated stimulating cells as compared to untreated (100%). ND = not detected.

References

1. Eckels DD, Hartzman RJ. Characterization of human T lymphocyte clones (TLCs) specific for HLA-region gene products. Immunogenetics 1982;16:117–128.
2. Myers LK, Ball EJ, Stastny P. Different specificities of an HLA-DRw6 haplotype detected by alloreactive T lymphocytes. Hum Immunol 1984;11:193–205.

Author Affiliation

Departments of Medicine and Laboratory Medicine, University of Alberta, Edmonton, Alberta, Canada.

An Alloreactive Clone that Recognizes an Epitope Only on Monocytes

C. Farrell, M.C. Honeyman, C.A. Hoadley, and M.C. Wolnizer

The clone B/LD2 (CD4+ve CD8-ve) was generated by limiting dilution from a ten day priming of cells from a Chinese individual, YKB (HLA-A 11,24; B 13,35; DR 2,w8; DQ w1,w3; D Syd2,blank) as responders and peripheral blood mononuclear cells (PBMC) from a Caucasian individual, LD2B (10th w/s 9083 HLA-A 3,–; B 7,–; DR 2,–; DQ w1,–; Dw2,–) as stimulators. The clone was expanded using PBMCs from LD2B as feeders and IL-2. Screening with an extensive panel of PBMCs from individuals of known HLA-A,B,DR,DQ,DP and D specificities, the clone was found to respond to about 70% of this panel but there was no correlation between a positive proliferative response and any known HLA class I or class II antigen. When the clone was subsequently stimulated with the 75 Tenth Workshop B-LCLs, no proliferation was seen with any cell. In addition, no proliferation was seen when the clone was stimulated with activated cells (PHA or alloactivation).

To investigate the response of the clone to the different cell types that are included in PBMCs, the following experiments were performed, using the specific stimulator cell LD2B throughout:

1. E-Rosetting; the clone only responded to the E-rosette negative population (non-T cell) and not to the E-rosette positive population (T cells).

Immunobiology of HLA
Volume II

Table 1. The Reactivity of Clone B/LD2 to Different Cell Populations

Sample No.	Treatment	Cell Type	CPM
(1)	PBMC	T+B+M	25090
(2)	Plastic NAD	T+B	4740
(3)	GAH NAD	T	2640
(4)	Plastic AD	M	22870
(5)	GAH AD	B	110
(6)	Plastic NAD + Plastic AD	T+B+M	25950

CPM, counts per minute; NAD, nonadherent; AD, adherent; GAH, goat anti-human Fab; T, T cells; B, B cells; M, monocytes.

2. Nycodenz Monocytes; a positive response was seen with cells from the Nycodenz interface (monocytes and platelets) but no response was detected to the cells in the pellet (lymphocytes and polymorphs). In a separate experiment, no response was obtained to a pure population of polymorphs or platelets.
3. Panning; in this experiment the monocytes were first removed by plastic adherence. The non-adherent cells were then incubated on a plate which was coated with goat anti-human Fab to remove the B cells. The adherent monocytes and B cells were removed by vigorous washing. The monocytes, B cells, plastic non-adherent, and goat anti-human non-adherent cells were used to stimulate the clone (Table 1). Proliferation was only seen when the monocytes were used as stimulators, and not when the B cells or the two non-adherent populations were used. When these monocytes were added back into the plastic non-adherent population, the responsiveness was restored to the level of the original cells. Three vascular endothelial cell lines (kindly provided by Dr. G. Russ, Queen Elizabeth hospital, South Australia) were also tested as stimulator cells but these did not elicit a response from the clone.

Table 2. Family Study

	HLA Phenotype	Monocyte Antigen
Mother	a,b	−,+
Father	c,d	−,+
Sibling 1	a,d	−,+
Sibling 2	a,c	−,−
Sibling 3	a,c	−,−
Sibling 4	b,c	+,−

These experiments indicate that the clone appears to recognize a monocyte antigen, and this is confirmed by the capacity of monocyte specific monoclonal antibodies FMC17 and FMC33, which recognize CDw14 (kindly provided by Dr. Heddy Zola, Flinders Medical Centre, Bedford Park, South Australia) to block the proliferative response. Further studies also indicate that this clone is not only proliferative but also cytotoxic.

Family studies were performed to investigate possible MHC linkage; in the one informative family, the antigen segregated with the MHC (Table 2) and in 14 out of 14 HLA-identical sibling pairs the results were concordant for the presence or absence of the antigen. From these results it would appear that clone is recognizing a polymorphic antigen present only on monocytes and which is MHC linked. The frequency of this antigen is 72% in the Caucasian population (56/78) and 7% in the small Chinese sample (1/15). This antigen is potentially important in transplantation as it is capable of initiating both proliferative and cytotoxic responses.

Author Affiliation

Tissue Typing Laboratory, NSW, Red Cross Blood Transfusion Service, Sydney, Australia.

HLA Class II Positive T-Cell Lines Lack Stimulatory Capability in the Mixed Leucocyte Reaction

Niels Ødum, Bo Hofmann, Niels Morling, Bodil K. Jakobsen, and Arne Svejgaard

There is conflicting evidence concerning the capability of activated, HLA class II positive T cells (Ta) to present alloantigen in the mixed leucocyte culture (MLC). Thus, Ta has been reported to be potent, poor, or unable to elicit allogeneic, proliferative responses.

We have investigated the capability of human alloactivated class II positive, CD4+ T-cell lines (1) to stimulate proliferative responses of naive T cells in the MLC and alloprimed T cells in the primed lymphocyte test (PLT) (2). None of nine Ta stimulated allospecific (e.g, DR1 or DPw1) PLT cells (median stimulation by Ta: 600 cpm). In the MLR, stimulation by Ta (irradiated 2400 rad) was only seen when the Ta were directed against the class II antigens on the responder (median stimulation of 15 experiments: 21,000 cpm), in contrast to other Ta (medium stimulation of 17 experiments: 1,000 cpm).

Immunobiology of HLA
Volume II

This difference was highly significant ($P < 0.001$) and thus, occasional "backstimulation" is likely to explain the above controversy.

The lack of stimulation by Ta was not due to a low expression of class II antigens since the surface densities on Ta of DR, DQ, and DP antigens were similar to those on peripheral non-T cells as judged by the reaction with monomorphic monoclonal antibodies (1) and because the Ta carried the DR specificities of the original T-cell donor.

Addition to the PLT of IL1, Ta autologous adherent cells, or interferon γ had no effect on the stimulatory capability of the Ta. In contrast, exogeneous recombinant IL2 reversed the lack of stimulation by Ta in the PLT as well in the MLR.

In conclusion, these findings suggest that class II positive Ta are capable of inducing IL2 responsiveness in naive T-responder cells, but they seem to lack the ability to induce the production of IL2 in these cells.

Acknowledgments. This paper was in part supported by the Danish Medical Research Council.

References

1. Ødum N, Hofmann B, Morling N, Platz P, Ryder LP, Tvede N, et al. Differences between primed allogeneic T-cell responses and the primary mixed leucocyte reaction (MLR): Primed T-cells acquire independence of the blocking effect of monoclonal antibodies against IL1 and the CD5, CD11a (LFA-1) and CD11C (p150,95) molecules. Scand J Immunol 1988;27:405.
2. Ødum N, Hartzman R, Jakobsen BK, Morling N, Platz P, Robbins F-M, et al. The HLA-DP polymorphism in Denmark investigated by local and international PLT reagents. Tissue Antigens 1986;28:105.

Author Affiliation

Tissue Typing Laboratory, Department of Clinical Immunology, State University Hospital (Rigshospitalet), Copenhagen, Denmark

Analysis of Defective Class II Allostimulation by γ-Interferon-Activated Fibroblasts*

David H. Maurer and Marilyn S. Pollack

Abstract: γ-Interferon (IFN-γ)-activated human dermal fibroblasts (FB_{IFN}) express HLA class II molecules (1–3) and specifically stimulate the proliferation of some allospecific, CD4-positive T-cell clones (2,3) or antigen-specific, class II-restricted T-cell clones (4,5). However, FB_{IFN} elicit little or no proliferation from naive T cells (2,3), resting, antigen-specific T cells (3,6), or certain T-cell clones (5). It has been proposed that particular T-cell populations may require accessory cell properties distinct from class II expression that cannot be provided by FB_{IFN} alone (3,6), but that can be restored by the addition of a competent accessory cell (7). We used primed lymphocyte techniques to demonstrate specific priming to lymphocyte-activating determinants (LAD) expressed by FB_{IFN} in cultures containing naive T cells, FB_{IFN}, and peripheral blood mononuclear (PBMC) accessory cells. Specific priming to FB LAD was dependent on two factors: allogenicity of the added accessory cells and prior treatment of FB with IFN-γ to induce HLA class II expression. These results suggest that under limited conditions, the HLA class II LAD expressed by FB_{IFN} can be recognized by naive T cells and that priming in this instance is mediated by bystander effects.

Preliminary experiments had indicated that although FB_{IFN} fail to stimulate primary allogeneic lymphoproliferation themselves, they could enhance proliferative response in a manner suggesting that they contribute to T-cell activation when added to ongoing allogeneic mixed leukocyte cultures (MLC) (Ref. 2 and unpublished observations). We reasoned that under such circumstances, specifically primed T cells with immunologic memory for LAD expressed by FB_{IFN} should be generated.

Table 1 illustrates typical results from experiments in which fresh responder PBMC were primed to stimulator PBMC in the presence of added FB third-party cells. Donors of stimulator PBMC were selected to minimize cross-reactive priming. After a 10-day bulk priming culture, recovered cells were assayed for secondary proliferative responses to fresh PBMC on day 2 when memory responses should be strongest and on day 6 when only primary responses would be expected. Addition of FB_{IFN} to allogeneic priming MLC elicited substantial (65%) secondary proliferative response to PBMC of the FB donor (C) on day 2 (Table 1, line 7). Addition of untreated FB to priming cultures elicited no more secondary proliferation to PBMC of the FB donor (line 6) than that observed with a control reagent primed in the absence of FB third-party cells (line 1). Parafor-

*This work was supported in part by NIH grants AI21289 and CA40552. Address correspondence to: Marilyn S. Pollack, Ph.D., The Methodist Hospital, F-501, 6565 Fannin, Houston, TX 77030.

Table 1. Priming Responses in Mixed Leukocyte Cultures Containing Third-Party Fibroblasts

	Primary Culture*			Secondary culture* Percent Relative Response‡ on Day 2				on Day 6			
				Stimulator PBMC							
Line	Resp PBMC	Stim PBMC	Accessory Cell§	A	B	C	D	A	B	C	D
1	A	B_x	–	0	100	35	4	<1	<1	20	<1
2	A	C_x	–	0	15	100	11	<1	128	7	2
3	A	B_x	C PBMC-x	0	100	135	10	<1	<1	<1	<1
4	A	A_x	C PBMC-x	0	33	100	9	<1	78	<1	<1
5	A	B_x	C PBMC-f	0	100	119	13	<1	<1	<1	<1
6	A	B_x	C FB-Ctrl-x	0	100	24	<1	<1	40	11	<1
7	A	B_x	C FB-IFN-x	0	100	65	14	<1	9	6	<1
8	A	B_x	C FB-Ctrl-f	0	100	34	<1	<1	<1	<1	<1
9	A	B_x	C FB-IFN-f	0	100	72	8	<1	<1	<1	<1
10	A	A_x	C FB-Ctrl-x	3	5	8	3	0	100	44	24
11	A	A_x	C FB-IFN-x	5	5	9	2	0	100	53	4
12	Unprimed A PBMC			<1	<1	<1	<1	0	100	83	52
13	Unprimed B PBMC			<1	<1	<1	<1	100	0	57	33
14	Anti-DPw2 T-cell line			56	0	100	<1	10	<1	4	<1

*10^6 responder PBMC were cultured with 10^6 irradiated (x) stimulator PBMC for 10 days in bulk priming cultures. Some primary cultures received additions of irradiated PBMC (3,000 rads) or FB (6,000 rads) accessory cells. Other cultures received viable or paraformaldeyhyde-fixed (f) PBMC or FB accessory cells.

†Responder PBMC recovered from primary culture were recultured in secondary restimulation assays with fresh, irradiated (3,000 R) stimulator PBMC for 48 hours and for 138 hours. Cultures were pulse labeled with methyl[^{3}H]thymidine for the final 8 hours of the culture period.

‡Secondary proliferative responses expressed as relative responses normalized to primary culture stimulator as 100% and primary culture responder as 0% except when no priming response occurred. Other responses were normalized to the maximal stimulator as 100% and the autologous stimulator as 0%.

A = Normal donor CC (DRw6,7,w52,w53; DQwl,2; DPw2,4); B = normal donor SR (DRw3,w52; DQw2; DPwl); C = normal donor RG (DR 1,4,w53; DQwl,3; DPw2); D = normal donor LA (DR 7, w8,w52,53; DQw2,3; DPw4).

maldehyde fixation of the added FB (lines 8 and 9) to prevent the shedding of FB-derived histocompatibility antigens did not affect the ability of FB_{IFN} to prime responder cells for memory responses. The specificity of priming under these conditions was indicated by the failure of reagents primed in the presence of FB_{IFN} to proliferate in secondary culture to irrelevant PBMC (donor D).

The mere presence of irradiated PBMC stimulator cells containing competent accessory cells in priming cultures was not sufficient to achieve priming to FB_{IFN} LAD. Responder cells primed to allogeneic FB_{IFN} in autologous MLC (lines 10 and 11) responded in secondary culture like unprimed PBMC (lines 12 and 13) (i.e., with maximal proliferation on day 6). Priming of naive T cells to FB LAD by this approach was clearly a function of the positive influence of a coextensive allogeneic MLC.

Because the experimental approach used in Table 1 was limited by the degree of cross-reactive or nonspecific priming observed with reagents primed in the absence of added FB (Table 1, line 1), we attempted to simplify the approach by using FB stimulator cells alone and by reconstituting putative bystander effects with recombinant IL2 (Table 2). Somewhat surprisingly, when paraformaldehyde-fixed FB were used to prime allogeneic responder PBMC, FB_{IFN}, but not untreated FB, specifically primed naive cells for memory responses to PBMC of the FB donor (lines 3 and 4). Although not tested in this experiment, viable FB had uniformly failed to prime T cells for specific responses to FB LAD (unpublished observations). Addition of recombinant IL2, although salutary for the priming response to FB_{IFN} LAD (line 6), was not required to elicit a specific priming response from naive allogeneic PBMC.

The specificity of the priming response to fixed FB_{IFN} stimulator cells under these conditions was indicated by the lack of a secondary response to unrelated PBMC (stimulators C and D), by the inability of responder PBMC cultured in the presence of recombinant IL2 alone to exhibit specific priming (line 9) and by the failure of the same FB to prime autologous responder PBMC (line 8). The fact that the unrelated stimulator D who shares the serological DR4 specificity with the donor of priming FB failed to elicit a secondary response suggests that the memory response elicited by paraformaldehyde-fixed FB_{IFN} in this case was not specific for LAD common to all DR4 molecules.

Previous findings of synergy between antigen-presenting FB_{IFN} and more conventional accessory cells like

Table 2. Priming Responses to Paraformaldehyde-Fixed FB Stimulator Cells

Primary Culture*				Secondary Culture†			
				Stimulator Cell			
Line	Resp PBMC	Stimulator Cell‡	Addition§	A	B	C	D
1	A	B PBMC-f	–	1,602±660‡	12,050±1,016	1,342±233	3,860±652
2	A	B PBMC-f	rIL2	805±151	5,575± 550	3,877±603	1,397±393
3	A	B $FB_{Ctrl\text{-}f}$	–	680± 58	2,595± 436	602± 99	180± 21
4	A	B $FB_{IFN\text{-}f}$	–	297± 41	10,697±1,163	800±138	737±202
5	A	B $FB_{Ctrl\text{-}f}$	rIL2	667±133	3,065± 607	1,635±359	1,030±311
6	A	B $FB_{IFN\text{-}f}$	rIL2	1,450±169	19,210± 173	3,852±630	2,890±360
7	B	B $FB_{Ctrl\text{-}f}$	rIL2	1,372±266	1,790± 260	5,437±839	723±216
8	B	B $FB_{IFN\text{-}f}$	rIL2	3,342±511	1,980± 218	3,115±313	1,287±260
9	A	–	rIL2	362± 60	1,100± 250	1,543±380	660±231
10	Anti-DPw2 T-cell line		–	335± 44	33,847±1,412	377±167	253± 79

*10^6 responder PBMC were cultured with 10^6 paraformaldehyde-fixed (f) PBMC or 0.5 × 10^6 fixed FB as described in Table 1. FB were either untreated (Ctrl) or precultured with IFN-γ (IFN) for 120 hours before fixation.

†Responder PBMC recovered from primary culture were recultured in secondary restimulation assays with fresh, irradiated (3,000 rads) stimulator PBMC for 48 hours. Cultures were pulse labeled with methyl[^{3}H]thymidine for the final 8 hours of the culture period.

§Some primary cultures received addition of 1.0 unit/ml of recombinant IL2 (Amgen).

‡Secondary proliferative responses expressed as mean cpm ± standard error.

A = Normal donor JM (B14,DR1/DR2,DQw1); B = normal donor RG (B14,DR1/DR4,DQw1/DQw3,DPw2); C = normal donor SR (DR3/DR3,DQw2/DQw2, DPw1/DPw1); D = normal donor DM (DR4/DR7,DQw3/DQw3, DPw4/DP-).

monocytes (6) and endothelial cells (7) in primary proliferative assays have been interpreted to suggest that nonantigen-specific signals necessary for specific activation of naive T cells may be delivered by adjacent cells that do not bear the stimulating antigen. Our experiments were designed to test this possibility directly using a response that can be generated entirely in vitro (i.e., the specific priming of naive T cells for memory responses to HLA-linked LAD in secondary culture). Our results suggest that the synergy observed in primary cultures with the admixture of nonstimulatory FB_{IFN} and cells possessing accessory functions may relate directly to the allogenicity of the additional cell type. Hence, so-called "bystander effects" mediated by the positive allogeneic effect of a coextensive MLC may permit naive T cells to proliferate and to be specifically primed for FB LAD. This represents a subtle, but important distinction from the capacity of a neutral accessory cell that is truly bystanding to T cell/FB interactions to deliver required activation signals to the T cell.

Nevertheless, our demonstration of the specific priming of naive T cells to LAD expressed by FB_{IFN} using two different experimental designs suggests that the histocompatibility antigens expressed by FB under the influence of IFN-γ are functional and can be recognized by naive T lymphocytes. The mechanism by which FB_{IFN} might prime T cells for memory proliferative responses but at the same time fail to elicit primary alloproliferation is unclear. One may speculate that in mixed leukocyte/FB_{IFN} cultures, a few naive, allospecific T cells possess high enough avidity for FB_{IFN} LAD to be activated without need of accessory activation pathways (8) or adhesion-strengthening interaction molecules (9) that FB may lack. However, because the number of responding clones is small and/or because viable FB mediate effects otherwise inhibitory to T-cell proliferation/priming, no response to FB_{IFN} is observed in primary cultures. Inhibitory effects mediated by class II-positive FB could include the absorption and internalization of IL1 via specific receptors (10) or the elaboration of prostaglandins (11) or other metabolites (3) that interfere with the expansion of activated T cells. These putative inhibitory effects could be surmounted by the positive allogeneic effects of a simultaneous, bystander MLC or by simple metabolic inactivation of FB_{IFN} stimulator cells thus permitting the expansion of a small number of responding naive T-cell clones to proceed. Further studies examining the allostimulatory capacity of IFN-γ-activated FB and other cultured cells are in progress.

References

1. Collins T, Korman AJ, Wake CT, Boss JM, Kappes DJ, Fiers W, et al. Immune interferon activates multiple class II major histocompatibility complex genes and the associated invariant chain gene in human endothelial cells and dermal fibroblasts. Proc Natl Acad Sci USA 1984;81:4917.
2. Maurer DH, Collins WE, Hanke JH, Van M, Rich RR, Pollack MS. Class II positive human dermal fibroblasts restimulate cloned allospecific T cells but fail to stimulate primary allogeneic lymphoproliferation. Hum Immunol 1985;14:245.
3. Geppert TD, Lipsky PE. Antigen presentation by interferon-gamma-treated endothelial cells and fibroblasts: Differential ability to function as antigen presenting cells despite comparable Ia expression. J Immunol 1985;135:3750.
4. Umetsu DT, Pober JS, Jabara HH, Fiers W, Yunis EJ, Burakoff SJ, et al. Human dermal fibroblasts present teta-

nus toxoid antigen to antigen-specific T cell clones. J Clin Invest 1985;76:254.
5. Maurer DH, Hanke JH, Mickelson E, Rich RR, Pollack MS. Differential presentation of HLA-DR, DQ and DP restriction elements by interferon-gamma-treated dermal fibroblasts. J Immunol 1987;139:715.
6. Umetsu DT, Katzen D, Jabara HH, Geha RS. Antigen presentation by human dermal fibroblasts: Activation of resting T lymphocytes. J Immunol 1986;136:440.
7. Geppert TD, Lipsky PE. Dissection of defective antigen presentation by interferon-gamma-treated fibroblasts. J Immunol 1987;138:385.
8. Yang SY, Chouaib S, Dupont B. A common pathway for T lymphocyte activation involving both the CD3-Ti complex and CD2 sheep erythrocyte receptor determinants. J Immunol 1986;137:1097.
9. Martz E. LFA-1 and other accessory molecules functioning in adhesions of T and B lymphocytes. Hum Immunol 1987; 18:3.
10. Dower SK, Kronheim SR, Hopp TP, Cantrell M, Deeley M, Gillis S, et al. The cell surface receptors for interleukin-1 alpha and interleukin-1 beta are identical. Nature 1986;324:266.
11. Korn JH. Modulation of lymphocyte mitogen responses by cocultured fibroblasts. Cell Immunol 1981;63:374.

Author Affiliation

Department of Microbiology and Immunology, Baylor College of Medicine, Houston, TX 77030, USA.

Pretreatment of Stimulator/Target Cells with Xyloside Inhibits Synthesis of the Class II Associated Proteoglycan, but Potentiates an MLR and Antigen Presentation

Susan Rosamond, Tara Rumbarger, Larry Brown, Carlen Gomez, Thomas J. Braciale, and Benjamin D. Schwartz

HLA class II molecules are associated with a number of invariant chain glycoproteins including γ_1 or I_i, the ~31 Kd predominant form of invariant chain; the ~33 Kd species γ_2 and γ_3; the 41 Kd species p41, the 25 Kd species, p25; and a number of acidic forms. In addition, we demonstrated that the class II molecules are also associated with a 40–70 Kd chondroitin sulfate proteoglycan molecule (CSPG) (1), and further demonstrated that the core protein of the CSPG is I_i (2).

Despite a fairly complete biochemical and molecular biological characterization of the invariant chain in both its glycoprotein and proteoglycan forms, relatively little is known regarding the physiologic role of these molecules. I_i appears not to be necessary for class II $\alpha\beta$ heterodimer formation or for cell surface expression. Computer modeling has led to the suggestion that I_i or the CSPG may protect the peptide-binding site of class II molecules until they contact foreign peptide. Biochemical parallels with transferrin receptor have suggested physiologic parallels, and raised the possibility that I_i and/or the CSPG may be involved in recycling of class II molecules.

To more precisely define the role of the CSPG in class II biology, we have used p-nitrophenyl β-D-xyloside, a specific inhibitor of proteoglycan synthesis. This substance competes with the natural substrate for the glycosaminoglycan. We have previously demonstrated that xyloside completely inhibits CSPG synthesis, but only marginally inhibits protein synthesis including that of the invariant chain (3). We have also shown that xyloside appears to inhibit antigen processing/presentation. Class II restricted influenza virus-specific human T-cell clones have been isolated which will lyse target cells bearing the appropriate complex of class II molecules and processed viral antigen. One of these clones, V1, recognizes influenza hemagglutinin (HA) and specifically HA peptide 126–145 in the context of a DR5 haplotype class II molecule. When $DR5^+$ lymphoblastoid cell lines were preincubated with xyloside, then pulsed with and allowed to process UV-irradiated virions in the presence of xyloside, the killing of the target cells by clone V1 was markedly inhibited compared to controls (3). Although several interpretations of these results are possible, the results are consistent with the class II associated CSPG playing a role in antigen processing.

In this report, we show that pretreatment with xyloside does not interfere with antigen presentation, but in fact potentiates antigen presentation and an MLR. For antigen presentation, $DR5^+$ cells were pretreated with xyloside, then incubated with HA peptide 126–146 and assayed for their ability to be lysed by clone V1 (Table 1, top). For the MLR, stimulator cells in the secondary MLR were pretreated with xyloside for 24 hours, then fixed with 0.1% glutaraldehyde, and assayed for their ability to stimulate a bulk population of primed T cells (Table 1, bottom). In both cases, pretreatment with xyloside increased the stimulation of the responding T cells.

Immunobiology of HLA
Volume II

Table 1. Xyloside Pretreatment Potentiates T Cell Responses

Stimulator Cells	Xyloside Pretreatment	Antigen	Cytotoxic Cells	% Cytotoxicity
TB-LCL's*	0	Intact virions**	V1†	59
TB-LCL's	5 mM	Intact virions	V1	38
TB-LCL's	0	HA peptide 125–146**	V1	61
TB-LCL's	5 mM	HA peptide 126–146	V1	76

Stimulator Cells	Xyloside Pretreatment	Responder Cells	^{3}H-dTR Incorporation
TB-LCL's‡	0	Bulk primed T cells	58,000
TB-LCL's	5 mM	Bulk primed T cells	72,000

*TB-LCL's are a DR5 haplotype positive B lymphoblastoid cell line.
**TB-LCL's were pulsed with UV-irradiated influenza virions at ~15 HA units/10^4 cells or 10 μg/ml HA peptide/10^4 cells.
†Effector:target = 10:1.
‡TB-LCL's were fixed with 0.1% glutaraldehyde after xyloside pretreatment.

These results indicate first that xyloside interferes with antigen processing but not presentation, and second, that xyloside in fact potentiates antigen presentation and an MLR. These findings are consistent with the interpretation that loss of the proteoglycan form of invariant chain induces a conformational change in class II molecules which makes them more accessible to foreign antigenic peptides and/or antigen-specific and allogeneic T cells.

Acknowledgments. The authors wish to thank Ms. Pat Stewart and Mrs. Lorraine Whiteley for expert preparation of the manuscript.

Supported in part by NIH grants AI-18925, AI-15608, AI-15322, AI-15353, and ML-33391, and by research support from Brown and Williamson Tobacco Corp, Phillip Morris, Inc., and the United States Tobacco Co.

References

1. Sant AJ, Cullen SE, Schwartz BD. Identification of a sulfate-bearing molecule associated with HLA class II molecules. Proc Natl Acad Sci USA 1984;81:1534.
2. Giacoletto KS, Sant AJ, Bono C, Gorka J, O'Sullivan DM, Quaranta V, et al. The human invariant chain is the core protein of the human class II associated proteoglycan. J Exp Med 1986;164:1422.
3. Rosamond S, Brown L, Gomez C, Braciale TJ, Schwartz BD. Xyloside inhibits synthesis of the class II associated chondroitin sulfate proteoglycan presentation events. J Immunol 1987;139:1946.

Author Affiliations

Division of Rheumatology, Departments of Medicine, Pathology, and Microbiology and Immunology, Howard Hughes Medical Institute, Washington University School of Medicine, St. Louis, MO 63110, USA

Roles of Accessory Molecules in Processing and Presentation of Foreign Antigens*

Shan Lu, Minzhen Xu, L. Thomas Chin, Quoc V. Nguyen, Patricia S. Reisert, Victor E. Reyes, Christopher H. Sorli, Lawrence J. Thomas, Xiao-rui Yao, Jacqueline Anderson, John E. Mole, Takeshi Sairenji, and Robert E. Humphreys

The I_i sequence Phe_{146}-Val_{164} was hypothesized to coil as an amphipathic, α helix in the desetope of class II MHC antigens until release in an acidic, foreign antigen-containing endosome to catalyze charging of the desetope with a structurally similar foreign peptide (1). A serum from one of four rabbits injected with a KLH-conjugated, synthetic peptide of I_i sequence 146–169, substituting Tyr for Phe_{146}, immunoprecipitated a 67-kD protein from Raji cells after 3 hours [^{35}S]methionine labeling and 69- and 67 kD-proteins after 5 or 10

*This work was supported by the American Cancer Society institutional grant IM-129 to Q.V.N., by NIH grant AI-23061 to T.S., by NIH grant AI/19337 to J.E.M., by NIH grant CA-37645 to R.E.H., and by NIH fellowship AI 07272 to Q.V.N. L.T.C. was a Betty Lee Stone fellow of the American Cancer Society, Massachusetts Division and was supported by NIH training grant CA 36762. Address correspondence to: R.E. Humphreys, Department of Pharmacology, University of Massachusetts Medical School, 55 Lake Avenue North, Worcester MA 01655, USA.

hours of labeling, respectively. The 67-kD protein was not sensitive to endoglycosidase F treatment or tunicamycin and had a pI about 5.5. p67 was not surface expressed as judged by immunofluorescence analyses with the antiserum and by immunoprecipitation of surface-biotinylated proteins. These human molecules might correspond to the murine proteins p72/74, described by Lakey et al. (2) to have a potential role in antigen presentation.

T cells recognize amphipathic helical peptides. For the development of T-cell help to B cells to produce antibodies against a superficial epitope on a protein antigen, a second, "carrier" determinant must also be presented to the helper T cell. This carrier determinant might be presented either by monocytic cells to expand the specific T_H population, or by specific B cells requiring T-cell help. Membranal immunoglobulin (mIg) concentrates antigen up to 10^4-fold, and selection of T-cell-presented determinants is defined by their structural linkage to the superficial epitopes seen by the mIg. In studies to characterize T-cell-presented, "carrier" determinants, only a few peptides digested from antigenic proteins were found to be T cell presented. Analyses of these peptides, in part derived from Kaiser's studies on the general, amphiphilic character of many peptide hormones (3), led DeLisi and Berzofsky to propose that potential amphipathic folding of digested peptides was a first requirement for T-cell presentation (4). From the finding of an amphipathic α helix in I_i, the electrophoretically invariant protein associated noncovalently with class II MHC antigens from synthesis until endosomal dissociation, Elliott et al. proposed that helix blocked the foreign antigen presenting site (desetope) of class II MHC antigens until catalyzing charging of that site by a concerted mechanism of I_i dissociation upon insertion of a foreign, structurally similar amphipathic peptide (1). An algorithm was developed to rank proline-free sequences which could fold into α helices with a moderately hydrophobic strip-of-helix and otherwise a generally hydrophilic surface (5). This algorithm predicted both class I and class II MHC-presented sequences and led to the views that 1) a given T-cell-presented peptide could be presented by both a class I allele and by a class II allele, and 2) class II and class I desetopes were structurally similar (V. Reyes et al., this volume).

Accessory molecules might catalyze or regulate MHC binding and presentation of peptides. An essential requirement for T-cell presentation of an amphipathic peptide was its binding to some allelic form of a class II MHC gene product on the antigen-presenting cell (APC) (6,7). However, such binding to a class II MHC complex per se did not guarantee stimulation of some T-cell clone in the immunized mouse (8). In studies of the binding of synthetic foreign peptides to immunopurified class II MHC antigens, Buss et al. found the association rate to be relatively slow (1.2 $M^{-1}s^{-1}$) (7). One might suspect that in vivo accessory proteins might catalyze or regulate antigen digestion, the binding of peptides to class II complexes, and the trafficking of those complexes to the cell surface.

Roles for at least three proteins can be proposed in catalyzing or regulating the processing of foreign antigen and its presentation by class II MHC molecules. The first is I_i, which is associated with class II molecules from their time of synthesis. A 25,000-dalton, COOH-terminal portion of I_i is released by proteolysis in an endosomal compartment about 20 to 40 minutes after synthesis (Q.V. Nguyen et al., this volume). That p25 was the COOH-terminal portion of I_i was demonstrated by immunoprecipitation of urea-denatured, resolubilized I_i, immunopurified with an anti-I_i monoclonal antibody (MAb), with two rabbit sera to two COOH-terminal I_i peptides but not with an NH_2-terminal peptide. Cleavage and release of I_i could be associated with charging of the desetope with structurally similar amphipathic helical peptides, if release of I_i and peptide charging were to occur by a concerted mechanism. Alternately, I_i might not bind to class II molecules through the desetope but at another region retarding class II molecules from bulk flow to the cell surface until "completion" of the class II complex with a foreign peptide. Other complexes of class II molecules with I_i, p70, p80, and CS-I_i are described (Sorli & Humphreys, this volume).

Proteases would be a second type of molecule regulating peptide charging of class II complexes. By cleavage of sites in I_i or the class II molecules themselves, proteases could promote or inhibit peptide binding. Cleavage at the basic dipeptides flanking the third hypervariable region of the β chain, Lys_{71}-Arg_{72}, and Arg_{93}-Arg_{94}, could lead to the displacement of that region and binding of structurally similar foreign peptides. The fact that the sequences of some foreign peptides closely resemble sequences of the hypervariable regions of their presenting class II MHC alleles (8) could reflect their substitution for such regions in proteolytically cleaved class II molecules.

A third candidate for some role in antigen processing and presentation is the p72/74 pair of proteins isolated by Lakey et al. (2). In a search for such accessory molecules that might bind T-cell-presented peptides, Lakey et al. prepared an affinity column with Sepharose-coupled cytochrome c peptide (81–104), which is T cell presented in H-2^k mice (2). This column specifically adsorbed p72/74. Rabbit anti-p72/74 serum reacted with a cell surface determinant and inhibited antigen presentation by B cells to a cloned T-cell line specific for the cytochrome c peptide.

In seeking to test the hypothesis that I_i might bind to class II complexes through a potential amphipathic helix, a peptide was synthesized with I_i sequence Phe_{146}-Met_{169}, substituting Tyr for Phe_{146}. One of four rabbits, immunized with this KLH-conjugated peptide, raised antibodies that immunoprecipitated a 67-kD protein from 3-hour [^{35}S]methionine-labeled Raji cells. After longer labeling times (5 or 10 hours) a 69-kD protein was also recognized and later bleeds from the same rabbit recognized in addition an 18-kD protein. The 69-kD protein, which appeared after substantial labeling times, could have been 1) a late-processed derivative of p67, 2) a protein that joined p67 after sequestration in some cellular compartment for a long time, or 3) a protein that joined p67 at its time of synthesis but from a larger pool that must become appreciably radiolabeled before seeing the p69 in anti-p67 immunoprecipitates. Two-dimensional gel electrophoresis showed p67 had a pI about 5.5 and no microheterogeneity. It was not sensitive to treatment with tunicamycin or endoglycosidase F.

The cellular distribution of p67 was assessed. p67 could be immunoprecipitated from Raji cells, EBV-transformed B cells, PHA-activated T cells, and hairy leukemic cells. Fluorescein-labeled $F(ab)_2$ anti-p67 antibody did not surface stain peripheral blood lymphocytes, Raji cells, CEM cells, and PHA-activated T cells any greater than did a parallel preparation of a preimmunization serum. Surface biotinylated preparations of these cells were immunoprecipitated, electrophoresed, blotted to nitrocellulose, and developed with ^{125}I-avidin. No p67 or p69 was observed while class II MHC α and β chains were well visualized. We concluded that the epitope of p67 recognized with our antiserum was not surface expressed.

The expression of p67 and p25 was decreased in hairy leukemic cells. Class II MHC-positive, hairy leukemic cells, which can constitute an appreciable mass of antigen-presenting cells, might have down-regulated mechanisms of antigen processing and presentation. To assess roles of accessory proteins in such cells, the levels of p67, I_i, p25 derived from I_i, and class II MHC antigens were determined in hairy leukemic cells, freshly transformed B lymphoblastoid cells, and in Raji lymphoblastoid cells. In hairy leukemic cells, the level of the antigen recognized with the antiserum to the peptide analog of I_i peptide (146–169) was appreciably diminished compared to B lymphoblastoid or Raji cells. Likewise, the level of p25 was decreased. p25 has been demonstrated to be derived in an endosomal compartment from the COOH-terminal exomembrane portion of I_i appearing about 20 to 40 minutes after synthesis (Q.V. Nguyen et al., this volume). The cleavage of I_i to p25 might reflect removal of I_i from class II MHC molecules and their charging with peptides structurally similar to I_i (Phe_{146}-Val_{164}). The decrease in p25 and p69 in hairy leukemic cells could reflect a fundamental alteration in antigen processing mechanisms in these cells.

In our future work we plan to raise a second antibody to immunopurified p67/69 to test functions described for murine p72/74, and to determine whether the I_i helix 146–164 is bound to the class II MHC desetope.

References

1. Elliott WL, Stille CJ, Thomas LJ, Humphreys RE. An hypothesis on the binding of an amphipathic, alpha helical sequence in Ii to the desetope of class II antigens. J Immunol 1987;138:2949.
2. Lakey EK, Margoliash E, Pierce SK. Identification of a peptide binding protein that plays a role in antigen presentation. Proc Natl Acad Sci USA 1987;84:1659.
3. Kaiser ET, Kezdy FJ. Amphiphilic secondary structure: Design of peptide hormones. Science 1984;223:249–255.
4. DeLisi C, Berzofsky JA. T-cell antigenic sites tend to be amphipathic structures. Proc Natl Acad Sci USA 1985; 82:7048.
5. Stille CJ, Thomas LJ, Reyes VE, Humphreys RE. Hydrophobic strip-of-helix algorithm for selection of T cell-presented peptides. Mol Immunol 1987;24:1021.
6. Babbitt BP, Allen PM, Matseuda G, Haber E, Unanue ER. Binding of immunogenic peptides to Ia histocompatibility molecules. Nature 1985;317:359.
7. Buss S, Selle A, Colon SM, Jenis DM, Grey HM. Isolation and characterization of antigen-Ia complexes involved in T cell recognition. Cell 1986;47:1071.
8. Guillet JG, Lai MZ, Briner TJ, Buss S, Sette A, Grey HM, et al. Immunological self, nonself discrimination. Science 1987;235:865.

Author Affiliation

University of Massachusetts Medical School, 55 Lake Avenue North, Worcester, MA 01655, USA.

Attempts to Understand the Effects of Anti-Class II Monoclonal Antibodies on Responder T Cells

P. Gladstone and C. Spooner

One method for characterizing class II molecules as antigen restriction elements uses anti-class II monoclonal antibodies (MAbs) to block HLA-restricted T-cell proliferation in response to nominal antigens or alloantigens (1). It is generally assumed that the antibodies block the stimulator's class II antigens. The effects of anti-class II MAb on responder cells has been less explored. While the role of Ia antigens on stimulator cells is essentially that of a passive carrier, on responder cells, Ia might serve to transduce signals intracellularly. For example, in murine B cells, anti-Ia antibodies are known to affect the subcellular localization of protein kinase C by elevation of cAMP levels (2). Our studies of MAb-induced functional changes in peripheral blood mononuclear cells (PBMCs) suggest that class II antigens can act as signal transducers and affect responder T cells via two

Immunobiology of HLA
Volume II

Table 1.

A: Inhibition of PA by HLA and Non-HLA Antibodies					
Donor 1			Donor 2		
MAb (Specificity)	PA	% Inhibition	MAb (Specificity)	PA	% Inhibition
–	1.70 ±0.004	–	–	.59±0.11	
HB10a (DR+DP)	1.17 ±07	31	HB10a (DR+DP)	.31±0.03	48
			G17-2 (CD4)	0.61±0.01	−3
VI.15 (DR)	0.484±0.02	72	F13 (CD14)	0.82±0.04	−39
P15.1 (DR)	1.53 ±0.05	10	60.3 (CD18)	0.62±0.02	−5
L243 (DR)	0.709±0.058	58	9.3 (CD28)	0.55±0.04	7
33.1 (DQ)	1.12 ±0.05	33	D8 (anti-PA)	0.03±0.01	95

B: Inhibition of PA by HB10a				
	Donor	No Ab	HB10a	% Inhibition
1	SAz	1.51±0.02	0.80±0.024	47
2	BKe	0.59±0.04	0.57±0.04	5
3	DSm	1.64±0.056	0.86±0.113	47
4	FDo	1.69±1.31	0.67±0.026	60

Immediately prior to assay, cells were washed twice with HB103 to remove secreted enzyme and inhibitors. For PA assay, cells were suspended in 0.3 ml HB103 plus 50 μl reaction buffer (60 mM Tris, pH 7.8, plus saturating amounts (5 to 10 μg of plasminogen). Plasminogen was purified from plasma on a lysine sepharose affinity column. The mixture was incubated at 37°C for 1.5 hours. The plasmin generated was assayed by addition of 250 μl of the synthetic chromogenic substrate S-2251 (1 mg/ml in 120 mM lysine buffer, pH 8.9) for 2 hours. The reaction was stopped by chilling and addition of 3% FBS; OD_{405} was read immediately on a Beckman spectrophotometer. A blank consisting of all reactants except cells was subtracted.

components – a cell surface protease and interleukin-1 (IL-1). Our investigations began as an attempt to better understand the relationship between HLA class I and II antigens and class III genes. Class III products are soluble serum proteins; C2 and factor B are complement serine proteases. Given the highly conserved linkage between class III genes and the other MHC genes, a study of a potential role for serine proteases in immune cell function was begun. One particular serine protease, plasminogen activator (PA), is a product of monocytes/macrophages (3). Using our anti-PA MAb to block the proteases' enzymatic activity, we found that anti-CD3-mediated proliferation was inhibited by 40 to 90% (4). This implies that PA is required for some forms of T-cell proliferation. A cell surface receptor for PA has been detected on monocytes, and we have found PA activity on T cells as well (5). To explore the relationship between membrane-bound PA and other surface molecules, an enzyme assay was used to measure PA activity on cells coated with MAb against various known surface antigens. The only antibodies that significantly reduced surface PA activity were anti-HLA antibodies. Because previous work by us had suggested a possible relationship between PA and IL-1, the effect of this same MAb on IL-1 release was explored, and striking effects have been seen. Thus, we report here two possibly related roles of Ia antigens – control of IL-1 release, and regulation of cell PA activity. The involvement of Ia antigens in IL-1 release has been previously suggested in murine and human cells (6,7).

Peripheral blood mononuclear cells were isolated from blood and assayed for PA as described in Table 1. Using this procedure, cell-associated PA activity is readily detectable on 2.5 × 10^6 PBMCs, although treatments were usually performed at 5 × 10^6/well. Cells incubated with purified MAb against HLA class II antigens showed dramatic alterations in cell-associated PA activity. A variety of other MAbs have been tested for inhibition of cell surface PA on PBMCs. These include antibodies against T-cell antigens – CD3(G19-4), CD4(G17-2, G19-2), CD5(10.2), and CD28(9.3) – and against B cell antigens – membrane immunoglobulin and CD20. Two antibodies from the LFA-1 family – 60.3 (anti-CD18) and 60.1 (anti-CD11) – were tested, as well as F13, an antibody against the monocyte antigen CD14. All antibodies were previously characterized (8). None of these non-HLA antibodies showed any significant inhibition of PA; the CD14 MAb frequently showed some increased PA activity. By contrast, most anti-class II MAbs as well as one of four class I MAbs (H1DE) produced significant PA inhibition. Representative results with several antibodies in two individuals are shown (Table 1A). Among the inhibitory monomorphic class II antibodies, are HB10a, which reacts with separated β chains of DR plus DP and 33.1 (Marti, G.E.) which reacts with DQ molecules. The inhibition is, therefore, not restricted to MAbs against any single class II molecule.

The degree of inhibition varies with the donor, as shown in Table 1B, with reductions in PA activity ranging from 5 to 60%. Over a 2 year period, PA inhibition studies have been performed on at least 50 individuals, 25 of whom have been tested twice; 10 of these have been tested three or more times. Complete results will be published elsewhere (Gladstone et al., in preparation). The MAb HB10a has different effects on different individuals, ranging from 93% inhibition to approximately 40% stimulation. Repeated testing of a single individual shows similar results when tests are repeated within 2 to 3 months of each other; tests separated by 6 to 9 months, however, may vary considerably. The susceptibility to anti-class II inhibition, therefore, is a property that can change in an individual, but only slowly. In general, females are more sensitive to inhibition than males, although the distributions do overlap. One explanation for these interindividual differences in response to DR MAb was tested, namely the possibility that these antibodies were not actually reacting with all individuals was assayed. Total PBMCs from 10 donors were stained with FITC-conjugated HB10a and VI.15. Several of the donors were insensitive to anti-Ia-dependent inhibition, but all showed positive staining reactions in both monocyte- and lymphocyte-sized populations (data not shown). The sensitivity to inhibition is also not merely a function of the percentage of monocytes among PBMCs.

One possible explanation for the interindividual variability in Ia-inhibitability is that there exist several different, independently varying, cell populations bearing PA, which differ in their response to HLA antibodies. Both monocytes/macrophages and lymphocytes do bear PA (3,9). Figure 1A shows simultaneous experiments on two individuals previously identified as showing distinctly different levels of Ia-inhibitability. The tightly adherent cells from each individual were separated by one round of plastic adherence, incubated with antibodies, and assayed. Neither adherent population shows Ia inhibition (Fig. 1B). However, the nonadherent populations of both individuals show significant Ia sensitivity. Thus, one reason for differences in sensitivity is the relative amount of PA activity derived from nonadherent cells in different individuals. As just mentioned, some individuals show stimulation of PA activity, particularly by VI.15 and L243. This stimulation is abolished when monocytes are completely removed. It is likely that this stimulation is Fc-dependent and is similar to the PA stimulation by nonspecific immune complexes previously reported (11). Consistent with this, three to four-fold stimulation by L243 seen in three individuals is lost when $F(ab)'_2$ fragments are used.

In parallel with the above experiments, cells were studied for the effects of anti-class II MAbs on secretion of IL-1. These experiments were performed because of our indications of a possible relation between PA and IL-1. In previous experiments on T-cell proliferation, we had found that inhibition of PA by anti-PA MAbs altered levels of IL-1 secretion. These results suggested that the inhibition of PA by class II MAbs might be mediated by effects on IL-1. Accordingly, IL-1 release was measured in nonactivated cells treated with anti-class II MAbs. In some individuals, certain antibodies (VI.15) caused strong stimulation of IL-1 release (e.g., donor DH tested and retested 6 months later) (Table 2A). As shown in Table 2B, the MAb HB10a shows stimulation only as an $F(ab)'_2$ and not as an intact IgG. This same difference is seen for the effects of HB10a on PA inhibition. While individual differences (such as those shown in Table 2A) do not correlate with DR type, it is possible they are due to polymorphisms for Fc receptors. The previous reports of anti-class II antibody effects on IL-1 release had

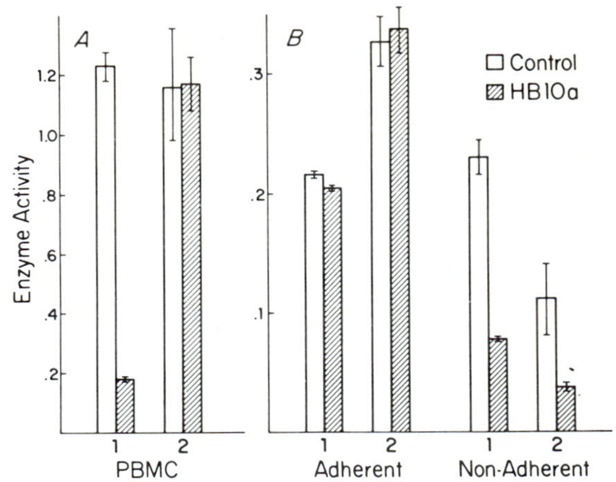

Figure 1. Adherent cells were separated by one to two cycles of adherence on plastic dishes. Adherent and nonadherent populations were assayed for enzyme activity. Adherent cells were assayed in situ to avoid artifacts caused by detachment procedures. Nonadherent cells were not to be used until two cycles of adherence were done. Cells were washed two times in Hank's solution, counted, resuspended in the serum-free medium (HB103; Hana Biologics, California), and distributed at concentrations of 2.5 to 7.5 × 10^6/well in 24-well Costar tissue culture plates. Cells were incubated with purified anti-class II antibodies described previously (10). MAbs were stored in PBS without azide at approximately 1.0 mg/ml. Antibodies were added to cells in 1 ml HB103 at final concentrations of 1.25 μg/ml and incubated for 15 to 18 hours at 37°C.

shown inhibition in murine cells (6) and stimulation in human cells (7); the present results show either positive or negative effects on IL-1 release, depending on the donor used. This may reconcile the conflicting findings in other published studies (6,7).

In summary, anti-HLA antibodies affect the activity of cell surface PA, a required participant in monocyte-dependent T-cell proliferation. These antibodies also influence, perhaps by a related mechanism, the release of IL-1 from a population other than tightly adherent monocytes. Some of these effects, particularly stimulation of PA by some MAb stimulation, are Fc-dependent, probably by interaction with Fc receptors. In light of our results, in vitro blocking experiments with anti-class II antibodies should be interpreted with caution. This caution is also important in interpreting results of in vivo use of antibodies. Reports of induction of suppressor cells in mice by anti-Ia MAbs (13) may also be explained by the data presented here.

Acknowledgment. We thank Susan Engardt and Mary Gallagher for helping to perform the experiments; Drs. D. Michael Strong, Jeffrey Ledbetter, Theta Tsu, Karen Nelson and Susan Radka for helpful advice; and Holly Chase for secretarial assistance.

Table 2. Results of ^{3}H-Thymidine Incorporation in a Standard Mouse Thymocyte Assay for IL-1 (12)*

A: Induction of IL-1 Release				
Donor	No Ab	HIDE†	HB10a	V1.15
DH	1.9±4	2.9±2	2.9±0.1	23.2±3.2
FJ	6.9±0.9	4.5±0.4	2.3±0.3	5.5±0.9
DH	22.1±2.3	–	32.0±5.8	70.2±7.4

B: IL-1 Release Induced by IgG and F(ab)'$_2$				
Donor	No Ab	HB10a IgG	HB-F'$_2$‡	V1.15
TS	28.6±4.2	30.9±4.1	127± 6.3	79.8±6.5
SK	77.6±9.5	59.3±10.2	180±10.7	164 ±4.5

*Initially, parallel cultures of cell in media with 5% human Ab serum were used as a source of secreted IL-1. In more recent experiments, Il-1 was assayed from the same serum-free supernatants of cells used for PA assay and comparable results were seen.
†HIDE is a class I monomorphic MAb. The others are DR monomorphic MAbs.
‡HB-F'2 refers to F(ab)'$_2$ of HB10a prepared by pepsin digestion at pH 4.5 and purification on a protein A-Sepharose column to remove Fc fragments.

References

1. Bergholtz BO, Thorsby E. HLA-D restriction of the macrophage dependent response of immune human T lymphocytes to PPD in vitro: inhibition of anti-HLA DR antisera. Scand J Immunol 1978;8:63.
2. Chen ZZ, Cambier J. Transmembrane signaling through B cell MHC Class II molecules: Anti-IA antibodies induce PK-C translocation. J Immunol 1986:in press.
3. Unkeless JC, Gordon S, Reich E. Secretion of plasminogen activator by stimulated macrophages. J Exp Med 1974; 139:834.
4. Gladstone P, Leahy M, Spooner C. The role of plasminogen activator in regulating T cell proliferation and γ IFN production. In: Proceedings of the 18th Leukocyte Culture Conference, Berlin: Walter de Gruyter, 1988, p 279.
5. Vassalli J-D, Baccino D, Belin D. A cellular binding site for the M_r 55,000 form of the human plasminogen activator, urokinase. J Cell Biol 1985;100:86.
6. Duram SK, Schmidt JA, Oppenheim JJ. Interleukin-1: an immunological perspective. Ann Rev Immunol 1985;3:263.
7. Palacios R. Monoclonal antibodies against human Ia antigens stimulate monocytes to secrete interleukin 1. Proc Natl Acad Sci USA 1985;82:6652.
8. Reinherz EL, Haynes BF, Nadler LM, Bernstein ID (eds). Leukocyte Typing II. Volumes 1 and 2. New York: Springer Verlag, 1986.
9. Goldfarb RH, Timonen T, Herberman RB. Production of plasminogen activator by human natural killer cells. J Exp Med 1984;159:935.
10. Gladstone P, Fueresz L, Pious D. Gene dosage and gene expression in the HLA region: evidence from deletion variants. Proc Natl Acad Sci USA 1982;79:1235.
11. Takemura R, Werb Z. Regulation of elastase and plasminogen activator secretion in resident and inflammatory macrophages by receptors for the Fc domain of immunoglobulin G. J Exp Med 1984;167:152.
12. Mizel SB, Oppenheim JJ, Rosenstreich DL. Characterization of lymphocyte-activating factor (LAF) produced by the macrophage cell line, P388D. J Immunol 1978;120:1497.
13. Perry LL, Green MI. Monoclonal anti-IA antibodies induce suppressor cells in mice. J Exp Med 1982;156:480.

Author Affiliation

Genetic Systems and Oncogen, Seattle, WA, USA

Inhibition by Anti-HLA Class II Monoclonal Antibodies of OKT3-Induced T Cell Proliferation: Analysis of the Mechanism

G. Ruggiero, C. Manzo, S. Fontana, E. Cosentini, G. Scala, G. Pirozzi, S. Ferrone, and S. Zappacosta

HLA class II molecules play a crucial role in immune responses, mainly in the mechanisms of antigen-specific interactions between T cells and monocytes. Recent reports have shown that monoclonal antibodies (mAbs) directed against the CD3-antigen receptor complex on T cells lead to T cell proliferation. This mitogenic effect on T cells is accessory cell dependent and requires an interaction of the anti-CD3 mAb (OKT3) with Fc receptors on monocytes, secretion of interleukins, and coordinate expression of interleukin receptors (1). It has been previously reported that mAbs to HLA-DR and -DP molecules can inhibit the OKT3-induced T lymphocyte proliferation (2). In this study, we have used a panel of mAbs to monomorphic or polymorphic determinants of

Immunobiology of HLA
Volume II

HLA class II molecules to clarify the molecular mechanisms underlying the inhibitory activity of anti-HLA class II mAbs on T cell proliferation.

In preliminary experiments, we observed that the anti-HLA-DR, -DP Q5/13 and CR10-343 mAbs displayed a high inhibitory activity on the proliferation of unfractionated peripheral blood leukocytes (PBL) stimulated with OKT3 mAb (2ng/ml). This inhibitory activity was also shared by other mAbs such as the anti-HLA-DQw1 mAb KS11, the anti HLA-DR,-DP mAb CR12-356, the anti-HLA-DQw3 mAb KS13 and the anti-HLA-DP1 mAb B7/21. The observed inhibition was neither mediated by a cytotoxicity effect nor by an interaction of the tested mAbs with the Fc receptors on monocytes, since $F(ab')_2$ fragments were equally effective. Kinetics experiments showed that a nearly complete inhibition was achieved when the antibodies were added within eight hours from the beginning of the cultures with the inhibition falling sharply thereafter. This suggested that the anti-HLA class II mAbs might interfere with early events of T cell activation and proliferation. To test this possibility, we first assessed that both the monocyte accessory cell population and the purified T lymphocytes were targets of the inhibitory activity of the anti-HLA class II mAbs tested. In fact, purified monocytes, previously pretreated with the anti-class II mAbs, were unable to function as accessory cells for the OKT3-induced T cell proliferation. Moreover preactivated T cells, which expressed class II molecules, were also inhibited when treated with the Q5/13 or the CR10-343 mAbs.

Next, we tested whether the anti-class II mAbs could inhibit the secretion of IL1 by the monocytic populations. Purified monocytes were cultured in the presence of LPS (10 μg/ml) and the anti-class II mAbs or in the presence of an anti-arsonate control mAb (R16.7). Results from these experiments showed that all the mAbs tested, with the exception of the CR12-356 mAb, inhibited the IL1 secretion by monocytes (Table 1). In further experiments, we tested whether the secretion of IL2 by T lymphocytes was also affected by the anti-class II mAbs. Peripheral blood leukocytes were cultured in the presence of OKT3 mAb (2 ng/ml) and anti-class II mAbs for 14 hours and the supernatant assayed for IL2 activity. We observed a nearly complete inhibition of the IL2 secretion of T cells by the anti-class II mAbs. Moreover, addition of IL2 reversed the inhibition of the OKT3-induced T cell proliferation exerted by the anti-class II mAbs, while exogeneous IL1 was only partially effective. The inhibition by anti-class II mAbs might also be the result of a reduced expression of the IL2 receptor on responding T cells, a critical event in the proliferation of T cells (3). To test this possibility, PBL were cultured for three days in the presence of OKT3 and optimal concentrations of the inhibitory mAbs. Cells were then stained with FITC-conjugated anti-IL2R mAb. A consistent decrease in the IL2R expression was observed when cells were cultured in the presence of the anti-class II mAbs.

Several reports have recently indicated that class II molecules regulate the activation and proliferation of T lymphocytes induced by antigens. However, the molecular mechanisms of this regulatory role are poorly understood. To address this issue, we have used a well-characterized panel of mAbs directed against monomorphic or polymorphic determinants of HLA class II molecules and inhibitory of the pathway of T cell proliferation operating through the CD3-T cell receptor complex. The above mentioned results suggest that HLA class II molecules regulate T cell proliferation by acting on both monocyte accessory cells and responder T cells. In particular, our results indicate that HLA class II molecules regulate IL1 as well as IL2 secretion and the expression of IL2R, further arguing in favor of the central role of HLA class II molecules in immune regulation.

Table 1. Effect of Anti-HLA Class II mAb on IL1 Production by Macrophages

LPS	mAb (ng/ml)	IL1 Units/ml	IL1 % Inhibition
–	–	0.1	–
+	–	11.5	–
+	Q5/13 (200)	6.8	48.4
+	Q5/6 (200)	4.3	67.2
+	CR12-356 (200)	15.7	0
+	CR10-343 (200)	7.7	41.0
+	$F(ab')_2$ of Q5/13 (800)	4.4	67.2
+	345.134 (200)	16.0	0
+	R16.7 (200)	13.1	0

Acknowledgment. Supported in part by CNR Grants (Progetti Finalizzati "Oncologia" and "Ingegneria Genetica"), by a Grant from Associazione Italiana per la Ricerca sul Cancro and by NIH Grant AI21384.

References

1. Meuer SC, Hogdon JC, Hussey RE, Protentis JP, Schlossman SF, Reinherz EL. Antigen-like effects of monoclonal antibodies directed at receptors on human T cell clones. J Exp Med 1983;158:988–993.
2. Akijama Y, Zicht R, Ferrone S, Bonnard GD, Herberman RB. Effect of monoclonal antibodies (MoAb) to class I and class II HLA antigens on lectin and MoAb OKT3-induced lymphocyte proliferation. Cell Immunol 1985;91: 477–491.
3. Robb RJ, Munk A, Smith KA. T cell growth factor receptors. Quantitation, specificity and biological relevance. J Exp Med 1981;154:1455–1474.

Author Affiliations

G. Ruggiero, S. Fontana, E. Cosentini, G. Scala, S. Zappacosta, Medical School, University of Naples; *C. Manzo, G. Pirozzi*, Cancer Institute "Pascale", Naples, Italy; *S. Ferrone*, New York Medical College, Valhalla, NY, USA

Change in the Fine Specificity of a Human Cytotoxic T-Cell Clone Under α-Interferon Treatment

Jean-Marc Zini, Marie-Véronique Clément, Patrick Haddad, Alain Haziot, Annie Soulié, Dominique Charron, Laurent Degos, Danièle Mathieu-Mahul, and Marilyne Sasportes

Abstract: We report in this work the case of a helper-independent CD4+, cytotoxic T-cell clone, produced within a family, which could recognize a new type of hybrid molecule (DRβ 3-DQα) on the specific target cell. Moreover, when treated by recombinant α-interferon (rIFN), the T-cell clone, called BJ1, was found to change its reactivity towards the family cells, used as targets in a cell-mediated cytotoxic assay (CML). Coincident with that event, it appeared that rIFN-α treatment of BJ1, induced a new T-cell receptor (TCR-α) gene rearrangement. This is the first observation of a modification in the recognition of the target molecule by a specific cytotoxic T-cell clone under interferon treatment.

Table 1. Cytolytic Activity of the T-Cell Clones BJ1 Before and After Interferon Treatment Within the Family Per

Family Per	HLA Haplotypes	% of Specific Lysis of BJ1	% of Specific Lysis of BJ1a
F.	a/b	2%	23%
M.	c/d	40%	35%
S.1	a/c	37%	33%
S.2	a/c	48%	45%
S.3	a/R1	3.5%	2%
S.4	a/d	−2%	0%
S.5	a/d	1%	3%
S.6	b/d	8%	4%
S.7	b/d	5%	2.5%
S.8	a/R2	82%	75%

a=A11;B38;Cw–;Dw9;DR14;DRw52; DQw1;DP1
b=A29;B44;Cw–;Dw7;DR7;DRw53; DQw2;DP–
c=A2;Bw51;Cw–;Dw7;DR7;DRw53; DQw2;DP1
d=A26;B8;Cw7;Dw3;DR3;DRw52; DQw2;DP4
R1=A2;Bw51;Cw–;Dw3;DR3;DRw52; DQw2;DP4
R2=A26;Bw51;Cw7;Dw7;DR7; DRw53; DQw2;DP1

BJ1 combination: Responder: S4 Stimulator: S8

Method: Peripheral blood mononuclear cells (PBMC) from donor S4 were sensitized in a mixed lymphocyte culture for 6 days with PBMC from his sibling S8 and restimulated 6 days in the presence of stimulating cells. The primed cells were then cloned by micromanipulation. Clonal cell population was tested for cell-mediated cytotoxicity using a standard chromium release assay, in which cloned cells (15,000) and ^{51}Cr-labeled targets (500) were mixed in a final volume of 0.15 ml. All the target cells were EBV-transformed cells line (LCL). For recombinant interferon-α treatment, cells were removed from the crude IL2 containing medium and cocultured with irradiated S8 LCL in RPMi 1640 medium containing 15% pooled human AB serum, recombinant IL2 (5 U/ml) (Glaxo, Geneva, Switzerland), and rIFN-α (Unicet, Levallois, France). The cells from the same clone, in the control experiment were cultured in parallel but without rIFN-α. a = A11,B38,Cw-,Dw9, DR14,DRw52,DQw1,DP1; b = A29,B44,Cw-,Dw7,DR7, DRw53,DQw2,DP-, c = A2,Bw51,Cw-,Dw7,DR7,DRw53, DQw2,DP1; d = A26,B8,Cw7,Dw3,DR3,DRw52,DQw2, DP4; R1 = A2,Bw51,Cw-,Dw3,DR3,DRw52,DQw2,DP4; R2 = A26,Bw51,Cw7,Dw7,DR7,DRw53,DQw2,DP1; BJ1 combination: Responder: S4; Stimulator: S8.

The mature T-cell clone BJ1 was found to proliferate and kill the specific target (1). The study of BJ1 cytotoxic activity was extended to the family members cells and 72 Epstein-Barr virus (EBV)-transformed lymphoblastoid B-cell lines, from the Tenth International Histocompatibility Workshop (Tables 1 and 2). BJ1 was able to kill all DR7+ cells, except the father cells and the sibling cells bearing the b HLA paternal haplotype, although the same class II products (DR7, DRw53, DQw2) were shared by the b paternal haplotype and the sensitizing maternal c haplotype, according to serology. One should notice that the DR4+ lymphoblastoid cell lines were not lysed by BJ1. Blocking experiments were also performed with the previously mentioned targets, and the workshop anti-class II monoclonal antibodies. Monomorphic anti-DR, and polymorphic anti-DR7 and anti-DRw53 monoclonal antibodies were able to inhibit the cytotoxic activity of BJ1 (Table 3).

In order to further determine the specificity recognized by BJ1 on the specific target cells, comparative two-dimensional polyacrylamide gel electrophoresis profiles (Fig. 1) from both parental cells were analyzed, using the monoclonal antibody TAL1B5 to immunoprecipitate class II molecules. There was no detectable expression of the DRβ3 chain in the father cells, compared to the mother cells.

The CML and biochemical data led us to consider that a hybrid molecule (DRβ3-DQw2α), present on all DR7+ cells and absent from the paternal and DR4+ cells as expected, could be the target structure recognized by BJ1 (2,3). The characterization of this molecule is now in progress.

We have previously shown that addition of recombinant interferon α or γ to precursor cytotoxic T-cell clones (CTL-p) resulted in acquisition of specific lytic activity (4). We have also treated the mature cytotoxic T-cell clone BJ1 by interferon. As previously reported (1,2), the same TCR-β rearrangement was found in BJ1 cells before and after IFN treatment, which indicated that they were true clones.

Moreover, we observed that, in addition to one rearranged TCR-α allele, recombinant interferon treatment

Immunobiology of HLA
Volume II

appeared to have induced rearrangement of the second TCR-α allele (Fig. 2). The question raised, was, therefore, to determine if the IFN-treated T-cell clone had maintained its original antigen specificity. For that purpose the protocol described above (CML, blocking experiments) was also carried out with the IFN-treated clone (Tables 1–3).

No difference could be noted with the panel of B-cell lines whether or not BJ1 was treated by rIFN. By contrast, when tested within the family, BJ1 treated by rIFN became able to lyse the father cells and the sibling cells bearing the b paternal haplotype, which demonstrated a modification in the pattern of its reactivity (Table 1).

Taken together, these data raise different questions. The first one concerns the binding affinity of the T-cell clone for its target, which could be increased under IFN treatment, following an increase in the expression of either IFN receptors or differentiation antigens like CD4, or adhesive molecules like LFA-1, or class II molecules (6). Preliminary experiments performed with precursors of CTLs and IFN-induced CTL do not exclude this possibility. The second concerns the rearranged TCR-α gene apparently induced under rIFN treatment. For that purpose, biochemical experiments using isoelectrofocusing and appropriate anti-TCR antibodies could be useful to distinguish between different functional TCR-α chains (7). Another approach, has been initiated to establish cDNA libraries from IFN-treated and untreated cytotoxic T-cell clones. This attempt will use a constant segment of TCR-α gene as a probe to screen the libraries and identify different functional TCR-α gene transcripts.

Table 2. Cytotoxicity of the T-Cell Clone BJ1 Untreated and Treated by rIFN on a Panel of 72 EBV Cell Lines and Percentage of Specific Lysis at the Effector/Target Cell Ratio of 30:1

Homozygous HLA DR EBV Cell Lines	Number of Cells	% of Specific Lysis of BJ1	% of Specific Lysis of BJ1 Treated by IFN
DR 1	7	<5%	<5%
DR 2	10	<5%	<5%
DR 3	6	<5%	<5%
DR 4	11	<5%	<5%
DR 5	11	<5%	<5%
DR w6	11	<5%	<5%
DR w8	8	<5%	<5%
DR 9	2	<5%	<5%
DR7 BM		37%	35%
DR7 PLH		27%	38%
DR7 LBUF		53%	52%
DR7 IBW 9		78%	53%
DR7 MOU		23%	34%
DR7 PITOUT		53%	51%
DR7 DBB		33%	39%
DR7 BER		42%	37%
DR7 WIN		40%	27%
DR7 LBUF		38%	35%

SD <10% of maximal lysis. Only DR7+ EBV cells were lysed whether BJ1 was treated by rIFN or not.

References

1. Chen LK, Tourvielle B, Burns GF, Bach FH, Mathieu-Mahul D, Sasportes M, et al. Interferon: A cytotoxic T lymphocyte differentiation signal. Eur J Immunol 1986;16:767.
2. Lotteau V, Teyton L, Burroughs D, Charron D. A novel HLA class II molecule (DR alpha-DQ Bêta) created by mismatched isotype pairing. Nature 1987;329:339.
3. Young JAT, Wilkinson D, Bodmer WF, Trowsdale. Sequence and evolution of HLA-DR7 and DR-w53 associated chain genes. Proc Natl Acad Sci USA 1987;84:4929.
4. Chen LK, Mathieu-Mahul D, Bach FH, Dausset J, Bensussan A, Sasportes M. Recombinant interferon can induce rearrangement of T-cell antigen receptor alpha chain genes and maturation to cytotoxicity in T-lymphocyte clones in vitro. Proc Natl Acad Sci USA 1986;83:4887.
5. Sim GK, Yague J, Nelson J, Marrak P, Palmer E, Augustin A, et al. Nature 1984;312:771.
6. Reinherz EL, Mener S, Shlossman SF. 1982. Antigen recognition by human T lymphocytes is linked surface expression of the T3 molecular complex. Cell 1982;19:821.
7. Borst J, De Vries E, Spits H, De Vries J, Boylston AW, Matthews EA. Complexity of T cell receptor recognition sites for defined alloantigens. J Immunol 1987;139:1952.

Author Affiliations

Jean-Marc Zini, *Marie-Véronique Clément*, *Patrick Haddad*, *Alain Haziot*, *Annie Soulié*, *Laurent Degos*, and *Marilyne Sasportes*, Unité Inserm U 93, Hôpital Saint-Louis, 75475 Paris, France; *Dominique Charron*, Institut des Cordeliers, 75006 Paris, France; and *Danièle Mathieu-Mahul*, Unité Inserm 301, Hôpital Saint-Louis, 75475 Paris, France

Table 3. Blocking of the Cytolytic Activity of T-Cell Clone BJ1 Before and After Interferon Treatment by Monoclonal Antibodies

Antibodies	Specificity	% of Lysis Inhibition of S8 by BJ1	% of Lysis Inhibition of S8 by BJ1a	% of Lysis Inhibition of F. by BJ1a
SFR 16 DR7M	DR7	48%	58%	54%
PL 3	DRw53	100%	100%	100%
GSP 4-1	DR	78%	71%	80%
B 7/21	DP	0%	0%	0%
I A3	DQ	0%	0%	0%

Methods: ^{51}Cr-labeled S8-LCL and father-LCL were used as targets. Monoclonal antibodies used at the dilution 1:100 were added to the target cells 30 min before mixing them with the cytotoxic clones. Percentage of specific lysis is given at the 30:1 effector/target ratio (SD <10% of maximal lysis).

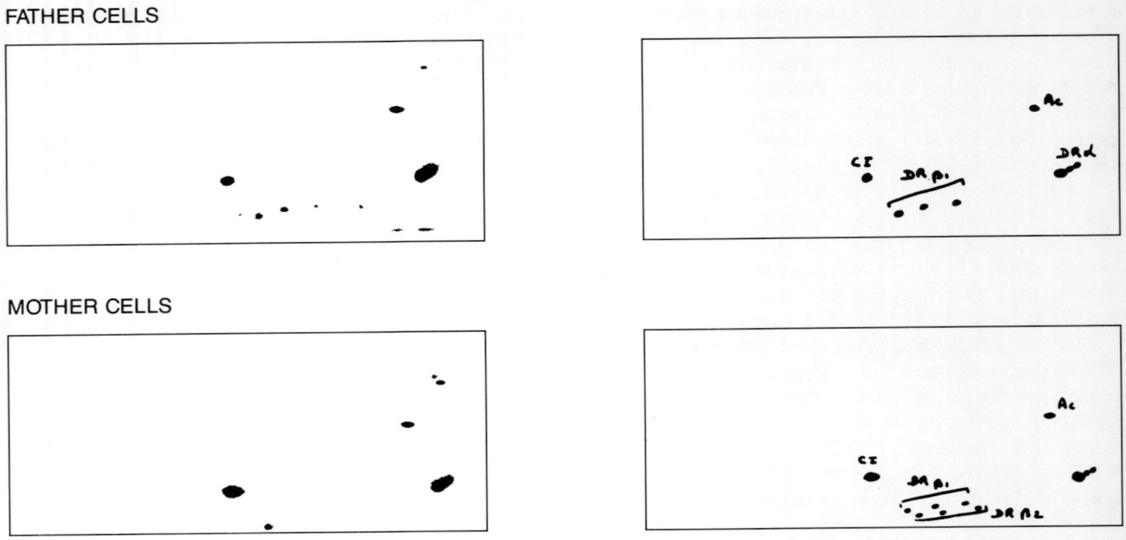

Figure 1. Biochemical studies of the father and the mother cells. Methods: The proteins are immunoprecipitated by TAL1B5, a monoclonal antibody that recognizes the HLA DR alpha chain. 10^7 cells were cultured with 0.1 mCi of 35S-methionine for 5 h. The membranes were extracted with 0.5% Nonidet P-40. The two-step precipitates were run in 2D-PAGE with NEPHGE for the first dimension and SDS-Page (10% acrylamide gel) in the second dimension.

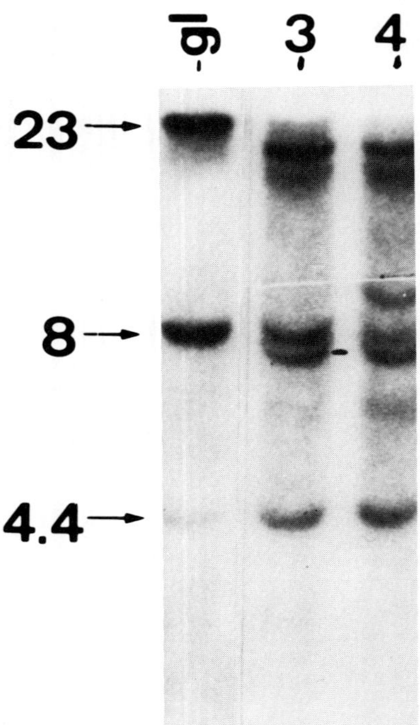

◄

Figure 2. The rearrangement of TCR-α sequences in BJ1 treated and untreated by rIFN. Methods: 10 μg of EcoR1-digested genomic DNA from an EBV-transformed B-cell line as germline control (gl), from BJ1 (3) and BJ1 treated by rIFN (BJ1a) (4) were electrophoresed on a 0.6% agarose gel and blotted onto nylon membranes (Compass, GENOFIT); The blot was hybridized under stringent conditions, with probes radiolabeled by hexamer priming to a specific activity of 10^9 cpm/μg. Hybridization with the pGA5, a cDNA corresponding to the α chain of the T-cell receptor expressed in Jurkat cells (from E. Palmer, Ref. 5). The 8-kb EcoR1 band corresponds to the variable region of the TCR-α; on BJ1 two bands are present: one of 8-kb germline and another of 7 kb; on BJ1a a new band of 11 kb appears coincident with the decrease of the 8 kb germline; on the other hand the 7 kb band is maintained.

Section XIV
HLA and Disease Associations

Allelic Sequence Variation in the HLA-DQ Region: Implications for Localization of Serological Specificities and for Susceptibility to Autoimmune Disease

Glenn T. Horn, Teodorica L. Bugawan, Christopher M. Long, and Henry A. Erlich

Abstract: The pattern of nucleotide sequence variation in the HLA-DQ genes has been analyzed with respect to the known serologically defined HLA types and to susceptibility to insulin-dependent diabetes (IDDM). Thus far, 8 allelic variants have been found for the DQα gene, and 12 alleles have been seen for DQβ. The DQw1 specificity correlates best with a related set of DQα variants, whereas the DQw2, DQw3, and DQ(blank) specificities correlate with alleles of DQβ. A pattern of amino acid substitutions at position 57 in the DQβ protein appears to reflect the degree of susceptibility to IDDM for the different HLA haplotypes. This portion of DQβ also shows significant homology with several human viruses, suggesting that it may be a preferred site for molecular mimicry.

The class II loci of the human major histocompatibility complex encode surface glycoproteins which present antigens to the helper T lymphocytes of the immune system (1). The genes for the α and β subunits of these proteins are clustered in three regions of the genome, designated HLA-DR, DQ, and DP. These HLA proteins are highly polymorphic, possibly as a result of their central role in the defense against pathogenic organisms (2). Detailed analysis of this polymorphism could assist in the understanding of histocompatibility, antigen presentation, immune response, and the autoimmune diseases (such as IDDM) that exhibit linkage to certain HLA types (3).

The development of the polymerase chain reaction (PCR) technique (4) has greatly reduced the effort required to determine allelic sequence variation (5). We applied this method to analyze the extent of variation in the HLA region. We examined alleles derived from the more frequent HLA haplotypes, from rare haplotypes, and from patients with autoimmune disease. This report summarizes our results for the HLA DQα and DQβ alleles, and compares allelic variation for all of the published sequences. A full description of this work will appear elsewhere (6).

Based on sequences determined from over 20 sources, there appear to be eight allelic variants of the DQα gene (Fig. 1A). The DQA1 alleles are closely related to each other and are more divergent from the other, DQα alleles. As a group, the DQA2, DQA3, and DQA4 alleles also appear to be most closely related to each other, and less homologous to the DQA1 alleles. The major variation in DQα tends to cluster in a region around codon 50.

The DQβ genes from over 50 sources have been examined, and grouped into 12 different alleles (Fig. 1B). Again, the alleles derived from cells typed as DQw1 (and designated DQB1) show the most homology to each other, particularly near positions 53 and 86 of the mature protein. The DQB2 allele from the DR3 and DR7 haplotypes appears to be the most distinct. Overall, sequence variation within the DQβ alleles is more dispersed than in DQα.

There are three major serologically defined types for the DQ proteins, designated DQw1, DQw2, and DQw3. Also, a fraction of homozygous cell lines type as DQ(blank). The segments of the DQα or DQβ subunits that best correlate with these types might be derived either from a direct comparison of the alleles, or by the analysis of haplotypes that have less frequent combinations of alleles. By sequencing alleles from a number of these "recombinant" haplotypes (6), we hypothesize that the DQB2 allele confers the DQw2 type and that the DQB3 alleles confer the DQw3 type. The DQα alleles do not correlate as well with the serology. For example, the DQA4.1 allele has been found both in cell lines typed as DQw2 or as DQw3. All four of the DQ(blank) haplotypes thus far analyzed contain the same distinctive DQB4 allele, suggesting it may be conferring this specificity. Since all of the DQw1 haplotypes analyzed contain both a DQA1 and DQB1 allele, we are not able to use this method to map the DQw1 serological specificity. However, the close homology of the DQA1 alleles suggests that they may confer the DQw1 type, although segments of DQβ near positions 53 and 85 also correlate with this specificity. These hypotheses are discussed in detail in reference 6.

The HLA-D region as a whole shows genetic linkage to susceptibility to IDDM (3), with the DR3 and DR4 types being increased in patients and the DR2 type being decreased. More recently it was shown that within DR4, the DQβ subtypes 3.1 and 3.2, which differ by four amino acid residues, are associated with differential susceptibility, with the DR4-DQw3.2 haplotype found most frequently in patients (7). The alleles associated with the highest susceptibility to IDDM (i.e., DQB3.2 and DQB2) contain an alanine at position 57, whereas the least susceptible alleles (i.e. DQB3.1 and DQB1.4) contain an aspartate. This pattern of variation, where the lack of aspartate correlates with IDDM susceptibility, also extends to many of the other DQβ alleles (6,8) as well as to some DRβ alleles (9) and to the diabetes-susceptible NOD mouse (10). This correlation does not, however, explain why the DQB2 allele found in many DR7 haplotypes (and identical to that found in DR3) does not confer increased susceptibility.

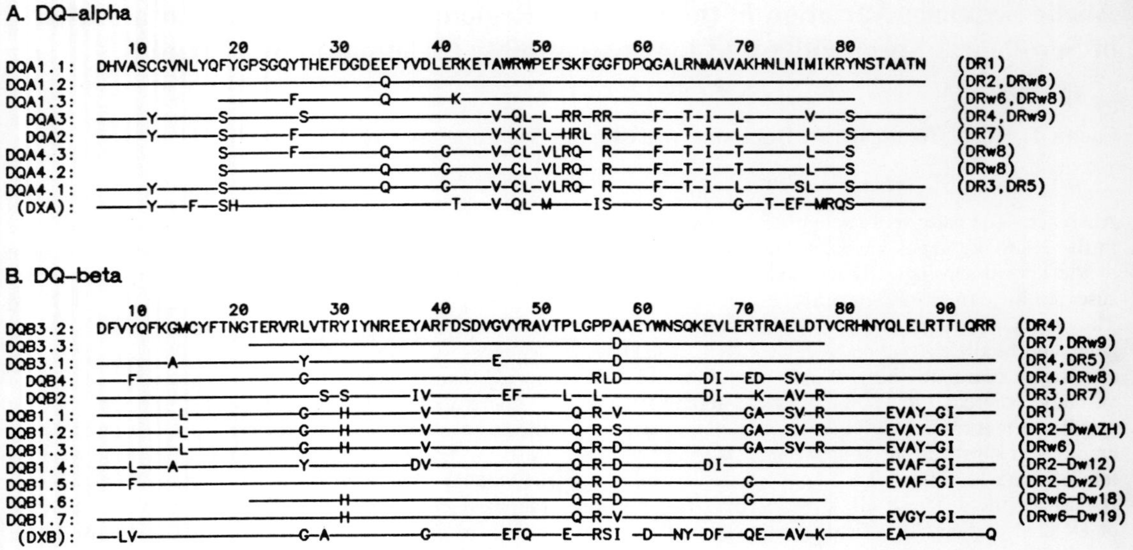

Figure 1. Allelic variation at the HLA-DQα and DQβ loci. DNA sequences were converted to one-letter amino acid codes and compared. The top line of each group shows a prototypic sequence, labeled with positions in the mature protein subunit. For the remaining alleles, only those residues that differ from the prototype are shown, with a dash indicating identity with the prototype. The alleles are listed in an order that groups the more homologous sequences together. The designation of the alleles, shown on the left, was chosen when possible to reflect the serological types. Closely related alleles are designated as subtypes. The HLA-DR haplotypes(s) from which the allele was most frequently isolated is indicated on the right. The majority of these sequences have been published elsewhere, with various designations. A full description of the sources(s) of each allelic sequence can be found in Ref. 6. The homologous segment of DNA from the DX region is shown at the bottom of each group. A; Variation of the DQα alleles. The sequence from the DR1 homozygous cell line LG-2 is used as a prototype. The analysis of PCR-amplified DQα alleles only determines sequences from codons 18 through 80 (5). A space has been inserted at position 55 of some alleles to maintain alignment. Note the clustering of the major variation around position 50, an area used to target oligonucleotide probes to discriminate the different alleles (4). B; Variation of the DQβ alleles. In this case, the allele from the DR4-DQw3.2 haplotype is used for comparison. The PCR amplification and cloning of DQβ only determines codons 21 through 77. Note that the variation here is more dispersed than in DQα, but exhibits some clustering near positions 55,71, and 86.

The correlation of the amino acid at position 57 with disease susceptibility suggests that this region of the protein may play an important role in the generation or suppression of an autoimmune response. Although this role could involve the actual process of antigen presentation, another mechanism invokes molecular mimicry of vital HLA epitopes by invading pathogens (11). To investigate this possibility, we have searched for segments of identity between this region of the DQβ alleles and the complete translations of the entire genome of the Epstein-Barr virus (EBV) (12). Table 2 demonstrates that EBV contains significantly more homology to the IDDM-susceptible DQB2 and DQB3.2 alleles than it does to the less

Table 1. Correlation of DQ Serotypes with DQα or DQβ Alleles

Specificity	Allele(s)
DQw1	DQγ1*
DQw2	DQβ2†
DQw3	DQβ3†
DQ(blank)	DQβ4†

*Based on allelic homology.
†Based on rare haplotypes.

Table 2. Homology between Selected HLA-DQβ Alleles and Epstein-Barr Virus Genome

Allele	Haplotype	Segment	Matches
DQB1.4	(DR2)	RPDAE	1
DQB2	(DR3)	GLPAA	5
		PAAEY	1
DQB3.2	(DR4)	GPPAA	20
		PPAAEY	1

The full sequence of the EBV genome was translated in all six phases and searched for the identity of five or more amino acids with the region of DQβ near position 57. The sequence found in common to the virus and the HLA gene is shown, as is the number of times it was found in the EBV genome.

susceptible DQB1.4 allele. Other viruses, some of which have been implicated in triggering IDDM, have also been found to contain segments homologous to the DQB3.2 allele (6). This finding suggests that the segment of HLA-DQβ near position 57 may be a preferred target for molecular mimicry by pathogenic viruses, and that the pattern of polymorphism seen in DQβ may be a result of selective pressure to avoid this mimicry.

References

1. Giles RC, Capra JD. Structure, function, and genetics of human class II molecules. Adv Immunol 1985;37:1.
2. Bodmer WF. Evolutionary significance of the HLA system. Nature 1972;237:139.
3. Svejgaard A, Platz P, Ryder LP. HLA and disease 1982–A survey. Immunol Rev 1983;70:193.
4. Saiki RK, Bugawan TL, Horn GT, Mullis KB, Erlich HA. Analysis of enzymatically amplified beta-globin and HLA-DQα DNA with allele specific probes. Nature 1986;324:163.
5. Scharf SJ, Horn GT, Erlich HA. Direct cloning and sequence analysis of enzymatically amplified genomic sequences. Science 1986;233:1076.
6. Horn GT, Bugawan TL, Long CM, Erlich HA. Allelic sequence variation of HLA-DQ genes; Relationship to serology and disease susceptibility. Manuscript submitted for publication.
7. Monos DS, Spielman RS, Gogolin KJ, Radka SF, Baker L, Zmijewski CM et al. HLA-DQw3.2 allele of the DR4 haplotype is associated with IDDM: Correlation between DQβ restriction fragments and DQβ chain variation. Immunogenetics 1987;26:299.
8. Todd JA, Bell JI, McDevitt HO. HLA-DQβ gene contributes to susceptibility and resistance to IDDM. Nature 1987;329:599.
9. Erlich HA, Horn GT, Bugawan TL, Scharf S. The molecular analysis of HLA class II gene polymorphism. In: Schacter et al. (eds): Molecular and Cellular Biology of Histocompatibility Antigens. 1987, pp 93–109.
10. Acha-Orbea H, McDevitt HO. The first eternal domain of the nonobese diabetic mouse Class II I-A beta chain is unique. Proc Natl Acad Sci USA 1987;84:2435.
11. Adams DD. Protection from autoimmune disease as the third function of the major histocompatibility gene complex. Lancet 1987;ii:245.
12. Baer R, Bankier AT, Biggin MD, Deininger PL, Farrell PJ, Gibson TJ, et al. DNA sequence and expression of the B95-8 Epstein-Barr virus genome. Nature 1984;310:207.

Author Affiliation

Department of Human Genetics, Cetus Corporation, Emeryville, CA 94608, USA.

The HLA-DQB1 Gene Contributes to the Genetic Susceptibility to Insulin-Dependent Diabetes Mellitus

John A. Todd, John I. Bell, and Hugh O. McDevitt

Over half of the genetic predisposition to autoimmune insulin-dependent diabetes mellitus (IDDM) is determined by one or more genes within the major histocompatibility complex (MHC) (1). Serologic, cellular, and DNA (restriction fragment length polymorphisms; RFLP) HLA typing methods have previously suggested that an IDDM susceptibility is closely linked to the HLA-DQ subregion (2). By direct sequencing of HLA genes from IDDM patients, we have established a correlation between a polymorphic residue of the DQB1-chain (position 57) and IDDM susceptibility/resistance that indicates a direct role for the DQ molecule in disease susceptibility (3).

Susceptibility to the autoimmune disease of IDDM is strongly associated with the MHC of the rat, mouse, and man (4). In particular, class II gene polymorphisms have been shown to be more strongly associated with IDDM than other MHC gene polymorphisms. The aims of this study were to establish which class II gene may be responsible for the genetic predisposition to IDDM encoded within the MHC. The only definitive approach was to directly sequence the major expressed class II genes (DRB1, DRB3, DQA1, and DQB1) isolated from IDDM patients. A comparison of the DQB1 chain and N-terminal domain amino acid sequences obtained from patients and several healthy control subjects revealed a strong correlation between the identity of amino acid 57 and the association of a particular DQ molecule with susceptibility or resistance to IDDM.

Table 1. Association of HLA-DQ_β with Insulin-Dependent Diabetes Mellitus

DR	DQ	IDDM Association	Residue 57
1	w5	Positive	Val
2	w6	Negative	Asp
2	w1.12	Neutral/negative	Asp
2	w1.AZH	Positive	Ser
3	w2	Positive	Ala
4	w7	Neutral/negative	Asp
4	w8	Positive	Ala
5	w7	Neutral/negative	Asp
6(w13)	w1.18	Neutral/negative	Asp
6(w13)	w1.19	Positive	Val
6(w14)	w1.9	Neutral/negative	Asp
7	w2	Neutral	Ala
8	w4	Neutral	Asp
9	w9	Neutral	Asp

Data are taken from several sources (2,4,5,7–9). First domain sequences of *DRBl; DQAl,* and *DQBl*-chains (see ref. 3 for complete first domain amino acid sequences) were obtained from diabetics and normals as previously described using the method of in vitro DNA amplification (10).

Previous studies, using serologic typing methods, have shown that each DR antigen (DR1 to DRw14) has a characteristic and reproducible association with IDDM susceptibility (Table 1). Most DR antigens, due to linkage disequilibrium, are usually associated with one particular DQ antigen (DQw1-9, Table 1). However, three DR antigens (DR2, DR4 and DRw6) are commonly associated with more than one DQ molecule. A careful comparison of the DQB1 chain sequences (obtained as part of this study and previously published sequences) within each of these three DR haplotypes revealed that DQB1 chains rarely found in IDDM patients all possessed Asp at position 57 and DQB1 chains commonly found in diabetics had Ala, Val, or Ser at residue 57 (Table 1). These correlations are statistically significant (5) and provide a simple explanation for the association (either positive, negative, or neutral) of most of the DR haplotypes with IDDM susceptibility.

A major exception to this correlation occurs in the DR7 haplotype, which is almost always associated with the DQw2 antigen. This DR7 DQw2β chain is identical to that found on DR3 haplotypes and has Ala at residue 57. The DR7 haplotype is not, however, positively associated with IDDM. One possible explanation is that the DR7 DQα chain, which has a unique structure, affects the overall structure of the DQ molecule, resulting in a "nonsusceptibility" molecule. DQα chain residues that may be important in this interaction include those in the second allelic hypervariable region (47 to 56).

It is likely that the overall conformation and shape of the DQ molecule will dictate T-cell recognition and its function in the immune response. By analogy with the class I crystal structure, a model for class II structure places β chain amino acid 57 on the molecule's surface pointing into the putative peptide binding site. Furthermore, it is possible that Asp57 may form a salt bridge with a conserved Arg residue at position 79 of the α chain (6). If such an interaction actually exists, the pleiotropic effect of residue 57 on the structure and function of the class II molecule could be more easily rationalized.

The disease susceptibility or resistance imposed by different DQ molecules is clearly not determined by residue 57 alone. There are multiple differences in both the α and β chains between each DQ molecule, and these presumably account for strengths of the associations of the different DQ molecules with IDDM (positive–DQw8,DQw2 > DQw5,DQwAZH > DQw4,DQw9 > DQw7 > DQw6,DQw1.12– negative).

Although we did not find any significant associations of DRB1 molecules with IDDM, we cannot rule out a role for DR and other, as yet unidentified, linked genes that

Immunobiology of HLA
Volume II

may influence disease susceptibility either directly or perhaps through modulation of the level of expression of class II molecules. Furthermore, the difference between the disease concordance rate for haploidentical siblings in diabetic families and that for monozygotic twins indicates that other genes, outside the MHC and on different chromosomes, are necessary for disease development (1). The discordance between monozygotic twins (25 to 50%) suggests that environmental factors may also be involved (1).

The residue 57 correlation extends to the nonobese diabetic (NOD) mouse. The NOD mouse develops a spontaneous form of IDDM that has many immunologic and pathologic features in common with the human disease. An IDDM susceptibility gene has also been mapped to the mouse MHC (11). The murine homologue of the DQB1 gene (Aβ) was previously isolated and sequenced (12). Compared with other mouse alleles, the NOD Aβ gene is unique at codon 57, which codes for Ser. Aβ chains from nondiabetic mouse strains, including the related nonobese normal (NON) strain (5) which has been commonly used in studies of genetic crosses with NOD, have Asp at position 57. Because the I-A heterodimer is the only major class II molecule expressed by the NOD MHC, it seems likely that the NOD Aβ chain is responsible for the autoimmune response.

The role of "Asp-positive" and "Asp-negative" DQ molecules in IDDM is unknown. Certain DQ molecules may influence the repertoire of T cells present in the immune system. DQ-restricted T-cell suppression and/or clonal deletion via DQ recognition by certain T-cell receptors during T-cell repertoire formation are two possible mechanisms currently under investigation.

Acknowledgment. Research supported by NIH grants CA 39069 and 33880 and by fellowships from the Science and Engineering Research Council, UK, to J.A.T. and the Alberta Heritage Fund to J.I.B.

References

1. Rotter JI, Landaw EM. Measuring the genetic contribution of a single locus to a multilocus disease. Clin Genetics 1984;26:529.
2. Kim SJ, Holbeck SL, Nisperos B, Hansen JA, Maeda H, Nepom GT. Identification of a polymorphic variant associated with HLA-DQw3 and characterized by specific restriction sites within the DQ β-chain gene. Proc Natl Acad Sci USA 1985;82:139.
3. Todd JA, Bell JI, McDevitt HO. HLA-DQβ gene contributes to susceptibility and resistance to insulin-dependent diabetes mellitus. Nature 329:599.
4. Eisenbarth GS. Type I diabetes mellitus. N Engl J Med 1986;314:1360.
5. Todd JA, Acha-Orbea H, Bell JI, Chao N, Fronek Z, Jacob CO, McDermott M, Sinha AA, Timmerman L, Steinman L, McDevitt HO. A molecular basis for MHC class II-associated autoimmunity. Science 1988;240:1003.
6. Brown JH, Jardetsky T, Saper MA, Samraovi B, Bjorkman, P, Wiley DC. Hypothetical model of the foreign antigen binding site of class II histocompatibility molecules. Nature 1988;332:845.
7. Thomson B. HLA DR antigens and susceptibility to insulin-dependent diabetes mellitus. Am J Hum Genet 1984;36: 1309.
8. Kohonen-Corish MRJ, Serjeanston SW, Lee HK, Ximmek P. Insulin-dependent diabetes mellitus: HLA-DR and DQ genotyping in three ethnic groups. Disease Markers 1987; 5:153.
9. Rosenshine S, Cascino I, Zeevi A, Duquesnoy RJ, Trucco M. DQα and β RFLP reveals the composition of the DQ molecule recognised by T-cell clones. Immunogenetics 1986;23:187.
10. Saiki RK, Scharf S, Faloona F, Mullis KB, Horn GT, Erlish HA, Arnheim N. Enzymatic amplification of β-globin genomic sequences and restriction site analysis for diagnosis of sickle cell anemia. Science 1985;230:1350.
11. Prochaska M, Leiter EH, Serreze DV, Coleman DL. Three recessive loci required for insulin-dependent diabetes in nonobese diabetic mice. Science 1987;237:286.
12. Acha-Orbea H, McDevitt HO. The first external domain of the nonobese diabetic mouse class II I-A β chain is unique. Proc Natl Sci USA 1987;84:2435.

Author Affiliations

John A. Todd, John I. Bell, Nuffield Departments of Clinical Medicine and Surgery, John Radcliffe Hospital, Headington, Oxford OX3 9DU, UK; *Hugh O. McDevitt*, Departments of Medical Microbiology and Medicine, Stanford University, Stanford, CA 94305, USA

HLA-DR4-Associated Disease: Oligonucleotide Probes Identify Specific Class II Susceptibility Genes in Type I Diabetes and Rheumatoid Arthritis

Gerald T. Nepom, Christine Seyfried, Susan Holbeck, Patricia Byers, Kenneth Wilske, Jerry Palmer, David Robinson, and Barbara Nepom

Distinct patterns of linked DRβ (HLA-DRB1) and DQβ (HLA-DQB1) genes are found on at least seven different haplotypes that carry the HLA-DR4 specificity (1,2). We have constructed 20 and 21-base synthetic oligonucleotide DNA probes capable of distinguishing between different DRβ alleles and between different DQβ alleles expressed on these DR4-positive haplotypes. Hybridization analysis indicates striking differences in the distribution of individual DQβ and DRβ alleles in DR4-positive patients with rheumatoid arthritis (RA) and DR4-positive patients with insulin-dependent diabetes mellitus (IDDM, type I diabetes). Probes identifying Dw14 or Dw4 DRβ genes hybridized to greater than 93% of patients with seropositive rheumatoid arthritis, consistent with the concept that these two alleles account for the DR4 association with this disease. No difference was observed between the distribution of different DQβ alleles in rheumatoid arthritis as compared to normal individuals. In contrast, DQβ 3.2 specific oligonucleotide probes hybridized to greater than 96% of DR4-positive IDDM patients, regardless of the linked DR4-positive DRβ gene.

HLA-DR4, like other HLA typing specificities, is a broad, public serologic specificity present on many different molecules (1-4). At least five different allelic molecules encoded by HLA-DRβ genes all carry the HLA-DR4 specificity. These five different alleles are closely related: they differ by between one and four amino acids clustered in the first DRβ polypeptide domain (5). Different DR4 alleles correspond to different HLA-D specificities and serve as genetic markers on different DR4-positive haplotypes, four of which are illustrated in Table1.

Also shown in Table 1 are the closely linked genes of the HLA-DQ locus. Two DQβ alleles, DQ3.1 and DQ3.2, carry the DQw3 typing specificity associated with different DR4 haplotypes (6). Previous studies have implicated the Dw4 and Dw14 DRβ genes, present on haplotypes 1 through 3 of Figure 1, as contributing to the HLA-associated susceptibility to rheumatoid arthritis (7,8); in contrast, the DQ3.2 gene has been shown to account for the HLA-associated susceptibility to type I diabetes (9,10). In order to directly compare differences between these distinct patterns of genetic susceptibility, we performed oligonucleotide hybridization analysis to distinguish among the key disease susceptibility genes present on different DR4-positive haplotypes.

Oligonucleotide probes for Dw4 and Dw14 have been previously described (7). They consist of 20-base oligomers that detect nucleotide substitutions at codon 71 and codon 86, the two substitutions which distinguish the Dw4 gene from the Dw14 gene. Oligonucleotide probes for DQ3.1 (5′-TCTGGTCACATAACGCACGCG-3′) and DQ3.2 (5′-CGCCCGATACACCCCCACGTC-3′) distinguish DQ3.1 from DQ3.2 in two of the four codons that differ between these two alleles. Detailed procedures for the analysis of genomic DNA using allele-specific probes, without amplification, are described elsewhere (7). Briefly, gels containing digests of genomic DNA or nitrocellulose slot blots were hybridized for three hours at 55° with 10^7 CPM/ml of the oligonucleotide probe, washed with 3.2 M tetramethyl ammonium chloride containing 0.5% SDS at 61°, and then analyzed by autoradiography.

Twenty-three Caucasian DR4-positive patients who fulfilled the clinical criteria for type I (juvenile onset) insulin-dependent diabetes, 43 Caucasian patients with seropositive erosive "classic" rheumatoid arthritis, and 34 clinically normal DR4-positive adult Caucasian controls were analyzed. Genomic DNA was prepared directly from patients' peripheral blood lymphocytes, or from autologous EBV-transformed B-cell lines. Hybridization analysis for DRβ, using Dw4 and Dw14 probes, was performed on Eco-R1 digested DNA samples. Hybridization analysis for DQβ was performed on Taq-1 digested genomic DNA. As previously described, these procedures provide a rapid and definitive identification of the individual DRβ and DQβ alleles present on different DR4 positive haplotypes.

Table 1. DR4-Positive Caucasoid Haplotypes in IDDM and RA

DQβ	$DR\beta_I$	$DR\beta_{III}$	HLA-B
DQ3.1	*Dw4(DR4)*	DRw53	(B44)
DQ3.2	*Dw4(DR4)*	DRw53	(Bw62)
DQ3.2	*Dw14(DR4)*	DRw53	(Bw60)
DQ3.2	Dw10(DR4)	DRw53	(B38)

The **DQ3.2** gene (**bold**) accounts for the association of HLA-DR4 with **IDDM**, the Dw4, and Dw14 genes (*italics*) for the association with *RA*.

Rheumatoid Arthritis

DNA from 40 of 43 (93%) patients with seropositive RA hybridized to the Dw4 or Dw14 DRβ-specific oligonucleotides. Thirty-one of these (72%) were identified by Dw4 oligonucleotide hybridization to an 11.0 kb band following Eco-R1 digestion, the pattern characteristic for Dw4-associated DRβ genes. Twelve (28%) were positive for Dw14 probe hybridization to an 11.0 kb band, confirming previous associations of both Dw4 and Dw14 with RA. Seven patients were heterozygous for

Immunobiology of HLA
Volume II

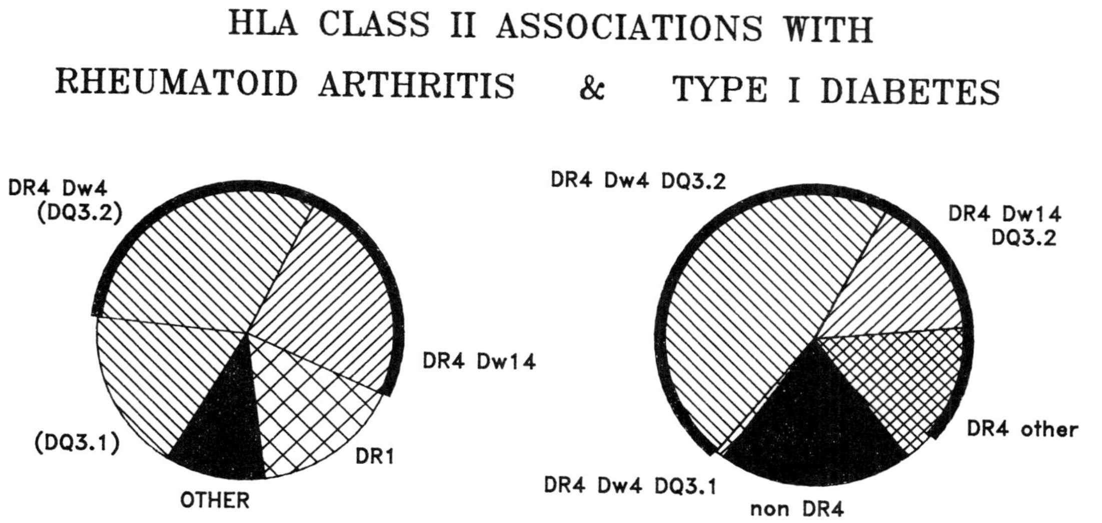

Figure 1. Relative frequencies of different class II haplotypes are shown for RA (left) and IDDM (right); HLA-DQ3.2-positive haplotypes are indicated by ▬▬ . Dw4/DQ3.1 haplotypes are prevalent in RA but not in IDDM.

Dw4 and Dw14. Eleven patients (26%) were characterized by a 3.8 kb band hybridizing to the Dw14 probe, a hybridization pattern that identifies an additional susceptibility gene associated with rheumatoid arthritis, characteristic of DR1-associated DRβ genes (11) found in most RA patients who lack Dw4 or Dw14. In summary, these oligonucleotide probes identify susceptible haplotypes because of the Dw4 or Dw14 DRβ genes, and identify DR1 susceptible haplotypes by virtue of a homologous nucleotide sequence in Dw1-associated DRβ genes.

As seen in Table 1, the Dw4 and Dw14 genes are present on three of the DR4 haplotypes shown. These three RA susceptibility haplotypes can differ for their linked DQβ alleles (Table1) in that some Dw4 haplotypes carry a DQ3.1 gene and some Dw4 haplotypes carry a DQ3.2 gene. This distribution of these two haplotypes in control Dw4 individuals was 35% and 65%; a similar distribution, 40% and 60%, was present among Dw4 RA patients. Therefore, we conclude that DQ allelic variation does not contribute to the disease susceptibility, which appears to be best characterized by specific alleles of the DRβ gene.

Type I Diabetes

A markedly different result is seen in the analysis of DR4-associated type I diabetes. Using RFLP analysis, the DQ3.2 allele appears to be responsible for the DR4 associations in IDDM (9,10,12,13). Oligonucleotide probe analysis distinguishing DQ3.1 from DQ3.2 confirms this previous result. Of 23 DR4 IDDM patients tested, all carried the DQ3.2 gene. Three of these individuals also carried a DQ3.1 gene; all three were homozygous for DR4 and heterozygous for DQ3.2 and DQ3.1.

Of the 23 DR4/DQ3.2 positive haplotypes in the IDDM patients, 17 (74%) carried a Dw4 gene, four carried a Dw14 gene, and two carried other DRβ genes. In other words, several different DR4 positive haplotypes were represented in the IDDM population, but all patients carried the DQ3.2 DQβ allele. We therefore conclude that the role of the DRβ locus in diabetes susceptibility is minimal, or secondary to a permissive function of the DQ3.2 gene.

Implications

It is now possible, using oligonucleotide probes, to detect individual DR4-associated HLA susceptibility genes implicated in rheumatoid arthritis and in Type I diabetes, with a high degree of precision and sensitivity, summarized in Table 2. The observation that the susceptibility genes in these two diseases are distinct provides

Table 2. Specific Class II Genes Associated with IDDM and RA

	No. of Patients	[+/+]	[+/−]	% Positive
DR4-positive IDDM vs DQ3.2	23	23	0	100%
RA vs Dw4		31	12	72%
vs Dw14	43	12	31	28%
vs 4+4+1*		40	3	93%

*DRβ_1 susceptibility genes identified with Dw4 and Dw14-specific oligonucleotide probes (see text).

a probable explanation for sporadic reports of families in which both RA and IDDM segregate together, and families in which these two diseases do not segregate together.

A key element of this analysis is the observation that susceptibility alleles differ from their nonsusceptibility counterparts by only a few amino acid substitutions. Oligonucleotide probe analysis provides a simple and precise method for distinguishing among these similar genes. Furthermore, it is probable that such subtle changes are sufficient to account for major differences in immune recognition contributing to disease susceptibility. Precise characterization of the molecular events involved in recognition of the susceptibility gene product will be a key element in advancing our understanding of the pathogenesis of autoimmune disease.

Acknowledgments. This work was supported by grant AR37296 from the National Institutes of Health and by a grant from the Juvenile Diabetes foundation.

References

1. Nepom BS, Nepom GT, Mickelson E, Antonelli P, Hansen JA. Electrophoretic analysis of human HLA-DR antigens from HLA-Dr4 homozygous cell lines: Correlation between beta-chain diversity and HLA-D. Proc Natl Acad Sci USA 1983;80:6962-6966.
2. Nepom GT. Immunogenetics of HLA associated diseases. In: Cruse J (ed): Concepts in Immunopathology, Vol. 5. Basel, Karger, pp 80–105, 1988.
3. Nepom GT, Nepom BS, Antonelli P, Mickelson E, Silver J, Goyert SM, et al. The HLA-DR4 family of haplotypes consists of a series of distinct DR and DS molecules. J Exp Med 1983;159:394-404.
4. Groner JP, Watson AJ, Bach FH. Dw/LD-related molecular polymorphism of DR4 beta-chains. J Exp Med 1983;157: 1687-1691.
5. Gregersen PK, Shen M, Song Q-L, Merryman P, Degar S, Seki T, et al. Molecular diversity of HLA-DR4 haplotypes. Proc Natl Acad Sci USA 1986;83:2642-2646.
6. Kim SJ, Holbeck SL, Nisperos B, Hansen JA, Maeda H, Nepom GT. Identification of a polymorphic variant associated with HLA-DQw3 and characterized by specific restriction sites within the DQ beta-chain gene. Proc Natl Acad Sci USA 1985;82:8139.
7. Nepom GT, Seyfried CE, Holbeck SL, Wilske KR, Nepom BS. Identification of HLA-Dw14 genes in DR4+ rheumatoid arthritis. Lancet 1986;ii (8514):1002-5.
8. Nepom B Nepom GT, Schaller J, Michelson E, Antonelli P, Hansen J. Characterization of specific HLA-DR4-associated histocompatibility molecules in patients with juvenile rheumatoid arthritis. J Clin Invest 1984;74:287-291.
9. Nepom BS, Palmer J, Kim S-J, Hansen JA, Holbeck SL, Nepom GT. Specific genomic markers for the HLA-DQ subregion discriminate between DR4+ insulin-dependent diabetes mellitus and DR4+ seropositive juvenile rheumatoid arthritis. J Exp Med 1986;164:345-350.
10. Nepom GT, Palmer J, Nepom B. Specific HLA class II variants associated with IDDM. In: Jaworski M, et al (eds): Immunology of Diabetes Mellitus. Amsterdam, Elsevier Publishers, pp 9–20, 1986.
11. Seyfried C, Mickelson E, Hansen Ja, Nepom GT. A specific nucleotide sequence defines a functional T-cell recognition epitope shared by diverse HLA-DR specificities. Human Immunol 1988;21:289–299.
12. Schreuder GM, Tilanus MG, Bontrop RE, Gruining GJ, Giphart MJ, van Rood JJ, et al. HLA-DQ polymorphism associated with resistance to type I diabetes detected with monoclonal antibodies, isoelectric point differences, and restriction fragment polymorphism. J Exp Med 1986;164: 938-43.
13. Michelsen B, Lernmark Å. Molecular cloning of a polymorphic DNA endonuclease fragment associates insulin-dependent diabetes mellitus with HLA DQ. J Clin Invest 1987;79:1144-1152.

Author Affiliations

Immunology Program, Virginia Mason Research Center, Seattle, Washington, and Diabetes Research Center, University of Washington School of Medicine, Seattle, WA, 98101, USA

Transfer and Expression of an IDDM Susceptibility Gene into Lymphoblastoid Cell Lines by Retroviral Vectors

William W. Kwok, Philip S. Thurtle, and Gerald T. Nepom

Analysis of structural variation of DQβ (HLA-DQB1) genes has established that one specific allele, the DQ3.2β gene, is highly correlated with IDDM (1–3) and accounts for the association of DR4 with that disease. We have expressed the cDNA of this HLA class II gene in human lymphoblastoid cell lines as a first step to investigate the functional contribution of this gene to immune responsiveness and IDDM. Here, we report the use of retroviral vectors to introduce the cDNA of DQ3.2β into an EBV transformed B-lymphoblastoid cell line MAT (DR3/ DQw2). Successful transfer and expression of the DQ3.2 cDNA in DQw2 cell lines were verified by means of specific oligonucleotide probes on DNA and RNA blots. DQ3.2β molecules were also detected on cell surfaces by flow cytometry, and immunoprecipitation experiments indicated that the introduced DQ3.2β chain

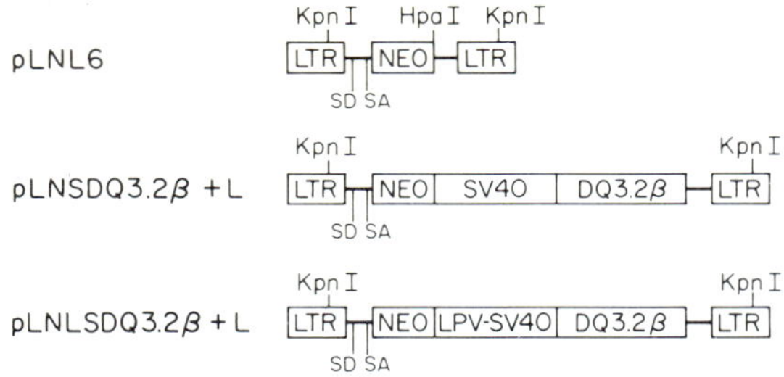

Figure 1. Schematic structure of pLNL6c8 and HLA vectors. SD is splice donor. SA is splice acceptor.

Table 1. Virus Titers of Retroviral Producer Cell Lines

Cell Line	Virus Titer (cfu/ml)
PA317LNL6c8*	2×10^6
PA317LNSDQ3.2β+Lc1b	4×10^4
PA317LNSDQ3.2β+Lc1	2×10^5
PA317LNSDQ3.2β+Lc2	1×10^3
PA317LNSDQ3.2β+Lc3	6×10^5
PA317LNLSDQ3.2β+Lc5	9×10^5
PA317LNLSDQ3.2β+Lc7	2×10^6

*PA317LNL6c8 is a line that produces no neo + viruses without the HLA cDNA insert. Cloned virus producing lines with the construct pLNSDQ3.2β+L were called PA317LNSDQ3.2β+Lc1, PA317LNSDQ3.2β+Lc2, etc. Cfu/ml were measured on NIH-3T3-TK$^-$ cells.

paired with an endogenous DQ2α chain to form an intact class II heterodimer.

Retroviral vectors have become versatile tools to introduce exogenous genes into different varieties of cell types (4). We used a retroviral vector, pLNL6, which has a 5′ LTR from Mo-MSV, a neomycin resistant gene, and a 3′ LTR from Mo-MLV as a parental vector for our recombinant vectors. A full-length DQ3.2β cDNA was obtained from the DR4/DQw3.2 cell line TAS. This cDNA, which included the 5′ ATG, but excluded the 3′ polyadenylation signal was linked to either the SV-40 promoter or the LPV enhanced SV-40 promoter (i.e., LPV enhancer linked to an SV-40 promoter without the 72 bp repeats) to form a transcriptional unit. The LPV enhancing element was also chosen because previous reports suggested that it should be a strong enhancer in lymphoid cells (5). These HLA transcriptional units were ligated to a unique HpaI site on the 3′ end of the neomycin resistant gene in pLNL6. Constructs pLNLSDQ3.2β+L and pLNSDQ3.1β+L have the DQ3.2β cDNA under the control of the LPV enhanced SV-40 promoter and the SV-40 promoter respectively (see Fig. 1). An infection and transfection protocol described by A.D. Miller et al, and the use of Psi-2 and PA317 amphotrophic packaging cell lines (6) were used to obtain numerous HLA virus producing cell lines. The virus titers of these cell lines are listed in Table 1. The virus titer ranged from $<1\times10^3$ cfu/ml to 2×10^6 cfu/ml. These virus titers were indicated by the number of G418 resistant NIH-3T3-TK$^-$ colonies; the method of the assay has been described previously (7). It should be noted that although our vector construct had a 5′ LTR from Mo-MSV, the retrovirus will replace this Mo-MSV LTR with the Mo-MLV LTR when it replicates. Thus all the integrated retroviral DNA will have both LTR from Mo-MLV (8).

A DR3/DQw2 homozygous cell line, MAT, was chosen as the recipient target for our gene transfer experiments, using co-cultivation infection (7). Selection with 1 mg/ml of G418 was applied 24 hours after the infection. G418-resistant B cells were grown up in four weeks and ready for analysis. The G418-resistant MAT line derived by infection with PA317LNL6c8 was called MATLNL6c8; the cell line derived by infection with PA317LNSDQ3.2β+Lc3 was called MATLNSDQ3.2β+Lc3 and so on. Experiments described below were carried out with these infected lines without further cloning.

A ^{32}P labelled oligomer corresponding to nucleotide sequences from codon 43 to codon 49 of the β1 exon of DQ3.2 was used as a probe for hybridization analysis, as previously described (3). Using restriction enzyme KpnI (which cuts once in both LTR), a 4.0 kb band was detected in all infected lines with the SV-40 transcriptional unit and a 4.1 kb band was detected in all cell lines with the LPV transcriptional unit. These results corresponded to successful transfer of the HLA cDNA; the intensity of the hybridization signals suggested that low copies of retroviral DNA, estimated to be less than three copies per cell, were present in all infected lines. RNA blots were also performed. Four transcripts with fragment sizes of 4.2 kb, 3.5 kb, 1.8 kb, and 1.4 kb were detected in MATLNSDQ3.2β+L. The 4.2 kb unspliced LTR-LTR transcript is the most prominent transcript. The 3.5 kb transcript is generated by utilization of the splice sites 5′ of the neo gene. This spliced transcript is necessary for synthesis of the neo gene product. The 1.8 kb transcript corresponds to initiation from the internal promoter and termination at the 3′ LTR. This should be the transcript from which the exogenous HLA molecules were synthesized. The 1.4 kb transcripts were endogenous DQ2β mRNA present in the MAT line, which cross-hybridized with our probe at low stringency. A 1.4 kb transcript, which is the endogenous DQ3.2β mRNA, could also be detected from a DR4, DQw3.2 (DQw8) EBV transformed cell line, Priess. The signal intensity of the 1.4 kb endogenous transcript from Priess was two to fivefold higher than the 1.8 kb transcript originated from the SV-40 promoter. The 1.8 kb transcript from the LPV enhanced SV-40 promoter of MATLNLSDQ3.2β+Lc7 was very weak. However, it was also observed that the DNA copy number of the transferred gene in this line was the lowest in our infected lines.

We then analyzed class II expression in the infected MAT cell lines. MAbs IVD12 (anti-DQw3) and 200.1 (anti-DQw3.2β (DQw8)) were used in FACS analyses, illustrated in Figure 2 and Table 2. Though uninfected MAT and MATLNL6c8 could not be stained by these two MAbs, the other infected lines stained positively. The level of surface expression of the DQ3.2β molecules in

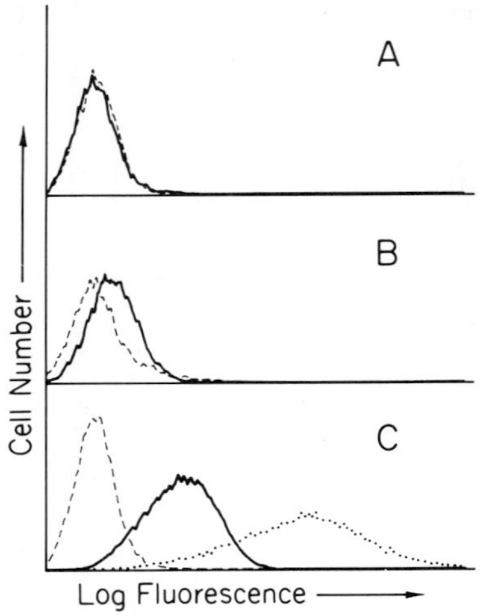

Figure 2. FACS analyses: Panel A, MAT;
Panel B, MATLNLSDQ3.2β+Lc7;
Panel C, Priess and MATLNSDQ3.2β+Lc3.
F9, an anti-leutinizing hormone-releasing hormone MAb, is used as a negative control. --- F9, — 200.1.
In panel C, · · · is MAT stained with MAb GSP 11.1 (anti-DR).

MATLNSDQ3.2β+Lc3 was twofold lower than the expression of DQ3.2β in Priess. MATLNLSDQ3.2β +Lc7 gave a much weaker staining with the MAbs. These results were consistent with the RNA blot data suggesting higher levels of SV40-initiated transcription compared to LPV-initiated transcription. The results also indicated that these anti-DQw3 MAbs recognize the β chain of the DQ3 molecules.

The FACS analyses suggested the DQ3.2β molecules could successfully pair with endogenous α molecules to form cell surface heterodimers. Immunoprecipitation experiments were carried out with other infected DQ3.2β MAT cell lines to determine whether the newly introduced DQ3.2β could be paired with the DQ2α and/or the DR3α to form these heterodimers. For these experiments, cells were labeled with ^{3}H-leucine, and cell lysates were pre-cleared with MAb GSP11.1 (anti-DR) before immunoprecipitation with MAb IVD12. Two-dimensional PAGE showed the presence of DQ2α molecules coprecipitated with the DQ3.2β chains, using MAb IVD12. These immunoprecipitation data agree with our previous finding that transcomplementation between DQ2α and DQ3β molecules can occur (9) in heterozygous DR3/DR4 individuals. Recently, A. Korman et al reported the use of retrovirus vectors to introduce HLA class II gene into mouse fibroblasts (10). The results discussed here show that it is also possible to introduce exogenous HLA cDNA into human B-cell lines by means of retroviral infection. Stable transfer and expression of the transduced genes are observed. These studies suggested that retroviral vectors can become a powerful tool to manipulate HLA class II genes for further experimental analyses.

Table 2. Reactivities of MAbs IVD12 and 200.1 on MAT, Infected MAT, and Priess Cell Lines

Cell Line	MAb Reactivity	
	IVD12	200.1
MAT	–	–
MATLNL6c8	ND	–
MATLNLSDQ3.2β+Lc7	ND	+
MATLNSDQ3.2β+Lc1b	ND	++
MATLNSDQ3.2β+Lc3	+++	+++
PRIESS	++++	++++

The level of expression of DQ3.2β on Priess is designated as ++++. +, ++ and +++ represent <10%, 10 to 25%, and 25 to 50% expression relative to Priess respectively. – = negative; ND = not done. (All cell lines were also stained with MAb GSP 11.1 (anti-DR) as a positive control.)

Acknowledgments. The vector pLNL6c8 and the cell line PA317LNL6c8 were kindly provided by Dr. R. Hock. We thank P. Gregersen for the DQ3.2 cDNA from TAS.

This work was supported by grant AR 37296 from the National Institutes of Health.

References

1. Nepom BS, Palmer J, Kim S-J, Hansen JA, Holbeck SL, Nepom GT. Specific genomic markers for the HLA-DQ subregion discriminate between DR4 positive from IDDM and DR4 positive JRA. J Exp Med 1986;164:345.
2. Todd JA, Bell JI, McDevitt HO. HLA-DQβ gene contributes to susceptibility and resistance to insulin-dependent diabetes mellitus. Nature 1987;325:599.
3. Holbeck S, Nepom GT. Analysis of allelic variation within HLA-DQβ genes using synthetic oligonucleotide probes. Immunogenetics 1986;24:251.
4. Coffin J. Genome structure. In: Weiss R, Teich N, Varmus H, Coffin J, (eds): RNA Tumor Viruses, Vol. 2. Cold Spring Harbor, NY, Cold Spring Laboratory, 1985, pp 17-74.
5. Mosthaf L, Pawlita M, Gruss P. A viral enhancer element specifically active in human haematopoietic cells. Nature 1985;315:597.
6. Miller AD, Buttimore C. Redesign of retrovirus packaging cell lines to avoid recombination leading to helper virus production. Mol Cell Biol 1986;6:2895.
7. Kwok WW, Schuening F, Stead RB, Miller AD. Retroviral transfer of genes into canine hemopoietic cells in culture: A model for human gene therapy. Proc Natl Acad Sci USA 1986;83:4552.
8. Varmus HE. Form and function of retroviral proviruses. Science 1982;216:812.
9. Nepom B, Schwarz D, Palmer J, Nepom GT. Transcomplementation of HLA genes in IDDM. Diabetes 1987; 36:114.
10. Korman AJ, Frantz DJ, Strominger JL, Mulligan RC. Expression of human class II major histocompatibility complex antigens using retrovirus vectors. Proc Natl Acad Sci USA 1987;84:2150.

Author Affiliations

Immunology Program, Virginia Mason Research Center, Seattle Washington, and Diabetes Research Center, University of Washington School of Medicine, Seattle, WA 98101, USA

DPA Polymorphisms in Normals, Type I Diabetics, and Patients with Graves' Disease

Klaus W.J. Badenhoop, Valerie Lewis, Varina Drummond, Gisele Schwarz, John Trowsdale, and Gian Franco Bottazzo

Among other organ-specific autoimmune diseases, Graves' disease is generally associated with the serological DR3 specificity. However, the associations shown so far have not been as clear as with other autoimmune diseases and the relative risks have been calculated to be around 3 to 5. Furthermore, only 50 to 60% of patients possess the DR3 specificity. Therefore the actual disease susceptibility gene(s) are thought to be in linkage disequilibrium with genes in the human leukocyte antigen (HLA) region and might lie in the DP subregion or another on the short arm of chromosome 6. To investigate this association further at the genomic level we studied the structure of the HLA class II genes in patients with Graves' disease, control subjects, diabetics, and families using restriction fragment length polymorphism (RFLP) analysis as well as serology.

Fifty-six Caucasoid patients with Graves' disease (thyrotoxicosis accompanied by diffuse radioactive iodine uptake, ophthalmopathy, and/or thyroid-stimulating immunoglobulins (TSI) were studied. These included a panel of 10 patients from West-Berlin (kindly provided by Prof. Horst Schleusener), originally typed as DR3 homozygous, who had been part of the German multicenter prospective study on the course of Graves' disease. The West-Berlin control panel did not differ in serological types from the London controls which consisted of 46 healthy Caucasoids. Forty-three type I diabetics and 12 families with one affected member were also studied. HLA specificities were assigned by ninth Histocompatibility Workshop definitions.

RFLP analysis was performed by standard procedures. Briefly, high molecular weight DNA was extracted from peripheral blood and 10 μg were digested with the restriction enzyme BglII to completion. DNA was separated in 0.8% agarose gel electrophoresis at 35 volts for 18 to 22 hours and blotted onto Biotrace membranes using the alkaline transfer method. Filters were prehybridized in 50% formamide, 5 × SSPE, 0.1% Denhardt's, 1% dextran sulfate, 0.1% sodium dodecyl-sulfate (SDS), 100 μg/ml of denatured salmon sperm DNA, and hybridized in prehybridization mix with 5 to 10 × 10^5 cpm/ml of oligolabeled probe, which had been labeled to a specific activity of 1 × 10^9cpm/μg. Washing and dehybridization were performed as in the tenth Histocompatibility Workshop Southern Blot protocol. A cDNA probe for the DPα chain gene was an 0.9-kb EcoRI fragment in pBR322 (1).

Twenty-six out of 47(55%) DR-typed Graves' patients possessed the DR3 specificity versus 9 out of 43 (21%) controls (p corrected = 0.005) This was reflected in the DQ serology, where 30/45 (67%) Graves' patients were DQ2, but 12/40 (30%) controls (p corrected = 0.005).

Digestion of genomic DNA revealed a 3.7-kb polymorphism which was more frequent in patients with Graves' disease (28/56,50%) versus 15/46 (33%) controls. This polymorphism was not DR3 linked (when DR3 homozygous patients were analyzed), although it was present in 15/25 (60%) DR3 Graves' patients versus 4/9 (44%) DR3 controls. Out of 43 type I diabetics, 12 (28%) showed the 3.7-kb polymorphism.

We then investigated the segregation of that polymorphism in families. Twelve families with one diabetic member were studied. The 3.7-kb band was detected in four families, where it segregated with the HLA haplotype (segregation of one family shown in Table 1). The other eight families did not show the polymorphism. Interestingly, the polymorphism segregated with DR9 in one family, where the diabetic proband was DR9/1, and his nondiabetic siblings (DR6/4) had islet-cell antibodies (ICA) and not the 3.7-kb band. Out of 5 non-DR3/non-DR4 type I diabetics, two showed the 3.7-kb polymorphism.

Analysis of RFLP with DP probes enables the investigation of DP polymorphisms at the genomic level (2). Using the restriction enzyme BglII and the DPα probe, we have detected a 3.7-kb polymorphism, which segregated with the HLA haplotype in families, was slightly increased in Graves' disease patients, and was also observed in Non-DR3/non-DR4 diabetics. Our results suggest that DP or genes in the vicinity (between DQ and DP) may play a role in the disease susceptibility to Graves' disease and may also be involved in non-DR3/non-DR4-associated type I diabetes.

Table 1. Segregation of the 3.7-kb BglII DPA Polymorphism in a Diabetic Family Related to the HLA-DR Haplotypes

Father	ab	DR6/9	+	
Mother	cd	DR1/4	−	
Sibling	ad	DR6/4	−	ICA +ve
Sibling	ad	DR6/4	−	ICA +ve
Sibling	bc	DR9/1	+	diabetic proband

+ = presence.

*K.B. was supported by the German Academic Exchange Service (DAAD).

References

1. Trowsdale J, Young JAT, Kelly AP, Austin PJ, Carson S, Meunier H, et al. Structure, sequence and polymorphism in the HLA-D region. Immunol Rev 1985;85:5.
2. Hyldig-Nielsen JJ, Morling N, Odum N, Ryder LP, Platz P, Jacobsen B, et al. Restriction fragment length polymorphism of the HLA-DP subregion and correlations to HLA-DP phenotypes. Proc Natl Acad Sci USA 1987;84:1644.

Author Affiliations

Klaus W.J. Badenhoop, Valerie Lewis, Varina Drummond, and *Gisele Schwarz*, Department of Diabetes and Immunogenetics, St. Bartholomews Hospital, London; *John Trowsdale*, Imperial Cancer Research Fund (ICRF), London; *Gian Franco Bottazzo*, Department of Immunology, Middlesex Hospital, London, UK

Restriction Fragment Length Polymorphism (RFLP) Defines a Graves' Disease Associated DR3 Subtype: Involvement of the DR-DQ Subregion

G. Semana, H. Allanic, F. Quillivic, B. Genetet, and R. Fauchet

The present study proposes to investigate, through Southern blot analysis, the MHC class II gene polymorphism in Graves' disease (GD) in order to define the HLA-related genetic susceptibility. Considering the preferential link between GD and the HLA DR3 antigen (1), 42 HLA DR3 GD patients were studied and compared with 42 HLA DR matched controls.

Southern's method was applied with three HLA-specific probes (DRβ, DQα, DQβ) and two restriction enzymes (Taq I, Bam HI).

The hybridization with DQβ probe is not informative when comparing patients with healthy controls.

The hybridization with DQα probe of the Taq I digested DNA reveals five fragments correlated with DR specificities and two 2.1Kb and 1.9Kb fragments which Spielman (2) has called respectively U and L, and which have been interpreted as alleles of the DXα gene. Genotype UU frequency is increased in GD patients as compared with controls (40.4% vs 21.4%) because of the increase in U allele in patients (65% vs 50%, $p < 0.05$). But, when the whole phenotype of patients and controls is considered, especially when comparing B8+DR3+ patients ($n = 34$) with B8+DR3+ controls ($n = 25$), the U frequency of the two groups balances (69% vs 64%) (Table 1). When the U frequency of B8+DR3+ patients ($n = 8$) is compared with that of B8+DR3+ controls ($n = 17$) the frequency balances but drops in both cases (31.2% vs 29.4%). Finally, comparison of nine non-DR3 patients with nine DR-matched healthy subjects reveals a low U frequency (22% vs 33%) which compares with that of DR3 non-B8 patients.

The hybridization with DRβ probe of Taq I digested DNA provides a definition of the involved DR3 subtype. Indeed there is a DR3 DRw52 associated polymorphism with 2 subtypes: DR3-1 Dw25 (14.5Kb+) and DR3-2 Dw24(12Kb+). Dw24 subtype comes out higher in GD patients to the distribution of the B8 DR3 haplotype in both groups (83.3% in patients vs 59.5% in controls).

This variant DR3 might play an important role in the presentation of the self antigen against the lymphocyte clones implicated in the autoimmune response. The increase of gene U frequency on the same B8 DR3(Dw24) haplotype makes it possible to extend to the DXα gene the chromosome 6 region involved in the susceptibility to the disease. Otherwise, such an over-representation of the allele U of the DXα gene, associated with a special DRβIII gene (Dw24 variant) raises the hypothesis of the existence in GD of hybrid molecules formed by the association of two chains (α and β) coded by genes from different loci (3). Such a hypothesis would imply the expression of a chain coded by the DXα gene on the cell surface; but the product of that gene has not yet been isolated.

In any case, these results assess the existence of a genetic susceptibility born by HLA in GD.

Table 1. Taq I/DQα-U Frequency

	DR3+		B8−DR3+		B8+ DR3+	
	GD	C	GD	C	GD	C
U	65.5%	50.0%	31.2%	29.4%	69.0%	64.0%

GD = Graves' disease; C = controls.

References

1. Allannic H, Fauchet R, Lorcy Y, Hein J, Guéguen M, Leguerrier AM, et al. HLA and Graves disease: an association with HLA-DRw3. J Clin Endocrinol Metab 1980;51:863.
2. Spielman RS, Lee J, Bodmer WF, Bodmer JG, Throwsdale J. Six HLA-D region α chain genes on human chromosome 6: polymorphisms and associations of DCα-related sequences with DR types. PNAS 1984;81:3461.
3. Lotteau V, Teyton L, Burroughs D, Charron D. A novel HLA Class II molecule (DRα DQβ) created by mismatched isotype pairing. Nature (in press).

Author Affiliations

G. Semana, F. Quillivic, B. Genetet, R. Fauchet, Laboraroire d'Histocompatibilité, Centre Régional de Transfusion Sanguine, Rennes; Clinique Médicale F, CHU - Rennes, France

Polymorphisms of the Human T-Cell Receptor α and β Chain Genes and Their Relationship to Insulin Dependent Diabetes Mellitus

Marie L. Hoover, Kay E. Black, Edward Ball, Marilyn Alford, James Marks, Peter Stastny, and J. Donald Capra

Abstract: The association of certain HLA-D alleles, with insulin-dependent diabetes mellitus (IDDM) is well known. However, several analyses have suggested that a second locus, unlinked to the HLA complex may be involved in this disease. The availability of genetic probes for the T-cell receptor (TCR) genes has provided an opportunity to test the hypothesis that the second locus might be one or more of the TCR genes. The human TCR is a heterodimer consisting of two glycosylated polypeptide chains, α and β (1–3). Restriction fragment length polymorphisms (RFLP) of both human genes have been described (4–11). Previously, we described an association between one of these RFLPs and IDDM (12). This report extends that analysis by examining over 150 additional patients for the associated marker and additional TCR polymorphisms. These studies document an association between the TCRβ chain polymorphism and IDDM and provide yet an additional risk factor particularly among individuals who are DR3/X (where X ≠ 4).

All the patients examined in this study were HLA-DR3 and/or DR4 Caucasians who fulfilled the criteria set by the National Diabetes Data Group. The controls were randomly selected, unrelated, normal Caucasians. Both populations were tested for the presence or absence of the polymorphic forms of the TCR genes and the allelic forms were named based on the corresponding molecular size of the hybridizing fragments. The initial screen consisted of testing 131 IDDM patients for the TCR α Bgl II polymorphism (4). There was no significant difference in the distribution of the three genotypes in the IDDM patients as compared to controls (Table 1). The discovery of an additional polymorphism within the TCR α gene, which is revealed utilizing the enzyme Taq I, enabled a more complete analysis of this locus (8). The Taq I polymorphism has been localized to a site between the most 3′ joining exon and the most 5′ constant region exon and is approximately 7 kb from the previously described Bgl II polymorphism. As these markers are not in linkage disequilibrium, we hypothesized that this Taq I marker might be associated. Seventy-four IDDM patients were screened and no association was found (Table 1). Furthermore, when the extended haplotype was determined in those individuals who were homozygous for either the Bgl II or Taq I marker, no significant difference was observed when compared to normal controls (Table 1). These results suggest that differences in the TCR α chain gene in the region from the most 3′ joining segment to the 3′ untranslated segment is not associated with IDDM. Additional studies will be required to determine if differences within the variable or joining segments of the TCR α chain gene are associated with IDDM.

Table 1. T-Cell Receptor α Chain Gene RFLPs in Normals and IDDM Patients

Phenotype or Genotype	Normal	IDDM
Bgl II 3.2/-	100 (78.1%)	114 (87.0%)
Bgl II 2.9/-	1 (0.8%)	1 (0.8%)
Bgl II 3.2/2.9	27 (21.1%)	16 (12.2%)
Taq I 7.8/-	4 (13.3%)	11 (14.9%)
Taq I 2.2/-	13 (43.3%)	35 (47.3%)
Taq I 7.8/2.2	13 (43.3%)	28 (37.8%)
Bgl II 3.2, Taq I 7.8	16 (30.7%)	44 (31.9%)
Bgl II 3.2, Taq I 2.2	34 (65.4%)	88 (63.8%)
Bgl II 2.9, Taq I 7.8	1 (1.9%)	1 (0.7%)
Bgl II 2.9, Taq I 2.2	1 (1.9%)	5 (3.6%)

In order to extend our previous observation of an association between the TCRβ chain gene and IDDM, we tested an additional 151 IDDM patients for the Bgl II TCRβ markers. The results of 256 patients and 74 HLA-DR-typed controls are depicted in Table 2. In the expanded analysis there is no association between the TCRβ chain gene and the total IDDM population, however when the patients were grouped based on DR type, there was a significant association between the TCRβ 9.3/8.6 genotype and DR3/X IDDM patients (where X ≠ 4, Table 3). Sixty percent of the DR3/X patients were heterozygous for the TCRβ constant segment markers resulting in a relative risk of 2.08. The control population was also subjected to the same analysis and there was no statistically significant association between DR type and TCRβ type in the control population (data not shown). The results presented here are largely in agreement with our previous findings, however the analysis of the larger patient population did result in the conclusion that there is no association between the 8.6 allele and IDDM in DR3/4 patients, as we previously reported (12).

The RFLP markers used in this study reflect changes in restriction sites within introns of both T α and T β.

Table 2. Distribution of T-Cell Receptor β Chain Constant Region RFLP in Normals and IDDM Patients

Genotype	Normal (n=74)	IDDM (n=256)
9.3/-	21 (28.4%)	57 (22.3%)
9.3/8.6	31 (41.9%)	132 (51.6%)
8.6/-	22 (29.7%)	67 (26.1%)

Table 3. Association of Tβ Polymorphism in DR3/X IDDM Patients

	Normals		DR 3/X IDDM
9.3/8.6			
Positive	31		42
Negative	43		28
Frequency	41.9%		60.0%
p value		0.045	
Relative risk		2.08	

Therefore, the association of the T β polymorphic alleles with IDDM most likely reflect additional mutational changes in or closely linked to the T β locus that are important in susceptibility to IDDM. These RFLP detectable differences may reflect variations within the V region genes of these two T-cell β chain alleles. The differences could result from changes, deletions, or insertion of nucleotides or entire deletions or duplications of V regions. To test this hypothesis, IDDM patients and controls were examined utilizing two different variable region probes (Vβ1 and Vβ8) which detect polymorphic differences within the variable region of the TCRβ gene locus using Taq I and Bam H1, respectively [unpublished results, (9)]. There was no significant difference in the Vβ1 genotypes of 75 IDDM patients and 52 controls as can be seen in Table 4. Similar results were obtained when 89 IDDM patients and 52 controls were tested for the presence or absence of the 2.2 kb Vβ8 hybridizing fragment. In unpublished studies conducted in conjunction with the above analysis, we have not found any of the known Vβ gene polymorphisms to be in linkage disequilibrium with the constant β gene polymorphism, therefore the lack of association between the variable region polymorphisms and IDDM is not unexpected. This suggests that differences closer to the Bg1 II polymorphism, such as in the diversity joining (Dβ2) or joining (Jβ2) genes which are important in generating diversity, may be necessary for the development of IDDM.

Although the role of the genes associated with the TCRβ chain polymorphisms described herein are still unclear, the use of the Bg1 II TCRβ RFLP marker for a gene encoded on chromosome 7 as well as similar markers for the HLA-DR and DQ loci on chromosome 6 should aid in predicting high-risk individuals in diabetic families. Furthermore, the ability to type individuals for two unlinked loci involved in susceptibility to IDDM should significantly enhance our understanding of the genetic and environmental factors involved in this disease.

Table 4. Distribution of T-Cell Receptor β Chain Variable Region RFLPs in Normals and IDDM Patients

Phenotype or Genotype	Normal	IDDM
vβ1		
3.9/–	3 (5.8%)	4 (5.3%)
3.4/–	33 (63.4%)	57 (76.0%)
3.9/3.4	16 (30.8%)	14 (18.7%)
vβ8		
2.2+	30 (57.7%)	65 (73.0%)*
2.2–	22 (42.3%)	24 (27.0%)

*Not significant.

Acknowledgments. Work supported by the Juvenile Diabetes Foundation (185574), American Cancer Society (IM413F), and the Greenwall Foundation. MLH is a Juvenile Diabetes Foundation Fellow.

References

1. Allison JP, McIntyre BW, Bloch D. Tumor-specific antigen of murine T-lymphoma defined with monoclonal antibody. J Immunol 1982;129:2293.
2. Haskins K, Kubo R, White J, Pigeon M, Kappler J, Marrack P. The major histocompatibility complex-restricted antigen receptor on T cells. I. Isolation with a monoclonal antibody. J Exp Med 1983;157:1149.
3. Meuer SC, Fitzgerald KA, Hussey RE, Hodgdon JC, Schlossman SF, Reinherz EL. Clonotypic structures involved in antigen-specific human T cell function. Relationship to the T3 molecular complex. J Exp Med 1983;157:705.
4. Hoover MH, Marks J, Chipman J, Palmer E, Stastny P, Capra JD. Restriction fragment length polymorphism of the gene encoding the alpha chain of the human T cell receptor. J Exp Med 1985;162:1087.
5. Robinson MA, Kindt TJ. Segregation of polymorphic T cell receptor genes in human families. Proc Natl Acad Sci USA 1985;82:3804.
6. Robinson MA, Kindt TJ. Molecular determination of T-cell receptor alpha and beta chain genotypes in human families. Human Immunol 1985;14:195.
7. Robinson MA, Kindt TJ. Molecular genotyping of human T-cell antigen receptor variable gene segments. Immunogenetics 1986;24:259.
8. Ball EJ, Dombrausky L, Hoover ML, Capra JD, Stastny P. Restriction fragment length polymorphism of the human T cell receptor alpha gene. I. Two polymorphic restriction sites localized to different regions of the gene. Immunogenetics 1987;26:48.
9. Flug F, Pelicci PG, Bonetti F, Knowles DM, Dalla-Favera R. T cell receptor gene rearrangements as markers of lineage and clonality in T-cell neoplasms. Proc Natl Acad Sci USA 1985;82:3460.
10. Tillinghast JP, Behlke MA, Loh DY. Structure and diversity of the human T-cell receptor β-chain variable region genes. Science 1986;233:879.
11. Siu G, Strauss EC, Lai E, Hood LE. Analysis of a human Vβ gene subfamily. J Exp Med 1986;164:1600.
12. Hoover ML, Angelini G, Ball E, Stastny P, Marks J, Rosenstock J, et al. HLA-DQ and T-cell receptor genes in insulin-dependent diabetes mellitus. Cold Spring Harbor Symp Quant Biol 1986;51:803.

Author Affiliations

Marie L. Hoover, Kay E. Black, J. Donald Capra, Department of Microbiology and the Center for Diabetes Research; *Edward Ball, Peter Stastny*, Department of Internal Medicine; *Marilyn Alford, James Marks*; Department of Pediatrics and the Center for Diabetes Research, The University of Texas Health Science Center at Dallas, Dallas TX, USA

HLA Haplotype Analysis in Danish HLA-DR3,4-Positive Insulin-Dependent Diabetics and Controls

M. Thomsen, J. Molvig, A. Zerbib, M. Abbal, C. de Preval, A. Cambon-Thomsen, A. Svejgaard, and J. Nerup

The association of insulin-dependent diabetes mellitus (IDDM) with HLA-DR3 and 4 is well established, and the increased risk of DR3,4 heterozygotes compared with DR3 or 4 homozygotes has been confirmed by numerous studies. Some reports have indicated that HLA factors other than DR might confer an increased risk for the disease, and that genetic factors not coded by the genes of the HLA region might be associated with IDDM. We wanted to apply a number of techniques to compare the distribution of HLA and non-HLA markers in a sample of DR3,4 heterozygous IDDM patients and controls from a homogeneous population. We chose to study a sample of unrelated HLA-DR3,4 heterozygous individuals from the homogeneous Danish population: 27 unrelated IDDM patients who had been HLA-A,B and DR typed in Copenhagen (included in the study by Svejgaard et al. (1), and 19 unrelated controls typed in Copenhagen and Aarhus. (The contribution of normal DR3,4-positive individuals by Dr. L.U. Lamm and Dr. L.S. Nielsen is gratefully acknowledged.) None of the controls had a family history of diabetes or other autoimmune disease.

The following markers were studied in details: 1) HLA-DR and DQ by serologic typing (two color fluorescence) with local and exchange sera and monoclonal antibodies including the specificity TA10, which splits DQw3; 2) the complement allotypes BF and C4 by electrophoresis and immunofixation; 3) the Dw specificities by use of local homozygous typing cells defining Dw1-8 and Dw14; 4) the DPw1-6 specificities by primed lymphocyte typing using local reagents and expansions of reagents obtained from R. Hartzman after the 9th Histocompatibility Workshop; and 5) restriction fragment length polymorphisms (RFLP) using informative restriction enzymes and the DQα, DQβ, Tcα, Tcβ, and ins310 probes.

The HLA region markers that are significantly more frequent in the IDDM patients than in the controls are shown in Table1. These are C4A3, C4B3, and DQw8. The frequency of B15 and Dw4 is also slightly elevated in diabetics. These markers are probably coded for by the DR4-carrying haplotype and are found together in 10 of the diabetics versus only 1 of the controls. The C4A3 and DQw8 were present in all of the diabetics, and it is of interest to note that the absence of these factors seems to be independent in the controls (only in one case is there a simultaneous absence of these two factors, i.e., 11 of the controls lack one of those factors versus none of the diabetics; $p=6\times10^{-6}$). The distribution of DP antigens showed no difference between diabetics and controls.

Three sets of markers are often found together in this material and constitute probably haplotypic associations in accordance with previous publications on family studies:

B15,BFS,C4A3B3,Dw4,DQw8;
B8,BFs,C4AQOB1,Dw3,DQw2;
and *B18,BFF1,C4A3BQO,Dw3,DQw2*.

Only the DR4-associated haplotype seems to be characteristic for the IDDM patients. The DR3-associated characters were not significantly more frequent in the patients than in the controls.

The DQα probe showed a limited allelic polymorphism displayed by presence of a 2.3 kb band (A) and/or a 2.15 kb band (B) after TaqI digestion. This polymorphism is probably caused by the DXα gene, and the B allele seems to be associated at the same time with the B18-positive DR3 haplotype mentioned above and the DR4,DQw7 haplotype found in the controls (2).

Our study thus shows that the DR4-associated markers exhibit a considerable heterogeneity, and high and low risk factors can be defined. In contrast, the DR3-associated markers show an even distribution in the DR3,4 heterozygous patients and controls. As the association with DQ and C4 markers with the susceptibility to diabetes seems to be equally strong, we might postulate that the DR4-positive haplotype in DR3,4-positive diabetics carries two diabetes suscept-

Table 1. Presence of Various Markers in HLA-DR3,4-Positive Diabetics and Controls

Marker	% Positive IDDM (n=27)	% Positive Controls (n=19)	Significance p
B8	74	74	
B18	15	5	
B15	52	32	
BFF1	15	5	
C4A3	100	68	0.003
C4B3	52	16	0.01
DQw8	100	68	0.003
Dw4	89	79	
Dxα2.3/2.3	78	58	
Tcα 7.9	60	50	
Insulin class I/I	69	53	

The p values are calculated by Fishers exact test, one-sided and without corrections for number of markers tested.

Immunobiology of HLA
Volume II

ibility genes, which both need to be present for the diabetes to develop.

The non-HLA markers studied by RFLP showed no significant association with IDDM, but it should be mentioned that the increased frequency of insulin class I (short) alleles in diabetics also has been found by others. Cells from all individuals have been established as lymphoblastoid lines, and the search for non-HLA markers associated with diabetes continues in this material with special attention to genes coding for variable regions of the T-cell receptor. In addition, a large collaborative study of more than 200 diabetic and normal families has been undertaken in a well defined area with a homogeneous population in the western Pyrenees. The application of different techniques in the search for genetic markers and the comparison of significant diabetes-associated factors in this population with those found in the present material will perhaps allow us to define more precisely the genetic risk factors for diabetes.

References

1. Svejgaard A, Jakobsen BK, Platz P, Ryder LP, Nerup J, Christy M, Borch-Johnsen K, Parving H-H, Deckert T, Mölsted-Pedersen L, Kühl C, Buschard K, Green A. HLA associations in insulin-dependent diabetes: search for heterogeneity in different groups of patients from a homogeneous population. Tissue Antigens 1986;28:237–244.
2. Zerbib A, Mölvig J, Thomsen M, Coppin H, Cambon-Thomsen A, Sommer E, Nerup J, de Preval C. Restriction fragment length polymorphism of HLA and non HLA genes in DR3/DR4 heterozygous Danish insulin-dependent diabetic patients and healthy individuals. Disease Markers 1988, in press.

Author Affiliations

M. Thomsen, A. Zerbib, M. Abbal, C. de Preval, A. Cambon-Thomsen, Inserm U100, CHU Purpan, 31052 Toulouse Cedex, France; *J. Molvig, J. Nerup*, Steno Memorial Hospital, Gentofte, Denmark; *A. Svejgaard*, Tissue Typing Laboratory, University Hospital, Copenhagen, Denmark

DR4 Associated DQw3 Subdivision in Insulin-Dependent Diabetes with Polyendocrinopathies: Serology and RFLP

G. Semana, H. Allanic, F. Quillivic, B. Genetet, and R. Fauchet

Numerous studies have revealed an association between HLA DR antigens and insulin dependent diabetes mellitus (IDDM). The highest risk of the occurrence of the disease is related to the HLA-DR3-DR4 phenotype (1). The aim of this study is to investigate the mechanisms of genetic transmission in patients with IDDM associated with another autoimmune endocrinopathy (IDDM-AE). For that purpose, 73 members of ten IDDM-AE families were studied, each family including a proband with IDDM and another AE (Graves' disease = 6, myxoedema = 3, anterior pituitary gland deficiency = 1). Moreover, six family members had only IDDM, and two others had another isolated AE (1 GD, 1 myxoedema). Using Southern blot analysis, the DNA from those 73 subjects were digested with Taq I and Bam HI restriction enzymes and hybridized with HLA-specific probes DRβ, DQα, and DQβ. Restriction patterns were compared to those of a panel of cells obtained from healthy unrelated controls (20 DR4 cells and 20 DR1 to DRw9 homozygous cells). All the subjects were HLA typed for A,B,DR and DQ loci (including TA 10 DQw3 subdivision). RFLP were correlated with serologic typing. Hybridization with DQα and DRβ specific probes did not yield significantly different patterns between family members and controls.

Hybridization with the DQβ specific probe separated DR4 DQw3 controls into two variants. This DQw3 polymorphism is defined as follows: DQw3-2 (Taq I:1,8 Kb; Bam HI: 12 Kb) present in 60% of DR4 DQw3 cells and DQw3-1 (Taq I:4,5 Kb; Bam HI:6,4 Kb,3,3 Kb) in the other 40% (2).

The distribution and segregation of the DQw3 subtypes in IDDM-AE families were analyzed. All the DR4 IDDM-AE probands (n = 7) carried the DQw3-2 variant. When other family members presented an isolated IDDM (n = 6), those patients also carried the DQw3-2 subtype (Table 1). In order to correlate RFLP results and DQw3 TA 10 subtypes, serological analysis was carried out on 16 DR4 cells for which RFLP had been performed (6 patients and 10 controls). In each case, there was perfect correlation between DQw3 TA 10− and DQw3-2 (DQw8) and between DQw3 TA 10+ and DQw3-1 (DQw7).

As a conclusion, this study suggests that in IDDM-AE, the susceptibility to the disease involves the HLA DQ subregion genes rather than the HLA DR genes themselves. The DQw8 variant (3) may play a prominent

Table 1. DQw3 Subtypes Distribution

	DR4 Controls (n=30)	DR4 IDDM-AE (n=7)	DR4 IDDM (n=6)
DQw8/DQw3.2	60%	100%	100%
DQw7/DQw3.1	40%	0%	0%

Immunobiology of HLA
Volume II

role in the occurrence of the disease. In the studied families, when other family members had an isolated IDDM, in all cases (n = 6), they also carried the DR4 DQw8 haplotype. On the whole, the combination of serological and RFLP studies provides a better definition of high risk haplotypes involved in IDDM.

References

1. Svejgaard A, Platz P, Ryder LP. Insulin dependent diabetes mellitus. In: Terasaki PI(ed): Histocompatibility Testing 1981, pp 287-300.
2. Nepom BS, Palmer J, Kim SJ, Hansen JA, Holbek S, Nepom GT. Specific genomic markers for the HLA-DQ subregion discriminate between DR4+ IDDM and DR4+ JRA. J Exp Med 1986;164:345-50.
3. Radka SF, Scott RG, Stewart SJ. Molecular complexity of HLA-DQw3: The Ta10 determinant is located on a subset of DQw3 B chains. Hum Immun 1987;18:287-300.

Author Affiliations

G. Semana, F. Quillivic, B. Genetet, R. Fauchet, Laboratoire d'Histocompatibilité, Centre Régional de Transfusion Sanguine, Rennes; *H. Allannic*, Clinique médicale F-CHU, Rennes, France

Polymorphism of HLA-DR4 Associated Gene Products in Rheumatoid Arthritis

Jerry S.S. Lanchbury, Lazaros I. Sakkas, Steven G.E. Marsh, Lorna J. Kennedy, Julia G. Bodmer, and Gabriel S. Panayi

Rheumatoid arthritis (RA) is associated with the public allospecificity HLA DR4 in Caucasian and other populations (1). Since DR4 and its linked DQw3 gene products are heterogeneous (2,3), molecular studies may be useful in discriminating gene product polymorphisms more closely associated with RA. We have examined DRβ, DQβ, and DQα polymorphisms in a panel of DR4 HTC and report data on a group of DR4 homozygous RA patients.

HLA class II gene products DR, DQ, and DP were identified by 2-DE (Tenth Histocompatibility Workshop conditions and reagents) using subregion specific MAbs L243 (DR), Tü22 (DQ), and B7/21 (DP) on a panel of normal DR4 HTC. L243 immunoprecipitated families of DR molecules corresponding to the variably glycosylated products of DRBl and DRB4 genes. Identity of products corresponding to the DRw53 antigen was confirmed by use of MAb PL3 (4). In contrast to the lack of variability apparent in the DRβ4 product, DRβl molecules from DR4 HTC of various Dw types were structurally polymorphic (Table 1).

Although Tü22 patterns contained minor DRβ-related components, DQβ polymorphism was evident among HTC with two common allelic electrophoretic variants 3.1 (acidic) and 3.2 (basic). Presence of either β-chain variant correlated with the serologic DQw7/IIB3 polymorphism (5). β3.1 molecules were associated with DQw7 positivity and the β3.2 variant with the IIB3 specificity. DQα electrophoretic polymorphism was also detected with two alleles, α3.1 and α3.2. Haplotypes

Table 1. Summary of Relationship Between HLA Class II Polymorphisms Defined by Serologic, Cellular, and Biochemical Methods on a Panel of DR4 HTCs

Name	10W*	Dw	DRβ1 2-DE	DQβ 2-DE	DQα 2-DE	DQw7	IIB3	DQw4
BSM	9032	4	4/14	3.2	3.2	–	+	–
AS		4	4/14	3.2	3.1	–	+	–
BOLETH	9031	4	4/14	3.2	3.2	–	+	–
DEU	9025	4	4/14	3.1	3.1	+	–	–
1296S		4	4/14	3.1/3.2	NT	+	+	–
MTF		14	4/14	3.2	NT	–	+	–
YAR	9026	10	10	3.2	NT	–	+	–
JHAF	9030	13	13	3.1	3.1	+	–	–
KT2		KT2	13	3.2	NT	–	+	–
KT17	9024	KT2	13	3.2	3.2	–	+	–
SJAH		15	15	WA	NT	–	wk+	+

*10th Histocompatibility Workshop cell line identity number.
†DRβ_1 biochemical nomenclature reflects Dw polymorphism associated with each electrophoretic variant.

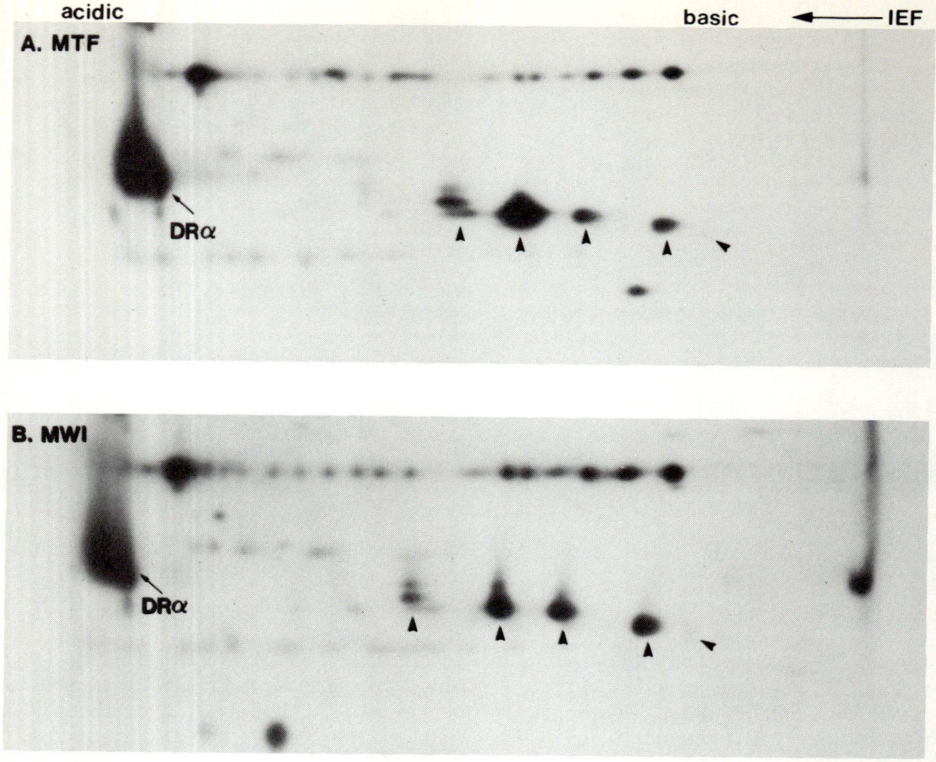

Figure 1. 2-DE gel profiles of HLA class II molecules immunoprecipitated with MAb L243. Identity of DRβ molecules from **A**, MTF, a normal DR4 Dw14 HTC; and **B**, MWI, a homozygous DR4 rheumatoid arthritis patient. Vertical arrowheads indicate DRβI glycoproteins and the oblique arrowhead the major DRw53β product.

exist such that α3.1 is generally associated with β3.1 and α3.2 with β3.2. Exceptionally, the HTC AS expresses α3.1 in association with β3.2 and may represent the result of an intra DQ region recombination event. DQ molecules from HTC SJAH (Dw15) typed DQw3 negative, DQw4 positive, weakly IIB3 positive, and were relatively unreactive with Tü22. They reflect additional DQ polymorphism associated with DR4. A summary of data relating to DQ polymorphisms is given in Table 1.

Epstein-Barr virus B lymphoblastoid cell lines were established from RA patients after serologic and RFLP screening. Genomic DNA digested with TaqI restriction endonuclease was hybridized with DRB and DQB cDNA probes. Patients positive for only DR4-associated DRB fragments (14.5, 6, 5.6, and 2.6 kb) and DQw3-associated DQB fragments (2 and 4.5 kb) were serologically typed for class II antigens. RA patients who typed for DR4, DRw53, and DQw3 antigens only, were then investigated for biochemical HLA class II polymorphisms. DR4 RA patients (n = 8) expressed only a single DRβ1 pattern, previously shown associated with Dw4 or Dw14 HTC (see Fig. 1 for an example). They thus represent individuals homozygous or heterozygous for either Dw4 or Dw14-associated products. At the DQβ level, however, patients may express either DQβ 3.1, 3.2 or both variants.

We have identified molecular polymorphisms of HLA-D region gene products DRβ, DQβ, and DQα that exist within DR4 haplotypes and have demonstrated the potential application of the technique to the dissection and identification of susceptibility determinants more specific to RA than DR4 alone.

Acknowledgments. This work was supported by the Arthritis and Rheumatism Council of Great Britain.

References

1. Stasny P, Joint report: Rheumatoid arthritis. In: Terasaki PI(ed): Histocompatibility Testing 1980. Los Angeles, UCLA Tissue Typing Press, 1980, pp 681-686.
2. Nepom GT, Nepom BS, Antonelli P, Mickelson E, Silver J, Goyert SM, et al. The HLA-DR4 family of haplotypes consists of a series of distinct DR and DS molecules. J Exp Med 1983;159:394.
3. Bontrop RE, Schreuder GMT, Mikulski EMA, van Milten-

burg RT, Giphart MJ. Polymorphisms within the HLA-DR4 haplotypes. Various DQ subtypes detected with monoclonal antibodies. Tissue Antigens, 1986;27:22.
4. Knowles RW, Flomenberg N, Horibe K, Winchester R, Radka SF, Dupont B. Complexity of the supertypic HLA-DRw53 specificity: Two distinct epitopes differentially expressed on one or all of the DR β-chains depending on the HLA-DR allotype. J Immunol 1986;137:2618.
5. Schreuder GMT, Tilanus MGJ, Bontrop RE, Bruining GJ, Giphart MJ, van Rood JJ, et al. HLA-DQ polymorphism associated with resistance to type I diabetes detected with monoclonal antibodies, isoelectric point differences, and restriction fragment length polymorphism. J Exp Med 1986;164:938.

Author Affiliations

Jerry S.S. Lanchbury, Lazaros I. Sakkas, Gabriel S. Panayi, Molecular Immunogenetics Laboratory, Rheumatology Unit, Department of Medicine, United Medical Schools of Guy's & St. Thomas' Hospitals, Guy's Hospital, London; *Steven G.E. Marsh, Lorna J. Kennedy, Julia G. Bodmer*, Tissue Antigen Laboratory, Imperial Cancer Research Fund, London, UK

New HLA DNA Polymorphisms Associated with Rheumatoid Arthritis

Hilliard Festenstein, Shirley D. Cutbush, William E. Ollier, Juliette Awad, Chryssa Pachoula Papasteriadis, and Michael J. Garlepp

Restriction fragment analysis of DNA from two DR4/Dw4 consanguineous homozygous typing cells (HTC) using a DQβ cDNA probe has allowed definition of two distinct patterns (1). These have been designated Φ (represented by HTC MCF) and Ω (HTC BM14) (Table 1). The two HTCs could also be distinguished using a DQα probe after Taq 1 digestion. BM14 had a characteristic 2.1 Kb fragment (U) and MCF a 1.9 Kb fragment (L). These are thought to represent alleles of the DXα gene (1). A DRβ probe gave identical patterns for all the enzymes used (Table 1).

The monoclonal antibody A-10-83 allowed further distinction. TA10 positivity was strongly, although not absolutely, associated with the Φ pattern (1). Other groups (2–4) have independently distinguished between these two subtypes of DR4/Dw4-DQw3 (Table 1). The DQβ restriction fragment patterns, the DQα RFLPs, and TA10 reactivity segregated together with MHC haplotypes in family analyses. Further, when data on unrelated controls (n = 29), patients with rheumatoid arthritis (RA) (n = 39) and insulin dependent diabetes mellitus (n = 46) were analyzed, they were strongly associated with each other and with specific HLA determinants. This allowed the construction of two preferential allelic associations (PAA), one containing Ω/U and the other Φ/L;

PAA 1 = DXαU-DQβΩTA10 negative-DQw3-Dw4-DR4---Bw62-Bw6-Cw3-A2
PAA2 = DXαL-DQβΦTA10 positive-DQw3-Dw4-DR4---B44 -Bw4-Cw3-A2.

These two PAAs contribute a major portion of the DR4 seen in our population of RA patients. The frequencies of these restriction fragment patterns were not statistically different in DR4-positive RA patients and DR4-positive controls. This would argue against the concept that DR4/DQw3 in RA is distinct from that in healthy individuals. Because RA is clinically heterogeneous, we subdivided our DR4 positive patients according to clinical criteria to determine whether a particular PAA, marked by either of these restriction fragment patterns, is associated with any particular patient subgroup (Table 2).

Table 1. DQβ Restriction Fragment Patterns in DR4/Dw4 Bearing PAAs

Designation		EcoRI	Pst I	Pvu II	Taq I
Ω	[1]	19[a]	12.5	3.3	3, 2
DQ3.2	[2]	NA[b]	NA	3.8	1.9, 2.6
DQR4	[4]	21	NA	3.7	2.3, 2.0
DQ3.DR4*	[3]	NA	14	NA	NA
Φ	[1]	13	4.4	2.6	2.2, 1.8
DQ3.1	[2]	NA	NA	2.6	1.5
DQR5	[4]	12	NA	2.1	2.1
DQ3.DR4†	[3]	NA	4.2	NA	NA

*Kilobase pairs; †information not available.

Table 2. Comparison of DQβΩ/Φ Frequencies in DR4 Positive British Caucasoid RA Patients and Controls

–	*n*	DQβΩ	DQβΦ
RA (total)	39	46%	67%
EA Positive	22	59%[a]	55%
EA Negative	17	29%[b]	82%
NER Positive	21	33%	86%[c]
NER Negative	18	61%	44%[d]
Controls	29	31%	72%

a vs. b; $p<0.05$; c vs. d; $p<0.008$; NER = Nodules and erosions; EA = extraarticular.

Immunobiology of HLA
Volume II

The Ω pattern was seen more frequently in patients with extraarticular manifestations (EA), e.g., vasculitis, neuropathy, Sjogren's syndrome, than in those without, whereas the Φ pattern was seen more frequently in those with nodules and erosions (NER).

Of further interest was the observation that a 6 kb Taq1 DQβ hybridizing fragment was demonstrated to be increased in frequency in RA patients (51%) compared to controls (20%). This increased frequency was particularly evident in females (59% compared to 29% in males). This fragment was seen in a number of DR types but was most strongly associated with DR1 and DR4.

Taken together these data suggest an important role for the DQ region of the MHC in the pathogenesis of RA. The fact that the Φ pattern includes TA10 but that the association is not absolute suggests further diversity in the region. It remains to be determined whether these polymorphic restriction sites represent markers for other genes in linkage disequilibrium or represent functionally different alleles of genes that predispose to RA and which help determine disease presentation. It is notable that in Greek patients with RA DR4 frequency is not increased (5) and that when DR4 is present, neither the complete Ω not Φ patterns are seen although components of each are represented. The Φ pattern includes two bands at 4.6 and 4.4 kb, whereas Ω has only the 4.4 kb band, but possesses a band at 2 kb with is not seen in Φ. The DR4 Greek patients studied are lacking both the 4.6 kb and 2 kb bands. This provides further indication that the PAAs represented by these patterns are important in RA and may provide clues as to the most critical sequences in the development of RA and its complications.

Acknowledgments. This work was supported by the Arthritis & Rheumatism Council of Great Britain & The Commonwealth (H.F., S.D.C., W.E.O.).

References

1. Festenstein H, Awad J, Cutbush S, Groves AV, Cassell P, Ollier W, et al. New HLA DNA polymorphism associated with autoimmune diseases. Nature 1986;322:64.
2. Holbeck SL, Nepom GT. Exon-specific oligonucleotide probes localise HLA DQβ allelic polymorphisms. Immunogenetics 1986;24:251.
3. Henson V, Maclaren N, Riley W, Wakeland EK. Polymorphisms of DQβ genes in HLA-DR4 haplotypes from healthy and diabetic individuals. Immunogenetics 1987;25:152.
4. Cohen-Haguenaur O, Robbins E, Massart C, Busson M, Deschamps I, Hors J. A systematic study of HLA Class II-β DNA restriction fragments in insulin-dependent diabetes mellitus. Proc Natl Acad Sci USA 1985;82:3335.
5. Papsteriadis CA, Kappou ID, Skopouli FN, Barla MN, Fostiropoulos GA, Moutsopoulis HM. Lack of HLA-antigen association in Greek rheumatoid arthritis patients. Rheumatol Int 1985;5:201.

Author Affiliations

Hilliard Festenstein, Shirley D. Cutbush, William E. Ollier, Juliette Awad, Michael J. Garlepp, Department of Immunology, The London Hospital Medical College, Turner Street, London El 2AD, UK; *Chryssa Pachoula Papasteriadis*, Department of Immunology, Evangelismos Hospital, Athens, Greece

DR4 and DR1 Define Different Subsets of Patients with Rheumatoid Arthritis

Peter Stastny, Nancy Olsen, Theodore Pincus, Muhammad A. Khan, and Edward J. Ball

The association of DR4 with rheumatoid arthritis (RA) is well established (1,2). However, it is known that a substantial number of subjects develop RA without having DR4. DR1 has also been observed to be increased in some studies. In a new series of patients, DR1 was found in 53.8% of DR4-negative RA patients and in 25% of DR4-negative normal controls (RR = 3.5, $p < 0.0001$). The relationship between DR4-associated and DR1-associated risk factors for development of RA is not well understood. DR1,DR4 heterozygotes occurred with increased frequency but not in excess of the expected from the frequency of the genes. Thus, having both DR1 and DR4 did not result in a much larger risk. It is not clear however, whether DR4 and DR1-associated risk factors are identical or whether they confer susceptibility for similar yet distinct RA-like syndromes.

The differences observed include rheumatoid-factor production, severity of x-ray changes, development of Felty's syndrome and vasculitis, and the possible interaction with risk factors related to an allelic form of the T-cell receptor β gene.

Rheumatoid factor production was again highly associated with DR4, although patients without rheumatoid factor did not have a significant association with this antigen (Table 1). In contrast, DR1 conferred risk for both groups of RA. Similarly, severity of radiographic changes by x-ray score, was associated with DR4 and not with DR1 (Table 2). Felty syndrome has been reported to be associated with a higher frequency of DR4 (3). A T-cell receptor V-β-8 allele has been reported to be increased predominantly in DR4-positive RA patients (4).

Immunobiology of HLA
Volume II

Table 1. HLA Antigens in RA with and Without Rheumatoid Factor

Class II Antigens	Normal (n=180) %	RF-Positive (n=116) %	RF-Positive RR	RF-Negative (n=23) %	RF-Negative RR
DR4	33.3	67.2	4.1‡	43.1	1.5 NS
DR1	18.3	31.9	2.1†	39.1	2.9*

*$p < 0.05$; †$p < 0.01$; ‡$p < 0.0001$; NS = not significant.

These results suggest that DR4 and DR1 are associated with distinct subsets of patients. It is possible that different immune responses involving either DR4 or DR1 restriction determinants are involved in susceptibility for clinically similar, yet distinguishable diseases classified as definite RA by fulfillment of five or more ARA criteria.

Table 2. HLA and Radiographic Changes in RA

Class II Antigens	Number Patients	X-Ray Score	p
DR4+	100	1.4	
DR4−	54	1.15	0.03
DR1+	23	1.25	
DR1−	29	1.08	0.35

References

1. Stastny P. Mixed lymphocyte cultures in rheumatoid arthritis. J Clin Invest 1978;57:1148-1157.
2. Stastny P. Association of the B-cell alloantigen DRw4 with rheumatoid arthritis. N Eng J Med 1978;298:869-871.
3. Dinant HJ, Muller WH, van den Berg-Loonen EM, Nijenhius LE, Engelfriet CP. HLA DRw4 in Felty's syndrome. Arthritis Rheum 1980;23:1336.
4. Ball EJ, Dombrausky L, Hoover M, Pincus T, Olsen N, Khan MA, et al. Rheumatoid arthritis is associated with a polymorphic restriction fragment of the human T-cell receptor beta gene. Abstract. American Rheumatism Association 51st Annual Scientific Meeting, 1987.

Author Affiliations

Department of Internal Medicine, The University of Texas Health Science Center at Dallas, Southwestern Medical School, Dallas, TX 75235; Division of Rheumatology and Immunology, Vanderbilt University, Nashville, TN; Cleveland Metropolitan General Hospital, Cleveland, OH, USA

Familial Rheumatoid Arthritis: Genetic Epidemiology of 75 White Multicase Families

T.S. Yamashita, M.A. Khan, T.L. Reynolds, F. Wolfe, and P. Stastny

The known association between rheumatoid arthritis (RA) and HLA-DR alleles DR4 and DR1 supports the view that genetic susceptibility plays an important role in the pathogenesis of this disease (1,2). Khan et al. (3) reported the first study of HLA antigens in familial RA. We have since extended that study to further define the basis for inheritance of genetic susceptibility to RA, and now report our results based on a study of 75 unrelated Caucasian families from the Midwest, 22 from Wichita, Kansas, and almost all of the remaining from Cleveland, Ohio. Each family had two or more cases of adult-onset definite or classical RA (by ARA criteria). HLA typing for alleles of HLA-A,-B,-C, and -DR loci was performed by the standard microlymphocytotoxicity test; specificities were assigned according to the Ninth Histocompatibility Workshop.

There were 385 males (58 with RA, 24 of them probands) and 419 females (127 with RA, 51 of them probands) (4). The proportion affected among females was higher than among males (31% versus 15%, $p < 0.0005$). The age of onset did not differ between the sexes (40.2 years for males and 41.3 for females). Eighty percent of the patients had serum rheumatoid factor (RF) (titer ≥ 1:40) by latex agglutination method, 60% had antinuclear antibodies (ANA) (titer ≥ 1:10) using rat kidney as substrates, and 75% had circulating immune complexes (CIC) (> 10% binding) by Clq binding assay. There were no differences in these serological features between male and female patients.

Among the 138 RA patients typed for HLA-DR alleles, 69% had DR4, 24% had DR1, 86% had either DR4 or DR1, or both, and 53% of the DR4(−) patients had DR1. Among the 289 unaffected family members, the corresponding values were 54%, 22%, 70%, and 33%. DR4 frequency in patients does not differ among males (71%) and females (68%). Among unaffected individuals, 60% of males and 49% of females carried DR4 ($p < .053$). The observed frequencies of DR4 genotypes (homozygous, heterozygous, and those lacking it) among affected and unaffected individuals differed significantly ($p < .005$). However, the observed genotype frequencies among affected and unaffected members were in agreement with the expected values under the Hardy-Weinberg equilibrium (Table 1). RF and ANA were more common among DR4(+) than DR4(−) individuals ($p < .03$). There was a direct correlation between the

Table 1. Frequencies of HLA-DR4 Genotypes

HLA-DR4 Genotype	Unaffected		Affected		Percent Affected	
	Obs	Exp	Obs	Exp	Females	Males
+/+	36	32.2	30	28.3	53%(23/ 43)	30%(7/23)
+/−	121	128.6	65	68.4	43%(42/ 98)	26%(23/88)
−/−	132	128.2	43	41.3	28%(31/110)	18%(12/65)

number of DR4 haplotypes carried (2,1, or 0) and the percent affected (53%, 43%, and 28% in females ($p < .026$), and 30%, 26%, and 18% for males ($p < .052$) (Table 1).

Affected sib-pair analysis in 37 families established linkage between HLA and RA since 18 affected sib-pairs shared both HLA haplotypes, 15 shared one, and 4 shared none, significantly different ($p < .01$) from 9.25, 18.5, and 9.25 expected if the disease and HLA were independent. The degrees of disease penetrance for males and females were 21% and 45%, respectively, under a recessive model; and 19% and 41%, respectively, under a dominant model. However, there was strong evidence of heterogeneity among pedigrees with respect to genetic models ($p < .0005$), and some pedigrees seemed to fit better with a recessive model than a dominant model. The log likelihood was maximum when the recombination fraction was 0 for both males and females. The log score was 2.42 under a dominant and 0.62 under a recessive model. The overall fit with a dominant linkage model was better than with a recessive model ($p < .0005$).

Acknowledgments. This work was supported by the National Institute of Arthritis, Diabetes, and Digestive and Kidney Diseases through grant AM 20618 to the Northeast Ohio Multipurpose Arthritis Center, and by the Irma Bender Arthritis Research Fund.

References

1. Stastny P. 1979. Association of the B-cell alloantigen DRw4 with rheumatoid arthritis. N Eng J Med, 1979;298: 869–871.
2. Stastny P, Ball EJ, Khan MA, Olsen NJ, Pinculs T, Gao XJ. HLA-DR4 and other genetic markers in rheumatoid arthritis. Br J Rheumatol 1988;27(suppl II):132–138.
3. Khan MA, Kushner I, Ballou SP, Braun WE. Familial rheumatoid arthritis and HLA-DRw4. Lancet 1979;1:921–922.
4. Khan MA, Yamashita TS, Reynolds TL, Wolfe F, Khan MK. HLA-DR4 genotype frequency and gender effect in familial rheumatoid arthritis. Tissue Antigens 1988;31:254–258.

Author Affiliations

T.S. Yamashita, Departments of Pediatrics and Epidemiology & Biostatistics, Case Western Reserve University, Cleveland, OH 44106, USA; *M.A. Khan* and *T.L. Reynolds*, Department of Medicine, Case Western Reserve University at Cleveland Metropolitan General Hospital, Cleveland, OH 44109 USA; *F. Wolfe*, Arthritis Center, Wichita, KA 67214, USA; and *P. Stastny*, Department of Internal Medicine, University of Texas, Dallas, TX 75235, USA

An HLA DRw52 Split Defined by Restriction Fragment Length Polymorphism: Population Genetics in Normals and in an HLA DRw52-Associated Autoimmune Disease

May K. Chatila, Lorie Luyrink, James McEleney, Jack L. Strominger, and David N. Glass

Using an HLA DR-specific repeated element probe (1) and the RFLP technique, two HLA DRw52-associated alleles, A and a, were defined with BamH I and five other restriction endonucleases, the two BamH I bands being 5.8 and 2.9 Kb (Fig. 1). The gene frequencies of A/a were 0.16 and 0.20, respectively, in 50 normal controls. The distribution relative to 0 (non-A/a alleles) met Hardy Weinberg equilibrium conditions ($G^2 = 1.4$, df = 3, $p > 0.5$) (Table 1). In DNA from 36 HTCs and the 50 normal controls, allele A correlated with HLA DRw52 in HLA DR5, HLA DR3 (non-HLA A1, B8) and some HLA DRw6 haplotypes; whereas, allele a correlated with HLA DRw8, HLA DRw3 (HLA A1, B8) and the remaining HLA DRw6 haplotypes. The RFLPs designated as A/a appear to correlate with variant HLA DRw52b and RFLP HLA DRw52a, respectively, described by others (2).

These RFLPs were studied in 33 patients with HLA DRw52 associated juvenile rheumatoid arthritis (JRA) and in 46 HLA DRw52+ controls; 75.8% of HLA

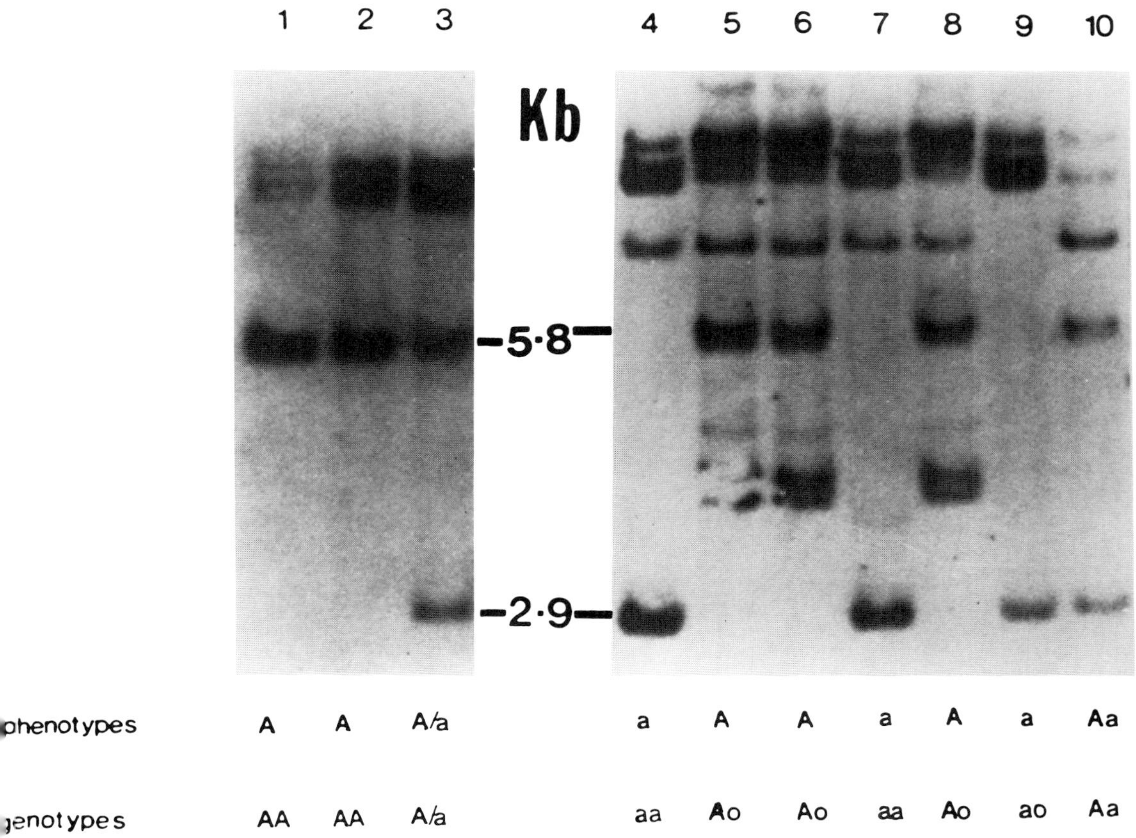

Figure 1.

Table 1. Hardy-Weinberg Equilibrium Conditions in Unselected Controls

Genotypes	Observed	Expected
AA	1	1.3
Ao	12	10.2
aa	3	2.0
ao	12	12.8
Aa	2	3.2
oo	20	20.4
Totals	50	50

Total $G^2 = 1.4$, df = 3, $p > 0.5$.

Table 2. Distribution of Genotypes in HLA-DRw52⁺ Patients and HLA-DRw52⁺ Controls

	Patients		Controls		Row Contribution	
Genotypes	No.	%	No.	%	to G^2*	RR†
AA	5	15.2	5	10.9	0.3	4.3
Ao	5	15.2	19	41.3	4.7	1.1
aa	10	30.3	4	8.7	5.0	10.8
ao	3	9.0	13	28.2	3.8	1.0
Aa	10	30.3	5	10.9	3.8	8.7
Totals	33	100.0	46	100.0		

*Total $G^2 = 17.65$, df = 4, p0 = 0.0014; †RR = relative risk (versus ao reference group).

DRw52+ patients but only 30.5% of controls were homozygous (AA, aa) or double heterozygotes (A/a) for the two alleles ($G^2 = 17.65$, df = 4, $p = 0.0014$) (Table 2). A and a contributed in an approximately equal manner to risk for this disease. In contrast, no more than 20% of patients were homozygous or double heterozygous for HLA DR5 and DRw8, the pair of HLA DRβ I coded alleles most strongly associated with this type of JRA (2). These findings and those of others suggest that both the HLA DRβ III and HLA DRβ I genes and their products will prove to play a role in disease pathogenesis perhaps on the basis of shared epitopes.

References

1. Spies T, Sorrentino R, Boss J, Strominger JL, Structural

organization of the DR subregion of the human major histocompatibility complex. Proc Natl Acad Sci USA 1985;82: 5165.
2. Gorski J, Mach B. Polymorphism of human Ia antigens: Gene conversion between two DR beta loci results in a new HLA-D/DR specificity. Nature 1986;322:67.
3. Howard JF, Sigsbee A, Glass DN. HLA genetics and inherited predisposition to JRA. J Rheumatol 1985;12:7.

Author Affiliations

May K. Chatila, Lorie Luyrink, James McEleney, David N. Glass, Department of Rheumatology and Immunology, Brigham and Women's Hospital, Boston; *Jack L. Strominger*, Dana Farber Cancer Institute, and Department of Molecular Biology and Biochemistry, Harvard University, Boston, MA02115, USA

Seropositive Rheumatoid Arthritis Is Associated with T-Cell Epitopes Encoded Within the Third Hypervariable Region of the HLA-DR1 and the HLA-DR4 Haplotype

C.M. Weyand and J. Goronzy

Major histocompatibility complex (MHC) class II molecules, which function to restrict the antigenic recognition of T lymphocytes, are important genetic factors in the susceptibility to seropositive rheumatoid arthritis (RA) (1–4). By utilizing alloreactive human T-cell clones we have defined disease-associated T-cell epitopes encoded within the HLA-DR subregion of patients with RA (2). Interestingly, such T-cell clones can recognize these T-cell epitopes outside of the HLA-DR4 haplotype, partially overcoming the incomplete genetic linkage between DR4 and the disease. We have now generated a panel of T-cell clones that identify multiple T-cell recognition sites shared among the DR molecules of HLA-DR4 and DR1. A segment of 15 amino acids within the third hypervariable region of the HLA-DRβ1 chain is of primary importance for the formation of these clustered T-cell epitopes. Within HLA-DR1 the T-cell clones identify a cassette of polymorphic sites identical to the third hypervariable region of a DR4 allele. The finding that the cross-reactive T-cell determinants are frequently expressed on antigen-presenting cells of RA patients suggests that these polymorphic structures are of particular relevance for T-cell-HLA interaction in RA.

To analyze the function and polymorphism of class II molecules associated with rheumatoid arthritis, we have established alloreactive, proliferative human T-cell

Table 1. Reactivity Pattern of Five Different T-Cell Clones on Homozygous Typing Lines

	HLA-DR	HLA-Dw	F3-29 (cpm)	GD2-30 (cpm)	C3-4 (cpm)	JA4-3 (cpm)	E1-10 (cpm)
BVR	1		3,200	0	4,151	9,244	9,244
MVL	1		26,084	0	6,577	4,492	8,259
OOS	1		30,730	6,862	3,883	13,417	19,229
BOD	4	4,4	0	0	438	378	22,108
BAR	4	4,4	76	102	n.t.	347	27,688
BSM	4	4,4	111	0	520	95	7,667
JAH	4	4,4	388	311	0	295	19,481
AL10	4	10,10	656	532	504	181	51
TS10	4	10,10	134	107	160	195	299
SST	4	13,13	30,311	0	n.t.	1,478	0
JHA13	4	13,13	22,297	0	67	2,051	690
KT2	4	13,13	21,312	0	302	12,265	811
LS40	4	14,14	34,041	3,643	11,250	5,078	2,044
THO	4	14,14	11,203	6,508	13,203	5,314	1,059
BIN40	4	14,14	20,957	15,344	16,887	6,255	662
ISA	4	14,14	30,279	7,071	10,448	13,398	2,510
HAS15	4	15,15	105	192	8,772	0	821

1×10^4 T cells were cocultured with 5×10^4 irradiated (10,000 rad) stimulator cells. Microcultures were pulsed with 1 μCi of [^{3}H]thymidine after 48 h and harvested after 60 h. Results are expressed as cpm representing thymidine incorporation of the clone in the presence of stimulator cells minus cpm incorporated by stimulators alone.

Immunobiology of HLA
Volume II

Table 2. Expression of Different T-Cell Epitopes on Cells of Patients with Seropositive Rheumatoid Arthritis

PAT	HLA-DR	F3-29 (cpm)	GD2-30 (cpm)	C3-4 (cpm)	JA4-3 (cpm)	E1-10 (cpm)
1	1,4	36,325	8,037	8,275	11,776	15,719
2	4,w9	28,890	0	4,897	0	0
3	1,4	0	0	7,599	0	6,106
4	4,w10	10,171	0	363	0	26,171
5	1,4	27,970	4,314	5,910	8,832	17,905
6	4,-	25,956	6,296	0	10,084	30,871
7	4,w10	27,918	9,074	5,910	8,324	12,210
8	1,4	28,201	4,390	10,808	9,280	16,951
9	4,-	33,843	0	10,132	0	0
10	4,5	29,786	3,176	5,066	4,710	329
11	4,w8	29,762	8,136	6,079	5,290	0
12	4,5	30,037	0	3,377	0	0
13	4,-	28,937	0	13,003	0	0
14	1,1	23,974	3,629	4,053	5,184	24,079
15	1,5	25,064	3,357	3,540	4,593	20,684

All patients tested had severe form of seropositive RA according to the American Rheumatism Association criteria. B cells of patients were transformed with Epstein-Barr virus. 5×10^4 irradiated B lymphoblasts were used to stimulate the different T-cell clones. Proliferation of T-cell clones was measured as described in Table 1.

clones (5,6). Clones were generated from peripheral blood lymphocytes of normal donors which were initially stimulated with B lymphoblasts of HLA-DR4$^+$ RA patients. To characterize the fine specificity of expanded clones, they were tested with homozygous typing lines representing the different subtypes of the HLA-DR4 family. We established 10 independently derived alloreactive T-cell clones, all of which were specific for T-cell epitopes within the DR4 haplotype (7). At least seven different reactivity patterns allowed the definition of multiple T-cell determinants and demonstrated an extensive polymorphism of T-cell stimulatory structures. Some of the clones were specific for a certain allele; most of the clones, however, reacted with determinants shared among different alleles. Surprisingly, 5 of the 10 clones tested cross-reacted with HLA-DR1$^+$ stimulator cells (Table 1). The specificity for a DR4-encoded T-cell epitope was not predictive for the reactivity of the different clones towards the DR1 haplotype. T-cell epitopes usually expressed by Dw14$^+$, Dw13$^+$, Dw15$^+$, and Dw4$^+$ stimulator cells were identified within the DR1 haplotype by using the described panel of alloreactive clones.

The clones utilized to define the DR1-DR4 shared T-cell recognition sites were originally stimulated against cells of RA patients. Thus, we tested the expression of these T-cell epitopes in a group of patients with severe seropositive rheumatoid arthritis. B lymphoblasts of the patients were used as stimulator cells for the different clones; the results of the coculture experiments are shown in Table 2. All patients tested carried T-cell epitopes described by the five different clones. Most of the patients were characterized by a concomitant expression of multiple of these epitopes. Thus, these clones can be used to study class II molecules in RA patients and to type antigen-presenting cells of patients for the presence of the defined T-cell sites.

To determine the molecular basis of these T-cell epitopes we attempted to localize the different determinants recognized within the DR4 as well in the DR1 haplotype. Blocking studies with monoclonal antibodies and correlation of reactivity pattern with allelic sequence polymorphism have demonstrated that the five cross-reactive clones were stimulated by a cluster of overlapping T-cell interaction sites encoded within the third hypervariable region of the HLA-DR4β1 chain (7). Which molecules within HLA-DR1 harbor the recognition sites seen by these clones? The proliferative response of the different clones to DR1$^+$ stimulators was measured in the presence of monoclonal antibodies specific for monomorphic determinants on HLA class II molecules. As shown in Table 3, all clones were reactive against DR-encoded molecules. Thus, a cluster of cross-stimulatory T-cell epitopes is expressed within the β1 chain of HLA-DR molecules of the DR1 as well as DR4 haplotype.

Sequence comparison reveals that the third hypervariable region of the HLA-DR1β1 chain is identical to one of the HLA-DR4 subtypes, Dw15 (8, 9) (Table 4). The cross-reactivity of the alloreactive clones, however, could not be correlated with their specificity for Dw15 within the DR4 family. Clones specific for Dw4 as well as the Dw14 subtype could interact with the β1 chain of the HLA-DR1 haplotype. The different subtypes of DR4 differ by limited sequence polymorphism with exchanges limited to residues within the third hypervariable region. Amino acid residues 71, 74, and 86 distinguish "Dw4-ness," "Dw14-ness," and "Dw15-ness." The sequence specific for Dw15 resembles positions 71 and 74 of Dw14 and position 86 of Dw4. The finding that T-cell clones identify a shared stretch of amino acids expressed within DR1 and DR4 proves the value of T-cell clones to detect disease-associated MHC structures. However, clones that clearly distinguished between Dw4 and Dw15 or Dw14 and Dw15 on DR4$^+$ stimulators were

Table 3. Inhibition Studies with Monoclonal Anti-HLA-DR Antibodies

Clone	Stimulator	Monoclonal Antibody L227	L203	L243	P4.1	B7/21	IV D 12	Genox3.53
E1-10	OOS	+	+++	+++	+++	–	n.t.	–
	JAH	–	+++	+++	+++	–	–	n.t.
GD3-30	OOS	+++	+++	+++	+++	–	n.t.	–
	ISA	+	+++	+++	+++	–	–	n.t.
F3-29	OOS	+	+++	+++	+++	–	n.t.	–
	LS40	+	+++	+++	+++	–	–	n.t.
C3-4	OOS	+	+++	+++	+++	–	n.t.	–
	LS40	–	+++	+++	+++	–	–	n.t.

T-cell clones were cocultured with HLA-DR1$^+$ stimulator cells as described in Table 1. At the initiation of the coculture, increasing concentrations of four different monoclonal antibodies (MAb) were included. MAb antibody L227, L203, L243, and P4.1 are reactive to monomorphic determinants on HLA-DR molecules. MAb IVD12 and Genox3.53 are specific for HLA-DQ-encoded molecules and B7/21 defines molecules encoded within the HLA-DP subregion. The inhibitory effect was ranked as +++ = > 70% inhibition, + = 30 to 70% inhibition, and 0 = < 30% inhibition of the proliferative response.

stimulated by Dw15 specific sequences within the DR1 backbone. Thus, the framework of the DR1β1 chain can keep the third hypervariable region of Dw15 in such a conformation that it allows a variety of T-cell receptor molecules (specific for very similar structures within DR4) to interact.

The advances in molecular biology of the MHC have provided sequence data for most disease-associated alleles. Up to now a disease-specific sequence for RA has not been identified. The finding that alloreactive T-cell clones define a cluster of T-cell epitopes created by similar but not identical sequences suggests that sequence analysis is of limited value in predicting T-cell function. A cluster of T-cell epitopes formed by the third hypervariable region of the DRβ1 chain is preferentially expressed by RA patients, consistent with the hypothesis that susceptibility to RA is inherited within a segment of 15 amino acids between positions 71 and 86 of the β1 chain. The RA-associated formation of that third hypervariable region is not restricted to a definite sequence. It includes subtypes Dw4, Dw13, Dw14, and Dw15, while Dw10 appears not to increase disease susceptibility. The linkage of that third hypervariable region to RA persists within the framework of a different HLA haplotype. The function of the third hypervariable region in maintaining pathological immune responses leading to RA remains to be elucidated. One interesting aspect of polymorphic HLA determinants is the selective interaction with fragments of foreign antigens (10,11). Alternatively, the function of the third hypervariable region in the selection of the T-cell repertoire during thymic T-cell education could turn out to be important (12).

Acknowledgments. We are grateful for the assistance of Ms. K.R. Jurutka and Ms. B. Saur in the preparation of this manuscript. We highly appreciate the technical support that was provided by the German Cancer Research Center, Heidelberg. We are thankful to G.C. Fathman who initiated our interest in HLA determinants.

Table 4. Sequence Polymorphism of the HLA-DRβ_1 Chain Comparing the Five Different Alleles of HLA-DR4 with HLA-DR1

Codon	57	67	69	70	71	74	86
HLA-DR1	Asp GAT	Leu CTC	Glu GAG	Glu CAG	Arg AGG	Ala GCG	Gly GGT
HLA-Dw4	–	–	–	–	Lys AAG	–	–
HLA-Dw14	–	–	–	–	–	–	Val GTG
HLA-Dw15	Ser AGC	–	–	–	–	–	–
HLA-Dw13	–	–	–	–	–	Glu GAG	Val GTG
HLA-Dw10	–	Ile ATC	– GAA	Asp GAC	Glu GAG	–	Val GTG

The HLA-DR4 family is characterized by a limited sequence polymorphism residing in the third hypervariable region of the first domain of the β_1 chain. The amino acid segment 71–86 of HLA-DR1 and HLA-Dw15 is identical. Sequence differences between HLA-DR4 and HLA-DR1 outside of the third hypervariable region are not shown.

References

1. Stastny P. Association of the B-cell alloantigen DRw4 with rheumatoid arthritis. N Engl J Med 1978;298:869.
2. Goronzy J, Weyand CM, Fathman CG. Shared T-cell recognition sites on human histocompatibility leucocyte antigen class II molecules of patients with seropositive rheumatoid arthritis. J Clin Invest 1986;77:1042.
3. Schiff B, Mizrachi Y, Orgad S, Yaren M, Gazit E. Association of HLA-Aw31 and HLA-DR1 with adult rheumatoid arthritis. Ann Rheum Dis 1982;41:403.
4. Duquesnoy RJ, Marrari M, Hackbarth S, Zeevi A. Serological and cellular definition of a new HLA-DR associated determinant, MC1, and its association with rheumatoid arthritis. Hum Immunol 1984;10:165.
5. Weyand CM, Goronzy J, Fathman CG. Human T-cell clones used to define functional epitopes on HLA-Class II molecules. Proc Natl Acad Sci USA 1986;83:762.
6. Goronzy J, Weyand CM, Fathman CG. Cloning of Human alloreactive T-cells. In Sabato G(ed):Methods in Enzymology: Immunochemical Techniques. New York, Academic Press, in press.
7. Goronzy J, Weyand CM. Single amino acid substitutions within the HLA-DR4 β_1-chain create multiple T-cell recognition sites. In: Dupont B(ed): Immunology of HLA: Vol. 2. Immunogenetics and Histocompatibility. New York, Springer-Verlag, 1987.
8. Gregerson PK, Shen M, Song Q, Merryman P, Degar S, Seki T, et al. Molecular diversity of HLA-DR4 haplotypes. Proc Natl Acad Sci USA 1986;83:2642.
9. Bell J, Esters P, John TSt, Saiki R, Watling PL, Erlich HA, et al. DNA sequence and characterization of human class II major histocompatibility complex β chains from the HLA-DR1 haplotype. Proc Natl Acad Sci USA 1985;82:3405.
10. Babbit B, Allen P, Matsueda G, Haber E, Unanue E. Binding of immunogenic peptides to Ia histocompatibility molecules. Nature 1985;317:359.
11. Guillet JG, Lai M-Z, Brenir TJ, Buus S, Sette A, Grey HM, et al. Immunological self, nonself discrimination. Science 1987;235:865.
12. Kappler JW, Wade T, White J, Kushnir E, Blackman M, Bill J, et al. A T-cell receptor V_β segment that imparts reactivity to a class II major histocompatibility product. Cell 1987;49:263.

Author Affiliation

Medizinische Klinik and Poliklinik V (Chairman: Professor W. Hunstein, M.D.), University of Heidelberg, 6900 Heidelberg, Federal Republic of Germany

In Vivo Expression of a Soluble Form of HLA Class I Antigens in Rheumatoid Synovial Cells

L. Teyton, V. Lotteau, and D. Charron

Human histocompatibility class I antigens are integral cell surface glycoproteins. Soluble forms could be generated from full-length mRNAs by an alternative RNA splicing and were secreted in vitro (1), the transmembrane spanning region of class I molecules being removed. It has previously been shown that HLA molecules are expressed at a high level by synovial cells in rheumatoid arthritis (RA). We have investigated the biochemistry of HLA class I molecules synthesized by synovial cells (SC) from RA and normal individuals by immunoprecipitation and two-dimensional (2D) gel electrophoresis.

In all cases tested (23 cases: 12 RA, 11 normal) SC from RA synthesized a high level of class I molecules, while SC from healthy donors expressed these molecules at a low level. Whatever their origin, SC loose HLA class I expression in in vitro culture and it could be easily reinduced by r-IFN-γ.

The most important point concerns the qualitative expression. Compared with those of homologous mononuclear cells from peripheral blood, 2D gel patterns of W6-32 immunoprecipitates of SC from RA exhibited additional class I spots which migrated around 39 kD. These molecules are expressed both in vivo and in vitro after r-IFN-γ treatment. They could not be detected in vivo in normal SC, perhaps because of their low amount in physiological conditions, but they are also easily induced in these cells by r-IFN-γ. These 39-kD molecules were identified as HLA class I antigens, HLA-A and -B, related as demonstrated by allele-specific antibodies (4E, 4B), associated with β2-microglobulin (binding with BBM1, and anti-β2-microglobulin monoclonal antibody [MAb]), and were normally glycosylated (1N-glycosylation). They were not expressed at the cell surface as assessed by W6-32 precipitates from iodinated cells (Fig. 1c). Furthermore they were recovered from the aqueous phases of Triton X-114 cellular extracts and culture supernatants (Fig. 1d). These water-soluble 39-kD molecules were also isolated from joint effusions from RA patients.

We describe for the first time in human cells that synthesize in vivo 39-kd soluble class I molecules. The questions now are: First—Is this phenomenon particular to synovium or to RA? However, the extra amount in RA synovium and joint effusions should probably interfere with immunological functions of local T cells and macrophages. The second question is: What is the function of the soluble forms of HLA class I molecules in physiological and pathological conditions?

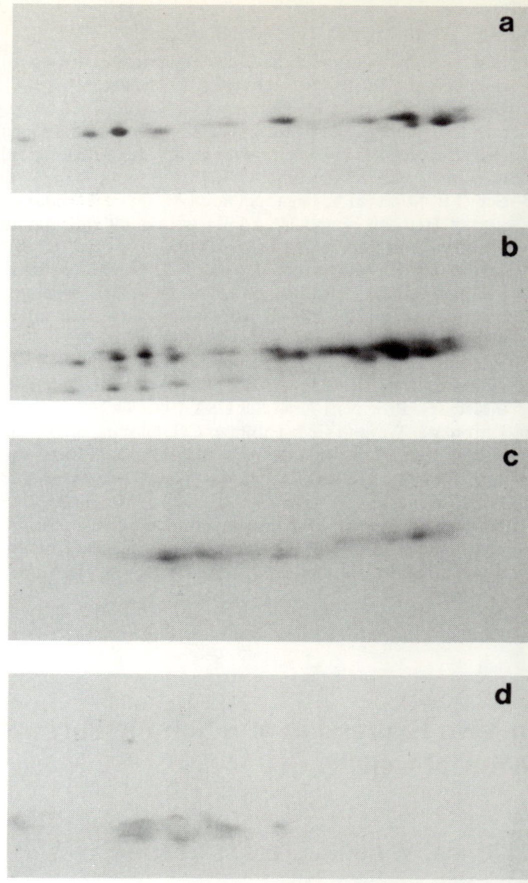

Figure 1. Secretion of soluble HLA class I molecules by synovial cells from RA patients. **a**: Peripheral blood mononuclear cells ([^{35}S]methionine labeling). **b**: Adherent synovial cells ([^{35}S]methionine labeling). **c**: Adherent synovial cells ([^{125}I]iodide labeling). **d**: Supernatant of adherent SC (unlabeled material, Western blot with HC-10 MAb). W6-32 precipitations of **a**), **b**), **c**), and *d* were analyzed by 2D gel electrophoresis (IEF in the first dimension) and run at the same time. The soluble 39-kD molecules were indicated by arrows in **c** and **d**.

References

1. Krangel MS. Secretion of HLA-A and B antigens via an alternative RNA splicing pathway. J Exp Med 1986;163:1173.
2. Teyton L, Lotteau V, Turmel P, Arenzana-Seisdedos R, Virelizier JL, Pujol JP, et al. HLA-DR, DQ, and DP antigen expression in rheumatoid synovial cells: A biochemical and quantitative study. J Immunol 1987;138:1730.

Author Affiliation

Institut Biomédical des Cordeliers, 75006 Paris, France

Characterization of HLA-DR$_{\beta}$ and HLA-DQ$_{\beta}$ Alleles Associated with Pemphigus Vulgaris

Animesh A. Sinha, Chaim Brautbar, Fanny Szafer, Adam Friedmann, Eli Tzfoni, John A. Todd, Lawrence Steinman, and Hugh O. McDevitt

The genetic association between human autoimmune diseases and particular class II alleles of the major histocompatibility complex (HLA) is well known (1) but poorly understood. Pemphigus vulgaris (PV), an autoimmune skin disease, is strongly associated with HLA-DR4 and HLA-DRw6. Recently, five of us (F.S., L.B., A.F., E.T., and L.S.) have identified two DQ$_{\beta}$ restriction fragment length polymorphisms (RFLPs) present in nearly all DR4 and DRw6 Israeli PV patients respectively, but few healthy class II-matched Israeli controls (2). Furthermore, it has been suggested that a shared stretch of amino acids, or epitope, in the third allelic hypervariable regions of the DR4$_{\beta I}$ and DRw6$_{\beta I}$ alleles may explain the increased susceptibility to PV associated with these two haplotypes (3). To address the molecular basis of autoimmunity, we have characterized the DQ$_{\beta}$ and DR$_{\beta}$ alleles of PV patients by nucleotide sequence analysis and dot-blot analysis using allele-specific oligonucleotide probes.

Pemphigus vulgaris is a severe autoimmune disease of the skin that presents most commonly in females over the age of 30 (4). Individuals afflicted with PV suffer from chronic, intraepidermal blistering, often covering the trunk, scalp, and mucous membranes. This is a rare

Immunobiology of HLA
Volume II

```
           67                                      76
R4(Dw10)   ATC CTG|GAA GAC GAG CGG GCC GCG|GTG GAC
R4(Dw4)    C-- --- --G C-G A-- --- --- --- --- ---
R4(Dw13)   C-- --- --G C-G AG- --- --- -A- --- ---
R4(Dw14)   C-- --- --G C-G AG- --- --- --- --- ---
R4(Dw15)   C-- --- --G C-G AG- --- --- --- --- ---
R5a        --- --- --- --- --- --- --- --- --- ---
R5b        T-- --- --- --- AG- --- --- --- --- ---
Rw6a       --- --- --- --- --- --- --- --- --- ---
Rw6b       C-- --- --G CGG AG- --- --- -A- --- ---
```

Figure 1. Comparison of $DR_{\beta I}$ third allelic hypervariable region nucleotide sequences in DR4, DR5, and DRw6 haplotypes. The DR4(Dw10) sequence is from a DR4 PV patient and is identical to the DR4(Dw10) sequence of Gregersen, et al. (15). (The other DR4 sequences are also from Gregersen, et al. (15); the DR5a sequence is from Bell, et al (8); and the DR5b sequence is from Didier et al (17). The DRw6a and DRw6b sequences are from Gorski and Mach (9).) Note that DR4(Dw10), DR5a and DRw6a are identical in this region. The sequence of the Dw10-specific oligonucleotide probe is boxed.

but potentially fatal disease. PV is mediated by autoantibodies directed against an antigen present on the surface of epidermal cells (5). Although PV occurs worldwide, the incidence is highest in people of Jewish descent (6). As with many autoimmune diseases, PV is strongly associated with particular alleles of the human class II. HLA-DR4 is highly over-represented in Ashkenazi Jews, whereas both HLA-DR4 and HLA-DRw6 are increased in non-Ashkenazi Jews and European PV patients (2). Overall, greater than 95% of PV bear DR4 or DRw6 serologic specificities. Virtually all DR4 Israeli PV patients are of the Dw10 cellular subtype (7).

From an examination of the nucleotide sequences of the highly polymorphic DR_β, DQ_α, and DQ_β loci, it is apparent that there are epitopes shared between alleles (8). Shuffling of epitopes by gene-conversion-like events is one mechanism via which the extensive polymorphism may have been generated (9). Such epitopes (which can be recognized by T-cells and antibodies) may form the basis of disease susceptibility in individuals of differing HLA types (10,11). By oligonucleotide dot-blot analysis, we assessed whether the increased susceptibility to PV in DR4 and DRw6 individuals results from the sharing of a common sequence in the third allelic hypervariable region of the $DR_{\beta I}$ first domain. Genomic DNA from several DR4 and DRw6 PV patients was amplified in vitro and hybridized to an oligonucleotide probe specific for the Dw10 epitope (present in DR4(Dw10), DRw6a, and some DR5 (here designated DR5a) $DR_{\beta I}$ alleles; (see Fig. 1 and refs. 8,3). As expected, we found that all DR4$^+$/DRw6$^-$ PV

Table 1. Frequency of Dw10 $DR_{\beta I}$ Epitope and DRw6 PV DQβ Allele in Israeli PV Patients and Controls

	Number Positive for			
	Dw10($DR_{\beta I}$)*		PV6(DQ_β)*	
Haplotype	PV†	Controls†	PV	Controls
DR4$^-$/DRw6$^+$	0/7	7/8	7/7	0/10
DR4$^+$/DRw6$^+$	ND	ND	6/6	1/13
DR4$^+$/DRw6$^-$	19/19	10/12	0/19	0/13
DR5/DR7	0/1	0/1	0/1	0/1

*DR_β of DQ_β loci were amplified in vitro as described (12,14). The amplified DNA was dot-blotted onto nitrocellulose, probed with ^{32}P-labeled oligonucleotide probes (boxed in Figs. 1 and 2), and washed as described (12,14).

†Patient and control groups include Ashkenazi and non-Ashkenazi Jews.

patients were positive for the Dw10 epitope (Table 1). However, none of the seven DR4$^-$/DRw6$^+$ PV patients were positive in this analysis (Table 1). Preliminary results from dot-blot hybridization using allele-specific probes suggest that most of these patients are of the DRw6b type (12). We also found one DR5/DR7 patient to be negative for the Dw10 epitope. Therefore, although there is some evidence for the epitope hypothesis in other HLA-linked diseases (10,11), our results clearly indicate that the DW10 $DR_{\beta I}$ epitope is not the only determinant of susceptibility to PV.

The presence of highly disease-specific DQ_β RFLPs in DR4 and DRw6 PV patients (2) prompted us to characterize the DQ_β alleles associated with PV. We used in vitro amplification (13) from a cDNA substrate (14) to rapidly obtain the nucleotide sequence of the first, and most polymorphic domain of DQβ loci from a DR4/DRw6 PV patient. One sequence obtained was identical to a previously published DQ_β sequence from a DR4(Dw10) homozygous typing cell (15) (Fig. 1). Therefore, the DQ_β RFLP found in DR4 PV patients does not reflect changes in the DQ_β first domain coding sequence. Currently, we are sequencing the second domain of DQ_β as well as DQ_α to further characterize the DQ loci of DR4 PV patients.

We found the predicted amino acid sequence of the other DQ_β allele from the DR4/DRw6 PV patient to be identical to the DQw1.1_β allele found in DR1 and some DRw6 haplotypes, except at codon 57, where an aspartic acid codon (GAC) replaces a valine codon (GTT) (Fig. 2). Comparison with other published DQ_β sequences reveals that DRw6 PV and DR2,DQw1.2 are identical for a 44 bp stretch that includes codon 57. Thus, DR2,DQw1.12 may have served as a donor sequence in the generation of DRw6 PV from an ancestral DQw1.1_β by a gene conversion-like event.

An allele-specific oligonucleotide probe ($PV6_\beta$) was designed to assess the frequency of the DRw6 PV DQ_β allele in DRw6 PV patients and controls. In vitro amplified DNA was hybridized to the $PV6_\beta$ probe in a dot-blot format. Thirteen out of 13 DRw6 Israeli PV patients possessed this variant DQ_β allele versus one in 13 Israeli

```
                                   50                                 60
DR1 DQw1.1      GGG GTG TAC CGG GCA GTG ACG CCG CAG GGG CGG CCT GTT GCC GAG TAC TGG AAC AGC CAG AAG
DR2 DQw1.2      --- --- --- --C --G --- --- --- --- --- --- --- -A- --- --- --- --- --- --- --- ---
DR2 DQw1.12     --- --- --T --- --G --- --C --- --- --- --- --- -AC --- --- --- --- --- --- --- ---
DR2 DQw1.AZH    --- --- --- --- --G --- --- --- --- --- --- --- AGC --- --- --- --- --- --- --- ---
DRw6 DQw1.1     --- --- --- --- --- --- --- --- --- --- --- --- --- --- --- --- --- --- --- --- ---
DRw6 DQw1.19    --T --- --- --- --G --- --- --- --- --- --- --- --- --- --- --- --- --- --- --- ---
DRw6 PV         --- --- --- --- --- --- --- --- --- [--- --- --- -AC --- ---] --- --- --- --- --- ---
```

Figure 2. Comparison of DQw1 DQ_{β} nucleotide sequences, codons 45-65. (The sources of the sequences are: DR1, DQw1.1, Bell et al (18); DR2,DQw1.2, DR2,DQw1.12, and DR2,DQw1.AZH, Lee et al (19); DRw6,DQw1.1, Schenning et al (20); and DRw6,DQw1.19, Turco et al (21).) The sequence of the $PV6DQ_{\beta}$ oligonucleotide probe is boxed.

DRw6 controls (Table 1). These results suggest that the DRw6 PV DQ_{β} allele is strongly disease-associated and highly disease-restricted. However, the sequence is not entirely disease-unique since it is present in a small proportion of healthy individuals and in the non-Israeli DRw6 HTC WT52 (data not shown). Our results with PV, and recent findings in insulin-dependent diabetes (16) suggest that residue 57 of the DQ_{β} molecule may play a critical role in disease susceptibility.

The frequency of this new DQ_{β} allele is currently being assessed in non-Israeli PV patients and controls. Further genetic studies will also be needed to determine the relative contributions of DQ_{β} and DR_{β} to disease susceptibility. This goal will be further aided when the PV autoantigen becomes available in suitable quantities for molecular studies.

Acknowledgments. Animesh A. Sinha is Centennial Fellow of the Medical Research Council of Canada and is a recipient of an Independent Research Allowance from the Alberta Heritage Foundation for Medical Research. We thank M. Trucco, P. Patek, B. Lee, P. Travers, Z. Fronek, K. Moody and K. Karagianes. This work was supported by the N.I.H.

References

1. Svejgaard A, Platz P, Ryder LP. Immunol Rev 1983;70: 193–218.
2. Szafer F, et al. Proc Natl Acad Sci USA 1987;84:6542–6545.
3. Gregersen PK, Silver J, Winchester RJ. Arthritis and Rheumatism (in press).
4. Singer KH, et al. Ann Rev Immunol 1985;3:87–108.
5. Stanley JR, et al. J Clin Invest 1982;70:281–288.
6. Ahmed AR. In: Ahmed AR(ed): Clinics in Dermatology–Pemphigus Vulgaris. 1983, pp 13–19.
7. Amar A, et al. Tissue Antigens 1984;23:17–22.
8. Bell J, et al. Proc Natl Acad Sci USA 1987;84:6234–6238.
9. Gorski J, Mach B. Nature 1986;322:67–70.
10. Silver J, Goyert SM. In: Southern BG, Moller E, Ferrone S (eds): HLA Class II Antigens. Berlin: Springer-Verlag, 1986:32–48.
11. Fathman CG, Goronzy J, Weyand C. Ann NY Acad Sci 1986;475:24–31.
12. Sinha AA, et al. Submitted for publication.
13. Saiki RK, et al. Nature (London) 1986;324:163–166.
14. Sinha AA, et al. Science (submitted).
15. Gregersen PK, et al. Proc Natl Acad Sci USA 1986;83: 2642–2646.
16. Todd JA, Bell JI, McDevitt HO. Nature (London) 1977; 329:599.
17. Didier DK, et al. J Immunol 1986;137:2627.
18. Bell JI, et al. Proc Natl Acad Sci USA 1985;82:3405.
19. Lee BSM, et al. Immunogenetics (in press).
20. Schenning L, et al. EMBO J 1984;3:447.
21. Turco E, et al. Immunogenetics (in press).

Author Affiliations

Animesh A. Sinha, John A. Todd, Department of Medical Microbiology, Stanford University, Stanford, CA 94305 USA; *Chaim Brautbar*, Tissue Typing Unit, Department of Clinical Microbiology, Hadassah Hospital, Hebrew University, Jerusalem, Israel, and Department of Medicine, Stanford University, Stanford, CA 94305, USA; *Fanny Szafer, Adam Friedmann*, Department of Genetics, Haddassah Hospital, Hebrew University, Jerusalem, Israel, and Department of Medicine, Stanford University, Stanford, CA 94305, USA; *Eli Tzfoni*, Department of Dermatology, Hadassah Hospital, Hebrew University, Jerusalem, Israel; *Hugh O. McDevitt*, Department of Genetics, Hadassah Hospital, Hebrew University, Jerusalem, Israel, and Department of Medical Microbiology, Stanford University, Stanford, CA 94305, USA

HLA Class II Allelic Variation and Susceptibility to Pemphigus Vulgaris

Stephen J. Scharf, Adam Friedmann, Chaim Brautbar, Fanny Szafer, Lawrence Steinman, Glenn Horn, and Henry A. Erlich

Abstract: Pemphigus vulgaris (PV) is a dermatologic autoimmune disease associated with DR4 and DRw6. The distribution of DRβ and DQβ alleles in PV patients and matched controls was compared using sequence-specific oligonucleotide probes and the polymerase chain reaction method of in vitro amplification. The DR4 susceptibility was strongly associated with the $DR\beta_I$ Dw10 allele. The DRw6 $DR\beta_I$ allele that shares an epitope in the third hypervariable region was not associated with PV. A DQβ allele, termed DQβ1.3, was found in virtually all DRw6 patients, but only 1 of 13 DRw6controls. The sequences of the disease-associated alleles suggest that disease susceptibility is conferred by the allele and not by an epitope.

Pemphigus vulgaris (PV) is an autoimmune dermatologic disease mediated by autoantibodies to an, as yet, uncharacterized epidermal membrane protein (1). In population studies, it is strongly associated with the serotypes HLA-DR4 and HLA-DRw6(2,3), with less than 5% of the patients possessing neither marker. Disease associations with two different HLA haplotypes can be interpreted to mean that either these two haplotypes share a common allele or epitope or, alternatively, that different alleles on the two haplotypes are capable of conferring disease susceptibility. Either model is consistent with the observed increase in DR4/DRw6 individuals among patients (3).

The analysis of class II haplotypes by cellular, biochemical, and molecular techniques has revealed extensive genetic heterogeneity *within* the serologically defined DR types. Sequence analysis of class II genes from various haplotypes demonstrated that a DRβI epitope at position 68-72 (termed here "IDE") was shared between a subset of DR4 haplotypes ("Dw10") (4) and a subset of DRw6 haplotypes (DRw13) (5, unpublished). This observation suggests the possibility that these shared polymorphic residues of the DRβI chain, which determine the DR specificity, may contribute to PV autoimmunity and account for both the DR4 and DRw6-associated susceptibility. However, a recent report by Szafer et al (3) demonstrated that specific RFLPs detected with a DQβ cDNA probe subdivided the DR4 and DRw6 haplotypes and showed significantly higher associations with PV than did the serologic markers. A *Bam*HI 2.5 kb fragment was present in 100% of the DR6$^+$ patients and 5.8% of DR6$^+$ controls while a *Pvu*II 6.9 kb fragment was present in 73% of DR4$^+$ patients and 19% of controls. These RFLP data were interpreted to mean that one particular DQβ allele in linkage disequilibrium with the DR4 serotype confers susceptibility as does another DQβ allele in linkage disequilibrium with DRw6. The analysis of class II allelic coding sequence variation is required to identify the putative class II alleles of epitopes that contribute to PV.

The sequence analysis of DRβI, DRβIII, and DQβ alleles from three U.S. DR4$^+$ PV patients (DR 4/4, 4/5, and 4/6) revealed that all the patients contained the "Dw10" DRβI allele (5a) found in only 10% or U.S. DR4$^+$ controls (6). Furthermore, three of the four DR4 haplotypes had the DQβ3.2 allele which is also present on 60 to 80% of control DR4 haplotypes. These sequence data, although obtained from only three patients, suggest that, for the DR4 susceptibility to PV, the DRβI Dw10 allele may contribute to autoimmunity but that the DQβ allele does not. These data were generated using the polymerase chain reaction (PCR) method for the in vitro amplification of specific DNA segment (7–9).

Here, we report the use of sequence-specific oligonucleotide (SSO) probes to determine the distribution of DR4, DRβI, and DQβ alleles in a panel of 52 PV patients and 56 controls. The frequency of the "IDE" sequence in patient and control DRw6 haplotypes was also determined to test the "shared epitope" model of PV susceptibility. In addition, we describe the identification of a novel DRw6 DQβ allele found in two PV patients and highly associated with PV. The frequency of this DQβ allele in DRw6$^+$ patients and DRw6$^+$ controls was, once again, determined using a set of DQβ sequence-specific oligonucleotide probes.

The analysis of PCR-amplified DNA from a panel of PV patients and controls with a set of DRβI and DQβ sequence-specific oligonucleotide probes, revealed that virtually all DR4$^+$ haplotypes contained the "Dw10" DRβI allele (Table 1). The frequency of the Dw10 subtype defined by MLC typing, among U.S. DR4$^+$ controls, is estimated to be about 10% (6). The frequency determined by oligonucleotide probe analysis in our control panel was 5% for non-Israeli DR4$^+$ controls and 60% for Israeli DR4$^+$ controls (Table 1). Sequence analysis of the DQβ alleles in 3 DR4$^+$ PV patients (5a) as well as the oligonucleotide probe analysis of a large panel of patients and controls, revealed that most DR4$^+$ haplotypes from patients contained the DQβ3.2 allele, as did most DR4$^+$ control haplotypes. Thus, the observed distribution of DRβI and DQβ alleles among patients and controls suggests that it is the DRβI allele ("Dw10") on the DR4 haplotype that is disease-associated and that may confer PV susceptibility.

The DRw6 susceptibility to PV, however, cannot be accounted for by the "shared epitope" at position 68-72 of the DRβ1 chain since only 29% of DRw6$^+$ patient haplotypes and 62% of control DRw6 haplotype contained this "IDE" sequence, as determined by oligonucleotide probe analysis (Table1). In the case of DR4/DRw6 samples, specific hybridization of the "IDE"

Immunobiology of HLA
Volume II

Table 1. DRβ I Oligonucleotide Probe Analysis

DR Haplotype	GH78 "DW10" PV	GH78 "DW10" Controls	GH101 "DW4" PV	GH101 "DW4" Controls	GH100 "Dw14" PV	GH100 "Dw14" Controls
Israeli						
DR4	24/24	15/25	0/24	1/25	6/24	10/25
% positive	100%	60%	0%	4%	25%	40%
	p=0.001					
DR6	4/14	8/13	–	–	–	–
% positive	29%	62%	–	–	–	–
	p=0.035					
Non-Israeli						
DR4	10/14*	1/19	1/14	6/19	3/14	10/19
% positive	71%	5%	7%	30%	21%	55%
	p=0.001					
DR6	2/10	7/7	–	–	–	–
% positive	20%	100%	–	–	–	–
	p=0.02					
*DR4 exceptions:	1) DR4̂/4 2) DR4/6̂					

ˆ: denotes susceptibility allele.

DNA from PV patients or controls was amplified for DRβ by PCR (7,8) for analysis with sequence specific oligonucleotide (SSO) probes (9). These SSOs hybridize specifically to their respective subtypes: GH78 (GACATCCTGGAAGACGAGC) to DR4 Dw10; GH101 (GGAGCAGAAGCGGGCCGCG) to DR4 Dw4; and GH100 (GGAGCAGAGGCGGGCCGCG) to DR4 Dw14. Some DRw6 DRβI alleles also contain a sequence encoding the "IDE" epitope complementary to the GH78 probe.

probe (GH78) to the PCR amplified DNA could be attributed to either haplotype. However, the unambiguous assignment of the probe-defined epitope to a specific haplotype could be accomplished by restriction endonuclease digestion of the PCR-amplified DNA with an enzyme that cleaves one of the alleles and Southern blot analysis with the GH78 probe (10).

Since the shared "IDE" epitope on the DRβI chain cannot account for the DRw6 susceptibility, the DQβ alleles of two DR5/DRw6 PV patients were determined. Both patient DRw6 haplotypes contained a previously unreported DQβ allele (Fig. 1) which we have designated DQβ1.3 (10,11) and is associated with the Dw9 subtype of DRw6, based on analysis of HTC's (unpublished). It differs from the DR1 DQβ allele, DQβ1.1, by only a Val to Asp substitution at amino acid position 57. At the nucleotide level, it is identical in this region to the DR2 Dw12 allele, DQβ1.4, and to the most common DRw6 DQβ allele, DQβ1.6, associated with the Dw18 subtype (10,11).

To determine the frequency of the DQβ1.3 allele among patient and control DRw6 haplotypes, we used a panel of four oligonucleotide probes to distinguish the DR1-like DQβ framework sequence (in DQβ1.1 and 1.3) from the DR6-like framework sequence (in DQβ1.6 and 1.7) and to identify the sequence around codon 57. These four oligonucleotide probes generate a hybridization pattern corresponding to the 4 DQβ alleles found on DRw6, DRw1, haplotypes (Fig. 1 and Fig. 2). The probe reactivity and the DQβ RFLP pattern for each Israeli patient and control sample is shown in Figure 2. All 13 DRw6⁺ PV patients were positive with the "DR1-like" probe GH69; reactivity to this probe correlates with the presence of the *Bam*HI 2.5 kb fragment. Eleven of 13 patients reacted with the "QRD" probe, CRX03. Two patients reacted with neither the "QRD" or the "QRV"

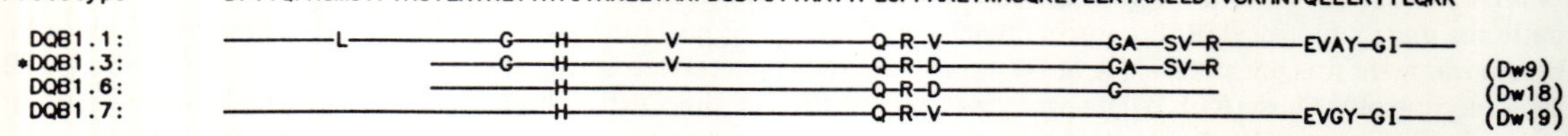

Figure 1. Protein sequences inferred from the nucleotide sequences for the DQβ alleles of DRw6, DQw1 haplotypes. The asterisk marks the allele (DQβ1.3) associated with PV. The DQβ1.3 allele is associated with the Dw9 subtype, the DQβ1.6 allele with the Dw18 subtype, and the DQβ1.7 subtype with the Dw19 subtype.

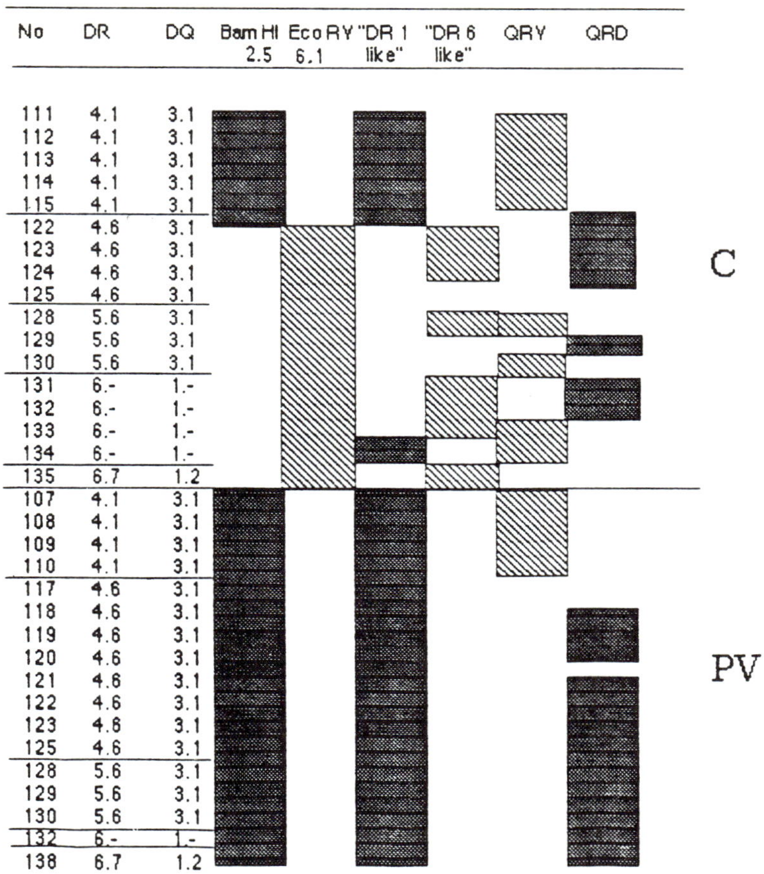

Figure 2. Oligonucleotide probe analysis of DQw1 PV patients and controls for DQβ allelic variation. "BamHI 2.5" and "EcoRV 6.9" designate a 2.5 kb BamHI genomic fragment and a 6.9 kb EcoRV genomic fragment detected with a DQβ cDNA probe (3). "DR1 like" refers to the SSO GH69 (GATGTGTCTGGTCACACCCCG) that hybridizes to the DQβ1.1 and DQβ1.3 alleles. "DR6 like" refers to the SSO GH80 (TCTTGTAACCAGACACATC), which hybridizes specifically to the DQβ1.6 and DQβ1.7 alleles (see Fig. 1). "QRV" refers to the SSO CRX02 (TCGGCAACAGGCCGCCCCT) which hybridizes specifically to the sequence encoding amino acid residues 53 to 57 in DQβ1.1 and DQβ1.7 alleles. "QRD" refers to the SSO CRXO3 (TCGGCGTCAGGCCGCCCCT) which hybridizes to the sequences encoding amino acid residues 53 to 57 in DQβ1.3 and DQβ1.6 alleles. Hybridization with the GH69 and CRX02 SSO probes corresponds to the DQβ1.1allele; with the GH69 and CRX03 probes to the DQβ1.3 allele; with the GH80 and CRX02 probes to the DQβ1.7 allele; and with the GH80 and CRX03 probes to the DQβ1.6 allele.

(CRX02) probes, due to a nucleotide polymorphism in this region. Only one of 13 DRw6$^+$ controls has the probe reactivity pattern associated with DQβ1.3 allele. These results suggest that the DRw6 susceptibility is strongly associated with a rare DQβ allele (DQβ1.3) that differs from a common DQβ allele (DQβ1.1) by only one residue. It may be relevant that a similar methodologic approach to the analysis of IDDM susceptibility (11,12) also found that the charge of the polymorphic residue at position 57 of the DQβ chain was correlated with genetic predisposition. In the case of PV, however, it is clear that it is the *allele*, not simply the *epitope* around position 57, that is disease associated since the most common DRw6 DQβ allele (DQβ1.6) contains the same sequence around position 57 but in a different framework (10). Thus, this novel DQβ allele appears to account for the DRw6 susceptibility to PV, consistent with the implications of the DQβ RFLP analysis(3). To evaluate the potential contribution of other class II alleles on this DRw6 haplotype, DRβ, DQα, and DPβ sequences are also being determined.

As noted above, sequence analysis of the DR4 susceptibility suggests that DRβI allelic variation and not DQβ variation are associated with PV in contrast to the predictions of the RFLP analysis. This apparent contradiction may be resolved by supposing that other class II sequences (e.g., DPβ sequences which cross-hybridize to the DQβ probe) on the DR4, D10 haplotype may also contribute to PV susceptibility. DPβ sequences from patient and control haplotypes are currently being determined from PCR clones. The association of the DR4, Dw10 haplotype with PV reported here is consistent with the ethnic distribution of this disease. The frequency of PV and the frequency of DR4 Dw10 haplotype are both increased among Ashkenazi Jews (3) (Table 1).

References

1. Razzaqui I, Ahmed AR, 1983 In: Ahmed AR (ed): Clinics in Dermatology--Pemphigus Vulgaris.
2. Brautbar C, Moscovitz M, Livshits T, Haim S, Hacham-Zadeh S, Cohen HA, et al. HLA-DRw4 in *Pemphigus vulgaris* patients in Israel. Tissue Antigens 1985;16:238.
3. Szafer F, Brautbar C, Tzfoni E, Frankel G, Sherman L, Cohen I, et al. Detection of disease specific restriction fragment length polymorphisms in *Phemphigus vulgaris* linked to the DQw1 and DQw3 alleles of the HLA-D region. Proc Natl Acad Sci 1987;84:6542.
4. Gregerson PK, Shen M, Song Q-L, Merryman P, Degar S, Seki T, et al. Molecular diversity of HLA-DR4 haplotypes. Proc Natl Acad Sci USA 1986;83:2642.
5. Gorski J, Mach B. Polymorphism of human Ia antigens: Gene conversion between two DR-beta loci results in a new HLA-D/DR specificity. Nature 1986;322:67.
6. Hansen JA, Beatty PG, Anasetti C, Martin PJ, Mickelson E, Thomas ED. Transplantation of hematopoietic stem cells (HSC). Brit Med Bull 1987;43:203.
7. Saiki RK, Scharf S, Faloona F, Mullis KB, Horn GT, Erlich HA, et al. Enzymatic amplification of β-globin genomic sequences and restriction site analysis for diagnosis of sickle cell anemia. Science 1985;230:1350.
8. Scharf SJ, Horn GT, Erlich HA. Direct cloning and sequence analysis of enzymatically amplified genomic sequences. Science 1986;233:1076.
9. Saiki RK, Bugawan TL, Horn GT, Mullis KB, Erlich HA. Analysis of enzymatically amplified β-globin and HLA-DQα DNA with allele-specific oligonucleotide probes. Nature 1986;324:163.
10. Scharf SJ, Long C, Erlich HA. Sequence analysis of the HLA-DRβ and HLA DQβ from three Pemphigus vulgaris patients.
11. Horn GT, Bugawan TL, Long CM, Erlich HA. Allelic sequence variation of the HLA-DQ loci: Relationship to serology and to insulin-dependent diabetes susceptibility. Proc Natl Acad Sci.
12. Todd JA, Bell JI, McDevitt HO. HLA-DQβ gene contributes to susceptibility and resistance to insulin-dependent diabetes mellitus. Nature 1987;329:599.

Author Affilitations

Stephen J. Scharf, Glenn Horn, Henry A. Erlich, Department of Human Genetics, Cetus Corporation, Emeryville, CA; *Lawrence Steinman*, Department of Neurology and Pediatrics, Stanford University, Stanford CA, USA; *Adam Friedmann, Fanny Szafer*, Department of Genetics, Institute for Life Sciences, Jerusalem; *Chaim Brautbar*, Hadassah Medical School, The Tissue Typing Unit and Department of Clinical Microbiology, and the Lautenberg Center for General and Tumor Immunology, Jerusalem, Israel.

DRβ Chain Polymorphism in Pemphigus Vulgaris

P.K. Gregersen, K.M. Barr, C. DeFeo, C. Bystryn, A. Franks, B. Sanchez, J. Silver, R.J. Winchester, and A. Nunez-Roldan

Previous studies by Amar et al (1) indicated that the risk for pemphigus vulgaris (PV) is preferentially associated with the Dw10 subtype of DR4, as well as with DRw6 in the Israeli population. This raised the possibility that third hypervariable (HVIII) region sequences, which are identical in DR4, Dw10 (2), and DRw6a (3) DRβ1 chains, might constitute a disease susceptibility epitope.

We have studied DRβ chain polymorphisms in 23 patients with PV. DRβ chains were immunoprecipitated with MoAb L243 and analyzed using two-dimensional gel electrophoresis with the kind assistance of Santiago Rodriquez de Cordoba (4). This technique readily distinguishes the Dw10 subtype of DR4. Selected individuals were also studied by DNA sequence analysis of DRβ clones generated using the polymerase chain reaction method of gene amplification from mRNA (5).

The results of these studies are summarized in Table 1. Twenty-one of 23 (91%) of pemphigus vulgaris patients were DR4+. Of these DR4+ patients, 19 were studied by two-dimensional gel analysis, and all but 1 of these individuals carried the Dw10 allele. The presence of the Dw10 allele was confirmed in several individuals by DNA sequence analysis. One DR4+ patient (MOH) carried the Dw14 allele. Of interest, this patient was the only non-Jewish DR4+ patient studied.

Two non-DR4 patients were identified. Both of these individuals typed as DRw6, consistent with previous reports (1). One Jewish patient (MIC) typed as DRw6,w8, and a North American black patient (WIT) typed as DR5,w6. DNA sequence analysis revealed the presence in both individuals of the DR6b allele. Unlike the DR6a DRβ chain, the DR6b allele has no sequence similarity

Immunobiology of HLA
Volume II

Table 1. Study Results of DRβ Chain Polymorphism in 23 Patients with Pemphigus Vulgaris

Patient	DR Type	Ethnic Origin	Allelic Subtype of DR4 or DRw6	Method of Analysis
ZAG	4,1	J	–	–
CMI	4,1	J	–	–
WIT	w6,5	B	DR6b	2D-gel/PCR
MOH	4,5	NJC	Dw14	2D-gel/PCR
FER	4,2	J	Dw10	2D-gel/PCR
LIS	4,w6	J	Dw10,DR6a	2D-gel/PCR
FLE	4,1	J	Dw10	2D-gel
STE	4,3	J	Dw10	2D-gel
MIC	w6,w8	J	DR6b	2D-gel/PCR
EIG	4,5	J	Dw10	2D-gel
BAR	4,-	J	Dw10	2D-gel
WAT	4,w6	J	Dw10	2D-gel
BRO	4,w8	J	Dw10	2D-gel
BEN	4,7	J	Dw10	2D-gel
LEV	4,w6	J	Dw10	2D-gel
GRO	4,2	J	Dw10	2D-gel
CAM	4,3	J	Dw10	2D-gel
LAI	4,3	J	Dw10	2D-gel
SHE	4,1	J	Dw10	2D-gel
ORL	4,-	J	Dw10	2D-gel
COH	4,w6	J	Dw10	2D-gel
GLU	4,1	J	Dw10	2D-gel
GUE	4,-	J	Dw10	2D-gel

J = Jewish; NJC = non-Jewish Caucasian; B = Black; 2D-gel = two-dimensional gel analysis; PCR = polymerase chain reaction; Dw14, Dw10 and DR6a sequences are identical to previous reports (2,3).

in the third hypervariable region with DR4 Dw10 β chains.

These studies indicate that the molecular basis for genetic susceptibility to PV is complex. Specifically, it appears that the dual association of pemphigus vulgaris with both DR4 and DRw6 cannot be simply explained by sharing of Dw10-like third hypervariable region sequences by the DRβ1 chains of these haplotypes. A more extended analysis of DRβ polymorphism in non-Jewish and non-DR4 pemphigus patients will be required to understand the complex relationship between class II polymorphisms and susceptibility to this disease.

References

1. Amar A, Rubinstein N, Hacham-Zadeh S, Cohen O, Cohen T, Brautbar C. Is predisposition to pemphigus vulgaris in Jewish patients mediated by HLA-Dw10 and DR4? Tissue Antigens 1984;23:17–22.
2. Gregersen PK, Shen M, Song Q, et al. Molecular diversity of HLA-DR4 haplotypes. Proc Natl Acad Sci USA 1986;83: 2642–2646.
3. Gorski J, Mach B. Polymorphism of human Ia antigens: gene conversion between two DR β loci results in a new HLA-D/DR specificity. Nature 1986;322:67–70.
4. Rodriquez de Cordoba S, Nunez-Roldan A, Winchester R, et al. Molecular characterization by high resolution isoelectric focusing of the products encoded by the class II region loci of the major histocompatibility complex in humans. I. DR and DQ variants. Hum Immunol 1987;20:71–93.
5. Saiki RK, Bugawan TL, Horn GT, Mullis KP, Erlich HA. Genetic analysis of enzymatically amplified β globin and HLA-DRα genomic DNA with allele specific oligonucleotide probes. Nature 1986;324:193.

Author Affiliations

P.K. Gregersen, K.M. Barr, B. Sanchez, J. Silver, R.J. Winchester and *A. Nunez-Roldan*, Molecular Biology Unit, Department of Rheumatic Diseases, Hospital for Joint Disease, New York, NY 10003, USA; *C. DeFeo, C. Bystryn*, and *A. Franks*, Department of Dermatology, New York University Medical Center, New York, NY 10003, USA

Immune Responses of Patients with Myasthenia Gravis to Peptides Representing Different Regions of the Human Acetylcholine Receptor

Edna Mozes, Chaim Brautbar, Oded Abramsky, Lawrence Steinman, Jonathan Rothbard, and Stefan Brocke

Abstract: Myasthenia gravis (MG) is an autoimmune disease characterized by the production of autoantibodies to the acetylcholine receptor (AChR). To find out which parts of the AChR are involved in the initiation and development of the disease, peptides representing different sequences of the human AChR α-subunit were synthesized. The latter were tested for their ability to stimulate T cells of myasthenic patients and healthy controls in proliferation assays and to bind to antibodies in the sera of these donors. Three of the peptides tested discriminated significantly between the two groups in the proliferation assay as well as in their binding to serum antibodies. HLA phenotyping revealed an increase in the joint expression of HLA-B8 and DR3 in the Israeli MG patients. HLA-DR3 and DR5 were associated with proliferative responses to two peptides. Thus, AChR epitopes that play a role in MG were suggested.

The higher frequency of certain histocompatibility antigens (HLA-B8,DR3) in patients with MG (1) suggests an immunoregulatory defect that might be expressed on the level of T cells. For the elucidation of the role of regulatory T cells in the pathogenesis of MG, it is important to study the immune response potential of MG patients to the relevant antigen, namely human AChR. However, due to minute amounts of AChR in human tissue and complicated purification methodologies, it is not possible to obtain AChR in sufficient quantities. It appeared that the employment of synthetic peptides from various regions in the human AChR molecule should alleviate the problem. To this end peptides representing different regions of the human AChR α-subunit were synthesized (2) and peripheral blood lymphocytes (PBL) of MG patients and of healthy controls were tested for their ability to proliferate (3) in the presence of an in vitro stimulus with these peptides. Three of the peptides tested (numbers 3,4, and 5 with sequences 195-212, 257-269, and 310-327, respectively) discriminated significantly between MG patients and healthy controls. The latter peptides were chosen to conform to a pattern present in many T-cell epitopes (4). Table 1 summarizes the results of the proliferative assays specific to peptides 3-5. As can be seen, the numbers of MG patients that responded to these peptides were significantly higher in comparison to controls. The differences between MG patients and healthy controls were the most significant when the responses to peptide 3 were measured. These results suggest certain sequences in the AChR molecule that might be critical in the pathogenesis of MG.

It was of interest to find out whether sera of patients with MG possess antibodies that react with the peptides representing sequences of the human AChR α-subunit especially because the detection of antibodies to AChR receptors in the serum of MG patients provides an approach to diagnosing and monitoring MG patients (5.6). Therefore, sera of MG patients and of healthy controls were tested by radioimmunoassay (3) to detect antibody titers specific to peptides 3-5. As demonstrated in Table 1, the number of myasthenic sera with elevated antibody titers to these peptides are significantly higher than those of the controls. In agreement with the proliferative responses, peptide 3 was the most discriminative with 79.5% of the patients sera reacting with it in contrast to 6.5% of positive sera in the group of healthy controls (Table 1), thus strengthening the evidence for a role that this sequence might have in the pathogenesis of the disease.

The significance of a certain frequency of positive proliferation responses in healthy controls and of a few sera with binding activity to peptides 3-5 in the latter group is unclear. A possible cross-reaction between sequences of the AChR α-subunit and certain bacteria

Table 1. Proliferative Responses and Antibody Titers to Peptides Representing Sequences of the Human AChR α-Subunit

		Proliferative Responses*			Serum Antibody†		
Peptide	Region	Patients	Controls	p Value	Patients	Controls	p Value
3	195-212	19/26 (73%)	6/29 (20.6%)	< 0.001	31/39 (79.5%)	2/31 (6.5%)	< 0.0001
4	257-269	13/23 (56.5%)	4/22 (18%)	< 0.02	19/40 (47.5%)	1/30 (3.3%)	< 0.001
5	310-327	10/25 (40%)	5/28 (17.8%)	< 0.02	21/43 (49%)	2/31 (6.5%)	< 0.001

*PBL (2×10^5/well) of MG patients and healthy controls were cultured in flat-bottom microtiter plates with different doses of peptides 3, 4, and 5. Cultures were incubated for 6 days. For the last 16 h (^{3}H)-thymidine (0.5 μc) was added to the culture. Thereafter cells were harvested and radioactivity was counted.

†Sera of MG patients and healthy controls were tested for their antibody titers against peptides 3, 4, and 5 in a solid phase radioimmunoassay. Microtiter plates were coated with the peptides (100 μl of 50 μg/ml) dissolved in phosphate buffered saline containing 0.2% glutaraldehyde. ^{125}I-labeled protein A was used to determine the amount of antibodies bound.

Immunobiology of HLA
Volume II

Table 2. Proliferative Responses to Peptides 3 and 4 in Individuals Possessing HLA-DR3 and HLA-DR5

(A)	Patients		(B) Patients		Controls	
	Responders	Nonresponders	Responders	Nonresponders	Responders	Nonresponders
I. Peptide 4 HLA-DR3	7/13	0/10*	7	0	1	0
II. Peptide 3 HLA-DR5	10/19	2/7	10	2	3	11†

*p < 0.005; †p < 0.007.

and viruses might explain these observations. Indeed, shared antigenic determinants between the AChR α-subunit and membrane proteins of several bacteria have been recently reported (7).

The HLA-A,B,C,DR, and DQ phenotypes of 45 MG patients were tested (8,9). Fifteen patients (33%) possessed HLA-B8. Fourteen of the 45 patients (31%) possessed HLA-DR3. In 12 cases of the latter DR3 occurred jointly with HLA-B8. It should be noted that both HLA-B8 and DR3 are present in a relatively low frequency of 9.4% and 10.4%, respectively, in the Israeli population. Moreover, the joint occurrence of B8 and DR3 in the healthy Israeli population is rather low (5.7%) but significantly higher ($p < 0.001$) in the Israeli MG patients.

It was also tested whether a possible correlation exists between the proliferative capacity of PBL of MG patients and controls to the various sequences of the AChR receptor and HLA-DR type. As shown in Table 2A, of the 13 HLA typed patients responding by proliferation to peptide 4, seven possessed HLA-DR3 whereas none of the nonresponding MG patients carried HLA-DR3. The difference between the groups of responder and nonresponder MG patients to peptide 4 is statistically significant ($p < 0.005$). It should be noted that in the group of healthy controls that was tested for proliferative responses to peptide 4, the only HLA-DR3 donor responded to the peptide (Table 2B). It is also shown in Table 2A that 10 of 19 (52.6%) of the myasthenic responders to peptide 3 carry the HLA-DR5 haplotype as compared to 2 of 7 (28.5%) of DR5 carriers in the group of nonresponder MG patients. If proliferative responses of patients and controls who are HLA-DR5 carriers are compared (Table 2B), a significant association is observed. Thus, 10 out of 12 patients who possess HLA-DR5 responded to peptide 3. In contrast, only 3 HLA-DR5 carriers in the control group responded and the remaining 11 healthy HLA-DR5 carriers were nonresponders to peptide 3 ($p < 0.007$). No correlation could be observed between proliferative responses to peptide 5 and any HLA-DR antigen. Antibody activity to peptides 3, 4, and 5 in the sera of MG patients was not associated with any particular HLA type.

MG has been shown to be linked to class II major histocompatibility genes in man and in its experimental counterpart experimental autoimmune myasthenia gravis (EAMG) in mice (10-12). The present study indicates that certain peptides of the AChR are particularly immunogenic for certain HLA-DR types. HLA-DR5 individuals respond better than healthy controls to peptide 3. Moreover, all HLA-DR3 patients respond to peptide 4. It is possible that peptide 3 binds more readily with an MG-associated class II product, whereas peptide 4 binds preferentially to HLA-DR3.

Finally, these studies suggest HLA-associated determinants of the AChR molecule that might be involved in the aberrant immune response in myasthenia gravis in humans.

Acknowledgments. Supported in part by the Los Angeles and California Chapters of the Myasthenia Gravis Foundation.

References

1. Compston D, Vincent A, Newsom-Davis J, Batchelor J. Clinical, pathological, HLA antigen and immunological evidence for disease heterogeneity in myasthenia gravis. Brain 1981;103:579.
2. Zamvil S, Mitchell G, Moore A, Kitamura K, Steinman L, Rothbard J. T-cell epitope of the autoantigen myelin basic protein that induces encephalomyelitis. Nature 1986; 324:258.
3. Katz D, Bentwich Z, Eshhar N, Löwy, Mozes E. Immune response potential to poly(Tyr,Glu)-poly(*DL*Ala) – poly(Lys) of human T cells of different donors. Proc Natl Acad Sci USA 1981;78:4505.
4. Rothbard JB. Peptides and the cellular immune response. Ann de l'Institut Pasteur 1986;137:518.
5. Lindstrom JM, Seybold ME, Lennon VA, Whittingham S, Duane D. Antibody to acetylcholine receptor in myasthenia gravis: Prevalence, clinical correlates and diagnostic value. Neurology 1976;26:1054.
6. Limburg P, The T, Hummel-Tappel E, Ousterhuis H. Anti-acetylcholine receptor antibodies in myasthenia gravis. I. Relation to clinical parameters in 250 patients. J Neurol Sci 1983;58:357.
7. Stefansson K, Dieperink ME, Richman DP, Gomez CM, Martan LS. Sharing of antigenic determinants between the nicotinic acetylcholine receptor and proteins in *Escherichia coli, Proteus vulgaris* and *K. Pneumoniae*. N Eng J Med 1985;312:224.
8. Mittal KK, Mickey MR, Singal DP, Terasaki PI. Refinement of microdroplet lymphocyte cytotoxicity test. Transplantation 1968;69:913.
9. Terasaki PI, Bernoco D, Park MS, Ozturk G, Iwaki Y. Microdroplet testing for HLA-A,B,C and D antigens. Am J Clin Pathol 1978;69:103.
10. Bell J, Smoot S, Newby C, Toyka K, Rassenti L, Smith K, et al. HLA-DQ beta chain polymorphism linked to myasthenia gravis. Lancet 1986;ii:1058.

11. Christadoss P, Lindstrom J, Melvold R, Talal N. I-A subregion mutation prevents immune EAMG. Immunogenetics 1985;21:33.
12. Waldor M, Sriram S, McDevitt HO, Steinman L. In vivo therapy with monoclonal anti I-A antibody suppresses immune response to AChR. Proc Natl Acad Sci USA 1983;80:2713.

Author Affiliations

Edna Mozes, Stefan Brocke, Department of Chemical Immunology, The Weizmann Institute of Science, Rehovot 76100, Israel, *Chaim Brautbar*, Tissue Typing Unit, Department of Clinical Microbiology and Lautenberg Center for General and Tumor Immunology, The Hebrew University, Hadassah Medical Center, Jerusalem, Israel, *Oded Abramsky*, Neuroimmunoloy Unit, The Hebrew University, Hadassah Medical Center, Jerusalem, Israel; *Lawrence Steinman*, Departments of Pediatrics, Neurology and Genetics, Stanford University School of Medicine, Stanford, CA 94305, USA; *Jonathan Rothbard*, Imperial Cancer Research Fund, London, UK

HLA Antigens in Japanese Patients with Childhood-Onset Myasthenia Gravis

Kazumasa Matsuki, Jiro Numaga, Masaru Takamizawa, Hiroo Maeda, Takeo Juji, Yoshiko Nomura, and Masaya Segawa

Myasthenia gravis (MG) is an autoimmune disorder that affects neuromuscular junctions. In Japanese, the age of onset of MG shows a bimodal distribution, with the first and highest peak before the age of two and the second peak in the third decade (1). This is in contrast with the distribution in Caucasians: in the latter population, it shows only a single adulthood peak. In this study, we have examined HLA antigens in 56 Japanese patients that developed MG before the age of six. DQB restriction fragment length polymorphism was also examined.

Table 1. HLA Antigen Frequencies in Childhood-Onset MG

Antigen	Patients (n=56)	Controls (n=70)	Relative Risk
Aw33	24 (42.9%)	11 (15.7%)	4.02*
B44	25 (44.6%)	9 (12.9%)	5.47†
DR1	4 (7.1%)	10 (14.3%)	
DR2	4 (7.1%)	24 (34.3%)	0.15*
DR4	7 (12.5%)	28 (40.0%)	0.21*
DRw8	16 (28.6%)	19 (27.1%)	
DR9	43 (76.8%)	19 (21.1%)	8.88‡
DRw10	0 (0.0%)	1 (1.4%)	
DRw11	1 (1.8%)	5 (7.1%)	
DRw12	6 (10.7%)	9 (12.9%)	
DRw13	28 (50.0%)	12 (17.1%)	4.83†
DRw14	1 (1.8%)	2 (2.9%)	
DRw52	47 (83.9%)	40 (57.1%)	
DRw53	49 (87.5%)	43 (61.4%)	
DQw1	45 (80.4%)	48 (68.6%)	
DQw3	55 (98.2%)	36 (51.4%)	51.94‡
DQW4	3 (5.4%)	14 (20.0%)	
DQw7	4 (7.1%)	13 (18.6%)	
DQw8/9	51 (91.1%)	22 (31.4%)	22.25§

*$Pc<0.05$; †$Pc<0.005$; ‡$Pc<0.00005$; §$Pc<0.000005$.

All the patients were outpatients at Segawa Neurological Clinic for Children. The diagnosis was made clinically before HLA typing. Patients with neonatal or juvenile MG were excluded from this study. None of the patients had thymoma.

Both the frequencies of DR9 and DRw13 were increased significantly (Table 1). Among them, 22 (39%) patients were DRw9/DRw13 heterozygotes (Table 2). The relative risk was highest for the DRw9/DRw13 heterozygotes. As for the DQ antigens, all except one patient (98%) expressed DQw3. The number of DQw3/DQw1 heterozygotes was also increased (79%). Increased frequencies of Aw33 and B44 may be due to the close linkage of these antigens to DRw13 in Japanese.

Assuming a single disease-predisposing allele in linkage disequilibrium with both DR9 and DRw13, we calculated the expected phenotype frequencies using the method of Thomson (2) (Table 2). Both recessive and dominant models were rejected. Instead, a synergistic model was favored, in which the DR9-associated disease-predisposing allele was dominant and the DRw13-associated allele was recessive in the absence of the other (3).

An RFLP analysis using the DQB probe showed an increased frequency of an EcoRI 20kb fragment (96% vs 39% in controls, $Pc<0.0001$) and a 100% presence of a TaqI 1.9kb fragment (including the DQw3 negative case) ($Pc<0.03$).

This is the first report that demonstrates an increase of heterozygotes for HLA class II antigens in MG. Our findings show that the "childhood-onset" myasthenia gravis has certain immunogenetically different characteristics when compared with adult-onset MG.

Table 2. DRw9 and DRw13 Phenotype Frequencies

	N of Phenotypes						
	9/9	9/w13	w13/w13	9/x	w13/x	x/x	
Observed	1	22	0	20	6	7	
Control	2	2	1	16	9	40	
Relative risk	0.9	32	0	2.9	1.2	0.2	
Single disease allele model							
Recessive	8.7	11.0	3.5	15.7	10.0	7.14	$\chi^2_4 = 21.8$
k1 = 0.393							$p<0.002$
k2 = 0.250							
Dominant	4.7	5.1	1.4	25.7	12.8	0.11	$\chi^2_3 = 61.7$
k1 = 0.570							$p<0.0000001$
k2 = 0.282							

Recessive and dominant models were rejected at $p<0.002$ and $p<0.0000001$, respectively.

Acknowledgment. The authors are grateful to Ms. M. Okuyama, Ms. A. Yu, and Ms. K. Osawa for their excellent technical assistance. K.M. thanks DR. G. Thomson for her helpful suggestions regarding the genetic analysis of the data.

References

1. Segawa M. Treatment of myasthenia gravis. Chiryo 1986; 68:1359.
2. Thomson G. Investigation of the mode of inheritance of the HLA associated diseases by the method of antigen genotype frequencies among the diseases individuals. Tissue Antigens 1983;21:81.
3. Louis EJ, Thomson G. Three-allele synergistic mixed model for insulin-dependent diabetes mellitus. Diabetes 1986; 35:958.

Author Affiliations

Kazumasa Matsuki, Masaru Takamizawa, Takeo Juji, Blood Transfusion Service; *Jiro Numaga*, Department of Ophthalmology, Tokyo University Hospital; *Hiroo Maeda*, Blood Transfusion Service, Saitama Medical School, Saitama; *Yoshiko Nomura, Masaya Segawa*, Segawa Neurological Clinic for Children, Tokyo, Japan

Association Between Myasthenia Gravis and HLA Bw46, BfS, C4A4, C4B2, DRw9 in Chinese

R.L. Dawkins, B.R. Hawkins, W. Gin, W.J. Zhang, E. Martin, and F.T. Christiansen

Adult onset myasthenia gravis (MG) is associated with A1,B8,DR3 in Caucasians (1). Complement allotyping has shown that patients with A1,B8,DR3 also have BfS,C4Q0,C4B1 indicating that the susceptibility gene(s) are marked by this particular supratype. To date no clear supratype associations have been described for any other racial groups. Weak associations between MG and Bw46 (2) and also DRw9 (3) have been reported in the Chinese. In an attempt to localize the crucial genes for MG, we have compared the typing of Caucasians and Chinese patients.

The study group consisted of 166 Hong Kong Chinese patients with local HLA typing. MG was diagnosed on the basis of a detailed neurologic examination and in most instances, a positive Tensilon test. Patients with the congenital form of the disease or with other causes of muscle weakness were excluded. Complement allotypes were determined using a standard electrophoretic and immunofixation technique and alleles were assigned in accordance with international agreement (4).

Overall increases in the frequencies of Bw46 (47%) and DRw9 (60.8%) but in no other antigens, were observed in the patients compared to healthy controls (29% and 21%, respectively). A variant of C4A3 (C4A3F) was found to be common but difficult to assign in the presence of C4A3 and/or C4A4. Although it was not possible to exclude heterozygosity for C4A3,4 and C4A3F,4 in a number of cases, nevertheless, BfS,C4A4,C4B2 seemed to be present in the majority of those with Bw46 and DRw9. It was noteworthy that the combination of Bw46,BfS,C4A4,C4B2,DRw9 (46,S,4,2,9) was increased in the younger patients who also generally had absent or low titers of anti-acetylcholine receptor antibodies (anti-AChR) and no detectable anti-striational antibodies. In fact the complete 46,S,4,2,9 supratype was found to be present in 21 of 29 (72%) patients in whom the age of onset of disease was known to be less than 12 years. None of these particular patients had antithyroid antibodies or a past history of thyroid disease; this was important because of a previously reported association between Bw46 and Graves Disease in the Chinese (5).

These findings indicate that there is a distinct form of MG that occurs in Chinese children and for which the disease susceptibility gene(s) are marked by 46,S,4,2,9. This is in contrast with the 1,8,S,0,1,3 supratype associated with the adult-onset MG and elevated anti-AChR titers in Caucasians. An autoimmune etiology cannot be excluded because anti-nuclear antibodies were detected in some of the pediatric cases. It is possible that the pediatric form of the disease is mediated by an antibody not detected by current assays.

Molecular characterization of the relevant supratype is in progress; to date no similarities with the 1,8,S,0,1,3 supratype have been detected within either class II or III regions.

References

1. Dawkins RL. Myasthenia gravis. In: Terasaki PI (ed): Histocompatibility Testing 1980. Los Angeles, UCLA Tissue Typing Laboratory, 1980, p 662.
2. Hawkins BR, Chan-Lui WY, Choi EKK, Ho AY. Strong association of HLA Bw46 with juvenile onset myasthenia gravis in Hong Kong Chinese. J Neurol Neurosurg Psychiatry 1984;47:555.
3. Thajeb P, Chee CY, Huang CC. The distribution of HLA A,B,DR antigens in Chinese myasthenia gravis. Tissue Antigens 1987;29:273.
4. Mauff G, Alper CA, Awdeh Z, et al. Statement on the nomenclature of human C4 allotypes. Immunobiol 1983; 164:184.
5. Chan SH, Yeo PPB, Lui KF, et al. HLA and thyotoxicosis (Graves Disease) in Chinese. Tissue Antigens 1978;12:109.

Author Affiliations

R.L. Dawkins, W. Gin, W.J. Zhang, E. Martin, F.T. Christiansen, Departments of Clinical Immunology, Royal Perth Hospital, Queen Elizabeth II Medical Centre and University of Western Australia, Perth, Western Australia; *B.R. Hawkins*, University of Hong Kong, Hong Kong

HLA Antigens in 206 Japanese Patients with Narcolepsy and 46 Patients with Essential Hypersomnia

Kazumasa Matsuki, Yutaka Honda, and Takeo Juji

Abstract: Narcolepsy is a relatively rare disease that affects sleep itself (1). Although there is agreement that narcolepsy has the strongest HLA association (2–8), controversy remains as to whether the association with HLA-DR2 is complete (9–13). In this study, we HLA-typed a total of 252 somnolent patients and examined if there was a case of DR2-negative narcolepsy in Japanese. Even though we have HLA-typed the largest number of the patients to date, we still have not found a single case of narcolepsy without DR2. This finding lent skepticism concerning the presence of DR2-negative narcolepsy at least in Japanese. However, employment of various diagnostic criteria other than ours led to apparent finding of DR2-negative "narcolepsy."

The subjects were 206 Japanese patients with narcolepsy and 46 patients with essential hypersomnia. All the previously reported cases were included (2–5). The diagnosis was made clinically before HLA typing. Our diagnostic criteria for narcolepsy are: a) recurrent daytime naps and/or lapses into sleep that occur almost every day over a period of at least six months; b) clinical confirmation of cataplexy in the patient's history. Cataplexy is defined here as the sudden bilateral loss of tone in skeletal muscles triggered by strong emotion. We do not consider attacks of muscle weakness as cataplexy if they lack intense emotional triggers. Nor do we consider them as cataplexy if consciousness is lost or memory is impaired during the attacks. Essential hypersomnia is defined as: a) excessive daytime somnolence that occurs almost daily over a period of at least six months; b) absence of cataplexy; c) not due to other known sleep disorders such as the sleep-apnea syndrome. Narcolepsy and essential hypersomnia both describe chronic somnolence of unknown etiology. The most important difference between the two categories is whether cataplexy is present or not.

HLA class II antigen frequencies in 206 Japanese patients with narcolepsy are shown in Table 1. All the patients with narcolepsy had DR2 and DQw1. The DR2 frequency in essential hypersomnia was also increased (57%) as compared with that in controls ($P<0.005$).

The patients with narcolepsy and essential hypersomnia were subclassified into 16 groups according to combinations of the four major narcoleptic symptoms (cataplexy, sleep paralysis, hypnagogic hallucinations,

Immunobiology of HLA
Volume II

Table 1. HLA Class II Antigen Frequencies in 206 Japanese Narcoleptics

Antigen	Patient (n=206)	Control (n=310)	χ^2	Relative Risk
DR1	5 (2.4%)	45 (14.5%)	−20.7	0.2†
DR2	206 (100.0%)	104 (33.5%)	225.1	816.1‖
DR4	69 (33.5%)	125 (40.3%)		
DRw8	18 (8.7%)	84 (27.1%)	−26.3	0.3†
DR9	23 (11.2%)	85 (27.4%)	−19.8	0.3†
DRw10	1 (0.5%)	2 (0.6%)		
DRw11	8 (3.9%)	20 (6.5%)		
DRw12	18 (8.7%)	41 (13.2%)		
DRw13	12 (5.8%)	51 (16.5%)	−13.0	0.3*
DRw14	5 (2.4%)	8 (2.6%)		
DRw52	61 (29.6%)	178 (57.4%)	−38.5	0.3‡
DRw53	92 (44.7%)	189 (61.0%)	−13.3	0.5*
DQw1	206 (100.0%)	213 (68.7%)	77.3	188.6§
DQw3	77 (37.4%)	157 (50.6%)	−8.8	0.6*

*$p<0.01$; †$p<0.0001$; ‡$p<10^{-9}$; §$p<10^{-17}$; ‖$p<10^{-40}$.

Table 3. Seven Sets of Diagnostic Criteria and DR2 Frequencies

Criteria	Equivalents	DR2 Frequencies
Sleep attacks	Roth	91%
Cataplexy	Honda	100%
REM sleep abnormalities	ASDC-APSS	95%
Cataplexy and/or sleep paralysis	Hypothetical	95%
Cataplexy and/or hypnagogic hallucinations	Hypothetical	96%
Sleep paralysis and/or hypnagogic hallucinations	Hypothetical	98%
Sleep-onset REM	REM-narcolepsy	98%

In criteria 1, the diagnosis of "narcolepsy" was made for the somnolent patients with sleep attacks; in criteria 2, for those with cataplexy; in criteria 3, for those with at least one REM sleep abnormality; in criteria 4, for those with either cataplexy or sleep paralysis or both; in criteria 7, for those with one or more sleep-onset REM periods.

and sleep-onset rapid eye movement (REM) periods) (Table 2). The DR2 frequencies in the patients with cataplexy were 100%, whereas those in the patients without cataplexy were less than 100%. As for specific signs and symptoms, the DR2 frequency was 100% in patients with cataplexy, 94.7% with sleep paralysis, 95.4% with hypnagogic hallucinations, and 98.6% with sleep-onset REM periods.

The right columns in Table 2 show the range of "narcolepsies" as defined by three different definitions.

Table 2. DR2 Frequencies in Somnolent Patients with Various Combinations of Narcoleptic Symptoms

Symptoms*				N of Patients HLA-DR2			Diagnostic Criteria		
C	HH	SP	REM	+	−	%DR2	Honda	ASDC-APSS	Roth
+	+	+	+	115	0	100%	*	*	*
+	+	+	−	28	0	100%	*	*	*
+	+	−	+	9	0	100%	*	*	*
+	+	−	−	3	0	100%	*	*	*
+	−	+	+	4	0	100%	*	*	*
+	−	+	−	1	0	100%	*	*	*
+	−	−	+	7	0	100%	*	*	*
+	−	−	−	6	0	100%	*	*	*
−	+	+	+	5	1	83%		*	*
−	+	+	−	6	7	46%		*	*
−	+	−	+	0	0	−		*	*
−	+	−	−	0	0	−		*	*
−	−	+	+	0	0	−		*	*
−	−	+	−	1	1	50%		*	*
−	−	−	+	2	1	67%		*	*
−	−	−	−	11	9	55%			*
DR2 frequencies							100%	95%	91%

*C = Cataplexy; HH = hypnagogic hallucinations; SP = sleep paralysis; REM = sleep-onset REM periods.

According to our criteria as described by Honda and colleagues (4,13), only somnolent patients with cataplexy were diagnosed as narcoleptics. The Association of Sleep Disorders Centers and the Association for the Psychophysiological Study of Sleep (ASDC-APSS) in 1979 (14) defined narcolepsy by the presence of REM sleep abnormalities. Thus, all the patients except in the last category were diagnosed as narcoleptics according to the 1979 ASDC-APSS classification. Roth (15) defines narcolepsy by the presence of sleep attacks. Honda's criteria is the most stringent of the three, and only this criteria gave the 100% DR2 frequency.

Table 3 shows seven sets of diagnostic criteria and DR2 frequencies. Only criteria 2 yielded the 100% DR2 frequency. Criteria 2 defines narcolepsy by the presence of cataplexy. All the other criteria do not require cataplexy as a requisite item for diagnosis. They reduced the DR2 frequency and led to the apparent finding of cases of DR2-negative "narcolepsy." These results suggest that the employment of certain diagnostic criteria will lead to the false finding of DR2-negative narcolepsy. These data also indicate that the criteria that include cataplexy as a sine qua non should be used in the study of the HLA-narcolepsy association.

There are historical controversies over the definition of narcolepsy. Depending on the investigator, narcolepsy may mean: 1) a symptom of irresistible sleep attacks, like in "narcolepsy-cataplexy"; 2) a syndrome of excessive daytime sleepiness and abnormal REM manifestations (ASDC-APSS, 1979) (14); 3) a disease entity characterized by excessive daytime sleepiness and cataplexy (1,4,13); 4) a pathophysiologic condition defined by the multiple sleep latency tests (16). Each of these definitions describes a similar, yet slightly different, pathology.

Our findings show that not all the somnolent patients are DR2 positive. However, according to our criteria, even in this, the largest scale study to date, we could not find any narcoleptic patients without DR2. This is in contrast with other reports that found DR2-negative narcoleptics in relatively small scale studies (9–12). Because we have HLA-typed the largest number of narcoleptic patients to date and still could not find a single case of DR2-negative narcolepsy, we remain skeptical about the presence of DR2-negative narcolepsy, at least in the Japanese. But, before any discussion on DR2-negative narcolepsy is productive, the definition of narcolepsy and clear and operational diagnostic criteria for narcolepsy should be established.

Acknowledgment. We are grateful to Ms. Machiko Okuyama, Ms. Angela Yu, and Ms. Keiko Osawa for their excellent technical assistance, Mrs. Mitsuko Matsuyama for her secretarial work, and Dr. Christopher J. Blackwell for his helpful discussions.

References

1. Honda Y, Asaka A, Tanimura M, Furusho T. A genetic study of narcolepsy and excessive daytime sleepiness in 308 families with a narcolepsy of hypersomnia proband. In: Guilleminault C, Lugaresi E (eds):Sleep/Wake Disorders. New York, Raven Press, 1983, p 198.
2. Honda Y, Doi Y, Juji T, Satake M. Narcolepsy and HLA: Positive DR2 as a prerequisite for the development of narcolepsy. Folia Psychiatr Neurol Jpn 1984;38:360.
3. Juji T, Satake M, Honda Y, Doi Y. HLA antigens in Japanese patients with narcolepsy. All the patients were DR2 positive. Tissue Antigens 1984;24:316.
4. Honda Y, Juji T, Matsuki K, Naohara T, Satake M, Inoko H, et al. HLA-DR2 and Dw2 in narcolepsy and in other disorders of excessive somnolence without cataplexy. Sleep 1986;9:133.
5. Matsuki K, Juji T, Tokunaga K, Naohara T, Satake M, Honda Y. HLA haplotype frequencies estimated from the data on HLA class I, II, and III antigens in 111 Japanese narcoleptics. J Clin Invest 1985;76:2078.
6. Langdon N, Welsh KI, Dam MV, Vaughan RW, Parkes D. Genetic markers in narcolepsy. Lancet 1984(ii):1178.
7. Billiard M, Seignalet J. Extraordinary association between HLA-DR2 and narcolepsy. Lancet 1985(i):226.
8. Poirier G, Montplaisir J, Decary F, Momege D, Lebrun A. HLA antigens in narcolepsy and idiopathic central nervous system hypersomnolence. Sleep 1986;9:153.
9. Andreas-Zietz A, Keller E, Scholz S, Albert ED, Roth B, Nevsimalova S, et al. DR2 negative narcolepsy. Lancet 1986(ii):684.
10. Guilleminault C, Grumet C. HLA-DR2 and narcolepsy: Not all narcoleptic-cataplectic patients are DR2. Human Immunology 1986;17:1.
11. Mueller-Eckhardt G, Meier-Ewert K, Schendel DJ, Reinecker FB, Mueller-Eckhardt C. HLA and narcolepsy in a German population. Tissue Antigens 1986;28:163.
12. Langdon N, Lock C, Welsh K, Vergani D, Dorow R, Wachtel H, et al. Immune factors in narcolepsy. Sleep 1986; 9:143.
13. Matsuki K, Honda Y, Juji T. Diagnostic criteria for narcolepsy and HLA-DR2 frequencies. Tissue Antigens (in press).
14. Association of Sleep Disorders Centers and The Association for the Psychophysiological Study of Sleep. Diagnostic classification of sleep and arousal disorders. Sleep 1979; 2:5.
15. Roth B. Narcolepsy and Hypersomnia. Basel, S. Karger, 1980.
16. Mittler M, van der Hoed Jr M, Caskadon MA. REM sleep episodes during the multiple sleep tests in narcoleptic patients. Electroencepharogr Clin Neurophysiol 1980; 46:479.

Author Affiliations

Kazumasa Matsuki, and Takeo Juji, Blood Transfusion Service, Tokyo University Hospital; *Yutaka Honda*, Seiwa Hospital, Neuropsychiatric Research Institute, Tokyo, Japan

DR2 and DQw1 Molecules in Narcolepsy

Hiroo Maeda, Ranko Hirata, Kazumasa Matsuki, Takeo Juji, and Yutaka Honda

Abstract: Narcolepsy is a clinical disease entity characterized by excessive daytime sleepiness and cataplexy. The cause and genetic basis for the disease has been unknown until 1984, when an extraordinary association between HLA-DR2 and narcolepsy was reported (1,2). This association was confirmed to be complete by the HTC typing (Dw2) (3,4) and by molecular genetics (DQR2.6) (4,5) and was confined to a specific DR2-carrying haplotype, HLA-DR2-Dw2-DQw1 (6). We have isolated DR2 and DQw1 molecules from nine DR2-positive Japanese patients with narcolepsy using anti-DR and anti-DQw1 related monoclonal antibodies and analyzed them by one- and two-dimensional IEF gel electrophoresis. The results indicated that both DQw1 α and β chains focused identically among patients and DR2-Dw2-positive normals but clearly distinct from those of DR2-Dw12 normal Japanese. DR2 α and β chains were identical among patients and DR2-Dw12 as well ad DR2-Dw2 normals. A second DR2 β chain isolated from one DR2 homozygous patient was identical with those of normal DR2-Dw2 individuals but distinct from DR2-Dw12 normals. The data strongly suggest that both DR2 and DQw1 molecules are identical among patients and normals.

Immunobiology of HLA
Volume II

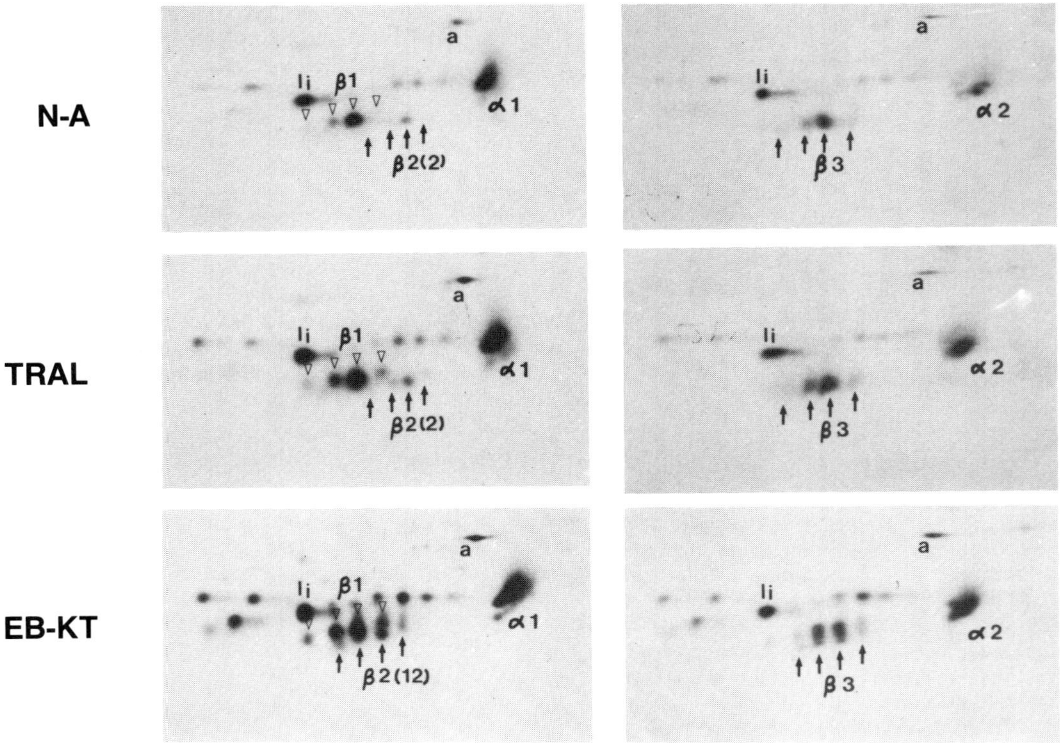

Figure 1. DR2 and DQw1 molecules were isolated from a B-cell line (NA) established from DR2-homozygous patient with narcolepsy and were compared with those from DR2-Dw2 (Tral) and DR2-Dw12 (EB-KT) B-cell lines by two-dimensional IEF gel electrophoresis (10). These molecules were separated by isoelectric focusing in the first dimension and by SDS-PAGE in the second dimension. In the figure, the acidic side is positioned on the right and the basic side on the left. The DR molecules isolated by anti-DR MAb Sh34 consist of one α (α1) and two β (β1 and β2) chains in all three cell lines (on the left columus). The DQ molecules isolated by MAb PLM13 consist of one α (α2) and one β (β 3) chain (on the right columus). All β chains are composed of a series of four spots. a = Actin; Ii = an invariant chain.

Narcolepsy is now a clinical disease entity characterized by excessive daytime sleepiness and cataplexy. The precise etiology of narcolepsy is not known, although genetic factors appear to play a role in the pathogenesis of the disease. We first described a strong association between HLA-DR2 and narcolepsy (1). The association has been complete in our series where DR2 was found in 100% of more than 200 narcoleptic patients. The same association was confirmed in British (2) and French (7) patients, although some investigators have reported that not all the patients are DR2-positive (8,9), suggesting that there are subsets in this disease or that DR2 is an associated marker rather than the disease susceptibility gene itself.

One of the approaches to test the second possibility is to compare DR2 and DQw1 molecules from patients and normals. Thus far no information on the DR2 and DQw1 molecules from patients with narcolepsy have been available. We have isolated both DR2 and DQw1 molecules from a patient, NA, who was HLA-typed for DR2/, and compared them with those from normal DR2-Dw2 (Tral) and DR2-Dw12 (EB-KT) homozygous typing cells by two-dimensional IEF gel electrophoresis (Fig. 1). Two DR molecules (α1β1 and α1 β2) isolated from NA using anti-DR framework MAb Sh34 focused identically with those from Tral, but the sole DR molecule (α1β1) was identical to that of EB-KT, whereas the other DR molecule (α1β2) was distinct from each other. The DQ molecule (α2β3) was isolated from NA by anti-DQw1 related MAb PLM13 (11) and was identical to that from Tral but distinct from that of EB-KT. Thus, two DR2 and one DQw1 molecules from the DR2-positive Japanese patient were indistinguishable from normal Caucasian DR2-Dw2 homozygous typing cells.

To further confirm the result obtained from the DR2 homozygous patient NA, we isolated DR2 and DQw1 molecules from eight B-lymphoid cell lines established from DR2-heterozygous patients (Table 1) and directly compared them by one-dimensional IEF gel electrophoresis. This method has several advantages over 2-D gel electrophoresis; a larger number of smaller quantity of samples can be separated in a single gel for the shorter time. Thus, multiple samples can be directly compared

Table 1. HLA Types of B-Cell Lines from Narcolepsy

No.	Name	A	B	DR/Dw	DQ
1	Na1	24,26	35,37	2,w10	w1,-
2	Na2	2,26	35,w55	2,4	w1,w3
3	Na4	2,24	w60,-	2,w14	w1,-
4	Na5	2,24	51,-	2,w11	w1,w3(TA10)
5	Na8	24,-	w61,-	2,4	w1,WA
6	Na9	2,-	35,51	2,HR8	w1,w3
7	Na10	24,26	35,-	2,w8	w1,w3
8	Na11	2,-	35,w54	2,4	w1,WA
9	CMG	3,3	7,7	2,2(w2)	w1,w1
10	Tra1	3,3	7,7	2,2(w2)	w1,w1
11	EB-KT	24,24	w52,w52	2,2(w12)	w1,w1
12	NA	24,26	35,51	2,-	w1,-

side by side. One disadvantage of this gel system is that multiple bands of a single molecule are so complex that it requires reduction of the number of bands for the exact interpretation. As a single DR or DQ molecule generates four different spots depending on the amounts of the sialic acid (see Fig. 1), we improved the methods by treating the isolated molecules with neuraminidase and then electrophoresed them in the range of pH 3.5-10, by which we succeeded in reducing the number of the bands or even in obtaining a single band of a molecule.

The gel patterns of DR and DQw1 molecules from eight DR2-heterozygous patients as well as DR2-Dw2 and DR2-Dw12 homozygous cell lines are demonstrated in Figure 2. In the gel for DR molecules isolated with anti-DR framework MAb Sh35, two DR β chains were isolated from heterozygous as well as homozygous cell lines. One DR β chain (β1) from all the DR2-heterozygous patients (No.1-No.8) was identical to each other and to one DR chain (β1) from DR2 homozygous cell lines (No.9-No.11). Another DR β chain from patients varied with the DR types on the opposite haplotypes, while that from DR2 homozygous cell lines, although much less faint, was a second DR β chain (β 2) from the same haplotype. The data only indicate that one DR2 β chain (β 1) was identical among all patients and DR2-positive normals and a second DR2 β chain from patients was not demonstrated, possibly because of its small quantity. One DQ β chain was isolated from DR2-heterozygous patients as well as homozygous normals. MAb PLM13 used for isolation of DQw1 molecule showed a stronger binding to DQw1 molecules from DR2-carrying haplotypes but much less binding to those from DR1 and DRw14 and thus immunoprecipitated one DQw1 molecule (α2β3) from each of the patients, which were identical among patients and DR2-Dw2 HTCs (No.9,10) but distinct from that of DR2-Dw12 HTC (No.11).

Two interpretations for the results are possible: 1) both DR2 (α1β2) and DQw1 (α2β3) molecules are identical among patients and normals; 2) DR2 or DQw1 molecules in patients carry very minute variations that do not effect the total charge of both molecules. The IEF gel analysis is the most discriminating technique available today concerning structural variations of DR and DQ molecules, thus we favored the first possibility although this is not a proof of identity at the amino acid sequence level. Even if DR2 and DQw1 molecules are identical among patients and normals, there still exist two possibilities: 1) DR2 or its linked HLA genes is not the disease susceptibility gene itself but just tightly linked to it; or 2) DR2 or its linked HLA genes is one of disease susceptibility genes. In this case, the sole presence of DR2 or DQw1 is not enough for the development of the disease but other factors, including a gene(s) outside the major histocompatibility complex, are required as well. Both lines of possibilities should be taken into consideration for future genetic studies on narcolepsy.

References

1. Juji T, Satake M, Honda Y, Doi Y. HLA antigens in Japanese patients with narcolepsy. Tissue Antigens 1984;24:316.
2. Langdon N, Welsh KI, Dam MV, Voughan RW, Parkes D. Genetic markers in narcolepsy. 1984;ii:1178.
3. Honda Y, Juji T, Matsuki K, Naohara T, Satake M, Inoko H, et al. HLA-DR2 and Dw2 in narcolepsy and in other disorders of excessive somnolence without cataplexy. Sleep 1986;9:133.
4. Marcadet A, Gebuhrer L, Betuel H, Seignalet J, Freidel AC, Confavreux C, et al. DNA polymorphism related to HLA-DR2 Dw2 in patients with narcolepsy. Immunogenetics 1985;22:679.
5. Inoko H, Ando A, Tsuji K, Matsuki K, Juji T, Honda H. HLA-DQ chain DNA restriction fragments can differentiate between healthy and narcoleptic individuals with HLA-DR2. Immunogenetics 1986;23:126.
6. Matsuki K, Juji T, Tokunaga K, Naohara T, Satake M, Honda Y. HLA haplotype frequencies estimated from the data on HLA class I, II, III antigens in 111 Japanese narcoleptics. J Clin Inv 1985;76:2078.
7. Billiard M, Seignalet J. Extraordinary association between HLA-DR2 and narcolepsy. Lancet 1985;i:226.
8. Guilleminault C, Grumet C. HLA-DR2 and narcolepsy: Not all narcoleptic-cataplectic patients are DR2. Human Immunol 1986;17:1.
9. Andreas-Zietz A, Keller E, Scholz S, Albert ED, et al. DR2-negative narcolepsy. Lancet 1986;ii:684.
10. Maeda H, Hirata R, Okuyama M, Thompson A, Tohyama H. Two-dimensional gel analysis of a second family of class II molecules by polymorphic HLA-DR4, 5, and w9 monoclonal antibodies. J Immunol 1984;132:2478.
11. Maeda H, Hirata R, Kambayashi H, Koning F, Schreuder GMT. Multiple epitopes on a single DQ molecule from the DQw3-carrying haplotypes. Tissue Antigens 1986; 28:136.

Author Affiliations

Hiroo Maeda, Ranko Hirata, Blood Transfusion Service, Saitama Medical Center, Saitama Medical School, Kawagoe; *Kazumasa Matsuki, Takeo Juji*, Blood Transfusion Service, Tokyo University Hospital, Tokyo; *Yutaka Honda*, Seiwa Hospital, Neuropsychiatric Research Foundation, Tokyo, Japan

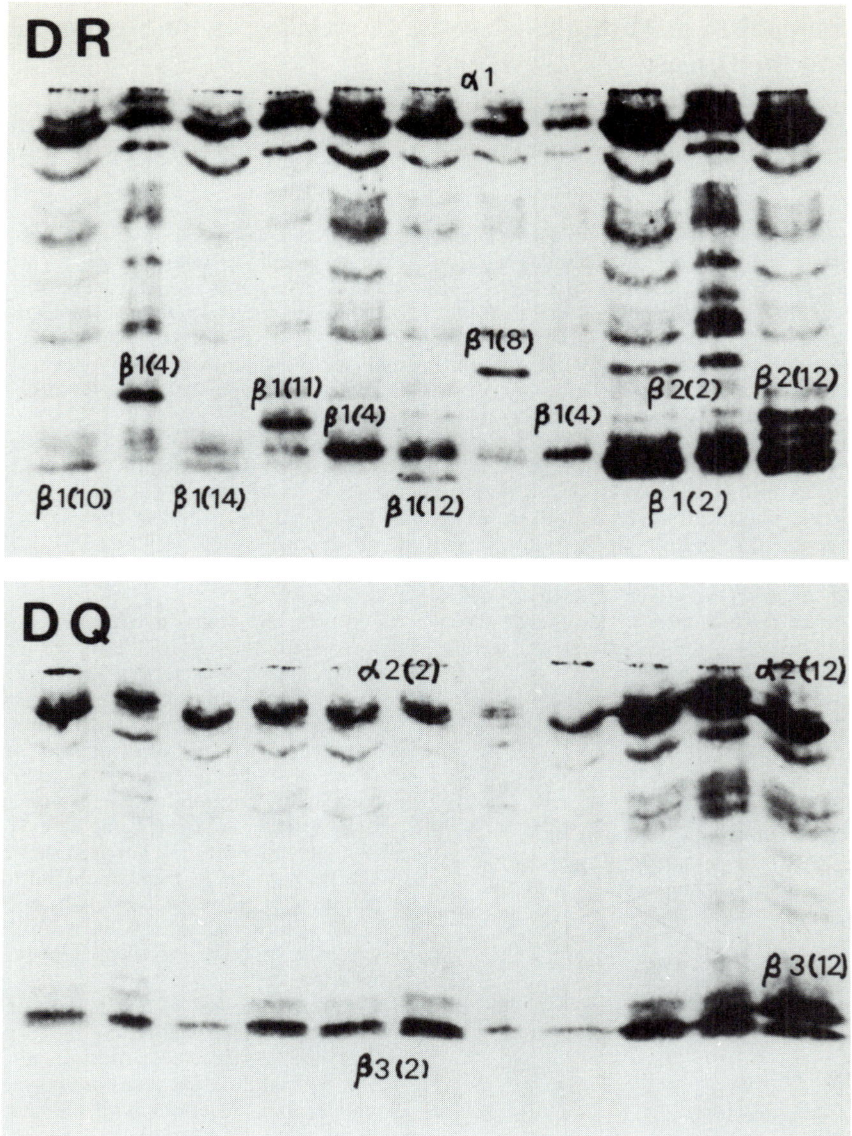

Figure 2. DR2 and DQw1 molecules were isolated from B-cell lines established from DR2-heterozygous patients with narcolepsy and separated by one-dimensional IEF gel electrophoresis. The isolated radiolabelled molecules were treated with neuraminidase and then applied on the acidic side (top). The gel was run at 600V for 7 hours and at 1000V for 1 hour using an LKB 2117 Multifor flat-bed gel apparatus. All α and β nomenclatures followed that in Figure 1. The numbers in parenthesis indicate the DR types of which the cells originated. In the case of DR2, the numbers in parenthesis following α2, β2, and β3 indicates the Dw types of which the DR2 cells are originated. The cell number in the bottom is listed in Table1.

DNA Polymorphism in Multiple Sclerosis, Correlations with Susceptibility and Resistance to the Disease

Ana Dos Santos, Alan Thompson, Juliette Awad, Michael J. Garlepp, and Hilliard Festenstein

Significant associations between certain HLA haplotypes and multiple sclerosis have been demonstrated by population and family studies (1–3). In the present, study blood samples from members of MS multiplex families, unrelated MS patients, and healthy English individuals were collected and typed for HLA class II antigens. DNA restriction fragment length polymorphisms (RFLP) were also analyzed using Southern blots. The enzymes and probes used were: EcoRI, Taq I and DQB, DQA.

Previously described HLA associations within the British population indicate that there is likely to be one or more genes in the HLA haplotype associated with DR2, predisposing to MS. They could also be associated with another DR antigen (4) such as DR4 as has been described in Jordanians. In this study the percentage of both DR2+ and DR4+ patients was greater than the controls.

We investigated the linked associations between DR2, DR4, and the RFLPs associated with the DQ and the DR genes. Normal population controls and patients were examined and the following preferential allelic associations (PAAs) were found to be significantly increased in the patients: PAA1: DR2-DQw1-DQB-2.1 kb; PAA2: DR4-DQw3-DQB-2.1 kb. Both the PAAs have a common RFLP which suggest that the 2.1 kb band may indicate the existence of a recessive genetic predisposition to MS.

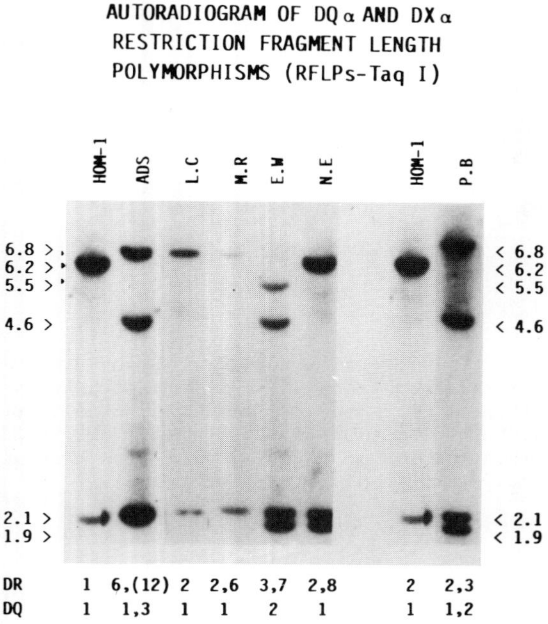

Figure 1. Southern blot obtained after hybridization of the DQα genomic DNA probe to genomic DNA isolated from MS patients and controls.

Furthermore, on examining DR2 patients and normal controls, DNA digested with Taq I and hybridized with the DQA probe gave a 6.2 kb band whicn is present at a reduced frequency in the patients compared to normal controls (50% vs 90.6%: $p=0.02$). These findings suggest that this band may be associated with MS resistance factors.

There are important exceptions to the classic associations of MS with HLA-DR2/Dw2. In view of these discrepancies we postulate that the interaction between genes within chromosome 6 and/or genes outside this region, plus environmental factors are very strong in groups where the classic association with DR2/Dw2 does not exist. The alternative would be that the genes making up the DR2 preferential allelic associations are present despite DR2 itself being absent. They could be associated with other DR antigens such as DR4.

The study of HLA distribution and MS in different populations therefore provides evidence that the balance between genetic background and environmental factors determines the relative susceptibility/resistance to this disease (5).

With respect to the HLA complex, we conclude that there is more than a single gene and/or gene product in each of the PAAs involved in the susceptibility/resistance to MS. Several different genes or molecules, possessing complementary functions with hierarchial as well as additive effects may be implicated in the pathogenesis and development of the disease.

Acknowledgments. These studies were supported by the Multiple Sclerosis Society of Great Britain (A.D.S, M.G), The London Hospital (A.T, S.A), and The Nuffield Foundation (H.F.).

We wish to thank P. Peterson and W. Bodmer for providing the DNA probes used in this study.

References

1. Ebbers G, Bulman D, Sadovnick A, Paty D, Warren S. A population-based study of multiple sclerosis in twins. N Engl J Med 1986;315;1639.

Immunobiology of HLA
Volume II

2. Batchelor R. Immunological and pathological aspects of MS. In: McApine's Multiple Sclerosis. London, Churchill Livingstone, 1985, pp 281–300.
3. McDonald W. Epidemiology and HLA associations. In: Multiple Sclerosis. Ann N Y Acad Sci 1984;436:109–117.
4. Elian M, Alonso A, Awad J, Dean G, Festenstein H. HLA associations with multiple sclerosis in Sicily and Malta. Dis Markers 1987;5:89–99.
5. Festenstein H, Awad J, Hitman G, Cassell P, Sachs J. New HLA DNA polymorphisms associated with autoimmune disease. Nature 1986;322:6074,64–67.

Author Affiliations

Ana Dos Santos, Juliette Awad, Michael J. Garlepp, Hilliard Festenstein, Department of Immunology; *Alan Thompson*, Department of Neurology, The London Hospital Medical College, Turner Street, London E1 2AD, UK

Chronic Progressive Multiple Sclerosis Is Associated with a Specific HLA-DQB Restriction Fragment Pattern

Olle Olerup, Sten Fredrikson, and Thomas Olsson

Multiple sclerosis (MS) is an inflammatory demyelinating disease of the central nervous system with a genetically determined susceptibility (1). Immunologic mechanisms are supposed to be involved in the pathogenesis. MS is associated with several HLA alleles, including A3, B7, and Dw2/DR2 (2). In Caucasians the strongest association has been found with the class II antigen DR2, with a relative risk of about three (2).

The majority of patients with MS exhibit a disease course characterized by relapses and remissions. However, many will eventually enter a progressive phase. The remaining 10 to 30% of MS patients have a primarily chronic progressive disease form, that is they show progressive evolution of disease from the onset. Chronic progressive and relapsing/remitting MS may represent two distinct disease entities (3), since they differ in respect to epidemiology, age at onset, initial symptoms, prognosis, and response to immunosuppression. Efforts to differentiate between chronic progressive and relapsing/remitting MS by serologic HLA class II typing have given divergent results.

We have studied 46 unrelated Caucasian patients with clinically definite MS and 40 healthy controls. Twenty patients had chronic progressive MS. Twenty-six patients had relapsing/remitting MS; this latter group included patients who showed a progressive evolution of the disease following initial relapses and remissions.

Patients and controls were HLA class II typed by restriction fragment length polymorphism (RFLP) analysis using almost full-length DRB, DQA, and DQB cDNA probes. The restriction enzyme used was *Taq I* (4).

Eleven of the 20 patients with primarily chronic progressive MS had a specific heterozygous DQB restriction fragment pattern, characterized by restriction fragments of 1.8+2.3+2.5 and 1.7+2.0+2.2 kilobases, whereas none of the 26 patients with relapsing/remitting MS ($p_{corr} < 0.0001$; Fisher's exact test) and only four of the 40 healthy controls ($p_{corr} < 0.005$; Fisher's exact test) exhibited this specific DQB band pattern. The relative risk of this DQB restriction fragment pattern was 11. Relapsing/remitting MS patients did not differ significantly from the healthy controls in this respect.

In the present study we have found an immunogenetically based heterogeneity (HLA-DQB) between patients with chronic progressive and with relapsing/remitting MS. This finding may favor the concept of MS as being two distinct disease entities and provide a starting point for molecular studies on the pathogenesis of chronic progressive MS. Furthermore, the possibility to subgroup MS patients at a genomic level might be used prognostically and to select homogeneous groups of patients for clinical trials.

Acknowledgments. This work was supported by the Medical Research Council of the Swedish Life Insurance Companies.

References

1. MacFarland HF, Greenstein J, McFarlin DE, Elridge R, Xu XH, Krebs H. Family and twin studies in multiple sclerosis. Annals of The New York Academy of Sciences 1984;436: 118–124.
2. Tiwari JL, Terasaki PI. Multiple sclerosis. In: Tiwari JL, Terasaki PI (eds): HLA and Disease Associations. New York, Springer-Verlag, 1985, pp 152–167.
3. Larsen JP, Kvaale G, Riise T, Nyland H, Aali J. Acta Neurol Scand 1985;72:145–150.
4. Carlsson B, Wallin J, Böhme J, Möller E. HLA-DR-DQ haplotypes defined by restriction fragment analysis. Correlation to serology. Human Immunology 1987;20:95–113.

Author Affiliations

Olle Olerup, Center for Biotechnology and Department of Clinical Immunology, Karolinska Institute at Huddinge Hospital, Stockholm; *Sten Fredrikson, Thomas Olsson*, Department of Neurology, Karolinska Institute at Huddinge Hospital, Stockholm, Sweden

Polymorphic Restriction Fragments Detected with HLA-DQα Discriminate Between DR2-Positive Multiple Sclerosis Patients and Healthy Controls

R.N. Heard, W.I. McDonald, J.R. Batchelor, and R.I. Lechler

Multiple sclerosis (MS) is one of several diseases of probable autoimmune pathogenesis that are associated with products of the MHC. In Caucasians, the most commonly seen extended haplotype is HLA-A3,-B7,-DR2, -DQw1, -C4*3, -C4B*1, -Bf*S, and an analysis of parental haplotypes in Scottish families suggests strongly that a susceptibility gene exists within this haplotype (1). An important observation has been that in some populations no association, or a negative association, with DR2 is observed. Furthermore, there is now some serological evidence that MS is more strongly associated with DQw1 than DR2 (2), and for these reasons HLA-DQ region genes may be important in influencing MS susceptibility.

Two complementary approaches to a more detailed genetic analysis of MHC-disease associations are restriction fragment length polymorphism (RFLP) studies and nucleotide sequence analysis. One of the advantages of RFLP analysis is its potential ability to detect polymorphisms in functionally important flanking sequences of genes, polymorphisms that would not be detected by sequencing cDNA clones. For this reason we have undertaken an extensive study of MHC Class II RFLP in MS.

The study comprised three phases. In the first phase, DNA pools were constructed (3) with DNA samples from 33 MS patients and 58 healthy controls from northeast Scotland, an area of high disease prevalence. The patients and controls were divided into DR2+ and DR2− groups, making four pools in all. These pools were digested with a panel of 14 4-, 5-, and 6-base cutters, and probed with full-length clones of DPα, DPβ, DQα, DQβ, and DRβ. Of the 70 enzyme/probe combinations studied, the majority failed to discriminate between MS and control pools. However, with a DQα probe, striking differences were observed following MspI digestion, and with DQβ further possible differences were seen using *BamH*I, *Bgl*II, and *Taq*I.

Table 1. DQα Fragments Generated by MspI in MS Patients and Healthy Unrelated Controls from Northeast Scotland

	Fragment (kb)	Patients	Controls	χ^2	p
All		(n = 32)	(n = 48)		
	4.90	9	33	12.707	<0.001
	3.25	10	2	11.046	<0.001
	2.50	10	2	11.046	<0.001
DR2+		(n = 23)	(n = 24)		
	4.90	7	15	4.850	<0.05
	3.25	7	1	5.738	<0.05
	2.50	7	1	5.738	<0.05

In the second part of the study, analysis of the individual DNA samples was then carried out using the enzyme/probe combinations identified in this way as being potentially discriminatory. Following MspI digestion and hybridization with DQα, three fragments were identified that showed highly significantly skewed distributions. A 4.9-kb fragment was negatively associated with MS and a 3.25-kb and 2.50-kb pair were positively associated. When only DR2-positive individuals were included in the analysis, the distribution of these fragments was unaltered (Table 1). A small number of other MspI/DQα fragments also discriminated between MS patients and controls, but at low levels of significance. None of the DQβ fragments showed significant disease association.

The third part of this study has been to test the generality of these observations in other northern European Caucasoid populations. Using the same enzymes and full-length DQα, DQβ, and DRβ probes, 109 patients and 112 controls from five European centers (Aberdeen, Belfast, London, Nijmegen, Rennes) have now been studied. An analysis of several DQα and DQβ fragments using provisional Tenth Workshop locus assignments shows that the extended haplotype DR2, Dw2,DQw6 is present equally in both the patients and controls studied. The 3.25 + 2.50 kb MspI/DQα cluster, which is provisionally assigned to DQw8/w9, is once again seen in a significantly greater number of patients (22.7% versus 6.4%, $p<0.01$). Taken together, these data suggest that DQ genes on both haplotypes may contribute to disease susceptibility. The possibility of a gene complementation effect is being investigated.

References

1. Francis DA, Batchelor JR, McDonald WI, Dodi IA, Hing SN, Hern JEC, et al. HLA genetic determinants in familial MS. Tissue Antigens 1987;29:7–12.
2. Francis DA, Batchelor JR, McDonald WI, Hing SN, Dodi IA, Fielder AHL, et al. Multiple sclerosis in Northeast Scotland: An association with HLA-DQw1. Brain 1987;110: 181–196.
3. Arnheim N, Strange C, Erlich H. Use of pooled DNA samples to detect linkage disequilibrium of polymorphic restriction fragments and human disease: Studies of the HLA class II loci. Proc Natl Acad Sci USA 1985;82:6970–6974.

Author Affiliations

R.N. Heard, J.R. Batchelor, and R.I. Lechler, Department of Immunology, Royal Postgraduate Medical School, Hammersmith Hospital, Du Cane Road, London W12.0HS UK; *W.I. McDonald*, Institute of Neurology, National Hospital for Nervous Diseases, 33 Queen Square, London WC1, UK

Immunobiology of HLA
Volume II

HLA Class III Complement Genes and Susceptibility to Systemic Lupus Erythematosus (SLE)

J. Richard Batchelor, Angela H.L. Fielder, and Mark J. Walport

Susceptibility to systemic lupus erythema (SLE) is partly governed by genetic factors. Among those implicated are genes within the HLA region. Originally, the strongest association between susceptibility to SLE and HLA gene products was with the class II HLA antigens, particularly DR3 and DR2 in Caucasoid populations. However, recent evidence indicates that the classical pathway complement genes located in the class III region (C2,C4A,C4B) are more closely associated. Among the polymorphic variants of these genes are so-called null or silent alleles, which result in non-expression of the relevant protein; some, but not all, null alleles are due to gene deletions. There is strong linkage disequilibrium between the class III alleles and those of HLA-B and DR. Therefore, an association between an HLA-B or DR antigen and SLE may well be secondary to one with a C4 or C2 polymorphism.

A variety of inherited, homozygous complement factor deficiencies are strongly associated with SLE, however, the majority of idiopathic SLE cases do not have homozygous deficiencies. This led us to suggest that heterozygous deficiency, due to a null allele, might be present in a substantial proportion of cases. We therefore carried out an immunogenetic study of 29 SLE patients and their first degree relatives to determine the frequency of C4 and C2 null alleles. Null alleles were found in 24 of the 29 patients (83%) compared with 18 of the 42 normal subjects (43%) (1). Nineteen of the 24 haplotypes carrying a null allele also carried DR3, thus it was impossible to distinguish their relative contributions to disease susceptibility. To clarify the position, the following studies were done

1. We determined the frequencies of C4 null alleles in 30 HLA genotyped Caucasoid SLE patients selected for absence of DR3; these were compared with a genotyped panel of normal subjects from which DR3 positive individuals were excluded.
2. We confirmed, in unselected SLE patients, using C4 and 21-OH gene probes, the earlier findings of a high frequency of C4 null alleles demonstrated by immunochemical methods.

Results as to the frequency of C4 null alleles in DR3-negative SLE patients showed that 18 of 30 patients (60%) had a C4 null allele. In contrast, 22 of 68 (37%) DR3 negative normal subjects had a C4 null allele (x^2=4.41, 0.05>P>0.01). The extended HLA haplotypes of these patients and the controls are given in full elsewhere (2). C4 and 21-OH gene probing of 60 unselected, Caucasoid SLE patients revealed that 39 of 60 patients (65%) had a deleted C4 gene. In contrast, 19 or 65 unselected normal subjects (29%) showed the same deletion (x^2=16.0, 0.001>P). In all subjects where a C4 deletion was seen, there was an accompanying deletion of the 21-OHA gene. No deletions of 21-OHB genes were observed.

This study shows that DR3 negative Caucasoid SLE patients have a significantly raised frequency of C4 null alleles, and indicates strongly that susceptibility to SLE is linked to the C4 rather than to DR polymorphism.

This study has also confirmed the significant excess of C4 null alleles in unselected cases of SLE by demonstrating a substantially higher frequency of deleted C4 (and 21-OHA) genes in the SLE cases compared with a group of normal subjects.

From our evidence, we conclude that the C4 null alleles, even in a single gene dose, are themselves directly responsible for conferring susceptibility to SLE.

Acknowledgments. We are grateful to the M.R.C. and A.R.C. for financial support.

References

1. Fielder AHL, Walport MJ, Batchelor RJ, Rynes RI, Black CM, Dodi IA, et al. Family study of the major histocompatibility complex in patients with systemic lupus erythematosus: Importance of null alleles of C4A and C4B in determining disease susceptibility. Brit Med J 1983;286:425.
2. Batchelor JR, Fielder AHL, Walport MJ, David J, Lord DK, Davey, et al. Family study of the major histocompatibility complex in HLA DR3 negative patients with systemic lupus erythematosus. Clin & Exp Immunol 1987;70.

Author Affiliations

J. Richard Batchelor, Angela H.L. Fielder, Departments of Immunology; *Mark J. Walport,* Department of Medicine, Royal Postgraduate Medical School, Hammersmith Hospital, London W12 OHS, UK

Autoantibodies to Procollagen Type VII in EBA and Bullous Eruption of SLE Are HLA-DR2 Associated

E.R. Heise, W.R. Gammon, W.A. Burke, J.D. Fine, D.T. Woodley, and R.A. Briggaman

Epidermolysis bullosa acquisita (EBA) is a chronic, acquired blistering disease characterized by circulating and tissue-bound IgG autoantibodies to the basement membrane zone of stratified squamous epithelium. In a preliminary study we found EBA to be associated with HLA-DR2 (1). Recently, EBA antibodies were found to react with epitopes present on a collagenase resistant domain of Type VII procollagen, a basement membrane zone (BMZ) matrix protein unique to stratified squamous epithelium (2). This feature distinguishes EBA antibodies from IgG anti-BMZ autoantibodies in all other diseases except the rare disorder, bullous eruption of systemic lupus erythematosus (BE-SLE). To determine whether HLA phenotype is a risk factor for the development of anti-Type VII procollagen, we typed 29 EBA patients (18 black, 11 white) from throughout the United States for 14 HLA-DR,DQ antigens. Twenty-five of the 29 patients were also typed for 39 HLA-A,B,C antigens. Six BE-SLE patients (all black), with anti-Type VII procollagen antibodies, also were typed for HLA antigens. The frequency of HLA antigens was compared with race matched U.S. population controls using the Fishers' Exact test (3). The frequency of HLA-DR2 was increased significantly in both white EBA patients ($P = 0.0006$, $RR = 13.05$) and black EBA patients ($P = 0.0013$, $RR = 4.81$). HLA-DR2 also was present in five of six black BE-SLE patients ($P = 0.0095$, $RR = 100.7$). Our results indicate that individuals with anti-Type VII procollagen antibodies and a diagnosis of EBA or BE-SLE have an increased frequency of HLA-DR2. Published studies indicate that systemic lupus erythematosus is associated with HLA-DR2 and B8,DR3 in whites but not in blacks. Goodpasture's disease, characterized by antibodies to Type IV collagen, is reported to be HLA-DR2 associated in whites. We conclude that HLA-DR2 or a closely linked gene constitutes an important risk factor for the development of anti-Type VII procollagen, EBA and BE-SLE.

Acknowledgments. This research was supported by NIH grant AR30475.

References

1. Gammon WR, Heise ER, Burke WA, Dole KC, Woodley DT, Briggaman RA. Production of epidermolysis bullosa acquisita autoantibodies is associated with HLA-DR2 phenotype (Abstr.). J Invest Dermatol 1986;86:476.
2. Woodley DT, Burgeson RE, Lunstrum GP, Reese MJ, Bruckner-Tuderman L, Gammon WR, et al. The epidermolysis bullosa acquisita antigen is type VII procollagen (Abstr.). Clin Res 1987;35:726A.
3. Tiwara JL, Terasaki PI. HLA and Disease Associations. New York, Springer-Verlag, New York, 1985, p 11.

Author Affiliations

W.R. Gammon, W.A. Burke, J.D. Fine, D.T. Woodley, R.A. Briggaman, Department of Dermatology, University of North Carolina School of Medicine, Chapel Hill; *E.R. Heise*, Department of Microbiology and Immunology, Bowman Gray School of Medicine of Wake Forest University, Winston-Salem, NC, USA

Absence of Polymorphism Between DR and DQ Sequences Isolated from Celiac Disease Patients and Normals

C.O. Jacob and H.O. McDevitt

Gluten-sensitive enteropathy or celiac disease is characterized by small intestinal mucosal injury and malabsorption. The disease is activated by dietary exposure to gluten and similar proteins. More that 90% of individuals with celiac disease express the DR3 or DR7 and DQw2 serologic specificities (Table 1). Two mechanisms could explain the association between class II MHC gene products and celiac disease: either a gene in linkage disequilibrium with class II genes predisposes to the disease or one of the class II genes itself is involved in celiac disease. This last hypothesis predicts that one of the genes, DR3/7 or DQw2α/β, in individuals with celiac disease might be different from those genes in individuals free of disease. Neither hypothesis excludes the role of other genes, unlinked to MHC, in the pathogenesis of the disease. To determine if class II MHC coding sequences

Table 1. Celiac Disease and HLA

HLA	Affected	Relative Risk	Disease Incidence
A1 B8 DR3 and/or DR7, DQw2	70 to 100%	11.6	0.2%
B7 DR2	↓	< 1	–

Identical twins (n=24): 75% concordance.
Affected sib pair sharing: 28% share 2 haplotype, 70 to 80% share 1 haplotype.

from celiac patients show differences that might relate to the etiology of the disease, a cDNA library in λ gt10 was constructed from a DR3/7 heterozygote celiac patient. The polymorphic DRβ and DQ α and β chains were cloned and sequenced by the dideoxy technique using synthetic oligonucleotide primers or the M13 primer. The resulting sequences of the β chains of DR3 and DR7 and DQ α and β chains show no differences between the celiac patient and published normal sequences. Moreover, using restriction fragment length polymorphism (RFLP) analysis and DRβ, DQα, and DQβ cDNA probes we have studied 21 celiac patients. The results of this analysis show no polymorphism associated with DRβ and DQ genotypes in celiac patients, compared to DR3/7 normal controls.

The absence of differences between class II gene sequences in the celiac patient and normals does not mean that class II gene products are not involved in the pathogenesis of the disease. Indeed, class II genes from only one patient were sequenced. However, the fact that no polymorphism could be demonstrated by RFLP would suggest that there are no primary sequence differences. There is still the possibility that differences in the transmembrane or intracytoplasmatic region or in the levels of expression are playing a role in the development of the disease. If another gene linked to a class II gene is responsible for the disease, the linkage disequilibrium between the two genes has to be extremely strong to have been conserved under selective pressure.

More probably, the disease may be caused by a multifactorial etiology where "normal" class II genes interact with other factors both genetic and environmental.

Author Affiliation

C.O. Jacob, H.O. McDevitt, Department of Medical Microbiology, Stanford University, Stanford, CA, USA

HLA-DP Region Is Relevant to Celiac Disease Susceptibility

M.J. Niven, C. Caffrey, J.A. Sachs, P.G. Cassell, R. Gallagher, P. Kumar, H. Festenstein, and G.A. Hitman

Genes important in determining susceptibility to celiac disease (CD) are located within or near the HLA region. Associations were first described with A1 and B8, and subsequently with DR3 and DR7. More recently strong associations have been shown with: (i) the serologically determined DQ antigen DQw2, present in 90% of CD patients (1); (ii) RFLPs of the DQ subregion, in particular a 2.1kb TaqI DQA2 fragment (2) and a 4.0kb RsaI DQB1 fragment (3). In this study we have examined a DPA gene-related RFLP in relation to susceptibility to CD, in order to investigate the role of genes centromeric to those already studied.

We studied 65 CD patients and 83 healthy controls. All were of British caucasoid descent and unrelated. Most had been previously HLA-DR and -DQ serotyped and DRA, DQA1, and DQA2 genotyped. The study included a small number of homozygous typing cells and healthy individuals who had also been DP-typed. The polymorphism (4), detected by BglII digestion and hybridization with a cDNA DPA probe (provided by J Trowsdale, ICRF, London), consists of two fragments, sized 3.5 and 2.2kb.

The 2.2kb fragment was common, occurring in all CD patients and in 80 of 83 (96%) of controls. In contrast the 3.5kb fragment was found in 29 out of 83 (35%) of healthy individuals compared to 51 out of 65 (78%) of patients with CD. The relative risk of the 3.5kb fragment for CD, calculated by the method of Woolf, is 6.8 (95% confidence limits 3.9–9.7); $\chi^2=25.4$, $p < 0.001$. Neither of the fragments was preferentially associated with any of the serologic or RFLP markers in the DR or DQ regions that we had previously studied, consistent with looser linkage between DP and DQ, compared to DQ and DR, as previously reported. This also indicates that the association found between CD and the DPA-related RFLP is independent of that with DR and DQ. The results in the DP-typed subjects are shown in Table 1. The 3.5kb fragment appeared to correlate with at least DPw2 and the 2.2kb fragment with other types. This is in contrast to the findings of Hyldig-Nielsen et al (4), who found the 3.5kb fragment to correlate with DPw1, DPw5, and the CDP-HEI specificity.

We would therefore suggest that genes within, or close to, the DP region play a role in determining suscept-

Table 1. Correlation of DP Types and DPA RFLP

	DP TYPE*	DPA RFLP	
1.	DPw2	3.5	–
2.	DPw3	–	2.2
3.	DPw4	–	2.2
4.	DPw4	–	2.2
5.	DPw4	–	2.2
6.	DPwl/w4	–	2.2
7.	DPwl/w4	–	2.2
8.	DPw2/w3	3.5	2.2
9.	DPw2/w4	3.5	2.2
10.	DPw3/X	–	2.2
11.	DPw3/w4	–	2.2
12.	DPw3/w4	–	2.2
13.	DPw4/X[†]	–	2.2
14.	DPw6/X	3.5	2.2
15.	DPw4/X	3.5	2.2

*1–5 are homozygous typing cells; 6–15 are healthy controls; [†]X = not determined.

ibility to CD, either in addition to or rather than genes in the DR and DQ regions.

Acknowledgments. We would like to thank Dr. C. Feighery for his assistance.

This work was supported by grants from the Wellcome Trust, the Coeliac Trust, and the Medical Research Council of Ireland.

References

1. Tosi R, Vismara D, Tanigaki N, Ferrara GB. Cicimarra F, Buffolana W, et al. Evidence that celiac disease is primarily associated with a DC locus allelic specificity. Clin Immunol Immunopathol 1983;28:395–404.
2. Hitman GA, Niven MJ, Festenstein H, Cassell PG, Awad J, Walker-Smith JP, et al. HLA class II Alpha chain gene polymorphisms in patients with insulin-dependent diabetes mellitus, dermatitis herpetifirmis and celiac disease. J Clin Invest 1987;79:609–615.
3. Howell MD, Austin RK, Kelleher D, Nepom GT, Kagnoff MF. An HLA-D region restriction fragment length polymorphism associated with celiac disease. J Exp Med 1986;164:333–338.
4. Hyldig-Nielsen JJ, Morling IV, Odum N, Ryder LP, Platz P, Jakobsen B, et al. Restriction fragment length polymorphism of the HLA-DP subregion and correlations to HLA-DR phenotypes. Proc Natl Acad Sci USA 1987;84:1644–48.

Author Affiliation

M.J. Niven, C. Caffrey, G.A. Hitman, Medical Unit, *H. Festenstein*, Department of Immunology, *P.G. Cassell*, Medical Unit and Bone and Joint Research Unit, *J.A. Sachs*, Department of Immunology and Bone and Joint Research Unit, The London Hospital Medical College, London, England; *R. Gallagher*, Department of Immunology, St. James' Hospital, Dublin, Ireland; *P. Kumar*, Department of Gastroenterology, St. Bartholomew's Hospital, London, UK

The Role of Lymphotoxin in the Pathogenesis of Celiac Disease

Nelson Fernandez, Graham A. Hitman, Hilliard Festenstein, Louise Garde, Mario O. Labeta, Cliff Stevens, John A. Walker-Smith, Anthony Meager, and John A. Sachs

Celiac disease (CD), a gluten-sensitive enteropathy, has an increased frequency of the serologically defined HLA class II antigens DR3 (90%), DR7 (50%), and DQw2 (95%). The use of Southern blot analysis with DQ and DR α and β probes has not further contributed to isolating the HLA-associated disease susceptibility genes. An alternative strategy is to look for variations in the structural patterns of class II-related gene products.

Lymphotoxin, a glycoprotein with a molecular weight of 25 kD, is a product of a gene located between the HLA-B and -DR region (1,2). We have demonstrated that DR monoclonal antibodies (MAbs) precipitate lymphotoxin secreted by Epstein-Barr virus (EBV)-transformed lymphoblastoid cell lines (LCLs) (3). We now report on the different two-dimensional SDS PAGE patterns that arise following DR immunoprecipitation of cell lines derived from HTCs and patients with CD.

The actin spot, invariant chain, DRα and DRβ chains did not appreciably differ in patients or the HTCs except for the inherent variability associated with the particular DR specificity. However, three different patterns relating to the 25 to 28 kD products were discernible. The 25 kD spot at pI 6.5 was present in *all* LCLs from HTCs (more than 30) and one of eight LCLs from patients with celiac disease (Fig. 1A). In the second pattern, found only in patients with CD (four of eight), the 25 kD spot was absent (Fig. 1B) even after prolonged incubation of the autoradiograph. In the third pattern, also found only in cell lines from patients with CD (three of eight), the 25 kD spot as well as a novel spot of 28 kD at pI 7.5 was observed (Fig. 1C).

Additional immuno-cytohistochemical studies indicated that lymphotoxin, identified by anti-lymphotoxin MAb 238/8 (4), is present in the enterocytes of jejunal biopsies from two patients with CD, but is absent from the biopsies of two controls.

These findings, together with the knowledge that lymphotoxin influences the cytotoxic reactions to virus-

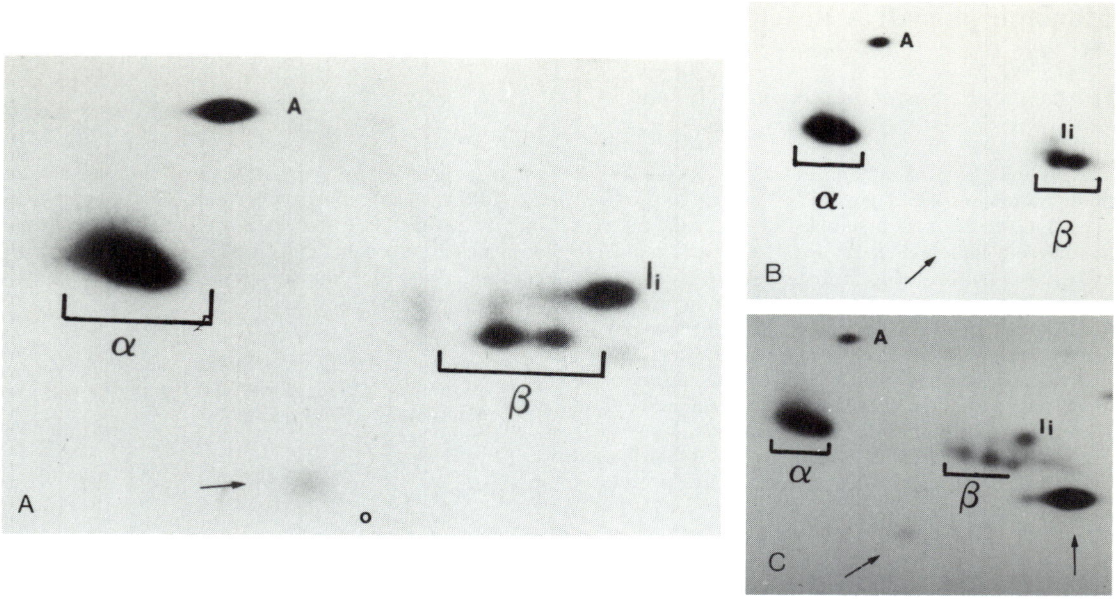

Figure 1. Autoradiograph of two-dimensional gels obtained after immunoprecipitation of LCLs with the DR monomorphic antibody L243. The two-dimensional gel bands corresponding to the α chain (a) and β chin (B), actin (A), and invariant chain (I) are indicated in all three panels. Panel A: The presence of the band designated g25. Panel B: The absence of the g25 band. Panel C: The presence of g25 and g28 bands.

infected cells and the increased expression of class II antigens in pancreatic cells (5), point to a possible regulatory role for lymphotoxin in target cell destruction in celiac disease.

Acknowledgments. The work was funded by a grant from the Coeliac Trust.

References

1. Gray P, Aggerual B, Benton C, Bringman TS, Henzel WJ, Jarrett JA, et al. Cloning and expression of cDNA for human lymphotoxin, a lymphokine with tumour necrosis activity. Nature 1984;312:721.
2. Spies T, Martin C, Nedopasov S, Fiers W, Pious D, Strominger J. Genes for the two necrosis factors α and β are linked to the major histocompatibility complex. PNAS 1986; 83:8702.
3. Sachs JA, Fernandez N, Kurpisz M, Labeta M, Cabrera T, Razak K, et al. Class II HLA-DR coprecipitate. In: Dupont JA (ed): Immunology of HLA. New York, Springer-Verlag, 1988.
4. Meager A, Parti S, Leung H, Woolley J, Peil E, Sidhu S, et al. A two-site sandwich immunoradiometric assay of human lymphotoxin with monoclonal antibodies and its application. J Immunol Methods (in press).
5. Pujol-Borrell R, Todd I, Doshin M, Bottazzo GF, Sutton R, Gray P, et al. HLA class II induction in human islet cells by interferon γ plus tumour necrosis factor or lymphotoxin. Nature 1987;326:304.

Author Affiliations

Nelson Fernandez, Hilliard Festenstein, Mario O. Labeta, Department of Immunology; *Graham A. Hitman, Louise Garde*, Medical Unit; *Cliff Stevens*, Bone & Joint Research Unit; *John A. Sachs*, Department of Immunology and Bone & Joint Research Unit, The London Hospital Medical College, London; *John A. Walker-Smith*, Queen Elizabeth Hospital for Children, London. *Anthony Meager*, National Institute for Biological Standards & Control, South Mimms, UK

Monomorphic HLA Reactivity in Heroin Addicts and Hydatidosis Patients

F. Ameglio, R. Sorrentino, N. Tanigaki, and R. Tosi

The presence of HLA antibodies was tested by a double precipitation test with purified ^{125}I-labeled class I or class II molecules. One hundred sixty-two of the 210 sera from heroin addicts (Fig. 1a) and 31 of the 32 sera from hydatidosis patients (Fig. 1c) showed significant reactivity against the class II molecules from the KT13 (DR4,4) cell line. Only one out of 100 male blood donors was positive (Fig. 1b). No positivities were found in patients with malaria, acute schistosomiasis, invasive amebiasis, as well as in HBsAg-positive subjects and in HIV-positive, non-heroin addict male homosexuals (Fig. 1 d–g, respectively). Similar reactivity was found against class I and class II molecules irrespective of the HLA phenotype. No reactivity was observed against β2m, BSA, and several glycoproteins. This phenomenon may be associated to repeated or continuous antigenic stimulation and could, conceivably, represent a mechanism underlying the immune hyporeactivity of the two groups of subjects considered.

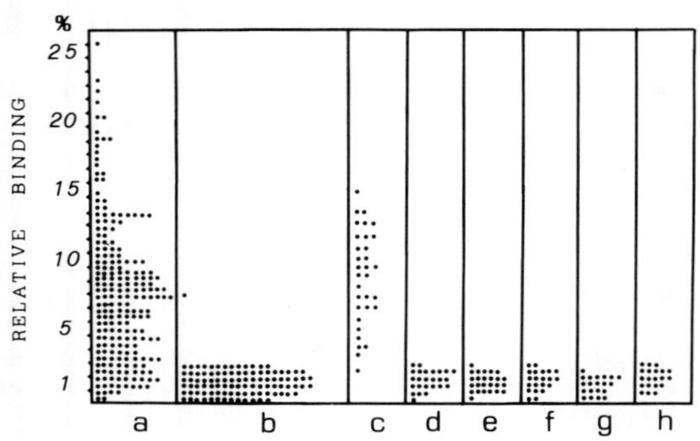

Figure 1. Binding of ^{125}I-labeled class II preparation KT13 (DR4,4) by different groups of sera (explained in text). Values calculated as proportion of binding given by rabbit anti-human class II serum 7147.

Acknowledgment. This investigation was supported by NIH Grant AI-20251 and a NATO grant to R.S. by Progetto Bilaterale CNR 860026504.

References

1. Ameglio F, Saba F, Bitti A, Aceti A, Tanigaki N, Sorrentino R, et al. Antibody reactivity to HLA classes I and II in sera from patients with hydatidosis. J Infect Dis 1987;156: 673–676.
2. Ameglio F, Dolei A, Benedetto A, Sorrentino R, Tanigaki N, Tosi R. Antibodies reactive with non-polymorphic epitopes on HLA molecules interferes in screening test for human immunodeficiency virus. J Infect Dis (in press).
3. Ameglio F, Benedetto A, Marotta P, Centis D, Sorrentino R, Tanigaki N, et al. A high proportion of sera of heroin addicts possesses anti-HLA class I and class II reactivity. Clin Immunol Immunopathol 1988;46:328–334.

Author Affiliations

F. Ameglio, Regina Elena Institute for Cancer Research, Rome; *R. Sorrentino*, Department of Cell Biology and Development, Universita' La Sapienza, Rome; *N. Tanigaki*, Roswell Park Memorial Institute, Buffalo, NY, USA; *R. Tosi*, Institute of Cell Biology, CNR, Rome, Italy

HLA Studies in Southern Chinese Suggest Hypotheses for Associations with Autoimmune Disease

B.R. Hawkins

Studies of HLA and autoimmune disease in Caucasians show a clear role for a gene or genes associated with B8 and DR3 in predisposing to autoimmunity. Autoimmune disease in Chinese occurs with a comparable prevalence to that in Caucasians but B8 and DR3 are rare, and DR3 usually occurs in linkage disequilibrium with B17 rather than B8. HLA studies of a number of autoimmune diseases among the predominantly southern Chinese population of Hong Kong have shown striking parallels to the findings in Caucasians.

Full details of the patients and controls are provided elsewhere (2-7).

Major differences in antigen frequencies between patients and controls are summarized in Table 1. Aw33, B17, and DR3 (which are in strong linkage disequilibrium in Chinese) are strongly associated with

Table 1. Percent of Selected HLA Frequencies in Patients and Normal Controls

	Controls (n=100)	Thyrotoxicosis (n=132)	Hashimoto's Thyroiditis (n=53)	Myasthenia Gravis Ocular (n=89)	Myasthenia Gravis Generalized (n=105)	IDDM (n=75)	SLE (n=100)
Aw33	14.0	15.2	5.5	11.2	16.2	28.0[1]	20.0
B17	17.0	7.6	7.5	10.1	11.4	33.3[2]	17.0
Bw46	31.0	54.6[3]	52.8[4]	66.3[5]	31.4	18.7	29.0
DR3	8.0	8.8	9.4	5.6	6.7	41.3[6]	9.0
DRw9	21.0	29.4	54.7[7]	79.8[8]	42.9[9]	34.7	33.0
DR2	38.0	44.1	18.9	29.2	37.1	32.0	62.0[10]

Comparisons With Normal Controls

	χ^2	p	Relative Risk		χ^2	p	Relative Risk
1	4.41	<.05	2.38	6	25.61	<.0001	8.10
2	5.40	<.025	2.44	7	16.42	<.0001	4.54
3	18.45	<.0001	2.67	8	62.84	<.0001	14.84
4	6.08	<.025	2.49	9	10.23	<.0025	2.82
5	22.12	<.0001	4.37	10	10.58	<.0025	2.66

insulin-dependent diabetes mellitus (IDDM). Bw46 and DRw9 (also in strong linkage disequilibrium) are strongly associated with Hashimoto's thyroiditis and myasthenia gravis, and Bw46 is associated with Graves' disease. DRw9 has an increased frequency in Graves' disease, IDDM, and systemic lupus erythematosus (SLE), but not at a statistically significant level. DR2 is strongly associated with SLE. In IDDM and generalized myasthenia gravis the strongest association is clearly with early onset disease (Table 2).

Hypotheses and Conclusions

The findings lead to the following hypotheses:

1. The Bw46, DRw9 haplotype in Chinese is a marker for a gene or genes that confer a general predisposition to "organ-specific" autoimmune disease (e.g., thyrotoxic Graves' disease, autoimmune thyroiditis, myasthenia gravis, and IDDM).
2. Predisposition to SLE, which is not an organ specific disease, is determined by other genes marked by DR2 rather than DRw9 and Bw46.
3. Specific predisposition to IDDM is conferred by a gene or genes marked by the haplotype Aw33,B17, DR3. Individuals who are already predisposed to autoimmune disease through possession of DRw9 and/or Bw46 and who are also predisposed to IDDM through possession of Aw33,B17,DR3 are at greatest risk. These people develop IDDM early in life, which causes an excess of DR3/DRw9 heterozygosity in patients presenting before the age of 20.

These hypotheses are strongly supported by findings in Caucasians. Organ-specific autoimmune diseases, such as Graves' disease, Addison's disease, and myasthenia gravis, are strongly associated with B8 and DR3. There is no single, convincing HLA association with SLE in Caucasians. Certainly, the strong B8,DR3 association seen in other autoimmune diseases does not exist.

Table 2. HLA Frequencies (%) and Age at Onset of Disease

	Age at Onset (years) Below 10	10–20	20–35	Over 35
IDDM	(n = 19)	(n = 34)	(n = 22)	
Aw33	42	29	14	
B17	47	41	14	
DR3	63	44	18	
DRw9	58	32	18	
DR3 + DRw9	47	15	0	
Ocular MG	(n = 44)	(n = 16)	(n = 14)	(n = 15)
Bw46	68	75	64	53
DRw9	91	81	79	47
Generalized MG	(n = 13)	(n = 25)	(n = 31)	(n = 36)
Bw46	62	56	26	8
DRw9	62	64	29	33

Instead there is evidence for an association with complement null alleles (1). An association with C4A null (which is in linkage disequilibrium with DR2) was also seen in the Hong Kong Chinese patients. Thus, predisposition to SLE is probably determined by genes within the MHC which have strongest linkage disequilibrium with complement null alleles in Caucasians, and with DR2 in Chinese. "Generalized" predisposition to autoimmune disease, which is conferred by genes related to B8,DR3 in Caucasians and Bw46,DRw9 in Chinese, may play an additional role in increasing susceptibility to SLE as seen by weak associations with the respective antigens in Caucasians and Chinese.

The strongest HLA association with IDDM in Caucasians is conferred by DR3/DR4 heterozygosity. To parallel the hypothesis proposed for Chinese it is suggested that generalized predisposition to autoimmunity is conferred by DR3-related genes, and specific predisposition to IDDM is conferred by DR4-related genes. As in Chinese, persons with both genes are at greatest risk and develop the disease early in life. This would imply that the association of IDDM with DR3 in Caucasians and Chinese is coincidental because the DR3 gene is associated with predisposition to autoimmunity in Caucasians and predisposition to IDDM in Chinese.

Acknowledgments. I am indebted to my colleagues in Hong Kong for providing the clinical material on which this paper is based.

References

1. Dunckley H, Gatenby PA, Hawkins BR, Naito S, Serjeantson SW. Partial deficiency of C4A is a genetic determinant of systemic lupus erythematosus in three ethnic groups. Journal of Immunogenetics 1987;14:209–218.
2. Hawkins BR, Lam KSL, Ma JTC, Low LCK, Cheung PT, Serjeantson SW, et al. Strong association of HLA-DR3/DRw9 heterozygosity with insulin-dependent diabetes mellitus of early onset in Chinese. Diabetes 1987;36:1297–1300.
3. Hawkins BR, Lam KSL, Ma JTC, Wang C, Yeung RTT. Strong association between HLA DRw9 and Hashimoto's thyroiditis in southern Chinese. Acta Endocrinologica 1987;114:543–546.
4. Hawkins BR, Ma JTC, Lam KSL, Wang CCL, Yeung RTT. Association of HLA antigens with thyrotoxic Graves' disease and periodic paralysis in Hong Kong Chinese. Clinical Endocrinology 1985;23:245–252.
5. Hawkins BR, Serjeantson SW, Higgins DA. Distribution and co-occurrence of MHC Class I, II, and III markers in Southern Chinese: implications for autoimmune disease. Disease Markers 1988 (in press).
6. Hawkins BR, Wong KL, Wong RWS, Chan KH, Dunckley H, Serjeantson SW. Strong association between the major histocompatibility complex and system lupus erythematosus in southern Chinese. Journal of Rheumatology 1987;14:1128–1131.
7. Hawkins BR, Yu YL, Wong V, Woo E, Ip, M, Dawkins RL. Possible evidence for a variant of myasthenia gravis based on HLA and acetylcholine receptor antibody patterns in Chinese patients. Quarterly Journal of Medicine (in press).

Author Affiliation

B.R. Hawkins, Department of Pathology, University of Hong Kong, Queen Mary Hospital, Hong Kong

MHC Gene Arrangements on Disease Associated Supratypes Revealed by Pulsed Field Gel Electrophoresis

R.L. Dawkins, Katsushi Tokunaga, G. Saueracker, P.H. Kay, F.T. Christiansen, and R. Anand

The major histocompatibility complex (MHC) represents one of the best characterized segments of human chromosome. The most striking feature of the MHC is the remarkable degree of polymorphism. Surprisingly, however, some 70% of Caucasoids possess at least one of only some 20 particular sets of MHC alleles or supratypes. Such supratypes serve as markers of unique chromosomal segments of approximately 2 megabases and it is likely that each supratype contains its own unique set of new genes, deletions, duplications, insertions, and other major arrangements.

The technique of pulsed field gel electrophoresis (PFGE), allows separation of large DNA molecules obtained by infrequently cutting restriction endonucleases (1). Recently, these techniques have been applied for long-range restriction mapping of the human MHC (2,3). However, the reported maps did not relate to particular supratypes. Therefore, we have mapped six different MHC supratypes using PFGE in combination with the rare cutting enzymes Mlu I, Not I, and Pvu I.

High molecular weight DNA was isolated from Epstein-Barr virus-transformed cell lines from subjects who were, in most cases, homozygous for the following supratypes:

HLA A3,B7,BfS,C4A3,C4B1,DR2 (3,7,2);
HLA A1,B8,BfS,C4AQO,C4B1,DR3 (1,8,3);
HLA B18,BfF1,C4A3,C4BQO,DR3(18,F1,3);
HLA Bw65,BfS, C4A2,C4B1+2,DR1 (65,1+2,1);
HLA Bw57,BfS,C4A6,C4B1,DR7 (57,6,7);
HLA A2,Bw46,BfS,C4A4,C4B2,DRw9 (2,46,9).

At least three different cell lines from unrelated individuals were examined for each supratype. Each of these supratypes has been reported to be associated with certain autoimmune diseases. DNA preparation, restriction enzyme digestion, and PFGE were performed as described (1). Southern blotting was carried out by standard methods for conventional agarose gels. The restriction fragments transferred onto a nylon membrane were sequentially hybridized with a series of probes specific for MHC genes.

Marked differences in restriction fragment length and band intensity were found for each supratype. The hybridization patterns were consistent within multiple samples of each supratype, except for some fragments within the class I region. These data suggest that each supratype has a unique and fixed structure throughout the entire HLA-B to DQ region of more that one megabase.

Compared with the 3,7,2 supratype, a number of previously undescribed extensive deletions and insertion were predicted in the other supratypes as follows. 1,8,3: some 40kb deletion close to the DR region, a 20kb deletion close to class III region, and a 40kb deletion near the TNF and class I genes. 18,F1,3: some 50kb deletion close to the DR region and a 10kb deletion close to the class III region. 65,1+2,1: a 20kb deletion close to the class III region. By contrast, an insertion of some 70kb was detected near the DQ or DR region in both 57,6,7 and 2,46,9. Furthermore, several restriction site variations were found between supratypes. For example, a Pvu I site within the DR region was found only in the 3,7,2, 1,8,3 and 65,1+2,1 supratypes, whereas only 1,8,3 and 18,F1,3 have a Pvu I site between DQ and DR genes.

These results indicate that the genomic organization of the MHC region differs extensively between various supratypes in terms of gene copy number, deletions, and insertions/duplications, as well as restriction site polymorphisms. Moreover, these data suggest that each supratype marks a unique set of genomic rearrangements not only within genes but between genes. The relationship between these newly recognized rearrangements and disease susceptibility is under investigation.

Acknowledgments. This work was funded by grants from the National Health and Medical Research Council of Australia and is further supported by the Immunogenetics Research Foundation.

References

1. Anand R. Pulsed field gel electrophoresis: a technique for fractionating large DNA molecules. Trends Genet 1986;23: 278-283.
2. Hardy DA, Bell JI, Long ED, Lindsten T, McDevitt HO. Mapping of the class II region of the human major histocompatibility complex by pulsed-field gel electrophoresis. Nature 1986;323:453-455.
3. Lawrance SK, Smith CL, Srivastava R, Cantor CR, Weissman SM. Megabase-scale mapping of the HLA gene complex by pulsed field gel electrophoresis. Science 1987;235: 1387-1390.

Author Affiliations

R.L. Dawkins, Katsushi Tokunaga, G. Saueracker, P.H. Kay, F.T. Christiansen, Departments of Clinical Immunology, Royal Perth Hospital, Queen Elizabeth II Medical Centre and University of Western Australia, Perth, Australia: *R. Anand*, Department of Biochemistry, Oxford University, Oxford, UK

Polymorphism of the HLA-Linked Steroid 21-Hydroxylase Genes

Nanda R. Rodrigues, Gillian Rumsby, and R. Duncan Campbell

Deficiency of the cytochrome P-450 steroid 21-hydroxylase (21-OHase), which causes congenital adrenal hyperplasia (CAH), is a monogenic autosomal recessive disorder that is linked to HLA (1). There are two 21-OHase genes in man, A and B, and they map to the HLA class III region ~ 3 kb 3′ to the complement genes C4A and C4B, respectively (2–4). The two 21-OHase genes have been isolated and characterized (5–7). Comparison of the two genes has shown that although they are highly conserved, there are three deleterious mutations in the 21-OHase A gene which cause frameshifts and introduce inphase premature termination codons. Thus the 21-OHase A gene is a pseudogene. A 21-OHase B gene from a patient with CAH has been characterized and sequenced (7). A G → C transversion in the 21-OHase B gene from the patient at codon 269 introduces a new NcoI restriction site into the gene. This restriction fragment length polymorphism (RFLP) was used to study other patients with CAH and normal individuals. The NcoI RFLP was found not to be confined to the 21-OHase B gene, but was also present in some 21-OHase A genes.

The presence of allelic variants of 21-OHase at a single locus is strongly suggested by the four different clinical manifestations of CAH. The 21-OHase genes lie between a region of the HLA which is highly polymorphic. A comparison of 21-OHase gene sequences (7) with sequences from other groups (5,6,8,) suggests polymorphism in 21-OHase genes. In the sequences compared, 79 differences were found. Most of the

Table 1. HLA Haplotypes of Patients with CAH (A) and Normal Individuals (B) Showing the NcoI Polymorphism

A

Proband	HLA Haplotype				TaqI (kb)		NcoI (kb)	
					3.2	3.7	3.1	2.3
JB	A9,32	B44,13	Cw4	DR1,7	+	−	½	½
CB	A3	Bw47,62	C6,1	DR7,5	+	½	+	½
MP	A2,3	B40,w3	Cw6	DR?	+	½	+	½
BB	A3,23	B39,44		DR2,7	+	−	−	+
RW	A1,29	B44,40/8	Cw5	DR3,7 DRw52	½	+	+	½
DM	A3,2	B7,44	C5		+	½	+	½
SM	A3,2	B7,44	C5		+	½	+	½
CBu	A2,11	Bw47,55	C6,3	DR7,4	+	½	½	+

B

Name	HLA Haplotype				NcoI (kb)	
					3.1	2.3
JH	A29,3	B44,7	Cw-	DR5,2	+	½
NJ	A2,11	B51(or blank),51		DR7	+	½
WG	A28,3	B35,14	Cw3,4	DRw52,53	+	½
CG	Not Available				+	½
MeP	A29,(29)	B44,(44)		DR7,2	+	½
AA	A2,19,(29)	Bw48,12,(44)	C-	DR8,7	+	½

Symbols: + indicates the presence of a band; − indicates the absence of a band; ½ indicates half intensity of a band compared with +.

differences may represent accumulation of point mutations in the pseudogene. Of the 17 differences in the 21-OHase B gene sequence that lie in exons, seven cause a codon change. The leucine at codon 6 (CTG) was absent in the other two published 21-OHase B sequences, but present in the 21-OHase A genes. A comparison of 21-OHase sequences from other species with human 21-OHase reveals a codon for leucine at the same position (7). This may indicate that there is polymorphism in the size of 21-OHase with either 494 or 495 amino acids.

The NcoI RFLP was first detected in the 21-OHase B gene of individual II.I who has CAH (7). A G → C transversion at position 1650 introduced the new NcoI restriction site (CCATGG) into the gene. This new NcoI site cuts down a 3.1 kb fragment to a 2.3 kb fragment and a 0.8 kb fragment. However, this polymorphism was also found in the 21-OHase A genes of individual II.I. The NcoI RFLP was used to study 20 other patients with CAH and a number of normal individuals from a random population. Table 1 shows the haplotypes and the band intensities with TaqI and NcoI of eight of the patients. Only two of the 20 patients showed a homozygous deletion of the 21-OHase B gene. The NcoI polymorphism was found in eight patients, but since the RFLP could be present in any one of the four 21-OHase genes, the frequency of the NcoI RFLP in patients would be ten out of a total of 62 genes, that is 16%.

It was observed that the Bw47, DR7 haplotype in the patients consistently had the NcoI fragment of 2.3 kb as seen from the banding patterns of the patients CB,AB,CBu and II.I As the Bw47 haplotype has a deleted 21-OHase B gene (2), the NcoI RFLP is in the 21-OHase A gene.

A study of normal individuals in a random population was carried out to find the frequency of occurrence of the NcoI polymorphism. Out of 17 individuals studied, six individuals showed the NcoI polymorphism (Table 1). This suggested that the NcoI RFLP was present at a substantially high frequency in a random population although it did not give any information as to which of the haplotypes and genes had the RFLP.

To differentiate between the 21-OHase A and B genes around the NcoI RFLP, a restriction endonuclease that gives distinct unique bands was selected. A single nucleotide substitution at position 1380 introduces a new AvaII site into the 21-OHase A gene which would yield two fragments of 147 bp and 328 bp instead of a single fragment of 475 bp which the 21-OHase B gene yields (Figure 1). If there is a NcoI RFLP in the 21-OHase A gene, the AvaII digest followed by a NcoI digest would yield fragments of 147 bp and 274 bp, whereas a 21-OHase B gene with the NcoI RFLP would yield a 420 bp fragment (Figure 1).

The AvaII/NcoI polymorphism in individual BB was studied in detail. BB shows no 3.7 kb TaqI fragment and hence was presumed to have a homozygous deletion of the 21-OHase B genes. The AvaII digest of BB, however, yields fragments of 475, 328, and 147 bp (Figure 1). In the AvaII/NcoI digest, both fragments of 475 bp and 328 bp are cut down to 420 bp and 274 bp, respectively. Thus all the 21-OHase genes in BB have the NcoI RFLP. The presence of the 475 bp AvaII fragment, however, suggests that one of the 21-OHase genes characterized by the 3.2 kb TaqI fragment is 21-OHase B-like. Thus one of the genes in BB may be a hybrid that contains the 5′ end of a 21-OHase A gene and the 3′ end of a 21-OHase B gene. Alternatively, a mutation introduced into the 5′

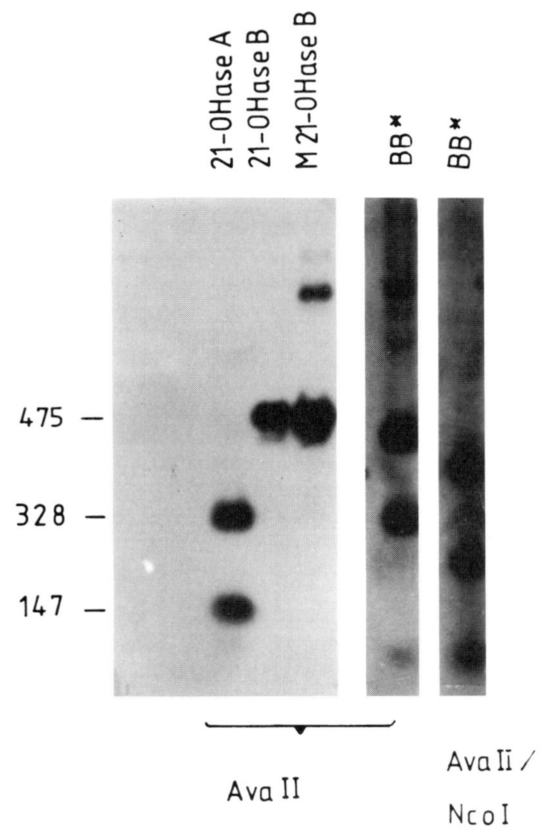

Figure 1. Left panel: Southern blot analysis of plasmid DNA containing 21-OHase A, 21-OHase B, and the 21-OHase B gene from the individual II.I digested with AvaII. Right panel: Southern blot analysis of genomic DNA from individual BB digested with AvaII and AvaII/NcoI. The 147 bp fragment in the genomic DNA has migrated further than the corresponding fragment in the plasmid DNA. The size of the 147 bp fragment was confirmed on the basis of markers run on the same gel. The probe used was AN which was obtained by isolating a 420 bp fragment from an AvaII/NcoI digest of the 3.7 kb TaqI subclone of the defective 21-OHase B gene from individual II.I.

region of a 21-OHase B gene may have resulted in the generation of an extra TaqI site which yields the 3.2 kb TaqI fragment. Thus the AvaII digest was able to distinguish 21-OHase A and 21-OHase B genes in BB, but did not identify which haplotype to which they belonged. Obviously, family studies are required to analyze the inheritance of the AvaII/NcoI restriction pattern so that the full implication of these results can be understood. The AvaII polymorphism can be used to study haplotypes in greater detail and extends the study to detect limits of deletion of the 21-OHase genes. The NcoI RFLP is interesting because of the substantially high frequency at which it occurs in a random population and use of an AvaII/NcoI double digest for detecting its presence in affected 21-OHase B genes may help in the detection of carriers.

A recent report (8) of a restriction polymorphism analysis of patients with CAH has shown that the lack of the 3.7 kb TaqI fragments does not necessarily indicate 21-OHase B gene deletion, as other restriction fragments characteristic of the 21-OHase B gene are present. Results obtained from the restriction analysis of individual BB in the present study support the same hypothesis. Thus more than one RFLP will have to be used in the study of patients to identify the nature of the genes on the affected haplotypes.

References

1. Dupont B, Oberfield SE, Smithwick EM, Lee TD, Levine LS. Close genetic linkage between HLA and congenital adrenal hyperplasia (21-hydroxylase deficiency). Lancet 1977;2:1309.
2. White PC, New M, Dupont B. HLA-linked congenital adrenal hyperplasia results from a defective gene encoding a cytochrome P-450 specific for steroid 21-hydroxylation. Proc Natl Acad Sci USA 1984;81:7505.
3. Carroll MC, Campbell RD, Porter RR. Mapping of steroid 21-hydroxylase genes adjacent to complement C4 genes in HLA, the major histocompatibility complex in man. Proc Natl Acad Sci USA 1985;82:521.
4. White PC, Grossberger D, Onufer BJ, Chaplin DD, New MI, Dupont B, et al. Two genes encoding steroid 21-hydroxylase are located near the genes encoding the fourth component of complement in man. Proc Natl Acad Sci USA 1985;82:1089.
5. Higashi Y, Yoshioka H, Gotoh O, Fujii-Kuriyama Y. Complete nucleotide sequence of two steroid 21-hydroxylase genes tandemly arranged in human chromosome: A pseudogene and a genuine gene. Proc Natl Acad Sci USA 1986; 83:2841.
6. White PC, New MI, Dupont B. Structure of human steroid 21-hydroxylase genes. Proc Natl Acad Sci USA 1986;83: 5111.
7. Rodrigues NR, Dunham I, Yu CY, Carroll MC, Porter RR, Campbell RD. Molecular characterization of the HLA-linked steroid 21-hydroxylase B gene from an individual with congenital adrenal hyperplasia. EMBO 1987;6:1653.
8. Matteson KJ, Phillips III JA, Miller WL, Chung B, Orlando PJ, Frisch H, et al. P450XXI (steroid 21-hydroxylase) gene deletions are not found in family studies of congenital adrenal hyperplasia. Proc Natl Acad Sci USA 1987;84:5858.

Author Affiliations

Nanda R. Rodrigues, R. Duncan Campbell, Department of Biochemistry, MRC Immunochemistry Unit, South Parks Road, Oxford; *Gillian Rumsby*, The Royal Marsden Hospital, Fulham Road, London, UK

HLA, Complotypes, and RFLP Analysis of 21-Hydroxylase (21-OH) and C4 Genes in Finnish Families with a 21-OH Deficient Patient

J. Partanen, S. Koskimies, I. Sipila, K. Simola, and A. Hakulinen

The 21-OHB gene encoding the enzyme 21-OH of steroid biosynthesis, and a closely-related pseudogene 21-OHA, are located in the MHC, adjacent to the C4 genes. We have performed HLA, Bf, and C4 typings, and RFLP analysis of the 21-OH and C4 genes in ten families with a patient having a 21-OH deficiency.

The phenotypes and Southern analysis were done using standard methods. The probes (pAT-A and p21-K4) were provided by DR. M.C. Carroll, Boston, USA. Six of the ten families had a patient with simple virilizing form of the disease, and four had salt-wasting form.

In the 20 chromosomes with a 21-OH deficiency, the frequencies of two HLA-B antigens were clearly increased: 50% (10/20) of them were HLA-B40 (9% in the reference population), and 10% (2/20) were HLA-Bw47 (0.4% in the reference population). The C4 phenotyping revealed that 20% (4/20) of the chromosomes had a rare C4A1 allotype (less than 0.4% in the reference population).

The 21-OH and C4 genes were studied using restriction enzymes Taq I, Kpn I, EcoRI, and Bgl II. When these results were combined with the phenotypic data, four major haplotypes, covering 80% of the chromosomes, were found (Table 1). *Type 1* (having a frequency of 30%) had a phenotype HLA A2/3 B40(w60) S30 DR1, and it always lacked the C4B and 21-OHB specific RFLP markers (see Table 1) suggesting their deletion. *Type 2* (20%) was HLA A24 B40 S01 DRw8 with the normal RFLP patterns indicating an absence of deletion. *Type 3* (20%) represented the F10 complotype, which was confirmed to lack the C4B and 21-OHB specific fragments. These haplotypes had HLA-DR7 in common but had different HLA A and B antigens. Using Bgl II and the C4 probe, a diagnostic 4.5 kb band for this complotype was observed. *The fourth type* (10%) had the normal C4 RFLP markers. It, however, lacked the Taq I and Kpn I fragments of the 21-OHB gene but the Bgl II pattern was normal, − 12 kb and 11 kb bands. We thus suggest that in this case, there is a 21-OHA gene instead of the 21-OHB, located adjacent to the C4B gene. These two chromosomes had the same MHC phenotypes. The remaining four haplotypes had different phenotypes.

In the salt-wasting form seven of the eight chromosomes appeared to carry a 21-OHB deletion, whereas only five of the twelve chromosomes in the simple virilizing form had the 21-OHB deletion.

Our results show that 1) there are at least four major haplotypes with a 21-OH deficiency in the Finnish population; 2) each of these haplotypes seems to have a characteristic RFLP pattern; 3) there is a diagnostic Bgl II band for the F10 complotype; and 4) the genetic basis of 21-OH deficiency is heterogeneous.

Author Affiliations

J. Partanen, S. Koskimies, Finnish Red Cross Blood Transfusion Service, Helsinki; *I. Sipila, K. Simola*, University of Helsinki; *A. Hakulinen*, University of Kuopio, Finland

Table 1. MHC Phenotypes, RFLP Patterns (in kb), and Suggested Gene Organizations of the C4 and 21-OH Genes in Four Major Haplotypes with a 21-OH Deficiency Observed in the Finnish Population

Phenotypes					RFLP				
	A	B	Complotype Genes	DR	21OH Taq I	Kpn I	Bgl II	C4 Taq I	Bgl II
Type 1									
	2/3	w60	S30 C4A;21OHA	1	3.2	4	11	7	14 1
Type 2									
	24	40	S01 C4A;21OHA;C4B:21OHB	w8	3.2 3.7	4 3	11 12	7 6	14 1
Type 3									
	Various		F10 C4A;21OHA	7	3.2	4	11	7	14 4.5
Type 4									
	2	18	S31 C4A;21OHA;C4B;21OHA	4	3.2	4	11 12	7 6	14 1

Immunobiology of HLA
Volume II

Section XV
HLA and Infectious Disease

Selection by Typhoid and Yellow Fever Epidemics Witnessed by the HLA-DR Locus

René R.P. de Vries, Geziena M.Th. Schreuder, Albert Naipal, Joe D'Amaro, and Jon J. van Rood

Several years ago we presented evidence for an HLA-linked control of survival from typhoid fever and yellow fever epidemics (1). At that time we were only able to type for HLA-A, -B, and -C antigens. Because we were interested to know whether these epidemics might have left even stronger marks on -DR or -DQ genes, we recently revisited the survivors and also performed HLA-DR and -DQ typing. The results indicate that most of the observed heterogeneity may be explained by a primary association of HLA-DR2 with mortality and DR4 + DRw13 with survival from typhoid and/or yellow fever. We cannot exclude one or two primary associations with HLA-B alleles, but no primary association was observed with HLA-A, -DRw52/53, or -DQ alleles. These data suggest that gene products on which the typhoid bacillus and/or the yellow fever virus could exert their apparent selective pressure are coded by genes located on or very near to the DRβ1 locus.

In May 1845, 50 Dutchmen and their families sailed to Surinam (the former Dutch Guyana) to start a new life as farmers. However, 2 weeks after their arrival an epidemic of typhoid fever broke out that raged for 4 months and killed 180 of the 367 or almost 50% of the settlers (2). Six years later an epidemic of yellow fever killed another 20% of those who survived the typhoid fever epidemic, so that total mortality due to these two epidemics was 60%. Between and after these epidemics the mortality was quite low. Most of the survivors and their descendants stayed in Surinam and intermarried until today. Thus we were able to trace and investigate 64 descendants of the survivors and to compare the frequencies of 26 polymorphic genes with those of a Dutch control group (1). Analysis of the variation in gene frequencies indicated that the differences observed for 4 out of the 26 polymorphisms tested were unlikely to be due to genetic drift, inbreeding, or admixture with non-Dutch genes. Therefore these differences strongly suggested selection by typhoid and/or yellow fever through genetic control of survival. These 4 polymorphisms were C3, Gm, GLO, and HLA-B (1).

Further analysis of the deviating HLA alleles suggested that at least part of the observed differences might be secondary to differences in frequencies of HLA-DR alleles. In that case perhaps stronger associations with survival might be observed. Because we also considered it important to map as precisely as possible the genes on which these infectious diseases might have exerted their selective pressure, we went back to Surinam, rebled (most of) the same individuals, and performed HLA-DR and -DQ typing.

A summary of the results is shown in Table 1. The first striking observation was that DQ was the only locus that did not deviate significantly from the controls. This observation either argues against an important role of DQ products in resistance to typhoid and/or yellow fever, or may indicate that polymorphic sites other than presently defined with serology are biologically relevant. The most significant differences were observed for DR and B alleles. Let us first concentrate on the alleles that showed a significantly decreased frequency among the survivors of the epidemics, indicating a possible association with mortality from one or both of the epidemics. The most significant decrease was observed for DR2, which has a gene frequency of 0.16 in the controls (see Fig. 1), but was absent in the postepidemic sample. The decreased frequencies of B7 and Cw7 are clearly secondary to this (3). The complete absence of DR2 would suggest a dominant susceptibility gene (either DR2 or in strong linkage disequilibrium with it), which certainly was not expected by us because that does not generate polymorphism so easily. B44 and Cw7 (which are in strong positive linkage disequilibrium [3]) also

Table 1. Comparison of HLA Gene Frequencies Between Survivors and Dutch Controls

Locus	p Value*	Alleles with Significantly Different Frequencies: Increased†		Decreased†
DQw1,2,3	n.s.			
DQ TA10/2B3	n.s.			
DRw52,53	.003		w53	
DR1-w14	.000004	4,w13	7	2
B	.000004	13,38(16), w57(17)	w58(17)	7, B44 (12)
C	.0004	w3		w7, w5
A	.0005		30	

We rebled 70 descendants from survivors, from which we deduced 114 haplotypes, the majority originating from survivors but diluted with about 10% from Dutchmen who arrived after the epidemics. Although we had selected these haplotypes so as to minimize double counting of genes, we could calculate that only about 65 haplotypes were not identical by descent. HLA typing was performed using routine procedures. Gene frequencies from "the survivors" were calculated by gene counting and from 502 random Dutch controls typed with the same reagents with a maximum likelihood method. Numbers of genes from the postepidemic group ("the survivors") were compared for each allele with the corresponding numbers of genes from the Dutch control group by means of chi-squaare test . For further details see ref. 1.

*From the sum of chi-squares for each allele, degrees of freedom equals number of alleles for that locus. n.s. = not significant.

†Among survivors as compared to controls. Underlining indicates possibly due to admixture with Negroid genes.

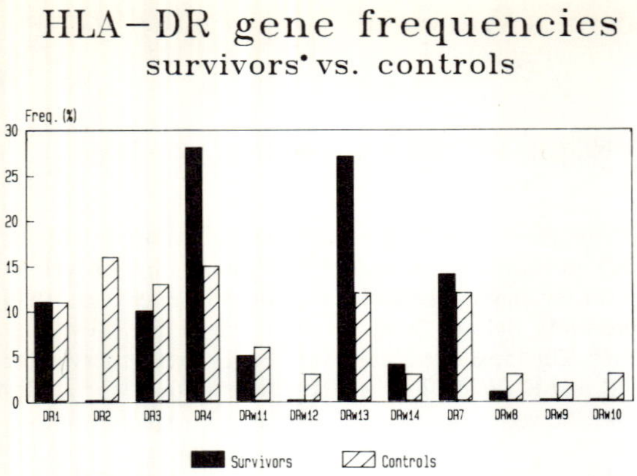

Figure 1. HLA-DR gene frequencies of survivors versus controls, after excluding six Bw58-DR7 haplotypes of the survivors, which may have been of Negroid origin.

showed a significantly decreased frequency, which in this case seemed to be a primary association.

More than 10 alleles at 5 loci were significantly increased in the postepidemic sample, indicating a possible association with resistance to these two infectious diseases. In order to define the primary association(s), we performed an analysis of A-C-B-DR haplotypes containing the alleles with increased frequencies. Part of the results is shown in Table 2. Not shown in this table are the DR7 haplotypes, because 6 of the 21 DR7 haplotypes contained Bw58(17), and are therefore quite possibly from Negroid origin. If we exclude these 6 haplotypes from the analysis, the associations of DRw53, DR7, Bw58(17), and Aw30 with survival lose their significance. Most of the other alleles with increased frequencies appear in Table 2. Inspection of this table and consultation of the reference tables in Baur et al. (3) taught us that the primary associations are with DRw13 and DR4, and that most of the increased frequencies of B alleles may be explained as being secondary to the observed associations of DR4 and DRw13 with survival from these typhoid and/or yellow fever epidemics (Fig. 1).

Assuming that the DR4 and DRw13 molecules might have conferred resistance and the DR2 molecule susceptibility, we inspected the DRβ1 and in the case of DR2 the βIII (4) sequences for a possible structural correlation (5). A remarkable correlation of the AA residue at position 71 in the third hypervariable region with survival was observed: all the DR alleles sequenced have a basic residue (arginine or lysine) at this position, except for DR2 (alanine) and DRw13 + some DR4 molecules (glutamic acid). This observation is at least compatible with the notion that this amino acid may control survival from typhoid and/or yellow fever!

Acknowledgments. The isolation and freezing of mononuclear cells from the descendants of the Dutch emigrants in Surinam was made possible with the help of Drs. R. Lai A Fat, W. van Kanten, and E. Li Fo Sjoe of the Medical Scientific Institute and University Hospital, Paramaribo, Surinam. S. Naipal-van den Berge, T. Prins, and T. van Westerop provided expert technical assistance.

Table 2. A-C-B-DR Haplotypes of Survivors Carrying DRw13 or DR4

DR	B	C	A	n*
w13	– w62(15)	– w3	– 2	3
w13	– w62(15)	– w3	– 3	5
w13	– w62(15)	– w3	– 29	5
w13	– 38(16)	– Blank	– 2	4
w13	– 38(16)	– Blank	– 26	4
w13	– w60(40)	– w3	– 2	1
w13	– w60(40)	– w3	– 3	2
w13	– w60(40)	– w3	– 11	2
w13	– 8	– w7	– 1	1
w13	– 8	– Blank	– 1	2
4	– w60(40)	– w3	– 2	1
4	– w60(40)	– w3	– 3	8
4	– w60(40)	– w3	– 11	1
4	– w61(40)	– w2	– 11	2
4	– w57(17)	– w6	– 1	9
4	– 7	– w7	– 24	3
4	– 27	– w2	– 3	1
4	– 27	– w2	– Blank	1
4	– 51(5)	– w8	– 11	2
4	– 44(12)	– w5	– 24	1
4	– w50(21)	– Blank	– w31	1

*Number observed among 114 haplotypes of descendants from survivors.

References

1. De Vries RRP, Meera Khan P, Bernini LF, van Loghem E, van Rood JJ. Genetic control of survival in epidemics. J Immunogenet 1979;6:271–287.
2. De Vries RRP. De epidemie te Voorzorg. Ned Tijdschr Geneeskd 1978;122:1851–1853.
3. Baur MP, Neugebauer M, Albert ED. Reference tables of two-locus haplotype frequencies for all MHC marker loci. In: Albert ED, Baur MP, Mayr WR (eds): Histocompatibility Testing 1984. Berlin, Springer-Verlag, 1984, p 677.
4. Lee BS, Rust NA, McMichael AJ, McDevitt HO. HLA-DR2 subtypes form an additional supertypic family of DRβ alleles. PNAS 1987;84:4591–4595.
5. Bosch ML. Molecular analysis of a human immune response gene: HLA-DRw6. Thesis, University of Leiden. Dordrecht, ICG Printing, 1987, pp 81–91.
6. Todd JA, Bell JI, McDevitt HO. HLA-DQβ gene contributes to susceptibility and resistance to insulin-dependent diabetes mellitus. Nature 1987;329:599–604.

Author Affiliation

Department of Immunohaematology and Blood Bank, University Hospital, Leiden, The Netherlands

HLA-DR Molecules Are Both Ir- and Is-Gene Products for T-Cell Responsiveness Against *M. Leprae*

René R.P. de Vries, Tom H.M. Ottenhoff, Diënne G. Elferink, and Li Shuguang

The type of leprosy that develops upon infection with *Mycobacterium leprae* (*M. leprae*) correlates with T-cell immune responsiveness against the bacillus (1,2) and is associated with human leukocyte antigen (HLA) (3). At one pole of the leprosy spectrum we find polar tuberculoid (TT) leprosy, characterized by high helper T-cell reactivity and associated with HLA-DR3 in a population from Surinam. At the other end of the spectrum is lepromatous leprosy, characterized by an *M. leprae*-specific T-cell unresponsiveness and associated with HLA-DQw1. In this paper, data are presented indicating that different HLA-DR alleles present different *M. leprae* epitopes to helper T cells and that presentation by the DR3 molecule induces particularly strong helper T-cell proliferation. DR molecules also present *M. leprae* antigens to suppressor T cells, but allelic differences control whether helper or suppressor cells are activated. These data may explain the association of HLA class II alleles with leprosy type (Table 1).

Table 1. HLA Class II Ir- and Is-Genes and Leprosy Type

Leprosy Type	Immune Response	Association	Mechanism?
Tuberculoid (TT)	T_H	DR3	Ir-gene
Lepromatous	T_S?	DQw1	Is-gene

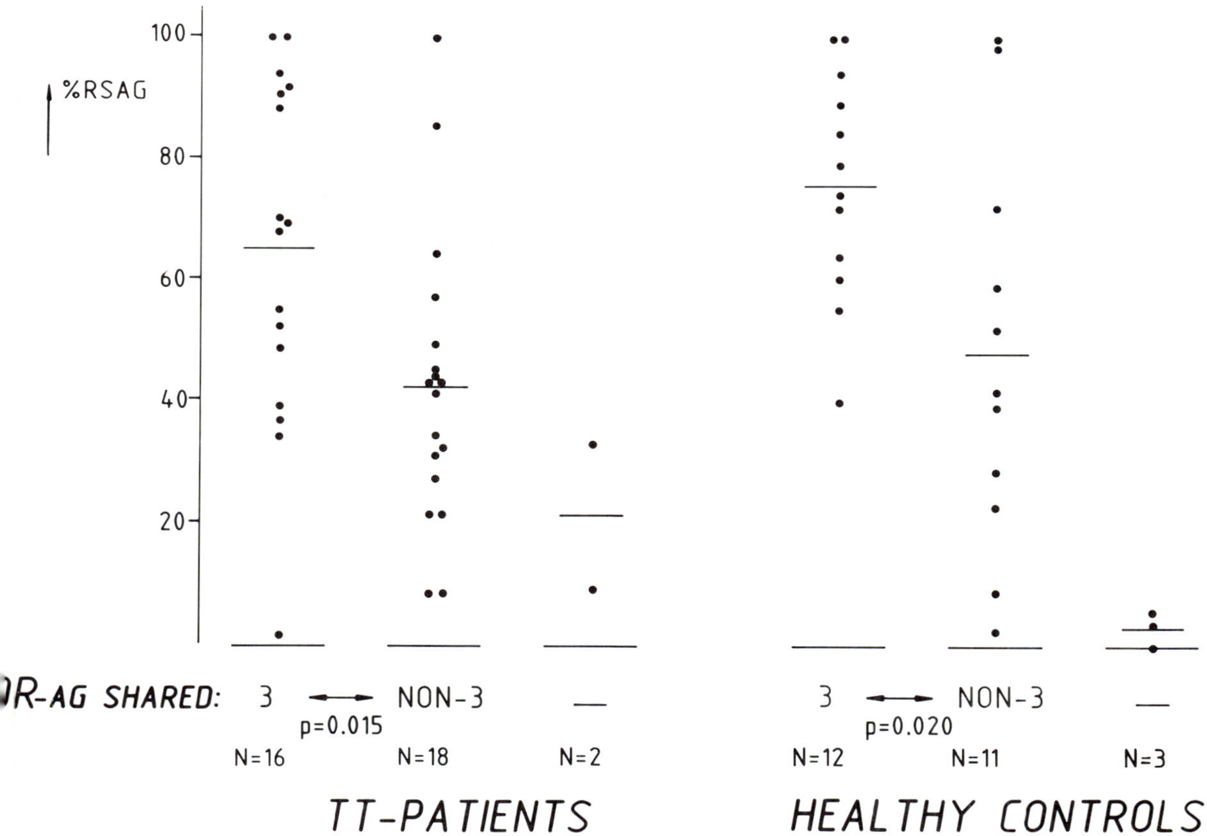

Figure 1. Preferential HLA-DR3 restriction of PPD-induced proliferation of TCL_{ppd} from three TT patients and three healthy individuals. The results are expressed in percent RSAG. Mean values for the T-cell-APC combination sharing respectively 2, 1, or 0 DR antigens are indicated by horizontal lines. The significance of the differences in the distribution of T-cell responses between the 2, 1, and 0 DR shared groups are indicated.

Table 2. Preferential Restriction of T_S Clones

	Presentation by Unrelated DR-/Dw-Matched APC	
	+	−
T_H clones	9	2
T_S clones	0	14

$p = 0.00003$.

M. leprae reactive helper T (Th) cells are probably responsible not only for acquired protective immunity to this mycobacterium but also delayed type hypersensitivity which may result in immunopathology (1). No detectable Th-cell reactivity against *M. leprae* is present in (polar) lepromatous leprosy patients. This unresponsiveness is *M. leprae* specific and may be a consequence of *M. leprae*-reactive suppressor T (Ts) cells (2). In order to better characterize these Th and Ts cells and the HLA molecules involved in the presentation of *M. leprae* antigen to them, we have cloned *M. leprae* reactive Th and Ts cells (4,5). The *M. leprae*-reactive Th clones were OKT3 and OKT4 positive, showed *M. leprae*-specific proliferation and γ-interferon production and did not suppress other *M. leprae* reactive T cells (4,6). With the use of monoclonal antibodies directed against different HLA class I and II molecules we could show that nearly always *M. leprae* antigens were presented to these Th clones by HLA-DR class II molecules (6). In order to test the hypothesis whether the above-mentioned association of HLA-DR3 with TT leprosy might be due to a HLA-DR immune response (Ir) gene, we investigated whether HLA-DR3 molecules might be better presenters than other DR molecules of *M. leprae* and other mycobacterial antigens to Th cells. This appeared to be the case indeed (Fig. 1).

We could generalize this observation by studying the helper T-cell response to the 65-kD proteins of *M. tuberculosis* and *M. leprae* which show 95% homology (7). With the use of deletion mutants (J.E.R. Thole et al., Infect Immunity 1988;56:1633–1640) and synthetic peptides (D.C. Anderson and T.M. Buchanan, Seattle), we were able to define five additional T-cell epitopes, apart from the two recently reported by Lamb et al. (8). It appeared that the recognition of each T-cell epitope was exclusively restricted by one HLA-class II locus or allele. These data, if confirmed on the population level, indicate HLA class II Ir-gene regulation of epitope recognition by helper T cells.

Recently we have succeeded in cloning from the peripheral blood of a borderline lepromatous leprosy patient T cells that specifically suppressed autologous Th cells reactive with mycobacteria (5). These Ts clones are also activated by *M. leprae* antigens presented by HLA class II molecules, which by blocking studies with monoclonal antibodies were shown also to be DR molecules. Interestingly, all 14 Ts clones raised from this patient only saw *M. leprae* when presented by autologous cells or those of an haplo-identical sibling, in contrast to Th clones from this patient (Table 2) to which *M. leprae* could be presented by DR/Dw matched unrelated APCs in 9 out of 11 cases. These data suggest that allelic differences between DR molecules may control whether certain epitopes are presented to helper or suppressor T cells. Thus far no DQw1-associated Is-gene and/or product has yet been demonstrated convincingly that might explain the association of HLA-DQ1 with lepromatous leprosy. However, in some instances *M. leprae*-specific suppression could be abolished with a monoclonal antibody directed against DQw molecules (own observations, and T. Sasazuki, personal communication).

Acknowledgment. Financial support for this study was obtained from the Immunology of Leprosy (IMMLEP) component of the UNDP/World Bank/WHO Special Program for Research and Training in Tropical Diseases, the Netherlands Leprosy Relief Association (NSL), the Dutch Foundation for Medical Research (MEDIGON grant 900-509-099) and the J.A. Cohen Institute for Radiopathology and Radiation Protection (IRS).

References

1. Godal T, Myrvang B, Stanford JL, Samuel DR. Recent advances in the immunology of leprosy with special reference to new aspects in immunoprophylaxis. Bull Inst Pasteur 1976;72:247–310.
2. Bloom BK, Mehra V. Immunological unresponsiveness in leprosy. Immunol Rev 1984;80:5–28.
3. De Vries RRP, van Eden W, Ottenhoff THM. HLA class II immune response genes and products in leprosy. Prog. Allergy 1985;36:95–113.
4. Ottenhoff THM, Klatser PR, Ivanyi J, Elferink BG, de Wit MYL, de Vries RRP. *Mycobacterium leprae* specific protein antigens defined by cloned human helper T cells. Nature 1986;319:66–68.
5. Ottenhoff THM, Elferink BG, Klatser PR, de Vries RRP. Cloned suppressor T cells from a lepromatous leprosy patient suppress *Mycobacterium leprae* reactive helper T cells. Nature 1986;322:462–464.
6. Ottenhoff THM, Neuteboom S, Elferink BG, de Vries RRP. Molecular localisation and polymorphism of HLA class II restriction determinants defined by *Mycobacterium leprae* reactive helper T cell clones from leprosy patients. J Exp Med 1986;164:1923–1939.
7. Thole JER, Keulen WJ, Kolk AHJ, Groothuis DG, Berwald LG, Tiesjema RH, et al. Characterization, sequence determination, and immunogenicity of a 64-kilodalton protein of *Mycobacterium bovis* BCG expressed in *Escherichia coli* K-12. Infect Immun 1987;55:1466–1475.
8. Lamb JR, Ivanyi J, Rees ADM, Rothbard JB, Howland K, Young RA, et al. Mapping of T cell epitopes using recombinant antigens and synthetic peptides. EMBO J 1987;6: 1245–1249.

Author Affiliation

Department of Immunohaematology and Blood Bank, University Hospital, Leiden, the Netherlands

B35 and CW4 Are Risk Factors for Development of AIDS

M. Jeannet, N.A. Carpentier, R. Sztajzel, P.J. Male, and B. Hirschel

Studies of the HLA system in different groups of patients with the acquired immune deficiency syndrome (AIDS) suggest that genetic factors associated with HLA phenotypes may predispose to the development of the disease. Among the HLA associations claimed to be significant, the most significant is the increased frequency of the HLA-DR5 allele in patients with AIDS and Kaposi's sarcoma (1,2). In addition, a recent Italian study suggests that HLA-B35 and CW4 represent factors of susceptibility to AIDS (3).

One hundred twenty-four individuals with antibodies to HIV, including 38 patients with AIDS (24 suffering from opportunistic infections and 14 from Kaposi's sarcoma or lymphoma) were typed for HLA-A,B,C and DR antigens. Eighty-six HIV positive individuals without AIDS were intravenous drug users, whereas among the 38 AIDS patients, 22 were homosexual men, and 16 were heroin addicts. As controls, 151 healthy blood donors and a group of 39 heroin addicts, all seronegative for HIV, were also studied. HLA typing was performed by the standard microlymphocytotoxicity assay using a large battery of antisera to all the presently well-defined HLA specificities. Statistical analysis was done by the χ-square with Yates correction.

The frequencies of B35 and CW4 were significantly higher than in the controls in AIDS patients with secondary infectious diseases and secondary cancers (Table 1) confirming a previous study (3); however, we could not confirm the decreased frequency of B44 and CW5 found in that study (3). Interestingly, DR2 was significantly increased in our patients, but only in those with opportunistic infections. On the other hand, DR5 was increased, although not significantly in patients with secondary cancers, as reported by others (1,2). The frequency of B8 and DR3 was low in both subgroups of AIDS patients, also confirming earlier studies (1). In 86 HIV positive drug abusers with and without lymphadenopathy syndrome, who had not progressed to AIDS, and in 39 HIV negative drug abusers, the frequencies of B35, CW4 and DR5 were not increased.

The present study confirms HLA associations already reported in AIDS. It provides strong evidence for the existence of HLA-linked factors of susceptibility to AIDS. The data also suggest that some HLA alleles which are clearly associated with the clinical patterns of HIV infection may serve as markers for predicting the rapid evolution of HIV infection to AIDS. The presence of B35 (in linkage disequilibrium with CW4) in an HIV-infected person appears to predispose to the development of AIDS. While the HLA-DR2 phenotype confers a higher risk of AIDS with opportunistic infection, HLA-DR5 is clearly associated with the development of Kaposi's sarcoma. In contrast the under-representation of HLA alleles such as B8 and DR3 in AIDS patients suggest that these antigens, or associated genes, operate as protective factors against the progression of HIV infection.

References

1. Pollack MS, Safai B, Myskowski PL, et al. Frequencies of HLA and Gm immunogenetic markers in Kaposi's sarcoma. Tissue Antigens 1983;21:1–8.
2. Prince HE, Schroff RW, Ayoub G, et al. HLA studies in acquired immunodeficiency syndrome patients with Kaposi's sarcoma. J Clin Immunol 1984;4:242–245.
3. Scorza Smeraldi R, Fabio G, Lazzarin A, et al. HLA-associated susceptibility to acquired immunodeficiency syndrome in Italian patients with human-immunodeficiency virus infection. Lancet 1986;ii:1187–1189.

Author Affiliations

M. Jeannet, N.A. Carpentier, R. Sztajzel, Transplantation Immunology Unit; *P.J. Male, B. Hirschel*, Department of Medicine, Hôpital Cantonal Universitaire, Geneva, Switzerland

Table 1. Distribution (%) of Selected HLA Antigens in 38 Patients with AIDS

HLA Antigen	Controls (n = 151)	AIDS (n = 38)	AIDS Group IVc (secondary infectious disease) (n = 24)	AIDS Group IVd (secondary cancers) (n = 14)
B8	25	11	13	7
B35	15	50*	42*	64*
CW4	14	45*	42*	50*
DR2	25	44	54*	25
DR3	25	14	21	0
DR5	30	42	33	58

*$p < 0.01$.

HLA and Tuberculosis in Chinese Patients

Y.P. Sun, J.B. An, Z.F. Jiang, C.Q. Dai, C. Li, and W. Chen

The association between tuberculosis and HLA-Bw35 in Chinese (rr = 7.4) was initially found in 1983 (1). In 1986, we studied 12 families with tuberculosis who had one parent and at least two siblings affected. The haplotype sharing between affected siblings significantly differed from the expected random distribution suggesting the existence of an HLA-linked gene for susceptibility to tuberculosis (2).

We subsequently examined 14 families with tuberculosis who had one parent and at least one child affected. Eight of these families with one affected parent and two affected siblings could also be used for sibling-pair analysis. Therefore, altogether 20 families were available for analysis.

Analysis of Segregation

Significant deviation from the random segregation of haplotypes in affected offspring of 20 families was found ($\chi^2 = 20.56$, $p < 0.001$). The calculation of χ^2 was done according to the formula:

$$\chi^2 = \frac{(\Sigma F_i - f_i \Sigma n)^2}{\Sigma V(fi)}$$

where Fi is the deviation of haplotype inheritance existing in affected offspring observed and fi is the expected deviation, V(fi) is the variance, and n is the coefficient bionominal distribution.

Table 2. Model of Inheritance

Dominant	Shared Haplotypes			
	(2)	(1)	(0)	X^2
Observed	32	14	0	
Expected	21.16	22.77	2.07	11.0
χ^2	5.55	3.38	2.07	
Recessive				
Observed	32	14	0	
Expected	31.92	12.79	1.29	1.4
χ^2	0.002	0.11	1.29	

Haplotype-Sharing Analysis Among Siblings

Significant deviation from the random distribution was also found in affected siblings. The haplotype-sharing distribution of all 46 affected siblings was shown in Table 1. Further genetic analysis using Thomson and Bodmer's method (3) revealed a strong correlation with autosomal recessive mode of inheritance ($\chi^2 = 1.42$, $p > 0.05$) (Table 2).

Table 1. HLA Haplotype Sharing Among Affected Siblings

No. of Pedigree	No. of Affected Siblings	Haplotype	No. of Siblings Sharing Haplotypes		
			(2)	(1)	(0)
1	2	ac ac	2	0	0
2	2	bd bc	0	2	0
3	3	ac ac ac	3	0	0
4	2	bd bd	2	0	0
5	3	ac bc ac	2	1	0
6	3	ac ac bc	2	1	0
7	2	ac ac	2	0	0
8	2	ac ac	2	0	0
9	2	ac ac	2	0	0
10	2	ac ad	0	2	0
11	3	ac bc bc	2	1	0
12	2	ac ac	2	0	0
13	2	ac bc	0	2	0
14	2	ac ad	0	2	0
15	2	ac ac	2	0	0
16	2	ac ad	0	2	0
17	2	ad ad	2	0	0
18	2	bc bc	2	0	0
19	4	ac ac ac ad	3	1	0
20	2	bc bc	2	0	0
Observed	46		32 (70%)	14 (30%)	0
Expected			(25%)	(50%)	(25%)

Immunobiology of HLA
Volume II

References

1. Jiang ZF, An JB, Sun YP, Mittal KK, Lee TD. Association of HLA-BW 35 with tuberculosis in the Chinese. Tissue Antigens 1983;22:86–88.
2. An JB, Sun YP, Li C, Chen WC, Lu LY, Xu LM, et al. A study on tuberculosis linked with HLA. Chinese Journal of Tuberculosis and Respiratory Diseases 1986;9(5): 260–263.
3. Bodmer WF. The HLA system and disease. Oliver Sharpey lecture, 1979. J Royal Col Physicians, London 1979;14: 43–59.

Author Affiliations

Y.P. Sun, W. Chen, Department of Immunology China-Japan Friendship Hospital; *Z.F. Jiang, J.B. An*, Beijing Research Institute of Pediatrics; *C.Q. Dai*, Beijing Tuberculosis Institute; *C. Li*, Ben Fu Railway Hospital, Beijing, People's Republic of China

Section XVI
Histocompatibility Antigens and Neoplastic Disease

AKR (H-2^k) Leukemias Express H-$2D^d$-Like Molecules

Mario O. Labeta, Nelson Fernandez, and Hilliard Festenstein

Our original finding on the expression of an H-$2D^d$-like 'allospecificity' in AKR-derived spontaneous leukemic line K.36 (1), has now been extended to the study of other AKR leukemic lines. Five AKR tumors were analyzed by cell-binding radioimmunoassay (CB.RIA), sodium dodecylsulfate polyacrylamide gel electrophoresis (SDS-PAGE), and two-dimensional (2-D) gel electrophoresis. Strongly positive binding by CB.RIA were obtained with the well-characterized and affinity purified anti-H-$2D^d$ monoclonal antibody (MAb) 34.5.8 with all the tumors (Figure 1). In some tumors (i.e., 424, K36.16, 369) the detection of the H-$2D^d$-reactive epitopes was facilitated by enzymic digestion of the whole cell with a sialidase (i.e., neuraminidase, Vibrio Cholerae). In the case of the 424 tumor, the enzymic digestion proved to be critical for the detection of the H-$2D^d$-like antigen. Direct immunoprecipitations of 46 kD and 12 kD bands consistent with the classical class I pattern by SDS-PAGE were obtained with 34.5.8 MAb using metabolically labeled NP-40 solubilized glycoproteins from 369,365, and K36.16 tumors. The same class I pattern was obtained with the NP-40 solubilized 424 glycoproteins but only after "preclearing" of the syngeneic H-$2K^k$ and H-$2D^k$specificities. The use of the nondenaturing, zwitterionic detergent 'CHAPS' to solubilize the 424 glycoproteins permitted the direct detection of the novel H-$2D^d$-like antigen by SDS-PAGE using the 34.5.8 MAbs. Thus, we speculate that the nonionic detergent may allow the formation of soluble protein complexes in which the novel antigen is associated with at least one H-2 class I specificity and not accessible to the specific MAb. The anti-H-$2K^k$ and H-$2D^k$ MoAb (15.1.5) may disrupt such complexes. Thus, after "preclearing" the syngeneic antigens, the H-$2D^d$-reactive epitopes can be detected. The zwitterionic detergent therefore, disaggregate such protein-protein complexes, permitting direct detection of the novel molecule.

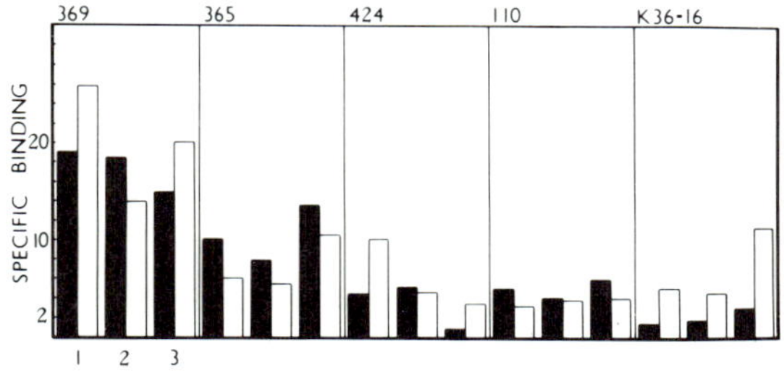

Figure 1. CB.RIA before (solid bars) and after (open bars) neuraminidase treatment of the AKR leukaemias. The MAbs used were (1) 15.1.5, anti-H-$2K^k$ and H-$2D^k$, (2) H100.5, anti-H-$2K^k$ and (3) 34.5.8, anti-H-$2D^d$. The specific binding is expressed as times the background values (250 cpm).

The 2-D gel electrophoresis analysis showed that the H-$2D^d$-like antigen has a molecular pattern clearly distinguishable from the classical H-$2D^d$ and also from the syngeneic H-$2K^k$ and H-$2D^k$ determinants.

The findings reported here suggest that (i) The expression of the H-$2D^d$-like allospecificity is a general phenomenon in the AKR leukemias; and (ii) the aberrant specificity could not readily be detected before because it was obscured either by a heavy sialylation pattern on this tumor or by close association with the syngeneic determinants.

The genetic basis of the expression of this extra specificity has not yet been established but possible explanations include the expression of Qa/TLa gene products on the leukaemic cells, which are not present in the normal counterpart, and the expression of an H-$2D^d$-like gene product by Qa/TLa genes, normally silent, which become activated by MHC gene rearrangement by retroviruses. The role of this molecule in tumor progression is being investigated.

Acknowledgment. This work was supported by The Cancer Research Campaign and The Medical Research Council of U.K.

References

1. Festenstein H, Schmidt W. Variation in MHC antigenic profiles of tumour cells and its biological effects. Immunological Rev 1981;60:85.

Author Affiliations

Department of Immunology, The London Hospital Medical College London, UK

MHC Class I Antigen Expression on Human Tumors

F. Ruiz-Cabello, A. Ferrón, M.A. López Nevot, S. Pedrinaci, J.M. Conejo, F. Esteban, and F. Garrido

Viral or tumoral antigens are recognized by cytotoxic T lymphocytes in association with histocompatibility class I antigens. Alterations in MHC class I antigen expression by tumor cell can abrogate cytotoxic T lymphocytes mediated lysis and may play a role in oncogenesis and metastasis (1). The absence of class I histocompatibility antigens has been correlated to greater tumor aggressiveness. Furthermore, the role of these molecules in tumor reaction has been confirmed in gene transfection experiments (2).

We have investigated HLA-class I antigen expression in 325 human neoplasias, comprising 234 primitive tumors and 91 metastases with immunohistologic techniques including peroxidase, alkaline phosphatase, and immunofluorescence. In these studies we used MAbs directed against class I heavy chain (W6/32), B2 microglobulin (GRH1), two locus-specific HLA-B MAbs (Joan 1 and α HLA-B), and several polymorphic MAbs (anti-A2, anti-A11, anti-B5, anti-B14, anti-B44, and anti-Bw4).

We found that HLA-class I loss occurs in nearly all types of tumor studied, although its frequency varied depending on the histologic type and the primitive or metastatic nature of the tumor. Class I negative tumors were observed with higher frequency on primary sarcomas, skin basal carcinomas, and breast ductal infiltrating carcinomas (Table 1). On laryngeal epidermoid carcinomas, class I antigen loss was correlated with a low degree of differentiation and in melanomas with more invasiveness.

Table 1. Class I Expression in Primary and Metastatic Tumors

Tumors	Primary	Metastatic
Melanoma	15/19*	15/29
Skin epidermoidal	7/8	1/1
Skin basal carcinoma	17/26	0/0
Colon adenocarcinoma	29/34	13/13
Stomach adenocarcinoma	28/31	27/30
Larynx carcinoma	27/34	9/14
Breast ductal carcinoma	30/39	0/1
Renal carcinoma	12/13	1/1
Sarcoma	0/4	0/2

*Class I positive/tested. A tumor was considered positive for class I antigen when more than 25% of cells were stained with W6/32 MAb.

Table 2. Selective Losses of HLA Class I Antigen in Human Tumors

	Lymphocytic Typing	Tumor Typing
Colon	A2,-/B5,B21	A2,-/-,-
Larynx	A2,-/B5,B44	A2,-/-,-
Stomach	A11,-/B14,B44	-,-/-,-

Lymphocytes were typed by standard microcytotoxicity test and tumors by immunohistologic techniques using polymorphic MAbs.

When we compared class I expression between primary tumor and autologous metastasis we found a high grade of heterogeneity. The highest frequency of class I negative metastatic tumors was observed on melanomas (Table 1).

The use of isotipic anti-HLA-B MAbs allows us to find selective loses of HLA-B locus products on colon, stomach, and laryngeal carcinomas. These tumors were W6/32 positive and HLA-B negative. This phenomenon was observed on 8.8% of colorectal carcinomas, 3.4% of stomach carcinomas, and 5.8% of laryngeal carcinomas. These results were confirmed with polymorphic MAbs. In addition we found that one stomach tumor was positive with W6/32 MAb and negative with polymorphic anti-HLA-A and HLA-B MAbs (Table 2).

We conclude that selective losses can be detected on human tumors with adecuated reagents and could be analyzed in relation to anti-tumoral immune responses.

References

1. Garrido A, Perez M, Delgado C, Garrido ML, Rojano J, Algarra I, et al. Influence of class I H-2 gene expression on local tumor growth. Clin and Exp Immunogenetics 1986; 3:98–110.
2. Hui K, Grosveld F, Festenstein H. Rejection of transplantable AKR leukemia cells following MHC-DNA mediated cell transformation. Nature 1984;31:750–752.

Author Affiliation

Servicio Analisis Clinicos, Hospital Virgen de las Nieves, Granada, Spain

Immunobiology of HLA
Volume II

Study of Class I HLA Genes in Human Tumors

M.A. Lopez Nevot, C. Romero, F. Ruiz-Cabello, S. Pedrinaci, J. Gutierrez, M.R. Oliva, C. Huelin, and F. Garrido

Within the area of immunology dealing with histocompatibility antigens in tumors, a topic of great interest is the correlation of phenotypic alterations with the presence (or absence) of MHC gene rearrangements (1). Data obtained in murine tumors has related the loss of class I expression or the appearance of aberrant molecules to structural changes in MHC class I genes (2).

The purpose of the present study is to use Southern blotting analyses to compare class I genes in human tumors and normal autologous cells. The tumors analyzed were 10 colon adenocarcinomas (7 cases class I positive, one case with selective losses of HLA-B antigens, and two cases negative, Fig. 1), four melanomas (two cases positive and two negative), two cases of stomach adenocarcinomas (both negative), and two larynx carcinomas (one showing selective losses of HLA-B antigens and one class I negative). High molecular weight DNA was isolated from the tumors; normal autologous mucosa and autologous peripheral lymphocytes .10 ug aliquots of DNA were digested with 150–300 units of different restriction enzymes (EcoRI, HindIII, BamHI, PstI, PvuII, MspI or HpaII) for 4 hours according to the supplier's specifications (Amersham). DNA fragments were separated by electrophoresis in 0.8% agarose gels for 16 hours and transferred to nitrocellulose membranes (Amersham) (3). A 1.4 kbp cDNA coding for HLA-B7 (pDP-001) was used as a probe (4). Membranes were treated for hybridization with nick-translated P32-labeled pDP-001 in 50% formamide solution over 36 to 40 hours at 42 C. Membranes were washed at 65 C for 30 min in two changes of 2xSSC/0.1% SDS. After washing, the membranes were dried and autoradiographed at −70 C until the signals were visible.

We found no alterations in class I RFLPs in tumors when compared to normal autologous cells. The findings suggest that in most tumors the loss of expression of class I molecules is the result of the action of regulatory mechanisms.

Selective losses of class I antigens can be explained by the existence of independent mechanism of control for the different class I loci.

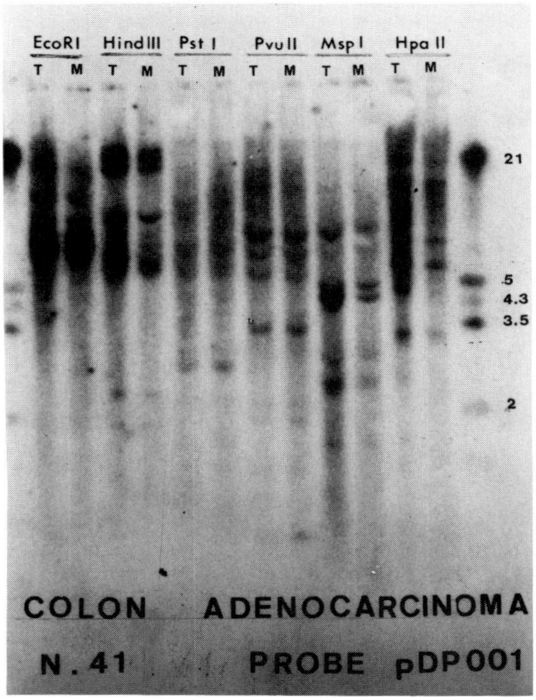

Figure 1. Southern blot analysis of class I genes from a class I negative colon adenocarcinoma (C-41). Tumor DNA (T) was compared with DNA from autologous normal mucosa (M). No differences on RFLPs were observed.

References

1. Garrido F. H-2 gene complex. In: David C (ed): Genes, molecules, functions. Bar Harbor, Maine, Plenum Corporation, NATO ASi series, Vol. 144, p 623–639, 1987.
2. Festenstein H, Garrido F. MHC antigens and malignancy Nature 1986;322:502–503.
3. Southern E. Detection of specific sequences among DNA fragments separated by gel electrophoresis. J Mol Biol 1975;98:503–518.
4. Sood AK, Pereira D, Weissman SM. Isolation and partial nucleotide sequence of a cDNA clone for human histocompatibility antigen HLA-B by use of oligodeoxynucleotide primer. Proc Natl Acad Sci USA 1981;78:616.

Author Affiliation

Servicio Analisis Clinicos, Hospital Virgen de las Nieves, Granada, Spain

Poor Prognosis in Childhood Acute Lymphoblastic Leukemia (ALL) Is Associated with HLA-A11

Shlomit Orgad, Ian J. Cohen, Yoram Neumann, Ruth Vogel, George Kende, Bracha Ramot, Rina Zaizov, and Ephraim Gazit

Acute childhood lymphoblastic leukemias (ALL) consists of several biologically distinct malignancies derived from different ancestral cells. A number of clinical and hematologic parameters are currently considered to be indicators of poor prognosis, i.e., male sex, high whiteblood cell counts, age below 2 or over 10 years, the presence of certain karyotypic abnormalities, and cell surface markers (1–3).

Table 1. Frequency of HLA-A11 in All Patients

	Number Studied	Frequency of -A11 Expected	Frequency of -A11 Found
PSG total	97	10	9
Dead	45	4	8
Alive	52	5	1
LTS	47	5	0

$p < 0.001$ for the difference between LTS vs. PSG dead; LTS = long-term survivors; PSG = prospective study group.

A possible association between HLA antigens, susceptibility or resistance to leukemia, and responsiveness to treatment has been studied in 144 ALL patients and compared to the other prognostic factors. The group studied included 37 patients classified as T-ALL, 41 CALLA+, 27 NULL, 12 B and pre-B, and 27 unclassified patients (those who were diagnosed before 1980). Eighty-six were Jews and 58 were Arabs. All the 97 newly diagnosed children (who were admitted between 1979–1984) comprised the group for the prospective study (PSG) and were followed for 6 years. Forty-five patients of the PSG died during the follow-up period. Forty-seven patients designated "long-term survivors" (LTS) have been followed 6 to 20 years after diagnosis, having completed a 3 to 5 year course of anti-leukemia therapy, remaining disease-free thereafter.

Poor prognosis was associated with high WBC counts at diagnosis, T-cell surface markers, ethnic origin, and specific HLA antigens: 1) a significant increase in HLA-A30 and decrease in HLA-B14 in the PSG Jewish

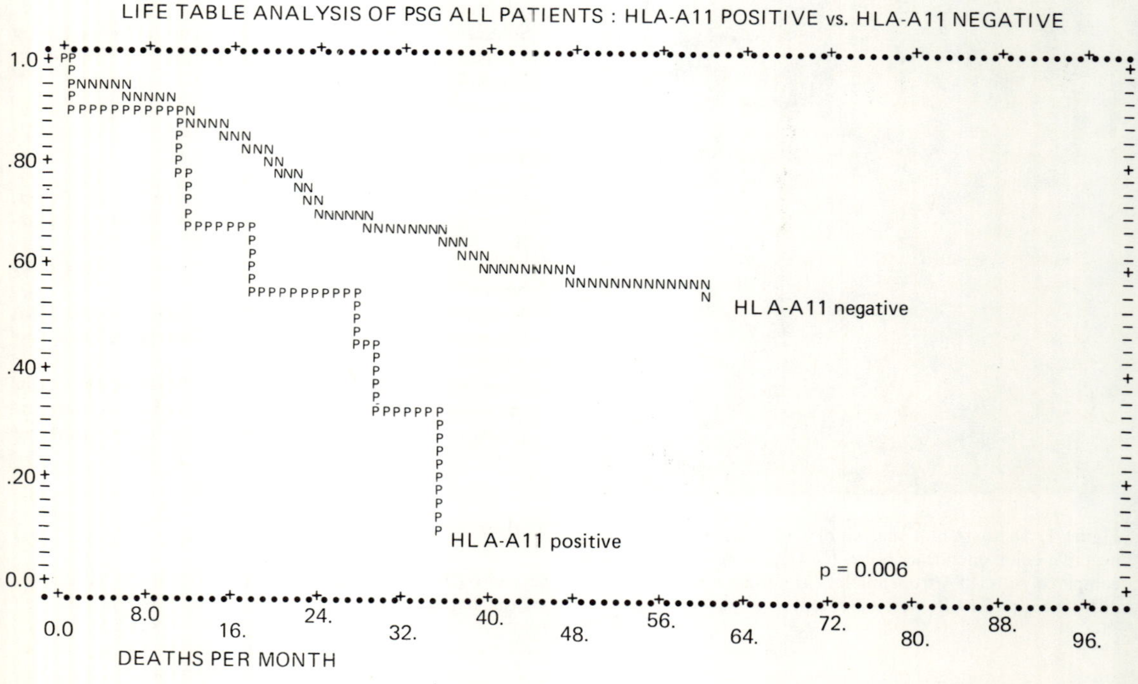

Figure 1. Life table analysis of 9 HLA-A11 positive and 86 HLA-A11 negative patients. $p<0.006$ for the difference between survival of HLA-A11 positive vs. negative patients. (*) Subsequent to the analysis, this only surviving patient has relapsed.

Immunobiology of HLA
Volume II

patients; 2) a complete absence of HLA-A11 in LTS. In the PSG, eight of nine HLA-A11 positive patients died during the follow-up period. (Table 1, Fig. 1). The fact that HLA-A11 was absent in the LTS group, suggests an association with poor prognosis and/or unfavorable response to treatment.

Recently, a specific resistance of Burkitt Lymphoma (BL) cell lines to HLA-A11 restricted lysis was described as due to down-regulation or altered presentation of the HLA-A11 epitope on the leukemic and EBV-infected B cells, which were isolated from the same BL patient (4).

In summary, HLA typing may be a useful prognostic marker of ALL. Because bone marrow transplantation is now a therapeutic alternative in poor risk leukemia, it seems justified to consider HLA-A11 positive patients as poor risk patients and therefore as potential candidates for this type of therapy, regardless of their leukemia subclass.

References

1. Greaves MF, Janossy G, Petto J, Humphrey K. Immunologically defined subclasses of acute lymphoblastic leukemia in children: Their relationship to presentation features and prognosis. Br J Haemat 1981;48:179.
2. Sallan SE, Ritz J, Pesando JM, Gelber R, O'Brian C, Hitchcok S, Coral F, Schlossman SF. Cell surface antigens prognostic implications in childhood acute lymphoblastic leukemia. Blood 1980;55:395.
3. Ramot B, Ben-Bassat I, Many A, Kende G, Neuman Y, Brok-Simon F, et al. Acute lymphoblastic leukemia subtypes in Israel: The Sheba Medical Center experience. Leukemia Res 1982;6:669.
4. Torsteinsdottir S, Masucci MG, Ehlin-Henriksson B, Brautbar C, Ben-Bassat H, Klein G, et al. Differentiation-dependent sensitivity of human B cell derived lines to major histocompatibility complex-restricted T-cell cytotoxicity. Proc Natl Acad Sci USA 1986;83:5620.

Author Affiliations

Shlomit Orgad, *Ephraim Gazit*, Division of Transplantation Immunology, Sheba Medical Center, Tel-Hashomer 52621, Israel; *Ian J. Cohen*, *Ruth Vogel*, *Rina Zaizov*, Sambur Institute of Pediatric Hematology & Oncology, Beilinson Medical Center, Petach-Tiqua, Israel; *Yoram Neumann*, *George Kende*, *Bracha Ramot*, Institute of Hematology, Sheba Medical Center, Tel-Hasomer, Israel

The Functional Heterogeneity of HLA Class II Determinants

H. Festenstein, L. Fainboim, C. Navarrete, D. Jaraquemada, and M. Bagnara

The original discovery by Navarrete et al. (1) that HLA-DQ and HLA-DR specificities are differentially expressed on leukemic cells and particularly that HLA-DQ is absent from acute myeloid leukemic cells, led us to investigate whether HLA-DQ has a specific unique role in the immune response. If so, the absence of HLA-DQ on the tumor cells may be advantageous for the development of leukemia. This concept was supported by the experiments of Corte et al. (2), which showed that HLA-DQ antibody was able to prevent the in vitro generation of cytotoxic T cells. These findings were subsequently confirmed in our laboratory (3). Previous genetic experiments in the mouse in which recombinant congenic mouse strains were employed also suggested that H-2 E (the functional counterpart of HLA-DQ) is important in the generation of cytotoxic T cells.

HLA-DQ also has a regulatory role in immune homeostasis. In a particular combination, two HLA identical siblings' lymphocytes were stimulated by the homozygous typing cell FPA(F) (HLA-BW35 homozygous and also HLA-DQ1). One of these HLA identical siblings' cells proliferated in response to the stimulation, while the other sibling's lymphocytes did not, indicating a non-HLA responder gene. When the siblings' lymphocytes were mixed and stimulated by FPA, there was no response. This suppression could be abrogated by the removal of the CD8-positive cells or by adding anti-DQ monoclonal antibodies to the cell culture (Fig. 1). When the phenotype of the activated T cells were examined, approximately 18% of the cells were found to be HLA-DQ positive, CD8 positive. Although the number of cells with this phenotype is relatively small, the potency of suppression by cells with this phenotype seems to be extremely high. Similar contrasuppressor effects by anti-DQ monoclonal antibodies have been reported by Sasazuki, in which the suppressor effects by streptococcal cell wall antigen and schistosomal antigen were also abrogated by the addition of anti-HLA-DQ monoclonal antibody. Recent studies from our laboratory have also shown that more than 80% of patients with AIDS have a significant number of CD8-DQ-positive cells, suggesting that immune deficiency in AIDS is not only dependent on the depletion of CD4 cells, but that there may also be an active suppressor process involving HLA-DQ.

However, it is not only HLA-DQ/CD8-positive cells that appear to play a role in immune homeostasis. Previous experiments from our laboratory have shown that polyclonally PHA-activated T cells and activated T cells in mixed lymphocyte cultures become suppressor cells once they express HLA class II determinants on the activated T cells. The active molecule in these instances has not been established. However, recent experiments in

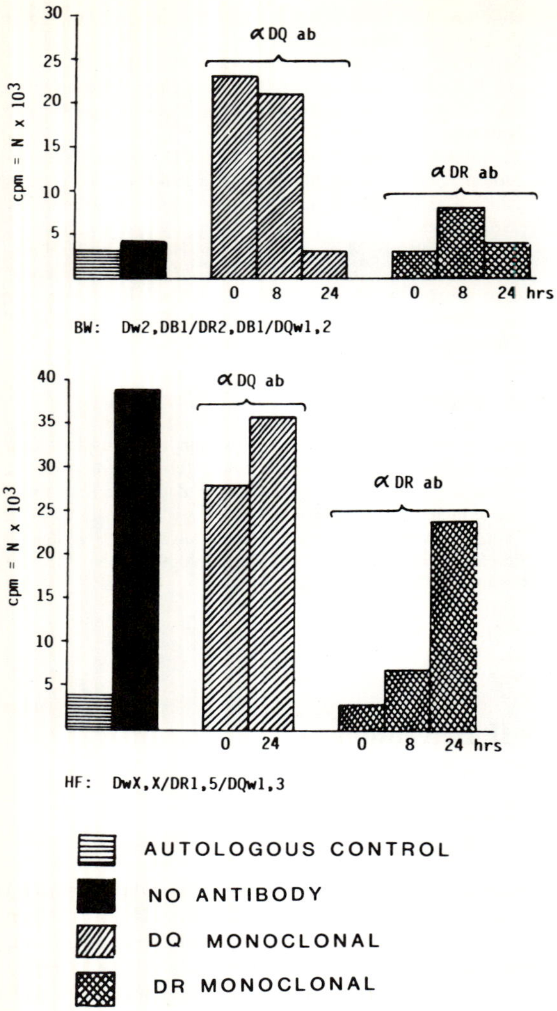

Figure 1. DQ1 antibody (Genox 3.5.1) added at 0.8 and 24 hours after commencement of mixed lymphocyte culture. The suppressed response of BW to FPA is abrogated by adding the anti-DQ1 antibody. H.F. is the control responder, in whom no effect was observed after adding anti-DQ1 antibody.

collaboration with Pawelec and Wernet and others in our own laboratory suggest that novel monomorphic HLA class II determinants other than HLA-DR, DQ, and DP may be the essential homeostatic molecules in this context—once they are expressed on activated T cells, the immune response is down regulated. This effect can be specifically abrogated by using the specific monoclonal antibodies Tu 39 and Edu 1. The former antibody has been shown to react with a subset of CD 4, "DY"-positive T cells.

Acknowledgments. This work has been supported by the Medical Research Council and Leukaemia Research Fund of Great Britain.

References

1. Navarrete C, Fainboim L, Festenstein H, Lister M, Amess J, Greaves MF. Selective expression of HLA-DR and HLA-MB antigens on acute leukemia. In: Knapp W (ed): Leukaemia markers. p 65. London: Academic Press, 1981.
2. Corte G, Moretta A, Cosulich ME, Ramarli D, Bargellesi A. A monoclonal anti-DC1 antibody selectively inhibits the generation of effector T cells mediating specific cytolytic activity. J Exp Med 1982;156:1539.
3. Pawelec GP, Fernandez N, Schneider E, Festenstein H, Brooker T, Wernet P. DY determinants, possibly associated with novel class II molecules, stimulate autoreactive, CD4+ T cells with suppressive activity.
4. Fernandez N, Labeta ML, Festenstein H. A non-HLA-DR, DP, DQ antigen structure. In: Dupont JA (ed): Immunology of HLA. New York, Springer Verlag, 1989.
5. Fainboim L, Navarrete C, Festenstein H. Precursor and effector phenotypes of activated human T lymphocytes. Nature 1980;288:391.

Author Affiliations

H. Festenstein, C. Navarrete, D. Jaraquemada, M. Bagnara, Department of Immunology, The London Hospital Medical College, London E1 2AD, UK; *L. Fainboim,* Fundacion CIMAE, Buenos Aires, Argentina and Department of Immunology, The London Hospital Medical College, London, England

Patterns of MHC Class II Molecules and mRNA Expression in Acute Myelogenous Leukemias

Teresa Improta and Robert J. Winchester

The occurrence of class II molecules on early normal and leukemic myelogenous cells suggests a novel function for these structures distinct from antigen presentation (1). Of particular interest the cell surface expression of DR, DP, and DQ molecules on hematopoietic cells is discordant. The particular pattern of expression is a marker of differentiation stage and lineage (2,3). HLA DR molecules are on myeloid and erythroid progenitors but, with the exception of monocytes and macrophages, not on their mature progency. DP is reported to be on a proportion of progenitor cells, whereas DQ is expressed on monocyte-macrophage cells but at very low levels, if at all, on myeloid progenitors.

Because of their clonality hemopoietic malignancies are a useful model for the study of myeloid lineage differentiation. There is general agreement that the class II phenotype of the acute myeloid leukemias is heterogeneous, indicative of the diversity of lineage and differentiation stage at which the myeloid stem cell becomes transformed. The interpretation of the results of these studies conducted on a small number of myelogenous leukemias was complicated by the lack of a specific monomorphic DQ antibody and, more importantly, primary reliance on surface membrane expression of the molecules to evaluate gene expression.

In the present study, we examined 20 acute myeloid leukemias representing different stages of cell differentiation, including the FAB subtypes from M1 to M5, defined by immunophenotyping and cytochemistry. The myeloid cell lines K562, KG1A, KG1, HL60, and U937 were also studied.

Direct expression of class II genes was determined by Northern analysis, using class II specific cDNA probes derived from the less conserved 3′ untranslated regions of β chain cDNA, and from part of the first and all of the second extracellular domains of DQα cDNA. These probes were shown not to exhibit class cross-hybridization (5). Quantitation of surface antigen expression has been performed using MAbs specific for HLA DR(SG157), DQ (1a3, a monomorphic reagent (4), Leu 10, IVD12, SG465), and DP(B7/21), measuring specific binding in indirect immunofluorescence by FACS analysis.

As shown in Figure 1, all the class II transcripts appear to be full length regardless of the level of expression. In general, the levels of mRNA for all class II loci were highly variable. However, coordination among the products of each subregion was demonstrated by the finding that DQ and/or DP mRNA were neither expressed in the absence of DR, nor did their levels exceed the amount of DR mRNA.

All the M1 and M2 leukemias had low or intermediate amounts of DR mRNA, compared to all other leukemias. Six of eight M1 and M2 leukemias lacked DQ and four of eight lacked DP mRNA.

The M3 leukemias and cell lines HL60, representing a promyelocytic population, were distinguished by the absence of any evidence of class II expression. The same result has been found for cell line K562, representative of a very early stage of myeloid differentiation. KG1A, KG1, and U937, respectively representing early myeloblast, myeloblast, and monocytic stages of differentiation, exhibited DR mRNA at low or intermediate levels, while the DP transcript was in low amount in KG1 and U937. Only KG1 expressed DQ mRNA.

The M4 and M5 leukemias had high or intermediate levels of DR mRNA. However, of importance, eight out of nine exhibited abundant DQ and DP mRNA, distinguishing them as a group from the M1-M2 AMLs.

In most, but not all leukemias the surface expression and level of subregion specific mRNA were parallel. Of particular interest for the leukemias LE(M2), EF and DE(M4), and RY(M5), we found high levels of DQ or DP mRNA and no (DE, LE, EF) or little (RY) protein

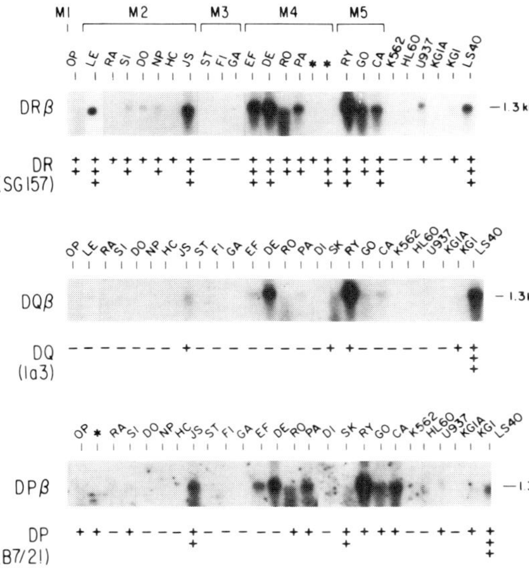

Figure 1. Northern blot analysis of HLA class II β mRNA in AMLs and human cell lines compared with the results of surface staining with SG157, 1a3, B7/12 moAbs. Total RNA from 20 patients and five cell lines were electrophoresed in duplicate, one part of the gel stained with ethidium bromide and the other blotted onto nylon membrane. Blots were probed, stripped, and reprobed successively with β tubulin as quantitative/qualitative control. LS40 = B lymphocyte cell line; * = no RNA.

antigen detectable in either cytoplasm or on the cell membrane. The quantity of DQα and β chain transcripts, determined by Northern analysis, was abundant, coordinated, and paralleled that of DR, excluding a gross defect in chain expression as an explanation.

Our results provide further evidence for the complexity in the regulation of MHC class II molecule expression. There is an overall coordination of regulation of the α and β chains of each locus in that the level of DR expression is always the greatest. At a second level, there appear to be DQ and DP subregion-specific regulatory signals. Thirdly, evidence was also obtained for a post-transcriptional control step for DQ and DP mRNA, although the interesting possibility cannot be excluded that DQ mRNA is not detectable, and that, instead, the probe cross-reacts with the product of a class II gene for which we do not have a specific antibody. Taken together, the differential expression we observed, involving the lack of DQ and the presence of DR and DP, particularly in the "early" cell types, suggests that the ontogenic sequence of gene activation may be DR, DP, and then DQ. The combined use of MAbs and class II specific probes represents a very sensitive method for studying gene expression at both mRNA and protein levels, allowing us to detect in some instances divergent expression of mRNA and surface antigens.

Acknowledgments. Dr. T. Improta is a recipient of a fellowship from Associazione Italiana per la Ricerca sul Cancro. This work was supported by United States Public Health Service Grants AM35404 and AI19411.

References

1. Winchester RJ, Ross GD, Jarowski CI, Wang CY, Halper J, Broxmeyer HE. Proc Nat Acad Sci USA 1977;74:9:4012–4016.
2. Torok-Storb B, Symington FW. In HLA Class II Antigens. A Comprehensive Review of Structure and Function. Berlin, Springer-Verlag, 1986.
3. Sparrow RL, Williams N. The pattern of HLA-DR and HLA-DQ antigen expression on clonable subpopulations of human myeloid progenitor cells. Blood 1986;2:379.
4. Shookster L, Matsuyama T, Burmester G, Winchester RJ. Monoclonal antibody 1a3 recognizes a monomorphic epitope unique to DQ molecules. Hum Immunol 1987;20:59.
5. Patier-Tonneau D, Turmel P, Auffrey C, Charron D. Construction of chain- and locus-specific HLA-class II DNA probes study of HLA-class II transcripts in leukemias. J Immunol 1986;137:2050.

Author Affiliation

Hospital for Joint Diseases of the New York University School of Medicine, New York, NY, USA

SIB-Pair Analyses in the Assessment of MHC Involvement in Germ Cell Tumor Pathogenesis

R.T.D. Oliver, M.S. Pollack, E.R. Heise, B. Dupont, D. Forman, S.G.E. Marsh, J. Bodmer, and W. Bodmer

It is paradoxical that despite the observation of an association of mouse MHC antigens with murine leukemia in the early 1960s (1), evidence for the involvement of HLA in human malignancy or infectious disease resistance/susceptibility is still very tenuous compared with the well-established associations seen between HLA antigens and autoimmune diseases (2).

Hodgkin's disease, the first malignancy thought to show an HLA association (3), illustrates the problem particularly well. Despite Amiel's original observation in 1967 (3), the 1972 HLA Workshop failed to demonstrate an unequivocal association despite typing more than 500 unrelated individuals with the disease (4). In contrast, Duasset et al. (5) only needed to review typing on 17 pairs of siblings affected by the disease and were able to demonstrate a significant distortion of the expected Mendelian distribution, with 10 of 17 (59%) of the siblings being HLA identical with the index case compared with 25% expected.

Today, there is increasing evidence that a similar association may be present in patients with testicular germ cell tumors. This paper reviews the evidence supporting this contention, with the aim to encourage the establishment of a registry to record HLA typing plus clinical and histopathologic information on the rare instances of affected sib pairs (including identical twins) to provide substantive evidence for the hypothesis.

Evidence for Genetic Experiment in Testicular Germ Cell Etiology

In contrast to the large amount of epidemiologic data collected on familial incidence of Hodgkin's disease showing a seven-fold increased risk among siblings of affected individuals (6), there are relatively few comparable data for patients with testis tumor. Recognizing this, there is now increased effort to rectify this defect, and several authors have already begun to notice a sig-

Table 1. Familial Testicular Tumors

	No. of Cases Reviewed	No. of Familial Cases Identified
Retrospective series*	3023	7 (0.2%)
Prospective series†	872	12 (1.3%)
Control population 10 year risk	10,000	7 (0.07%)

*Pugh 1976 (14), Oliver et al. 1986 (10); †Oliver, Forman, and Dielman unpublished series.

nificant excess in prospectively studied cases compared with that recorded in large retrospective series or expected on the basis of frequency in unrelated individuals (Table 1). The data suggest that there is a three-fold increased risk in the retrospective series, but a 19-fold increased risk in the prospective series.

Possible evidence that the genetic susceptibility does not just influence the occurrence, but may in fact affect which pathologic subtype of tumor occurs, comes from comparing the type of tumor seen in twin sib pairs with that in other familial pairs. Identical twins are significantly more likely to have the same histology than are other sibling case pairs (80% vs. 33%; $p<0.05$).

HLA Studies in Testicular Germ Cell Tumors

Possible data for HLA involvement come from four studies, which have investigated DR antigen frequency

Table 2. Testicular Germ Cell Tumors DR7/DW7* Frequency

	Patient No./%	Controls No./%
Oliver et al. 1985 (10)	101 34%	390 26%
Pollack et al. 1983 (8)	103 32%	176 28%
DeWolf et al. 1979* (7)	53 28%	150 9%
Aiginger et al. 1982 (9)	69 29%	160 28%
Total	328 32%†	876 24%†

†$p < 0.05$ RR.

Table 3. HLA Genotype of Family Members with Germ Cell Tumors

Family No.	Relationship	Genotype
1	Patient	A1,B8/AX,BX
	Uncle	A1,B8/A2,B7
2	Patient	A3,BW44/AW30,B70
	Nephew	A3,BW44/AW26,BW57
3	Patient	A32,B27/A2,B13
	Brother	A32,B27/A29,B7
	Brother	A32,B27/A2,B13
4	Patient	A2,B12/A28,B13
	Non-identical	A2,B12/A28,B13
5	Patient	A3,B7/A2,B17
	Brother	A3,B7/A28,B14
6	Patient	A3,BW35/A24,B7
	Brother	A3,BW35/A28,B44
7	Patient	A3,B7/A28,B44
	Brother	A3,B7/A2,B40

in unrelated patients (7–10). The marginal increase in DR/DW7 antigen frequency recorded (Table 2) is similar to the increase in incidence of A-1 seen in patients with Hodgkin's disease. However, were it not for the familial incidence of testicular tumor, these results would be of no real interest. Despite the small number of families tested (Table 3), there is already a suggestive association equal to that seen in the limited Hodgkin's disease families type (Table 4).

Discussion

These observations, even if confirmed by more substantial studies, would be of no particular interest unless they produced either improved insight into the actual cause of these tumors or offered some practical therapeutic advantage. Increasing evidence that mumps virus induce testicular atrophy may be an extremely important etiologic factor (11), and reports of increased incidence of

Table 4. Familial HLA Studies in Malignancy

	No. of Pairs of Family Cases	Two Haplotypes Identical	One Haplotype in Common	Two Haplotypes Different
Hors et al. 1980 (16) (Hodgkins sib pairs)	7	6	1	0
Hors et al. 1983 (17) Hodgkins sib pairs)	14	6	6	2
Hors et al. 1984 (18) (Hodgkins sib pairs)	13	7	3	3
Total	34	19	10	5
This series (testis tumors sib pairs	6	2	4	0
This series (testis tumors 2nd-degree relatives)	2	0	2	0

Table 5. Testicular Germ Cell Tumours DR3/DR5 Frequency

	Seminoma		Control	
	N		N	
Oliver et al. (1985) DR5	31	36%*	390	19
Pollack et al. (1983) DR5	15	54%**	176	25%
Oliver et al. (1985) DR3	31	13%	390	23%
Pollack et al. (1983) DR3	15	0%†	176	22%

* v control $p < 0.04$ RR = 0.11
** v control $p < 0.05$ RR = 3.38
† v control $p < 0.05$ RR = 2.38

herpes type 2 in these patients (12) open up new avenues to explore.

Possible avenues might include population studies of influence of HLA phenotype on outcome of infection with these viruses, as has been done for herpes type 1 (13), and looking for evidence of cellular immunity to these viruses in patients with testicular germ cell tumors and correlating this with HLA phenotype. However, before doing so, more extensive family studies will be required to prove the HLA association suggested by the preliminary data summarized in Table 3.

In particular, efforts should be made to focus on other genetic polymorphism within the HLA region because gene linkage is presumably the explanation for the weak association seen in unrelated cases and the stronger association seen in familial cases.

At the same time, attempts should be made to try to firm up the observation in identical twins, which suggests that the genes can influence the actual pathologic subtype as well as the actual occurrence of germ cell tumors. That HLA-associated genes may be involved is suggested by the observation that two (7,9) of the four studies of unrelated patients reported an excess of DR5 and deficiency of DR3 in patients with seminoma (Table 5) and two (6,9) also observed an excess of DR/DW7 in patients with terato-carcinoma. Of passing interest in this respect is family 3 (Table 3), in which the two HLA-identical siblings both had seminoma, while the third haplo-matched individual had a terato-carcinoma.

Conclusion

Sib pair analyses provide a powerful tool for preliminary screening of diseases where results from HLA typing of unrelated individuals show weak associations of uncertain significance. The relative rarity of testicular tumors makes international collaboration to study affected sib pairs (including identical twins) and informative second-degree relatives highly desirable. This should include freezing of samples for later DNA polymorphism analysis and central registration of information, particularly in view of the intriguing suggestion that HLA genes may influence which pathologic subtype of tumor that develops.

References

1. Lilly F, Boyse EA, Oldham J. Genetic basis of susceptibility to viral leukaemogenesis. Lancet 1964;ii:1207.
2. Editorial. New views on HLA and disease. Lancet 1985; i:559–561.
3. Amiel JL, Curtoni ES, Mateus PL, Tosi RM (eds). Study of leucocyte phenotypes in Hodgkins disease. In: Histocompatibility Testing 1967. Copenhagen, Munksgaard, 1967; 75–81.
4. Morris PJ, Lawler S, Oliver RTD. HLA and Hodgkin's disease. Joint report of combined analysis. In: Dausset J, Colombani J, Histocompatibility Testing 1972. Copenhagen: Munksgaard, 1972:669–677.
5. Dausset J, Colombani, Hors J. Major histocompatibility complex and cancer, with special reference to human familial tumours (Hodgkin's disease and other malignancies). Cancer Surveys 1982;1:119–147.
6. Grufferman S, Cole P, Smith PG, Lukes RJ. Hodgkin's disease in siblings. N Engl J Med 1977;296:248–253.
7. DeWolf WC, Lange PH, Einaison ME, Yunus EJ. HLA and testicular cancer. Nature 1979;277:216–217.
8. Pollak MS, Vugrin D, Hennessey W, Herr HW, et al. HLA antigens in patients with germ cell cancers of the testis. Cancer Res 1982;42:2470–2473.
9. Aiginger P, Schwartz HP, Kuzmits R, et al. HLA-A, B, C and Dr-antigens and testicular cancer: a prospective study. Proc of Am Assoc Cancer Res 1983;24:190.
10. Oliver RTD, Stephenson CA, Parkinson MC, Forman D, et al. Germ cell tumours of the testicle a model for MHC influence on human malignancy. J Immunogenetics 1986; 13:85–92.
11. Beard CM, Benson RC Jr, Kelalis PP, Elveback LR, et al. The incidence and outcome of mumps orchitis in Rochester. Minnesota 1935–1974. Mayo Clinic Proc 1977;52(1): 3–7.
12. Oliver RTD, Blandy JP, Hope-Stone HF. Current controversies in the management of testicular germcell tumours. In: Pavone Macaluso M, et al. (eds). Testicular Tumours. New York: Plenum, 1985:403–409.
13. Blackwelder WC, Dolin R, Mittal KK, McNamara PM, et al. A population study of herpesvirus infections and HLA antigens. Am J Epid 1982;115:569–576.
14. Pugh RCB. Testicular tumours. In: Pugh RCB (ed). Pathology of the Testis. Blackwell Scientific Publications, Oxford: 1976:139–159.
15. Weissbach L, Widman T. Familial tumour of the testis. Eur J Urol 1986;12:104–108.
16. Hors J, Steinberg G, Andrieu JM, Jacquillat C, et al. HLA genotypes in familial Hodgkin's disease. Eur J Cancer 1980;16:809–815.
17. Hors J, Dausset J. HLA and susceptibility to Hodgkin's disease. Immunol Rev 1983;70:167–192.
18. Hors J, Bonaiti-Pellie C, D'Agay MF, Rappaport H, et al. Hodgkin's disease. In: Albert ED, Baur MP, Mayr WR (eds). Histocompatibility Testing 1984. Berlin: Springer-Verlag, 1984;411–414.

Author Affiliations

R.T.D. Oliver, The London Hospital Medical College; *M.S. Pollack*, Baylor College of Medicine; *E.R. Heese*, Bowman Gray School NC; *B. Dupont*, Memorial Sloan-Kettering; *D. Forman*, ICRF Oxford; *S.G.E. Marsh*, *J. Bodmer*, and *W. Bodmer*, ICRF London

Section XVII
Transplantation Immunology

Induction of Specific Immunosuppression in the Mouse Using Cells of Recipient Origin Transfected with Genes from the Organ Donor

Kathryn J. Wood, Joren C. Madsen, Riccardo A. Superina, and Peter J. Morris

Abstract: Preoperative transfusion with blood from randomly selected donors has been shown by many centers to have a beneficial effect on renal allograft survival in clinical transplantation (1). This observation raises the question of the immunologic specificity of this phenomenon. In this study we have investigated the blood transfusion effect in the mouse by using a novel approach to determine if the sharing of a single major histocompatibility complex (MHC) class I antigen by the blood donor and the organ donor is sufficient to induce specific unresponsiveness to a fully allogeneic organ graft. MHC class I genes, H-2K or H-2D, isolated from the genome of the organ donor C57BL/10 (H-2^b) were expressed in cells of recipient origin, C3H.He (H-2^k) by DNA-mediated gene transfer. These recipient cells expressing donor (as well as recipient) class I gene products were used to treat C3H.He recipients at various times before transplantation of a C57BL/10 heart. We show that class I antigens in the absence of class II or minor histocompatibility antigens can induce specific unresponsiveness to fully allogeneic vascularized organ grafts.

Donor-specific transfusions have been demonstrated to have a marked beneficial effect on allograft survival in clinical and experimental transplantation (2,3). In addition, transfusions from randomly selected blood donors, unrelated to the organ donor, improve graft survival in clinical renal transplantation (1). The immunologic basis for these observations remains unclear. Previous attempts to explain these observations by determining the role of donor class I, class II, and/or minor histocompatibility antigens in the induction of the blood transfusion effect have produced contradictory findings (4–7). For this study, H-2D and H-2K class I heavy chain genes cloned from the genome of the C57BL/10 (H-2^b) mouse were transfected into L cells, a fibroblast cell line derived from C3H.He (H-2^k) mice. Transfected cells expressing high levels of the H-$2K^b$ or H-$2D^b$ antigen were sorted and cloned using an Ortho cytofluorograf system 50L after labeling with a polymorphic monoclonal antibody specific for H-$2D^b$ (27.11.13S) (8) or H-$2K^b$ (B8-24-3) (9) molecules, respectively. The number of H-$2K^b$ or H-$2D^b$ molecules expressed by each clone was determined by direct saturation binding of the appropriate ^{125}I-labelled polymorphic monoclonal antibody. The L-D^b clone R1, expressed $1.0 \pm 0.1 \times 10^{-5}$ (mean $\pm$ SD) D^b molecules/cell and the L-K^b clone G, expressed $2.2 \pm 0.3 \times 10^{-5}$ K^b molecules/cell. C3H.He (H-2^k) recipients were treated intravenously (IV) with either L-D^b or L-K^b transfectants at various times before transplantation of a fully vascularized C57BL/10 (H-2^b) heart graft (10). This approach allowed us to evaluate the ability of isolated donor class I MHC locus products to induce specific unresponsiveness, without contamination by donor class II MHC or minor histocompatibility antigens.

C57BL/10 hearts transplanted into untreated C3H.He mice were rejected within 10 days (mean survival time – MST = 9.2 days, range 8–10 days). When C3H.He recipients were transfused with C57BL/10 blood (0.25ml) IV, 14 days before transplantation, the survival of C57BL/10 heart grafts was significantly prolonged (MST=34.2 days, range 13–72 days). This transfusion effect was donor specific, only the survival of grafts sharing H-2 antigens with the blood donor being prolonged. For example, Balb/c (H-2^d) hearts transplanted into C3H.He recipients transfused with C57BL/10 blood were rejected in the same time as Balb/c hearts transplanted into untreated control animals (MST=9.7 days, range 9–11 days).

To evaluate the contribution of H-2D locus class I antigens in the induction of the transfusion effect in this strain combination, C3H.He recipients were treated IV with 10^7 L-D^b R-1 cells, 14 days before transplantation of a fully allogeneic C57BL/10 heart. This resulted in modest, but significant prolongation of the C57BL/10 heart grafts (MST=14.3 days, range 9–18 days) compared to untreated control animals or animals treated with 10^7 parental, untransfected L cells using the same protocol (Fig. 1A). Dose-response studies showed that this modest prolongation of survival was only obtained when recipients were pretreated with 10^7 L-D^b cells (Fig. 1B). This effect was donor (H-2^b) specific, for Balb/c hearts transplanted into C3H.He recipients treated with 10^7 L-D^b cells were rejected within 11 days (MST=9.6 days, range 9–11 days). The optimal time interval between treatment with L-D^b cells and transplantation was found to be 14 days (data not shown). These results show that pretreatment with a single class I antigen shared with the organ donor is sufficient to induce prolonged survival of a fully allogeneic heart.

In the second part of this study, C3H.He recipients were treated with L cells transfected with the H-$2K^b$ gene to investigate the effect of sharing isolated H-2K locus class I antigens with the organ donor on the induction of immunologic unresponsiveness. In the first experiments 10^7 L-K^b cells, clone G, were used to treat C3H.He recipients IV, 14 days before transplantation. This resulted in prolongation of the survival of C57BL/10 hearts to 14.4 days. The effect of pretreating recipients with 5×10^6 L-K^b cells was also investigated, as clone G expressed approximately twice as many H-$2K^b$ molecules/cell as H-$2D^b$ molecules expressed by the L-D^b clone R1 (2.2×10^5 vs 1.0×10^5). This improved the survival of C57BL/10 hearts transplanted

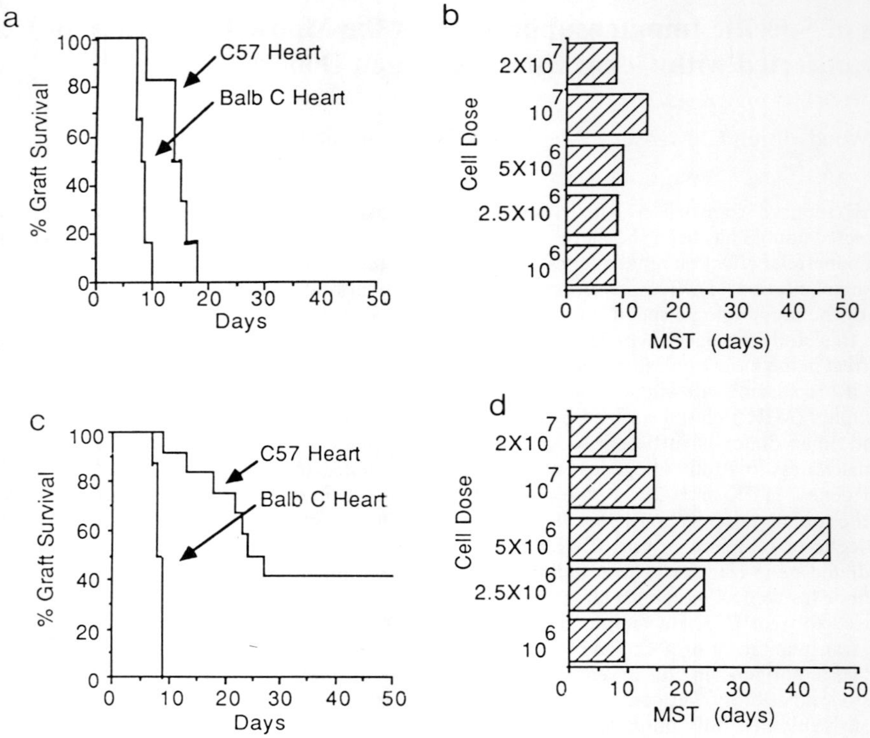

Figure 1. Survival curves of C57BL/10 (H-2^b) cardiac allografts transplanted into C3H/He (H-2^k) mice pretreated with L cell (H2^k) transfectants expressing H-$2K^b$ or H-$2D^b$ gene products. Parts A and C plot the survival time of C57BL/10 hearts or third-party Balb/c (H-2^d) hearts transplanted into C3H.He recipients treated with an optimal IV dose of L-D^b (A) or L-K^b transfectants (C), 10^7 and 5×10^6 cells respectively, 14 days before transplantation. Treatment with 10^7 L-D^b cells resulted in MST of 14.3 days, range 9 – 18 days (n=6), $P<0.05$, and with 5×10^6 L-K^b cells resulted in a MST of 45.2 days, range 9 to 100 days (n=7), $p<0.05$ compared to the survival of third party control hearts. Plots B and D show the dose response analysis for L-D^b and L-K^b transfectants, respectively.

into these recipients to 45.2 days (Fig. 1C), and three of the 12 grafts transplanted in this group survived indefinitely (>100 days). It should be noted that this treatment, 5×10^6 L-K^b cells, IV, 14 days before transplantation, where only a single donor class I antigen is shared with organ donor, proved more effective in inducing prolonged survival of a C57BL/10 heart graft than preoperative donor-specific blood transfusion in this strain combination. A full dose response analysis showed that the survival of C57BL/10 hearts was also significantly prolonged when C3H.He recipients were treated with 2.5×10^6 L-K^b cells, although doses of $<10^6$ or $>2 \times 10^7$ were ineffective (Fig. 1D). The state of unresponsiveness induced by preoperative treatment with recipient cells expressing donor class I K locus molecules was again specific, for C3H.He recipients treated with 5×10^6 L-K^b cells rejected a Balb/c heart graft (MST 9.3 days, range 9–11). Just as for the induction of unresponsiveness using recipient cells transfected with D locus class I genes, the optimal time interval between pretreatment with L-K^b cells and transplantation was found to be 14 days (data not shown).

These data clearly demonstrate that pretreatment with a single donor-specific class I molecule, in this case either H-2K or D locus products, can induce specific unresponsiveness and prolonged survival of a fully allogeneic cardiac allograft. The induction of the unresponsive state was dependent upon: (i) The antigen load—for both K and D locus products, maximal prolongation of graft survival was induced when recipients were treated with approximately 10^{12} molecules of donor class I antigen 14 days before transplantation. In the case of the L-D^b transfectants, (clone R1), pretreatment with 10^7 cells (10^5 molecules/cell) was required and for the L-K^b transfectants 5×10^6 cells induced optimal prolongation of graft survival (2×10^5 molecules/cell). (ii) The intrinsic immunogenicity of the antigen—at the same antigen load, H-2K molecules were more effective than H-2D class I antigens in this strain combination. Other studies have shown K locus antigens to be more immunogenic than D locus products (e.g., 11).

By using this model, where class I genes isolated from the genome of the organ donor are transfected into cells of recipient origin, we have also demonstrated that donor class I antigens can induce specific unresponsiveness to a fully allogeneic graft in the absence of donor class II MHC or minor histocompatibility antigens. This is in contrast to a previous report suggesting that donor class I antigens can only induce prolonged graft survival when they are expressed in the presence of donor minor histocompatibility antigens (5).

The aim of these studies was to determine if the sharing of single locus products by the blood donor and the organ donor was sufficient to induce prolonged survival of a graft mismatched for all other histocompatibility antigens, both major and minor antigens. We felt that this might provide an explanation for the observation that preoperative transfusions from randomly selected blood donors unrelated to the organ donor, has a beneficial effect on graft survival in clinical transplantation (1). As we can demonstrate that pretreatment with cells expressing a single class I antigen of donor origin is sufficient to prolong the survival of a fully allogeneic cardiac allografts, this strongly suggests that the beneficial effect of preoperative transfusion in clinical transplantation may arise from fortuitous sharing of single MHC antigens by the blood donors and the subsequent organ donor. Unfortunately the donors who provide blood for transfusion of potential transplant recipients are not routinely HLA typed. Thus at present, it is not possible to confirm this prediction by examining the clinical data. However, these findings suggest that if the HLA phenotype of the blood donors was known, in some instances it might be possible to select the recipient of a donated organ such that the maximum number of HLA antigens were shared by the blood donors and the organ donor in order to increase the possibility of inducing a specific unresponsive state in the recipient.

Acknowledgments. We would like to thank Andy Mellor for providing H-2K^b and H-2D^b genes.

This work was supported by grants from the Medical Research Council, UK and the British Heart Foundation.

References

1. Opelz G. Current relevance of the transfusion effect in renal transplantation. Transplant Proc 1985;17:1015.
2. Salvatierra O, Vinceti F, Amend W, Potter D, Iwaki Y, Opelz G, et al. Deliberate donor specific blood transfusions prior to living related renal transplantation. Ann Surg 1980;192:543.
3. Fabre JW, Morris PJ. The effect of donor strain blood pretreatment on renal allograft rejection in rats. Transplantation 1972;14:608.
4. Wood KJ, Evins J, Morris PJ. Suppression of renal allograft rejection in the rat by class I antigens on purified erythrocytes. Transplantation 1985;39:56.
5. Majoor GD, van Breda Vriesman PJC. Requirement for both MHC and non-MHC antigens residing on the same erythrocyte for donor erythrocyte-mediated prolongation of rat renal allograft survival. Transplantation 1986;41:92.
6. Soulillou JP, Blandin F, Gunther E, Lemoine V. Genetics of the blood transfusion effect on heart allografts in rats. Transplantation 1984;38:63.
7. Wood ML, Monaco AP. Induction of unresponsiveness to skin allografts in adult mice disparate at defined regions of the H-2 complex. II. Effect of pre-graft donor-specific blood transfusions in ALS-treated mice. Transplantation 1984;37:39.
8. Ozato K, Sachs DH. Monoclonal antibodies to mouse MHC antigens. III. Hybridoma antibodies reacting to antigens of the H-2b haplotype reveal genetic control of isotype expression. J Immunol 1981;126:317.
9. Kohler G, Fischer Lindahl K, Heusser C. Characterisation of a monoclonal anti-H-2Kb antibody. Immune Syst 1981; 2:202.
10. Superina RA, Peugh WN, Wood KJ, Morris PJ. Assessment of primarily vascularised cardiac allografts in mice. Transplantation 1986;42:227.
11. Corry RJ, Winn HJ, Russell PS. Primarily vascularized allografts of hearts in mice. The role of H-2D,H-2K and non-H-2 antigens in rejection. Transplantation 1973;16: 343.

Author Affiliations

Kathryn J. Wood, Joren C. Madsen, Peter J. Morris, Nuffield Department of Surgery, University of Oxford, John Rad-Cliffe Hospital, Oxford, England; *Riccardo A. Superina*, Department of Surgery, Hospital for Sick Children, Toronto, Canada

Donor T Cells Are Responsible for the Blood Transfusion Effect Across an H-2 Class I Disparity

Leo P. de Waal, Esther van Twuyver, Marieke Verveld, W. Martin Kast, and Cees J.M. Melief

Abstract: Donor-specific blood transfusion has a favorable effect on the survival of renal transplants. However, the mechanism underlying this effect is not known. We have adopted a mouse model to study the influence of donor-specific spleen cell transfusion on T-cell reactivity measured in vivo by skin graft rejection and DTH-reaction, and in vitro by the determination of cytotoxic T-lymphocyte precursor (CTLp) frequencies. The results indicated that the induction of donor-specific suppression in C57B1/6 (B6) recipient mice by spleen cell transfusion depends on the H2 mismatches between transfusion donor and recipient (manuscript in preparation). Donor-specific spleen cell transfusion across a class I H2 mutant difference (bm1, bm11) resulted in a significant prolongation of skin allograft survival and in a severe depression of the alloantigen-specific CTLp frequency. The same result was observed across a class II H2 mutant difference (bm12). However, donor-recipient mismatching for class I and class II H2 antigens completely abolished the immunosuppressive effects. Furthermore, we observed that prolongation of skin allograft survival always correlated with a marked decrease in CTLp frequency whereas no correlation was found with DTH suppression.

In the present study we have analyzed the requirements of the donor spleen cell inoculum necessary to induce the transfusion effect across the bm1 class I mutant difference. B6 recipient mice received three intravenous (IV) injections with 5.10^7 donor spleen cells with one-week intervals. A week after the last IV injection, the B6 recipient mice were hemisplenectomized and after one further week, skin grafting was performed. The results confirmed our previous findings that spleen cell transfusion across a class I H2 difference induces skin allograft prolongation and a concomitant decrease in CTLp frequency (Table 1, line 1–4) (1,2). Irradation of the donor spleen cell inoculum with 2000 rad completely abolished these effects (Table 1, line 5). Separation of the donor spleen cell inoculum in a T-cell enriched (nylon wool purification) and a T-cell depleted (anti-Thyl + C) fraction revealed that the induction of the transfusion effect resided in the T-cell fraction (Table 1, line 6–7). These results are in agreement with the data of Rammensee et al (3,4) who showed that the IV injection of spleen cells into H2 class I incompatible recipients leads to a functional clonal deletion of donor reactive CTL in the recipient. These data are explained by assuming that CTLp of recipient origin can be inactivated when they recognize antigens on the membrane of donor-derived veto cells (3,4).

In agreement with the veto cell concept, the IV injection of (B6*bml)F1 spleen cells into B6 recipients induces prolongation of skin allograft survival and a concomitant decrease in CTLp frequency for donor alloantigens (data not shown). Thus, we show for the first time that veto cell suppression can induce skin allograft enhancement.

References

1. Van Twuyver E, Kast WM, Mooijaart RJD, Melief CJM, de Waal LP. Transfusion induced skin allograft enhancement across an H-2 class-I mismatch is caused by a clonal deletion of donor specific cytotoxic T-lymphocyte precursors within the allograft. Transplant Proc (in press).
2. Kast WM, Van Twuyver E, Mooijaart RJD, Verveld M, Kampluis ASA, Melief CJM, de Waal LP. Mechanism of skin allograft enhancement across an H-2 class I mutant difference. Eur J Immunol (in press).
3. Rammensee HG, Fink PJ, Bevan MJ. Functional clonal deletion of class I-specific cytotoxic T lymphocytes by veto cells that express antigen. J Immunol 1984;133:2390–2396.
4. Rammensee HG, Bevan MJ. Mutual tolerization of histoincompatible lymphocytes. Eur J Immunol 1987;17:893–895.

Author Affiliations

Central Laboratory of the Netherlands Red Cross Blood Transfusion Service, and Laboratory for Experimental and Clinical Immunology, University of Amsterdam, Amsterdam, The Netherlands

Table 1. Effect of Donor-Specific Spleen Cell Transfusion Across a Class I Mutant Difference on Skin Allograft Survival and CTLp Frequency

Donor*	Recipient*	Graft†	CTLp Frequency†	Survival (days)
1 –	B6	bm11	1:2000	17,17,24,24,25,26,32,33
2 –	B6	bm1	1:2400	19,21,25,27,28,28,32,37,39
3 bm11	B6	bm11	1:42.200	35,37, 8x > 100
4 bml	B6	bm1	1:29.900	24,44,44,46,48, 5x > 130
5 ⚡ bml	B6	bm1	1:2700	17,18,21,21,27,21,35
6 bml T cells	B6	bm1	1:22.800	27,31,31,33,37, 3x > 62
7 bml non-T cells	B6	bm1	1:4900	17,19,19,20,21,23,23,26,26,32

*Recipient mice received three IV injections of $5 \cdot 10^7$ donor spleen cells with 1-week interval.
†One week after the last IV injection, recipient mice were hemisplenectomized and after 1 further week, skin grafting was performed.

Immunobiology of HLA
Volume II

T Cells Marked by the 2H4 Antigen Function in Allosuppression

Eleanor L. Ramos, Laurence A. Turka, John E. Leggat, Edgar L. Milford, and Charles B. Carpenter

A mouse monoclonal antibody (MAb), anti-2H4 (IgG1 subclass), has recently been described (1) that recognizes 200/220 kD glycoproteins (2) of the leukocyte common antigen/T200 family. The 2H4 antigen is found on 42% of unfractionated human T cells, 41% of CD4+ lymphocytes, 54% of CD8+ lymphocytes, and over 30% of both peripheral blood B cells and null cells. In a pokeweed mitogen system that measures B cell immunoglobulin production, the CD4+2H4+ cells were found to be inducers of suppression (1) and CD4+2H4− cells to be inducers of help (3). In the autologous mixed lymphocyte response (AMLR), CD8+2H4+ cells were found to have suppressor effector function (4). Additionally, in a concanavalin A− activated system, suppressor cell activity belonged to the 2H4+ subset of T cells (2). As the role of these subsets have not been established in alloimmunity, we studied the proliferative response and generation of suppressor cells in an allogeneic MLR using 2H4 enriched or depleted cells as the responding population.

Peripheral blood mononuclear cells from normal human volunteers were T cell enriched on nylon wool columns. The T cells were then further fractionated using an immunoaffinity column. The nonadherent fraction (2H4− cells) had 20% contamination, but these were weakly staining cells. The adherent fraction (2H4+ cells) contained 77% brightly staining cells. Proliferation (^{3}HTdR incorporation) was measured on day 6. Generation of suppressor cells was measured by addition of lightly irradiated day 10 responders (modulators or putative suppressor cells) to a fresh test MLR. Percent suppression was defined as [1-(experimental cpm-spontaneous cpm)/(control cpm-spontaneous cpm)] × 100. Experimental and control cpms were from modulated and control cultures, respectively. Spontaneous cpms were from MLRs with autologous irradiated responders and no modulators.

Because 2H4+ cells were positively selected, we confirmed that anti-2H4 MAb coated cells responded in the same fashion as noncoated cells in an MLR or suppressor generating cultures (data not shown). As seen in Table 1, in each of the five experiments, the 2H4+ cells proliferated less than 2H4− cells; the CPM ratio of 2H4+/2H4− varied between 0.63 and 0.84. When cells generated from bulk cultures consisting of 2H4+ or 2H4− responder cells were used as modulators in a fresh test MLR, modulator-generating cultures utilizing 2H4+ responders produced suppressor cells that were twice as potent as those produced by their 2H4− counterparts. This effect was observed in all three experiments performed.

These results are consistent with the hypothesis that the 2H4 antigen marks T cells active in the suppressor limb of the alloimmune response.

References

1. Morimoto C, Letvin NL, Distaso JA, Aldrich WR, Schlossman SF. The isolation and characterization of the human suppressor inducer T cell subset. J Immunol 1986;134:1508.
2. Morimoto C, Letvin NL, Rudd CE, Hagan M, Takeuchi T, Schlossman SF. The role of the 2H4 molecule in the generation of suppressor function in Con A-activated T cells. J Immunol 1986;137:3247.
3. Morimoto C, Letvin NL, Boyd AW, Hagan M, Brown HM, Kornacki MM, et al. The isolation and characterization of the human helper inducer T cell subset. J Immunol 1985;134:3762.
4. Takeuchi T, Rudd CE, Schlossman SF, Morimoto C. Functional characterization of the 2H4 molecule in T8+ cells in the AMLR system (submitted).

Author Affiliation

Laboratory of Immunogenetics and Transplantation, Brigham and Women's Hospital, Boston, MA 02115, USA

Table 1. Functional In Vitro Studies on 2H4 Enriched (2H4+) or Depleted (2H4−) Normal Peripheral Blood*

	MLR (CPM)				% Suppression†		
Experiment	2H4+	2H4−	2H4+/2H4−	Experiment	2H4+	2H4−	2H4+/2H4−
1	154850	245193	0.63	6	71.4	38.3	1.9
2	77473	116935	0.66	7	38.6	18.1	2.1
3	29881	35576	0.84	8	39.4	16.9	2.3
4	60242	87345	0.69				
5	92857	130733	0.71				

*Responder cells enriched or depleted for 2H4+ cells.
†Products of MLRs generated with 2H4-enriched or depleted T cells were tested for ability to suppress proliferation in an MLR using whole mononuclear cells from the same responder and stimulator.

Human Non-MHC-Restricted Immunosuppression by Alloantigen and IL2-Activated Cells*

Donna D. Kostyu, Manar Ghanayem, and Cindy Brissette-Storkus

We have detected and characterized a small population of human non-MHC restricted suppressor cells which arise following alloantigen and IL2 stimulation and which are potent inhibitors of T-cell activation.

Activated cell cultures containing suppressor cells were produced by mixed lymphocyte culture and expanded in 10% partially purified IL2. The ability of these activated cell cultures to inhibit phytohemagglutinin (PHA) responses was measured by coculturing them with mononuclear cells and PHA. Controls monitored effects of cell density and alloresponsiveness in the absence of PHA. All activated cells were irradiated (3,000 rads) prior to use.

A variety of alloantigen-activated lymphocyte cultures were tested for their ability to suppress normal PHA responses (see Table 1). Suppression appeared related to early exposure to IL2. Of numerous cultures primed to alloantigen alone (no IL2) for 11 to 18 days (e.g., Max'LS$_x$d11), none were suppressive. However, several cultures primed to alloantigen and expanded with IL2 beginning around d4-d7 of culture were potent inhibitors of PHA responses. Suppression was always nonspecific and non-MHC restricted. Donor-specific differences are suggested since some activated cells (e.g., MC'LuS$_x$) were highly suppressive, while others such as the CMH'EdB$_x$ were not, even though activated and expanded under similar conditions.

Table 1. Varied Ability of Alloantigen, IL2-Activated Lymphocyte Cultures to Suppress Normal PHA Responses

Cells	PHA	Added Activated Cells$_x$		Cpm (±)	SD
DK	−	DK$_x$		211 ±	105
DK	+	DK$_x$		25,228 ±	2,735
DK	+	LuS'DaHo$_x$d18	(IL2)*	31,381 ±	14,404
DK	+	LuS'EdS$_\Delta$d21	(IL2)	272 ±	141
DK	+	MC'LuS$_x$d14	(IL2)	483 ±	29
JaS	−	JaS$_x$		1,314 ±	287
JaS	+	JaS$_x$		43,914 ±	n.d.
JaS	+	CMH'EdB$_x$d12	(IL2)	105,583 ±	50,774
JaS	+	MC'LuS$_x$d11	(IL2)	540 ±	212
JaS	+	CMB'S$_x$d13	(IL2)	887 ±	394
DK	−	DK$_x$		1,930 ±	1,162
DK	+	DK$_x$		32,923 ±	4,733
DK	+	JBG'JAK$_x$d13		29,022 ±	9,172
DK	+	MAX'LS$_x$d11		22,842 ±	3,561
DK	+	ADS'MLA$_x$d11		36,070 ±	2,830
DK	+	MC'LuS$_x$d11	(IL2)	524 ±	236
JAK	−	JAK$_x$		703 ±	313
JAK	+	JAK$_x$		47,287 ±	9,504
JAK	+	MC'LuS$_x$d12	(IL2)	506 ±	217
JAK	+	DK'RSt$_x$d12	(IL2)	323 ±	61

*(IL2) denotes cultures exposed to exogenous IL2 early in culture activation.

The MC'LuS cells were characterized in some detail. In titration experiments, the addition of 3,000 activated MC'LuS cells to 50,000 normal peripheral blood lymphocytes (PBL) was still sufficient for >50% suppression. The actual suppressor cell did not appear to be a classic CD3+ cell or a macrophage, as depletion of OKT3$^+$ cells and of LeuM3$^+$ cells did not abrogate suppression. It is possible that these suppressor cells are similar to the non-T, HNK1$^+$ cells identified in human bone marrow and murine spleen (1–3). Suppression was not reversible with IL2, and was not due to cytolytic cells or shifts in kinetics of proliferation. A suppressor factor identified in supernatants of MC'LuS-activated cells was capable of suppressing PHA, Con A, PWM, and MLR responses.

These nontraditional, non-T suppressor cells may be of interest not only because of their broad and potent inhibitory activity, but also because of their induction with alloantigen and IL2.

References

1. Tilden AB, Abo T, Balch CM. Suppressor cell function of human granular lymphocytes identified by the HNK-1 (Leu7) monoclonal antibody. J Immunol 1983;130:1171.
2. Mortari F, Bains MA, Singhal SK. Immunoregulatory activity of human bone marrow. Identification of suppressor cells possessing OKM1, SSEA-1 and HNK-1 antigens. J Immunol 1986;137:1133.
3. Stober S, Palathumpat V, Schwadron R, Hertel-Wulff B. Cloned natural suppressor cells prevent lethal graft-versus-host disease. J Immunol 1987;138:699.

Author Affiliation

Department of Microbiology and Immunology, Duke Medical Center, Durham, NC 27710, USA

*This work was supported in part by NIH grant AI-08897 and NICHHD grant 21244.

Immunobiology of HLA
Volume II

A Human Lymphoblastoid Cell Line Secreting Antiidiotypic Antibodies

D.P. Singal, M.A. Blajchman, F.A. Ofosu, S. Joseph, B. Roberge, E.K.M. Smith, and D. Ludwin

Abstract: It has been postulated that antiidiotypic antibodies play an important role in maintaining tolerance in vivo, thereby enhancing allograft survival in renal transplant recipients (1–3). These antibodies have been shown to be present in recipients following blood transfusions and in recipients of long-term functioning renal allografts (4,5). It has been possible to transform human B lymphocytes with Epstein-Barr Virus (EBV) to obtain stable lymphoblastoid cell lines that will produce monoclonal antibodies of the desired specificity (6). Using this technique of lymphoblastoid transformation with EBV, we have established a B cell line from a recipient of a long-term functioning renal allograft. This B cell line secreted antiidiotypic antibodies, which specifically inhibited (a) the lymphocytotoxic activity of the kidney donor alloantigen-specific anti-HLA reference sera, and (b) proliferative responses in MLC of recipient's lymphocytes and of third-party cells sharing HLA-B,DR antigens with the recipient, against stimulator cells bearing HLA-B/DR antigens present in the kidney donor.

Transplant recipient PS (HLA-A1,30; B14,18; DR1,w8; DRw52; DQw1) received a kidney from a cadaver donor (HLA-A1,10; B8) in March, 1970. The recipient had received 46 pretransplant blood transfusions and is currently receiving 50 mg of azathioprine per day and 15 mg of prednisone on alternate days. An enriched B lymphocyte population obtained from peripheral blood lymphocytes was incubated with EBV-containing supernatant derived from the marmoset cell line B95-8 in flat-bottomed tissue culture plates. Cells from positive cultures were transferred as individual clusters and grown in 3 ml MEM medium in polystyrene tubes. B cell line "A" was selected after picking colonies successively three times by the cluster picking technique. Supernatant from cell line A, henceforth called Supernatant A, had 19.24 μg/ml IgM. Supernatant A was concentrated 25-fold by Amicon and tested for antiidiotypic antibodies.

Supernatant A was tested for inhibition of lymphocytotoxic activity of antisera defining various HLA-A (1,2,3,10,11,24,25,26,29), B (5,7,8,13,14,15,16,17, 18,35,44), and DR (1,2,3,4,5,7) specificities. In these studies, at least two antisera defining each of these HLA specificities were utilized. Inhibition of lymphocytotoxic activity of antisera defining some HLA specificities by the Supernatant A is given in Table 1. Of the HLA-A,B specificities, maximal (100%) inhibition was observed with anti-B8 sera. Significant inhibition (61%) was also observed for sera defining B5. On the other hand, no inhibition was observed for sera directed against the remaining HLA-A,B specificities. Of the HLA-DR specificities, anti-DR3 sera showed maximal (100%) inhibition. In contrast, no inhibition (12 to 38%) was observed for sera defining DR1,2,4,5, and 7.

Initial MLC-inhibition experiments were carried out with responder cells from the recipient PS and two unrelated normal healthy panel members. Supernatant A resulted in significant inhibition (47% and 53%) of responses of PS to both stimulators in MLC (results not shown). In subsequent experiments, cells from different responders and stimulators were utilized. The results from one such experiment utilizing responder cells from a normal healthy panel member SP sharing B14,DR1, DQw1 with recipient PS are shown in Figure 1. In this experiment the stimulator cells were obtained from five unrelated panel members, of which three (CS,GS,ES) are positive for B8,DR3, and the remaining two (SJ,MD) did not carry these antigens. It can be seen that Supernatant A caused significant inhibition (31 to 42%) of proliferative responses of SP to stimulator cells from B8,DR3 donors. In contrast, no inhibition (12 to 16%) was observed with stimulator cells from SJ and MD.

Next, we utilized responder lymphocytes from panel members who did not share any HLA-B,DR antigens with the recipient. In these experiments stimulator cells were obtained from panel members positive or negative for the B,DR(?) antigens present in the kidney donor. Supernatant A did not inhibit responses of these responder cells to stimulator lymphocytes in MLC (Table 2).

Results from immunodepletion experiments are shown in Table 3. It is evident that the lymphocytotoxic-inhibiting activity and the MLC-inhibiting activity in the supernatant was found in the fraction containing IgM. For example, the IgM-depleted fraction caused 36% inhibition of anti-B8 reference serum. In contrast, fraction associated with IgM resulted in 100% inhibition

Table 1. Inhibition of Lymphocytotoxic Activity of Anti-HLA Sera by Supernatant A

	Anti-HLA Sera								
Percent Inhibition	A10	B5	B7	B8	B44	DR2	DR3	DR4	DR5
	0	61	25	100	0	20	100	25	12

The percent inhibition of lymphocytotoxic activity of anti-HLA sera by Supernatant A was calculated by the formula:

$$\frac{\text{\% Killing of cells by anti-HLA serum diluted with medium} - \text{\% Killing of cells by anti-HLA serum diluted with Supernatant A}}{\text{\% Killing of cells by anti-HLA serum diluted with medium}}$$

Each value represents mean percent inhibition of at least two HLA antisera. Anti-HLA sera that were inhibited by 59% or more (99% confidence limit) were considered as being inhibited by Supernatant A.

RESPONDER: SP=HLA-A11, 33; B14 44; Cwl; DRl; DQw1

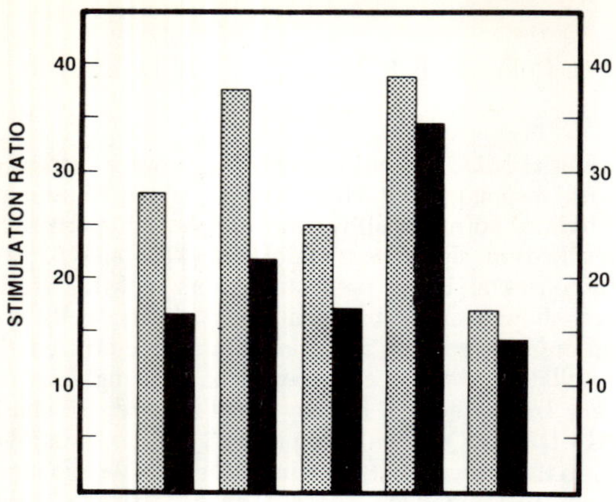

STIMULATORS	CS	GS	ES	SJ	MD
HLA-A	1,3	1,24	1	2,24	2,23
-B	8,44	8	8,35	22,61	18,44
-C	5	-	4	1	-
-DR	3,4	3,4	3,6	2	7
-DRw	52,53	52	52	-	53
-DQw	2,3	2,3	1,2	1	2

Figure 1. Inhibition of proliferative responses in MLC by Supernatant A. MLC-inhibition test was carried out using 100×10^3 responder cells, 100×10^3 x-irradiated stimulator lymphocytes, and 10 μl of medium MEM (▨) or supernatant A(■). The cultures were incubated at 37°C in water saturated 5% CO_2-in-air atmosphere and labeled with 0.5 μCi of ^{3}H-thymidine for 18 hr. Total culture time, including thymidine pulse, was 120 hr. Percent inhibition was calculated by comparing stimulation ratio (SR) values in control (medium) and experimental (Supernatant A) cultures. MLC combinations which showed inhibition of 22% or more (99% confidence limit) were scored as inhibited.

of anti-B8 serum. In MLC-inhibition experiments, fraction associated with IgM caused 72% and 78% inhibition of responses. On the other hand, IgM-depleted fraction did not cause inhibition in MLC.

Supernatant A was absorbed either by anti-HLA (anti-B8, anti-DR3) Ig column or by MLC-activated lymphoblasts. The absorbed supernatant was tested for inhibition of lymphocytotoxic activity of anti-HLA sera and of proliferative responses in MLC. Absorption of supernatant by anti-B8 or anti-DR3 Ig column removed lymphocytotoxic inhibitory activity to both anti-B8 and anti-DR3 sera. However, supernatant absorbed by anti-B8 and anti-DR3 Ig column still caused 82% and 50% inhibition, respectively, of proliferative responses of lymphocytes from SP to stimulator cells from two donors, GS and CS, in MLC. Supernatant absorbed by MLC-activated lymphocytes failed to inhibit proliferative responses in MLC. In contrast, the MLC-activated lymphocyte-absorbed supernatant still caused 100% inhibition of lymphocytotoxic activity of anti-B8 and anti-DR3 sera.

The results in the present study demonstrate that antibody(ies) secreted by an EBV-transformed B cell line from a recipient with a long-term functioning renal allograft could specifically inhibit both the lymphocytotoxic activity of kidney donor-specific anti-HLA reference sera and the proliferative responses of recipient's lymphocytes, and of third-party unrelated healthy panel members, sharing B14,DR1,DQw1 with the recipient, to x-irradiated stimulator cells from normal healthy subjects carrying B8,DR3. Results from immunodepletion experiments demonstrated that the inhibitory activity in Supernatant A was associated with the IgM fraction. Absorption experiments suggested that different antibodies may be responsible for inhibition of lymphocytotoxic activity of anti-HLA sera and of proliferative responses in MLC.

Table 2. Inhibition of Proliferative Responses in MLC by Supernatant A

Responders DS = A2,24;B7,44;Cw3;DR2,w8;DRw52;DQw1					LB = A1;B7,8;Cw7;DR3,5;DRw52;DQw2,3				
Stimulators			Medium	Sup A	Stimulators			Medium	Sup A
I.D.	B8	DR3			I.D.	B8	DR3		
CL	+	–	8.2	10.9	AH	+	+	8.5	8.5
PE	+	+	9.9	11.6	SJ	–	–	18.1	16.6
SE	–	–	10.2	11.6	MD	–	–	23.5	19.9

Results are given as stimulation ratios (SR). Percent inhibition was calculated by comparing SR values in control (medium) and experimental (Supernatant A) cultures.

Table 3. Immunoglobulin Characteristics of the Inhibitory Factor in Supernatant A

	Lymphocytotoxic Activity of Anti-HLA-B8 Serum			Proliferative Responses in MLC			
Titer	Medium	IgM	IgM-depleted	Stimulator	Medium	IgM	IgM-depleted
1:4	100	0	100	GS	20.1	5.7	27.3
1:8	70	0	45	CS	32.9	7.4	35.1
1:16	45	0	35				

Immunoglobulin characteristics of the inhibitory factor in Supernatant A was investigated by passing supernatant through a DEAE AFFIL-GEL blue (BioRad, Richmond, CA) column. Both immunoglobulin depleted and the eluted fraction containing IgM were tested for inhibition of lymphocytotoxic activity of anti-HLA-B8 serum and for proliferative responses in MLC.
For lymphocytotoxic activity of anti-HLA-B8 serum, the results are given as percent cells killed, and for proliferative responses in MLC, the data are given as stimulation ratios (SR).

In summary, we have generated antiidiotypic antibodies in vitro by EBV-transformation of B lymphocytes from a recipient with a long-term functioning renal allograft. This methodology will permit the production of such antibodies in large quantities, which could provide antisera for immunochemical characterization of idiotypic receptors on immunoglobulins and T lymphocytes.

Acknowledgment. This work was supported by a grant from the Canadian Red Cross Society, Blood Transfusion Service.

References

1. Singal DP, Ludwin D, Blajchman MA. Annotation: Blood transfusion and transplantation. Brit J Haemat 1985;61:595.
2. Singal DP, Fagnilli L, Joseph S. Blood transfusions induce antiidiotypic antibodies in renal transplant patients. Transplant Proc 1983;15:1005.
3. Singal DP, Joseph S, Ali M, Blajchman MA, Smith EKM, Ludwin D. Antiidiotypic antibodies and suppressor cells in patients receiving pretransplant donor-specific blood transfusions. Transplant Proc 1987;19:3399.
4. Singal DP, Joseph S. Role of blood transfusions on the induction of antibodies against recognition sites on T lymphocytes in renal transplant patients. Hum Immunol 1982;4:93.
5. Miyajima T, Higuchi R, Kashiwabara H, Yokoyama T, Fujimoto S. Anti-idiotypic antibodies in a patient with functioning graft. Nature 1980;283:306.
6. Steinitz M, Klein G, Koskimies S, Makela O. EB virus-induced B lymphocyte cell lines producing specific antibody. Nature 1977;269:420.

Author Affiliations

D.P. Singal, S. Joseph, B. Roberge, Department of Pathology; *E.K.M. Smith, D. Ludwin*, Department of Medicine, McMaster University; *M.A. Blajchman*, Departments of Pathology and Medicine, McMaster University and Canadian Red Cross Society, BTS Service; *F.A. Ofosu*, Department of Pathology, McMaster University and Canadian Red Cross Society, BTS Service, Hamilton, Ontario, Canada

Inhibitory Effects of FK506 in Combination with Cyclosporine A on Human Lymphocyte Alloproliferative Responses

G. Eiras, A. Zeevi, S. Todo, L. Makowka, T.E. Starzl, and R.J. Duquesnoy

FK506 (FK) is a novel immunosuppressive drug isolated from *Streptomyces tsukubaensis*. Its in vitro effect on lymphocyte proliferation is several hundred-fold greater than that of Cyclosporine A (CsA). Studies were conducted on the inhibitory effect of combinations of FK and CsA on lymphocyte proliferation in vitro. The secondary proliferative response (PLT) of alloreactive T cells was significantly inhibited by a combination of 0.06 ng/ml of FK and 1 ng/ml of CsA. At these low concentrations, these drugs were ineffective. The synergistic effect between FK and CsA was also demonstrated through sequential exposure of alloreactive lymphocytes to FK followed by CsA. Pretreatment of IL2-expanded alloreactive cells with FK yielded cell cultures whose PLT activity was considerably more sensitive to low doses of CsA than control cells or CsA-pretreated cells. These findings provide evidence for a synergism between the immunosuppressive effects of FK and CsA. The combination of low-dose FK plus CsA immunosuppression may benefit organ transplant outcome.

Recent studies have described a new, potent immunosuppressive drug, FK 506, which induces considerable prolongation of allograft survival (1–3). FK is isolated from *Streptomyces tsukubaensis* in Osaka, Japan. Its molecular structure resembles the macrolide family with a molecular weight of 822 daltons (4). The immunosuppressive effect of FK can be demonstrated in primary mixed lymphocyte reaction (MLR), and in secondary proliferation of alloreactive T cells generated from MLR or propagated from organ transplant biopsies (5,6).

Studies on lymphocyte proliferation in primary MLR have previously shown that inhibition was achieved with doses of FK that were several hundred-fold lower than those of CsA (5,6). An example of the difference between FK and CsA is illustrated in Table 1. Significant inhibition (>70%) of the alloreactive response, measured in a primed lymphocyte test (PLT), of a liver biopsy propagated lymphocyte culture was achieved with a low dose of 0.25 ng/ml of FK. On the contrary, a 200-fold higher concentration of CsA was necessary to obtain the same level of inhibition. In vivo studies have shown a similarly potent effect of FK on the prolongation of transplant survival in experimental animals (8–10).

Polypharmaceutical therapy is currently widely used in organ transplantation to avoid side effects associated with high doses of a single agent. Studies have been conducted to determine in vitro synergism between FK and CsA. Our first approach was to add to lymphocyte cultures low doses of these drug combinations. As shown in Figure 1, the combination of 0.06 ng/ml of FK and 10 ng/ml of CsA induced a significant inhibition (40%) of the PLT reactivity of liver biopsy-grown cells. On the other hand neither drug was effective at these low doses. In addition, a 10-fold lower dose of CsA (1 ng/ml) and

Table 1. Dose Effect of FK506 and CsA on the PLT Reactivity of Liver Biopsy-Grown Cells

	% Inhibition of PLT Response*	
Drug (ng/ml)	CsA	FK506
500	87	nd
100	85	nd
50	70	95
10	10	93
1	2	85
0.25	nd	75
0.12	nd	20
0.06	nd	5

Alloreactive lymphocyte cultures were propagated from human liver transplant biopsy in the presence of interleukin-2 (IL2) as previously reported (7). Secondary proliferation was assessed in a 3-day primed lymphocyte test (PLT) whereby 10^4 responder cells were incubated with 10^5 irradiated (2,000 rads) stimulator cells. Lymphocyte proliferation was determined by [^{3}H]thymidine incorporation. The inhibitory effect was measured at different concentrations of drugs and the results are expressed as percent inhibition.

$$\% \text{ inhibition} = \left(1 - \frac{\text{cpm with drug}}{\text{cpm without drug}}\right) \times 100$$

*The PLT activity of liver biopsy (LB) cells against the donor irradiated cells was 15,241 ± 1,421 cpm and the background was 76 ± 15 cpm.

0.06 ng/ml of FK induced a significant PLT inhibition indicating that FK acts synergistically with a relatively wide dose range of CsA (Fig. 1). In contrast, a 2-fold reduction of the FK dose (to 0.03 ng/ml) added to 10 ng/ml of CsA greatly diminished the synergistic inhibitory effect between these drugs (11,12).

The second approach to study the synergistic effect between these two drugs was through sequential exposure of alloreactive lymphocytes to FK followed by CsA. Because FK and CsA have no effect on IL2-induced

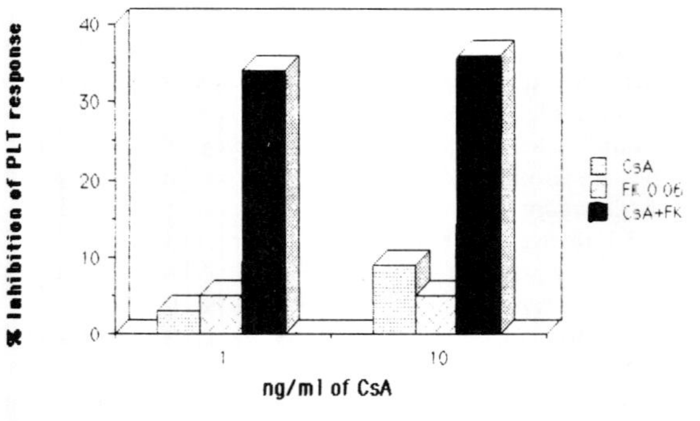

Figure 1. Synergism between the immunosuppressive effects of FK506 and CsA. The donor-specific PLT reactivity of liver biopsy-propagated lymphocyte culture was tested in the presence of combinations of low doses of FK (0.06 ng/ml) and CsA (10 ng/ml or 1 ng/ml).

Table 2. Enhanced CsA Sensitivity of Heart Biopsy Lymphocyte Cultures Pretreated with FK506

	% Inhibition of PLT Response of Pretreated Cultures*			
CsA (μg/ml)	Control	CsA (10 μg/ml)	FK (0.1 μg/ml)	FK (0.01 μg/ml)
0.5	80	80	90	85
0.1	20	20	82	60
0.05	2	2	78	40
0.01	0	0	25	20
0.005	0	0	24	20

*Lymphocyte cultures propagated from a heart biopsy were expanded in IL2 in the absence (control) or presence of various concentrations of FK (0.1 μg/ml or 0.01 μg/ml) and CsA (10 μg/ml). After 4 days of incubation, the cell cultures were washed, counted, and tested for donor-specific PLT responses in the presence of various doses of CsA.

proliferation of activated T cells (11), experiments were set up whereby heart transplant biopsy grown lymphocytes were expanded in IL2 in the presence of various concentrations of FK (0.1 μg/ml and 0.01 μg/ml) or CsA (10 μg/ml). After 4 days in culture, these pretreated cells showed similar PLT reactivity and IL2 responsiveness as untreated control cultures. On the other hand, FK-pretreated cultures showed increased CsA sensitivity when tested in PLT (Table 2). For instance, the PLT response of alloreactive T cells pretreated with 0.1 ng/ml of FK was inhibited by 78% with a low dose of 0.05 ng/ml of CsA. On the other hand, pretreatment with 10 ng/ml of CsA did not effect the CsA sensitivity of a subsequent PLT (Table 2).

These findings of a synergism between FK and CsA have an important clinical application in that low doses of these drug combinations may have less adverse side effects but are still effective in increasing transplant survival. Indeed, Todo and his colleagues, using the experimental models of heterotopic heart transplantation in rats and renal allografts in dogs, have recently demonstrated that the most efficient immunosuppressive protocols are combinations of low doses of FK with subtherapeutical doses of CsA and prednisone (8).

The mechanism of the synergism between FK and CsA is unknown. Preliminary observations at our institution have shown that FK enhances CsA uptake by peripheral blood lymphocytes (13). It is possible that the synergism between FK and CsA might be related to lymphocyte membrane binding of these drugs.

Acknowledgment. This work was supported by NIH grants AI23467 and HL36416. Address correspondence to: Adriana Zeevi, Ph.D., University of Pittsburgh, Division of Clinical Immunopathology, Room 5711, One Children's Place, Pittsburgh, PA 15213-3417, USA.

References

1. Ochiai T, Nakajima K, Kagale M, Nagata M, Suzuki T, Asano T, et al. Effect of a new immunosuppressive agent,

FK506 on heterotopic cardiac allotransplantation in the rat. Transplant Proc 1987;19:1284–1286.
2. Lee P, Murase N, Todo S, Makowka L, Starzl T. The immunosuppressive effects of FR 900506 in rats receiving heterotopic cardiac allografts. Surg Res Commun 1987;1:325.
3. Todo S, Makowka L, Starzl T. Immunosuppression with FR 900506 for canine kidney and liver transplants. Surg Res Commun 1987;1:330.
4. Goto T, Kino T, Hatanaka H, Nishiyama M, Okuhara M, Kohsaka M, et al. Discovery of FK506, a novel immunosuppressant isolated from Streptomyces Tsukubaensis. Transplant Proc 1987;19(suppl 6):4.
5. Kino T, Inamura N, Sakai F, Nakahara K, Goto T, Ohuhara M, et al. Effect of FK506 on human mixed lymphocyte reactions in vitro. Transplant Proc 1987;19(suppl 6):36.
6. Zeevi A, Duquesnoy R, Eiras G, Todo S, Makowka L, Starzl TE. In vitro immunosuppressive effects of FR900506 on human T lymphocyte activation. Surg Res Commun 1987;1:315.
7. Zeevi A, Fung JJ, Zerbe T, Kaufman C, Rabin B, Griffith B, et al. Allospecificity of activated T cells grown from endomyocardial biopsies from heart transplant patients. Transplantation 1986;41:620.
8. Todo S, Demetrius AJ, Veda Y, Imventarza O, Okuda K, Casavilla A, et al. Canine kidney transplantation with FK506 alone or in combination with cyclosporine and steroids. Transplant Proc 1987;19(suppl 6):57.
9. Ochiai T, Nagata M, Nakajima K, Sakamoto K, Asano T, Isono K. Prolongation of canine renal allograft survival by treatment with FK506. Transplant Proc 1987;19(suppl 6):53.
10. Todo S, Podesta L, Chapchap P, Kahn D, Pan CE, Veda Y, et al. Orthotopic liver transplantation in dogs receiving FK506. Transplant Proc 1987;19(suppl 6):64.
11. Zeevi A, Duquesnoy R, Eiras G, Rabinowich H, Todo S, Makowka L, et al. Immunosuppressive effect of FK506 on in vitro lymphocyte activation: Synergism with cyclosporine A. Transplant Proc 1987;19(suppl 6):40.
12. Zeevi A, Duquesnoy R, Eiras G, Rabinowich H, Todo S, Makowka L, et al. In vitro immunosuppressive effects of FK506 in combination with other drugs. Transplant Proc, T Proceed 20:220; Supplement 1988.
13. Sanghvi A, Warty V, Zeevi A, Diven W, Duquesnoy R, Makowka L, et al. FK506 enhances cyclosporine uptake by peripheral blood lymphocytes. Transplant Proc 1987; 19(suppl 6):45.

Author Affiliations

G. Eiras, *A. Zeevi*, and *R.J. Duquesnoy*, Department of Pathology; *S. Todo*, *L. Makowka*, and *T.E. Starzl*, Department of Surgery, University of Pittsburgh, Pittsburgh, PA 15213-3417, USA

Allogeneic Rat T-Suppressor Inducer Cells with In Vivo and In Vitro Activity

David M. Rothstein, Andrew H. Frankel, Cheng Kwok, Edgar L. Milford, and Charles B. Carpenter

The presence of allogeneic Ag-specific T-suppressor inducer cells with in vivo activity has been demonstrated by Hutchinson (1), using a rat renal allograft model. We now demonstrate the presence of in vivo and in vitro T-suppressor inducer activity in the W3/25+ (CD4+) subset of lymphocytes from WF strain rats rejecting LEW strain heterotopic cardiac allografts transplanted 5 days previously. Adoptive transfer of 2 to 4 $\times$ 10^7 of these immune day 5 W3/25+ [d5 W3/25+] splenocytes, or their sonicates 7 days prior to transplantation, prolongs allograft survival two- to threefold in unmodified or irradiated hosts (200 to 300 rads). This prolongation is specific for the donor strain (Table 1).

Analogous experiments in vitro using immune day 5 W3/25+ lymph node cells or splenocytes, co-incubated with syngeneic naive OX8+ (CD8+) [nOX8+] lymphocytes for 7 days, give rise to modulators capable of potent suppression of a test MLR (as measured by 3HTdR uptake) (Fig. 1). Unlike the situation in vivo, this suppression is not specific for the original donor strain when tested in an MLR with third party stimulators. Neither nOX8+ nor d5 W3/25+ lymphocytes incubated alone for 7 days will suppress an MLR. Furthermore, neither cell type, alone or in combination will suppress an MLR prior to incubation.

In further experiments, the modulators were incubated in Transwell™ dual-chambered culture wells separated by a semipermeable membrane, preventing direct cell to cell contact. The modulators incubated in each chamber were then tested independently for suppressor activity in an MLR. When d5 W3/25+ and nOX8+ cells were incubated in the bottom and top chambers of the Transwell, respectively, neither modulator was

Table 1.

Graft	LEW	LEW	BN
Irrad. (d=0)	−	+	+
Control	5.5±0.8(6)*	7.1±0.9(4)	14.3±5.9(3)
d5W3/25+spl	11.0±1.6(4)	17.0±5.7(4)	5.0(1)
Sonicate	10.5±2.1(2)	13.0±1.4(2)	5.5±.5(2)

*Graft survival in days (n).

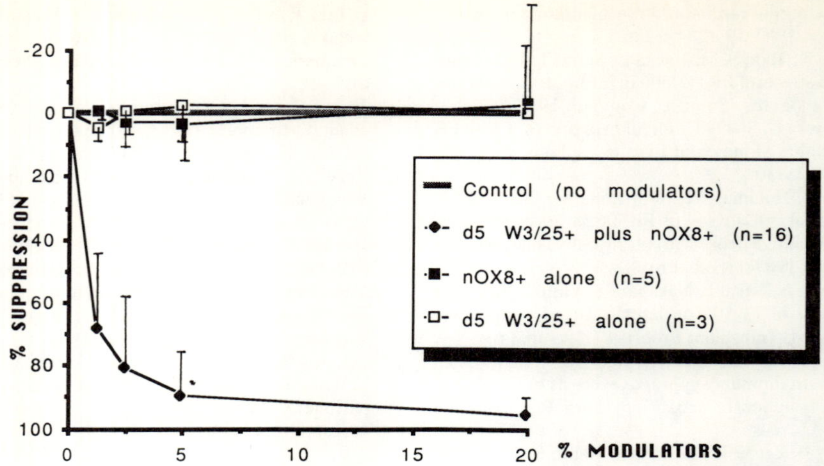

Figure 1. Modulators consisting of WF naive OX8+ (nOX8+), WF day 5 rejecting W 3/25+ (d5 W3/25+), or combination of both cell-types, were incubated for 7 days and then added in various concentrations to a fresh WF X LEW* MLR.

capable of suppressing a test MLR. When d5 W3/25+ plus nOX8+ lymphocytes were placed together in the bottom chamber, with either nOX8 or d5 W3/25+ in the top chamber, and incubated for 7 days, the combination of d5 W3/25+ plus nOX8+ lymphocytes in the bottom chamber was capable of potent suppression of a test MLR. Furthermore, the direct interaction between these two cell types was capable of inducing potent suppressor effectors from a population of cells in either the nOX8+ or the d5 W3/25+ cells in the top chamber.

In conclusion, these experiments imply that a CD4+ T-suppressor inducer is present within the d5-rejecting spleen or lymph node. Direct cell to cell contact appears to be required between the d5 CD4+ inducers and the nOX8+ (CD8+) cells to trigger the induction of Ts effectors. Once the above direct interaction has occurred, factors are released capable inducing a population of cells in either the nOX8+ or the d5 W3/25+ lymphocytes to become T-suppressor effectors.

Comment

After 6 months of entirely consistent results in repeated experiments, we are now unable to reproduce either the in vivo or in vitro phenomena. The reason for this abrupt change is completely obscure. Further evaluation is in progress.

References

1. Hutchinson IV. Suppressor T cells in allogeneic models. Transplantation 1986;41(5):547.

Author Affiliation

Laboratory of Immunogenetics and Transplantation Brigham and Women's Hospital, Boston, MA 02115, USA

Differential Effects of Helper and Cytotoxic T-Lymphocyte Clones on Macrophage Function*

Mark L. Jordan, Ann Carlson, Rosemary A. Hoffman, and Richard L. Simmons

Although it has been well documented that immune dysfunction increases host susceptibility to infection, the precise mechanisms defining this relationship are not well understood (1). Indeed, although allograft recipients are at increased risk of infection presumably because of pharmacological immunosuppression, whether the process of allosensitization in and of itself is capable of altering this risk is unknown. Within the context of the allograft response, closer examination of allosensitized lymphocyte-macrophage interactions occurring in response to a septic stimulus could help to answer this question.

The present studies were undertaken to determine whether allosensitized T cells with defined functional capacities are capable of modifying macrophage responses and whether these responses differ in the presence of a septic stimulus. We therefore studied the effects of cloned allosensitized murine helper or cytotoxic T cells (derived by limiting dilution analysis of a C57BL/6 anti-DBA/2 MLC) on both interleukin (IL1, assessed by the enhancement of BALB/c thymocyte proliferation to a suboptimal dose of phytohemagglutinin) and prostaglandin E_2 (PGE_2, determined by radioimmunoassay) production by syngeneic resident murine peritoneal macrophages (MO). In order to study macrophage responses to lymphocytes, a coculture system was used whereby macrophages were pretreated with endotoxin for 24 hours, washed, and then exposed to resting allosensitized cloned lymphocytes for a further 48 hours. We chose to examine IL1 and PGE_2 production as parameters of macrophage function since 1) both IL1 and PGE_2 are synthesized by macrophages, 2) both macrophage IL1 and PGE_2 synthesis can be augmented by exposure to agents (such as LPS) that mimic circulating endotoxin in vivo (2), 3) IL1 and PGE_2 have distinct functional effects on both macrophages and lymphocytes, and 4) we have found that these allosensitized T-cell clones do not produce detectable levels of either IL1 or PGE_2.

Our initial experiments have confirmed that both macrophage IL1 and PGE_2 production could be stimulated by the same dose (10 μg/ml) of LPS that has previously been found to stimulate other in vitro macrophage functions. Furthermore, IL1 and PGE_2 production plateaued at 48 hours following endotoxin pretreatment, allowing us to study both responses simultaneously. This plateau effect of IL1 and PGE_2 production has been observed by others and is in keeping with the reciprocal regulatory effects of PGE_2 and IL1 on their endogenous production (2).

Without prior exposure to endotoxin, macrophage production of IL1 was completely unaltered by the presence of either cloned helper or cytotoxic T lymphocytes (CTL) (Table 1). PGE_2 production, on the other hand, was slightly increased after coculture with CTL, but was unaffected by the presence of helper cells. Macrophage pretreatment with endotoxin caused significant elevations in both IL1 and PGE_2 production, although no further increase was observed after 48 hours. Addition of cloned helper cells to LPS-pretreated macrophages resulted in both a significant rise in macrophage IL1 production and a reduction in PGE_2 synthesis. Similar patterns have been observed after macrophage treatment with indomethacin (2) and raise the possibility that certain T-cell products may affect macrophage arachidonate metabolism. Another possibility is that LPS-treated macrophages themselves may induce helper T-cell IFN-γ production, which in turn has been shown to augment macrophage IL1 synthesis. In contrast, directly opposite effects were observed after coculture of LPS-treated macrophages with cloned CTL. IL1 production was significantly decreased, whereas PGE_2 release was augmented. It has been hypothesized that one mechanism by which immunological "suppression" occurs is via macrophage-derived PGE_2 either inducing a subset of T-suppressor lymphocytes or by directly inhibiting T-cell function. We have previously observed that cloned helper, but *not* CTL, are functionally inhibited by PGE_2 (3). It thus appears that cloned CTL may be inherently *resistant* to the inhibitory effects of PGE_2. Taken together with the data from the current studies, this would suggest that these cloned CTL may in fact possess "suppressor" activity by virtue of their ability to induce macrophage PGE_2 and suppress IL1 production while being themselves unaffected by PGE_2. Alternatively, these CTL may be inducing a subpopulation of macrophages to preferentially secrete PGE_2 while reducing IL1 production. A further possibility is that endotoxin induces the synthesis of a specific monokine to which the cloned CTL respond with release of a potent suppressor factor. It should be emphasized that in the above studies, *syngeneic* macrophages and lymphocytes were used to avoid de novo lymphocyte allospecific stimulation. Whether the above responses are observed with allogeneic macrophage-lymphocyte interactions will require further study.

These experiments suggest that septic stimuli may render the macrophage susceptible to specific functional modulation by activated T lymphocytes. Allosensitized helper T cells, consistent with their nomenclature and identification as IL2-producing cells, appear to alter macrophage function so that the overall effect is that of enhancement (↑IL1, ↓PGE_2) of immune function. On the other hand, allosensitized CTL appear to exert "suppressive" effects (↓IL1, ↑PGE_2) on these macrophage functions. Evaluation of other specific macrophage

Table 1. Production of IL1 and PGE_2 by Cocultures of Mouse Peritoneal Macrophages and Allosensitized Helper or Cytotoxic T-Cell Clones*

T Cells	MØ	LPS Pretreatment	IL 1 Activity (cpm ± SEM)	PGE_2 (ng/10^6 MØSEM)
None	+	−	10,731 ± 1,211	0.62 ± 0.1
	+	+	45,261 ± 5,635†	12.50 ± 2.5†
Helper clone	−	−	0	0
(199-11)	−	+	421 ± 35	0
	+	−	10,117 ± 1,004	1.05 ± 0.2
	+	+	73,853 ± 8,473†‡	9.06 ± 0.5†‡
Cytotoxic clone	−	−	0	0
(199-4)	−	+	370 ± 28‡	0
	+	−	8,326 ± 749‡	1.95 ± 0.2
	+	+	25,910 ± 2,117†‡	17.10 ± 0.5†‡

*1 × 10^6 freshly harvested C57BL/6 MØwere incubated for 24 hr in the presence or absence of endotoxin (10 μg/ml LPS). After removing the LPS by washing each well three times with fresh medium (RPMI-1640) either 1 × 10^6 cloned helper (199-11) or cytotoxic (199-4) T cells (both clones C57BL/6 anti-DBA /2) were added to the MØ monolayers. Following 48 hr of coculture, the resulting conditioned medium (CM) was rendered cell free by centrifugation for 10 min at 2,000 rpm, and then assayed for both IL1 activity (in a 72-hr thymocyte proliferation assay) and PGE_2 (by radioimmunoassay). CM obtained from control cultures of helper or cytotoxic clones without MØ, with or without LPS pretreatment, as well as medium (RPMI-1640) alone with or without added LPS (10 μg/ml), contained neither IL1 nor PGE_2. Data are expressed as mean ± SEM of the results from three experiments.
†$p < 0.01$ vs. MØ alone.
‡$p < 0.01$ vs. MØ (LPS pretreated).

interactions with allosensitized cloned T cells, such as modulation of macrophage phagocytic and microbicidal functions, as well as release of other secretory molecules (leukotrienes, TNF, procoagulant activity) with potentially important roles in host defense will require further study. Nevertheless, these initial data suggest for the first time that functionally defined allosensitized T-cell subsets differentially modulate the secretory function of endotoxin-exposed macrophages. Further studies are being undertaken to determine whether these responses are unique to certain T-cell clones and/or macrophage functions.

References

1. Matas AJ, Simmons RL, Najarian JS. Sepsis following kidney transplantation. In: Hardy JD (ed): Critical Surgical Illness. W.B. Saunders, Philadelphia, PA, 1980, p 552.
2. Kunkel SL, Chensue SW, Phan SH. Prostaglandins as endogenous mediators of interleukin 1 production. J Immunol 1986;136:186.
3. Jordan ML, Hoffman RA, Debe EF, Simmons RL. In vitro locomotion of allosensitized T lymphocyte clones in response to metabolites of arachidonic acid is subset specific. J Immunol 1986;137:661.

Author Affiliations

Mark L. Jordan, Division of Urology, Toronto Western Hospital, University of Toronto, Suite 8-304, 399 Bathurst St., Toronto, Ontario, Canada M5T 2S8; *Ann Carlson*, *Rosemary A. Hoffman*, and *Richard L. Simmons*, Department of Surgery, University of Pittsburgh, Pittsburgh, PA 15213-3417, USA

Alloreactivity Studies of Lymphocytes in Bronchoalveolar Lavages from Canine Lung Allografts*

H. Rabinowich, A. Zeevi, R. Dal-Col, D. Hurlan, B.P. Griffith, R.L. Hardesty, I.L. Paradis, J.H. Dauber, and R.J. Duquesnoy

Although heart-lung transplantation is increasingly being recognized as a therapeutic modality to treat end-stage lung disease, its success rate is still considerably lower than other organ transplants. The transplanted lung appears more susceptible to rejection than heart, liver, and kidney transplants (1). The diagnosis of lung allograft rejection is difficult and there is limited value in measurements of nonimmunological parameters such as radiographic pulmonary infiltrates, arterial blood gases, and clinical status. The bronchoalveolar lavage (BAL) of pulmonary allografts has been proven useful in the diagnosis of rejection and infection (2–4). This approach enables access to lymphocytes directly involved in lung allograft rejection. It also provides a unique model of in situ transplant immunity (2,3). In order to better understand and monitor the cellular events taking place during rejection episodes, an experimental model of canine unilateral lung transplant has been developed (5). In this dog model, rejection was allowed to occur by maintaining the dog on subtherapeutic immunosuppression. Sequential BAL procedures were utilized to probe the development of in situ immune response to the lung allograft. In this paper we describe our findings on donor-specific alloreactivity of immunocompetent cells retrieved by the BAL procedure during the posttransplant period.

Orthotopic left unilateral lung allotransplantations were performed on beagle dogs utilizing size-matched unrelated donors. During the first 4 days posttransplant allograft recipients were treated with Cyclosporine A (CsA) (15 mg/kg/day) and azathioprine (25 mg/day). Thereafter, the CsA therapy was withdrawn and the dogs were maintained only on the suboptimal dose of azathioprine. This protocol of reduced immunosuppression led to rejection, which usually was demonstrable clinically 3 to 4 weeks posttransplant. Sequential BAL procedures were performed on the lung allograft and the contralateral control lung on a weekly basis during the 1st month posttransplantation (4). Rejection was assessed clinically by weekly chest radiographs. Acute pulmonary rejection was confirmed by histopathological evaluation of open lung biopsies. BAL lymphocytes (3) from the lung allograft were tested for donor alloreactivity while BAL lymphocytes from the contralateral normal lung and peripheral blood lymphocytes (PBLs) served as controls.

BAL from grafted and control lungs yielded cells significantly different in number and functions. A typical example of cell counts in these BAL is shown in Table 1. The total number of BAL cells was significantly higher in the allograft than the control lung throughout the experimental period. Fifteen days posttransplant the predominant cell type of the BAL fluid from the allograft changed from macrophages to polymorphonuclear cells (PMNs). In the later posttransplantation

Table 1. Cell Content of BAL Fluid of Canine Lung Allograft Recipient No. R7

Day Post TX	Lung	Differential Cell Count (%)* M	Lymph	PMN	Total Cell Count ($\times 10^{-5}$/ml BAL Fluid)
0	Control	74	22	4	1.2
8	Allograft	24	8	60	4
	Control	86	2	12	1.5
15	Allograft	13	5	81	11.5
	Control	81	4	14	1.6
28	Allograft	39	39	17	8.5
	Control	90	7	2	1.1

*BAL cells were washed two times with Hank's balanced salt solution and analyzed immediately for total and differential cell counts in hemocytometer using 0.1% crystal violet stain.

Table 2. Donor-Specific PLT Reactivity of BAL Lymphocytes from Rejecting Canine Lung Transplanted Recipient

Dog No.	Day Post-transplantation	Culture	[³H]Thymidine (cpm) BAL Allograft	BAL Control
K7	19	Background	365	757
		PLT	2,214	842
K13	19	Background	305	114
		PLT	1,458	171
R5	30	Background	831	629
		PLT	2,577	1,115
R7	28	Background	680	335
		PLT	2,453	421

The ability of the allograft and control BAL lymphocytes to proliferate in the presence of donor antigens was evaluated in 72-hour PLT (3). BAL lymphocytes (5×10^4) were incubated in the presence of 10^5 irradiated donor cells. BAL lymphocytes incubated without stimulator cells served as background controls. The proliferative response was measured as [^{3}H]thymidine incorporation into DNA during the final 20 hours of the incubation.

*These studies were supported in part by grants HL-36416 and AI-23467 from the National Institutes of Health.

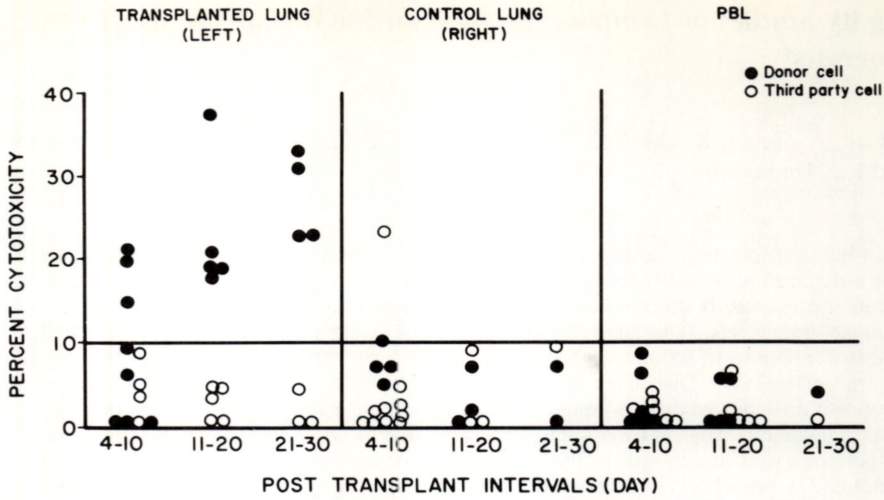

Figure 1. Cytotoxic activity of BAL and peripheral blood lymphocytes of canine lung allograft recipients. PBL and BAL lymphocytes were expanded for 4 to 7 days in the presence of human recombinant IL2 (50 U/ml). The cytotoxic activity of the expanded lymphocytes was determined in a standard 4-hour ^{51}Cr release assay (6). ^{51}Cr-labeled target cells were prepared from phytohemagglutinin-stimulated lymph node cells from the donor or a third party. These target cells (5×10^3/well) were suspended in RPMI 1640 + 5% dog serum in round-bottomed microwells. The effector/target cell ratios used were 40:1 to 10:1.

period (20 to 30 days), the percentage of PMNs decreased concomitant with a significant increase in the percentage of lymphocytes. Lavages of the right control lung yielded similar cell numbers and differentials as lavages obtained preoperatively. Increased proportions of PMNs were found exclusively in the lavage fluid of the allograft but not in the contralateral control lung or in autotransplanted lungs (5). The presence of PMNs in the rejecting canine lung is concordant with the results reported for lavage fluid of rejecting human heart-lung allografts (4).

The alloreactivity of the allograft-derived BAL lymphocytes was assayed in a primed lymphocyte test (PLT) using donor-derived stimulator cells (3). Table 2 presents the PLT response of allograft BAL lymphocytes from four different dogs. No significant proliferation occurred with lymphocytes from the control lungs or with PBLs (data not shown). The donor-specific PLT responses presented in Table 2 appeared associated with acute pulmonary rejection. This was confirmed by open lung biopsies obtained on each of the indicated days posttransplantation. Significant PLT responses could be demonstrated only when the proportion of PMNs in the lavage fluid was low. Elevated numbers of PMNs in the lavage fluid of the allograft in the early posttransplantation period were found to interfere with the in vitro PLT proliferative responses.

The cytotoxic potential of BAL cells was assessed in cell-mediated lympholysis (CML) assays. The lymphoid cells in BAL were expanded in the presence of human rIL2 and then examined for their cytotoxic activity against phytohemagglutinin (PHA) blasts from the donor or third-party unrelated donors (6). Figure 1 illustrates the significant donor-specific cytotoxicity of allograft BAL lymphocytes. This lytic activity was demonstrated as early as 1 week posttransplantation. Neither BAL cells from the right control lung nor PBLs showed CML activity against donor targets. The CML activity of allograft BAL lymphocytes was donor specific; no significant lysis of third-party allogeneic PHA blasts was observed.

Functional testing of allograft BAL lymphocytes provide opportunity to elucidate cellular mechanisms operating in vivo during lung allograft rejection. Norin et al. have demonstrated increased Con A-dependent cytotoxicity of BAL lymphocytes against unrelated target cells during lung allograft rejection (7). In this study we have presented evidence that during rejection the allograft BAL lymphocyte exhibited two properties of donor-specific alloreactive T cells: secondary proliferation and cytotoxicity. These findings extend our previous observation on human heart-lung (3) and human heart transplants (8).

The functional analysis of serially obtained BAL lymphocytes enables us to investigate the dynamics of immunological mechanisms associated with the rejection of canine lung transplants. In a previous report we have shown that the accumulation of donor-specific cytotoxic cells in the graft may occur well before any radiographic changes appear in the transplanted lung (6). The functional analysis of BAL cells offers a new means of immunological monitoring and diagnosis of lung transplant rejection.

References

1. Griffith BP, Hardesty RL, Trento A, Paradis IL, Duquesnoy RJ, Zeevi A, et al. Heart-lung transplantation: Lessons learned and future hopes. Ann Thorac Surg 1987;43:6.
2. Fung JJ, Zeevi A, Kaufman C, Paradis IL, Dauber JH, Hardesty RL, et al. Interactions between bronchoalveolar lymphocytes and macrophages in heart-lung transplant recipients. Hum Immunol 1985;14:281.
3. Zeevi A, Rabinowich H, Paradis I, Gryzan S, Dauber JH, Hardesty RL, et al. Lymphocyte activation in bronchoalveolar lavages from heart-lung recipients. Transplant Proc 1988;20:273.
4. Gryzan S, Paradis IL, Hardesty RL, Griffith BP, Dauber JH. Bronchoalveolar lavage in heart-lung transplantation. Heart Transplant 1985;4:414.
5. Herlan D, Kormos R, Zeevi A, Paradis IL, Wei L, Nalesnik M, et al. Dynamics of bronchoalveolar lavage in the canine lung transplant. Transplant Proc 1988;20:832 Suppl 1.
6. Rabinowich H, Zeevi A, Herlan D, Griffith BP, Hardesty RL, Kormos R, et al. Functional studies on lymphocytes in bronchoalveolar lavages from canine lung allografts. Transplant Proc 1988;20:836 Suppl 1.
7. Norin AJ, Kamhotz SL, Pinsker KL, Emerson EE, Veith FJ. Con A-dependent cell-mediated cytotoxicity in bronchoalveolar lavage fluid. Transplantation 1986;42:466.
8. Kaufman C, Zeevi A, Weber T, Zerbe T, Griffith B, Kormos R, et al. Correlation of prophylactic therapy protocol with functional studies of lymphocytes propagated from heart transplant biopsies. Immunogenetics and Histocompatibility, Vol 2, in press.

Author Affiliations

H. Rabinowich, A. Zeevi, I.L. Paradis, J.H. Dauber, and *R.J. Duquesnoy,* Departments of Pathology and Medicine; *R. Dal-Col, D. Hurlan, B.P. Griffith,* and *R.L. Hardesty,* Department of Surgery, University of Pittsburgh, School of Medicine, Pittsburgh, PA 15213-3417, USA

Selective Detection of In Vivo-Activated:, Cytotoxic T Lymphocytes by Limiting Dilution Analysis: Murine CTL Distribution After Allosensitization with Sponge Matrix Allografts

Charles G. Orosz, D. Keith Bishop, Barbara Horstemeyer, and Ronald M. Ferguson

Abstract: We have modified an existing limiting dilution analysis (LDA) technique to detect only in vivo-activated cytotoxic T lymphocytes (CTL) with specificity for graft alloantigens. Using this technique, we demonstrate in sponge allograft recipients that graft sites accumulate donor-reactive CTL, most of which are activated. Prior to transplantation, draining lymph nodes and peripheral blood always contain modest numbers of alloreactive CTL, none of which are activated. At the peak of rejection, only a small fraction of the lymph node CTL, and none of the peripheral blood CTL demonstrate evidence of prior activation. These observations reveal the graft-induced distribution of alloactivated, donor-reactive CTL, and raise questions regarding the use of peripheral blood to monitor immune responses associated with allograft rejection.

Limiting dilution analysis (LDA) is a quantitative immunologic technique that has been used increasingly in both human and murine studies to monitor the behavior of functional T-cell subpopulations in various lymphoid compartments. In previous studies, we used LDA to examine the accumulation of donor-reactive CTL in sponge matrix allografts (1). We found that sponge allografts accumulate donor-reactive CTL steadily from day 5 to day 10 after implantation. Thereafter, the CTL population in sponge allografts remains stable at about 0.1% of the cellular infiltrate for at least seven more days. However, when we examined the donor-reactive CTL population in regional lymph nodes and peripheral blood, we found that these sites constitutively maintain high levels (about 0.1%) of these CTL, and that graft implantation causes minimal changes in these baseline frequencies that are difficult to detect, even by LDA (Orosz et al, manuscript in preparation).

We reasoned that many of the graft-reactive CTL in the allograft recipient may not encounter graft alloantigens, and therefore do not participate directly in graft rejection. However, conventional LDA techniques are not designed to discriminate between T cells at various stages of differentiation (i.e., naive vs. activated vs. memory T cells). If the size of the participating donor-reactive CTL subpopulation is relatively small, then alloantigen-induced activation and clonal expansion of the participating cells would be difficult to detect against the background of non-participating, donor-reactive CTL, using conventional LDA techniques. Nonetheless, it is precisely this participating subpopulation that we would want to monitor as an index of ongoing alloreactivity. Consequently, we developed an LDA technique that detects only in vivo-alloactivated CTL.

The key to this technique was provided by our experiments on alloantigen specificity of the CTL that infiltrate sponge allografts. We initially found that sponge

Immunobiology of HLA
Volume II

Table 1. Alloantigen Specificity of CTL in Sponge Infiltrates of C57B1/6 Recipients of DBA/2 Sponge Allografts

LDA Target		LDA Stimulator DBA/2	C57B1/6	C3H/HeN
DBA/2	f^{-1}	1,012	2,538	967
	range	790 – 1,409	2,055 – 3,320	791 – 1,243
	p	0.9129	0.3574	0.3229
C3H/HeN	f^{-1}	27,433	96,478	5,010
	range	19,911 – 44,085	89,181 – 200,911	4,024 – 6,636
	p	0.7432	0.8734	0.1199

Three C57B1/6 (H-2^b) mice received DBA/2 (H-2^d) sponge allografts (1). After 10 days the sponge infiltrates were recovered, pooled, and added in various dilutions to replicate limiting dilution microcultures containing 2% EL4 SN (as a source of growth factors) plus 5×10^5 irradiated DBA/2, C57B1/6, or C3H/HeN (H-2^k) splenocytes (1,2). Seven days later, all microwells were split and tested for lytic activity against ^{51}Cr-labeled DBA/2 or C3H/HeN, Con-A activated splenocytes. The pattern of cytolytic activity was analyzed for CTL frequency by the method of χ^2 minimization (3), which yields a frequency estimate (f^{-1}), a frequency range representing the 95% confidence limits of the estimate (range), and χ^2 analysis of frequency estimate probability (p), which is acceptable if it is greater than 0.05.

allografts acquire large numbers of LDA-detectable CTL with irrelevant alloreactivity, along with the donor-reactive CTL. In split well experiments designed to determine if these irrelevant CTL were cross-reactive with donor alloantigens, we unexpectedly found that cells placed in LDA microcultures containing non-donor allogeneic stimuli could develop donor-reactive cytotoxic activity, exclusively (2). Apparently, LDA microculture conditions can promote both the function and the clonal expansion of donor-reactive CTL from sponge allografts, without the need for further donor-alloantigen stimulation. Presumably, the activating alloantigenic stimulus was delivered in vivo, and is no longer required for further in vitro function.

Table 2. Distribution of DBA/2-Reactive CTL in Recipients of Sponge Isografts (Day 10)

Responder Cells	LDA Stimulator	Anti-DBA/2 CTL f^{-1}	% Active CTL
LNC	DBA/2	1,873	
LNC	C57Bl/6	<385,529	<0.4%
PBC	DBA/2	3,410	
PBC	C57Bl/6	<385,529	<0.9%
SIC	DBA/2	<385,529	
SIC	C57Bl/6	<385,529	NC

Three C57BL/6 mice received C57BL/6 sponge isografts. After 10 days, the regional lymph node cells (LNC), peripheral blood cells (PBC), and sponge infiltrating cells (SIC) were retrieved, and appropriately pooled cells were added in various dilutions to replicate, and appropriately pooled cells were added in various dilutions to replicate limiting dilution microcultures containing 2% EL4 SN plus 5×10^5 irradiated DBA/2 splenocytes or 5×10^5 irradiated C57Bl/6 splenocytes. Seven days later the microwells were tested for lytic activity against ^{51}Cr-labeled DBA/2, Con-A activated splenocytes. The pattern of cytolytic activity was analyzed for CTL frequency by the method of χ^2 minimization. NC = not able to calculate.

In view of this, we substituted syngeneic feeder cells for allogeneic stimulator cells in LD microcultures, and used this modified LDA technique, along with the conventional LDA technique, to analyze the donor-reactive CTL frequency in sponge allograft infiltrates. As shown in Table 1, most of the DBA/2-reactive CTL in DBA/2 sponge allografts of C57BL/6 mice were detected without the need for DBA/2 alloantigens in the LDA. In contrast, only a small fraction of the C3H/HeN-reactive CTL that are recovered from DBA/2 sponge allografts in the same mice can be detected in the LDAs that lack the specific C3H/HeN stimuli. We interpret this to mean that most of the donor-reactive CTL in sponge allografts are activated, whereas most of the irrelevant CTL that accumulate in sponge allografts are inactive.

We have expanded this analysis to include the donor-reactive CTL in regional lymph nodes and peripheral blood. As shown in Table 2, if mice receive sponge isografts, they accumulate no LDA-detectable, DBA/2-reactive CTL at the graft site, and all of the DBA/2-reactive CTL in the regional lymph nodes and peripheral blood show no evidence of prior contact with alloantigen (i.e., remain undetectable in LDAs lacking the DBA/2 stimulus). In contrast, C57B1/6 mice that receive DBA/2 sponge allografts accumulate many donor-reactive CTL at the graft site, most of which are activated (Tables 1 and 3). Interestingly, only a small fraction of the DBA/2-reactive CTL in the regional lymph nodes, and none of the CTL in peripheral blood, show evidence of in vivo alloactivation during the period of peak CTL responses in the sponge graft (Table 3). Kinetic studies have indicated that in vivo alloactivated CTL are rare in the regional lymph node and absent from the peripheral blood throughout the initial ten-day period when activated, donor-reactive CTL are accumulating rapidly at the graft site (data not shown).

These data demonstrate that a modified LDA technique can be used to differentially detect and quantify

Table 3. Distribution of DBA/2 Reactive CTL in Recipients of DBA/2 Sponge Allografts (Day 10)

Cell Source	LDA Stimulator	Anti-DBA/2 CTL f^{-1}	% Active CTL
LNC	DBA/2	814	
LNC	C57Bl/6	57,208	1.4%
PBC	DBA/2	2,902	
PBC	C57Bl/6	<385,529	<0.7%
SIC	DBA/2	3,599	
SIC	C57Bl/6	4,385	82.1%

Three C57BL/6 mice received DBA/2 sponge allografts. After 10 days, the appropriate lymphoid populations were retrieved, pooled, and added in various dilutions to replicate limiting dilution microcultures containing growth factor plus irradiated DBA/2 or C57Bl/6 splenocytes. Seven days later the microwells were tested for lytic activity against ^{51}Cr-labeled DBA/2, Con-A activated splenocytes and the pattern of cytolytic activity was analyzed for CTL frequency by the method of Chi square minimization.

the in vivo alloactivated subpopulation of donor-reactive CTL in various lymphoid compartments of murine sponge allograft recipients. We propose that only this CTL subpopulation should be monitored as an index of ongoing allosensitization because LDA monitoring of the total donor-reactive CTL population is relatively uninformative, except at the graft site. During our studies, we found that lymphoid compartments differ in their content of alloactivated CTL during allosensitization. Based on our experimental data, the priority for the monitoring of lymphoid compartments for activated CTL should be graft > regional lymph node > peripheral blood. Indeed, our observations raise serious questions regarding the use of peripheral blood to monitor cellular immune responses associated with allograft rejection.

Acknowledgments. We wish to thank Ms. Cindy Narcross for her excellent secretarial assistance and Ms. Teri Bailey for her helpful administrative assistance in association with this manuscript.

This is paper #38 from The Ohio State University, Therapeutic Immunology Program, Division of Transplantation. This project was supported by NIH grant 5 R01 AM34774-03, by the Medical Research and Development Fund of the Dept. of Surgery, and by PHS grant number P 30 CA 1605814 awarded by the National Cancer Institute, DHHS.

References

1. Orosz C, Zinn N, Sirinek L, Ferguson R. *In vivo* mechanisms of alloreactivity. I. Frequency of donor-reactive cytotoxic T lymphocytes in sponge matrix allografts. Transplantation 1986;41:75.
2. Orosz C, Zinn N, Sirinek L, Ferguson R. *In vivo* mechanisms of alloreactivity. II. Allospecificity of cytotoxic T lymphocytes in sponge matrix allografts as determined by limiting dilution analysis. Transplantation 1986;41:84.
3. Taswell C. Limiting dilution analysis for the determination of immunocompetent cell frequencies. J Immunol 1981; 126:1614.

Author Affiliation

Therapeutic Immunology Laboratories, Div. of Transplantation, Dept. of Surgery, The Ohio State University College of Medicine, Columbus, Ohio 43210

Alloreactive Lymphocyte Adherence to Human Arterial Endothelial Cell Monolayers

Y.L. Colson, B.H. Markus, A. Zeevi, and R.J. Duquesnoy

Abstract: The early events of the cellular infiltration of allografts involve interactions between lymphocytes and vascular endothelium. To investigate the role of HLA antigens in the recognition and adherence of recipient lymphocytes to donor epithelium, an in vitro adherence assay has been developed. Donor-primed lymphocytes (DPL) exhibited a nearly two-fold increase in adherence during coculture with donor human arterial endothelial cells (HAEC) as compared to third-party primed lymphocytes (TPL). The degree of DPL adherence correlated well with the extent of donor HLA antigen expression on HAEC. DPL adherence to HAEC expressing class I HLA antigens was blocked by class I-specific monoclonal antibodies (mAbs). Following treatment with gamma-interferon (Y-IFN) HAEC express class I and class II HLA antigens, and DPL adherence is blocked by class I or class II-specific mAbs. In contrast, the adherence of TPL to HAEC was unaffected by these mAbs. These studies suggest that, in the allograft situation, allorecognition may play a role in the adherence of recipient lymphocytes to the allograft endothelium.

Activated T lymphocytes infiltrating human cardiac and hepatic allografts can be propagated in invitro cultures supplemented with interleukin-2. These cells exhibit donor specific alloreactivity towards class I and/or class

Immunobiology of HLA
Volume II

Table 1. HLA Antigen Expression on HAEC Monolayers

HAEC	Untreated	Y-IFN treated
% of HAEC staining (±SD)		
Class I antigen	86.6 ± 3.7	84.6 ± 3.8
Class II antigen	9.7 ± 1.6	73.4 ± 2.8
Mean channel fluorescence (±SD)		
Class I antigen	61.1 ± 2.5	361.3 ± 2.0
Class II antigen	37.5 ± 1.8	>90

Human arterial endothelial cells were isolated and cultured from donor aorta specimens obtained at the time of organ procurement as previously described.† Verification of endothelial cells origin was obtained by fluorscein isothiocyanate conjugated (FITC)-labelled rabbit antihuman Factor VIII staining (Atlantic Antibodies, Scarborough, ME) and analysis by flowcytometry or fluorescent microscopy. Greater than 95% of the cells in all HAEC cultures stained positive for Factor VIII.

HLA antigen expression was determined by indirect immunofluorescence using an unspecific mouse IgG antibody (Cooperbiochemical) for control binding, anti-β 2 microglobulin (DAKO) or PA2.6 (P. Parham, Palo Alto, CA) for class I HLA antigen assessment, anti-class II HLA mABs [TU35 (A. Ziegler, Tubingen, W. Germany) and CA2.206 (Charron, Paris, France)], and an FITC-labelled goat anti-mouse $F(ab')_2$ antibody (Tago) as a secondary antibody. For preparation of HAEC suspensions, cells were removed from the tissue culture plates by incubation with 0.125% trypsin/0.05% EDTA (GIBCO, Grand Island, NY) at 37°C for 10 to 15 minutes and then washed twice in phosphate buffered saline with 0.04% sodium azide (PBS-azide). HAEC were then resuspended in 50 μl of PBS-azide at a final concentration of 5×10^4 cells/ml and incubated with the various mAb at 4°C for 60 min. HAEC were washed with PBS-azide after the different staining procedures. Flow cytometric studies were done on FACSTAR (Becton-Dickinson). Fluorescence seen with nonspecific antibody controls or autofluorescence was subtracted from alloantigen fluorescence to determine the degree of alloantigen-specific staining reported here. Y-IFN treated HAEC monolayers were exposed to 300 U/ml Y-IFN (Cellular Products) for 3 days prior to flowcytometric analysis.

II HLA antigens (1,2). Biopsies obtained during the early post-transplant period more frequently yield class I HLA-specific T cells, whereas later biopsies generally grow class II HLA-specific lymphocytes. This has led to the postulate of a sequential appearance of class I- followed by class II-specific cells infiltrating the allograft (3). Although the presence of alloreactive lymphocytes in cellular infiltrates of human organ transplants undergoing acute rejection has been well documented, the interactions occurring between lymphocytes and the vascular endothelium are not well understood. These interactions consist of various phases including recognition, adherence, activation, increased permeability, and eventually penetration of lymphocytes through the endothelial barrier. Since the infiltration of these class I- and class II-specific T cells results from a penetration of the lymphocytes through the vascular endothelium, an in vitro model was developed to study the interactions

Table 2. Comparison of Alloreactive Lymphocyte Adherence to Donor HAEC Monolayers*

HAEC	% Lymphocytes Adherent (±SE) DPL	TPL	
Untreated	20.1 ± 1.7	12.5 ± 1.4	p<0.02
Y-IFN treated	42.3 ± 3.2	21.6 ± 1.6	p<0.002

*Y-IFN treated HAEC monolayers were exposed to 300 U/ml Y-IFN for 3 days prior to performance of the adherence assay as outlined in the text. The number of adherent lymphocytes was assessed by a three field count at 200x using a Leitz Diavert inverted microscope. Only those well in which HAEC monolayers remained intact were included for data analysis. Data from triplicate assays (n=3) were pooled and the results were expressed as the mean percentage ± SE lymphocytes added initially that remained adherent to the HAEC monolayers. Statistical significance between DPL and TPL populations was analyzed by the Mann–Whitney Rank Sum test.

between alloreactive lymphocytes and human arterial endothelial cell (HAEC)* cultures derived from transplant donors.

HAEC cultures were established and HLA antigen expression on "resting" and on Y-IFN treated HAEC monolayers was studied using flowcytometric analysis.† Table 1 shows the results of these experiments. Greater than 80% of HAEC constitutively express class I HLA antigens, while class II antigen expression is minimal. Pretreatment of HAEC monolayers with Y-IFN resulted in enhanced expression of class I HLA antigens, as shown by the increase in mean channel fluorescent intensity, and in the induction of class II HLA antigen expression on greater than 70% of HAEC. These findings have led us to postulate that the sequential infiltration of class I- followed by class II-specific T cells depends on an HLA-specific interaction between the donor organ endothelium and circulating recipient lymphocytes and relates to the differential expression of HLA antigens on vascular endothelial cells (2,4).

An in vitro adherence assay was developed to analyze HLA-specific lymphocyte-endothelial interactions by directly quantitating lymphocyte adherence to HAEC monolayers (5).‡ Lymphocytes specific for the donor (DPL) or unrelated (TPL) HLA antigens were generated in 10-day mixed lymphocyte cultures of fresh peripheral blood lymphocytes and irradiated donor or third-party splenocytes. DPL or TPL were fluorescently labelled

---

*Abbreviations: DPL, donor-primed lymphocytes; Y-IFN, gamma interferon; HAEC, human arterial endothelial cells; mAb(s), monoclonal antibody(ies); TPL, third party-primed lymphocytes.

†Markus BH, Colson YL, Fung JJ, Zeevi A, and Duquesnoy RJ. HLA antigen expression on cultured human arterial endothelial cells. Tissue Antigens 1988 (in press).

‡Colson YL, Markus BH, Zeevi A, Duquesnoy RJ. Adherence of alloreactive lymphocytes to human arterial endothelial cell monolayers. Clin Exp Immunol 1988 (in press).

Table 3. Monoclonal Antibody Inhibition of DPL Adherence to HAEC Monolayers*

HAEC Treatment	Lymphocyte Specificity	% Lymphocytes adherent to HAEC monolayer Control	Mouse IgG	Anti-Class I	Anti-Class II
None	DPL	20.1 ± 1.7	21.4 ± 1.5†	9.8 ± 1.0‡	18.9 ± 1.9
	TPL	12.5 ± 1.4	14.6 ± 0.8	16.4 ± 1.0	14.6 ± 0.8
Y-IFN	DPL	42.3 ± 3.2	39.9 ± 2.6†	30.2 ± 2.1‡	27.8 ± 1.1§
	TPL	21.6 ± 1.6	26.3 ± 1.9	32.2 ± 2.2	30.4 ± 2.1

*The adherence assay was performed as outlined previously in the text. HAEC monolayers were preincubated with mouse IgG, anti-class I [W6/32 (Pelfreeze, Rogers, AR)], or anti-class II HLA monoclonal antibody for 30 minutes before lymphocyte exposure. Final antibody dilution was 1:40. Data from triplicate assays was pooled and presented as means ± SE. Statistical significance between DPL and TPL populations was analyzed by the Mann–Whitney Rank Sum test; †DPL mouse IgG vs. TPL mouse IgG: $p<0.02$; ‡DPL + anti-class I mAb vs. DPL mouse IgG: $p<0.02$; §DPL + anti-class II mAb vs. DPL mouse IgG: $p<0.002$.

with hydroethidine before being cocultured on donor HAEC monolayers at 10^5/ml for 18 hours. Cocultures were then inverted and centrifuged at 7.5 g for 1 minute before the numbers of adherent lymphocytes were counted by fluorescent microscopy. The fluorescent label does not affect the ability of the lymphocytes to adhere to the HAEC monolayers and it is important in differentiating lymphocytes from other cell types, especially when the lymphocytes undergo blastogenesis and migrate through the endothelial cell monolayer. Monoclonal antibody (mAb) blocking studies were performed as outlined above following a 30-minute preincubation of HAEC monolayers with a 1:40 final dilution of the designated antibody.

Initial studies were designed to determine if recognition of donor class I HLA antigens would alter lymphocyte adherence to HAEC monolayers. The number of DPL adherent to HAEC monolayers expressing donor class I HLA antigens was nearly twice the number of adherent TPL (Table 2). Although we found similar levels of adherence in the TPL population to those reported in the literature (6,7,8). DPL adherence was significantly greater, indicating the validity of this in vitro model in the study of lymphocyte–endothelial interactions unique to the allograft situation. To confirm that the differences in adherence were related to the allorecognition of HLA antigens, monoclonal antibody blocking studies were performed. The increase in DPL adherence, as compared to TPL adherence, was abolished in the presence of an anti-class I HLA mAb, but remained unaffected by mouse IgG or anti-class II mAbs (Table 3).

To study class II HLA antigen recognition in lymphocyte adherence to endothelium, HAEC monolayers were pretreated with Y-IFN and the above adherence assay was repeated. Although both DPL and TPL adherence to Y-IFN treated HAEC was significantly greater than the adherence to untreated HAEC, the DPL population continued to have a greater number of adherent cells than did the TPL population (Table 2). DPL adherence is presumably related to the allorecognition of HLA antigens on donor endothelium by alloreactive lymphocytes. Therefore, mAbs which prevent allorecognition should also interfere with any HLA-specific interaction required for the increased adherence seen in the DPL population. Since both class I and class II HLA antigens are expressed on HAEC that have been pretreated with Y-IFN, lymphocyte allorecognition may be directed against either the class I or class II antigenic determinants. The results of these monoclonal antibody blocking studies are given in Table 3. Although TPL adherence remained unaffected by the various mAbs used in the study, DPL adherence was prevented from surpassing TPL adherence in the presence of anti-class I or anti-class II mAbs. These findings suggest that HLA-specific allorecognition may be important in the adherence of donor-specific lymphocytes to allograft endothelium.

In the normal physiological situation, infiltrating lymphocytes adhere to autologous endothelium. However, in the allograft situation, lymphocytes adhere to an endothelium that displays allogeneic HLA antigens capable of triggering proliferative or cytotoxic lymphocyte responses (2).% In this paper, we have investigated this unique lymphocyte–endothelial interaction in vitro. In addition to demonstrating the differential expression of HLA antigens on HAEC, we have shown that alloreactive lymphocyte adherence is increased in the context of allorecognition and may, in fact, be related to an HLA-specific lymphocyte–endothelial cell interaction. These findings demonstrate the usefulness of this in vitro model to study the interactions between the vascular endothelium and lymphocytes involved in allograft immunity.

Acknowledgments. The authors would like to thank the Pittsburgh Transplant Foundation for their aid in donor sample procurement. We also wish to thank Drs. Bartley Griffith, Robert Hardesty, Carlos Esquivel, Shunzaboro Iwatsuki, Alfredo Trento, Robert Gordon, Andrei Steiber and Thomas Starzl for their cooperation. Utilization of the flowcytometer in the Pittsburgh Cancer Institute under supervision of Dr. Theresa Whiteside and the expert technical assistance of John Meharra was greatly appreciated.

Supported by the American Association of University Women Educational Foundation American Fellowship

Program (to Y. Colson), the Deutsche Forschungsgemeinschaft Research Fellowship Program (to B. Markus) and by grants HL36416 and A123467 from National Institute of Health and American Heart Association, Western Pennsylvania Chapter.

References

1. Zeevi A, Fung J, Zerbe TR, Kaufman C, Rabin BS, Griffith B, et al. Allospecificity of activated T cells grown from endomyocardial biopsies from heart transplant patients. Transplantation 1986;41:620.
2. Fung JJ, Zeevi A, Markus B, Zerbe TR, Duquesnoy RJ. Dynamics of allospecific T lymphocyte infiltration in vascularized human allografts. Immunol Res 1986;5:149.
3. Duquesnoy RJ, Zeevi A, Fung JJ, Kaufman C, Zerbe TR, Griffith B, et al. Sequential infiltration of class I and class II specific alloreactive T cells in human cardiac allografts. Transp. Proc. XIX. 1987;2560.
4. Colson YL, Markus BH, Zeevi A, Duquesnoy RJ. Interactions between endothelial cells and alloreactive T cells involved in allograft immunity. Transp Proc XX(2), 1988;273.
5. Colson YL, Markus BH, Zeevi A, Duquesnoy RJ. Adherence of allogeneic lymphocytes to human arterial endothelial cell monolayers. Transp Proc XX(1), 1988;54.
6. Yu CL, Haskard DO, Cavender D, Johnson A, Ziff M. Human gamma interferon increases the binding of T lymphocytes to endothelial cells. Clin Exp Immunol 1985; 62:554.
7. Haskard D, Cavender D, Ziff M. Phorbol ester stimulated T lymphocytes show enhanced adhesion to human endothelial cell monolayers. J Immunol 1986;137:1429.
8. Damle NK, Doyle LV, Bender JR, Bradley EC. Interleukin 2 activated human lymphocytes exhibit enhanced adhesion to normal vascular endothelial cells and cause their lysis. J Immunol 1987;138:1779.

Author Affiliation

Division of Clinical Immunopathology, University of Pittsburgh, Pittsburgh, PA 15213-3417, USA

In Vitro Lymphocyte Priming, Clonal Selection, and Response to Third-Party Lymphocytes by VG01⁺ and VG01⁻ T-Cell Fractions

Jorge F. Denegri and Jeanne Peterson

In this study we describe the selection of alloreactive primed lymphocytes by their expression of VG01 activation antigen. We also report that in vitro primed T lymphocytes (PTL) depleted of the VG01 positive cell fraction maintain their primary MLC proliferative capability suggesting that $VG01^+$ primed T cells also include the suppressor lymphocytes generated during MLC.

Peripheral blood lymphocytes were obtained from normal donors. After Ficoll-Hypaque separation the lymphocytes were stimulated in vitro with irradiated allogeneic mononuclear cells for 10 to 14 days. At the end of the priming period the responder lymphocytes were labeled with anti-VG01 and GAM-FITC and sorted into $VG01^+$ and $VG01^-$ cell fractions in a FACS 420 (Becton & Dickinson). Their proliferative response to the original stimulators and to third-party unrelated mononuclear cells was measured at 48, 72, and 120 hours, and compared to unseparated antibody treated and nontreated control PTL. The purity of the sorted VG01 depleted and enriched subpopulations was $> 95\%$.

$VG01^+$ PTL were responsible for the accelerated (48 to 72 h) proliferative response to alloantigens seen in the secondary MLC (Fig. 1). This finding was consistently found in six independent experiments. We also demonstrated a significant enrichment in the proliferative response of the $VG01^+$ PTL when compared to control unseparated PTL (paired t test, $p > 0.02$). The $VG01^-$ PTL were highly depleted of allospecific responding cells at 48 hours. The kinetics of response of $VG01^+$ primed cells was the one seen in a secondary MLC and similar to the kinetics of unseparated PTL.

The proliferative response of the $VG01^-$ PTL is shown in Figure 1. The kinetics of proliferation of this enriched cell fraction to third-party unrelated lymphocytes and also to the original stimulators did not show an early peak and was similar to that seen in a primary MLC. In contrast, the control unseparated PTL, containing both VG01 cell fractions, and the $VG01^+$ cell fraction did not proliferate to third party unrelated lymphocytes (Fig. 1). These preliminary observations together with our initial report on VG01[1] suggest:

1. Upon in vitro activation with allogeneic cells, specific alloreactive T cells express VG01. Thus, using anti-VG01 antibodies it could be theoretically possible to enrich or deplete recently alloactivated T lymphocytes. Resting mature lymphocytes are probably $VG01^-$. This cell subpopulation maintains the proliferative response to allogeneic cells after depletion of activated $VG01^+$ lymphocytes. It could be theoretically possible to selectively deplete recently activated T cells while sparing immune-competent resting lymphocytes.

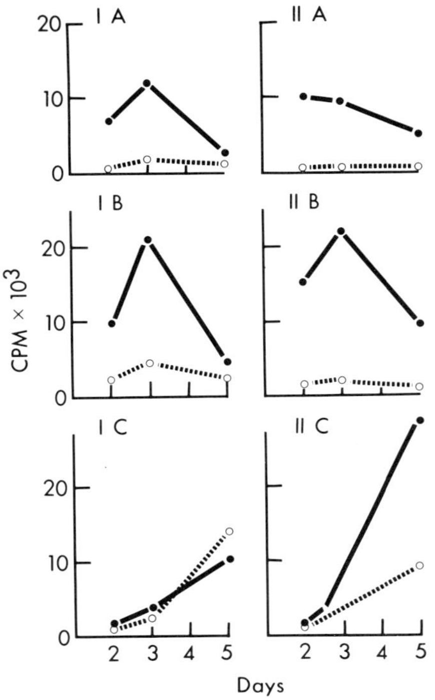

Figure 1. Proliferative response of VG01⁺ and VG01⁻ PTL. I, II, two independent experiments. A controlled unseparated PTL, B VG01⁺ PTL, C VG01⁻ PTL. ●———● Proliferation to original stimulators. ○ • • • • ○ Proliferation to third-party lymphocytes.

2. It is known that T lymphocytes activated in a mixed lymphocyte culture suppress the allogeneic response of autologous T cells and that this activity could be MHC specific and non-specific (2). The proliferative response observed in the VG01⁻ T cells, after depletion of the VG01⁺ PTL, suggests that the PTL suppressor activity resides within the VG01⁺ subpopulation.
3. Once the activated VG01⁺ PTL have been removed, the VG01⁻ PTL proliferate to the original stimulators with a primary MLC kinetics. This observations suggests that during in vitro priming not all the specific alloreactive clones are activated. It may be equivalent with the isolation during different rejection episodes of T-cell clones with different specificities against donor antigens.

References

1. Denegri JF, Gilks CB, Altin M, Peterson J, Lafleur L. A novel lymphocyte activation antigen present on allospecific primed lymphocytes and defined by the monoclonal antibody VG01. Transplantation (in press).
2. Odum N, Hofmann B, Jakobsen BK, Langhoff E, Morling N, Platz P, et al. HLA-DP related suppression of mixed lymphocyte reaction with alloactivated lymphocytes. Tissue Antigens 1986;27:32–43.

Author Affiliations

Jeanne Peterson, Immunology Laboratory, Division of Hematopathology, Department of Pathology, Vancouver General Hospital, Vancouver; *Jorge F. Denegri*, Immunology Laboratory, Division of Hematopathology, Department of Pathology, Vancouver General Hospital, and University of British Columbia, Vancouver, British Columbia, Canada

Association of Allograft Rejection and Inducible Interleukin-2 Receptor Expression on CD8 Lymphocytes

J.E. Anderson, M. Kelly, S. Jensik, J. Williams, and H. Gebel

Acute rejection of the transplanted kidney is a leading cause of graft failure in renal transplant recipients. Early clinical diagnosis of rejection in these patients is difficult because other processes such as drug toxicity and infection can diminish graft function, elevate serum creatinine, and mimic rejection. Recently, we had begun testing a new peripheral blood assay for the diagnosis of rejection among solid organ allograft recipients. A significant increase in inducible interleukin-2 receptor (IL2R) expression on CD8 lymphocytes from hepatic allograft recipients was observed during periods of rejection versus quiescence (37 ± 5 vs 9 ± 2; $p < 0.001$) documented by serial liver biopsies (1). In this study, we obtained peripheral blood from ten renal allograft recipients at least bi-weekly post-transplant to determine the predictive value of monitoring inducible IL2R on CD8 lymphocytes.

Ten randomly selected renal allografts (7 cadaveric and 3 living related) recipients were entered into the study over an 8-month period. For each recipient, there was at least a two antigen HLA-A,B,DR donor match. Routine immunosuppressive therapy consisted of antilymphocyte globulin for cadaveric recipients until the creatinine fell below 3.0 mg/dl or for 14 days, cortico-

Table 1. Peripheral Blood CD8 Lymphocytes Expressing Inducible IL2R in Hepatic and Renal Allograft Recipients

		Samples	
	Subjects	Tested	%CD8/CD25‡
Normals	10	43	19±2 (mean±SE)
Hepatic (no rejection)*	8	29	10±2
Hepatic (rejecting)*	3	6	37±5 (p<0.001)
Renal (stable)†	8	43	8±2
Renal (subsequent rejection)†	2	9	45±7 (p<0.001)

*Samples were obtained at the time of biopsy before the a.m. immunosuppressive medication dosages (corticosteroids and cyclosporine). Biopsies were routinely performed bi-weekly during the posttransplant hospitalization. Periods of nonrejection were determined by a biopsy with a histologic diagnosis of "no evidence of rejection" if the previous and subsequent biopsies also had "no evidence of rejection." Periods of rejection were defined by a histologic diagnosis of "rejection." Samples obtained after adjustment of immunosuppressive therapy to treat rejection were not included.

†Samples were obtained during the first two weeks posttransplant. Two patients subsequently rejected at 21 and 22 days posttransplant. Biopsies were not obtained on the renal patients during the first two weeks posttransplant.

‡Percentage of high density CD8 positive lymphocytes expressing IL2R.

steroids and cyclosporine (5–10 mg/kg/day) when the patient's creatinine fell below 3.0 mg/dl. Rejection was diagnosed based on clinical criteria consisting of deteriorating renal function, graft tenderness, fever, hypertension, and weight gain. Biopsies were obtained only if there was clinical suspicion of rejection. Peripheral blood samples were obtained prior to the a.m. immunosuppressive medication dosages. Mononuclear cells were isolated by density gradient centrifugation. A portion of the cells were incubated with a FITC conjugated anti-CD4 or anti-CD8, and PE conjugated anti-CD25 (Becton–Dickinson, California). The remaining cells were first cultured with phytohemagglutinin and then incubated with the monoclonal antibodies. Two-color flow cytofluorometry was performed with a Coulter EPICS V equipped with an Argon laser emitting 400 mW at 488 nm. Quadrant analysis was performed and the percentage of CD4 and CD8 cells expressing CD25 were calculated. In that low density CD8-positive cells may be NK cells (2,3), only high density CD8-positive cells were used for the calculations.

All patients achieved successful graft function postoperatively. Two patients had a rejection episode diagnosed approximately three weeks postoperatively (day 21 and 22, respectively). The first was a living related recipient receiving corticosteroids and cyclosporine. The second was a cadaveric recipient who received four days of anti-lymphocyte globulin, corticosteroids, and cyclosporine beginning on day 5.

The renal allograft recipients experiencing a rejection episode had a significantly increased percentage of CD8 lymphocytes expressing inducible IL2R during the first two weeks after transplantation (Table 1). For the patients who did not experience a subsequent rejection episode, there was a lack of activation among the CD8 lymphocytes during the two weeks after transplantation. Earlier studies of hepatic allograft recipients showed a correlation between an increased percentage of CD8 lymphocytes expressing inducible IL2R and histologic evidence of rejection (Table 1). These preliminary results indicate that, among renal and hepatic transplant patients, monitoring inducible IL2R expression on CD8 lymphocytes appears to be a useful, non-invasive method facilitating detection of rejection episodes.

References

1. Anderson J, Kelly M, Economou S, Jensik S, Williams J, Gebel H. Rejection of transplanted organs: A new method of detection. Fed Proc 1987;46:1357.
2. Lanier LL, Loken MR. Human lymphocyte subpopulations identified by using three-color immunofluorescence and flow cytometry analysis: Correlation of leu-2, leu-3, leu-7, leu-8, and leu-11 cell surface antigen expression. J Immunol 1984;132:151–156.
3. Gebel HM, Anderson JE, Gottschalk LR, Bray RA. Determination of helper:suppressor T-cell ratios. N Eng J Med 1987;316:113.

Author Affiliations

Departments of General Surgery and Immunology/Microbiology, Rush University, Chicago, IL 60612, USA

An In Vitro Model to Investigate the Importance for Renal Transplant Survival of Mismatches for Specific Histocompatibility Antigens

Joseph M. MacSween, James A. Collicut, Stuart L. Eastwood, and Allan D. Cohen

The importance of specific donor and recipient HLA antigens for transplant survival has been difficult to estimate from clinical outcomes because of constraints on experimentation imposed by transplantation. As an alternative approach we investigated the feasibility of inducing sensitization and effector responses in vitro using assays for macrophage migration inhibition (MIF) and stimulation (MStF) factors. Supernatants were assayed for MIF and MStF activity in triplicate by the capillary tube method using the human monocytoid cell line U937 as target cells (1). The ratio of migration areas produced by supernatants of alloantigen-stimulated cells to those of unstimulated cells was multiplied by 100 to give a migration index with values of < 91, 91–109, and > 109 taken as MIF, negative, or MStF responses respectively. These responses were determined for recipient mononuclear cells stimulated by an extract of donor spleen or peripheral blood mononuclear cells at 3, 7, 10, and 14 days after renal transplantation from cadaver donors (2). The responses of cells sensitized in vitro in mixed lymphocyte reactions (MLR) were also determined using responder cells from renal transplant recipients obtained prior to transplantation and donor cells as stimulators. Lymphokine responses in MLR were also evaluated using cells from healthy individuals as stimulators and responders. Stimulator cells were pulsed with puromycin 5 µg/ml for two hours prior to culture and responder cells were stimulated with a soluble extract from the stimulator cells after 72 hours culture. Typing for HLA A,B,C, and DR antigens was performed with tissue typing trays obtained from the Canadian Red Cross.

Post transplant responses to donor cell extracts in recipients treated with prednisone and cyclosporine correlated with one-year graft survival; MIF responders had 11/21 (52%) functioning grafts compared to 23/25 (92%) for MStF responders (X^2 =9.3, p = 0.0027). In patients treated with prednisone and azathioprine, MIF responders had 4/12 (33%) functioning grafts compared to 11/15 (73%) for MStF responders (X^2 = 4.32, p = 0.035). After sensitization by donor cells in MLR there was a highly significant correlation between 30 pretransplant responses and the corresponding posttransplant responses (r = 0.80, t = 6.6, p = < 0.0001). There was also a significant correlation with pretransplant responses in MLR and graft survival at 100 days in patients treated with prednisone and cyclosporine in that MIF responders had 5/8 (63%) functioning kidneys whereas MStF responders had 21/21 (100%) functioning kidneys (X^2 = 8.78, p = 0.003).

We propose to correlate HLA B and DR phenotypes with these lymphokine responses since matching for these antigens has been linked to improved transplant survival. In order to discern whether the phenotypes of the stimulator or responder cells were more important in influencing the lymphokine responses, we reviewed the results of 360 assays using cells typed for HLA B antigens and 346 assays using cells typed for HLA DR antigens. When the frequencies of the eight commonest HLA B antigens and DR 1–8 in our study population were compared to those published for a normal Caucasian population (3), the greatest divergence from the norm occurred when the DR antigens of the stimulator cells were separated on the basis of the MIF/MStF responses (Table 1).

These results demonstrate that MIF/MStF responses to donor cell extracts after sensitization in vivo following transplantation or in vitro in MLR correlate with renal transplant survival. These in vitro responses appear therefore to be a valid model reflecting immunologic factors that influence renal transplant survival suggesting that they can be used to investigate the effect of specific HLA phenotypes on graft survival. The observation that the frequency of individual HLA DR antigens in stimulator cells separated on the basis of MIF or MStF responses differed maximally from that published for North American Caucasians suggests that certain stimulating DR antigens selectively associated with either MIF or MStF, presumably by influencing the responses. The HLA B antigens and the DR antigens of responder cells did not show this segregation suggesting they were less important in influencing the responses.

Acknowledgment. This work was supported in part by the Kidney Foundation of Canada and by the Dalhousie University Internal Medicine Research Foundation.

Table 1. Comparison of the Observed Frequency of 8 HLA B and 8 HLA DR Antigens with that Expected in the Normal Population

Antigens	Population Analyzed	Comparison of Observed and Expected Antigen Frequency: Stimulators	Responders
HLA B	Unseparated	$\chi^2 = 16.7$	$\chi^2 = 29.0$
	MIF and MStF Subpopulations	$\chi^2 = 26.5$	$\chi^2 = 31.6$
HLA DR	Unseparated	$\chi^2 = 30.2$	$\chi^2 = 28.3$
	MIF and MStF Subpopulations	$\chi^2 = 55.9$	$\chi^2 = 36.5$

References

1. MacSween JM, Rajaraman K, Rajaraman R, Fox RA. Macrophage migration inhibition factor (MIF): Reducing the variables. J Immunol Methods 1982;52:127–136.
2. MacSween JM, Cohen AD, Rajaraman K. Donor-specific cellular responses and preexisting nonspecific responsiveness as interrelated determinants of renal transplant survival. Transplant Proc 1984;16:1410–1412.
3. Baur MP, Neugebaur M, Albert ED. Reference tables of two-locus haplotype frequencies for all MHC marker loci. In: Albert ED et al (eds): Histocompatibility Testing 1984. Berlin, Springer-Verlag, 1984, pp 692–694.

Author Affiliations

Department of Medicine, Dalhousie University, Camp Hill Hospital and Transplant Unit, Victoria General Hospital, Halifax, Nova Scotia, Canada

Characterization of γ:δ T-Cell Receptor Expressed on Peripheral Blood Lymphocytes After Allogeneic Bone Marrow Transplantation

E. Vilmer, M. Schumpp, F. Triebel, F. Sigaux, and A. Bensussan

Abstract: Two kinds of T-cell antigen receptor (TCRs) α:β and γ:δ, have so far been identified as independently expressed on the cell surface of distinct subsets of T cells in association with the CD3 (T3) complex (1,3). The surface expression of these TCR was systematically analyzed in peripheral lymphocytes (PBL) from 27 patients (pts) having received allogeneic bone marrow transplantation for leukemia. A predominant emergence of circulating CD3+ lymphocytes– 40% of PBL– lacking α:β heterodimer was demonstrated in 3 pts. In two pts, we showed that these cells expressed γ chains associated to CD3. In one pt, functional analyses indicated that this γ:δ subset representing 40% of PBL may have played a role in a profound immunodeficiency state which occurred six months after grafting. Molecular studies of γ:δ cells indicated that these circulating cells preferentially expressed a single V_γ segment of the T-cell receptor γ gene (TRG): $V_\gamma5$ for pt 1 and $V_\gamma9$ for pt 2.

Three and six months after grafting, cytofluorographic analyses of PBL were performed in 2 groups of pts grafted for leukemia with an HLA-matched donor. The former group (11 pts) received T-cell depleted bone marrow, and the latter (16 pts) unmanipulated marrow. We used anti CD3, CD4, CD8 monoclonal antibodies (mAb), anti WT31, mAb which recognizes a constant part of α:β heterodimer (4), and anti TiγA mAb directed towards the products of a Vγ9-JγP rearrangement (5,6). Previously we confirmed that, in normal blood, 5% of cells (range 0 to 10%) were CD3+ WT31- and that 3% of PBL reacted with anti TiγA. In 3 out of 27 pts, the presence of peripheral CD3+ WT31-cells attaining 40% PBL was detected (Table 1). Anti TiγA reacted with 33 to 40% of PBL from pt 2, but did not with PBL from pt 1 and 3. The CD3+ WT31-population was characterized in pt 1 and pt 2. Pt 1 was grafted with T-cell depleted bone marrow, for chronic myelogenous leukemia. He

Table 1. Longitudinal Phenotypic Analysis of PBL from the 3 Patients

Patient No.	Time After Grafting (min)	CD3	Membrane Antigens*			
			WT31	CD4	CD8	TiγA
1	3	37	35	17	37	0
	6	70	29	5	28	0
	12	75	25	4	23	0
2	3	86	45	9	36	40
	6	74	36	14	34	33
3	6	95	56	18	82	0

*Percentage of positive PBL.

Table 2. Functional Analysis of PBL from Pt 1 before and after Sorting with Anti WT31 or Anti CD3 mAb

PBL	Medium	T3 Beads	Allogeneic Unrelated Cells
Donor cells	135	7850	5430
Unfractionated patient cells	50	1237	150
WT31+ sorted patient cells	40	5491	1283
WT31– sorted patient cells	20	620	45
CD3+ sorted patient cells	105	930	210
Donor cells and WT31+ sorted cells	150	8632	ND
Donor cells and WT31– sorted cells	112	730	ND

These results are expressed in incorporation of (^{3}H) thymidine, cpm.

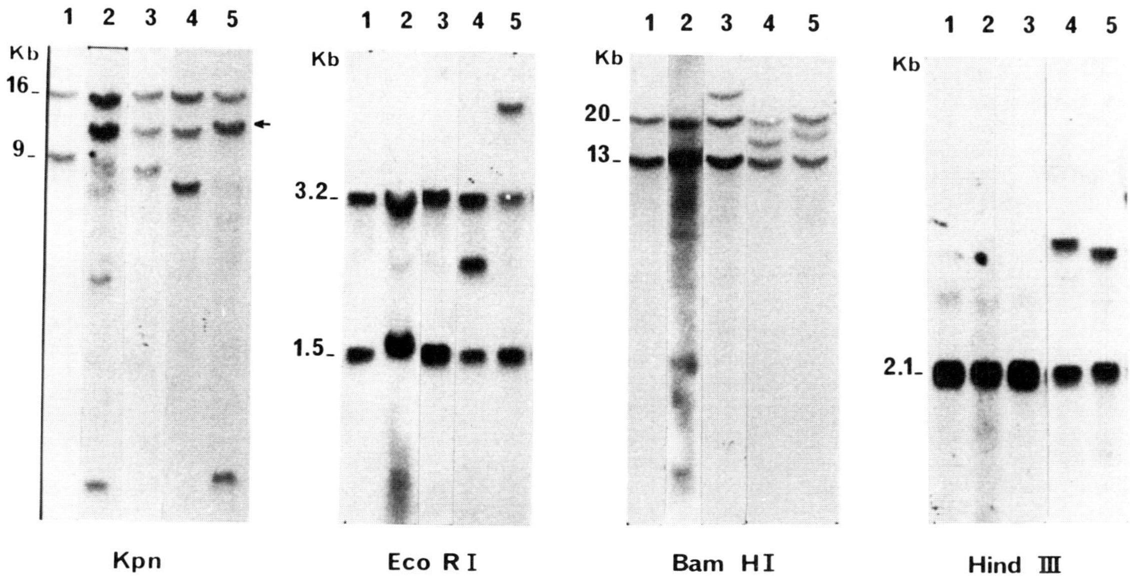

Figure 1. Analysis of TRGγ by southern blotting from clones and cell lines derived from pt 2 PBL. Clones and cell lines were CD3+WT31−TiγA+. Hybridizations were performed using Jγ probe (Ph 60) (6). Lane 1 indicates the germ line control, lane 2, the cell line and lane 3, 4, 5, different TiγA + clones. In KpnI digested DNA, clones and cell line demonstrated the apperance of a 12-kb rearranged band and the loss of 9- kb germ line band consistent with a Vγ9-Jγp rearrangement (6). On the second Vγ locus allele three different patterns of Vγ rearrangements were shown in each clone (lanes 3,4,5).

developed a severe cytomegalovirus (CMV) retinitis in the 24th week post grafting which evoked an unusual and persistent immunodeficient state which was unexpected. Pt 2 was grafted for acute lymphoblastic leukemia in second remission with unmanipulated bone marrow. In the second month, a CMV infection with cytopenia and hepatitis spontaneously resolved. As of the third month post grafting, no new complications were noted.

The biochemical structure of the TCR was studied using CD3+ WT31- clones derived from pt 1 and pt 2 PBL according to methods already described (7). In a one-dimensional SDS PAGE, a disulfide-linked structure of 90 kd was immunoprecipitated with the CD3 complex in the clones derived from pt 1 and pt 2. Upon reducing conditions, the 90-kd structure dissociated into 3 subunits of 45, 40, and 37 kd in pt 1 and into 2 subunits of 44 and 41 kd in pt 2. According to procedures previously detailed, sequential immunoprecipitations were performed with a hetero antiserum anti Cγ and we showed that 40, 37 kd band from pt 1 clones and the 44, 41 kd bands from pt 2 clones were γ chains. Thus, the upper band of 45 kd in pt 1 clone has to correspond to the δ chain.

In an attempt to understand the relationship between the emergence of this γ:δ subset and the immunodeficiency after grafting, functional studies were performed from PBL and from γ:δ clones (7). PBL from pt 1 did not proliferate to any stimulus (Table 2) 6 months after grafting. Such unresponsiveness was generally not observed in other patients studied (data not shown). To study the function of purified subsets, cell sorting experiments of PBL from pt 1 were performed using anti WT31 and anti CD3 mAb. Unseparated PBL, or the CD3 sorted cells, did not respond to stimuli whereas the WT31+pt PBL, like donor PBL, exhibited a positive response to the stimuli tested. These results suggest that WT31-cells may have an inhibitory effect. To test this hypothesis, we used the bone marrow donor cells, genotypically identical to the pt cells. The WT31−sorted cells and not the WT31+ cells were able to inhibit the proliferative response of donor cells induced by antiCD3 beads (Table 2). Taken together, these data suggest that the CD3+WT31- subset may mediate the unresponsiveness of PBL from pt 1. In contrast, PBL from pt 2 exhibited a normal proliferative response to mitogens and to anti CD3 beads and the immune reconstitution occurred according to the classical timing observed in grafted pts. To explain this difference between both pts, we speculate that CD3+WT31−cells from pt 1 were specifically activated in vivo to induce this suppressive effect. We next assessed the ability to activate the TCR/CD3 complex in these γ:δ clones from pt 1 and pt 2 as described previously (8,9). When clones DS6 and DS8 from pt 1 were triggered with anti-CD3 beads, no proliferative response was observed. However CD3 triggering enhanced exogenous rIL2 response. In contrast, several Ti gamma A positive clones from pt 2 could be directly induced to proliferate by the same stimuli, without adding rIL2.

We sought to find out whether this circulating γ:δ subset was clonal and/or expressed a preferential Vγ gene segment. DNA from PBL, cell lines, and clones were digested with EcoR1, BamHI, Hind III, and KpnI enzymes and assayed by Southern blot analysis using Cβ or Jγ probe (6,10). For pt 1 and pt 2, no rearranged TCRβ band was detected in the PBL analysis. In all the γ:δ clones studied, the TCRβ chain genes were found in germ line configuration. For pt 1, the size of rearranged fragments detected with the Jγ probe, directly in PBL or in γ:δ clones, was consistent with the involvement of Vγ5 and Jγ1 segments in TRGγ rearrangement. The monoallelic characteristic of this Vγ5 rearrangement in the two clones studied and the sequence analysis of the Vγ5 segment (11) have indicated that the rearrangement of Vγ5 was functional. In pt 2, the circulating γ cells being anti-TiγA positive we have sought a Vγ9-JγP rearrangement which is detectable after KpnI DNA digestion (10) (Fig. 1). As expected γ expressing cell line and clones demonstrated the appearance of a rearranged band typical of this Vγ9-JγP rearrangement. Nevertheless, these data do not allow us to determine the clonality of this subset without the characterization of Vγ segment rearranged on the second Vγ locus allele. In the TiγA positive cell line and in each of the 6 clones tested, different patterns of Vγ rearrangements were observed suggesting that the TiγA positive PBL subset was polyclonal as was the minor TiγA-positive population normally residing in the blood (5,6).

References

1. Bank I, Delinho RA, Brenner MB, Cassimeris J, Alt FW, Chess L. A functional T3 molecule associated with a novel heterodimer on the surface of immature human thymocytes. Nature 1986;322:179.
2. Brenner MB, McLean DP, Dyalynas DP, Strominger JL, Smith JA, Owen FL, et al. Identification of putative second T cell receptor. Nature 1986;322:145.
3. Weiss A, Newton M, Crommie D. Expression of T3 in association with a molecule distinct from the T cell antigen receptor heterodimer. Proc Natl Acad Sci USA 1986;83: 6998.
4. Spits H, Borst J, Tax W, Capel PJA, Terhost C, Devries JE. Characteristics of a monoclonal antibody (WT31) that recognizes a common epitope on a human T cell receptor for antigen. J Immunol 1985;135:1922.
5. Jitsukawa S, Faure F, Lipinski M, Triebel F, Hercend T. A novel subset of human lymphocytes with a TCR gamma complex. J Exp Med 1987;166:1192–1197.
6. Triebel F, Faure F, Graziani M, Jitsukawa S, Lefranc MP, Hercend T. A unique Vγ9 rearrangement is expressed in a majority of CD3+TCR alpha-beta-circulating lymphocytes. J Exp Med 1988;167:694.
7. Vilmer E, Guglielmi P, David V, Leca G, Rabian C, Degos L, et al. Predominant expression of circulating CD3+ lymphocytes bearing gamma T cell receptor in a prolonged immunodeficiency after allogeneic bone marrow transplantation. J Clinical Invest 1988;82:755.
8. Bensussan A, Tourvieille B, Chen LK, Dausset J, Sasportes M. Phorbol ester induced a differential effector function of human allospecific CTL and NK clones. Proc Natl Acad Sci USA 1985;82:6642.
9. Bensussan A, Acuto O, Hussey RE, Milanese C, Reinherz EL. T3-Ti receptor triggering of T8+ suppressor T cells beads to unresponsiveness to interleukin. 2 Nature 1984; 311:565.
10. Vilmer E, Triebel F, Rabian C, David V, Schumpp M, Leca G, et al. Prominent expansion of circulating lymphocytes bearing gamma T cell receptor with preferential expression of variable gamma genes, after allogeneic bone marrow transplantation. Blood 1988;72:841.
11. Bories JC, Guglielmi P, Sigaux F, Bensussan A. Nucleotide sequence of a CDNA corrresponding to a new human variable segment of a functionally rearranged T cell receptor gamma chain gene. Nucleic Acid Research 1987;15:10059.
12. Huck S, Lefranc MP. Rearrangements to the JP1 JP and JP2 segments in the human T-cell rearranging gamma gene (TRG) locus. FEBS letter 1987;224:291.

Author Affiliations

E. Vilmer, M. Schumpp, A. Bensussan, INSERM 93; *F. Sigaux*, INSERM, 301, Hôpital Saint-Louis, Paris; *F. Triebel*, Laboratoire d'Immunulogie Institut Gustave Roussy, Villejuif, France

TCA and MLCs in Bone Marrow Transplant Patients

A. van Leeuwen, A.A. Biegel, F.P.M.J. van Bree, R. Willemze, W. Fibbe, A. Termijtelen, and J.J. van Rood

The TCA system in man is a di-allelic system of HLA-like antigens, encoded by genes located about 15 cM telomoric to HLA-A. The TCA antigens are detected by complement-dependent cytotoxicity assays using ethidium bromide as indicator. In normal persons, T_G lymphocyte enriched suspensions are used for typing purposes (1).

In T-ALL, AML, and CML patients, using nonenriched suspensions, an increased level of TCA-positive cells appears in the peripheral blood, including lymphoid blasts. In general, those high levels decrease during treatment with anti-leukemic agents and, in some cases, increase again upon relapse. TCA expression is not elevated in patients with CLL, HCL, and non-T, non-B ALL (2).

Immunobiology of HLA
Volume II

Increased TCA expression after BMT in autologous and allogeneic situations.

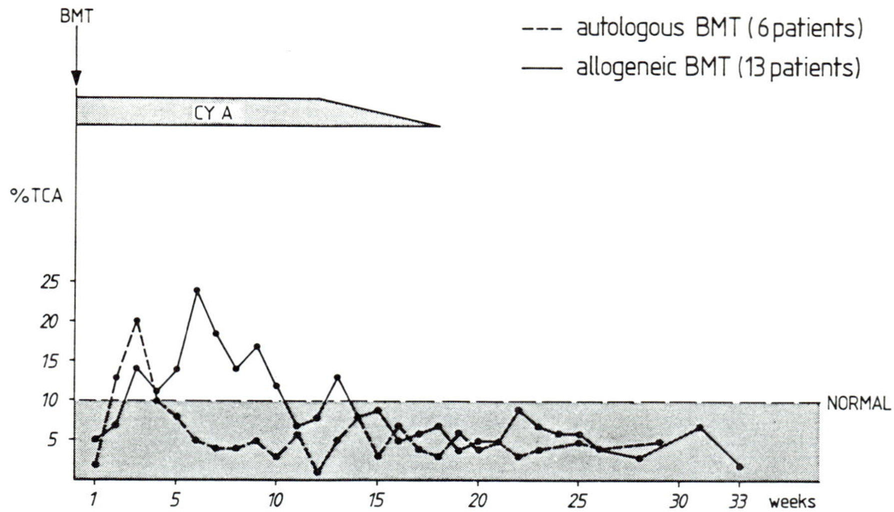

Figure 1. The cyclosporin A (Cy A) therapy indicated in this figure applies only to the allogeneic BMTs.

Since some melanoma cell lines and a teratocarcinoma cell line also carry an elevated TCA level, this marker is not limited to lymphoid cells alone. Its biological function is as yet unknown. However, it may be a differentiation marker, since TCA positive cells are also increased upon the repopulation of the peripheral blood of patients after bone marrow transplantation (BMT). In this situation, the appearance of TCA-positive cells is slightly delayed in allogeneic T-cell-depleted BM transplants as compared to autologous, nondepleted situations. The comparison is set out in Figure 1 where it is shown that the peak of maximum TCA expression in the autologous BM transplants (4 AML, 1 NHL, and 1 ALL) is between two and three weeks after transplantation, while in the allogeneic transplants (7 AML, 3 CML, 2 NHL, and 1 ALL), the peak appears between three and ten weeks after transplantation. All patients showed a normal TCA expression before BMT. Note that the allogeneic BMT patients were treated with cyclosporin A for about four months in contrast to the autologous BMT patients who did not receive any cyclosporin A.

Our preliminary findings so far do not support the notion that the increased expression of TCA after BMT correlate with increased expression of TCA in the bone marrow transplant.

Taken together, we studied 15 allogeneic BMT recipients. All patients were transplanted with an HLA-identical sibling, except one. This patient received BM from his phenotypically identical mother. In this last case, the MLC reaction was positive before BMT. Two AML patients did not show any increase in TCA expression after BMT. They were negative in MLC reactivity before transplantation. The remaining 12 patients showed slight stimulation of the MLC reactivity.

This observation suggests an influence of donor cells that causes an increased expression of TCA. The effect of this increased TCA expression on the outcome of the graft is under study.

Acknowledgments. Supported in part by the Dutch Foundation for Medical and Health Research (MEDIGON) and the J.A. Cohen Institute for Radiopathology and Radiation Protection (IRS).

We thank Mrs. Ellen van der Willik for typing the manuscript.

References

1. Van Leeuwen A, Giphart MJ, De Groot G, Morolli B, Festenstein H, Nijenhuis LE, et al. Two different T-cell systems in humans, one of which is probably equivalent to Qa or Tla in mice. Human Immunol 1985;12:235–246.
2. Van Leeuwen A, Schrier PI, Giphart MJ, Noordermeer IA, Ruiter DJ, Rubinstein P, et al. TCA—a polymorphic genetic marker in leukemias and melanoma cell lines. Blood 1986; 67:1139–1142.

Author Affiliations

A. van Leeuwen, *F.P.M.J. van Bree*, *A. Termijtelen*, *J.J. van Rood*, Department of Immunohematology/Blood Bank; *R. Willemze*, *W. Fibbe*, Department of Hematology, University Hospital, Leiden, The Netherlands; *A.A. Biegel*, Department of Medicine, Indiana University Medical College, Indianapolis, IN, USA

Section XVIII
Clinical Transplantation

Marrow Transplantation from Unrelated HLA-A,B,DR-Matched Volunteer Donors

Patrick G. Beatty, John A. Hansen, Charlotte Atcher, and E. Donnall Thomas

Abstract: The Seattle Marrow Transplant Team has conducted searches for unrelated donors for 318 patients. HLA-A,B,DR-matched MLC compatible donors were found for 18 (6%), with an average search time of 165 days (range 25-598). Of these 18 patients transplanted, plus four with minor HLA incompatibilities, nine are currently alive. Acute graft-versus-host disease (AGVHD) prophylaxis with cyclosporine and methotrexate appears to have had a substantial beneficial impact upon both incidence of severe AGVHD and survival. The current challenge is to expand the number of available HLA-typed volunteer donors and to explore strategies for defining minimal acceptable limits of HLA incompatibility for unrelated donors.

Bone marrow transplantation has become a standard form of therapy for several hematologic diseases. However, less than 40% of patients will have a genotypically matched sibling (1), with another 10% having an acceptable partially matched relative (2). For this reason, several transplant centers have begun exploring the use of HLA-matched unrelated donors when there is no available family donor.

In 1979, the Seattle Bone Marrow Transplant team performed a successful marrow transplant for a 10-year-old leukemic patient from an HLA-A,B,DR-matched unrelated donor (3). The potential for additional unrelated donor transplants, however, was seemingly unapproachable because HLA-matched unrelated donors could not be found for more than a small fraction of patients. To expand the base for potential donor searches, a collaborative effort was initiated between the Blood Center of Southeastern Wisconsin in Milwaukee, the Portland Red Cross, the Puget Sound Blood Center in Seattle, the St. Paul Red Cross, and the Anthony Nolan Foundation of London. The combined database was approximately 20,000 in the United States and 50,000 in the United Kingdom, most of whom were typed for HLA-A,B only.

Table 1. HLA-A,B,DR MLC Nonreactive Match Search

Stage of Search	# of Patients (%)	Avg Time to Complete (Range)
Begin donor search	318 (100%)	–
Find HLA-A,B match	242 (76%)	28d (0-737d)
Find HLA-A,B,DR match	49 (15%)	113d (7-547d)
Find HLA-A,B,DR,D (MLC-)	31 (10%)	165d (25-598d)
Transplanted	18 (6%)	205d (89-576d)

Table 2. AGVHD Prophylaxis Impact on Transplant Patients

AGVHD Prophylaxis	# Severe AGVHD (%)	# Alive
Methotrexate (MTX)	4/6 (67%)	0/6 (0%)
MTX + cyclosporin	1/16 (5%)	9/16 (56%)

From 1980 to May 1987, 318 patients in Seattle met protocol criteria for transplantation, of whom 31 patients (10%) found HLA-A,B,DR MLC nonreactive matches (see Table 1). Seven patients elected not to proceed, one donor elected not to proceed, two patients had other types of transplant, two patients had transplants at other centers, and one patient died before transplant.

Twenty-two patients have been transplanted as of mid-August 1987. The first six patients received standard methotrexate (MTX) for AGVHD prophylaxis. Of these six, four developed severe Grade III or IV AGVHD, and none of these six have survived. The subsequent 22 received a combination of MTX and cyclosporin. Of these, only one experienced severe AGVHD, and nine are alive, engrafted, and free of significant GVHD from 80 to 914 days after transplant (see Table 2).

The favorable results in the most recent group of 16 patients indicates the efficacy of transplants from HLA-A,B,DR-matched unrelated donors and suggests that survival may approximate the results of transplants from HLA genotypically identical siblings. In order to bring this potentially curative therapy to more patients, it will be necessary to both expand the currently available number of volunteer donors and improve the efficiency of the search process so donors can be located in a timely way before the clinical condition of the patient begins to deteriorate. However, even with registry sizes of 100,000, we would expect to find HLA-A,B,DR-matches for only 30% of patients (4). For the remainder, it will be necessary to develop strategies for choosing partially compatible unrelated volunteer donors with minimal HLA disparities that are least likely to lead to graft rejection or severe GVHD.

Acknowledgments. Supported in part by grants CA 18029 from the National Cancer Institute and HL 33775 from the National Heart, Lung and Blood Institute, National Institutes of Health. Dr. Thomas is a recipient of a Research Career Award (AI 02425) from the National Institute of Allergy and Infectious Diseases.

References

1. Appelbaum FR, Hewlett J, Kopecky K, Angle C, Coltman C. Prospective comparative study of bone marrow transplanta-

Immunobiology of HLA
Volume II

tion versus continued chemotherapy for adult acute nonlymphocytic leukemic: A Southwest Oncology Group Study. Blood 1986;68:216a.
2. Beatty PG, Clift RA, Mickelson EM, Nisperos B, Flournoy N, Martin PJ, et al. Marrow transplantation from related donors other than HLA identical siblings. N Engl J Med 1985;313:765.
3. Hansen JA, Clift RA, Thomas ED, Buckner CD, Storb R, Giblett ER. Transplantation of marrow from an unrelated donor to a patient with acute leukemia. N Engl J Med 1980; 303:565.
4. Beatty PG, Dahlberg S, Mickelson EM, Nisperos B, Opelz G, Martin JP, et al. Probability of finding HLA matched unrelated marrow donors. Transplantation 1988;45:714.

Author Affiliations

Patrick G. Beatty, John A, Hansen, Fred Hutchinson Cancer Research Center, The Puget Sound Blood Center, and Department of Medicine, University of Washington; *Charlotte Atcher*, The Puget Sound Blood Center; *E. Donnall Thomas*, Fred Hutchinson Cancer Research Center and Department of Medicine, University of Washington, Seattle, WA, USA

Risk Factors for Graft Rejection of Marrow Transplants From HLA-Haploidentical Donors

C. Anasetti, D. Amos, P.G. Beatty, P.J. Martin, E.D. Thomas, and J.A. Hansen

Marrow transplantation from an HLA-identical sibling can be lifesaving for patients suffering from a variety of hematologic malignancies, but only 40% of the patients in the United States have such a donor. On the other hand, nearly all patients have an HLA-haploidentical relative, such as a parent or a child, who could serve as a donor. We reported previously an increased incidence of acute graft-versus-host disease in recipients of an HLA haploidentical marrow (1). In this study, we determined the incidence of graft failure in patients with hematologic malignancies transplanted from family members who share one HLA haplotype but differ to a variable degree for the HLA-A,B, and D antigens of the nonshared haplotype. Furthermore, we identified the risk factors for graft failure by using a multivariate analysis.

The relevance of HLA compatibility to achievement of sustained marrow engraftment was analyzed in 269 patients transplanted from HLA haploidentical family members. Results were compared to those for 930 patients transplanted from an HLA genotypically identical sibling donor. All patients had a hematologic malignancy and were treated with cyclophosphamide and total body irradiation followed by the infusion of unmodified marrow cells. Recipients of HLA haploidentical marrow had a 14.6% probability of graft failure versus 2.2% for the controls ($p < 0.001$).

A multivariate proportional hazards regression analysis was used to identify the risk factors associated with graft failure among patients transplanted with HLA-haploidentical marrow. Patients with a positive anti-donor lymphocytotoxic crossmatch, reactive against either T cells or B cells, had a higher rater of graft failure than patients with a negative test (49.7% versus 12.1%, $p < 0.01$) (Table 1). Marrow grafts incompatible for HLA-B and -D had a 33.7% probability of failure versus 9.8% ($p < 0.001$) for all other combinations of incompatibility (Table 2). Patients with leukemia in relapse had a higher rate of graft failure than those in remission (19.0% versus 9.5%, $p < 0.05$).

Residual host lymphocytes were detected in 11 of 14 patients with graft failure who had informative markers, by either HLA typing or karyotype analysis, sug-

Table 1. Failure to Achieve Sustained Engraftment of Haploidentical Marrow Transplants

Crossmatch	No. of Patients	Cases of Graft Failure
Anti-donor T ± B cells		
Positive	7	3
Negative	262	30
Anti-donor B cells		
Positive	11	4
Negative	258	29

Table 2. Failure to Achieve Sustained Engraftment of Haploidentical Marrow Transplants

Host vs. Graft Incompatibility	No. of Patients	Cases of Graft Failure
Zero locus	43	3
One locus		
HLA-A only	47	4
B only	20	1
D only	54	6
Two loci		
HLA-A+B	15	2
A+D	18	1
B+D	53	15
Three loci	19	1

Immunobiology of HLA
Volume II

gesting that immune rejection was the cause of graft failure.

These results indicate that donor incompatibility at the major histocompatibility complex increases the probability of marrow graft rejection, particularly in patients sensitized specifically to their donor.

Acknowledgments. Supported in part by grants CA 18029, CA09515, and CA 15704 from the National Cancer Institute and HL 33775 and HL 36444 from the National Heart, Lung and Blood Institute, National Institutes of Health. Dr. Thomas is a recipient of a Research Career Award (AI 02425) from the National Institute of Allergy and Infectious Diseases.

References

1. Beatty PG, Clift RA, Mickelson EM, Nisperos B, Flournoy N, Martin PJ, et al. Marrow transplantation from related donors other than HLA identical siblings. N Engl J Med 1985;313:765–771.

Author Affiliations

C. Anasetti, D. Amos, Fred Hutchinson Cancer Research Center; *P.J. Martin, E.D. Thomas*, Fred Hutchinson Cancer Research Center and Department of Medicine, University of Washington; *P.G., Beatty, J.A. Hansen*, Fred Hutchinson Cancer Research Center, The Puget Sound Blood Center, and Department of Medicine, University of Washington, Seattle, WA 98104, USA

HLA-DP and Acute Graft Versus Host Disease

Niels Ødum, Claus Munk Petersen, Bodil K. Jakobsen, Niels Jacobsen, Niels Morling, Lars Lamm, and Arne Svejgaard

Abstract: In vitro studies have demonstrated that HLA-DP antigens can elicit allogeneic proliferative and cytotoxic responses (1), induce allospecific (2) and nonspecific suppression (3), and act as restriction elements in antigen presentation to cloned T cells (4). Nevertheless, only little is presently known about the immunological role of HLA-DP antigens in vivo. Recently, HLA-DP antigens have been shown to contribute to the susceptibility to juvenile chronic arthritis (5,6), probably multiple sclerosis (7,8), and perhaps severe aplastic anemia (9). To study the possible role of HLA-DP antigens in transplantation, we DP-typed 15 recipients successfully engrafted with HLA-haploidentical, DR-compatible bone marrow (BM) and their BM donors using primed lymphocyte typing (PLT) (10). Severe acute graft versus host disease (GVHD) ($\geq$ grade 2) developed within 3 months after BM transplantation in all recipients of HLA-DP-incompatible BM (in the "GVH-direction"), but in only one of seven recipients of HLA-DP-compatible BM. This difference was very significant ($p= 0.0014$, Fisher's exact test). Incompatibility for HLA class I antigens did not correlate with acute GVHD or with incompatibility for HLA-DP antigens. Thus, HLA-DP determinants play a role as transplantation antigens.

Results and Discussion. The strength of the one-way donor-to-recipient mixed leucocyte culture (MLC) reaction (MLR) has usually been considered predictive for the risk of developing acute GVHD after allogeneic BM transplantation (BMT): a high MLR indicates an increased risk and a low MLR indicates a decreased risk of developing severe acute GVHD. Recently, the validity of the MLR as a predictive test in relation to GVHD has been questioned (11). Thus, de Gast et al. (11) found no correlation between the MLR and either the incidence or the severity of acute GVHD in recipients of allogeneic BM. We have analyzed the correlation between the donor-to-recipient one-way MLR and the occurrence of severe acute GVHD ($\geq$ grade 2) in 15 recipients of haploidentical, DR-compatible BM. Figure 1 shows the relative responses (RR) obtained from these MLR combinations and from 38 MLRs between healthy haploidentical , DR-compatible family members. The RR values obtained from the donor-to recipient MLRs seemed to be higher in combinations, with recipients later developing severe acute GVHD (median RR= 11%, range=0-32%) than in combinations without severe GVHD (median RR= 4%, range=0-11%), but this difference was, however, not significant, and it was not possible from the RR value to predict whether or not a given recipient would develop severe acute GVHD. However, all three BMT recipients stimulating their donors to RR values above 20% developed severe acute GVHD.

In this small sample, there was no significant difference between the MLRs obtained from DP-compatible (median RR= 5%, range=0–26%) and DP-incompatible donor-recipient combinations (median RR= 8%, range=0–32%), whereas the MLRs between 11 haploidentical, DR-compatible, DP-incompatible healthy individuals were significantly ($p<$ k0.01) higher (median RR= 17%, range =5-48%, Fig 1) than those between 27 haploidentical, DR- and DP-compatible family members (RR= 6%, range=0-39%). The RR values seen between DP-incompatible but otherwise

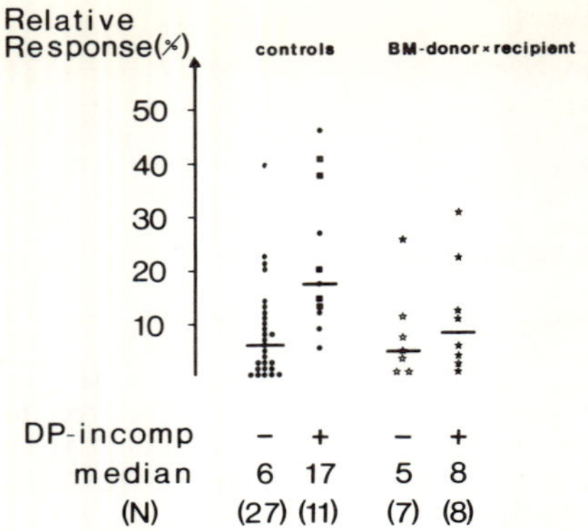

Figure 1. Influence of HLA-DP incompatibility on the one-way MLR between BM donors and their recipients (open and closed asterixes indicate absence and presence, respectively, of severe acute GVHD (≥ grade 2) in BM recipients within 3 months after BMT) and between healthy, haploidentical, DR-compatible family members (•), some of whom were DP-incompatible but otherwise HLA-identical due to HLA-DR/DP recombinations (■).

HLA-identical siblings obtained from three HLA-DR/DP recombinant families (10,12, and unpublished) ranged between 13 and 41%, indicating that DP antigens or antigens in close linkage disequilibrium with DP may be more potent stimulators of the primary MLR than suggested previously (1,13). Nevertheless, our data stress that the low MLRs are of little value in determining DP incompatibility in individual combinations. Moreover, our findings like those of others (11), indicate that the MLR is of limited value in predicting which recipients of allogeneic BM will develop severe acute GVHD. In contrast, incompatibility for DP antigens (in the "GVH-direction," i.e., recipient possesses DP antigen lacked by donor) seems to be highly predictive for severe acute GVHD because all eight recipients of haploidentical, DR-compatible, DP-incompatible BM developed severe acute GVHD (≤ grade 2), whereas only one of seven recipients of haploidentical, DR- and DP-compatible, BM developed severe acute GVHD within 3 months after BMT (p= 0.0014, Table 1).

A few isolated cases of severe acute GVHD following BMT with HLA-DP mismatched BM have been reported (14,15), two of which were recipients of DP-incompatible but otherwise HLA-identical BM (due to HLA-DR/DP recombinatorion in either recipient or donor) (14,15). We DP-typed eight recipients of HLA-identical BM who developed severe acute GVHD and the corresponding donors and all donor-recipient pairs were identical for DP (data not shown).

Pawelec et al. (16) also failed to demonstrate DP incompatibility as a cause of GVHD in recipients of HLA-identical BM. Thus, these cases of severe GVHD (as in one of the recipients of DR- and DP-compatible BM [Table 1]) are most probably due to the cumulative effects of disparity for multiple non-HLA minor histo-

Table 1. Acute Graft Versus Host Disease in Recipients of HLA-Identical and Haploidentical Bone Marrow

Recipient/Donor†	GVHD (≥ grade 2)	
	+	−
(a) Haploidentical, DR-compatible, DP-incompatible*	8	0
(b) Haploidentical, DR- and DP-compatible*	1	6
(c) HLA-identical	23	63

*In the "GVHD direction"; †a vs. c: p = 0.000079, Fisher's exact test; b vs. c: p = NS.

compatibility antigens (17,18). When comparing the incidence of severe acute GVHD between recipients of HLA-identical BM and recipients of haploidentical BM, we found no significant difference between recipients of HLA-identical BM and recipients of haploidentical, DR- and DP-compatible BM (Table 1). In contrast, the incidence of severe GVHD was significantly higher in recipients of haploidentical, DR-compatible, DP-incompatible marrow as compared with recipients of HLA-identical BM (p= 0.000079, Table 1). Incompatibility for HLA-class I antigens did not correlate with acute GVHD or with incompatibility for HLA-DP antigens. Taken together, these data provide strong evidence that HLA-DP antigens play a role as transplantation antigens.

Acknowledgments. This paper was supported in part by the Danish Cancer Society (Kræftens Bekæmpelse) and the Danish Medical Research Council.

References

1. Shaw S, Johnson AM, Shearer GM. Evidence for a new segregant series of B cell antigens that are encoded in the HLA-D region and that stimulate secondary allogeneic proliferative and cytotoxic responses. J Exp Med 1980; 152:65.
2. Ødum N, Hofmann B, Jakobsen BK, Langhoff E, Morling N, Platz P, et al. HLA-DP related suppression of mixed lymphocyte reaction with alloactivated lymphocytes. Tissue Antigens 1986;27:32.
3. Pawelec G, Wernet P, Rosenlund R, Blaurock M, Schneider EM. Strong lymphoproliferative suppressive function of PLT clones specific for SB-like antigens. Hum Immunol 1984;9:145.
4. Eckels DD, Lake P, Lamb J, Johnson AH, Shaw S, Woody JN, et al. SB restricted presentation of influenza and herpes virus antigen to human T-lymphocyte clones. Nature 1983;301:716.
5. Hoffman RW, Shaw S, Frencis LC, Larson MG, Petersen RA, Chylack LT, et al. HLA-DP antigens in patients with pauciarticular juvenile rheumatoid arthritis. arthritis Rheum 1986;29:1057.
6. Ødum N, Morling N, Friis J, Heilmann C, Hyldig-Nielsen W, Jakobsen BK, et al. Increased frequency of HLA-DPw2 in pauciarticular onset juvenile chronic arthritis. Tissue Antigens 1986;28:244.

7. Moen T, Stein R, Bratlie A, Bondevik E. Distribution of HLA-SB antigens in multiple sclerosis. Tissue Antigens 1984;245:126.
8. Ødum N, Hyldig-Nielsen JJ, Morling N, Sandberg-Wollheim M, Platz P, Svejgaard A. HLA-DP antigens involved in the susceptibility to multiple sclerosis. Tissue Antigens 1988;31:235.
9. Ødum N, Platz P, Morling N, Jacobsen N, Ryder LP, Svejgaard A. Increased frequency of HLA-DPw3 in severe aplastic anemia. Tissue Antigens 1986;28:184.
10. Ødum N, Hartzman RJ, Jakobsen BK, Morling N, Platz P, Robbins FM, et al. The HLA-DP polymorphism in Denmark investigated by local and international PLT reagents. Definition of two "new" DP antigens. Tissue Antigens 1986;28:105.
11. de Gast GC, Mickelson EM, Beatty PG, Amos D, Schock HG, Thomas Ed. Effect of mixed leucocyte culture reactivity on GVHD and relapse in HLA-matched BMT. Bone Marrow Transplant 1987;2:84.
12. Ødum N, Hartzman R, Jakobsen BK, Morling N, Platz P, Robbins FM, et al. An HLA-DR/DP recombinant family involving DPw6. Evidence for cross-reactivity between DPw6 and GNN 2B. Tissue Antigens 1986;27:44.
13. Termijtelen A, Naipal- van den Berge SJ, van Rood JJ. The influence of matching for SB on MLC typing is significant but marginal. Scand J Immunol 1984;19:265.
14. Møller E, Carlsson B, Wallin J. Distinct or overlapping functions of human class II genes. Scand J Immunol 1984; 22:483.
15. Amar A, Nepon GT, Mickelson E, Erlich H, Hansen JA. HLA-DP and HLA-DO genes in presumptive HLA-identical siblings: Structural and functional identification of allelic variation. J Immunol 1987;138:1947.
16. Pawelec G, Rhninger G, Smidt H, Wernet P. HLA-DP matching and graft-versus-host disease in allogeneic bone marrow transplantation. Transplantation 1986;42:558.
17. Korngold R, Sprent J. Lethal GVHD across minor histocompatibility barriers: Nature of effector cells and role of the H-2 complex. Immunological Rev 1983;71:5.
18. Goulmy E. Class I restricted human cytotoxic T lymphocytes directed against minor transplantation antigens and their possible role in organ transplantation. Prog Allergy 1985;36:44.

Author Affiliations

Niels Ødum, Bodil K. Jakobsen, Niels Morling, Arne Svejgaard, Tissue Typing Laboratory; *Niels Jacobsen*, Department of Infectious Diseases, State University Hospital (Rigshospitalet), Copenhagen; *Lars Lamm*, Tissue Typing Laboratory, Kommune Hospitalet, Arhus, Denmark

Frequent GvHR-Like Syndromes Following Syngeneic Bone Marrow Transplantation Suggest Inappropriately Controlled Autoreactivity

Claudia Müller, Hermann Einsele, Gerhard Krüger, Angelika Vallbracht, Roland Dopfer, Helmuth Schmidt, and Gerhard Ehninger

Development of graft versus host disease is the rule in major histocompatibility complex mismatched bone marrow transplants, but with variable incidence it is also observed in recipients of HLA-identical grafts (1) and even reported to occur in rare cases after syngeneic or autologous bone marrow transplant (BMT) (2), suggesting that differences in histocompatibility antigens are not an absolute requirement for the induction of GvHD.

In this study acute GvHR-like syndromes have been observed in seven of nine patients with BMT from identical twin donors. Genetic identity of donor-recipient twin pairs was verified from typing of different red cell and HLA class I and II antigens, as well as from data on serum protein and enzyme polymorphisms. Beside signs of mild hepatic disorder, increase in temperature, and changes in blood lymphocyte counts, eight of these patients developed an erythematous rash 7 to 11 days after BMT involving 25 to 100% of the total body surface. In one patient, the rash could be related to allergic reactions during antibiotic treatment. Another patient simultaneously suffered from extensive diarrhea. Clinical features and course of dermal and intestinal manifestations were indistinguishable from acute GvHD grade I-III (3) of patients after allogenic MHC-matched BMT. Due to the severity of the cutaneous lesions, three patients were treated with immunosuppressive drugs. The clinical diagnosis was confirmed by histologic as well as immunohistologic criteria of skin biopsies, which revealed abnormal expression of HLA-DP and DR antigens on keratinocytes, reduction of Langerhans cells, and dermal infiltrates of $CD4^+$ or CD 8^+ T lymphocytes similar to that in dermal manifestations of allogenic GvHR or of autoimmune diseases. In addition, peripheral blood lymphocytes revealed an early shift of the $CD4^+/CD8^+$ T-cell ratio below 1 in correlation with clinical symptoms before day 20 after BMT. In at least three of these patients, GvHR-like symptoms appeared to be related to reactivation of cytomegalovirus (CMV) or herpes simplex virus infection or to the development of antiviral immunity against hepatitis B infection. In one patient skin rashes could be observed on two occasions following clinical signs of herpes labialis. Another patient with reactivation CMV infection during severe marrow aplasia before transplantation developed extended skin lesions only after a successful second transplantation with high CMV-antibody titers several months later.

Complications related to conditioning irradiation or chemotherapy were unlikely to underlie the observed syndromes due to their association with graft take, the time lag between treatment and development of the disease, as well as the clinical pattern of different organ manifestations. There was no evidence for allergic reactions or transfusion-induced alloreactivity. Thus GvHR-like syndromes appear to occur more frequently after syngeneic BMT than expected previously (2) and as in animal models (4), may represent auto- rather than alloreactivity. In these patients viruses, as well as conditioning treatment with cyclosporin A or cyclophosphamide are suggested to trigger regulation failure of autoreactivity after human bone marrow grafting. Investigation of specificity and immunoregulatory features of cellular infiltrates in organ manifestations of GvHR-like syndromes in syngeneic and autologous BMT could lead to a better understanding of the development and control of autoreactivity in man.

Acknowledgments. Supported by DFG Sonderforschungsbereich 120, Project B1, B5.

References

1. Gale R. Graft-versus-host disease. Immunol Rev 1985; 88:193.
2. Reinherz EL, Parkman R, Rappeport J, Rosen FS, Schlossman FS. Aberrations in suppressor T-cells in human graft-versus-host disease. N Engl J Med 1979;300:1061.
3. Thomas ED, Storb R, Clift RA. Bone marrow transplantation. N Engl J Med 1975;292:823–825.
4. Hess AD, Horwitz L, Beschorner WE, Santos GW. Development of graft-versus-host disease-like syndrome in cyclosporin-treated rats after syngeneic bone marrow transplantation. J Exp Med 1985;161:718.

Author Affiliations

Claudia Müller, Hermann Einsele, Helmuth Schmidt, Gerhard Ehninger, Medical University Clinic, Tübingen; *Gerhard Krüger*, Institute of Pathology, University of Cologne, Cologne; *Angelika Vallbracht*, Institute of Hygiene, Department of Virology, Tübingen, Federal Republic of Germany

Use of Synthetic Oligonucleotides for the Study of Hypervariable Regions of Human DNA and Their Clinical Applications

L. Ugozzoli, P. Briata, C. Delfini, G. Lucarelli, B. Wallace, and G.B. Ferrara

Allogenic marrow transplantation has an established role in the therapy of thalassemia. However, about 15% of patients do not engraft subsequently or reject the transplant.

Routine engraftment markers include red cell antigens phenotyping, studies of hemoglobin synthesis, and cytogenetic analysis. These markers of engraftment are not reliably informative because leukopenia, transfusion dependence, severe immunosuppression, and sex match are factors that may limit their utility, particularly in the early posttransplant period.

Recently, the utility of synthetic oligonucleotides to harness repetitive sequences of human DNA for documentation of engraftment and mixed hematopoietic chimerism following bone marrow transplantation (BMT) has been demonstrated (1,2). These variable regions consist of tandem repeats of a short sequences (mini-satellite) and polymorphism resulting from allelic differences in the number of repeats (3). The resulting mini-satellite length variation can be detected using any restriction endonuclease that does not cleave the repeat unit, thus giving a set of stable, inherited genetic markers when they exhibit multiple alleles.

The aim of our study was to use these hypervariable sequences as markers of engraftment in BMT. For this study, three oligonucleotides (0-YNH-24, 0-AY-27, and 0-33.6) were synthesized and employed as probes to serially assess the engraftment in 19 thalassemic patients undergoing BMT.

DNA from peripheral blood specimens from patients before and after transplantation and from their respective donors was analyzed. The purified DNA was digested with Hinf I, and the resultant fragments were then fractionated by electrophoresis. The gel was then dried and directly hybridized with each of the oligonucleotides. Our results demonstrate that:

1. by using these three oligonucleotides in familial segregation analysis, the mendelian inheritance of these genetic markers occurs independently from the HLA system;
2. these markers can easily distinguish between two HLA-identical siblings and therefore can be employed to document engraftment of donor cells after allogenic BMT;
3. with the association of the three probes, we were able to discriminate 100% of HLA-identical donor/recipient sibling pairs;
4. the association of the three probes showed that 18 of 19 HLA-identical sibling pairs had both donor-specific and recipient-specific fragments and 1 of 19 had a donor-specific fragment (Table 1);
5. it is possible to get the result within 15 days after the BMT;

Table 1. DNA Analysis from 19 HLA-Identical Patient/Donor Pairs

UPN	0-YNH-24	0-AY-27	0-33.6
321	N.I	D-R	D-R
325	N.I	D-R	R
324	D-R	D-R	N.I
328	D-R	N.I	N.I
329	N.I	N.I	D
331	N.I	D-R	D-R
332	D-R	D-R	D-R
330	D-R	D-R	N.I
358	N.I	D-R	N.I
333	D-R	D-R	D-R
345	D-R	N.I	R
323	N.I	D-R	D-R
371	D-R	D-R	D-R
382	D-R	D-R	D-R
377	D-R	–	–
391	D-R	D-R	N.I
390	D-R	D-R	D-R
318	D-R	D-R	D-R
389	D-R	N.I	D-R

D = Informative fragment for the donor; D-R = informative fragments for both recipient and donor; N.I = not informative; R = informative fragment for the recipient; UPN = unique patient number. – = not done.

6. by studying BMT in 19 thalassemic patients, we found a 100% correlation between the data obtained by clinical follow-up (or by other methods performed later) and the results shown by our methodology.

The analysis of different restriction patterns derived from the 19 sibling pairs after transplantation allowed us to establish 15 cases of engraftment, one transient mixed hematopoietic chimerism followed by complete engraftment, and three cases of rejection.

References

1. Wallace RB, Petz LD, Yam PY. Application of synthetic DNA probes to the analysis of DNA sequence variants in man. Cold Spring Harbor Symp Quant Biol 1986;LI:257.
2. Yam PY, Petz LD, Ali S, Dean Stock A, Wallace RB. Development of a single probe for documentation of chimerism following bone marrow transplantation. Am J Hum Genet (in press).
3. Jeffereys AJ, Wilson V, Thein SL. Hypervariable "minisatellite" regions in human DNA. Nature 1985;314:67.

Author Affiliations

L. Ugozzoli, P. Briata, Instituto Nazionale per la Ricerca sul Cancro, Genova, Italy; *C. Delfini, G. Lucarelli*, Centro Trapianto Midollo osseo, Pesaro, Italy; *B. Wallace*, City of Hope Medical Center, Duarte, CA, USA; *G.B. Ferrara*, Instituto Nazionale per la Ricerca sul Cancro, Genova, Catterdra di Genetica Umana, I Facoltà di Medicina, Università di Napoli, and Centro ricerche Immunoematologiche AVIS, Bergamo, Italy.

A DRw13-Specific DR Antigen Structure Implicated in Poor Kidney Graft Survival

Marnix L. Bosch, Emma L. Lagaay, Annemarie Termijtelen, Jon J. van Rood, and Marius J. Giphart

Abstract: The presence of the major histocompatibility antigen HLA-DRw6 in the recipient of a kidney allograft has been shown to influence graft survival. We demonstrate that this effect (impaired graft survival) resides in the DRw13 subgroup of DRw6 only. Comparison of available DR β sequences identified two DRw13-specific residues in the third hypervariable region of the DR β I chain. We postulate that these negatively charged residues create a bias in the T-cell repertoire of DRw13+ individuals, resulting in relatively strong alloreactivity and poor graft survival.

Products of the HLA class II region play a vital role in the immunologic processes that mediate allograft rejection: class II antigens expressed on the thymic epithelium of the recipient are involved in the generation of an allodirected T-cell repertoire (1) and class II antigens expressed on kidney allografts serve as targets for immunologic graft rejection (2). Indeed it has been demonstrated that HLA class II disparity between kidney donor and -recipient negatively influences graft survival (3,4) and it has been possible to isolate anti-donor class II directed CD4+ host-derived cytotoxic T cells from rejected kidney grafts (5).

A putative immune response gene (Ir gene) phenomenon has been described in kidney transplantation, implicating DRw6+ patients as high responders against alloantigens (6). We have analyzed kidney graft survival data from patients transplanted in the Netherlands after having received two to five pretransplant blood transfusions, as recent data have shown that the DRw6 recipient effect is the most clear-cut in this group (7). Both the influence of the presence of DRw6 and of the DRw6 subtypes DRw13 and DRw14 (8) were analyzed. The results of these analyses are shown in Figure 1. As can be judged from Figure 1, a DRw6 recipient effect is detectable in this patient group: kidneys transplanted into DRw6 recipients do less well than those transplanted into nega-

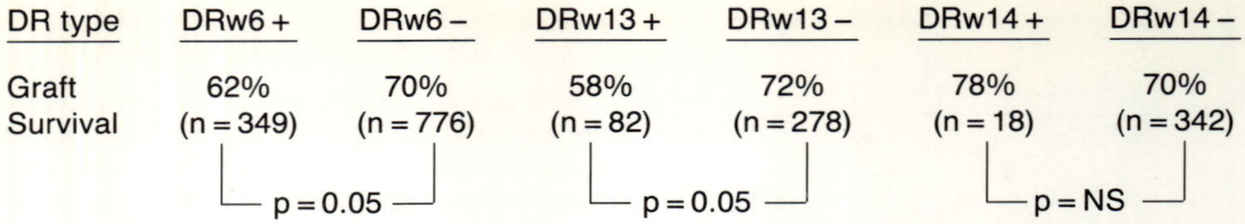

Figure 1. Two-year kidney allograft survival in patients transplanted in the Netherlands typed for HLA-DRw6 subtypes and who received two to five pretransplant blood transfusions.

tive recipients. The same result is found in DRw13+ recipients, but not in DRw14+ recipients.

These results urged us to look for a DRw13-specific structure. The makeup and subdivision of the HLA-DRw6 haplotype is depicted in Figure 2. As can be seen from this figure all DRw13 cells express the so-called DR β I gene, independent of the HLA-Dw specificities of the individual cells (9,10). The other HLA class II antigens expressed by these cells, both DR- and DQ-derived differ between the different HLA-Dw splits of DRw13, Dw18, and Dw19 in isoelectric point (9,11) and amino-acid sequence (M.L. Bosch, unpublished observations). This makes the DR β I chain the only common trait of all DRw13+ cells. As the known sequences of the DR beta III genes of DRw13 and DRw14 cells are identical (10), it seems likely that the functional differences between DRw13 and DRw14 reside on the DR β I chains.

To identify a DRw13-specific antigen structure we compared the DR β I chain sequences of DRw13 and DRw14 haplotypes (10) as shown in Figure 3. The only difference between these two sequences imparting DRw13 specificity is printed boldface (at amino acids 70 and 71); all other amino acid differences are also found in other haplotypes and are therefore not DRw13-specific. Figure 4 compares the amino acids at positions 70 and 71 of all DR β genes of which the sequence is known. The two "DRw13-specific" residues also occur in the DR4 subtype HLA-Dw10.

The two DRw13- (and Dw10) specific residues at positions 70 and 71 of the DR β I chain impart a strongly negative charge to this region of the chain. All DR β chains expressed by other haplotypes (as well as the DRw13-DR beta III chain) are either neutral or positively charged in that region (Fig. 4).

The third hypervariable region of HLA-DR antigens, containing amino acids 70 and 71, is probably important in the recognition of these antigens by T cells (12), implying that it may also be important in the generation of the T-cell repertoire in the thymus. We propose that the molecular makeup of this region may be responsible for a bias in the T-cell repertoire. The T cells in DRw13+ individual would be relatively unresponsive to class II antigens containing acidic residues in the third hypervariable region, and relatively strongly responsive to allo class II antigens that are neutral or positively charged in this region. As all non-DRw13 haplotypes (except Dw10) express only neutral or positively charged HLA-DR β chains, DRw13+ individuals will strongly reject allografts, except when the graft also expresses DRw13. In other words, kidney transplantation across a charge difference in the third hypervariable region of the DR antigens of donor and recipient results in poor graft survival.

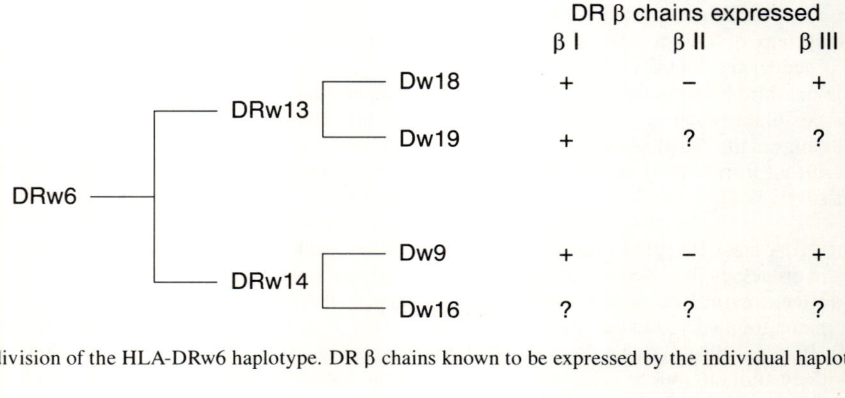

Figure 2. Subdivision of the HLA-DRw6 haplotype. DR β chains known to be expressed by the individual haplotypes are shown at right.

```
                  1 0                 2 0                 3 0                 4 0                 5 0
consensus P R F L E Q x K x E C H F F N G T E R V R F L x R Y F Y N Q E E Y V R F D S D V G E Y R A V
DRw13-I   . . . . . Y S T S . . . . . . . . . . . . . . D . . . H . . . . N . . . . . . . . . P . . .
DRw14-I   . . . . . Y S T S . . . . . . . . . . . . . . D . . . H . . . . . . . . . . . . . . . . . .

                          6 0                 7 0                 8 0                   9 0
consensus T E L G R P D A E Y W N S Q K D L L E Q x R A A V D T Y C R H N Y G V G E S F T V Q R R
DRw13-I   . . . . . . . . . . . . . . . I . . . D E . . . . . . . . . . . . . . . . . . . . . . .
DRw14-I   . . . . . . A . D D . . . . . . . . . R R . . E . . . . . . . . . . . . . . . . . . . .
```

Figure 3. Predicted amino acid sequences of the first domain of DR β I chains of the DRw13 and DRw14 haplotypes related to a DR β consensus sequence (as deduced from the references 13–18).

Acknowledgments. We gratefully acknowledge the technical assistance of Jacqueline Anholts and the help of Peter de Lange with the computer analyses. This work would have been impossible without the help of the Eurotransplant transplantation centers.

References

1. von Boehmer H. The selection of the alpha-beta heterodimeric T-cell receptor for antigen. Immunol Today 1986; 7:333.
2. Hall BM, Duggin GG, Phillips J, Bishop GA, Horvath JS, Tiller DJ. Increased expression of HLA-DR antigens on renal tubular cells in renal transplants: Relevance to the rejection response. Lancet 1984;2:247.
3. Persijn GG, Gabb BW, van Leeuwen A, Nagtegaal J, van Rood JJ. The effect of matching for HLA-A, -B and -DR antigens on kidney graft survival in Eurotransplant. Lancet 1978;1:495.
4. Morris PJ, Ting A. Study of HLA-DR with relevance to renal transplantation. Immunol Rev 1982;66:103.
5. Asher NL, Hoffman R, Hants DW, Simmons RL. Cellular events within the rejecting allograft. Transplantation 1983; 35:193.
6. Hendriks GFJ, Schreuder GMTH, Claas FHJ, D'Amaro J, Persijn GG, Cohen et al. HLA-DRw6 and renal allograft rejection. Brit Med J 1983;8:85
7. Lagaay EL, Persijn GG, Termijtelen A, Hendriks GFJ, de Lange P, van Rood JJ. The difference in kidney graft survival in DRw6+ and DRw6− recipients may be explained by a blood transfusion policy that is disadvantageous for DRw6+ recipients. Transplantation (in press).
8. Schreuder GMT, Parlevliet J, Termijtelen A, van Rood JJ. Reanalysis of the HLA-DRw6 complex. Tissue antigens 1983;21:62.
9. Bosch ML, Termijtelen A, Gerrets R, Schreuder GMT, Bontrop RE, Giphart MJ. Polymorphisms within the HLA-DRw6 haplotype. II. Protein charge heterogeneity reflects MLC subtyping of HLA-DRw6 homozygous cells. Immunogenetics 1985;22:23.
10. Gorski J, Mach B. Polymorphism of human Ia antigens: Gene conversion between two DR beta loci results in an new HLA-D/DR specificity. Nature 1986;322:67.
11. Bosch ML, Fei HM, Bontrop RE, Gerrets R, Tilanus MGJ, Termijtelen A, et al. Polymorphisms within the HLA-DRw6 haplotype. III. DQ alpha and DQ beta polymorphism, associated with HLA-D. Human Immunol 1987;19:91.
12. Ronchese F, Schwartz RH, Germain RN. Functionally distinct subsites on a class II major histocompatibility complex molecule. Nature 1987;329:256.
13. Didier DK, Schiffenbauer J, Shuman S, Abruzzini LF, Gorski J, Watling DL, et al. Characterization of two distinct DR beta chain alleles at the beta III locus of the DR 5 haplotype: Beta III alleles are highly conserved. J Immunol 1986; 137:2627.
14. Gregersen PK, Shen M, Song QL, Merryman P, Degar S, Seki T, et al. Molecular diversity of HLA-DR4 haplotypes. Proc Natl Acad Sci USA 1986;83:2642.
15. Gregersen PK, Moriuchi T, Karr RW, Obata F, Moriuchi J,

Amino Acid 70	Amino Acid 71	Haplotype/DR β Chain
GLN	LYS	DR3/I;DRw52a/III;DRw52b/III;DR4Dw4/I
GLN	ALA	DR2a/III;DR2b/I;DR2/I
GLN	ARG	DR4Dw13/I;DR4Dw14/I;DR4Dw15/I;DR1
ARG	ARG	DRw6Dw9/I;DRw52/III;DRw6Dw19/I!
GLU	ARG	DR5/I;DR7/I;DR2a/I;DR2b/III;DR2/III
GLU	ASP	DRw6DRw13/I;DR4Dw10/I

Figure 4. Comparison of the amino acids 70 and 71 of DR β chains of various haplotypes. (Sequences were deduced from refs. 13–18.)

Maccari J, et al. Polymorphism of HLA-DR beta chains in DR 4, -7 and -9 haplotypes: Implications for the mechanisms of allelic variation. Proc Natl Acad Sci USA 1986; 83:9149.
16. Cains JS, Curtsinger JM, Dahl CA, Freeman S, Alter BJ, Back FH. Sequence polymorphism of HLA-DR beta I alleles relating to T-cell-recognized determinants. Nature 1985;317:166.
17. Bell JI, Estess P, St. John T, Saiki R, Watling DL, Ehrlich HA, et al. DNA sequence and characterization of human class II major histocompatibility complex beta chains from the DR1 haplotype. Proc Natl Acad Sci USA 1985;82:3405.
18. Wu S, Saunders TL, Bach FH. Polymorphism of human Ia antigens: Reciprocal intergenic exchange between two DR beta loci. Nature 1986;324:676.

Author Affiliations

Marnix L. Bosch, Department of Immunobiology, RIVM, BILTHOVEN, and Department of Immunohaematology and Bloodbank, University Hospital, Leiden, The Netherlands; *Emma L. Lagaay*, Eurotransplant Foundation, University Hospital; *Annemarie Termijtelen, Jon J. van Rood, Marius J. Giphart*, Department of Immunohaematology and Bloodbank, University Hospital, Leiden, The Netherlands

DNA Typing of HLA Class II Antigens in Kidney Transplantation

Hitoshi Shimoyama, Tadashi Horimi, Takahiro Okabayashi, Isao Takeda, Michihiro Kaneda, Shintaro Chono, Izuru Tanaka, Noriyuki Miyake, Dofu Hayashi, Kunzo Orita, Hidetoshi Inoko, Asako Ando, and Kimiyoshi Tsuji

Southern hybridization analysis with HLA class II cDNA probes allows us the rapid investigation of the class II region of human MHC on population samples by detection of restriction fragment length polymorphisms (RFLPs) (1). This new typing method, called DNA typing, can be very helpful in the detection of the polymorphism of HLA class II haplotypes which may escape detection by conventional methodologies. As indicated above, HLA-DNA typing is expected to be a sharper typing than conventional HLA typing. In this report, we demonstrate the preliminary application of DNA typing in clinical kidney transplantation. The subjects were 13 cases of cadaveric kidney transplantation and eight cases of living-related kidney transplantation with established HLA-A,-B and -DR. Probes were full length cDNA of DRβ chain (pDRβ102), DQα chain (pDCα107), DQβ chain (pDCβ101), and DPβ chain (pDAβ5) (2). Two different endonucleases, EcoRI and PstI, were used for digestion of DNA. RFLP bands were investigated and compared between donors and recipients. The relationship between the number of mismatch bands of the each typing and the post-transplant clinical course was investigated. Statistical significance was estimated by the Fisher exact test and the χ^2 test with Yate's correlation.

As for cadaveric kidney transplantation, surviving graft showed five or less mismatch bands in DQα (EcoR + PstI) typing ($p<0.05$), and also showed three or less mismatch bands in DQβ (EcoRI) typing ($p<0.05$). Least association was observed between compatibility in DRβ typing and graft survival, but surviving graft showed five or less mismatch bands in DRβ(PstI)-DQα(PstI) typing ($p<0.01$) and also in DRβ(PstI)-DQβ(EcoRI) typing ($p<0.01$).

As for 21 cases of kidney transplantation, surviving graft showed five or less mismatch bands in DQα(EcoRI + PstI) typing ($p<0.01$), and also showed three or less mismatch bands in DQβ(EcoRI) typing ($p<0.05$). No or least association was observed between compatibility in DPβ typing and graft survival.

The results of our research indicate the implication of DQ typing in clinical kidney transplantation. The degree of association between DNA typing and graft survival is strongly related to the degree of polymorphism of each probe. Recent studies revealed that DQβ genes are most polymorphic and DPβ genes are least polymorphic in β-chain genes. DQα genes are also highly polymorphic as much as β-chain genes (3). These findings support our results that allograft survival seems to be associated with the compatibility of DNA typing with DQα probe and DQβ probe.

Acknowledgments. We thank Ms. Y. Iwabe for her expert secretarial assistance.

References

1. Inoko H, Ando A, Ito M, Tsuji K. Southern hybridization analysis of DNA polymorphism in the HLA-D region. Hum Immunol 1986;16:304.
2. Inoko H, Ando A, Kimura M, Tsuji K. Isolation and characterization of the cDNA clones and genomic clones of the HLA class II antigens. In: Tsujik (ed): Present and Future Development of Transplantation. Tokyo, Tokai University Press, 1985, p 188.
3. Auffray C, Strominger JL. Molecular genetics of the human major histocompatibility complex. In: Harris H, Hirschhorn K (eds):Advances in Human Genetics, Vol. 15. New York, Plenum Publishing Co, 1986, p 197.

Immunobiology of HLA
Volume II

Author Affiliations

Hitoshi Shimoyama, Takahiro Okabayashi, Michihiro Kaneda, Shintaro Chono, Izuru Tanaka, Noriyuki Miyake, Dofu Hayashi, Kunzo Orita, Department of Surgery, Okayama University Medical School, Okayama 700, Japan; *Tadashi Horimi, Isao Takeda*, Department of Surgery, Okayama University Medical School, Okayama and Section of General Surgery and Renal Transplantation Center, Kochi Prefectural Central Hospital, Kochi 780, Japan; *Hidetoshi Inoko, Asako Ando, Kimiyoshi Tsuji*, Department of Transplantation, Tokai University School of Medicine, Isehara 259-11, Japan

Donor Selection for Highly Immunized Patients Based on Acceptable HLA-A and -B Mismatches

Frans H.J. Claas, Y. Gijbels, J.J. van der Velden-de Munck, L.P. de Waal, Joe D'Amaro, Frits F.J. Hendriks, Guido G. Persijn, and Jon J. van Rood

The accumulation of highly immunized patients on renal transplant waiting lists is a universal problem. These patients have alloantibodies against many HLA antigens due to previous graft failures, pregnancies, and blood transfusions. Because of these high levels of alloantibodies, it is very difficult to find a cross match-negative donor for them. Therefore, several strategies have been developed to increase the chance of finding a cross match-negative kidney for such highly sensitized patients. One possibility is to select HLA-A and -B identical or compatible donors, but the chance of finding such a donor is often very low owing to the enormous polymorphism of the HLA system. Other schemes involve the distribution of the sera from these patients to many tissue-typing laboratories. Each potential donor is tested against these sera, and donor-recipient combinations with negative crossmatches are identified by trial and error (1,2). We describe here a scheme that does not need the distribution of patient sera and can be operated from a central organ-sharing office. The scheme is based on data of Eurotransplant, which show that in the case of a negative crossmatch, HLA-A and -B mismatches are less important for graft survival in highly sensitized patients, whereas DR matching has a strong influence especially in retransplantation (3). The aim of the scheme is an exact definition of the HLA-A and -B incompatibilities that will not result in a positive crossmatch. Therefore, the sera of highly immunized patients are screened against a panel of lymphocyte donors who are selected on the basis of having only one HLA-A or -B mismatch with the specific patient (Table 1). Using this approach, HLA-A and -B antigens towards which the patient did not develop antibodies can be detected. Kidney donor selection takes place by implementing the HLA-A, -B, and -DR antigens of every potential donor in the central computer, which will select crossmatch-negative recipients on the basis of their own HLA-A, -B, and -DR antigens in combination with the acceptable HLA-A and -B mismatches. In this way, the donor is always optimally matched for HLA-DR but may have several HLA-A and -B mismatches.

In the past 2 years, the sera from 73 highly sensitized patients were tested for the absence of specific alloantibodies against HLA-A and -B antigens using the protocol indicated in Table 1. All these patients showed antibody reactivity against more than 85% of the panel donors (panel size = 53 donors), which was not due to the presence of autoantibodies. In patients with less than 100% panel reactivity, a first indication concerning these acceptable HLA-A and -B mismatches can be deduced from the HLA-group of the negative panel donors in the screening. Nevertheless, using our approach, patients with 100% panel reactivity were found to have acceptable mismatches also.

In 59 of the 73 patients tested, it was possible to define one or more acceptable HLA-A or -B mismatches. Addition of these antigens to the patients own HLA antigens with regard to kidney donor selection significantly increased the chance for finding a crossmatch-negative donor. In the meantime, 34 of these patients have received a HLA-DR compatible crossmatch-negative kidney graft following this scheme with a 6-month graft survival of 86%. This scheme will be used in the near future for all highly sensitized patients in The Netherlands.

Table 1. Principle of Detection of Acceptable HLA-A and -B Mismatches in Highly Sensitized Patients

HLA type of the patients: HLA-A1 -A2 -B7 -B8

Selected Panel Donors				Cross-Match With Patient Serum
A1	A2	B7	Bw60	+
A1	A2	B7	B44	–
A1	A2	B7	B35	+
A1	A3	B7	B8	–

Acceptable mismatches: A3 and B44.

References

1. Bradley BA, Klouda PT, Ray TC, Gore SM. Negative crossmatch selection of kidney for highly sensitized patients. Transpl Proc 1985;17:2465.
2. Schäfer AJ, Hasert K, Opelz G. Collaborative transplant study crossmatch and antibody project. Transpl Proc 1985;17:2469.
3. Hendriks GFJ, de Lange P, D'Amaro J, Schreuder GMTh, Claas FHJ, Persijn GG, et al. Eurotransplant experience with highly sensitized patients. Scand J Urol & Nephrol 1985;92:81.

Author Affiliations

Frans H.J. Claas, Y. Gijbels, J.J. van der Velden-de Munck, Joe D'Amaro, Jon J. van Rood, Department of Immunohaematology & Blood Bank, University Hospital, Leiden; *L.P. de Waal*, The Central Laboratory of the Netherlands Red Cross Blood Transfusion Service, Amsterdam; *Frits F.J. Hendriks*, Dijkzigt, University Hospital, Rotterdam; *Guido G. Persijn*, Eurotransplant Foundation, Leiden, The Netherlands

Influence of HLA Matching in Cardiac Allograft Recipients Receiving Cyclosporin A and Azathioprine

Hilliard Festenstein, Magdi Yacoub, Daniel McCloskey, Juliette Awad, Marina Martin, Amando Gamba, Asghar Khaghani, Julian Holmes, and John Smith

In 1986, we reported the effect of HLA matching and other immunologic factors on the outcome of cardiac transplantation. The data presented here are those derived from 352 patients who received cyclosporin A and azathioprine following cardiac transplantation (see Table 1). These data were analyzed using the method of Peto et al. (3). Their ages varied from 6 to 63 years (mean 52). The ABO blood grouping was the only criteria used for donor/recipient selection. No attempt was made to match for HLA antigens. The overall graft survival was 74% at 1 year and 72% at 2 years. The effect of matching for class I (HLA-A,-B), class II DR, DRw52/DRw53, and DQ on graft survival were determined. Matching for class II antigens appeared to have an influence on the 2-year graft survival (78% for one DR mismatch and 68% for two DR mismatched, p=NS). Similar influence of HLA-DRw52/53 matching on a 2-year survival was observed (76% for no mismatch and 70% for one mismatch). Matching for class I antigens alone did not significantly influence the outcome. However, when the combination of class I (HLA-B) and class II (HLA-DR) were analyzed, there was a significant influence on the

Table 1. HLA Matching Outcome in Cardiac Transplantation

Factor		No.	% Survival 300 Days	% Survival 750 Days	O/E
Overall survival		352	74	72	
HLA-A	0MM	21	82	82	1.07
	1MM	140	86	86	0.92
	2MM	155	84	82	1.06
HLA-B	1MM	99	84	84	0.97
	2MM	200	84	84	0.99
HLA-Bw4,w6	0MM	185	72	72	1.07
	1MM	127	78	76	0.90
HLA-DR	1MM	111	80	78	0.81
	2MM	109	70	68	1.20
HLA-DRw52,w53	0MM	47	76	76	0.94
	1MM	113	72	70	1.09
HLA-DQ	0MM	70	78	76	0.85
	1MM	90	72	72	1.11

Table 1. *Continued.*

Factor		No.	% Survival 300 Days	% Survival 750 Days	O/E
Overall survival		352	74	72	
HLA-B + DR	2MM	53	74	74	1.01
	3MM	94	80	80	0.75
	4MM	74	68	66	1.27
	2/3MM	147	78	78	0.86
	4MM	74	68	66	1.30
HLA-B + w52, w53	2MM	63	76	76	0.89
	3MM	99	80	76	0.84
	4MM	50	70	70	1.14
Crossmatch result	+ve	33	56	52*	1.87
	−ve	223	76	76	0.88
HLA-DRw6 in recipient	−ve	179	78	76	0.90
	+ve	66	68	68	1.29
HLA-DRw6 in donor	−ve	235	76	74	0.95
	+ve	82	72	72	1.15
HLA-A2 in donor & not in recip	Yes	78	78	76	0.84
	No	248	72	72	1.05
Donor sex	♂	223	76	76†	0.84
	♀	116	68	64	1.32

*p = 0.0086; †p = 0.0351.

Immunobiology of HLA
Volume II

2-year graft survival suggesting a synergistic effect of class I and class II matching. The lymphocytotoxic crossmatch result had a significant affect on graft survival (52% positive crossmatch vs. 76% negative crossmatch, p = 0.0086). The sex of the donor had a significant affect on 2-year survival (64% female donors vs. 76% male donors, p = 0.0035). The sex of the recipient did not influence the 2-year survival. Due to the low number of female recipients, effect of donor gender cannot be investigated. It is concluded that HLA matching and the crossmatch result can have an important effect on the medium term result of cardiac transplantation.

References

1. Yacoub M, Festenstein H, Doyle P, Martin M, McCloskey D, Awad J, et al. The influence of HLA matching in cardiac allograft recipients receiving cyclosporine A and Azathioprine. Transpl Proc 1987;19:2487.
2. Yacoub M, Alivizatos P, Khaghani A, Mitchell A. The use of cyclosporine azathioprine and antithymocyte globulin with or without low dose steroid for immunosuppression of cardiac transplant patients. Transpl Proc 1985; 17:221.
3. Peto R, Pike MC, Armitage P, Breslow NE, Cox DR, Howard SV, et al. Design and analysis of randomized clinical trials requiring prolonged observation of each patient. II. Analysis and examples. Br J Cancer 1977:35:1–39.

Author Affiliations

Hilliard Festenstein, Daniel McCloskey, Juliette Awad, Julian Holmes, John Smith, Department of Immunology, London Hospital Medical College, London; *Magdi Yacoub, Marina Martin, Armando Gamba, Asghar Khaghani,* Harefield Hospital, Harefield, UK

HLA Antigen Expression in Cardiac Transplant Biopsies: Association with Rejection and Propagation of Infiltrating Alloreactive Lymphocytes

Linda T. White, Tony R. Zerbe, Adriana Zeevi, Robert L. Hardesty, Bartley P. Griffith, and Rene J. Duquesnoy

Abstract: The expression of class I and class II HLA antigens in situ was studied in serial endomyocardial biopsies of human cardiac allografts. The presence and intensity of this expression were assessed in relation to other parameters of cellular rejection: histological evidence of cellular infiltration, the IL2 facilitated growth of lymphocytes from the biopsy tissue, and donor alloreactivity of biopsy-grown cells.

All tissues expressed class II-DR and -DQ antigens on vascular endothelium and interstitial structures, but class II expression on myocardium was never demonstrated. Expression of class I antigen on myocytes correlated with an increased histological grade of rejection in the biopsy. No significant association of intensity of HLA expression on vascular endothelium with histological evidence of rejection was established. However, a biopsy with myocardial or strong endothelial staining of HLA antigens was significantly more likely to be associated with successful growth of graft-infiltrating lymphocytes than was a biopsy with weak staining; such cells were also significantly more likely to be donor allospecific.

Immunostaining for HLA antigens was performed on 133 endomyocardial biopsies obtained from 26 cardiac transplant recipients during the first 4 months after transplant. Donor tissue obtained at the time of transplantation from 13 of these patients was also available for staining. Immunosuppression protocols for all patients included Cyclosporine A and prednisone.

Indirect immunoperoxidase staining (1) was performed on 4-μ frozen sections using the following monoclonal antibodies: w6/32 (HLA-ABC),L243(HLA-DR), and Leu-10(HLA-DQ). A dilution of each antibody was chosen so that quantitative differences in endothelial staining among the specimens were maximized. Vascular endothelial staining was graded semiquantitatively; staining of the myocytes was judged to be negative or positive at this dilution. All readings were made by a single individual (L.W.) independently of histological grading for rejection.

Three or four concurrently obtained biopsy fragments, as well as the frozen biopsy fragment used in the immunostaining, were stained with hematoxylin and eosin and graded (by T.Z.) using histologic criteria for cellular rejection after Billingham (2).

For 25 of the 26 transplant recipients, concurrent myocardial fragments were also cultured in the laboratory with irradiated autologous stimulator cells in medium containing interleukin-2 (IL2). Lymphocytes obtained by this expansion were then tested in a primed lymphocyte test (PLT) to evaluate donor alloreactivity (3).

The results of the immunostaining were compared to histological evidence of cellular rejection. All donor tissues taken at transplantation and posttransplant biopsies expressed class I class II-DR and -DQ antigens on the

Table 1. HLA Antigen Staining on Vascular Endothelium of Cardiac Transplant Biopsies: No correlation with Histologic Grade of Rejection

Rejection		Class I			DR			DQ		
Grade	N*	Weak	Moderate	Strong	Weak	Moderate	Strong	Weak	Moderate	Strong
0	26	28.0%†	60.0%	12.0%	30.8%	46.2%	23.1%	24.0%	48.0%	28.0%
1	61	42.6%	42.6%	14.8%	28.8%	44.1%	27.1%	13.8%	62.1%	24.1%
2	20	31.6%	52.6%	15.8%	30.0%	35.0%	35.0%	20.0%	70.0%	10.0%
3+	26	26.1%	39.1%	34.8%	19.2%	50.0%	30.8%	15.4%	50.0%	34.6%
			NS‡			NS			NS	

*Number of biopsies; †percent of biopsies (each rejection grade); ‡not significant; $p > 0.05$ by chi-square analysis.

vascular endothelium and interstitial cells. However, a significant correlation between the staining intensity of the endothelium and the grade of rejection was not apparent (Table1). Myocardium, in contrast to endothelium, never stained with class II antibodies. Class I expression on myocytes was demonstrated only in posttransplant biopsy specimens; this expression was significantly associated with the degree of cellular rejection (Table 2). Also, myocyte staining was associated with a diffuse pattern of rejection in which inflammatory infiltrate was seen in more than one concurrent biopsy fragment rather than with a focal pattern in which only one concurrent fragment contained a cellular infiltrate ($p = 0.01$). The class I myocyte staining was therefore associated with both a higher grade of cellular rejection and with the involvement of a greater proportion of the heart tissue.

The intensity of immunostaining correlated with the degree of the lymphocyte growth from posttransplant biopsies (Fig.1). This was observed for both class I and class II (DR and DQ) antigens on endothelium. Also, class I expression on myocytes was associated with a higher yield of lymphocyte cultures propagated from biopsies. Statistical analysis of the combined data showed that myocyte staining and/or strong endothelial expression of any of the three MHC gene products was significantly associated with increased biopsy growth of infiltrating lymphocytes ($\chi^2 = 5.97$; $p < 0.05$).

Table 2. Correlation between Class I HLA Antigen Staining on Myocytes and the Histologic Rejection Grades of Heart Transplant Biopsies

Rejection		Class I Staining on Myocytes	
Grade	N*	Negative	Positive
0	26	92.0%	8.0%
1	61	80.3%	19.7%
2	20	57.9%	42.1%
3+	26	53.8%	46.1%
		$\chi^2 = 13.35$; $p < 0.005$	

*Number of biopsies; †percent of biopsies (each rejection grade).

Many cultures of biopsy-grown cells exhibited donor alloreactivity as determined by PLT. The percentage of lymphocyte cultures with PLT reactivity correlated with the intensity of HLA antigen expression on endothelium (class I, DR and DQ) and myocytes in concurrent biopsies (Fig. 1).

Cellular rejection of allografts involves the infiltration of recipient cells from the circulation into the graft tissue. Upregulation of donor major histocompatibility complex (MHC) expression on vascular endothelium may enhance the primary immune response (3) and increase infiltration of donor-allospecific T cells (4). Cytokines released by such activated cells may then further modulate MHC expression in the graft (4,5). In addition, donor-specific cytotoxic T cells may directly damage cells of the graft displaying increased or aberrant levels of class I or II antigens. Increased expression of MHC antigens on the graft tissues therefore could both precipitate cellular infiltration and occur in response to such infiltration.

In this study, the strong correlation of *increased* class II expression with the demonstration of donor-alloreactive cells in the graft suggests that such upregulation occurs in response to the cytokines produced and induced by activated T cells. Class II expression was not solely a response to the products of infiltrating cells, however; expression of class II MHC gene products on endothelium preceded cellular infiltration, as in the donor tissue obtained at transplantation. Although myocardial expression of class I antigens was related to significant cellular rejection, such antigen expression did not invariably accompany infiltration and often persisted beyond the period of active infiltration, as others have found (6). Immunostaining for HLA antigens in endomyocardial biopsies would appear to have limited prognostic value in the monitoring of allograft rejection; however, the HLA expression is clearly consistent with other parameters of rejection, especially the infiltration by alloreactive T cells which can be propagated in vitro from biopsies.

Acknowledgment. This work was supported by NIH grant AI-23467 and a grant in aid from the Western Pennsylvania chapter of the American Heart Association.

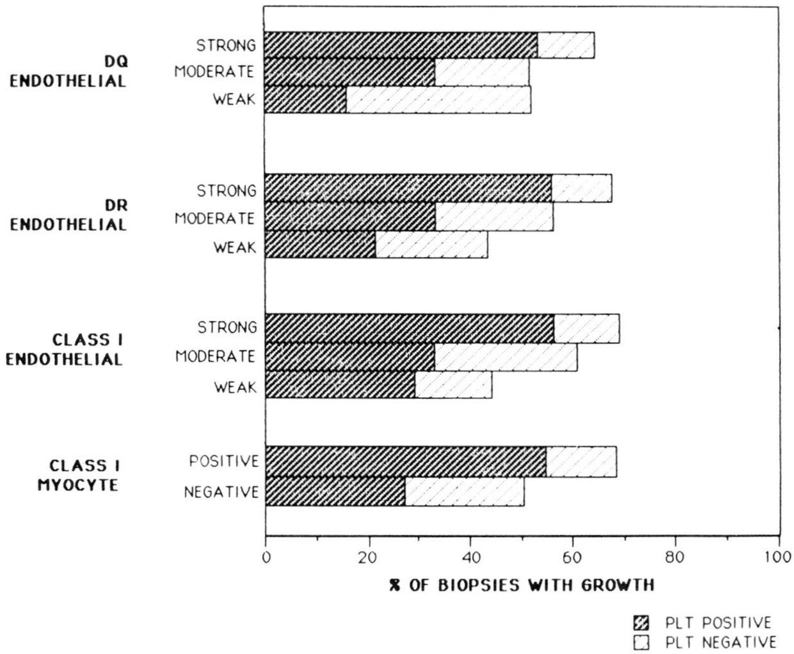

Figure 1. Correlation of growth of infiltrating cells and their donor reactivity with the intensity of immunostaining (107 biopsies).

References

1. Poppema S, Brocker EB, DeLeij L, Terbrack D, Visscher T, Terhaar A, et al. In situ analysis of the mononuclear cell infiltrate in primary malignant melanoma of the skin. Clin Exp Immunol 1983;51:77–82.
2. Billingham ME. Diagnosis of cardiac rejection by endomyocardial biopsy. Heart Transplant 1981;1:25–30.
3. Zeevi A, Fung J, Zerbe T, Kaufman C, Rabin B, Griffith B, et al. Allospecificity of activated T cells grown from endomyocardial biopsies from heart transplant patients. Transplantation 1986;41:620–626.
4. Pober JS, Collins T, Gimbrone MA, Cotran R, Gitlin J, Fiersi W, et al. Lymphocytes recognize human vascular endothelial and dermal fibroblast Ia antigen induced by recombinant interferon. Nature 1983;305:726–729.
5. Pober JS, Gimbrone M, Cotran R, Reiss C, Burakoff S, Fiers W, et al. Ia expression by vascular endothelium in inducible by activated T cells and by human interferon. J Exp Med 1983;157:1339–1353.
6. Suitters A, Rose M, Higgins A, Yacoub MH. MHC antigen expression in sequential biopsies from cardiac transplant patients—correlation with rejection. Clin Exp Immunol 1987;69:575–583.

Author Affiliations

Linda T. White, Tony R. Zerbe, Adriana Zeevi, and *Rene J. Duquesnoy*, Department of Pathology, University of Pittsburgh School of Medicine, Pittsburgh, PA 15261, USA; *Robert L. Hardesty* and *Bartley P. Griffith*, Department of Surgery, University of Pittsburgh School of Medicine, Pittsburgh, PA 15261, USA

Alloreactive Lymphocyte Propagation from Endomyocardial Biopsies from Heart Transplant Patients on Different Prophylactic Immunosuppressive Treatment Protocols

Christina Kaufman, Adriana Zeevi, Thomas Weber, Tony Zerbe, Bartley Griffith, Robert Kormos, Robert Hardesty, and Rene Duquesnoy

Cyclosporin is currently the preferred immunosuppressive drug to treat heart transplant recipients. During the early posttransplant period, an efficient immunosuppressive therapy is essential to reduce allograft rejection and thereby to increase transplant survival. However, infection remains a major complication of immunosuppression. Various prophylactic treatment protocols have been investigated to optimize heart transplant survival. They are based on tapered doses of cyclosporin in combination with prednisone (protocol A) supplemented with azathioprine (protocol B). Two protocols include additional antilymphocyte antibodies, namely rabbit antithymocyte globulin (RATG) (1.5 mg/kg/day) (protocol C) or murine anti-T-lymphocyte monoclonal antibody (OKT3; 5 mg/day) (protocol D).

In this study, we compared these four prophylactic treatment protocols for the relation to the histologic evidence of rejection, the propagation of lymphocyte cultures from endomyocardial biopsies, and the alloreactivity of such cultured lymphocytes as measured by primed lymphocyte testing (PLT) and cell-mediated lympholysis (CML).

Histologic evidence of rejection was graded according to Billingham's criteria regarding the degree of cellular infiltration in the perivascular areas and the extent of myocyte necrosis (1). Endomyocardial biopsies with no or minimal cellular inflammatory infiltrates (and absence of myocyte injury) received histologic grades 0 or 1, respectively. A mild cellular infiltrate with no myocyte injury received a grade 2, whereas an inflammatory infiltrate with myocyte necrosis was assigned grade 3. Table 1 shows the distribution of histologic grades among biopsies from heart transplant patients on different immunosuppression protocols. There were no significant differences in biopsy histology of patients on protocols A and B. It was noted that patients on protocol C, which included RATG treatment, showed a lower frequency of biopsies with significant cellular infiltrates. On the other hand, a high incidence of histologic rejection was observed in patients on protocol D, which included OKT3 prophylaxis.

Infiltrating lymphocytes can be propagated in vitro by culturing heart biopsy specimens with interleukin-2 and autologous irradiated feeder cells (2). The success rate of lymphocyte propagation correlates with the histologic rejection grade (3) and such cultured cells frequently show donor-specific alloreactivity as measured in PLT and CML assays. Lymphocyte growth from biopsies with grades 2 and 3 is the highest (about 70 to 80%) during the first month after transplantation and gradually decreases thereafter (4). The success rate of lymphocyte culturing from biopsies with grades 0 and 1 is about 30% throughout the posttransplant period of up to 6 months (3). These findings demonstrated that biopsies with no or minimal histologic evidence of cellular infiltration often enough yield lymphocyte cultures with donor reactivity. Recent studies (5) have shown that during the early posttransplant period, lymphocyte growth from grade 0 and

Table 1. Distribution of Histological Grade in Prophylactic Treatment Protocols

Drugs*	Protocol	No. of Patients	No. of Biopsies†	Percent of Biopsies with Grade 0	1	2	3
CsA+Pred	A	18	127	21	43	16	20
CsA+Pred+ Aza	B	15	107	29	32	23	16
CsA+Pred+ Aza+RATG	C	24	148	47	39	14	8
CsA+Pred+ Aza+OKT3	D	8	50	22	28	28	22

*CsA = Cyclosporin A; Pred = prednisone; Aza = azathioprine; RATG = rabbit antithymocyte globulin; OKT3 = murine anti-CD3 monoclonal antibody (Orthoclone).
†Biopsies taken in the first 90 days after transplantation.

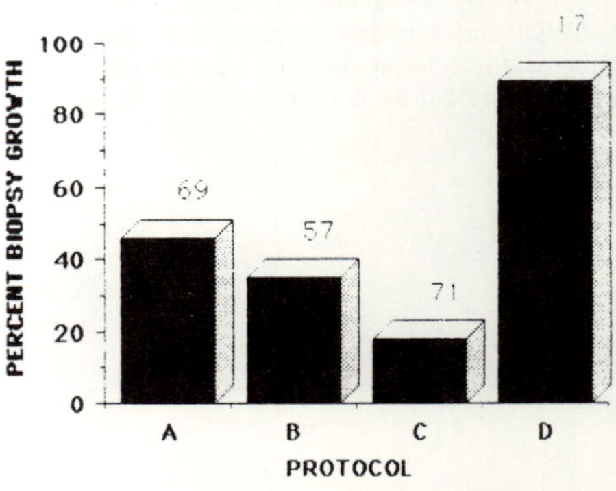

Figure 1. Lymphocyte growth from endomyocardial biopsies from heart transplant patients on different protocols of immunosuppressive prophylaxis. The drug combinations used in protocols A, B, C, and D are shown in Table 1.

Immunobiology of HLA
Volume II

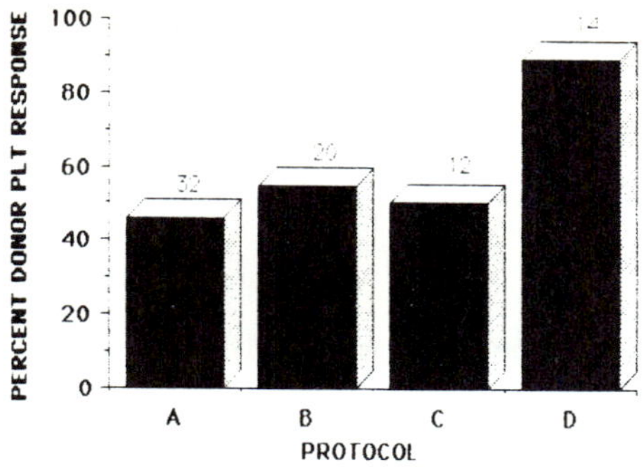

Figure 2. Donor-specific PLT reactivity of biopsy-grown lymphocytes derived from heart transplant patients on different immunosuppressive protocols. In the PLT test, 10^4 responder cells were incubated for 3 days with 10^5 irradiated (2,000 rads) donor cells and lymphocyte proliferation was determined by $[^3H]$thymidine incorporation (2).

1 biopsies is significantly associated with an upcoming rejection episode.

A comparative analysis was performed of lymphocyte growth rates from biopsies from patients on different immunosuppressive protocols. The biopsies were grouped according to histologic grade. Most informative were the biopsies with no or minimal cellular infiltrations (grades 0 and 1). Figure 1 shows a very high lymphocyte growth rate from grade 0 and 1 biopsies from patients treated under the OKT3 protocol D. These lymphocyte cultures also exhibited a high incidence of donor-specific PLT reactivity (Fig.2). In contrast, patients on protocols A and B and especially C (RATG) showed considerably lower frequencies of lymphocyte growth from grade 0-1 biopsies (Fig. 1). The cultured cells also displayed lower frequency of donor alloreactivity (Fig. 2). No significant differences were observed between grade 2-3 biopsies in the different protocol groups of patients. The frequency of lymphocyte growth was in the 70 to 90% range and 65 to 75% of the cultures had donor alloreactivity (data not shown).

The most notable observation in this study was that grade 0-1 biopsies from patients on the OKT3 prophylactic treatment protocol D frequently yielded lymphocytes with donor-specific PLT and also CML activity (data not shown). This means that in this group of patients the cardiac allograft was infiltrated by significant numbers of alloreactive lymphocytes before histologic changes indicative of rejection could be seen. Since the outgrowth of lymphocytes from biopsies with no or minimal cellular infiltration is associated with a higher incidence of subsequent rejection episodes (5), the biopsy findings suggest that patients on OKT3 prophylaxis may be at higher risk of rejection. Indeed, recent clinical studies by Kormos et al. (6) have demonstrated that the actuarial rejection-free survival after 3 months was only 35% in the OKT3 group, significantly less than the 60% value for the group of patients on RATG prophylaxis. The incidence of infection was similar in both groups.

Consistent with the clinical findings, the present results of the histologic analysis and the lymphocyte culture studies of endomyocardial biopsies suggest that OKT3 prophylactic treatment may have an adverse effect on cardiac transplant outcome by increasing the incidence of rejection.

Acknowledgments. This work was supported by grants AI-23467 and HL-35416 from the National Institutes of Health and a grant from the American Heart Association, Western Pennsylvania Chapter.

References

1. Billingham ME. Some recent advances in cardiac pathology. Hum Pathol 1979;10:367.
2. Zeevi A, Fung JJ, Zerbe T, Kaufman C, Rabin B, Griffith B, et al. Allospecificity of activated T cells grown from endomyocardial biopsies from heart transplant patients. Transplantation 1986;41:620.
3. Fung JJ, Zeevi A, Markus B, Zerbe T, Duquesnoy RJ. Dynamics of allospecific T lymphocyte infiltration in vascularized human allografts. Immunol Res, 1986;5:149.
4. Weber T, Kaufman C, Zeevi A, Zerbe T, Hardesty R, Kormos K, et al. Propagation of lymphocytes from human heart transplant biopsies: Methodological considerations. Transplant Proc, in press.
5. Duquesnoy RJ, Weber T, Zeevi A, Kaufman C, Zerbe TR, Hardesty RJ, et al. Propagation of lymphocytes infiltrating endomyocardial biopsies from heart transplant patients. Transplant Proc, in press.
6. Kormos RL, Herlan DB, Curran M, Hardesty RL, Armitage J, Griffith B. Monoclonal vs. polyclonal antibody therapy for prophylaxis against rejection following cardiac transplantation. Submitted for publication.

Author Affiliations

Christina Kaufman, Adriana Zeevi, Thomas Weber, Tony Zerbe, and *Rene Duquesnoy*, Division of Clinical Immunopathology, Department of Pathology, University of Pittsburgh, Pittsburgh, PA 15261, USA; *Bartley Griffith, Robert Kormos*, and *Robert Hardesty*, Department of Surgery, University of Pittsburgh, Pittsburgh, PA 15261, USA

Autolymphocytotoxins in Cardiac Transplantation: Serology, Significance, and Association with HLA-A1 and HLA-A3

C. McNamara, W.E. Braun, A.A. Zachary, L. Klingman, R.W. Stewart, C.E. Pippenger, G.A. Teresi, and G. Rincon

The presence of an autolymphocytotoxin (ALCT) in any potential organ recipient, and particularly cardiac recipients who often receive procainamide, can easily be confused with human leukocyte antigen (HLA) antibodies when serum screening and donor cross matching are done unless autocross matching and appropriate absorptions are performed. This has been a routine procedure in our pretransplant serum testing for cardiac transplant candidates and has revealed five patients with an ALCT.

The sera of 99 potential cardiac transplant recipients were screened against a T-lymphocyte panel for the presence of antibody. The sera of 46 patients reacted with 0% of the panel cells (group 1), 34 with less than 10% (group 2), 12 with 10 to 50% (group 3), and 7 with >50% (group 4). None of groups 1, 2, or 3, but 5 of 7 in group 4 had an ALCT as well as ANA titers of 1:320 or more (Table 1). An ALCT may be suspected by typically universal reactivity with a normal lymphocyte panel, autocytotoxicity, its IgM class and increased cytotoxicity at 5°C, its removal by absorption with autologous lymphocytes as well as by the wash steps in certain cytotoxicity tests such as the antiglobulin method, and its association with the use of procainamide and high ANA titers.

Of the five patients with an ALCT, all five had received procainamide, two were parous females, and only one had been transfused. Two of these patients have been successfully transplanted (one after plasmapheresis of the IgM antibody, and one after it had spontaneously disappeared), and two patients are still waiting for transplantation. One patient with newly discovered systemic lupus erythematosus (SLE) had plasmapheresis removal of his IgM antibody but died before a cardiac donor could be found. The effectiveness of plasmapheresis in removing IgM autocytotoxins was most likely related to the fact that 80% of IgM is contained in the intravascular compartment compared to only about 45% for IgG.

Of interest was the fact that all five patients with an ALCT had either HLA-A1 or A3 compared with only 38 or 82 without an ALCT (Fisher's exact test p uncorrected = 0.026). The relative risk for formation of an ALCT following procainamide in an individual with HLA-A1 or A3 was 5.8 (Table 2). Because a slow rate of acetylation has been related to procainamide sensitization, it may be that either an epitope shared by HLA-A1 and A3 is associated in some way with slow acetylation, or slow acetylation and an HLA-A1/A3 epitope may be independent factors in the development of an ALCT (1). In addition to genetic factors, more recent studies have shown that procainamide, but not its metabolite *N*-acetylprocainamide (NAPA), inhibits the generation of certain suppressor T lymphocytes and by that mechanism may permit autoantibody formation (2). A parallel situation has been described by Batchelor et al. who reported that patients with hydralazine-induced systemic lupus typically were slow acetylators and also possessed HLA-DR4 (3). DR4 was present in three of our five patients, and DQw3 was present in three of four tested, the only negative being the patient with newly discovered systemic lupus.

Table 1. Serological Characteristics of the Five Patients with Autolymphocytotoxins

				Serum Screening			
				Unabsorbed		Autoabsorbed	
Patient	Procainamide	ANA Titer	Anti-DNA	Auto XM	Prelim. Screen	Auto XM	Prelim. Screen
W.G.	Yes	>1:320	–	+	100%	–	61%
M.C.	Yes, but not since 1984	>1:320	–	+	100%	2X autoabsorbed (incomplete)	
M.K.	Yes	1:320	–	+	100%	No absorptions done	
A.M.	Yes	>1:320	+ 26% (Crithidia assay)*	+	100%⁺	–	20%
M.J.L.	Yes	>1:320	–	+	100%⁺	–	18% nonspecific†

*For antibody to double-stranded DNA.
⁺In these cases, column separation of the patient's serum showed that the ALCT was IgM.
†Tested by the basic technique with no wash at 22°C; 1 month later a nonautoabsorbed sample had 24% PRA with A29 and B8 specificity.

Immunobiology of HLA
Volume II

Table 2.

	HLA-A1 or A3	
	+	–
ALCT		
+	5	0
–	38	44
Relative risk = 5.8		

The 5 patients with ALCTs all had either HLA-A1 or A3 compared with just 38 of 82 potential cardiac transplant recipients without ALCTs (Fisher's exact test p uncorrected = 0.026). There were 12 ALCT-negative patients who were not HLA typed.

In a separate investigation of 30 cardiac patients whose sera were obtained simply for a procainamide level, 10 reacted with one or more cells of the normal lymphocyte panel and only two of these reacted with >50% of the panel (60% and 88%). Pregnancy and/or transfusions were not involved in their sensitization. Further studies of procainamide and NAPA in 10 different ratios that were consistent with dosage levels obtained in humans failed to show any direct cytotoxic effect on lymphocytes.

The natural history of such antibodies in cardiac transplant recipients is unknown, but information from rheumatologic patients has suggested that the titer of the ALCT should fall promptly with titers reduced by approximately 75% within 3 to 16 weeks, although ANA may persist for months (4). It is also unclear whether an ALCT would have an adverse effect on the outcome of a cardiac transplant. However, the fact that hyperacutely rejected cardiac allografts have had deposition of IgM and fibrin suggests that such antibodies might be relevant to allograft injury (5).

These data suggest that potential cardiac transplant recipients may have a disproportionately high ALCT frequency that is most likely due to procainamide, and individuals with HLA-A1 or A3 may be at greater risk for the formation of an ALCT when receiving this drug (relative risk = 5.8).

References

1. Woosley RL, Drayer DE, Reidenberg MM, Nies AS, Carr K, Oates JA. Effect of acetylator phenotype on the rate at which procainamide induces antinuclear antibodies in the lupus syndrome. N Engl J Med 1978;298:1157–1159.
2. Ochi T, Goldings EA, Lipsky PE, Ziff M. Immunomodulatory effect of procainamide in man: Inhibition of human suppressor T-cell activity in vitro. J Clin Invest 1983;71:36–45.
3. Batchelor JR, Tinoco RM, Hughes GRV, Naish PF, Bing RF, Welsh KI, et al. Hydralazine-induced systemic lupus erythematosus: Influence of HLA-DR and sex on susceptibility. Lancet 1980;1:1107–1109.
4. Bluestein HG, Zvaifler NJ, Weisman NH, Shapiro RF. Lymphocyte alterations by procainamide: Relation to drug-induced lupus erythematosus syndrome. Lancet 1979;2: 816–819.
5. Singh G, Thompson N, Griffith B, Bernstein R, Rabin BS, Hardesty R, et al. Histocompatibility in cardiac transplantation with particular reference to immunopathology of positive serologic crossmatch. Clin Immunol Immunopathol 1983;28:56–66.

Author Affiliations

Histocompatibility and Immunogenetics Laboratory, Cardiac Transplantation and Cardiothoracic Surgery, Cardiology, and Biochemistry, Cleveland Clinic Foundation, Cleveland, OH 44106, USA.

Role of HLA Class II-Specific Alloreactive T Cells in Biliary Epithelium Injury Associated with Liver Transplant Rejection

S. Saidman, B. Markus, A.J. Demetris, J. Fung, A. Zeevi, T. Starzl, and R. Duquesnoy

Abstract: In this study, we report the establishment of lymphocyte cultures from liver allograft biopsies, and correlate their primed lymphocyte testing (PLT) specificity towards HLA class I and/or class II donor antigens with clinicopathological findings. Early posttransplant biopsies generally yielded more class I-specific cells than later biopsies. There appears to be no correlation of hepatocellular enzyme levels (SGOT or SGPT) with PLT specificity. However, the levels of GGTP and AP, which are biliary enzymes, were increased in patients whose biopsies yielded class II-specific rather than class I-specific cells. There was also a trend towards more damage to bile duct epithelium in biopsies yielding class II-specific cells. These results are probably related to the interaction of class II-specific T cells with bile duct epithelium, which has been shown to express class II HLA antigens.

Although liver transplantation has become an accepted treatment for a variety of end-stage liver diseases, the mechanisms of liver transplant rejection are still obscure. A novel approach to learn more about the immunopathology of transplant rejection is the propagation and characterization of lymphocytes directly from the

Table 1. HLA-Associated Alloreactivity of Liver Biopsy-Grown Lymphocyte Cultures

Specificity	No. of Cultures	Posttransplant Day of Biopsy Median	Range
Class I	8	8.5*	4–48
Class I + II	15	17.0	2–237
Class II	12	29.0	5–753
	35[†]		

*Class I versus class II differences for all biopsies ($n = 20$), p = 0.076; for biopsies with rejection ($n = 16$), p = 0.003.
[†]Thirty-five of 67 biopsy-grown lymphocyte cultures showed class I and/or class II HLA-specific reactivity.

allograft. Previous studies have shown that activated T-cell cultures can be generated from liver allograft biopsies by incubation with recombinant IL2 (1,2). Such cultured lymphocytes frequently exhibit alloreactivity towards donor cells as measured by primed lymphocyte testing (PLT) and by cell-mediated lympholysis (CML). Often they demonstrate restricted specificity patterns against a limited number of donor HLA antigens and this alloreactivity can be blocked with HLA-specific monoclonal antibodies against appropriate determinants. The aim of this study was to evaluate the association of the HLA class I and class II allospecificity of lymphocytes grown from liver allograft biopsies with clinical, biochemical, and histopathologic findings.

T lymphocytes were propagated from portions of percutaneous liver biopsies taken at times of liver allograft dysfunction, or from removed allografts. The tissue was divided into small segments and cultured in microculture wells in the presence of 200 µl of recombinant IL2 as described previously (1). After approximately 2 weeks, sufficient cells were obtained for PLT assays against cryopreserved donor splenocytes. Class I- and/or class II-specific alloreactivity was determined by testing an informative panel of HLA-typed lymphocytes and/or blocking with anti-class I and class II monoclonal antibodies (3).

Liver tissue from the same biopsy core or from the same region of the liver (in failed allografts) was submitted for histopathologic review with special emphasis on the presence of a mononuclear portal inflammatory infiltrate associated with ductual epithelial and venular endothelial damage, findings considered to be important in hepatic rejection (4). Information about the clinical and biochemical data was obtained from the medical charts and included serum levels of total and direct bilirubin, glutamate oxaloacetic transaminase (SGOT), glutamate pyruvate transaminase (SGPT), alkaline phosphatase (AP), and γ-glutamyl transpeptidase (GGTP). All values were recorded from the day prior to obtaining the biopsy. Statistical analysis of relevant data was performed by the nonparametric statistical tests of Mann-Whitney and Kruskal-Wallis using the SPSSPC software package (5).

For 35 out of 67 lymphocyte cultures established we could assess allospecificity towards HLA class I ($n = 8$), mixed class I and II ($n = 15$) or HLA class II antigens

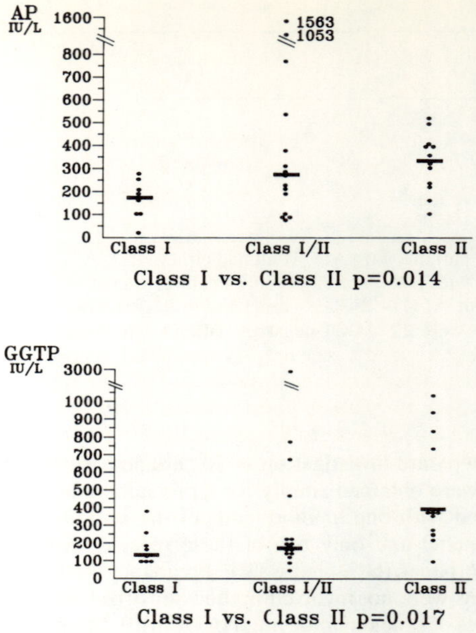

Figure 1. Serum levels of the biliary enzymes alkaline phosphatase (AP) and γ-glutamyl transpeptidase (GGTP) from liver transplant recipients on the day prior to liver biopsy. Patients are grouped according to whether their biopsies grew lymphocytes reactive against only class I antigens, a combination of class I and II antigens, or only class II antigens. Bars indicate the median levels for each group. The p values are for the comparison of median levels in the group yielding only class I-reactive cells versus median levels in the group yielding only class II-reactive cells, using the Mann-Whitney rank sum test.

($n = 12$). As shown in Table 1, class I-specific cells were generated from biopsies obtained earlier during the posttransplant time period than those yielding class II-specific cells. This is concordant with previous data from heart transplant biopsies showing that class I-specific cells are more predominant in earlier biopsies whereas class II-specific cells are generally found in later biopsies (6).

No statistically significant differences were found in the bilirubin, SGOT, and SGPT values between the groups yielding class I- and/or class II-specific cells, although a trend towards higher levels in the group of biopsies yielding class II-specific cells was noted. In contrast, significant differences were found for levels of AP and GGTP, both of which are biliary enzymes. Figure 1 shows that the median levels of serum AP were almost twice as high in the class II group as in the class I group (332 versus 170 IU/ml, normal range < 100 IU/ml). Intermediate levels of AP were found in the group of biopsies yielding a mixture of class I- and II-specific cells. Similarly, the median GGTP levels were approximately 2.5 times higher in the class II group than in the class I group (365 versus 142 UI/ml, normal range males <44; females <32 IU/ml).

Other studies have shown that increased GGTP levels reflect injury to the biliary epithelium (7). Thus greater elevations of GGTP (and probably also AP) levels in the class II group may suggest that a cellular infiltrate of class II-specific cells in the liver allograft is associated with more injury to the biliary epithelium. Normal biliary epithelium expresses class I HLA antigens, but no or very low levels of class II antigens (8,9). In contrast, increased expression of class II HLA antigens is frequently observed in liver allografts, especially during rejection (10). The expression of class II antigens on biliary epithelial cells can be induced by lymphokines released by activated T cells, such as γ interferon (11). During rejection, the biliary epithelium is often surrounded by infiltrating lymphocytes (12,13). Our studies have shown a marked trend towards increased bile duct damage in biopsies yielding class II-specific lymphocyte cultures as compared to biopsies growing class I-specific cells (3).

These data suggest that a cellular infiltrate of class II-specific alloreactive T cells may lead to injury of the bile duct epithelium. At the same time, an increased expression of class II HLA antigens may render the biliary epithelium a preferred target in the cellular rejection of the liver allograft.

Acknowledgment. This work was supported by NIH grants AI 23467 and HL 36416. B. Markus was a recipient of a Fellowship Grant from the Deutsche Forschungsgemeinschaft.

References

1. Fung JJ, Zeevi A, Starzl TE, Demetris AJ, Iwatsuki S, Duquesnoy RJ. Functional characterization of infiltrating T lymphocytes in human hepatic allografts. Hum Immunol 1986;16:182.
2. Markus BH, Fung JJ, Zeevi A, Starzl TE, Demetris AJ, Duquesnoy RJ. Analysis of T lymphocytes infiltrating human hepatic allografts. Transplant Proc 1987;19:2470.
3. Markus BH, Demetris AJ, Saidman SL, Fung JJ, Zeevi A, et al. Alloreactive T lymphocytes cultured from liver transplant biopsies: Associations of HLA specificity with clinicopathological findings. Clin Transplant 1988; 2:70.
4. Demetris AJ, Jaffe R, Starzl TE. Pathol Annu, in press.
5. Norusis MJ. SPSS/PC+ for the IBM PC/XT/AT. SPSS Inc., Chicago, 1986.
6. Zeevi A, Fung J, Zerbe T, Kaufman C, Rabin B, Griffith B, et al. Allospecificity of activated T cells grown from endomyocardial biopsies from heart transplant patients. Transplantation 1986;41:620.
7. Leonard TB, Neptun DA, Popp JA. Serum gamma glutamyl transferase as a specific indicator of bile duct lesions in the rat liver. Am J Pathol 1984;116:262.
8. Lautenschlager I, Hayry P. Expression of the major histocompatibility complex antigens on different liver cellular components in rat and man. Scand J Immunol 1981; 14:421.
9. Daar AS, Fuggle SV, Fabre JW, Ting A, Morris PJ. The detailed distribution of MHC Class II antigens in normal human organs. Transplantation 1984;38:293.
10. Demetris AJ, Lasky S, van Thiel DH, Starzl TE, Whiteside T. Induction of DR/Ia antigens in human liver allografts. Transplantation 1985;5:504.
11. Demetris AJ, Markus B, Saidman S, Fung J, Nalesnik M, Makowka L, et al. Establishment of primary cultures of human biliary epithelium and induction of class II MHC antigens by interferon gamma. Transplant Proc, in press.
12. Porter KA. Pathology of the orthotopic homograft and heterograft. In: Starzl TE (ed): Experience in Hepatic Transplantation. Philadelphia, WB Saunders, 1969, p 422.
13. Fennell RH. Ductular damage in liver transplant rejection: Its similarity to that of primary biliary cirrhosis and graft versus host disease. Pathol Annu 1981;16(2):289.

Author Affiliations

S. Saidman, A.J. Demetris, A. Zeevi, and *R. Duquesnoy*, Department of Pathology, University of Pittsburgh, Pittsburgh, PA, USA; *B. Markus, J. Fung*, and *T. Starzl*, Department of Surgery, University of Pittsburgh, Pittsburgh, PA, USA

Cultured Epidermis as Long-Term Allografts Without Anti-HLA Sensitization

H. Betuel, G. Mauduit, L. Gebuhrer, M. Faure, and J. Thivolet

Epidermal cells (keratinocytes) obtained from a skin biopsy can be grown in vitro and yield a thin sheet of epithelium suitable for grafting as permanent autografts or as allografts (1). During culture, Langerhans cells (LC) and other class II-bearing cells are loosed, whereas the expression of class I antigens on the cultured keratinocytes is retained (2). These epidermal allografts (EAG) devoid of LC are not rejected clinically or histologically and do not induce cellular response in nonimmunosuppressed patients (1–3). In the present report, we monitored the humoral response of 19 recipients directed against the HLA antigens present on the EAG.

Methodology

Obtention of cultured epidermis. Epidermal cells were obtained from small skin specimens of donors undergoing plastic surgery and cultured as described (1) to produce epithelial sheets.

Patients. There were 19 adult recipients, 12 women and 7 men, whose mean age was 69. Nine females had had pregnancies and three had been transfused. Two male recipients had received transfusions. HLA A,B,C DR typing of the recipients and of their donors was performed by the standard NIH test. Screening for anti-HLA was performed every 15 days for 2 months.

Grafting. The recipients of allografts suffered from chronic leg ulcers. A 120-cm^2 area split-thickness skin graft from the forearm was excised, meshed to expand the surface, and placed as autograft on the leg ulcer (1). The EAG from cultured epidermis was placed on the graft removal site of the forearm. The epithelial grafts were layered on a gauze and placed with the basal cells layers directly in contact with the prepared sites. The gauzes were removed on day 5. Dressing was changed daily for 10 days, after which the grafts were left exposed.

Results

Immunization. In 14 pairs of HLA-typed donors and recipients, the mean number of A,B DR mismatches was respectively 1.6, 1.9, and 1.2. Crossmatches performed before graft were negative. None of the seven male patients developed anti-HLA antibodies. Among the 12 female patients, 2 had preexisting anti-IgM antibodies that did not increase and persisted after EAG. In only two cases were antibodies detected after EAG. In the first case, the recipient typed as HLA A28 A11 B51 B14 CW3 DR2 DQw1 and the donor as A2 A9 B7 B38 DR2 DRW13 DQW1. The antibody was an IgG anti-A2; still detectable though weak after 5 months, the graft was not rejected. No family typing was possible. In the second case, the recipient typed as A2 A3 B49 B7 CW3 DRW13 DRW14 DQW1 and the donor as A11 A28 B51 B27 CW2 DRW13 DRW11 DQW3. On day 15, strong anti-DR1, DR9, DR10 activity was detected. The donor was not DR1. Family inquiry revealed that the son was DR1, inherited from the father. In short, 2 of 19 recipients developed anti-HLA antibodies after EAG; direct involvement of donor's cells was disputable in one case and could be excluded in the other.

Fate of Epidermal Allografts

After culture, the epithelial sheets are composed of a basal cell layer and a few suprabasal cell layers of keratinocytes. HLA class I antigens were present on basal cells only (2,3). On skin biopsies taken after and beyond 28 days, class I antigens were present on basal cells and on keratinocytes of the epidermis (2,3). Neither melanocytes nor Langerhans cells could be observed by histologic and electronmicroscopic examination of the cultured epithelium prior to grafting (1). During the next few weeks, the grafts spread outward and merged with each other and with the surrounding skin. No inflammatory mononuclear infiltrates were present. No clinical or histologic sign of rejection could be observed during the follow-up (1,2). Beyond 2 months, there was progressive replacement of donor type keratinocytes by epithelial cells from the recipient (3).

Discussion

Grafting of cultured epithelium devoid of LC did not lead to anti-HLA sensitization in 15 of 17 patients. This contrasts with the strong immunogenicity by the full-thickness skin graft. Antibodies were detected in two multiparous recipients. In one instance, the specificity—anti-DR1—did not correspond to the antigens present on the donor's lymphocytes—DRW13 DRW11—; furthermore Langerhans cells are not encountered in EAG as documented previously (1,2). In the other case, anti-HLA.A2 was present and the donor was A2; however, this patient had had four pregnancies and it cannot be excluded that the husband was A2 on account of the high frequency of this antigen. Thus, in both cases, it can be suggested that antibody formation occurred by a secondary response to a nonspecific challenge brought by the cells present in the cultured epidermis. The lack of LC in cultured epithelial cell suspensions results in an absence of stimulation of peripheral blood lymphocytes. It has also been verified that recipient cells failed to proliferate when cultured with epidermal cells (lacking LC) taken from sequential biopsies of EAG (1). Donor keratinocytes do not induce anti-HLA immunization during their existence as allografts; they are not rejected, but are progressively colonized by epithelial cells originating from the recipient. Another example of the lack of sensitization in the absence of class II antigens is represented by platelet transfusions. Platelet express class I but not class II antigens. Transfusions of purified platelets cleared of lymphocytes does not lead to anti-HLA immunization in more than 20 recipients, as we have shown in a protocol of purified platelets transfusions prior to kidney transplantation. (4). These two models illustrated the crucial role of class II antigens in allorecognition and subsequent immunologic response.

References

1. Thivolet J, Faure M, Demiden A, Mauduit G. Long-term survival and immunological tolerance of human epidermal allografts produced in culture. Transplantation 1986;42:275.
2. Mauduit G, Vincent C, Gielen V, Faure M, Demiden A, Thivolet J. Expression of class I MHC antigens by cultured human epidermis and epidermal allografts. Tissue Antigens 1987;29:65.
3. Gielen V, Faure M, Mauduit G, Thivolet J. Progressive replacement of human cultured epithelial allografts by recipient cells as evidenced by HLA class I antigens expression. Dermatologica 1987;175:166.
4. Betuel H, Cantarovitch D, Robert F, Gebuhrer L, Touraine JL. Platelet transfusions preparative for kidney transplantation. Transplant Proc 1985;17:2335.

Author Affiliations

H. Betuel, L. Gebuhrer, Histocompatibility Laboratory, Blood Transfusion Centre; *G. Mauduit, M. Faure, J. Thivolet,* Clinique Dermatologique, Hôpital Edoard Herriot, Lyon, France.

Section XIX
In Vivo Applications of MHC and Anti-MHC Products

The Effect of Tumor Necrosis Factor (TNF) on (NZB × NZW)F_1 Lupus Nephritis

Chaim O. Jacob and Hugh O. McDevitt

The (NZB × NZW)F_1 hybrid mouse develops severe autoimmune disease characterized by fatal glomerulonephritis, antibody to nuclear antigens, and female predominance, remarkably similar to systemic lupus erythematosus in humans. (NZB × NZW)F_1 × NZB backcross studies by Kotzin and Palmer have shown that the major genetic contribution from NZW mice maps within the H-2 region but may be separate from the class II loci (1). Since the I-A^z β chain from the NZW parental strain has an identical sequence to the normal I-A^u haplotype (Acha-Orbea and McDevitt) and it was postulated that levels of Ia expression in affected tissues may be involved in autoimmune pathogenesis (2), the regulatory role of cytokines was tested in vivo. We present evidence suggesting the immunoregulation by interferon gamma (IFN-γ) and tumor necrosis factor (TNF) may be involved in the pathogenesis of lupus nephritis in the (NZB × NZW) model system.

Two groups of 25 female NZB/W F_1 mice, 4 months old, received rIFN-γ or equivalent volumes of PBS over a period of 3 months. As an additional control group, 16 age- and sex-matched mice of the parental strain NZW, were similarly treated with IFN-γ. Death occurred at an earlier age in the NZB/W F_1 group that received IFN-γ compared with the PBS control mice (Fig. 1). The difference in survival between treated and control mice was statistically significant ($p \leq 0.001$).

While PBS-treated NZB/W F_1 control mice began to die at ~8 months of age (50% survival, 9.5 months), in the IFN-γ-treated group, 75 to 80% of NZB/W F_1 were dead by 8 months. The life span of the NZW control group was not affected by the IFN-γ treatment (not shown). High grade proteinuria was detected significantly earlier in the IFN-γ-treated NZB/W F_1 group compared with the control group. The onset of death correlated closely with the onset and amount of proteinuria. Similarly, anti-DNA antibodies reached peak levels earlier in the IFN-γ-treated group than in the control group, although no significant differences in the peak levels of anti-DNA antibodies were observed between the two groups as a whole (not shown).

Given these results we asked whether we could block disease development by using monoclonal antibody to IFN-γ. Two groups of 20 female NZB/W F_1 mice received purified anti-IFN-γ mAB (DB-1), and as controls, 10 mice received an irrelevant monoclonal IgG_1 antibody (TE33) and 20 animals received PBS. All mice were 4 months old at the beginning of the experiment. Treatment was given for a period of three months. The results of these experiments show that monoclonal anti IFN-γ antibodies can delay significantly the disease development. At the age of 11 months, 80 of both control groups were dead, while 95% of mice were alive in the DB-1-treated group (3).

Since TNF synergises with IFN-γ in a number of activities (4), we have tested the in vivo effects of TNF-α in this murine model.

A group of 20 female (NZB × NZW)F_1 mice was treated with 10 μg recombinant murine TNF-α, three times per week intraperitoneally and as controls 20 mice received PBS. All mice were 4 months old at the beginning of the treatment. We have chosen 4 months of age

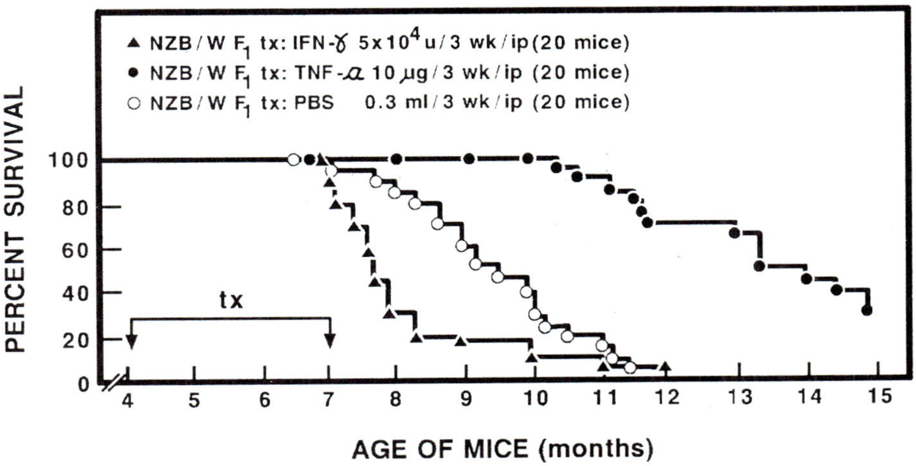

Figure 1. Survival of (NZB × NZW)F_1 female mice treated with IFN-γ, TNF-α or PBS. Arrows indicate the time period of treatment.

© Springer-Verlag New York 1989
Immunobiology of HLA
Volume II

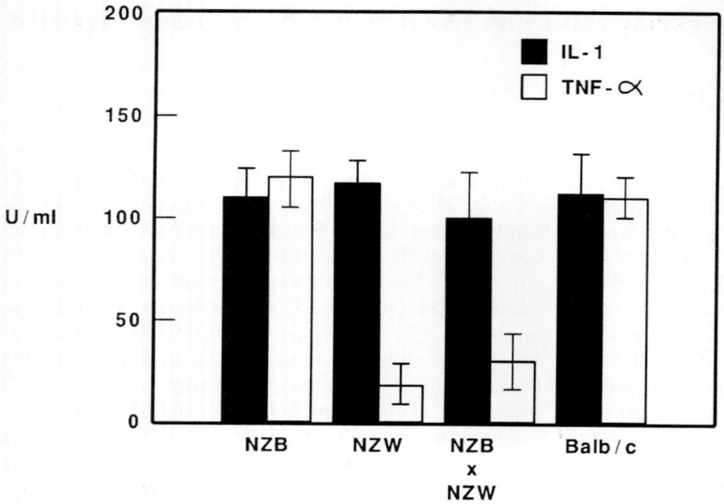

Figure 2. TNF-α and IL-1 production by peritoneal exudate cells in various mouse strains.

for treatment since at this age F_1 mice show only minor disease. Treatment was given for a period of 3 months.

Figure 1 shows the improved survival rate of mice treated with TNF-α. At the age of 11 months 90% of treated animals were alive, while 80 to 85% of controls were dead. Even at 15 months of age (8 months after the treatment was stopped) about 30% of treated mice were still alive, while all control mice were dead by 12 months of age. In parallel with the dramatically prolonged survival of TNF-α-treated mice, the development of proteinuria was significantly delayed. By 9 months, 65% of control mice have high grade proteinuria, while only 10 to 15% of the treated mice show proteinuria of this degree. No weight loss was observed in the treated group versus controls (not shown).

Based on these results and taking into consideration recent cytogenetic studies which have shown that the genes for human and mouse TNF-α and lymphotoxin (TNF-β) are located tandemly within the major histocompatibility complex (5,6), we further tested the hypothesis that a TNF-gene is involved in (NZB × NZW)F_1lupus nephritis.

A PvuII 300 bp fragment derived from exon 4 of the murine TNF-α gene was used as a probe to search for polymorphic restriction fragments in various mouse strains. NZB mice, similar to Balb/c mice, show a BamHI fragment of about 10 kb, while the NZW mice show an 11.5 kb band. This polymorphism can also be documented by a BamHI and HincII double digestion. Thus, NZW mice show a nonpolymorphic 4.3 kb band and polymorphic 2.7 kb band, in contrast to NZB mice, which possess a 1.2 kb polymorphic band (not shown). No polymorphism could be demonstrated using an 871 bp KpnI-XBaI fragment probe from murine TNF-β cDNA with 12 different restriction enzymes. Interestingly, the classical autoimmune mouse strains MRL−+/+, MRL-1pr, BXSB, and NOD show the same TNF-α RFLP pattern as the NZW mice, but some nonautoimmune strains have this pattern as well (7).

Peritoneal exudate cells obtained from NZB, NZW, (NZB × NZW)F_1 and Balb/c mice 4 to 5 days after thioglycollate administration were tested for TNF-α production using an in vitro bioassay which quantitates the cytolytic activity of TNF-α on L-M murine connective tissue cells (8). Adherent cells were activated for TNF-α production with IFN-γ, LPS or IFN-γ and LPS in various doses for 16 hours and cell-free supernatants were assayed for TNF-α cytolytic activity. As shown in Figure 2, NZB and Balb/c mice show very similar activities in all experiments. By contrast, NZW mice show very low TNF-α biological activity regardless of the method used for activation. The F_1 mice show an intermediate level but in most experiments TNF-α levels were much closer to the NZW parent than to the NZB parent. Moreover, the supernatants of the peritoneal exudate cells from the various strains were assayed for the presence of IL-1. In contrast to the major differences in production of TNF-α, all strains tested produce similar quantities of IL-1 (Fig. 2).

Taken together, the data suggest that the NZW parent may contribute to the multigenic lupus-like autoimmune disease in (NZB × NZW)F_1 mice by donating a TNF-α gene capable of only a low level of TNF-α production. Thus, the restriction fragment length polymorphism coincides with the levels of inducibility of TNF-α in the mice tested. Most importantly, replacement therapy with recombinant TNF-α induces a very significant delay in the development of the disease, suggesting that such treatment is supplementing the low TNF production by a possible abnormal TNF-α gene contributed by the NZW parent. While we favor the hypothesis that IFN-γ and TNF-α modulate the immune response by regulating MHC class II antigen expression, the mechanism of activity of these cytokines in this in vivo system remains to be determined.

Acknowledgments. We are very grateful to Dr. H.M. Shepard for advice, help, and interest throughout these studies. We thank Dr. Gail Lewis and the bioassay service group in Genentech for their contributions, May Koo for excellent technical assistance, and Karen Moody for preparation of the manuscript.

References

1. Kotzin BL, Palmer E. The contribution of NZW genes to lupus-like disease in (NZB × NZW)F_1 mice. J Exp Med 1986;165:1237.
2. Bottazzo GR, Pujol-Borrell R, Hanafusa T, Feldman M. Role of aberrant HLA-DR expression and antigen presentation in induction of endocrine autoimmunity. Lancet 1983;ii:1115.
3. Jacob CO, van Der Meide PH, McDevitt HO. In vivo treatment of (NZB × NZW)F_1 lupus-like nephritis with monoclonal antibody to γ interferon. J Exp Med 1987; 166:584.
4. Beutler B, Cerami A. Cachectin and tumor necrosis factor as two sides of the same biological coin. Nature 1986; 320:584.
5. Muller U, Jongeneel CV, Nedospasov SA, Fischer-Lindahl K, Steinmetz M. Tumor necrosis factor and lymphotoxin genes map close to the H-2D in the mouse major histocompatibility complex. Nature 1987;325:265.
6. Dunham I, Sargent CA, Trowsdale J, Campbell RD. Molecular mapping of the human major histocompatibility complex by pulse field gel electrophoresis. Proc Natl Acad Sci USA 1987;84:7237.
7. Gardner SM, Mock BA, Hilgers J, Huppi KE, Roeder WD. Mouse lymphotoxin and tumor necrosis factor: Structural analysis of the cloned genes, physical linkage and chromosomal position. J Immunol 1987;139:476.
8. Mukavitz-Kramer S, Carver ME. Serum free in vitro bioassay for the detection of tumor necrosis factor. J Immonol Methods 1986;93:201.

Author Affiliation

Department of Medical Microbiology, Stanford University, Stanford, CA 94305, USA

Treatment of EAE by Anti-MHC Class II-Specific Monoclonal Antibodies in Rhesus Monkeys

M. Jonker and L. Steinman

Abstract: Major histocompatibility complex (MHC) class II antigens play a crucial role in the immune response. T lymphocytes recognize antigens in the context of MHC class II antigens on antigen-presenting cells. It has been shown that class II antigens are expressed on nonlymphoid cells in tissues that are attacked by the immune response in allografts or autoimmune diseases. The presence of these class II antigens may provide an essential extra stimulus to the immune system to reject an allograft or to keep an autoimmune disease going. Masking these MHC class II antigens by monoclonal antibodies (MAb) may therefore suppress allo- or autoreactivity. The feasibility of this was tested by treating monkeys that had experimental allergic encephalomyelitis (EAE). While untreated monkeys invariably die within 1 to 3 days after the onset of EAE, anti-class II treatment, started as soon as the first clinical signs of EAE became apparent, improved the disease course in two monkeys and cured two other monkeys while only one of the five treated monkeys died within 3 days. This shows that anti-MHC class II monoclonal antibodies can modify the immune response in autoimmunity and underlines the role of MHC class II in autoimmune disease.

Results and Discussion. Experimental allergic encephalomyelitis (EAE) is an autoimmune disease that is T cell-mediated. EAE is induced in rhesus monkeys by intradermal immunization with central nervous system tissue from calf brain emulsified in Freund's complete adjuvant (1). Untreated monkeys died within 24 to 72 hours after the onset of clinical signs. All of these animals had large hemorrhagic lesions in the white matter and sometimes in the grey matter of cerebrum and cerebellum.

The monkeys that proved to be susceptible for EAE were treated with MAb at the onset of clinical signs. Three monkeys received a mixture of GM11 and FN16, both reactive with rhesus monkey monomorphic MHC class II determinants (2). These monkeys received a daily dose of 2 mg/kg of GM11 (IgG2a) and 1 mg/kg of FN16 (IgG1) for a maximum of 10 days. One of the monkeys died within 48 hours after the onset of EAE signs. One monkey had to be sacrificed 7 days after the onset of EAE signs because of its uncoordinated movements which did not improve. Both of these monkeys did not have macroscopically visible lesions in the brain (like the untreated monkeys) but many microscopically visible smaller lesions perivascularly. The third monkey had much milder neurological signs (some

Table 1. Reactivity of Genox353 on Rhesus Monkey Cells

Monkey Identification	Rhesus D/DR Type	Genox353 Reactivity
1WM	1,1	+
2D	1,1	−
2F	1,1	−
OUA	2,2	−
OYE	2,2	−
2BR	3,3	−
1IR	3,3	+
OKD	5,5	−
1FV	5,5	−
OJU	8,8	+
2B	8,8	−
2C	8,8	−
2G	8,8	−
2V	8,8	−
2BI	101,101	+
OME	101,−	+
1RF	101,−	+
ONZ	9,9	−
2AK	9,9	−
2I	9,9	−

Genox353 reactivity was tested by cytotoxicity or by indirect immunofluorescence and fluorescent-activated cell sorter (FACS) analysis. Most animals tested were homozygous for the RhLA system.

Table 2. EAE in Rhesus Monkeys

Monkey	DR Type	Onset of EAE Day Post-immunization	Treatment	Survival After EAE (Days)
3983	1,8	22	−	1
1JZ	3,8	54	−	1
3988	2,2	29	−	2
1NK	8,101	47	−	2
1JI	2,2	14	−	3
1PS	1,8	38	Genox353	22
1VJ	1,8	73	Genox353	>200
1RU	2,4	17	GM11+ FN16	2
1WB	7,8	51	GM11+ FN16	6
1RC	3,5	41	GM11+ FN16	>200

EAE was induced by injecting rhesus monkeys with 2 ml of 0.5 mg of brain homogenate in complete Freund's adjuvant. The animals were inspected three times a day for EAE signs; MAb treatment was started as soon as the first neurological signs became apparent.

motor weakness) when these became apparent. This animal improved significantly after the first few days of MAb treatment and had normal motoric function by the end of treatment. However, the CSF still contained many lymphocytes. Ten days after discontinuation of MAb treatment this animal showed new EAE signs (uncoordinated movements). These symptoms disappeared, however, without further treatment and this animal remained symptom free for nearly a year. It was then sacrificed for histological examination.

Two monkeys received Genox353, a MAb specific for a polymorphic determinant of the human HLA-DQwl antigen (3). This antibody cross-reacts well with rhesus monkeys and also detects a polymorphic determinant in this species (Table 1). The Genox353 preparation was a mixture of the original IgG1 and an isotype switch variant (IgG2a) in a 1:2 ratio. The monkeys received 2 mg/kg of this mixture daily for 10 days. Both monkeys survived much longer than untreated animals. One of the monkeys had to be sacrificed 22 days after the onset of EAE because of uncoordinated movements and hind limb paralysis which did not improve. The other monkey clearly improved after the start of MAb therapy. Two weeks later this animal presented new EAE signs. Since these symptoms did not improve this animal was given a course of CD4 specific MAb (1 mg/kg of a mixture of OKT4+4A), which was previously shown to be also effective in curing EAE signs in monkeys (4). This animal then slowly improved and remained free of symptoms for more than 1½ years.

The results presented here clearly show the beneficial influence of MHC class II-specific MAb on the clinical course of EAE in rhesus monkeys (Table 2). Both a mixture of two MAb that recognize a monomorphic structure of a MHC class II antigen as well as MAb reactive with a polymorphic structure of a MHC class II antigen were effective. The course of events in EAE may be described as follows. After immunization, autoreactive T cells are activated, which leads to cellular infiltrates in the brain. The inflammatory reaction causes the release of interleukins and this may subsequently stimulate MHC class II expression on many cells in the brain and also may attract granulocytes to the lesions. The activated autoreactive T cells are most likely $CD4^+$ and MHC class II positive. Previous experiments demonstrated that anti-CD4 MAb have a beneficial influence on EAE and that this treatment may prevent initiation of new lesions as well as the blocking of interleukin production by these cells (4). MHC class II MAb could act similarly. By interacting with activated ($CD4^+$) MHC class II expressing T cells, these cells may be prevented from further lesion formation, while they may also prevent a further activation of the T cells in the brain by blocking the antigen-presenting function of the MHC class II molecules. The fact that Genox353, specific for a polymorphic structure of a class II gene product in rhesus monkeys, indicates that haplotype specific suppression is feasible in primates. The results are in line with similar results obtained in mice where EAE can be modulated with class II-specific MAb (5) and in F1 animals with MAb specific for the susceptible haplotype (6). Although murine MAb specific for MHC class II products may not be the therapeutic agents for long-lasting therapy in autoimmune diseases (because they readily elicit antibodies to them [7]), these primate experiments clearly show the involvement of MHC class II gene products in the autoimmune process. Other ways should be investigated to inactivate the MHC class II antigens, for instance by using peptides that block MHC class II function in a specific manner.

Acknowledgment. This work was supported in part by Medigon-ZWO, the National Multiple Sclerosis Society, grant RG 1709-A-1, and NIH contract PR 4692.

References

1. van Lambalgen R, Jonker M. Experimental allergic encephalomyelitis in rhesus monkeys. I. Immunological parameters in EAE resistant and susceptible rhesus monkeys. Clin Exp Immunol 1987;68:100.
2. Nooij FJM, Borst JG, van Meurs GJE, Jonker M. Differentiation antigens of rhesus monkey lymphocytes. I. Identification of T cells bearing CD3 and CD8 and a subset of CD8 bearing cells. Eur J Immunol 1986;16:975.
3. Brodsky FM, Parham P, Barnstable CJ, Crumpton MJ, Bodmer WF. Monoclonal antibodies for analysis of the HLA system. Immunol Rev. 1979;47:3.
4. van Lambalgen R, Jonker M. Experimental allergic encephalomyelitis in rhesus monkeys. II. Treatment of EAE with anti-T lymphocyte subset monoclonal antibodies. Clin Exp Immunol 1987;68:305.
5. Sriram S, Steinman L. Anti-IA antibody suppresses in vivo encephalomyelitis; treatment model for disease linked to IR genes. J Exp Med 1983;158:1362.
6. Sriram S, Topham DJ, Carroll L. Haplotype-specific suppression of experimental allergic encephalomyelitis with anti-IA antibodies. J Immunol 1987;139:1485.
7. Jonker M, Nooij FJM, den Butter G, van Lambalgen R, Fuccello AJ. Side effects and immunogenicity of murine lymphocyte specific monoclonal antibodies in subhuman primates. Transplantation, in press.

Author Affiliations

M. Jonker, Primate Center TNO, 151 Lange Kleiweg, 2288, GJ Rijswijk, The Netherlands; *L. Steinman*, Stanford University Medical Center, Departments of Neurology and Pediatrics, Stanford, CA 94305, USA.

Section XX
T-Lymphocyte Differentiation Activation and Regulation

In Vitro Differentiation of a Human Thymic Clone Coexpressing CD4 and CD8 Molecules

Orly Azogui, Danièle Mathieu-Mahul, Martine Amiot, Hélène Dastot, Laurent Degos, and Laurence Boumsell

Abstract: In the thymus, T-cell differentiation results in T-cell receptor (TCR) gene rearrangement and the generation of two distinct subsets of T lymphocytes distinguishable by the expression of CD4 and CD8 glycoproteins (1). It is not obvious whether CD4+8+ thymocytes are or include the intermediates between CD4−8− thymocytes and mature T-cell subsets (2). some authors have suggested that double-positive CD4+8+ cells evolve into more mature single-positive cells which then leave the thymus (3), while others suggested that the double-positive cells die because of some selection pressure (4). The data presented here provide evidence that CD4+8+ cloned thymocytes are capable of giving rise to a CD4+8− subclone in vitro. The experiments performed on the TCR gene rearrangement of the two cell populations showed that the loss of CD8 molecules within the CD4+8+ cloned cell population is followed by a new TCR-α gene rearrangement, suggesting that CD4+8− cells differentiate from CD4+8+ thymic precursors.

To generate a continuous cell line, normal human thymocytes were cultured with phytohemagglutinin (PHA) and recombinant IL2. Amplification of the thymocytes resulted in a high proportion of CD8+ (80%) and CD4+ (40%) cells with a population of CD4+8+ cells (20%). After a short culture period, thymocytes were cloned by limiting dilution, using irradiated allogeneic peripheral blood lymphocytes (PBL) as a feeder layer. Several clones were obtained that retained the characteristics

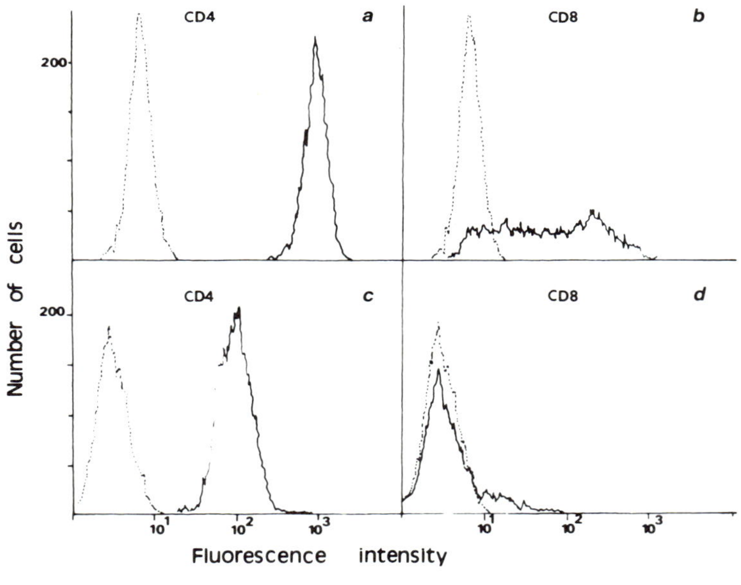

Figure 1. Expression of CD4 and CD8 molecules on the C44A (a and b) and its subclone C44B (c and d). The cells were stained with CD4 and CD8 MAbs, or a negative ascite control (dotted lines). Method: Thymus was obtained from a 2-year-old child at the time of cardiac surgery. Thymocytes were cultured at 1×10^6 cells ml^{-1} in the presence of 1/600 phytohemagglutinin (Gibco) and recombinant IL2 (30 U mo^{-1}; Biogen, Geneva) in a final volume of 1 ml in 24-well tissue culture plates (Falcon). The culture medium consisted of RPMI 1640 (Gibco) supplemented with 1% penicillin-streptomycin and 20% inactivated pooled human serum. The cells were grown in a humidified atmosphere with 5% CO^2 at 37°C. After 5 days of bulk culture, thymocytes were cloned by limiting dilution in 96-well V-bottom plates with fresh allogeneic irradiated PBL as feeder cells and expanded under conditions similar to those used in the initial bulk culture. For cytofluorographic analysis, 2×10^5 cloned cells were incubated at room temperature with 1/100 dilutionof the appropriate antibodies for 30 minutes. The cells were then washed twice and incubated for an additional 30 minutes at 4°C with fluorescein-conjugated goat anti-mouse immunoglobulin (Meloy) and washed twice. Cell analysis was performed on a model Facstar (Becton Dickinson, Mountain View, CA). The results shown represent analysis of 10^4 cells for each population.

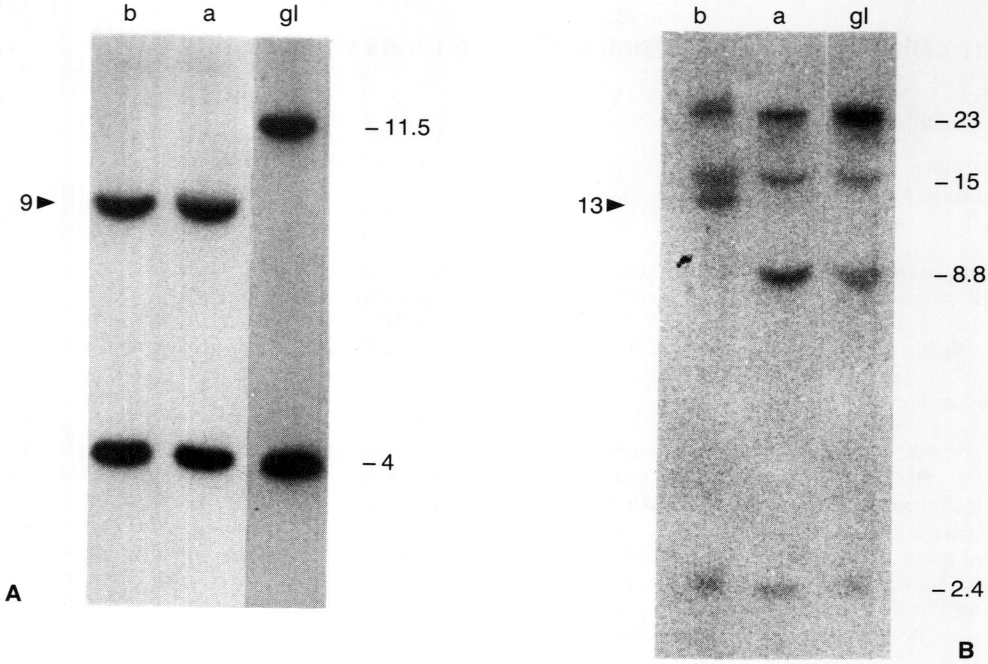

Figure 2. The rearrangement of TCR-β (A) and TCR-α (B) sequences in C44A and C44B thymic clones. Method: Aliquots of 10 μg of EcoR1-digested genomic DNAs from an EBV-transformed B-cell line as a germline control (gl), from C44A (a), and C44B (b), were electrophoresed on a 0.6% agarose gel and blotted onto nylon membranes (Compass, GENOFIT). The blot was hybridized under stringent conditions, with probes radiolabeled by hexamer priming to a specific activity of 10^9 cpm/μg. A, Hybridization with the 0.4 kb *Bgl*II fragment from pDo-β1 clone (kindly provided by Dr. Strominger's laboratory) containing all Cβ coding sequences; the 11.5-kb EcoRl Cβ1 germline band is deleted and replaced by a unique 9-kb band (indicated by a triangle) in C44A and C44B cells. B, Hybridization with the 1.2-kb insert from PY14 TCR-α cDNA clone[6]. C44A DNA contains all the germline EcoRl fragments. On the other hand, in C44B DNA, the 8.8-kb germline fragment has been deleted and replaced by a single 13-kb band (as indicated by a triangle).

of IL2 and feeder-dependent cell lines. Cytofluorographic analysis has shown that 12 of 14 clones expressed only CD4 molecules, while 2 clones coexpressed CD4 and CD8 molecules CD4+8+. One CD4+8+ clone (C44A) was maintained in long-term culture for further study (Fig 1, a and b). After 2 months of culture, the percentage of double-positive cells decreased dramatically, leading to a large majority of cells expressing only CD4 antigens (C44B) (Fig. 1, c and d).

Clonality of C44A and C44B was studied by Southern blot analysis of TCR-β gene configuration. EcoR1-digested DNAs from C44A and C44B cells hybridized with a Cβ TCR probe revealed the same and unique 9-kb rearranged band (Fig. 2A). This observation was corroborated by the use of other restriction enzymes. This confirmed the clonality of those cells and indicated that C44B cells were originating from C44A cells. Although C44A was of thymic origin, it was reactive with CD3 and WT31 monoclonal antibody (MAb) as found by flow-cytometry and biochemical analysis, thus indicating that CD3 molecules were associated with a TCR-αβ heterodimer at the cell membrane. In addition, sequential immunoprecipitations performed with CD3 and WT31 antibodies showed that all the CD3 associated molecules bore the WT31 determinant (data not shown). In addition, these results were in agreement with the detection of one TCR-α allele rearrangement, using pGA5 and some Jα sequence probes. This rearrangement indicated the deletion of Vα sequences (data not shown). Because C44A lost CD8 molecules after 2 months in culture, it was important to compare the TCR-α gene configuration of C44A and C44B cells. Using Southern blot analysis, EcoR1-digested DNAs from both clones were hybridized with another cDNA probe PY14 (5); while C44A cell DNA (Fig. 2B, lane a) was in germline configuration with this probe, in C44B cell DNA (Fig. 2B, lane b) the 8.8-kb germline band was deleted and replaced by a single 13-kb rearranged band. These data suggested that the rearrangement process was not complete in C44A cells but continued during the maturation of C44A into C44B cells, possibly (but not necessarily) representing the second allele rearrangement. Serial studies of surface phenotype and of TCR-α sequence configuration at different culture times of C44A and C44B have defined sequential events: the new TCR-α gene rearrangement occurred only after the loss of CD8 molecules. It has to be noted that this modality was specific to the C44B clone. Other thymic clones (CD4+) maintained under the same culture conditions kept their original TCR-α gene rearrangement.

Taken together, our data strongly suggest that a human thymic CD4+8+ clone differentiates in vitro into a CD4+8− T-cell clone. This is in agreement with recent evidence which reports that in vivo depletion of the murine CD8+ T cells prevents the appearance of CD4+ cell populations (6). The differentiative pathways of the CD4+8+ T-cell clone that gives rise to a CD4+8− T-cell subclone could be either a subclonal expansion of CD4+8− cells coming out from the CD4+8+ population, or the disappearance of CD8 antigens from host cells. An extrathymic environment, rather than the thymic one shown to be necessary for β-chain gene rearrangements (7,8) was sufficient to promote the differentiation process. Such factors as HLA class II antigens (9), IFN α (10), and IL1 (11) (alone or in combination), provided by feeder cells could be responsible for the maturation from C44A to C44B.

Considering the recent data about a physical link or interaction between TCR and CD4 or CD8 molecules (12) and the second Vα-TCR gene rearrangement observed after the loss of CD8 molecules in C44B, one is led to speculate that these two events are causally related. Two different reports that show that the cytotoxic function is linked to the antigen receptor-CD8 interactions (13,14) could sustain this hypothesis. As the association between α and β chain has been shown to determine both MHC restriction and antigenic specificity (15,16), the loss of CD8 molecules on C44A, followed by a second TCR-α gene rearrangement, could suggest strongly a further stage in the change of the specificity and/or a definition of the MHC restriction controlled by the TCR. Moreover, it has been recently reported that CD4+8+ thymic cells with a unique Vβ gene segment might be the target for clonal deletion of self-reactive T cells in certain mouse strains (17). The change of TCR-α gene rearrangement in C44B is consistent with the possibility that the development of a CD4+8− subclone from CD4+8− precursors, with a supposedly different specificity for antigen and/or MHC restriction reflects a mechanism for clonal selection.

References

1. Reinherz EL, Schlossman SF. The differentiation and function of human T lymphocytes. Cell, 1980;19:821.
2. Mathieson BJ, Fowlkes BJ. Cell surface antigen expression on thymocytes: Development and phenotypic differentiation of intrathymic subsets. Immunol Rev, 1984;82:141.
3. Ceredig R, Dialynas DP, Fitch FW, MacDonald HR. Factor producing cell in the thymus: Ontogeny, frequency and precursors of T cell growth. Quantitative recovery in a subpopulation of phenotypically mature thymocytes defined by monoclonal antibody G.K.1.5. J Exp Med 1983; 158:1654.
4. Scollay R, Bartlett P, Shortman K. T cell development in the adult murine thymus: Changes in the expression of the surface antigens LyT2, L3T4 and B2A2 during development from early precursor cells to emigrants. Immunol Rev, 1982;82:79.
5. Yanagi Y, Chan A, Chin B, Minden M, Mak TW. Analysis of cDNA clones specific for human T cells and the α and β chain of the T-cell-receptor genes. Proc Natl Acad Sci USA 1985;82:3430.
6. Smith L. CD4+ murine T cells develop from CD8+ precursors *in vivo*. Nature, 1987;326:798.
7. Williams GT, Kingston R, Owen MJ, Jenkinson EJ, Owen JJT. A single micromanipulated stem cell gives rise to multiple T-cell receptor gene rearrangements in the thymus in vitro. Nature, 1986;324:63.
8. Stutman O. Intrathymic and extrathymic T cell maturation Immunol Rev, 1978;42:138.
9. Blue ML, Daley JF, Levine H, Schlossman SF. Class II major histocompatibility complex molecules regulate the development of the T4+T8− inducer phenotype of cultured human thymocytes. Proc Natl Acad Sci USA 1985; 82:8178.
10. Chen LK, Mathieu-Mahul D, Back FH, Dausset J, Bensussan A, Sasportes M. Recombinant interferon α can induce rearrangement of T-cell antigen receptor α-chain genes and maturation to cytotoxicity in T-lymphocyte clones in vitro. Proc Natl Acad Sci USA 1986;83:4887.
11. Rock KL, Benacerraf B. Thymic T cells are driven to expand upon interaction with self-class II major histocompatibility complex gene products on accessory cells. Proc Natl Acad Sci USA 1984;81:1221.
12. Saizawa K, Rojo J, Janeway CA Jr. Evidence for a physical association of CD4 and CD3: α:β T-cell receptor. Nature 1987;328:260.
13. Havran WL, Fitch FW. Characterization of murine cytolytic-helper hybrid T cell clones. Nature 1987; 325:65.
14. Gabert J, Langlet C, Zamoyska R, Parnes JR, Schmitt-Verhulst AM, Malissen B. Reconstitution of MHC Class I specificity by transfer of the T cell receptor and LyT-2 genes. Cell 1987;50:545.
15. Dembic Z, Haas W, Weiss S, McCubrey J, Kiefer H, Von Boehmer H, et al. Transfer of specificity by murine α and β T-cell receptor genes. Nature 1986;320:232.
16. Blackman M, Yagüe J, Kubo R, Gay D, Coleclough C, Palmer E, et al. The T cell repertoire may be biased in favor of MHC recognition. Cell 1986;47:349.
17. Kappler JW, Roehm M, Marrack P. T cell tolerance by clonal elimination in the thymus. Cell, 1987;49:273.

Author Affilations

Orly Azogui, *Martine Amiot*, *Hélène Dastot*, *Laurent Degos*, and *Laurence Boumsell*, Unité de recherches sur l'Immunogénétique de la transplantation 93, Hôpital Saint-Louis, 1 rue Claude Vellefaux 75475, Paris, Cedex 10, France; *Danièle Mathieu-Mahul*, Unité 301, de l'Institut National de la santé et de la recherche médicale, Hôpital Saint-Louis, 1 rue Claude Vellefaux, 75475 Paris, Cedex 10, France

Functional Characterization of T-cell Clones Using Related T-Cell Receptor V Gene Segments

Edward J. Ball, Kay Black, Marie Hoover, J. Donald Capra, and Peter Stastny

T lymphocytes play a critical role in immune responses involved in defense against pathogens, immune surveillance, and allograft rejection. Their participation is dependent on specific recognition of foreign and self antigens through the T-cell receptor for antigen (TCR). It is clear that this receptor is able to recognize a vast array of antigens. The discovery of the genes encoding the TCR α and β chain genes has allowed the investigation of how the TCR repertoire is generated (reviewed in ref. 1). Receptor diversity is generated by several mechanisms: 1) germline diversity due to a modest number of variable (V), diversity (D), and joining (J) gene segments; 2) combinatorial joining of V D, and J segments; 3) junctional diversity; 4) N-region diversity; and 5) combinatorial association of TCR α and β chains. The role of somatic mutation in modification of the repertoire appears to be minimal for the TCR in contrast to the immunoglobulins.

We have begun to study the contribution of the V-β-8 gene segment family to the human T-cell repertoire. This family consists of five members (2), two of which are clearly pseudogenes. A third member of the family may lack a promotor region and thus also be a pseudogene. The two members of the family that are expressed (Vb 8.1 and Vb8.2) are 98% homologous. Acuto and colleagues (3,4) described a monoclonal antibody (Ti3A) that reacts with T cells expressing V-β-8 gene-encoded segments. Further, they have shown that the Vb8 family can be used to construct a number of specificities.

In our experiments, peripheral blood T cells positive for Vb8 TCR were selected using a panning technique with the Ti3A monoclonal antibody (gift of Dr. E. Reinherz, Boston, MA) and expanded with IL2. After establishment, lines were selected multiple times by stimulation with irradiated allogeneic mononuclear cells in the absence of IL2. In such a manner, bulk T-cell lines specific for DR1-DRw9 haplotypes were generated. These lines were > 95% Ti3A-positive in FACS analysis (compared with 4.2 $\pm$ 1.7% in peripheral blood). Cloned T-cell lines were generated from these bulk lines by limiting dilution. Analysis of the panel reactivity of these T-cell clones revealed that each DR-haplotype-specific bulk line in fact contained T-cell clones recognizing several different related specificities.

A summary of Vb8+ T-cell clones recognizing two different DR2-associated antigens is shown in Table 1. Each of the clones was >95% Ti3A+ in FACS analysis. Both of the DR2 cells missed by CO2 and CO3 (AKI and PBO) were serologically typed as DR2-short (missed by 10W1031 and 1032 antisera). CO1 and C12 T-cell clones also missed the DR2-short cells, but they also missed three other DR2 stimulators that were serologically "long" (OBA, MSM, VGU).

Southern blot analysis of DNA extracted from Ti3A-positive clones confirmed that the V-β-8 genes had been rearranged. Comparison of EcoRI Vb8 and C-β fragments to published genomic sequences (2) allows one to determine which V-β-8 and C-β segments were used to construct the receptor. From such an analysis CO1 was inferred to have used Vb8.2 and C-β-1, whereas CO2 and CO3 used Vb8.1 and C-β-2.

The data show that the repertoire constructed from the Vb8 gene family is quite large. Further analysis of Vb8+ clones will be useful in determining structure/specificity relationships of the TCR.

Table 1. Panel Testing of DR2-Specific VB-8 T-Cell Clones

Stimulator Cell			T1005.UBO.03 Clones (% RR)			
ID	B	DR	CO3	CO2	CO1	C12
LDO	7,44	2,11	68.8	54.6	41.7	69.4
DHU	7,44	2,7	70.4	77.1	53.9	48.2
DBU	7,44	2,9	68.2	nt	32.0	nt
MGA	7,44	2,–	78.4	63.9	nt	61.9
DJE	7,51	2,7	68.8	74.0	88.9	56.7
IDO	7,52	2,1	75.4	59.8	nt	50.2
GNU	18,44	2,7	71.2	nt	99.9	nt
LBU	18,44	2,4	100.0	nt	41.8	nt
TWI	18,38	2,14	91.6	84.1	nt	59.2
BKE	51,35	2,1	83.2	nt	95.2	100.0
RWI	44,–	2,4	55.3	nt	41.0	nt
JKR	27,60	2,4	65.8	nt	42.0	nt
SSI	8,53	2,9	82.7	100.0	nt	77.1
KBR	37,62	2,3	66.5	31.8	26.0	62.0
OBA	37,42	2,3	59.5	54.4	4.8	11.2
MSM	18,–	2,7	60.8	20.3	3.7	7.6
VGU	35,61	2,14	56.4	41.3	1.8	10.0
AKI	45,–	2,3	1.2	nt	2.6	2.0
PBO	35,38	2,4	3.1	5.0	nt	8.0
KZO	7,51	4,–	0.6	nt	1.4	2.1
JDE	7,44	4,4	0.0	nt	0.0	nt
LMY	7,60	4,6	2.9	nt	6.1	1.1
RDA	7,58	12,14	3.2	9.5	4.6	15.6
MMA	7,62	4,6	1.8	nt	2.3	1.3
SCL	14,62	4,6	3.6	4.7	0.6	5.2
MKR	27,44	5,6	1.2	nt	2.9	2.0
BBR	22,62	6,7	2.4	nt	1.6	2.4
DSP	44,57	7,7	1.1	nt	11.3	4.2

References

1. Kronenberg M, Siu G, Hood L, Shastri N. The molecular genetics of the T-cell antigen receptor and T-cell antigen recognition. Ann Rev Immunol 1986;4:529–591.
2. Siu G, Strauss E, Lai E, Hood L. Analysis of a human V beta gene subfamily. J Exp Med 1986;164:1600–1614.

Immunobiology of HLA
Volume II

3. Acuto O, Campen T, Royer H, Hussey R, Poole C, Reinherz E. Molecular analysis of T cell receptor (Ti) variable region (V) gene expression. Evidence that a single Ti beta V gene family can be used in formation of V domains on phenotypically and functionally diverse T cell populations. J Exp Med 1985;161:1326–1343.
4. Acuto O, Hussey R, Reinherz E. Multiple class I and class II major histocompatibility complex allospecificities are generated with T cell receptor variable (V) domains created by a single Ti beta V gene family. J Exp Med 1985;162:1387–1392.

Author Affiliations

University of Texas Health Science Center at Dallas, Southwestern Medical School, Dallas, TX 75235, USA

TCR γ Chain Expression on Human Peripheral Blood T Lymphocytes

Frits Koning, Rafick P. Sekaly, Erwin Tschachler, Roberto Biassoni, Marvin S. Reitz Jr., Eric O. Long, and John E. Coligan

Abstract: Expression of the human T-cell receptor (TCR) γ chain has been detected on subpopulations of peripheral T cells (1–5), a thymus-derived clone (6), and the T-cell line PEER (7). At least three forms of γ chains are known to exist in γδ TCRs (Table 1): [1] disulfide-linked Cγ1 chains have been detected on T-cell clones derived from the periphery (2-5,8); [2] a 55 kD Cγ2 chain, nondisulfide linked to a 40 kD δ chain is expressed on a cell line derived from an immunodeficiency patient (1) and on the PEER cell line (7); and [3] a second nondisulfide-linked ~41 kD Cγ2 chain has been found on a peripheral blood-derived cell line (8) and on a thymic clone (6). This latter receptor is poorly characterized and, furthermore, it is unknown if all three types of γ chains can be expressed by a single individual. Here we report that the 41 kD nondisulfide-linked γ chain can be separated into two components that are indistinguishable in charge, but differ slightly in size. This type of γ chain associates with a 36 kD putative δ chain. We show that all three forms of γδ TCRs can be expressed on $CD3^{+}WT31^{-}$ lymphocytes from one individual. However, the 55 kD γ chain was not expressed by all individuals, which most likely reflects the absence of one of the Cγ2 alleles (9,10).

The human peripheral blood-derived, HTLV-I-transformed cell line ET702 expresses CD3 but lacks expression of the epitope detected by WT31 (not shown), which indicated that this cell line might express a γδ TCR (1,2). Iodine-125 labeled cell lysates from this cell line were immunoprecipitated with anti-γ and anti-CD3δ reagents (11), followed by SDS-PAGE analysis (Fig. 1A). In addition to CD3 proteins, anti-CD3δ precipitated nondisulfide-linked (compare nonreduced and reduced gels) 41 kD and 36 kD TCR chains from digitonin solubilized cells (Fig. 1A). Using NP40 solubilized cells, the anti-γ serum precipitated only the 41 kD chain. Thus, the 41 kD protein is the γ chain, whereas the 36 kD protein is a putative δ chain. Removal of N-linked carbohydrate revealed that the γ chain had a protein backbone of ~34 kD, whereas the putative δ chain protein backbone was found to be 31 kD (not shown). In several experiments, it was observed that the ~41 kD γ-chain protein band appeared to consist of two components which were poorly separated by SDS-PAGE. Two-dimensional IEF/SDS-PAGE analysis revealed that these two γ-chain proteins are indistinguishable by charge (pI ~7), but differ slightly in molecular weight (not shown). These molecular weight differences most likely reflect differences in glycosylation. Similar molecular weight differences have previously been observed for the disulfide-linked γ chain (2) and may, therefore, reflect similar patterns of glycosylation of these two γ chains.

Nondisulfide-linked ~40k kD γ chains have also been detected on a thymus-derived clone (6) and on a polyclonal cell line derived from peripheral blood (8). On the

Table 1. Characteristics of Human γδ T-Cell Receptors

Type*	Molecular Weights γ/δ	Protein Backbones γ/δ	Disulfide Binding	Probable Cγ Locus	References
I	36-40/40	31/34	+	Cγ1	2,3,8, this study
I	36-40/43-45	31/?	+	Cγ1	2,4
II	55/40	40-43/31	−	Cγ2	1,7,8,11,14, this study
III	44/62	?/?	−	Cγ2	6
III	35-45/?	?/?	−	Cγ2	8
III	41/36	34/31	−	Cγ2	This study

*Based on approximate size and disulfide linkage of the γ chains.

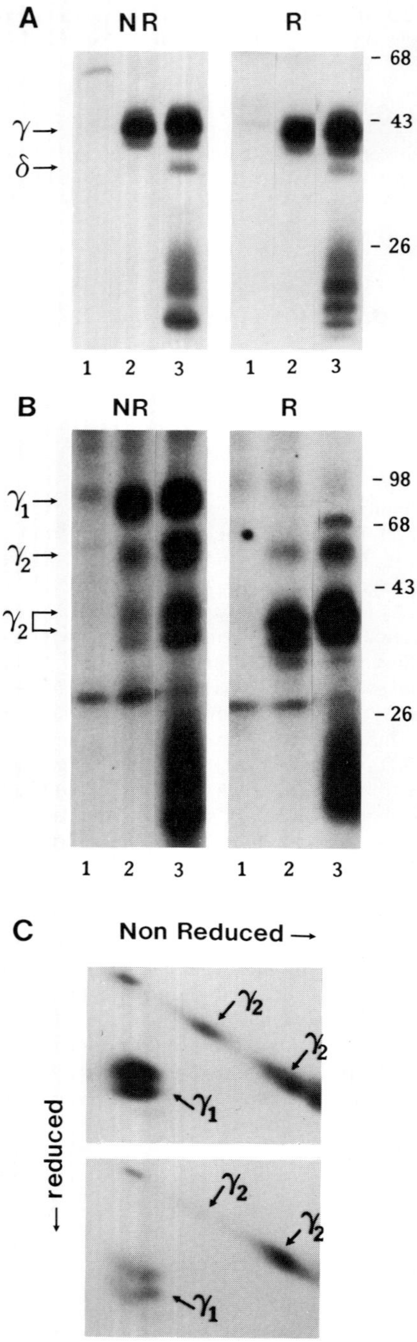

◄
Figure 1. *A* and *B*: analysis of TCR γ-chain expression on ^{125}I-labeled ET702 cells (**A**) and CD3$^+$WT31$^-$ cultured peripheral blood lymphocytes (**B**). Immunoprecipitations were carried out in detergent lysates using control serum (lanes 1), anti-γ serum (lanes 2), and anti-CE3δ serum (lanes 3) and analyzed under either nonreducing (NR) or reducing (R) conditions as indicated. *C*: Two-dimensional nonreduced/reduced SDS-PAGE analysis of TCR γ-chain expression on CD3$^+$WT31$^-$ lymphocytes from two unrelated individuals. Precipitations were carried out with anti-CD3δ and analyzed by nonreduced SDS-PAGE in the first dimension, followed by reduced SDS-PAGE in the second.

thymic clone, the γ chain was associated with a putative 62 kD δ chain (6). As shown in Figure 1, the 41kD nondisulfide-linked γ chain expressed on the ET702 cell line is expressed with a putative ~36 kD δ chain. A similar γδ complex was observed on CD3$^+$WT31$^-$ cells derived from an unrelated individual (not shown). It thus appears that this γ chain can be expressed with two different putative δ chains (6; this study). further studies will be needed to establish the molecular basis for these observations.

Next we investigated if all three γ chains found in γδ TCR complexes can be expressed on peripheral blood lymphocytes from a single donor. CD3$^+$WT31$^-$ T lymphocytes were isolated by FACS sorting. The resulting cell population was activated by PHA (1%) stimulation and grown in IL-2 (100 U/ml)-containing media for 2 weeks. Subsequently TCR expression was analyzed by immunoprecipitation followed by SDS-PAGE. Three CD3-associated TCR γ chains could be identified (Fig. 1B). All three γ chains were also precipitated by the anti-γ serum (Fig. 1B), but not by antisera specific for TCR α and β chains (not shown). Whereas two of the detected γ chains have a similar electrophoretic mobility under both nonreducing and reducing conditions, the third γ chain containing TCR complex (80 kD nonreduced) could be reduced to subunits of ~40 kD. Isolation of these three γ chains (by electroelution from gel slices) followed by removal of N-linked glycosylation and SDS-PAGE analysis confirmed that three distinct γ chains (protein backbones of 43, 34, and 31 kD, see also Table 1) were present in those complexes (not shown). Thus, both the Cγ1-derived, disulfide-linked γ chain and the two Cγ2-derived, non-disulfide-linked γ chains (9, 10, 12, 13) are expressed on these CD3$^+$WT31$^-$ lymphocytes. Subsequent analysis of TCR-γ-chain expression on CD3$^+$WT31$^-$ lymphocites of two additional individuals revealed that one individual lacked expression of the 55 kD Cγ2 chain (Fig. 1C, bottom), while the other expressed all three γ chains (Fig, 1C, top).

Although three forms of γ chins have been described on human T cells (1–7), T-cell lines derived from peripheral blood appeared to express almost exclusively the disulfide-linked form (2–4). This has led to speculations that expression of nondisulfide-linked γ chains might be restricted (4). The results in this study clearly demonstrate that on CD3$^+$WT31$^-$, peripheral T cells isolated from two of three individuals express both the disulfide-linked and the two nondisulfide-linked γ chains. However, prolonged culture time of these cells resulted in a predominant expression of the disulfide-linked γ chain on these cells (not shown). The culture conditions used, therefore may favor outgrowth of cells expressing the disulfide-linked γ chain, which in part may explain the predominant expression of this type of γ chain on cloned T cells. It may furthermore indicate that cells expressing nondisulfide-linked γ chains have different growth requirements compared with cells expressing disulfide-linked γ chains.

Analysis of TCR γ-chain expression on $CD3^+WT31^-$ peripheral T cells from a number of individuals indicated the apparent lack of expression of the 55 kD γ chain by some individuals (Fig. 1C and not shown). Recent analysis of the genes coding for the human TCR γ chains indicate that the Cγ1 locus codes for the disulfide-linked γ chain, whereas two allelic forms of the Cγ2 locus exist, one coding for the ~55 kD non-disulfide-linked γ chain and the other for the ~41 kD nondisulfide-linked γ chain (8,9,11,12). Lack of expression of the 55 kD γ chain, therefore, most likely reflects the absence of the allele coding for this γ chain indicating homozygosity for the γ loci, whereas individuals expressing all three γ chains are heterozygous.

In summary, our studies indicate that the 41 kD non-disulfide-linked γ chain can form a heterodimer with a putative 36 kd δ chain. This γδ TCR, as well as a non-disulfide-linked 55/40 γδ TCR and a disulfide-linked 40/40 kD γδ TCR can be detected on $CD3^+WT31^-$ peripheral lymphocytes. Lack of expression of the 55 kD γ chain by some individuals most likely reflects the absence of one of the Cγ2 alleles.

References

1. Brenner MB, McLean J, Dialynas DP, Strominger JL, Smith JA, Owen FL, et al. Identification of a putative second T-cell receptor. Nature 1986;322:145.
2. Borst J, van de Griend RJ, van Oostveen JW, Ang S-L, Melief CL, Seidman JG, et al. A T-cell receptor γ/CD3 complex found on cloned functional lymphocytes. Nature 1987;325:683.
3. Moingeon P, Ythier A, Goubin G, Faure F, Nowill A, Delmon L, et al. A unique T-cell receptor complex expressed on human fetal lymphocytes displaying natural-killer-like activity. Nature 1986;323:638.
4. Lanier LL, Federspiel NA, Ruitenberg JJ, Phillips JH, Allison JP, Littman D, et al. The T cell antigen receptor complex expressed on normal peripheral blood $CD4^-$, $CD8^-$ T lymphocytes. J Exp Med 1987;165:1076.
5. Ang SL, Seidman JG, Peterman GM, Duby AD, Benjamin D, Lee SJ, et al. Functional γ chain-associated T cell receptors on cerebrospinal fluid-derived natural killer-like T cell clones. J Exp Med 1987;165:1453.
6. Bank I, DePinho RA, Brenner MB, Cassimeris J, Alt FW, Chess L. A funtional T3 molecule associated with a novel heterodimer on the surface of immature human thymocytes. Nature 1986;322:179.
7. Weiss A, Newton M, Crommie D. Expression of T3 in association with a molecular distinct from the T-cell antigen receptor heterodimer. Proc Natl Acad Sci USA 1986; 83:6998.
8. Brenner MB, McLean J, Scheft H, Riberdy J, Ang S-L, Seidman JG, et al. Two forms of the T-cell receptor γ protein found on peripheral blood cytotoxic T lymphocytes. Nature 1987;325:689.
9. Lefranc M-P, Forster A, Rabbitts THE. Genetic polymorphism and exon changes of the constant regions of the human T-cell rearranging gene γ. Proc Natl Acad Sci USA 1986;83:9596.
10. Pelicci PG, Subar M, Weiss A, Dalla-Favera R, Littman DR. Molecular diversity of the human T-gamma constant region genes. Science 1987;237:1051.
11. Koning F, Maloy WL, Cohen D, Coligan JE. Independent association of T cell receptor β and γ chains with CD3 in the same cell. J Exp Med 1987;166:595.
12. Littman DR, Newton M, Crommie D, Ang S-L, Seidman JG, Gettner SN, et al. Characterization of an expressed CD3-associated Ti γ-chain reveals Cγ domain polymorphism. Nature 1987;326:85.
13. Krangel MS, Band H, Hata S, McLean J, Brenner MB. Structurally divergent human T cell receptor γ proteins encoded by distinct Cγ genes. Science 1987;237:64.
14. Hata S, Brenner MB, Krangel MS. Identification of putative human T cell receptor δ complementary DNA clones. Science 1987;238:678.

Author Affiliations

Frits Koning, John E. Coligan, Biological Resources Branch; *Rafick P. Sekaly, Roberto Biassoni, Eric O. Long*, Laboratory of Immunogenetics, National Institute of Allergy and Infectious Diseases; *Erwin Tschachler, Marvin S. Reitz Jr.*, Laboratory of Tumor Cell Biology, National Cancer Institute, National Institutes of Health, Bethesda, MD 20892, USA

Molecular and Functional Characterization of Human T-Cell Clones Expressing TCR-$\gamma\delta$ Antigen Receptor

Armand Bensussan, Catherine Grillot-Courvalin, and Etienne Vilmer

The T-cell antigen receptor (TCR) expressed on most T lymphocytes is a 90-kD disulfide-linked heterodimer composed of α and β glycoprotein subunits. The Tiα/β heterodimer is noncovalently associated with the CD3 glycoprotein complex; Ti$\alpha\beta$ molecule alone determines both antigen recognition and MHC restriction whereas the CD3 complex is implicated in signal transduction. It has been suggested that the presence of Ti$\alpha\beta$ was mandatory for CD3 expression at the cell surface. However, expression of a CD3-associated γ product has been described recently on CD3+, CD4-, CD8- T cells that failed to react with the monoclonal antibody WT31, which recognizes a framework determinant present on all $\alpha\beta$ heterodimers (1).

We have cloned CD2+, CD3+, WT 31− lymphocytes from peripheral blood mononuclear cells of a patient presenting with prolonged immunodeficiency following allogeneic bone marrow transplantation. The expression of T-cell γ genes in a cloned cell line CD4− and CD8− named DS6 was studied at the protein and the DNA level; we have demonstrated that DS6 cells express a CD3/Ti$\gamma\delta$ complex and we have shown that the γ protein results from a productive rearrangement involving the Vγ5 and Jγ1 segments (2).

The analysis of DS6 cytolytic activity have revealed that DS6 exhibited a non-MHC- restricted cytotoxic activity against a large panel of NK-sensitive target cell lines as well as against phytohemagglutinin (PHA) blast cells. While a mitogenic combination of anti-CD2 antibodies or sepharose coupled anti-CD3 antibodies resulted in an increased response to rIL2, no proliferation was observed in the absence of exogeneous IL2. These findings suggested that DS6 cells failed to produce IL2 upon activation. To determine whether the CD3+ Ti$\gamma\delta$ DS6 cloned cells were capable of suppressing B-lymphocyte differentiation, the T-cell clone was cultured with allogeneic splenic lymphocytes stimulated with PWM. The results indicated that addition of CD3 Ti$\gamma\delta$ cells resulted in a profound decrease in plasma cell generation as well as IgM and IgG secretion (Table 1), whereas the cell recovery was similar at the end of cultures. In parallel, the same CD3 Ti$\gamma\delta$ cells did not alter allogeneic proliferative response.

Taken together these functional results suggested that the suppression of B-cell differentiation induced by the DS6 clone might not be totally mediated through cell-mediated cytotoxicity.

Table 1. Clone DS6 Induces a Marked Suppression of PWM-Induced B-Cell Differentiation

Cells in Culture*	nb ($\times 10^6$)† (cells/ culture)	% cIg†	IgM† (ng/ml)	IgG† (ng/ml)
A + medium	0.5	1	–	–
A + 1.10⁵ DS6	0.5	1	–	–
A + PWM	1	24	650	1,000
A + 1.10⁵ DS6 + PWM	1	4	70	200

*1.10⁶ spleen lymphocytes were cultured in RPMI + 10% FCS.
†Measured on day 7 of culture. The generation of Ig-containing cells (cIg) was estimated by direct fluorescence on fixed cells using fluorescein-conjugated polyvalent goat IgG to human Ig. The amount of IgM or IgG secreted in cell-free culture supernatants was determined by an enzyme-linked immunosorbent assay.

References

1. Brenner M, McLean J, Scheft H, Riberdy J, Ang SL, Seidman JG, et al. Two forms of the T cell receptor γ protein found on peripheral blood cytotoxic T lymphocytes. Nature 1987;325:689.
2. Bories JC, Guglielmi P, Sigaux F, Bensussan A. Nucleotide sequence of a cDNA corresponding to a new human variable segment of functionally rearranged T cell receptor gamma chain gene. Nucleic Acid Res 1987;15:10059.

Author Affiliations

Inserm U 93 and U 108, Hôpital Saint-Louis, 1 Avenue Claude Vellefaux 75475, Paris Cedex 10, France

Immunobiology of HLA
Volume II

A Novel Polyclonal T-Cell Receptor Expression Defect Probably Not Linked to HLA

J.R. Regueiro, J.L. Vicario, M. López Botet, M.O. de Landázuri, J. Alcami, A. Corell, J.M. Martin-Villa, J. Martínez-Laso, P. Pérez-Aciego, E. Paz, and A. Arnaiz-Villena

Knowledge of the immune system has expanded rapidly as a result of experimental work in animals and the study of human patients with immunodeficiency (ID). We describe here a polyclonal T-cell receptor complex (TcR) expression defect occurring in two children (D and V) within an otherwise healthy Spanish family. Sib V survived vaccination with attenuated poliomyelitis virus, but was clinically an ID with autoimmune features (1). Sib D has been normally vaccinated and symptomless throughout life. Although both sibs had normal levels of CD1, CD2, CD4, CD8, CD11, CD14, CD16, CD18, CD20, CD25, and HLA-DR-positive peripheral blood mononuclear cells (PBMC), only about 6% of the CD2+ T cells expressed surface antigen-specific T-cell receptor (Ti was detected with WT31 and BMA031 monoclonals), and only about 23% weakly showed surface CD3 (2). This abnormal phenotype was stable *in vitro*. Because CD3 expression was always higher than Ti expression, the defect on CD3 expression is very likely secondary to the defect on Ti expression. The functional data confirmed previous phenotype findings: CD3-mediated T-cell activation and Ti-mediated responses to allogenic and tetanus toxoid antigens were severely depressed, whereas activation through CD2 was normal in both sibs (2).

Family clustering suggests a genetic pathogenesis, but linkage to HLA is unlikely (Fig. 1). However, although the defective phenotype has been stable since 1985, we do not know whether it is (or was) permanent. Indeed, the older sib S is 7 years old, symptomless to date, and unaffected but he is HLA-different from his other brothers and sister, whereas C is still 1 year old (the defect was detected in V when 2 years old and in D when 4). We are following the family to rule out that it is a late-onset HLA-linked defect, although no other living relative tested so far presents it (including both parents, four paternal aunts, two paternal uncles, one maternal aunt, and the maternal grandmother). Linkage to other blood group markers is also excluded (see Fig. 1, footnote).

At the DNA level, the genes encoding for Ti-α, -β and -γ, and CD3-δ chains were present and apparently normal by Southern analysis, and polyclonal Ti-β chain genes rearrangements were detected in both children's T-cell blasts (2). Specific RNA transcripts for α, β Ti and ε, δ CD3 were detected in both sibs' PBMC, but an abnormal α/β Ti dimer has been immunoprecipitated from their cytoplasms (3). It seems thus likely that a small natural mutation in Ti genes or regulatory elements is giving rise to an abnormal TcR which is unable to reach the T-cell surface in these individuals. Sib V's late autoimmune disease suggests that immune systems with an incomplete T-cell repertoire can cope better with foreign than with self antigens, stressing the role of the thymus in achieving self-tolerance rather than in generating a repertoire for foreign antigens.

Lastly, it has been shown that CD2-mediated T-cell activation is regulated by the modulation of TcR structures (4). In contrast, our polyclonal mutant CD2+Ti− T-cells are not activated *in vivo* (normal CD25 levels were recorded, see above), suggesting that regulation of the CD2 activation pathway does not require TcR surface expression.

Acknowledgments. Supported in part by FIS, NATO, and CAICYT.

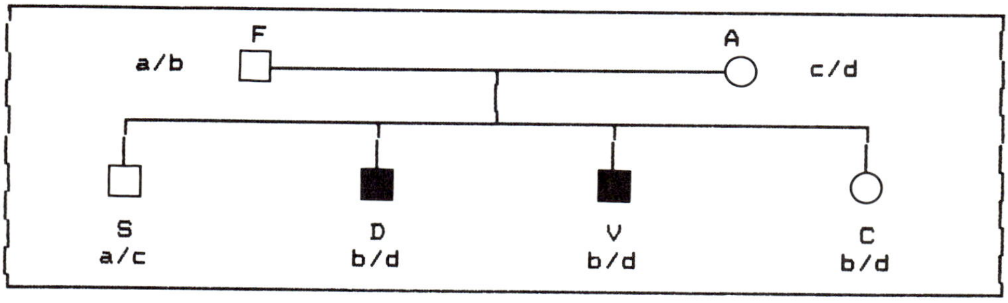

Figure 1. Family pedigree showing HLA haplotypes. Shaded symbols denote TcR-defective phenotype, square symbols a male subject, and circles a female subject. Capitals identify the different individuals. Other tested blood groups which did not segregate with the defect were the following: ABO, Rh, Fy2/Fyb, K/k, MNS, JKa/JKb 2nd C3.2 = HLA-A32 − Cwl − Bw55 − Bw6 C2C BfF C4A3B1 − DR3 − DRw52 − DQw2; b = HLA-A1 − Cw6 − B17 − Bw4 C2C BfS C4A6B1 − DR4 − DRw53 − DQw3; c = HLA-A26 − B18 − Bw6 C2C BfF1 C4A3BQO − DR3 − DRw52 − DQw2; d = HLA-A32 − B51 − Bw4 C2C BfS C4A3B1 − DR2 − DQw1.

References

1. Regueiro JR, Arnaiz-Villena A, de Landázuri MO, Martin-Villa JM, Vicario JL, Pascual-Ruiz V, et al. Familial defect of CD3 (T3) expression by T cells associated with rare gut epithelial cell autoantibodies. Lancet 1986;I:1274.
2. Regueiro JR, López-Botet M, de Landázuri MO, Alcami J, Corell A, Martin-Villa JM, et al. An in vivo functional immune system lacking polyclonal T-cell surface expression of the CD3/Ti(WT31) complex. Scand J Immunol 1987;26:699.
3. Alarcon B, Regueiro JR, Arnaiz-Villena A, Terhorst C. Familial defect in the surface expression of the T-cell receptor-CD3 complex. New Eng J Med 1988;319:1203.
4. Reinherz EL. A molecular basis for thymic selection: regulation of T11 induced thymocyte expansion by the T3-Ti antigen / MHC receptor pathway. Immunology Today 1985; 6:75.

Author Affiliations

J.R. Regueiro, J.L. Vicario, J. Alcami, A. Corell, J.M. Martin-Villa, J. Martinez-Laso, P. Pérez-Aciego, E. Paz, A. Arnaiz-Villena, Immunología, Hospital 12 de Octubre; *M. López Botet, M.O. de Landázuri*, Hospital de la Princesa, Madrid, Spain.

CDw29 Molecules from the Monocyte Surface Are Required for CD3-Induced T-Cell Activation

Martine Amiot, Stéphane Huet, Orly Azogui, Hélène Dastot, Alain Bernard, and Laurence Boumsell

Abstract: We previously described a MAb termed K20 recognizing in humans a glycoprotien complex not restricted to hematopoïetic lineages, but preferentially expressed on early hematopoïetic cells, cells of the granulocytic, monocytic, and T-cell lineages (1). Moreover K20 and 4B4 MAb showed very similar tissue distribution during the blind analysis performed during the course of the Third International Workshop on Human Leucocyte Differentiation Antigens (2), thus forming CDw29 (CD, cluster of differentiation). We show here that CDw29 recognize the same molecular complexes on human thymus cells that a MAb recognizes the β constant 130 kD subunit of the very late antigen (VLA) family (3). The VLA family is part of a super family, collectively referred to as the integrin receptors. They are involved in cell-matrix and cell-cell adhesion in many physiologically important processes (4–6). We show that CDw29, in conjunction with MAb UCHL1, defines three subsets of CD4 cells. We observed a complete inhibition by CDw29"K20" of the T-cell activation induced by soluble or fixed CD3 in co-cultures of purified T-cells and fixed monocytes. In contrast, under the same coculture conditions, K20 did not inhibit, phytohemagglutin (PHA) and partially inhibited CD2-induced mitosis. Moreover, we show that the molecular complexes recognized by K20 on the monocyte surface are necessary to trigger T-cell activation via CD3, but not via CD2 or PHA.

To investigate more precisely the protein constituting the family of molecular complexes immunoprecipitated from T-cells by K20, immunoprecipitations were carried out with ^{125}I-labeled lysates of thymocytes and analyzed on a sodium dodecyl sulfate (SDS) 7% polyacrylamide gel electrophoresis (PAGE). K20 immunoprecipitated four predominant protein bands of 160, 130, 85, and 70 kD and a weaker protein band of 200 kD (Fig. 1, lane C). A similar pattern was obtained under nonreducing and reducing conditions, indicating that these complexes are made of five noncovalently associated subunits (results not shown). In addition, it has been previously demonstrated that MAb A1A5 recognizes the 130 kD constant subunit of the VLA family formed by the molecular association of heterodimers, which are composed of a same β chain of 130 kD and different α chains on T-cells (210 kD in VLA1, 165 kD in VLA2, 150 kD in VLA4) (3,4,7,8,). To compare the molecular complexes defined by K20 or 4B4 MAb (9) with the VLA complexes, comparative and sequential immunoprecipitation experiments were performed on thymocytes (Fig. 1). The molecular complexes immunoprecipitated by K20, 4B4, and A1A5 showed identical protein composition on thymocytes. The results of sequential immunoprecipitation clearly show that K20, 4B4, and A1A5 recognized the same molecular complexes on thymocytes (Fig. 1, lanes D and E).

Next, to analyze the expression of these family of molecules during T-cell activation and differentiation, the molecular complexes detected by K20 are compared on T cells before and after activation by PHA. K20 immunoprecipitates from TPBL, thymocytes, and activated thymocytes (PHA or clone) the same five protein bands (results not shown). However, although the level of expression of each protein band is similar on thymocytes and TPBL, variations in the intensities of 70, 85, 160, and 200 kD protein bands are observed after the activation of thymocytes. On PHA-activated thymocytes, the relative expression of the 210-130kD heterodimer increases, whereas the relative intensity of 70, 85 and 160 kD protein bands decreases. These variations

Immunobiology of HLA
Volume II

Figure 1. Comparison of the protein complexes precipitated by 4B4, A1A5 or K20 and sequential immunoprecipitations on thymocyte lysates. Cells were surface labeled with ^{125}I then lysed in standard lysis buffer containing 1% nonidet P40 (1). (The MAb A1A5 and 4B4 were kindly provided by Dr. Hemler and Morimoto, respectively.) A thymocyte lysate was immunoprecipitated with MAb "4B4" (lane A), with MAb "A1A5," precleared by several cycles of immunoprecipitations of the molecules reacting with A1A5 (lane D) or K20 (lane E), then immunoprecipitated with K20 (lane D) or A1A5 (lane E). ►

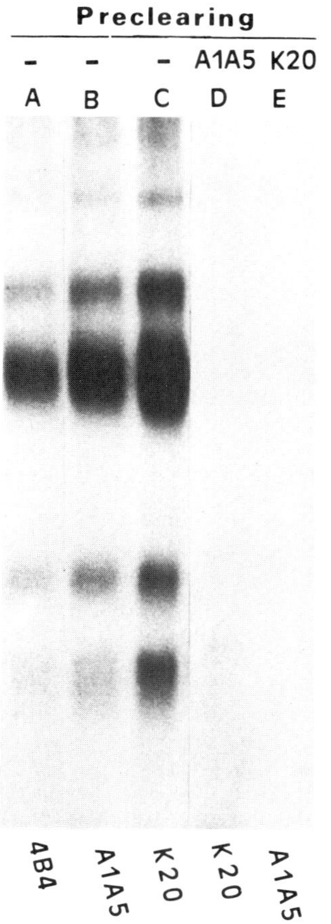

in expression, which are observed on a bulk culture of PHA-activated thymocytes, are still more obvious on two clones of thymic cells. The differences observed in the relative expression of each protein during T-cell activation and differentiation could be explained either by a modification of the expression of the different subunits, or by a modification of the association between the different glycoproteins themselves or finally by a selection of a subpopulation during T-cell activation.

The reactivity of various T-cell populations was next analyzed by immunofluorescence flow cytometry. All the thymocytes were stained by K20 with a low antigenic density, whereas only 60 to 70% of TPBL were stained by K20. Double-labeling experiments were carried out to determine the relationship of the $K20^{+}$ lymphocyte subset to CD4 and CD8 population. K20 similar to 4B4, recognizes a major population of CD8 T-lymphocytes and, furthermore, it appears to be important for the delineation of T-lymphocyte subpopulations by subdividing $CD4^{+}$ cells. In the $CD4^{+}$ subset, it has been shown that MAb UCHL1 (10), similar to MAb 4B4 (9), defines a subpopulation of T-cells that proliferates maximally to soluble antigen and provides help for PWM-

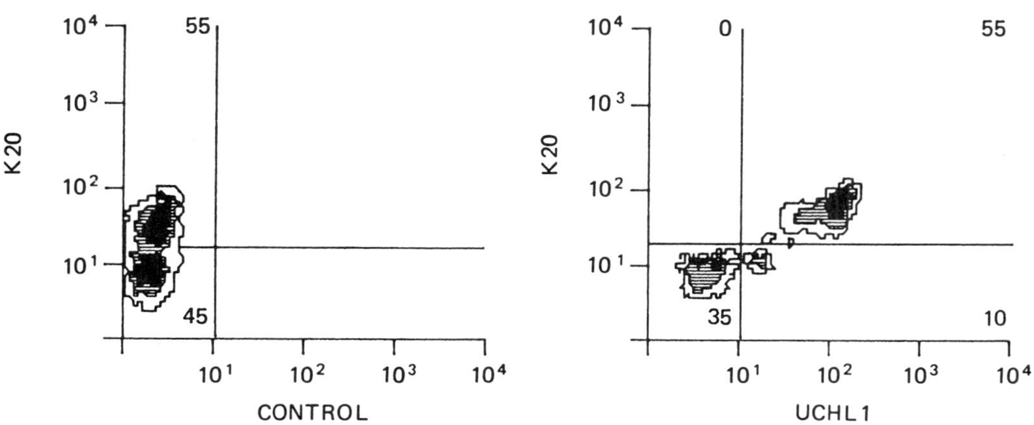

Figure 2. Two-color immunofluorescence reveals the existence of three subsets of CD4 cells in peripheral blood. CD4 cells were obtained by negative selection using a cytotoxic CD8 MAb and complement-dependant lysis. Next, the cells were labeled using MAb UCHL1 or an isotype-matched control ascitis and a fluorescein isothiocyanate conjugated-Fab fragment of a goat anti-mouse Ig. Finally, biotinylated K20 and PE-conjugated steptavidin were added. Ten thousand cells were analyzed per sample, using a Facstar microfluorimeter. The numbers represent the percentages of cells in the respective quadrants. (UCHL1 was provided by Dr. Smith.)

Table 1. The Molecular Complexes Recognized by CDw29"K20" MAb Are Involved in T-Cell Activation via CD3 or via CD2

Responding Cell Populations	PHA	Anti-CD3 Soluble	Anti-CD3 Sepharose	Anti-CD2 Pairs MAb D66 + T11.1	Anti-CD2 Pairs MAb GT2 + T11.1
Experiment #1					
T + fixed monocytes	66*	56	143	183	147
T + fixed monocytes + K20	70	0.5	0.3	81	98
T + fixed monocytes + anti-DR	66	60	125	NT	132
Experiment #2					
T	NT†	0.2	0.2	16.1	0.3
T + fixed monocytes	NT	15	122	150	183
T + fixed (monocytes + K20)	NT	1.2	17	140	160

*cpm × 10^{-3}; †NT = not tested.

Effect of K20 MAb on proliferative responses of T cells induced by PHA, CD3, or CD2 molecules. Cells cultures were performed in 96 well culture plates in a final volume of 200 μl 5 × 10^4 cells in RPMI 1640. Ten percent FCS were seeded per well. After a 3 day (activation by PHA or via CD3) or a 4 day (via CD2) incubation, cultures were pulsed with 1 μCi of (^{3}H) thymidine during the final 15 hours of incubation. T cells and monocytes used in proliferative responses were obtained using several steps of purification as described (12). Monocytes were fixed by treatment with 1% paraformaldehyde before or after treatment with MAb (12).

stimulated immunoglobulin synthesis. The cellular distribution of UCHL1 antigen on leukocytes is very similar to that of CDw29, and in the CD4$^+$ subset, the comparison of the subset defined by K20 and UCHL1 indicated that the K20+CD4+ subset is included in the subset UCHL1+CD4+ and that only 5 to 10% of cells are UCHL1+K20− (Fig. 2). Thus it will be interesting to determine whether the CD4+ UCHL1+K20− subset can also proliferate maximally to soluble antigens and provide help for PWM stimulated immunoglobulin synthesis.

Recently Huet (submitted) and others (11) have shown that among CD4+ subset, only the CD4+ CDw29+ cells can proliferate after stimulation through the CD3 molecule in the presence of accessory cells. Therefore, we investigated whether the CDw29 molecule was directly involved in this activation process. We compared the effect of K20 MAb on the activation of T cells induced either with PHA or via CD3 or CD2 molecules. The results, confirming our pervious observations (1), show that K20 does not inhibit the proliferative response of purified T cells induced by PHA in the presence of fixed monocytes (Table 1). In contrast, K20 totally inhibits the proliferative response of purified T-cells induced via CD3 molecules in the presence of fixed monocytes. To show that this inhibition is not due to Fc-R saturation, the inhibition of the proliferation induced by soluble CD3 or CD3 MAb coated on sepharose beads was compared, and the results were similar, as shown in Table 1. To test the effect of K20 on the activation of T-cells induced via CD2 molecules, two different pairs of mitogenic CD2 MAb "GT2 + T11" and "D66 + T 11" were used, as we have previously shown that these two combination of CD2 MAb deliver distinct signals to T-cells (12). K20 MAb partially inhibits the proliferative response of T-cells to these two CD2 pairs of MAb (Table 1).

In our previous work (13,14), we had shown the plurality of the monocyte surface molecules necessary to complement the primary signals directly delivered to T-cells by CD3 or CD2 MAb, which involves Fc receptors, CD11b, and HLA class I molecules. Therefore, we investigated whether the inhibitory effect of K20 on CD3 or CD2 induced T-cell activation was at least partly due to the monocyte surface molecule recognized by K20. Table 1 shows that under conditions where K20 is only fixed to the monocyte culture, a complete inhibition of T-cell proliferation induced by CD3 sepharose beads was still observed. In contrast, under the same conditions, K20 did not inhibit CD2-induced T-cell proliferation.

These results show that the molecular complexes recognized by K20 on the monocyte surface are necessary to trigger the activation of resting T cells via CD3, by not via CD2 or PHA. We do not know whether the molecular complexes recognized by K20 on the surface of resting T cells are also necessary to trigger the activation of these cells via CD3, while we show that they are involved in the activation of resting T cells via CD2. However, we know that the activation of clonal population of activated thymocytes via CD2 or CD3 is not inhibited by K20 (results not shown). Thus, this indicates that the molecular complexes recognized by K20 would be critical in mediating cell interaction, as exemplified by the ability of CDw29-bearing cells to proliferate maximally to soluble antigen. In contrast, the molecular complexes recognized by K20 would not be involved at later stages of T-cell functions, when cell interactions are no longer required for autocrine proliferation.

Acknowledgments. This work was supported by INSERM, ARC, and Fondation pour la rechérche sur les leucémies.

References

1. Amiot M, Bernard A, Tran HC, Leca G, Kanellopoulos J, Boumsell L. The human cell surface glycoprotein complex (gp 120,200) recognized by monoclonal antibody K20 is a

component binding to phytohemagglutinin (PHA) on T-cell. Scand J Immunol 1986;23:109.

2. McMichael AJ, Gotch F. T-cell antigens: New and previously defined clusters. In: McMichael AJ (ed): Leucocyte typing III. New York, Oxford University Press, 1987, p 31.
3. Hemler ME, Ware CF, Strominger JL. Characterization of a novel differentiation antigen complex recognized by a monoclonal antibody (A-1A5): Unique activation-specific molecular forms on stimulated T cells. J Immunol 1983; 131:334.
4. Hemler ME, Huang C, Schwarz L. The VLA protein family. J Biol Chem 1987;262:3300.
5. Takada Y, Huang C, Hemler ME. Fibronectin receptor structures in the VLA family of heterodimers. Nature 1987;326:607.
6. Hynes RO. Integrins: A family of cell surface receptors. Cell 1987;48:549.
7. Hemler ME, Jacobson JG, Strominger JL. Biochemical characterization of VLA-1 and VLA-2. J Biol Chem 1985;260:15246.
8. Hemler ME, Huang C, Takada Y, Schwarz L, Strominger JL, Clabby ML. Characterization of the cell surface heterodimer VLA-4 and related peptides. J Biol Chem 1987; 262:11478.
9. Morimoto C, Letvin NL, Boyd AW, Hagan M, Brown HM, Kornacki MM, et al. The isolation and characterization of the human helper inducer T cell subset. J Immunol 1985; 134:3762.
10. Smith SH, Brown MH, Rowe D, Callards RE, Beverley PCL. Functional subsets of human helper-inducer cells defined by a new monoclonal antibody, UCHL1. Immunology 1986;58:63.
11. Fox EJ, Lewis DE, Deemer KP, Masry MNE, Rich RR. T suppressor cell growth factor and anti-CD3 antibodies stimulate reciprocal subsets of T lymphocytes. J Exp Med 1987;166:404.
12. Brottier P, Boumsell L, Gelin C, Bernard A. T-cell activation via CD2T, gp50 molecules. Accessory cells are required to trigger T-cell activation via CD2-D66 plus CD2-9.6/T11$_1$ epitopes. J Immunol 1985;135:1624.
13. Huet S, Boumsell L, Raynal B, Degos L, Dausset J, Bernard A. Role in T-cell activation for HLA-class I molecules from accessory cells. Further distinction between activation signals delivered to T-cells via CD2 and CD3 molecules. Proc Natl Acad Sci USA 1987;84:7222.
14. Huet S, Wakasugi H, Sterkers G, Gilmour J, Tursz T, Boumsell L, et al. T-cell activation via CD2 T, gp50: The role of accessory cells in activating resting T-cells via CD2. J Immunol 1986;137:1420.

Author Affiliations

Martine Amiot, Orly Azogui, Hélène Dastot, Laurence Boumsell INSERM U93, Institut de recherche sur les maladies du sang, Hôpital Saint-Louis, 75475 Paris; *Stéphane Huet, Alain Bernard*, Laboratoire d'Immunologie des tumeurs de l'enfant, Institut Gustave Roussy, 94800 Villejuif, France

Differential Effects of an Anti-IL-2 Receptor Antibody on Cytotoxic and Suppressor Cells

Laurence A. Turka, Charles B. Carpenter, Edmond J. Yunis, and Edgar L. Milford

Mitogen or antigen activation of resting CD4+ T cells provides a signal for the secretion of interleukin 2 (IL-2) and the development of IL-2 receptors (IL-2Rs) on their surface (1). The released IL-2 stimulates activated CD4+ helper cells in an autocrine fashion and, also exerts a paracrine effect on non-IL-2 secreting cells such as cytotoxic T lymphocytes. The specific requirement of CD8+ suppressor cells for IL-2 is less clear. We therefore studied the in vitro effects of SA36.6G, an anti-IL-2R monoclonal antibody directed to an epitope on the 55 kD chain of the IL-2R, on allostimulated proliferation, cytotoxic and suppressor cell generation, and the phenotype of activated cells.

When present, SA36.6G was added at culture initiation. Proliferation (^{3}HTdR uptake) was measured on day 6, and cytotoxicity (lysis of ^{51}Cr-labeled stimulator cells-CTL) on day 6 or day 11. Generation of T suppressors was measured by the ability of lightly irradiated day 10 responders (modulators) to suppress proliferation when added to a fresh test mixed lymphocyte response (MLR).

SA36.6G inhibited thymidine incorporation in the MLR in a dose-dependent fashion, but even at maximal concentrations (1:10 dilution of ascites) did not fully suppress proliferation. In several different MLR combinations, SA36.6G (1:1000, a saturation dilution by flow cytometry) inhibited proliferation by an average of 44% (101,297 cpm vs. 57,888 cpm; $p < 0.002$).

The generation of cytotoxic T cells on culture day 6 was markedly inhibited by SA36.6G(39.4% lysis vs. 14.6% lysis; $p < 0.01$). In contrast to its effect at day 6, SA36.6G did not reduce CTL when the assay was performed on day 11 (19.2% lysis vs. 22.8% lysis; p = NS). This occurred despite the continued presence of SA36.6G in the culture media, as measured by binding to PHA activated lymphoblasts.

In eight separate experiments, SA36.6G did not block the generation of suppressor cells in culture (66.3% suppression vs. 85.0% suppression; p = NS). When modulators from control and SA36.6G blocked cultures were tested for specificity against a panel of 12 third-party stimulators, the patterns of suppression were identical,

Table 1. Effect of SA36.6G on Blast Generation and Phenotype

	% Blasts	% Blasts Positive for			
		IL-2R	2H4	CD4;2H4	CD8;2H4
Medium control	49 ± 15	78 ± 15	27 ± 9	14 ± 6	16 ± 9
+ SA36.6G	22 ± 18	29 ± 25	53 ± 13	28 ± 9	34 ± 13

All p values < 0.02 by Wilcoxon signed rank test.

and the overall mean percent suppressions for the two modulators were not significantly different.

As shown in Table 1, SA36.6G reduced the percentage of cultured cells that were activated blasts, but within the blast population, doubled the number of cells that were 2H4+. The 2H4 surface antigen is part of the T200 family of molecules and identifies suppressor cells in the control of B-cell activation. Early work suggests a similar role in the response to soluble and alloantigens. Increases in 2H4+ cells were seen in both the CD4+ and CD8+ subsets.

Staining of blasts for IL-2R revealed an over 60% decline in IL-2R+ cells in SA36.6G containing cultures, and those cells that were IL-2R+ bore a smaller number of receptors as indicated by staining intensity. When cells from day 6 control cultures were sorted into IL-2R+ and IL-2R− fractions, virtually all of the cytotoxicity and suppressor activity found in the unsorted population was accounted for solely by the IL-2R+ fraction. In contrast, when cells from cultures blocked with SA36.6G were sorted on day 11 (since this represented peak cytotoxicity), the bulk of the cytotoxicity and suppression were seen in the IL-2R negative fraction (data not shown).

These results demonstrate that an anti-IL-2R monoclonal antibody can partially inhibit allo-MLR proliferation and early cytotoxic cell generation while sparing specific suppressor cells. The subsequent escape of cytotoxic cells may be mediated via IL-2 independent cell growth, or utilization of an IL-2R not recognized by SA36.6G, such as the 75 kD chain of the IL-2R (2).

References

1. Gullberg M, Smith KA. Regulation of T cell autocrine growth. J Exp Med 1986;163:270.
2. Bich-Thuy LT, Dukovich M, Peffer NJ, Fauci AS, Kehrl JH, Greene WC. Direct activation of human resting T cells by IL 2: The role of an IL 2 receptor distinct from the Tac protein. J Immunol 1987;139:1550.

Author Affiliations

Laurence A. Turka, Department of Nephrology, University of Michigan Medical Center, Ann Arbor, MI 48109, USA; *Charles B. Carpenter, Edgar L. Milford*, Laboratory of Immunogenetics and Transplantation, Brigham and Women's Hospital, Boston, MA 02115; *Edmond J. Yunis*, Immunogenetics Laboratory, Dana-Farber Cancer Institute, Boston, MA 02115, USA

Coordinate-Enhanced Expression of Three Adhesion Molecules (LFA-3, CD2, and LFA-1) and Three Other Molecules (4B4, UCHL1, and Pgp-1) Defines a Human T-Cell Subset Containing Memory Cells and Characterized by Enhanced γ Interferon Production

Martin E. Sanders, Malegapuru W. Makgoba, Susan O. Sharrow, David Stephany, Timothy A. Springer, Howard A. Young, and Stephen Shaw

Abstract: The present report summarizes our experimental findings regarding phenotypic and functional differences between two reciprocal subsets of human peripheral blood T cells: memory cells versus naive cells. Human memory T cells are phenotypically distinguishable from naive T cells by increased expression of six cell-surface molecules (LFA-3, CD2, LFA-1, 4B4, UCHL1, and Pgp-1). Neonatal cord blood consists almost exclusively of naive cells expressing lower levels of these markers, which increase in expression following phytohemagglutin (PHA) activation. As expected, only the high-expressing T cells in adults proliferate in response to recall antigens. Memory cells produce greater than five times more γ interferon with PHA activation than naive cells, although they produce equivalent amount of IL2. Our results integrate findings reported from multiple laboratories regarding different markers (e.g., 4B4, 2H4, HB-11, UCHL1, and Pgp-1),

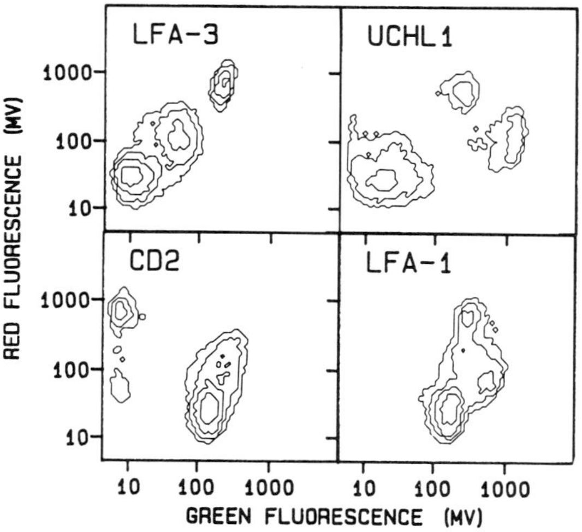

Figure 1. Two-color flow cytometric analysis of adult peripheral blood mononuclear cells for expression of various markers (x axis) vs. 4B4 (axis). Cells were stained with the x-axis MAb, followed by FITC-conjugated goat anti-mouse IgG, followed by blocking with mouse IgG, and subsequent staining with phycoerythrin-(PE) conjugated 4B4. In each panel, the cell population with the highest level of 4B4 represents monocytes.

Table 1. Proliferative Response and Lymphokine Production of LFA-3 Subsets of T Cells

	Proliferation (CPM)		Lymphokine Production with PHA (Units)	
	PHA	Tetanus Toxoid	IL2	IFN-γ
Donor 1				
LFA-3+	15483	2398	67	>50
LFA-3−	39500	0	119	8
Donor 2				
LFA-3+	7977	3401	9	>50
LFA-3−	59616	163	7	8
Donor 3				
LFA-3+	22351	5370	ND	ND
LFA-3−	66124	280	ND	ND

Data previously published in part in reference 10. Cells were electronically sorted after staining with the LFA-3 MAb TS2/9 followed by fluorescein isothiocynate- (FITC) conjugated goat anti-mouse IgG, with >95% purity for each population. Proliferation data are from 3-day cultures for PHA and 6-day cultures for tetanus toxoid. Irradiated autologous monocytes not treated with LFA-3 MAb were used as antigen-presenting cells. Cultures were pulsed with tritiated thymidine overnight prior to harvest. Data represent geometric means of counts per minute minus media control. Lymphokine data are from supernatants collected at 48 hours from PHA-activated cultures of T cells electronically sorted for LFA-3 expression. IL2 was determined using the CTLL-2 asay and IFNγ was measured using a radioimmunoassay (Centocor, Malvern, PA). ND indicates not done.

and demonstrate that these subsets represent different maturational stages of T cells rather than different T-cell lineages.

Antigen-independent adhesion is an important event in T-cell recognition (1,2). We have shown that such adhesion is mediated by at least two independent pathways; one pathway involving T-cell CD2 interaction with target cell LFA-3, and a second pathway involving T-cell LFA-1 interaction with target cell ICAM-1 and other as yet unidentified ligands (2–4). In addition, three of the molecules involved in antigen-independent adhesion, LFA-3, CD2, and LFA-1, each have been demonstrated to have signaling functions potentially of relevance to T-cell activation. LFA-3 interaction with CD2 augments PHA activation of T cells, and MAb to LFA-3 can induce IL1 secretion by monocytes and thymic epithelium (5–7). Certain pairs of MAb to CD2 activate T cells (8), and at least one MAb to LFA-1 has been reported to activate murine T cells (9). Thus, increased expression of these molecules on a subset of T cells could contribute to functional differences of that subset through a variety of mechanisms including enhanced adhesion properties and enhanced transmembrane signaling functions.

We have found enhanced expression of LFA-3, CD2, and LFA-1 on a single subset of T lymphocytes (10,11) and showed that this same subset of T lymphocytes also expresses enhanced levels of three molecules (4B4 or CDw29, UCHL1, and Pgp-1) previously shown by others to identify T-cell subsets that have memory function (12–15). Results of two-color flow cytometric analysis for five of these markers are shown in Figure 1, in which peripheral blood mononuclear cells from a normal donor are stained in each panel with 4B4 on the red axis and one of the other coordinately expressed markers on the green axis. In each panel, the population with highest expression of 4B4 is made up of monocytes, as this population stains with the Leu M3 MAb (10). However, each panel clearly demonstrates two major subsets of lymphoid cells differing in quantity of expression of 4B4. The lymphoid subset with higher expression of 4B4 also shows an enhanced mean expression of LFA-3 (approximately 8-fold), UCHL1 (approximately 30fold), CD2 (approximately 2.8-fold), and LFA-1 (approximately 2.4-fold). Other experiments have shown this same subset has enhanced expression of the human analogue of Pgp-1 and low expression of another molecule, 2H4 or CD45R (10).

If the subset with enhanced expression of LFA-3 and the other coordinately expressed molecules represents memory cells, there should be very few cells of this phenotype in neonatal blood because the fetus has limited antigenic exposure. Flow cytometric analysis of human cord blood showed that essentially all of the T cells were of the low-expression phenotype (10). Furthermore, if the low-expression subset represents naive cells, activation of that subset should lead to phenotypic conversion into the subset with high expression. Activation of cord blood T cells with PHA led to nearly uniform enhancement of expression of LFA-3, CD2, LFA-1, 4B4, and UCHL1, and a slight decrease in CD45R expression in 3 days (10). Likewise, activation of adult T cells with PHA led to uniform expression of LFA-3 and enhanced expression of CD2 on T cells by 3 days. This expression persisted at levels equivalent to that seen on the high expression subset in peripheral blood for at least 12 days, after the T cells had lost >85% of the peak expression of IL2 receptor (data not shown).

Proliferation studies performed with adult T cells electronically sorted for LFA-3 expression showed that proliferative response to the memory antigen tetanus toxoid was restricted to the high-expression subset (Table 1). Proliferative response to PHA was consistently better in the low-expression subset. However, despite more vigorous proliferation to PHA, the low-expression subset produced five-fold less γ interferon than the high-expression subset, whereas IL2 production was nearly equivalent or slightly greater in the low-expressing subset (Table 1). These proliferation data are consistent with those reported for cells sorted for expression of 4B4 (12), or UCHL1 (13), or reciprocally sorted for CD45R (16,17). The lymphokine data extend to the human species the finding of enhanced γ interferon production by memory cells, as previously reported for murine T cells sorted for Pgp-1 expression (18).

Acknowledgments. The authors thank Dr. Yoji Shimizu and Dr. Gene Shearer for helpful discussion of the manuscript, and the clinical testing laboratory of Program Resources Inc., NCI/FCRF, Frederick, MD, for assistance with the lymphokine assays.

References

1. Spits H, Van Schooten W, Keizer H, Van Seventer G, Van de Rijn M, Terhorst C, De Vries JE. Alloantigen recognition is preceded by nonspecific adhesion of cytotoxic T cells and target cells. Science 1986;232:403.
2. Shaw S, Luce GEG, Quinones R, Gress RE, Springer TA, Sanders ME. Two antigen-independent adhesion pathways used by human cytotoxic T cell clones. Nature 1986; 323:262.
3. Selvaraj P, Plunkett ML, Dustin M, Sanders ME, Shaw S, Springer TA. The T lymphocyte glycoprotein CD2 (LFA-2/T11/E-rosette receptor) binds the cell surface ligand LFA-3. Nature 1987;326:400.
4. Dustin ML, Sanders ME, Shaw S, Springer TA. Purified lymphocyte function associated antigen-3 (LFA-3) binds to CD2 and mediates T lymphocyte adhesion. J Exp Med 1987;165:677 .
5. Hunig T, Tiefenthaler G, Meyer zum Buschenfelde KH, Meuer SC. Alternative pathway activation of T cells by binding of CD2 to its cell-surface ligand. Nature 1987; 326:298.
6. Bierer BE, Peterson A, Takai Y, Greenstein J, Herrmann S, Seed B, et al. Evidence that LFA-3 activates T cells via the CD2 receptor. Fed Proc 1987;46:1498.
7. Le P, Denning S, Springer T, Haynes B, Singer K. Anti-LFA-3 monoclonal antibody induces interleukin 1 (IL1) release by thymic epithelial (TE) cells and monocytes. Fed Proc 1987;46:447A.
8. Meuer SC, Hussey RE, Fabbi M, Fox D, Acuto O, Fitzgerald KA, et al. An alternative pathway of T-cell activation: A functional role for the 50 kd T11 sheep erythrocyte receptor protein. Cell 1984;36:897.
9. Pircher H, Groscurth P, Baumhutter S, Aguet M, Zinkernagel RM, Hengartner H. A monoclonal antibody against altered LFA-1 induces proliferation and lymphokine release of cloned T cells. Eur J Immunol 1986;16:172.
10. Sanders ME, Makgoba MW, Sharrow SO, Stephany D, Springer TA, Young HA, et al. Three adhesion molecules (LFA-3, CD2, LFA-1) and three other molecules (UCHL1, 4B4, Pgp-1) show increased expression on a human T cell subset containing memory cells and characterized by enhanced gamma interferon production. J Immunol 1988; 140:1401.
11. Sanders ME, Makgoba MW, Luce GE, Shaw S. T cell adhesion receptors LFA-1 and CD2 and their ligands ICAM-1 and LFA-3: Analysis in adhesion, cell mediated lysis and as markers of T cell subsets. In: Kappler J, Davis M (eds): UCLA Symposia on Molecular and Cellular Biology. New Series. vol. 73 New York, Alan R. Liss, Inc., 1987.
12. Morimoto C, Letvin NL, Boyd AW, Hagan M, Brown HM, Kornacki MM, et al. The isolation and characterization of the human helper inducer T cell subset. J Immunol 1985; 134:3762.
13. Smith SH, Brown MH, Rowe D, Callard RE, Beverley PC. Functional subsets of human helper-inducer cells defined by a new monoclonal antibody, UCHL1. Immunology 1986; 58:63.
14. Budd RC, Cerottini JC, MacDonald HR. Phenotypic identification of memory cytolytic T lymphocytes in a subset of Lyt-2+ cells. J Immunol 1987;138:1009.
15. Budd RC, Cerottini JC, Horvath C, Bron C, Pedrazzini T, Howe RC, et al. Distinction of virgin and memory T lymphocytes. Stable acquisition of the Pgp-1 glycoprotein concomitant with antigenic stimulation. J Immunol 1987; 138:3120.
16. Morimoto C, Letvin NL, Distaso JA, Aldrich WR, Schlossman SF. The isolation and characterization of the human suppressor inducer T cell subset. J Immunol 1985; 134:1508.
17. Tedder TF, Cooper MD, Clement LT. Human lymphocyte differentiation antigens HB-10 and HB-11. II. Differential production of B cell growth and differentiation factors by distinct helper T cell subpopulations. J Immunol 1985; 134:2989.
18. Budd RC, Cerottini JC, MacDonald HR. Selectively increased production of interferon-gamma by subsets of Lyt-2+ and L3T4+ T cells identified by expression of Pgp-1. J Immunol 1987;138:3583.

Author Affiliations

Martin E. Sanders, Malegapuru W. Makgoba, Susan O. Sharrow, David Stephany, Stephen Shaw, the Immunology Branch, National Cancer Institute, Bethesda, MD 20892; *Timothy A. Springer,* the Dana Farber Cancer Institute, Boston, MA 02115; *Howard A. Young,* the Biological Response Modifiers Program, NCI/Frederick Cancer Research Facility, Frederick, MD 21701,USA

VG01 T-Cell Activation Antigen Selects for Antigen Reactive Cells

Jorge F. Denegri and Jeanne Peterson

Upon immune stimulation, mature T lymphocytes express new membrane proteins known as activation antigens. These antigens, their function mostly unknown, could be either transiently expressed as with IL-2, 4F2, TLiSA, A-1A5, or could be permanently acquired representing a further differentiation step as with Pgp-1 and possibly Ta_1 (1,2). We recently described VG01, a novel lymphocyte activation antigen (3), expressed by the majority of CD16 lymphocytes, a small proportion of circulating T lymphocytes, and monocytes. Upon mitogen and allostimulation in vitro, VG01 is newly expressed by the activated T cells, and its expression persists for a prolonged period of time. Alloactivated T cells generated in vitro and selected by their expression of VG01 are responsible for the specific proliferative response observed in secondary MLC(3).

In this study, we show that the small fraction of circulating VG01$^+$ T cells are responsible for the in vitro lymphocyte proliferative response to PPD and Candida in known sensitized subjects. Thus, VG01 appears to be also expressed upon in vivo activation, and the VG01 T lymphocytes may include memory T cells with a rather stable expression of this marker.

Peripheral blood lymphocytes were obtained from normal donors and separated by a Ficoll-Hypaque density gradient. Proliferative responses to soluble antigens, PHA, and primed lymphocyte testing (PLT) was performed with VG01-enriched and -depleted lymphocytes (>95% pure) obtained by cell sorting in a FACS 420 (Becton & Dickinson), and compared with the proliferative response of control unseparated T cells. a constant number of irradiated antigen presenting cells (M1 +) was added to each well in the microtiter plates.

The VG01$^+$ mononuclear cells sorted after in vitro priming were responsible for the accelerated proliferative response to alloantigens seen in PLT. The proliferative response of the VG01$^+$ lymphocytes (secondary memory) was allospecific, since they proliferated to the original stimulators and not to third party-unrelated lymphocytes (Table1).

To assess whether VG01 is a marker of activation and memory, we investigated the lymphocyte in vitro proliferative response to PPD and Candida in known reactive normal donors (Table 1). The proliferative response to PPD and Candida was almost entirely found in the VG01$^+$ lymphocytes, while the VG01$^-$ cell fraction was almost devoid of responding cells. There were no significant differences when the response of the VG01$^+$ cells was compared with the PPD and Candida response in control unseparated mononuclear cells. The average proliferative response to these two antigens in four independent experiments (Table 1) was similar for the unseparated and the VG01$^+$ mononuclear cells, and was several times greater than the proliferative response of the VG01$^-$ subpopulation. In contrast to the second results, PHA stimulation performed simultaneously demonstrated that the average proliferative response of VG01$^-$ lymphocytes to this mitogen was significantly higher (paired t test, $p > 0.05$) than the response of unseparated and VG01$^+$ mononuclear cells. In subsequent experiments, we proved that the difference in the proliferative response between the VG01-enriched and -depleted cell fractions was neither due to their kinetics nor to the effect of anti-VG01.

The proliferative response to PPD and Candida, as described in this report, suggests that VG01 is acquired in vitro and in vivo upon antigen stimulation. The in vitro proliferation of circulating VG01$^+$ cells to the soluble antigens was not due to the effect of the antibody on this membrane protein. VG01$^-$ lymphocytes have a greater response to PHA (Table 1) and alloantigens in a primary MLC (3). Thus, the lack of proliferative response to PPD and Candida by the VG01-depleted subpopulation is unlikely due to suppressor activity. The lack of apparent enrichment for antigen reactive cells may be due to the simultaneous enrichment of NK cells in the VG01$^+$ cells and complex cell to cell interactions. To assess this possibility, it would require to deplete the VG01$^+$ subpopulation of CD16 cells or to measure the enrichment of antigen reactive cells by limiting dilution.

Recently, Pgp-1 has been suggested as a useful marker to differentiate virgin from memory lymphocytes (1). The 85 kD protein is expressed by memory T cells in a stable fashion and, similar to VG01, it is not expressed on thymocytes. Our data suggest that VG01 is expressed in the differentiation process following immune activation and that VG01$^+$ cells include memory T lymphocytes. VG01 could be useful as a marker to separate virgin from memory T cells in the peripheral blood of

Table 1. Proliferative Response of VGO1-Enriched and -Depleted T Cells

	T Lymphocytes		
	Unseparated	VGO1 +	VGO1 −
PPD*	167 ± 53	106 ± 25	9 ± 5
Candida*	128 ± 73	148 ± 52	6 ± 2
PHA*	200 ± 64	155 ± 55	417 ± 180
PLT†	54 ± 28%	100%	8.6 ± 7.5

*Results expressed as cpm × 10^3, $\bar{x}$, ±, SD.
†Results normalized within each experiment to percent maximum response, always found in the VGO1+ cells, and expressed as $\bar{x}$ ± SD.
For all results, average of four independent experiments.

Immunobiology of HLA
Volume II

normal individuals and has the potential of selecting for specific alloreactive cells in sensitized subjects.

References

1. Budd RC, Cerottini JC, Horvath C, Bron C, Pedrazzini T, Howe RC, et al. Distinction of virgin and memory T lymphocytes. J Immunol 1987;138:3120–3129.
2. Hafler DA, Fox DA, Benjamin D, Wiener HL. Antigen reactive memory T cells are defined by Ta1. J Immunol 1986; 137:414–418.
3. Denegri JF, Gilks CB, Altin M, Peterson J, Lafleur L. A novel lymphocyte activation antigen present on allospecific primed lymphocytes and defined by the monoclonal antibody VG01[1]. Transplantation (in press).

Author Affiliations

Jeanne Peterson, Immunology Laboratory, Division of Hematopathology, Department of Pathology, Vancouver General Hospital, Vancouver; *Jorge F. Denegri*, Immunology Laboratory, Division of Hemapathology, Department of Pathology, Vancouver General Hospital, and University of British Columbia, Vancouver V52 1M9, British Columbia, Canada

Identification of Autoreactive Human Bone Marrow and Peripheral Blood CD3–, CD4–, CD8–Prothymocytes

Brigitte Autran, Guy Gorochov, Djavad Mossalayi, and Patrice Debre

Abstract: T-lymphocyte colonies (TLC) can be generated from CD3-/bone marrow cells, but the developmental stage of their precursors (CFU-TL), considered as prothymocytes, is not yet elucidated (1–3, D. Mossalayi et al*). We present a study of the cell surface phenotype and of the function of TLC cells.

Colonies derived from human peripheral blood and bone marrow CFU-TL were mostly composed of CD3+ CD4+ cells and of rare CD8+ cells. About 5% of cells, though they were CD2+ and CD7+, displayed no CD3, CD4, or CD8 antigens. When plated in agar cultures, those cells gave rise to mature TLC.

The CD3−, CD4−, CD8− cells were auto- and alloreactive. This self-reactivity, actively suppressed by mature T cells, could depend on recognition of the DR molecule. These data suggest that the self-reactive CD2+ CD7+ CD3− CD4− CD8− cells could represent an early stage in the T-cell lineage.

T-cell colonies were driven from normal human peripheral blood (PB) and bone marrow (BM) mononuclear cells in semi-solid agar cultures containing PHA and lymphocyte-conditioned medium (1–3). On day 8 of in vitro cultures, PB- or BM-derived TLC were harvested, pooled, and tested for their cell surface markers in an immunofluorescence analysis. The same results were obtained for TLC of both PB- and BM-origins. As shown in Table 1, most of the colony cells had the phenotype of mature T cells composed of CD2+ and CD7+ cells (98 to 99%) and CD3+ cells (95%). Most of these cells were CD4+ (72 to 78%) and a few were CD8+ (16 to 22%). They displayed the α/B T-cell receptor (TCR) by their large positivity for the WT31 monoclonal antibody (moAb) and a lower positivity with a panel of other anti-TCR moAbs: 2D4, 30/3D6, 31/162, 42/161, and 65 which detect subsets of normal, α/B TCR-bearing PBL (4). About 35 to 50% of these TLC displayed the class II MHC molecules (DR, DP, DQ). None of them were double-positive CD4+ CD8+ cells, nor did they display any cell-surface antigen of the killer or natural killer cells, B cells, or of the myelo-monocytic lineage.

A small contingent (about 5%) of cells was CD3−. We, therefore, tried to enrich in CD3− cells the TLC cell suspensions after a complement-mediated cytolysis using three anti-CD3, CD4, CD5 moAbs.

The residual viable cells from both PB- and BM-derived TLC were then analyzed for their T-cell markers and were shown to be negative for the following antigens: CD1, CD3, CD4, CD5, CD6, and CD8 and for the TCR. They were then referred to as phenotypically immature T lymphocytes (ITL). Half of them displayed the MHC class II molecules (DR, DP, DQ) (Table 1).

Since phenotypically immature T lymphocytes could be individualized among mature TLC cells, we then tried to characterize the proliferative responses of those cells in mixed lymphocyte reactions (MLR). The target cells used in those MLR assays were 25 grays irradiated, peripheral blood lymphocytes from autologous or allogeneic donors. Table 2 shows the results of the MLR assays. Both PB- and BM-derived TLC mature T cells were able to proliferate against allogeneic target cells; however, they were unresponsive to the autologous stimulating PBL. In contrast, the ITL obtained after depletion of the mature T-cell colonies of several donor origin could proliferate against autologous resting PBL. Treating TLC cells with rabbit complement alone could not induce an autoreactivity of the nondepleted cells (data

*D. Mossalayi, J.C. Lecron, P. Goube De Laforest et al. Characterization of prothymocytes with cloning (CFU-TL) in human bone marrow.

Table 1. Immunofluorescence Analysis of Peripheral Blood Derived TLC : Existence of a CD3-, CD4-, CD8-Subset

Monoclonal Antibodies: Clusters of Differentiation	Antibody	Untreated TLC Cells†	T-Depleted TLC Cells‡
CD1	OKT6	–§	–
CD2	OKT11	98	95
CD3	OKT3	94	–
CD4	IOT4	78	–
CD5	T101	90	–
CD6	MBG6	89	–
CD7	I21	99	96
CD8	IOT8	16	–
TCR	WT31	90	–
	2D4	1	–
	30/3D6	1	–
	31/161	10	–
	42/161	4	–
	65	5	–
Class II MHC			
DR	D1-12	49	48
DP	B7-21	47	52
DQ	Leu 10	47	47
CD11	OKM1	–	–
CD16	Leu 11b	–	–
CD19	B4	–	–

*As proposed by the three first Workshops on Human Leucocyte Differentiation Antigens (1982, 1984, 1986).
†T-lymphocyte colonies were harvested after culture in semi-solid agar medium in Petri dishes. The lower layer was composed of RPMI 1640 medium supplemented with 10% human serum, 1% glutamine, 1% PHA-M, 2% LCM, and 5% agar. Cells were plated in the upper layer, composed of supplemented RPMI containing no PHA or LCM and 3% agar. Cells were harvested at day 8, pooled, desagregated, and extensively washed before being tested in an indirect immunofluorescence (IF) assay using the defined moAbs as first-step reagents and an FITC anti-F(ab)'2 antiserum. Results of IF were read under a microscope and verified by cytofluorometric analysis.
‡TLC cells were, after harvest, submitted to a complement-dependent cytolysis using three moAbs: IOT3 (CD3), IOT4 (CD4), and T101 (CD5) and a pretested baby rabbit complement. Cells were incubated 30 minutes at 4°C with the moAbs, then were twice incubated 40 minutes at 37°C with the complement. After washings, cells were tested in the same indirect IF analysis as above. Ethidium bromide was added before reading to exclude the residual dead cells.
%Percentages of positive cells.

Table 2. Autologous and Allogeneic MLR of the Peripheral Blood- and Bone Marrow-Derived TLC

Responding Cells*	Stimulating Cells†: None	Autologous PBL	Allogeneic PBL
PB-TLC			
Total TLC cells	958‡	1511	29237
	(137)	(189)	(3409)
T-depleted TLC cells	1319	15746	23068
	(221)	(2224)	(3271)
BM-TLC			
Total TLC cells	2140	3145	14878
	(134)	(581)	(1570)
T-depleted TLC cells	2190	14879	6944
	(287)	(1570)	(1246)

*Responding cells from the PB- and BM-derived TLC were prepared as described in Table 1. Briefly, after harvest and washings, TLC were either directly used as effectors in MLR assays, either submitted to the predefined complement-dependent cytolysis and used as effectors after extensive washings at a 1 × 10^6 cells/ml concentration in RPMI 1640 medium supplemented with 1% L-glutamine, 1% sodium pyruvate, and 10% human serum.
†Autologous or allogeneic PBL were irradiated (25 grays) and mixed with the responding cells at a 1 × 10^6 cells/ml concentration in 96-well culture plates, and incubated for 5 days. Tritiated thymidine was incorporated during the last 18 hours of culture.
‡Results of proliferative responses are given as mean cpm ± SD of triplicate cultures.

not shown). An alloreactivity of the CD3– CD4– CD8– cells was regularly detected in association with the autoreactivity of these phenotypically immature T cells (Table 2).

To examine which antigenic structures were involved during the self-recognition process of these CD_3– CD4– CD8– cells, we investigated the inhibiting effects of an anti-DR monoclonal antibody, D1-12. A dose-response curve was obtained, and an 80% inhibition of the autologous MLR (A-MLR) could be regularly obtained (Table 3) to the same extent as for the allogeneic MLR.

Since no self-reactivity could be detected among the mature nondepleted TLC, though they were composed of up to 5% of ITL, we then looked for the suppressive effect of the mature T cells on the self-reactivity of ITL. The A-MLR of the ITL could also be extincted in a dose-dependant fashion by adding increasing amounts of autologous nonirradiated PBL. A 90% inhibition of the self-reactivity of the TLC cells depleted in CD3– CD4– CD8– cells could be obtained by adding 50 × 10^3 autologous PBL to the mixed lymphocyte culture.

The data presented demonstrated that a weak subpopulation of phenotypically immature T lymphocytes present in the TLC is able, when isolated, to proliferate against autologous PBL. The phenotype of these allo- and auto-reactive T cells is in accordance with that of prethymic cells as demonstrated in leukemic T cells (5–12) since they are both CD2 and CD7 positive, but do not display the TCR nor the CD1, CD3, CD4, CD5, CD6, CD8 antigens. Half of these ITL expressed the MHC class II molecules.

Mature CD3+ CD4+ autoreactive cells have been shown, in both human and murine species, to recognize self antigens through a CD3-linked TCR (13–15). However, since the first report on the self-reactivity of neonatal murine thymocytes (16), dull Thy1+, Lyt1–, CD4–, CD8– precursors of autoreactive T cells have been isolated in the murine spleen (17). The autoreactivity of these phenotypically immature T cells coexists

Table 3. Inhibition Patterns of the Autologous MLR Reactivities of the T-Depleted PB-Derived T-Lymphocyte Colonies

Stimulating Cells*	D1.12 Ascitic Dilution†			Inhibiting Factors		PBL Added × 10³/Well‡		
	None	0.6 × 10-3	2.5 × 10-2	1 × 10-2	None	3	12	50
Allogeneic PBL	0%§	1.5%	60%	95%	NT			
Autologous PBL	0%	2%	50%	85%	0%	35%	80%	92%

*Twenty-five grays irradiated stimulating PBL were mixed above with the T-depleted TLC cells derived from PBL prepared as described above.
†An anti-DR molecule monoclonal antibody (D1-12) in ascitic fluid was added to the MLR at the beginning of the cultures in serial dilutions.
‡Nonirradiated autologous PBL was added, at the beginning of the cultures, in serial dilutions to the MLR.
§Percentages of the inhibition were calculated as: $1 - \frac{\text{inhibited cpm} - \text{control cpm}}{\text{noninhibited cpm} - \text{control cpm}}$.

with an alloreactivity, even at the clonal level. Anti-self-reacting cytotoxic T cells have also been shown to differentiate from Thy1- normal murine BM precursors (18) and could be enriched, as prethymic cells, after a density gradient centrifugation (19). In all these models, the self-reactivity of the precursor cells was MHC-restricted.

In our study, the T-depletion of TLC resulted in an enrichment of allo- and self-reactive CD_3-, CD_4-, CD_8- T cells. Both the allo- and auto-reactivity seemed to depend on recognition of HLA-DR molecules. Altogether, these data suggest either the presence of two distinct subsets of allo- and auto-reactive cells among the ITL or that the ITL react against monomorphic determinants displayed on both autologous and allogeneic class II molecules.

References

1. Triebel F, Robinson WA, Hayward AR, Goube de la Forest P. Blood 1981;58:911.
2. Mossalayi MD, Goube de la Forest P, Guilhot F, Lemaire M, Larroque V, Tanzer J. Int J Cell Clon 1984;2:34.
3. Mossalayi MD, Ahmad M, Tanzer J, Janossy G, Goube de la Forest P. Exp Hematol 1986.
4. Boylston AW, Borst J, Yssel H, Blanchard D, Spits H, de Vries JE. J Immunol 1986;137:741.
5. Reinherz EL, Kung PC, Goldstein G, Levey RH, Schlossman SF. Proc Natl Acad Sci USA 1980;77:1588.
6. Janossy G, Tidman N, Papageorigiou ES, Kung PC, Goldstein GJ. Immunol 1981;126:1608.
7. Snodgrass HR, Dembic Z, Steinmetz M, Von Boehmer H, Nature 1985;315:232.
8. Collins MKL, Ranigawa G, Kissonerghis AM, Ritter M, Price KM, Tonegawa S, Owen MJ. Proc Natl Acad Sci USA 1985;82:4503.
9. Sangster RN, Minowada J, Suciu-Foca N, Minden M, Tak W Mak. J Exp Med 1986;163:1491.
10. Furley AJ, Mizutani S, Weilbaecher K, Dhaliwal HS, Ford AM, Chan LC, et al. Cell 1986;46:75.
11. Vodinelich L, Tax W, Bai Y, Pegram S, Caper P, Greaves MF. Blood 1983;62:1108.
12. Sutherland DR, Rudd CE, Greaves MF. J Immunol 1984;133:327.
13. Altman A, Katz DH. J Immunol 1980;125:1536.
14. Bensussan A, Meuer SC, Schlossmann SF, Reinherz EI. J Exp med 184;159:559.
15. Tilkin AF, Michon J, Kayibanda M, Henin Y, Sterkers G, Betuel H, Levy JP. J Immunol 1987;137:674.
16. Howe ML, Golstein AL, Battisto JR. Proc Natl Acad Sci USA 1970;67:613.
17. de Talance A, Regnier D, Spinella S, Morisset J, Seman M. J Immunol 1986;67:613.
18. Benveniste P, Miller RG. J Immunol 1986;136:4399.
19. Hurme M, Shivola M. J Immunol 1985;135:1108.

Author Affiliation

Laboratoire d'Immunologie Cellulaire et Tissulaire, UA CNRS 625, CHU Pitié-Salpétrière, Paris, France

A Novel Lymphokine Is Required for the Regulation of Autoaggressive Responses

Karen Rosenkrantz, Karim Bhimani, Karl Welte, Jochen Buch, James DiSanto, Esther Levi, Ulrich Hammerling, Bo Dupont, and Neal Flomenberg

Abstract: Limiting dilution studies indicate that cells with the potential to lyse autologous target cells exist in the peripheral blood of all normal individuals. These cells are down regulated by a second less frequent cell population. When recombinant IL-2 is substituted for crude lymphocyte-conditioned medium in these experiments, autocytotoxicity develops normally. Under these conditions, however, the autocytotoxic response is not down regulated. Mixing crude lymphocyte-conditioned medium with recombinant IL-2 restores the regulation

LIMITING DILUTION ANALYSES OF CELL MEDIATED CYTOTOXIC RESPONSES

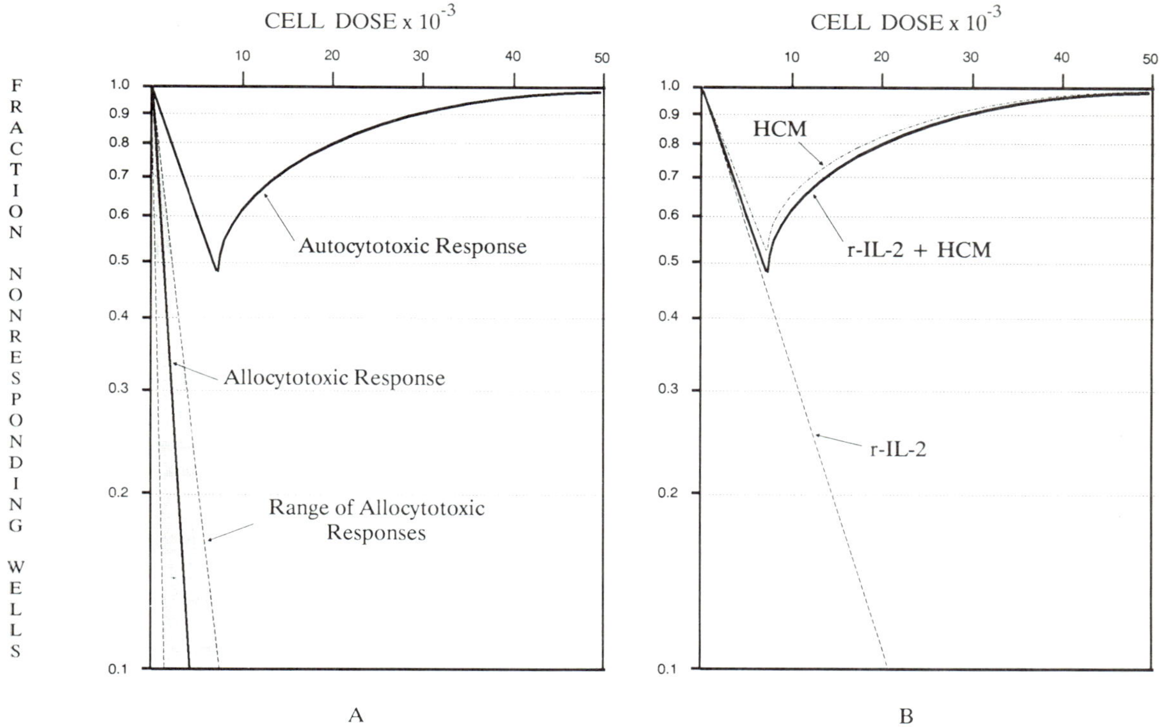

Figure 1. Limiting dilution assays were performed in V-bottom microtiter plates (Flow Laboratories) in 0.2 ml of RPMI-1640 medium (GIBCO) with 10% autologous serum and an IL-2 source. Wells contained graded concentrations of responder peripheral blood mononuclear cells (PBMC) and 5×10^4 irradiated (4000 rad) stimulator (autologous or HLA-mismatched) PBMC. On day 7, 100 μl of culture medium was removed and replaced with fresh medium. On day 10, individual wells were assayed for cytotoxicity in a chromium release assay by adding 3×10^3 ^{51}Cr-labeled PHA-activated lymphoblasts to each well. The supernatants were harvested after 4 hours. The cytolytic activity of an individual well was scored as positive when the release of ^{51}Cr exceeded the mean of the control plate containing stimulator cells alone by 3 standard deviations (1). Human lymphocyte-conditioned medium (HCM) was produced from PBMC that had been pulsed for 2 hours with purified phytohemagglutinin (PHA-HA-16, Burroughs Wellcome) and phorbol myristic acetate (Sigma). Recombinant IL-2 (Amgen) was utilized at a concentration of 50 U/ml.

of autocytotoxicity normally seen at high-responder cell dose. These findings indicate that a second soluble factor present in the conditioned medium is necessary for either the activation, growth, or differentiation of the regulatory cell population. Gel filtration studies indicate that the molecular weight of this factor is between 60 and 80 kD. This factor appears to be distinct from known immunologically active cytokines. It is conceivable that deficiencies of this factor may be relevant to the pathogenesis of autoimmune diseases or GVH responses.

Limiting dilution studies of allocytotoxic responses toward major histocompatibility complex gene products in mouse and man have demonstrated that these responses conform to a traditional single hit kinetic model. In unsensitized humans, the frequency of T cells capable of lysing cells bearing allogeneic MHC products is such that if 5000 responder cells are seeded per microculture, all microcultures will exhibit detectable allocytotoxic responses. If the cell dose is further increased to 50,000 cells per microculture, strong cytotoxicity persists in all microcultures. The range of responses typically observed is indicated by the shaded area in Figure 1A.

Table 1. Cytokines Tested for Autoregulatory Activity

IL-1 α	TNF α
IL-1 β	TNF β
IL-2	α-Interferon
IL-3	γ-Interferon
IL-4	G-CSF
IL-5	GM-CSF

Although cells with autocytolytic potential are not readily demonstrated in bulk culture studies of human peripheral blood, such cells are readily demonstrated in all normal adults under limiting dilution conditions (1). In contrast to allocytotoxic responses, the autocytotoxic response is byphasic. Strong autocytotoxic responses are observed at responder cell doses of 2500 to 10,000 cells per microculture. However, as responder cell dose is further increased, the autocytotoxic response returns to baseline (1). A consensus curve of autocytolytic activity is shown in Figure 1A. Biphasic limiting dilution dose-response curves of this sort reflect the activities of two counterpoised cell populations. One population, with autocytolytic potential, is present at a higher frequency and its effects are, therefore, seen at low-responder cell dose. The second population, which is capable of down-regulating this activity, is present at lower frequency. The effects of this regulatory population are, therefore, most readily seen at high-responder cell dose. Both cell populations have been characterized previously and appear to be T lymphocytes (2).

When autologous limiting dilution cultures are performed in recombinant IL-2 rather than crude lymphocyte-conditioned medium, the autocytotoxic response develops normally. However, under these conditions, this response is not down regulated and remains strongly positive at cell doses of up to 50,000 per microculture. Mixing recombinant IL-2 with crude lymphocyte-conditioned medium restores the regulatory component of the biphasic curve. Representative examples are shown in Figure 1B. The ability to restore the regulation of autoaggressive responses in these mixing experiments indicates that a soluble factor distinct from IL-2 is necessary for the development of autoregulation. This factor does not directly inhibit autocytotoxic responses, since inhibition is seen only at high-responder cell doses. Rather, the observed results indicate that this soluble factor is necessary for the activation, growth, or differentiation of the regulatory cell population. Fractionation of conditioned medium by gel filtration chromatography suggests that this substance has a molecular weight of between 60 and 80 kd. A variety of recombinant or highly purified cytokines (listed in Table 1.) have been tested for their ability to complement recombinant IL-2 in this assay. None has demonstrated any activity. Residual phorbol ester in the conditioned medium has also been excluded as the responsible moiety.

These studies thus indicate that a novel cytokine with a molecular weight between 60 and 80 kD is necessary for the control of autoaggressive responses in normal peripheral lymphoid populations. Preliminary studies in patients undergoing therapeutic marrow transplantation suggest that this autoregulatory network may play a role in the establishment of donor–host tolerance and prevention of graft versus host disease (3). Therefore, it is conceivable that qualitative or quantitative deficiencies of this factor may contribute to the pathogenesis of autoimmune diseases or graft versus host responses.

References

1. Rosenkrantz K, Dupont B, Flomenberg N. Generation and regulation of autocytotoxicity in mixed lymphocyte cultures: evidence for active suppression of autocytotoxic cells. Proc Natl Acad Sci USA 1985;82:4508–12.
2. Rosenkrantz K, Dupont B, Williams D, Flomenberg N. Autocytotoxic and autosuppressor T-cell lines generated from autologous lymphocyte cultures. Human Immunol 1987; 19:189–203.
3. Rosenkrantz K, Keever C, Kirsch J, Horvath A, Bhimani K, O'Reilly R, Dupont B, Flomenberg N. In vitro correlates of graft-host tolerance after HLA-matched and mismatched marrow transplants: suggestions from limiting dilution analysis. Transplant Proc 1987;19:98–103.

Author Affiliation

Memorial Sloan-Kettering Cancer Center, New York, N.Y. 10021, USA

Immunoregulatory Circuits Initiated by Autoreactive T-Cell Clones: I. Generation of T-Helper, T-Amplifier, and T-Suppressor Clones

Niciole Suciu-Foca, Elaine Reed, Elena Cristea, and Donald W. King

Abstract: The idiotype network postulates that T- and B-cell clones are interconnected through idiotypic determinants borne by the immune receptor. Because the antigen receptor of T cells, like the Ig receptor of B cells, is clonally distributed, the idiotypes characterizing one clone can stimulate the proliferation of another clone bearing complementary anti-idiotypic receptors. In previous studies (1–3), we provided evidence supporting this concept by demonstrating that (allo) antigen-specific T cells trigger the production of autoantibodies and of autoreactive (i.e., AMLC-reactive) anti-idiotypic T cells. We now document the existence of anti-idiotypic T-cell clones with helper, amplifier, suppressor, and suppressor/amplifier function.

Autoreactive T-cell clones (AMLC-TCC) were established by stimulating T lymphocytes with irradiated T lymphoblasts from an autologous alloreactive T-cell line (MLC-TCL). Responding T lymphocytes were obtained from a DR3,5 heterozygous individual. T lymphoblasts used for AMLC priming, derived from an MLC-TCL sensitized against allogeneic cells with the DR3,4 phenotype. After tertiary priming, T-cell clones were generated by limiting dilution at 0.3 cell/well. Clones were expanded in medium containing IL-2 and irradiated stimulating cells, consisting of mixtures of autologous PBL and MLC-TCL (at a 1 to 1 ratio) (4). Twenty-six TCC exhibiting blastogenic responses when stimulated with MLC-TCL were characterized with respect to their cognitive specificity and ability to induce, suppress, or amplify T-cell proliferation and Ig secretion by autologous B cells.

Testing of the ability of these TCCs to proliferate in response to allogeneic PBL or to MLC-TCL different from the original sensitizer indicated that significant stimulation (proliferation index > 10) occurred only when the clones were challenged with the MLC-TCL used for priming. Although the autologous EBV-transformed B cells induced no significant blastogenic responses, as ascertained by increased rates of tritiated thymidine incorporation, they stimulated clonal expansion when used for feeding the TCC. Allogeneic LBCL, failed to stimulate growth. This suggests that the AMLC-TCC recognize shared or cross-reactive idiotypes on the Ig receptors of autologous B cells and/or that this response is self-Ia restricted.

Attempts to characterize the function of the AMLC-TCC included testing their ability to secrete B-cell growth and differentiation factors upon stimulation with autologous B cells. The assay was performed by coculturing highly purified B-cell suspensions with AMLC-TCC in the presence of Pokeweed mitogen. After 6 days of incubation, the amount of ^{3}H-Td R incorporated by the cells and the amount of Ig released in the supernatant was quantitated. Of 26 clones tested, 5 exhibited helper activity, i.e., they stimulated the growth and maturation of autologous B cells (Table1). The nature of the factor produced by these clones is still uncertain. However, at least one of these five clones seems to secrete BCGF-12 kD, as indicated by the inhibitory effect displayed by the anti-BCGF-12 kD receptor antibody (MAb NDA$_3$) on the PWM-stimulated response of B cells grown with the autologous AMLC-TCC.

Table 1. Stimulation of B-Cell Growth and Maturation by AMLC-TCC with Helper Function

Cultures Containing: B	TCC	PWM	^{3}H-TdR (Mean cpm)	IgG (µg/ml)	IgM (µg/ml)
+	–	+	1,084	1	3
+	N3Ai2	+	12,672	10	16
+	N5B10	+	8,976	9	18
+	N6E7	+	14,628	12	15
+	N8E4	+	8,167	6	10
+	N2C6	+	7,430	8	1

B cells (10^5) were cultured with or without TCC (10^4/cells in flat-bottomed microwells in a volume of 0.2 ml of medium containing PWM (10 µg/ml). All reactions were done in triplicates. After 6 days of incubation at 37°C, 100 microliter aliquots of supernatants were collected from each well and the amount of Ig was quantitated by ELISA (3). Cells were labeled with ^{3}H-TdR, and incubated for an additional 8 hours . Radiolabel incorporation was measured by liquid scintillation counting.

Table 2. Amplification of T-Helper Activity by AMLC-TCC

Cultures Containing B cells + PWM +	^{3}H-TdR Incorporation (cpm)	Ig Synthesis IgG (µg/ml)	Ig Synthesis IgM (µg/ml)
–	1,450	2	4
T cells	22,540	18	24
T cells (1000r)	12,560	10	12
AMLC-TCC	2,810	3	4
AMLC-TCC + T cells	32,690	28	34
AMLC-TCC + T (1000r)	25,780	19	25
AMLC-TCC (1000r) + T cells (1000r)	10,407	11	10

Cultures were set up in the autologus system using 10^4 AMLC-TCC, 10^4 purified T cells and 10^5 purified B cells in medium containing PWM. After 6 days of incubation at 37°C, ^{3}H-TdR incorporation was measured by liquid scintillation counting, and Ig synthesis was determined by ELISA.

Immunobiology of HLA
Volume II

Table 3. Inhibition of MLC Responses and B-Cell Differentiation by AMLC-TCC with Suppressor Activity

MLC Responses ^{3}H-TdR (Autologous + Allogeneic PBL)		PWM Responses (Autologous T + B Cells) ^{3}H-TdR	Ig (μg/ml) G	Ig (μg/ml) M
NO AMLC-TCC	48,642	35,730	16	24
+ AMLC-TCC				
12F6	19,510	18,420	6	14
13B8	18,540	16,570	9	10
17C1	10,260	10,210	8	6

MLC were set up using 5 × 10^4 responding cells and 5 × 10^4 irradiated stimulating cells. AMLC-TCC were tested at 1 × 10^4 cells/culture. PWM stimulated cultures contained 1 × 10^4 T cells, 1 × 10^5, and 1 × 10^4 cells from the AMLC-TCC.

As opposed to these 5 clones, 12 other clones failed to stimulate B-cell proliferation and Ig synthesis when added to purified B-cell suspensions, yet acted as amplifiers of the helper activity exhibited by autologous T cells. This was demonstrated in experiments in which the TCC were cultured together with autologous T and B cells in the presence of PWM. As illustrated in Table 2, proliferation and Ig synthesis were significantly higher when the TCC was added to T- and B-cell mixtures than when B cells were grown only with TCC or only with T cells. The amplification effect was slightly diminished when T cells were irradiated with 1000 rads, yet it was entirely abolished by irradiating the clone (Table 2). This suggests that the AMLC-TCC produced an amplification factor that potentiates the production of B-cell growth and differentiation factors by radioresistant T-helper cells. An additional seven AMLC-TCC were characterized as suppressors. These TCC inhibited the ability of autologous T cells to respond in MLC to the specific stimulator (used for priming the MLC-TCL) or to other stimulators sharing the same HLA-DR antigen. When added to cultures of autologous T and B lymphocytes, the suppressor AMLC-TCC inhibited by 50% or more the process of PWM-induced proliferation and Ig synthesis (Table 3).

Testing of the remaining three AMLC-TCC for their effect on the MLC reactivity of autologous T cells and for their ability to produce helper factors, yielded results that appeared to be rather conflicting. These three TCC suppressed the MLC response of autologous peripheral blood lymphocytes, yet augmented the ability of T cells to induce B-cell proliferation and Ig synthesis in the PWM system.

The paradoxical occurrence of a dual suppressor and amplification effect may reflect the capacity of such AMLC-TCC to induce the production of autoantibodies. Such antibodies may in turn inhibit the MLC reactivity of autologous T cells. To explore this possibility, supernatants from cultures containing AMLC-TCC, autologous T and B cells, were tested for binding to resting or to MLC-stimulated T lymphocytes. Indirect immunofluorescence studies on an Ortho-spectrum 3 cytofluorograph revealed significant binding to autologous but not to allogeneic T lymphocytes (Table 4).

To determine whether such anti-T-cell autoantibodies recognize structures associated with the T-cell receptor, we preincubated resting and MLC-activated T lymphocytes for 18 hours with MoAb OKT3 (at 0.3 mg/ml). After incubation, cells were washed, reincubated with supernatants containing the putative autoantibodies, and stained with fluorescinated anti-human Ig. Replicate samples were stained with fluorescinated MoAb OKT3. There was almost complete inhibition of binding of supernatant Ig as well as of MoAb OKT3. These comodulation experiments suggest that the structure recognized by supernatant Ig on the membrane of the cell is associated with the T3 antigen that is part of the T-cell receptor complex.

Thus, AMLC-reactive clones, which react as suppressors of MLC and as amplifiers of Ig secretion in the autologous system, seem to induce the production of anti-T-cell autoantibodies.

Analysis of T-cell receptor rearrangement and expression in the 26 AMLC-TCC showed that all the clones that we have characterized functionally, including the suppressors, utilize the α and the β chain of the T-cell receptor for antigens. This confirms our previous conclusion that there are no structural differences between the T-cell receptor utilized by suppressor cells and helper or cytotoxic cells, and that T-cell function is not encoded by the antigen receptor (5). Our data indicate the existence of regulatory circuits comprising MHC-restricted T cells that activate T-helper and suppressor cells, which in turn act on autologous B lymphocytes.

Table 4. Characterization of AMLC-TCC Which Suppress MLC Responses and Amplify B-Cell Function

Cultures				% Cells Reacting with Ig from MLC Supernatant			
				Autologous T		Allogenic T	
Cells Added	MLC (cpm)	PWM (cpm)	IgM	Small	MLC Blast	Small	MLC Blast
None	24,870	28,890	14	0.0	0.0	0.0	0.0
TCC	8,312	42,370	22	5.8	36.0	0.4	0.6

AMLC-TCC (10^4 cells) were added to MLC or to PWM cultures. Cultures were harvested after 6 days. Supernatants from MLCs with or without AMLC-TCC were collected and tested on resting or activated T cells by indirect immunofluorescence using an Ortho spectrum III cytofluorograph.

Acknowledgments. This work was supported by NIH grants AL 25210 and HL 36581.

References

1. Suciu-Foca N, Reed E, Rohowsky C, Kung P, King DW. Anti-idiotypic antibodies to anti-HLA receptors induced by pregnancy. Proc Natl Acad Sci USA 1983;80:830–834.
2. Suciu-Foca N, Rohowsky C, Kung P, and King DW. Idiotypic-like determinants on human T lymphocytes alloactivated in mixed lymphocyte culture (MLC). J Exp Med 1982;156:283–288.
3. Rohowsky C, Suciu-Foca N, Kung P, Tang TF, Reed E, King DW. Suppressor T cells generated in autologous MLR with activated T lymphoblasts. Transplant Proc 1983;15(1): 765–770.
4. Suciu-Foca N, Rohowsky C, Kung P, King DW. Idiotypic receptors for soluble antigens on human T-lymphocytes clones. Human Immunol 1984;93:34–47.
5. Yoshikai Y, Yanagi Y, Suciu-Foca N, Mak TW. Presence of T cell receptor mRNA in functionally distinct T cells and elevation during intrathymic differentiation. Nature 1984; 310:506–508.

Author Affiliations

Nicole Suciu-Foca, Elaine Reed, Elena Cristea, Department of Pathology, College of Physicians and Surgeons of Columbia University, New York, NY; *Donald W. King*, Department of Pathology, University of Chicago, Chicago, IL, USA

Down-Regulatory Role of CD8 Molecules in CD2 MAb and CD3 MAb Induced Nonspecific Cytotoxicity of Cytotoxic T-Lymphocyte Clones

G.A. van Seventer, R.A.W. van Lier, H. Spits, K.C. Kuijpers, and C.J.M. Melief

The CD8 (T8) antigen is thought to function as an additional stabilizer of the interaction of the cytotoxic T lymphocyte (CTL) with the target cell by binding to monomorphic determinants on the HLA class I molecules. Recent evidence, however, indicates that the CD8 molecules do not merely serve as passive adhesion structures, but also exert regulating effects on T-cell activity (1). Therefore, we decided to reinvestigate the role of the CD8 antigen in CD8⁺ HLA class I and class II allospecific human CTL clones.

An HLA class I-specific CD8⁺ CTL clone (KOR-18) and an HLA class II-specific CD8⁺ CTL clone (JR-2-2) were tested for blocking of antigen-specific cytotoxicity with a panel of CD8 MAbs. Only the class I-specific clone KOR-18 was found to be inhibited by the CD8 MAbs, whereas the class II-specific clone JR-2-2 was not influenced at all (Table 1).

MAbs directed against the CD3 antigen have been described to induce CTL clones to exhibit nonspecific cytotoxicity, which is mediated by Fc receptors on the target cells. CD3 MAb-induced cytotoxicity of CTL clones is reported to be inhibited by CD8 MAbs, even when the target cells lack HLA class I antigens, the putative natural ligands of the CD8 antigen (1). This inhibition by CD8 MAbs, however, is only found when suboptimal concentrations of CD3 MAb were used to induce nonspecific cytotoxicity. Therefore, we investigated whether the class II-specific clone JR-2-2 could be inhibited by CD8 MAbs in this assay. As shown in Table 1, both clones (KOR-18 and JR-2-2) were specifically blocked by CD8 MAb when suboptimal concentrations of CD3 MAb were used. This indicates the involvement of CD8 antigens in the cytotoxic activity of both types of CTL clones.

Besides T-cell receptor/CD3 complex-mediated T-cell activation, several alternative T-cell activation pathways are described. One of them is the CD2 antigen activation pathway. Combinations of CD2 MAb can induce resting T cells to proliferate and CTL to exert nonspecific cytotoxicity (2). To further analyze the functional role of CD8 molecules, we performed experiments in which CD2 MAbs were used to induce nonspecific cytotoxicity. As indicated in Table 1 CD8 MAbs could inhibit CD2 MAb-induced cytotoxicity of both CTL clones. However, the inhibition of the clone JR-2-2 was less effective because only one of the tested CD8 MAbs could inhibit significantly. The differential effects of the CD8 MAbs on antigen-specific cytotoxicity on one hand and in both CD2 MAb and CD3 MAb-induced cytotoxicity on the other hand may reflect differences in functionally relevant epitopes on the CD8 molecule, as has been described before (1). These combined findings suggest a down-regulatory function of CD8 not only in TCR/CD3 activation, but also in TCR/CD3 controlled alternative activation routes such as the CD2 activation pathway. The absence of evidence for CD8 involvement in the antigen-specific recognition of HLA class II antigens by the CD8⁺ CTL clone JR-2-2 may be explained by assuming a high affinity interaction of the TCR with (allo)antigen. This higher affinity interaction provides an activation signal of sufficient strength to be able to overrule any down regulation by the CD8 molecules. Moreover, clone JR-2-2 may be highly sensitive for triggering of cytotoxic activity, a result of which it is hard to

Table 1.

CTL KOR-18	Antigen-Specific Cytotoxicity	CD3 MAb-Induced Cytotoxicity	CD2 MAb-Induced Cytotoxicity
	59%	43%	29%
CD8 MAb 1 (IgG1)	25%	6%	0%
CD8 MAb 2 (IgG2a)	6%	12%	0%
CD8 MAb 3 (IgG2b)	1%	3%	0%
CD8 MAb 4 (IgG1)	47%	19%	17%
Control MAb (IgG1)	60%	40%	27%
Background lysis		2%	2%
CTL JR-2-2			
	35%	45%	58%
CD8 MAb 1	33%	10%	23%
CD8 MAb 2	28%	15%	40%
CD8 MAb 3	34%	20%	52%
CD8 MAb 4	37%	38%	54%
Control MAb	32%	38%	58%
Background lysis		12%	12%

The CD3 MAb (CLB-T3/4.1, IgG1) was used in a dilution of 10^{-6} of ascites fluid in the experiments with clone KOR-18 and in a dilution of 10^{-7} with clone JR-2-2. The CD2 MAbs (CLB-T11.1/2 and GT2, both IgG1) were used in dilutions of 4.10^{-2} and 10^{-4}, respectively. All CD8 MAbs and the control MAb (directed against CD6) were tested in a dilution of 4×10^{-2} of ascitic fluid. E:T ratio was 10:1, specific cytotoxicity was tested on cells of the cell line JY, nonspecific cytotoxicity was tested on cells of the cell line Daudi. In all experimental settings the mean of six-fold determinations was taken. The results are given as the percentage specific ^{51}Cr-release as determined in a 4 h assay.

down regulate it by CD8 MAbs not only in TCR/CD3-mediated activation, but in CD2-mediated activation as well. This could make this clone independent of the adhesion-like function of CD8 (or CD4) molecules. Thus, the effectiveness of a down-regulatory role of CD8 in CTL activation is thought to be inversely correlated with the strength of the activation signal. In this view, CD8 mediated regulation of T-cell activation can prevent nonspecific triggering of cytotoxicity by interactions of insufficient affinity and may, therefore, play an important role in the maintenance of self-tolerance.

Acknowledgments. This study was supported by a grant from the Netherlands Kidney Foundation (grant no. 84502).

References

1. Van Seventer GA, van Lier RAW, Spits H, Ivanyi P, Melief CJM. Evidence for a regulatory role of the T8(CD8) antigen in antigen-specific and anti-T3-(CD3)-induced lytic activity of allospecific cytotoxic T lymphocyte clones. Eur J Immunol 1986;16:1363.
2. Siliciano RS, Pratt JC, Schmidt RE, Reinherz EL. Activation of cytolytic T lymphocyte and natural killer cell function through the T11 sheep erythrocytes binding protein. Nature 1986;317:428.

Author Affiliations

G.A. van Seventer, R.A.W. van Lier, K.C. Kuijpers, Central Laboratory of the Netherlands Red Cross Blood Transfusion Service, Amsterdam; *C.J.M. Melief*, Netherlands Cancer Institute, Amsterdam, The Netherlands; *H. Spits*, UNICET, Dardilly, France.

Inhibitors of Cytotoxic T Lymphocyte Galactosyltransferase Inhibit Cytotoxic T Lymphocyte Interaction with Their Target Cells*

S.M. Tatum, E.A. Kurt-Jones, B.D. Shur, and R.R. Lindquist

We previously suggested that cytotoxic T lymphocyte (CTL) surface galactosyltransferase (Gal-Tase) may participate in the adhesion between CTL and target cells (1). Gal-Tase catalyzes the transfer of galactose from UDP-GAl to *N*-acetylglucosamine on the specific oligosaccharide chain of the acceptor molecule. Gal-Tase exhibits specificity for both sugar donor and acceptor molecules. On transferring the donor sugar to the acceptor molecule, the newly glycosylated product exhibits poor affinity for the acceptor binding site of the enzyme and Gal-Tase disassociates from its product. Since sugar nucleotides are not normally present in the extracellular fluids, the availability of extracellular donor sugar nucleotide frequently limits glycosyltransferase catalysis. Hence acceptor glycoproteins may interact with the glycosyltransferase acceptor binding site and form a

*This work was supported in part by grants from the National Cancer Institute and the American Cancer Society. Address correspondence to: R.R. Lindquist, M.D., Pathology Department, University of Connecticut, Farmington, CT 06032, USA.

Table 1. Effect of UDP-Dialdehyde on Galactosyltransferase and on CTL Activity

UDP-Dialdehyde Concentration (moles)	Gal-Tase Activity (% Inhib)	CTL Killing Activity (% Inhib)	CTL Conjugate Formation (% Inhib)
10^{-4}	35	11	12
5×10^{-4}	60	51	52
10^{-3}	75	107	73

Control Gal-Tase activity was 11.7 ± 3.7 pmoles/million cells, control CTL killing activity was 50.4 ± 3.7%, and control CTL conjugate formation was 22 ± 5.0%.

Table 2. Effect of α-Lactalbumin on Galactosyltransferase, Lactose Synthetase, and CTL Activities

α-Lactalbumin Concentration (g%)	Gal-Tase Activity (% Inhib)	Lactose Synthesis Activity (% Max Inc)	CTL Killing Activity (% Inhib)
0.01	92	35	73
0.05	73	71	42
0.10	63	84	29
0.50	55	100	34
1.00	55	89	19

stable enzyme-substrate complex which is only reversible after the donor sugar-nucleotide becomes available and initiates catalysis. If this suggestion is true, we would expect that inhibitors of Gal-Tase would inhibit CTLs from adhering to their specific target cells and from killing the target cells.

We now report the results of experiments using two inhibitors of Gal-Tase, galactosyl UDP-dialdehyde and α-lactalbumin. We determined the effect of these two Gal-Tase inhibitors on secondary $H2^b$ anti-$H2^d$ CTL (1-1). Nucleotide-dialdehyde forms a Schiff's base with nucleotide-binding enzymes such as glycosyltransferase and thus inhibits their binding and catalytic activity. UDP-dialdehyde was synthesized from equimolar concentrations of UDP and sodium metaperiodate. Varying concentrations of UDP-dialdehyde were added to mixtures of 1-1 CTLs and 50,000 specific target cells and the effect of UDP-dialdehyde on Gal-Tase and alloimmune cytolysis was determined. The results of the experiments are presented in the Table 1. From these results, we calculated that the ID50 of UDP-dialdehyde for cell-surface Gal-Tase was 2.5×10^{-4}M and the ID50 for CTL cytolytic activity was 2.2×10^{-4}M. Thus, as predicted by our prior suggestion, UDP-dialdehyde inhibits CTL cytolytic activity and the inhibition parallels ($r = 0.9760$) UDP-dialdehyde-induced inhibition of CTL surface Gal-Tase. Also shown in Table 1 is the effect of UDP-dialdehyde on CTL-target cell conjugate formation, and from the data it can be seen that UDP-dialdehyde inhibits CTL-target cell interaction.

α-lactalbumin, unlike nucleotide-dialdehyde, is not a general inhibitor of glycosyltransferases. It only inhibits Gal-Tase. Strictly speaking, α-lactalbumin is not an inhibitor of Gal-Tase. α-Lactalbumin is a modifier of Gal-Tase and changes the acceptor specificity. α-Lactalbumin after binding to Gal-Tase modifies the specific acceptor for galactose from *N*-acetylglucosamine to glucose. Therefore, rather than adding galactose to terminal *N*-acetylglucosamine it adds it to free glucose to make lactose (galactose-glucose). Thus, although α-lactalbumin does not actually inhibit Gal-Tase activity, it modifies the enzyme so that terminal *N*-acetylglucosamine of glycoproteins on target cells can no longer occupy the acceptor binding site on the enzyme with high affinity. Therefore, if our suggestion is correct, we would anticipate the α-lactalbumin would inhibit CTL-mediated specific target cell killing. The results of our experiment with α-lactalbumin are presented in Table 2. From Table 2 it is seen that α-lactalbumin increases lactose synthesis and concomitantly decreases Gal-Tase activity, thus demonstrating that α-lactalbumin changes the specificity of the enzyme. Table 2 also indicates that along with this change in enzyme specificity, α-lactalbumin inhibits specific alloimmune cytolysis. Not shown because of space limitations are the data from dextrose dispersion experiments that indicate that α-lactalbumin inhibits conjugate formation and disrupts preformed conjugates.

Thus, these experiments further support the view that CTL Gal-Tase acts as an adhesion molecule on CTL. This view has several attractive features; foremost is that it provides a mechanism explaining CTL recycling. It has long been known that CTL after delivering the lethal hit releases the target cell and goes on to adhere to and kill new target cells even before the hit target cell disintegrates. This recycling poses a dilemma, for Tatum and Lindquist (2) found recently that CTL very tenaciously adheres to target cells. A very considerable force of 3.33×10^{-5} dynes is necessary to dislodge 50% of CTL from target cells to which they adhere. Yet in spite of this tenacious adhesion, on recycling CTL releases the target cell. Thus the CTL must have a mechanism for adhesion that is easily reversible. Gal-Tase fulfills this requirement. Gal-Tase binds acceptor sugars with high affinity; but on adding the donor sugar to the acceptor sugar, Gal-Tase no longer binds the acceptor with high affinity. Hence CTL-target adhesion mediated by Gal-Tase would be a firm adhesion, until UDP-donor sugar becomes available and is added to the acceptor sugar moiety on the target cell. On adding the donor sugar to the acceptor sugar on the target cell, CTL cell surface Gal-Tase no longer binds the acceptor on target cell and the two cells separate.

References

1. Kurt EA, Shur BD, Lindquist RR. Cytolytic T lymphocyte galactosyl-transferase activity. Fed Proc 1981;40:1150.
2. Tatum SM, Lindquist RR. Measurement of adhesive strength between CTL and target cell. Leukocyte Biol 1987;42:330.

Author Affiliations

University of Connecticut Medical School, Farmington, CT, USA. Present address: *S.M. Tatum*, Connecticut Cancer Institute; *E.A. Kurt-Jones*, Harvard Medical School; *B.D. Shur*, M.D. Anderson Hospital & Tumor Institute

Molecular Cloning of an Inducible Cytotoxic T-Lymphocyte-Associated Gene (Hu-CTLA-1) and Gene Localization to Human Chromosome 14

Patrick Haddad, Jean-Marc Zini, Marie-Veronique Clement, Jean-François Brunet, Marie-Genevieve Mattei, Pierre Golstein, Laurent Degos, Christian-Jacques Larsen, Daniele Mathieu-Mahul, and Marilyne Sasportes

Abstract: Cytotoxic T lymphocytes (CTL) recognize the antigen in association with MHC products or the allogenic MHC proteins on the target cells through the T-cell receptor (TcR). TcR triggering could be followed by secretory processes involved in target cell lysis. CTLs have, indeed, cytolytic granules that contain the pore-forming protein perforin and serine esterases (granzymes). These proteases and perforin might form a cytolytic cascade as in complement-mediated cytotoxicity. Recently, a murine cDNA clone, named CTLA-1 (or C.1.1.), whose coding sequences are strongly related to the serine esterase, granzyme B, G, or H, has been isolated; this CTLA-1 gene is transcribed at high levels in murine CTLs. For this reason, we were interested in

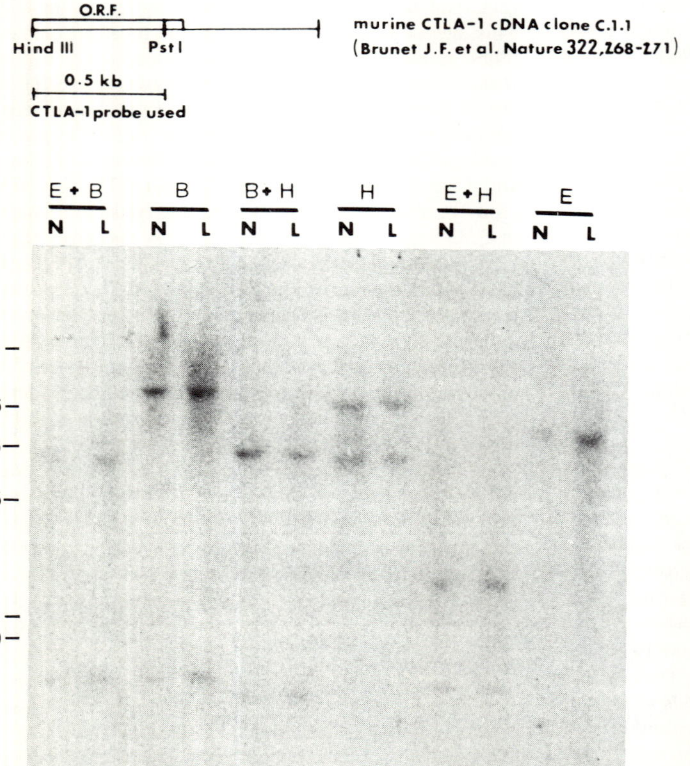

Figure 1. Southern blot analysis of CTLA-1-related sequences in human genomic DNA with the murine cDNA probe. DNA was isolated from PBL of a healthy donor (N) or from acute T-cell leukemia cells (L). It was digested with restriction enzymes and analyzed by Southern blot hybridization with the HindIII-PstI 0.5 kb fragment of a CTLA-1 cDNA clone (see ref. 13). Hybridization was performed under nonstringent conditions (washes at 65°C in 0.5 SSC). E = EcoRI; B = BamHI; H = HindIII.

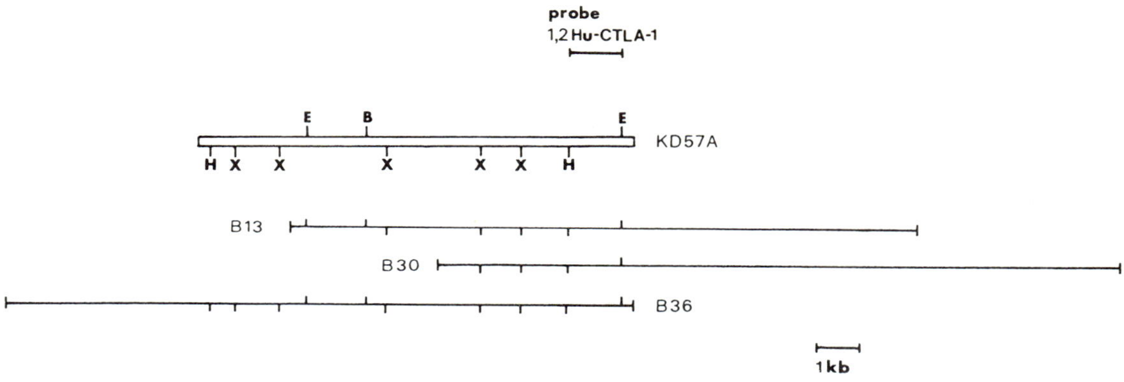

Figure 2. Restriction map of recombinant phage clone detected by a CTLA-1 cDNA probe. Clone KD57A was isolated from a human genomic DNA library established from T-ALL cells. Screening of this library was performed with a murine CTLA-1 cDNA probe consisting of a HindIII-PstI 0.5 kb fragment. A Hind III-EcoRI 1.2 kb fragment from genomic clone KD57A was subcloned into the pGEM plasmid and used as a probe to screen another human genomic DNA library (see text) from which four clones, designated B13, B30, B34, and B36, were recovered. Only clones B13, B30, and B36 overlapped the KD57A insert. E = EcoRI; H = HindIII; B = BamHI; X = XbaI.

cloning the human homologue of CTLA-1 gene. We report the isolation of human sequences of CTLA-1 gene, the localization of the gene on human chromosome 14, and the co-inducibility of Hu-CTLA-1 gene expression and cytotoxicity in CTLs.

Considerable data have shown that pore formation plays an important role in cell killing mediated by cytotoxic T lymphocytes (CTLs) or natural killer cells (NK) (1–3). CTLs and NK could lyse appropriate cells by secreting the pore-forming protein, perforin (4,5), which inserts and polymerizes in the target cell membrane. In addition, cytolytic granules contain serine esterases (6,7) or granzymes (8) which could activate perforin. Because of the existence of structural and functional homologies between perforin and the ninth component of complement (C9) (9,10), perforin and the granzymes have been proposed to constitute a novel T-cytotoxic cascade. Nevertheless, this process was not found to be systematically required for cell-mediated lysis either in vitro or in vivo. (11,12). These apparently contradictory results suggest that different mechanisms might be responsible for target cell lysis.

Two murine cDNA clones, CTLA-1 (13) and CCP-1 (14), have been isolated from two independent cDNA libraries established from murine CTLs. These clones have been shown to be specific for T-cell mediated cytotoxicity. Interestingly, the nucleotide sequence of these two clones revealed high homology with the serine esterase granzyme B, G, or H (8). Moreover, the murine CTLA-1 gene has been localized (13) on the D segment of chromosome 14 in the same chromosomal region to which the TcRα locus has been mapped. Recently, we described the generation of alloreactive T-cell clones following α-interferon treatment (15). In view of these results, it seemed extremely relevant that precursor T cells rearrange the TCRα gene at the same time they acquire their specific lytic function (16,17). All these observations incited us to isolate a human equivalent of the CTLA-1 gene.

By Southern blot analysis, we first analyzed the organization of the CTLA-1 related sequences in human genomic DNA using the HindIII-PstI 0.5 kb fragment from a murine CTLA-1 clone (clone C.1.1 in ref. 13) that contains most of the coding sequences (see Fig. 1). Under non-stringent conditions, the murine CTLA-1 cDNA probe was able to recognize human specific sequences. The systematic presence of two distinct bands with all restriction enzymes (except EcoRI) was observed suggesting that the murine cDNA recognized either at least two exons in the same human gene or two distinct homologous loci in the human genome.

To isolate human CTLA-1-related sequences, we screened a human genomic DNA library established from an acute T-cell leukemia (18) with the HindIII-PstI 0.5 kb fragment. One positive clone, KD57A, containing a 10 kb insert, was isolated; restriction analysis showed that sequences homologous to murine coding sequences of murine CTLA-1 gene were located in a (1.2 kb EcoRI-HindIII fragment 1.2 Hu-CTLA-1), as indicated at the top of Figure 2. In addition, this fragment did not contain any repetitive sequences. It was subcloned into the pGEM plasmid in order to provide a human CTLA-1 probe that was used to perform Southern blot analysis of human DNA. Two distinct bands were detected with almost all of the restriction enzymes including HindIII enzyme, which obviously did not cut within the 1.2 kb genomic probe (data not shown). Moreover, for each restriction enzyme digest, one of these bands was more intense, suggesting that the human genome might contain at least two distinct CTLA-1 loci. To confirm this hypothesis, another human DNA library established from a pre-B leukemia

Table 1. Expression of the Hu-CTLA-1 Gene in Different Human Leukocytes

Cells	Cytotoxicity	Detection of the 1 kb Hu-CTLA-1 Transcript
CTL clone Bj1α	++	++
CTL clone PIK6	++	++
Rex T-cell line	–	–
Precursor CTL clone BJ4	–	–
CTL clone BJ4α (treated by rIFNα)	++	++
PBL treated for 3 days by:		
PHA alone	ND	+
PHA + rIL2	ND	+
IL2 alone	ND	+/–
IL2 + rIFNα	ND	+/–

Clonal cell populations were tested for cell-mediated cytotoxicity using a standard chromium release assay. Northern blots were hybridized with the 1.2 kb Hu-CTLA-1 probe. A single band running at 1.0 kb was detected in cells with a positive expression (stringent conditions were used). ND = Not determined; PBL = peripheral blood leucocytes; PHA = phytohemagglutinin A; rIL2 = recombinant interleukin 2; rIFNα = recombinant interferon α.

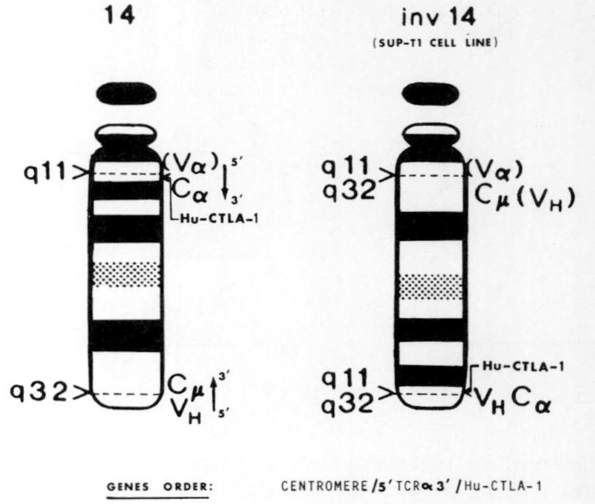

Figure 3. Localization of the Hu-CTLA-1 gene to human chromosome 14 by in situ hybridization with the 1.2 kb Hu-CTLA-1 probe. Only one specific site was found, in band q11-q12—interestingly, in a region to which the TcR α-chain gene has been mapped. The gene order on chromosome 14 was determined by in situ hybridization with the 1.2 kb Hu-CTLA-1 probe on SUP-T1 cell line metaphase spreads (see text).

(19) was screened with the 1.2 Hu-CTLA-1 probe: four positive recombinant phages were analyzed (see Fig 2); three clones overlapped the KD57A insert. On the other hand, the fourth clone (B34 clone, not presented here) had a distinct restriction map and might contain CTLA-1 sequences from the second locus.

To determine the coding capacity of the 1.2 Hu-CTLA-1 probe, Northern blot analysis was performed with cytotoxic and non-cytotoxic T-cell clones or lines. The results are summarized in Table 1. A 1.0 kb transcript was detected only in cytotoxic T-cell clones, suggesting the association of Hu-CTLA-1 gene expression and cytotoxic function, as reported in the murine system.

Importantly, the expression of the murine CTLA-1 gene was induced under conditions leading the differentiation of non-cytotoxic cells into cytotoxic cells. This effect was clearly shown with murine PC60 hybridoma cells, constitutive for growth but inducible for cytotoxicity in the presence of interleukin 2: CTLA-1 gene transcript was detected only in cytotoxic PC60 cells (13). A very similar phenomenon was observed with human precursor T-cell clones, which became cytotoxic after α- or γ-interferon treatment and expressed an Hu-CTLA-1 transcript (see T-cell clone BJ 4 on Table 1). In parallel, the conditions of Hu-CTLA-1 gene expression were established for normal PBL, as summarized in Table 1. The Hu-CTLA-1 gene was specifically expressed in PBL activated by PHA. Further experiments using human T-cell clones, either precursors or cytotoxic after IFN treatment, and mature CTLs treated by IFN, are in progress to establish the positive relationship between Hu-CTLA-1 gene expression, IFN treatment, and cytotoxicity.

As previously mentioned, the murine CTLA-1 gene has been located on the same chromosomal region as the TcRα gene locus. By in situ hybridization with the 1.2 kb human probe, the Hu-CTLA-1 gene was located on band q11 of chromosome 14 in the human genome. Therefore, in mouse and man, TCRα and CTLA-1 genes map close to each other. Next, the gene order on chromosome 14 with respect to the centromere was established by analyzing the tumoral SUP-T1 cell line. These cells are characterized by a chromosome 14 inversion: the breakpoints of the abnormal chromosome 14 are respectively in band q11, between the V and C regions of TCRα gene, and, in band q32, in the heavy chain immunoglobulin locus (20) (see Fig. 3). A positive signal was found in band q11 on normal chromosome 14, while it was detected in band q32 on the inverted chromosome 14. This was consistent with the following gene order: centromere/5′ TCRα 3′/Hu-CTLA-1. Although the distance between these two genes has not been determined, an attractive possibility might be that the Hu-CTLA-1 gene is cis-regulated by the TcRα gene.

References

1. Dourmashkin RR, Deteix P, Simone CB, Henkart P. Electron microscopic demonstration of lesion in target cell membranes associated with antibody-dependent cellular cytotoxicity. Clin Exp Immunol 1980;42:554.
2. Dennert G, Podack ER. Cytolysis by H2 specific T-killer cells: Assembly of tubular complexes during the lytic reaction. J Exp Med 1982;157:1483.

3. Podack ER, Dennert G. Cell mediated cytolysis: Assembly of two types of tubules with putative cytolytic function by cloned natural killer cells. Nature 1983;302:442.
4. Henkart PA. Mechanism of lymphocyte-mediated cytotoxicity. Ann Rev Immunol 1985;3:31
5. Podack ER, Königsberg PJ. Cytolytic T cell granules. Isolation, structural, biochemical and functional characterization. J Exp Med 1984;160:695.
6. Pasternak MS, Eisen HN. A novel serine esterase expressed by cytotoxic T lymphocytes. Nature 1985;314:743.
7. Pasternak MS, Verret CR, Liu MA, Eisen HN. Serine esterase in cytolytic T lymphocytes. Nature 1986;322:740.
8. Tschopp J, Masson D. A family of serine esterases in lytic granules of cytolytic T lymphocytes. Cell 1987;49:697.
9. Young JD-E, Cohn ZA, Podack ER. The ninth component of complement and the pore-forming protein (perforin 1) from cytotoxic T cells: Structural, immunological and functional similarities. Science 1986;233:184.
10. Tschopp J, Masson D, Stanley K. Structural/functional similarity between proteins involved in complement-and cytotoxic T-lymphocyte-mediated cytolysis. Nature 1986; 322:831.
11. Ostergaard HL, Kane KP, Mescher MF, Clark WR. Cytotoxic T lymphocyte mediated lysis without release of serine esterase. Nature 1987;330:71.
12. Trenn G, Takayama H, Sitkovsky MV. Exocytosis of cytolytic granules may not be required for target cell lysis by cytotoxic T-Lymphocytes. Nature 1987;330:72.
13. Brunet JF, Dosseto M, Denizot F, Mattei M-G, Clark WR, Haqqi TM, et al. The inducible cytotoxic-T-lymphocyte-associated gene transcript CTLA-1 sequence and gene localization to mouse chromosome 14. Nature 1986; 322:268.
14. Lobe CG, Finlay BB, Paranchych W, Paetkau VH, Bleackley RC. Novel serine proteases encoded by two cytotoxic T lymphocyte-specific genes. Science 1986;232:858.
15. Chen LK, Tourvieille B, Burns GF, Back FH, Mathieu-Mahul D, Sasportes M, et al. Interferon: A cytotoxic T lymphocyte differentiation signal. Eur J Immunol 1986;16:767.
16. Mathieu-Mahul C, Chen, LK, Larsen CJ, Bensussan A. Detection of T-cell receptor alpha locus rearrangements n human antigen-specific T lymphocytes clones with a DNA probe specific for the chromosome region 14q11. J Exp Med 1986;163:1308.
17 .Chen LK, Mathieu-Mahul D, Bach FH, Dausset J, Bensussan A, Sasportes M. Recombinant interferon alpha can induce rearrangement of T-cell antigen receptor alpha and maturation to cytotoxicity in T-lymphocyte clone in vitro. Proc Natl Acad Sci USA 1986;83:4887.
18. Mathieu-Mahul D, Bernheim A, Sigaux F, Daniel MT, Larsen CJ, Berger R. Translocation t(1;14) and rearrangement of the alpha chain T cell receptor gene in a T cell acute lymphoblastic leukemia. CR Acad Sc Paris 1986;302:525.
19. Chen Z, Le Pasuer D, Dausset J, Degos L, Flandrin G, Cohen D, et al. Human T-cell rearranging genes gamma are very frequently rearranged in B-lineage acute lymphoblastic leukemias but not in chronic B-cell proliferations. J Exp Med 1987;165:1000.
20. Baer R, Forster A, Rabbitts TH. The mechanism of chromosome 14 inversion in a human T cell lymphoma. Cell 1987;50:97.

Author Affiliations

Patrick Haddad, Jean-Marc Zini, Marie-Veronique Clement, Laurent Degos, Marilyne Sasportes, Inserm Unit 93; *Christian-Jacques Larsen, Daniele Mathieu-Mahul*, Inserm Unit 301, Hôpital Saint-Louis, Paris; *Jean-François Brunet, Pierre Golstein*, Inserm Unit 136, Campus de Luminy, 13288 Marseille; *Marie-Genevieve Mattei* Inserm Unit 242, Hôpital d'Enfants de la Timone, Marseille Cedex 5 France

A Cluster of Antibodies (RR1/1, LB-2 and 84H10) That Inhibit LFA-1-Dependent Lymphoid and Myeloid Cell Adhesion Bind Intercellular Adhesion Molecule-1 (ICAM-1)

M.W. Makgoba, M.E. Sanders, G.E. Ginther Luce, M.L. Dustin, P. Mannoni, E.A. Clark, T.A. Springer, and S. Shaw

Abstract: Recent functional studies indicate that intercellular adhesion molecule-1 (ICAM-1), recognized by monoclonal antibody (MAb) RR1/1, is a ligand for lymphocyte-associated antigen (LFA)-1 in several models of antigen-independent B- and T-cell adhesion and in cytotoxic T-cell recognition. Independent reports have identified other MAbs (LB-2 and 84H10) that inhibit LFA-1-dependent adhesion by B cells or myeloid cells. We analyzed the relationships of these three MAbs. The results show that: 1) all three MAbs inhibit LFA-1-dependent conjugate formation with particular targets and that mixes of these MAbs do not show additive inhibition; 2) LB-2 and 84H10 each inhibit the binding of the other, whereas binding of RR1/1 neither inhibits nor is inhibited by binding of LB-2 or 84H10; 3) all three MAbs bind purified 84H10 protein; and 4) binding of these three MAbs is concordant on 12 different cell types and is similarly enhanced on monocytes with activation. The simplest interpretation of these data is that a single protein, ICAM-1, is recognized by the MAbs RR1/1, LB-2, and 84H10. This finding unifies and extends previous observations; ICAM-1 functions as an adhesion molecule in a variety of leukocyte interactions, apparently as a ligand for LFA-1. Furthermore,

Table 1. MAb Inhibition of LFA-1-Dependent Conjugate Formation

Mab	HeLa	U937	U266
TS2/9	38	24	27
MHM23	0	2	2
LB-2	ND	1	23
RR1/1	2	2	21
84H10	3	1	23
84H10 + LB-2	ND	1	24
84H10 + RR1/1	2	4	21
RR1/1 + LB-2	ND	2	24

Conjugate formation was measured between the cytotoxic T-cell clone 8.2 and three antigen-negative targets HeLa, U937, and U266. Enumeration of conjugates was by two-color microfluorometry as described (10,19). The results are expressed as a % of targets that end up in conjugates. All assays were done in the continuous presence of saturating concentrations of the LFA-3 MAb TS2/9(17) to isolate the LFA-1 pathway. MAbs were continuously present during the assay at 100 μg/ml purified IgG except for MHM23, the LFA-1 Mab (20) and TS2/9 which were present at 300 μg/ml of Fab fragment. ND = not determined.

Table 2. MAb Binding to Purified ICAM-1 Protein

MAb	OD
TS2/9	75
84H10	850
LB-2	750
RR1/1	800

ICAM-1 was purified from L428 cells by an 84H10 immunochromatography column as described (11). The purified material was applied to glutaraldehyde/poly-L-lysine treated microtiter plates as described before (11). MAbs binding to the immobilized antigen was measured by ELISA using purified MAb (IgG) at 50 μg/ml as the first step, β-galactosidase-conjugated $F(ab')_2$ sheep anti-mouse IgG (H+L) as the second step, and measuring the rate of conversion of p-nitrophenyl-β-galactoside to a colored product (using the Bethesda Research Laboratories Hybridoma Screening Reagent according to the manufacturers directions). Data are reported as the average of the OD 405 nm of duplicate wells after subtraction of the OD in the absence of first step MAb.

MAb binding to either of two distinct ICAM-1 epitopes inhibits ICAM-1-dependent adhesion.

The LFA-1 antigen was identified 5 years ago in studies of monoclonal inhibition of antigen-specific cell-mediated cytotoxicity in both man and mouse (1,2). It has since been shown to be an adhesion receptor for all types of leucocytes (3,4) in a variety of such interactions. LFA-1 is a member of the cytoadhesin family of adhesion receptors (5–8). Homology between the members of this family extend to the position-specific protein (PSP) found in *Drosophila* (7,8), indicating that the cytoadhesins form a primitive cell-cell recognition system in both mammalian and nonmammalian species. Intercellular adhesion molecule-1 (ICAM-1) was postulated as a ligand for LFA-1 based on studies of monoclonal inhibition of homotypic B-cell aggregation and heterotypic adhesion of T lymphoblasts to fibroblasts (9,10). These findings have been extended and confirmed in studies of antigen-independent conjugate formation with T-cell clones (11). Recently, the ICAM-1 monoclonal antibody RR1/1 was shown to inhibit certain but not all types of antigen-specific CML interaction, raising the possibility of other alternative molecules that may act as LFA-1 ligands (12,13). Other MAbs had been described that inhibited leukocyte adhesion in a variety of models systems and suggested that they may act as ligands. Are these alternative LFA-1 ligands? The two relevant MAbs are 84H10 and LB-2. The MAb 84H10 was identified by preferential binding to myeloid leukemic cells and subsequently found to inhibit such cells to bone marrow stromal cells (unpublished observation). More recently, the 84H10 MAb has been shown to inhibit LFA-1-dependent homotypic B-cell aggregates (14). The MAb LB-2 was identified as a B-cell activation antigen (15) and subsequently shown to inhibit both myeloid and B-cell homotypic aggregates (16).

The initial studies compared the ability of three MAbs (RR1/1, LB-2, and 84H10) to inhibit LFA-1-dependent antigen-independent conjugate formation between a T-cell clone and targets of three different cell lineages. The LFA-1 pathway of adhesion was isolated by complete inhibition of the CD2/LFA-3 pathway using saturating amounts of the LFA-3 mab TS2/9 Fab (17). The results (Table 1) demonstrated a similar degree and pattern of inhibition by all three MAbs on the three targets. Furthermore, there was no additive inhibition when mixes were used as compared with single MAbs alone. The result suggested that the three MAbs act in the same pathway, either by binding to the same epitope or to different epitopes. The lack of inhibition on the target U266 raised the possibility of alternative molecules that may act as LFA-1 ligands, as has been previously suggested (9,12,13).

To determine whether all three MAbs recognize the same antigen, binding studies were carried on an 84H10 immunopurified protein bound to microtiter plates by the gluteraldehyde/poly-L-lysine method (11). The immunopurified 84H10 protein from L428 cells ran as a single band of 95 kD on SDS-PAGE (11). All three MAbs bound to the purified ICAM-1 (Table 2). The specificity of binding was demonstrated by the lack of binding of an irrelevant LFA-3 MAb TS2/9. This finding unifies a number of observations regarding the function, structure, and genetics of ICAM-1 (see below).

Because these three MAbs bind ICAM-1 and inhibit LFA-1-dependent adhesion, it was evaluated whether they bind to the same epitope. LB-2, 84H10, and RR1/1 MAbs were radiolabeled and the binding of each to L428 cells was measured in the presence of an excess of each MAb (Fig. 1). LB-2 and 84H10 appear to bind to the same epitope since they compete with each other for binding. RR1/1 binds to a different epitope since no

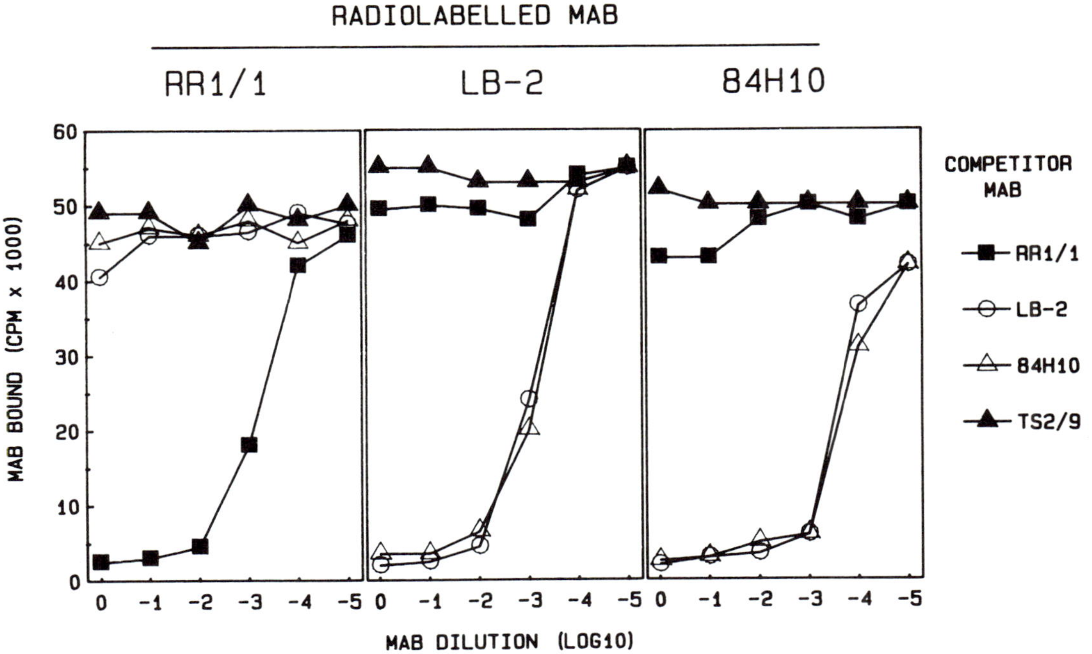

Figure 1. Competitive binding assay. MAbs were radiolabeled with ^{125}I by the iodo-bead method according to the manufacturer's protocol (Pierce Chemical Co., Rockford, IL). Usually 50% of the MAb retained activity by serial absorption studies. Serial 10-fold dilutions of the unlabeled MAbs were prepared in 0.5% human serum albumen in phosphate buffered saline starting at a concentration of 5 mg/ml. L428 cells at 2×10^5 per tube in 100 μl of 0.5% human serum albumen/phosphate buffered saline were preincubated with 30 μl of unlabeled MAb at 4°C for 15 minutes. After this, 20 μl of radiolabeled MAb equivalent to 0.1 μg was added and further incubated for another 60 minutes. The samples were washed three times with buffer and counted for radioactivity.

inhibition was seen in either direction between LB-2 or 84H10 and RR1/1. No nonspecific inhibition was observed since the TS2/9 MAb which binds to LFA-3 did not inhibit the binding of radiolabeled LB-2, 84H10, and RR1/1. Thus, ICAM-1 MAb that inhibit LFA-1-dependent adhesion can do so by binding to at least two distinct epitopes on ICAM-1.

If all three MAbs bind the same protein, a simple testable hypothesis is to study the tissue distribution of each epitope with the prediction that their expression would be concordant. Our studies using 12 different cell types show an identical pattern for the RR1/1 and 84H10/LB-2 epitopes (Table 3). Other studies demonstrated staining of a variety of cells on tissue sections, particularly endothelium for RR1/1 epitope (9,10). This pattern of staining is consistent with that reported for the 84H10 epitope in the Third International Workshop on Leukocyte Differentiation Antigens (18) or expression of the LB-2 epitope (16). Binding is coordinately enhanced for both epitopes following activation of monocytes (MWM in preparation).

The above studies demonstrate that the molecules defined by the MAbs LB-2, 84H10 are identical to the antigen ICAM-1 defined by the MAb RR1/1. There is

Table 3. Expression of the RR1/1 and 84H10/LB-2 Epitopes on Human Cells and Cell Lines*

Cell Type	Description	RR1/1	84H10
F2B	B cell line	813	1076
DAUDI	Burkitt's lymphoma	918	820
U266	Plasmacytoid	2066	1323
Jurkat	T cell line	64	64
8.2	T cell clone	224	189
PBT	Peripheral blood T cells	4	4
PBA	Peripheral blood	483	288
	adherent cell	205	175
U937	Promomocytoid cell line	2441	3094
L428	Reed-Sternberg cell line		
K562	Erythroleukaemia cell line	672	577
RBC	Human erythrocytes	0	0
HeLa	Cervical carcinoma cell line	700	1258

*Expression of the RR1/1 and 84H10 epitopes was measured by flow microfluorometry as described (21) on 12 indicated cell types. Data were collected on 50,000 cells by using a logarithmic gain. Fluorescence intensity is expressed in the linear units of millivolts. Expression on all cell types was unimodal. Values above 5 mv are positive.

variation between cell types in ICAM-1 molecular weight as described for RR1/1 ranging from 90 to 114 kD (9), LB-2 antigen of 76 kD (15), or 84 kD (16), and the 84H10 antigen of 105 kD (18). This finding is consistent with the interpretation that cell-type-dependent differences in carbohydrate processing account for the variation in molecular weight. With regard to genetics, the gene coding for ICAM-1 maps to chromosome 19 based on prior studies with the LB-2 MAb (15). Finally, studies of cell adhesion using purified ICAM-1 have shown that ICAM-1 mediates LFA-1 dependent adhesion of B, T, and myeloid cells (11,12). The adhesion is inhibited by each of the three ICAM-1 MAbs and the LFA-1 MAb (11). This provides biochemical evidence that ICAM-1 is a ligand for LFA-1 in a variety of leukocyte interactions, and furthermore, that adhesion can be inhibited by each of two distinct epitopes on the ICAM-1 protein.

Acknowledgments. T.A.S. is supported by NIH Grant CA 31798. We thank Dr. J.E.K. Hildreth for the MHM23 MAb; and Drs. A. Singer and A. Rosenberg for constructive criticism.

References

1. Kurzinger K, Reynolds T, Germain RN, Davignon D, Martz E, Springer TA. A novel lymphocyte function-associated antigen (LFA-1): Cellular distribution, quantitative expression, and structure. J Immunol 1981;127:596.
2. Hildreth JE, Gotch FM, Hildreth PD, McMichael AJ. A human lymphocyte-associated antigen involved in cell-mediated lympholysis. Eur J Immunol 1983;13:202.
3. Springer TA, Dustin ML, Kishimoto TK, Marlin SD. The lymphocyte function-associated LFA-1, CD2, and LFA-3 molecules: Cell adhesion receptors of the immune system. Annu Rev Immunol 1987;5:223.
4. Martz E. LFA-1 and other accessory molecules functioning in adhesions of T and B lymphocytes. Hum Immunol 1987; 18:3.
5. Law SKA, Gagnon J, Hildreth JEK, Wells CE, Willis AC, Wong AJ. The primary structure of the β-subunit of the cell surface adhesion glycoproteins LFA-1, CR3 and P150, 95 and its relationship to the fibronectin receptor. EMBO J 1987;6:915.
6. Kishimoto TK, O'Connor K, Lee A, Roberts TM, Springer TA. Cloning of the beta subunit of the leukocyte adhesion proteins: Homology to an extracellular matrix receptor defines a novel supergene family. Cell 1987;48: 681.
7. Takada Y, Strominger JL, Hemler ME. The very late antigen family of heterodimers is part of a superfamily of molecules involved in adhesion and embroyogenesis. Proc Natl Acad Sci USA 1987;84:3239.
8. Hynes RO. Integrins: A family of cell surface receptors. Cell 1987;48:549.
9. Rothlein R, Dustin ML, Marlin SD, Springer TA. A human intercellular adhesion molecule (ICAM-1) distinct from LFA-1. J Immunol 1986;137:1270.
10. Dustin ML, Rothlein R, Bhan AK, Dinarello CA, Springer TA. Induction by IL 1 and Interferon-gamma: Tissue distribution, biochemistry, and function of a natural adherence molecule (ICAM-1). J Immunol 1986;137:245.
11. Makgoba MW, Sanders ME, Luce GE, Marlin SD, Springer TA, Clark EA, et al. ICAM-1: A ligand for LFA-1 in B, T, and myeloid cell adhesion. Nature 1988;331:86.
12. Makgoba MW, Sanders ME, Luce GE, Gugel EA, Dustin ML, Springer TA, et al. Intercellular adhesion molecule-1 (ICAM-1) monoclonal antibody inhibits cytotoxic T lymphocyte recognition. Ann N Y Acad Sci 1988;532:427.
13. Makgoba MW, Sanders ME, Luce GE, Gugel EA, Dustin ML, Springer TA, et al. Functional evidence that intercellular adhesion molecule (ICAM-1) is a ligand for LFA-1 in cytotoxic T cell recognition. Eur J Immunol 1988;18:86.
14. Marlin SD, Miller LM, Springer TA. Inhibition of phorbol ester-induced homotypic aggregation by monoclonal antibodies (Abstract). In: McMichael et al (eds): Leucocyte Typing III. White Cell Differentiation Antigens. New York, Oxford University Press, 1987; p 832.
15. Clark EA, Ledbetter JA, Holly RC, Dinndorf PA, Shu G. Polypeptides on human B lymphocytes associated with cell activation. Hum Immunol 1986;16:100.
16. Patarroyo M, Clark EA, Prieto J, Kantor C, Gahmberg CG. Identification of a novel adhesion molecule in human leukocytes by monoclonal antibody LB-2. FEBS Lett 1987;210:127.
17. Sanchez-Madrid F, Krensky AM, Ware CF, Robbins E, Strominger JL, Burakoff SJ, et al. Three distinct antigens associated with human T-lymphocyte-mediated cytolysis: LFA-1, LFA-2, LFA-3. Proc Natl Acad Sci USA 1982; 79:7489.
18. Hogg N, Horton MA. Joint report of the myeloid section of the III International Workshop on Human Leukocyte Differentiation Antigens. In McMichael AJ et al(eds): Leucocyte Typing III. White Cell Differentiation Antigens. New York, Oxford University Press, 1987.
19. Shaw S, Luce GEG, Quinones R, Gress RE, Springer TA, Sanders ME. Two antigen-independent adhesion pathways used by human cytotoxic T cell clones. Nature 1986; 323:262.
20. Hildreth JE, August JT. The human lymphocyte function-associated (HLFA) antigen and a related macrophage differentiation antigen (HMac-1): Functional effects of subunit-specific monoclonal antibodies. J Immunol 1985; 134:3272.
21. Makgoba MW, Shaw S, Gugel EA, Sanders ME. Human T cell rosetting is mediated by LFA-3 on autologous erythrocytes. J Immunol 1987;138:3587.
22. Marlin SD, Springer TA. Purified Intercellular adhesion molecule-1(ICAM-1) is a ligand for lymphocyte function-associated antigen 1 (LFA-1). Cell 1988;51:813.

Author Affiliations

M.W. Makgoba, M.E. Sanders, G.E. Ginther Luce, S. Shaw, Experimental Immunology Branch, NCI, NIH, Bethesda, MD 20892; *M.L. Dustin, T.A. Springer*, Dana Farber Cancer Institute, Boston, MA 02115; *E.A. Clark*, Regional Primate Center, Washington University, Seattle, WA 98195, USA; *P. Mannoni*, Inserm Unit 119, Marseille, 13009, France

Intrathymic Injection of Fetal and Adult CD4⁻8⁻ Thymocytes: Analysis of Thymic and Peripheral Progeny

C. Guidos, J.L. Weissman, and B. Adkins

We have examined the development of mature peripheral T cells from intrathymically injected fetal or adult $CD4^-8^-$ thymic precursors. Thymic and peripheral progeny of FACS-purified adult or day 14 fetal $CD4^-8^-$ thymocytes were characterized by three-color FACS analysis at various time points following injection into sublethally irradiated, Thy-1 congenic adult mice. The intrathymic progeny of $10^5 - 3 \times 10^5$ adult $CD4^-8^-$ thymocytes represented up to 20% of the total thymocytes at 1 week post-injection, but declined rapidly thereafter. Within the thymus, donor-derived cells rapidly acquired CD4 and CD8 expression; by 1 week post-injection greater than 97% were $CD4^+8^+$. Reconstituted thymic lobes analyzed at 2 weeks contained normal proportions of $CD4^-8^-$, $CD4^+8^-$, $CD4^-8^+$, and $CD4^+8^+$ donor-derived thymocytes. At later time points (3 to 5 weeks), donor-derived cells of the $CD4^+8^+$ phenotype declined in proportion with a concomitant increase in the proportions of $CD4^+8^-$ and $CD4^-8^+$ cells. These results suggest that adult $CD4^-8^-$ cells transiently colonize the thymus, giving rise primarily to $CD4^+8^+$ cells that show a limited intrathymic life span.

Extrathymic progeny of adult $CD4^-8^-$ cells were barely detectable in the peripheral blood and lymph nodes at 1 week post-injection, but increased rapidly to maximal levels (20 to 30% of total T cells) by 3 to 4 weeks. This was followed by a steady decline from 4 to 12 weeks, after which the proportion of donor-derived T-cells (1 to 3% of total T cells) was stably maintained for several months. The peripheral T-cell progeny of $CD4^-8^-$ precursors included $CD4^+8^-$ and $CD4^-8^+$ cells in proportions equivalent to those found among the indigenous host T-cell population in the same animals. In addition to the major $CD4^+8^-$ and $CD4^-8^+$ subsets, a small proportion of donor-derived peripheral T cells were $CD4^-8^-$ or $CD4^+8^+$. The presence of donor-derived $CD4^-8^-$ and $CD4^+8^+$ cells in peripheral lymphoid tissues did not appear to be an artifact of the irradiation or intrathymic injection procedures since 1) similar levels of $CD4^-8^-$ and $CD4^+8^+$ cells were found among the host peripheral T cells in the same animals, and 2) $CD4^-8^-$ and $CD4^+8^+$ T-cells were also detected in the peripheral blood and lymph nodes of normal untreated animals. The peripheral $CD4^-8^-$ cells were all $CD5^{hi}$, and 20 to 30% expressed a subset of T-cell receptor alpha-beta heterodimers reactive with monoclonal antibody F23.1. Donor-derived peripheral T-cells of all phenotypes expressed mature levels of the T-cell receptor-associated CD3 complex. Furthermore, the proportion of $F23.1^+$ donor-derived cells was equivalent to the proportion found among the host T-cell population. Following intrathymic injection, therefore, adult $CD4^-8^-$ thymocytes appear to differentiate normally within the thymus and to export the correct proportions of phenotypically mature progeny.

Similar to adult $CD4^-8^-$ thymocytes, we have also found that day 14 fetal $CD4^-8^-$ thymocytes proliferate, differentiate, and export mature (CD3 and F23.1 positive) progeny to the periphery. However, we observed two major differences in the recolonization patterns of fetal and adult $CD4^-8^-$ thymocytes. First, in the majority of cases (eight of nine), the thymic progeny of fetal $CD4^-8^-$ thymocytes consisted primarily of $CD4^+8^-$ and $CD4^-8^+$ cells at 2 weeks post-injection. Since this phenotypic profile was not observed among progeny of adult $CD4^-8^-$ cells until 3 weeks after injection, it is possible that fetal $CD4^-8^-$ thymocytes differentiate more rapidly than do adult $CD4^-8^-$ cells. Second, in contrast to the normal $CD4^+8^-$:$CD4^-8^+$ ratios (2–5:1) observed among peripheral progeny of adult $CD4^-8^-$ cells, fetal thymocyte-derived peripheral T cells showed abnormal ratios (0.5–1.3:1). Since the $CD4^+8^-$:$CD4^-8^+$ ratios of intrathymic fetal thymocyte progeny were normal, the abnormal extrathymic ratios were likely due to selective export or selective proliferation and/or survival of $CD4^-8^+$ progeny in the periphery, or deceased export, proliferation, or survival of $CD4^+8^-$ cells. These data indicate that there may be a fundamental difference between adult and fetal thymic precursors in the colonization of an adult host.

Author Affiliation

Department of Pathology, Stanford University School of Medicine, Stanford, California, USA

Section XXI
Cell Surface Antigens Other than HLA

Molecular Characterization of Endothelial Monocyte Antigens

Ronald E. Bontrop, Frans H.J. Claas, Wink Baldwin, Nel Otting, Jon J. van Rood, and Marius J. Giphart

Abstract: Previous reports have demonstrated the existence of an alloantigenic system not detectable on resting T and B lymphocytes, which is expressed on renal capillary endothelium and monocytes (1,2). In contrast to resting B lymphocytes, EBV-transformed B cells appeared to express EM antigens as shown by absorption studies. These endothelium monocyte (EM) antigens appear to be important in transplantation immunology since EM antibodies are present in the circulation during acute graft rejection and can be eluted from biopsies taken from a graft that is being rejected (2–6). Although the genetics is not completely clear, EM antigens have shown to be polymorphic (7). However, the molecular localization of these antigens has been a question of debate. Therefore, this was studied by analyzing immunoprecipitates obtained of EBV-transformed B cells using EM alloantisera. Gel electrophoretic techniques showed that EM alloantigenic determinants are located on a subset of HLA-DR molecules.

Microcytotoxicity testing with platelet-absorbed EM alloantisera demonstrated that EBV-transformed B cells express EM antigens (Table 1), whereas resting B cells do not. For that reason, an attempt was made to characterize these EM antigens by use of immunoprecipitation assays and gel electrophoretic techniques. To this purpose, the cell line LUY (HLA-DRw8-Dw8-DQw7) was radiolabeled with ^{35}S-methionine, after which cell extracts were prepared in a lysisbuffer containing NP-40 as a detergent. Subsequent immunoprecipitation and SDS-PAGE analysis demonstrated that both the SC and FO alloantiserum isolated 29 and 34-kD polypeptide chains from the cell extract. Such a profile resembles the SDS-PAGE characteristics of HLA class II molecules.

Serum SC appeared to isolate the highest amount of EM antigen and was therefore selected for further study. To investigate whether EM antigenic determinants are located on HLA class II molecules, and if so, on which type of HLA class II molecules, sequential immunoprecipitation assays were conducted. For this set of experiments, radiolabeled cell extracts were divided into four equal aliquots. As is shown in Figure 1, from the first aliquot, HLA-DR molecules were depleted by extensive immunoprecipitation with monoclonal antibody B8.11.2 (Fig. 1, lanes 1 to 4). Subsequent immunoprecipitation with the SC serum demonstrated that this aliquot still did contain EM alloantigens (Fig.1, lane 5). The second aliquot was depleted for HLA-DQ molecules (lanes 6 to 8), whereas the third aliquot was extensively depleted for HLA-DP molecules (Fig. 1, lands 10 and 11). Also, after removal of HLA-DQ and HLA-DP molecules, both aliquot two (Fig. 1, lane 9) and three (Fig. 1, lane 12) still contained molecules bearing EM allodeterminants. Direct immunoprecipitation with serum SC on the fourth aliquot demonstrated that the amount of EM antigens isolated in this way (Fig. 1, lane 13) equals the amount of EM antigens that can be isolated after removal of HLA-DR (Fig. 1, lane 5), HLA-DQ (Fig. 1, lane 9) or HLA-DP molecules (Fig. 1, lane 12). These experiments suggest that EM alloantigens are not located on HLA-DR, -DQ, or -DP molecules.

The amount of EM alloantigens expressed on EBV-BLCL seems to be a minor amount as compared with HLA-DR molecules (Fig. 1, lanes 1 and 5). Control experiments did demonstrate that EM antigens could be isolated from cell extracts that have been precleared by 1) a mixture of different HLA-DR monoclonal antibodies, 2) by HLA-DR alloantisera, and 3) by a mixture of HLA-DR + DQ + DP locus specific monoclonal antibodies. The only monoclonal antibody tested that seems to be reactive with EM alloantigens, is PdV 5.2, which is directed against a HLA class II framework structure. The EM alloantiserum SC could not isolate EM alloantigens from OOS (HLA-DR1-Dw1-DQw5) and EKR (HLA-DR7-Dw7-DQw2) cell extracts. This is what one should expect since these EBV cell lines do not give a positive reaction in complement-dependent cytotoxicity with this serum (Table 1).

To test which type of HLA class II molecules bear EM alloantigens, two-dimensional gel panel studies were performed. To this purpose EM alloantigens, HLA-DR, -DQ, and -DP molecules were isolated from the cell lines listed in Table 1 and analyzed by two-dimensional gel electrophoresis. The two-dimensional gel profiles of the HLA class II molecules expressed by these cell lines have been described in detail (8–10).

Table 1. Reactivity Pattern of Three EM Alloantisera* with a Panel of EBV-Transformed B-Cell Lines

Reference Cell	HLA Phenotype	EM Sera SC	FO	KR
OOS	DR1-Dw1-DQw5	–	–	–
IWB	DRw15-Dw2-DQw6	5	–	NT
SAS	DRw15-Dw12-DQw6	3	2	2
HAR	DRw17-Dw3-DQw2	3	3	–
QBL	DRw17-Dw3-DQw2	5	2	2
BSM	DR4-Dw4-DQw8	4	2	–
ATH	DRw11-Dw5-DQw7	5	4	4
KRA	DRw13-Dw19-DQw6	5	4	4
BRU	DRw14-Dw9-DQw5	4	3	3
EKR	DR7-Dw7-DQw2	–	3	2
LUY	DRw8-Dw8-DQw7	5	5	NT

*Reaction scores: No lysis; 1–2 = 20 to 40% lysis; 3–4, 60 to 80% lysis; 5 = 100% lysis.

Immunobiology of HLA
Volume II

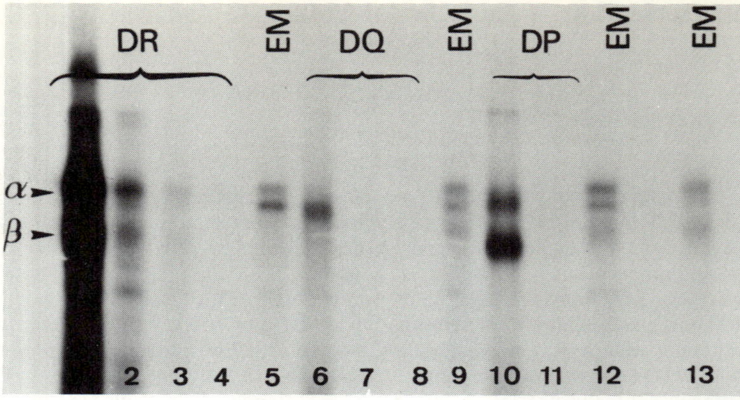

Figure 1. SDS-PAGE analysis of HLA class II and EM antigens isolate from the cell line LUY (HLA-DRw8-Dw8-DQw7). The cell extract was divided into four equal aliquots and HLA class II molecules were isolated by immunoprecipitation or have been described in detail (8–10). The first aliquot was depleted for HLA-DR molecules (lanes 1, 2, 3, and 4), after which EM antigens could be isolated (lane 5). HLA-DQ molecules (lanes 6, 7, and 8) and HLA-DP molecules (lanes 10, 11, and 12) were isolated from the second and third aliquot, respectively. After removal of these molecules, EM antigens could be isolated from both the second (lane 9) and third (lane 12) aliquot. Lane 13 shows the instant isolation of EM alloantigens from the fourth aliquot.

The two-dimensional profiles of the EM alloantigens did not match with the two-dimensional profiles of HLA-DQ and -DP molecules that were isolated from the same cell line. Consequently, it can be concluded that EM allodeterminants detected with the serum SC are not located on HLA-DQ or -DP molecules. Gel electrophoresis did demonstrate that EM antigens are composed of an apparently invariant α chain and a polymorphic β chain. The EM alloantigen two-dimensional profiles perfectly matched with the two dimensional profiles of HLA-DR molecules isolated from the same cell line. This indicates that the EM antigens detected with the serum SC are located on HLA-DR or HLA-DR-like molecules. The fact that HLA-DR alloantisera did not remove EM alloantigens from the cell extracts may be explained by the low titer of these sera. The SC serum was not able to discriminate between DRβI or DRβIII molecules and only small amounts of HLA-DR or DR-like molecules were isolated with the serum. This suggests that not all HLA-DR molecules bear EM allodeterminants. Therefore, the EM allodeterminant could be a determinant on HLA-DR molecules that is not expressed constitutively or which is present on a small subset of HLA-DR molecules. EM alloantigens are expressed by endothelial monocytes, renal capillary endothelium, and EBV-transformed B-cell lines, but not by resting B or T lymphocytes. Therefore, EBV transformation of B cells seems to induce the expression of these EM alloantigens. At present, we do not know whether the EM antigenic structure is a protein or, for example, a sugar determinant. If the latter is true, then the differential tissue distribution could be explained by a difference in posttranslational modification of HLA class II polypeptides. Different cell types are known to express different sorts of glycosilation enzymes and, therefore, postranslational modification may be responsible for the induction of extra heterogeneity in the HLA class II system giving raise for instance to EM antigens.

Acknowledgments. This work was in part supported by the Dutch Foundation for Medical Research (MEDIGON), which is subsidized by the Dutch Organization for the Advancement of Pure Research Protection (zwo) and the J.A. Cohen Institute for Radiopathology and Radiation Protection (IRS).

References

1. Claas FHJ, Paul LC, van Es LA, van Rood JJ. Antibodies against donor antigens on endothelial cells and monocytes in eluates of rejected kidney allografts. Tissue Antigens 1980;15:19.
2. Paul LC, Claas FHJ, van Es LA, Kalff MW, de Graeff J. Accelerated rejection of a renal allograft associated with pretransplantation antibodies directed against donor antigen on endothelium and monocytes. N Engl J Med 1979;300:1258.
3. Baldwin III WM, Soulilou JP, Claas FHJ, Peyrat MA, van Es LA, van Rood JJ. Antibodies to endothelial antigens in eluates of 88 human kidneys: Correlation with graft survival and presence of T- and B-cell antibodies. Transpl Proc 1981;XIII:1547.
4. Cerilli J, Brasile L, Galouzis LT, de Francis ME. Clinical significance of antimonocyte antibody in kidney transplant recipients. Transplantation 1981;32:495.
5. Stastny P. Endothelial-monocyte antigens. Transpl Proc 1980;12:(suppl.1):32.
6. Thompson JS, Overlin VL, Herbick JM, Severson CD, Claas FHJ, D'Amaro J, et al. New granulocyte antigens

demonstrated by microgranulocytotoxicity assay. J Clin Invest 1980;65:1431.
7. Moraes JR Stastny P. Eight groups of human endothelial cell alloantigens. Tissue Antigens 1976;8:273.
8. Bontrop RE, Schreuder GMTh, Elferink DG, Mikulski MMA, Geerse R, Giphart MJ. Molecular, serologic and functional evidence for an apparent HLA-DR triplet. J Immunol 1986;137:211.
9. Bontrop RE, Baas EJ, Otting N, Schreuder GMTh, Giphart MJ. Molecular diversity of HLA-DQ. DQ alpha and beta chain isoelectric point differences and their relation to serologically defined HLA-DQ allospecificities. Immunogenetics 1987;25:305.
10. Baas EJ, Bontrop RE, Otting N, Giphart MJ. Molecular analysis of HLA-DP specificities HLA-DPw1, -DPw2 and -DPw4: DP beta chain heterogeneity correlates with PLT subtyping. Tissue Antigens 1988;31:5.

Author Affiliations

Ronald E. Bontrop, Frans H.J. Claas, Nel Otting, Jon J. van Rood, Marius J. Giphart, Department of Immunohematology and Blood Bank, University Hospital Leiden, 2300 RC Leiden, The Netherlands; *Wink Baldwin*, Pathology Department, Duke University Medical Center, Durham, NC 27710, USA

Stimulation of Growth and Differentiation of EBV-Transformed Cell Lines by Use of Monoclonal Antibodies to New Differentiation Antigens

Nicole Suciu-Foca, Simon Chen, Elaine Reed, Ji Dong Cai, Pablo Rubinstein, and Donald W. King

The commitment of hematopoetic stem cells to the B-cell lineage and the development of Ig-producing cells involves the rearrangement of Ig genes and the translation and membrane expression of their product. Resting Ig-expressing B lymphocytes are activated by cross-linking of membrane Ig by antigen or anti-Ig (isotype, allotype, or idiotype-specific) antibodies and triggered to proliferate and differentiate by a variety of cytokines (1,2).

Activated T and B lymphocytes express on their surface new differentiation antigens that are absent from the membrane of resting cells. Some of these antigens serve as receptors for hormones, lymphokines, monokines, and cell-cell interaction signals which regulate the growth, differentiation, and network of communications within and between the T-cell and B-cell compartment. To investigate the nature of such interactions we developed a series of murine monoclonal antibodies (MAb) that react with activated, but not with resting T and B lymphocytes. Two of these reagents, MAb NDA_3 and MAb NDA_4, identify activation antigens that are expressed by human T lymphocytes relatively late following activation and are also present on normal activated B cells and Epstein-Barr virus (EBV)-transformed B cells (3,4).

The antigen recognized by MAb NDA_3 on the surface of activated B cells is a 36 kD glycoprotein, whereas the antigen detected by MAb NDA_4 is a 46 kD molecule. To determine whether MAb NDA_3 and MAb NDA_4 recognize differentiation antigens involved in the growth and maturation of activated B cells, the effect of these antibodies on the rate of growth and Ig synthesis exhibited by EBV-transformed lymphoblastoid cell lines (LBCL) was tested. Cells from different LBCL were suspended in culture medium containing different concentrations of MAb (1 to 10 μg/ml) and seeded at low density (2.5×10^3 cells/culture) in 96 well trays. After 24, 48, and 72 hours, cells were labeled with 3H-Td R, and the amount of radiolabel incorporation was measured by liquid scintillation counting. At all time intervals, the rate of proliferation was significantly higher in cultures grown with MAb NDA_3 or MAb NDA_4 compared with cultures grown in medium only or with irrelevant MAb used as isotype controls. The stimulating effect of both MAb on cell proliferation was dose-related over the range of 1 to 10 μg/ml with a plateau reached at higher concentrations. Quantitation of the amount of Ig in culture supernatant by ELISA also showed a significant increase in secreted Ig.

We next compared LBCL treated with MAb NDA_3 or NDA_4 and control populations for growth parameters. A striking difference between treated and control cells involved a change in the pattern of growth in liquid cultures. In the presence of MAb NDA_3 or MAb NDA_4, cells grew in fine clumps or as individual cells, whereas control cultures maintained the characteristic pattern of growth consisting in the formation of large, microscopically detectable clumps. This difference is significant because it indicates a decreased level of dependency on locally produced growth factors, as also observed in the presence of exogenous BCGF. Futhermore, when purified Ig of MAb NDA_3 or MAb NDA_4 was added to the cultures (at 1 to 5 μg/ml), the amount of fetal calf serum required to sustain growth could be lowered to 1%. Next we investigated the in vitro cloning efficiency of cells propagated in the presence of MAb NDA_3 and NDA_4. There was a 7- to 10-fold increase in cloning efficiency when LBCL were seeded at <1 cell/well in medium containing MAb NDA_3 or MAb NDA_4.

Certain lymphokines such as IL-4 stimulate Ig heavy chain class switching. To determine whether MAb NDA_3 and NDA_4 have a similar effect, we analyzed B-cell clones propagated in the presence of these MAb with respect to the isotype of membrane and secreted Ig. The line used in these experiments was SKW 6.4 which produces only membrane IgM. Cytofluorographic analysis of the Ig phenotype showed that 12% of the subclones grown with NDA_3 and 15% of clones grown with NDA_4 express both membrane IgM and IgG. Cytoplasmic staining revealed the presence of IgG secreting plasma cells within these clones and ELISA determinations showed that supernatants from some clones contained high levels of IgG.

To further analyze the nature of the changes induced by MAb NDA_3 and MAb NDA_4 in LBCL we studied, by Northern blots, the level of Ig heavy chain transcription in LBCL grown with and without MAb. For this, 20 μg of total RNA isolated from cells incubated with or without MAb NDA_3 or NDA_4 were denatured in formaldehyde and subjected to electrophoresis in 1% or 1.4% agarose/formaldehyde gels. After electrophoresis, gels were stained with ethidium bromide to identify the 18S and 28S ribosomal (r) RNA. The staining of 18S (2.0 kb) and 28S (4.7 kb) rRNA subunits was used as an internal standard for the quantities of total RNA as well as for molecular weight markers. The RNA was transferred to nitrocellulose filters and hybridized to 32P-labeled nick-translated probes. The probes used were an EcoR1 fragment of clone CH4-51 which contains the Ig μ-chain gene; a Stul-Pvul fragment of about 300 basepairs, which is specific for the Ig-γ 1 gene; and BglII-Pvul fragment of about 660 basepairs which is specific for the Eg γ 3 gene.

Northern blot analysis using RNA extracted from SKW 6.4 LBCL incubated with MAb NDA_3 or with MAb NDA_4 for 24 hours showed 300 to 400% increase in the rate of mRNA transcription.

Analysis of two other B-cell lines showed, however, that while the level of u transcription was not significantly changed, the level of γ 1 transcription was increased by 310 and 340%, respectively. There were no changes in the level of γ 2 and γ 3 expression or in the transcription of RNA complementary to the cDNA probe for actine and for the α chain of HLA-DR that were used as controls.

Taken together these data indicate that MAb NDA_3 and MAb NDA_4 stimulate the proliferation and maturation of EBV-transformed lymphoblastoid cells line.

Acknowledgment. This work was supported by NIH grants AI 25210 and HL 36581.

References

1. Zola H. The surface antigens of human B lymphocytes. Immunology Today 1987;8:308.
2. Gordon J, Guy GR. The molecules controlling B lymphocytes. Immunology Today 1987;8:339.
3. Suciu-Foca N, Rohowsky-Kochan C, Rosochacki SJ, Cristea E, Cai J, Reed E, et al. NDA_3: A differentiation antigen associated with the receptor for B cell growth factor. Immunol Res 1986:165.
4. Suciu-Foca N, Cai J, Gutierrez C, Reed E, Morrision S, Pernis B, et al. New differentiation antigens associated with the growth and maturation of B lymphocytes. J Immunol 1988;140;395–403.

Author Affiliations

Nicole Suciu-Foca, Simon Chen, Elaine Reed, J. Dong Cai, Pablo Rubinstein, Department of Pathology, College of Physicians and Surgeons of Columbia University, New York, NY; *Donald W. King*, Department of Pathology, University of Chicago, Chicago, IL, USA

Characterization of a Human B-Lymphocyte Antigen Associated with the Receptor for BCGF-12 kD

Nicole Suciu-Foca, Simon Chen, Garbor Hirka, Elaine Reed, Pablo Rubinstein, and Donald W. King

Abstract: The transition of B cells from the resting to the activated state is accompanied by the expression of receptors for growth and maturation factors. Some of these factors are produced by activated monocytes (e.g., IL-1, TNF-2), whereas others are products of activated T cells (e.g., IL-2, IL4, IL5, IL6 interferon γ human BCGF-12kd), or of activated B cells (1–5). Although the molecular structure of most of these cytokines and of their corresponding genes have been elucidated, little is known about the structure of cytokine receptors other than IL-2. In an attempt to unravel the structure of receptors for growth and maturation factors, we developed a series of murine monoclonal antibodies that detect new differentiation antigens (NDA) expressed by T and/or by B lymphocytes following activation (6–9). We now present evidence that a 36 kD glycoprotein recognized by one of these monoclonal antibodies named "MAb NDA_3" is the receptor for BCGF-12 kD. The evidence includes the demonstration that MAb NDA_3: 1) stimulates the growth of normal activated B cells and of B cells transformed by EBV infection, mimicking the effect of BCGF; 2) acts as a competitive antagonist of BCGF-12 kD stimulated B-cell proliferation; 3) is prevented from binding to the membrane of activated B cells by BCGF-12 kD; and 4) competitively inhibits ^{125}I-BCGF-12 kD binding to the cell membrane.

Immunobiology of HLA
Volume II

Table 1. Growth-Promoting Activity of MAb NDA_3

Responding B Blasts	^{3}H-TdR Incorporation (Mean CPM) in Cultures Containing: Medium	BCGF	MAb NDA_3	MAb NDA_3
SAC-activated	22,570	64,720	76,380	20,430
LBCL-SKW 6.4	18,520	120,480	125,426	20,640

PBL was obtained from fresh peripheral blood by Ficoll-Hypague gradient centrifugation, followed by monocyte depletion through adherence to plastic for 16 hours at 37°C. B cells were purified on a nylon wood column and high density cells were obtained by centrifugation on a discontinuous Percoll gradient (30, 50, and 100%) at the interphase between the 50 and 100% layer. The purity of the B-cell suspension was >96% as determined by immunofluorescent analysis of surface immunoglobulin staining. Resting high density B cells were suspended in RPMI-1640 medium supplemented with glutamine, antibiotics, and 10% fetal calf serum (10% FCS) and plated at 2×10^6 cells per well in 24 flat-bottom wells. SAC activation was induced by adding to the cultures 1 ml of 0.1% SAC. Forty-eight hours later, blasts were collected on a Percoll gradient, washed and plated at 1×10^5 cells/well. Semipurified commercial BCGF was obtained from Cellular Products, Buffalo, N.Y. and used at a final concentration of 5 or 10% (v/v). MAb NDA_3 and MAb NDA_9 were used at a final concentration of 10 μg/ml. Purified Ig from MAb was free of mycoplasma as determined by hybridization with 3H-DNA probe homologous to Mycoplasma and Acholeplasma ribosomal RNA (Mycoplasma TC Detection Kit, Gen-Probe, San Diego, CA). There were no endotoxins in the Ig solution as measured by use of the Limulus Amebocyte Lysate Pyrotell test (Associates of Cape Cod, Inc., Woods Hole, MA).

All cultures were labeled with 3H-TdR after 48 h and harvested 18 hours later. Results are expressed as mean cpm of triplicate reactions. The difference between triplicates was less than 10%.

Table 2. Inhibition of NDA_3 Expression on the Membrane of LBCL-IC by MAb NDA_3

LBCL IC Preincubated in:	Mean Fluorescence Intensity NDA_3	Ia	Transferrin Receptor
PBS	5,207	12,165	2,860
MAb NDA_3	1,297	12,268	2,936
BCGF-12kd	1,544	11,221	3,048
BCGF-SP	1,656	13,432	2,876
IL_2	4,883	12,559	2,754

LBCL-IC cells were incubated for 1 hour at 4°C with MAb NDA_3 (20 μg/ml), BCGF-12 kd (8 μg/ml), Semipurified (SP) commercial BCGF (15%) or interleukin 2 (15%). Cells were stained with FITC-conjugated MAb NDA_3, MAb L_{243}, or MAb NDA_9. Mean fluorescence intensity was calculated from the product between the percent positive cells and the mean channel of fluorescence minus the values obtained for the negative control.

The cell surface distribution of the antigen recognized by MAb NDA_3 was investigated by direct immunofluorescence using an Ortho spectrum III cytofluorometer. This antigen is not present on the membrane of resting T cells, B cells, or monocytes obtained from fresh peripheral blood, lymph nodes, spleen, or tonsils (7,9). In vitro activation of T cells with plant mitogens, allogeneic HLA-DR, or soluble antigens (tetanus toxoid or purified protein derivative of Mycobacterium tuberculosis) induced the expression of NDA_3 on a fraction (10 to 40%) of the population of responding T lymphocytes. Similarly, B cells stimulated with SAC, PWM, anti-μ, and BCGF or with EBV express NDA_3 at the peak of the blastogenic response.

MAb NDA_3 had no significant effect on blastogenic responses of T cells to mitogens, soluble antigens, or allogeneic stimulating cells. This antibody inhibits, however, the response of resting, high density B cells to SAC and to BCGF.

In contrast to resting B cells, SAC-activated B blast cells or EBV-transformed B cells are stimulated to proliferate when MAb NDA_3 is added to the cultures. The magnitude of this growth-enhancing effect is of the same order as that of BCGF (Table 1). Thus, similar to certain MAb reacting with hormone receptors, MAb NDA_3 mimics the effect of the ligand by stimulating cell proliferation.

To determine whether MAb NDA_3 competes with BCGF- for binding to the same receptor site, B cells from the monoclonal LBCL SKW 6.4 were seeded at low density (2.5×10^3 cells/ml) and grown for 18 hours in medium containing different concentrations of BCGF- and MAb NDA_3. There was significant stimulation of growth in the presence of BCGF. The response to BCGF was dose related with a plateau reached at 5% (v/v). Similarly, the proliferative response triggered by MAb NDA_3, increased over a dose of 1 to 5 μg/ml reaching a plateau at higher concentrations. When BCGF and MAb NDA_3 were added together to the cultures, at optimal concentrations, there was no additive effect.

Furthermore, cells preincubated with MAb NDA_3 (at 10 μg/ml) for 1 hour at 4°C showed no enhanced proliferation when grown in medium containing BCGF, suggesting that BCGF and NDA_3 compete for the same receptor site.

The growth-promoting activity of the semi-purified, commercial BCGF has been reported to reside in its 12 kD fraction. The possibility that the BCGF-12 kD fraction of commercial BCGF blocks binding of MAb NDA_3 to its epitope was explored by competitive binding studies.

Cytofluorometric analysis of cells from LBCL-IC showed that the percent of the cells reacting with FITC-conjugated MAb NDA_3 and the mean channel of fluorescence decreased significantly (70%) when the cells were preincubated for 2 hours at 4°C or for 16 hours at 37°C with BCGF-12 kD (2 μg/ml) semipurified commercial BCGF (5%) or "cold," i.e., non-fluorescinated MAb NDA_3 (20 μ/ml). Interleukin 2 and interferon α, β, and γ, used as specificity controls did not inhibit binding of MAb NDA_3 to the cell membrane (9). Similarly, BCGF-12 kD did not alter the cell surface expression of the transferrin receptor and of Ia antigens, as determined

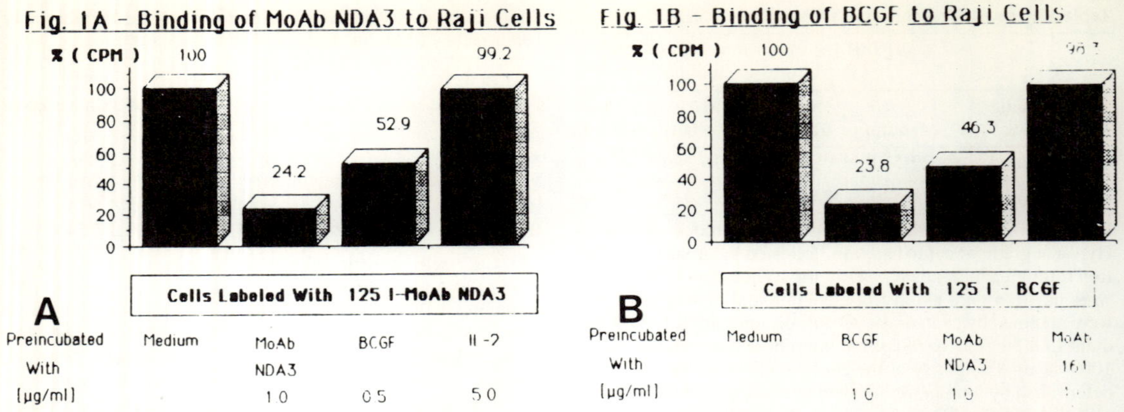

Figure 1. *A*. Binding of ^{125}-I-labeled MAb NDA_3 to LBCL and blocking by human BCGF-12 kD. *B* Binding of ^{125}-I-labeled BCGF-12 kD and blocking by MAb NDA_3. MAb NDA_3 was purified from culture supernatants by gel filtration through sephacryl S-300 superfine (Pharmacia). The flow-through fraction was tested by ELISA using anti-mouse u to determine the concentration of IgM. The fraction of NDA_3 that may be inhibited from binding by an excess of "cold" NDA_3 was determined by adding to a standard number of lymphoblasts (10^6) a standard quantity of labeled antibody 0.06 μg, containing 5×10^5 CPM alone or in the presence of 1 or 10μ of cold antibody. The inhibitable or "specific" fraction of the labeled reagent was 50% of the total CPM bound. Inhibition by cold antibody, by BCGF, and by an unrelated monoclonal antibody of the same (IgM) isotype, MAb 161, is expressed as percent binding (CPM) compared to cells preincubated in medium only. For these experiments, MAb NDA_3 was purified as above and BCGF was separated by FPLC on a superose 12 column (Pharmacia) from a stock kindly provided by Dr. Mehta (5). Purity of the active fraction, $M_r = 12$ kd, was monitored by SDS-PAGE (12.5% gels) and its concentration by spectroflurometry (OD=280) (5). MAb NDA_3 and BCGF were radiolabeled using the Enzymobead radioiodination reagent (Bio-Rad, Richmond, VA) essentially following the manufacturer's specifications. Binding of labeled NDA_3 or BCGF was conducted using 5×10^5 cells in the presence (or absence) of the inhibitor indicated. Volume was kept constant (1 ml) with RPMI-1640 containing lyzoyme (0.1%). All experiments were conducted in triplicate and results are expressed as the respective means. All binding and inhibition experiments were conducted at 4°C and each incubation phase was 2 hours.

by use of FITC-conjugated MAb NDA_9 and MAb L243, respectively (Table 2).

The possibility that the antigen recognized by MAb NDA_3 is structurally identical with or at least sterically related to the receptor for human BCGF-12 kD was further tested by incubating B cells from the RAJI cell lines with either BCGF or with MAb NDA_3 and examining their ability to bind radiolabeled MAb NDA_3 or radiolabeled BCGF. The results (Fig. 1) are consistent with this hypothesis, since the binding of the labeled MAb NDA_3 was significantly reduced in the presence of BCGF-12 kD. Cold MAb NDA_3 and MAb 161, used as an isotype control (IgM) had inhibitory and neutral effects, respectively, on the binding of ^{125}I-MAb NDA_3. Similarly, IL-2 was not inhibitory (data not shown). From the specific activity of the labeled antibody (6.9×10^6 cpm/μg) and the amount specifically bound, we calculated the number of sites to be $2.5 \pm 0.2 \times 10^4$. Scatchard analyses (not shown) yielded a remarkably close estimate, 27,000 sites per cell, for the Raji cell line. These estimates are higher than those obtained by Mehta (5) for the high-affinity receptors (4 to 8×10^3) using the natural ligand BCGF-12 kD. Low-affinity receptors were found to be more numerous, however. Our estimates are based on an assumed antibody valence of 1, which may not be numerically accurate given the pentameric nature of IgM: the proximity of the receptor sites on the cell surface will determine whether they lie within bridging distance by IgM. Additionally, the number of BCGF-12 kD receptor sites varies considerably within the same B-cell line in a manner apparently unrelated to the cell cycle (data not shown). These variations may depend on the autocrine production of growth factors.

Thus, the binding of MAb NDA_3 to its target molecules can be easily and specifically blocked by BCGF in both native solutions and purified forms. This indicates that the NDA_3-binding molecule is the putative BCGF receptor. To further characterize this receptor, we determined whether binding of ^{125}I-BCGF to the membrane of Raji cells is blocked by MAb NDA_3. Results obtained in this set of investigations, showed that MAb NDA_3 blocks the binding of BCGF-12 kD. There was no inhibition of BCGF-12 kD binding when cells were preincubated with anti-Ia or with anti-transferrin receptor antibodies. Taken together, our results indicated that the 36kd molecule identified by MAb NDA_3 is the human B-cell receptor for BCGF-12 kD. a factor that stimulated B-cell proliferation.

Acknowledgment. This work was supported by NIH grants AI 25210 and HL 36581.

References

1. Kehrl JH, MIller A, Fauci AS. Effect of tumor necrosis factor a on mitogen-activated human B cells. J Exp Med 1987;166:786–791.
2. Taga T, Kawanish Y, Hardy RR, Hirano T, Kishimoto T. Receptors for B cell stimulatory factor a. Quantitation, specificity, distribution and regulation of their expression. J Ex Med 1987;166:967–981.
3. Buck J, Hammerling U, Hoffman MK, Levi E, Welte K. Purification and biochemical characterization of a human autocrine growth factor produced by Epstein-Barr virus transformed B cells. J Immunol 1987;138:2923–2928.
4. Gordon J, Guy GR. The molecules controlling B lymphocytes. Immunology Today 1987;8:339.
5. Mehta SR, Grant SR, Maizel AL. Characterization of the cell surface receptors for human B cell growth factor of 1,000 molecular weight.
6. Suciu-Foca N, Reed E, Rubinstein P, Mackenzie W, Ng WA, King DW. A late differentiation antigen associated with the helper inducer function of human T cells. Nature 1985; 138:465.
7. Suciu-Foca N, Rosochacki C, Cai J, Reed E, Rubinstein P, King DW. NDA_3: a differentiation antigen associated with the receptor for B cell growth factor. Immunol Res 1986;5:16.
8. Suciu-Foca N, Rosochacki SJ, Cai J, Reed E, Rubinstein P, King DW. Immunological and biochemical characterization of an epitope of the transferrin receptor involved in the production of gamma interferon and B cell growth factor. Cellular Immunol 1987;110:265–281.
9. Suciu-Foca N, Cai J, Gutierrez C, Reed E, Morrison S, Pernis B, et al. New differentiation antigens associated with the growth and maturation of B lymphocytes. J Immunol 1988;140:395–403.

Author Affiliations

Nicole Suciu-Foca, Simon Chen, Garbor Hirka, Elaine Reed, Pablo Rubinstein, Department of Pathology, College of Physicians and Surgeons of Columbia University; New York, NY; *Donald W. King*, Department of Pathology, University of Chicago, Chicago, IL, USA

Section XXII
Histocompatibility Antigens in Nonhuman Species

Evolutionary Analysis of MHC Genes in Nonhuman Primates

Hajime Takata, Mie Haranaka, Osamu Takenaka, Nobukatsu Miwa, Yuichi Murayama, and Hisami Iri

Restriction enzyme-cleaved genomic DNAs from nonhuman primates were analyzed by the Southern hybridization with 12 probes encoding HLA class II and III proteins. These probes were contributed to this workshop, in addition to the class I probe (pC250) from Cw3 genomic clone described by T. Strachan et al. (1). The latter seems to be HLA-C specific in high stringency condition (2).

Fifteen species were examined in addition to prosimians: common tupai (treeshrew) (*Tupaia glis*) and ring-tailed lemur (*Lemur catta*); New World monkeys consisting of common marmoset (*Callithrix jacchus*), cotton-topped tamarin (*Saguinus oedipus*), owl monkey (*Aotus trivirgatus*), squirrel monkey (*Saimiri sciureus*) and white-throated monkey (*Cebus capucinus*); Old World monkeys consisting of rhesus macaque (*M. mulatta*), pig-tailed macaque (*M. nemestrina*), bonnet macaque (*M. Radiata*), crab-eating macaque (*M. fascicularis*), Japanese macaque (*M. fuscata*), hamadryas baboon (*Papio hamadryas*); and hominoids consisting of orangutan (*Pongo pygmaeus*) and chimpanzee (*Pan troglodytes*).

Enzyme-cleaved genomic DNAs from all species tested, except prosimians, were strongly hybridized with all probes used, suggesting high homology among HLA and their MHCs. For common tupai and ring-tailed lemur, however, the faint bands were observed in the high stringency hybridization condition with class I probe (Table 1).

On the other hand, the class III probe showed the different hybridization patterns among prosimians, New World's, Old World's, hominoids, and humans, which appeared to reflect the difference of class III gene evolution in the primates.

Genomic DNAs from the harems of Japanese macaque were digested with EcoRI, Pst I, BamHI, HindIII, and Taq I and hybridized to class II probes. When DRβ, DQα, DQβ, DPα, and DPβ were used as probe, the polymorphism, which appeared to correlate with haplotypes of their MHC gene and resemble HLA genes in the pattern, were detected. In the case of small families, segregated haplotypes were identified on the DNA digested by EcoRI, Pst I, and Taq I using DQβ, DPα, and DPβ probes.

These results suggest that human MHC genes are well conserved in the species of nonhuman primates above prosimians, and, at least in Japanese macaque, MHC gene structure is quite similar to that of humans.

References

1. Strachan T, Dodge AB, Smillie D, Dyer PA, Sodyer R, Jordan BR, et al. An HLA-C-specific DNA probe. Immunogenetics 1986;23:115.
2. Smeaton I, Summers CW, Harris R, Strachan T. Restriction fragment length polymorphism at the HLA-C locus. Immunogenetics 1987;25:179.

Author Affiliations

Hajime Takata, Mie Haranaka, Hisami Iri, Department of Clinical Laboratories, Keio University School of Medicine, Tokyo; *Osamu Takenaka, Nobukatsu Miwa, Yuichi Murayama*, Department of Biochemistry, Primate Research Institute, Kyoto University, Inuyama, Japan

Table 1. Summary of the Southern Hybridization of the DNA from Primates with Human MHC Class I, II and III Probes

Primates (number)	Class I_α	DR_α	DR_β	DQ_α	DQ_β	DP_α	DP_β	DO_α	DZ_β	C2	C4	Bf	21-OH
Hominoids													
Chimpanzee (3)	+	+	+	+	+	+	+	+	+	+	+	+	+
Orang-utan (1)	+	+	+	+	+	+	+	+	+	+	+	+	+
Old World's													
Baboon (2)	+	+	+	+	+	+	+	+	+	+	+	+	+
Japanese macaque (2)	+	+	+	+	+	+	+	+	+	+	+	+	+
Crab-eating macaque (2)	+	+	+	+	+	+	+	+	+	+	+	+	+
Bonnet macaque (2)	+	+	+	+	+	+	+	+	+	+	+	+	+
Pig-tailed macaque (2)	+	+	+	+	+	+	+	+	+	+	+	+	+
Rhesus macaque (2)	+	+	+	+	+	+	+	+	+	+	+	+	+
New World's													
White-throated monkey (1)	+	+	+	+	+	+	+	+	+	+	+	+	+
Squirrel monkey (1)	+	+	+	+	+	+	+	+	+	+	+	+	+
Owl monkey (3)	+	+	+	+	+	+	+	+	+	+	+	+	+
Cotton-topped tamarin (1)	+	+	+	+	+	+	+	+	+	+	+	+	+
Common marmoset (1)	+	+	+	+	+	+	+	+	+	+	+	+	+
Prosimians													
Ring-tailed lemur (1)	∓	+	+	+	+	+	+	±	+	+	+	+	+
Common tupai (1)	∓	+	+	±	+	+	+	±	±	+	+	+	+

+ = Strongly hybridized; ± = hybridized weakly; ∓ = hybridized, but very weakly.

The Detection of Feline Class I and Class II MHC Homologue Molecules and Genes by Biochemical and Molecular Genetic Techniques

Marilyn S. Pollack, Richard G. Cook, Audrey A. Hayes, Samatha Mooney, Neil C. Pedersen, J. David Holcomb, and J. Fielding Hejtmancik

Although we previously reported that cats apparently have MHC systems with limited polymorphism because they have relatively low MLC responses and fail to develop lymphocytotoxic antibodies in response to transfusion or pregnancy that can be used for conventional MHC typing (1), the presence on feline cells of class I and class II I-E-type MHC homologues has now been demonstrated using cross-reacting monoclonal antibodies (MoAb) and DNA probes.

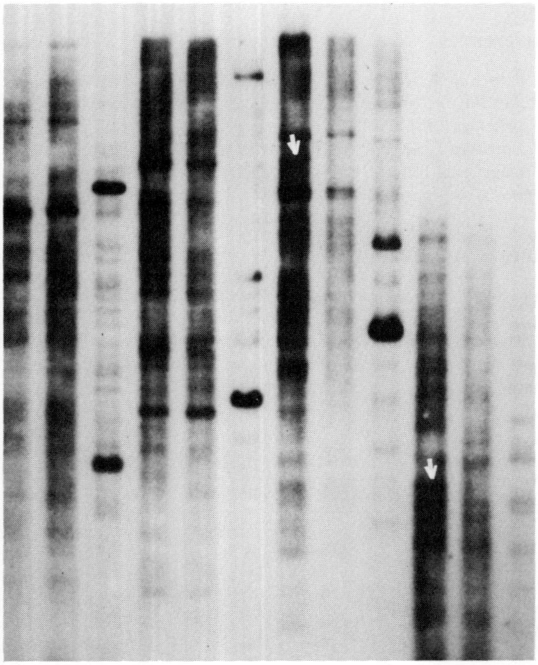

Figure 1. Autoradiograph showing four different restriction enzyme digestion fragments of DNA hybridizing with a human DR α probe. Lanes 1, 2, and 3 = Pst I; lanes 4, 5, and 6 = Hind III; lanes 7, 8, and 9 = EcoR I; lanes 10, 11, and 12 =Taq I. Lanes 1, 4, 7, and 10 have digested DNA extracted from a feline lung fibroblast cell line; lanes 2, 5, 8, and 11 have DNA extracted from an unrelated feline tongue fibroblast cell line; and lanes 3, 6, 9, and 12 have DNA from a human (control) cell line. The cat lines' DNA bands differ somewhat from those found for the human DNA, but also show some clear differences in pattern from each other (examples marked with arrows).

The feline class I antigen homologues were detected with both immunofluorescence (IF) and one- and two-dimensional gel electrophoresis techniques using the anti-human class I MAb W6/32 on in vitro cultured feline fibroblasts and lymphoid cells, but not on fresh lymphoid cells. The association of bovine β-2 microglobulin with feline class I heavy chains apparently generates the determinant recognized by the MAb W6/32 (e.g., [2]) because a long chase period was required to see biosynthetically labeled class I heavy chains with an *absence* of labeled β-2 microglobulin, but a definite 12K band was seen in surface iodinated preparations. Neefjes et al. (3), who also demonstrated the presence of class I MHC homologues on feline cells with the same W6/32 antibody, may have incorrectly concluded that there is a relative scarcity of class I antigens on nonstimulated cat PBL because they did not appear to realize that β-2 microglobulin exchange was required and did not allow enough time.

Feline class II I-E-like molecules were detected with IF techniques using the species cross-reactive anti-mouse I-E antibody 40D (courtesy D. Sachs) only when peripheral blood mononuclear cells were activated, for example with mitogens. Although the report by Neefjes et al. (3) indicates that resting feline cells express large amounts of conventional class II MHC molecules, they used antibodies that would also react with I-A-like molecules and the difference between their studies and ours may relate to a relatively high expression of I-A-like versus I-E-like class II molecules on resting feline cells. Limited polymorphism and/or this unusual regulation of expression of different class II molecules may explain precious studies by ourselves (see above) and others (e.g., the relative susceptibility of cats to viral leukemia and other infections, and the failure of cats to develop cytotoxic lymphocytes after immunization with virally infected allogenic cells [4]) indicating the relative lack of functional polymorphism of the feline MHC. Our human cell studies of γ interferon induced fibroblast class II antigens indicate that these cells have DR (I-E-like molecules) but frequently lack expression of DQ (I-A-like molecules) and that this apparently affects their functional properties (e.g., [5]). Cells lacking I-E-like molecules may have similar functional deficits.

Both class I and class II feline MHC determinants have now also been demonstrated with cross-reactive DNA probes. Initial studies were performed using four different enzymes digest: Pst I, Hind III, EcoR I, and Taq I, and hybridization bands were detected in all cases. Although no polymorphic bands, different between two

Immunobiology of HLA
Volume II

unrelated cats' fibroblast DNA preparations, were identified for the single class I probe used to date, DNA probe HG28 (pJY 150-RI-(HLA-B7)) (data not shown) bands in different positions were identified using the class II probes as illustrated in Figure 1. These cross hybridization techniques can thus be used to identify polymorphic feline MHC DNA segments and continuation and expansion of these studies is currently in progress.

Acknowledgments. This work was supported in part by the Robert H. Winn Foundation for Cat Research, The Houston Cat Club, and by NIH grants CA 40552 and AI 17897.

References

1. Pollack MS, Mastrota F, Chin-Louie J, Mooney S, Hayes A. Preliminary studies of the feline histocompatibility system. Immunogenetics 1982;16:339.
2. Kahn-Perles B, Boyer C, Arnold B, Sanderson AR, Ferrier P, Lemmonnier FA. Acquisition of HLA class I W6/32 defined antigenic determinant by heavy chains from different species following association with bovine beta β_2-microglobulin. J Immunol 1987;138:2190.
3. Neefjes JJ, Hensen EJ, deKroon TIP, Ploegh HD. A biochemical characterization of feline MHC products: Unusually high expression of class II antigens on peripheral blood lymphocytes. Immunogenetics 1986;23:341–347.
4. McCarthy J, Grant CK. Cytotoxic effector cell populations in cats. In: Hardy Jr. WD, Essex M, McClelland AJ; (eds): International Feline Leukemia Virus Meeting. New York, Elsevier/North Holland, 1980, pp 213.
5. Maurer DH, Hanke JH, Mickelson E, Rich RR, Pollack MS. Differential presentation of HLA-DR, DQ, and DP restriction elements by interferon-γ-treated dermal fibroblasts. J Immunol 1987;139:715–723.

Author Affiliations

Marilyn S. Pollack, J. David Halcomb, J. Fielding Hejtmancik, Baylor College of Medicine; *Richard C. Cook,* Baylor College of Medicine and Howard Hughes Medical Institute, Houston, TX; *Audrey A. Hayes, Samatha Mooney,* The Animal Medical Center, New York, NY; *Neil C. Pedersen,* School of Veterinary Medicine, University of California, Davis, CA

Section XXIII
New Techniques

New Strategies for the Production of Monoclonal Antibodies Against MHC-Encoded Antigens

Barbara Zimmer, Ilias Doxiadis, Jutta Trenkmann, and Hans Grosse-Wilde

In the past, monoclonal antibodies (Mabs) against MHC-encoded antigens have been produced by immunization of mice with whole lymphoid cells or purified antigens (1). Here, we present a simple and very effective immunization technique, yielding highly specific MAbs reacting against MHC products under denaturated conditions.

The technique is based on the immunization of mice with immobilized proteins bound on nitrocellulose paper. To this, 10×10^6 cells of the human lymphoblastoid cell line 9W0404 SUD (HLA-A2, B44, Cw5, DR4, DRw53, DQw3, Dw4)[2] (2) were lysed in Triton-X 114-containing buffer, and separated by sodium dodecyl sulfate polyacrylamide gel electrophoresis (SDS-PAGE). After transfer of the proteins onto nitrocellulose paper (NC), areas of interest within the MHC products—determined by their molecular weight—were marked on each side of the blot, at approximately 44 kd for the heavy chain of class I antigens, and at approximately 28 kd for the beta-chains of class II products. The antigen-bearing NC-bands were cut out, pulverized mechanically in liquid nitrogen, resuspended in phosphate-buffered saline, and injected intraperitoneally into Balb/c mice. Fusion of the removed mouse spleen cells with the mouse myeloma cell line 8653 was performed according to standard protocols.

Seventy-five percent of the growing clones showed specific antibody production against the MHC-encoded antigens, as shown by various screening procedures (e.g., enzyme linked immunosorbent assays [ELISA], dot blotting technique, SDS-PAGE, and one- and two-dimensional isoelectric focusing gels [1D-IEF, 2D-gels]).

After subcloning, four stable hybridomas were established: ZIM-1 through ZIM-3 against monomorphic MHC class I antigens (HLA-A, B, and C loci), and ZIM-4 recognizing MHC class II antigens. The specificity of these antibodies is shown in Table 1.

We have demonstrated that this new and simple immunization technique is a very reliable method for the production of Mabs against MHC-encoded antigens. Furthermore, nitrocellulose paper does not interfere with antibody production against proteins bound on it.

This work was supported in part by the Deutsche Forschungsgemeinschaft through SFB102 TPE4 and Do 313-1.

Table 1. Characterization of Mabs ZIM-1 Through ZIM-4

Tests	ZIM-1, ZIM-2 ZIM-3	ZIM-4
ELISA (whole lysate)*	+	+
Complement dependent micro-lymphocytotoxicity assay		
Against T cell	–	–
Against B cell	–	+
Immunofluorescence	Weak	Strong
Dot-blotting technique*	+	+
SDS-PAGE		
Blotting at 44 kd (class I)	+	–
Blotting at 28 kd (class II β)	–	+
1D-IEF		
Class I/II antigens	+	+
2D-gels		
Class I/II antigens	+	+
Immunoprecipitation		
Class I heavy chain and β-2-microglobulin	+	–
Class II a and β chains	–	+

*Triton-X 114 lysed and phase-separated lysate of B-LCLs.

On the contrary, the high rate of specific antibody secretion of the established hybridomas (75% of all growing clones) supports the argument that NC acts as an effective adjuvant.

In summary, this technique of antigen separation through standard SDS-PAGE and identification of the immobilized proteins as immunogens on nitrocellulose paper appears to be very useful for the production of Mabs against any protein with a known molecular weight.

References

1. Brodsky FM, Parham P, Barnstable CJ, Crumpton MJ, Bodmer WF. Monoclonal antibodies for analysis of the HLA system. Immunol Rev 1979;47:3.
2. Grosse-Wilde H, Doxiadis I, Brandt H. Definition of HLA-D with HTC. In: E.D. Albert, M.P. Baur, and W.R. Mayr (eds). Histocompatibility Testing 1984. Berlin, Heidelberg, New York, Tokyo, Springer-Verlag, 1984:249.

Author Affiliation

Institute of Immunogenetics, University Hospital of Essen, F.R.G.

HLA Typing Using Soluble Antigens from Serum or Plasma

Ilias Doxiadis and Hans Grosse-Wilde

In 1970, van Rood et al. (1) claimed the existence of HLA-A2 in the serum of healthy individuals. Later, attempts were made to utilize serum as a source for HLA typing by an ELISA-like assay (2). The soluble class I antigens were captured by a monoclonal antibody (Mab) recognizing a polymorphic region, and development with a second antibody directed against beta-2-microglobulin (β-2m). This antibody was radioactive labeled. However, this method is restricted until Mab towards HLA antigens are available. To date, this is the case for few alloantigens only.

In this report, we present a simple nonradioactive technique for the detection of HLA class I molecules in serum or plasma. The method is based on immunoprecipitation of class I molecules with the Mab w6/32, which recognizes class I molecules in association with β-2m. This step is followed by a neuraminidase treatment and a denaturation of the precipitates. The complexity of the soluble class I determinants was analyzed by two-dimensional gel electrophoresis, where three distinct molecules with a molecular weight of 45, 39, and 33 kd, respectively, were identified. The latter molecules revealed spots at the same isoelectric point (pI) range as found for the 45 kd molecule, which represents presumably native class I antigens. Comparative experiments using peripheral blood lymphocytes of the same individual showed spots on the same pI position for the 45 kd molecule, whereas, as expected, the 39 and 33 kd molecules were missing. These results encouraged us to apply one-dimensional isoelectric focusing (1D-IEF) followed by immunoblotting for the characterization of soluble class I determinants. This methodology is already established for the identification of cell surface-bound class I antigens (3). Our results, including the segregation analysis in a family (Fig. 1), can be summarized as follows:

1. Soluble 45 kd molecules recovered from serum or plasma of different individuals represent the native HLA class I molecules usually found on the cell surface.
2. All HLA class I molecules present in serum or plasma are associated with β-2m.
3. The 1D-IEF banding pattern of soluble class I gene products parallels that of cell lysates.

The data presented here should open new avenues for serology of HLA class I antigens.

References

1. van Rood JJ, van Leeuwen A, van Santen MCT. Anti-HL-A2 inhibitor in normal human serum cytotoxins. Nature 1970; 226:366.
2. Russo C, Fotino M, Antonelli P, Carbonara AO, Ferrone S. In: E.D. Albert, M.P. Baur, and W.R. Mayr (eds.). Berlin, Heidelberg, New York, Tokyo: Springer-Verlag, 1984:447.
3. Neefjes JJ, Doxiadis I, Stam NJ, Beckers CJ, Ploegh HL. An analysis of class I antigens of man and other species by one-dimensional IEF and immunoblotting. Immunogenetics 1986;23:164.

Author Affiliation

Institute of Immunogenetics, University Hospital of Essen, F.R.G.

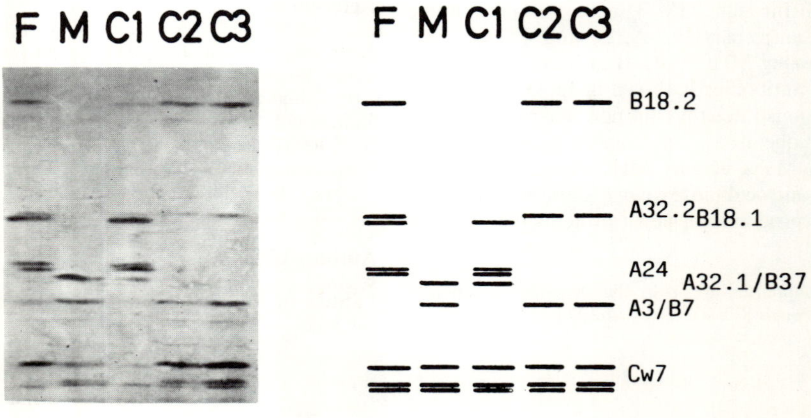

Figure 1. Segregation of soluble class I molecules recovered from serum.

Immunobiology of HLA
Volume II

Index to Volume II